MECHANICS OF SOLIDS AND MATERIALS

Mechanics of Solids and Materials intends to provide a modern and integrated treatment of the foundations of solid mechanics as applied to the mathematical description of material behavior. The book blends both innovative topics (*e.g.*, large strain, strain rate, temperature, time-dependent deformation and localized plastic deformation in crystalline solids, and deformation of biological networks) and traditional topics (*e.g.*, elastic theory of torsion, elastic beam and plate theories, and contact mechanics) in a coherent theoretical framework. This, and the extensive use of transform methods to generate solutions, makes the book of interest to structural, mechanical, materials, and aerospace engineers. Plasticity theories, micromechanics, crystal plasticity, thin films, energetics of elastic systems, and an overall review of continuum mechanics and thermodynamics are also covered in the book.

Robert J. Asaro was awarded his PhD in materials science with distinction from Stanford University in 1972. He was a professor of engineering at Brown University from 1975 to 1989, and has been a professor of engineering at the University of California, San Diego since 1989. Dr. Asaro has led programs involved with the design, fabrication, and full-scale structural testing of large composite structures, including high-performance ships and marine civil structures. His list of publications includes more than 170 research papers in the leading professional journals and conference proceedings. He received the NSF Special Creativity Award for his research in 1983 and 1987. Dr. Asaro also received the TMS Champion H. Mathewson Gold Medal in 1991. He has made fundamental contributions to the theory of crystal plasticity, the analysis of surface instabilities, and dislocation theory. He served as a founding member of the Advisory Committee for NSF's Office of Advanced Computing that founded the Supercomputer Program in the United States. He has also served on the NSF Materials Advisory Committee. He has been an affiliate with Los Alamos National Laboratory for more than 20 years and has served as consultant to Sandia National Laboratory. Dr. Asaro has been recognized by ISI as a highly cited author in materials science.

Vlado A. Lubarda received his PhD in mechanical engineering from Stanford University in 1980. He was a professor at the University of Montenegro from 1980 to 1989, Fulbright fellow and a visiting associate professor at Brown University from 1989 to 1991, and a visiting professor at Arizona State University from 1992 to 1997. Since 1998 he has been an adjunct professor of applied mechanics at the University of California, San Diego. Dr. Lubarda has made significant contributions to phenomenological theories of large deformation elastoplasticity, dislocation theory, damage mechanics, and micromechanics. He is the author of more than 100 journal and conference publications and two books: *Strength of Materials* (1985) and *Elastoplasticity Theory* (2002). He has served as a research panelist for NSF and as a reviewer to numerous international journals of mechanics, materials science, and applied mathematics. In 2000 Dr. Lubarda was elected to the Montenegrin Academy of Sciences and Arts. He is also recipient of the 2004 Distinguished Teaching Award from the University of California.

Mechanics of Solids and Materials

ROBERT J. ASARO
University of California, San Diego

VLADO A. LUBARDA
University of California, San Diego

CAMBRIDGE UNIVERSITY PRESS
Cambridge, New York, Melbourne, Madrid, Cape Town, Singapore, São Paulo

Cambridge University Press
40 West 20th Street, New York, NY 10011-4211, USA

www.cambridge.org
Information on this title: www.cambridge.org/9780521859790

© Robert Asaro and Vlado Lubarda 2006

This publication is in copyright. Subject to statutory exception
and to the provisions of relevant collective licensing agreements,
no reproduction of any part may take place without
the written permission of Cambridge University Press.

First published 2006

Printed in the United States of America

A catalog record for this publication is available from the British Library.

Library of Congress Cataloging in Publication Data

Asaro, Robert J.
Mechanics of solids and materials / Robert J. Asaro and Vlado A. Lubarda.
    p.   cm.
Includes bibliographical references and index.
ISBN 0-521-85979-4 (hardback)
1. Mechanics, Applied.   I. Lubarda, Vlado A.   II. Title.
TA350.A735   2006
620.1 – dc22   2005025722

ISBN-13   978-0-521-85979-0 hardback
ISBN-10   0-521-85979-4 hardback

Cambridge University Press has no responsibility for
the persistence or accuracy of URLs for external or
third-party Internet Web sites referred to in this publication
and does not guarantee that any content on such
Web sites is, or will remain, accurate or appropriate.

Contents

Preface		*page* xix
PART 1: MATHEMATICAL PRELIMINARIES		
1 Vectors and Tensors		1
1.1.	Vector Algebra	1
1.2.	Coordinate Transformation: Rotation of Axes	4
1.3.	Second-Rank Tensors	5
1.4.	Symmetric and Antisymmetric Tensors	5
1.5.	Prelude to Invariants of Tensors	6
1.6.	Inverse of a Tensor	7
1.7.	Additional Proofs	7
1.8.	Additional Lemmas for Vectors	8
1.9.	Coordinate Transformation of Tensors	9
1.10.	Some Identities with Indices	10
1.11.	Tensor Product	10
1.12.	Orthonormal Basis	11
1.13.	Eigenvectors and Eigenvalues	12
1.14.	Symmetric Tensors	14
1.15.	Positive Definiteness of a Tensor	14
1.16.	Antisymmetric Tensors	15
1.16.1.	Eigenvectors of W	15
1.17.	Orthogonal Tensors	17
1.18.	Polar Decomposition Theorem	19
1.19.	Polar Decomposition: Physical Approach	20
1.19.1.	Left and Right Stretch Tensors	21
1.19.2.	Principal Stretches	21
1.20.	The Cayley–Hamilton Theorem	22
1.21.	Additional Lemmas for Tensors	23
1.22.	Identities and Relations Involving ∇ Operator	23
1.23.	Suggested Reading	25

2 Basic Integral Theorems — 26

- 2.1. Gauss and Stokes's Theorems — 26
  - 2.1.1. Applications of Divergence Theorem — 27
- 2.2. Vector and Tensor Fields: Physical Approach — 27
- 2.3. Surface Integrals: Gauss Law — 28
- 2.4. Evaluating Surface Integrals — 29
  - 2.4.1. Application of the Concept of Flux — 31
- 2.5. The Divergence — 31
- 2.6. Divergence Theorem: Relation of Surface to Volume Integrals — 33
- 2.7. More on Divergence Theorem — 34
- 2.8. Suggested Reading — 35

3 Fourier Series and Fourier Integrals — 36

- 3.1. Fourier Series — 36
- 3.2. Double Fourier Series — 37
  - 3.2.1. Double Trigonometric Series — 38
- 3.3. Integral Transforms — 39
- 3.4. Dirichlet's Conditions — 42
- 3.5. Integral Theorems — 46
- 3.6. Convolution Integrals — 48
  - 3.6.1. Evaluation of Integrals by Use of Convolution Theorems — 49
- 3.7. Fourier Transforms of Derivatives of $f(x)$ — 49
- 3.8. Fourier Integrals as Limiting Cases of Fourier Series — 50
- 3.9. Dirac Delta Function — 51
- 3.10. Suggested Reading — 52

PART 2: CONTINUUM MECHANICS

4 Kinematics of Continuum — 55

- 4.1. Preliminaries — 55
- 4.2. Uniaxial Strain — 56
- 4.3. Deformation Gradient — 57
- 4.4. Strain Tensor — 58
- 4.5. Stretch and Normal Strains — 60
- 4.6. Angle Change and Shear Strains — 60
- 4.7. Infinitesimal Strains — 61
- 4.8. Principal Stretches — 62
- 4.9. Eigenvectors and Eigenvalues of Deformation Tensors — 63
- 4.10. Volume Changes — 63
- 4.11. Area Changes — 64
- 4.12. Area Changes: Alternative Approach — 65
- 4.13. Simple Shear of a Thick Plate with a Central Hole — 66
- 4.14. Finite *vs.* Small Deformations — 68
- 4.15. Reference *vs.* Current Configuration — 69
- 4.16. Material Derivatives and Velocity — 71
- 4.17. Velocity Gradient — 71

4.18.	Deformation Rate and Spin	74
4.19.	Rate of Stretching and Shearing	75
4.20.	Material Derivatives of Strain Tensors: $\dot{\mathbf{E}}$ vs. $\mathbf{D}$	76
4.21.	Rate of F in Terms of Principal Stretches	78
4.21.1.	Spins of Lagrangian and Eulerian Triads	81
4.22.	Additional Connections Between Current and Reference State Representations	82
4.23.	Transport Formulae	83
4.24.	Material Derivatives of Volume, Area, and Surface Integrals: Transport Formulae Revisited	84
4.25.	Analysis of Simple Shearing	85
4.26.	Examples of Particle and Plane Motion	87
4.27.	Rigid Body Motions	88
4.28.	Behavior under Superposed Rotation	89
4.29.	Suggested Reading	90

5 Kinetics of Continuum — 92

5.1.	Traction Vector and Stress Tensor	92
5.2.	Equations of Equilibrium	94
5.3.	Balance of Angular Momentum: Symmetry of σ	95
5.4.	Principal Values of Cauchy Stress	96
5.5.	Maximum Shear Stresses	97
5.6.	Nominal Stress	98
5.7.	Equilibrium in the Reference State	99
5.8.	Work Conjugate Connections	100
5.9.	Stress Deviator	102
5.10.	Frame Indifference	102
5.11.	Continuity Equation and Equations of Motion	107
5.12.	Stress Power	108
5.13.	The Principle of Virtual Work	109
5.14.	Generalized Clapeyron's Formula	111
5.15.	Suggested Reading	111

6 Thermodynamics of Continuum — 113

6.1.	First Law of Thermodynamics: Energy Equation	113
6.2.	Second Law of Thermodynamics: Clausius–Duhem Inequality	114
6.3.	Reversible Thermodynamics	116
6.3.1.	Thermodynamic Potentials	116
6.3.2.	Specific and Latent Heats	118
6.3.3.	Coupled Heat Equation	119
6.4.	Thermodynamic Relationships with p, V, T, and s	120
6.4.1.	Specific and Latent Heats	121
6.4.2.	Coefficients of Thermal Expansion and Compressibility	122
6.5.	Theoretical Calculations of Heat Capacity	123
6.6.	Third Law of Thermodynamics	125
6.7.	Irreversible Thermodynamics	127
6.7.1.	Evolution of Internal Variables	129

6.8.	Gibbs Conditions of Thermodynamic Equilibrium	129
6.9.	Linear Thermoelasticity	130
6.10.	Thermodynamic Potentials in Linear Thermoelasticity	132
	6.10.1. Internal Energy	132
	6.10.2. Helmholtz Free Energy	133
	6.10.3. Gibbs Energy	134
	6.10.4. Enthalpy Function	135
6.11.	Uniaxial Loading and Thermoelastic Effect	136
6.12.	Thermodynamics of Open Systems: Chemical Potentials	139
6.13.	Gibbs–Duhem Equation	141
6.14.	Chemical Potentials for Binary Systems	142
6.15.	Configurational Entropy	143
6.16.	Ideal Solutions	144
6.17.	Regular Solutions for Binary Alloys	145
6.18.	Suggested Reading	147

7 Nonlinear Elasticity — 148

7.1.	Green Elasticity	148
7.2.	Isotropic Green Elasticity	150
7.3.	Constitutive Equations in Terms of B	151
7.4.	Constitutive Equations in Terms of Principal Stretches	152
7.5.	Incompressible Isotropic Elastic Materials	153
7.6.	Elastic Moduli Tensors	153
7.7.	Instantaneous Elastic Moduli	155
7.8.	Elastic Pseudomoduli	155
7.9.	Elastic Moduli of Isotropic Elasticity	156
7.10.	Elastic Moduli in Terms of Principal Stretches	157
7.11.	Suggested Reading	158

PART 3: LINEAR ELASTICITY

8 Governing Equations of Linear Elasticity — 161

8.1.	Elementary Theory of Isotropic Linear Elasticity	161
8.2.	Elastic Energy in Linear Elasticity	163
8.3.	Restrictions on the Elastic Constants	164
	8.3.1. Material Symmetry	164
	8.3.2. Restrictions on the Elastic Constants	168
8.4.	Compatibility Relations	169
8.5.	Compatibility Conditions: Cesàro Integrals	170
8.6.	Beltrami–Michell Compatibility Equations	172
8.7.	Navier Equations of Motion	172
8.8.	Uniqueness of Solution to Linear Elastic Boundary Value Problem	174
	8.8.1. Statement of the Boundary Value Problem	174
	8.8.2. Uniqueness of the Solution	174
8.9.	Potential Energy and Variational Principle	175
	8.9.1. Uniqueness of the Strain Field	177

Contents ix

	8.10. Betti's Theorem of Linear Elasticity	177
	8.11. Plane Strain	178
	8.11.1. Plane Stress	179
	8.12. Governing Equations of Plane Elasticity	180
	8.13. Thermal Distortion of a Simple Beam	180
	8.14. Suggested Reading	182

9 Elastic Beam Problems   184

    9.1. A Simple 2D Beam Problem   184
    9.2. Polynomial Solutions to $\nabla^4 \phi = 0$ 185
    9.3. A Simple Beam Problem Continued   186
        9.3.1. Strains and Displacements for 2D Beams   187
    9.4. Beam Problems with Body Force Potentials   188
    9.5. Beam under Fourier Loading   190
    9.6. Complete Boundary Value Problems for Beams   193
        9.6.1. Displacement Calculations   196
    9.7. Suggested Reading   198

10 Solutions in Polar Coordinates   199

    10.1. Polar Components of Stress and Strain   199
    10.2. Plate with Circular Hole   201
        10.2.1. Far Field Shear   201
        10.2.2. Far Field Tension   203
    10.3. Degenerate Cases of Solution in Polar Coordinates   204
    10.4. Curved Beams: Plane Stress   206
        10.4.1. Pressurized Cylinder   209
        10.4.2. Bending of a Curved Beam   210
    10.5. Axisymmetric Deformations   211
    10.6. Suggested Reading   213

11 Torsion and Bending of Prismatic Rods   214

    11.1. Torsion of Prismatic Rods   214
    11.2. Elastic Energy of Torsion   216
    11.3. Torsion of a Rod with Rectangular Cross Section   217
    11.4. Torsion of a Rod with Elliptical Cross Section   221
    11.5. Torsion of a Rod with Multiply Connected Cross Sections   222
        11.5.1. Hollow Elliptical Cross Section   224
    11.6. Bending of a Cantilever   225
    11.7. Elliptical Cross Section   227
    11.8. Suggested Reading   228

12 Semi-Infinite Media   229

    12.1. Fourier Transform of Biharmonic Equation   229
    12.2. Loading on a Half-Plane   230
    12.3. Half-Plane Loading: Special Case   232
    12.4. Symmetric Half-Plane Loading   234
    12.5. Half-Plane Loading: Alternative Approach   235
    12.6. Additional Half-Plane Solutions   237

	12.6.1. Displacement Fields in Half-Spaces	238
	12.6.2. Boundary Value Problem	239
	12.6.3. Specific Example	240
12.7.	Infinite Strip	242
	12.7.1. Uniform Loading: $-a \leq x \leq a$	243
	12.7.2. Symmetrical Point Loads	244
12.8.	Suggested Reading	245

13 Isotropic 3D Solutions — 246

13.1.	Displacement-Based Equations of Equilibrium	246
13.2.	Boussinesq–Papkovitch Solutions	247
13.3.	Spherically Symmetrical Geometries	248
	13.3.1. Internally Pressurized Sphere	249
13.4.	Pressurized Sphere: Stress-Based Solution	251
	13.4.1. Pressurized Rigid Inclusion	252
	13.4.2. Disk with Circumferential Shear	253
	13.4.3. Sphere Subject to Temperature Gradients	254
13.5.	Spherical Indentation	254
	13.5.1. Displacement-Based Equilibrium	255
	13.5.2. Strain Potentials	256
	13.5.3. Point Force on a Half-Plane	257
	13.5.4. Hemispherical Load Distribution	258
	13.5.5. Indentation by a Spherical Ball	259
13.6.	Point Forces on Elastic Half-Space	261
13.7.	Suggested Reading	263

14 Anisotropic 3D Solutions — 264

14.1.	Point Force	264
14.2.	Green's Function	264
14.3.	Isotropic Green's Function	268
14.4.	Suggested Reading	270

15 Plane Contact Problems — 271

15.1.	Wedge Problem	271
15.2.	Distributed Contact Forces	274
	15.2.1. Uniform Contact Pressure	275
	15.2.2. Uniform Tangential Force	277
15.3.	Displacement-Based Contact: Rigid Flat Punch	277
15.4.	Suggested Reading	279

16 Deformation of Plates — 280

16.1.	Stresses and Strains of Bent Plates	280
16.2.	Energy of Bent Plates	281
16.3.	Equilibrium Equations for a Plate	282
16.4.	Shear Forces and Bending and Twisting Moments	285
16.5.	Examples of Plate Deformation	287
	16.5.1. Clamped Circular Plate	287
	16.5.2. Circular Plate with Simply Supported Edges	288

Contents

	16.5.3. Circular Plate with Concentrated Force	288
	16.5.4. Peeled Surface Layer	288
16.6.	Rectangular Plates	289
	16.6.1. Uniformly Loaded Rectangular Plate	290
16.7.	Suggested Reading	291

PART 4: MICROMECHANICS

17 Dislocations and Cracks: Elementary Treatment 293

17.1.	Dislocations	293
	17.1.1. Derivation of the Displacement Field	294
17.2.	Tensile Cracks	295
17.3.	Suggested Reading	298

18 Dislocations in Anisotropic Media 299

18.1.	Dislocation Character and Geometry	299
18.2.	Dislocations in Isotropic Media	302
	18.2.1. Infinitely Long Screw Dislocations	302
	18.2.2. Infinitely Long Edge Dislocations	303
	18.2.3. Infinitely Long Mixed Segments	303
18.3.	Planar Geometric Theorem	305
18.4.	Applications of the Planar Geometric Theorem	308
	18.4.1. Angular Dislocations	311
18.5.	A 3D Geometrical Theorem	312
18.6.	Suggested Reading	314

19 Cracks in Anisotropic Media 315

19.1.	Dislocation Mechanics: Reviewed	315
19.2.	Freely Slipping Crack	316
19.3.	Crack Extension Force	319
19.4.	Crack Faces Loaded by Tractions	320
19.5.	Stress Intensity Factors and Crack Extension Force	322
	19.5.1. Computation of the Crack Extension Force	323
19.6.	Crack Tip Opening Displacement	325
19.7.	Dislocation Energy Factor Matrix	325
19.8.	Inversion of a Singular Integral Equation	328
19.9.	2D Anisotropic Elasticity – Stroh Formalism	329
	19.9.1. Barnett–Lothe Tensors	332
19.10.	Suggested Reading	334

20 The Inclusion Problem 335

20.1.	The Problem	335
20.2.	Eshelby's Solution Setup	336
20.3.	Calculation of the Constrained Fields: u^c, e^c, and σ^c	338
20.4.	Components of the Eshelby Tensor for Ellipsoidal Inclusion	341
20.5.	Elastic Energy of an Inclusion	343
20.6.	Inhomogeneous Inclusion: Uniform Transformation Strain	343
20.7.	Nonuniform Transformation Strain Inclusion Problem	345
	20.7.1. The Cases $M = 0, 1$	349

20.8.	Inclusions in Isotropic Media	350
	20.8.1. Constrained Elastic Field	350
	20.8.2. Field in the Matrix	351
	20.8.3. Field at the Interface	352
	20.8.4. Isotropic Spherical Inclusion	353
20.9.	Suggested Reading	354

21 Forces and Energy in Elastic Systems — 355

21.1.	Free Energy and Mechanical Potential Energy	355
21.2.	Forces of Translation	357
	21.2.1. Force on an Interface	359
	21.2.2. Finite Deformation Energy Momentum Tensor	360
21.3.	Interaction Between Defects and Loading Mechanisms	362
	21.3.1. Interaction Between Dislocations and Inclusions	364
	21.3.2. Force on a Dislocation Segment	365
21.4.	Elastic Energy of a Dislocation	366
21.5.	In-Plane Stresses of Straight Dislocation Lines	367
21.6.	Chemical Potential	369
	21.6.1. Force on a Defect due to a Free Surface	371
21.7.	Applications of the J Integral	372
	21.7.1. Force on a Clamped Crack	372
	21.7.2. Application of the Interface Force to Precipitation	372
21.8.	Suggested Reading	374

22 Micropolar Elasticity — 375

22.1.	Introduction	375
22.2.	Basic Equations of Couple-Stress Elasticity	376
22.3.	Displacement Equations of Equilibrium	377
22.4.	Correspondence Theorem of Couple-Stress Elasticity	378
22.5.	Plane Strain Problems of Couple-Stress Elasticity	379
	22.5.1. Mindlin's Stress Functions	380
22.6.	Edge Dislocation in Couple-Stress Elasticity	381
	22.6.1. Strain Energy	382
22.7.	Edge Dislocation in a Hollow Cylinder	384
22.8.	Governing Equations for Antiplane Strain	386
	22.8.1. Expressions in Polar Coordinates	388
	22.8.2. Correspondence Theorem for Antiplane Strain	389
22.9.	Antiplane Shear of Circular Annulus	390
22.10.	Screw Dislocation in Couple-Stress Elasticity	391
	22.10.1. Strain Energy	392
22.11.	Configurational Forces in Couple-Stress Elasticity	392
	22.11.1. Reciprocal Properties	393
	22.11.2. Energy due to Internal Sources of Stress	394
	22.11.3. Energy due to Internal and External Sources of Stress	394
	22.11.4. The Force on an Elastic Singularity	395

Contents

22.12.	Energy-Momentum Tensor of a Couple-Stress Field	396
22.13.	Basic Equations of Micropolar Elasticity	398
22.14.	Noether's Theorem of Micropolar Elasticity	400
22.15.	Conservation Integrals in Micropolar Elasticity	403
22.16.	Conservation Laws for Plane Strain Micropolar Elasticity	404
22.17.	M Integral of Micropolar Elasticity	404
22.18.	Suggested Reading	406

PART 5: THIN FILMS AND INTERFACES

23 Dislocations in Bimaterials — 407

- 23.1. Introduction — 407
- 23.2. Screw Dislocation Near a Bimaterial Interface — 407
  - 23.2.1. Interface Screw Dislocation — 409
  - 23.2.2. Screw Dislocation in a Homogeneous Medium — 409
  - 23.2.3. Screw Dislocation Near a Free Surface — 409
  - 23.2.4. Screw Dislocation Near a Rigid Boundary — 410
- 23.3. Edge Dislocation (b_x) Near a Bimaterial Interface — 410
  - 23.3.1. Interface Edge Dislocation — 415
  - 23.3.2. Edge Dislocation in an Infinite Medium — 417
  - 23.3.3. Edge Dislocation Near a Free Surface — 417
  - 23.3.4. Edge Dislocation Near a Rigid Boundary — 418
- 23.4. Edge Dislocation (b_y) Near a Bimaterial Interface — 419
  - 23.4.1. Interface Edge Dislocation — 420
  - 23.4.2. Edge Dislocation in an Infinite Medium — 422
  - 23.4.3. Edge Dislocation Near a Free Surface — 422
  - 23.4.4. Edge Dislocation Near a Rigid Boundary — 423
- 23.5. Strain Energy of a Dislocation Near a Bimaterial Interface — 423
  - 23.5.1. Strain Energy of a Dislocation Near a Free Surface — 426
- 23.6. Suggested Reading — 427

24 Strain Relaxation in Thin Films — 428

- 24.1. Dislocation Array Beneath the Free Surface — 428
- 24.2. Energy of a Dislocation Array — 430
- 24.3. Strained-Layer Epitaxy — 431
- 24.4. Conditions for Dislocation Array Formation — 432
- 24.5. Frank and van der Merwe Energy Criterion — 434
- 24.6. Gradual Strain Relaxation — 436
- 24.7. Stability of Array Configurations — 439
- 24.8. Stronger Stability Criteria — 439
- 24.9. Further Stability Bounds — 441
  - 24.9.1. Lower Bound — 441
  - 24.9.2. Upper Bound — 443
- 24.10. Suggested Reading — 446

25 Stability of Planar Interfaces — 447

- 25.1. Stressed Surface Problem — 447
- 25.2. Chemical Potential — 449

25.3.	Surface Diffusion and Interface Stability	450
25.4.	Volume Diffusion and Interface Stability	451
25.5.	2D Surface Profiles and Surface Stability	455
25.6.	Asymptotic Stresses for 1D Surface Profiles	457
25.7.	Suggested Reading	459

PART 6: PLASTICITY AND VISCOPLASTICITY

26 Phenomenological Plasticity — 461

26.1.	Yield Criteria for Multiaxial Stress States	462
26.2.	Von Mises Yield Criterion	463
26.3.	Tresca Yield Criterion	465
26.4.	Mohr–Coulomb Yield Criterion	467
26.4.1.	Drucker–Prager Yield Criterion	468
26.5.	Gurson Yield Criterion for Porous Metals	470
26.6.	Anisotropic Yield Criteria	470
26.7.	Elastic-Plastic Constitutive Equations	471
26.8.	Isotropic Hardening	473
26.8.1.	J_2 Flow Theory of Plasticity	474
26.9.	Kinematic Hardening	475
26.9.1.	Linear and Nonlinear Kinematic Hardening	477
26.10.	Constitutive Equations for Pressure-Dependent Plasticity	478
26.11.	Nonassociative Plasticity	480
26.12.	Plastic Potential for Geomaterials	480
26.13.	Rate-Dependent Plasticity	482
26.14.	Deformation Theory of Plasticity	484
26.14.1.	Rate-Type Formulation of Deformation Theory	485
26.14.2.	Application beyond Proportional Loading	486
26.15.	J_2 Corner Theory	487
26.16.	Rate-Dependent Flow Theory	489
26.16.1.	Multiplicative Decomposition $\mathbf{F} = \mathbf{F}^e \cdot \mathbf{F}^p$	489
26.17.	Elastic and Plastic Constitutive Contributions	491
26.17.1.	Rate-Dependent J_2 Flow Theory	492
26.18.	A Rate Tangent Integration	493
26.19.	Plastic Void Growth	495
26.19.1.	Ideally Plastic Material	497
26.19.2.	Incompressible Linearly Hardening Material	498
26.20.	Suggested Reading	501

27 Micromechanics of Crystallographic Slip — 502

27.1.	Early Observations	502
27.2.	Dislocations	508
27.2.1.	Some Basic Properties of Dislocations in Crystals	511
27.2.2.	Strain Hardening, Dislocation Interactions, and Dislocation Multiplication	514
27.3.	Other Strengthening Mechanisms	517
27.4.	Measurements of Latent Hardening	519
27.5.	Observations of Slip in Single Crystals and Polycrystals at Modest Strains	523

Contents

27.6.	Deformation Mechanisms in Nanocrystalline Grains	525
	27.6.1. Background: AKK Model	530
	27.6.2. Perspective on Discreteness	535
	27.6.3. Dislocation and Partial Dislocation Slip Systems	535
27.7.	Suggested Reading	537

28 Crystal Plasticity 538

28.1.	Basic Kinematics	538
28.2.	Stress and Stress Rates	541
	28.2.1. Resolved Shear Stress	542
	28.2.2. Rate-Independent Strain Hardening	544
28.3.	Convected Elasticity	545
28.4.	Rate-Dependent Slip	547
	28.4.1. A Rate Tangent Modulus	548
28.5.	Crystalline Component Forms	550
	28.5.1. Additional Crystalline Forms	553
	28.5.2. Component Forms on Laboratory Axes	555
28.6.	Suggested Reading	555

29 The Nature of Crystalline Deformation: Localized Plastic Deformation 557

29.1.	Perspectives on Nonuniform and Localized Plastic Flow	557
	29.1.1. Coarse Slip Bands and Macroscopic Shear Bands in Simple Crystals	558
	29.1.2. Coarse Slip Bands and Macroscopic Shear Bands in Ordered Crystals	559
29.2.	Localized Deformation in Single Slip	560
	29.2.1. Constitutive Law for the Single Slip Crystal	560
	29.2.2. Plastic Shearing with Non-Schmid Effects	560
	29.2.3. Conditions for Localization	563
	29.2.4. Expansion to the Order of σ	565
	29.2.5. Perturbations about the Slip and Kink Plane Orientations	567
	29.2.6. Isotropic Elastic Moduli	570
	29.2.7. Particular Cases for Localization	571
29.3.	Localization in Multiple Slip	576
	29.3.1. Double Slip Model	576
	29.3.2. Constitutive Law for the Double Slip Crystal	576
29.4.	Numerical Results for Crystalline Deformation	580
	29.4.1. Additional Experimental Observations	580
	29.4.2. Numerical Observations	582
29.5.	Suggested Reading	584

30 Polycrystal Plasticity 586

30.1.	Perspectives on Polycrystalline Modeling and Texture Development	586
30.2.	Polycrystal Model	588
30.3.	Extended Taylor Model	590

	30.4.	Model Calculational Procedure	592
		30.4.1. Texture Determinations	593
		30.4.2. Yield Surface Determination	594
	30.5.	Deformation Theories and Path-Dependent Response	596
		30.5.1. Specific Model Forms	597
		30.5.2. Alternative Approach to a Deformation Theory	598
		30.5.3. Nonproportional Loading	598
	30.6.	Suggested Reading	600
31	Laminate Plasticity		601
	31.1.	Laminate Model	601
	31.2.	Additional Kinematical Perspective	604
	31.3.	Final Constitutive Forms	604
		31.3.1. Rigid-Plastic Laminate in Single Slip	605
	31.4.	Suggested Reading	607

PART 7: BIOMECHANICS

32	Mechanics of a Growing Mass		609
	32.1.	Introduction	609
	32.2.	Continuity Equation	610
		32.2.1. Material Form of Continuity Equation	610
		32.2.2. Quantities per Unit Initial and Current Mass	612
	32.3.	Reynolds Transport Theorem	612
	32.4.	Momentum Principles	614
		32.4.1. Rate-Type Equations of Motion	615
	32.5.	Energy Equation	615
		32.5.1. Material Form of Energy Equation	616
	32.6.	Entropy Equation	617
		32.6.1. Material Form of Entropy Equation	618
		32.6.2. Combined Energy and Entropy Equations	619
	32.7.	General Constitutive Framework	619
		32.7.1. Thermodynamic Potentials per Unit Initial Mass	620
		32.7.2. Equivalence of the Constitutive Structures	621
	32.8.	Multiplicative Decomposition of Deformation Gradient	622
		32.8.1. Strain and Strain-Rate Measures	623
	32.9.	Density Expressions	624
	32.10.	Elastic Stress Response	625
	32.11.	Partition of the Rate of Deformation	626
	32.12.	Elastic Moduli Tensor	627
		32.12.1. Elastic Moduli Coefficients	629
	32.13.	Elastic Strain Energy Representation	629
	32.14.	Evolution Equation for Stretch Ratio	630
	32.15.	Suggested Reading	631
33	Constitutive Relations for Membranes		633
	33.1.	Biological Membranes	633
	33.2.	Membrane Kinematics	634

	33.3.	Constitutive Laws for Membranes	637
	33.4.	Limited Area Compressibility	638
	33.5.	Simple Triangular Networks	639
	33.6.	Suggested Reading	640

PART 8: SOLVED PROBLEMS

34 Solved Problems for Chapters 1–33   641

Bibliography   833
Index   853

Preface

This book is written for graduate students in solid mechanics and materials science and should also be useful to researchers in these fields. The book consists of eight parts. Part 1 covers the mathematical preliminaries used in later chapters. It includes an introduction to vectors and tensors, basic integral theorems, and Fourier series and integrals. The second part is an introduction to nonlinear continuum mechanics. This incorporates kinematics, kinetics, and thermodynamics of a continuum and an application to nonlinear elasticity. Part 3 is devoted to linear elasticity. The governing equations of the three-dimensional elasticity with appropriate specifications for the two-dimensional plane stress and plane strain problems are given. The applications include the analyses of bending of beams and plates, torsion of prismatic rods, contact problems, semi-infinite media, and three-dimensional isotropic and anisotropic elastic problems. Part 4 is concerned with micromechanics, which includes the analyses of dislocations and cracks in isotropic and anisotropic media, the well-known Eshelby elastic inclusion problem, energy analyses of imperfections and configurational forces, and micropolar elasticity. In Part 5 we analyze dislocations in bimaterials and thin films, with an application to the study of strain relaxation in thin films and stability of planar interfaces. Part 6 is devoted to mathematical and physical theories of plasticity and viscoplasticity. The phenomenological or continuum theory of plasticity, single crystal, polycrystalline, and laminate plasticity are presented. The micromechanics of crystallographic slip is addressed in detail, with an analysis of the nature of crystalline deformation, embedded in its tendency toward localized plastic deformation. Part 7 is an introduction to biomechanics, particularly the formulation of governing equations of the mechanics of solids with a growing mass and constitutive relations for biological membranes. Part 8 is a collection of 180 solved problems covering all chapters of the book. This is included to provide additional development of the basic theory and to further illustrate its application.

The book is transcribed from lecture notes we have used for various courses in solid mechanics and materials science, as well as from our own published work. We have also consulted and used major contributions by other authors, their research work and written books, as cited in the various sections. As such, this book can be used as a textbook for a sequence of solid mechanics courses at the graduate level within mechanical, structural, aerospace, and materials science engineering programs. In particular, it can be used for the introduction to continuum mechanics, linear and nonlinear elasticities, theory of dislocations, fracture mechanics, theory of plasticity, and selected topics from thin films and biomechanics. At the end of each chapter we offer a list of recommended references for additional reading, which aid further study and mastering of the particular subject.

Standard notations and conventions are used throughout the text. Symbols in bold, both Latin and Greek, denote tensors or vectors, the order of which is indicated by the context. Typically the magnitude of a vector will be indicated by the name symbol unbolded. Thus, for example, **a** or **b** indicate two vectors or tensors. If **a** and **b** are vectors, then the scalar product, *i.e.*, the dot product between them is indicated by a single dot, as **a** · **b**. Since **a** and **b** are vectors in this context, the scalar product is also $ab\cos\theta$, where θ is the angle between them. If **A** is a higher order tensor, say second-order, then the dot product of **A** and **a** produces another vector, *viz.*, **A** · **a** = **b**. In the index notation this is expressed as $A_{ij}a_j = b_i$. Unless explicitly stated otherwise, the summation convention is adopted whereby a repeated index implies summation over its full range. This means, accordingly, that the scalar product of two vectors as written above can also be expressed as $a_j b_j = \phi$, where ϕ is the scalar result. Two additional operations are introduced and defined in the text involving double dot products. For example, if **A** and **B** are two second-rank tensors, then **A** : **B** = $A_{ij}B_{ij}$ and **A** ·· **B** = $A_{ij}B_{ji}$. For higher order tensors, similar principles apply. If **C** is a fourth-rank tensor, then **C** : **e** $\Rightarrow C_{ijkl}e_{kl} = \{...\}_{ij}$..

In finite vector spaces we assume the existence of a convenient set of *basis vectors*. Most commonly these are taken to be orthogonal and such that an arbitrary vector, say **a**, can be expressed *wrt* its components along these base vectors as **a** = $a_1\mathbf{e}_1 + a_2\mathbf{e}_2 + a_3\mathbf{e}_3$, where $\{\mathbf{e}_1, \mathbf{e}_2, \mathbf{e}_3\}$ are the orthogonal set of base vectors in question. Other more or less standard notations are used, *e.g.*, the left- or right-hand side of an equation is referred to as the *lhs*, or *rhs*, respectively. The commonly used phrase with respect is abbreviated as *wrt*, and so on.

We are grateful to many colleagues and students who have influenced and contributed to our work in solid mechanics and materials science over a long period of time and thus directly or indirectly contributed to our writing of this book. Specifically our experiences at Stanford University, Brown University, UCSD, Ford Motor Company (RJA), Ohio State University (RJA), University of Montenegro (VAL), and Arizona State University (VAL) have involved collaborations that have been of great professional value to us. Research funding by NSF, the U.S. Army, the U.S. Air Force, the U.S. Navy, DARPA, the U.S. DOE, Alcoa Corp., and Ford Motor Co. over the past several decades has greatly facilitated our research in solid mechanics and materials science. We are also most grateful to our families and friends for their support during the writing of this book

La Jolla, California   Robert J. Asaro
July, 2005   Vlado A. Lubarda

PART 1: MATHEMATICAL PRELIMINARIES

1 Vectors and Tensors

This chapter and the next are concerned with establishing some basic properties of vectors and tensors in real spaces. The first of these is specifically concerned with vector algebra and introduces the notion of tensors; the next chapter continues the discussion of tensor algebra and introduces Gauss and Stokes's integral theorems. The discussion in both chapters is focused on laying out the algebraic methods needed in developing the concepts that follow throughout the book. It is, therefore, selective and thus far from inclusive of all vector and tensor algebra. Selected reading is recommended for additional study as it is for all subsequent chapters. Chapter 3 is an introduction to Fourier series and Fourier integrals, added to facilitate the derivation of certain elasticity solutions in later chapters of the book.

1.1 Vector Algebra

We consider three-dimensional *Euclidean vector spaces*, $\mathcal{E}$, for which to each vector such as **a** or **b** there exists a scalar formed by a *scalar product* $\mathbf{a} \cdot \mathbf{b}$ such that $\mathbf{a} \cdot \mathbf{b} = $ *a real number in* $\mathcal{R}$ and a *vector product* that is another vector such that $\mathbf{a} \times \mathbf{b} = \mathbf{c}$. Note the definitions *via* the operations of the symbols, $\cdot$ and $\times$, respectively. Connections to common geometric interpretations will be considered shortly.

With α and β being scalars, the properties of these operations are as follows

$$\mathbf{a} \cdot \mathbf{b} = \mathbf{b} \cdot \mathbf{a}, \quad \forall \, \mathbf{a}, \mathbf{b} \in \mathcal{E}, \tag{1.1}$$

$$(\alpha \mathbf{a} + \beta \mathbf{b}) \cdot \mathbf{c} = \alpha (\mathbf{a} \cdot \mathbf{c}) + \beta (\mathbf{b} \cdot \mathbf{c}), \quad \forall \, \mathbf{a}, \mathbf{b}, \mathbf{c} \in \mathcal{E}, \tag{1.2}$$

$$\mathbf{a} \cdot \mathbf{a} \geq 0, \quad \text{with} \quad \mathbf{a} \cdot \mathbf{a} = 0 \quad \text{iff} \quad \mathbf{a} = \mathbf{0}, \tag{1.3}$$

$$\mathbf{a} \times \mathbf{b} = -\mathbf{b} \times \mathbf{a}, \tag{1.4}$$

$$(\alpha \mathbf{a} + \beta \mathbf{b}) \times \mathbf{c} = \alpha (\mathbf{a} \times \mathbf{c}) + \beta (\mathbf{b} \times \mathbf{c}). \tag{1.5}$$

Also,

$$\mathbf{a} \cdot (\mathbf{a} \times \mathbf{b}) = 0, \tag{1.6}$$

$$(\mathbf{a} \times \mathbf{b}) \cdot (\mathbf{a} \times \mathbf{b}) = (\mathbf{a} \cdot \mathbf{a})(\mathbf{b} \cdot \mathbf{b}) - (\mathbf{a} \cdot \mathbf{b})^2. \tag{1.7}$$

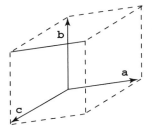

Figure 1.1. Geometric meaning of a vector triple product. The triple product is equal to the volume of the parallelepiped formed from the three defining vectors, **a**, **b**, and **c**.

The magnitude of **a** is
$$|\mathbf{a}| \equiv a = (\mathbf{a} \cdot \mathbf{a})^{1/2}. \tag{1.8}$$

Two vectors are *orthogonal* if
$$\mathbf{a} \cdot \mathbf{b} = 0. \tag{1.9}$$

From the above expressions it follows that if $\mathbf{a} \times \mathbf{b} = \mathbf{0}$, then **a** and **b** are *linearly dependent*, *i.e.*, $\mathbf{a} = \alpha \mathbf{b}$ where α is any scalar.

A *triple product* is defined as
$$[\mathbf{a}, \mathbf{b}, \mathbf{c}] \equiv \mathbf{a} \cdot (\mathbf{b} \times \mathbf{c}). \tag{1.10}$$

It is evident from simple geometry that the triple product is equal to the volume enclosed by the parallelepiped constructed from the vectors **a**, **b**, **c**. This is depicted in Fig. 1.1. Here, again, the listed vector properties allow us to write
$$\begin{aligned}[\mathbf{a}, \mathbf{b}, \mathbf{c}] &= [\mathbf{b}, \mathbf{c}, \mathbf{a}] = [\mathbf{c}, \mathbf{a}, \mathbf{b}] \\ &= -[\mathbf{b}, \mathbf{a}, \mathbf{c}] = -[\mathbf{a}, \mathbf{c}, \mathbf{b}] \\ &= -[\mathbf{c}, \mathbf{b}, \mathbf{a}]\end{aligned} \tag{1.11}$$

and
$$[\alpha \mathbf{a} + \beta \mathbf{b}, \mathbf{c}, \mathbf{d}] = \alpha[\mathbf{a}, \mathbf{c}, \mathbf{d}] + \beta[\mathbf{b}, \mathbf{c}, \mathbf{d}]. \tag{1.12}$$

Furthermore,
$$[\mathbf{a}, \mathbf{b}, \mathbf{c}] = 0 \tag{1.13}$$

iff **a**, **b**, **c** are linearly dependent.

Because of the first of the properties (1.3), we can establish an *orthonormal basis* (Fig. 1.2) that we designate as $\{\mathbf{e}_1, \mathbf{e}_2, \mathbf{e}_3\}$, such that
$$\mathbf{e}_i \cdot \mathbf{e}_j = \delta_{ij} = \begin{cases} 1, & \text{if } i = j, \\ 0, & \text{otherwise.} \end{cases} \tag{1.14}$$

The δ_{ij} is referred to as the Kronecker delta. Using the basis $\{\mathbf{e}_i\}$, an arbitrary vector, say **a**, can be expressed as
$$\mathbf{a} = a_1 \mathbf{e}_1 + a_2 \mathbf{e}_2 + a_3 \mathbf{e}_3 \tag{1.15}$$

or
$$\mathbf{a} = a_i \mathbf{e}_i, \tag{1.16}$$

1.1. Vector Algebra

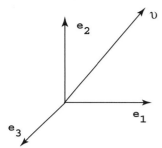

Figure 1.2. A vector v in an orthonormal basis.

where the repeated index i implies summation, i.e.,

$$\mathbf{a} = a_i \mathbf{e}_i = \sum_{i=1}^{3} a_i \mathbf{e}_i. \qquad (1.17)$$

We can use (1.14) to show that

$$a_i = \mathbf{a} \cdot \mathbf{e}_i = a_r \mathbf{e}_r \cdot \mathbf{e}_i = a_r \delta_{ri}. \qquad (1.18)$$

The properties listed previously allow us to write

$$\mathbf{e}_1 = \mathbf{e}_2 \times \mathbf{e}_3, \quad \mathbf{e}_2 = \mathbf{e}_3 \times \mathbf{e}_1, \quad \mathbf{e}_3 = \mathbf{e}_1 \times \mathbf{e}_2. \qquad (1.19)$$

We note that these relations can be expressed as

$$\mathbf{e}_i \times \mathbf{e}_j = \epsilon_{ijk} \mathbf{e}_k, \qquad (1.20)$$

where the *permutation tensor* is defined as

$$\epsilon_{ijk} = \begin{cases} +1, & \text{if } i, j, k \text{ are an even permutation of } 1, 2, 3, \\ -1, & \text{if } i, j, k \text{ are an odd permutation of } 1, 2, 3, \\ 0, & \text{if any of } i, j, k \text{ are the same.} \end{cases} \qquad (1.21)$$

Some useful results follow. Let $\mathbf{a} = a_p \mathbf{e}_p$ and $\mathbf{b} = b_r \mathbf{e}_r$. Then,

$$\mathbf{a} \cdot \mathbf{b} = (a_p \mathbf{e}_p) \cdot (b_r \mathbf{e}_r) = a_p b_r (\mathbf{e}_p \cdot \mathbf{e}_r) = a_p b_r \delta_{pr}. \qquad (1.22)$$

Thus, the scalar product is

$$\mathbf{a} \cdot \mathbf{b} = a_p b_p = a_r b_r. \qquad (1.23)$$

Similarly, the vector product is

$$\mathbf{a} \times \mathbf{b} = a_p \mathbf{e}_p \times b_r \mathbf{e}_r = a_p b_r \mathbf{e}_p \times \mathbf{e}_r = a_p b_r \epsilon_{pri} \mathbf{e}_i = \epsilon_{ipr} (a_p b_r) \mathbf{e}_i. \qquad (1.24)$$

Finally, the component form of the triple product,

$$[\mathbf{a}, \mathbf{b}, \mathbf{c}] = [\mathbf{c}, \mathbf{a}, \mathbf{b}] = \mathbf{c} \cdot (\mathbf{a} \times \mathbf{b}), \qquad (1.25)$$

is

$$\mathbf{c} \cdot (\epsilon_{ipr} a_p b_r \mathbf{e}_i) = \epsilon_{ipr} a_p b_r c_i = \epsilon_{pri} a_p b_r c_i = \epsilon_{ijk} a_i b_j c_k. \qquad (1.26)$$

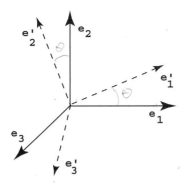

Figure 1.3. Transformation *via* rotation of basis.

1.2 Coordinate Transformation: Rotation of Axes

Let v be a vector referred to two sets of basis vectors, $\{\mathbf{e}_i\}$ and $\{\mathbf{e}'_i\}$, *i.e.*,

$$v = v_i \mathbf{e}_i = v'_i \mathbf{e}'_i. \tag{1.27}$$

We seek to relationship of the v_i to the v'_i. Let the transformation between two bases (Fig. 1.3) be given by

$$\mathbf{e}'_i = \alpha_{ij} \mathbf{e}_j. \tag{1.28}$$

Then

$$\mathbf{e}'_i \cdot \mathbf{e}_j = \alpha_{is} \mathbf{e}_s \cdot \mathbf{e}_j = \alpha_{is} \delta_{sj} = \alpha_{ij}. \tag{1.29}$$

It follows that

$$v'_s = v \cdot \mathbf{e}'_s = v \cdot \alpha_{sp} \mathbf{e}_p = v_p \alpha_{sp} = \alpha_{sp} v_p$$

and thus

$$v'_i = \alpha_{ij} v_j. \tag{1.30}$$

For example, in the two-dimensional case, we have

$$\begin{aligned}\mathbf{e}'_1 &= \cos\theta \mathbf{e}_1 + \sin\theta \mathbf{e}_2, \\ \mathbf{e}'_2 &= -\sin\theta \mathbf{e}_1 + \cos\theta \mathbf{e}_2,\end{aligned} \tag{1.31}$$

with the corresponding transformation matrix

$$\alpha = \begin{bmatrix} \cos\theta & \sin\theta \\ -\sin\theta & \cos\theta \end{bmatrix}. \tag{1.32}$$

Another way to describe the transformation in (1.28) is to set

$$\mathbf{e}' = \beta \cdot \mathbf{e}. \tag{1.33}$$

A straightforward manipulation, however, shows that β and α are related by

$$\beta = \alpha^T, \tag{1.34}$$

where the *transpose*, α^T, is defined in the sequel.

1.3 Second-Rank Tensors

A vector assigns to each direction a scalar, *viz.*, the magnitude of the vector. A second-rank tensor assigns to each vector another (unique) vector, *via* the operation

$$\mathbf{A} \cdot \mathbf{a} = \mathbf{b}. \tag{1.35}$$

More generally,

$$\mathbf{A} \cdot (\alpha \mathbf{a} + \beta \mathbf{b}) = \alpha \mathbf{A} \cdot \mathbf{a} + \beta \mathbf{A} \cdot \mathbf{b}. \tag{1.36}$$

Second-rank tensors obey the following additional rules

$$\begin{aligned} (\mathbf{A} + \mathbf{B}) \cdot \mathbf{a} &= \mathbf{A} \cdot \mathbf{a} + \mathbf{B} \cdot \mathbf{a}, \\ (\alpha \mathbf{A}) \cdot \mathbf{a} &= \alpha \mathbf{A} \cdot \mathbf{a}, \\ (\mathbf{A} \cdot \mathbf{B}) \cdot \mathbf{a} &= \mathbf{A} \cdot (\mathbf{B} \cdot \mathbf{a}), \\ \mathbf{A} + \mathbf{B} &= \mathbf{B} + \mathbf{A}, \\ \alpha (\mathbf{A} \cdot \mathbf{B}) &= (\alpha \mathbf{A}) \cdot \mathbf{B}, \\ \mathbf{A} \cdot (\mathbf{B} + \mathbf{C}) &= \mathbf{A} \cdot \mathbf{B} + \mathbf{A} \cdot \mathbf{C}, \\ \mathbf{A} \cdot (\mathbf{B} \cdot \mathbf{C}) &= (\mathbf{A} \cdot \mathbf{B}) \cdot \mathbf{C}. \end{aligned} \tag{1.37}$$

Each tensor, $\mathbf{A}$, has a unique transpose, $\mathbf{A}^T$, defined such that

$$\mathbf{a} \cdot (\mathbf{A}^T \cdot \mathbf{b}) = \mathbf{b} \cdot (\mathbf{A} \cdot \mathbf{a}). \tag{1.38}$$

Because of (1.36)–(1.38), we can write

$$(\alpha \mathbf{A} + \beta \mathbf{B})^T = \alpha \mathbf{A}^T + \beta \mathbf{B}^T, \tag{1.39}$$

and

$$(\mathbf{A} \cdot \mathbf{B})^T = \mathbf{B}^T \cdot \mathbf{A}^T. \tag{1.40}$$

1.4 Symmetric and Antisymmetric Tensors

We call the tensor $\mathbf{A}$ symmetric if $\mathbf{A} = \mathbf{A}^T$. $\mathbf{A}$ is said to be antisymmetric if $\mathbf{A} = -\mathbf{A}^T$. An arbitrary tensor, $\mathbf{A}$, can be expressed (or decomposed) in terms of its symmetric and antisymmetric parts, *via*

$$\mathbf{A} = \frac{1}{2}(\mathbf{A} + \mathbf{A}^T) + \frac{1}{2}(\mathbf{A} - \mathbf{A}^T), \tag{1.41}$$

where

$$\begin{aligned} \text{sym}(\mathbf{A}) &\equiv \frac{1}{2}(\mathbf{A} + \mathbf{A}^T), \\ \text{skew}(\mathbf{A}) &\equiv \frac{1}{2}(\mathbf{A} - \mathbf{A}^T). \end{aligned} \tag{1.42}$$

1.5 Prelude to Invariants of Tensors

Let $\{\mathbf{f}, \mathbf{g}, \mathbf{h}\}$ and $\{\mathbf{l}, \mathbf{m}, \mathbf{n}\}$ be two arbitrary bases of $\mathcal{E}$. Then it can be shown that

$$\chi_1 = ([\mathbf{A} \cdot \mathbf{f}, \mathbf{g}, \mathbf{h}] + [\mathbf{f}, \mathbf{A} \cdot \mathbf{g}, \mathbf{h}] + [\mathbf{f}, \mathbf{g}, \mathbf{A} \cdot \mathbf{h}]) / [\mathbf{f}, \mathbf{g}, \mathbf{h}]$$
$$= ([\mathbf{A} \cdot \mathbf{l}, \mathbf{m}, \mathbf{n}] + [\mathbf{l}, \mathbf{A} \cdot \mathbf{m}, \mathbf{n}] + [\mathbf{l}, \mathbf{m}, \mathbf{A} \cdot \mathbf{n}]) / [\mathbf{l}, \mathbf{m}, \mathbf{n}],$$

$$\chi_2 = ([\mathbf{A} \cdot \mathbf{f}, \mathbf{A} \cdot \mathbf{g}, \mathbf{h}] + [\mathbf{f}, \mathbf{A} \cdot \mathbf{g}, \mathbf{A} \cdot \mathbf{h}] + [\mathbf{A} \cdot \mathbf{f}, \mathbf{g}, \mathbf{A} \cdot \mathbf{h}]) / [\mathbf{f}, \mathbf{g}, \mathbf{h}]$$
$$= ([\mathbf{A} \cdot \mathbf{l}, \mathbf{A} \cdot \mathbf{m}, \mathbf{n}] + [\mathbf{l}, \mathbf{A} \cdot \mathbf{m}, \mathbf{A} \cdot \mathbf{n}] + [\mathbf{A} \cdot \mathbf{l}, \mathbf{m}, \mathbf{A} \cdot \mathbf{n}]) / [\mathbf{l}, \mathbf{m}, \mathbf{n}],$$

$$\chi_3 = [\mathbf{A} \cdot \mathbf{f}, \mathbf{A} \cdot \mathbf{g}, \mathbf{A} \cdot \mathbf{h}]/[\mathbf{f}, \mathbf{g}, \mathbf{h}] = [\mathbf{A} \cdot \mathbf{l}, \mathbf{A} \cdot \mathbf{m}, \mathbf{A} \cdot \mathbf{n}]/[\mathbf{l}, \mathbf{m}, \mathbf{n}].$$

In proof of the first of the above, consider the first part of the *lhs*,

$$[\mathbf{A} \cdot \mathbf{f}, \mathbf{g}, \mathbf{h}] = [\mathbf{A} \cdot (f_p \mathbf{e}_p), g_q \mathbf{e}_q, h_r \mathbf{e}_r] \tag{1.43}$$
$$= [f_p (\mathbf{A} \cdot \mathbf{e}_p), g_q \mathbf{e}_q, h_r \mathbf{e}_r] = f_p g_q h_r [\mathbf{A} \cdot \mathbf{e}_p, \mathbf{e}_q, \mathbf{e}_r].$$

Thus the entire expression for χ_1 becomes

$$\chi_1 = \frac{f_p g_q h_r}{[\mathbf{f}, \mathbf{g}, \mathbf{h}]} ([\mathbf{A} \cdot \mathbf{e}_p, \mathbf{e}_q, \mathbf{e}_r] + [\mathbf{e}_p, \mathbf{A} \cdot \mathbf{e}_q, \mathbf{e}_r] + [\mathbf{e}_p, \mathbf{e}_q, \mathbf{A} \cdot \mathbf{e}_r]). \tag{1.44}$$

The term in $(\ldots)$ remains unchanged if p, q, r undergo an even permutation of 1, 2, 3; it reverses sign if p, q, r undergo an odd permutation, and is equal to 0 if any of p, q, r are made equal. Thus set $p = 1$, $q = 2$, $r = 3$, and multiply the result by ϵ_{pqr} to take care of the changes in sign or the null results just described. The full expression becomes

$$\frac{(f_p g_q h_r)\epsilon_{pqr}}{[\mathbf{f}, \mathbf{g}, \mathbf{h}]} ([\mathbf{A} \cdot \mathbf{e}_1, \mathbf{e}_2, \mathbf{e}_3] + [\mathbf{e}_1, \mathbf{A} \cdot \mathbf{e}_2, \mathbf{e}_3] + [\mathbf{e}_1, \mathbf{e}_2, \mathbf{A} \cdot \mathbf{e}_3]) \tag{1.45}$$
$$= [\mathbf{A} \cdot \mathbf{e}_1, \mathbf{e}_2, \mathbf{e}_3] + [\mathbf{e}_1, \mathbf{A} \cdot \mathbf{e}_2, \mathbf{e}_3] + [\mathbf{e}_1, \mathbf{e}_2, \mathbf{A} \cdot \mathbf{e}_3].$$

Since $[\mathbf{e}_1, \mathbf{e}_2, \mathbf{e}_3] = +1$, the quantity

$$([\mathbf{A} \cdot \mathbf{f}, \mathbf{g}, \mathbf{h}] + [\mathbf{f}, \mathbf{A} \cdot \mathbf{g}, \mathbf{h}] + [\mathbf{f}, \mathbf{g}, \mathbf{A} \cdot \mathbf{h}]) / [\mathbf{f}, \mathbf{g}, \mathbf{h}] \tag{1.46}$$

is invariant to changes of the basis $\{\mathbf{f}, \mathbf{g}, \mathbf{h}\}$.

Given the validity of the expressions for χ_1, χ_2, and χ_3, we thereby discover three invariants of the tensor $\mathbf{A}$, *viz.*,

$$I_A = ([\mathbf{A} \cdot \mathbf{f}, \mathbf{g}, \mathbf{h}] + [\mathbf{f}, \mathbf{A} \cdot \mathbf{g}, \mathbf{h}] + [\mathbf{f}, \mathbf{g}, \mathbf{A} \cdot \mathbf{h}]) / [\mathbf{f}, \mathbf{g}, \mathbf{h}],$$

$$II_A = ([\mathbf{A} \cdot \mathbf{f}, \mathbf{A} \cdot \mathbf{g}, \mathbf{h}] + [\mathbf{f}, \mathbf{A} \cdot \mathbf{g}, \mathbf{A} \cdot \mathbf{h}] + [\mathbf{A} \cdot \mathbf{f}, \mathbf{g}, \mathbf{A} \cdot \mathbf{h}]) / [\mathbf{f}, \mathbf{g}, \mathbf{h}], \tag{1.47}$$

$$III_A = [\mathbf{A} \cdot \mathbf{f}, \mathbf{A} \cdot \mathbf{g}, \mathbf{A} \cdot \mathbf{h}]/[\mathbf{f}, \mathbf{g}, \mathbf{h}].$$

The commonly held descriptors for two of these are

$$I_A = \text{trace of } \mathbf{A} = \text{tr}(\mathbf{A}),$$

$$III_A = \text{determinant of } \mathbf{A} = \det(\mathbf{A}) = |\mathbf{A}|.$$

1.6 Inverse of a Tensor

If $|\mathbf{A}| \neq 0$, $\mathbf{A}$ has an inverse, $\mathbf{A}^{-1}$, such that

$$\mathbf{A} \cdot \mathbf{A}^{-1} = \mathbf{A}^{-1} \cdot \mathbf{A} = \mathbf{I}, \tag{1.48}$$

where $\mathbf{I}$, *the identity tensor*, is defined *via* the relations

$$\mathbf{a} = \mathbf{I} \cdot \mathbf{a} = \mathbf{a} \cdot \mathbf{I}. \tag{1.49}$$

Useful relations that follow from the above are

$$|\alpha \mathbf{A}| = \alpha^3 |\mathbf{A}|,$$
$$|\mathbf{A} \cdot \mathbf{B}| = |\mathbf{A}||\mathbf{B}|. \tag{1.50}$$

Thus, it follows that

$$|\mathbf{A} \cdot \mathbf{A}^{-1}| = |\mathbf{I}| = 1 = |\mathbf{A}||\mathbf{A}^{-1}|,$$
$$|\mathbf{A}^{-1}| = \frac{1}{|\mathbf{A}|} = |\mathbf{A}|^{-1}. \tag{1.51}$$

1.7 Additional Proofs

We deferred formal proofs of several lemmas until now in the interest of presentation. We provide the proofs at this time.

LEMMA 1.1: *If $\mathbf{a}$ and $\mathbf{b}$ are two vectors, $\mathbf{a} \times \mathbf{b} = \mathbf{0}$ iff $\mathbf{a}$ and $\mathbf{b}$ are linearly dependent.*

Proof: If $\mathbf{a}$ and $\mathbf{b}$ are linearly dependent then there is a scalar such that $\mathbf{b} = \alpha \mathbf{a}$. In this case, if we express the vector product $\mathbf{a} \times \mathbf{b} = \mathbf{c}$ in component form, we find that $c_i = \epsilon_{ijk} a_j \alpha a_k = \alpha \epsilon_{ijk} a_j a_k$. But the summations over the indices j and k will produce pairs of multiples of $a_\beta a_\gamma$, and then again $a_\gamma a_\beta$, for which the permutator tensor alternates algebraic sign, thus causing such pairs to cancel. Thus, in this case $\mathbf{a} \times \mathbf{b} = \mathbf{0}$.

Conversely, if $\mathbf{a} \times \mathbf{b} = \mathbf{0}$, we find from (1.3) to (1.8) that $\mathbf{a} \times \mathbf{b} = \pm |\mathbf{a}||\mathbf{b}|$. If the plus signs holds, we have from the second of (1.3)

$$(|\mathbf{b}|\mathbf{a} - |\mathbf{a}|\mathbf{b}) \cdot (|\mathbf{b}|\mathbf{a} - |\mathbf{a}|\mathbf{b}) = 2|\mathbf{a}|^2|\mathbf{b}|^2 - 2|\mathbf{a}||\mathbf{b}|\mathbf{a} \cdot \mathbf{b} = 0. \tag{1.52}$$

Because of the third property in (1.3) this means that $|\mathbf{b}|\mathbf{a} = |\mathbf{a}|\mathbf{b}$. When the minus sign holds, we find that $|\mathbf{b}|\mathbf{a} = -|\mathbf{a}|\mathbf{b}$. In either case this leads to the conclusion that $\mathbf{b} = \alpha \mathbf{a}$.

Next we examine the relations defining properties of the triple product when pairs of the vectors are interchanged. Use (1.26) to calculate the triple product. This yields $[\mathbf{a}, \mathbf{b}, \mathbf{c}] = \epsilon_{ijk} a_i b_j c_k$. Next imagine interchanging, say $\mathbf{a}$ with $\mathbf{b}$; we obtain $[\mathbf{b}, \mathbf{a}, \mathbf{c}] = \epsilon_{ijk} b_i a_j c_k = \epsilon_{ijk} a_j b_i c_k = -\epsilon_{jik} a_j b_i c_k = -\epsilon_{ijk} a_i b_j c_k$, where the last term involved merely a reassignment of summation indices. Thus $[\mathbf{a}, \mathbf{b}, \mathbf{c}] = -[\mathbf{b}, \mathbf{a}, \mathbf{c}]$. Proceeding this way all members of (1.11) are generated.

We now examine the triple product property expressed in (1.12).

LEMMA 1.2: *If $\mathbf{a}, \mathbf{b}, \mathbf{c}$, and $\mathbf{d}$ are arbitrary vectors, and α and β arbitrary scalars, then*

$$[\alpha \mathbf{a} + \beta \mathbf{b}, \mathbf{c}, \mathbf{d}] = \alpha[\mathbf{a}, \mathbf{c}, \mathbf{d}] + \beta[\mathbf{b}, \mathbf{c}, \mathbf{d}], \quad \forall \mathbf{a}, \mathbf{b}, \mathbf{c}, \mathbf{d} \in \mathcal{E}, \quad \alpha, \beta \in \mathcal{R}. \tag{1.53}$$

Proof: Begin with the property of scalar products between vectors expressed in (1.2) and replace $\mathbf{c}$ with $\mathbf{c} \times \mathbf{d}$. Then,

$$(\alpha \mathbf{a} + \beta \mathbf{b}) \cdot (\mathbf{c} \times \mathbf{d}) = \alpha \mathbf{a} \cdot (\mathbf{c} \times \mathbf{d}) + \beta \mathbf{b} \cdot (\mathbf{c} \times \mathbf{d}) = \alpha[\mathbf{a}, \mathbf{c}, \mathbf{d}] + \beta[\mathbf{b}, \mathbf{c}, \mathbf{d}]. \qquad (1.54)$$

Of course, the first term in the above is the triple product expressed on the *lhs* of the lemma.

1.8 Additional Lemmas for Vectors

LEMMA 1.3: *If*

$$\mathbf{v} = \alpha \mathbf{a} + \beta \mathbf{b} + \gamma \mathbf{c}, \qquad (1.55)$$

where $\mathbf{a}, \mathbf{b}, \mathbf{c}, \mathbf{v}$ *are all vectors, then*

$$\alpha = \frac{\epsilon_{ijk} v_i b_j c_k}{\epsilon_{pqr} a_p b_q c_r}, \quad \beta = \frac{\epsilon_{ijk} a_i v_j c_k}{\epsilon_{pqr} a_p b_q c_r}, \quad \gamma = \frac{\epsilon_{ijk} a_i b_j v_k}{\epsilon_{pqr} a_p b_q c_r}. \qquad (1.56)$$

Proof: The three relations that express the connections are

$$\begin{aligned} v_1 &= \alpha a_1 + \beta b_1 + \gamma c_1, \\ v_2 &= \alpha a_2 + \beta b_2 + \gamma c_2, \\ v_3 &= \alpha a_3 + \beta b_3 + \gamma c_3. \end{aligned} \qquad (1.57)$$

By Cramer's rule

$$\alpha = \frac{\begin{vmatrix} v_1 & b_1 & c_1 \\ v_2 & b_2 & c_2 \\ v_3 & b_3 & c_3 \end{vmatrix}}{\begin{vmatrix} a_1 & b_1 & c_1 \\ a_2 & b_2 & c_2 \\ a_3 & b_3 & c_3 \end{vmatrix}}. \qquad (1.58)$$

Thus, the lemma is proved once the two determinants are expressed using the permutation tensor.

LEMMA 1.4: *Given a vector* $\mathbf{a}$, *then for arbitrary vector* $\mathbf{x}$,

$$\mathbf{a} \times \mathbf{x} = \mathbf{a} \text{ iff } \mathbf{a} = \mathbf{0}. \qquad (1.59)$$

Proof: Express the ith component of $\mathbf{a} \times \mathbf{x}$ as

$$\epsilon_{ijk} a_j x_k, \qquad (1.60)$$

and then form the product $\mathbf{a} \cdot \mathbf{a}$ to obtain

$$\epsilon_{ijk} a_j x_k \epsilon_{irs} a_r x_s = (\delta_{jr}\delta_{ks} - \delta_{js}\delta_{kr}) a_j x_k a_r a_s = a^2 x^2 - (\mathbf{a} \cdot \mathbf{x})^2 = a^2.$$

The expression just generated is zero as may be seen, for example, by letting $\mathbf{x}$ be equal to $\mathbf{e}_1, \mathbf{e}_2, \mathbf{e}_3$, respectively. Note that the third equation of (1.70) below has been used.

LEMMA 1.5: *Suppose that for any vector* $\mathbf{p}$, $\mathbf{p} \cdot \mathbf{q} = \mathbf{p} \cdot \mathbf{t}$, *then we have*

$$\mathbf{p} \cdot \mathbf{q} = \mathbf{p} \cdot \mathbf{t} \Rightarrow \mathbf{q} = \mathbf{t}. \tag{1.61}$$

Proof: The relation $\mathbf{p} \cdot \mathbf{q} = \mathbf{p} \cdot \mathbf{t}$ can be rewritten as

$$\mathbf{p} \cdot (\mathbf{q} - \mathbf{t}) = p_1(q_1 - t_1) + p_2(q_2 - t_2) + p_3(q_3 - t_3) = 0. \tag{1.62}$$

As in the previous lemma, letting $\mathbf{p}$ be systematically equal to $\mathbf{e}_1, \mathbf{e}_2, \mathbf{e}_3$ shows that $p_1 = p_2 = p_3 = 0$.

We reexamine now the operation of the cross product between vectors to develop two additional lemmas of interest.

LEMMA 1.6: *Given the vectors* $\mathbf{p}, \mathbf{q}, \mathbf{r}$ *we have*

$$\mathbf{p} \times (\mathbf{q} \times \mathbf{r}) = \mathbf{q}(\mathbf{p} \cdot \mathbf{r}) - \mathbf{r}(\mathbf{p} \cdot \mathbf{q}). \tag{1.63}$$

Proof: The proof is most readily done by expressing the above in component form, *i.e.*,

$$\epsilon_{rsi} p_s \epsilon_{ijk} q_j r_k = \epsilon_{rsi} \epsilon_{ijk} p_s q_j r_k = \epsilon_{irs} \epsilon_{ijk} p_s q_j r_k. \tag{1.64}$$

Use the identity given by the third equation of (1.70) and write

$$\epsilon_{irs} \epsilon_{ijk} p_s q_j r_k = q_r(p_s r_s) - r_r(p_s q_s) \tag{1.65}$$

to complete the proof.

A simple extension of the last lemma is that

$$(\mathbf{p} \times \mathbf{q}) \times \mathbf{r} = \mathbf{q}(\mathbf{p} \cdot \mathbf{r}) - \mathbf{p}(\mathbf{r} \cdot \mathbf{q}). \tag{1.66}$$

The proof is left as an exercise.

1.9 Coordinate Transformation of Tensors

Consider coordinate transformations prescribed by (1.28). A tensor $\mathbf{A}$ can be written alternatively as

$$\mathbf{A} = A_{ij} \mathbf{e}_i \mathbf{e}_j = A'_{ij} \mathbf{e}'_i \mathbf{e}'_j = A'_{ij} \alpha_{ir} \mathbf{e}_r \alpha_{js} \mathbf{e}_s. \tag{1.67}$$

Since $\mathbf{e}'_p \cdot \mathbf{A} \cdot \mathbf{e}'_q = A'_{pq}$, performing this operation on (1.67) gives

$$A'_{pq} = \mathbf{e}'_p \cdot A_{ij} \mathbf{e}_i \mathbf{e}_j \cdot \mathbf{e}'_q = \alpha_{pi} \alpha_{qj} A_{ij}. \tag{1.68}$$

Transformation of higher order tensors can be handled in an identical manner.

1.10 Some Identities with Indices

The following identities involving the Kronecker delta are useful and are easily verified by direct expansion

$$\begin{aligned}\delta_{ii} &= 3, \\ \delta_{ij}\delta_{ij} &= 3, \\ \delta_{ij}\delta_{ik}\delta_{jk} &= 3, \\ \delta_{ij}\delta_{jk} &= \delta_{ik}, \\ \delta_{ij}A_{ik} &= A_{jk}.\end{aligned} \quad (1.69)$$

Useful identities involving the permutation tensor are

$$\begin{aligned}\epsilon_{ijk}\epsilon_{kpq} &= \delta_{ip}\delta_{jq} - \delta_{iq}\delta_{jp}, \\ \epsilon_{pqs}\epsilon_{sqr} &= -2\delta_{pr}, \\ \epsilon_{ijk}\epsilon_{ijk} &= 6.\end{aligned} \quad (1.70)$$

1.11 Tensor Product

Let $\mathbf{u}$ and $\mathbf{v}$ be two vectors; then there is a tensor $\mathbf{B} = \mathbf{uv}$ defined *via* its action on an arbitrary vector $\mathbf{a}$, such that

$$(\mathbf{uv}) \cdot \mathbf{a} = (\mathbf{v} \cdot \mathbf{a})\mathbf{u}. \quad (1.71)$$

Note that there is the commutative property that follows, *viz.*,

$$\begin{aligned}(\alpha\mathbf{u} + \beta\mathbf{v})\mathbf{w} &= \alpha\mathbf{uw} + \beta\mathbf{vw}, \\ \mathbf{u}(\alpha\mathbf{v} + \beta\mathbf{w}) &= \alpha\mathbf{uv} + \beta\mathbf{uw}.\end{aligned} \quad (1.72)$$

By the definition of the transpose as given previously, we also have

$$(\mathbf{uv})^T = \mathbf{vu}. \quad (1.73)$$

The identity tensor, $\mathbf{I}$, can be expressed as

$$\mathbf{I} = \mathbf{e}_p\mathbf{e}_p, \quad (1.74)$$

if $\mathbf{e}_1, \mathbf{e}_2, \mathbf{e}_3$ are orthonormal. Indeed,

$$(\mathbf{e}_p\mathbf{e}_p) \cdot \mathbf{a} = (\mathbf{a} \cdot \mathbf{e}_p)\mathbf{e}_p = a_p\mathbf{e}_p = \mathbf{a} = \mathbf{I} \cdot \mathbf{a}.$$

LEMMA 1.7: *If $\mathbf{u}$ and $\mathbf{v}$ are arbitrary vectors, then*

$$|\mathbf{uv}| = 0, \quad \text{and} \quad \text{tr}(\mathbf{uv}) = \mathbf{u} \cdot \mathbf{v}. \quad (1.75)$$

Proof: Replace $\mathbf{A}$ in (1.47) with $\mathbf{uv}$, and use $\{\mathbf{a}, \mathbf{b}, \mathbf{c}\}$ as a basis of $\mathcal{E}$. Then the third equation from (1.47) becomes

$$[(\mathbf{uv}) \cdot \mathbf{a}, (\mathbf{uv}) \cdot \mathbf{b}, (\mathbf{uv}) \cdot \mathbf{c}] = III_{uv}[\mathbf{a}, \mathbf{b}, \mathbf{c}],$$

1.12. Orthonormal Basis

where
$$III_{uv} = |\mathbf{uv}|.$$

But,
$$(\mathbf{uv}) \cdot \mathbf{a} = (\mathbf{v} \cdot \mathbf{a})\mathbf{u} \parallel \mathbf{u},$$

and similarly
$$(\mathbf{uv}) \cdot \mathbf{b} = (\mathbf{v} \cdot \mathbf{b})\mathbf{u} \parallel \mathbf{u}, \quad (\mathbf{uv}) \cdot \mathbf{c} = (\mathbf{v} \cdot \mathbf{c})\mathbf{u} \parallel \mathbf{u}.$$

Consequently,
$$|\mathbf{uv}| = 0. \tag{1.76}$$

Next, we note that the first equation from (1.47), with $\mathbf{A} = \mathbf{uv}$, leads to
$$[(\mathbf{uv}) \cdot \mathbf{a}, \mathbf{b}, \mathbf{c}] + [\mathbf{a}, (\mathbf{uv}) \cdot \mathbf{b}, \mathbf{c}] + [\mathbf{a}, \mathbf{b}, (\mathbf{uv}) \cdot \mathbf{c}] = I_{uv}[\mathbf{a}, \mathbf{b}, \mathbf{c}], \tag{1.77}$$

where
$$I_{uv} = \mathrm{tr}\,(\mathbf{uv}).$$

But the *lhs* of the relation (1.77) can be rearranged as
$$\mathbf{v} \cdot \mathbf{a}[\mathbf{u}, \mathbf{b}, \mathbf{c}] + \mathbf{v} \cdot \mathbf{b}[\mathbf{a}, \mathbf{u}, \mathbf{c}] + \mathbf{v} \cdot \mathbf{c}[\mathbf{a}, \mathbf{b}, \mathbf{u}].$$

Since $\{\mathbf{a}, \mathbf{b}, \mathbf{c}\}$ is a basis of $\mathcal{E}$, $\mathbf{u}$ can be expressed as
$$\mathbf{u} = \alpha\mathbf{a} + \beta\mathbf{b} + \gamma\mathbf{c}, \tag{1.78}$$

which, when substituted into the above, yields for the various bracketed terms
$$\begin{aligned}[\alpha\mathbf{a} + \beta\mathbf{b} + \gamma\mathbf{c}, \mathbf{b}, \mathbf{c}] &= \alpha[\mathbf{a}, \mathbf{b}, \mathbf{c}], \\ [\mathbf{a}, \alpha\mathbf{a} + \beta\mathbf{b} + \gamma\mathbf{c}, \mathbf{c}] &= \beta[\mathbf{a}, \mathbf{b}, \mathbf{c}], \\ [\mathbf{a}, \mathbf{b}, \alpha\mathbf{a} + \beta\mathbf{b} + \gamma\mathbf{c}] &= \gamma[\mathbf{a}, \mathbf{b}, \mathbf{c}]. \end{aligned} \tag{1.79}$$

But, the first of (1.47) gives
$$(\alpha\mathbf{a} \cdot \mathbf{v} + \beta\mathbf{b} \cdot \mathbf{v} + \gamma\mathbf{c} \cdot \mathbf{v})\,[\mathbf{a}, \mathbf{b}, \mathbf{c}] = \mathrm{tr}\,(\mathbf{uv})[\mathbf{a}, \mathbf{b}, \mathbf{c}],$$

so that
$$\mathrm{tr}\,(\mathbf{uv}) = \mathbf{u} \cdot \mathbf{v}. \tag{1.80}$$

1.12 Orthonormal Basis

Let us now refer the tensor $\mathbf{A}$ to an orthonormal basis, $\{\mathbf{e}_1, \mathbf{e}_2, \mathbf{e}_3\}$. The $\mathbf{e}_i$ are unit vectors in this context. Let A_{ij} be the components of $\mathbf{A}$ relative to this basis. Then
$$\mathbf{A} \cdot \mathbf{e}_j = A_{pj}\mathbf{e}_p \quad \text{and} \quad A_{ij} = \mathbf{e}_i \cdot \mathbf{A} \cdot \mathbf{e}_j. \tag{1.81}$$

Now form $\mathbf{A} = A_{pq}\mathbf{e}_p\mathbf{e}_q$ and look at its operation on a vector $\mathbf{a} = a_r\mathbf{e}_r$. We have
$$(A_{pq}\mathbf{e}_p\mathbf{e}_q) \cdot a_r\mathbf{e}_r = A_{pq}a_r\mathbf{e}_p\mathbf{e}_q \cdot \mathbf{e}_r = A_{pr}a_r\mathbf{e}_p = a_r A_{pr}\mathbf{e}_p.$$

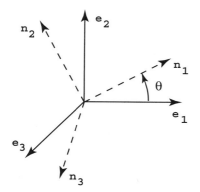

Figure 1.4. Coordinate system (axes) of the principal directions.

But,

$$\mathbf{A} \cdot \mathbf{a} = \mathbf{A} \cdot a_r \mathbf{e}_r = a_r \mathbf{A} \cdot \mathbf{e}_r = a_r A_{pr} \mathbf{e}_p. \tag{1.82}$$

Thus, $\mathbf{A}$ can be expressed as

$$\mathbf{A} = A_{pq} \mathbf{e}_p \mathbf{e}_q, \quad \text{with} \quad A_{pq} = \mathbf{e}_p \cdot \mathbf{A} \cdot \mathbf{e}_q. \tag{1.83}$$

Note if $\mathbf{A} = \mathbf{uv}$, then $A_{ij} = u_i v_j$, because $\mathbf{A} = u_i v_j \mathbf{e}_i \mathbf{e}_j$ and $\mathbf{u} = u_i \mathbf{e}_i$, $\mathbf{v} = v_j \mathbf{e}_j$.

By using an orthonormal basis $\{\mathbf{e}_i\}$, the invariants of a second-rank tensor $\mathbf{A}$ can be expressed from (1.47) as follows. First, consider

$$I_A = [\mathbf{A} \cdot \mathbf{e}_1, \mathbf{e}_2, \mathbf{e}_3] + [\mathbf{e}_1, \mathbf{A} \cdot \mathbf{e}_2, \mathbf{e}_3] + [\mathbf{e}_1, \mathbf{e}_2, \mathbf{A} \cdot \mathbf{e}_3]$$
$$= [A_{p1} \mathbf{e}_p, \mathbf{e}_2, \mathbf{e}_3] + [\mathbf{e}_1, A_{r2} \mathbf{e}_r, \mathbf{e}_3] + [\mathbf{e}_1, \mathbf{e}_2, A_{s3} \mathbf{e}_s]. \tag{1.84}$$

To evaluate the above, use $A_{p1} \mathbf{e}_p = A_{11} \mathbf{e}_1 + A_{21} \mathbf{e}_2 + A_{31} \mathbf{e}_3$ and, for example,

$$[A_{11} \mathbf{e}_1 + A_{21} \mathbf{e}_2 + A_{31} \mathbf{e}_3, \mathbf{e}_2, \mathbf{e}_3] = [A_{11} \mathbf{e}_1, \mathbf{e}_2, \mathbf{e}_3] = A_{11} [\mathbf{e}_1, \mathbf{e}_2, \mathbf{e}_3] = A_{11}.$$

Likewise,

$$[\mathbf{e}_1, A_{r2} \mathbf{e}_r, \mathbf{e}_3] = A_{22}, \quad [\mathbf{e}_1, \mathbf{e}_2, A_{s3} \mathbf{e}_s] = A_{33}.$$

Thus,

$$I_A = A_{11} + A_{22} + A_{33} = \text{tr } \mathbf{A}. \tag{1.85}$$

Similar manipulations yield

$$II_A = \frac{1}{2}(A_{pp}^2 - A_{pq} A_{qp}), \tag{1.86}$$

$$III_A = \det \mathbf{A} = |\mathbf{A}| = \epsilon_{pqr} A_{p1} A_{q2} A_{r3}. \tag{1.87}$$

1.13 Eigenvectors and Eigenvalues

Let $\boldsymbol{\sigma}$ be a symmetric tensor, *i.e.*, $\sigma_{ij} = \sigma_{ji}$ in any orthonormal basis, $\{\mathbf{e}_i\}$. Examine the "normal components" of $\boldsymbol{\sigma}$, *e.g.*, $\sigma_{nn} = \mathbf{n} \cdot \boldsymbol{\sigma} \cdot \mathbf{n} = n_i \sigma_{ij} n_j$. Look for extremum values for

1.13. Eigenvectors and Eigenvalues

σ_{nn} *wrt* the orientation of **n**. Let θ be the angle between **n** and $\mathbf{e}_1$ (Fig. 1.4). We require that

$$\partial \sigma_{nn}/\partial\theta = 0 = \partial n_i/\partial\theta(\sigma_{ij}n_j) + n_i \partial(\sigma_{ij}n_j)/\partial\theta$$

$$= \partial n_i/\partial\theta \sigma_{ij}n_j + n_i\sigma_{ij}\partial n_j/\partial\theta$$

$$= 2\partial n_i/\partial\theta(\sigma_{ij}n_j) = 2(\partial \mathbf{n}/\partial\theta) \cdot \mathbf{T}^{(n)} = 0,$$

where $\mathbf{T}^{(n)} = \boldsymbol{\sigma} \cdot \mathbf{n}$ and $T_i^{(n)} = \sigma_{ij}n_j$. Since $\partial \mathbf{n}/\partial\theta$ is orthogonal to **n**, we conclude that $\mathbf{T}^{(n)} = \boldsymbol{\sigma} \cdot \mathbf{n}$ must be codirectional with **n**. Hence, $\mathbf{T}^{(n)} = \boldsymbol{\sigma} \cdot \mathbf{n} = \lambda^{(n)}\mathbf{n}$. This leads to the homogeneous set of equations

$$(\sigma_{ij} - \lambda^{(n)}\delta_{ij})n_j = 0. \tag{1.88}$$

In dyadic notation these are

$$\boldsymbol{\sigma} \cdot \mathbf{n} - \lambda^{(n)}\mathbf{n} = \mathbf{0} \tag{1.89}$$

or

$$\mathbf{A} \cdot \mathbf{n} = \mathbf{0}, \quad \mathbf{A} = \boldsymbol{\sigma} - \lambda \mathbf{I}. \tag{1.90}$$

Conditions need to be sought whereby (1.89) can have nontrivial solutions.

LEMMA 1.8: *Recall* (1.5) *viz.* $\det \mathbf{A} = [\mathbf{A} \cdot \mathbf{f}, \mathbf{A} \cdot \mathbf{g}, \mathbf{A} \cdot \mathbf{h}]$, *where* $\{\mathbf{f}, \mathbf{g}, \mathbf{h}\}$ *is an arbitrary basis of* $\mathcal{E}$. *If* $[\mathbf{A} \cdot \mathbf{f}, \mathbf{A} \cdot \mathbf{g}, \mathbf{A} \cdot \mathbf{h}] = 0$, *then* $\{\mathbf{A} \cdot \mathbf{f}, \mathbf{A} \cdot \mathbf{g}, \mathbf{A} \cdot \mathbf{h}\}$ *must be linearly dependent. That is,* $[\mathbf{a}, \mathbf{b}, \mathbf{c}] = 0$ *iff* $\{\mathbf{a}, \mathbf{b}, \mathbf{c}\}$ *are linearly dependent.*

Proof: If one of $\{\mathbf{p}, \mathbf{q}, \mathbf{r}\}$ are zero, $[\mathbf{p}, \mathbf{q}, \mathbf{r}] = \mathbf{p} \cdot (\mathbf{q} \times \mathbf{r}) = 0$. Next, if $\{\mathbf{a}, \mathbf{b}, \mathbf{c}\}$ are linearly dependent, there exist α, β, γ (not all zero) such that

$$\alpha \mathbf{a} + \beta \mathbf{b} + \gamma \mathbf{c} = \mathbf{0}. \tag{1.91}$$

But, such triple products are

$$[\alpha \mathbf{a} + \beta \mathbf{b} + \gamma \mathbf{c}, \mathbf{b}, \mathbf{c}] = 0 = \alpha[\mathbf{a}, \mathbf{b}, \mathbf{c}]. \tag{1.92}$$

The converse result follows from the fact that $\mathbf{a} \times \mathbf{b} = \mathbf{0}$ iff **a** and **b** are linearly dependent.

Return now to the possible solution of the equation $\mathbf{A} \cdot \mathbf{n} = \mathbf{0}$. Suppose $|\mathbf{A}| = 0$, then if $\{\mathbf{f}, \mathbf{g}, \mathbf{h}\}$ form a basis of $\mathcal{E}$, they are linearly dependent, *i.e.*,

$$\alpha(\mathbf{A} \cdot \mathbf{f}) + \beta(\mathbf{A} \cdot \mathbf{g}) + \gamma(\mathbf{A} \cdot \mathbf{h}) = \mathbf{0},$$

which leads to

$$\mathbf{A} \cdot (\alpha \mathbf{f} + \beta \mathbf{g} + \gamma \mathbf{h}) = \mathbf{0}.$$

Thus,

$$\mathbf{n} = \alpha \mathbf{f} + \beta \mathbf{g} + \gamma \mathbf{h}. \tag{1.93}$$

Conversely, if **n** satisfies $\mathbf{A} \cdot \mathbf{n} = \mathbf{0}$, then we can choose two vectors **l** and **m** that together with **n** form a basis of $\mathcal{E}$. Then

$$\det \mathbf{A} = [\mathbf{A} \cdot \mathbf{l}, \mathbf{A} \cdot \mathbf{m}, \mathbf{A} \cdot \mathbf{n}] = 0,$$

because $\mathbf{A} \cdot \mathbf{n} = \mathbf{0}$.

Thus, if $\mathbf{A} \cdot \mathbf{n} = \lambda^{(n)}\mathbf{n}$ is to have a solution for $\mathbf{n}$ with $\lambda^{(n)}$, then

$$\det(\mathbf{A} - \lambda^{(n)}\mathbf{I}) = 0. \tag{1.94}$$

Using an arbitrary basis $\{\mathbf{a}, \mathbf{b}, \mathbf{c}\}$, we obtain

$$[\mathbf{A} \cdot \mathbf{a} - \lambda^{(n)}\mathbf{a}, \mathbf{A} \cdot \mathbf{b} - \lambda^{(n)}\mathbf{b}, \mathbf{A} \cdot \mathbf{c} - \lambda^{(n)}\mathbf{c}] = 0, \tag{1.95}$$

which becomes

$$\lambda^3 - I_\mathbf{A}\lambda^2 + II_\mathbf{A}\lambda - III_\mathbf{A} = 0. \tag{1.96}$$

Equation (1.96), referred to as a characteristic equation, has three solutions.

1.14 Symmetric Tensors

Symmetric tensors, *e.g.*, $\mathbf{S}$, possess real eigenvalues and corresponding eigenvectors, $\{\lambda_1, \lambda_2, \lambda_3\}$ and $\{\mathbf{p}_1, \mathbf{p}_2, \mathbf{p}_3\}$, respectively. We may write $\mathbf{S}$ in the various forms such as $\mathbf{S} = \mathbf{S} \cdot \mathbf{I} = \mathbf{S} \cdot (\mathbf{p}_r\mathbf{p}_r) = \mathbf{S} \cdot \mathbf{p}_r\mathbf{p}_r = \lambda^{(r)}\mathbf{p}_r\mathbf{p}_r$. Thus the *spectral representation* of $\mathbf{S}$ is

$$\mathbf{S} = \lambda^{(r)}\mathbf{p}_r\mathbf{p}_r \quad (\text{sum on } r). \tag{1.97}$$

The invariants of $\mathbf{S}$ are

$$\begin{aligned} III_S &= \lambda_1\lambda_2\lambda_3, \\ II_S &= \lambda_1\lambda_2 + \lambda_2\lambda_3 + \lambda_1\lambda_3, \\ I_S &= \lambda_1 + \lambda_2 + \lambda_3. \end{aligned} \tag{1.98}$$

1.15 Positive Definiteness of a Tensor

If for any arbitrary vector $\mathbf{a}$, $\mathbf{a} \cdot \mathbf{A} \cdot \mathbf{a} \geq 0$, the tensor $\mathbf{A}$ is said to be *positive semidefinite*. If $\mathbf{a} \cdot \mathbf{A} \cdot \mathbf{a} > 0$, $\mathbf{A}$ is said to be *positive definite*.

Let $\mathbf{S}$ be a symmetric, positive semidefinite tensor with the associated eigenvectors and eigenvalues, $\mathbf{p}_i$ and λ_i. Then, as before,

$$\mathbf{S} = \lambda_r\mathbf{p}_r\mathbf{p}_r \quad (\text{sum on } r). \tag{1.99}$$

Now form the double products indicated above and note that

$$\mathbf{a} \cdot \mathbf{S} \cdot \mathbf{a} = \mathbf{a} \cdot \lambda_r\mathbf{p}_r\mathbf{p}_r \cdot \mathbf{a} = \lambda_r(\mathbf{a} \cdot \mathbf{p}_r)^2 \geq 0. \tag{1.100}$$

Since $\mathbf{a}$ is arbitrary, $\lambda_r \geq 0$ for all $r = 1, 2, 3$. It follows that

$$\mathbf{S}^{1/2} = \lambda_r^{1/2}\mathbf{p}_r\mathbf{p}_r. \tag{1.101}$$

Also,

$$\mathbf{S}^{-1} = \lambda_r^{-1}\mathbf{p}_r\mathbf{p}_r, \tag{1.102}$$

1.16. Antisymmetric Tensors

which is readily verified *via*

$$\mathbf{S} \cdot \mathbf{S}^{-1} = \lambda_r \mathbf{p}_r \mathbf{p}_r \cdot \lambda_s^{-1} \mathbf{p}_s \mathbf{p}_s \quad \text{(sum on r,s)}$$
$$= \lambda_r \lambda_s^{-1} \mathbf{p}_r \mathbf{p}_s \mathbf{p}_r \cdot \mathbf{p}_s = \lambda_r \lambda_s^{-1} \mathbf{p}_r \mathbf{p}_s \delta_{rs}$$
$$= \mathbf{p}_r \mathbf{p}_r = \mathbf{I}.$$

LEMMA 1.9: *If $\mathbf{A}$ is an arbitrary tensor, then $\mathbf{A}^T \cdot \mathbf{A}$ and $\mathbf{A} \cdot \mathbf{A}^T$ are positive semidefinite.*

Proof: Clearly,

$$(\mathbf{A}^T \cdot \mathbf{A})^T = \mathbf{A}^T \cdot (\mathbf{A}^T)^T = \mathbf{A}^T \cdot \mathbf{A} = \text{ symmetric,}$$
$$(\mathbf{A} \cdot \mathbf{A}^T)^T = (\mathbf{A}^T)^T \cdot \mathbf{A}^T = \mathbf{A} \cdot \mathbf{A}^T = \text{ symmetric.}$$

Thus,

$$\mathbf{a} \cdot \left[(\mathbf{A}^T \cdot \mathbf{A}) \cdot \mathbf{a}\right] = \mathbf{a} \cdot \left[\mathbf{A}^T \cdot (\mathbf{A} \cdot \mathbf{a})\right] = (\mathbf{a} \cdot \mathbf{A}^T) \cdot (\mathbf{A} \cdot \mathbf{a})$$
$$= (\mathbf{A} \cdot \mathbf{a}) \cdot (\mathbf{A} \cdot \mathbf{a}) \geq 0, \quad (1.103)$$

and

$$\mathbf{a} \cdot \left[(\mathbf{A} \cdot \mathbf{A}^T) \cdot \mathbf{a}\right] = (\mathbf{a} \cdot \mathbf{A}) \cdot (\mathbf{A}^T \cdot \mathbf{a}) = (\mathbf{a} \cdot \mathbf{A}) \cdot (\mathbf{a} \cdot \mathbf{A}) \geq 0. \quad (1.104)$$

1.16 Antisymmetric Tensors

If

$$\mathbf{W}^T = -\mathbf{W}, \quad (1.105)$$

the tensor $\mathbf{W}$ is said to be *antisymmetric*.

Let $\mathbf{a}$ and $\mathbf{b}$ be arbitrary vectors, then

$$\mathbf{b} \cdot (\mathbf{W} \cdot \mathbf{a}) = \mathbf{a} \cdot (\mathbf{W}^T \cdot \mathbf{b}) = -\mathbf{a} \cdot (\mathbf{W} \cdot \mathbf{b}). \quad (1.106)$$

Thus, for example, if $\mathbf{a} = \mathbf{b}$,

$$\mathbf{a} \cdot \mathbf{W} \cdot \mathbf{a} = 0. \quad (1.107)$$

Furthermore,

$$W_{ij} = \mathbf{e}_i \cdot \mathbf{W} \cdot \mathbf{e}_j = \begin{cases} 0, & \text{if } i = j, \\ W_{ij}, & \text{if } i \neq j, \\ -W_{ji}, & \text{if } i \neq j. \end{cases} \quad (1.108)$$

1.16.1 Eigenvectors of W

Examine

$$\mathbf{W} \cdot \mathbf{p} = \lambda \mathbf{p}, \quad (1.109)$$

where $\mathbf{p}$ is a unit vector. Form the product

$$\mathbf{p} \cdot \mathbf{W} \cdot \mathbf{p} = \lambda = 0. \quad (1.110)$$

Thus $\lambda = 0$ and $\mathbf{W} \cdot \mathbf{p} = \mathbf{0}$.

Let $\{\mathbf{q}, \mathbf{r}, \mathbf{p}\}$ be a unit orthonormal basis; then

$$\mathbf{p} = \mathbf{q} \times \mathbf{r}, \qquad \mathbf{q} = \mathbf{r} \times \mathbf{p},$$
$$\mathbf{r} = \mathbf{p} \times \mathbf{q}, \qquad [\mathbf{p}, \mathbf{q}, \mathbf{r}] = 1. \tag{1.111}$$

Recall that if $\{\mathbf{i}, \mathbf{j}\}$ is a pair of unit vectors from the set $\{\mathbf{q}, \mathbf{r}\}$, then

$$W_{ij} = \mathbf{i} \cdot \mathbf{W} \cdot \mathbf{j}, \quad \mathbf{W} = W_{ij}\mathbf{ij}. \tag{1.112}$$

Thus

$$\mathbf{W} = \omega(\mathbf{rq} - \mathbf{qr}), \tag{1.113}$$

which is readily verified *via*

$$\mathbf{W} \cdot \mathbf{p} = \omega(\mathbf{rq} - \mathbf{qr}) \cdot \mathbf{p} = \mathbf{0}. \tag{1.114}$$

The scalar ω is then obtained as

$$\omega = \mathbf{r} \cdot \mathbf{W} \cdot \mathbf{q} = -\mathbf{q} \cdot \mathbf{W} \cdot \mathbf{r}. \tag{1.115}$$

Now set $\mathbf{w} = \omega\mathbf{p}$, and let $\mathbf{a}$ be an arbitrary vector. We have

$$\mathbf{W} \cdot \mathbf{a} - \mathbf{w} \times \mathbf{a} = \omega(\mathbf{rq} - \mathbf{qr}) \cdot \mathbf{a} - \omega\mathbf{p} \times \mathbf{a}. \tag{1.116}$$

Next, write

$$\mathbf{a} = (\mathbf{a} \cdot \mathbf{p})\mathbf{p} + (\mathbf{a} \cdot \mathbf{q})\mathbf{q} + (\mathbf{a} \cdot \mathbf{r})\mathbf{r}, \tag{1.117}$$

and form

$$\mathbf{W} \cdot \mathbf{a} - \mathbf{w} \times \mathbf{a}$$
$$= \omega\left[(\mathbf{rq}) \cdot \mathbf{a} - (\mathbf{qr}) \cdot \mathbf{a} - \mathbf{p} \times (\mathbf{a} \cdot \mathbf{p})\mathbf{p} - \mathbf{p} \times (\mathbf{a} \cdot \mathbf{q})\mathbf{q} - \mathbf{p} \times (\mathbf{a} \cdot \mathbf{r})\mathbf{r}\right]$$
$$= \omega\left[(\mathbf{q} \cdot \mathbf{a})\{\mathbf{r} - \mathbf{p} \times \mathbf{q}\} - (\mathbf{r} \cdot \mathbf{a})\{\mathbf{q} + \mathbf{p} \times \mathbf{r}\}\right].$$

But the two terms in the $\{\ldots\}$ above are equal to zero, and so

$$\mathbf{W} \cdot \mathbf{a} - \mathbf{w} \times \mathbf{a} = \mathbf{0}. \tag{1.118}$$

So associated with $\mathbf{W}$ there is an *axial vector* $\mathbf{w} \equiv \omega\mathbf{p}$ such that

$$\mathbf{W} \cdot \mathbf{a} = \mathbf{w} \times \mathbf{a}, \tag{1.119}$$

where

$$\mathbf{W} \cdot \mathbf{p} = \mathbf{0}. \tag{1.120}$$

It is readily deduced from the above that

$$I_W = \mathrm{tr}\,\mathbf{W} = 0,$$
$$II_W = \omega^2, \tag{1.121}$$
$$III_W = \det \mathbf{W} = 0.$$

Also, if $\mathbf{u}$ and $\mathbf{v}$ are arbitrary vectors, then the tensor $\mathbf{L}$ defined as

$$\mathbf{L} \equiv \mathbf{uv} - \mathbf{vu} \tag{1.122}$$

1.17 Orthogonal Tensors

is an antisymmetric tensor and

$$\mathbf{w} = \mathbf{v} \times \mathbf{u} \tag{1.123}$$

is its axial vector.

1.17 Orthogonal Tensors

An orthogonal tensor $\mathbf{Q}$ has the property of preserving scalar products, *i.e.*,

$$(\mathbf{Q} \cdot \mathbf{a}) \cdot (\mathbf{Q} \cdot \mathbf{b}) = \mathbf{a} \cdot \mathbf{b}. \tag{1.124}$$

This property has the following effect

$$(\mathbf{Q} \cdot \mathbf{a}) \cdot (\mathbf{Q} \cdot \mathbf{b}) = (\mathbf{a} \cdot \mathbf{Q}^T) \cdot (\mathbf{Q} \cdot \mathbf{b}) = \mathbf{a} \cdot (\mathbf{Q}^T \cdot \mathbf{Q}) \cdot \mathbf{b} = \mathbf{a} \cdot \mathbf{b}.$$

Thus, we deduce that

$$\mathbf{Q}^T \cdot \mathbf{Q} = \mathbf{I}, \quad \text{or} \quad \mathbf{Q}^{-1} = \mathbf{Q}^T. \tag{1.125}$$

Note that

$$\mathbf{Q}^T \cdot (\mathbf{Q} - \mathbf{I}) = -(\mathbf{Q} - \mathbf{I})^T. \tag{1.126}$$

Since $\mathbf{Q}^T \cdot \mathbf{Q} = \mathbf{I}$ and $\det \mathbf{Q}^T = \det \mathbf{Q}$, we have

$$(\det \mathbf{Q})^2 = 1. \tag{1.127}$$

Thus, for a proper orthogonal (rotation) tensor,

$$\det \mathbf{Q} = 1. \tag{1.128}$$

The proper orthogonal tensor has one real eigenvalue, which is equal to 1. The corresponding eigenvector, $\mathbf{p}$, is parallel to the axis of rotation associated with $\mathbf{Q}$, *i.e.*,

$$\mathbf{Q} \cdot \mathbf{p} - 1\mathbf{p} = \mathbf{0}. \tag{1.129}$$

Now introduce $\mathbf{q}$ and $\mathbf{r}$ as before, *viz.*,

$$\mathbf{p} = \mathbf{q} \times \mathbf{r}, \quad \mathbf{q} = \mathbf{r} \times \mathbf{p}, \quad \mathbf{r} = \mathbf{p} \times \mathbf{q}. \tag{1.130}$$

Then,

$$\mathbf{Q} \cdot \mathbf{p} = \mathbf{p} = \mathbf{Q}^T \cdot \mathbf{p}, \quad (\text{because } \mathbf{Q}^T = \mathbf{Q}^{-1}). \tag{1.131}$$

In addition, we may deduce

$$\mathbf{q} \cdot (\mathbf{Q} \cdot \mathbf{p}) = 0 = \mathbf{r} \cdot (\mathbf{Q} \cdot \mathbf{p}) = \mathbf{p} \cdot (\mathbf{Q} \cdot \mathbf{q}) = \mathbf{p} \cdot (\mathbf{Q} \cdot \mathbf{r}) \tag{1.132}$$

and

$$\mathbf{p} \cdot \mathbf{Q} \cdot \mathbf{p} = 1. \tag{1.133}$$

Some other results that may be straightforwardly deduced are

$$(\mathbf{Q} \cdot \mathbf{q}) \cdot (\mathbf{Q} \cdot \mathbf{r}) = \mathbf{q} \cdot \mathbf{r},$$
$$(\mathbf{Q} \cdot \mathbf{q}) \cdot (\mathbf{Q} \cdot \mathbf{q}) = (\mathbf{q} \cdot \mathbf{Q}^T) \cdot (\mathbf{Q} \cdot \mathbf{q}) = \mathbf{q} \cdot \mathbf{q} = 1, \tag{1.134}$$
$$|\mathbf{Q} \cdot \mathbf{q}| = |\mathbf{Q} \cdot \mathbf{r}| = 1.$$

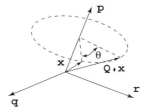

Figure 1.5. Geometric interpretation of an orthogonal tensor in terms of the rotation of a material fiber.

Thus, the pairs $(\mathbf{q}, \mathbf{r})$ and $(\mathbf{Q} \cdot \mathbf{q}, \mathbf{Q} \cdot \mathbf{r})$ are orthogonal to $\mathbf{p}$. Because of this property, we then can write

$$\mathbf{Q} \cdot \mathbf{q} = \alpha \mathbf{q} + \beta \mathbf{r}, \quad \mathbf{Q} \cdot \mathbf{r} = \gamma \mathbf{q} + \delta \mathbf{r}, \tag{1.135}$$

and

$$\alpha^2 + \beta^2 = 1, \quad \gamma^2 + \delta^2 = 1. \tag{1.136}$$

The determinant of $\mathbf{Q}$ is

$$\det(\mathbf{Q}) = [\mathbf{Q} \cdot \mathbf{p}, \mathbf{Q} \cdot \mathbf{q}, \mathbf{Q} \cdot \mathbf{r}] = [\mathbf{p}, \alpha \mathbf{q} + \beta \mathbf{r}, \gamma \mathbf{q} + \delta \mathbf{r}]$$
$$= \alpha \delta - \beta \gamma = 1. \tag{1.137}$$

Note also, since $(\mathbf{Q} \cdot \mathbf{q}) \cdot (\mathbf{Q} \cdot \mathbf{r}) = 0$, that

$$\alpha \gamma + \beta \delta = 0. \tag{1.138}$$

In addition,

$$\begin{aligned} \alpha^2 + \beta^2 = 1, \quad & \gamma^2 + \delta^2 = 1, \\ \alpha \delta - \beta \gamma = 1, \quad & \alpha \gamma + \beta \delta = 0. \end{aligned} \tag{1.139}$$

These are satisfied by the following assignments

$$\alpha = \delta = \cos \theta, \quad \beta = -\gamma = \sin \theta. \tag{1.140}$$

Therefore,

$$\begin{aligned} -\mathbf{q} \cdot \mathbf{Q} \cdot \mathbf{r} = \mathbf{r} \cdot \mathbf{Q} \cdot \mathbf{q} &= \sin \theta, \\ \mathbf{q} \cdot \mathbf{Q} \cdot \mathbf{q} = \mathbf{r} \cdot \mathbf{Q} \cdot \mathbf{r} &= \cos \theta, \end{aligned} \tag{1.141}$$

and thus

$$\mathbf{Q} = \mathbf{p}\mathbf{p} + \cos \theta (\mathbf{q}\mathbf{q} + \mathbf{r}\mathbf{r}) - \sin \theta (\mathbf{q}\mathbf{r} - \mathbf{r}\mathbf{q}). \tag{1.142}$$

This last result is depicted in Fig. 1.5, which shows the result of operating on a typical vector, $\mathbf{x}$. As shown, the effect of the tensor operation is to rotate $\mathbf{x}$ about the eigenvector $\mathbf{p}$ by the angle θ. That is, the result of operating with $\mathbf{Q}$ is to produce a rotated vector, $\mathbf{x}' = \mathbf{Q} \cdot \mathbf{x}$, as shown in the figure.

As expected, orthogonal tensors enter the discussion of material motion prominently with respect to describing rotations of bodies and material fibers. This will, for example, appear explicitly in our consideration of the the polar decomposition theorem introduced in the next section.

1.18. Polar Decomposition

As an example, with respect to the basis $\mathbf{e} = \{\mathbf{e}_1, \mathbf{e}_2, \mathbf{e}_3\}$, let $Q_{11} = Q_{22} = \cos\theta$, $Q_{12} = -Q_{21} = -\sin\theta$, $Q_{33} = 1$, and $Q_{ij} = 0$ otherwise. Equations (1.129) then become

$$\cos\theta p_1 + \sin\theta p_2 = p_1,$$
$$-\sin\theta p_1 + \cos\theta p_2 = p_2, \quad (1.143)$$
$$p_3 = p_3,$$

with $\mathbf{p} = p_i \mathbf{e}_i$. The solution to this set is trivially $p_1 = p_2 = 0$ and $p_3 = 1$; thus $\mathbf{p} = \mathbf{e}_3$.

Next choose $\mathbf{q} = \mathbf{e}_1$ and $\mathbf{r} = \mathbf{e}_2$ to satisfy the requirement of a right-handed triad basis, $\{\mathbf{p}, \mathbf{q}, \mathbf{r}\}$. Clearly then $Q_{qq} = \mathbf{q} \cdot \mathbf{Q} \cdot \mathbf{q} = \cos\theta = \mathbf{r} \cdot \mathbf{Q} \cdot \mathbf{r} = Q_{rr}$, $Q_{rq} = \mathbf{r} \cdot \mathbf{Q} \cdot \mathbf{q} = \sin\theta = -(\mathbf{q} \cdot \mathbf{Q} \cdot \mathbf{r}) = -Q_{qr}$, and hence

$$\mathbf{Q} = \mathbf{pp} + \cos\theta(\mathbf{qq} + \mathbf{rr}) - \sin\theta(\mathbf{qr} - \mathbf{rq}). \quad (1.144)$$

But $\mathbf{x} = x_s \mathbf{e}_s = (\mathbf{x} \cdot \boldsymbol{\xi}_s)\boldsymbol{\xi}_s$, where $\boldsymbol{\xi}_s = \{\mathbf{p}, \mathbf{q}, \mathbf{r}\}$ and where $x_1 = \cos\alpha$, $x_2 = \sin\alpha$ and $x_3 = 0$. The relationship $\mathbf{x}' = \mathbf{Q} \cdot \mathbf{x}$ leads to

$$\mathbf{x}' = [\mathbf{pp} + \cos\theta(\mathbf{qq} + \mathbf{rr}) - \sin\theta(\mathbf{qr} - \mathbf{rq})] \cdot (\cos\alpha \mathbf{q} + \sin\alpha \mathbf{r})$$
$$= \cos(\theta + \alpha)\mathbf{q} + \sin(\theta + \alpha)\mathbf{r}.$$

The result is exactly what was expected, namely that the fiber $\mathbf{x}$ inclined by θ to the $\mathbf{e}_1$ base vector is now rotated by α so that its total inclination is $\theta + \alpha$.

1.18 Polar Decomposition Theorem

Let $\mathbf{A}$ be an arbitrary tensor that possesses an inverse $\mathbf{A}^{-1}$. The following theorem, known as the *polar decomposition theorem*, will be useful in the analysis of finite deformations and the development of constitutive relations.

THEOREM 1.1: *An invertible second-order tensor $\mathbf{A}$ can be uniquely decomposed as*

$$\mathbf{A} = \mathbf{Q} \cdot \mathbf{U} = \mathbf{V} \cdot \mathbf{Q}, \quad (1.145)$$

where $\mathbf{Q}$ is an orthogonal tensor and $\mathbf{U}$ and $\mathbf{V}$ are symmetric, positive definite tensors.

Proof: Recall that the forms $\mathbf{A}^T \cdot \mathbf{A}$ and $\mathbf{A} \cdot \mathbf{A}^T$ are positive semidefinite, symmetric tensors. If $\mathbf{A}$ is invertible, *i.e.*, $\mathbf{A}^{-1}$ exists, then det $\mathbf{A} \neq 0$ and $\mathbf{A} \cdot \mathbf{n} \neq \mathbf{0}$ if $\mathbf{n} \neq \mathbf{0}$; also by this $\mathbf{A}^T \cdot \mathbf{n} \neq \mathbf{0}$. Recall also, that $\mathbf{A}^T \cdot \mathbf{A}$ and $\mathbf{A} \cdot \mathbf{A}^T$ have unique, positive square roots.

Let $\mathbf{U}$ be the square root of $\mathbf{A}^T \cdot \mathbf{A}$ and $\mathbf{V}$ be the square root of $\mathbf{A} \cdot \mathbf{A}^T$. But then $\mathbf{U}$ and $\mathbf{V}$ have unique inverses, $\mathbf{U}^{-1}$ and $\mathbf{V}^{-1}$. Consequently, if

$$\mathbf{A} = \mathbf{Q} \cdot \mathbf{U}, \quad \text{then} \quad \mathbf{Q} = \mathbf{A} \cdot \mathbf{U}^{-1}, \quad (1.146)$$

and if

$$\mathbf{A} = \mathbf{V} \cdot \mathbf{R}, \quad \text{then} \quad \mathbf{R} = \mathbf{V}^{-1} \cdot \mathbf{A}. \quad (1.147)$$

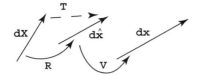

Figure 1.6. Decomposition of motions into a rotation and translation followed by a deformation.

Now,
$$\mathbf{Q}^T \cdot \mathbf{Q} = (\mathbf{A} \cdot \mathbf{U}^{-1})^T \cdot (\mathbf{A} \cdot \mathbf{U}^{-1})$$
$$= (\mathbf{U}^{-1})^T \cdot \mathbf{A}^T \cdot \mathbf{A} \cdot \mathbf{U}^{-1}$$
$$= \mathbf{U}^{-1} \cdot \mathbf{U}^2 \cdot \mathbf{U}^{-1}$$
$$= \mathbf{I} \cdot \mathbf{I} = \mathbf{I}.$$

Similarly, we find that $\mathbf{R}^T \cdot \mathbf{R} = \mathbf{I}$, and thus that $\mathbf{Q}$ and $\mathbf{R}$ are orthogonal tensors. With $\mathbf{U} = (\mathbf{A}^T \cdot \mathbf{A})^{1/2}$ and $\mathbf{V} = (\mathbf{A} \cdot \mathbf{A}^T)^{1/2}$ we have established that

$$\mathbf{A} = \mathbf{Q} \cdot \mathbf{U} = \mathbf{V} \cdot \mathbf{R}. \tag{1.148}$$

The question is now, are $\mathbf{Q}$ and $\mathbf{R}$ unique? Suppose we proposed another decomposition, $\mathbf{A} = \mathbf{Q}' \cdot \mathbf{U}'$. Then

$$\mathbf{A}^T \cdot \mathbf{A} = (\mathbf{Q}' \cdot \mathbf{U}')^T \cdot (\mathbf{Q}' \cdot \mathbf{U}') = (\mathbf{U}' \cdot \mathbf{Q}'^T) \cdot (\mathbf{Q}' \cdot \mathbf{U}') = (\mathbf{U}')^2.$$

But $(\mathbf{U}')^2 = \mathbf{U}^2$ and thus $\mathbf{U}' = \mathbf{U}$. This means that

$$\mathbf{Q}' = \mathbf{A} \cdot \mathbf{U}^{-1} = \mathbf{Q} \tag{1.149}$$

is unique! A similar consideration establishes the uniqueness of $\mathbf{R}$.

Finally, we ask if $\mathbf{Q} = \mathbf{R}$? To address this, write

$$\mathbf{A} = \mathbf{Q} \cdot \mathbf{U} = (\mathbf{R} \cdot \mathbf{R}^T) \cdot (\mathbf{V} \cdot \mathbf{R}) = \mathbf{R} \cdot (\mathbf{R}^T \cdot \mathbf{V} \cdot \mathbf{R})$$
$$= \mathbf{R} \cdot \left[(\mathbf{V}^{1/2} \cdot \mathbf{R})^T \cdot (\mathbf{V}^{1/2} \cdot \mathbf{R}) \right] = \mathbf{R} \cdot (\mathbf{R}^T \cdot \mathbf{V} \cdot \mathbf{R})$$
$$= \mathbf{Q} \cdot \mathbf{U}.$$

Therefore, $\mathbf{R} = \mathbf{Q}$ and $\mathbf{U} = \mathbf{R}^T \cdot \mathbf{V} \cdot \mathbf{R}$.

COROLLARY 1.1: *If $\mathbf{A}$ is an invertible tensor and $\mathbf{A} = \mathbf{Q} \cdot \mathbf{U} = \mathbf{V} \cdot \mathbf{Q}$, then*

$\mathbf{U}$ has eigenvectors $\mathbf{p}_i$ and eigenvalues λ_i,

$\mathbf{V}$ has eigenvectors $\mathbf{q}_i$ and eigenvalues μ_i,

where

$$\lambda_i = \mu_i \quad \text{and} \quad \mathbf{q}_i = \mathbf{Q} \cdot \mathbf{p}_i.$$

1.19 Polar Decomposition: Physical Approach

It is illustrative to view the motions involved in the polar decomposition discussed above as a sequential set of motions. This is schematically shown in Fig. 1.6.

1.19. Polar Decomposition: Physical Approach

1.19.1 Left and Right Stretch Tensors

In the first motion the element $d\mathbf{X} \to d\hat{\mathbf{x}}$, such that

$$d\hat{\mathbf{x}} = \mathbf{R} \cdot d\mathbf{X} + \mathbf{T}, \qquad (1.150)$$

where $\mathbf{T}$ is a translation vector and $\mathbf{R}$ is an orthogonal tensor representing a rotation. This gives for a deformation gradient,

$$\hat{\mathbf{F}} = \partial \hat{\mathbf{x}} / \partial \mathbf{X} = \mathbf{R}. \qquad (1.151)$$

Since there is no deformation in this first step, we have

$$\hat{\mathbf{C}} = \hat{\mathbf{F}}^T \cdot \hat{\mathbf{F}} = \mathbf{I} = \mathbf{R}^T \cdot \mathbf{R} \Rightarrow \mathbf{R} \text{ is orthogonal.} \qquad (1.152)$$

Next, let $d\hat{\mathbf{x}}$ be deformed into the element $d\mathbf{x}$ *via* a pure deformation,

$$d\mathbf{x} = \mathbf{V} \cdot d\hat{\mathbf{x}} = \mathbf{V} \cdot (\mathbf{R} \cdot d\mathbf{X} + \mathbf{T}) = (\mathbf{V} \cdot \mathbf{R}) \cdot d\mathbf{X} + \mathbf{V} \cdot \mathbf{T},$$

which yields

$$\mathbf{F} = \mathbf{V} \cdot \mathbf{R}. \qquad (1.153)$$

This is the left form of the polar decomposition. The tensor $\mathbf{V}$ is the left stretch tensor. Note that

$$\mathbf{C} = \mathbf{F}^T \cdot \mathbf{F} = (\mathbf{V} \cdot \mathbf{R})^T \cdot \mathbf{V} \cdot \mathbf{R} = \mathbf{R}^T \cdot \mathbf{V}^T \cdot \mathbf{V} \cdot \mathbf{R} = \mathbf{R}^T \cdot \mathbf{V}^2 \cdot \mathbf{R}. \qquad (1.154)$$

Alternatively, the total motion may be viewed as occurring first *via* a pure deformation given so that

$$d\hat{\mathbf{x}}' = \mathbf{U} \cdot d\mathbf{X}, \qquad (1.155)$$

followed by a rotation, $\mathbf{R}$, and a translation, $\mathbf{T}$. Theses motions result in

$$d\mathbf{x} = \mathbf{R} \cdot d\hat{\mathbf{x}}' + \mathbf{T} = (\mathbf{R} \cdot \mathbf{U}) \cdot d\mathbf{X} + \mathbf{T}, \qquad (1.156)$$

and

$$\mathbf{F} = \mathbf{R} \cdot \mathbf{U}. \qquad (1.157)$$

The tensor $\mathbf{U}$ is the right stretch tensor.

1.19.2 Principal Stretches

Write, again,

$$d\mathbf{x} = \mathbf{F} \cdot d\mathbf{X}, \quad d\mathbf{X} = \mathbf{N} dS, \quad d\mathbf{x} = \mathbf{n} ds, \qquad (1.158)$$

where all quantities should have obvious meanings. Then the principal stretches can be expressed as

$$\Lambda(\mathbf{N})\mathbf{n} = \mathbf{F} \cdot \mathbf{N} \quad \text{or} \quad \lambda(\mathbf{n})\mathbf{n} = \mathbf{F} \cdot \mathbf{N}, \qquad (1.159)$$

where $\Lambda(\mathbf{N}) = \lambda(\mathbf{n}) = ds/dS$. Since $\mathbf{N}$ and $\mathbf{n}$ are unit vectors, $\mathbf{N}$ is transformed into $\mathbf{n}$ *via*

$$\mathbf{n} = \mathbf{R} \cdot \mathbf{N}. \qquad (1.160)$$

This relation implies that

$$\mathbf{R} = \mathbf{nN}. \tag{1.161}$$

Thus, noting that $\mathbf{F} = \mathbf{R} \cdot \mathbf{U} = \mathbf{V} \cdot \mathbf{R}$, we have

$$\Lambda(\mathbf{N})\mathbf{R} \cdot \mathbf{N} = (\mathbf{R} \cdot \mathbf{U}) \cdot \mathbf{N} = \mathbf{F} \cdot \mathbf{N}$$

or

$$\mathbf{R} \cdot [\mathbf{U} - \Lambda(\mathbf{N})\mathbf{I}] \cdot \mathbf{N} = \mathbf{0}. \tag{1.162}$$

For arbitrary $\mathbf{R}$ this leads to

$$[\mathbf{U} - \Lambda(\mathbf{N})\mathbf{I}] \cdot \mathbf{N} = \mathbf{0} \tag{1.163}$$

for the principal stretch ratios, $\Lambda(\mathbf{N})$, in the *unrotated* principal directions, $\mathbf{N}$.

Similarly, we find that

$$\lambda(\mathbf{n})\mathbf{n} = (\mathbf{V} \cdot \mathbf{R}) \cdot \mathbf{N} = \mathbf{V} \cdot \mathbf{n}, \tag{1.164}$$

which yields

$$[\mathbf{V} - \lambda(\mathbf{n})\mathbf{I}] \cdot \mathbf{n} = \mathbf{0} \tag{1.165}$$

for the principal stretch ratios in the *rotated* principal directions, $\mathbf{n}$.

1.20 The Cayley–Hamilton Theorem

LEMMA 1.10: *Let $f(\lambda)$ be a real polynomial and $\mathbf{A}$ an arbitrary tensor, and let λ be an eigenvalue of $\mathbf{A}$. Then if $f(\mathbf{A})$ is the tensor obtained from the polynomial function of $\mathbf{A}$ constructed from the appropriate multiples (i.e., dot products) of $\mathbf{A}$, then $f(\lambda)$ is an eigenvalue of $f(\mathbf{A})$. Also an eigenvector of $f(\mathbf{A})$ associated with eigenvalue $f(\lambda)$ is an eigenvector of $\mathbf{A}$ associated with λ.*

Proof: The proof is obtained using simple induction. Let $\mathbf{p}$ be an eigenvector of $\mathbf{A}$ associated with λ *via* the equation $\mathbf{A} \cdot \mathbf{p} = \lambda \mathbf{p}$. Then,

$$\mathbf{A}^r \cdot \mathbf{p} = \lambda^r \mathbf{p} \tag{1.166}$$

holds for $r = 1$. Assume that this relation holds for $r = n$. If we prove that it also holds for $r = n + 1$, we would have shown that it holds by induction for all r. Therefore,

$$\mathbf{A}^{n+1} \cdot \mathbf{p} = \mathbf{A} \cdot (\mathbf{A}^n \cdot \mathbf{p}) = \mathbf{A} \cdot (\lambda^n \mathbf{p}) = \lambda^{n+1} \mathbf{p}, \tag{1.167}$$

and thus the relation holds for all r. Further, since $f(\mathbf{A})$ is a linear combination of powers of $\mathbf{A}$, it follows that $f(\lambda)$ will be an eigenvalue of $f(\mathbf{A})$ and $\mathbf{p}$ will be its associated eigenvector.

THEOREM 1.2: *Let A be an arbitrary tensor and the set of three λ's be its associated eigenvalues. Then, since the λ's satisfy (1.96), we have*

$$\lambda^3 - I_A \lambda^2 + II_A \lambda - III_A = 0. \tag{1.168}$$

1.22. Relations Involving ∇ Operator

Apply Lemma 1.10 to this characteristic polynomial. This leads to the result that the tensor $\mathbf{A}^3 - I_A \mathbf{A}^2 + II_A \mathbf{A} - III_A \mathbf{I} = \mathbf{0}$ has three eigenvalues each equal to 0. The Cayley–Hamilton theorem states that for arbitrary $\mathbf{A}$ this tensor is $\mathbf{0}$. This means that the tensor, $\mathbf{A}$, satisfies its own characteristic equation, i.e., (1.96).

1.21 Additional Lemmas for Tensors

LEMMA 1.11: *Consider the quadratic form*

$$\lambda = \mathbf{A} : (\mathbf{xx}) = A_{ij} x_i x_j. \tag{1.169}$$

Then

$$\begin{aligned} \partial \lambda / \partial \mathbf{x} &= 2 \operatorname{sym}(\mathbf{A}) \cdot \mathbf{x}, \\ \partial^2 \lambda / \partial \mathbf{x} \, \partial \mathbf{x} &= 2 \operatorname{sym}(\mathbf{A}), \end{aligned} \tag{1.170}$$

or, in component form,

$$\begin{aligned} \partial \lambda / \partial x_i &= (A_{ij} + A_{ji}) x_j, \\ \partial^2 \lambda / \partial x_i \partial x_j &= A_{ij} + A_{ji}. \end{aligned} \tag{1.171}$$

Proof: Consider

$$\partial \lambda / \partial x_k = A_{ij} (\partial x_i / \partial x_k) x_j + A_{ij} x_i (\partial x_j / \partial x_k). \tag{1.172}$$

Since $\partial x_i / \partial x_k = \delta_{ik}$, the substitution into the above, proves the first assertion. The rest is proved simply by continuing the argument.

LEMMA 1.12: *If $r^2 = \mathbf{x} \cdot \mathbf{x} = x_i x_i$ and $f(r)$ is an arbitrary function of r, then*

$$\nabla f(r) = f'(r) \mathbf{x} / r. \tag{1.173}$$

Proof: The components of $\nabla f(r)$ are $\partial f / \partial x_i$. Thus,

$$\frac{\partial f}{\partial x_i} = \frac{\partial f}{\partial r} \frac{\partial r}{\partial x_i} \tag{1.174}$$

and since $r^2 = x_j x_j$ and $\partial r^2 / \partial x_j = 2r \partial r / \partial x_j = 2 x_j$, it follows that $\partial r / \partial x_j = x_j / r$. Thus,

$$\frac{\partial f(r)}{\partial x_i} = \frac{\partial f}{\partial r} \frac{\partial r}{\partial x_i} = \frac{\partial f}{\partial r} \frac{x_i}{r}. \tag{1.175}$$

1.22 Identities and Relations Involving ∇ Operator

In this section several useful and illustrative results involving the gradient, and divergence, operation are given. They are listed as a collection of lemmas as follows.

LEMMA 1.13: *Let* **r** *be a position vector and thus* $\mathbf{r} = x_1 \mathbf{e}_1 + x_2 \mathbf{e}_2 + x_3 \mathbf{e}_3$. *Then*

$$\nabla \cdot \mathbf{r} = 3,$$
$$\nabla \times \mathbf{r} = \mathbf{0}, \qquad (1.176)$$
$$\nabla \cdot \left(\frac{\mathbf{r}}{r^3}\right) = 0.$$

Proof: The proof of the first assertion is straightforward, since $\partial x_i / \partial x_j = \delta_{ij}$ and

$$\nabla \cdot \mathbf{r} = \frac{\partial x_i}{\partial x_i} = 3. \qquad (1.177)$$

The curl of **r** is taken so that its ith component is $\epsilon_{ijk} \partial x_k / \partial x_j$. But $\partial x_k / \partial x_j = \delta_{kj}$ and this leads to $\epsilon_{ijk} \delta_{kj} = 0$, because $\epsilon_{ijk} = -\epsilon_{ikj}$ and $\delta_{kj} = \delta_{jk}$. This proves the second assertion. The third equation is expressed as

$$\nabla \cdot \left(\frac{\mathbf{r}}{r^3}\right) = \frac{\partial}{\partial x_i} \frac{x_i}{r^3} = 3\frac{1}{r^3} + x_i \frac{\partial r^{-3}}{\partial x_i}. \qquad (1.178)$$

Since $r^2 = x_i x_i$, we find

$$r \frac{\partial r}{\partial x_i} = x_i, \quad \frac{\partial r^{-3}}{\partial x_i} = -3r^{-4} \frac{\partial r}{\partial x_i} = -3r^{-5} x_i. \qquad (1.179)$$

Substituting this result in the expression for $\nabla \cdot (\mathbf{r}/r^3)$ given above proves the lemma.

LEMMA 1.14: *Let* $f(u, v)$ *be a scalar function of u and v, where $u = u(\mathbf{x})$ and $v = v(\mathbf{x})$. Then the gradient of f may be expressed as*

$$\nabla f(\mathbf{x}) = \frac{\partial f}{\partial u} \nabla u + \frac{\partial f}{\partial v} \nabla v. \qquad (1.180)$$

Proof: First write the gradient of f as

$$\nabla f = \frac{\partial f}{\partial x_i} \mathbf{e}_i. \qquad (1.181)$$

By the chain rule of calculus,

$$\frac{\partial f}{\partial x_i} \mathbf{e}_i = \left(\frac{\partial f}{\partial u} \frac{\partial u}{\partial x_i} + \frac{\partial f}{\partial v} \frac{\partial v}{\partial x_i}\right) \mathbf{e}_i. \qquad (1.182)$$

When this expression is expanded, and its terms reassembled, it is found that

$$\nabla f(\mathbf{x}) = \frac{\partial f}{\partial u} \nabla u + \frac{\partial f}{\partial v} \nabla v, \qquad (1.183)$$

as desired.

LEMMA 1.15: *Let* $f(\mathbf{x})$ *be a scalar field and* $\mathcal{G}$ *a vector field. Then*

$$\nabla \cdot (f\mathcal{G}) = \nabla f \cdot \mathcal{G} + f \nabla \cdot \mathcal{G}. \qquad (1.184)$$

Proof: We again use the chain rule of calculus to write

$$\nabla \cdot (f\mathcal{G}) = \frac{\partial (f\mathcal{G}_i)}{\partial x_i} = \frac{\partial f}{\partial x_i}\mathcal{G}_i + f\frac{\partial \mathcal{G}_i}{\partial x_i} \qquad (1.185)$$
$$= \nabla f \cdot \mathcal{G} + f \nabla \cdot \mathcal{G}.$$

LEMMA 1.16: *If **u** is a vector field then,*

$$\nabla \cdot (\nabla \times \mathbf{u}) = 0, \qquad (1.186)$$

or, in other words, the divergence of the curl vanishes.

Proof: When expressed in component form, this is

$$\frac{\partial}{\partial x_i} \epsilon_{ijk} \frac{\partial u_k}{\partial x_j} = \epsilon_{ijk} \frac{\partial^2 u_k}{\partial x_i \partial x_j}. \qquad (1.187)$$

But this will vanish because the second derivatives of **u** are symmetric, whereas ϵ_{ijk} is antisymmetric in ij, so that $\epsilon_{ijk} u_{k,ij} = 0$.

1.23 Suggested Reading

There are a number of excellent texts and reference books concerned with vector and tensor algebra in physics and applied mathematics. Those that are specifically concerned with or provide a framework directly relevant to applications to solid mechanics include Synge and Shild (1949), Ericksen (1960), Brillouin (1964), Wrede (1972), and Boehler (1987). The reader is also directed to Chadwick (1999) for an excellent summary of such results as applied to the analysis of deformation and forces. The following is a list of books recommended for additional reading.

Boehler, J. P. (1987), *Application of Tensor Functions in Solid Mechanics*, Springer-Verlag, Wien.
Brillouin, L. (1964), *Tensors in Mechanics and Elasticity*, Academic Press, New York.
Chadwick, P. (1999), *Continuum Mechanics: Concise Theory and Problems*, Dover Publications, Mineola, New York.
Ericksen, J. L. (1960), Tensor Fields. In *Hanbuch der Physik* (S. Flugge, ed.), Band III/1, Springer-Verlag, Berlin.
Eringen, A. C. (1971), Tensor Analysis. In *Continuum Physics* (A. C. Eringen, ed.), Vol. 1, Academic Press, New York.
Spencer, A. J. M. (1971), Theory of Invariants. In *Continuum Physics* (A. C. Eringen, ed.), Vol. 1, Academic Press, New York.
Synge, J. L., and Schild, A. (1949), *Tensor Calculus*, University Press, Toronto.
Wrede, R. C. (1972), *Introduction to Vector and Tensor Analysis*, Dover Publications, New York.

2 Basic Integral Theorems

2.1 Gauss and Stokes's Theorems

The divergence theorem of Gauss may be expressed as follows:

THEOREM 2.1: *If V is a volume bounded by the closed surface S and $\mathbf{A}$ is a vector field that possesses continuous derivatives (and is singled valued in V), then*

$$\int_V \nabla \cdot \mathbf{A}\, dV = \int_S \mathbf{A} \cdot \mathbf{n}\, dS = \int_S \mathbf{A} \cdot d\mathbf{S}, \tag{2.1}$$

where $\mathbf{n}$ is the outward pointing unit normal vector to S. Note that we may extend this result to the case where $\mathbf{A}$ is a tensor field with the same proviso's. The basic theorem is proven below.

Stokes's theorem takes the following form:

THEOREM 2.2: *If S is an open, two-sided surface bounded by a closed, nonintersecting curve C, and if $\mathbf{A}$ again has continuous derivatives, then*

$$\oint_C \mathbf{A} \cdot d\mathbf{r} = \int_S (\nabla \times \mathbf{A}) \cdot \mathbf{n}\, dS = \int_S (\nabla \times \mathbf{A}) \cdot d\mathbf{S}, \tag{2.2}$$

where C is measured positive if the motion along it is counterclockwise (then S would be on the left).

A special case of Stokes's theorem for the plane is

THEOREM 2.3: *If S is a closed region of the $x - y$ plane bounded by a closed curve S, and if $M(x, y)$ and $N(x, y)$ are two continuous functions having continuous derivatives in S, then*

$$\oint_C M(x, y)\, dx + N(x, y)\, dy = \int_S \left(\frac{\partial N}{\partial x} - \frac{\partial M}{\partial y} \right) dS, \tag{2.3}$$

where C is traversed in the same positive direction.

2.2. Vector and Tensor Fields

2.1.1 Applications of Divergence Theorem

A useful application of the divergence theorem involves the computation of the volume of a solid. Consider the integral

$$\int_S \mathbf{r} \cdot \mathbf{n} \, dS, \tag{2.4}$$

where $\mathbf{r}$ is the position vector and all other quantities have their obvious meanings as above. The divergence theorem states that

$$\int_S \mathbf{r} \cdot \mathbf{n} \, dS = \int_V \nabla \cdot \mathbf{r} \, dV$$
$$= \int_V \nabla \cdot (x_i \mathbf{e}_i) \, dV = \int_V \frac{\partial x_i}{\partial x_i} \, dV = 3V. \tag{2.5}$$

Next, consider the integral

$$\int_V \nabla \varphi \, dV, \tag{2.6}$$

where φ is a scalar function of position. To interpret this, define a vector $\mathbf{a} = \varphi \mathbf{b}_0$ where $\mathbf{b}_0$ is an arbitrary *constant* vector. Then,

$$\int_V \nabla \cdot (\varphi \mathbf{b}_0) \, dV = \int_S \varphi \mathbf{b}_0 \cdot \mathbf{n} \, dS$$
$$= \int_V \mathbf{b}_0 \cdot \nabla \varphi \, dV \tag{2.7}$$
$$= \int_S \mathbf{b}_0 \cdot (\varphi \mathbf{n}) \, dS.$$

Now take the constant vector, $\mathbf{b}_0$, out of the integrals and note that it was indeed arbitrary. Thus, what is established is

$$\int_V \nabla \varphi \, dV = \int_S \varphi \mathbf{n} \, dS. \tag{2.8}$$

Clearly other useful integral lemmas may be derived by similar manipulations of the Gauss or Stokes theorems.

2.2 Vector and Tensor Fields: Physical Approach

Consider, as an example, fields that arise from electrostatics and Coulomb's law. If $\mathbf{f}$ is the force between two charged points at distance r with charges q and q_0, its representation is given as

$$\mathbf{f} = \frac{1}{4\pi \epsilon_0} \frac{q q_0}{r^2} \hat{\mathbf{u}}. \tag{2.9}$$

The permeability of the vacuum is ϵ_0. As shown in Fig. 2.1, $\hat{\mathbf{u}}$ is the unit vector from charge q to charge q_0 and r is the distance between them.

Now introduce a *vector field* $\mathbf{E}(x_1, x_2, x_3)$ such that

$$\mathbf{E}(x_1, x_2, x_3) = \mathbf{E}(\mathbf{r}) = \frac{1}{4\pi \epsilon_0} \frac{q}{r^2} \hat{\mathbf{u}}. \tag{2.10}$$

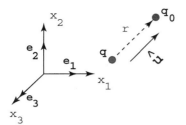

Figure 2.1. Interaction of point charges.

Thus the force exerted on charge q_0 from the *field* of charge q can be expressed as

$$\mathbf{f} = \mathbf{E} q_0. \tag{2.11}$$

Now let there be as many as n charges, each with a net charge q_i. The force they exert on the test charge q_0 is given by the linear superposition,

$$\mathbf{f}(\mathbf{r}) = \left(\sum_{i=1}^{n} \frac{q_i}{4\pi \epsilon_0 |\mathbf{r} - \mathbf{r}_i|^2} \hat{\mathbf{u}} \right) q_0. \tag{2.12}$$

Thus, we define the field arising from this distribution of charges as

$$\mathbf{E}(\mathbf{r}) = \sum_{i=1}^{n} \frac{q_i}{4\pi \epsilon_0 |\mathbf{r} - \mathbf{r}_i|^2} \hat{\mathbf{u}}. \tag{2.13}$$

Finally, if we define a continuous *charge density* by the limit $\rho = \lim_{\Delta v \to 0} \frac{\Delta q}{\Delta v}$, the total charge in volume V is

$$Q = \int_V \rho(x_1, x_2, x_3) \, dV. \tag{2.14}$$

The total field arising from this continuous distribution of charge is, therefore,

$$\mathbf{E}(\mathbf{r}) = \frac{1}{4\pi \epsilon_0} \int_V \frac{\rho(x_1, x_2, x_3) \hat{\mathbf{u}}}{|\mathbf{r} - \mathbf{r}'|^2} \, dV. \tag{2.15}$$

Such is an example of a simple vector field.

2.3 Surface Integrals: Gauss Law

We will be interested in integrals evaluated over a surface of the general form, such as arises in the Gauss law

$$\int_S \mathbf{E}(\mathbf{x}) \cdot \mathbf{n} \, dS = q/\epsilon, \tag{2.16}$$

where $\mathbf{n}$ is the *outward pointing* unit normal vector to the surface S. The normal to the surface can be defined as

$$\mathbf{n} = \mathbf{u} \times \mathbf{v}, \tag{2.17}$$

where $\mathbf{u}$ and $\mathbf{v}$ are locally unit vectors, tangent to the surface element as sketched in Fig. 2.2. If $\mathbf{u}$ and $\mathbf{v}$ are not unit vectors, then

$$\mathbf{n} = \frac{\mathbf{u} \times \mathbf{v}}{|\mathbf{u} \times \mathbf{v}|}. \tag{2.18}$$

2.4. Evaluating Surface Integrals

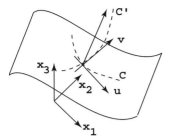

Figure 2.2. Defining the local normal to a surface.

Let the surface be given by

$$g(x_1, x_2, x_3) = 0 \quad \Rightarrow \quad x_3 = f(x_1, x_2). \tag{2.19}$$

Imagine following the curve over an arc C, along x_1. The corresponding displacement undergone would be

$$\mathbf{u} = u_1 \mathbf{e}_1 + (\partial f / \partial x_1) u_1 \mathbf{e}_3. \tag{2.20}$$

Similarly, on an arc C', along x_2, we obtain

$$\mathbf{v} = v_2 \mathbf{e}_2 + (\partial f / \partial x_2) v_2 \mathbf{e}_3. \tag{2.21}$$

Thus,

$$\mathbf{N} = \mathbf{u} \times \mathbf{v} = (-\partial f / \partial x_1 \mathbf{e}_1 - \partial f / \partial x_2 \mathbf{e}_2 + \mathbf{e}_3) u_1 v_2,$$

$$\mathbf{n} = \frac{\mathbf{u} \times \mathbf{v}}{|\mathbf{u} \times \mathbf{v}|} = \frac{-\partial f / \partial x_1 \mathbf{e}_1 - \partial f / \partial x_2 \mathbf{e}_2 + \mathbf{e}_3}{[1 + (\partial f / \partial x_1)^2 + (\partial f / \partial x_2)^2]^{1/2}}. \tag{2.22}$$

2.4 Evaluating Surface Integrals

Examine the integral

$$\int_S \mathbf{G}(x_1, x_2, x_3) \, dS, \tag{2.23}$$

where $\mathbf{G}$ may be either a scalar or tensor field. Such integrals may be considered as the limit of a summation over an infinite number of infinitesimal surface elements (Fig. 2.3), *i.e.*,

$$\int_S \mathbf{G}(x_1, x_2, x_3) \, dS = \lim_{\substack{N \to \infty \\ \Delta S_i \to 0}} \sum_{i=1}^{N} G(x_1^i, x_2^i, x_3^i) \Delta S_i. \tag{2.24}$$

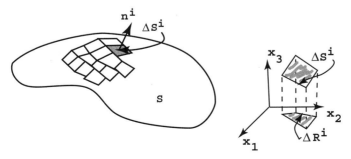

Figure 2.3. Limiting approach to evaluating surface integrals.

We want to relate these surface elements to the projections they make on the $x_1 - x_2$ coordinate plane. Note that if ΔR_i is the projection of ΔS_i, then

$$\Delta S_i = \Delta R_i/(\mathbf{n} \cdot \mathbf{e}_3). \tag{2.25}$$

Thus,

$$\int_S \mathbf{G}(x_1, x_2, x_3)\,dS = \lim_{\substack{N \to \infty \\ \Delta R_i \to 0}} \sum_{i=1}^N G(x_1^i, x_2^i, x_3^i)\Delta R_i/(\mathbf{n} \cdot \mathbf{e}_3), \tag{2.26}$$

which in the limit becomes

$$\int\int_R \frac{G(x_1, x_2, x_3)}{\mathbf{n} \cdot \mathbf{e}_3}\,dx_1\,dx_2. \tag{2.27}$$

Since on S, $x_3 = f(x_1, x_2)$, and recalling the result from (2.22) for $\mathbf{n}$, (2.26) becomes

$$\int_S G(x_1, x_2, x_3)\,dS = \int\int_R G(x_1, x_2, f)\left[1 + (\partial f/\partial x_1)^2 + (\partial f/\partial x_2)^2\right]^{1/2} dx_1\,dx_2. \tag{2.28}$$

If, instead, we had used projections on the $x_1 - x_3$ plane, we would have

$$\int_S G(x_1, x_2, x_3)\,dS = \int\int_R G[x_1, g(x_1, x_2), x_3]\left[1 + (\partial g/\partial x_1)^2 + (\partial g/\partial x_2)^2\right]^{1/2} dx_1\,dx_3, \tag{2.29}$$

where

$$x_2 = g(x_1, x_3). \tag{2.30}$$

Alternatively, on the $x_2 - x_3$ plane,

$$\int_S G(x_1, x_2, x_3)\,dS = \int\int_R G[h(x_2, x_3), x_2, x_3]\left[1 + (\partial h/\partial x_2)^2 + (\partial h/\partial x_3)^2\right]^{1/2} dx_2\,dx_3, \tag{2.31}$$

where

$$x = h(x_2, x_3). \tag{2.32}$$

We are, however, even more interested in integrals of the form,

$$\int_S \mathbf{F} \cdot \mathbf{n}\,dS, \tag{2.33}$$

such as appears in (2.16). Replace $\mathbf{G}$ with $\mathbf{F} \cdot \mathbf{n}$ in (2.28) to obtain

$$\int_S \mathbf{F} \cdot \mathbf{n}\,dS = \int\int_R \mathbf{F} \cdot \mathbf{n}\left[1 + (\partial f/\partial x_1)^2 + (\partial f/\partial x_2)^2\right]^{1/2} dx_1\,dx_2. \tag{2.34}$$

The substitution of (2.22) into the above gives

$$\int_S \mathbf{F} \cdot \mathbf{n}\,dS = \int\int_R \{-F_1(\partial f/\partial x_1) - F_2(\partial f/\partial x_2) + F_3[x_1, x_2, f(x_1, x_2)]\}\,dx_1\,dx_2,$$

or

$$\int_S \mathbf{F} \cdot \mathbf{n}\,dS = \int\int_R \{-F_1[x_1, x_2, f(x_1, x_2)](\partial f/\partial x_1) - F_2[x_1, x_2, f(x_1, x_2)](\partial f/\partial x_2)$$
$$+ F_3[x_1, x_2, f(x_1, x_2)]\}\,dx_1\,dx_2,$$

2.5. The Divergence

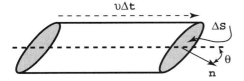

Figure 2.4. Flux of fluid through a cylinder.

where
$$\mathbf{F} = F_1 \mathbf{e}_1 + F_2 \mathbf{e}_2 + F_3 \mathbf{e}_3. \tag{2.35}$$

2.4.1 Application of the Concept of Flux

The flux of the vector field $\mathbf{F}$ through the surface S is defined by

$$\int_S \mathbf{F} \cdot \mathbf{n} \, dS, \tag{2.36}$$

where $\mathbf{n}$ is a unit outward normal to S. For example, consider the flow of fluid through a cylinder, during a time increment Δt, as depicted in Fig. 2.4. With v as the velocity, the distance travelled in time Δt is $v\Delta t$, where v is the magnitude of v. The volume of fluid flowing through the surface inclined as shown is equal to $v\Delta t \cos\theta$. The mass of fluid flowing through the inclined surface is, with ρ as the mass density, $\rho v \Delta t \cos\theta = \rho \Delta t v \cdot \mathbf{n}$, and thus the rate of mass flow through the inclined surface is $\rho v \cdot \mathbf{n} \Delta S$. As shown in the sketch, ΔS is the area of the inclined surface element. Thus, if we define

$$\mathbf{F} \equiv \rho v(x_1, x_2, x_3), \tag{2.37}$$

we find that

$$\text{Rate of mass flow through } S = \int_S \mathbf{F} \cdot \mathbf{n} \, dS. \tag{2.38}$$

2.5 The Divergence

Recall the Gauss law

$$\int_S \mathbf{E}(\mathbf{x}) \cdot \mathbf{n} \, dS = q/\epsilon_0. \tag{2.39}$$

Let $q = \bar{\rho}_{\Delta V} \Delta V$, where $\bar{\rho}_{\Delta V}$ is the average charge density in ΔV. Then,

$$\int_S \mathbf{E}(\mathbf{x}) \cdot \mathbf{n} \, dS = \bar{\rho}_{\Delta V} \Delta V / \epsilon_0,$$
$$\frac{1}{\Delta V} \int_S \mathbf{E}(\mathbf{x}) \cdot \mathbf{n} \, dS = \bar{\rho}_{\Delta V} / \epsilon_0. \tag{2.40}$$

Taking an appropriate limit, there follows

$$\lim_{\Delta V \to 0} \frac{1}{\Delta V} \int_S \mathbf{E}(\mathbf{x}) \cdot \mathbf{n} \, dS = \rho(x_1, x_2, x_3)/\epsilon_0. \tag{2.41}$$

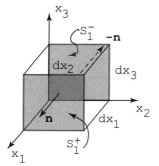

Figure 2.5. A cuboid with edges aligned with the coordinate axes.

Now, define the *divergence* of **F** to be

$$\operatorname{div} \mathbf{F} \equiv \lim_{\Delta V \to 0} \frac{1}{\Delta V} \int_S \mathbf{F} \cdot \mathbf{n} \, dS. \tag{2.42}$$

The Gauss law then becomes

$$\operatorname{div} \mathbf{F} = \rho/\epsilon_0. \tag{2.43}$$

We next calculate div **F** over a cuboid whose volume shrinks to zero (Fig. 2.5). Calculate first

$$\int_{S_1^\pm} \mathbf{F} \cdot \mathbf{n} \, dS, \tag{2.44}$$

on S_1^+, where $\mathbf{F} \cdot \mathbf{n} = \mathbf{F} \cdot \mathbf{e}_1 = F_1(x_1, x_2, x_3)$. Take the centroid of the cuboid at (x_1, x_2, x_3). On S_1^+ we have

$$\mathbf{F} \cdot \mathbf{n} = F_1 \approx F_1(x_1, x_2, x_3) + (\partial F_1/\partial x_1) dx_1/2, \tag{2.45}$$

while on S_1^-,

$$\mathbf{F} \cdot \mathbf{n} = -F_1 \approx -F_1(x_1, x_2, x_3) + (\partial F_1/\partial x_1) dx_1/2. \tag{2.46}$$

Thus, we obtain

$$\int_{S_1^+} + \int_{S_1^-} = [\partial F_1(x_1, x_2, x_3)/\partial x_1] \, dx_1 \, dx_2 \, dx_3 = [\partial F_1(x_1, x_2, x_3)/\partial x_1] dx_1 \Delta S_1^\pm,$$

where $\Delta S_1^\pm = dx_2 \, dx_3$. This also shows that

$$\lim_{\Delta V \to 0} \frac{1}{\Delta V} \int_{S_1^\pm} \mathbf{F} \cdot \mathbf{n} \, dS = \partial F_1/\partial x_1. \tag{2.47}$$

By the same sort of argument we have

$$\lim_{\Delta V \to 0} \frac{1}{\Delta V} \int_{S_2^\pm} \mathbf{F} \cdot \mathbf{n} \, dS = \partial F_2/\partial x_2, \tag{2.48}$$

and

$$\lim_{\Delta V \to 0} \frac{1}{\Delta V} \int_{S_3^\pm} \mathbf{F} \cdot \mathbf{n} \, dS = \partial F_3/\partial x_3. \tag{2.49}$$

2.6. Divergence Theorem

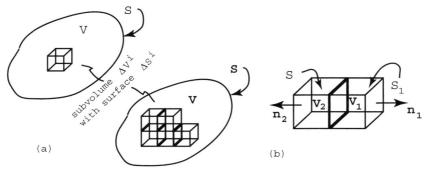

Figure 2.6. Limiting approach to evaluating surface integrals.

Consequently,
$$\text{div}(\mathbf{F}) = \partial F_1/\partial x_1 + \partial F_2/\partial x_2 + \partial F_3/\partial x_3. \tag{2.50}$$

Letting
$$\nabla \equiv \partial/\partial x_1 \mathbf{e}_1 + \partial/\partial x_2 \mathbf{e}_2 + \partial/\partial x_3 \mathbf{e}_3, \tag{2.51}$$

we can write
$$\nabla \cdot \mathbf{F} = (\partial/\partial x_1 \mathbf{e}_1 + \partial/\partial x_2 \mathbf{e}_2 + \partial/\partial x_3 \mathbf{e}_3) \cdot (F_1 \mathbf{e}_1 + F_2 \mathbf{e}_2 + F_3 \mathbf{e}_3)$$
$$= \partial F_1/\partial x_1 + \partial F_2/\partial x_2 + \partial F_3/\partial x_3. \tag{2.52}$$

This yields another expression of the Gauss law, namely
$$\nabla \cdot \mathbf{E} = \rho/\epsilon_0. \tag{2.53}$$

2.6 Divergence Theorem: Relation of Surface to Volume Integrals

Assume that
$$\int_S \mathbf{F} \cdot \mathbf{n} \, dS = \sum_{i=1}^{N} \int_{S_i} \mathbf{F} \cdot \mathbf{n} \, dS, \tag{2.54}$$

as sketched in Fig. 2.6. The two subsurface integrals over S_1 and S_2 include the integrals over the subsurface facet common to both small subelements. The value of $\mathbf{F}$ is the same at each point on this common facet (if $\mathbf{F}$ is continuous). But the outward pointing normals $\mathbf{n}$ on such a common facet are such that at each point, $\mathbf{n}_1 = -\mathbf{n}_2$. Thus, the integrals over the facet cancel. With this in mind, rewrite (2.54) as

$$\int_S \mathbf{F} \cdot \mathbf{n} \, dS = \sum_{i=1}^{N} \left(\frac{1}{\Delta V_i} \int_{S_i} \mathbf{F} \cdot \mathbf{n} \, dS \right) \Delta V_i. \tag{2.55}$$

Taking the limit as $N \to \infty$, or as $\Delta V_i \to 0$, we have

$$\int_S \mathbf{F} \cdot \mathbf{n} \, dS = \lim_{\substack{N \to \infty \\ \Delta V_i \to 0}} \sum_{i=1}^{N} \left(\frac{1}{\Delta V_i} \int_{S_i} \mathbf{F} \cdot \mathbf{n} \, dS \right) \Delta V_i. \tag{2.56}$$

But the term within (...) in (2.56) is in fact $\nabla \cdot \mathbf{F}$ and thus

$$\int_S \mathbf{F} \cdot \mathbf{n} \, dS = \int_V \nabla \cdot \mathbf{F} \, dV. \tag{2.57}$$

This is the Gauss divergence theorem.

To illustrate its application, consider the rate of flow of mass through a surface S, i.e.,

$$\text{flow rate} = \int_S \rho v \cdot \mathbf{n} \, dS. \tag{2.58}$$

Let the surface S be closed. Then we can write

$$\text{amount of mass in V} = \int_V \rho \, dV, \tag{2.59}$$

and

$$\frac{\text{change of mass within V}}{\text{unit time}} = \int_V \partial \rho / \partial t \, dV. \tag{2.60}$$

The conservation of mass requires that

$$\int_V \partial \rho / \partial t \, dV = -\int_S \rho v \cdot \mathbf{n} \, dS. \tag{2.61}$$

Applying the divergence theorem to the second integral in (2.61), we obtain

$$\int_S (\rho v) \cdot \mathbf{n} \, dS = \int_V \nabla \cdot (\rho v) \, dV. \tag{2.62}$$

Consequently,

$$\int_V \partial \rho / \partial t \, dV = -\int_V \nabla \cdot (\rho v) \, dV. \tag{2.63}$$

Since this is true for all V, we conclude that

$$\partial \rho / \partial t = -\nabla \cdot (\rho v), \tag{2.64}$$

which is known as a *continuity equation*.

2.7 More on Divergence Theorem

Another form of the divergence theorem of interest for later application can be stated as follows.

LEMMA 2.1: *If S is the surface bounding the volume V, and $\mathbf{n}$ is its unit outward pointing normal, and $\mathbf{u}(\mathbf{x})$ and $\mathbf{T}(\mathbf{x})$ are arbitrary vector and tensor fields in V, then*

$$\int_S \mathbf{u} \left(\mathbf{T}^T \cdot \mathbf{n} \right) dS = \int_V [\mathbf{u} \nabla \cdot \mathbf{T} + (\nabla \mathbf{u}) \cdot \mathbf{T}] \, dV. \tag{2.65}$$

Some special cases immediately follow. For example, if $\mathbf{u}$ is a fixed vector, then

$$\mathbf{u} \int_S \mathbf{T}^T \cdot \mathbf{n} \, dS = \mathbf{u} \int_V \nabla \cdot \mathbf{T} \, dV. \tag{2.66}$$

Because **u** is arbitrary, (2.66) implies that

$$\int_S \mathbf{T}^T \cdot \mathbf{n} \, dS = \int_V \nabla \cdot \mathbf{T} \, dV. \tag{2.67}$$

Suppose further that $\mathbf{T} = \mathbf{I}$. Then (2.66) becomes

$$\int_S \mathbf{u}\mathbf{n} \, dS = \int_V \nabla \mathbf{u} \, dV. \tag{2.68}$$

By taking the trace of both sides of (2.68), we obtain

$$\int_S \mathbf{u} \cdot \mathbf{n} \, dS = \int_V \nabla \cdot \mathbf{u} \, dV. \tag{2.69}$$

It should be noted that (2.65) can, in fact, be derived from (2.69) by setting **u** to be $(\mathbf{u} \cdot \mathbf{a})(\mathbf{T} \cdot \mathbf{b})$, where **a** and **b** are arbitrary vectors.

2.8 Suggested Reading

Malvern, L. E. (1969), *Introduction to the Mechanics of a Continuous Medium*, Prentice Hall, Englewood Cliffs, New Jersey.

Marsden, J. E., and Tromba, A. J. (2003), *Vector Calculus*, 5th ed., W. H. Freeman and Company, New York.

Truesdell, C., and Toupin, R. (1960), The Classical Field Theories. In *Handbuch der Physik* (S. Flügge, ed.), Band III/1, Springer-Verlag, Berlin.

Wrede, R. C. (1972), *Introduction to Vector and Tensor Analysis*, Dover, New York.

[Handwritten note at top: SINCE f(x) THAT WE'LL BE USING IS CONTINUOUS ON [-c, c] BUT ACTUAL LOAD IS ONLY CONTINUOUS ON [0, +c], WE CAN DO WHATEVER WE WANT ON [-c, 0]. USUALLY IT IS MOST CONVENIENT TO PROVIDE EITHER SYMMETRY OR ANTISYMMETRY]

3 Fourier Series and Fourier Integrals

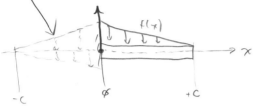

3.1 Fourier Series

Let $f(x)$ be a continuous, integrable function defined on the interval $[-c, c]$. Consider the Fourier series of $f(x)$, viz.,

[Handwritten: ARBITRARY / JUST MAKES IT CONVENIENTLY SYMMETRIC]

$$f(x) = (a_0/2) + \sum_{k=1}^{\infty}[a_k \cos(k\pi x/c) + b_k \sin(k\pi x/c)]. \tag{3.1}$$

The coefficients, a_k and b_k, indexed by the integers k, can be identified as follows. Multiply each side of (3.1) by $\cos(n\pi x/c)$, n being an integer, and integrate over $[-c, c]$ to obtain

$$\int_{-c}^{c} f(x) \cos(n\pi x/c)\, dx = \int_{-c}^{c} (a_0/2) \cos(n\pi x/c)\, dx$$
$$+ \int_{-c}^{c} \sum_{k=1}^{\infty}[a_k \cos(k\pi x/c) \cos(n\pi x/c)$$
$$+ b_k \sin(k\pi x/c) \cos(n\pi x/c)]\, dx.$$

To proceed, we note that

$$\int_{-c}^{c} \cos(k\pi x/c) \cos(n\pi x/c)\, dx = c/\pi \int_{-\pi}^{\pi} \cos(k\lambda) \cos(n\lambda)\, d\lambda, \tag{3.2}$$

where

$$\lambda = \pi x/c, \quad d\lambda = (\pi/c)\, dx. \tag{3.3}$$

Thus,

$$\int_{-\pi}^{\pi} \cos(k\lambda) \cos(n\lambda)\, d\lambda = \begin{cases} 0, & k \neq n, \\ \pi, & k = n, \\ 2\pi, & k = n = 0, \end{cases} \tag{3.4}$$

and

$$\int_{-\pi}^{\pi} \sin(k\lambda) \cos(n\lambda)\, d\lambda = 0. \tag{3.5}$$

3.2. Double Fourier Series

Similarly,

$$\int_{-\pi}^{\pi} \sin(k\lambda)\sin(n\lambda)\,d\lambda = \begin{cases} 0, & k \neq n, \\ \pi, & k = n, \neq \emptyset \end{cases} \quad (3.6)$$

as long as $k \neq 0$ and $n \neq 0$. Consequently,

$$\int_{-c}^{c} \cos(k\pi x/c)\cos(n\pi x/c)\,dx = \begin{cases} 0, & k \neq n, \\ c, & k = n, \\ 2c, & k = n = 0, \end{cases} \quad (3.7)$$

and

$$\int_{-c}^{c} \cos(k\pi x/c)\sin(n\pi x/c)\,dx = 0. \quad (3.8)$$

Therefore, we obtain

$$a_n = \frac{1}{c}\int_{-c}^{c} f(x)\cos(n\pi x/c)\,dx. \quad n = 0, 1, 2, \ldots. \quad (3.9)$$

Likewise, if (3.1) were multiplied by $\sin(n\pi x/c)$ and integrated over the interval $[-c, c]$, it would be found that

$$b_n = \frac{1}{c}\int_{-c}^{c} f(x)\sin(n\pi x/c)\,dx, \quad n = 1, 2, \ldots. \quad (3.10)$$

It is noted in passing that if $f(x)$ were an odd function of x, such that $f(-x) = -f(x)$, then $a_n = 0$ for all n. Likewise if $f(-x) = f(x)$, that is if $f(x)$ were an even function of x, then $b_n = 0$.

3.2 Double Fourier Series

Let R be a rectangle defined by the region $-a \leq \alpha \leq a$, $-b \leq \beta \leq b$, and let $\varphi_n(\alpha, \beta)$ be a set of continuous functions, none of which vanishes identically in R. Such a set is said to be *orthogonal* if

$$\iint_R \varphi_n(\alpha, \beta)\varphi_m(\alpha, \beta)\,d\alpha\,d\beta = 0, \quad \text{if } n \neq m. \quad (3.11)$$

The number

$$\|\varphi_n\| = \left[\iint_R \varphi_n^2(\alpha, \beta)\,d\alpha\,d\beta\right]^{1/2} \quad (3.12)$$

is called the norm of φ_n. The set is said to be *normalized* if $\|\varphi_n\| = 1$, for $n = 0, 1, 2, \ldots$. It is equivalent for normalization that

$$\iint_R \varphi_n^2(\alpha, \beta)\,d\alpha\,d\beta = 1, \quad \text{for } n = 0, 1, 2, \ldots. \quad (3.13)$$

Of course, it is always possible to define constants, say μ_n, such that

$$\mu_n = \frac{1}{\|\varphi_n\|}, \quad (3.14)$$

which can be used to normalize the members of the set φ_n.

As was done for a function of a single variable, it is possible to expand a function of two variables in terms of these orthogonal functions as

$$f(\alpha, \beta) = c_0 \varphi_0(\alpha, \beta) + c_1 \varphi_1(\alpha, \beta) + \ldots + c_n \varphi_n(\alpha, \beta) + \ldots . \tag{3.15}$$

The coefficients are obtained by using the orthogonality property. This gives

$$c_n = \frac{\iint_R f(\alpha, \beta) \varphi_n(\alpha, \beta) \, d\alpha \, d\beta}{\iint_R \varphi^2(\alpha, \beta) \, d\alpha \, d\beta} . \tag{3.16}$$

If it is assumed that the equality holds in (3.15), the series will converge uniformly.

3.2.1 Double Trigonometric Series

Consider the orthogonal set consisting of the functions

$$1, \cos(mx), \sin(mx), \cos(ny), \sin(ny),$$
$$\cos(mx)\cos(ny), \sin(mx)\cos(ny), \tag{3.17}$$
$$\cos(mx)\sin(ny), \sin(mx)\sin(ny), \ldots, (n, m = 1, 2, 3, \ldots).$$

These are clearly orthogonal on the square defined by $-\pi \leq x \leq \pi$ and $-\pi \leq y \leq \pi$. For reference we note that the norms are

$$\|1\| = 2\pi, \quad \|\cos(mx)\| = \|\sin(mx)\| = \sqrt{2\pi},$$
$$\|\cos(mx)\cos(ny)\| = \|\sin(mx)\sin(ny)\| = \|\cos(mx)\sin(ny)\| = \pi. \tag{3.18}$$

The above leads to the following system

$$a_{mn} = \frac{1}{\pi^2} \iint_R f(x, y) \cos(mx) \cos(ny) \, dx \, dy,$$

$$b_{mn} = \frac{1}{\pi^2} \iint_R f(x, y) \sin(mx) \cos(ny) \, dx \, dy,$$

$$c_{mn} = \frac{1}{\pi^2} \iint_R f(x, y) \cos(mx) \sin(ny) \, dx \, dy, \tag{3.19}$$

$$d_{mn} = \frac{1}{\pi^2} \iint_R f(x, y) \sin(mx) \sin(ny) \, dx \, dy,$$

for $m, n = 1, 2, \ldots$. For the cases where either $m = 0$ or $n = 0$, we have

$$A_{m0} = \frac{\iint_R f(x, y) \cos(mx) \, dx \, dy}{\|\cos(mx)\|^2}$$

$$= \frac{1}{2\pi^2} \iint_R f(x, y) \cos(mx) \, dx \, dy, \quad (m = 1, 2, \ldots), \tag{3.20}$$

3.3. Integral Transforms

$$A_{0n} = \frac{\iint_R f(x,y)\cos(ny)\,dx\,dy}{\|\cos(ny)\|^2}$$

$$= \frac{1}{2\pi^2} \iint_R f(x,y)\cos(ny)\,dx\,dy, \quad (n = 1, 2, \ldots), \tag{3.21}$$

$$B_{m0} = \frac{\iint_R f(x,y)\sin(mx)\,dx\,dy}{\|\sin(mx)\|^2}$$

$$= \frac{1}{2\pi^2} \iint_R f(x,y)\sin(mx)\,dx\,dy, \quad (m = 1, 2, \ldots), \tag{3.22}$$

$$B_{0n} = \frac{\iint_R f(x,y)\sin(ny)\,dx\,dy}{\|\sin(ny)\|^2}$$

$$= \frac{1}{2\pi^2} \iint_R f(x,y)\sin(ny)\,dx\,dy, \quad (n = 1, 2, \ldots). \tag{3.23}$$

Note that for symmetry of expression we may define $A_{m0} = \frac{1}{2}a_{m0}$, with similar definitions for the other A and B terms. Finally note that

$$A_{00} = \frac{\iint_R f(x,y)\,dx\,dy}{\|1\|} = \frac{1}{4\pi^2} \iint_R f(x,y)\,dx\,dy, \tag{3.24}$$

whereby we may define $A_{00} = \frac{1}{4} a_{00}$.

When this is assembled, the series expansion may be written as

$$f(x,y) = \sum_{m,n=0}^{\infty} \lambda_{mn}[a_{mn}\cos(mx)\cos(ny) + b_{mn}\sin(mx)\cos(ny) \tag{3.25}$$

$$+ c_{mn}\cos(mx)\sin(ny) + d_{mn}\sin(mx)\sin(ny)],$$

with

$$\lambda_{mn} = \frac{1}{4}, \quad \text{for } m = n = 0,$$

$$\lambda_{mn} = \frac{1}{2}, \quad \text{for } m > 0,\ n = 0 \text{ or } m = 0,\ n > 0, \tag{3.26}$$

$$\lambda_{mn} = 1, \quad \text{for } m, n > 0.$$

The rectangular domain of interest is easily transformed into the area defined by $-a \leq \alpha \leq a$ and $-b \leq \beta \leq b$ by the change in variables, $x = \pi\alpha/a$, $y = \pi\beta/b$.

3.3 Integral Transforms

Let $f(x)$ be a function that has a convergent integral over the domain $[0, \infty]$. Also let $K(\alpha x)$ define a function of (α, x). Then, if the integral

$$I_f(\alpha) = \int_0^{\infty} f(x) K(\alpha x)\,dx \neq \infty \tag{3.27}$$

3. Fourier Series and Fourier Integrals

is convergent, then $I_f(\alpha)$ is the *integral transform* of $f(x)$ by the *kernel* $K(\alpha x)$. If for each $I_f(\alpha)$, (3.27) is satisfied by only one $f(x)$, then (3.27) has an inverse

$$f(x) = \int_0^\infty I_f(\alpha) H(\alpha x)\, d\alpha. \tag{3.28}$$

If $H(\alpha x) = K(\alpha x)$, then $K(\alpha x)$ is a *Fourier kernel*.

A Mellin transform of $K(x)$ is defined as

$$K(s) = \int_0^\infty K(x) x^{s-1}\, dx. \tag{3.29}$$

LEMMA 3.1: If $K(\alpha x)$ is a Fourier kernel then $K(s)K(1-s) = 1$.

Proof: By definition,

$$K(s) = \int_0^\infty K(x) x^{s-1}\, dx, \tag{3.30}$$

and thus, by implication,

$$\int_0^\infty \alpha^{s-1} I_f(\alpha)\, d\alpha = \int_0^\infty \alpha^{s-1} \int_0^\infty f(x) K(\alpha x)\, dx\, d\alpha$$
$$= \int_0^\infty f(x)\, dx \int_0^\infty \alpha^{s-1} K(\alpha x)\, d\alpha. \tag{3.31}$$

Now let $\eta \equiv \alpha x$, and write

$$\int_0^\infty K(\alpha x) \alpha^{s-1}\, d\alpha = x^{-s} \int_0^\infty K(\eta) \eta^{s-1}\, d\eta = x^{-s} K(s). \tag{3.32}$$

Then,

$$\int_0^\infty \alpha^{s-1} I_f(\alpha)\, d\alpha = I(s) = \int_0^\infty x^{-s} f(x)\, dx\, K(s) \tag{3.33}$$

$$I(s) = K(s) F(1-s).$$

Furthermore, if

$$f(x) = \int_0^\infty I_f(\alpha) K(\alpha x)\, d\alpha, \tag{3.34}$$

we have

$$\int_0^\infty f(x) x^{s-1}\, dx = \int_0^\infty x^{s-1} \int_0^\infty I_f(\alpha) K(\alpha x)\, d\alpha\, dx$$
$$= \int_0^\infty I_f(\alpha)\, d\alpha \int_0^\infty x^{s-1} K(\alpha x)\, dx. \tag{3.35}$$

Thus, when compared with (3.33),

$$F(s) = \int_0^\infty I_f(\alpha) \alpha^{-s}\, d\alpha \int_0^\infty \eta^{s-1} K(\eta)\, d\eta = I(1-s) K(s), \tag{3.36}$$

which shows that

$$F(1-s) = I(s) K(1-s), \tag{3.37}$$

3.3. Integral Transforms

and, therefore,
$$K(s)K(1-s) = 1, \tag{3.38}$$
as desired.

For example, if
$$K(\alpha x) = \sqrt{2/\pi}\cos(\alpha x), \tag{3.39}$$
the Fourier cosine transform of a function $f(x)$ is
$$F_c(\alpha) = \sqrt{2/\pi}\int_0^\infty f(x)\cos(\alpha x)\,dx, \tag{3.40}$$
with its inverse
$$f(x) = \sqrt{2/\pi}\int_0^\infty F_c(\alpha)\cos(\alpha x)\,d\alpha. \tag{3.41}$$

As another example, let $K(\alpha x)$ be defined as
$$K(\alpha x) = \sqrt{2/\pi}\sin(\alpha x). \tag{3.42}$$
The corresponding set of transform and inverse is
$$F_s(\alpha) = \sqrt{2/\pi}\int_0^\infty f(x)\sin(\alpha x)\,dx, \tag{3.43}$$
and
$$f(x) = \sqrt{2/\pi}\int_0^\infty F_s(\alpha)\sin(\alpha x)\,d\alpha. \tag{3.44}$$

By the lemma proved above, if
$$I_f(\alpha) = \int_0^\infty f(x)K(\alpha x)\,dx, \tag{3.45}$$
then
$$f(x) = \int_0^\infty I_f(\alpha)H(\alpha x)\,d\alpha, \tag{3.46}$$
and
$$K(s)H(1-s) = 1, \tag{3.47}$$
where $K(s)$ and $H(s)$ are the Mellin transforms of $K(x)$ and $H(x)$, respectively.

In short, what is desired is to show that if $f(x)$ is defined on $[0, \infty]$, then it is possible that
$$f(x) = \frac{2}{\pi}\int_0^\infty d\alpha \int_0^\infty f(\eta)\cos(\alpha\eta)\cos(\alpha x)\,d\eta, \tag{3.48}$$
or, more generally,
$$f(x) = \frac{1}{\pi}\int_{-\infty}^\infty d\alpha \int_{-\infty}^\infty f(\eta)\cos(\alpha\eta)\cos(\alpha x)\,d\eta. \tag{3.49}$$

Before exploring this, it is necessary to establish some integrability properties of $f(x)$. This is done next.

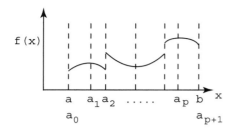

Figure 3.1. Function with finite number of extrema and discontinuities.

3.4 Dirichlet's Conditions

We say that the function $f(x)$ satisfies Dirichlet's conditions if

(1) $f(x)$ has only a finite number of extrema in an interval $[a, b]$, and
(2) $f(x)$ has only a finite number of discontinuities in the interval $[a, b]$ and no infinite discontinuities.

Let $a_0 = a$, $a_{p+1} = b$ and a_i, $i = 1, 2, \ldots, p$ be the points where $f(x)$ has either an extrema or a discontinuity, as depicted in Fig. 3.1. The satisfaction of Dirichelet's conditions leads to the following lemmas.

LEMMA 3.2: *If $f(x)$ satisfies Dirichlet's conditions in the interval $[a, b]$, then*

$$\lim_{\omega \to \infty} \int_a^b f(x) \sin(\omega x)\, dx = 0,$$
$$\lim_{\omega \to \infty} \int_a^b f(x) \cos(\omega x)\, dx = 0.$$
(3.50)

Proof: From the definition of the a_i above we may rewrite the first of the integrals as

$$\int_a^b f(x) \sin(\omega x)\, dx = \sum_{k=0}^{p} \int_{a_k}^{a_{k+1}} f(x) \sin(\omega x)\, dx.$$
(3.51)

[PROVIDED s.t. on EACH SUB-INTERVAL THERE ARE NO EXTREMA OR DISCONTINUITIES]

In each subinterval $f(x)$ is monotone, either increasing or decreasing, and by the second mean value theorem of calculus we have

$$\int_{a_k}^{a_{k+1}} f(x) \sin(\omega x)\, dx = f(a_k + 0) \int_{a_k}^{\zeta} \sin(\omega x)\, dx$$
$$+ f(a_{k+1} - 0) \int_{\zeta}^{a_{k+1}} \sin(\omega x)\, dx,$$
(3.52)

[i.e., THERE IS A ZETA (ζ) BETWEEN a_k AND a_{k+1} s.t. $f(\zeta)$ IS EXACT - WE DON'T NEED TO FIND IT]

where ζ is a point in the subinterval $[a_k, a_{k+1}]$. When the integrations are performed, there results

$$\int_{a_k}^{a_{k+1}} f(x) \sin(\omega x)\, dx = f(a_k + 0) \frac{\cos(\omega a_k) - \cos(\omega \zeta)}{\omega}$$
$$+ f(a_{k+1} - 0) \frac{\cos(\omega \zeta) - \cos(\omega a_{k+1})}{\omega}.$$
(3.53)

3.4. Dirichlet's Conditions

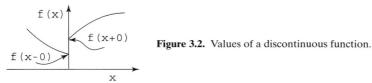

Figure 3.2. Values of a discontinuous function.

Thus,

$$\lim_{\omega \to \infty} \int_a^b f(x) \sin(\omega x) \, dx = 0. \tag{3.54}$$

Since p is finite, we obtain

$$\lim_{\omega \to \infty} \int_a^b f(x) \sin(\omega x) \, dx = \sum_{k=0}^{p} \lim_{\omega \to \infty} \int_{a_k}^{a_{k+1}} f(x) \sin(\omega x) \, dx = 0, \tag{3.55}$$

as proposed above. The proposal in the second of (3.50) is proved by a similar procedure.

The same considerations lead to the following additional results that are proved in standard textbooks on integral transforms.

LEMMA 3.3: *If $f(x)$ satisfies the Dirichlet conditions on the interval $[a, b]$, then*

$$\lim_{\omega \to \infty} \int_a^b f(x) \frac{\sin(\omega x)}{x} \, dx = \begin{cases} 0, & \text{if } a > 0, \\ \frac{1}{2}\pi f(0^+), & \text{if } a = 0. \end{cases} \tag{3.56}$$

More generally,

LEMMA 3.4: *If $f(x)$ satisfies Dirichlet's conditions on the interval $[a, b]$, then*

$$\lim_{\omega \to \infty} \frac{2}{\pi} \int_a^b f(x+u) \frac{\sin(\omega u)}{u} \, du = \begin{cases} f(x+0) + f(x-0), & \text{if } a < 0 < b, \\ f(x+0), & \text{if } a = 0 < b, \\ f(x-0), & \text{if } a < 0 = b, \\ 0, & \text{if } 0 < a < b \text{ or } a < b < 0. \end{cases}$$

Proof: The proof of this lemma requires the following identity, which is demonstrated first, *viz.*,

$$\lim_{n \to \infty} \frac{1}{\pi} \int_{-\pi}^{\pi} f(x+u) \frac{\sin(n+1/2)u}{2\sin(u/2)} \, du = \frac{f(x+0) + f(x-0)}{2}. \tag{3.57}$$

It is, for this purpose, sufficient to show that (Fig. 3.2)

$$\lim_{n \to \infty} \frac{1}{\pi} \int_0^\pi f(x+u) \frac{\sin(n+1/2)u}{2\sin(u/2)} \, du = \frac{f(x+0)}{2}, \tag{3.58}$$

$$\lim_{n \to \infty} \frac{1}{\pi} \int_{-\pi}^0 f(x+u) \frac{\sin(n+1/2)u}{2\sin(u/2)} \, du = \frac{f(x-0)}{2}. \tag{3.59}$$

Focus on the demonstration of the first of these, as the proof of the second follows similar lines. It is readily shown that

$$\frac{1}{2} f(x+0) = \frac{1}{\pi} \int_0^\pi f(x+0) \frac{\sin(n+1/2)u}{2\sin(u/2)} \, du, \qquad (3.60)$$

which leads to the strategy of proving that

$$\lim_{n \to \infty} \frac{1}{\pi} \int_0^\pi [f(x+u) - f(x+0)] \frac{\sin(n+1/2)u}{2\sin(u/2)} \, du = 0. \qquad (3.61)$$

Consider the function of u defined by

$$\varphi(u) = \frac{f(x+u) - f(x+0)}{2\sin(u/2)} = \frac{f(x+u) - f(x+0)}{u} \cdot \frac{u}{2\sin(u/2)}. \qquad (3.62)$$

Since this function has a derivative at $x+0$, the ratio $f(x+u) - f(x+0)/u$ exists as $u \to 0$. This means that this same ratio is absolutely integrable on the interval $0 \le x \le \pi$. But the function $u/2\sin(u/2)$ is bounded, and so the function defined as $\varphi(u)$ is integrable on $0 \le x \le \pi$. We furthermore have that

$$\int_0^\pi [f(x+u) - f(x+0)] \frac{\sin(n+1/2)u}{2\sin(u/2)} \, du = \int_0^\pi \varphi(u) \sin(n+1/2)u \, du. \qquad (3.63)$$

To continue with the proof, note that for any $\epsilon > 0$, the inequality

$$\frac{1}{\pi} \int_{-\delta}^\delta |f(x+u)| \, du < \frac{1}{2} \epsilon \qquad (3.64)$$

holds for sufficiently small δ. The function $(1/u) f(x+u)$ is absolutely integrable over $-\infty < u \le -\delta$ and $\delta \le u < \infty$. Therefore, by the previous lemma we have

$$\lim_{\ell \to \infty} \frac{1}{\pi} \int_\delta^\infty f(x+u) \frac{\sin(\ell u)}{u} \, du = \lim_{\ell \to \infty} \frac{1}{\pi} \int_{-\infty}^{-\delta} f(x+u) \frac{\sin(\ell u)}{u} \, du = 0.$$

Next consider the equality

$$\lim_{m \to \infty} \frac{1}{\pi} \int_{-\pi}^\pi f(x+u) \frac{\sin(mu)}{2\sin(u/2)} \, du = \frac{f(x+0) + f(x-0)}{2}, \qquad (3.65)$$

which has already been demonstrated, and where $m = n + 1/2$ for integer n. This can be rewritten as

$$\lim_{m \to \infty} \frac{1}{\pi} \int_{-\delta}^\delta f(x+u) \frac{\sin(mu)}{2\sin(u/2)} \, du = \frac{f(x+0) + f(x-0)}{2}, \qquad (3.66)$$

because the integrals on the intervals $-\pi \le u \le -\delta$ and $\delta \le u \le \pi$ vanish as $m \to \infty$.

Note that the integral on the *lhs* of (3.66) differs from

$$\frac{1}{\pi} \int_{-\delta}^\delta f(x+u) \frac{\sin(mu)}{u} \, du \qquad (3.67)$$

by the quantity

$$\frac{1}{\pi} \int_{-\delta}^\delta f(x+u) \left[\frac{1}{2\sin(u/2)} - \frac{1}{u} \right] \sin(mu) \, du, \qquad (3.68)$$

3.4. Dirichlet's Conditions

where the function in [...] is continuous and where it is seen to be zero at $u = 0$ (by appropriate limits). But the integral in (3.68) vanishes in the limit as $m \to \infty$, and (3.66) can be replaced by

$$\lim_{m \to \infty} \frac{1}{\pi} \int_{-\delta}^{\delta} f(x+u) \frac{\sin(mu)}{u} \, du = \frac{f(x+0) + f(x-0)}{2}. \tag{3.69}$$

Now, let $m \leq \ell < m+1$, so that $\ell = m + \xi$, where $0 \leq \xi < 1$. The mean value theorem of calculus gives

$$\frac{\sin(\ell u) - \sin(mu)}{u} = (\ell - m)\cos(hu), \tag{3.70}$$

where $m \leq h \leq \ell$. Therefore, for any ℓ,

$$\frac{1}{\pi} \left| \int_{-\delta}^{\delta} f(x+u) \frac{\sin(\ell u)}{u} \, du - \int_{-\delta}^{\delta} f(x+u) \frac{\sin(mu)}{u} \, du \right|$$

$$= \frac{1}{\pi} \left| \int_{-\delta}^{\delta} f(x+u) \xi \cos(hu) \, du \right| \tag{3.71}$$

$$\leq \frac{1}{\pi} \int_{-\delta}^{\delta} |f(x+u)| \, du < \frac{\epsilon}{2}.$$

For large ℓ and m, we have

$$\left| \frac{f(x+0) + f(x-0)}{2} - \frac{1}{\pi} \int_{-\delta}^{\delta} f(x+u) \frac{\sin(mu)}{u} \, du \right| < \frac{\epsilon}{2}. \tag{3.72}$$

Thus, combining the last two inequalities, we find that

$$\left| \frac{f(x+0) + f(x-0)}{2} - \frac{1}{\pi} \int_{-\delta}^{\delta} f(x+u) \frac{\sin(\ell u)}{u} \, du \right| < \frac{\epsilon}{2}, \tag{3.73}$$

for all sufficiently large ℓ. Finally, by (3.4),

$$\left| \frac{f(x+0) + f(x-0)}{2} - \frac{1}{\pi} \int_{-\infty}^{\infty} f(x+u) \frac{\sin(\ell u)}{u} \, du \right| < \frac{\epsilon}{2}, \tag{3.74}$$

for sufficiently large ℓ. This proves the lemma in question.

LEMMA 3.5: *If $f(x)$ satisfies the Dirichlet conditions on the interval $[a, b]$, then*

NOTE: COULD BE DISCONTINUOUS AT x

$$\frac{1}{2}[f(x+0) + f(x-0)] = \frac{1}{\pi} \int_0^{\infty} d\alpha \int_{-\infty}^{\infty} f(\eta) \cos[\alpha(\eta - x)] \, d\eta, \tag{3.75}$$

and, if $f(x)$ is continuous,

$$f(x) = \frac{1}{\pi} \int_0^{\infty} d\alpha \int_{-\infty}^{\infty} f(\eta) \cos[\alpha(\eta - x)] \, d\eta. \tag{3.76}$$

This last lemma will be used below to demonstrate an integral inversion theorem.

Proof: We again suppose that $f(x)$ is absolutely integrable on the entire x axis. Then, by the definition of an improper integral, we have

$$\frac{1}{\pi}\int_0^\infty d\lambda \int_{-\infty}^\infty f(u)\cos\lambda(u-x)\,du = \lim_{\ell\to\infty}\frac{1}{\pi}\int_0^\ell d\lambda \int_{-\infty}^\infty f(u)\cos\lambda(u-x)\,du. \tag{3.77}$$

The integral

$$\int_{-\infty}^\infty f(u)\cos\lambda(u-x)\,du \tag{3.78}$$

is convergent for $-\infty \le \lambda \le \infty$, because

$$|f(u)\cos\lambda(u-x)| \le |f(u)|, \tag{3.79}$$

and $f(u)$ is, as already stipulated, absolutely integrable on the entire axis. It follows from the previous lemmas that

$$\int_0^\ell d\lambda \int_{-\infty}^\infty f(u)\cos\lambda(u-x)\,du = \int_{-\infty}^\infty du \int_0^\ell f(u)\cos\lambda(u-x)\,d\lambda$$
$$= \int_{-\infty}^\infty f(u)\frac{\sin\ell(u-x)}{u-x}\,du = \int_{-\infty}^\infty f(x+u)\frac{\sin(\ell u)}{u}\,du, \tag{3.80}$$

where an obvious substitution of variables has been made. But, from (3.77) we have

$$\frac{1}{\pi}\int_0^\infty d\lambda \int_{-\infty}^\infty f(u)\cos\lambda(u-x)\,du = \lim_{\ell\to\infty}\frac{1}{\pi}\int_{-\infty}^\infty f(x+u)\frac{\sin(\ell u)}{u}\,du. \tag{3.81}$$

If the function $f(x)$ has left-hand and right-hand derivatives at the point x, then the limit on the *rhs* exists and is equal to $\frac{1}{2}[f(x+0)+f(x-0)]$. Thus, the integral on the *lhs* exists and is

$$\frac{1}{\pi}\int_0^\infty d\lambda \int_{-\infty}^\infty f(u)\cos\lambda(u-x)\,du = \frac{1}{2}[f(x+0)+f(x-0)], \tag{3.82}$$

which proves the lemma.

3.5 Integral Theorems

Let $f(x)$ have a convergent integral over the interval $[0,\infty]$ and satisfy the Dirichlet conditions described above. As $f(x)$ is defined for $x \ge 0$, define $f(x)$ in the interval $[-\infty, 0]$ as $f(-x) = f(x)$, i.e., make $f(x)$ an even function. Then we may construct the integral

$$\frac{1}{\pi}\int_0^\infty d\alpha \int_{-\infty}^\infty f(\eta)\cos[\alpha(\eta-x)]\,d\eta = \frac{1}{\pi}\int_0^\infty d\alpha \int_0^\infty f(\eta)\cos[\alpha(\eta-x)]\,d\eta$$
$$+ \frac{1}{\pi}\int_0^\infty d\alpha \int_{-\infty}^0 f(\eta)\cos[\alpha(\eta-x)]\,d\eta. \tag{3.83}$$

3.5. Integral Theorems

But,

$$\int_{-\infty}^{0} f(\eta)\cos[\alpha(\eta - x)]\,d\eta = \int_{0}^{\infty} f(-\eta)\cos[\alpha(-\eta - x)]\,d\eta \qquad (3.84)$$
$$= \int_{0}^{\infty} f(\eta)\cos[\alpha(\eta + x)]\,d\eta.$$

Consequently,

$$\frac{1}{\pi}\int_{0}^{\infty} d\alpha \int_{-\infty}^{\infty} f(\eta)\cos[\alpha(\eta - x)]\,d\eta$$
$$= \frac{1}{\pi}\int_{0}^{\infty} d\alpha \int_{0}^{\infty} f(\eta)\{\cos[\alpha(\eta - x)] + \cos[\alpha(\eta + x)]\}\,d\eta$$
$$= \frac{2}{\pi}\int_{0}^{\infty} \cos(\alpha x)\,d\alpha \int_{0}^{\infty} f(\eta)\cos(\alpha\eta)\,d\eta.$$

Therefore, with appeal to (3.76), if

$$F_c(\alpha) = \sqrt{2/\pi}\int_{0}^{\infty} f(\eta)\cos(\alpha\eta)\,d\eta, \qquad (3.85)$$

then

$$f(x) = \sqrt{2/\pi}\int_{0}^{\infty} F_c(\alpha)\cos(\alpha x)\,d\alpha. \qquad (3.86)$$

On the other hand, if $f(x)$ is defined in the interval $[-\infty, 0]$ as $f(-x) = -f(x)$, then

$$F_s(\alpha) = \sqrt{2/\pi}\int_{0}^{\infty} f(\eta)\sin(\alpha\eta)\,d\eta \qquad (3.87)$$

implies

$$f(x) = \sqrt{2/\pi}\int_{0}^{\infty} F_s(\alpha)\sin(\alpha x)\,d\alpha. \qquad (3.88)$$

Noting that

$$\int_{-m}^{m}\cos[\alpha(\eta - x)]\,d\alpha = 2\int_{0}^{m}\cos[\alpha(\eta - x)]\,d\alpha,$$
$$\int_{-m}^{m}\sin[\alpha(\eta - x)]\,d\alpha = 0, \qquad (3.89)$$

we have, as before,

$$f(x) = \lim_{m\to\infty}\frac{1}{\pi}\int_{-\infty}^{\infty} f(\eta)\,d\eta \int_{0}^{m}\cos[\alpha(\eta - x)]\,d\alpha$$
$$= \lim_{m\to\infty}\frac{1}{\pi}\int_{-\infty}^{\infty} f(\eta)\,d\eta\,\frac{1}{2}\int_{-m}^{m} e^{i\alpha(\eta - x)}\,d\alpha,$$

or

$$f(x) = \frac{1}{2\pi}\int_{-\infty}^{\infty} e^{-i\alpha x}\,d\alpha \int_{-\infty}^{\infty} f(\eta)e^{i\alpha\eta}\,d\eta. \qquad (3.90)$$

Thus,
$$F(\alpha) = \frac{1}{\sqrt{2\pi}} \int_{-\infty}^{\infty} f(x) e^{i\alpha x} \, dx \tag{3.91}$$

implies
$$f(x) = \frac{1}{\sqrt{2\pi}} \int_{-\infty}^{\infty} F(\alpha) e^{-i\alpha x} \, d\alpha. \tag{3.92}$$

3.6 Convolution Integrals

The convolution of $f(x)$ and $g(x)$ is defined as
$$f \star g = \frac{1}{\sqrt{2\pi}} \int_{-\infty}^{\infty} g(\eta) f(x-\eta) \, d\eta. \tag{3.93}$$

Since
$$\int_{-\infty}^{\infty} g(\eta) f(x-\eta) \, d\eta = \frac{1}{\sqrt{2\pi}} \int_{-\infty}^{\infty} g(\eta) \, d\eta \int_{-\infty}^{\infty} F(t) e^{-it(x-\eta)} \, dt,$$

where
$$F(t) = \frac{1}{\sqrt{2\pi}} \int_{-\infty}^{\infty} f(x) e^{itx} \, dx, \tag{3.94}$$

we have
$$\frac{1}{\sqrt{2\pi}} \int_{-\infty}^{\infty} g(\eta) \, d\eta \int_{-\infty}^{\infty} F(t) e^{-it(x-\eta)} \, dt = \frac{1}{\sqrt{2\pi}} \int_{-\infty}^{\infty} F(t) e^{-itx} \, dt \int_{-\infty}^{\infty} g(\eta) e^{it\eta} \, d\eta$$
$$= \int_{-\infty}^{\infty} F(t) G(t) e^{-itx} \, dt,$$

with
$$G(t) = \frac{1}{\sqrt{2\pi}} \int_{-\infty}^{\infty} g(x) e^{itx} \, dx. \tag{3.95}$$

Therefore,
$$\int_{-\infty}^{\infty} F(t) G(t) e^{-itx} \, dt = \int_{-\infty}^{\infty} g(\eta) f(x-\eta) \, d\eta. \tag{3.96}$$

In a special case where $x = 0$, (3.96) becomes
$$\int_{-\infty}^{\infty} F(t) G(t) \, dt = \int_{-\infty}^{\infty} g(\eta) f(-\eta) \, d\eta. \tag{3.97}$$

If we replace $F(t)$ with $F_c(t)$, i.e., if $f(x)$ is an even function, $f(-x) = f(x)$, and if the same is done for $g(x)$, the above becomes
$$\int_{0}^{\infty} F_c(t) G_c(t) \, dt = \int_{0}^{\infty} f(\eta) g(\eta) \, d\eta. \tag{3.98}$$

3.7. Transforms of Derivatives

3.6.1 Evaluation of Integrals by Use of Convolution Theorems

The above results may be used to evaluate integrals. For example, let

$$f(x) = e^{-bx}, \tag{3.99}$$

and let us evaluate two integrals, *viz.*,

$$I_1 = \int_0^\infty e^{-bx} \cos(\alpha x) \, dx,$$
$$I_2 = \int_0^\infty e^{-bx} \sin(\alpha x) \, dx. \tag{3.100}$$

We find

$$I_1 = \left[-\frac{1}{b} e^{-bx} \cos(\alpha x)\right]_0^\infty - \frac{\alpha}{b} \int_0^\infty e^{-bx} \sin(\alpha x) \, dx = \frac{1}{b} - \frac{\alpha}{b} I_2,$$
$$I_2 = \left[-\frac{1}{b} e^{-bx} \sin(\alpha x)\right]_0^\infty + \frac{\alpha}{b} \int_0^\infty e^{-bx} \cos(\alpha x) \, dx = \frac{\alpha}{b} I_1. \tag{3.101}$$

When the above are solved, we obtain

$$I_1 = \frac{b}{\alpha^2 + b^2} \Rightarrow F_c(\alpha) = \sqrt{2/\pi} \frac{b}{\alpha^2 + b^2},$$
$$I_2 = \frac{\alpha}{\alpha^2 + b^2} \Rightarrow F_s(\alpha) = \sqrt{2/\pi} \frac{\alpha}{\alpha^2 + b^2}. \tag{3.102}$$

But, from the definition of the Fourier cosine transform,

$$f(x) = \sqrt{2/\pi} \int_0^\infty F_c(\alpha) \cos(\alpha x) \, d\alpha = \frac{2}{\pi} \int_0^\infty \frac{b}{\alpha^2 + b^2} \cos(\alpha x) \, d\alpha$$
$$= \frac{2b}{\pi} \int_0^\infty \frac{\cos(\alpha x)}{\alpha^2 + b^2} \, d\alpha = e^{-bx}.$$

Imagine the same development for a companion function, $g(x) = e^{-ax}$ and substitute both results into (3.99). First, it is found that

$$F_c(\alpha) = \sqrt{2/\pi} \frac{b}{\alpha^2 + b^2} \quad \text{and} \quad G_c(\alpha) = \sqrt{2/\pi} \frac{a}{\alpha^2 + a^2}, \tag{3.103}$$

and then, after the substitution,

$$\frac{2ab}{\pi} \int_0^\infty \frac{d\alpha}{(\alpha^2 + b^2)(\alpha^2 + a^2)} = \int_0^\infty e^{-(a+b)x} \, dx = \frac{1}{a+b}. \tag{3.104}$$

3.7 Fourier Transforms of Derivatives of $f(x)$

To obtain formulae for the transformations of derivatives of functions, first define

$$F^{(r)}(\alpha) = \frac{1}{\sqrt{2\pi}} \int_{-\infty}^\infty \frac{d^r f(x)}{dx^r} e^{i\alpha x} \, dx \tag{3.105}$$

and then integrate by parts to obtain

$$F^{(r)}(\alpha) = \left[\frac{1}{\sqrt{2\pi}} \frac{d^{r-1} f}{dx^{r-1}} e^{i\alpha x}\right]_{-\infty}^{\infty} - \frac{1}{\sqrt{2\pi}} \int_{-\infty}^{\infty} \frac{d^{r-1} f}{dx^{r-1}} (i\alpha) e^{i\alpha x} \, dx \quad (3.106)$$

$$= -(i\alpha) F^{(r-1)}(\alpha), \quad \text{if} \quad \frac{d^{r-1} f}{dx^{r-1}} \to 0 \quad \text{as} \quad x \to \pm\infty.$$

In general, by repetitive application of this procedure, it is found that

$$F^{(r)}(\alpha) = (-i\alpha)^r F(\alpha). \quad (3.107)$$

3.8 Fourier Integrals as Limiting Cases of Fourier Series

It is useful to explore the existence of the above Fourier integrals as limiting cases of the Fourier series considered at in previous sections. Although the following is not as rigorous as the proof of the Fourier theorem given earlier, the discussion of the Fourier integral as a limiting case of a Fourier series is valuable as a connection between the two.

As before, let $f(x)$ be defined for all real x, and let it be piecewise smooth (with a possible finite number of discontinuities) on every finite interval $-\ell \leq x \leq \ell$. Then $f(x)$ can be expanded as a Fourier series

$$f(x) = \frac{a_0}{2} + \sum_{n=1}^{\infty} \left(a_n \cos \frac{n\pi x}{\ell} + b_n \sin \frac{n\pi x}{\ell}\right), \quad (3.108)$$

where

$$a_n = \frac{1}{\ell} \int_{-\ell}^{\ell} f(u) \cos \frac{n\pi x}{\ell} \, du, \quad n = 0, 1, 2, \ldots,$$

$$b_n = \frac{1}{\ell} \int_{-\ell}^{\ell} f(u) \sin \frac{n\pi x}{\ell} \, du, \quad n = 1, 2, 3, \ldots. \quad (3.109)$$

We note that at points where $f(x)$ is discontinuous we must interpret the value as $\frac{1}{2}[f(x+0) + f(x-0)]$. Now, substitute (3.109) into the above to obtain

$$f(x) = \frac{1}{2\ell} \int_{-\ell}^{\ell} f(u) \, du + \sum_{n=1}^{\infty} \frac{1}{\ell} \int_{-\ell}^{\ell} f(u) \cos \frac{n\pi}{\ell} (u - x) \, du. \quad (3.110)$$

Suppose that $f(x)$ is absolutely integrable on the entire x axis, that is suppose that the integral

$$\int_{-\infty}^{\infty} |f(x)| \, dx \quad (3.111)$$

exists. Then, as $\ell \to \infty$, (3.110) becomes

$$f(x) = \lim_{\ell \to \infty} \sum_{n=1}^{\infty} \frac{1}{\ell} \int_{-\ell}^{\ell} f(u) \cos \frac{n\pi}{\ell} (u - x) \, du. \quad (3.112)$$

Setting

$$\lambda_1 = \frac{\pi}{\ell}, \quad \lambda_2 = \frac{2\pi}{\ell}, \quad \lambda_2 = \frac{3\pi}{\ell}, \ldots,$$

$$\Delta \lambda_n = \lambda_{n+1} - \lambda_n = \frac{\pi}{\ell}, \quad (3.113)$$

3.9. Dirac Delta Function

and using this in the above, the sum takes the form

$$\frac{1}{\pi} \sum_{n=1}^{\infty} \Delta \lambda_n \int_{-\ell}^{\ell} f(u) \cos \lambda_n(u-x)\, du. \tag{3.114}$$

But for fixed x this looks like the sum that would define the integral of the function

$$\zeta(\lambda) = \frac{1}{\pi} \int_{-\infty}^{\infty} f(u) \cos \lambda(u-x)\, du, \tag{3.115}$$

and, therefore, it is natural to conclude that, as $\ell \to \infty$,

$$f(x) = \frac{1}{\pi} \int_0^{\infty} d\lambda \int_{-\infty}^{\infty} f(u) \cos \lambda(u-x)\, du. \tag{3.116}$$

3.9 Dirac Delta Function

Consider the function

$$\Delta_v(s) \equiv \frac{1}{\pi} \frac{\sin(vs)}{s}. \tag{3.117}$$

It can be verified that

$$\int_{-\infty}^{\infty} \Delta_v(s)\, ds = 1, \tag{3.118}$$

and

$$\lim_{v \to 0} \int_{-\infty}^{\infty} \Delta_v(s) f(s)\, ds = f(0). \tag{3.119}$$

To prove the latter, expand $f(s)$ about the point $s = 0$, i.e.,

$$f(s) = f(0) + (\partial f/\partial s)_0 s + 1/2!(\partial^2 f/\partial s^2)_0 s^2 + \ldots. \tag{3.120}$$

Then,

$$\lim_{v \to 0} \int_{-\infty}^{\infty} \Delta_v(s) f(s)\, ds = \lim_{v \to 0} \int_{-\infty}^{\infty} \frac{1}{\pi} f(0) \frac{\sin vs}{s}\, ds + \lim_{v \to 0} \int_{-\infty}^{\infty} \frac{1}{\pi} (\partial f/\partial s)_0 \sin vs\, ds$$

$$+ \lim_{v \to 0} \int_{-\infty}^{\infty} \frac{1}{2\pi} (\partial^2 f/\partial s^2)_0 s \sin vs\, ds$$

$$= f(0).$$

We also explore the function, $\eta(s-a)$, where $\eta(s)$ is shown in Fig. 3.3. Clearly,

$$\int_{-\infty}^{\infty} \eta(s)\, ds = 2 \frac{1}{2} \epsilon \frac{1}{\epsilon} = 1. \tag{3.121}$$

Then, by expanding $f(s)$ about the point $s = a$, we deduce that

$$\lim_{\epsilon \to 0} \int_{-\infty}^{\infty} \eta(s) f(s)\, ds = \lim_{\epsilon \to 0} \int_{a-1/2\epsilon}^{a+1/2\epsilon} f(a) \frac{1}{\epsilon}\, ds. \tag{3.122}$$

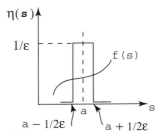

Figure 3.3. A step function as a generalized function.

The $f(a)$ here is, of course, just the first term of a Taylor expansion of $f(s)$ about the point $s = a$. The remaining terms in the Taylor expansion produce null results; for example, the next term in the expansion yields

$$\lim_{\epsilon \to 0} \int_{a-1/2\epsilon}^{a} (s-a)\frac{1}{\epsilon}\,ds + \lim_{\epsilon \to 0} \int_{a}^{a-1/2\epsilon} (s-a)\frac{1}{\epsilon}\,ds = 0, \qquad (3.123)$$

and so on. Thus,

$$\lim_{\epsilon \to 0} \int_{-\infty}^{\infty} \eta(s) f(s)\,ds = f(a). \qquad (3.124)$$

We note that

$$\lim_{\epsilon \to 0} \eta(s-a) = \infty, \quad \text{at } s = a, \qquad (3.125)$$

and

$$\lim_{\epsilon \to 0} \int_{-\infty}^{\infty} \eta(s)\,ds = 1. \qquad (3.126)$$

Consequently, we define

$$\lim_{\epsilon \to 0} \eta(s-a) \equiv \delta(s-a), \qquad (3.127)$$

and

$$\lim_{v \to 0} \Delta_v(s) \equiv \delta(s). \qquad (3.128)$$

The so defined Dirac delta function $\delta(s)$ is such that

$$\int_{-\infty}^{\infty} \delta(s)\,ds = 1, \quad \delta(s) = \begin{cases} \infty, & \text{if } s = 0, \\ 0, & \text{if } s \neq 0, \end{cases} \qquad (3.129)$$

and

$$\int_{-\infty}^{\infty} f(s)\delta(s-a)\,ds = f(a). \qquad (3.130)$$

3.10 Suggested Reading

Brown, J. W., and Churchill, R. V. (2001), *Fourier Series and Boundary Value Problems*, 6th ed., McGraw-Hill, Boston.

Goldberg, R. R. (1961), *Fourier Transforms*, Cambridge University Press, Cambridge, UK.

Little, R. W. (1973), *Elasticity*, Prentice Hall, Englewood Cliffs, New Jersey.

3.10. Suggested Reading

Papoulis, A. (1962), *The Fourier Integral and Its Application*, McGraw-Hill, New York.

Pinkus, A., and Samy, Z. (1997), *Fourier Series and Integral Transforms*, Cambridge University Press, New York.

Sneddon, I. N. (1951), *Fourier Transforms*, McGraw-Hill, New York.

Sneddon, I. N. (1961), *Fourier Series*, Routledge and Paul, London.

Sneddon, I. N. (1972), *The Use of Integral Transforms*, McGraw-Hill, New York.

Wolf, K. B. (1979), *Integral Transforms in Science and Engineering*, Plenum, New York.

PART 2: CONTINUUM MECHANICS

4 Kinematics of Continuum

4.1 Preliminaries

Let $\phi(\mathbf{x})$ be a scalar field which is a function of x_1, x_2, x_3 (Fig. 4.1), *i.e.*,

$$\phi = \phi(x_1, x_2, x_2). \tag{4.1}$$

The function ϕ is continuous if

$$\lim_{\alpha \to 0} |\phi(\mathbf{x} + \alpha \mathbf{a}) - \phi(\mathbf{x})| = 0, \quad \forall\, \mathbf{x} \in \mathcal{D},\ \mathbf{a} \in \mathcal{E}. \tag{4.2}$$

The field ϕ is differentiable within $\mathcal{D}$ if there is a vector field, $\mathbf{w}$, such that

$$\lim_{\alpha \to 0} |\mathbf{w} \cdot \mathbf{a} - \alpha^{-1}[\phi(\mathbf{x} + \alpha \mathbf{a}) - \phi(\mathbf{x})]| = 0, \quad \forall\, \mathbf{x} \in \mathcal{D},\ \mathbf{a} \in \mathcal{E}. \tag{4.3}$$

The vector field $\mathbf{w}$ is unique and is the *gradient* of ϕ, $\mathbf{w} = \operatorname{grad} \phi$.

If $\mathbf{u}$ is a vector field, $\mathbf{u}(\mathbf{x})$, then its gradient is defined by

$$[\operatorname{grad} \mathbf{u}(\mathbf{x})]^T \cdot \mathbf{a} = \operatorname{grad}[\mathbf{u}(\mathbf{x}) \cdot \mathbf{a}], \quad \forall\, \mathbf{x} \in \mathcal{D},\ \mathbf{a} \in \mathcal{E}. \tag{4.4}$$

The divergence of $\mathbf{u}(\mathbf{x})$, $\operatorname{div} \mathbf{u}$, or $\nabla \cdot \mathbf{u}$, is defined by

$$\nabla \cdot \mathbf{u} = \operatorname{tr}[\operatorname{grad} \mathbf{u}(\mathbf{x})]. \tag{4.5}$$

The curl is defined by

$$[\operatorname{curl} \mathbf{u}(\mathbf{x})] \cdot \mathbf{a} = \nabla \cdot [\mathbf{u}(\mathbf{x}) \times \mathbf{a}], \quad \forall\, \mathbf{x} \in \mathcal{D},\ \mathbf{a} \in \mathcal{E}. \tag{4.6}$$

The tensor field, $\mathbf{T}(\mathbf{x})$, if differentiable, has its divergence defined as

$$\operatorname{div} \mathbf{T}(\mathbf{x}) \cdot \mathbf{a} = \nabla \cdot \mathbf{T}(\mathbf{x}) \cdot \mathbf{a} = \operatorname{div}[\mathbf{T}(\mathbf{x}) \cdot \mathbf{a}]. \tag{4.7}$$

It is readily shown that if ϕ, $\mathbf{u}$, and $\mathbf{T}$ are scalar, vector, and tensor fields, respectively, then with reference to rectangular Cartesian unit base vectors $\{\mathbf{e}_1, \mathbf{e}_2, \mathbf{e}_3\}$, we can write

$$\phi = \phi(\mathbf{x}),$$
$$\mathbf{u}(\mathbf{x}) = u_p(\mathbf{x})\mathbf{e}_p, \tag{4.8}$$
$$\mathbf{T}(\mathbf{x}) = T_{pq}(\mathbf{x})\mathbf{e}_p\mathbf{e}_q.$$

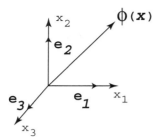

Figure 4.1. A scalar field, $\phi(\mathbf{x})$.

It is also readily shown from the discussion and definitions given above that, *inter alia*,

$$\text{grad } \phi(\mathbf{x}) = \nabla \phi = \partial \phi(\mathbf{x})/\partial x_p \mathbf{e}_p,$$
$$\text{grad } \mathbf{u}(\mathbf{x}) = \nabla \mathbf{u}(\mathbf{x}) = \partial u_p/\partial x_q \mathbf{e}_p \mathbf{e}_q,$$
$$\text{div } \mathbf{u}(\mathbf{x}) = \nabla \cdot \mathbf{u}(\mathbf{x}) = \partial u_p/\partial x_p, \quad (4.9)$$
$$\text{curl } \mathbf{u}(\mathbf{x}) = \epsilon_{pqr} \partial u_r/\partial x_q \mathbf{e}_p,$$
$$\text{div } \mathbf{T}(\mathbf{x}) = \nabla \cdot \mathbf{T}(\mathbf{x}) = \partial T_{pq}/\partial x_p \mathbf{e}_q.$$

4.2 Uniaxial Strain

A uniaxial *measure of strain* should, at the minimum, quantitatively describe changes in length as depicted in Fig. 4.2. There are, however, an infinite number of ways to do this. For example,

$$\textit{nominal strain:} \quad e = \Delta \ell/\ell_0 = (\ell - \ell_0)/\ell_0 = \ell/\ell_0 - 1. \quad (4.10)$$

The *stretch* associated with this change in length is

$$\lambda \equiv \ell/\ell_0 \quad \rightarrow \quad e = \lambda - 1. \quad (4.11)$$

Other definitions of uniaxial strain follow, and as examples we define

$$\textit{natural strain:} \quad \eta = \frac{\ell - \ell_0}{\ell} = 1 - 1/\lambda, \quad (4.12)$$

$$\textit{Lagrangian strain:} \quad E = \frac{1}{2}\frac{\ell^2 - \ell_0^2}{\ell_0^2} = \frac{1}{2}(\lambda^2 - 1), \quad (4.13)$$

$$\textit{Eulerian strain:} \quad \mathcal{E} = \frac{1}{2}\frac{\ell^2 - \ell_0^2}{\ell^2} = \frac{1}{2}(1 - 1/\lambda^2). \quad (4.14)$$

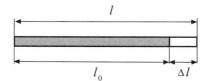

Figure 4.2. Uniaxial stretch.

4.3. Deformation Gradient

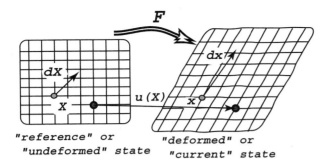

Figure 4.3. Kinematics of deformation, describing the motion of material particles from the "reference state" to the "deformed state."

Still another measure is defined by

$$\int_0^\epsilon d\epsilon = \int_{\ell_0}^\ell \frac{d\ell}{\ell} = \epsilon = \ln \frac{\ell}{\ell_0} = \ln \lambda. \qquad (4.15)$$

Accordingly, this measure of strain is referred to as the *logarithmic strain* or *true strain*.

4.3 Deformation Gradient

To provide a precise description of changes in the shape, size, and orientation of a solid body, we introduce various tensors that describe deformation of a body. The displacement field $\mathbf{u}(\mathbf{x})$ describes the change in position, relative to a convenient coordinate system, of all points in the body. We shall assume this field to be single valued and continuous for the present. The property of being single valued implies that there are no holes, gaps, or interpenetrations of matter in the body. The displacement field is nonuniform if the body is deformed and is uniform only if the body has undergone rigid translations.

To describe the picture above we need to describe how each "material point" is displaced, *i.e.*, we want to associate a displacement, $\mathbf{u}$, with each material point (Fig. 4.3). We identify material points by their positions in the reference state, $\mathbf{X}$. Thus,

        $\mathbf{X}$ is position in the reference state,

        $\mathbf{x}$ is position in the current (deformed) state, and

        $\mathbf{x}$ is function of $\mathbf{X}$, $\mathbf{x} = \mathbf{x}(\mathbf{X})$.

The gradient of $\mathbf{x}(\mathbf{X})$ is defined as

$$\mathbf{F} = \partial \mathbf{x}/\partial \mathbf{X}. \qquad (4.16)$$

If $\{\mathbf{e}_1, \mathbf{e}_2, \mathbf{e}_3\}$ is a set of convenient base vectors in the current configuration and $\{\mathbf{E}_1, \mathbf{E}_2, \mathbf{E}_3\}$ are base vectors in the reference configuration, then an explicit, and most natural, representation of $\mathbf{F}$ is

$$\mathbf{F} = \partial x_i(\mathbf{X})/\partial X_j \, \mathbf{e}_i \mathbf{E}_j. \qquad (4.17)$$

This is a so-called *two-point tensor*, which relates $d\mathbf{x}$ and $d\mathbf{X}$, such that

$$d\mathbf{x} = \mathbf{F} \cdot d\mathbf{X}. \qquad (4.18)$$

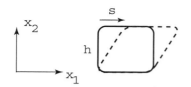

Figure 4.4. Case of 2D simple shear.

Since **F** is invertible,

$$d\mathbf{X} = \mathbf{F}^{-1} \cdot d\mathbf{x}. \tag{4.19}$$

In terms of displacement gradients we may express the above as follows. Since

$$\mathbf{x} = \mathbf{X} + \mathbf{u}, \tag{4.20}$$

the displacement gradient is

$$\partial \mathbf{u}/\partial \mathbf{X} = \partial \mathbf{x}/\partial \mathbf{X} - \mathbf{I} = \mathbf{F} - \mathbf{I}, \tag{4.21}$$

or, in the current state,

$$\partial \mathbf{u}/\partial \mathbf{x} = \mathbf{I} - \partial \mathbf{X}/\partial \mathbf{x} = \mathbf{I} - \mathbf{F}^{-1}. \tag{4.22}$$

For example, consider a simple shear deformation. Define a shear strain as $\gamma = s/h$, as depicted in Fig. 4.4. Then, the deformation mapping is

$$x_1 = X_1 + \gamma X_2, \quad x_2 = X_2. \tag{4.23}$$

The in-plane deformation gradient has the components

$$\mathbf{F} = \begin{bmatrix} 1 & \gamma \\ 0 & 1 \end{bmatrix}, \tag{4.24}$$

whereas the displacement gradient is

$$\frac{\partial \mathbf{u}}{\partial \mathbf{X}} = \begin{bmatrix} 0 & \gamma \\ 0 & 0 \end{bmatrix}. \tag{4.25}$$

4.4 Strain Tensor

Let $\mathcal{R}_0$ and $\mathcal{R}$ be regions within the body in the reference (undeformed) configuration and current (deformed) configuration, respectively. Further, let **N** be a unit vector embedded within the body in the reference configuration; the deformation transforms **N** to **n** in the deformed state (Fig. 4.5). The square of the stretch of **N**, $\lambda(\mathbf{N})$, can be calculated as

$$\lambda^2(\mathbf{N}) = (\mathbf{F} \cdot \mathbf{N}) \cdot (\mathbf{F} \cdot \mathbf{N}) = \mathbf{N} \cdot \mathbf{F}^T \cdot \mathbf{F} \cdot \mathbf{N}. \tag{4.26}$$

Define $\mathbf{C} \equiv \mathbf{F}^T \cdot \mathbf{F}$ as the right *Cauchy–Green deformation tensor*. Then the stretch, squared, is also given by

$$\lambda^2(\mathbf{N}) = \mathbf{N} \cdot \mathbf{C} \cdot \mathbf{N}. \tag{4.27}$$

By using the polar decomposition theorem (1.145), we obtain

$$\mathbf{F} = \mathbf{R} \cdot \mathbf{U}, \quad \mathbf{C} = \mathbf{F}^T \cdot \mathbf{F} = \mathbf{U}^2, \quad \lambda^2(\mathbf{N}) = \mathbf{N} \cdot \mathbf{U}^2 \cdot \mathbf{N}. \tag{4.28}$$

4.4. Strain Tensor

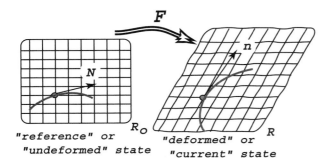

Figure 4.5. Stretch of an arbitrary fiber **N** in the reference state to **n** in the current, deformed state.

The tensor **U** is called the right stretch tensor. If we now define the strain in the direction **N** as

$$E(\mathbf{N}) = \frac{1}{2}\left[\lambda^2(\mathbf{N}) - 1\right] = \mathbf{N} \cdot \mathbf{E} \cdot \mathbf{N}, \tag{4.29}$$

we deduce the strain tensor

$$\mathbf{E} = \frac{1}{2}(\mathbf{C} - \mathbf{I}), \tag{4.30}$$

which is known as the *Green strain* (or *Lagrangian strain*) tensor.

An alternative approach to define this strain tensor is as follows. Let

$$\mathrm{d}S = \text{length of } \mathrm{d}\mathbf{X},$$
$$\mathrm{d}s = \text{length of } \mathrm{d}\mathbf{x}.$$

Then,

$$\mathrm{d}S^2 = \mathrm{d}\mathbf{X} \cdot \mathrm{d}\mathbf{X} = \mathrm{d}\mathbf{X} \cdot \mathbf{I} \cdot \mathrm{d}\mathbf{X},$$
$$\mathrm{d}s^2 = \mathrm{d}\mathbf{x} \cdot \mathrm{d}\mathbf{x} = (\mathbf{F} \cdot \mathrm{d}\mathbf{X}) \cdot (\mathbf{F} \cdot \mathrm{d}\mathbf{X}) = \mathrm{d}\mathbf{X} \cdot \mathbf{F}^T \cdot \mathbf{F} \cdot \mathrm{d}\mathbf{X}.$$

Combining the last two results, we have

$$\mathrm{d}s^2 - \mathrm{d}S^2 = \mathrm{d}\mathbf{X} \cdot (\mathbf{F}^T \cdot \mathbf{F} - \mathbf{I}) \cdot \mathrm{d}\mathbf{X}. \tag{4.31}$$

This leads to the definition of *Green strain* as

$$\mathbf{E} = \frac{1}{2}(\mathbf{F}^T \cdot \mathbf{F} - \mathbf{I}) = \frac{1}{2}(\mathbf{C} - \mathbf{I}),$$

because then

$$\mathrm{d}s^2 - \mathrm{d}S^2 = 2\mathrm{d}\mathbf{X} \cdot \mathbf{E} \cdot \mathrm{d}\mathbf{X}. \tag{4.32}$$

The Green strain is symmetric as may be seen *via* the manipulations

$$E_{ij} = \frac{1}{2}(F^T_{is}F_{sj} - \delta_{ij}),$$

$$E_{ji} = \frac{1}{2}(F^T_{js}F_{si} - \delta_{ji})$$
$$= \frac{1}{2}(F_{sj}F^T_{is} - \delta_{ij}) = \frac{1}{2}(F^T_{is}F_{sj} - \delta_{ij}).$$

This is also obvious from (4.4), because $\mathbf{F}^T \cdot \mathbf{F}$ and **I** are symmetric.

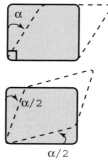

Figure 4.6. Angle change during shearing deformation.

4.5 Stretch and Normal Strains

Consider an infinitesimal vector, $d\mathbf{X}$, which is deformed by $\mathbf{F}$ into the vector element, $d\mathbf{x}$. The magnitudes of the lengths of these vectors are, respectively, dS and ds. Then,

$$ds^2 = d\mathbf{x} \cdot d\mathbf{x} = d\mathbf{X} \cdot \mathbf{F}^T \cdot \mathbf{F} \cdot d\mathbf{X} = d\mathbf{X} \cdot \mathbf{C} \cdot d\mathbf{X}. \tag{4.33}$$

By defining the unit vector, $\hat{\mathbf{N}} = d\mathbf{X}/dS$, we can write

$$ds^2/dS^2 = \hat{\mathbf{N}} \cdot \mathbf{C} \cdot \hat{\mathbf{N}} = \lambda^2(\hat{\mathbf{N}}). \tag{4.34}$$

Thus, $\lambda(\hat{\mathbf{N}})$ is the *stretch* of a fiber, initially of unit length, which was lying in the direction $\hat{\mathbf{N}}$ in the reference state.

For example, consider a uniaxial stretching, and take $\hat{\mathbf{N}} = \mathbf{e}_1$. Then,

$$\lambda^2(\mathbf{e}_1) = C_{11} = 2E_{11} + 1. \tag{4.35}$$

Also,

$$\frac{ds(\mathbf{e}_1) - dS(\mathbf{e}_1)}{dS} = \lambda(\mathbf{e}_1) - 1 = \sqrt{2E_{11} + 1} - 1. \tag{4.36}$$

If $E_{11} \ll 1$, we have

$$\frac{ds - dS}{dS} \approx E_{11}, \tag{4.37}$$

which is just the definition of a small uniaxial engineering (or nominal) strain measure. Clearly, by expanding $(1 + 2E_{11})^{1/2}$, we obtain

$$e_{11} = \frac{ds - dS}{dS} \sim E_{11} - \frac{1}{2} E_{11}^2 + \ldots. \tag{4.38}$$

4.6 Angle Change and Shear Strains

We begin by examining Fig. 4.6. Note that a convenient measure of the obvious distortion would be the change in the 90° angle made between the horizontal and vertical sides of the square section. This angular change is α as shown. Thus one measure of the distortion is simply $\cos(\pi/2 - \alpha)$, which is zero if $\alpha = 0$ and is nonzero if $\alpha \neq 0$, in which case $\cos(\pi/2 - \alpha) = \sin \alpha \neq 0$. We also note that the distortions in both figures are the same, and the displacements differ only by a rigid body rotation. Thus, $\cos(\pi/2 - \alpha)$ is an acceptable measure of strain. Now let us consider two infinitesimal vector elements, $d\mathbf{X}_1$

4.7. Infinitesimal Strains

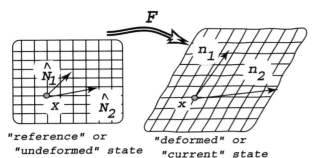

Figure 4.7. Angle change between two fibers.

and $d\mathbf{X}_2$, in the reference state; in the current state they become, respectively, $d\mathbf{x}_1$ and $d\mathbf{x}_2$ (Fig. 4.7). Unit vectors associated with these two can be defined as

$$\hat{\mathbf{N}}_1 = d\mathbf{X}_1/|d\mathbf{X}_1|, \quad \hat{\mathbf{n}}_1 = d\mathbf{x}_1/|d\mathbf{x}_1|, \tag{4.39}$$

$$\hat{\mathbf{N}}_2 = d\mathbf{X}_2/|d\mathbf{X}_2|, \quad \hat{\mathbf{n}}_2 = d\mathbf{x}_2/|d\mathbf{x}_2|. \tag{4.40}$$

We can now form the cosine of the angle between them as

$$\cos(\hat{\mathbf{n}}_1, \hat{\mathbf{n}}_2) = \hat{\mathbf{n}}_1 \cdot \hat{\mathbf{n}}_2 = \frac{d\mathbf{x}_1 \cdot d\mathbf{x}_2}{|d\mathbf{x}_1||d\mathbf{x}_2|}. \tag{4.41}$$

In view of the connections

$$d\mathbf{x}_1 = \mathbf{F} \cdot d\mathbf{X}_1, \quad d\mathbf{x}_2 = \mathbf{F} \cdot d\mathbf{X}_2,$$

$$d\mathbf{x}_1 = d\mathbf{X}_1 \cdot \mathbf{F}^T, \quad d\mathbf{x}_2 = d\mathbf{X}_2 \cdot \mathbf{F}^T,$$

the angle change can be expressed as

$$\cos(\hat{\mathbf{n}}_1, \hat{\mathbf{n}}_2) = \frac{d\mathbf{X}_1 \cdot \mathbf{F}^T \cdot \mathbf{F} \cdot d\mathbf{X}_2}{(d\mathbf{X}_1 \cdot \mathbf{C} \cdot d\mathbf{X}_1)^{1/2}(d\mathbf{X}_2 \cdot \mathbf{C} \cdot d\mathbf{X}_2)^{1/2}}. \tag{4.42}$$

But, $d\mathbf{X}_1 = \hat{\mathbf{N}}_1|d\mathbf{X}_1|$ and $d\mathbf{X}_2 = \hat{\mathbf{N}}_2|d\mathbf{X}_2|$, so that

$$\cos(\hat{\mathbf{n}}_1, \hat{\mathbf{n}}_2) = \frac{\hat{\mathbf{N}}_1 \cdot \mathbf{C} \cdot \hat{\mathbf{N}}_2}{\lambda(\hat{\mathbf{N}}_1)\lambda(\hat{\mathbf{N}}_2)}. \tag{4.43}$$

As an example, let $\hat{\mathbf{N}}_1 = \mathbf{e}_1$ and $\hat{\mathbf{N}}_2 = \mathbf{e}_2$. Then,

$$\cos(\hat{\mathbf{n}}_1, \hat{\mathbf{n}}_2) = \frac{C_{12}}{\lambda(\mathbf{e}_1)\lambda(\mathbf{e}_2)} = \frac{C_{12}}{\sqrt{C_{11}}\sqrt{C_{22}}}$$

$$= \frac{2E_{12}}{\sqrt{2E_{11}+1}\sqrt{2E_{22}+1}}. \tag{4.44}$$

If all strains are small, $|E_{ij}| \ll 1$, we have

$$\cos(\hat{\mathbf{n}}_1, \hat{\mathbf{n}}_2) \approx 2E_{12}. \tag{4.45}$$

4.7 Infinitesimal Strains

In general, the connection between positions in the reference and deformed states can be written as

$$\mathbf{x}(\mathbf{X}) = \mathbf{X} + \mathbf{u}(\mathbf{X}). \tag{4.46}$$

Consider a particular component of strain, say

$$E_{11} = \frac{1}{2}\left(F_{1s}^T F_{s1} - 1\right). \tag{4.47}$$

Since

$$F_{ij} = \delta_{ij} + \partial u_i/\partial X_j, \quad F_{ij}^T = \delta_{ij} + \partial u_j/\partial X_i, \tag{4.48}$$

we have

$$E_{11} = \frac{1}{2}\left[(\delta_{1s} + \partial u_s/\partial X_1)(\delta_{s1} + \partial u_s/\partial X_1) - 1\right],$$

i.e.,

$$E_{11} = \frac{1}{2}\left[(\partial u_1/\partial X_1 + \partial u_1/\partial X_1) + (\partial u_s/\partial X_1)(\partial u_s/\partial X_1)\right]. \tag{4.49}$$

In general, we obtain that

$$E_{ij} = \frac{1}{2}\left[(\partial u_i/\partial X_j + \partial u_j/\partial X_i) + (\partial u_s/\partial X_i)(\partial u_s/\partial X_j)\right]. \tag{4.50}$$

If $|\partial \mathbf{u}/\partial \mathbf{X}| \ll 1$, the E_{ij} reduce to the infinitesimal strain components,

$$E_{ij} \approx e_{ij} = \frac{1}{2}(\partial u_i/\partial X_j + \partial u_j/\partial X_i). \tag{4.51}$$

4.8 Principal Stretches

Consider the stretch of a fiber along the direction of the unit vector $\mathbf{n}$. We have

$$\lambda^2(\mathbf{n}) = \mathbf{n} \cdot \mathbf{C} \cdot \mathbf{n}. \tag{4.52}$$

We seek those $\mathbf{n}$ for which the stretch is extreme (principal stretch). To find the extremum introduce a Lagrange multiplier, μ, *via*

$$\mathcal{L} = \mathbf{n} \cdot \mathbf{C} \cdot \mathbf{n} - \mu(\mathbf{n} \cdot \mathbf{n} - 1). \tag{4.53}$$

Then,

$$\partial \mathcal{L}(\mathbf{n}, \mu)/\partial \mathbf{n} = 2(\mathbf{C} \cdot \mathbf{n} - \mu \mathbf{n}) = \mathbf{0},$$

which leads to

$$\mathbf{C} \cdot \mathbf{n} - \mu \mathbf{n} = \mathbf{0}, \tag{4.54}$$

where the μ's are the eigenvalues of $\mathbf{C}$. By using the results from Chapter 1, we have

$$\mu^3 - I_C \mu^2 + II_C \mu - III_C = 0, \tag{4.55}$$

where the invariants of $\mathbf{C}$ are

$$I_C = \mathrm{tr}\,(\mathbf{C}),$$
$$II_C = [\mathbf{e}_1, \mathbf{C} \cdot \mathbf{e}_2, \mathbf{C} \cdot \mathbf{e}_3] + [\mathbf{C} \cdot \mathbf{e}_1, \mathbf{e}_2, \mathbf{C} \cdot \mathbf{e}_3] + [\mathbf{C} \cdot \mathbf{e}_1, \mathbf{C} \cdot \mathbf{e}_2, \mathbf{e}_3], \tag{4.56}$$
$$III_C = [\mathbf{C} \cdot \mathbf{e}_1, \mathbf{C} \cdot \mathbf{e}_2, \mathbf{C} \cdot \mathbf{e}_3].$$

4.9 Eigenvectors and Eigenvalues of Deformation Tensors

The deformation gradient may be decomposed *via* the polar decomposition theorem as $\mathbf{F} = \mathbf{R} \cdot \mathbf{U}$, with $\mathbf{C} = \mathbf{U}^T \cdot \mathbf{U} = \mathbf{U}^2$. Thus, if

$$\mathbf{C} = \sum_{i=1}^{3} \mu_p \mathbf{n}_p \mathbf{n}_p, \tag{4.57}$$

then

$$\mathbf{U} = \sum_{i=1}^{3} \mu_p^{1/2} \mathbf{n}_p \mathbf{n}_p, \tag{4.58}$$

where the eigenvalues, μ_p, are associated with eigenvectors, $\mathbf{n}_p$, *via*

$$\mathbf{C} \cdot \mathbf{n}_p = \mu_p \mathbf{n}_p. \tag{4.59}$$

Consider the stretch of an eigenvector, $\lambda(\mathbf{n}_p)$. Its square is

$$\lambda^2(\mathbf{n}_p) = \mathbf{n}_p \cdot \mathbf{C} \cdot \mathbf{n}_p = \mu_p. \tag{4.60}$$

If $p \neq s$, then $\mathbf{n}_p \cdot \mathbf{n}_s = 0$. Indeed,

$$\mathbf{C} \cdot \mathbf{n}_p = \mu_p \mathbf{n}_p, \tag{4.61}$$

and thus, since $\mathbf{C}$ is symmetric,

$$\mathbf{n}_s \cdot \mathbf{C} \cdot \mathbf{n}_p - \mathbf{n}_p \cdot \mathbf{C} \cdot \mathbf{n}_s = 0. \tag{4.62}$$

Also,

$$\mathbf{n}_s \cdot \mu_p \mathbf{n}_p - \mathbf{n}_p \mu_s \mathbf{n}_s = 0,$$

i.e.,

$$(\mu_p - \mu_s) \mathbf{n}_s \cdot \mathbf{n}_p = 0. \tag{4.63}$$

Thus, if $\mu_p \neq \mu_s$, then $\mathbf{n}_p \cdot \mathbf{n}_s = 0$.

Now consider $\cos(\mathbf{F} \cdot \mathbf{n}_j, \mathbf{F} \cdot \mathbf{n}_k)$. We have

$$\frac{\mathbf{n}_j \cdot \mathbf{C} \cdot \mathbf{n}_k}{\sqrt{\mu_j}\sqrt{\mu_k}} = \frac{\mathbf{n}_j \cdot (\mu_k \mathbf{n}_k)}{\sqrt{\mu_j}\sqrt{\mu_k}} = \frac{\sqrt{\mu_k}}{\sqrt{\mu_j}} \mathbf{n}_j \cdot \mathbf{n}_k = 0. \tag{4.64}$$

Therefore, the eigenvectors of $\mathbf{C}$ undergo pure stretch, *i.e.*, they are not rotated by the deformation.

4.10 Volume Changes

Consider the volume element, δV as it is transformed into δv through $\mathbf{F}$ (Fig. 4.8). Clearly,

$$\delta V = |\mathrm{d}X_1||\mathrm{d}X_2||\mathrm{d}X_3|, \quad \text{and} \quad \delta v = |\mathrm{d}x_1||\mathrm{d}x_2||\mathrm{d}x_3|. \tag{4.65}$$

Also,

$$\delta v = [\mathbf{F} \cdot \mathrm{d}\mathbf{X}_1, \mathbf{F} \cdot \mathrm{d}\mathbf{X}_2, \mathbf{F} \cdot \mathrm{d}\mathbf{X}_3]. \tag{4.66}$$

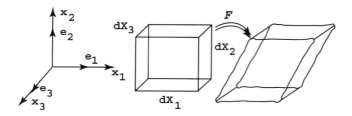

Figure 4.8. Changes in volume of a typical volume element.

Now, if $d\mathbf{X}_i = \mathbf{e}_i$, then $\delta V = 1$, and

$$\delta v = (\delta v/\delta V) = III_F = \det \mathbf{F} = |\mathbf{F}|. \tag{4.67}$$

4.11 Area Changes

The area element, bounded by the infinitesimals $d\mathbf{X}$ and $d\mathbf{Y}$ shown in Fig. 4.9, can clearly be described as $\hat{\mathbf{N}} dS = d\mathbf{X} \times d\mathbf{Y}$; dS is the magnitude of the area and $\hat{\mathbf{N}}$ is the unit normal to the surface "patch." The componental representation for the vectorial area of the undeformed surface patch is

$$\hat{N}_i dS = \epsilon_{ijk} dX_j dY_k. \tag{4.68}$$

Likewise, for the deformed surface patch,

$$\hat{n}_i ds = \epsilon_{ijk} dx_j dy_k, \tag{4.69}$$

where ds is the area of the deformed surface patch and $\hat{\mathbf{n}}$ is its unit normal. Since $d\mathbf{x} = \mathbf{F} \cdot d\mathbf{X}$, we have

$$\hat{n}_i ds = \epsilon_{ijk} F_{js} dX_s F_{kp} dY_p. \tag{4.70}$$

Now multiply both sides of (4.70) by F_{it} and perform the indicated summations to obtain

$$\hat{n}_i F_{it} ds = \epsilon_{ijk} F_{js} F_{kp} F_{it} dX_s dX_p. \tag{4.71}$$

Note that within the *rhs* we find the term $\epsilon_{tsp}(\det \mathbf{F}) \sim \epsilon_{ijk} F_{js} F_{kp} F_{it}$. This allows us to write

$$\hat{n}_i F_{it} ds = \epsilon_{tsp}(\det \mathbf{F}) dX_s dY_p = \hat{N}_t(\det \mathbf{F}) dS. \tag{4.72}$$

Taking a product with F_{tr}^{-1} gives

$$\hat{n}_i F_{it} F_{tr}^{-1} ds = \hat{N}_t F_{tr}^{-1}(\det \mathbf{F}) dS,$$

i.e.,

$$\hat{n}_r ds = (\det \mathbf{F}) \hat{N}_t F_{tr}^{-1} dS, \tag{4.73}$$

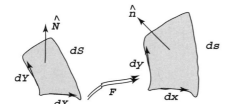

Figure 4.9. Area change of a surface patch (element) during a general deformation.

4.12. Area Changes: Alternative Approach

and thus

$$\hat{\mathbf{n}} ds = (\det \mathbf{F})\hat{\mathbf{N}} \cdot \mathbf{F}^{-1} dS = (\det \mathbf{F})\mathbf{F}^{-T} \cdot \hat{\mathbf{N}} dS. \tag{4.74}$$

This is known as Nanson's relation between the undeformed and deformed vectorial areas. By multiplying (4.74) with itself, and then taking a square root of the result, we obtain

$$ds = (\det \mathbf{F})(\hat{\mathbf{N}} \cdot \mathbf{C}^{-1} \cdot \hat{\mathbf{N}})^{1/2} \, dS, \tag{4.75}$$

where $\mathbf{C} = \mathbf{F}^T \cdot \mathbf{F}$ is the right Cauchy–Green deformation tensor. Substituting this back into (4.74), we obtain an expression for the unit normal $\hat{\mathbf{n}}$ in terms of the unit normal $\hat{\mathbf{N}}$, i.e.,

$$\hat{\mathbf{n}} = \frac{\mathbf{F}^{-T} \cdot \hat{\mathbf{N}}}{(\hat{\mathbf{N}} \cdot \mathbf{C}^{-1} \cdot \hat{\mathbf{N}})^{1/2}}. \tag{4.76}$$

Alternatively, if (4.74) is rewritten as

$$\hat{\mathbf{N}} dS = \frac{1}{\det \mathbf{F}} \mathbf{F}^T \cdot \hat{\mathbf{n}} \, ds = \frac{1}{\det \mathbf{F}} \hat{\mathbf{n}} \cdot \mathbf{F} \, ds, \tag{4.77}$$

the multiplication with itself gives

$$dS = \frac{1}{\det \mathbf{F}} (\hat{\mathbf{n}} \cdot \mathbf{B} \cdot \hat{\mathbf{n}})^{1/2} \, ds, \tag{4.78}$$

where $\mathbf{B} = \mathbf{F} \cdot \mathbf{F}^T$ is the left Cauchy–Green deformation tensor. When (4.78) is substituted back into (4.77), we obtain an expression for the unit normal $\hat{\mathbf{N}}$ in terms of the unit normal $\hat{\mathbf{n}}$, i.e.,

$$\hat{\mathbf{N}} = \frac{\mathbf{F}^T \cdot \hat{\mathbf{n}}}{(\hat{\mathbf{n}} \cdot \mathbf{B} \cdot \hat{\mathbf{n}})^{1/2}}. \tag{4.79}$$

Two identities can be observed from the above results, one for the ratio of the deformed and undeformed surface areas and another for the cosine of the angle between their unit normals. These are

$$\frac{ds}{dS} = (\det \mathbf{F})(\hat{\mathbf{n}} \cdot \mathbf{B} \cdot \hat{\mathbf{n}})^{-1/2} = (\det \mathbf{F})(\hat{\mathbf{N}} \cdot \mathbf{C}^{-1} \cdot \hat{\mathbf{N}})^{1/2}, \tag{4.80}$$

$$\hat{\mathbf{n}} \cdot \hat{\mathbf{N}} = \frac{\hat{\mathbf{n}} \cdot \mathbf{F}^T \cdot \hat{\mathbf{n}}}{(\hat{\mathbf{n}} \cdot \mathbf{B} \cdot \hat{\mathbf{n}})^{1/2}} = \frac{\hat{\mathbf{N}} \cdot \mathbf{F}^{-T} \cdot \hat{\mathbf{N}}}{(\hat{\mathbf{N}} \cdot \mathbf{C}^{-1} \cdot \hat{\mathbf{N}})^{1/2}}. \tag{4.81}$$

4.12 Area Changes: Alternative Approach

A second-order tensor $\mathbf{F}$ has a dual tensor $\mathbf{F}^*$, such that

$$\mathbf{F}^* \cdot (\mathbf{a} \times \mathbf{b}) = (\mathbf{F} \cdot \mathbf{a}) \times (\mathbf{F} \cdot \mathbf{b}), \tag{4.82}$$

where $\mathbf{a}$ and $\mathbf{b}$ are arbitrary vectors. To see this and identify $\mathbf{F}^*$, let $\mathbf{c}$ be still a third arbitrary vector. Then, by using the result from Problem 1.7, we obtain

$$[\mathbf{F}^* \cdot (\mathbf{a} \times \mathbf{b})] \cdot \mathbf{c} = [(\mathbf{F} \cdot \mathbf{a}) \times (\mathbf{F} \cdot \mathbf{b})] \cdot (\mathbf{F} \cdot \mathbf{F}^{-1} \cdot \mathbf{c}) = (\det \mathbf{F})(\mathbf{a} \times \mathbf{b}) \cdot \mathbf{F}^{-1} \cdot \mathbf{c}.$$

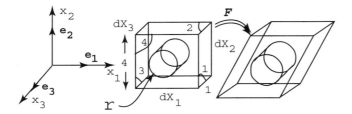

Figure 4.10. Block with circular hole subject to a deformation involving a simple shear.

Recalling that $\mathbf{F}^{-1} \cdot \mathbf{c} = \mathbf{c} \cdot \mathbf{F}^{-T}$, and by invoking cyclic property of triple product and commutative property of scalar product, the above becomes

$$\mathbf{c} \cdot \mathbf{F}^* \cdot (\mathbf{a} \times \mathbf{b}) = (\det \mathbf{F}) \mathbf{c} \cdot \mathbf{F}^{-T} \cdot (\mathbf{a} \times \mathbf{b}). \tag{4.83}$$

Since this holds for arbitrary trio of vectors **a**, **b**, **c**, we obtain

$$\mathbf{F}^* = (\det \mathbf{F}) \mathbf{F}^{-T}. \tag{4.84}$$

Applying this result to the computation of area, we find

$$d\mathbf{x} \times d\mathbf{y} = (\mathbf{F} \cdot d\mathbf{X}) \times (\mathbf{F} \cdot d\mathbf{Y}) = \mathbf{F}^* \cdot (d\mathbf{X} \times d\mathbf{Y}) = (\det \mathbf{F}) \mathbf{F}^{-T} \cdot \hat{\mathbf{N}} dS.$$

4.13 Simple Shear of a Thick Plate with a Central Hole

A three-dimensional state of simple shear, as depicted in Fig. 4.10, is prescribed by

$$\begin{aligned} x_1 &= X_1 + \gamma X_2, \\ x_2 &= X_2, \\ x_3 &= X_3. \end{aligned} \tag{4.85}$$

The corresponding deformation gradient and its inverse are

$$\mathbf{F} = \begin{bmatrix} 1 & \gamma & 0 \\ 0 & 1 & 0 \\ 0 & 0 & 1 \end{bmatrix}, \quad \mathbf{F}^{-1} = \begin{bmatrix} 1 & -\gamma & 0 \\ 0 & 1 & 0 \\ 0 & 0 & 1 \end{bmatrix}. \tag{4.86}$$

Furthermore,

$$(\mathbf{F}^{-1})^T = \mathbf{F}^{-T} = \begin{bmatrix} 1 & 0 & 0 \\ -\gamma & 1 & 0 \\ 0 & 0 & 1 \end{bmatrix}, \tag{4.87}$$

and

$$\hat{\mathbf{n}} ds = (\det \mathbf{F}) \mathbf{F}^{-T} \cdot \hat{\mathbf{N}} dS. \tag{4.88}$$

On side 1 we have $\hat{\mathbf{N}} = \mathbf{e}_1$, $dS = dX_3 dX_2$, and $\det \mathbf{F} = 1$. Thus,

$$\hat{\mathbf{n}} ds = \mathbf{F}^{-T} \cdot \mathbf{e}_1 dX_3 dX_2. \tag{4.89}$$

4.13. Simple Shear of a Hollow Plate

This means that

$$\begin{aligned}
dA_1^2 &= \hat{\mathbf{n}}ds \cdot \hat{\mathbf{n}}ds \\
&= \mathbf{F}^{-T} \cdot \mathbf{e}_1 (dX_3 dX_2)^2 \cdot (\mathbf{F}^{-T} \cdot \mathbf{e}_1) \\
&= \mathbf{e}_1 \cdot \mathbf{F}^{-1} \cdot \mathbf{F}^{-T} \cdot \mathbf{e}_1 (dX_3 dX_2)^2 \\
&= F_{1s}^{-1} F_{s1}^{-T} (dX_3 dX_2)^2 \\
&= (1 + \gamma^2)(dX_3 dX_2)^2,
\end{aligned}$$

that is,

$$dA_1 = \sqrt{1 + \gamma^2}\, dX_3 dX_2. \tag{4.90}$$

Thus, on all of side 1,

$$\int_{\text{side 1}} dA = \int_0^1 \int_0^4 \sqrt{1 + \gamma^2}\, dX_2\, dX_3 = 4\sqrt{1 + \gamma^2}. \tag{4.91}$$

Similarly, for side 3,

$$\int_{\text{side3}} dA = 4\sqrt{1 + \gamma^2}. \tag{4.92}$$

On side 2, we have $\hat{\mathbf{N}} = \mathbf{e}_2$, $dS = dX_1 dX_3$. Thus,

$$\hat{\mathbf{n}}ds = \mathbf{F}^{-T} \cdot \mathbf{e}_2 dX_1 dX_3, \tag{4.93}$$

and

$$\begin{aligned}
dA_2^2 &= \hat{\mathbf{n}}ds \cdot \hat{\mathbf{n}}ds \\
&= \mathbf{e}_2 \cdot \mathbf{F}^{-1} \cdot \mathbf{F}^{-T} \cdot \mathbf{e}_2 (dX_1 dX_3)^2 \\
&= F_{2s}^{-1} (F_{s2}^{-1})^T (dX_1 dX_3)^2 \\
&= 1(dX_1 dX_3)^2 = \text{no change!}
\end{aligned}$$

For side 4 we have the same result as for side 2.

Now consider the circular hole. Its radius in the reference state is r. With the origin at the hole's center, and using a polar coordinate system aligned with the $x_1 - x_2$ axes, we have that

$$\hat{\mathbf{N}} dS = (\cos\theta \mathbf{e}_1 + \sin\theta \mathbf{e}_2) r\, d\theta\, dX_3, \tag{4.94}$$

and (Fig. 4.11)

$$\begin{aligned}
\hat{\mathbf{n}}ds &= (\det \mathbf{F})\mathbf{F}^{-T} \cdot \hat{\mathbf{N}} dS \\
&= (F_{1s}^{-1})\hat{N}_s r\, d\theta\, dX_3 \mathbf{e}_1 + (F_{2s}^{-1})^T \hat{N}_s r\, d\theta\, dX_3 \mathbf{e}_2 \\
&= [\cos\theta \mathbf{e}_1 + (\sin\theta - \gamma \cos\theta)\mathbf{e}_2] r\, d\theta\, dX_3.
\end{aligned}$$

Consequently,

$$\begin{aligned}
dA^2 &= \hat{\mathbf{n}} \cdot ds \cdot \hat{\mathbf{n}} \cdot ds \\
&= [\cos\theta \mathbf{e}_1 + (\sin\theta - \gamma \cos\theta)\mathbf{e}_2] \cdot [\cos\theta \mathbf{e}_1 + (\sin\theta - \gamma \cos\theta)\mathbf{e}_2] r^2 d\theta^2 dX_3^2,
\end{aligned}$$

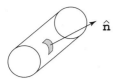

Figure 4.11. Normal to the hole's surface. See Fig. 4.10 for a complete perspective of the geometry.

or

$$dA^2 = [\cos^2\theta + (\sin\theta - \gamma\cos\theta)^2]r^2 d\theta^2 dX_3^2$$
$$= (\cos^2\theta + \sin^2\theta - 2\gamma\sin\theta\cos\theta + \gamma^2\cos^2\theta)r^2 d\theta^2 dX_3^2$$
$$= (1 - \gamma\sin 2\theta + \gamma^2\cos^2\theta)r^2 d\theta^2 dX_3^2.$$

Finally,

$$dA = \sqrt{1 - \gamma\sin 2\theta + \gamma^2\cos^2\theta}\, r\, d\theta\, dX_3, \tag{4.95}$$

which, when integrated, yields

$$A = r \int_0^1 \int_0^{2\pi} \sqrt{1 - \gamma\sin 2\theta + \gamma^2\cos^2\theta}\, d\theta\, dX_3. \tag{4.96}$$

4.14 Finite *vs.* Small Deformations

It is worthwhile to explore the differences between finite and small deformations for the purpose of understanding how even seemingly qualitative effects can arise in the transition from infinitesimal to finite deformation. We do this here *via* a simple example. Consider the deformation

$$\begin{aligned} x_1 &= X_1 - \gamma X_2 + \beta X_3, \\ x_2 &= \gamma X_1 + X_2 - \alpha X_3, \\ x_3 &= -\beta X_1 + \alpha X_2 + X_3. \end{aligned} \tag{4.97}$$

What we show is that this deformation involves only a rigid body rotation if the α, β, γ are "small," but a general deformation if they are "finite."

Indeed, the deformation gradient is

$$\mathbf{F} = \begin{bmatrix} 1 & -\gamma & \beta \\ \gamma & 1 & -\alpha \\ -\beta & \alpha & 1 \end{bmatrix}, \tag{4.98}$$

whereas the corresponding Green strain is

$$\mathbf{E} = \frac{1}{2}\begin{bmatrix} \beta^2 + \gamma^2 & -\alpha\beta & -\alpha\gamma \\ -\alpha\beta & \alpha^2 + \gamma^2 & -\beta\gamma \\ -\alpha\gamma & -\beta\gamma & \alpha^2 + \beta^2 \end{bmatrix}. \tag{4.99}$$

4.15. Reference vs. Current Configuration

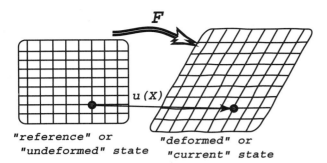

"reference" or "undeformed" state

"deformed" or "current" state

Figure 4.12. Reference and current coordinate systems.

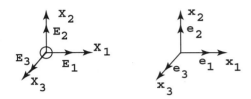

Clearly, the strains vanish to $\mathcal{O}(\xi\zeta)$ where $\xi, \zeta = \alpha, \beta, \gamma$. If, however, these quadratic terms are not negligible, the state is one that involves deformation. In any event, the rotations do not vanish. The rotation vector, $\mathbf{w}$, can be calculated as

$$w_i = -\frac{1}{2} \epsilon_{ijk} W_{jk}, \tag{4.100}$$

which yields $w_1 = \alpha$, $w_2 = \beta$, $w_3 = \gamma$.

4.15 Reference vs. Current Configuration

We again begin by defining the deformation commencing from the reference to the current states as shown in Fig. 4.12. We introduce two coordinate frames each belonging to one state or the other. In particular, the frame $\{\mathbf{E}_1, \mathbf{E}_2, \mathbf{E}_3\}$ is defined in the reference state, and $\{\mathbf{e}_1, \mathbf{e}_2, \mathbf{e}_3\}$ is defined in the current, *i.e.*, deformed state. For clarity we will use Greek subscripts to denote components referred to the reference state and Latin subscripts for the current state. Thus a vector such as $\mathbf{u}$ is expressed as either

$$\mathbf{u} = u_\pi \mathbf{E}_\pi, \quad \text{or} \quad \mathbf{u} = u_s \mathbf{e}_s. \tag{4.101}$$

Their components are formed as

$$u_\pi = \mathbf{u} \cdot \mathbf{E}_\pi, \quad \text{or} \quad u_s = \mathbf{u} \cdot \mathbf{e}_s. \tag{4.102}$$

Higher order tensors can be handled similarly. For example,

$$\boldsymbol{\sigma} = \sigma_{\alpha\beta} \mathbf{E}_\alpha \mathbf{E}_\beta = \sigma_{sr} \mathbf{e}_s \mathbf{e}_r, \tag{4.103}$$

where

$$\sigma_{\alpha\beta} = \mathbf{E}_\alpha \cdot \boldsymbol{\sigma} \cdot \mathbf{E}_\beta \quad \sigma_{sr} = \mathbf{e}_s \cdot \boldsymbol{\sigma} \cdot \mathbf{e}_r. \tag{4.104}$$

Mixed representations are also possible, *e.g.*,

$$\boldsymbol{\sigma} = \sigma_{\alpha i} \mathbf{E}_\alpha \mathbf{e}_i = \sigma_{s\beta} \mathbf{e}_s \mathbf{E}_\beta, \tag{4.105}$$

where

$$\sigma_{\alpha i} = \mathbf{E}_\alpha \cdot \boldsymbol{\sigma} \cdot \mathbf{e}_i, \tag{4.106}$$

$$\sigma_{s\beta} = \mathbf{e}_s \cdot \boldsymbol{\sigma} \cdot \mathbf{E}_\beta. \tag{4.107}$$

Some components are most naturally represented as mixed components, including those of the deformation gradient. Recall

$$\mathbf{x} = \mathbf{x}(\mathbf{X}, t), \tag{4.108}$$

where the possibility that the deformation mapping depends on time, t, has been explicitly included. Then the gradient is

$$\mathbf{F} = \partial \mathbf{x}/\partial \mathbf{X} = F_{i\alpha} \mathbf{e}_i \mathbf{E}_\alpha, \tag{4.109}$$

because $\mathbf{x}$ "lives" in the current configuration, and $\mathbf{X}$ in the reference state.

We note that $\mathbf{F}$ is invertible and so $\mathbf{F}^{-1}$ exists – this is physically demandable from the various pictures of the deformation presented to date, and from the statements that the displacements involved are single valued. Thus each material point is, at this juncture, taken to originate from a unique point in the reference configuration and is displaced to a unique point in the current configuration. Thus, the process is "conceptually invertible," and $\mathbf{F}^{-1}$ exists. Therefore, we can write

$$\mathbf{X} = \Psi(\mathbf{x}, t), \quad \mathbf{F}^{-1} = \partial \Psi(\mathbf{x}, t)/\partial \mathbf{x} = \partial \mathbf{X}/\partial \mathbf{x}. \tag{4.110}$$

For a component representation, we have

$$\mathbf{F}^{-1} = F^{-1}_{\alpha r} \mathbf{E}_\alpha \mathbf{e}_r, \quad F^{-1}_{\alpha r} = \partial X_\alpha/\partial x_r. \tag{4.111}$$

LEMMA 4.1: *If ϕ is a scalar field, then*

$$\mathrm{Grad}\, \phi = \partial \phi/\partial \mathbf{X} = \mathbf{F}^T \cdot \mathrm{grad}\, \phi, \tag{4.112}$$

where $\mathrm{grad}\, \phi = \partial \phi/\partial \mathbf{x}$.

Proof: First we write $\mathrm{Grad}\, \phi = \partial \phi/\partial \mathbf{X} = \partial \phi/\partial X_\alpha \mathbf{E}_\alpha$. Then,

$$\partial \phi/\partial \mathbf{X} = \partial \phi/\partial \mathbf{x} \cdot \partial \mathbf{x}/\partial \mathbf{X} = (\partial \phi/\partial x_i \mathbf{e}_i) \cdot (\partial x_p/\partial X_\alpha \mathbf{e}_p \mathbf{E}_\alpha)$$

$$= (\partial x_p/\partial X_\alpha \mathbf{E}_\alpha \mathbf{e}_p) \cdot (\partial \phi/\partial x_i \mathbf{e}_i)$$

$$= \mathbf{F}^T \cdot \mathrm{grad}\, \phi.$$

COROLLARY 4.1: *Similarly to Lemma 4.1 above, we find that for a vector field $\boldsymbol{u}$,*

$$\mathrm{Grad}\, \boldsymbol{u} = (\mathrm{grad}\, \boldsymbol{u}) \cdot \mathbf{F}, \tag{4.113}$$

i.e.,

$$\mathbf{u}\overleftarrow{\nabla}^0 = (\mathbf{u}\overleftarrow{\nabla}) \cdot \mathbf{F}, \quad \frac{\partial u_i}{\partial X_j} = \frac{\partial u_i}{\partial x_k} \frac{\partial x_k}{\partial X_j}. \tag{4.114}$$

For clarity, the arrow above the nabla operator is attached to indicate the direction in which the operator applies.

4.16 Material Derivatives and Velocity

Let t be a time like variable that increases monotonically with the deformation process. It is assumed that $\mathbf{F}$ evolves with t. Likewise all scalar or tensor fields may evolve with time. For a scalar field then

$$\phi = \text{scalar field} = \phi(\mathbf{X}, t). \tag{4.115}$$

We may define the "material derivative" of ϕ as

$$\dot{\phi}(\mathbf{X}, t) = \partial \phi(\mathbf{X}, t)/\partial t, \tag{4.116}$$

that is, a derivative taken at constant reference position, $\mathbf{X}$. Stated otherwise, this is a rate of change at a fixed material point. Clearly $\mathbf{X}$ does not depend on time. But the current position of a material point, $\mathbf{x}$, does depend on time, so that in the current configuration where we have the representation, $\phi(\mathbf{x}, t)$, the time rate of ϕ is

$$\dot{\phi}(\mathbf{x}, t) = \partial \phi(\mathbf{x}, t)/\partial t + (\partial \phi/\partial \mathbf{x}) \cdot (\partial \mathbf{x}/\partial t). \tag{4.117}$$

Defining the velocity as

$$v(\mathbf{x}, t) = \partial \mathbf{x}/\partial t, \tag{4.118}$$

the rate of change of the scalar field is

$$\dot{\phi}(\mathbf{x}, t) = \partial \phi(\mathbf{x}, t)/\partial t + \nabla \phi \cdot v. \tag{4.119}$$

But, $\nabla \phi = \text{grad } \phi$, and so

$$\dot{\phi}(\mathbf{x}, t) = \partial \phi(\mathbf{x}, t)/\partial t + (\text{grad } \phi) \cdot v. \tag{4.120}$$

4.17 Velocity Gradient

The velocity gradient is defined as

$$\mathbf{L} = \text{grad } v = \text{grad } \dot{\mathbf{x}}, \tag{4.121}$$

i.e.,

$$\mathbf{L} = v \overleftarrow{\nabla}, \quad L_{ij} = \frac{\partial v_i}{\partial x_j}. \tag{4.122}$$

By using (4.113), we obtain

$$\text{Grad } \dot{\mathbf{x}} = (\text{grad } \dot{\mathbf{x}}) \cdot \mathbf{F}, \tag{4.123}$$

or

$$\text{grad } \dot{\mathbf{x}} = \text{grad } v = \mathbf{L} = (\text{Grad } \dot{\mathbf{x}}) \cdot \mathbf{F}^{-1}. \tag{4.124}$$

But,

$$\text{Grad } \dot{\mathbf{x}} = \dot{\mathbf{F}} \tag{4.125}$$

and so

$$\mathbf{L} = \text{grad } v = \text{grad } \dot{\mathbf{x}} = \dot{\mathbf{F}} \cdot \mathbf{F}^{-1}. \tag{4.126}$$

Note that $\mathbf{F} \cdot \mathbf{F}^{-1} = \mathbf{I}$. Taking the time derivative of this identity, we obtain

$$\dot{\mathbf{F}} \cdot \mathbf{F}^{-1} + \mathbf{F} \cdot (\mathbf{F}^{-1})^{\cdot} = \mathbf{0} \quad \Rightarrow \quad (\mathbf{F}^{-1})^{\cdot} = -\mathbf{F}^{-1} \cdot \dot{\mathbf{F}} \cdot \mathbf{F}^{-1}. \tag{4.127}$$

Thus,

$$\mathbf{L} = \dot{\mathbf{F}} \cdot \mathbf{F}^{-1} = -\mathbf{F} \cdot (\mathbf{F}^{-1})^{\cdot}. \tag{4.128}$$

LEMMA 4.2: *If $\mathbf{F}$ is invertible and depends on a time like parameter, t, then*

$$\frac{d}{dt}(\det \mathbf{F}) \equiv \frac{d}{dt} J = (\det \mathbf{F}) \, \mathrm{tr}(\dot{\mathbf{F}} \cdot \mathbf{F}^{-1}). \tag{4.129}$$

Proof: Begin by writing

$$\det \mathbf{F} = [\mathbf{F} \cdot \mathbf{a}, \mathbf{F} \cdot \mathbf{b}, \mathbf{F} \cdot \mathbf{c}], \tag{4.130}$$

where $\{\mathbf{a}, \mathbf{b}, \mathbf{c}\}$ is a set of any convenient orthonormal unit vectors. Then,

$$\frac{d}{dt}(\det \mathbf{F}) = \left[\frac{d\mathbf{F}}{dt} \cdot \mathbf{a}, \mathbf{F} \cdot \mathbf{b}, \mathbf{F} \cdot \mathbf{c}\right] + \left[\mathbf{F} \cdot \mathbf{a}, \frac{d\mathbf{F}}{dt} \cdot \mathbf{b}, \mathbf{F} \cdot \mathbf{c}\right] + \left[\mathbf{F} \cdot \mathbf{a}, \mathbf{F} \cdot \mathbf{b}, \frac{d\mathbf{F}}{dt} \cdot \mathbf{c}\right].$$

Since

$$\mathbf{L} = \frac{d\mathbf{F}}{dt} \cdot \mathbf{F}^{-1}, \tag{4.131}$$

we have

$$\frac{d}{dt}(\det \mathbf{F}) = [\mathbf{L} \cdot \mathbf{F} \cdot \mathbf{a}, \mathbf{F} \cdot \mathbf{b}, \mathbf{F} \cdot \mathbf{c}] + [\mathbf{F} \cdot \mathbf{a}, \mathbf{L} \cdot \mathbf{F} \cdot \mathbf{b}, \mathbf{F} \cdot \mathbf{c}] + [\mathbf{F} \cdot \mathbf{a}, \mathbf{F} \cdot \mathbf{b}, \mathbf{L} \cdot \mathbf{F} \cdot \mathbf{c}].$$

Clearly, then,

$$\frac{d}{dt}(\det \mathbf{F}) = (\mathrm{tr}\,\mathbf{L})(\det \mathbf{F}). \tag{4.132}$$

Recalling that $(\delta v/\delta V) \equiv J = \det \mathbf{F}$, this means that

$$\dot{J} = dJ/dt = (\delta v/\delta V)^{\cdot} = J \,\mathrm{tr}\,(\dot{\mathbf{F}} \cdot \mathbf{F}^{-1}) = J(\mathrm{tr}\,\mathbf{L}). \tag{4.133}$$

But, $\mathbf{L} = \mathrm{grad}\,v = \partial v / \partial \mathbf{x}$, and we have

$$\mathrm{tr}\,\mathbf{L} = \mathrm{div}\,v = \nabla \cdot v, \tag{4.134}$$

or

$$\dot{J} = J \,\mathrm{div}\,(v). \tag{4.135}$$

Let us return now to the rate of change of an infinitesimal material fiber (Fig. 4.3), and recall that

$$d\mathbf{x} = \mathbf{F} \cdot d\mathbf{X}. \tag{4.136}$$

Then,

$$(d\mathbf{x})^{\cdot} = d\dot{\mathbf{x}} = \dot{\mathbf{F}} \cdot d\mathbf{X} = \mathbf{L} \cdot \mathbf{F} \cdot d\mathbf{X} = \mathbf{L} \cdot d\mathbf{x}. \tag{4.137}$$

4.17. Velocity Gradient

Consider
$$d\mathbf{x} = \boldsymbol{\ell} ds, \tag{4.138}$$
where

$\boldsymbol{\ell} =$ a unit vector $\parallel d\mathbf{x}$,

$ds =$ the arc length of $d\mathbf{x}$.

By taking the time derivative we obtain
$$d\dot{\mathbf{x}} = \dot{\boldsymbol{\ell}} ds + \boldsymbol{\ell} d\dot{s}. \tag{4.139}$$

Recalling that $\mathbf{L} = \dot{\mathbf{F}} \cdot \mathbf{F}^{-1}$ and $d\mathbf{X} = \mathbf{F}^{-1} \cdot d\mathbf{x}$, we obtain
$$d\dot{\mathbf{x}} = \dot{\boldsymbol{\ell}} ds + \boldsymbol{\ell} d\dot{s} = \mathbf{L} \cdot \boldsymbol{\ell} ds. \tag{4.140}$$

It is noted that
$$\boldsymbol{\ell} \cdot \boldsymbol{\ell} = 1 \quad \Rightarrow \quad \boldsymbol{\ell} \cdot \dot{\boldsymbol{\ell}} = 0.$$

Thus,
$$d\dot{s} = \boldsymbol{\ell} \cdot \mathbf{L} \cdot \boldsymbol{\ell} \, ds \tag{4.141}$$
and
$$\dot{\boldsymbol{\ell}} = \mathbf{L} \cdot \boldsymbol{\ell} - (\boldsymbol{\ell} \cdot \mathbf{L} \cdot \boldsymbol{\ell})\boldsymbol{\ell}. \tag{4.142}$$

Consequently, the change in length per unit length of the material fiber, $d\mathbf{x}$, is found to be
$$d\dot{s}/ds = \boldsymbol{\ell} \cdot \mathbf{L} \cdot \boldsymbol{\ell}. \tag{4.143}$$

If we decompose $\mathbf{L}$ into its symmetric and antisymmetric parts as
$$\mathbf{L} = \frac{1}{2}(\mathbf{L} + \mathbf{L}^T) + \frac{1}{2}(\mathbf{L} - \mathbf{L}^T), \tag{4.144}$$
and call
$$\mathbf{D} = \frac{1}{2}(\mathbf{L} + \mathbf{L}^T), \tag{4.145}$$
then
$$d\dot{s}/ds = \boldsymbol{\ell} \cdot \mathbf{D} \cdot \boldsymbol{\ell}. \tag{4.146}$$

The symmetric tensor $\mathbf{D}$ is called the *rate of deformation* or simply the *deformation rate*. The term velocity strain is also in use.

We next look at the rate of change of the included angle between two material fibers, *i.e.*, at rates of shear. Let $d\mathbf{x}$ and $d\mathbf{y}$ be two infinitesimal fibers in the current state (Fig. 4.13). Then,
$$d\mathbf{x} = \mathbf{F} \cdot d\mathbf{X} = \boldsymbol{\ell} ds^x, \quad d\mathbf{y} = \mathbf{F} \cdot d\mathbf{Y} = \mathbf{m} ds^y. \tag{4.147}$$

As for the angle between them, in the current state we have
$$\cos\theta = \boldsymbol{\ell} \cdot \mathbf{m} \tag{4.148}$$

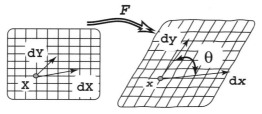

Figure 4.13. Stretch and angular change between arbitrary fibers.

"reference" or "undeformed" state

"deformed" or "current" state

and

$$\frac{d}{dt}(\cos\theta) = -\sin\theta\,\dot\theta = \dot{\boldsymbol{\ell}}\cdot\mathbf{m} + \boldsymbol{\ell}\cdot\dot{\mathbf{m}}. \tag{4.149}$$

Now, identify $\sin\theta = |\boldsymbol{\ell}\times\mathbf{m}|$, and use (4.142) to get

$$\dot\theta = \frac{1}{|\boldsymbol{\ell}\times\mathbf{m}|}\left[(\boldsymbol{\ell}\cdot\mathbf{L}\cdot\boldsymbol{\ell} + \mathbf{m}\cdot\mathbf{L}\cdot\mathbf{m})(\boldsymbol{\ell}\cdot\mathbf{m}) - \boldsymbol{\ell}\cdot(\mathbf{L}+\mathbf{L}^T)\cdot\mathbf{m}\right]. \tag{4.150}$$

4.18 Deformation Rate and Spin

Recall again the definitions of the symmetric and antisymmetric parts of **L**, *viz.*,

$$\mathbf{L} = \frac{1}{2}(\mathbf{L}+\mathbf{L}^T) + \frac{1}{2}(\mathbf{L}-\mathbf{L}^T). \tag{4.151}$$

We call **D** the symmetric part and **W** the antisymmetric part of **L**, *i.e.*,

$$\mathbf{D} = \frac{1}{2}(\mathbf{L}+\mathbf{L}^T), \tag{4.152}$$

$$\mathbf{W} = \frac{1}{2}(\mathbf{L}-\mathbf{L}^T). \tag{4.153}$$

From the polar decomposition theorem

$$\mathbf{F} = \mathbf{R}\cdot\mathbf{U} = \mathbf{V}\cdot\mathbf{R}, \tag{4.154}$$

it follows that

$$\dot{\mathbf{F}} = \dot{\mathbf{R}}\cdot\mathbf{U} + \mathbf{R}\cdot\dot{\mathbf{U}} = \dot{\mathbf{V}}\cdot\mathbf{R} + \mathbf{V}\cdot\dot{\mathbf{R}} \tag{4.155}$$

and

$$\mathbf{F}^{-1} = (\mathbf{R}\cdot\mathbf{U})^{-1} = \mathbf{U}^{-1}\cdot\mathbf{R}^T = (\mathbf{V}\cdot\mathbf{R})^{-1} = \mathbf{R}^T\cdot\mathbf{V}^{-1}. \tag{4.156}$$

Therefore,

$$\mathbf{L} = \dot{\mathbf{F}}\cdot\mathbf{F}^{-1} = \dot{\mathbf{R}}\cdot\mathbf{R}^T + \mathbf{R}\cdot\dot{\mathbf{U}}\cdot\mathbf{U}^{-1}\cdot\mathbf{R}^T \tag{4.157}$$

or

$$\mathbf{L} = \dot{\mathbf{V}}\cdot\mathbf{V}^{-1} + \mathbf{V}\cdot\dot{\mathbf{R}}\cdot\mathbf{R}^T\cdot\mathbf{V}^{-1}, \tag{4.158}$$

because

$$\dot{\mathbf{R}} \cdot \mathbf{R}^T = -(\dot{\mathbf{R}} \cdot \mathbf{R}^T)^T. \tag{4.159}$$

For $\mathbf{D}$ and $\mathbf{W}$, we obtain

$$\mathbf{D} = \frac{1}{2}\mathbf{R} \cdot (\dot{\mathbf{U}} \cdot \mathbf{U}^{-1} + \mathbf{U}^{-1} \cdot \dot{\mathbf{U}}) \cdot \mathbf{R}^T \tag{4.160}$$

and

$$\mathbf{W} = \frac{1}{2}\mathbf{R} \cdot (\dot{\mathbf{U}} \cdot \mathbf{U}^{-1} - \mathbf{U}^{-1} \cdot \dot{\mathbf{U}}) \cdot \mathbf{R}^T + \dot{\mathbf{R}} \cdot \mathbf{R}^T. \tag{4.161}$$

If the current state is taken as the reference state, *i.e.*, if $\mathbf{F} = \mathbf{I}$, $\mathbf{U} = \mathbf{I}$, and $\mathbf{R} = \mathbf{I}$, then instantaneously

$$\mathbf{D} = \frac{1}{2}(\dot{\mathbf{U}}_0 + \dot{\mathbf{U}}_0) = \dot{\mathbf{U}}_0 \tag{4.162}$$

and

$$\mathbf{W} = \dot{\mathbf{R}}_0. \tag{4.163}$$

The subscript 0 signifies the assignment of reference to current state.

4.19 Rate of Stretching and Shearing

Once again consider two infinitesimal fibers $d\mathbf{x}$ and $d\mathbf{y}$ in the current state, as sketched in Fig. 4.13. Recall the connections $d\mathbf{x} = \mathbf{F} \cdot d\mathbf{X}$ and $d\mathbf{y} = \mathbf{F} \cdot d\mathbf{Y}$, where $d\mathbf{X}$ and $d\mathbf{Y}$ are the same fibers in the reference state. Recall also the representation of $d\mathbf{x}$ and $d\mathbf{y}$ in terms of unit vectors along them and their respective lengths, *viz.*, $d\mathbf{x} = \boldsymbol{\ell}ds^x$ and $d\mathbf{y} = \mathbf{m}ds^y$. For the change in length per unit length of $d\mathbf{x}$, we have

$$d\dot{s}^x/ds^x = \boldsymbol{\ell} \cdot \mathbf{L} \cdot \boldsymbol{\ell} = \boldsymbol{\ell} \cdot \mathbf{D} \cdot \boldsymbol{\ell}. \tag{4.164}$$

Also, from (4.150),

$$\dot{\theta} = \frac{1}{|\boldsymbol{\ell} \times \mathbf{m}|}\left[(\boldsymbol{\ell} \cdot \mathbf{L} \cdot \boldsymbol{\ell} + \mathbf{m} \cdot \mathbf{L} \cdot \mathbf{m})(\boldsymbol{\ell} \cdot \mathbf{m}) - \boldsymbol{\ell} \cdot (\mathbf{L} + \mathbf{L}^T) \cdot \mathbf{m}\right]. \tag{4.165}$$

If it happened that, instantaneously, $\boldsymbol{\ell} \cdot \mathbf{m} = 0$ in the current configuration, then

$$\dot{\theta} = -\boldsymbol{\ell} \cdot (\mathbf{L} + \mathbf{L}^T) \cdot \mathbf{m} = -2\boldsymbol{\ell} \cdot \mathbf{D} \cdot \mathbf{m}. \tag{4.166}$$

This suggests a possible definition of shearing rate as

$$\dot{\gamma} = -\frac{1}{2}\dot{\theta} = \boldsymbol{\ell} \cdot \mathbf{D} \cdot \mathbf{m}. \tag{4.167}$$

Since $\mathbf{D}$ is symmetric, it can be represented *via* a spectral form

$$\mathbf{D} = \sum_{r=1}^{3} \alpha_r \mathbf{p}_r \mathbf{p}_r. \tag{4.168}$$

We also have the relations

$$\mathbf{p}_r \cdot \mathbf{p}_s = 0, \quad \text{if } r \neq s,$$

$$(d\dot{s}/ds)_{\mathbf{p}_r} = \alpha_r, \tag{4.169}$$

$$-\frac{1}{2}\dot{\theta}(\mathbf{p}_s, \mathbf{p}_q) = \mathbf{p}_s \cdot \sum_{r=1}^{3} \alpha_r \mathbf{p}_r \mathbf{p}_r \cdot \mathbf{p}_q.$$

Note a property of the last relation, *viz.*,

$$-\frac{1}{2}\dot{\theta}(\mathbf{p}_s, \mathbf{p}_q) = \mathbf{p}_s \cdot \sum_{r=1}^{3} \alpha_r \mathbf{p}_r \mathbf{p}_r \cdot \mathbf{p}_q = 0, \quad \text{if } s \neq q. \tag{4.170}$$

Thus the principal directions of **D** do not undergo relative angular changes and consequently undergo only rigid body rotations as a triad of vectors. Indeed, recall (4.142) for the time derivative of a unit vector along a fiber in the current state,

$$\dot{\boldsymbol{\ell}} = \mathbf{L} \cdot \boldsymbol{\ell} - (\boldsymbol{\ell} \cdot \mathbf{L} \cdot \boldsymbol{\ell})\boldsymbol{\ell}, \tag{4.171}$$

and apply it to a principal direction, $\mathbf{p}_i$. There follows

$$\dot{\mathbf{p}}_i = \mathbf{L} \cdot \mathbf{p}_i - (\mathbf{p}_i \cdot \mathbf{L} \cdot \mathbf{p}_i)\mathbf{p}_i. \tag{4.172}$$

This can be reduced to

$$\dot{\mathbf{p}}_i = \mathbf{D} \cdot \mathbf{p}_i + \mathbf{W} \cdot \mathbf{p}_i - (\mathbf{p}_i \cdot \mathbf{D} \cdot \mathbf{p}_i)\mathbf{p}_i \quad \text{(no sum on } i\text{)}$$
$$= \alpha_i \mathbf{p}_i + \mathbf{W} \cdot \mathbf{p}_i - \alpha_i \mathbf{p}_i$$
$$= \mathbf{W} \cdot \mathbf{p}_i \Rightarrow \text{ a rigid rotation.}$$

Thus **W** can be thought of as the instantaneous spin of the principal directions of **D**.

4.20 Material Derivatives of Strain Tensors: Ė vs. D

Recall that

$$\mathbf{E} = \frac{1}{2}\left(\mathbf{F}^T \cdot \mathbf{F} - \mathbf{I}\right) = \frac{1}{2}(\mathbf{C} - \mathbf{I}). \tag{4.173}$$

Taking the time derivative, we find

$$\dot{\mathbf{E}} = \frac{1}{2}\dot{\mathbf{C}} = \frac{1}{2}\left(\dot{\mathbf{F}}^T \cdot \mathbf{F} + \mathbf{F}^T \cdot \dot{\mathbf{F}}\right)$$
$$= \frac{1}{2}\left(\mathbf{F}^T \cdot \mathbf{L}^T \cdot \mathbf{F} + \mathbf{F}^T \cdot \mathbf{L} \cdot \mathbf{F}\right)$$
$$= \frac{1}{2}\left[\mathbf{F}^T \cdot (\mathbf{L} + \mathbf{L}^T) \cdot \mathbf{F}\right].$$

Thus,

$$\dot{\mathbf{E}} = \mathbf{F}^T \cdot \mathbf{D} \cdot \mathbf{F} = \frac{1}{2}\dot{\mathbf{C}}. \tag{4.174}$$

4.20. Material Derivative of Strain

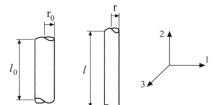

Figure 4.14. Geometry of uniaxial tension.

Only when the reference state is taken as the current state, *i.e.*, when $\mathbf{F} = \mathbf{I}$ instantaneously, do we have

$$\dot{\mathbf{E}}_0 = \mathbf{D}_0. \tag{4.175}$$

As an example, consider a homogeneous uniaxial tension or compression test. The reference gauge length of the uniaxial specimen is taken as ℓ_0, and its current length is ℓ (Fig. 4.14). The geometry of uniaxial tension, or compression, then dictates the following deformation map

$$\begin{aligned} x_1 &= (1 + \frac{r - r_0}{r_0})X_1 \\ x_2 &= (1 + \frac{\ell - \ell_0}{\ell_0})X_2 \\ x_3 &= (1 + \frac{r - r_0}{r_0})X_3. \end{aligned} \tag{4.176}$$

The stretches are defined as $\lambda \sim length/initial\ length$, and so for the three principal directions,

$$\lambda_1 = r/r_0, \quad \lambda_2 = \ell/\ell_0, \quad \lambda_3 = r/r_0. \tag{4.177}$$

The radial stretch in both directions is assumed to be the same, *e.g.*, as would be the case for an isotropic material. For the deformation gradient we obtain

$$\mathbf{F} = \begin{bmatrix} \lambda_1 & 0 & 0 \\ 0 & \lambda_2 & 0 \\ 0 & 0 & \lambda_3 \end{bmatrix}, \quad \mathbf{F}^{-1} = \begin{bmatrix} 1/\lambda_1 & 0 & 0 \\ 0 & 1/\lambda_2 & 0 \\ 0 & 0 & 1/\lambda_3 \end{bmatrix}. \tag{4.178}$$

Also,

$$\mathbf{F} = \mathbf{R} \cdot \mathbf{U} \Rightarrow \mathbf{U} = \begin{bmatrix} \lambda_1 & 0 & 0 \\ 0 & \lambda_2 & 0 \\ 0 & 0 & \lambda_3 \end{bmatrix}, \quad \mathbf{R} = \mathbf{I}. \tag{4.179}$$

Thus,

$$\mathbf{C} = \mathbf{F}^T \cdot \mathbf{F} = \mathbf{U}^2 = \begin{bmatrix} \lambda_1^2 & 0 & 0 \\ 0 & \lambda_2^2 & 0 \\ 0 & 0 & \lambda_3^2 \end{bmatrix}. \tag{4.180}$$

The strain tensor is

$$\mathbf{E} = \frac{1}{2}(\mathbf{F}^T \cdot \mathbf{F} - \mathbf{I}) = \frac{1}{2}(\mathbf{C} - \mathbf{I})$$

$$= \frac{1}{2}\begin{bmatrix} \lambda_1^2 - 1 & 0 & 0 \\ 0 & \lambda_2^2 - 1 & 0 \\ 0 & 0 & \lambda_3^2 - 1 \end{bmatrix}. \quad (4.181)$$

The velocity gradient is

$$\mathbf{L} = \dot{\mathbf{F}} \cdot \mathbf{F}^{-1} = \begin{bmatrix} \dot{\lambda}_1/\lambda_1 & 0 & 0 \\ 0 & \dot{\lambda}_2/\lambda_2 & 0 \\ 0 & 0 & \dot{\lambda}_3/\lambda_3 \end{bmatrix} = \begin{bmatrix} \dot{r}/r_0 & 0 & 0 \\ 0 & \dot{\ell}/\ell_0 & 0 \\ 0 & 0 & \dot{r}/r_0 \end{bmatrix}. \quad (4.182)$$

It is clear that

$$\mathbf{L} = \mathbf{D}, \quad \mathbf{W} = \mathbf{0}. \quad (4.183)$$

The rate of strain is

$$\dot{\mathbf{E}} = \frac{1}{2}\begin{bmatrix} 2\lambda_1\dot{\lambda}_1 & 0 & 0 \\ 0 & 2\lambda_2\dot{\lambda}_2 & 0 \\ 0 & 0 & 2\lambda_3\dot{\lambda}_3 \end{bmatrix}. \quad (4.184)$$

As a check on the above, form

$$\dot{\mathbf{E}} = \mathbf{F}^T \cdot \mathbf{D} \cdot \mathbf{F}$$

$$= \begin{bmatrix} \lambda_1 & 0 & 0 \\ 0 & \lambda_2 & 0 \\ 0 & 0 & \lambda_3 \end{bmatrix} \cdot \begin{bmatrix} \dot{\lambda}_1/\lambda_1 & 0 & 0 \\ 0 & \dot{\lambda}_2/\lambda_2 & 0 \\ 0 & 0 & \dot{\lambda}_3/\lambda_3 \end{bmatrix} \cdot \begin{bmatrix} \lambda_1 & 0 & 0 \\ 0 & \lambda_2 & 0 \\ 0 & 0 & \lambda_3 \end{bmatrix}$$

$$= \frac{1}{2}\begin{bmatrix} 2\lambda_1\dot{\lambda}_1 & 0 & 0 \\ 0 & 2\lambda_2\dot{\lambda}_2 & 0 \\ 0 & 0 & 2\lambda_3\dot{\lambda}_3 \end{bmatrix}, \quad (4.185)$$

in agreement with the previous result.

4.21 Rate of F in Terms of Principal Stretches

The right stretch tensor can be expressed in terms of its eigenvalues – principal stretches λ_i (assumed here to be different) and corresponding eigendirections $\mathbf{N}_i$ as

$$\mathbf{U} = \sum_{i=1}^{3} \lambda_i \mathbf{N}_i \mathbf{N}_i. \quad (4.186)$$

The rate of $\mathbf{U}$ is then

$$\dot{\mathbf{U}} = \sum_{i=1}^{3} \left[\dot{\lambda}_i \mathbf{N}_i \mathbf{N}_i + \lambda_i \left(\dot{\mathbf{N}}_i \mathbf{N}_i + \mathbf{N}_i \dot{\mathbf{N}}_i\right)\right]. \quad (4.187)$$

4.21. Rate of F in Terms of Principal Stretches

If $\mathbf{e}_i^0$ ($i = 1, 2, 3$) are the fixed reference unit vectors, the unit vectors $\mathbf{N}_i$ of the principal directions of $\mathbf{U}$ can be expressed as

$$\mathbf{N}_i = \mathcal{R}_0 \cdot \mathbf{e}_i^0, \tag{4.188}$$

where $\mathcal{R}_0$ is the rotation that carries the orthogonal triad $\{\mathbf{e}_i^0\}$ into the Lagrangian triad $\{\mathbf{N}_i\}$. Defining the spin of the Lagrangian triad by

$$\mathbf{\Omega}_0 = \dot{\mathcal{R}}_0 \cdot \mathcal{R}_0^{-1}, \tag{4.189}$$

it follows that

$$\dot{\mathbf{N}}_i = \dot{\mathcal{R}}_0 \cdot \mathbf{e}_i^0 = \mathbf{\Omega}_0 \cdot \mathbf{N}_i = -\mathbf{N}_i \cdot \mathbf{\Omega}_0, \tag{4.190}$$

and the substitution into (4.187) gives

$$\dot{\mathbf{U}} = \sum_{i=1}^{3} \dot{\lambda}_i \mathbf{N}_i \mathbf{N}_i + \mathbf{\Omega}_0 \cdot \mathbf{U} - \mathbf{U} \cdot \mathbf{\Omega}_0. \tag{4.191}$$

If the spin tensor $\mathbf{\Omega}_0$ is expressed on the axes of the Lagrangian triad as

$$\mathbf{\Omega}_0 = \sum_{i \neq j} \Omega_{ij}^0 \mathbf{N}_i \mathbf{N}_j, \tag{4.192}$$

it is readily found that

$$\mathbf{\Omega}_0 \cdot \mathbf{U} = \Omega_{12}^0 (\lambda_2 - \lambda_1) \mathbf{N}_1 \mathbf{N}_2 + \Omega_{23}^0 (\lambda_3 - \lambda_2) \mathbf{N}_2 \mathbf{N}_3 + \Omega_{31}^0 (\lambda_1 - \lambda_3) \mathbf{N}_3 \mathbf{N}_1. \tag{4.193}$$

Consequently,

$$\mathbf{\Omega}_0 \cdot \mathbf{U} - \mathbf{U} \cdot \mathbf{\Omega}_0 = \mathbf{\Omega}_0 \cdot \mathbf{U} + (\mathbf{\Omega}_0 \cdot \mathbf{U})^T = \sum_{i \neq j} \Omega_{ij}^0 (\lambda_j - \lambda_i) \mathbf{N}_i \mathbf{N}_j. \tag{4.194}$$

The substitution into (4.191) yields

$$\dot{\mathbf{U}} = \sum_{i=1}^{3} \dot{\lambda}_i \mathbf{N}_i \mathbf{N}_i + \sum_{i \neq j} \Omega_{ij}^0 (\lambda_j - \lambda_i) \mathbf{N}_i \mathbf{N}_j. \tag{4.195}$$

Similarly, the rate of the Lagrangian strain tensor is

$$\dot{\mathbf{E}} = \sum_{i=1}^{3} \lambda_i \dot{\lambda}_i \mathbf{N}_i \mathbf{N}_i + \sum_{i \neq j} \Omega_{ij}^0 \frac{\lambda_j^2 - \lambda_i^2}{2} \mathbf{N}_i \mathbf{N}_j. \tag{4.196}$$

The principal directions of the left stretch tensor $\mathbf{V}$, appearing in the spectral representation

$$\mathbf{V} = \sum_{i=1}^{3} \lambda_i \mathbf{n}_i \mathbf{n}_i, \tag{4.197}$$

are related to principal directions $\mathbf{N}_i$ of the right stretch tensor $\mathbf{U}$ by

$$\mathbf{n}_i = \mathbf{R} \cdot \mathbf{N}_i = \mathcal{R} \cdot \mathbf{e}_i^0, \quad \mathcal{R} = \mathbf{R} \cdot \mathcal{R}_0. \tag{4.198}$$

The rotation tensor **R** is the rotation from the polar decomposition of the the deformation gradient $\mathbf{F} = \mathbf{V} \cdot \mathbf{R} = \mathbf{R} \cdot \mathbf{U}$. By differentiating the above expression for $\mathbf{n}_i$, there follows

$$\dot{\mathbf{n}}_i = \mathbf{\Omega} \cdot \mathbf{n}_i, \tag{4.199}$$

where the spin of the Eulerian triad $\{\mathbf{n}_i\}$ is defined by

$$\mathbf{\Omega} = \dot{\mathcal{R}} \cdot \mathcal{R}^{-1} = \boldsymbol{\omega} + \mathbf{R} \cdot \mathbf{\Omega}_0 \cdot \mathbf{R}^T, \quad \boldsymbol{\omega} = \dot{\mathbf{R}} \cdot \mathbf{R}^{-1}. \tag{4.200}$$

On the axes $\mathbf{n}_i$, the spin $\mathbf{\Omega}$ can be decomposed as

$$\mathbf{\Omega} = \sum_{i \neq j} \Omega_{ij} \, \mathbf{n}_i \, \mathbf{n}_j. \tag{4.201}$$

By an analogous derivation, as used to obtain the rate $\dot{\mathbf{U}}$, it follows that

$$\dot{\mathbf{V}} = \sum_{i=1}^{3} \dot{\lambda}_i \, \mathbf{n}_i \, \mathbf{n}_i + \sum_{i \neq j} \Omega_{ij} \, (\lambda_j - \lambda_i) \, \mathbf{n}_i \, \mathbf{n}_j. \tag{4.202}$$

The rate of the rotation tensor

$$\mathbf{R} = \sum_{i=1}^{3} \mathbf{n}_i \, \mathbf{N}_i \tag{4.203}$$

is

$$\dot{\mathbf{R}} = \sum_{i=1}^{3} \left(\dot{\mathbf{n}}_i \, \mathbf{N}_i + \mathbf{n}_i \, \dot{\mathbf{N}}_i \right) = \mathbf{\Omega} \cdot \mathbf{R} - \mathbf{R} \cdot \mathbf{\Omega}_0, \tag{4.204}$$

or

$$\dot{\mathbf{R}} = \sum_{i \neq j} \left(\Omega_{ij} - \Omega_{ij}^0 \right) \mathbf{n}_i \, \mathbf{N}_j. \tag{4.205}$$

Finally, the rate of the deformation gradient

$$\mathbf{F} = \sum_{i=1}^{3} \lambda_i \, \mathbf{n}_i \, \mathbf{N}_i \tag{4.206}$$

is

$$\dot{\mathbf{F}} = \sum_{i=1}^{3} \left[\dot{\lambda}_i \, \mathbf{n}_i \, \mathbf{N}_i + \lambda_i \left(\dot{\mathbf{n}}_i \, \mathbf{N}_i + \mathbf{n}_i \, \dot{\mathbf{N}}_i \right) \right]. \tag{4.207}$$

Since $\dot{\mathbf{n}}_i = \mathbf{\Omega} \cdot \mathbf{n}_i$ and $\dot{\mathbf{N}}_i = \mathbf{\Omega}_0 \cdot \mathbf{N}_i$, it follows that

$$\dot{\mathbf{F}} = \sum_{i=1}^{3} \dot{\lambda}_i \, \mathbf{n}_i \, \mathbf{N}_i + \mathbf{\Omega} \cdot \mathbf{F} - \mathbf{F} \cdot \mathbf{\Omega}_0, \tag{4.208}$$

and

$$\dot{\mathbf{F}} = \sum_{i=1}^{3} \dot{\lambda}_i \, \mathbf{n}_i \, \mathbf{N}_i + \sum_{i \neq j} \left(\lambda_j \Omega_{ij} - \lambda_i \Omega_{ij}^0 \right) \mathbf{n}_i \, \mathbf{N}_j. \tag{4.209}$$

4.21. Rate of F in Terms of Principal Stretches

4.21.1 Spins of Lagrangian and Eulerian Triads

The inverse of the deformation gradient can be written in terms of the principal stretches as

$$\mathbf{F}^{-1} = \sum_{i=1}^{3} \frac{1}{\lambda_i} \mathbf{N}_i \, \mathbf{n}_i. \tag{4.210}$$

Using this and (4.209) we obtain an expression for the velocity gradient

$$\mathbf{L} = \dot{\mathbf{F}} \cdot \mathbf{F}^{-1} = \sum_{i=1}^{3} \frac{\dot{\lambda}_i}{\lambda_i} \mathbf{n}_i \, \mathbf{n}_i + \sum_{i \ne j} \left(\Omega_{ij} - \frac{\lambda_i}{\lambda_j} \Omega^0_{ij} \right) \mathbf{n}_i \, \mathbf{n}_j. \tag{4.211}$$

The symmetric part of this is the rate of deformation tensor,

$$\mathbf{D} = \sum_{i=1}^{3} \frac{\dot{\lambda}_i}{\lambda_i} \mathbf{n}_i \, \mathbf{n}_i + \sum_{i \ne j} \frac{\lambda_j^2 - \lambda_i^2}{2 \lambda_i \lambda_j} \Omega^0_{ij} \mathbf{n}_i \, \mathbf{n}_j, \tag{4.212}$$

whereas the antisymmetric part is the spin tensor

$$\mathbf{W} = \sum_{i \ne j} \left(\Omega_{ij} - \frac{\lambda_i^2 + \lambda_j^2}{2 \lambda_i \lambda_j} \Omega^0_{ij} \right) \mathbf{n}_i \, \mathbf{n}_j. \tag{4.213}$$

Evidently, for $i \ne j$ from (4.212) we have

$$\Omega^0_{ij} = \frac{2 \lambda_i \lambda_j}{\lambda_j^2 - \lambda_i^2} D_{ij}, \quad \lambda_i \ne \lambda_j, \tag{4.214}$$

which is an expression for the components of the Lagrangian spin $\mathbf{\Omega}_0$ in terms of the stretch ratios and the components of the rate of deformation tensor. Substituting (4.214) into (4.213) we obtain an expression for the components of the Eulerian spin $\mathbf{\Omega}$ in terms of the stretch ratios and the components of the rate of deformation and spin tensors, i.e.,

$$\Omega_{ij} = W_{ij} + \frac{\lambda_i^2 + \lambda_j^2}{\lambda_j^2 - \lambda_i^2} D_{ij}, \quad \lambda_i \ne \lambda_j. \tag{4.215}$$

The inverse of the rotation tensor $\mathbf{R}$ is

$$\mathbf{R}^{-1} = \sum_{i=1}^{3} \mathbf{N}_i \, \mathbf{n}_i, \tag{4.216}$$

so that, by virtue of (4.205), the spin $\boldsymbol{\omega}$ can be expressed as

$$\boldsymbol{\omega} = \dot{\mathbf{R}} \cdot \mathbf{R}^{-1} = \sum_{i \ne j} \left(\Omega_{ij} - \Omega^0_{ij} \right) \mathbf{n}_i \, \mathbf{n}_j. \tag{4.217}$$

Thus,

$$\omega_{ij} = \Omega_{ij} - \Omega^0_{ij}, \tag{4.218}$$

where Ω^0_{ij} are the components of $\mathbf{\Omega}_0$ on the Lagrangian triad $\{\mathbf{N}_i\}$, whereas Ω_{ij} are the components of $\mathbf{\Omega}$ on the Eulerian triad $\{\mathbf{n}_i\}$. When (4.214) and (4.215) are substituted into (4.218), we obtain an expression for the spin components ω_{ij} in terms of the stretch ratios and the components of the rate of deformation and spin tensors, which is

$$\omega_{ij} = W_{ij} + \frac{\lambda_j - \lambda_i}{\lambda_i + \lambda_j} D_{ij}. \qquad (4.219)$$

4.22 Additional Connections Between Current and Reference State Representations

LEMMA 4.3: *If* **u** *is a vector field and* **T** *is a tensor field representing properties of a deforming body, then*

$$\text{Div } \mathbf{u} = J \text{div}\,(J^{-1}\mathbf{F} \cdot \mathbf{u}), \quad \text{Div } \mathbf{T} = J \text{div}\,(J^{-1}\mathbf{F} \cdot \mathbf{T}). \qquad (4.220)$$

Proof: The above relations depend on the result

$$\frac{\partial}{\partial x_p}\left(J^{-1}F_{p\alpha}\right) = 0, \qquad (4.221)$$

which is justified from Lemma 4.1. In fact replacing t with X_α, we obtain

$$\partial J/\partial X_\alpha = J \,\text{tr}\left(\frac{\partial \mathbf{F}}{\partial X_\alpha} \cdot \mathbf{F}^{-1}\right) = J \frac{\partial F_{q\rho}}{\partial X_\alpha} F_{\rho q}^{-1} = J \frac{\partial^2 x_q}{\partial X_\alpha \partial X_\rho} F_{\rho q}^{-1}, \qquad (4.222)$$

where we note that $F_{\alpha i}^{-1} = \partial x_i/\partial X_\alpha$, and $J = \det \mathbf{F}$. Thus,

$$\frac{\partial}{\partial x_p}\left(J^{-1}F_{p\alpha}\right) = \frac{\partial}{\partial X_\pi}\left(J^{-1}F_{p\alpha}\right)\frac{\partial X_\pi}{\partial x_p}$$

$$= \left(J^{-1}\frac{\partial^2 x_p}{\partial X_\alpha \partial X_\pi} - J^{-1}\frac{\partial^2 x_q}{\partial X_\pi \partial X_\rho} F_{\rho q}^{-1} F_{p\alpha}\right) F_{\pi p}^{-1}$$

$$= J^{-1}\left(\frac{\partial^2 x_q}{\partial X_\alpha \partial X_\rho} F_{\rho q}^{-1} - \frac{\partial^2 x_q}{\partial X_\pi \partial X_\rho} \delta_{\pi\alpha} F_{\rho q}^{-1}\right) = 0.$$

Furthermore, we obtain

$$\text{Div } \mathbf{u} - J \text{div}\,(J^{-1}\mathbf{F} \cdot \mathbf{u}) = \frac{\partial u_\pi}{\partial X_\pi} - J\left(J^{-1}F_{p\pi}u_\pi\right)$$

$$= \frac{\partial u_\pi}{\partial X_\pi} - F_{p\pi}\frac{\partial u_\pi}{\partial x_p}$$

$$= \frac{\partial u_\pi}{\partial X_\pi} - \frac{\partial u_\pi}{\partial x_p}\frac{\partial x_p}{\partial X_\pi} = 0$$

and

$$\text{Div } \mathbf{T} - J \text{div}\,(J^{-1}\mathbf{F} \cdot \mathbf{T}) = \frac{\partial T_{\pi\rho}}{\partial X_\pi}\mathbf{E}_\rho - J\frac{\partial}{\partial x_p}(J^{-1}F_{p\pi}T_{\pi q})\mathbf{e}_q$$

$$= \frac{\partial T_{\pi\rho}}{\partial X_\pi}\mathbf{E}_\rho - F_{p\pi}\frac{\partial T_{\pi q}}{\partial x_p}\mathbf{e}_q$$

$$= \frac{\partial}{\partial X_\pi}(T_{\pi\rho}\mathbf{E}_\rho - T_{\pi q}\mathbf{e}_q) = \mathbf{0}.$$

4.23 Transport Formulae

We now turn our attention to the rates of change of integrals over material curves, surfaces, and volume elements. We focus attention to the area or volume elements in the current state.

LEMMA 4.4: *Let C, S, and V represent a material line (curve), a surface, and a volume element, respectively in the current state; as usual, $\phi(\mathbf{x})$ is a scalar field. Then,*

$$\frac{d}{dt}\int_C \phi\, d\mathbf{x} = \int_C (\dot\phi\, d\mathbf{x} + \phi\, \mathbf{L}\cdot d\mathbf{x}),$$

$$\frac{d}{dt}\int_S \phi\, \mathbf{n}\, da = \int_S [(\dot\phi + \phi\, \mathrm{tr}\,\mathbf{L})\mathbf{n} - \phi\, \mathbf{L}^T\cdot\mathbf{n}]\, da, \quad (4.223)$$

$$\frac{d}{dt}\int_V \phi\, dv = \int_V (\dot\phi + \phi\, \mathrm{tr}\,\mathbf{L})\, dv,$$

where $\mathbf{n}$ is the unit normal to the external surface, S, of the body.

If $\mathbf{u}(\mathbf{x})$ is a continuous and differentiable vector field, then the corresponding results are

$$\frac{d}{dt}\int_C \mathbf{u}\cdot d\mathbf{x} = \int_C (\dot{\mathbf{u}} + \mathbf{L}^T\cdot\mathbf{u})\cdot d\mathbf{x},$$

$$\frac{d}{dt}\int_S \mathbf{u}\cdot\mathbf{n}\, da = \int_S (\dot{\mathbf{u}} + \mathbf{u}\,\mathrm{tr}\,\mathbf{L} - \mathbf{L}\cdot\mathbf{u})\cdot\mathbf{n}\, da, \quad (4.224)$$

$$\frac{d}{dt}\int_V \mathbf{u}\, dv = \int_V (\dot{\mathbf{u}} + \mathbf{u}\,\mathrm{tr}\,\mathbf{L})\, dv.$$

Proof: We focus on the second of the above relations, as the other relations follow by direct analogy. We have

$$\frac{d}{dt}\int_S \mathbf{u}\cdot\mathbf{n}\, da = \frac{d}{dt}\int_{S_R} [\mathbf{u}\cdot(J\mathbf{F}^{-T}\cdot\mathbf{N}\, dA)]$$

$$= \int_{S_R} \frac{\partial}{\partial t}[\mathbf{u}\cdot(J\mathbf{F}^{-T}\cdot\mathbf{N}\, dA)]$$

$$= \int_{S_R} \frac{\partial}{\partial t}(J\mathbf{F}^{-1}\cdot\mathbf{u})\cdot\mathbf{N}\, dA$$

$$= \int_{S_R} [J\mathbf{F}^{-1}\cdot\dot{\mathbf{u}} + (J\,\mathrm{tr}\,\mathbf{L})\cdot\mathbf{F}^{-1}\cdot\mathbf{u} + J(-\mathbf{F}^{-1}\cdot\mathbf{L})\cdot\mathbf{u}]\cdot\mathbf{N}\, dA$$

$$= \int_{S_R} (\dot{\mathbf{u}} + \mathbf{u}\,\mathrm{tr}\,\mathbf{L} - \mathbf{L}\cdot\mathbf{u})\cdot(J\mathbf{F}^{-T}\cdot\mathbf{N}\, dA)$$

$$= \int_S (\dot{\mathbf{u}} + \mathbf{u}\,\mathrm{tr}\,\mathbf{L} - \mathbf{L}\cdot\mathbf{u})\cdot\mathbf{n}\, da.$$

In developing the above use was made of the following

$$(\mathbf{F}^{-1})^{\cdot} = -\mathbf{F}^{-1}\cdot\dot{\mathbf{F}}\cdot\mathbf{F}^{-1},$$

$$\mathbf{L} = \dot{\mathbf{F}}\cdot\mathbf{F}^{-1} = -\mathbf{F}\cdot(\mathbf{F}^{-1})^{\cdot},$$

and $\dot{J} = \text{tr } \mathbf{L}$. Also, if $\text{d}\mathbf{x} = \mathbf{F} \cdot \text{d}\mathbf{X}$ and $\text{d}\mathbf{y} = \mathbf{F} \cdot \text{d}\mathbf{Y}$ are vector elements in the current and reference states respectively, then $\mathbf{n}\text{d}a = (\mathbf{F} \cdot \text{d}\mathbf{X}) \times (\mathbf{F} \cdot \text{d}\mathbf{Y}) = J\mathbf{F}^{-T} \cdot \mathbf{N}\text{d}A$.

4.24 Material Derivatives of Volume, Area, and Surface Integrals: Transport Formulae Revisited

We recall the results for the rate of change of a volume element,

$$\frac{\text{d}}{\text{d}t}(\text{d}V) = \frac{\partial v_i}{\partial x_i}\,\text{d}V. \tag{4.225}$$

Examine integrals of the type

$$\mathcal{A}_{ij\ldots}(t) = \int_V A_{ij\ldots}(\mathbf{x}, t)\,\text{d}V, \tag{4.226}$$

and, more particularly, their time derivatives

$$\frac{\text{d}}{\text{d}t}\mathcal{A}_{ij\ldots}(t) = \frac{\text{d}}{\text{d}t}\int_V A_{ij\ldots}(\mathbf{x}, t)\,\text{d}V. \tag{4.227}$$

Since this integral is taken over a fixed amount of mass (*i.e.*, material), we may interchange the order of differentiation and integration. This leads to

$$\frac{\text{d}}{\text{d}t}\int_V A_{ij\ldots}(\mathbf{x}, t)\,\text{d}V = \int_V \frac{\text{d}}{\text{d}t}\left[A_{ij\ldots}(\mathbf{x}, t)\,\text{d}V\right], \tag{4.228}$$

which on using (4.225) leads to

$$\frac{\text{d}}{\text{d}t}\int_V A_{ij\ldots}(\mathbf{x}, t)\,\text{d}V = \int_V \left[\frac{\text{d}A_{ij\ldots}(\mathbf{x}, t)}{\text{d}t} + A_{ij\ldots}(\mathbf{x}, t)\frac{\partial v_p}{\partial x_p}\right]\text{d}V. \tag{4.229}$$

We next recall the relation for the material derivative operator, *viz.*,

$$\frac{\text{d}}{\text{d}t} = \frac{\partial}{\partial t} + v_p \frac{\partial}{\partial x_p}. \tag{4.230}$$

When this is used in the above it is found that

$$\frac{\text{d}}{\text{d}t}\int_v A_{ij\ldots}(\mathbf{x}, t)\,\text{d}V = \int_V \left\{\frac{\partial A_{ij\ldots}(\mathbf{x}, t)}{\partial t} + \frac{\partial}{\partial x_p}\left[v_p A_{ij\ldots}(\mathbf{x}, t)\right]\right\}\text{d}V. \tag{4.231}$$

The divergence theorem may be used on the second term on the *rhs* to obtain

$$\frac{\text{d}}{\text{d}t}\int_v A_{ij\ldots}(\mathbf{x}, t)\,\text{d}V = \int_V \frac{\partial A_{ij\ldots}(\mathbf{x}, t)}{\partial t}\,\text{d}V + \int_S v_p A_{ij\ldots}(\mathbf{x}, t)n_p\,\text{d}S, \tag{4.232}$$

where $\mathbf{n}$ is the outward pointing unit normal to the surface S that bounds V and n_p is its pth component.

We next examine surface integrals of the form

$$\mathcal{B}_{ij\ldots}(t) = \int_S B_{ij\ldots}(\mathbf{x}, t)n_p\,\text{d}S, \tag{4.233}$$

and derivatives of the type

$$\frac{\text{d}}{\text{d}t}\int_S B_{ij\ldots}(\mathbf{x}, t)n_p\,\text{d}S = \int_S \frac{\text{d}}{\text{d}t}\left[B_{ij\ldots}(\mathbf{x}, t)n_p\,\text{d}S\right]. \tag{4.234}$$

4.25. Analysis of Simple Shearing

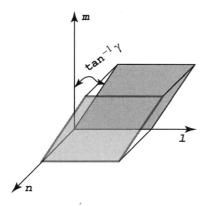

Figure 4.15. Geometry of simple shear.

From the solved Problem 4.11 of Chapter 34, with $dS_p \equiv n_p dS$, we have

$$\frac{dS_p}{dt} = (\partial v_q/\partial x_q)dS_p - (\partial v_q/\partial x_p)dS_q, \tag{4.235}$$

which describes the rate of change of surface area elements. Using this in (4.234), we obtain

$$\frac{d\mathcal{B}_{ij...}(t)}{dt} = \int_S \left[\frac{dB_{ij...}(\mathbf{x},t)}{dt} + \frac{\partial v_q}{\partial x_q} B_{ij...}(\mathbf{x},t)\right] dS_p - \int_S B_{ij...}(\mathbf{x},t)\frac{\partial v_p}{\partial x_q} dS_p.$$

Finally, consider line integrals of the type

$$\mathcal{L}_{ij...}(t) = \int_C L_{ij...}(\mathbf{x},t)\ell_p \, ds, \tag{4.236}$$

where ds is an element of arc length and $\boldsymbol{\ell}$ is the running tangent at each point on the line; ℓ_p is then the pth component of that tangent. Recall that $\ell_p ds = dx_p$ and

$$\frac{d}{dt}(dx_p) = \frac{\partial v_p}{\partial x_k} dx_k. \tag{4.237}$$

Then,

$$\frac{d}{dt}\int_C L_{ij...}(\mathbf{x},t)\, dx_p = \int_C \frac{d}{dt}[L_{ij...}(\mathbf{x},t)\, dx_p] \tag{4.238}$$

and

$$\frac{d\mathcal{L}_{ij...}(t)}{dt} = \int_C \frac{dL_{ij...}(\mathbf{x},t)}{dt} dx_p + \int_C \frac{\partial v_p}{\partial x_k}[L_{ij...}(\mathbf{x},t)] dx_k. \tag{4.239}$$

4.25 Analysis of Simple Shearing

Consider the motion depicted in Fig. 4.15. The motion is associated with the deformation gradient,

$$\mathbf{F} = \mathbf{I} + \gamma(t)\boldsymbol{\ell}\mathbf{m}. \tag{4.240}$$

Let $\mathbf{n}$ be a third unit vector that along with $\boldsymbol{\ell}$ and $\mathbf{m}$ form a right-handed triad and serve as a basis system. The deformation prescribed by the above $\mathbf{F}$ is volume preserving because

$$\det \mathbf{F} = [\mathbf{F}\cdot\boldsymbol{\ell}, \mathbf{F}\cdot\mathbf{m}, \mathbf{F}\cdot\mathbf{n}] = [\boldsymbol{\ell}, \mathbf{m} + \gamma\boldsymbol{\ell}, \mathbf{n}] = [\boldsymbol{\ell}, \mathbf{m}, \mathbf{n}] = 1. \tag{4.241}$$

We also note that

$$F^{-1} = I - \gamma \ell m,$$
$$F^T = I + \gamma m\ell, \quad (4.242)$$
$$I = \ell\ell + mm + nn.$$

Thus,

$$C = F^T \cdot F = \ell\ell + (1+\gamma^2)mm + nn + \gamma(\ell m + m\ell). \quad (4.243)$$

Since $C \cdot n = n$, the vector n is clearly an eigenvector with an eigenvalue of 1; let p_1 and p_2 be the other two eigenvectors of C. Then,

$$C = \sum_{i=1}^{3} \lambda_i^2 p_i p_i, \quad (4.244)$$

where formally $p_3 = n$. But since det $F = \lambda_1 \lambda_2 \lambda_3 = 1$, and $\lambda_3 = 1$, it is clear that $\lambda_1 = 1/\lambda_2$. Recall that $\det C = \det(F^T \cdot F) = (\det F)^2$. Thus, letting $\lambda = \lambda_1$, we write

$$C = \lambda^2 p_1 p_1 + \lambda^{-2} p_2 p_2 + nn. \quad (4.245)$$

Next, express the eigenvectors as

$$p_1 = \cos\theta\, \ell + \sin\theta\, m,$$
$$p_2 = -\sin\theta\, \ell + \cos\theta\, m. \quad (4.246)$$

We use these in (4.245), and then in (4.243), to form the equations

$$C \cdot p_1 = \lambda^2 p_1, \quad C \cdot p_2 = \lambda^{-2} p_2. \quad (4.247)$$

Upon expanding, we obtain the set

$$\lambda^2 \cos^2\theta + \lambda^{-2} \sin^2\theta = 1,$$
$$\lambda^2 \sin^2\theta + \lambda^{-2} \cos^2\theta = 1 + \gamma^2, \quad (4.248)$$
$$(\lambda^2 - \lambda^{-2}) \sin\theta \cos\theta = \gamma,$$

which has the solution

$$\lambda = (1 + \tfrac{1}{4}\gamma^2)^{1/2} + \tfrac{1}{2}\gamma = \cot\psi,$$
$$\theta = \tfrac{1}{2}\pi - \psi, \quad (4.249)$$
$$\psi = \tfrac{1}{2}\tan(2/\gamma).$$

Thus, the principal stretches are

$$\lambda_1 = \cot\psi, \quad \lambda_2 = \tan\psi, \quad \lambda_3 = 1, \quad (4.250)$$

with the corresponding principal directions

$$\mathbf{p}_1 = \sin\psi\,\boldsymbol{\ell} + \cos\psi\,\mathbf{m},$$
$$\mathbf{p}_2 = -\cos\psi\,\boldsymbol{\ell} + \sin\psi\,\mathbf{m}, \tag{4.251}$$
$$\mathbf{p}_3 = \mathbf{n}.$$

But γ is a function of time and we note that $\dot{\mathbf{F}} = \dot{\gamma}\,\boldsymbol{\ell}\mathbf{m}$. Thus, the velocity gradient becomes

$$\mathbf{L} = \dot{\gamma}\,\boldsymbol{\ell}\mathbf{m} \cdot (\mathbf{I} - \gamma\boldsymbol{\ell}\mathbf{m}) = \dot{\gamma}\,\boldsymbol{\ell}\mathbf{m}, \tag{4.252}$$

which, in turn, means that the velocity is

$$v = \dot{\gamma}(\boldsymbol{\ell}\mathbf{m}) \cdot \mathbf{x}. \tag{4.253}$$

The rate of deformation and spin rates become

$$\mathbf{D} = \frac{1}{2}(\boldsymbol{\ell}\mathbf{m} + \mathbf{m}\boldsymbol{\ell}),$$
$$\mathbf{W} = \frac{1}{2}(\boldsymbol{\ell}\mathbf{m} - \mathbf{m}\boldsymbol{\ell}). \tag{4.254}$$

Consider the effect of $\mathbf{W}$ on an arbitrary vector $\mathbf{a}$. We have

$$\mathbf{W} \cdot \mathbf{a} = \frac{1}{2}\dot{\gamma}[(\mathbf{a}\cdot\mathbf{m})\boldsymbol{\ell} - (\mathbf{a}\cdot\boldsymbol{\ell})\mathbf{m}] = -\frac{1}{2}\dot{\gamma}\mathbf{n}\times\mathbf{a}, \tag{4.255}$$

and thus $\frac{1}{2}\dot{\gamma}\mathbf{n}$ is the axial vector of $\mathbf{W}$.

4.26 Examples of Particle and Plane Motion

Consider a deformation map prescribed by

$$\begin{aligned}x_1 &= X_1 + kt\,X_3,\\ x_2 &= X_2 + kt\,X_3,\\ x_3 &= X_3 - kt(X_1 - X_2),\end{aligned} \tag{4.256}$$

where the geometry is sketched in Fig. 4.16. The parameter k is a constant and t is time. We will show that the motion of an arbitrary particle is along a straight line always orthogonal to $\mathbf{X}$, the vector from the origin to the initial position of the particle. Further, if there was a slab, as also depicted in the figure, lying with its faces perpendicular to the X_1 axis, this slab would reorient such that, in the limit as $t \to \infty$, the slab's faces would lie inclined to the X_1 axes at $\pi/4$.

From (4.256) we first form the components of the deformation gradient and the vector of material velocity. These are

$$\mathbf{F} = \begin{bmatrix} 1 & 0 & kt \\ 0 & 1 & kt \\ -kt & -kt & 1 \end{bmatrix}, \tag{4.257}$$

and

$$v = kX_3\mathbf{e}_1 + kX_3\mathbf{e}_2 - k(X_1 + X_2)\mathbf{e}_3. \tag{4.258}$$

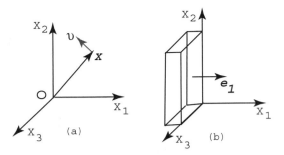

Figure 4.16. (a) Position vector, **X**, of an arbitrary particle. (b) Slab with its initial faces perpendicular to the X_1 axis.

Since the vector to the arbitrary material particle is **X**, the velocity is – and for all t is – orthogonal to **X**, because

$$v \cdot \mathbf{X} = [kX_3\mathbf{e}_1 + kX_3\mathbf{e}_2 - k(X_1 + X_2)\mathbf{e}_3] \cdot [X_1\mathbf{e}_1 + X_2\mathbf{e}_2 + X_3\mathbf{e}_3] = 0. \quad (4.259)$$

Next consider the two vectors that define the sides of the slab. In the current state they are

$$\mathbf{F} \cdot \mathbf{e}_2 = F_{q2}\mathbf{e}_q, \quad \mathbf{F} \cdot \mathbf{e}_3 = F_{r3}\mathbf{e}_r. \quad (4.260)$$

Thus, the normal to the slab's face is

$$\mathbf{n} = \frac{F_{q2}\mathbf{e}_2 \times F_{r3}\mathbf{e}_3}{|F_{q2}\mathbf{e}_2||F_{r3}\mathbf{e}_3|}. \quad (4.261)$$

A straightforward calculation shows that

$$\mathbf{n} \cdot \mathbf{e}_1 = \frac{1 - (kt)^2}{[1 + (kt)^2]^{1/2}[1 + 2(kt)^2]^{1/2}}, \quad (4.262)$$

which, in the limit, becomes

$$\lim_{t \to \infty} \mathbf{n} \cdot \mathbf{e}_1 = 1/\sqrt{2}, \quad (4.263)$$

thus demonstrating the proposition.

4.27 Rigid Body Motions

LEMMA 4.5: *Let $\{\mathbf{e}_1, \mathbf{e}_2, \mathbf{e}_3\}$ and $\{\mathbf{e}'_1, \mathbf{e}'_2, \mathbf{e}'_3\}$ be two orthonormal unit bases systems. Then*

$$\mathcal{P} = \mathbf{e}_p\mathbf{e}'_p \quad (4.264)$$

is an orthogonal tensor.

Proof: If **a** and **b** are two arbitrary vectors and if $\mathcal{P}$ is an orthogonal tensor, then

$$(\mathcal{P} \cdot \mathbf{a}) \cdot (\mathcal{P} \cdot \mathbf{b}) = \mathbf{a} \cdot \mathbf{b}. \quad (4.265)$$

To verify this, write

$$(\mathcal{P} \cdot \mathbf{a}) \cdot (\mathcal{P} \cdot \mathbf{b}) = (\mathbf{e}'_p\mathbf{e}_p \cdot \mathbf{a}) \cdot (\mathbf{e}'_p\mathbf{e}_p \cdot \mathbf{b})$$

$$= (\mathbf{a} \cdot \mathcal{P}^T) \cdot (\mathcal{P} \cdot \mathbf{b}) = (\mathbf{a} \cdot \mathbf{e}_p\mathbf{e}'_p) \cdot (\mathbf{e}'_s\mathbf{e}_s \cdot \mathbf{b}).$$

4.28. Behavior under Superposed Rotation

Figure 4.17. Two material fibers.

The last equality verifies the proposition since $\mathbf{e}'_p \cdot \mathbf{e}'_s = \delta_{ps}$ and, of course, $a_p = \mathbf{a} \cdot \mathbf{e}_p$ and $b_s = \mathbf{b} \cdot \mathbf{e}_s$. Thus $(\mathcal{P} \cdot \mathbf{a}) \cdot (\mathcal{P} \cdot \mathbf{b}) = \mathbf{a} \cdot \mathbf{b}$, as $\mathcal{P}$ defines an orthogonal tensor.

LEMMA 4.6: *If the distance between every pair of material points is the same following a motion, the motion is said to be rigid. For this to be the case the mapping must be of the form*

$$\mathbf{x}(t) = \mathbf{c}(t) + \mathbf{Q}(t) \cdot \mathbf{X}, \tag{4.266}$$

where $\mathbf{Q}$ is an orthogonal tensor.

Proof: Refer to Fig. 4.17 showing two material vectors, $\mathbf{x}$ and $\mathbf{y}$. First note that the term $\mathbf{c}(t)$ simply represents a uniform translation of the body. This clearly produces no changes in the length of any fibers or any changes in the relative angle between any two fibers. Thus, we are concerned only with the remaining part of the transformation, *i.e.*,

$$\mathbf{x} - \mathbf{c} = \mathbf{Q} \cdot \mathbf{X}. \tag{4.267}$$

We proceed as follows. Form the scalar product between the two vectors $\mathbf{x} - \mathbf{c}$ and $\mathbf{y} - \mathbf{c}$,

$$(\mathbf{x} - \mathbf{c}) \cdot (\mathbf{y} - \mathbf{c}) = (\mathbf{Q} \cdot \mathbf{X}) \cdot (\mathbf{Q} \cdot \mathbf{Y}) = (\mathbf{X} \cdot \mathbf{Q}^T) \cdot (\mathbf{Q} \cdot \mathbf{Y}) = \mathbf{X} \cdot \mathbf{Y},$$

which is constant for given $\mathbf{X}$ and $\mathbf{Y}$. Also, $|\mathbf{x} - \mathbf{y}| = |\mathbf{X} - \mathbf{Y}|$, because $\mathbf{Q}$ is orthogonal and $\mathbf{x} - \mathbf{y} = \mathbf{Q} \cdot (\mathbf{X} - \mathbf{Y})$. Furthermore, since $\mathbf{x} = \mathbf{c} + \mathbf{Q} \cdot \mathbf{X}$, the velocity is

$$\dot{\mathbf{x}} = v = d\mathbf{x}/dt = \dot{\mathbf{c}}(t) + \dot{\mathbf{Q}} \cdot \mathbf{X}. \tag{4.268}$$

But,

$$\mathbf{X} = \mathbf{Q}^T \cdot [\mathbf{x} - \mathbf{c}(t)], \tag{4.269}$$

and so

$$v = \dot{\mathbf{x}} = \dot{\mathbf{c}}(t) + \dot{\mathbf{Q}} \cdot \mathbf{Q}^T \cdot (\mathbf{x} - \mathbf{c}). \tag{4.270}$$

Conversely,

$$\mathbf{Q} \cdot \mathbf{Q}^T = \mathbf{I} \;\Rightarrow\; \dot{\mathbf{Q}} \cdot \mathbf{Q}^T + \mathbf{Q} \cdot \dot{\mathbf{Q}}^T = 0 \;\Rightarrow\; \dot{\mathbf{Q}} \cdot \mathbf{Q}^T = -\mathbf{Q} \cdot \dot{\mathbf{Q}}^T.$$

Thus, $\dot{\mathbf{Q}} \cdot \mathbf{Q}^T = \mathbf{W}$ is antisymmetric, and we so write

$$v = \dot{\mathbf{x}} = \dot{\mathbf{c}} + \mathbf{W} \cdot (\mathbf{x} - \mathbf{c}). \tag{4.271}$$

4.28 Behavior under Superposed Rotation

If a time-dependent rotation $\mathbf{Q}(t)$ is superposed to the deformed configuration at time t, an infinitesimal material line element $d\mathbf{x}$ becomes

$$d\mathbf{x}^* = \mathbf{Q} \cdot d\mathbf{x}, \tag{4.272}$$

whereas in the undeformed configuration

$$dX^* = dX. \tag{4.273}$$

Consequently, since $dx = F \cdot dX$, we have

$$F^* = Q \cdot F. \tag{4.274}$$

This implies that

$$U^* = U, \quad C^* = C, \quad E^* = E, \tag{4.275}$$

and

$$V^* = Q \cdot V \cdot Q^T, \quad B^* = Q \cdot B \cdot Q^T. \tag{4.276}$$

The objective rate of the deformation gradient F transforms according to

$$\overset{\triangledown}{F^*} = Q \cdot \overset{\triangledown}{F}, \quad \overset{\triangledown}{F} = \dot{F} - W \cdot F. \tag{4.277}$$

The rotation R becomes

$$R^* = Q \cdot R. \tag{4.278}$$

The spin $\omega = \dot{R} \cdot R^{-1}$ changes to

$$\omega^* = \Omega + Q \cdot \omega \cdot Q^T, \quad \Omega = \dot{Q} \cdot Q^{-1}. \tag{4.279}$$

The velocity gradient transforms as

$$L^* = \Omega + Q \cdot L \cdot Q^T, \tag{4.280}$$

whereas the rate of deformation and the spin tensor become

$$\begin{aligned} D^* &= Q \cdot D \cdot Q^T, \\ W^* &= \Omega + Q \cdot W \cdot Q^T. \end{aligned} \tag{4.281}$$

4.29 Suggested Reading

Chadwick, P. (1999), *Continuum Mechanics: Concise Theory and Problems*, Dover, New York.

Chung, T. J. (1996), *Applied Continuum Mechanics*, Cambridge University Press, Cambridge, UK.

Eringen, A. C. (1967), *Mechanics of Continua*, Wiley, New York.

Fung, Y.-C. (1965), *Foundations of Solid Mechanics*, Prentice Hall, Englewood Cliffs, New Jersey.

Hill, R. (1978), Aspects of Invariance in Solid Mechanics, *Adv. Appl. Mech.*, Vol. 18, pp. 1–75.

Hunter, S. C. (1983), *Mechanics of Continuous Media*, Ellis Horwood, Chichester, UK.

Jaunzemis, W. (1967), *Continuum Mechanics*, The Macmillan, New York.

Lai, W. M., Rubin, D., and Krempl, E. (1993), *Introduction to Continuum Mechanics*, Pergamon, New York.

Malvern, L. E. (1969), *Introduction to the Mechanics of a Continuous Medium*, Prentice Hall, Englewood Cliffs, New Jersey.

4.29. Suggested Reading

Ogden, R. W. (1984), *Non-Linear Elastic Deformations*, Ellis Horwood, Chichester, UK (2nd ed., Dover, 1997).

Prager, W. (1961), *Introduction of Mechanics of Continua*, Ginn and Company, Boston.

Sedov, L. I. (1966), *Foundations of the Non-Linear Mechanics of Continua*, Pergamon, Oxford, UK.

Spencer, A. J. M. (1992), *Continuum Mechanics*, Longman Scientific & Technical, London.

Truesdell, C., and Toupin, R. (1960), The Classical Field Theories. In *Handbuch der Physik* (S. Flügge, ed.), Band III/1, Springer-Verlag, Berlin.

Truesdell, C., and Noll, W. (1965), The Nonlinear Field Theories of Mechanics. In *Handbuch der Physik* (S. Flügge, ed.), Band III/3, Springer-Verlag, Berlin (2nd ed., 1992).

5 Kinetics of Continuum

5.1 Traction Vector and Stress Tensor

There are two general types of forces we consider in the mechanics of solid bodies, *viz.*, applied surface traction and body forces. Examples of body forces include, *inter alia*, gravitational forces (self-weight), electrostatic forces (in the case of charged bodies in electric fields), and magnetic forces. Body forces are usually described as a density of force and thus have units of newtons per cubic meter [N/m^3]. Surface traction has, as expected, units of newtons per square meter [N/m^2]. We then describe a body force density as $\mathbf{b}$, so that the total body force acting on a volume element δv is $\mathbf{b}\delta v$. Surface traction must be described in terms of the vector force involved and with respect to the surface element it acts on; this, in turn, is described by the unit normal, $\mathbf{n}$, to the surface element that the traction acts on. Let $\mathbf{T_n}$ be the traction vector measured per unit area of the surface element involved. Then the total force acting on an area element with normal $\mathbf{n}$ and area ds is $\mathbf{T_n}ds$. These are depicted in Fig. 5.1.

Consider a thin "wafer" of material whose lateral dimensions are of $\mathcal{O}(h)$ and whose thickness is of $\mathcal{O}(\epsilon)$; we will soon take the limit as $\lim_{\epsilon/h \to 0}$. In any event, $\epsilon/h \ll 1$. Equilibrium for the wafer requires that

$$\int_\Gamma \epsilon \mathbf{T}_\Gamma \, ds + \int_\Omega \mathbf{T_{n^+}} \, dA + \int_\Omega \mathbf{T_{n^-}} \, dA + \int_\Omega \epsilon \mathbf{b} \, dA = \mathbf{0}. \tag{5.1}$$

Here Ω is the total area of the wafer's top and bottom sides and Γ is the perimeter around its edge. The traction vectors, $\mathbf{T_{n^+}}$ and $\mathbf{T_{n^-}}$ act on the "top" and "bottom" sides of the wafer, respectively. The wafer's thickness is ϵ. If $\epsilon \to 0$, we have

$$\int_\Gamma \epsilon \mathbf{T}_\Gamma \, ds \to \mathbf{0}, \quad \int_\Omega \epsilon \mathbf{b} \, dA \to \mathbf{0}, \tag{5.2}$$

and we arrive at the notion of *traction continuity*

$$\int_\Omega \mathbf{T_{n^+}} \, dA + \int_\Omega \mathbf{T_{n^-}} \, dA = \mathbf{0}. \tag{5.3}$$

Since Ω is arbitrary, we must have

$$\mathbf{T_{n^+}} = -\mathbf{T_{n^-}}. \tag{5.4}$$

5.1. Traction Vector and Stress Tensor

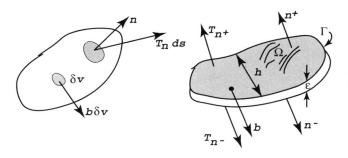

Figure 5.1. Traction vector over the surface element.

Consider next the tetrahedron shown in Fig. 5.2. This is known as the Cauchy tetrahedron. Let a_1, a_2, a_3 be the areas of the three faces having normals in the $\mathbf{e}_1, \mathbf{e}_2, \mathbf{e}_3$ directions, respectively. The volume of the tetrahedron is v. Note that the edges of the oblique face make intercepts with the axes at locations marked a, b, c. The distances along the axes of these intercepts are $\epsilon_1, \epsilon_2, \epsilon_3$. Note also that the three faces aligned with the coordinate axes have unit normals along $-\mathbf{e}_1, -\mathbf{e}_2, -\mathbf{e}_3$, respectively. Simple geometry reveals that

$$a_1 = \frac{1}{2}\epsilon_2\epsilon_3, \quad a_2 = \frac{1}{2}\epsilon_1\epsilon_3, \quad a_3 = \frac{1}{2}\epsilon_2\epsilon_1, \quad v = \frac{1}{6}\epsilon_1\epsilon_2\epsilon_3. \tag{5.5}$$

We need to calculate the area of the oblique face, and for this purpose we define two convenient vectors that define the oblique surface element. If we call a_n the magnitude of the area of this oblique surface element, and $\mathbf{n}$ its unit normal, its vector area will be $a_n\mathbf{n}$. Clearly,

$$a_n\mathbf{n} = \frac{1}{2}(\mathbf{b} - \mathbf{a}) \times (\mathbf{c} - \mathbf{a}), \tag{5.6}$$

where $\mathbf{a}, \mathbf{b}$, and $\mathbf{c}$ are the edge vectors comprising the sides of the inclined face of the Cauchy tetrahedron. In terms of the intercepts and unit base vectors, this becomes

$$\begin{aligned} a_n\mathbf{n} &= \frac{1}{2}(-\epsilon_1\mathbf{e}_1 + \epsilon_2\mathbf{e}_2) \times (-\epsilon_1\mathbf{e}_1 + \epsilon_3\mathbf{e}_3) \\ &= \frac{1}{2}(\epsilon_3\epsilon_1\mathbf{e}_2 + \epsilon_1\epsilon_2\mathbf{e}_3 + \epsilon_2\epsilon_3\mathbf{e}_1). \end{aligned} \tag{5.7}$$

Thus,

$$a_n\mathbf{n} = a_i\mathbf{e}_i \Rightarrow a_i = a_n(\mathbf{n} \cdot \mathbf{e}_i). \tag{5.8}$$

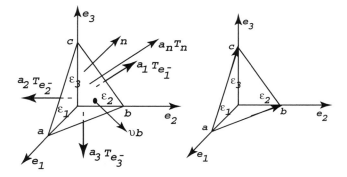

Figure 5.2. The Cauchy tetrahedron.

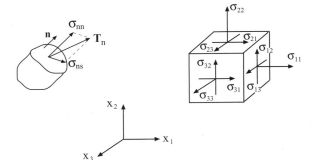

Figure 5.3. Rectangular stress components of Cauchy stress.

The equilibrium of the tetrahedron yields

$$a_n \mathbf{T_n} + a_1 \mathbf{T}_{\mathbf{e}_1^-} + a_2 \mathbf{T}_{\mathbf{e}_2^-} + a_3 \mathbf{T}_{\mathbf{e}_3^-} + v\mathbf{b} = \mathbf{0}, \tag{5.9}$$

or, more compactly,

$$\mathbf{T_n} + (\mathbf{n} \cdot \mathbf{e}_i)\mathbf{T}_{\mathbf{e}_i^-} + v/a_n \mathbf{b} = \mathbf{0}. \tag{5.10}$$

In the limit when $\epsilon_i \to 0$, or $v/a_n \to 0$, we obtain

$$\mathbf{T_n} = -(\mathbf{n} \cdot \mathbf{e}_i)\mathbf{T}_{\mathbf{e}_i^-} = (\mathbf{n} \cdot \mathbf{e}_i)\mathbf{T}_{\mathbf{e}_i}. \tag{5.11}$$

This suggests the definition

$$\sigma \equiv \mathbf{e}_i \mathbf{T}_{\mathbf{e}_i}, \tag{5.12}$$

such that

$$\mathbf{T_n} = \mathbf{n} \cdot \sigma. \tag{5.13}$$

But,

$$\mathbf{T}_{\mathbf{e}_i} = T_{\mathbf{e}_i j} \mathbf{e}_j, \tag{5.14}$$

which suggests a further definition

$$\sigma = T_{\mathbf{e}_i j} \mathbf{e}_i \mathbf{e}_j. \tag{5.15}$$

This yields

$$\sigma = \sigma_{ij} \mathbf{e}_i \mathbf{e}_j, \quad \sigma_{ij} = T_{\mathbf{e}_i j}. \tag{5.16}$$

With reference to Fig. 5.3, we can represent the normal and shear components of stress as

$$\begin{aligned}\sigma_{nn} &= \mathbf{n} \cdot \sigma \cdot \mathbf{n}, \\ \sigma_{ns} &= \mathbf{n} \cdot \sigma \cdot \mathbf{s}.\end{aligned} \tag{5.17}$$

5.2 Equations of Equilibrium

Consider a loading system on a solid body consisting of body forces and applied surface traction. Equilibrium requires that

$$\int_S \mathbf{T_n}\, dS + \int_V \mathbf{b}\, dV = \mathbf{0}, \tag{5.18}$$

5.3. Balance of Angular Momentum: Symmetry of σ

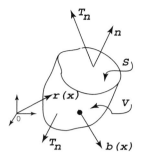

Figure 5.4. A portion of an equilibrated body under surface tractions and body forces.

where dS is a surface area element and S the body's surface as depicted in Fig. 5.4. Since $\mathbf{T_n} = \mathbf{n} \cdot \sigma$, (5.18) can be written as

$$\int_S \mathbf{n} \cdot \sigma \, dS + \int_V \mathbf{b} \, dV = \mathbf{0}. \tag{5.19}$$

The divergence theorem gives

$$\int_S \mathbf{n} \cdot \sigma \, dS = \int_V \nabla \cdot \sigma \, dV, \tag{5.20}$$

which, in turn, leads to

$$\int_v (\nabla \cdot \sigma + \mathbf{b}) \, dV = \mathbf{0}. \tag{5.21}$$

Since V is arbitrary, we find that at every point

$$\text{div } \sigma + \mathbf{b} = \nabla \cdot \sigma + \mathbf{b} = \mathbf{0}. \tag{5.22}$$

These are the equilibrium equations. Since

$$\sigma = \sigma_{ij} \mathbf{e}_i \mathbf{e}_j, \tag{5.23}$$

we can write, in component form, the three equilibrium equations as

$$\frac{\partial \sigma_{ji}}{\partial x_j} + b_i = 0, \quad i = 1, 2, 3. \tag{5.24}$$

The components of the body force per unit current volume are b_i.

5.3 Balance of Angular Momentum: Symmetry of σ

The requirement of vanishing net torque on the solid body requires that

$$\int_S \mathbf{r} \times \mathbf{T_n} \, dS + \int_V \mathbf{r} \times \mathbf{b} \, dV = \mathbf{0}, \tag{5.25}$$

where $\mathbf{r}$ is a position vector within the body measured from an arbitrary point (Fig. 5.4). Using the fact that $\mathbf{T_n} = \mathbf{n} \cdot \sigma = \sigma^T \cdot \mathbf{n}$, the above becomes

$$\int_S \mathbf{r} \times (\sigma^T \cdot \mathbf{n}) \, dS + \int_V \mathbf{r} \times \mathbf{b} \, dV = \mathbf{0}. \tag{5.26}$$

The corresponding equation in the index notation is

$$\int_S \epsilon_{ijk} x_j \sigma_{lk} n_l \, dS + \int_V \epsilon_{ijk} x_j b_k \, dV = 0. \tag{5.27}$$

By the Gauss divergence theorem

$$\int_S \epsilon_{ijk} x_j \sigma_{lk} n_l \, dS = \int_V \epsilon_{ijk} \frac{\partial}{\partial x_l}(x_j \sigma_{lk}) \, dV$$
$$= \int_V \epsilon_{ijk} \left(\sigma_{jk} + x_j \frac{\partial \sigma_{lk}}{\partial x_l} \right) dV. \tag{5.28}$$

Then, (5.27) becomes

$$\int_V \epsilon_{ijk} x_j \left(\frac{\partial \sigma_{lk}}{\partial x_l} + b_k \right) dV + \int_V \epsilon_{ijk} \sigma_{jk} \, dV = 0. \tag{5.29}$$

The first integral vanishes by equilibrium equations. Thus,

$$\int_V \epsilon_{ijk} \sigma_{jk} \, dV = 0. \tag{5.30}$$

This holds for the whole volume V, or any part of it, so that we must have at every point

$$\epsilon_{ijk} \sigma_{jk} = 0. \tag{5.31}$$

Since ϵ_{ijk} is antisymmetric in jk, the stress tensor must be symmetric, $\sigma_{jk} = \sigma_{kj}$. Alternatively, by multiplying (5.31) with ϵ_{mni} and by using $\epsilon - \delta$ relation

$$\epsilon_{mni} \epsilon_{jki} = \delta_{mj} \delta_{nk} - \delta_{mk} \delta_{nj}, \tag{5.32}$$

there follows

$$\epsilon_{mni} \epsilon_{ijk} \sigma_{jk} = \sigma_{mn} - \sigma_{nm} = 0. \tag{5.33}$$

Consequently,

$$\sigma_{mn} = \sigma_{nm}, \quad \text{or} \quad \boldsymbol{\sigma} = \boldsymbol{\sigma}^T, \tag{5.34}$$

establishing that the Cauchy stress tensor is a symmetric tensor.

5.4 Principal Values of Cauchy Stress

To explore principal values, σ_i, of the Cauchy stress tensor, we consider the equation

$$\boldsymbol{\sigma} \cdot \mathbf{n} = \sigma \mathbf{n}. \tag{5.35}$$

The eigenvalues

$$\sigma_i = \mathbf{n}_i \cdot \boldsymbol{\sigma} \cdot \mathbf{n}_i \tag{5.36}$$

are the principal stresses, and the eigenvectors $\mathbf{n}_i$ are the unit normals to principal planes of the stress tensor $\boldsymbol{\sigma}$. Since

$$\boldsymbol{\sigma} \cdot \mathbf{n}_i - \sigma_i \mathbf{n}_i = \mathbf{0}, \tag{5.37}$$

there is no shear stress on the principal planes. The principal stresses and principal planes are determined by solving the eigenvalue problem (5.35). If the three eigenvalues are

5.5. Maximum Shear Stresses

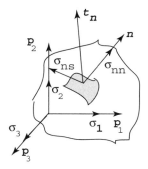

Figure 5.5. Coordinate system of the principal directions.

different, there are three mutually orthogonal principal directions. If two eigenvalues are equal, *e.g.*, $\sigma_1 = \sigma_2 \neq \sigma_3$, any direction in the plane orthogonal to $\mathbf{n}_3$ is the principal direction. If three eigenvalues are equal to each other, the stress state is said to be spherical; any direction is the principal direction, without shear stress on any plane.

5.5 Maximum Shear Stresses

Imagine that the principal stresses have been determined along with the corresponding eigenvectors, *i.e.*, the principal directions of the Cauchy stress tensor $\boldsymbol{\sigma}$. Call them $\{\mathbf{p}_1, \mathbf{p}_2, \mathbf{p}_3\}$. Call the three principal stresses $\{\sigma_1, \sigma_2, \sigma_3\}$, and assume without loss of generality that they are algebraically ordered as $\sigma_1 \geq \sigma_2 \geq \sigma_3$. Now consider an arbitrary area element within the body at a point with unit normal $\mathbf{n}$. The traction acting on this area element is $\mathbf{t_n}$. The components of $\mathbf{t_n}$ with respect to the principal directions are then

$$t_1^n = \sigma_1 n_1,$$
$$t_2^n = \sigma_2 n_2, \qquad (5.38)$$
$$t_3^n = \sigma_3 n_3,$$

where the components of the unit normal $\mathbf{n}$ are likewise referred to the principal directions of $\boldsymbol{\sigma}$. The normal stress on the area element is

$$\sigma_{nn} = \mathbf{n} \cdot \boldsymbol{\sigma} \cdot \mathbf{n} = \sigma_1 n_1^2 + \sigma_2 n_2^2 + \sigma_3 n_3^2. \qquad (5.39)$$

The stress component σ_{ns}, defined in Fig. 5.5 is the shear stress acting in the area element resulting from the tangential force resolved from $\mathbf{t_n}$. Then since the magnitude of $\mathbf{t_n}$ is

$$\mathbf{t_n} \cdot \mathbf{t_n} = \sigma_{nn}^2 + \sigma_{ns}^2, \qquad (5.40)$$

we have

$$\sigma_{ns}^2 = \mathbf{t_n} \cdot \mathbf{t_n} - \sigma_{nn}^2. \qquad (5.41)$$

With the obvious substitutions, we obtain

$$\sigma_{ns}^2 = \sigma_1^2 n_1^2 + \sigma_2^2 n_2^2 + \sigma_3^2 n_3^2 - (\sigma_1 n_1^2 + \sigma_2 n_2^2 + \sigma_3 n_3^2)^2. \qquad (5.42)$$

Extremum values for σ_{ns} are found using the Lagrangian multiplier. Let

$$\mathcal{L} = \sigma_{ns}^2 - \lambda(n_i n_i - 1), \qquad (5.43)$$

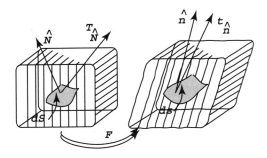

Figure 5.6. The traction vector in the reference and deformed states.

subject to the constraint that $n_i n_i = 1$. Then $\partial \mathcal{L}/\partial \lambda = 0$ yields the system

$$n_1 \left[\sigma_1^2 - 2\sigma_1(\sigma_1 n_1^2 + \sigma_2 n_2^2 + \sigma_3 n_3^2) + \lambda \right] = 0,$$
$$n_2 \left[\sigma_2^2 - 2\sigma_2(\sigma_1 n_1^2 + \sigma_2 n_2^2 + \sigma_3 n_3^2) + \lambda \right] = 0, \quad (5.44)$$
$$n_3 \left[\sigma_3^2 - 2\sigma_3(\sigma_1 n_1^2 + \sigma_2 n_2^2 + \sigma_3 n_3^2) + \lambda \right] = 0.$$

These equations are solved for λ and the components of **n**, subject to the constraint that $n_1^2 + n_2^2 + n_3^2 = 1$.

There are two sets of solutions possible. The first set is

$$n_1 = \pm 1, \quad n_2 = 0, \quad n_3 = 0, \quad \text{for which } \sigma_{ns} = 0,$$
$$n_1 = 0, \quad n_2 = \pm 1, \quad n_3 = 0, \quad \text{for which } \sigma_{ns} = 0,$$
$$n_1 = 0, \quad n_2 = 0, \quad n_3 = \pm 1, \quad \text{for which } \sigma_{ns} = 0.$$

The shear stresses are zero because these solutions for **n** are just to align it with a principal direction; this yields minimum values for σ_{ns}.

Another set of solutions is

$$n_1 = 0, \quad n_2 = \pm 1/\sqrt{2}, \quad n_3 = \pm 1/\sqrt{2}, \quad \text{for which } \sigma_{ns} = (\sigma_2 - \sigma_3)/2,$$
$$n_1 = \pm 1/\sqrt{2}, \quad n_2 = 0, \quad n_3 = \pm 1/\sqrt{2}, \quad \text{for which } \sigma_{ns} = (\sigma_1 - \sigma_3)/2,$$
$$n_1 = \pm 1/\sqrt{2}, \quad n_2 = \pm 1/\sqrt{2}, \quad n_3 = 0, \quad \text{for which } \sigma_{ns} = (\sigma_1 - \sigma_2)/2.$$

The second expression gives the maximum shear stress, which is equal to half the difference between the maximum and minimum principal stress. Also, the maximum shear stress acts in the plane that bisects the right angle between the directions of the maximum and minimum principal stresses.

5.6 Nominal Stress

Recall the connection between area elements in the deformed and the reference state,

$$\hat{\mathbf{n}} ds = (\det \mathbf{F}) \mathbf{F}^{-T} \cdot \hat{\mathbf{N}} dS. \quad (5.45)$$

Define the nominal traction $\mathbf{T}_{\hat{\mathbf{N}}}$ acting on the area element in the reference state (Fig. 5.6) by

$$\mathbf{T}_{\hat{\mathbf{N}}} dS = \mathbf{t}_{\hat{\mathbf{n}}} ds, \quad (5.46)$$

5.7. Equilibrium in the Reference State

where $\mathbf{t}_{\hat{\mathbf{n}}}$ is the true traction acting in the corresponding area element in the deformed state. Since $\mathbf{t}_{\hat{\mathbf{n}}} = \boldsymbol{\sigma} \cdot \hat{\mathbf{n}}$, we have

$$\mathbf{T}_{\hat{\mathbf{N}}} dS = \boldsymbol{\sigma} \cdot \hat{\mathbf{n}} ds = \boldsymbol{\sigma} \cdot (\det \mathbf{F}) \mathbf{F}^{-T} \cdot \hat{\mathbf{N}} dS. \tag{5.47}$$

This defines the nominal stress tensor, or the *first Piola–Kirchhoff stress* tensor,

$$\mathbf{P} = (\det \mathbf{F}) \boldsymbol{\sigma} \cdot \mathbf{F}^{-T}, \tag{5.48}$$

such that $\mathbf{T}_{\hat{\mathbf{N}}} = \mathbf{P} \cdot \hat{\mathbf{N}}$. If we decompose the nominal stress on the bases in the undeformed and deformed configuration, we have

$$\mathbf{P} = P_{ij} \mathbf{e}_i \mathbf{E}_j. \tag{5.49}$$

Thus, the nominal stress defined in this way has the following interpretation

$P_{ij} = \mathbf{e}_i \cdot \mathbf{P} \cdot \mathbf{E}_j = i^{\text{th}}$ component of force acting on an area element

that had its normal in the j direction

and had a unit area in the reference state.

The representation $\hat{\mathbf{P}}$ is also used in the literature, where $\hat{\mathbf{P}} = \mathbf{P}^T$. This is associated with the relationship to nominal traction, *viz.*,

$$\mathbf{T}_{\hat{\mathbf{N}}} = \hat{\mathbf{N}} \cdot \hat{\mathbf{P}}. \tag{5.50}$$

In this case

$$\hat{\mathbf{P}} = \hat{P}_{ij} \mathbf{E}_i \mathbf{e}_j, \tag{5.51}$$

so that

$\hat{P}_{ij} = \mathbf{E}_i \cdot \hat{\mathbf{P}} \cdot \mathbf{e}_j = j^{\text{th}}$ component of force acting on an area element

that had its normal in the i direction

and had a unit area in the reference state.

It is noted that first Piola–Kirchhoff stress is not symmetric. Since

$$\boldsymbol{\sigma} = \frac{1}{\det \mathbf{F}} \mathbf{P} \cdot \mathbf{F}^T, \tag{5.52}$$

the symmetry of the Cauchy stress ($\boldsymbol{\sigma} = \boldsymbol{\sigma}^T$) implies

$$\mathbf{P} \cdot \mathbf{F}^T = (\mathbf{P} \cdot \mathbf{F}^T)^T = \mathbf{F} \cdot \mathbf{P}^T,$$

or

$$\mathbf{P} \cdot \mathbf{F}^T = \mathbf{F} \cdot \mathbf{P}^T. \tag{5.53}$$

5.7 Equilibrium in the Reference State

We had already established the concept of nominal stress based on the idea of traction equality, *i.e.*,

$$\mathbf{T}_{\mathbf{N}} dS = \mathbf{t}_{\mathbf{n}} ds. \tag{5.54}$$

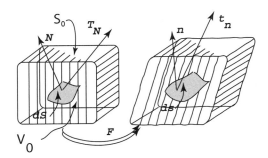

Figure 5.7. Nominal and true traction in the reference and current states.

To establish equilibrium in the reference state, we recall the definition of the nominal stress, *viz.*, $\mathbf{P} = (\det \mathbf{F})\boldsymbol{\sigma} \cdot \mathbf{F}^{-T}$. With this in mind, global equilibrium requires that

$$\int_{S_0} \mathbf{T_N}\, dS_0 + \int_{V_0} \mathbf{B}\, dV_0 = \mathbf{0}. \tag{5.55}$$

The body force per unit reference volume is $\mathbf{B}$. With the understanding that $\mathbf{T_n} = \mathbf{P} \cdot \mathbf{N}$, where $\mathbf{N}$ is the unit normal to the body's surface S_0 (see Fig. 5.7), (5.55) becomes

$$\int_{S_0} \mathbf{P} \cdot \mathbf{N}\, dS_0 + \int_{V_0} \mathbf{B}\, dV_0 = \mathbf{0}. \tag{5.56}$$

By the usual route, the divergence theorem leads to

$$\int_{V_0} (\text{Div } \mathbf{P}^T + \mathbf{B})\, dV_0 = \mathbf{0}, \tag{5.57}$$

i.e.,

$$\text{Div } \mathbf{P}^T + \mathbf{B} = \mathbf{0}. \tag{5.58}$$

By taking the time derivative of (5.56) we find that

$$\frac{d}{dt} \int_{V_0} (\text{Div } \mathbf{P}^T + \mathbf{B})\, dV_0 = \mathbf{0}, \tag{5.59}$$

where we note that $dV_0/dt = 0$. Thus, in rate form,

$$\text{Div } \dot{\mathbf{P}}^T + \dot{\mathbf{B}} = \mathbf{0}. \tag{5.60}$$

Such a simple rate form equation does not hold for the Cauchy stress and its rate.

In component form (5.58) becomes

$$\frac{\partial P_{ij}}{\partial X_j} + B_i = 0. \tag{5.61}$$

whereas, in rate form,

$$\frac{\partial \dot{P}_{ij}}{\partial X_j} + \dot{B}_i = 0. \tag{5.62}$$

5.8 Work Conjugate Connections

Consider an increment of work performed during an incremental displacement, *i.e.*,

$$\delta w = \int_{S_0} (\mathbf{P} \cdot \mathbf{N}) \cdot \delta \mathbf{u}\, dS_0. \tag{5.63}$$

5.8. Work Conjugate Connections

The rate of such working, at fixed imposed stress, is

$$\dot{w} = \int_{S_0} (\mathbf{P} \cdot \mathbf{N}) \cdot \dot{\mathbf{u}} \, dS_0. \tag{5.64}$$

But the connection between positions in the current and reference states is

$$\mathbf{x} = \mathbf{X} + \mathbf{u}, \quad \dot{\mathbf{x}} = \dot{\mathbf{u}}. \tag{5.65}$$

Recalling also that

$$\mathbf{F} = \partial \mathbf{x}/\partial \mathbf{X}, \quad \partial \dot{\mathbf{x}}/\partial \mathbf{X} = \partial \dot{\mathbf{u}}/\partial \mathbf{X} = \dot{\mathbf{F}}, \tag{5.66}$$

we can write

$$\dot{w} = \int_{S_0} (\dot{\mathbf{u}} \cdot \mathbf{P}) \cdot \mathbf{N} \, dS_0 = \int_{V_0} \text{Div}\,(\dot{\mathbf{u}} \cdot \mathbf{P}) \, dV_0 = \int_{V_0} \text{Div}\,(\mathbf{P}^T \cdot \dot{\mathbf{u}}) \, dV_0,$$

or

$$\dot{w} = \int_{V_0} (\dot{\mathbf{u}} \cdot \text{Div}\,\mathbf{P}^T + \mathbf{P} : \dot{\mathbf{F}}) \, dV_0. \tag{5.67}$$

Consider the case where $\mathbf{B} = \dot{\mathbf{B}} = \mathbf{0}$, $\text{Div}\,\mathbf{P}^T = \mathbf{0}$, and so

$$\dot{w} = \int_{V_0} \mathbf{P} : \dot{\mathbf{F}} \, dV_0, \tag{5.68}$$

where $\mathbf{P} : \dot{\mathbf{F}} = P_{ij} \dot{F}_{ij}$. Note, if $\hat{\mathbf{P}} = \mathbf{P}^T$ is used as a nominal stress, then

$$\mathbf{P} : \dot{\mathbf{F}} = P_{ij} \dot{F}_{ij} = \hat{P}_{ji} \dot{F}_{ij} = \hat{\mathbf{P}} \cdot \cdot \, \dot{\mathbf{F}}. \tag{5.69}$$

If $\mathcal{R}_0$ is the work rate per unit reference volume, then

$$\mathcal{R}_0 = \frac{\text{work rate}}{\text{unit reference volume}} = \mathbf{P} : \dot{\mathbf{F}} = P_{ij} \dot{F}_{ij}. \tag{5.70}$$

Since $\mathbf{P} = (\det \mathbf{F}) \boldsymbol{\sigma} \cdot \mathbf{F}^{-T}$, we have

$$\mathcal{R}_0 = (\det \mathbf{F})(\boldsymbol{\sigma} \cdot \mathbf{F}^{-T}) : \dot{\mathbf{F}}, \tag{5.71}$$

and, since $\boldsymbol{\sigma} = \boldsymbol{\sigma}^T$,

$$\mathcal{R}_0 = (\det \mathbf{F}) \, \boldsymbol{\sigma} : \frac{1}{2}\left[\dot{\mathbf{F}} \cdot \mathbf{F}^{-1} + (\dot{\mathbf{F}} \cdot \mathbf{F}^{-1})^T\right]. \tag{5.72}$$

This is is equivalent to

$$\mathcal{R}_0 = (\det \mathbf{F}) \, \boldsymbol{\sigma} : \mathbf{D} = \boldsymbol{\tau} : \mathbf{D}, \tag{5.73}$$

where we define the *Kirchhoff stress* as

$$\boldsymbol{\tau} = (\det \mathbf{F}) \, \boldsymbol{\sigma}. \tag{5.74}$$

Recalling that

$$\dot{\mathbf{E}} = \mathbf{F}^T \cdot \mathbf{D} \cdot \mathbf{F} \;\Rightarrow\; \mathbf{D} = \mathbf{F}^{-T} \cdot \dot{\mathbf{E}} \cdot \mathbf{F}^{-1}, \tag{5.75}$$

we can further write

$$\begin{aligned}
\mathcal{R}_0 &= (\det \mathbf{F})\,\boldsymbol{\sigma} : (\mathbf{F}^{-T} \cdot \dot{\mathbf{E}} \cdot \mathbf{F}^{-1}) \\
&= (\det \mathbf{F})\,\sigma_{ij} F_{js}^{-T} \dot{E}_{sp} F_{pi}^{-1} \\
&= (\det \mathbf{F})\,\sigma_{ij} F_{js}^{-T} F_{pi}^{-1} \dot{E}_{sp} \\
&= (\det \mathbf{F})\,F_{pi}^{-1} \sigma_{ij} F_{js}^{-T} \dot{E}_{sp}.
\end{aligned}$$

Thus,

$$\mathcal{R}_0 = (\det \mathbf{F})\,\mathbf{F}^{-1} \cdot \boldsymbol{\sigma} \cdot \mathbf{F}^{-T} : \dot{\mathbf{E}}, \tag{5.76}$$

from which we extract the stress measure

$$\mathbf{S} = (\det \mathbf{F})\,\mathbf{F}^{-1} \cdot \boldsymbol{\sigma} \cdot \mathbf{F}^{-T}, \tag{5.77}$$

which we call the *second Piola–Kirchhoff stress*. Thus, for the rate of working per unit initial volume we have

$$\mathcal{R}_0 = \mathbf{P} : \dot{\mathbf{F}} = \boldsymbol{\tau} : \mathbf{D} = \mathbf{S} : \dot{\mathbf{E}}. \tag{5.78}$$

To summarize, we have introduced three measures of stress, each *conjugate* to a particular deformation measure vis-à-vis the work rate per unit reference volume, *i.e.*,

$$\text{Nominal stress} = \mathbf{P} = (\det \mathbf{F})\,\boldsymbol{\sigma} \cdot \mathbf{F}^{-T},$$
$$\text{Kirchhoff stress} = \boldsymbol{\tau} = (\det \mathbf{F})\,\boldsymbol{\sigma}, \tag{5.79}$$
$$\text{2}^{\text{nd}}\text{ Piola–Kirchhoff stress} = \mathbf{S} = (\det \mathbf{F})\,\mathbf{F}^{-1} \cdot \boldsymbol{\sigma} \cdot \mathbf{F}^{-T}.$$

We can say that $\mathbf{P}$ is work conjugate to $\mathbf{F}$, whereas $\mathbf{S}$ is work conjugate to $\mathbf{E}$.

5.9 Stress Deviator

The general tensor, such as $\boldsymbol{\sigma}$, can be uniquely decomposed into a *deviatoric* part as

$$\boldsymbol{\sigma} = \boldsymbol{\sigma}' + \frac{1}{3}(\text{tr}\,\boldsymbol{\sigma})\mathbf{I}. \tag{5.80}$$

In proof, assume there were two such decompositions, *viz.*, $\sigma_{ij} = \lambda \delta_{ij} + \sigma'_{ij} = \lambda^* \delta_{ij} + \sigma'^{*}_{ij}$, with $\sigma'_{ii} = \sigma'^{*}_{ii} = 0$. Then, $\sigma_{ii} = 3\lambda = 3\lambda^*$ and from $\lambda \delta_{ij} + \sigma'_{ij} = \lambda \delta_{ij} + \sigma'^{*}_{ij}$ it follows that $\sigma'_{ij} = \sigma'^{*}_{ij}$. Note also that the principal directions of $\boldsymbol{\sigma}$ and $\boldsymbol{\sigma}'$ are coincident.

5.10 Frame Indifference

Consider the scenario depicted in Fig. 5.8, *i.e.*, a uniaxial tensile test with

$$\sigma_{22} = \sigma, \quad \sigma_{ij} = 0 \text{ otherwise.} \tag{5.81}$$

For the simple case of linear elasticity, with Young's modulus E, one is tempted to write

$$\dot{\epsilon}_{22} = \frac{1}{E}\dot{\sigma}_{22}. \tag{5.82}$$

5.10. Frame Indifference

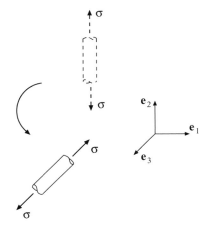

Figure 5.8. Rotating body with stress relative to a fixed reference frame.

This relation in rate form, on the surface, looks innocent enough! But, imagine that the rod shown in the figure were to rotate, with the loading system attached to it as indicated. With respect to a fixed basis such as $\{\mathbf{e}_1, \mathbf{e}_2, \mathbf{e}_3\}$ the components of σ would change. However, there are clearly no changes in stress state within the body that would cause any additional deformation. Thus a stress rate such as $\dot{\sigma}$ would clearly be inappropriate for use in a constitutive law.

We introduce the concept of an observer, *i.e.*, a *frame* of observation. We say that two such frames are equivalent if

(a) they measure the distance between any two arbitrary points to be the same;
(b) they measure the orientation between any two vectors to be the same;
(c) the time elapsed between any two events is the same; and
(d) the relative time between any two events is the same.

Let the two frames be called $\mathcal{F}$ and $\mathcal{F}'$, respectively. The two frames will satisfy the invariance conditions (a)–(d) above iff the spatial-time coordinates are connected *via*

$$\mathbf{x}' = \mathbf{c}(t) + \mathbf{Q}(t) \cdot \mathbf{x}, \quad t' = t - a, \tag{5.83}$$

where $\mathbf{Q}$ is an orthogonal tensor and a is a scalar constant (Fig. 5.9). Equation (5.83) can be inverted to yield

$$\mathbf{x} = \mathbf{Q}^T(t' + a)[\mathbf{x}' - \mathbf{c}(t' + a)], \quad t = t' + a. \tag{5.84}$$

If a scalar field ϕ, a vector field $\mathbf{u}$, and a tensor field $\mathbf{T}$ are to be objective, they must transform according to

$$\begin{aligned} \phi'(\mathbf{x}', t') &= \phi(\mathbf{x}, t), \\ \mathbf{u}'(\mathbf{x}', t') &= \mathbf{Q}(t) \cdot \mathbf{u}(\mathbf{x}, t), \\ \mathbf{T}'(\mathbf{x}', t') &= \mathbf{Q}(t) \cdot \mathbf{T}(\mathbf{x}, t) \cdot \mathbf{Q}^T(t). \end{aligned} \tag{5.85}$$

These transformation rules ensure that the directions of $\mathbf{u}$ and $\mathbf{T}$ are not altered by the transformation, relative to two rotated frames. To see this consider the two sets of

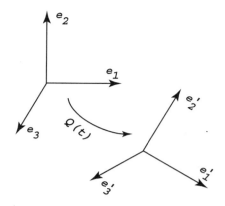

Figure 5.9. Rotated coordinate frame.

orthonormal bases, $\{e_1, e_2, e_3\}$ and $\{e'_1, e'_2, e'_3\}$, where the $\{e_i\}$ bases, defined in $\mathcal{F}$, coincide with the $\{e'_i\}$ bases defined in $\mathcal{F}'$, and where according to (5.85),

$$e'_i = \mathbf{Q}(t) \cdot e_i. \tag{5.86}$$

Thus,

$$\mathbf{Q}(t) = e'_p(t) e_p. \tag{5.87}$$

Now, form the components of **u** as

$$\mathbf{u}'(\mathbf{x}', t') \cdot e'_i = u'_i(\mathbf{x}', t') = \mathbf{Q}(t) \cdot \mathbf{u}(\mathbf{x}, t) \cdot e'_i = \mathbf{u}(\mathbf{x}, t) \cdot \mathbf{Q}^T(t) \cdot e'_i$$
$$= \mathbf{u}(\mathbf{x}, t) \cdot (e_p e'_p) \cdot e'_i = \mathbf{u}(\mathbf{x}, t) \cdot e_p \delta_{pi} = \mathbf{u}(\mathbf{x}, t) \cdot e_i = u_i(\mathbf{x}, t).$$

A similar manipulation confirms that

$$T'_{ij}(\mathbf{x}', t') = e'_i \cdot \mathbf{T}'(\mathbf{x}', t') \cdot e'_j = e_i \cdot \mathbf{T}(\mathbf{x}, t) \cdot e_j = T_{ij}(\mathbf{x}, t). \tag{5.88}$$

Let us next obtain the relations for the velocity v, acceleration **a**, the rate of deformation gradient $\dot{\mathbf{F}}$, and the velocity gradient **L**, under an observer transformation $e'_i = \mathbf{Q} \cdot e_i$. As before, $\mathbf{x}' = \mathbf{c}(t) + \mathbf{Q} \cdot \mathbf{x}$ and $t' = t - a$. In frame $\mathcal{F}$ the motion is given as

$$\mathbf{x} = \boldsymbol{\Psi}(\mathbf{X}, t), \tag{5.89}$$

whereas in frame $\mathcal{F}'$

$$\mathbf{x}' = \mathbf{c}(t' + a) + \mathbf{Q}(t' + a) \cdot \boldsymbol{\Psi}(\mathbf{X}, t' + a) = \boldsymbol{\Psi}'(\mathbf{X}, t'). \tag{5.90}$$

We obtain

$$v'(\mathbf{x}', t') = \frac{\partial \mathbf{x}'}{\partial t'} = \frac{\partial \boldsymbol{\Psi}'(\mathbf{X}, t')}{\partial t'}$$
$$= \dot{\mathbf{c}}(t' + a) + \dot{\mathbf{Q}}(t' + a) \cdot \boldsymbol{\Psi}(\mathbf{X}, t' + a)$$
$$+ \mathbf{Q}(t' + a) \cdot \frac{\partial \boldsymbol{\Psi}(\mathbf{X}, t' + a)}{\partial t}.$$

Since $\mathbf{x}' = \mathbf{c} + \mathbf{Q} \cdot \mathbf{x}$, there follows $\mathbf{x} = \mathbf{Q}^T \cdot (\mathbf{x}' - \mathbf{c})$, and

$$v'(\mathbf{x}', t') = \dot{\mathbf{c}} + \mathbf{Q} \cdot v + (\dot{\mathbf{Q}} \cdot \mathbf{Q}^T) \cdot (\mathbf{x}' - \mathbf{c}). \tag{5.91}$$

Furthermore,

$$\partial \mathbf{x}'/\partial \mathbf{X} = \mathbf{Q}(t' + a) \cdot \partial \boldsymbol{\Psi}/\partial \mathbf{X} = \mathbf{Q} \cdot \mathbf{F}(\mathbf{X}, t) = \mathbf{F}',$$

5.10. Frame Indifference

so that
$$\mathbf{F}' = \mathbf{Q} \cdot \mathbf{F}, \tag{5.92}$$

and
$$\dot{\mathbf{F}}'(\mathbf{X}, t) = \dot{\mathbf{Q}}(t) \cdot \mathbf{F}(\mathbf{X}, t) + \mathbf{Q} \cdot \dot{\mathbf{F}}(\mathbf{X}, t). \tag{5.93}$$

For the velocity gradient, we can write
$$\begin{aligned}\mathbf{L}' = \dot{\mathbf{F}}' \cdot \mathbf{F}'^{-1} &= (\mathbf{Q} \cdot \dot{\mathbf{F}} + \dot{\mathbf{Q}} \cdot \mathbf{F}) \cdot (\mathbf{Q} \cdot \mathbf{F})^{-1} \\ &= \mathbf{Q} \cdot \dot{\mathbf{F}} \cdot \mathbf{F}^{-1} \cdot \mathbf{Q}^T + \dot{\mathbf{Q}} \cdot \mathbf{Q}^T. \end{aligned} \tag{5.94}$$

Let
$$\mathbf{\Omega} \equiv \dot{\mathbf{Q}} \cdot \mathbf{Q}^T. \tag{5.95}$$

Since
$$\mathbf{Q} \cdot \mathbf{Q}^T = \mathbf{I} \Rightarrow (\mathbf{Q} \cdot \mathbf{Q}^T)^{\cdot} = \mathbf{0},$$

we have
$$\dot{\mathbf{Q}} \cdot \mathbf{Q}^T = -\mathbf{Q} \cdot \dot{\mathbf{Q}}^T = -(\dot{\mathbf{Q}} \cdot \mathbf{Q}^T)^T, \tag{5.96}$$

that is, $\dot{\mathbf{Q}} \cdot \mathbf{Q}^T$ is antisymmetric. Furthermore,
$$\mathbf{D}' = \frac{1}{2}\left(\mathbf{L}' + \mathbf{L}'^T\right), \tag{5.97}$$

and
$$\mathbf{D}' = \frac{1}{2}\left(\mathbf{Q} \cdot \mathbf{L} \cdot \mathbf{Q}^T + \mathbf{Q} \cdot \mathbf{L}^T \cdot \mathbf{Q}^T + \mathbf{\Omega} + \mathbf{\Omega}^T\right).$$

Thus,
$$\mathbf{D}' = \mathbf{Q} \cdot \mathbf{D} \cdot \mathbf{Q}^T. \tag{5.98}$$

Finally, we show that if a vector $\mathbf{u}$ and a tensor $\mathbf{T}$ are objective, their so-called convected rates
$$\overset{\triangle}{\mathbf{u}} = \dot{\mathbf{u}} + \mathbf{L}^T \cdot \mathbf{u}, \tag{5.99}$$

and
$$\overset{\triangle}{\mathbf{T}} = \dot{\mathbf{T}} + \mathbf{L}^T \cdot \mathbf{T} + \mathbf{T} \cdot \mathbf{L}. \tag{5.100}$$

are also objective. In frame $\mathcal{F}'$ we have
$$\begin{aligned}\overset{\triangle}{\mathbf{u}'} &= \dot{\mathbf{u}}' + \mathbf{L}'^T \cdot \mathbf{u}' \\ &= (\mathbf{Q} \cdot \mathbf{u})^{\cdot} + (\mathbf{Q} \cdot \mathbf{L}^T \cdot \mathbf{Q}^T - \mathbf{\Omega}) \cdot (\mathbf{Q} \cdot \mathbf{u}) \\ &= \mathbf{Q} \cdot (\dot{\mathbf{u}} + \mathbf{L}^T \cdot \mathbf{u}) + (\dot{\mathbf{Q}} \cdot \mathbf{Q}^T - \mathbf{\Omega}) \cdot (\mathbf{Q} \cdot \mathbf{u}) \\ &= \mathbf{Q} \cdot (\dot{\mathbf{u}} + \mathbf{L}^T \cdot \mathbf{u}) = \mathbf{Q} \cdot \overset{\triangle}{\mathbf{u}}.\end{aligned}$$

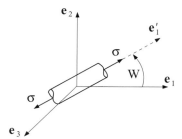

Figure 5.10. Components of stress formed on spinning axes.

Thus, $\overset{\triangle}{\mathbf{u}}$, as given by (5.99), is objective. Similarly, for the tensor $\mathbf{T}$,

$$\overset{\triangle}{\mathbf{T}'} = \dot{\mathbf{T}}' + \mathbf{L}'^T \cdot \mathbf{T}' + \mathbf{T}' \cdot \mathbf{L}'$$
$$= (\mathbf{Q} \cdot \mathbf{T} \cdot \mathbf{Q}^T)^{\cdot} + (\mathbf{Q} \cdot \mathbf{L}^T \cdot \mathbf{Q}^T - \boldsymbol{\Omega}) \cdot (\mathbf{Q} \cdot \mathbf{T} \cdot \mathbf{Q}^T)$$
$$+ (\mathbf{Q} \cdot \mathbf{T} \cdot \mathbf{Q}^T) \cdot (\mathbf{Q} \cdot \mathbf{L} \cdot \mathbf{Q}^T + \boldsymbol{\Omega}),$$

i.e.,

$$\overset{\triangle}{\mathbf{T}'} = \mathbf{Q} \cdot \overset{\triangle}{\mathbf{T}} \cdot \mathbf{Q}^T. \tag{5.101}$$

Thus, $\overset{\triangle}{\mathbf{T}}$, as given by (5.100), is objective.

Further results can be obtained by using the polar decomposition theorem, $\mathbf{F} = \mathbf{R} \cdot \mathbf{U}$. Then,

$$\mathbf{L} = \dot{\mathbf{F}} \cdot \mathbf{F}^{-1} = (\dot{\mathbf{R}} \cdot \mathbf{U} + \mathbf{R} \cdot \dot{\mathbf{U}}) \cdot (\mathbf{U}^{-1} \cdot \mathbf{R}^T)$$
$$= \dot{\mathbf{R}} \cdot \mathbf{R}^T + \mathbf{R} \cdot (\dot{\mathbf{U}} \cdot \mathbf{U}^{-1}) \cdot \mathbf{R}^T$$
$$= \boldsymbol{\Omega} + \mathbf{R} \cdot (\dot{\mathbf{U}} \cdot \mathbf{U}^{-1}) \cdot \mathbf{R}^T, \quad \boldsymbol{\Omega} = \dot{\mathbf{R}} \cdot \mathbf{R}^T. \tag{5.102}$$

Recall that the *spin* $\mathbf{W}$ is given by

$$\mathbf{W} = \frac{1}{2} \mathbf{R} \cdot (\dot{\mathbf{U}} \cdot \mathbf{U}^{-1} - \mathbf{U}^{-1} \cdot \dot{\mathbf{U}}) \cdot \mathbf{R}^T + \dot{\mathbf{R}} \cdot \mathbf{R}^T. \tag{5.103}$$

If $\dot{\mathbf{U}} = 0$, then $\mathbf{W} = \dot{\mathbf{R}} \cdot \mathbf{R}^T$, i.e., $\mathbf{W} = \boldsymbol{\Omega}$. Consequently, $\mathbf{L} = \dot{\mathbf{R}} \cdot \mathbf{R}^T$ and $\mathbf{W} = \mathbf{L} = \dot{\mathbf{R}} \cdot \mathbf{R}^T$, as well. It follows that

$$\overset{\triangledown}{\mathbf{T}} = \dot{\mathbf{T}} - \mathbf{W} \cdot \mathbf{T} + \mathbf{T} \cdot \mathbf{W} \tag{5.104}$$

is another objective rate of the tensor $\mathbf{T}$, viz., the Jaumann rate.

To better understand the Jaumann rate, consider the stress tensor $\boldsymbol{\sigma}$. Its representation in the basis $\{\mathbf{e}_i\}$ is

$$\boldsymbol{\sigma} = \sigma_{ij} \mathbf{e}_i \mathbf{e}_j. \tag{5.105}$$

5.11. Continuity Equation and Equations of Motion

We ask now about the precise character of the base vectors $\mathbf{e}_i$ that are used to form the components of $\boldsymbol{\sigma}$. Let the basis $\{\mathbf{e}_i\}$ spin with the rate $\mathbf{W}$ (material spin), as sketched in Fig. 5.10. Then,

$$\dot{\boldsymbol{\sigma}} = \dot{\sigma}_{ij}\mathbf{e}_i\mathbf{e}_j + \sigma_{ij}\dot{\mathbf{e}}_i\mathbf{e}_j + \sigma_{ij}\mathbf{e}_i\dot{\mathbf{e}}_j, \tag{5.106}$$

and

$$\dot{\mathbf{e}}_i = \mathbf{W} \cdot \mathbf{e}_i = W_{ki}\mathbf{e}_k. \tag{5.107}$$

Consequently,

$$\dot{\boldsymbol{\sigma}} = \dot{\sigma}_{ij}\mathbf{e}_i\mathbf{e}_j + W_{ik}\sigma_{kj}\mathbf{e}_i\mathbf{e}_j - \sigma_{ik}W_{kj}\mathbf{e}_i\mathbf{e}_j, \tag{5.108}$$

i.e.,

$$\dot{\boldsymbol{\sigma}} = \dot{\sigma}_{ij}\mathbf{e}_i\mathbf{e}_j + \mathbf{W} \cdot \boldsymbol{\sigma} - \boldsymbol{\sigma} \cdot \mathbf{W}. \tag{5.109}$$

This leads to definition of corotational stress rate, observed in the frame that instantaneously rotates with material spin $\mathbf{W}$,

$$\dot{\sigma}_{ij}\mathbf{e}_i\mathbf{e}_j = \dot{\boldsymbol{\sigma}} - \mathbf{W} \cdot \boldsymbol{\sigma} + \boldsymbol{\sigma} \cdot \mathbf{W}. \tag{5.110}$$

The stress rate

$$\overset{\triangledown}{\boldsymbol{\sigma}} = \dot{\boldsymbol{\sigma}} - \mathbf{W} \cdot \boldsymbol{\sigma} + \boldsymbol{\sigma} \cdot \mathbf{W} \tag{5.111}$$

is known as the *Jaumann stress rate*.

5.11 Continuity Equation and Equations of Motion

Consider an arbitrary material region that occupies V at time t. Let ρ be its mass density, and as such is a continuous, differentiable, scalar field; thus the third equation of (4.223) applies. If diffusion or chemical reactions are neglected, the total mass in this region is conserved during deformation processes that may cause V to change, *i.e.*,

$$\frac{d}{dt}\int_V \rho \, dV = 0. \tag{5.112}$$

When the third equation of (4.223) is applied, this gives

$$\int_V (\dot{\rho} + \rho \operatorname{div} \boldsymbol{v}) \, dV = 0. \tag{5.113}$$

Since the region V is arbitrary, we have the continuity equation

$$\dot{\rho} + \rho \operatorname{div} \boldsymbol{v} = 0, \tag{5.114}$$

at each point of the body.

Using (5.114), we may also show that

$$\frac{d}{dt}\int_V \rho\phi \, dV = \int_V \rho\dot{\phi} \, dV. \tag{5.115}$$

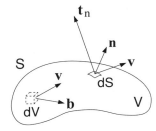

Figure 5.11. The volume V of the body bounded by closed surface S. The body force per unit mass is **b** and the surface traction over S is $\mathbf{t}_n$. The velocity vector at an arbitrary point is **v**.

Indeed,

$$\frac{d}{dt}\int_V \rho\phi\, dV = \int_V [(\rho\phi)^{\cdot} + \rho\phi\, \text{tr}\, \mathbf{L}]\, dV$$

$$= \int_V [\rho\dot\phi + (\dot\rho + \rho\, \text{div}\, v)\phi]\, dV$$

$$= \int_V \rho\dot\phi\, dV. \tag{5.116}$$

Let $\hat{\mathbf{b}}$ be the density of body force, measured per unit mass, as opposed to **b**, introduced earlier which was measured per unit current volume. Then conservation of linear momentum dictates that

$$\frac{d}{dt}\int_V \rho v\, dV = \int_V \rho\hat{\mathbf{b}}\, dV + \int_S \mathbf{t_n}\, dS, \tag{5.117}$$

where S is the bounding surface of the region in question, **n** is its outward pointing unit normal, and $\mathbf{t_n} = \boldsymbol{\sigma}\cdot\mathbf{n}$ is the traction vector. Now use (5.115) and the divergence theorem to obtain

$$\int_V \left[\rho\dot v - (\text{div}\,\boldsymbol{\sigma} + \rho\hat{\mathbf{b}})\right] dV = \mathbf{0}. \tag{5.118}$$

Because V is arbitrary, we obtain

$$\rho\mathbf{a} = \rho\dot v = \rho\ddot{\mathbf{x}} = \text{div}\,\boldsymbol{\sigma} + \rho\hat{\mathbf{b}}. \tag{5.119}$$

These are the equations of motion.

5.12 Stress Power

The rate at which external surface and body forces are doing work (the mechanical power input) on a body currently occupying the volume V bounded by the surface S is given by

$$\mathcal{P} = \int_S t_i^n v_i\, dS + \int_V \rho b_i v_i\, dV, \tag{5.120}$$

where b_i are the components of the body force (per unit mass), and t_i^n are the components of the traction vector over the surface element with unit normal **n** (Fig. 5.11). Since by the Cauchy relation $t_i^n = \sigma_{ij}n_j$, the surface integral can be expressed, with a help of the Gauss divergence theorem, as

$$\int_S t_i^n v_i\, dS = \int_S \sigma_{ij}v_i n_j\, dS = \int_V \frac{\partial}{\partial x_j}(\sigma_{ij}v_i)\, dV. \tag{5.121}$$

5.13. The Principle of Virtual Work

Thus, the mechanical power input is

$$\mathcal{P} = \int_V \left[\left(\frac{\partial \sigma_{ij}}{\partial x_j} + \rho b_i \right) v_i + \sigma_{ij} \frac{\partial v_i}{\partial x_j} \right] dV. \quad (5.122)$$

Since, by the equations of motion,

$$\frac{\partial \sigma_{ij}}{\partial x_j} + \rho b_i = \rho \frac{dv_i}{dt}, \quad (5.123)$$

the first integral in the expression for $\mathcal{P}$ becomes

$$\int_V \rho v_i \frac{dv_i}{dt} dV = \int_V \frac{1}{2} \rho \frac{d}{dt}(v_i v_i) dV = \frac{d}{dt} \int_V \frac{1}{2} \rho (v_i v_i) dV. \quad (5.124)$$

This is evidently the rate of the kinetic energy

$$K = \int_V \frac{1}{2} \rho (v_i v_i) dV, \quad (5.125)$$

so that the mechanical power input is

$$\mathcal{P} = \frac{dK}{dt} + \int_V \sigma_{ij} \frac{\partial v_i}{\partial x_j} dV. \quad (5.126)$$

But, the velocity gradient is the sum of the rate of deformation and spin tensors,

$$\frac{\partial v_i}{\partial x_j} = L_{ij} = D_{ij} + W_{ij}, \quad (5.127)$$

and because $\sigma_{ij} W_{ij} = 0$ (σ_{ij} being a symmetric and W_{ij} an antisymmetric tensor), we finally obtain

$$\mathcal{P} = \frac{dK}{dt} + \int_V \sigma_{ij} D_{ij} dV. \quad (5.128)$$

The second term is the so-called stress power. Thus, the mechanical power input goes into the change of kinetic energy of the body and the stress power associated with the deformation of the body. The quantity $\sigma_{ij} D_{ij}$ is the stress power per unit current volume. The stress power per unit initial volume is $\tau_{ij} D_{ij}$, where $\tau_{ij} = (\rho_0/\rho)\sigma_{ij}$ is the Kirchhoff stress. The stress power per unit mass is $(1/\rho)\sigma_{ij} D_{ij}$.

5.13 The Principle of Virtual Work

Consider a body that is in a state of static equilibrium under the action of a system of body forces **b** and surface traction $\mathbf{T_n}$; the unit vector **n** is again the normal to the body's surface S. The surface S is taken to be composed of two parts, S_T and S_u, where on S_T traction boundary conditions are imposed and on S_u displacement boundary conditions are imposed, as sketched in Fig. 5.12. Note that this assumes that a displacement field, **u**, exists that gives rise to the equilibrium state of stress, such that the stresses satisfy the equations of static equilibrium, viz., $\nabla \cdot \boldsymbol{\sigma} + \mathbf{b} = \mathbf{0}$. Now let us consider a displacement field $\delta \mathbf{u}$ consistent with the constraints imposed on the body. Thus $\delta \mathbf{u}$ must vanish on S_u, but is arbitrary on S_T. Additionally, we take $\delta \mathbf{u}$ to be continuous and differentiable (as needed) and to be infinitesimal. We assume that this *virtual displacement* does not affect static

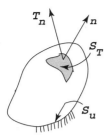

Figure 5.12. Virtual displacements atop boundary conditions.

equilibrium which continues to prevail during the imposition of $\delta \mathbf{u}$ on $\mathbf{u}$. The increment of work done through these displacements is then

$$\delta w = \int_V \mathbf{b} \cdot \delta \mathbf{u}\, dV + \int_S \mathbf{T_n} \cdot \delta \mathbf{u}\, dS, \tag{5.129}$$

or, in the component form,

$$\delta w = \int_V b_i \cdot \delta u_i\, dV + \int_S T_i \delta u_i\, dS. \tag{5.130}$$

But $\mathbf{T_n} = \boldsymbol{\sigma} \cdot \mathbf{n}$ or $T_i = \sigma_{ij} n_j$, and the second integral in (5.129) becomes, *via* the divergence theorem,

$$\int_S \mathbf{T_n} \cdot \delta \mathbf{u}\, dS = \int_S (\boldsymbol{\sigma} \cdot \mathbf{n}) \cdot \delta \mathbf{u}\, dS$$

$$= \int_S (\boldsymbol{\sigma} \cdot \delta \mathbf{u}) \cdot \mathbf{n}\, dS = \int_V \nabla \cdot (\boldsymbol{\sigma} \cdot \delta \mathbf{u})\, dV$$

$$= \int_V \nabla \cdot \boldsymbol{\sigma} \cdot \delta \mathbf{u}\, dV + \int_V \boldsymbol{\sigma} : \operatorname{grad} \delta \mathbf{u}\, dV. \tag{5.131}$$

The equilibrium conditions allow us to write

$$\int_V \nabla \cdot \boldsymbol{\sigma} \cdot \delta \mathbf{u}\, dV = -\int_V \mathbf{b} \cdot \delta \mathbf{u}\, dV. \tag{5.132}$$

Define a small strain measure corresponding to $\delta \mathbf{u}$ as

$$\delta \mathbf{e} = \frac{1}{2} \operatorname{sym}(\operatorname{grad} \delta \mathbf{u}). \tag{5.133}$$

Then, because of the symmetry of $\boldsymbol{\sigma}$, (5.129) becomes

$$\int_V \mathbf{b} \cdot \delta \mathbf{u}\, dV + \int_{S_T} \mathbf{T_n} \cdot \delta \mathbf{u}\, dS = \int_V \boldsymbol{\sigma} : \delta \mathbf{e}\, dV, \tag{5.134}$$

or, in the component form,

$$\int_V b_i \delta u_i\, dV + \int_{S_T} T_i \delta u_i\, dS = \int_V \sigma_{ij} \delta e_{ij}\, dV. \tag{5.135}$$

This is a principle of virtual work.

A similar development in terms of the nominal stress and an increment in $\delta \mathbf{F}$ yields the result

$$\int_{V_0} \mathbf{P} : \delta \mathbf{F}\, dV_0 = \int_{V_0} \mathbf{B} \cdot \delta \mathbf{u}\, dV_0 + \int_{S_0} \delta \mathbf{u} \cdot \mathbf{P} \cdot \mathbf{N}\, dS_0, \tag{5.136}$$

where V_0 and S_0 are the volume and surface of the body in the reference state.

5.14 Generalized Clapeyron's Formula

Let σ_{ij} be any statically admissible stress field, in equilibrium with body forces b_i prescribed within the volume V of the body and surface tractions T_i prescribed over the boundary S of V. Let $\hat{u}_i$ be any continuous and differentiable displacement field within V, giving rise to strain field $\hat{e}_{ij} = (\hat{u}_{i,j} + \hat{u}_{j,i})/2$. The notation for partial differentiation is used such that $\hat{u}_{i,j} = \partial \hat{u}_i / \partial x_j$. Then, since $T_i = \sigma_{ij} n_j$, where n_i defines the unit outward normal to S, the application of the Gauss divergence theorem gives

$$\int_S T_i \hat{u}_i \, dS = \int_S \sigma_{ij} \hat{u}_i n_j \, dS = \int_V (\sigma_{ij} \hat{u}_i)_{,j} \, dV \qquad (5.137)$$
$$= \int_V (\sigma_{ij,j} \hat{u}_i + \sigma_{ij} \hat{u}_{i,j}) \, dV.$$

But, $\sigma_{ij,j} = -b_i$ by equilibrium, and $\sigma_{ij} \hat{u}_{i,j} = \sigma_{ij} \hat{e}_{ij}$ by symmetry of σ_{ij}, and the substitution in above establishes a generalized Clapeyron's formula

$$\int_V \sigma_{ij} \hat{e}_{ij} \, dV = \int_S T_i \hat{u}_i \, dS + \int_V b_i \hat{u}_i \, dV. \qquad (5.138)$$

This remarkable formula clearly leads to the principle of virtual work. Suppose we take $\hat{u}_i$ to be the difference of any kinematically admissible displacement field u_i^k and the true displacement field u_i^t of the boundary value problem corresponding to prescribed body forces within V, surface tractions on S_T, and displacements on $S_u = S - S_T$ (if any). A kinematically admissible displacement field is continuous and differentiable, and it satisfies the prescribed displacement conditions on S_u. Call the displacement difference a virtual displacement field, i.e.,

$$\hat{u}_i = \delta u_i = u_i^k - u_i^t. \qquad (5.139)$$

Clearly, $\delta u_i = 0$ over S_u. The generalized Clapeyron's formula then becomes

$$\int_V \sigma_{ij} \delta e_{ij} \, dV = \int_S T_i \delta u_i \, dS + \int_V b_i \delta u_i \, dV. \qquad (5.140)$$

This is the virtual work principle, previously established in (5.135). If (5.140) holds for any kinematically admissible virtual displacement field δu_i, giving rise to virtual strain field δe_{ij} by usual relations, then the stress field σ_{ij} is in equilibrium with given body forces b_i in V and surface tractions T_i on S_T, i.e.,

$$\sigma_{ij,j} + b_i = 0 \quad \text{in } V, \quad \text{and} \quad \sigma_{ij} n_j = T_i \quad \text{on } S_T. \qquad (5.141)$$

5.15 Suggested Reading

Chadwick, P. (1999), *Continuum Mechanics: Concise Theory and Problems*, Dover, New York.
Chung, T. J. (1996), *Applied Continuum Mechanics*, Cambridge University Press, Cambridge.
Eringen, A. C. (1967), *Mechanics of Continua*, Wiley, New York.
Fung, Y.-C. (1965), *Foundations of Solid Mechanics*, Prentice Hall, Englewood Cliffs, New Jersey.
Gurtin, M. E. (1981), *An Introduction to Continuum Mechanics*, Academic Press, New York.

Hill, R. (1978), Aspects of Invariance in Solid Mechanics, *Adv. Appl. Mech.*, Vol. 18, pp. 1–75.

Malvern, L. E. (1969), *Introduction to the Mechanics of a Continuous Medium*, Prentice Hall, Englewood Cliffs, New Jersey.

Prager, W. (1961), *Introduction of Mechanics of Continua*, Ginn and Company, Boston.

Sedov, L. I. (1966), *Foundations of the Non-Linear Mechanics of Continua*, Pergamon, Oxford, UK.

Spencer, A. J. M. (1992), *Continuum Mechanics*, Longman Scientific & Technical, London.

Truesdell, C., and Toupin, R. (1960), The Classical Field Theories. In *Handbuch der Physik* (S. Flügge, ed.), Band III/1, Springer-Verlag, Berlin.

Truesdell, C., and Noll, W. (1965), The Nonlinear Field Theories of Mechanics. In *Handbuch der Physik* (S. Flügge, ed.), Band III/3, Springer-Verlag, Berlin (2nd ed., 1992).

6 Thermodynamics of Continuum

6.1 First Law of Thermodynamics: Energy Equation

A deforming body, or a given portion of it, can be considered to be a thermodynamic system in continuum mechanics. The first law of thermodynamics relates the mechanical work done on the system and the heat transferred into the system to the change in total energy of the system. The rate at which external surface and body forces are doing work on a body currently occupying the volume V bounded by the surface S is given by (5.128), i.e.,

$$\mathcal{P} = \frac{\mathrm{d}}{\mathrm{d}t} \int_V \frac{1}{2} \rho\, v_i v_i \,\mathrm{d}V + \int_V \sigma_{ij} D_{ij} \,\mathrm{d}V. \tag{6.1}$$

Let q_i be a vector whose magnitude gives the rate of heat flow by conduction across a unit area normal to q_i. The direction of q_i is the direction of heat flow, so that in time $\mathrm{d}t$ the heat amount $q_i \mathrm{d}t$ would flow through a unit area normal to q_i. If the area $\mathrm{d}S$ is oriented so that its normal n_i is not in the direction of q_i, the rate of outward heat flow through $\mathrm{d}S$ is $q_i n_i \mathrm{d}S$ (Fig. 6.1). Let a scalar r be the rate of heat input per unit mass due to distributed internal or external heat sources (*e.g.*, radiation and heating due to dissipation). The total heat input rate into the system is then

$$\mathcal{Q} = -\int_S q_i n_i \,\mathrm{d}S + \int_V \rho r \,\mathrm{d}V = \int_V \left(-\frac{\partial q_i}{\partial x_i} + \rho r \right) \mathrm{d}V. \tag{6.2}$$

First law of thermodynamics states that in any process the total energy of the system is conserved, if no work is done or heat transferred to the system from outside. Alternatively, the whole energy of the universe (system and its surrounding) is conserved. According to the first law of thermodynamics there exists a state function of a thermodynamic system, called the total energy of the system $\mathcal{E}_{\mathrm{tot}}$, such that its rate of change is

$$\dot{\mathcal{E}}_{\mathrm{tot}} = \mathcal{P} + \mathcal{Q}. \tag{6.3}$$

Neither $\mathcal{P}$ nor $\mathcal{Q}$ is in general the rate of any state function, but their sum is. The total energy of the system consists of the macroscopic kinetic energy and the internal energy of the system,

$$\mathcal{E}_{\mathrm{tot}} = \int_V \frac{1}{2} \rho\, v_i v_i \,\mathrm{d}V + \int_V \rho u \,\mathrm{d}V. \tag{6.4}$$

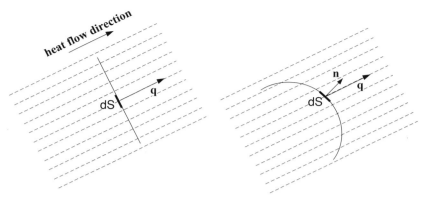

Figure 6.1. (a) Heat flow through the surface orthogonal to the direction of the heat flow. (b) The heat flow vector **q** through the surface element dS with a unit normal **n**.

The specific internal energy (internal energy per unit mass) is denoted by u. It includes the elastic strain energy and all other forms of energy that do not contribute to macroscopic kinetic energy (*e.g.*, latent strain energy around dislocations, phase-transition energy, twinning energy, and energy of random thermal motion of atoms).

Substituting (6.1), (6.2), and (6.4) into (6.3) and having in mind the general result for a scalar or tensor field A,

$$\frac{d}{dt}\int_V \rho A \, dV = \int_V \rho \frac{dA}{dt} \, dV, \qquad (6.5)$$

gives

$$\int_V \left(\rho \dot{u} - \sigma_{ij} D_{ij} + \frac{\partial q_i}{\partial x_i} - \rho r\right) dV = 0. \qquad (6.6)$$

This holds for the whole body and for any part of it, so that locally, at each point, we can write

$$\rho \dot{u} = \sigma_{ij} D_{ij} - \frac{\partial q_i}{\partial x_i} + \rho r. \qquad (6.7)$$

This is the energy equation in the deformed configuration.

6.2 Second Law of Thermodynamics: Clausius–Duhem Inequality

The first law of thermodynamics is a statement of the energy balance, which applies regardless of the direction in which the energy conversion between work and heat is assumed to occur. The second law of thermodynamics imposes restrictions on possible directions of thermodynamic processes. A state function, called the entropy of the system, is introduced as a measure of microstructural disorder of the system. The entropy can change by interaction of the system with its surroundings through the heat transfer, and by irreversible changes that take place inside the system due to local rearrangements of microstructure caused by deformation. The entropy input rate due to heat transfer is

$$-\int_S \frac{q_i n_i}{T} \, dS + \int_V \rho \frac{r}{T} \, dV = \int_V \left[-\frac{1}{\rho}\frac{\partial (q_i/T)}{\partial x_i} + \frac{r}{T}\right] \rho \, dV, \qquad (6.8)$$

6.2. Second Law of Thermodynamics

where $T > 0$ is the absolute temperature. The temperature is defined as a measure of the coldness or hotness. It appears in the denominators of the above integrands, because a given heat input causes more disorder (higher entropy change) at lower than at higher temperature (state at lower temperature being less disordered and thus more sensitive to the heat input).

An explicit expression for the rate of entropy change caused by irreversible microstructural changes inside the system depends on the type of deformation and constitution of the material. Denote this part of the rate of entropy change (per unit mass) by γ. The total rate of entropy change of the whole system is then

$$\int_V \rho \frac{ds}{dt} \, dV = \int_V \left[-\frac{1}{\rho} \frac{\partial (q_i/T)}{\partial x_i} + \frac{r}{T} + \gamma \right] \rho \, dV. \tag{6.9}$$

Locally, at each point of a deformed body, the rate of specific entropy is

$$\dot{s} = -\frac{1}{\rho} \frac{\partial (q_i/T)}{\partial x_i} + \frac{r}{T} + \gamma. \tag{6.10}$$

Because irreversible microstructural changes increase a disorder, they always contribute to an increase of the entropy. Thus, γ is always positive, and is referred to as the entropy production rate. The inequality

$$\gamma > 0 \tag{6.11}$$

is a statement of the second law of thermodynamics for irreversible processes. Therefore, from (6.10), we can write

$$\dot{s} \geq -\frac{1}{\rho} \frac{\partial (q_i/T)}{\partial x_i} + \frac{r}{T}. \tag{6.12}$$

The equality sign applies only to reversible processes ($\gamma = 0$). Inequality (6.12) is known as the Clausius–Duhem inequality.

Since

$$\frac{\partial (q_i/T)}{\partial x_i} = \frac{1}{T} \frac{\partial q_i}{\partial x_i} - \frac{1}{T^2} q_i \frac{\partial T}{\partial x_i}, \tag{6.13}$$

the inequality (6.12) can be rewritten as

$$\dot{s} \geq -\frac{1}{\rho T} \frac{\partial q_i}{\partial x_i} + \frac{r}{T} + \frac{1}{\rho T^2} q_i \frac{\partial T}{\partial x_i}. \tag{6.14}$$

The heat spontaneously flows in the direction from the hot to cold part of the body, so that $q_i(\partial T/\partial x_i) \leq 0$. Since $T > 0$, it follows that

$$\frac{1}{\rho T^2} q_i \frac{\partial T}{\partial x_i} \leq 0. \tag{6.15}$$

Thus, a stronger (more restrictive) form of the Clausius–Duhem inequality is

$$\dot{s} \geq -\frac{1}{\rho T} \frac{\partial q_i}{\partial x_i} + \frac{r}{T}. \tag{6.16}$$

Inequality (6.16) can alternatively be adopted if the temperature gradients are negligible or equal to zero. For the Carathéodory's formulation of the second law of thermodynamics and the resulting expression for the entropy production rate, see Boley and Wiener (1960) and Kestin (1979).

6.3 Reversible Thermodynamics

If deformation is such that there are no permanent microstructural rearrangements within the material (*e.g.*, thermoelastic deformation), the entropy production rate γ is equal to zero. The rate of entropy change is due to heat transfer only, and

$$T\dot{s} = -\frac{1}{\rho}\frac{\partial q_i}{\partial x_i} + r. \tag{6.17}$$

Since

$$\frac{1}{\rho}\sigma_{ij} D_{ij} = \frac{1}{\rho_0} S_{ij} \dot{E}_{ij}, \tag{6.18}$$

the energy equation (6.7) becomes

$$\dot{u} = \frac{1}{\rho_0} S_{ij} \dot{E}_{ij} + T\dot{s}. \tag{6.19}$$

Equation (6.19) shows that the internal energy is a thermodynamic potential for determining S_{ij} and T, when E_{ij} and s are considered to be independent state variables. Indeed, by partial differentiation of

$$u = u(E_{ij}, s), \tag{6.20}$$

we have

$$\dot{u} = \frac{\partial u}{\partial E_{ij}} \dot{E}_{ij} + \frac{\partial u}{\partial s} \dot{s}, \tag{6.21}$$

and comparison with (6.19) gives

$$S_{ij} = \rho_0 \frac{\partial u}{\partial E_{ij}}, \quad T = \frac{\partial u}{\partial s}. \tag{6.22}$$

In the theory of the so-called entropic elasticity, describing thermo-mechanical behavior of some elastomers, the internal energy depends only on temperature (*i.e.*, there is no change of internal energy due to deformation at constant temperature); see Chadwick (1974) and Holzapfel (2000).

6.3.1 Thermodynamic Potentials

The Helmholtz free energy is related to internal energy by

$$\phi = u - Ts. \tag{6.23}$$

By differentiating and incorporating (6.19), the rate of the Helmholtz free energy is

$$\dot{\phi} = \frac{1}{\rho_0} S_{ij} \dot{E}_{ij} - s\dot{T}. \tag{6.24}$$

This indicates that ϕ is the portion of internal energy u available for doing work at constant temperature ($\dot{T} = 0$), *i.e.*, the change of ϕ is the increment of work done at constant T. The Helmholtz free energy is a thermodynamic potential for S_{ij} and s, when E_{ij} and T are considered to be independent state variables. Indeed, by partial differentiation of

$$\phi = \phi(E_{ij}, T), \tag{6.25}$$

6.3. Reversible Thermodynamics

we have

$$\dot{\phi} = \frac{\partial \phi}{\partial E_{ij}} \dot{E}_{ij} + \frac{\partial \phi}{\partial T} \dot{T}, \tag{6.26}$$

and comparison with (6.24) gives

$$S_{ij} = \rho_0 \frac{\partial \phi}{\partial E_{ij}}, \quad s = -\frac{\partial \phi}{\partial T}. \tag{6.27}$$

The Gibbs energy is defined as a thermodynamic potential with stress and temperature as independent (controllable) variables, such that

$$g(S_{ij}, T) = \phi(E_{ij}, T) - \frac{1}{\rho_0} S_{ij} E_{ij}. \tag{6.28}$$

By differentiating (6.28) and using (6.24), it follows that

$$\dot{g} = \frac{\partial g}{\partial S_{ij}} \dot{S}_{ij} + \frac{\partial g}{\partial T} \dot{T} = -\frac{1}{\rho_0} E_{ij} \dot{S}_{ij} - s \dot{T}, \tag{6.29}$$

so that

$$E_{ij} = -\rho_0 \frac{\partial g}{\partial S_{ij}}, \quad s = -\frac{\partial g}{\partial T}. \tag{6.30}$$

Note that

$$u(E_{ij}, s) - g(S_{ij}, T) = \frac{1}{\rho_0} S_{ij} E_{ij} + T s. \tag{6.31}$$

Finally, the enthalpy function is introduced as a thermodynamic potential with stress and entropy as independent variables, such that

$$h(S_{ij}, s) = u(E_{ij}, s) - \frac{1}{\rho_0} S_{ij} E_{ij} = g(S_{ij}, T) + T s. \tag{6.32}$$

By either (6.19) or (6.29), the rate of change of enthalpy is

$$\dot{h} = \frac{\partial h}{\partial S_{ij}} \dot{S}_{ij} + \frac{\partial h}{\partial s} \dot{s} = -\frac{1}{\rho_0} E_{ij} \dot{S}_{ij} + T \dot{s}. \tag{6.33}$$

This demonstrates that the enthalpy is a portion of the internal energy that can be absorbed or released as heat, Tds, when stress S_{ij} is held constant. For example, if we compress the solid adiabatically, it warms up. If we then keep the stress constant, the amount of heat that is conducted and/or convected away is the enthalpy change. Furthermore, (6.33) yields

$$E_{ij} = -\rho_0 \frac{\partial h}{\partial S_{ij}}, \quad T = \frac{\partial h}{\partial s}. \tag{6.34}$$

The fourth-order tensors

$$\Lambda^e_{ijkl} = \left(\frac{\partial S_{ij}}{\partial E_{kl}}\right)_T = \frac{\partial^2 (\rho_0 \phi)}{\partial E_{ij} \partial E_{kl}}, \tag{6.35}$$

$$M^e_{ijkl} = \left(\frac{\partial E_{ij}}{\partial S_{kl}}\right)_T = \frac{\partial^2 (\rho_0 g)}{\partial S_{ij} \partial S_{kl}} \tag{6.36}$$

are the isothermal elastic stiffness and compliance tensors. The two fourth-order tensors are the inverse of each other. Being defined as the the second partial derivatives of $\rho_0 \phi$ and $\rho_0 g$ with respect to strain and stress, respectively, the tensors Λ_{ijkl} and M_{ijkl} possess reciprocal symmetries

$$\Lambda^e_{ijkl} = \Lambda^e_{klij}, \quad M^e_{ijkl} = M^e_{klij}. \tag{6.37}$$

The adiabatic elastic stiffness and compliance tensors are defined be the second derivatives of $\rho_0 u$ and $\rho_0 h$ with respect to strain and stress, respectively.

6.3.2 Specific and Latent Heats

The ratio of the absorbed amount of heat and the temperature increase is called the heat capacity. Because the increment of heat is not a perfect differential, the specific heat depends on the path of transformation. The two most important kinds of transformations are those taking place at constant stress (pressure) and constant strain (volume). Specific heats at constant strain and stress are thus defined by

$$c_E = T\left(\frac{\partial s}{\partial T}\right)_E, \quad c_S = T\left(\frac{\partial s}{\partial T}\right)_S, \tag{6.38}$$

where

$$s = \bar{s}(E_{ij}, T) = \hat{s}(S_{ij}, T). \tag{6.39}$$

The latent heats of change of strain and stress are the second-order tensors

$$l^E_{ij} = T\left(\frac{\partial s}{\partial E_{ij}}\right)_T, \quad l^S_{ij} = T\left(\frac{\partial s}{\partial S_{ij}}\right)_T. \tag{6.40}$$

In view of the reciprocal relations

$$\rho_0\left(\frac{\partial s}{\partial E_{ij}}\right)_T = -\left(\frac{\partial S_{ij}}{\partial T}\right)_E, \quad \rho_0\left(\frac{\partial s}{\partial S_{ij}}\right)_T = -\left(\frac{\partial E_{ij}}{\partial T}\right)_S, \tag{6.41}$$

the latent heats can also be expressed as

$$l^E_{ij} = -\frac{1}{\rho_0} T\left(\frac{\partial S_{ij}}{\partial T}\right)_E, \quad l^S_{ij} = \frac{1}{\rho_0} T\left(\frac{\partial E_{ij}}{\partial T}\right)_S. \tag{6.42}$$

The physical interpretation of the specific and latent heats follows from

$$ds = \left(\frac{\partial s}{\partial E_{ij}}\right)_T dE_{ij} + \left(\frac{\partial s}{\partial T}\right)_E dT = \frac{1}{T}\left(l^E_{ij} dE_{ij} + c_E dT\right), \tag{6.43}$$

$$ds = \left(\frac{\partial s}{\partial S_{ij}}\right)_T dS_{ij} + \left(\frac{\partial s}{\partial T}\right)_S dT = \frac{1}{T}\left(l^S_{ij} dS_{ij} + c_S dT\right). \tag{6.44}$$

Thus, the specific heat at constant strain c_E (often denoted by c_v) is the heat amount ($T\,ds$) required to increase the temperature of a unit mass for the amount dT at constant strain ($dE_{ij} = 0$). Similar interpretation holds for c_S (often denoted by c_p). The latent heat l^E_{ij} is the second-order tensor whose ij component represents the heat amount associated with a change of the corresponding strain component by dE_{ij}, at fixed temperature and fixed values of the remaining five strain components. Analogous interpretation applies to l^S_{ij}.

6.3. Reversible Thermodynamics

By partial differentiation, we have from (6.39)

$$\frac{\partial \hat{s}}{\partial T} = \frac{\partial \bar{s}}{\partial T} + \frac{\partial \bar{s}}{\partial E_{ij}} \frac{\partial E_{ij}}{\partial T}. \tag{6.45}$$

The multiplication by T and incorporation of (6.38)–(6.42) gives the relationship

$$c_S - c_E = \frac{\rho_0}{T} l^S_{ij} l^E_{ij}. \tag{6.46}$$

Furthermore, since

$$\frac{\partial \hat{s}}{\partial S_{ij}} = \frac{\partial \bar{s}}{\partial E_{kl}} M^e_{ijkl}, \tag{6.47}$$

it follows that

$$l^S_{ij} = M^e_{ijkl} l^E_{kl}. \tag{6.48}$$

When this is inserted into (6.46), we obtain

$$c_S - c_E = \frac{\rho_0}{T} M^e_{ijkl} l^E_{ij} l^E_{kl}. \tag{6.49}$$

For positive definite elastic compliance M^e_{ijkl}, it follows that

$$c_S > c_E. \tag{6.50}$$

The change in temperature caused by adiabatic straining dE_{ij}, or adiabatic stressing dS_{ij}, is obtained by setting $ds = 0$ in (6.43) and (6.44). This gives

$$dT = -\frac{1}{c_E} l^E_{ij} dE_{ij}, \quad dT = -\frac{1}{c_S} l^S_{ij} dS_{ij}. \tag{6.51}$$

6.3.3 Coupled Heat Equation

Suppose that the heat conduction is specified by a generalized Fourier law

$$q_i = -K_{ij} \frac{\partial T}{\partial x_j}, \quad K_{ij} = K_{ji}. \tag{6.52}$$

If the inequality applies in (6.15), the second-order tensor of conductivities K_{ij} must be positive-definite, i.e.,

$$\frac{\partial T}{\partial x_i} K_{ij} \frac{\partial T}{\partial x_j} > 0. \tag{6.53}$$

For simplicity it is assumed that K_{ij} is a constant tensor, although it could more generally depend on temperature and deformation. The nominal rate of heat flow is

$$q^0_i = -K^0_{ij} \frac{\partial T}{\partial X_j}, \quad K^0_{ij} = (\det \mathbf{F}) F^{-1}_{im} K_{mn} F^{-T}_{nj}, \tag{6.54}$$

which follows from

$$q^0_i n^0_i dS^0 = q_i n_i dS \quad \Rightarrow \quad q^0_i = (\det \mathbf{F}) F^{-1}_{ij} q_j, \tag{6.55}$$

and (6.52). Since

$$\frac{1}{\rho} \frac{\partial q_i}{\partial x_i} = \frac{1}{\rho^0} \frac{\partial q^0_i}{\partial X_i}, \tag{6.56}$$

we obtain from (6.17)

$$T\dot{s} = \frac{1}{\rho^0} K_{ij}^0 \frac{\partial^2 T}{\partial X_i \partial X_j} + r. \tag{6.57}$$

Combining this with (6.43) yields the heat equation

$$\frac{1}{\rho^0} K_{ij}^0 \frac{\partial^2 T}{\partial X_i \partial X_j} + r = l_{ij}^E \dot{E}_{ij} + c_E \dot{T}, \tag{6.58}$$

where

$$l_{ij}^E = T \frac{\partial \bar{s}}{\partial E_{ij}} = -\frac{1}{\rho^0} T \frac{\partial S_{ij}}{\partial T} = -T \frac{\partial^2 \phi}{\partial E_{ij} \partial T}, \tag{6.59}$$

$$c_E = T \frac{\partial \bar{s}}{\partial T} = -T \frac{\partial^2 \phi}{\partial T^2}. \tag{6.60}$$

Since (6.58) involves the rates of both temperature and strain, it is referred to as a coupled heat equation. The temperature and deformation fields cannot be determined separately, but simultaneously.

6.4 Thermodynamic Relationships with p, V, T, and s

In many thermodynamic considerations in materials science it is the pressure and volume that, together with temperature and entropy, appear in the analysis. We thus list in this section the corresponding thermodynamic expressions, reckoned per one mole of the substance (*i.e.*, $\rho_0 V_0$ is the mass of one mole). The energy equation is

$$du = -p dV + T ds. \tag{6.61}$$

Taking $u = u(V, s)$, we obtain

$$p = -\left(\frac{\partial u}{\partial V}\right)_s, \quad T = \left(\frac{\partial u}{\partial s}\right)_V. \tag{6.62}$$

The Helmholtz free energy is $\phi = \phi(V, T) = u(V, s) - Ts$, which gives

$$d\phi = -p dV - s dT, \tag{6.63}$$

and

$$p = -\left(\frac{\partial \phi}{\partial V}\right)_T, \quad s = -\left(\frac{\partial \phi}{\partial T}\right)_V. \tag{6.64}$$

The Gibbs energy is $g = g(p, T) = \phi(V, T) + pV$, and

$$dg = V dp - s dT, \tag{6.65}$$

indicating that s determines how fast g varies with T, and V how fast g varies with p. Furthermore,

$$V = \left(\frac{\partial g}{\partial p}\right)_T, \quad s = -\left(\frac{\partial g}{\partial T}\right)_p. \tag{6.66}$$

6.4. Thermodynamic Relations with p, V, T, and s

Finally, the enthalpy is $h = h(p, s) = u(V, s) + pV$, which gives

$$dh = V dp + T ds, \tag{6.67}$$

and

$$V = \left(\frac{\partial h}{\partial p}\right)_s, \quad T = \left(\frac{\partial h}{\partial s}\right)_p. \tag{6.68}$$

It is noted that

$$g = h - Ts, \quad u = \phi - g + h. \tag{6.69}$$

The Maxwell relations are

$$\left(\frac{\partial p}{\partial T}\right)_V = \left(\frac{\partial s}{\partial V}\right)_T, \quad \left(\frac{\partial p}{\partial s}\right)_V = -\left(\frac{\partial T}{\partial V}\right)_s, \tag{6.70}$$

and

$$\left(\frac{\partial V}{\partial T}\right)_p = -\left(\frac{\partial s}{\partial p}\right)_T, \quad \left(\frac{\partial V}{\partial s}\right)_p = \left(\frac{\partial T}{\partial p}\right)_s. \tag{6.71}$$

It may be noted that at constant V,

$$d\phi = -s dT, \quad du = T ds, \tag{6.72}$$

whereas at constant p,

$$dg = -s dT, \quad dh = T ds. \tag{6.73}$$

A physicist by "free energy" usually means the Helmholtz free energy, whereas a chemist usually means Gibbs free energy (because they study their reactions at constant p or T).

6.4.1 Specific and Latent Heats

The specific heats at constant volume and pressure are defined by

$$c_v = T\left(\frac{\partial s}{\partial T}\right)_V, \quad c_p = T\left(\frac{\partial s}{\partial T}\right)_p. \tag{6.74}$$

Similarly, the latent heats are defined by

$$l_v = T\left(\frac{\partial s}{\partial V}\right)_T, \quad l_p = T\left(\frac{\partial s}{\partial p}\right)_T. \tag{6.75}$$

Since $s = s(V, T)$, we have

$$ds = \left(\frac{\partial s}{\partial V}\right)_T dV + \left(\frac{\partial s}{\partial T}\right)_V dT, \tag{6.76}$$

and, upon multiplication with T,

$$T ds = l_v dV + c_v dT. \tag{6.77}$$

In particular, at $V = $ const., we have $du = T ds = c_v dT$. On the other hand, by writing $s = s(p, T)$, we obtain

$$ds = \left(\frac{\partial s}{\partial p}\right)_T dp + \left(\frac{\partial s}{\partial T}\right)_p dT \tag{6.78}$$

and, upon multiplication with T,

$$T\mathrm{d}s = l_p \mathrm{d}p + c_p \mathrm{d}T. \tag{6.79}$$

In particular, at $p = \text{const.}$, we have $\mathrm{d}h = T\mathrm{d}s = c_p \mathrm{d}T$.

The relationship between c_p and c_v can be derived by partial differentiation of $s = s[V(p,t), T]$, as follows

$$\left(\frac{\partial s}{\partial T}\right)_p = \left(\frac{\partial s}{\partial V}\right)_T \left(\frac{\partial V}{\partial T}\right)_p + \left(\frac{\partial s}{\partial T}\right)_V. \tag{6.80}$$

The multiplication with T establishes

$$c_p = l_v \left(\frac{\partial V}{\partial T}\right)_p + c_v. \tag{6.81}$$

Similarly, the relationship between l_p and l_v can be derived by partial differentiation of $s = s[p(V,t), T]$,

$$\left(\frac{\partial s}{\partial V}\right)_T = \left(\frac{\partial s}{\partial p}\right)_T \left(\frac{\partial p}{\partial V}\right)_T = -\left(\frac{\partial s}{\partial p}\right)_T \left(\frac{\partial^2 \phi}{\partial V^2}\right)_T. \tag{6.82}$$

The last transition is made by using the relationship $p = -(\partial \phi/\partial V)_T$. The multiplication with T then establishes

$$l_v = -\left(\frac{\partial^2 \phi}{\partial V^2}\right)_T l_p. \tag{6.83}$$

6.4.2 Coefficients of Thermal Expansion and Compressibility

Coefficient of volumetric thermal expansion and compressibility coefficient are defined by

$$\alpha = \frac{1}{V}\left(\frac{\partial V}{\partial T}\right)_p, \quad \beta = -\frac{1}{V}\left(\frac{\partial V}{\partial p}\right)_T. \tag{6.84}$$

In view of one of the Maxwell relations, we also have

$$\alpha = \frac{1}{V}\left(\frac{\partial V}{\partial T}\right)_p = -\frac{1}{V}\left(\frac{\partial s}{\partial p}\right)_T. \tag{6.85}$$

The coefficient of linear thermal expansion is $\alpha/3$, whereas the modulus of compressibility is $\kappa = 1/\beta$. The coefficients α and β appear in the expression for the increment of volumetric strain

$$\frac{\mathrm{d}V}{V} = \alpha \mathrm{d}T - \beta \mathrm{d}p. \tag{6.86}$$

This follows from the above definitions of α and β, and

$$V = V(T, p) \quad \Rightarrow \quad \mathrm{d}V = \left(\frac{\partial V}{\partial T}\right)_p \mathrm{d}T + \left(\frac{\partial V}{\partial p}\right)_T \mathrm{d}p. \tag{6.87}$$

An important relation for the pressure gradient of enthalpy at constant temperature can be derived by partial differentiation of $h = h[p, s(p, T)]$. This gives

$$\left(\frac{\partial h}{\partial p}\right)_T = \left(\frac{\partial h}{\partial p}\right)_s + \left(\frac{\partial h}{\partial s}\right)_p \left(\frac{\partial s}{\partial p}\right)_T = V + T\left(\frac{\partial s}{\partial p}\right)_T, \tag{6.88}$$

i.e.,
$$\left(\frac{\partial h}{\partial p}\right)_T = V(1 - \alpha T). \tag{6.89}$$

To establish additional useful relationships involving specific heats and coefficient of thermal expansion, we first recall the following result from calculus. If $z = z(x, y)$, then

$$\left(\frac{\partial z}{\partial x}\right)_y \left(\frac{\partial x}{\partial y}\right)_z \left(\frac{\partial y}{\partial z}\right)_x = -1. \tag{6.90}$$

Applying this to variables (T, V, p), we have

$$\left(\frac{\partial p}{\partial T}\right)_V \left(\frac{\partial T}{\partial V}\right)_p \left(\frac{\partial V}{\partial p}\right)_T = -1. \tag{6.91}$$

In view of the definitions (6.84) for α and β, this gives

$$\left(\frac{\partial p}{\partial T}\right)_V = \frac{\alpha}{\beta}, \tag{6.92}$$

or, recalling the Maxwell relation $(\partial p/\partial T)_V = (\partial s/\partial V)_T$,

$$\left(\frac{\partial s}{\partial V}\right)_T = \frac{\alpha}{\beta}. \tag{6.93}$$

Now, from (6.81) and (6.84),

$$c_p - c_v = \left(\frac{\partial V}{\partial T}\right)_p l_v = \alpha V l_v = \alpha V T \left(\frac{\partial s}{\partial V}\right)_T, \tag{6.94}$$

and, in view of (6.93),

$$c_p - c_v = \frac{\alpha^2}{\beta} VT. \tag{6.95}$$

Finally, we derive a useful expression for the entropy change. Since

$$l_p = T\left(\frac{\partial s}{\partial p}\right)_T = -T\left(\frac{\partial V}{\partial T}\right)_p = -T\alpha V, \tag{6.96}$$

we obtain, from (6.79),

$$ds = -\alpha V dp + c_p \frac{dT}{T}. \tag{6.97}$$

6.5 Theoretical Calculations of Heat Capacity

In this section we give a brief summary of theoretical calculations of heat capacity c_v from solid-state physics. We start with a monatomic gas. Each atom has three degrees of freedom, so that the mean kinetic energy of an atom is $3 \cdot \frac{1}{2}kT$, where $k = 1.38062 \times 10^{-23}$ J/K is the Boltzmann constant. Thus, the molar heat capacity c_v of a monatomic gas is $c_v = 3R/2$, where $R = N_A k = 8.314$ J/K is the universal gas constant, $N_A = 6.02217 \times 10^{23}$ mol^{-1} being the Avogadro's number. For diatomic molecule there are additional energy contributions, one vibrational along the axis joining two atoms, and two rotational around the axes normal to common axis. Each translational and rotational degree

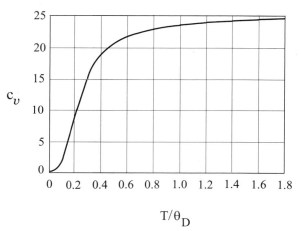

Figure 6.2. Specific heat c_v (J/mol K) of a solid as a function of temperature T (normalized by the Debye's temperature θ_D).

of freedom contributes on average an energy equal to $\frac{1}{2}kT$ per molecule (principle of equipartition of energy among the active degrees of freedom). This gives a theoretical value $c_v = 5R/2$. At higher temperature, one needs to include the vibrational (kinetic and potential) energy contributions and quantum mechanics effects. If this is done, it is found that

$$c_v = R\left[\frac{5}{2} + \left(\frac{\theta}{T}\right)^2 \frac{e^{\theta/T}}{(e^{\theta/T}-1)^2}\right], \quad \theta = \frac{h\nu}{k}. \tag{6.98}$$

The Planck's constant is $h = 6.62620 \times 10^{-34}$ J s, and ν is a characteristic vibrational frequency of the Einstein harmonic oscillator. The corresponding result for a solid is

$$c_v = 3R\left(\frac{\theta}{T}\right)^2 \frac{e^{\theta/T}}{(e^{\theta/T}-1)^2}. \tag{6.99}$$

At high temperature this gives a Dulong–Petit limit

$$\lim_{\theta/T \to 0} c_v = 3R. \tag{6.100}$$

The expression (6.99) is not satisfactory at low temperature, where it predicts an exponential decrease of c_v with temperature, whereas a dependence on T^3 is generally observed. This is accomplished by Debye's expression

$$c_v = 3R\left[12\left(\frac{\theta_D}{T}\right)^{-3} \int_0^{\theta_D/T} \frac{x^3 dx}{e^x - 1} - \frac{3\theta_D/T}{e^{\theta_D/T}-1}\right], \tag{6.101}$$

where θ_D is the Debye's temperature of a solid (the temperature of a crystal's highest mode of vibration). The values of θ_D are tabulated; for example, $\theta_D^{Cu} = 343$ K, whereas $\theta_D^{Al} = 428$ K. The plot of c_v/R vs. T/θ_D for a typical solid is shown in Fig. 6.2. For many solids for which the room temperature is greater than θ_D, the Dulong–Petit limit $c_v = 3R$ is practically attained already at 298 K.

6.6. Third Law of Thermodynamics

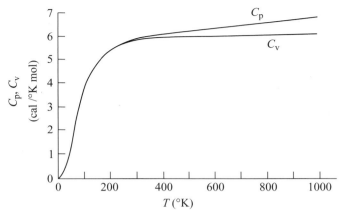

Figure 6.3. Specific heats c_p and c_v of copper vs. temperature T (from Lupis, 1983).

If the coefficients of thermal expansion and compressibility are experimentally determined, then the plot for c_p vs. temperature can be obtained from that for c_v and the relationship $c_p - c_v = (\alpha^2/\beta)VT$. Typically, one obtains a plot as sketched in Fig. 6.3. Knowing the variation $c_p(T)$ is important to determine the entropy change associated with the change of temperature (at constant pressure). Since, at constant pressure, $Tds = c_p dT$, we obtain

$$s - s_0 = \int_{T_0}^{T} c_p(T) dT. \tag{6.102}$$

In a certain temperature range, the approximation $c_p(T) = a + bT + cT^{-2}$ may be satisfactory. The coefficients a, b, and c are also tabulated for various materials and temperature ranges. For most metals, the entropy increases because of phase transformations (melting and boiling) are

$$\begin{aligned} \Delta s_m^0 &= 8.3 \div 12.6 \, \frac{J}{K \, mol} \quad \text{(Richard's rule)}, \\ \Delta s_b^0 &\approx 92 \, \frac{J}{K \, mol} \quad \text{(Trouton's rule)}. \end{aligned} \tag{6.103}$$

The superscript 0 indicates that the values are given at standard pressure of 1 atm. Better estimates of Δs_m are possible by including into considerations the crystallographic structure of solid phase. Usually,

$$\Delta s_m^0(\text{h.c.p.}) - \Delta s_m^0(\text{f.c.c.}) = \Delta s_m^0(\text{f.c.c.}) - \Delta s_m^0(\text{b.c.c.}) \approx 1 \, \frac{J}{K \, mol}.$$

6.6 Third Law of Thermodynamics

It is first assumed that there is a lower limit of temperature that the matter can exhibit. This is called the absolute zero of temperature. Now, if the entropy of each element in its perfect crystalline state (without vacancies, dislocations, or any disorder) is taken to be zero at the absolute zero of temperature, every substance composed of elements has a finite positive entropy, unless it is a perfect crystalline substance. All perfect substances have zero entropy at $T = 0$ K. This is the third law of thermodynamics.

This law simplifies the calculation of entropies. For example, if perfect iron and perfect carbon have zero entropies at 0 K, then the entropy of perfect cementite Fe_3C is also zero at 0 K. If we do not adopt the third law, we could take that the entropy of perfect iron is, say, 8 J/K mol, and of carbon 4 J/K mol, but then the entropy of the perfect cementite would be 28 J/K mol. Note also that, if we take that $s = 0$ at $T = 0$, then from $dg = dh - Tds - sdT$ we obtain that at the absolute zero of temperature $dg = dh$. Furthermore, from statistical thermodynamics, we have $s = k \ln w_{max}$, where w_{max} is the probability of the state with maximum number of quantum states (under given conditions of energy, volume, etc.). At the absolute zero of temperature, there would be a perfect order and only one state. Thus, $w_{max} = 1$ and $s = 0$.

With the adopted third law, in a test at constant pressure the entropy of a perfect substance at the temperature T can be calculated from

$$s = \int_0^T \frac{c_p}{T} dT. \tag{6.104}$$

Clearly, the ratio c_p/T must remain finite as $T \to 0$ K. Similarly, in a test at constant volume (leading to another state from that obtained in a test at constant pressure), the entropy of a perfect substance at the temperature T would be

$$s = \int_0^T \frac{c_v}{T} dT. \tag{6.105}$$

The integrand is clearly not divergent at $T = 0$ K, because $c_v \sim T^3$ near the absolute zero of temperature.

The entropy of a perfect substance at $T = 0$ K is equal to zero regardless of the pressure or volume there. Thus,

$$\left(\frac{\partial s}{\partial p}\right)_{T=0} = 0, \tag{6.106}$$

and since, by the Maxwell's relation and the definition of the coefficient of thermal expansion,

$$\left(\frac{\partial s}{\partial p}\right)_T = -\left(\frac{\partial V}{\partial T}\right)_p = -\alpha V, \tag{6.107}$$

we conclude that

$$(\alpha)_{T=0} = 0. \tag{6.108}$$

Furthermore, since

$$\left(\frac{\partial s}{\partial V}\right)_{T=0} = 0, \tag{6.109}$$

and

$$\left(\frac{\partial s}{\partial V}\right)_T = \left(\frac{\partial p}{\partial T}\right)_V = \frac{\alpha}{\beta}, \tag{6.110}$$

we conclude that

$$(\beta)_{T=0} \neq 0, \quad i.e., \quad (\kappa)_{T=0} \neq \infty. \tag{6.111}$$

The modulus of compressibility is $\kappa = 1/\beta$.

6.7 Irreversible Thermodynamics

For irreversible thermodynamic processes (*e.g.*, processes involving plastic or viscoelastic deformation) we can adopt a thermodynamics with internal state variables. A set of internal (structural) variables is introduced to describe, in some average sense, the essential features of microstructural changes that occurred at the considered place during the deformation process. These variables are denoted by ξ_j ($j = 1, 2, \ldots, n$). For simplicity, they are assumed to be scalars (extension to include tensorial internal variables is straightforward). Inelastic deformation is considered to be a sequence of constrained equilibrium states. These states are created by a conceptual constraining of internal variables at their current values through imposed thermodynamic forces f_j. The thermodynamic forces or constraints are defined such that the power dissipation (temperature times the entropy production rate) due to structural rearrangements can be expressed as

$$T\gamma = f_j \dot{\xi}_j. \tag{6.112}$$

The rates of internal variables $\dot{\xi}_j$ are called the fluxes, and the forces f_j are their affinities. By the second law, $\gamma > 0$, and therefore $f_j \dot{\xi}_j > 0$.

If various equilibrium states are considered, each corresponding to the same set of values of internal variables ξ_j, the neighboring states are related by the usual laws of reversible thermodynamics (thermoelasticity), such as (6.17) and (6.19). If neighboring constrained equilibrium states correspond to different values of internal variables, then

$$T\dot{s} = -\frac{1}{\rho}\frac{\partial q_i}{\partial x_i} + r + f_j \dot{\xi}_j. \tag{6.113}$$

Combining this with the energy equation gives

$$\dot{u} = \frac{1}{\rho_0} S_{ij} \dot{E}_{ij} + T\dot{s} - f_j \dot{\xi}_j. \tag{6.114}$$

Thus, the internal energy is a thermodynamic potential for determining S_{ij}, T and f_j, when E_{ij}, s and ξ_j are considered to be independent state variables. Indeed, after partial differentiation of

$$u = u(E_{ij}, s, \xi_j), \tag{6.115}$$

the comparison with (6.114) gives

$$S_{ij} = \rho_0 \frac{\partial u}{\partial E_{ij}}, \quad T = \frac{\partial u}{\partial s}, \quad f_j = -\frac{\partial u}{\partial \xi_j}. \tag{6.116}$$

The Helmholtz free energy

$$\phi = \phi(E_{ij}, T, \xi_j) = u(E_{ij}, s, \xi_j) - Ts \tag{6.117}$$

is a thermodynamic potential for determining S_{ij}, s and f_j, such that

$$S_{ij} = \rho_0 \frac{\partial \phi}{\partial E_{ij}}, \quad s = -\frac{\partial \phi}{\partial T}, \quad f_j = -\frac{\partial \phi}{\partial \xi_j}. \tag{6.118}$$

This clearly follows because

$$\dot{\phi} = \frac{1}{\rho_0} S_{ij} \dot{E}_{ij} - s \dot{T} - f_j \dot{\xi}_j. \tag{6.119}$$

If the Gibbs energy

$$g = g(S_{ij}, T, \xi_j) = \phi(E_{ij}, T, \xi_j) - \frac{1}{\rho_0} S_{ij} E_{ij} \tag{6.120}$$

is used, we have

$$E_{ij} = -\rho_0 \frac{\partial g}{\partial S_{ij}}, \quad s = -\frac{\partial g}{\partial T}, \quad f_j = -\frac{\partial g}{\partial \xi_j}. \tag{6.121}$$

This follows because

$$\dot{g} = -\frac{1}{\rho_0} E_{ij} \dot{S}_{ij} - s \dot{T} - f_j \dot{\xi}_j. \tag{6.122}$$

It is noted that in (6.118)

$$f_k = \bar{f}_k(E_{ij}, T, \xi_j), \tag{6.123}$$

whereas in (6.121)

$$f_k = \hat{f}_k(S_{ij}, T, \xi_j), \tag{6.124}$$

indicating different functional dependences of the respective arguments.

Finally, with the enthalpy

$$h = h(S_{ij}, s, \xi_j) = u(E_{ij}, s, \xi_j) - \frac{1}{\rho_0} S_{ij} E_{ij} \tag{6.125}$$

as a thermodynamic potential, it is found that

$$E_{ij} = -\rho_0 \frac{\partial h}{\partial S_{ij}}, \quad T = \frac{\partial h}{\partial s}, \quad f_j = -\frac{\partial h}{\partial \xi_j}, \tag{6.126}$$

because

$$\dot{h} = -\frac{1}{\rho_0} E_{ij} \dot{S}_{ij} + T \dot{s} - f_j \dot{\xi}_j. \tag{6.127}$$

By taking appropriate cross-derivatives of the previous expressions, we establish the Maxwell relations. For example,

$$\frac{\partial E_{kl}(S_{ij}, T, \xi_j)}{\partial T} = \rho_0 \frac{\partial \hat{s}(S_{ij}, T, \xi_j)}{\partial S_{kl}},$$
$$\frac{\partial S_{kl}(E_{ij}, T, \xi_j)}{\partial T} = -\rho_0 \frac{\partial \bar{s}(E_{ij}, T, \xi_j)}{\partial E_{kl}}, \tag{6.128}$$

6.8. Gibbs Conditions of Thermodynamic Equilibrium

and

$$\frac{\partial E_{kl}(S_{ij}, T, \xi_j)}{\partial \xi_m} = \rho_0 \frac{\partial \hat{f}_m(S_{ij}, T, \xi_j)}{\partial S_{kl}},$$
$$\frac{\partial S_{kl}(E_{ij}, T, \xi_j)}{\partial \xi_m} = -\rho_0 \frac{\partial \bar{f}_m(E_{ij}, T, \xi_j)}{\partial E_{kl}}. \tag{6.129}$$

6.7.1 Evolution of Internal Variables

The selection of appropriate internal variables is a difficult task, which depends on the material constitution and the type of deformation. Once internal variables are selected, it is necessary to construct evolution equations that govern their change during the deformation. For example, if the fluxes are assumed to be linearly dependent on the affinities, we may write

$$\dot{\xi}_i = \Lambda_{ij} f_j. \tag{6.130}$$

The coefficients Λ_{ij} obey the Onsager reciprocity relations if $\Lambda_{ij} = \Lambda_{ji}$.

For some materials and for some range of deformation, it may be appropriate to assume that at a given temperature T and given pattern of internal rearrangements ξ_j, each flux depends only on its own affinity, i.e.,

$$\dot{\xi}_i = \text{function}(f_i, T, \xi_j). \tag{6.131}$$

The flux dependence on the stress S_{ij} comes only through the fact that $f_i = \hat{f}_i(S_{kl}, T, \xi_k)$. This type of evolution equation is often adopted in metal plasticity, where it is assumed that the crystallographic slip on each slip system is governed by the resolved shear stress on that system (or, at the dislocation level, the motion of each dislocation segment is governed by the local Peach–Koehler force on that segment).

6.8 Gibbs Conditions of Thermodynamic Equilibrium

The system is in a thermodynamic equilibrium if its state variables do not spontaneously change with time.

Theorem 1: *In an isolated system, the equilibrium state is the state that has the maximum value of entropy.*

This is a direct consequence of the second law of thermodynamics. The increment of entropy is due to irreversible processes within the system, and its interaction with the surrounding, $ds = ds^{irr} + ds^{surr}$. By second law, $ds^{irr} > 0$. Thus, for an isolated system ($ds^{surr} = 0$), we must have

$$ds > 0, \tag{6.132}$$

i.e., the entropy of an isolated system can only increase during a thermodynamic process. If the process comes to equilibrium, the entropy is greater than the entropy in any nearby nonequilibrium state. This, for example, indicates that the entropy of the whole universe (which is an isolated system) always increases.

Theorem 2: *At constant stress and temperature, the direction of a spontaneous change of the thermodynamic state is in the direction of decreasing Gibbs energy.*

This clearly follows from (6.122), which, at constant stress and temperature, reduces to

$$\dot{g} = -f_j \dot{\xi}_j < 0. \tag{6.133}$$

The inequality follows from the second law, requiring $f_j \dot{\xi}_j > 0$. Thus, a transformation from the thermodynamic state can occur at constant stress and temerature only if it is associated with a decrease of the Gibbs energy. Because there is no spontaneous change from the equilibrium state, the Gibbs energy is minimum at the equilibrium state (relative to all neighboring states at constant stress and temperature).

Gibbs originally formulated his celebrated thermodynamic equilibrium condition as:

Any virtual variation from the equilibrium state at constant pressure and temperature which does not involve irreversible processes would give $\delta g \geq 0$; if the variation involves irreversible changes, then $\delta g < 0$.

If the substance can be in either *a* or *b* phase (structure), at given pressure and temperature, a stable equilibrium phase is one that corresponds to lower Gibbs energy.

Theorem 3: *Among all neighboring states with the same strain and entropy, the equilibrium state is one with the lowest internal energy.*

This follows from (6.114), which, at constant strain and temperature, reduces to

$$\dot{u} = -f_j \dot{\xi}_j < 0. \tag{6.134}$$

Thus, when the system undergoing a thermodynamic process at constant strain and entropy comes to rest at its equilibrium, its internal energy attains its minimum. (It is hard to control entropy in the experiment; conceptually one would need to extract the heat from the internal dissipation, such that $ds = 0$).

Theorem 4: *Among all neighboring states with the same strain and temperature, the equilibrium state is one with the lowest Helmholtz free energy.*

This follows from (6.119), which, at constant strain and temperature, reduces to

$$\dot{\phi} = -f_j \dot{\xi}_j < 0. \tag{6.135}$$

Theorem 5: *Among all neighboring states with the same stress and temperature, the equilibrium state is one with the lowest enthalpy.*

This follows from (6.127), which, at constant strain and temperature, reduces to

$$\dot{h} = -f_j \dot{\xi}_j < 0. \tag{6.136}$$

6.9 Linear Thermoelasticity

The structure of the constitutive equations relating the stress, strain, entropy and temperature in linear thermoelasticity is readily derived by assuming a quadratic representation of the Helmholtz free energy in terms of strain and temperature. The material parameters are specified in the accordance with observed isothermal elastic behavior and

6.9. Linear Thermoelasticity

measured coefficients of thermal expansion and the specific heat. For isotropic materials, this yields

$$\phi(e_{ij}, T) = \frac{1}{2}\lambda_T e_{kk}^2 + \mu e_{ij}e_{ij} - \kappa_T \alpha_0 (T - T_0) e_{kk} \\ - \frac{c_V^0}{2T_0}(T - T_0)^2 - s_0(T - T_0) + f_0, \quad (6.137)$$

where λ_T and μ are the isothermal Lamé elastic constants, $\kappa_T = \lambda_T + 2\mu/3$ is the isothermal bulk modulus, and α_0, c_V^0, and s_0 are, respectively, the coefficient of volumetric thermal expansion, the specific heat at constant strain, and the specific entropy (per unit volume), all in the reference state with temperature T_0. The infinitesimal strain is e_{ij}. The corresponding free energy (per unit volume) is $\phi(0, T_0) = \phi_0$. The stress and entropy in the deformed state are the gradients of f with respect to strain and temperature, which gives

$$\sigma_{ij} = \left(\frac{\partial \phi}{\partial e_{ij}}\right)_T = \lambda_T e_{kk}\delta_{ij} + 2\mu e_{ij} - \kappa_T \alpha_0 (T - T_0)\delta_{ij}, \quad (6.138)$$

$$s = -\left(\frac{\partial \phi}{\partial T}\right)_e = \kappa_T \alpha_0 e_{kk} + \frac{c_V^0}{T_0}(T - T_0) + s_0. \quad (6.139)$$

The specific heat at constant strain, associated with (6.137) is

$$c_V = T\left(\frac{\partial s}{\partial T}\right)_e = -T\left(\frac{\partial^2 \phi}{\partial T^2}\right)_e = c_V^0 \frac{T}{T_0}. \quad (6.140)$$

Once the Helmholtz free energy is specified as a function of strain and temperature, the internal energy $u = \phi + Ts$ can be expressed in terms of the same independent variables by simple substitution of (6.137) and the corresponding expression for the entropy. This yields

$$u(e_{ij}, T) = \frac{1}{2}\lambda_T e_{kk}^2 + \mu e_{ij}e_{ij} + \kappa_T \alpha_0 T_0 e_{kk} + \frac{c_V^0}{2T_0}(T^2 - T_0^2) + u_0. \quad (6.141)$$

In the sequel, it will be assumed that the internal energy vanishes in the reference state, so that

$$u_0 = 0, \quad \phi_0 = -T_0 s_0. \quad (6.142)$$

However, the internal energy is a thermodynamic potential whose natural independent state variables are strain and entropy, rather than strain and temperature. The desired representation $u = u(e_{ij}, s)$ can be obtained from $u = \phi + Ts$ by eliminating the temperature in terms of strain and entropy. The purely algebraic transition is simple, but little indicative of the underlying thermodynamics. An independent derivation, starting from the energy equation and utilizing the experimental data embedded in the Duhamel–Neumann extension of Hooke's law and the assumed specific heat behavior, is desirable. The systematic procedure to achieve this, and to derive the expressions for other thermodynamic potentials, is presented in next section.

6.10 Thermodynamic Potentials in Linear Thermoelasticity

The four thermodynamic potentials are derived in this section in terms of their natural independent state variables. The derivation is in each case based only on the Duhamel–Neumann extension of Hooke's law, and an assumed linear dependence of the specific heat on temperature.

6.10.1 Internal Energy

The increment of internal energy is expressed in terms of the increments of strain and entropy by the energy equation

$$du = \sigma_{ij} de_{ij} + T ds . \tag{6.143}$$

Because u is a state function, du is a perfect differential, and the Maxwell relation holds

$$\left(\frac{\partial \sigma_{ij}}{\partial s} \right)_e = \left(\frac{\partial T}{\partial e_{ij}} \right)_s . \tag{6.144}$$

The thermodynamic potential $u = u(e_{ij}, s)$ is sought corresponding to the Duhamel–Neumann expression

$$\sigma_{ij} = \lambda_T e_{kk} \delta_{ij} + 2\mu e_{ij} - \kappa_T \alpha_0 (T - T_0) \delta_{ij} , \tag{6.145}$$

and an assumed linear dependence of the specific heat on temperature

$$c_V = c_V^0 \frac{T}{T_0} . \tag{6.146}$$

By partial differentiation from (6.145) it follows that

$$\left(\frac{\partial \sigma_{ij}}{\partial s} \right)_e = \left(\frac{\partial \sigma_{ij}}{\partial T} \right)_e \left(\frac{\partial T}{\partial s} \right)_e = -\kappa_T \alpha_0 \left(\frac{\partial T}{\partial s} \right)_e \delta_{ij} , \tag{6.147}$$

so that the Maxwell relation (6.144) gives

$$\left(\frac{\partial T}{\partial e_{ij}} \right)_s = -\kappa_T \alpha_0 \left(\frac{\partial T}{\partial s} \right)_e \delta_{ij} . \tag{6.148}$$

The thermodynamic definition of the specific heat at constant strain is

$$c_V = T \left(\frac{\partial s}{\partial T} \right)_e , \tag{6.149}$$

which, in conjunction with (6.146), specifies the temperature gradient

$$\left(\frac{\partial T}{\partial s} \right)_e = \frac{T_0}{c_V^0} . \tag{6.150}$$

The substitution into (6.148) yields

$$\left(\frac{\partial T}{\partial e_{ij}} \right)_s = -\frac{\kappa_T \alpha_0 T_0}{c_V^0} \delta_{ij} . \tag{6.151}$$

The joint integration of the above two equations provides the temperature expression

$$T = -\frac{\kappa_T \alpha_0 T_0}{c_V^0} e_{kk} + \frac{T_0}{c_V^0} (s - s_0) + T_0 . \tag{6.152}$$

6.10. Thermodynamic Potentials in Thermoelasticity

When this is inserted into (6.145), we obtain an expression for the stress in terms of the strain and entropy,

$$\sigma_{ij} = \lambda_S e_{kk}\delta_{ij} + 2\mu e_{ij} - \frac{\kappa_T \alpha_0 T_0}{c_V^0}(s - s_0)\delta_{ij}. \tag{6.153}$$

The adiabatic (isentropic) Lamé constant λ_S is related to its isothermal counterpart λ_T by

$$\lambda_S = \lambda_T + \frac{\alpha_0^2 T_0}{c_V^0}\kappa_T^2. \tag{6.154}$$

By using (6.152) and (6.153), the joint integration of

$$\sigma_{ij} = \left(\frac{\partial u}{\partial e_{ij}}\right)_s, \quad T = \left(\frac{\partial u}{\partial s}\right)_e, \tag{6.155}$$

yields a desired expression for the internal energy in terms of its natural independent variables e_{ij} and s. This is

$$u(e_{ij}, s) = \frac{1}{2}\lambda_S e_{kk}^2 + \mu e_{ij}e_{ij} - \frac{\kappa_T \alpha_0 T_0}{c_V^0}(s - s_0)e_{kk}$$
$$+ \frac{T_0}{2c_V^0}(s - s_0)^2 + T_0(s - s_0). \tag{6.156}$$

6.10.2 Helmholtz Free Energy

An independent derivation of the Helmholtz free energy $\phi = \phi(e_{ij}, T)$ again begins with the pair of expressions (6.145) and (6.146). The increment of ϕ is

$$d\phi = \sigma_{ij}de_{ij} - s\,dT, \tag{6.157}$$

with the Maxwell relation

$$\left(\frac{\partial \sigma_{ij}}{\partial T}\right)_e = -\left(\frac{\partial s}{\partial e_{ij}}\right)_T. \tag{6.158}$$

By evaluating the temperature gradient of stress from (6.145), and by substituting the result into (6.158), we find

$$\left(\frac{\partial s}{\partial e_{ij}}\right)_T = \kappa_T \alpha_0 \delta_{ij}. \tag{6.159}$$

The integration of above, in conjunction with

$$\left(\frac{\partial s}{\partial T}\right)_e = \frac{c_V^0}{T_0}, \tag{6.160}$$

provides the entropy expression

$$s = \kappa_T \alpha_0 e_{kk} + \frac{c_V^0}{T_0}(T - T_0) + s_0. \tag{6.161}$$

By using (6.145) and (6.161), the joint integration of

$$\sigma_{ij} = \left(\frac{\partial \phi}{\partial e_{ij}}\right)_T, \quad s = -\left(\frac{\partial \phi}{\partial T}\right)_e, \tag{6.162}$$

yields a desired expression for the Helmholtz free energy in terms of its natural independent variables e_{ij} and T. This is

$$\phi(e_{ij}, T) = \frac{1}{2}\lambda_T e_{kk}^2 + \mu e_{ij} e_{ij} - \kappa_T \alpha_0 (T - T_0) e_{kk} \\ - \frac{c_V^0}{2T_0}(T - T_0)^2 - s_0 T. \tag{6.163}$$

6.10.3 Gibbs Energy

The increment of the Gibbs energy is

$$dg = -e_{ij} d\sigma_{ij} - s \, dT, \tag{6.164}$$

with the Maxwell relation

$$\left(\frac{\partial e_{ij}}{\partial T}\right)_\sigma = \left(\frac{\partial s}{\partial \sigma_{ij}}\right)_T. \tag{6.165}$$

To derive the function $g(\sigma_{ij}, T)$, independently of the connection $g = \phi - \sigma_{ij} e_{ij}$ and without tedious change of variables, we begin with the thermoelastic stress-strain relation and the expression for the specific heat,

$$e_{ij} = \frac{1}{2\mu}\left(\sigma_{ij} - \frac{\nu_T}{1+\nu_T}\sigma_{kk}\delta_{ij}\right) + \frac{\alpha_0}{3}(T - T_0)\delta_{ij}, \tag{6.166}$$

$$c_P(T) = c_P^0 \frac{T}{T_0}. \tag{6.167}$$

The first one is a simple extension of Hooke's law to include thermal strain, and the second one is the assumed linear dependence of the specific heat at constant stress on temperature. The thermodynamic definition of the specific heat c_P is

$$c_P = T\left(\frac{\partial s}{\partial T}\right)_\sigma. \tag{6.168}$$

By differentiating (6.166) to evaluate the temperature gradient of strain, and by substituting the result into the Maxwell relation (6.165), we find

$$\left(\frac{\partial s}{\partial \sigma_{ij}}\right)_T = \frac{\alpha_0}{3}\delta_{ij}. \tag{6.169}$$

The integration of this, in conjunction with

$$\left(\frac{\partial s}{\partial T}\right)_\sigma = \frac{c_P^0}{T_0}, \tag{6.170}$$

provides the entropy expression

$$s = \frac{\alpha_0}{3}\sigma_{kk} + \frac{c_P^0}{T_0}(T - T_0) + s_0. \tag{6.171}$$

Using (6.166) and (6.171), the joint integration of

$$e_{ij} = -\left(\frac{\partial g}{\partial \sigma_{ij}}\right)_T, \quad s = -\left(\frac{\partial g}{\partial T}\right)_\sigma, \tag{6.172}$$

6.10. Thermodynamic Potentials in Thermoelasticity

yields a desired expression for the Gibbs energy in terms of its natural independent variables σ_{ij} and T. This is

$$g(\sigma_{ij}, T) = -\frac{1}{4\mu}\left(\sigma_{ij}\sigma_{ij} - \frac{\nu_T}{1+\nu_T}\sigma_{kk}^2\right) - \frac{\alpha_0}{3}(T-T_0)\sigma_{kk} \qquad (6.173)$$
$$- \frac{c_P^0}{2T_0}(T-T_0)^2 - s_0 T .$$

The relationship between the specific heats c_P^0 and c_V^0 can be obtained in various ways. For example, by reconciling the entropy expressions (6.161) and (6.171), and by using the relationship

$$e_{kk} = \frac{1}{3\kappa_T}\sigma_{kk} + \alpha_0(T-T_0), \qquad (6.174)$$

following from (6.164), it is found that

$$c_P^0 - c_V^0 = \kappa_T \alpha_0^2 T_0 . \qquad (6.175)$$

6.10.4 Enthalpy Function

The increment of enthalpy is

$$dh = -e_{ij}d\sigma_{ij} + Tds , \qquad (6.176)$$

with the Maxwell relation

$$\left(\frac{\partial e_{ij}}{\partial s}\right)_\sigma = -\left(\frac{\partial T}{\partial \sigma_{ij}}\right)_s . \qquad (6.177)$$

To derive the function $h(\sigma_{ij}, s)$, we again begin with the expressions (6.166) and (6.167). By partial differentiation from (6.166) it follows that

$$\left(\frac{\partial e_{ij}}{\partial s}\right)_\sigma = \left(\frac{\partial e_{ij}}{\partial T}\right)_\sigma \left(\frac{\partial T}{\partial s}\right)_\sigma = \frac{\alpha_0}{3}\left(\frac{\partial T}{\partial s}\right)_\sigma \delta_{ij} . \qquad (6.178)$$

The substitution into the Maxwell relation (6.177) gives

$$\left(\frac{\partial T}{\partial \sigma_{ij}}\right)_s = -\frac{\alpha_0}{3}\left(\frac{\partial T}{\partial s}\right)_\sigma \delta_{ij} = -\frac{\alpha_0 T_0}{3c_P^0}\delta_{ij} . \qquad (6.179)$$

The definition (6.168), in conjunction with (6.167), was used in the last step. The joint integration of the above equation and

$$\left(\frac{\partial T}{\partial s}\right)_\sigma = \frac{T_0}{c_P^0}, \qquad (6.180)$$

provides the temperature expression

$$T = -\frac{\alpha_0 T_0}{3c_P^0}\sigma_{kk} + \frac{T_0}{c_P^0}(s-s_0) + T_0 . \qquad (6.181)$$

When this is substituted into (6.166), there follows

$$e_{ij} = \frac{1}{2\mu}\left(\sigma_{ij} - \frac{\nu_S}{1+\nu_S}\sigma_{kk}\delta_{ij}\right) + \frac{\alpha_0 T_0}{3c_P^0}(s-s_0)\delta_{ij} . \qquad (6.182)$$

The adiabatic Poisson's ratio v_S is related to its isothermal counterpart v_T by

$$v_S = \frac{v_T + 2\mu(1 + v_T)a}{1 - 2\mu(1 + v_T)a}, \quad v_T = \frac{v_S - 2\mu(1 + v_S)a}{1 + 2\mu(1 + v_S)a}, \qquad (6.183)$$

where

$$a = \frac{\alpha_0^2 T_0}{9 c_P^0}. \qquad (6.184)$$

The adiabatic and isothermal Young's moduli are related by

$$\frac{1}{E_T} - \frac{1}{E_S} = \frac{\alpha_0^2 T_0}{9 c_P^0}. \qquad (6.185)$$

A simple relationship is also recorded

$$\frac{c_P^0}{c_V^0} = \frac{\kappa_S}{\kappa_T}. \qquad (6.186)$$

This easily follows by noting, from (6.152) and (6.181), that for adiabatic loading

$$T_0 - T = \frac{\kappa_T \alpha_0 T_0}{c_V^0} e_{kk} = \frac{\alpha_0 T_0}{3 c_P^0} \sigma_{kk}. \qquad (6.187)$$

Since for adiabatic loading $\sigma_{kk} = 3\kappa_S e_{kk}$, the substitution into (6.187) yields (6.186).

Returning to the enthalpy function, by using (6.181) and (6.182), the joint integration of

$$e_{ij} = -\left(\frac{\partial h}{\partial \sigma_{ij}}\right)_s, \quad T = \left(\frac{\partial h}{\partial s}\right)_\sigma, \qquad (6.188)$$

yields the expression for the enthalpy in terms of its natural independent variables σ_{ij} and s. This is

$$h(\sigma_{ij}, s) = -\frac{1}{4\mu}\left(\sigma_{ij}\sigma_{ij} - \frac{v_S}{1 + v_S}\sigma_{kk}^2\right) - \frac{\alpha_0 T_0}{3 c_P^0}(s - s_0)\sigma_{kk}$$
$$+ \frac{T_0}{2 c_P^0}(s - s_0)^2 + T_0(s - s_0). \qquad (6.189)$$

6.11 Uniaxial Loading and Thermoelastic Effect

The derived representations of thermodynamic potentials for arbitrary three-dimensional states of stress and strain are greatly simplified in the case of uniaxial and spherical states of stress. The corresponding results are listed in Problems 6.1 and 6.2 of Chapter 34. To illustrate the use of some of the derived formulas, consider the uniaxial loading paths shown in Fig. 6.4. The path OAB is an adiabatic (fast loading) path, the path OC is an isothermal (slow loading) path, the path AC is a constant stress path, and the path BC is a constant longitudinal strain path. Along the adiabatic path OAB (see Problem 6.1)

$$u = -h = \frac{1}{2 E_S}\sigma^2, \qquad (6.190)$$

6.11. Uniaxial Loading and Thermoelastic Effect

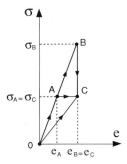

Figure 6.4. Uniaxial loading along isothermal path OC, and along adiabatic path OAB. The paths AC and BC are the constant stress and constant longitudinal strain paths, respectively.

whereas along the isothermal path OC [see the expressions for $\phi(\sigma, T)$ and $g(\sigma, T)$ from Problem 6.1]

$$\phi - \phi_0 = g_0 - g = \frac{1}{2E_T}\sigma^2, \quad \phi_0 = g_0 = -T_0 s_0. \tag{6.191}$$

The temperature drop along the adiabatic path is

$$T - T_0 = -\frac{\alpha_0 T_0}{3 c_P^0} \sigma, \tag{6.192}$$

in accord with Kelvin's formula describing Joule's thermoelastic effect. The entropy increase along the isothermal path is

$$s - s_0 = \frac{\alpha_0}{3}\sigma, \tag{6.193}$$

with the corresponding absorbed heat given by $T_0(s - s_0)$.

The heat absorbed along the constant stress path AC is equal to the enthalpy change

$$h_C - h_A = \frac{\alpha_0 T_0}{3}\sigma_A - \frac{\alpha_0^2 T_0}{18 c_P^0}\sigma_A^2. \tag{6.194}$$

This is in agreement with the result following from

$$\int_{T_A}^{T_C} c_P(T)\,dT = \frac{c_P^0}{2T_0}(T_0^2 - T_A^2). \tag{6.195}$$

The heat absorbed along the constant longitudinal strain path BC is

$$u_C - u_B = \left(\frac{1}{2E_T}\sigma_C^2 + \frac{\alpha_0 T_0}{3}\sigma_C\right) - \frac{1}{2E_S}\sigma_B^2, \tag{6.196}$$

which gives

$$u_C - u_B = \frac{\alpha_0 T_0}{3}\sigma_A - \frac{E_S}{E_T}\frac{\alpha_0^2 T_0}{18 c_P^0}\sigma_A^2. \tag{6.197}$$

This can be confirmed by integrating

$$\int_{s_B}^{s_C} T\,ds = \int_{s_B}^{s_C}\left[T_0 + \frac{T_0}{c_P^0}\left(s - s_0 - \frac{\alpha_0}{3}\sigma\right)\right]ds, \tag{6.198}$$

with the stress variation, along the path BC, given by

$$\sigma = \frac{E_S}{E_T}\sigma_A - \frac{\alpha_0 T_0}{3c_P^0} E_S(s - s_0). \tag{6.199}$$

For metals, the second term on the right-hand side of (6.197) is much smaller than the first term, being associated with small departures of c_P and c_V from their reference values c_P^0 and c_V^0, inherent in linear approximations $c_P = c_P^0 T/T_0$ and $c_V = c_V^0 T/T_0$, which are valid for sufficiently small temperature differences $(T - T_0)$.

An alternative derivation of (6.197) proceeds by noting that along the path BC, $d\sigma = -E_T\alpha_0 dT/3$ (because the longitudinal component of strain is fixed along that path). The corresponding increment of entropy is

$$ds = \frac{\alpha_0}{3} d\sigma + \frac{c_P^0}{T_0} dT = \frac{c_P^0}{T_0} \frac{E_T}{E_S} dT. \tag{6.200}$$

The relationship (6.185) between isothermal and adiabatic Young's moduli was used. Therefore,

$$\int_{T_B}^{T_C} T ds = \frac{c_P^0}{2T_0} \frac{E_T}{E_S} (T_0^2 - T_B^2). \tag{6.201}$$

The incorporation of (6.192) reproduces (6.197).

Yet another derivation is possible by starting from an expression for the heat increment in terms of the latent and specific heat, i.e.,

$$T ds = l_{ij}^e de_{ij} + c_V dT. \tag{6.202}$$

The components of the latent heat tensor at constant strain are defined by

$$l_{ij}^e = T \left(\frac{\partial s}{\partial e_{ij}}\right)_T = \kappa_T \alpha_0 T \delta_{ij}, \tag{6.203}$$

which gives

$$T ds = \kappa_T \alpha_0 T de_{kk} + c_V dT. \tag{6.204}$$

Because along the path BC,

$$de_{kk} = \frac{2}{3}(1 + \nu_T)\alpha_0 dT, \tag{6.205}$$

the substitution into (6.204), and integration from T_B to $T_C = T_0$, gives (6.197). This derivation is facilitated by noting that, in view of (6.183),

$$(1 + \nu_T)\frac{\alpha_0^2 T_0}{9c_P^0} = \frac{\nu_S - \nu_T}{E_S}. \tag{6.206}$$

The individual contributions of the latent and specific heat to the total heat absorbed along the path BC are

$$\int_B^C c_V dT = \frac{1 - 2\nu_S}{1 - 2\nu_T}(u_C - u_B), \tag{6.207}$$

$$\int_B^C l_{ij}^e de_{ij} = 2\frac{\nu_S - \nu_T}{1 - 2\nu_T}(u_C - u_B). \tag{6.208}$$

6.12 Thermodynamics of Open Systems: Chemical Potentials

The contribution given by (6.207) is smaller than $(u_C - u_B)$, because $v_S > v_T$. Since the lateral strain is not held constant along the path BC, there is a small but positive contribution to absorbed heat from the latent heat, and this is represented by (6.208). Both, (6.197) and (6.207) display in their structure simple combination of adiabatic and isothermal elastic constants, *via* the ratio terms E_S/E_T and

$$\frac{c_V^0}{c_P^0} \frac{E_S}{E_T} = \frac{1 - 2v_S}{1 - 2v_T}. \tag{6.209}$$

6.12 Thermodynamics of Open Systems: Chemical Potentials

In addition to heat and work transfer, an open thermodynamic system allows a mass transfer across its boundary. Consider a homogeneous system, consisting of one phase, made up from uniform mixture of k components. For example, an Al-Zn fcc phase consists of Al and Zn components. Let $n_1, n_2, \ldots, n_k$ be the numbers of moles of these components. Let U, V, and S be the internal energy, volume, and entropy within the whole system (extensive properties; if u is the internal energy per unit current mass, then $U = \rho V u$ and $S = \rho V s$, where ρ is the current mass density). Suppose

$$U = U(V, S, n_1, n_2, \ldots, n_k), \tag{6.210}$$

then

$$dU = \left(\frac{\partial U}{\partial V}\right)_{S,n_i} dV + \left(\frac{\partial U}{\partial S}\right)_{V,n_i} dS + \sum_{i=1}^{k} \left(\frac{\partial U}{\partial n_i}\right)_{S,V,n_{j\neq i}} dn_i. \tag{6.211}$$

Since, at constant composition (all n_i fixed), the relationships from the thermodynamics of closed system apply, we have

$$p = -\left(\frac{\partial U}{\partial V}\right)_{S,n_i}, \quad T = \left(\frac{\partial U}{\partial S}\right)_{V,n_i}, \tag{6.212}$$

where p and T are the pressure and temperature (intensive properties). If V and S are held constant, (6.211) indicates that the internal energy changes because of the changes in composition alone. Thus, we introduce the so-called chemical potential of the i-th component by

$$\mu_i = \left(\frac{\partial U}{\partial n_i}\right)_{V,S,n_{j\neq i}}. \tag{6.213}$$

Consequently, the overall increment of internal energy can be expressed as

$$dU = -p dV + T dS + \sum_{i=1}^{k} \mu_i dn_i. \tag{6.214}$$

This can be viewed as the energy equation for an open system. The term $-p dV$ is the work done to the system, $T dS$ is the heat transferred to the system, and each $\mu_i dn_i$ is the energy change due to an infinitesimal change of the component i at fixed V and S and fixed number of moles of other components. (For example, the addition of interstitial atoms changes the internal energy of the system, and this change is governed by the corresponding chemical potential).

The chemical potential can also be introduced as the gradient of other thermodynamic potentials, by holding their independent variables fixed. For example, considering the Helmholtz free energy

$$\Phi(V, T, n_1, n_2, \ldots, n_k) = U(V, S, n_1, n_2, \ldots, n_k) - TS, \qquad (6.215)$$

we have

$$d\Phi = -p\,dV - S\,dT + \sum_{i=1}^{k} \mu_i\,dn_i, \qquad (6.216)$$

where

$$p = -\left(\frac{\partial \Phi}{\partial V}\right)_{T,n_i}, \quad S = -\left(\frac{\partial \Phi}{\partial T}\right)_{V,n_i}, \quad \mu_i = \left(\frac{\partial \Phi}{\partial n_i}\right)_{V,T,n_{j\neq i}}. \qquad (6.217)$$

With the Gibbs energy

$$G(p, T, n_1, n_2, \ldots, n_k) = \Phi(V, T, n_1, n_2, \ldots, n_k) + pV \qquad (6.218)$$

as the thermodynamic potential, we obtain

$$dG = V\,dp - S\,dT + \sum_{i=1}^{k} \mu_i\,dn_i, \qquad (6.219)$$

where

$$V = \left(\frac{\partial G}{\partial p}\right)_{T,n_i}, \quad S = -\left(\frac{\partial G}{\partial T}\right)_{p,n_i}, \quad \mu_i = \left(\frac{\partial G}{\partial n_i}\right)_{p,T,n_{j\neq i}}. \qquad (6.220)$$

Finally, if the enthalpy of the system

$$H(p, S, n_1, n_2, \ldots, n_k) = U(V, S, n_1, n_2, \ldots, n_k) + pV \qquad (6.221)$$

is used, we have

$$dH = V\,dp + T\,dS + \sum_{i=1}^{k} \mu_i\,dn_i, \qquad (6.222)$$

with

$$V = \left(\frac{\partial H}{\partial p}\right)_{S,n_i}, \quad T = \left(\frac{\partial H}{\partial S}\right)_{p,n_i}, \quad \mu_i = \left(\frac{\partial H}{\partial n_i}\right)_{p,S,n_{j\neq i}}. \qquad (6.223)$$

If the stress state is not pure pressure, the stress and strain tensors are related by

$$S_{ij} = \frac{1}{V_0}\left(\frac{\partial \Phi}{\partial E_{ij}}\right)_{T,n_i}, \quad E_{ij} = \frac{1}{V_0}\left(\frac{\partial G}{\partial S_{ij}}\right)_{E_{ij},n_i}, \qquad (6.224)$$

where V_0 is the initial volume of the system. Similar relations hold if U and H are used as thermodynamic potentials.

One can easily establish the Maxwell's relations corresponding to each thermodynamic potential (or the set of independent state variables). For example, by using the Gibbs energy, i.e., p, T, and n_i as independent state variables, there follows

$$\left(\frac{\partial V}{\partial n_i}\right)_{p,T,n_{j\neq i}} = \left(\frac{\partial \mu_i}{\partial p}\right)_{T,n_j}, \qquad (6.225)$$

6.13. Gibbs–Duhem Equation

$$\left(\frac{\partial S}{\partial n_i}\right)_{p,T,n_{j\neq i}} = -\left(\frac{\partial \mu_i}{\partial T}\right)_{p,n_j}, \tag{6.226}$$

$$\left(\frac{\partial \mu_i}{\partial n_j}\right)_{p,T,n_{r\neq j}} = \left(\frac{\partial \mu_j}{\partial n_i}\right)_{p,T,n_{r\neq i}}. \tag{6.227}$$

For the two component system, $G = G(p, T, n_1, n_2)$, the last reciprocal relation reads

$$\left(\frac{\partial \mu_1}{\partial n_2}\right)_{p,T,n_1} = \left(\frac{\partial \mu_2}{\partial n_1}\right)_{p,T,n_2} = \left(\frac{\partial^2 G}{\partial n_1 \partial n_2}\right)_{p,T}. \tag{6.228}$$

6.13 Gibbs–Duhem Equation

The Gibbs energy

$$G = G(p, T, n_1, n_2, \ldots, n_k) \tag{6.229}$$

is a homogeneous function of degree 1 with respect to n_i, so that

$$G(p, T, \lambda n_1, \lambda n_2, \ldots, \lambda n_k) = \lambda G(p, T, n_1, n_2, \ldots, n_k), \tag{6.230}$$

and

$$\sum_{i=1}^{k} \left(\frac{\partial G}{\partial n_i}\right)_{p,T,n_{j\neq i}} n_i = G. \tag{6.231}$$

Recalling a definition of the chemical potential

$$\mu_i = \left(\frac{\partial G}{\partial n_i}\right)_{p,T,n_{j\neq i}}, \tag{6.232}$$

we conclude that

$$G = \sum_{i=1}^{k} \mu_i n_i. \tag{6.233}$$

Therefore, the Gibbs energy of the system is the weighted sum of the chemical potentials of its components. If the system consists of n_1 moles of only one component, then $G_1 = n_1 \mu_1$, i.e., the chemical potential of this component (associated with the change of energy of the system due to change of the amount of this component in the system at constant V and S) is $\mu_1 = G_1/n_1$ (molar Gibbs energy).

An additional important relationship can be derived by applying a total differential to (6.233). We obtain

$$dG = \sum_{i=1}^{k} \mu_i dn_i + \sum_{i=1}^{k} n_i d\mu_i. \tag{6.234}$$

By equating this to (6.219), i.e.,

$$dG = Vdp - SdT + \sum_{i=1}^{k} \mu_i dn_i, \tag{6.235}$$

there follows

$$-V dp + S dT + \sum_{i=1}^{k} n_i d\mu_i = 0. \tag{6.236}$$

This is known as the Gibbs–Duhem equation. In a test at constant pressure and temperature, it reduces to

$$\sum_{i=1}^{k} n_i d\mu_i = 0. \tag{6.237}$$

Having established the representation of the Gibbs energy in terms of chemical potentials, it is straightforward to derive the expressions for other thermodynamic potentials. Indeed, since $H = G + TS$, we obtain

$$H = TS + \sum_{i=1}^{k} \mu_i n_i. \tag{6.238}$$

Furthermore, since $U = H - pV$,

$$U = -pV + TS + \sum_{i=1}^{k} \mu_i n_i, \tag{6.239}$$

and, since $\Phi = U - TS$,

$$\Phi = -pV + \sum_{i=1}^{k} \mu_i n_i. \tag{6.240}$$

The last two can be generalized for the case of nonhydrostatic state of stress as

$$U = V_0 S_{ij} E_{ij} + TS + \sum_{i=1}^{k} \mu_i n_i, \quad \Phi = V_0 S_{ij} E_{ij} + \sum_{i=1}^{k} \mu_i n_i. \tag{6.241}$$

6.14 Chemical Potentials for Binary Systems

Consider a binary system consisting of n_1 moles of component 1 and n_2 moles of component 2. By (6.233), its Gibbs energy is

$$G = n_1 \mu_1 + n_2 \mu_2, \tag{6.242}$$

where μ_1 and μ_2 are the chemical potentials of two components in the system. The molar Gibbs energy is defined by

$$G_m = \frac{G}{n_1 + n_2}. \tag{6.243}$$

Denoting by

$$X_1 = \frac{n_1}{n_1 + n_2}, \quad X_2 = \frac{n_2}{n_1 + n_2} \tag{6.244}$$

the concentrations of two components, we can write

$$G_m = X_1 \mu_1 + X_2 \mu_2. \tag{6.245}$$

6.15. Configurational Entropy

Consider a process at constant p and T. The Gibbs–Duhem equation (6.237) then gives

$$n_1 d\mu_1 + n_2 d\mu_2 = 0 \quad \Rightarrow \quad X_1 d\mu_1 + X_2 d\mu_2 = 0. \tag{6.246}$$

Consequently, upon applying the differential to (6.245), we obtain

$$dG_m = \mu_1 dX_1 + \mu_2 dX_2. \tag{6.247}$$

Since $dX_1 = -dX_2$ (because $X_1 + X_2 = 1$), the above reduces to

$$dG_m = (\mu_2 - \mu_1) dX_2, \tag{6.248}$$

i.e.,

$$\mu_2 - \mu_1 = \frac{dG_m}{dX_2}. \tag{6.249}$$

Solving (6.245) and (6.249) for μ_1 and μ_2, there follows

$$\mu_1 = G_m - X_2 \frac{dG_m}{dX_2}, \tag{6.250}$$

$$\mu_2 = G_m + (1 - X_2) \frac{dG_m}{dX_2}, \tag{6.251}$$

both being expressed in terms of the molar Gibbs energy $G_m = G_m(X_2)$ and the concentration X_2.

6.15 Configurational Entropy

We derive in this section an expression for the configurational entropy that is often adopted in the thermodynamics of open systems and alloy solutions. The number of distinguished ways in which N_1 particles of type 1 and N_2 particles of type 2 can fill the $N_1 + N_2$ available sites is given by the well-known formula

$$\Omega = \frac{(N_1 + N_2)!}{N_1! N_2!}. \tag{6.252}$$

The configurational entropy is defined as

$$S^{\text{conf}} = k \ln \Omega = k \ln \frac{(N_1 + N_2)!}{N_1! N_2!}, \tag{6.253}$$

where k is the Boltzmann's constant. Recalling the Stirling's formula from calculus

$$m! = \sqrt{2\pi m}\, m^m e^{-m}, \tag{6.254}$$

we have

$$\ln m! = \frac{1}{2} \ln(2\pi m) + m \ln m - m. \tag{6.255}$$

For large integers m, this is approximately equal to

$$\ln m! \approx m \ln m - m. \tag{6.256}$$

Applying this approximation to (6.253), we now have

$$S^{\text{conf}} = k[(N_1 + N_2)\ln(N_1 + N_2) - (N_1 + N_2) \\ -(N_1 \ln N_1 - N_1) - (N_2 \ln N_2 - N_2)], \qquad (6.257)$$

i.e.,

$$S^{\text{conf}} = -k\left(N_1 \ln \frac{N_1}{N_1 + N_2} + N_2 \ln \frac{N_2}{N_1 + N_2}\right). \qquad (6.258)$$

If there is n_1 moles of particles 1 and n_2 moles of particles 2, then $N_1 = N_A n_1$ and $N_2 = N_A n_2$, where N_A is the Avogadro's number, so that

$$\frac{N_1}{N_1 + N_2} = \frac{n_1}{n_1 + n_2} = X_1, \quad \frac{N_2}{N_1 + N_2} = \frac{n_2}{n_1 + n_2} = X_2 \qquad (6.259)$$

are the concentration of two types of particles in the mixture. Therefore, by dividing (6.258) with the total number of moles $n_1 + n_2$, we obtain

$$S_m^{\text{conf}} = -R(X_1 \ln X_1 + X_2 \ln X_2), \qquad (6.260)$$

where $R = N_A k$ is the universal gas constant, and $S_m^{\text{conf}} = S^{\text{conf}}/(n_1 + n_2)$ is the molar configurational entropy.

6.16 Ideal Solutions

The above expression for the configurational entropy is adopted in the theory of solid solutions. The molar entropy of the solid solution is assumed to consists of two parts: a weighted sum of the entropies of pure components (at given p and T), and the entropy of the components' mixing, i.e.,

$$S_m = \sum_{i=1}^{k} X_i S_m^{(i)} + S_m^{\text{mix}} = \sum_{i=1}^{k} X_i S_m^i - R \sum_{i=1}^{k} X_i \ln X_i. \qquad (6.261)$$

The molar entropy of pure component i at given p and T is denoted by $S_m^{(i)}$. Similarly, the molar Gibbs energy is taken to be

$$G_m = \sum_{i=1}^{k} X_i G_m^{(i)} + G_m^{\text{mix}}, \quad G_m^{\text{mix}} = -T S_m^{\text{mix}} = RT \sum_{i=1}^{k} X_i \ln X_i. \qquad (6.262)$$

If these expressions are adopted, the solution model is referred to as an ideal solution. Note that for an ideal solution model, $H_m^{\text{mix}} = 0$ and $\Phi_m^{\text{mix}} = G_m^{\text{mix}}$. The chemical potentials of the components in an ideal solution can be readily calculated. For example, for a binary solution, (6.262) gives

$$G_m = X_1 G_m^{(1)} + X_2 G_m^{(2)} + RT(X_1 \ln X_1 + X_2 \ln X_2) \\ = (1 - X_2) G_m^{(1)} + X_2 G_m^{(2)} + RT[(1 - X_2)\ln(1 - X_2) + X_2 \ln X_2],$$

so that

$$\frac{dG_m}{dX_2} = G_m^{(2)} - G_m^{(1)} + RT \ln \frac{X_2}{1 - X_2}. \qquad (6.263)$$

6.17 Regular Solutions for Binary Alloys

The substitution of this into (6.250) and (6.251) then gives

$$\mu_1 = G_m^{(1)} + RT \ln X_1, \quad \mu_2 = G_m^{(2)} + RT \ln X_2. \quad (6.264)$$

The logarithmic terms are the mixing contributions to chemical potentials.

6.17 Regular Solutions for Binary Alloys

The regular solution model involves an additional parameter ω (representing in some average sense the relative energies of like and unlike bonds between the two components of the alloy), such that the molar enthalpy of mixing is

$$H_m = \omega X_1 X_2. \quad (6.265)$$

The entropy of mixing is taken to be as in an ideal solution,

$$S_m^{mix} = -R(X_1 \ln X_1 + X_2 \ln X_2), \quad (6.266)$$

so that the Gibbs energy of mixing ($G_m^{mix} = H_m^{mix} - T S_m^{mix}$) becomes

$$G_m^{mix} = \omega X_1 X_2 + RT(X_1 \ln X_1 + X_2 \ln X_2). \quad (6.267)$$

The molar Gibbs energy is accordingly

$$G_m = X_1 G_m^{(1)} + X_2 G_m^{(2)} + G_m^{mix}, \quad (6.268)$$

whereas the chemical potentials of two components in the solution are, from (6.250) and (6.251),

$$\mu_1 = G_m^{(1)} + \omega X_2^2 + RT \ln X_1,$$
$$\mu_2 = G_m^{(2)} + \omega X_1^2 + RT \ln X_2. \quad (6.269)$$

The most distinguished feature of a regular solution model is that it can lead to a miscibility gap. Namely, whereas an ideal solution implies complete solubility of two components throughout the concentration range, a regular solution allows a possibility that two components at low temperature may not mix for some concentration, but form a mixture of two solution phases with different concentrations. Assuming that $\omega > 0$, the temperature below which there may be a miscibility gap is obtained from the requirements

$$\frac{d^2 G_m}{d X_2^2} = \frac{d^3 G_m}{d X_2^3} = 0. \quad (6.270)$$

This specifies the critical temperature

$$T_{cr} = \frac{\omega}{2R}. \quad (6.271)$$

The miscibility boundary is defined by

$$\frac{dG_m}{dX_2} = G_m^{(2)} - G_m^{(1)}, \quad (6.272)$$

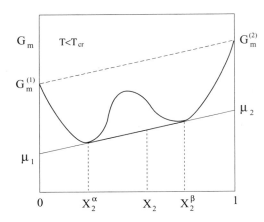

Figure 6.5. The molar Gibbs energy vs. the concentration X_2 below the critical temp T_{cr}. The mixture of α and β phases has lower Gibbs energy than the solution in the concentration range $X_2^\alpha < X_2 < X_2^\beta$.

or simply $dG_m^{mix}/dX_2 = 0$, which gives

$$\frac{T}{T_{cr}} = 2\frac{1 - 2X_2}{\ln\frac{1-X_2}{X_2}}. \tag{6.273}$$

This is clearly symmetric with respect to the midpoint of the concentration range $X_2 = 0.5$. It can be easily verified that for $T < T_{cr}$,

$$\frac{d^2 G_m}{dX_2^2} = -2\omega + \frac{RT}{X_1 X_2} < 0, \tag{6.274}$$

implying an unstable solution within the miscibility gap. The mixture of two solutions has a lower Gibbs energy in this range.

If the miscibilty gap at some $T < T_{cr}$ extends from X_2^α to X_2^β, the alloy of concentration $X_2^\alpha < X_2 < X_2^\beta$ is a mixture of two solutions, one of the concentration X_2^α and the other of the concentration X_2^β. This mixture has the Gibbs energy which is on the straight line tangent to the solution curve at both concentration points, X_2^α and X_2^β (Fig. 6.5), and thus given by

$$G_m(X_2) = G_m^\alpha + \frac{G_m^\beta - G_m^\alpha}{X_2^\beta - X_2^\alpha}(X_2 - X_2^\alpha), \tag{6.275}$$

where $G_m^\alpha = G_m(X_2^\alpha)$ and $G_m^\beta = G_m(X_2^\beta)$ are the Gibbs energies of regular solutions at the corresponding concentrations, calculated from (6.268). The two phases (α and β) in the mixture are in thermodynamic equilibrium, so that

$$\mu_1(X_2^\alpha) = \mu_1(X_2^\beta), \quad \mu_2(X_2^\alpha) = \mu_2(X_2^\beta). \tag{6.276}$$

The amounts of the α and β phases in the mixture are determined from the lever rule. There is $(X_2^\beta - X_2)/(X_2^\beta - X_2^\alpha)$ of the α phase, and $(X_2 - X_2^\alpha)/(X_2^\beta - X_2^\alpha)$ of the β phase in the mixture of overall concentration X_2. Note also that

$$\mu_2(X_2^\alpha) - \mu_1(X_2^\alpha) = \mu_2(X_2^\beta) - \mu_1(X_2^\beta) = G_m^{(2)} - G_m^{(1)}, \tag{6.277}$$

6.18 Suggested Reading

Suggested texts for thermodynamics of solids and materials are listed below. They include the books on continuum thermodynamics, thermoelasticity, metallurgical thermodynamics, thermodynamics of materials, and chemical thermodynamics.

Boley, B. A., and Weiner, J. H. (1960), *Theory of Thermal Stresses*, Wiley, New York.

Callen, H. B. (1960), *Thermodynamics*, Wiley, New York.

DeHoff, R. T. (1993), *Thermodynamics in Materials Science*, McGraw-Hill, New York.

Denbigh, K. (1981), *The Principles of Chemical Equilibrium*, 4th ed., Cambridge University Press, Cambridge.

Dickerson, R. E. (1969), *Molecular Thermodynamics*, The Benjamin/Cummings Publ., Menlo Park, CA.

Ericksen, J. L. (1991), *Introduction to the Thermodynamics of Solids*, Chapman and Hall, London.

Fung, Y.-C. (1965), *Foundations of Solid Mechanics*, Prentice Hall, Englewood Cliffs, New Jersey.

Gaskell, D. R. (2003), *Introduction to the Thermodynamics of Materials*, 4th ed., Taylor & Francis, New York.

Kestin, J. (1979), *A Course in Thermodynamics*, McGraw-Hill, New York.

Kovalenko, A. D. (1969), *Thermoelasticity*, Wolters–Noordhoff, Groningen, The Netherlands.

Lupis, C. H. P. (1983), *Chemical Thermodynamics of Materials*, Prentice Hall, Englewood Cliffs, New Jersey.

Malvern, L. E. (1969), *Introduction to the Mechanics of a Continuous Medium*, Prentice Hall, Englewood Cliffs, New Jersey.

McLellan, A. G. (1980), *The Classical Thermodynamics of Deformable Materials*, Cambridge University Press, Cambridge.

Müller, I. (1985), *Thermodynamics*, Pitman, Boston.

Noda, N., Hetnarski, R. B., and Tanigawa, Y. (2003), *Thermal Stresses*, Taylor & Francis, New York.

Ragone, D. V. (1995), *Thermodynamics of Materials*, Wiley, New York.

Sneddon, I. N. (1974), *The Linear Theory of Thermoelasticity*, CISM Udine, Springer-Verlag, Wien.

Swalin, R. A. (1972), *Thermodynamics of Solids*, Wiley, New York.

Ziegler, H. (1983), *An Introduction to Thermomechanics*, 2nd revised ed., North-Holland, Amsterdam.

7 Nonlinear Elasticity

In this chapter we give a concise treatment of nonlinear elasticity, which includes both geometrical and material nonlinearities (finite deformations and nonlinear constitutive equations). This is done to illustrate the application of the general framework of nonlinear continuum mechanics developed in previous chapters and to give an introduction to important subject of rubber elasticity. A detailed coverage of linear elasticity is presented in Part III of the book.

7.1 Green Elasticity

Elastic deformation is a reversible process which does not cause any permanent change of internal structure of the material. Experiments indicate that there is no net work left in a material upon any closed cycle of elastic strain, *i.e.*,

$$\oint \mathbf{S} : d\mathbf{E} = 0, \tag{7.1}$$

where $\mathbf{E}$ is the Green strain and $\mathbf{S}$ its conjugate symmetric Piola–Kirchhoff stress. This means that

$$\mathbf{S} : d\mathbf{E} = d\Phi \tag{7.2}$$

is a total differential, which leads to

$$\mathbf{S} = \frac{\partial \Phi}{\partial \mathbf{E}}, \quad \Phi = \Phi(\mathbf{E}). \tag{7.3}$$

The function $\Phi = \Phi(\mathbf{E})$ is the strain energy function per unit initial volume. It represents the work done to isothermally deform a unit of initial volume to the state of strain $\mathbf{E}$.

In view of the relationship between the Kirchhoff stress τ and the symmetric Piola–Kirchhoff stress $\mathbf{S}$, we have

$$\tau = (\det \mathbf{F})\sigma = \mathbf{F} \cdot \frac{\partial \Phi}{\partial \mathbf{E}} \cdot \mathbf{F}^T. \tag{7.4}$$

With a specified strain energy function for a given material, (7.4) defines the stress response corresponding to the deformation gradient $\mathbf{F}$. Because stress is derived from the strain

7.1. Green Elasticity

energy function, the equation is referred to as the constitutive equation of hyperelasticity or Green elasticity.

The nonsymmetric Piola–Kirchhoff stress (denoted by $\hat{\mathbf{P}}$ in Section 5.8) is

$$\mathbf{P} = \frac{\partial \Phi}{\partial \mathbf{F}}, \quad P_{ji} = \frac{\partial \Phi}{\partial F_{ij}}, \tag{7.5}$$

which follows from

$$d\Phi = \frac{\partial \Phi}{\partial F_{ij}} dF_{ij} = P_{ji} dF_{ij}. \tag{7.6}$$

Because Φ is unaffected by the rotation of the deformed configuration,

$$\Phi(\mathbf{F}) = \Phi(\mathbf{Q} \cdot \mathbf{F}). \tag{7.7}$$

By choosing $\mathbf{Q} = \mathbf{R}^T$, where $\mathbf{R}$ is the rotation tensor from the polar decomposition of the deformation gradient $\mathbf{F} = \mathbf{V} \cdot \mathbf{R} = \mathbf{R} \cdot \mathbf{U}$, it follows that Φ depends on $\mathbf{F}$ only through $\mathbf{U}$, or $\mathbf{C} = \mathbf{U}^2$, i.e.,

$$\Phi = \Phi(\mathbf{C}), \quad \mathbf{C} = \mathbf{F}^T \cdot \mathbf{F}. \tag{7.8}$$

The functional dependences of Φ on different tensor arguments such as $\mathbf{F}$, $\mathbf{U}$, or $\mathbf{C}$ are, of course, different.

Constitutive equations of nonlinear elasticity can be derived without assuming the existence of the strain energy function. Suppose that at any state of elastic deformation, the stress is a single-valued function of strain, regardless of the deformation path along which the state has been reached. Since no strain energy is assumed to exist, the work done by the stress could in general be different for different deformation paths. This type of elasticity is known as Cauchy elasticity, although experimental evidence does not indicate existence of any Cauchy-elastic material that is also not Green-elastic. In any case, we write

$$\mathbf{S} = \mathbf{f}(\mathbf{E}), \tag{7.9}$$

where $\mathbf{f}$ is a second-order tensor function, whose representation depends on the elastic properties of the material. Since

$$\mathbf{S} = (\det \mathbf{F}) \mathbf{F}^{-1} \cdot \boldsymbol{\sigma} \cdot \mathbf{F}^{-T}, \tag{7.10}$$

we have

$$\boldsymbol{\sigma} = \frac{1}{\det \mathbf{F}} \mathbf{F} \cdot \mathbf{g}(\mathbf{U}) \cdot \mathbf{F}^T, \tag{7.11}$$

where

$$\mathbf{g}(\mathbf{U}) = \mathbf{f}(\mathbf{E}), \quad \mathbf{E} = \frac{1}{2}(\mathbf{U}^2 - \mathbf{I}). \tag{7.12}$$

By using the polar decomposition $\mathbf{F} = \mathbf{R} \cdot \mathbf{U}$, (7.11) becomes

$$\boldsymbol{\sigma} = \frac{1}{\det \mathbf{U}} \mathbf{R} \cdot \mathbf{U} \cdot \mathbf{g}(\mathbf{U}) \cdot \mathbf{U} \cdot \mathbf{R}^T. \tag{7.13}$$

Thus, the stress response of the Cauchy elasticity can be put in the form

$$\boldsymbol{\sigma} = \mathbf{R} \cdot \mathbf{h}(\mathbf{U}) \cdot \mathbf{R}^T, \tag{7.14}$$

with the function **h** defined by

$$\mathbf{h}(\mathbf{U}) = \frac{1}{\det \mathbf{U}} \cdot \mathbf{U} \cdot \mathbf{g}(\mathbf{U}) \cdot \mathbf{U}. \tag{7.15}$$

7.2 Isotropic Green Elasticity

If the strain energy does not depend along which material directions the principal strains are applied, so that

$$\Phi\left(\mathbf{Q} \cdot \mathbf{E} \cdot \mathbf{Q}^T\right) = \Phi(\mathbf{E}) \tag{7.16}$$

for any rotation tensor $\mathbf{Q}$, the material is elastically isotropic. A scalar function which satisfies (7.16) is said to be an isotropic function of its second-order tensor argument. Such a function can be expressed in terms of the principal invariants of the strain tensor $\mathbf{E}$, *i.e.*,

$$\Phi = \Phi(I_E, II_E, III_E), \tag{7.17}$$

where

$$I_E = \operatorname{tr} \mathbf{E}, \quad II_E = \frac{1}{2}\left[\operatorname{tr}(\mathbf{E}^2) - (\operatorname{tr}\mathbf{E})^2\right], \quad III_E = \det \mathbf{E}. \tag{7.18}$$

Note that the definition for II_E used here differs in sign from the definition of second invariant used in Chapters 1 and 4. Since

$$\frac{\partial I_E}{\partial \mathbf{E}} = \mathbf{I}, \quad \frac{\partial II_E}{\partial \mathbf{E}} = \mathbf{E} - I_E \mathbf{I}, \quad \frac{\partial III_E}{\partial \mathbf{E}} = \mathbf{E}^2 - I_E \mathbf{E} - II_E \mathbf{I}, \tag{7.19}$$

(7.3) yields, upon partial differentiation,

$$\mathbf{S} = c_0 \mathbf{I} + c_1 \mathbf{E} + c_2 \mathbf{E}^2. \tag{7.20}$$

The parameters are

$$c_0 = \frac{\partial \Phi}{\partial I_E} - I_E \frac{\partial \Phi}{\partial II_E} - II_E \frac{\partial \Phi}{\partial III_E}, \quad c_1 = \frac{\partial \Phi}{\partial II_E} - I_E \frac{\partial \Phi}{\partial III_E},$$

$$c_2 = \frac{\partial \Phi}{\partial III_E}. \tag{7.21}$$

For example, if the Saint-Venant–Kirchhoff assumption is adopted,

$$\Phi = \frac{1}{2}(\lambda + 2\mu)I_E^2 + 2\mu II_E, \tag{7.22}$$

a generalized Hooke's law for finite strain is obtained,

$$\mathbf{S} = \lambda I_E \mathbf{I} + 2\mu \mathbf{E}. \tag{7.23}$$

The material constants are λ and μ.

If a cubic representation of Φ is assumed,

$$\Phi = \frac{1}{2}(\lambda + 2\mu)I_E^2 + 2\mu II_E + \frac{l + 2m}{3}I_E^3 + 2m I_E II_E + n III_E, \tag{7.24}$$

the stress response becomes

$$\mathbf{S} = [\lambda I_E + l I_E^2 + (2m - n)II_E]\mathbf{I} + [2\mu + (2m - n)I_E]\mathbf{E} + n\mathbf{E}^2. \tag{7.25}$$

The constants l, m and n are the Murnaghan's constants.

7.3 Constitutive Equations in Terms of B

The finite strain constitutive equations of isotropic elasticity are often expressed in terms of the left Cauchy–Green deformation tensor $\mathbf{B} = \mathbf{V}^2$. Since

$$\dot{\Phi} = \frac{\partial \Phi}{\partial \mathbf{B}} : \overset{\triangledown}{\mathbf{B}} = \boldsymbol{\tau} : \mathbf{D} \tag{7.26}$$

and

$$\overset{\triangledown}{\mathbf{B}} = \mathbf{B} \cdot \mathbf{D} + \mathbf{D} \cdot \mathbf{B}, \tag{7.27}$$

we obtain

$$\boldsymbol{\tau} = \mathbf{B} \cdot \frac{\partial \Phi}{\partial \mathbf{B}} + \frac{\partial \Phi}{\partial \mathbf{B}} \cdot \mathbf{B}, \tag{7.28}$$

written in a symmetrized form. The function $\Phi = \Phi(\mathbf{B})$ is an isotropic function of $\mathbf{B}$. Expressing the strain energy in terms of the invariants of $\mathbf{B}$,

$$\Phi = \Phi(I_B, II_B, III_B), \tag{7.29}$$

(7.28) gives

$$\boldsymbol{\tau} = 2\left[\left(III_B \frac{\partial \Phi}{\partial III_B}\right)\mathbf{I} + \left(\frac{\partial \Phi}{\partial I_B} - I_B \frac{\partial \Phi}{\partial II_B}\right)\mathbf{B} + \left(\frac{\partial \Phi}{\partial II_B}\right)\mathbf{B}^2\right]. \tag{7.30}$$

If $\mathbf{B}^2$ is eliminated by using the Cayley–Hamilton theorem, (7.30) can be rewritten as

$$\boldsymbol{\tau} = 2\left[\left(III_B \frac{\partial \Phi}{\partial III_B} + II_B \frac{\partial \Phi}{\partial II_B}\right)\mathbf{I} + \left(\frac{\partial \Phi}{\partial I_B}\right)\mathbf{B} + \left(III_B \frac{\partial \Phi}{\partial II_B}\right)\mathbf{B}^{-1}\right]. \tag{7.31}$$

Note that the invariants of $\mathbf{E}$ and $\mathbf{B}$ are related by

$$I_E = \frac{1}{2}(I_B - 3), \quad II_E = \frac{1}{4}II_B + \frac{1}{2}I_B - \frac{3}{4},$$

$$III_E = \frac{1}{8}(III_B + II_B + I_B - 1), \tag{7.32}$$

$$I_B = 2I_E + 3, \quad II_B = 4II_E - 4I_E - 3,$$

$$III_B = 8III_E - 4II_E + 2I_E + 1. \tag{7.33}$$

The constitutive equation of isotropic elastic material in terms of the nominal stress is

$$\mathbf{P} = \mathbf{F}^{-1} \cdot \boldsymbol{\tau} = \mathbf{F}^T \cdot \left(\frac{\partial \Phi}{\partial \mathbf{B}} + \mathbf{B}^{-1} \cdot \frac{\partial \Phi}{\partial \mathbf{B}} \cdot \mathbf{B}\right). \tag{7.34}$$

By using the strain energy representation (7.29), this becomes

$$\mathbf{P} = 2\mathbf{F}^T \cdot \left[\left(\frac{\partial \Phi}{\partial I_B} - I_B \frac{\partial \Phi}{\partial II_B}\right)\mathbf{I} + \left(\frac{\partial \Phi}{\partial II_B}\right)\mathbf{B} + \left(III_B \frac{\partial \Phi}{\partial III_B}\right)\mathbf{B}^{-1}\right]. \tag{7.35}$$

Different forms of the strain energy function were used in the literature. For example, Ogden (1984) used

$$\Phi = \frac{a}{2}(I_B - 3 - \ln III_B) + c\left(III_B^{1/2} - 1\right)^2, \tag{7.36}$$

where a and c are the material parameters. Blatz and Ko (1962) proposed an expression for the strain energy for compressible foamed elastomers. Other representations can be found in Beatty (1996) and Holzapfel (2000).

7.4 Constitutive Equations in Terms of Principal Stretches

The strain energy of an isotropic material can be conveniently expressed in terms of the principal stretches λ_i (the eigenvalues of $\mathbf{U}$ and $\mathbf{V}$, which are invariant quantities), *i.e.*,

$$\Phi = \Phi(\lambda_1, \lambda_2, \lambda_3). \tag{7.37}$$

Suppose that all principal stretches are different, and that $\mathbf{N}_i$ and $\mathbf{n}_i$ are the principal directions of the right and left stretch tensors $\mathbf{U}$ and $\mathbf{V}$, respectively, so that

$$\mathbf{U} = \sum_{i=1}^{3} \lambda_i \mathbf{N}_i \mathbf{N}_i, \quad \mathbf{E} = \sum_{i=1}^{3} \frac{1}{2} (\lambda_i^2 - 1) \mathbf{N}_i \mathbf{N}_i, \tag{7.38}$$

and

$$\mathbf{V} = \sum_{i=1}^{3} \lambda_i \mathbf{n}_i \mathbf{n}_i, \quad \mathbf{F} = \sum_{i=1}^{3} \lambda_i \mathbf{n}_i \mathbf{N}_i. \tag{7.39}$$

For an isotropic elastic material, the principal directions of the strain tensor $\mathbf{E}$ are parallel to those of its conjugate stress tensor $\mathbf{S}$, and we can write

$$\mathbf{S} = \sum_{i=1}^{3} S_i \mathbf{N}_i \mathbf{N}_i. \tag{7.40}$$

The principal Piola–Kirchhoff stresses are

$$S_i = \frac{\partial \Phi}{\partial E_i} = \frac{1}{\lambda_i} \frac{\partial \Phi}{\partial \lambda_i}, \tag{7.41}$$

with no sum on i. Recall that $\lambda_i^2 = 1 + 2E_i^2$.

The principal directions of the Kirchhoff stress $\boldsymbol{\tau}$ of an isotropic elastic material are parallel to those of $\mathbf{V}$, so that

$$\boldsymbol{\tau} = \sum_{i=1}^{3} \tau_i \mathbf{n}_i \mathbf{n}_i. \tag{7.42}$$

The corresponding principal components are

$$\tau_i = \lambda_i^2 S_i = \lambda_i \frac{\partial \Phi}{\partial \lambda_i}. \tag{7.43}$$

Finally, decomposing the nominal stress as

$$\mathbf{P} = \sum_{i=1}^{3} P_i \mathbf{n}_i \mathbf{N}_i, \tag{7.44}$$

we have

$$P_i = \lambda_i S_i = \frac{\partial \Phi}{\partial \lambda_i}. \tag{7.45}$$

7.5 Incompressible Isotropic Elastic Materials

For an incompressible material the deformation is necessarily isochoric, so that $\det \mathbf{F} = 1$. Only two invariants of $\mathbf{E}$ are independent, because

$$III_E = -\frac{1}{4}(I_E - 2II_E). \tag{7.46}$$

Thus, the strain energy can be expressed as

$$\Phi = \Phi(I_E, II_E). \tag{7.47}$$

If (7.30) is specialized to incompressible materials, there follows

$$\sigma = -p\mathbf{I} + 2\left[\left(\frac{\partial \Phi}{\partial I_B} - I_B \frac{\partial \Phi}{\partial II_B}\right)\mathbf{B} + \left(\frac{\partial \Phi}{\partial II_B}\right)\mathbf{B}^2\right], \tag{7.48}$$

where p is an arbitrary pressure. Similarly, (7.31) gives

$$\sigma = -p_0\mathbf{I} + 2\left[\left(\frac{\partial \Phi}{\partial I_B}\right)\mathbf{B} + \left(\frac{\partial \Phi}{\partial II_B}\right)\mathbf{B}^{-1}\right]. \tag{7.49}$$

Here, all terms proportional to $\mathbf{I}$ are absorbed in p_0.

Equation (7.48) can also be derived by viewing an incompressible material as a material with internal constraint

$$III_B - 1 = 0. \tag{7.50}$$

A Lagrangian multiplier $-p/2$ is then introduced, such that

$$\Phi = \Phi(I_B, II_B) - \frac{p}{2}(III_B - 1), \tag{7.51}$$

and (7.28) directly leads to (7.48).

For the Mooney–Rivlin rubber model, the strain energy is

$$\Phi = aI_E + bII_E = \frac{a+b}{2}(I_B - 3) + \frac{b}{4}(II_B + 3), \tag{7.52}$$

whereas for the neo-Hookean material

$$\Phi = \frac{a}{2}(I_B - 3). \tag{7.53}$$

The strain energy representation, suggested by Ogden (1982, 1984),

$$\Phi = \sum_{n=1}^{N} \frac{a_n}{\alpha_n}\left(\lambda_1^{\alpha_n} + \lambda_2^{\alpha_n} + \lambda_3^{\alpha_n} - 3\right) \tag{7.54}$$

may be used in some applications, where N is positive integer, but α_n need not be integers. The material parameters are a_n and α_n. Incompressibility constraint is $\lambda_1 \lambda_2 \lambda_3 = 1$. Other representations in terms of principal stretches λ_i have also been explored (Anand, 1986; Arruda and Boyce, 1993).

7.6 Elastic Moduli Tensors

The rate-type constitutive equation for finite deformation elasticity is obtained by differentiating (7.3) with respect to a time like monotonically increasing parameter t. This gives

$$\dot{\mathbf{S}} = \boldsymbol{\Lambda} : \dot{\mathbf{E}}, \quad \boldsymbol{\Lambda} = \frac{\partial^2 \Phi(\mathbf{E})}{\partial \mathbf{E}\, \partial \mathbf{E}}. \tag{7.55}$$

The fourth-order tensor Λ is the tensor of elastic moduli (or tensor of elasticities) associated with a conjugate pair of tensors $(\mathbf{E}, \mathbf{S})$. The rates of the conjugate tensors $\mathbf{E}$ and $\mathbf{S}$ are

$$\dot{\mathbf{E}} = \mathbf{F}^T \cdot \mathbf{D} \cdot \mathbf{F}, \quad \dot{\mathbf{S}} = \mathbf{F}^{-1} \cdot \overset{\circ}{\boldsymbol{\tau}} \cdot \mathbf{F}^{-T}, \qquad (7.56)$$

where

$$\overset{\circ}{\boldsymbol{\tau}} = \dot{\boldsymbol{\tau}} - \mathbf{L} \cdot \boldsymbol{\tau} - \boldsymbol{\tau} \cdot \mathbf{L}^T \qquad (7.57)$$

is the Oldroyd rate of the Kirchhoff stress $\boldsymbol{\tau}$ ($\mathbf{L}$ being the velocity gradient). The substitution of (7.56) into (7.55) gives $\overset{\circ}{\boldsymbol{\tau}}$ in terms of the rate of deformation $\mathbf{D}$,

$$\overset{\circ}{\boldsymbol{\tau}} = \mathcal{L} : \mathbf{D}. \qquad (7.58)$$

The corresponding elastic moduli tensor is

$$\mathcal{L} = \mathbf{F}\mathbf{F}\Lambda \mathbf{F}^T \mathbf{F}^T. \qquad (7.59)$$

The products are here such that the Cartesian components of the two tensors of elasticities are related by

$$\mathcal{L}_{ijkl} = F_{im} F_{jn} \Lambda_{mnpq} F_{kp} F_{lq}. \qquad (7.60)$$

The rate-type constitutive equation (7.58) can be rewritten in terms of the Jaumann rate $\overset{\triangledown}{\boldsymbol{\tau}}$ as

$$\overset{\triangledown}{\boldsymbol{\tau}} = \hat{\mathcal{L}} : \mathbf{D}, \qquad (7.61)$$

where

$$\hat{\mathcal{L}} = \mathcal{L} + 2H. \qquad (7.62)$$

This follows because of the relationships

$$\overset{\triangledown}{\boldsymbol{\tau}} = \overset{\circ}{\boldsymbol{\tau}} + \mathbf{D} \cdot \boldsymbol{\tau} + \boldsymbol{\tau} \cdot \mathbf{D} = \overset{\triangle}{\boldsymbol{\tau}} - \mathbf{D} \cdot \boldsymbol{\tau} - \boldsymbol{\tau} \cdot \mathbf{D}, \qquad (7.63)$$

where

$$\overset{\triangle}{\boldsymbol{\tau}} = \dot{\boldsymbol{\tau}} + \mathbf{L}^T \cdot \boldsymbol{\tau} + \boldsymbol{\tau} \cdot \mathbf{L} \qquad (7.64)$$

is the Cotter–Rivlin convected rate of the Kirchhoff stress. The Cartesian components of the fourth-order tensor H are

$$H_{ijkl} = \frac{1}{4} \left(\tau_{ik} \delta_{jl} + \tau_{jk} \delta_{il} + \tau_{il} \delta_{jk} + \tau_{jl} \delta_{ik} \right). \qquad (7.65)$$

For metals, $\hat{\mathcal{L}} \approx \mathcal{L}$. The elastic moduli tensors $\Lambda, \mathcal{L}$, and $\hat{\mathcal{L}}$ all possess the basic and reciprocal (major) symmetries, e.g.,

$$\mathcal{L}_{ijkl} = \mathcal{L}_{jikl} = \mathcal{L}_{ijlk}, \quad \mathcal{L}_{ijkl} = \mathcal{L}_{klij}. \qquad (7.66)$$

7.7 Instantaneous Elastic Moduli

The instantaneous elastic moduli relate the rates of conjugate stress and strain tensors, when these are evaluated at the current configuration as the reference (this being denoted in the sequence by a subscript or superscript o). Thus, since $\dot{\mathbf{E}}_\circ = \mathbf{D}$, we write

$$\dot{\mathbf{S}}_\circ = \mathbf{\Lambda}_\circ : \dot{\mathbf{E}}_\circ = \mathbf{\Lambda}_\circ : \mathbf{D}. \tag{7.67}$$

The tensor of instantaneous elastic moduli $\mathbf{\Lambda}_\circ$ can be related to the corresponding tensor of elastic moduli $\mathbf{\Lambda}$ by using the relationship between $\dot{\mathbf{E}}$ and $\dot{\mathbf{E}}_\circ$. Indeed, we have

$$\dot{\mathbf{S}} = (\det \mathbf{F})\, \mathbf{F}^{-1} \cdot \dot{\mathbf{S}}_\circ \cdot \mathbf{F}^{-T}, \quad \dot{\mathbf{E}} = \mathbf{F}^T \cdot \mathbf{D} \cdot \mathbf{F}. \tag{7.68}$$

The substitution into (7.55) gives

$$\dot{\mathbf{S}}_\circ = \mathbf{\Lambda}_\circ : \mathbf{D},$$
$$\mathbf{\Lambda}_\circ = (\det \mathbf{F})^{-1}\, \mathbf{F}\mathbf{F}\mathbf{\Lambda}\mathbf{F}^T\mathbf{F}^T = (\det \mathbf{F})^{-1}\, \mathcal{L}. \tag{7.69}$$

Since, from (7.56), $\dot{\mathbf{S}}_\circ = \overset{\circ}{\boldsymbol{\tau}}$, (7.67) becomes

$$\overset{\circ}{\boldsymbol{\tau}} = \mathcal{L}_\circ : \mathbf{D}, \quad \mathcal{L}_\circ = \mathbf{\Lambda}_\circ. \tag{7.70}$$

7.8 Elastic Pseudomoduli

The nonsymmetric nominal stress $\mathbf{P}$ is derived from the strain energy function as its gradient with respect to deformation gradient $\mathbf{F}$, such that

$$\mathbf{P} = \frac{\partial \Phi}{\partial \mathbf{F}}, \quad P_{ji} = \frac{\partial \Phi}{\partial F_{ij}}. \tag{7.71}$$

The rate of the nominal stress is, therefore,

$$\dot{\mathbf{P}} = \mathbf{K} \cdot\cdot\, \dot{\mathbf{F}} = \mathbf{K} \cdot\cdot\, (\mathbf{L} \cdot \mathbf{F}), \quad \mathbf{K} = \frac{\partial^2 \Phi}{\partial \mathbf{F}\, \partial \mathbf{F}}. \tag{7.72}$$

A two-point tensor of elastic pseudomoduli is denoted by $\mathbf{K}$. The Cartesian component representation of (7.72) is

$$\dot{P}_{ji} = K_{jilk} \dot{F}_{kl}, \quad K_{jilk} = \frac{\partial^2 \Phi}{\partial F_{ij}\, \partial F_{kl}}. \tag{7.73}$$

The elastic pseudomoduli K_{jilk} are not true moduli because they are partly associated with the material spin. They clearly possess the reciprocal symmetry

$$K_{jilk} = K_{lkji}. \tag{7.74}$$

In view of the connection

$$\mathbf{P} = \mathbf{S} \cdot \mathbf{F}^T, \tag{7.75}$$

the differentiation gives

$$\mathbf{K} \cdot\cdot\, \dot{\mathbf{F}} = \left(\mathbf{\Lambda} : \dot{\mathbf{E}}\right) \cdot \mathbf{F}^T + \mathbf{S} \cdot \dot{\mathbf{F}}^T. \tag{7.76}$$

Upon using

$$\dot{\mathbf{E}} = \frac{1}{2}\left(\dot{\mathbf{F}}^T \cdot \mathbf{F} + \mathbf{F}^T \cdot \dot{\mathbf{F}}\right), \tag{7.77}$$

(7.76) yields the connection between the elastic moduli $\mathbf{K}$ and Λ. Their Cartesian components are related by

$$K_{jilk} = \Lambda_{jmln} F_{im} F_{kn} + S_{jl}\delta_{ik}. \tag{7.78}$$

Since $\mathbf{F} \cdot \mathbf{P}$ is a symmetric tensor, i.e.,

$$F_{ik}P_{kj} = F_{jk}P_{ki}, \tag{7.79}$$

by differentiation and incorporation of (7.73) it follows that

$$F_{jm}K_{milk} - F_{im}K_{mjlk} = \delta_{ik}P_{lj} - \delta_{jk}P_{li}. \tag{7.80}$$

This corresponds to the symmetry in the leading pair of indices of the true elastic moduli

$$\Lambda_{ijkl} = \Lambda_{jikl}. \tag{7.81}$$

The tensor of elastic pseudomoduli Λ can be related to the tensor of instantaneous elastic moduli, appearing in the expression

$$\dot{\mathbf{P}}_\circ = \mathbf{K}_\circ \cdot \cdot \mathbf{L}, \tag{7.82}$$

by recalling the relationship

$$\dot{\mathbf{P}} = (\det \mathbf{F})\mathbf{F}^{-1} \cdot \dot{\mathbf{P}}_\circ. \tag{7.83}$$

This gives

$$\mathbf{K}_\circ = (\det \mathbf{F})^{-1}\mathbf{F}\mathbf{K}\mathbf{F}^T, \tag{7.84}$$

with the Cartesian component representation

$$K^\circ_{ijkl} = (\det \mathbf{F})^{-1} F_{im}K_{mjnk}F_{ln}. \tag{7.85}$$

In addition, from (7.78), we have

$$K^\circ_{jilk} = \Lambda_{jilk} + \sigma_{jl}\delta_{ik}. \tag{7.86}$$

7.9 Elastic Moduli of Isotropic Elasticity

The constitutive structure of isotropic elasticity is

$$\mathbf{S} = \frac{\partial \Phi}{\partial \mathbf{E}} = 2\frac{\partial \Phi}{\partial \mathbf{C}} = 2\left[\left(\frac{\partial \Phi}{\partial I_C} - I_C \frac{\partial \Phi}{\partial II_C}\right)\mathbf{I} + \left(\frac{\partial \Phi}{\partial II_C}\right)\mathbf{C} + \left(III_C \frac{\partial \Phi}{\partial III_C}\right)\mathbf{C}^{-1}\right]. \tag{7.87}$$

The strain energy function $\Phi = \Phi(I_C, II_C, III_C)$ is here expressed in terms of the principal invariants of the right Cauchy–Green deformation tensor $\mathbf{C} = \mathbf{F}^T \cdot \mathbf{F} = \mathbf{I} + 2\mathbf{E}$. The corresponding elastic moduli tensor is

$$\Lambda = \frac{\partial \mathbf{S}}{\partial \mathbf{E}} = \frac{\partial^2 \Phi}{\partial \mathbf{E} \partial \mathbf{E}} = 4\frac{\partial^2 \Phi}{\partial \mathbf{C} \partial \mathbf{C}}, \tag{7.88}$$

7.10. Elastic Moduli in Terms of Principal Stretches

which is thus defined by the fully symmetric tensor $\partial^2\Phi/(\partial\mathbf{C}\,\partial\mathbf{C})$. Since

$$\frac{\partial I_C}{\partial \mathbf{C}} = \mathbf{I}, \quad \frac{\partial II_C}{\partial \mathbf{C}} = \mathbf{C} - I_C\mathbf{I},$$

$$\frac{\partial III_C}{\partial \mathbf{C}} = \mathbf{C}^2 - I_C\mathbf{C} - II_C\mathbf{I} = III_C\mathbf{C}^{-1},$$

(7.89)

and in view of the symmetry $C_{ij} = C_{ji}$, we obtain

$$\frac{\partial^2 \Phi}{\partial C_{ij} \partial C_{kl}} = c_1 \delta_{ij}\delta_{kl} + c_2\left(\delta_{ij}C_{kl} + C_{ij}\delta_{kl}\right) + c_3 C_{ij} C_{kl}$$
$$+ c_4 \left(\delta_{ij}C_{kl}^{-1} + C_{ij}^{-1}\delta_{kl}\right) + c_5 \left(C_{ij}C_{kl}^{-1} + C_{ij}^{-1}C_{kl}\right)$$
$$+ c_6 C_{ij}^{-1} C_{kl}^{-1} + c_7 \left(C_{ik}^{-1}C_{jl}^{-1} + C_{il}^{-1}C_{jk}^{-1}\right)$$
$$+ c_8 \left(\delta_{ik}\delta_{jl} + \delta_{il}\delta_{jk}\right).$$

(7.90)

The parameters c_i ($i = 1, 2, \ldots, 8$) are

$$c_1 = \frac{\partial^2 \Phi}{\partial I_C^2} - 2I_C \frac{\partial^2 \Phi}{\partial I_C \partial II_C} + I_C^2 \frac{\partial^2 \Phi}{\partial II_C^2} - \frac{\partial \Phi}{\partial II_C},$$

$$c_2 = \frac{\partial^2 \Phi}{\partial I_C \partial II_C} - I_C \frac{\partial^2 \Phi}{\partial II_C^2},$$

$$c_3 = \frac{\partial^2 \Phi}{\partial II_C^2}, \quad c_5 = III_C \frac{\partial^2 \Phi}{\partial II_C \partial III_C},$$

(7.91)

$$c_4 = III_C \frac{\partial^2 \Phi}{\partial III_C \partial I_C} - III_C I_C \frac{\partial^2 \Phi}{\partial II_C \partial III_C},$$

$$c_6 = III_C^2 \frac{\partial^2 \Phi}{\partial III_C^2} + III_C \frac{\partial \Phi}{\partial III_C},$$

$$c_7 = -\frac{1}{2} III_C \frac{\partial \Phi}{\partial III_C}, \quad c_8 = \frac{1}{2} \frac{\partial \Phi}{\partial II_C}.$$

7.10 Elastic Moduli in Terms of Principal Stretches

For isotropic elastic material the principal directions $\mathbf{N}_i$ of the right Cauchy–Green deformation tensor

$$\mathbf{C} = \sum_{i=1}^{3} \lambda_i^2 \mathbf{N}_i \mathbf{N}_i, \quad C_i = \lambda_i^2,$$

(7.92)

where λ_i are the principal stretches, are parallel to those of the symmetric Piola–Kirchhoff stress **S**. Thus, the spectral representation of **S** is

$$\mathbf{S} = \sum_{i=1}^{3} S_i \, \mathbf{N}_i \, \mathbf{N}_i. \tag{7.93}$$

From the analysis presented in Section 4.21 it readily follows that

$$\dot{\mathbf{C}} = \sum_{i=1}^{3} 2\lambda_i \dot{\lambda}_i \, \mathbf{N}_i \, \mathbf{N}_i + \sum_{i \neq j} \Omega_{ij}^0 \left(\lambda_j^2 - \lambda_i^2 \right) \mathbf{N}_i \, \mathbf{N}_j, \tag{7.94}$$

and

$$\dot{\mathbf{S}} = \sum_{i=1}^{3} \dot{S}_i \, \mathbf{N}_i \, \mathbf{N}_i + \sum_{i \neq j} \Omega_{ij}^0 \left(S_j - S_i \right) \mathbf{N}_i \, \mathbf{N}_j. \tag{7.95}$$

For elastically isotropic material the strain energy can be expressed as a function of the principal stretches, $\Psi = \Psi(\lambda_1, \lambda_2, \lambda_3)$, so that

$$S_i = \frac{\partial \Phi}{\partial E_i} = \frac{1}{\lambda_i} \frac{\partial \Phi}{\partial \lambda_i}, \tag{7.96}$$

$$\dot{S}_i = \sum_{j=1}^{3} \frac{\partial S_i}{\partial \lambda_j} \dot{\lambda}_j, \quad \frac{\partial S_i}{\partial \lambda_j} = -\delta_{ij} \frac{1}{\lambda_i^2} \frac{\partial \Phi}{\partial \lambda_i} + \frac{1}{\lambda_i} \frac{\partial^2 \Phi}{\partial \lambda_i \partial \lambda_j}. \tag{7.97}$$

Thus, (7.95) can be rewritten as

$$\dot{\mathbf{S}} = \sum_{i,j=1}^{3} \frac{\partial S_i}{\partial \lambda_j} \dot{\lambda}_j \, \mathbf{N}_i \, \mathbf{N}_i + \sum_{i \neq j} \Omega_{ij}^0 \left(\lambda_j^2 - \lambda_i^2 \right) \frac{S_j - S_i}{\lambda_j^2 - \lambda_i^2} \mathbf{N}_i \, \mathbf{N}_j. \tag{7.98}$$

Since

$$\dot{\mathbf{S}} = \Lambda : \dot{\mathbf{E}} = \frac{1}{2} \Lambda : \dot{\mathbf{C}}, \tag{7.99}$$

we recognize from (7.94) and (7.98), by inspection, that

$$\Lambda = \sum_{i,j=1}^{3} \frac{1}{\lambda_j} \frac{\partial S_i}{\partial \lambda_j} \mathbf{N}_i \, \mathbf{N}_i \, \mathbf{N}_j \, \mathbf{N}_j + \sum_{i \neq j} \frac{S_j - S_i}{\lambda_j^2 - \lambda_i^2} \mathbf{N}_i \, \mathbf{N}_j \left(\mathbf{N}_i \, \mathbf{N}_j + \mathbf{N}_j \, \mathbf{N}_i \right). \tag{7.100}$$

7.11 Suggested Reading

Antman, S. S. (1995), *Nonlinear Problems of Elasticity*, Springer-Verlag, New York.
Beatty, M. F. (1996), Introduction to Nonlinear Elasticity. In *Nonlinear Effects in Fluids and Solids* (M. M. Carroll and M. A. Hayes, eds.), pp. 13–112, Plenum, New York.
Carlson, D. E., and Shield, R. T., eds. (1982), *Finite Elasticity*, Martinus Nijhoff, The Hague.
Green, A. E., and Adkins, J. E. (1960), *Large Elastic Deformations*, Oxford University Press, Oxford, UK.
Gurtin, M. E. (1981), *Topics in Finite Elasticity*, SIAM, Philadelphia.
Holzapfel, G. A. (2000), *Nonlinear Solid Mechanics*, Wiley, Chichester, UK.
Leigh, D. C. (1968), *Nonlinear Continuum Mechanics*, McGraw-Hill, New York.

7.11. Suggested Reading

Lurie, A. I. (1990), *Nonlinear Theory of Elasticity*, North-Holland, Amsterdam.

Malvern, L. E. (1969), *Introduction to the Mechanics of a Continuous Medium*, Prentice Hall, Englewood Cliffs, New Jersey.

Marsden, J. E., and Hughes, T. J. R. (1983), *Mathematical Foundations of Elasticity*, Prentice Hall, Englewood Cliffs, New Jersey.

Murnaghan, F. D. (1951), *Finite Deformation of an Elastic Solid*, Wiley, New York.

Ogden, R. W. (1984), *Non-Linear Elastic Deformations*, Ellis Horwood, Chichester, UK (2nd ed., Dover, 1997).

Rivlin, R. S. (1960), Some Topics in Finite Elasticity. In *Structural Mechanics* (J. N. Goodier and N. Hoff, eds.), pp. 169–198, Pergamon, New York.

Rivlin, R. S., ed. (1977), *Finite Elasticity*, ASME, AMD, Vol. 27, New York.

Treloar, L. R. G. (1975), *The Physics of Rubber Elasticity*, Clarendon Press, Oxford, UK.

Truesdell, C., and Noll, W. (1965), The Nonlinear Field Theories of Mechanics. In *Handbuch der Physik* (S. Flügge, ed.), Band III/3, Springer-Verlag, Berlin (2nd ed., 1992).

Wang, C.-C., and Truesdell, C. (1973), *Introduction to Rational Elasticity*, Noordhoff International, Leyden, The Netherlands.

PART 3: LINEAR ELASTICITY

8 Governing Equations of Linear Elasticity

8.1 Elementary Theory of Isotropic Linear Elasticity

Consider a bar of uniform cross section, composed of a homogeneous isotropic material, subject to uniaxial tension of magnitude σ. Assuming small strain, the response is linearly elastic if

$$\sigma = Ee, \qquad (8.1)$$

where E is *Young's modulus* and e is the longitudinal strain. Likewise if a homogeneous body is subject to a shear stress, τ, the linearly elastic response is

$$\tau = G\gamma = 2G(\gamma/2). \qquad (8.2)$$

The constant G is the elastic shear modulus, and $\gamma/2$ is the shear strain; γ is the so-called engineering shear strain.

For isotropic materials, equations (8.1) and (8.2) can be generalized to

$$\sigma'_{ij} = 2Ge'_{ij}, \quad p = -\frac{1}{3}\sigma_{kk} = -Ke_v, \qquad (8.3)$$

where

$$\sigma'_{ij} = \sigma_{ij} - \frac{1}{3}\sigma_{kk}\delta_{ij} = \sigma_{ij} + p\delta_{ij}, \qquad (8.4)$$

$$e'_{ij} = e_{ij} - \frac{1}{3}e_v\delta_{ij}, \quad e_v = e_{kk}. \qquad (8.5)$$

This is referred to as the generalized Hooke's law. The prime designates the deviatoric part. The volumetric strain is e_v. The bulk modulus, K, is related to E and G by

$$K = \frac{GE}{3(3G-E)}. \qquad (8.6)$$

The most general form of an isotropic forth-order tensor, **C**, *via* its components on the $\{e_i\}$ basis is

$$C_{ijkl} = \lambda\delta_{ij}\delta_{kl} + \mu(\delta_{ik}\delta_{jl} + \delta_{il}\delta_{jk}) + \upsilon(\delta_{ik}\delta_{jl} - \delta_{il}\delta_{jk}). \qquad (8.7)$$

This is the elastic moduli tensor which relates the stress and strain tensors,

$$\sigma_{ij} = C_{ijkl} e_{kl}. \tag{8.8}$$

The symmetries of the stress and strain tensors imply

$$e_{kl} = e_{lk} \Rightarrow C_{ijkl} = C_{ijlk}, \tag{8.9}$$

and

$$\sigma_{ij} = \sigma_{ji} \Rightarrow C_{ijkl} = C_{jikl}. \tag{8.10}$$

Thus, $\upsilon = 0$ in (8.7), and $C_{ijkl} = C_{klij}$. The end result takes the form

$$\sigma_{ij} = C_{ijkl} e_{kl},$$
$$C_{ijkl} = \lambda \delta_{ij} \delta_{kl} + \mu (\delta_{ik} \delta_{jl} + \delta_{il} \delta_{jk}). \tag{8.11}$$

These are the constitutive equations of isotropic linear elasticity. The constants λ and μ are known as the Lamé constants. Consistent with the previous relations in terms of the Young's and shear moduli, it is readily shown that

$$\mu \equiv G = \frac{E}{2(1+\nu)}, \quad \lambda = \frac{\nu E}{(1+\nu)(1-2\nu)}, \quad \nu = \frac{\lambda}{2(\lambda + \mu)}, \tag{8.12}$$

and

$$e_{ij} = -\frac{\nu}{E} \sigma_{kk} \delta_{ij} + \frac{1+\nu}{E} \sigma_{ij}. \tag{8.13}$$

The Poisson's ratio of lateral contraction ν is defined such that the lateral strain in uniaxial tension is $e_{\text{lat}} = -\nu e$, where e is the longitudinal strain. In the expanded form, the constitutive equations of linear elasticity read

$$e_{11} = \frac{1}{E}[\sigma_{11} - \nu(\sigma_{22} + \sigma_{33})],$$
$$e_{22} = \frac{1}{E}[\sigma_{22} - \nu(\sigma_{33} + \sigma_{11})], \tag{8.14}$$
$$e_{33} = \frac{1}{E}[\sigma_{33} - \nu(\sigma_{11} + \sigma_{22})],$$

and

$$e_{12} = \frac{1}{G} \sigma_{12}, \quad e_{23} = \frac{1}{G} \sigma_{23}, \quad e_{31} = \frac{1}{G} \sigma_{31}. \tag{8.15}$$

Note also that

$$e_{ij} = \frac{1}{2\mu} \sigma'_{ij} + \frac{1}{9K} \sigma_{kk} \delta_{ij}. \tag{8.16}$$

8.2. Elastic Energy in Linear Elasticity

The following is a useful list of the relationships among various elastic constants of isotropic material:

$$E = 2\mu(1+\nu) = 3\kappa(1-2\nu) = \frac{9\kappa\mu}{3\kappa+\mu} = \frac{\mu(3\lambda+2\mu)}{\lambda+\mu}$$

$$= \frac{\lambda(1+\nu)(1-2\nu)}{\nu} = \frac{9\kappa(\kappa-\lambda)}{3\kappa-\lambda},$$

$$\nu = \frac{E}{2\mu} - 1 = \frac{1}{2} - \frac{E}{6\kappa} = \frac{\lambda}{2(\lambda+\mu)} = \frac{3\kappa-2\mu}{2(3\kappa+\mu)} = \frac{\lambda}{3\kappa-\lambda},$$

$$\lambda = \frac{E\nu}{(1+\nu)(1-2\nu)} = \frac{2\mu\nu}{1-2\nu} = \frac{3\kappa\nu}{1+\nu} = \kappa - \frac{2}{3}\mu$$

$$= \frac{3\kappa(3\kappa-E)}{9\kappa-E} = \frac{\mu(E-2\mu)}{3\mu-E},$$

$$\mu = G = \frac{E}{2(1+\nu)} = \frac{3\kappa E}{9\kappa-E} = \frac{\lambda(1-2\nu)}{2\nu} = \frac{3\kappa(1-2\nu)}{2(1+\nu)} = \frac{3}{2}(\kappa-\lambda),$$

$$\kappa = \frac{E}{3(1-2\nu)} = \lambda + \frac{2}{3}\mu = \frac{\lambda(1+\nu)}{3\nu} = \frac{2\mu(1+\nu)}{3(1-2\nu)} = \frac{\mu E}{3(3\mu-E)}.$$

The expressions for the combinations of elastic moduli, expressed solely in terms of the Poisson's ratio, are also noted:

$$\frac{\mu}{\lambda+\mu} = 1 - 2\nu, \quad \frac{\lambda}{\lambda+2\mu} = \frac{\nu}{1-\nu},$$

$$\frac{\lambda+2\mu}{E} = \frac{1-\nu}{(1+\nu)(1-2\nu)}, \quad \frac{4\mu}{E}\frac{\lambda+\mu}{\lambda+2\mu} = \frac{1}{1-\nu^2}.$$

8.2 Elastic Energy in Linear Elasticity

For small strains, the elastic strain energy per unit volume is

$$\Phi = \frac{1}{2}\sigma_{ij}e_{ij} = \frac{1}{2}C_{ijkl}e_{ij}e_{kl}. \qquad (8.17)$$

The total elastic strain energy within the body of volume V is

$$\frac{1}{2}\int_V \sigma_{ij}e_{ij}\,dV = \frac{1}{2}\int_V \sigma_{ij}\frac{1}{2}\left(\frac{\partial u_i}{\partial x_j} + \frac{\partial u_j}{\partial x_i}\right)dV$$

$$= \frac{1}{2}\int_V \sigma_{ij}\frac{\partial u_i}{\partial x_j}\,dV. \qquad (8.18)$$

The symmetry of the Cauchy stress $\sigma_{ij} = \sigma_{ji}$ was used in the last step. Since

$$\frac{\partial}{\partial x_j}(\sigma_{ij} u_i) = \frac{\partial \sigma_{ij}}{\partial x_j} u_i + \sigma_{ij} \frac{\partial u_i}{\partial x_j} = -b_i u_i + \sigma_{ij} \frac{\partial u_i}{\partial x_j}, \tag{8.19}$$

where b_i are the body forces per unit volume ($\sigma_{ij,j} + b_i = 0$), we have from (8.18)

$$\frac{1}{2} \int_V \sigma_{ij} \frac{\partial u_i}{\partial x_j} \, dV = \frac{1}{2} \int_V \left[\frac{\partial}{\partial x_j}(\sigma_{ij} u_i) + b_i u_i \right] dV. \tag{8.20}$$

Upon applying the Gauss divergence theorem to the first integral on the *rhs*, we obtain

$$\begin{aligned}
\frac{1}{2} \int_V \sigma_{ij} \frac{\partial u_i}{\partial x_j} \, dV &= \frac{1}{2} \int_S \sigma_{ij} n_j u_i \, dS + \frac{1}{2} \int_V b_i u_i \, dV \\
&= \frac{1}{2} \int_S T_i u_i \, dS + \frac{1}{2} \int_V b_i u_i \, dV \\
&= \text{the work done.}
\end{aligned} \tag{8.21}$$

Thus, comparing (8.18) and (8.21), we have

$$\frac{1}{2} \int_V \sigma_{ij} e_{ij} \, dV = \frac{1}{2} \int_S T_i u_i \, dS + \frac{1}{2} \int_V b_i u_i \, dV. \tag{8.22}$$

We note that in later parts of this book $\Phi(\mathbf{e})$ is called $W(\mathbf{e})$, the strain energy density. We further take the strain energy density to be a strictly positive function, in the sense that

$$W(\mathbf{e}) = \frac{1}{2} C_{ijk\ell} e_{ij} e_{k\ell} \geq 0, \tag{8.23}$$

where the equality holds only if $\mathbf{e} = \mathbf{0}$.

8.3 Restrictions on the Elastic Constants

In this section we explore restrictions on the elastic moduli that limit the number of possible nonzero values and the range of values that elastic moduli may have.

8.3.1 Material Symmetry

We begin with the statement of basic symmetry

$$C_{ijk\ell} = C_{jik\ell} = C_{k\ell ij} = C_{ij\ell k}. \tag{8.24}$$

The symmetry in the first two indices follows from the symmetry of the stress tensor whereas the symmetry in the last two follows from the symmetry of the strain tensor. The reciprocal symmetry $C_{ijk\ell} = C_{k\ell ij}$ follows from (8.17), because

$$C_{ijk\ell} e_{ij} e_{k\ell} = C_{k\ell ij} e_{k\ell} e_{ij} \quad \Rightarrow \quad (C_{ijk\ell} - C_{k\ell ij}) e_{ij} e_{k\ell} = 0. \tag{8.25}$$

It is convenient to introduce here a contracted (Voigt) notation for the elastic constants, whereby

$$\begin{aligned}
&\hat{\sigma}_1 \leftarrow \sigma_{11}, \quad \hat{\sigma}_2 \leftarrow \sigma_{22}, \quad \hat{\sigma}_3 \leftarrow \sigma_{33} \\
&\hat{\sigma}_4 \leftarrow \sigma_{23}, \quad \hat{\sigma}_5 \leftarrow \sigma_{31}, \quad \hat{\sigma}_6 \leftarrow \sigma_{12}.
\end{aligned} \tag{8.26}$$

8.3. Elastic Constants

Similarly, for the strains,

$$\hat{e}_1 \leftarrow e_{11}, \quad \hat{e}_2 \leftarrow e_{22}, \quad \hat{e}_3 \leftarrow e_{33}$$
$$\hat{e}_4 \leftarrow 2e_{23}, \quad \hat{e}_5 \leftarrow 2e_{31}, \quad \hat{e}_6 \leftarrow 2e_{12}. \tag{8.27}$$

The use of the factor of 2 in the notation for the strain components ensures the symmetry in the expression

$$\hat{\sigma}_\alpha = \hat{C}_{\alpha\beta} e_\beta, \tag{8.28}$$

i.e., $\hat{C}_{\alpha\beta} = \hat{C}_{\beta\alpha}$. In defining the components of $\hat{\mathbf{C}}$ the same indicial notation is used on the index pairs ij and $k\ell$ in the tensor $\mathbf{C}$.

Consider a coordinate transformation specified by the orthogonal tensor, $\mathbf{Q}$,

$$\mathbf{e}'_\alpha = Q_{\alpha j} \mathbf{e}_j. \tag{8.29}$$

On the basis $\{\mathbf{e}'_i\}$ the components of $\mathbf{C}$ would be

$$C'_{ijk\ell} = Q_{ip} Q_{jq} Q_{kr} Q_{\ell t} C_{pqrt}. \tag{8.30}$$

If $\mathbf{Q}$ represents a symmetry operation, we have

$$C'_{ijk\ell} = C_{ijk\ell}. \tag{8.31}$$

As an example, consider the inversion process as specified by the *improper* orthogonal tensor with components

$$\mathbf{Q}^I = \begin{bmatrix} -1 & 0 & 0 \\ 0 & -1 & 0 \\ 0 & 0 & -1 \end{bmatrix}. \tag{8.32}$$

All materials will be invariant under the transformation given by $\mathbf{Q}^I$.

Now let the initial basis be aligned with the most natural set of axes in the material. For example, if the material was cubic these would be the cube axes of the unit cell. Then specify a rotation about the $\mathbf{e}_3$ axis as

$$\mathbf{Q}(\theta) = \begin{bmatrix} \cos\theta & \sin\theta & 0 \\ -\sin\theta & \cos\theta & 0 \\ 0 & 0 & 1 \end{bmatrix}. \tag{8.33}$$

Reflection, on the other hand say across a plane whose unit normal was $\mathbf{n}$, would be specified by

$$\mathbf{Q}^R = \mathbf{I} - 2\mathbf{n}\mathbf{n}. \tag{8.34}$$

Clearly, the effect of $\mathbf{Q}^R$ would be to leave any vector, such as $\mathbf{m}$, in the plane unchanged, because $\mathbf{m} \cdot \mathbf{n} = 0$, and would cause $\mathbf{n}$ to become $-\mathbf{n}$, i.e.,

$$\mathbf{Q}^R \cdot \mathbf{m} = \mathbf{m} \quad \text{and} \quad \mathbf{Q}^R \cdot \mathbf{n} = -\mathbf{n}. \tag{8.35}$$

Let one such reflection plane be

$$\mathbf{n} = \cos\theta \, \mathbf{e}_1 + \sin\theta \, \mathbf{e}_2 + 0 \, \mathbf{e}_3. \tag{8.36}$$

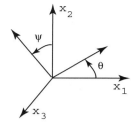

Figure 8.1. Angles θ and ψ defining two reflection planes.

In this case

$$\mathbf{Q}^R(\theta) = \begin{bmatrix} -\cos(2\theta) & -\sin(2\theta) & 0 \\ -\sin(2\theta) & \cos(2\theta) & 0 \\ 0 & 0 & 1 \end{bmatrix}. \tag{8.37}$$

The angle θ is defined in Fig. 8.1 along with a second angle ψ used to define a second reflection plane. In our example, if $\theta = 0$ then the operation is specified by

$$\mathbf{Q}^R(0) = \begin{bmatrix} -1 & 0 & 0 \\ 0 & 1 & 0 \\ 0 & 0 & 1 \end{bmatrix}, \tag{8.38}$$

and represents a reflection across the $x_2 - x_3$ plane, *i.e.*, through the x_1 axis.

To characterize the applicable symmetry operations we will use the form of the improper orthogonal tensor given in (8.37), which specifies a reflection through a plane whose normal is in the $x_1 - x_2$ plane, *and* one with respect to a plane whose normal is in the $x_2 - x_3$ plane; the latter is defined by the angle ψ in Fig. 8.1. This would, in fact, be associated with

$$\mathbf{Q}^R(\psi) = \begin{bmatrix} 1 & 0 & 0 \\ 0 & -\cos(2\psi) & -\sin(2\psi) \\ 0 & -\sin(2\psi) & \cos(2\psi) \end{bmatrix}. \tag{8.39}$$

Additional details can be found in the books by Musgrave (1970) and Ting (1996); here we list some results in two forms. First, the matrix $\hat{\mathbf{C}}$ is shown by using Greek letters to represent (positive) values of potentially nonzero elastic constants. Symmetry will dictate the number of independent moduli. Second, the values of θ and ψ will be given along with the symmetry type. The number of independent constants is called q. Note that $\hat{\mathbf{C}}$ is symmetric and that elements below the diagonal are not indicated.

I. Triclinic, no symmetry, $q = 21$:

$$\hat{\mathbf{C}} = \begin{bmatrix} \alpha & \beta & \gamma & \delta & \epsilon & \zeta \\ & \eta & \iota & \kappa & \xi & \pi \\ & & \rho & \tau & \chi & \phi \\ & & & \omega & \upsilon & \nu \\ & & & & \varphi & \varepsilon \\ & & & & & \varpi \end{bmatrix}. \tag{8.40}$$

8.3. Elastic Constants

II. Monoclinic, one symmetry plane, $x_1 = 0$, $\theta = 0$, $q = 13$:

$$\hat{\mathbf{C}} = \begin{bmatrix} \alpha & \beta & \gamma & \delta & 0 & 0 \\ \eta & \iota & \kappa & 0 & 0 & \\ & \rho & \tau & 0 & 0 & \\ & & \omega & 0 & 0 & \\ & & & \varphi & \varepsilon & \\ & & & & \varpi & \end{bmatrix}. \tag{8.41}$$

Note that the single symmetry plane can be alternatively taken at $x_2 = 0$, $\theta = \pi/2$ or $\psi = 0$, or at $x_3 = 0$, $\psi = 0$.

III. Orthotropic (or Rhombic), three symmetry planes at $\theta = 0, \pi/2$ and $\psi = \pi/2, q = 9$:

$$\hat{\mathbf{C}} = \begin{bmatrix} \alpha & \beta & \gamma & 0 & 0 & 0 \\ & \eta & \iota & 0 & 0 & 0 \\ & & \rho & 0 & 0 & 0 \\ & & & \omega & 0 & 0 \\ & & & & \varphi & 0 \\ & & & & & \varpi \end{bmatrix}. \tag{8.42}$$

IV. Tetragonal, five symmetry planes at $\theta = 0, \pm\pi/4, \pi/2$ and $\psi = \pi/2$, $q = 6$:

$$\hat{\mathbf{C}} = \begin{bmatrix} \alpha & \beta & \gamma & 0 & 0 & 0 \\ & \alpha & \gamma & 0 & 0 & 0 \\ & & \rho & 0 & 0 & 0 \\ & & & \omega & 0 & 0 \\ & & & & \omega & 0 \\ & & & & & \varpi \end{bmatrix}. \tag{8.43}$$

V. Transversely Isotropic (Hexagonal), the symmetry plane is the x_3 axis, $q = 5$:

$$\hat{\mathbf{C}} = \begin{bmatrix} \alpha & \beta & \gamma & 0 & 0 & 0 \\ & \eta & \gamma & 0 & 0 & 0 \\ & & \rho & 0 & 0 & 0 \\ & & & \omega & 0 & 0 \\ & & & & \omega & 0 \\ & & & & & \blacksquare \end{bmatrix}. \tag{8.44}$$

In this case $\blacksquare = \frac{1}{2}(\hat{C}_{11} - \hat{C}_{12}) = \frac{1}{2}(\alpha - \beta)$.

VI. Cubic, nine symmetry planes including the coordinate axes and those whose normals make angles of $\pm\pi/4$ with the coordinate axes, $q = 3$:

$$\hat{\mathbf{C}} = \begin{bmatrix} \alpha & \beta & \beta & 0 & 0 & 0 \\ & \alpha & \beta & 0 & 0 & 0 \\ & & \alpha & 0 & 0 & 0 \\ & & & \omega & 0 & 0 \\ & & & & \omega & 0 \\ & & & & & \omega \end{bmatrix}. \tag{8.45}$$

VII. Isotropic, any plane is a symmetry plane, $q = 2$:

$$\hat{\mathbf{C}} = \begin{bmatrix} \alpha & \beta & \beta & 0 & 0 & 0 \\ & \alpha & \beta & 0 & 0 & 0 \\ & & \alpha & 0 & 0 & 0 \\ & & & \blacksquare & 0 & 0 \\ & & & & \blacksquare & 0 \\ & & & & & \blacksquare \end{bmatrix}, \tag{8.46}$$

where $\blacksquare = \frac{1}{2}(\alpha - \beta)$. The above matrix is commonly rewritten as

$$\hat{\mathbf{C}} = \begin{bmatrix} \lambda + 2\mu & \lambda & \lambda & 0 & 0 & 0 \\ & \lambda + 2\mu & \lambda & 0 & 0 & 0 \\ & & \lambda + 2\mu & 0 & 0 & 0 \\ & & & \mu & 0 & 0 \\ & & & & \mu & 0 \\ & & & & & \mu \end{bmatrix}. \tag{8.47}$$

The constants λ and μ are the Lamé constants of the constitutive law (8.11).

8.3.2 Restrictions on the Elastic Constants

There are two conditions we impose on the elastic constants, namely the so-called *strong ellipticity* condition

$$\mathbf{ab} : \mathbf{C} : \mathbf{ab} = a_i b_j C_{ijk\ell} a_k b_\ell > 0, \tag{8.48}$$

and the *strong convexity* condition

$$\mathbf{e} : \mathbf{C} : \mathbf{e} = C_{ijk\ell} e_{ij} e_{k\ell} > 0. \tag{8.49}$$

The convexity condition is the more severe of the two, which may be readily seen by noting that the dyadic product **ab** would represent a strain tensor that is singular; yet (8.49) is to be true for *all* admissible strain tensors, including those that are not.

For isotropic materials (8.49) leads to

$$\begin{aligned} \lambda(e_{ii})^2 + 2\mu(e_{ij}e_{ij}) &> 0, \\ (\lambda + \frac{2}{3}\mu)(e_{ii})^2 + 2\mu[e_{ij}e_{ij} - \frac{1}{3}(e_{ii})^2] &> 0. \end{aligned} \tag{8.50}$$

These imply

$$\lambda + \frac{2}{3}\mu > 0, \quad \mu > 0. \tag{8.51}$$

Since the connections between the modulii (E, ν) and (λ, μ) are

$$\lambda = \frac{E\nu}{(1+\nu)(1-2\nu)}, \quad \mu = \frac{E}{2(1+\nu)}, \tag{8.52}$$

the conditions expressed in (8.50) become

$$E > 0, \quad -1 < \nu < 1/2. \tag{8.53}$$

8.4. Compatibility Relations

Another approach to establishing such restrictions, still based on the positive-definite nature of the elastic energy function, is as follows. If the elastic strain energy is assumed to be positive-definite function of strain e_{ij}, then

$$\sigma_{ij} e_{ij} = \lambda (e_{kk})^2 + 2\mu e_{ij} e_{ij} > 0. \tag{8.54}$$

Denoting by e'_{ij} the deviatoric part of strain tensor, such that

$$e_{ij} = e'_{ij} + \frac{1}{3} e_{kk} \delta_{ij}, \tag{8.55}$$

we have

$$e_{ij} e_{ij} = e'_{ij} e'_{ij} + \frac{1}{3} (e_{kk})^2. \tag{8.56}$$

Consequently, the inequality (8.54) can be rewritten as

$$2\mu e'_{ij} e'_{ij} + \kappa (e_{kk})^2 > 0, \tag{8.57}$$

where

$$\kappa = \lambda + \frac{2}{3} \mu = \frac{E}{3(1 - 2\nu)}, \quad \mu = \frac{E}{2(1 + \nu)}. \tag{8.58}$$

By taking $e_{kk} = 0$ in (8.57), we deduce that $\mu > 0$; by taking $e'_{ij} = 0$, we deduce that $\kappa > 0$. This implies from equations (8.58) that $E > 0$ and $-1 < \nu < 1/2$. Since experience does not reveal any isotropic elastic material with negative Poisson's ratio, the physical range of ν is $0 < \nu < 1/2$. Negative values of ν would imply negative values of λ, because $\lambda = 2\mu\nu/(1 - 2\nu)$. In the limit $\nu \to 1/2$, the material behaves as incompressible (λ and $\kappa \to \infty$), and $\mu = E/3$. If $\nu = 1/3$, we have $E = \kappa$ and $\lambda = 2\mu$; if $\nu = 1/4$, then $E = 3\kappa/2$ and $\lambda = \mu$.

8.4 Compatibility Relations

If the strains are derived from a continuous, single valued vector displacement field **u**, then they must satisfy a particular set of differential equations. Conversely, if this set of differential equations is satisfied by the components of strain, the corresponding displacement field is a continuous displacement field. We derive the compatibility equations below, using the following convenient notation for partial derivatives

$$\partial(..)/\partial x_\ell \equiv (..)_{,\ell}, \quad \partial^2(..)/\partial x_\ell \partial x_r \equiv (..)_{,\ell r}.$$

With this notation the components of the infinitesimal strain tensor become

$$e_{ij} = \frac{1}{2}(u_{i,j} + u_{j,i}). \tag{8.59}$$

Thus, by inspection, we observe that

$$e_{\ell\ell, mm} = u_{\ell,\ell mm}, \quad e_{mm,\ell\ell} = u_{\ell,m\ell\ell}, \tag{8.60}$$

and

$$e_{\ell m,\ell m} = \frac{1}{2}(u_{\ell,m\ell m} + u_{m,\ell\ell m}) = \frac{1}{2}(u_{\ell,\ell mm} + u_{m,m\ell\ell}). \tag{8.61}$$

Therefore,
$$e_{\ell\ell,mm} + e_{mm,\ell\ell} = 2e_{\ell m,\ell m}. \tag{8.62}$$

Similarly,
$$e_{\ell\ell,mn} = u_{\ell,\ell mn}, \quad e_{\ell m,\ell n} = \frac{1}{2}(u_{\ell,m\ell n} + u_{m,\ell\ell n}), \tag{8.63}$$

and
$$e_{\ell n,\ell m} = \frac{1}{2}(u_{\ell,n\ell m} + u_{n,\ell\ell m}), \quad e_{mn,\ell\ell} = \frac{1}{2}(u_{m,n\ell\ell} + u_{n,m\ell\ell}). \tag{8.64}$$

Consequently,
$$e_{\ell m,\ell n} + e_{\ell n,\ell m} - e_{mn,\ell\ell} = e_{\ell\ell,mn}. \tag{8.65}$$

In general, we may express six independent relations of this type as

$$\begin{aligned} e_{11,22} + e_{22,11} &= 2e_{12,12}, \\ e_{22,33} + e_{33,22} &= 2e_{23,23}, \\ e_{33,11} + e_{11,33} &= 2e_{31,31}, \\ e_{12,13} + e_{13,12} - e_{23,11} &= e_{11,23}, \\ e_{23,21} + e_{21,23} - e_{31,22} &= e_{22,31}, \\ e_{31,32} + e_{32,31} - e_{12,33} &= e_{33,12}, \end{aligned} \tag{8.66}$$

or, in more compact form,
$$e_{ij,kl} + e_{kl,ij} = e_{ik,jl} + e_{jl,ik}. \tag{8.67}$$

The above are called the Saint-Venant's compatibility equations for the strain components. In what immediately follows, we revisit the establishment of such equations from two different approaches.

8.5 Compatibility Conditions: Cesàro Integrals

Consider two points of the deformed body, A and B. Denoting their displacement components by u_i^A and u_i^B, we can write

$$u_i^B = u_i^A + \int_A^B du_i. \tag{8.68}$$

If displacement components u_i^B are to be single-valued in a simply connected region, the integral on the right-hand side of the above equation must be path-independent. By imposing this condition, we will again arrive at the Saint-Venant's compatibility equations. To demonstrate this, we proceed as follows. First, we have

$$u_i^B = u_i^A + \int_A^B du_i = u_i^A + \int_A^B \frac{\partial u_i}{\partial x_j} dx_j = u_i^A + \int_A^B e_{ij}\, dx_j + \int_A^B \omega_{ij}\, dx_j, \tag{8.69}$$

8.5. Cesàro Integrals

where e_{ij} and ω_{ij} are the infinitesimal strain and rotation tensors (symmetric and antisymmetric parts of the displacement gradient $u_{i,j}$). By using the integration by parts, we then write

$$\int_A^B \omega_{ij}\,dx_j = \int_A^B \omega_{ij}\,d(x_j - x_j^B) = (x_j^B - x_j^A)\omega_{ij}^A - \int_A^B (x_j - x_j^B)\frac{\partial \omega_{ij}}{\partial x_k}\,dx_k. \tag{8.70}$$

But, as can be easily verified by inspection,

$$\frac{\partial \omega_{ij}}{\partial x_k} = \frac{\partial e_{ik}}{\partial x_j} - \frac{\partial e_{jk}}{\partial x_i}. \tag{8.71}$$

Hence, by substituting (8.71) into (8.70), and then this into (8.69), we obtain

$$u_i^B = u_i^A + (x_j^B - x_j^A)\omega_{ij}^A + \int_A^B f_{ik}\,dx_k,$$

where

$$f_{ik} = e_{ik} - (x_j - x_j^B)\left(\frac{\partial e_{ik}}{\partial x_j} - \frac{\partial e_{jk}}{\partial x_i}\right). \tag{8.72}$$

The above integral has to be path independent. Thus, $f_{ik}\,dx_k = f_{ij}\,dx_j = dg_i$ has to be a perfect differential. The necessary and sufficient condition for this is that

$$\frac{\partial f_{ik}}{\partial x_j} = \frac{\partial f_{ij}}{\partial x_k}, \tag{8.73}$$

(both then being equal to $\partial^2 g_i/\partial x_j \partial x_k$). Substituting (8.72) into (8.73) leads to

$$\frac{\partial^2 e_{ij}}{\partial x_k \partial x_l} + \frac{\partial^2 e_{kl}}{\partial x_i \partial x_j} - \frac{\partial^2 e_{ik}}{\partial x_j \partial x_l} - \frac{\partial^2 e_{jl}}{\partial x_i \partial x_k} = 0. \tag{8.74}$$

There are 81 of these equations, but some of them are trivial identities, and some are repetitions due to symmetry in indices ij and kl. The contraction $k = l$ yields six linearly independent Saint-Venant's equations

$$\frac{\partial^2 e_{ij}}{\partial x_k \partial x_k} + \frac{\partial^2 e_{kk}}{\partial x_i \partial x_j} - \frac{\partial^2 e_{ik}}{\partial x_j \partial x_k} - \frac{\partial^2 e_{jk}}{\partial x_i \partial x_k} = 0. \tag{8.75}$$

They can be compactly rewritten as

$$\Xi_{ij} = \epsilon_{ikl}\epsilon_{jmn}\frac{\partial^2 e_{ln}}{\partial x_k \partial x_m} = 0. \tag{8.76}$$

The permutation tensor is denoted by ϵ_{ijk}. Thus, for a given strain field e_{ij}, the necessary and sufficient conditions for the existence of single-valued displacement field u_i within a simply connected region (apart from the rigid-body motion) are the six Saint-Venant's compatibility equations $\Xi_{ij} = 0$. These equations are linearly independent, but differentially related by three Bianchi conditions

$$\frac{\partial \Xi_{ij}}{\partial x_j} = 0. \tag{8.77}$$

These follow from the symmetry of the mixed partial derivative $\partial^2/\partial x_m \partial x_n$ and the skew-symmetry of the permutation tensor ϵ_{ijk}. If $\Xi_{ij} \neq 0$, the incompatibility tensor Ξ_{ij} represents a measure of the degree of strain incompatibility.

For $(N+1)$-tiply connected region, in addition to six Saint-Venant's compatibility equations (8.76), there are $3N$ Cesàro integral conditions for the existence of single-valued displacements. They are

$$\oint_{C_\alpha} f_{ij} \mathrm{d}x_j = 0, \tag{8.78}$$

where C_α ($\alpha = 1, 2, \ldots, N$) are any closed irreducible curves around N internal "cavities" and

$$f_{ij} = e_{ij} - (x_k - x_k^0)(e_{ij,k} - e_{jk,i}). \tag{8.79}$$

The coordinates of a selected reference point on C_α are x_k^0. Since this point can be selected arbitrarily on each C_α, equations (8.78) and (8.79) give rise to $6N$ integral conditions

$$\oint_{C_\alpha} (e_{ij} - x_k \epsilon_{lik} \epsilon_{lpq} e_{pj,q}) \mathrm{d}x_j = 0, \tag{8.80}$$

$$\oint_{C_\alpha} \epsilon_{ijl} e_{jk,l} \mathrm{d}x_k = 0. \tag{8.81}$$

8.6 Beltrami–Michell Compatibility Equations

By substituting the stress-strain relations

$$e_{ij} = -\frac{\nu}{E} \sigma_{kk} \delta_{ij} + \frac{1+\nu}{E} \sigma_{ij} \tag{8.82}$$

into the Saint-Venant's compatibility equations (8.75), and by using the equilibrium equations

$$\sigma_{ij,j} + b_i = 0, \tag{8.83}$$

we can deduce six linearly independent compatibility equations expressed in terms of stresses. These are the Beltrami–Michell compatibility equations

$$\sigma_{ij,kk} + \frac{1}{1+\nu} \sigma_{kk,ij} = -\frac{\nu}{1-\nu} b_{k,k} \delta_{ij} - b_{i,j} - b_{j,i}. \tag{8.84}$$

In particular, it follows that the hydrostatic stress satisfies the Poisson's equation

$$\Delta^2 \sigma_{kk} = -\frac{1+\nu}{1-\nu} b_{k,k}. \tag{8.85}$$

If there are no body forces, the Beltrami–Michell equation reduce to

$$\sigma_{ij,kk} + \frac{1}{1+\nu} \sigma_{kk,ij} = 0, \tag{8.86}$$

whereas the hydrostatic stress becomes a harmonic function, satisfying the Laplace's equation

$$\Delta^2 \sigma_{kk} = 0. \tag{8.87}$$

8.7 Navier Equations of Motion

The Cauchy equations of motion, in terms of stresses, are

$$\frac{\partial \sigma_{ij}}{\partial x_j} + b_i = \rho \frac{\partial^2 u_i}{\partial t^2}, \tag{8.88}$$

8.7. Navier Equations of Motion

where b_i are the components of the body force per unit volume. By substituting the stress-strain relations

$$\sigma_{ij} = 2\mu e_{ij} + \lambda e_{kk}\delta_{ij} = \mu\left(\frac{\partial u_i}{\partial x_j} + \frac{\partial u_j}{\partial x_i}\right) + \lambda \frac{\partial u_k}{\partial x_k}\delta_{ij} \qquad (8.89)$$

into (8.88), we obtain the Navier equations of motion, in terms of displacements,

$$(\lambda + \mu)\frac{\partial^2 u_j}{\partial x_i \partial x_j} + \mu \frac{\partial^2 u_i}{\partial x_j \partial x_j} + b_i = \rho \frac{\partial^2 u_i}{\partial t^2}. \qquad (8.90)$$

Since $T_i = \sigma_{ij} n_j$, the boundary conditions over the bounding surface of the body S_T, where tractions are externally prescribed, are

$$\lambda \frac{\partial u_k}{\partial x_k} n_i + \mu\left(\frac{\partial u_i}{\partial x_j} + \frac{\partial u_j}{\partial x_i}\right) n_j = T_i^{\text{ext}}, \qquad (8.91)$$

whereas

$$u_i = u_i^{\text{ext}}, \qquad (8.92)$$

where displacements are prescribed (S_u). In the vector notation, the Navier equations are

$$(\lambda + \mu)\nabla(\nabla \cdot \mathbf{u}) + \mu \nabla^2 \mathbf{u} + \mathbf{b} = \rho \frac{\partial^2 \mathbf{u}}{\partial t^2} \quad \text{in } V,$$

$$\lambda(\nabla \cdot \mathbf{u})\mathbf{n} + \mu(\mathbf{u}\nabla + \nabla\mathbf{u}) \cdot \mathbf{n} = \mathbf{T}^{\text{ext}} \quad \text{on } S_T, \qquad (8.93)$$

$$\mathbf{u} = \mathbf{u}^{\text{ext}} \quad \text{on } S_u.$$

The nabla operator expressed with respect to three typical sets of coordinates is

$$\nabla = \mathbf{e}_1 \frac{\partial}{\partial x_1} + \mathbf{e}_2 \frac{\partial}{\partial x_2} + \mathbf{e}_3 \frac{\partial}{\partial x_3},$$

$$\nabla = \mathbf{e}_r \frac{\partial}{\partial r} + \mathbf{e}_\theta \frac{1}{r}\frac{\partial}{\partial \theta} + \mathbf{e}_z \frac{\partial}{\partial z}, \qquad (8.94)$$

$$\nabla = \mathbf{e}_r \frac{\partial}{\partial r} + \mathbf{e}_\theta \frac{1}{r}\frac{\partial}{\partial \theta} + \mathbf{e}_\phi \frac{1}{r \sin\theta}\frac{\partial}{\partial \phi}.$$

The corresponding Laplacian operators are, respectively,

$$\nabla^2 = \frac{\partial^2}{\partial x_1^2} + \frac{\partial^2}{\partial x_2^2} + \frac{\partial^2}{\partial x_3^2},$$

$$\nabla^2 = \frac{1}{r}\frac{\partial}{\partial r}\left(r \frac{\partial}{\partial r}\right) + \frac{1}{r^2}\frac{\partial^2}{\partial \theta^2} + \frac{\partial^2}{\partial z^2}, \qquad (8.95)$$

$$\nabla^2 = \frac{1}{r^2}\frac{\partial}{\partial r}\left(r^2 \frac{\partial}{\partial r}\right) + \frac{1}{r^2 \sin\theta}\frac{\partial}{\partial \theta}\left(\sin\theta \frac{\partial}{\partial \theta}\right) + \frac{1}{r^2 \sin^2\theta}\frac{\partial^2}{\partial \phi^2}.$$

8.8 Uniqueness of Solution to Linear Elastic Boundary Value Problem

8.8.1 Statement of the Boundary Value Problem

The boundary value problem of linear elasticity is specified by the equations

$$\sigma_{ij,j} + b_i = 0,$$

$$e_{ij} = \frac{1}{2}(u_{i,j} + u_{j,i}), \tag{8.96}$$

$$e_{ij,kl} + e_{kl,ij} = e_{ik,jl} + e_{jl,ik},$$

and the boundary conditions

$$\begin{aligned} T_i &= \sigma_{ij} n_j \quad \text{prescribed on } S_T, \\ u_i &\quad \text{prescribed on } S_u. \end{aligned} \tag{8.97}$$

The outward pointing normal to the bounding surface of the body is n_i. Furthermore, if there exists a strain energy function $W = W(e_{ij})$, then

$$\sigma_{ij} = \frac{\partial W}{\partial e_{ij}}. \tag{8.98}$$

For linear elasticity,

$$\sigma_{ij} = C_{ijkl} e_{ij}, \quad W = \frac{1}{2}\sigma_{ij} e_{ij} = \frac{1}{2} C_{ijkl} e_{ij} e_{kl}. \tag{8.99}$$

If the elastic strain energy is assumed to be positive-definite function of strain, the elastic moduli tensor C_{ijkl} is a positive-definite tensor.

8.8.2 Uniqueness of the Solution

The so specified boundary value problem of linear elasticity has a unique solution for the stress and strain fields. The proof is based on the assumption of the positive-definiteness of the elastic strain energy. Suppose that two solutions exist that satisfy (8.8.1)–(8.99). Denote the corresponding displacement, strain, and stress fields by $u_i^{(1)}, e_{ij}^{(1)}, \sigma_{ij}^{(1)}$ and $u_i^{(2)}$, $e_{ij}^{(2)}, \sigma_{ij}^{(2)}$. Let

$$u_i^* = u_i^{(1)} - u_2^{(2)}, \quad e_{ij}^* = e_{ij}^{(1)} - e_{ij}^{(2)}, \quad \sigma_{ij}^* = \sigma_{ij}^{(1)} - \sigma_{ij}^{(2)}. \tag{8.100}$$

Since

$$\sigma_{ij,j}^{(1)} + b_i = 0, \quad \sigma_{ij,j}^{(2)} + b_i = 0, \tag{8.101}$$

we evidently have

$$\sigma_{ij,j}^* = 0. \tag{8.102}$$

Consequently, the fields u_i^*, e_{ij}^*, σ_{ij}^* represent a solution to the homogeneous boundary value problem

$$\sigma_{ij,j}^* = 0 \text{ and } b_i^* = 0 \text{ in } V, \quad T_i^* = \sigma_{ij}^* n_j = 0 \text{ on } S_T, \quad u_i^* = 0 \text{ on } S_u. \tag{8.103}$$

Clearly, at every point of the bounding surface S of the body the product $u_i^* T_i^* = 0$, because either u_i^* or T_i^* vanishes at the boundary. Thus, upon integration and the application of the Gauss divergence theorem,

$$0 = \int_S u_i^* T_i^* \, \mathrm{d}S = \int_S u_i^* \sigma_{ij}^* n_j \, \mathrm{d}S = \int_V (u_i^* \sigma_{ij}^*)_{,j} \, \mathrm{d}V. \tag{8.104}$$

Performing the partial differentiation within the integrand, and using $\sigma_{ij,j}^* = 0$ and $\sigma_{ij}^* u_{i,j}^* = \sigma_{ij}^* e_{ij}^*$, then gives

$$0 = \int_V \sigma_{ij}^* e_{ij}^* \, \mathrm{d}V = 2 \int_V W(e_{ij}^*) \, \mathrm{d}V. \tag{8.105}$$

But W is a positive-definite function of strain and the integral of $W(e_{ij}^*)$ over the volume V vanishes only if $W = 0$ at every point of the body. This is possible only if $e_{ij}^* = 0$ at every point, i.e.,

$$e_{ij}^{(1)} = e_{ij}^{(2)} \quad \text{and therefore} \quad \sigma_{ij}^{(1)} = \sigma_{ij}^{(2)}. \tag{8.106}$$

Thus the uniqueness of the solution to the boundary value problem of linear elasticity: there is only one stress and strain field corresponding to prescribed boundary conditions and given body forces.

8.9 Potential Energy and Variational Principle

Consider infinitesimal deformation of an elastic material. In this section, for the sake of generality, we do not actually restrict to linearly elastic material, but the material that is characterized by the strain energy that is an arbitrary single valued function of a small strain, such that

$$W = W(e_{mn}) = \int_0^{e_{mn}} \sigma_{ij} \, \mathrm{d}e_{ij}, \quad \sigma_{ij} = \frac{\partial W}{\partial e_{ij}}. \tag{8.107}$$

The principle of virtual work (5.140) then gives

$$\int_V \delta W \, \mathrm{d}V = \int_S T_i \delta u_i \, \mathrm{d}S + \int_V b_i \delta u_i \, \mathrm{d}V, \tag{8.108}$$

because $\delta W = \sigma_{ij} \delta e_{ij} = (\partial W / \partial e_{ij}) \delta e_{ij}$. With body forces b_i regarded as given in V and surface forces T_i given on S_T, and recalling that $\delta u_i = 0$ on $S_u = S - S_T$, (8.108) delivers a variational principle

$$\delta \Pi = 0, \quad \Pi = \int_V [W(e_{mn}) - b_i u_i] \, \mathrm{d}V - \int_{S_T} T_i u_i \, \mathrm{d}S. \tag{8.109}$$

The functional $\Pi = \Pi(u_i)$ is the potential energy of the elastic body. Since b_i and T_i are regarded as fixed, it is a functional of the displacement field. For the true equilibrium displacement field, Π has a stationary value, because $\delta \Pi = 0$ for any kinematically admissible variation δu_i.

We next show the equilibrium displacement field minimizes the potential energy among all kinematically admissible displacement fields. The following proof relies on the adoption of an incremental stability (stability in the small), according to which

$$d\sigma_{ij} de_{ij} > 0 \tag{8.110}$$

for any set of strain and corresponding stress increments. Indeed, let u_i^t be the true equilibrium displacement field and u_i^k an arbitrary kinematically admissible displacement field (continuous and differentiable, satisfying the displacement boundary conditions on S_u, if any). The difference in the potential energy associated with these two fields is

$$\Pi(u_i^k) - \Pi(u_i^t) = \int_V [W(e_{mn}^k) - W(e_{mn}^t) - b_i(u_i^k - u_i^t)] dV \\ - \int_S T_i(u_i^k - u_i^t) dS. \tag{8.111}$$

Note that $u_i^k - u_i^t = 0$ on S_u, so that the above surface integral can be extended from S_T to S. But the principle of virtual work gives

$$\int_V \sigma_{ij}^t (e_{ij}^k - e_{ij}^t) dV = \int_S T_i(u_i^k - u_i^t) dS + \int_V b_i(u_i^k - u_i^t) dV. \tag{8.112}$$

When this is substituted in (8.111), there follows

$$\Pi(u_i^k) - \Pi(u_i^t) = \int_V [W(e_{mn}^k) - W(e_{mn}^t) - \sigma_{ij}^t (e_{ij}^k - e_{ij}^t)] dV. \tag{8.113}$$

In view of (8.107), this can be rewritten as

$$\Pi(u_i^k) - \Pi(u_i^t) = \int_V \left[\int_0^{e_{mn}^k} (\sigma_{ij} - \sigma_{ij}^t) de_{ij} - \int_0^{e_{mn}^t} (\sigma_{ij} - \sigma_{ij}^t) de_{ij} \right] dV \\ = \int_V \left[\int_{e_{mn}^t}^{e_{mn}^k} (\sigma_{ij} - \sigma_{ij}^t) de_{ij} \right] dV.$$

The elastic deformation is path-independent, and the value of the integral

$$\int_{e_{mn}^t}^{e_{mn}^k} (\sigma_{ij} - \sigma_{ij}^t) de_{ij} \tag{8.114}$$

does not depend on the strain path between e_{mn}^t and e_{mn}^k. Thus, selecting a strain path that corresponds to a straight line in stress space from σ_{ij}^t to σ_{ij}^k, the stress increment $d\sigma_{ij}$ is codirectional with $\sigma_{ij} - \sigma_{ij}^t$. Consequently, if the material is incrementally stable, such that $d\sigma_{ij} de_{ij} > 0$, we have

$$\int_{e_{mn}^t}^{e_{mn}^k} (\sigma_{ij} - \sigma_{ij}^t) de_{ij} > 0, \tag{8.115}$$

and thus

$$\Pi(u_i^k) - \Pi(u_i^t) > 0. \tag{8.116}$$

Therefore, among all kinematically admissible displacement fields, the true (equilibrium) displacement field minimizes the potential energy of the body.

8.9.1 Uniqueness of the Strain Field

If two displacement fields $u_i^{(1)}$ and $u_i^{(2)}$ both minimize the potential energy, then

$$\int_V \left[\int_{e_{mn}^{(1)}}^{e_{mn}^{(2)}} (\sigma_{ij} - \sigma_{ij}^{(1)}) \mathrm{d}e_{ij} \right] \mathrm{d}V = 0. \tag{8.117}$$

But, the incremental stability $\mathrm{d}\sigma_{ij}\mathrm{d}e_{ij} > 0$ implies that $(\sigma_{ij} - \sigma_{ij}^{(1)})\mathrm{d}e_{ij} > 0$ along the strain path that corresponds to a straight line in stress space from $\sigma_{ij}^{(1)}$ to $\sigma_{ij}^{(2)}$. Consequently, the above integral can vanish if and only if $e_{mn}^{(2)} = e_{mn}^{(1)}$, which assures the uniqueness of the strain field (possibly, to within a rigid body displacement).

8.10 Betti's Theorem of Linear Elasticity

This important theorem of linear elasticity follows from the generalized Claopeyron's formula and the linearity between the stress and strain, $\sigma_{ij} = C_{ijkl}e_{kl}$, where C_{ijkl} are the elastic moduli obeying the reciprocal symmetry $C_{ijkl} = C_{klij}$. Indeed, the Clapeyron's formula [see (5.138) from Chapter 5] gives

$$\int_V \sigma_{ij}\hat{e}_{ij}\,\mathrm{d}V = \int_S T_i\hat{u}_i\,\mathrm{d}S + \int_V b_i\hat{u}_i\,\mathrm{d}V. \tag{8.118}$$

Let $\hat{\sigma}_{ij} = C_{ijkl}\hat{e}_{kl}$ be the true equilibrium stress field associated with the displacement field $\hat{u}_i$. Then,

$$\hat{\sigma}_{ij}e_{ij} = C_{ijkl}e_{ij}\hat{e}_{kl}, \tag{8.119}$$

and

$$\sigma_{ij}\hat{e}_{ij} = C_{ijkl}e_{kl}\hat{e}_{ij} = C_{klij}e_{ij}\hat{e}_{kl} = C_{ijkl}e_{ij}\hat{e}_{kl}. \tag{8.120}$$

Thus,

$$\hat{\sigma}_{ij}e_{ij} = \sigma_{ij}\hat{e}_{ij}. \tag{8.121}$$

Consequently, from (8.118) it follows that

$$\int_S T_i\hat{u}_i\,\mathrm{d}S + \int_V b_i\hat{u}_i\,\mathrm{d}V = \int_S \hat{T}_i u_i\,\mathrm{d}S + \int_V \hat{b}_i u_i\,\mathrm{d}V. \tag{8.122}$$

This is the Betti's reciprocal theorem of linear elasticity. The work of the first type of loading on the displacement due to the second type of loading, is equal to the work of the second type of loading on the displacement due to the first type of loading.

A classical illustration of the application of this theorem is as follows. Consider a uniform rod of length l and cross-sectional area A. The rod is made of isotropic elastic material, with the elastic modulus E and Poisson's ratio ν. Two types of loading applied to this rod are shown in Fig. 8.2. The first loading is a uniform tensile stress σ applied to the ends of the rod, giving rise to axial force $P = \sigma A$. The second loading is a pair of two equal but opposite transverse forces F applied to the lateral surface of the rod, anywhere along the length of the rod. The transverse distance between the points of the application of two forces is h. The Betti's theorem allows us to easily calculate the elongation of the rod due to this pair of forces, without solving the complicated boundary value problem. Indeed, the first type of loading causes the lateral contraction of the rod of magnitude

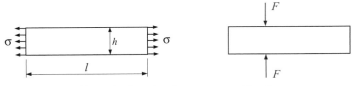

Figure 8.2. A uniform elastic rod under two types of loading.

$\Delta h = \nu(Ph/EA)$. By Betti's theorem, the work of F on the displacement due to $P = \sigma A$, must be equal to the work of P on the displacement due to F, i.e.,

$$Fv\frac{Ph}{EA} = P\Delta. \tag{8.123}$$

Thus, the elongation of the rod due to the pair of forces F is

$$\Delta = \nu\frac{Fh}{EA}. \tag{8.124}$$

8.11 Plane Strain

Plane deformations are defined such that

$$e_{33} = e_{13} = e_{23} = 0. \tag{8.125}$$

If we allow for the possibility of thermal strains, with the thermal expansion coefficient α, the corresponding thermoelastic constitutive equations are

$$\begin{aligned} e_{11} &= \frac{1}{E}[\sigma_{11} - \nu(\sigma_{22} + \sigma_{33})] + \alpha \Delta T, \\ e_{22} &= \frac{1}{E}[\sigma_{22} - \nu(\sigma_{11} + \sigma_{33})] + \alpha \Delta T, \\ e_{33} &= 0 \Rightarrow \sigma_{33} = \nu(\sigma_{11} + \sigma_{22}) - \alpha E \Delta T, \\ e_{12} &= e_{21} = \frac{1}{2G}\sigma_{12}. \end{aligned} \tag{8.126}$$

Suppose that we introduce body forces *via* a potential function, $V(x_1, x_2)$, such that

$$\mathbf{b} = -\partial V/\partial \mathbf{x}. \tag{8.127}$$

Using the compact notation for differentiation, the equations of equilibrium become

$$\begin{aligned} \sigma_{11,1} + \sigma_{12,2} - V_{,1} &= 0 \\ \sigma_{21,1} + \sigma_{22,2} - V_{,2} &= 0. \end{aligned} \tag{8.128}$$

All of the compatibility equations (8.66) are identically satisfied except the first one, which gives

$$e_{11,22} + e_{22,11} = 2e_{12,12}. \tag{8.129}$$

8.11. Plane Strain

Introduce the Airy stress function, $\phi(x_1, x_2)$, such that

$$\sigma_{11} = \phi_{,22} + V(x_1, x_2),$$
$$\sigma_{22} = \phi_{,11} + V(x_1, x_2), \qquad (8.130)$$
$$\sigma_{12} = -\phi_{,12}.$$

With such definitions the equilibrium equations (8.128) are identically satisfied. To determine the governing equations for ϕ we first substitute (8.130) into the constitutive expressions (8.126) and demand the satisfaction of the compatibility relation (8.129). Thus, we substitute

$$e_{11} = \frac{1}{E}\{(1-\nu^2)\phi_{,22} - \nu(1+\nu)\phi_{,11} + [1-\nu(1+2\nu)]V\} + (1+\nu)\alpha\Delta T,$$

$$e_{22} = \frac{1}{E}\{(1-\nu^2)\phi_{,11} - \nu(1+\nu)\phi_{,22} + [1-\nu(1+2\nu)]V\} + (1+\nu)\alpha\Delta T,$$

$$e_{12} = -\frac{1}{2G}\phi_{,12}$$

into (8.129) to get

$$\nabla^4 \phi = -\frac{(1-2\nu)}{(1-\nu)}\nabla^2 V - \frac{\alpha E}{(1-\nu)}\nabla^2 T. \qquad (8.131)$$

If there are no body forces and thermal gradients, or if $\nabla^2 V = 0$ and $\nabla^2 T = 0$, then ϕ satisfies the biharmonic equation

$$\nabla^4 \phi = 0, \quad \nabla^4 \phi \equiv \frac{\partial^4 \phi}{\partial x_1^4} + \frac{2\partial^4 \phi}{\partial x_1^2 \partial x_2^2} + \frac{\partial^4 \phi}{\partial x_2^4}. \qquad (8.132)$$

If the thermal field is such that $\nabla^2 T = 0$, there would be no effect on ϕ, or on the stresses; there would, of course, be an effect on the strains. An example in Section 8.13 will illustrate these effects. For this purpose we consider thermal strains alone – linear superposition allows the addition of the effects of applied loads or displacements.

8.11.1 Plane Stress

The governing equations for plane stress are similar. In that case we set $\sigma_{33} = \sigma_{13} = \sigma_{23} = 0$, and the constitutive equations are

$$e_{11} = \frac{1}{E}(\sigma_{11} - \nu\sigma_{22}), \quad e_{22} = \frac{1}{E}(\sigma_{22} - \nu\sigma_{11}),$$

$$e_{33} = -\frac{\nu}{E}(e_{11} + e_{22}), \quad e_{12} = \frac{1}{2G}\sigma_{12}, \qquad (8.133)$$

$$e_{13} = e_{23} = 0.$$

In the absence of thermal fields, the governing equation for the stress potential is

$$\nabla^4 \phi = -(1-\nu)\nabla^2 V(x_1, x_2). \qquad (8.134)$$

8.12 Governing Equations of Plane Elasticity

In this section we summarize the governing equations of plane elasticity, simultaneously for both plane stress and plane strain. The stress-strain relations are

$$e_{ij} = \frac{1}{2\mu}\left(\sigma_{ij} - \frac{3-\vartheta}{4}\sigma_{kk}\delta_{ij}\right), \quad (i,j = 1, 2), \tag{8.135}$$

where ϑ is the Kolosov constant defined by

$$\vartheta = \begin{cases} 3 - 4\nu, & \text{plane strain}, \\ \dfrac{3-\nu}{1+\nu}, & \text{plane stress}. \end{cases} \tag{8.136}$$

Because the physical range of the Poisson ratio is $0 \le \nu \le 1/2$, the Kolosov constant is restricted to $0 \le \vartheta \le 3$ for plane strain, and $5/3 \le \vartheta \le 3$ for plane stress. The inverted form of (8.135) is

$$\sigma_{ij} = 2\mu\left(e_{ij} + \frac{1}{2}\frac{3-\vartheta}{\vartheta-1}e_{kk}\delta_{ij}\right), \quad (i,j = 1, 2). \tag{8.137}$$

There are two Cauchy equations of motion

$$\frac{\partial \sigma_{ij}}{\partial x_j} + b_i = \rho \frac{\partial^2 u_i}{\partial t^2}, \tag{8.138}$$

and one Saint-Venant's compatibility equation

$$\frac{\partial^2 e_{11}}{\partial x_2^2} + \frac{\partial^2 e_{22}}{\partial x_1^2} = 2\frac{\partial^2 e_{12}}{\partial x_1 \partial x_2}. \tag{8.139}$$

The corresponding Beltrami–Michell comaptibility equation is

$$\nabla^2(\sigma_{11} + \sigma_{22}) = \frac{4}{1+\vartheta}\frac{\partial b_i}{\partial x_i}, \tag{8.140}$$

where ∇^2 is the two-dimensional Laplacian operator. Finally, the two Navier equations of motion are

$$\mu\nabla^2 u_i + \frac{2\mu}{\vartheta-1}\frac{\partial^2 u_j}{\partial x_i \partial x_j} + b_i = \rho \frac{\partial^2 u_i}{\partial t^2}, \quad (i,j=1,2). \tag{8.141}$$

8.13 Thermal Distortion of a Simple Beam

As an illustration of the application of derived equations, consider the following thermoelastic beam problem. Let $T(x_2)$ be the temperature variation specified across the beam shown in Fig. 8.3. Our objective is to calculate the displacements of the beam, and as needed the stresses and strains caused by what amounts to purely thermal loading. For this case, $\nabla^2 T = d^2 T/dx_2^2$, since the gradient is solely through the thickness of the beam, i.e., in its x_2 direction. Equation (8.131) becomes

$$\nabla^4 \phi = -\frac{\alpha E}{(1-\nu)}\frac{d^2 T}{dx_2^2}. \tag{8.142}$$

8.13. Thermal Distortion of a Simple Beam

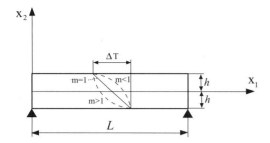

Figure 8.3. A simple beam subject to a thermal gradient.

Consistent with all the boundary conditions specified so far, we can set $\sigma_{22} = \sigma_{12} = 0$, which are typical assumptions of *elementary beam theory*. Then,

$$\partial^4 \phi / \partial x_1^4 = 0,$$
$$\frac{\partial^4 \phi}{\partial x_1^2 \partial x_2^2} = 0. \tag{8.143}$$

The integration of (8.143) is elementary, but illustrative. Toward that end, we write

$$\frac{\partial^2}{\partial x_2^2}\left(\frac{\partial^2 \phi}{\partial x_2^2}\right) = -\frac{\alpha E}{(1-\nu)}\frac{d^2 T}{dx_2^2}. \tag{8.144}$$

Since $\sigma_{11} = \phi_{,22}$, this reduce to

$$\frac{\partial^2}{\partial x_2^2}\left[\sigma_{11} + \frac{\alpha E}{(1-\nu)}T(x_2)\right] = 0. \tag{8.145}$$

Two integrations later, we obtain

$$\sigma_{11} + \frac{\alpha E}{(1-\nu)}T(x_2) = C_1 x_2 + C_2. \tag{8.146}$$

The integration constants, C_1 and C_2, are determined from the requirement of static equilibrium

$$\int_{-h}^{h} \sigma_{11}(x_2)\, dx_2 = 0,$$
$$\int_{-h}^{h} \sigma_{11}(x_2) x_2\, dx_2 = 0. \tag{8.147}$$

The integrations yield

$$\sigma_{11}(x_2) = \frac{\alpha E}{(1-\nu)}\left[-T(x_2) + \frac{1}{2h}\int_{-h}^{h} T(\eta)\, d\eta + \frac{x_2}{2/3h^3}\int_{-h}^{h} \eta T(\eta)\, d\eta\right].$$

The corresponding strains are

$$e_{11}(x_2) = \frac{1}{E}\sigma_{11} + \alpha T(x_2), \quad e_{22}(x_2) = -\frac{\nu}{E}\sigma_{11} + \alpha T(x_2), \quad e_{12} = 0.$$

We now analyze the resulting deflection of the beam. With the usual assumptions of elementary beam theory, we approximate the beam's curvature as

$$1/r \approx -e_{11}/x_2.$$

Thus,
$$(1/r)_{\text{top}} \approx -e_{11}(h)/h, \quad (1/r)_{\text{bottom}} \approx -e_{11}(-h)/(-h), \tag{8.148}$$

and so
$$1/r \approx \frac{e_{11}(-h) - e_{11}(h)}{2h} \approx \frac{\partial^2 u_2}{\partial x_1^2}. \tag{8.149}$$

As a specific example, take
$$T(x_2) = (1/2)^m \Delta T (1 - x_2/h)^m, \tag{8.150}$$

where ΔT is a given difference in temperature between the bottom and top sides of the beam. The analysis is straightforward. In particular, the maximum deflection of the beam is found to be
$$\delta_{\max} \approx \frac{L}{16} \alpha \Delta T (L/h) g(m), \tag{8.151}$$

with
$$g(m) \equiv \frac{6m}{(m+1)(m+2)}. \tag{8.152}$$

The result shows how the shape of the temperature variation is important in affecting the magnitude of the induced thermal distortion. For example, if $m < 1$, say $m \approx 1/3$, then $g \sim 0.6$, which represents an appreciable reduction in the maximum displacement at the midpoint of the beam. On the other hand, if $m \approx 3$, then $g \sim 0.9$, and the reduction in maximum displacement is far less. Thus manipulating the temperature profile may provide a mechanism to mitigate intolerably large thermally induced deflections.

8.14 Suggested Reading

Barber, J. R. (2002), *Elasticity*, 2nd ed., Kluwer, Dordrecht.
Filonenko-Borodich, M. (1964), *Theory of Elasticity*, Noordhoff, Groningen, Netherlands.
Fung, Y.-C. (1965), *Foundations of Solid Mechanics*, Prentice Hall, Englewood Cliffs, New Jersey.
Green, A. E., and Zerna, W. (1954), *Theoretical Elasticity*, Oxford University Press, London.
Landau, L. D., and Lifshitz, E. M. (1986), *Theory of Elasticity*, 3rd English ed., Pergamon, New York.
Little, R. W. (1973), *Elasticity*, Prentice Hall, Englewood Cliffs, New Jersey.
Love, A. E. H. (1944), *A Treatise on the Mathematical Theory of Elasticity*, Dover, New York.
Malvern, L. E. (1969), *Introduction to the Mechanics of a Continuous Medium*, Prentice Hall, Englewood Cliffs, New Jersey.
Musgrave, M. J. P. (1970), *Crystal Acoustics: Introduction to the Study of Elastic Waves and Vibrations in Crystals*, Holden-day, San Francisco.
Muskhelishvili, N. I. (1963), *Some Basic Problems of the Mathematical Theory of Elasticity*, Noordhoff, Groningen, Netherlands.

8.14. Suggested Reading

Shames, I. H., and Cozzarelli, F. A. (1997), *Elastic and Inelastic Stress Analysis*, Taylor & Francis, London.

Sokolnikoff, I. S. (1956), *Mathematical Theory of Elasticity*, 2nd ed., McGraw-Hill, New York.

Timoshenko, S. P., and Goodier, J. N. (1970), *Theory of Elasticity*, 3rd ed., McGraw-Hill, New York.

9 Elastic Beam Problems

9.1 A Simple 2D Beam Problem

Approximate solutions can be developed for two-dimensional boundary value problems in plane stress or plane strain by representing the Airy stress function as a polynomial in x_1 and x_2. This general methodology is developed *via* several specific examples.

To begin, recall the general form of the biharmonic equation for the case of plane strain,

$$\nabla^4 \phi = -\frac{1-2\nu}{1-\nu} \nabla^2 V, \tag{9.1}$$

or, for plane stress,

$$\nabla^4 \phi = -(1-\nu)\nabla^2 V. \tag{9.2}$$

Recall also the connections between ϕ and the stresses, *viz.*,

$$\begin{aligned}\sigma_{11} &= \phi_{,22} + V(x_1, x_2), \\ \sigma_{22} &= \phi_{,11} + V(x_1, x_2), \\ \sigma_{12} &= -\phi_{,12}.\end{aligned} \tag{9.3}$$

To illustrate how an approximate solution may be constructed consider the boundary value problem for the simple beam shown in Fig. 9.1. Here the boundary conditions are specified as

$$\begin{aligned}\sigma_{12} &= 0 \text{ on } x_2 = \pm b, \\ \sigma_{22} &= 0 \text{ on } x_2 = \pm b, \\ \sigma_{11} &= 0 \text{ on } x_1 = 0,\end{aligned} \tag{9.4}$$

and

$$\int_{-b}^{b} \sigma_{12} \, dx_2 = F \text{ on } x_1 = 0. \tag{9.5}$$

The last boundary condition is the so-called integral or global boundary condition, in contrast to point wise boundary conditions of (9.4).

9.2. Polynomial Solutions to $\nabla^4 \phi = 0$

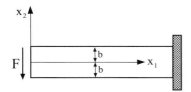

Figure 9.1. A cantilever beam under the end loading.

In the subsequent analysis it is assumed that no potentials act, so that $V = 0$. Also plane stress prevails so that the governing equation for ϕ is simply $\nabla^4 \phi = 0$. We then explore possible polynomial solutions to this biharmonic equation.

9.2 Polynomial Solutions to $\nabla^4 \phi = 0$

Consider the general n^{th} order polynomial

$$p_n(x_1, x_2) = a_0 x_1^n + a_1 x_1^{n-1} x_2 + a_2 x_1^{n-2} x_2^2 + \cdots + a_n x_2^n$$
$$= \sum_{i=0}^{n} a_i x_1^{n-i} x_2^i. \qquad (9.6)$$

If (9.6) is substituted into (9.2), with $V = 0$, a polynomial of the form

$$q_{n-4}(x_1, x_2) = \nabla^4 p_n(x_1, x_2) = \sum_{i=0}^{n-4} b_i x_1^{n-4-i} x_2^i \qquad (9.7)$$

is generated. For example,

$$b_0 = n(n-1)(n-2)(n-3)a_0 + 4(n-2)(n-3)a_2 + 24 a_4, \qquad (9.8)$$

and so on. For (9.6) to be a solution to (9.2), all the $q_n(x_1, x_2)$ must be zero. This requires that

$$b_i = 0, \quad i = 0, 1, \ldots, n-4. \qquad (9.9)$$

Thus, (9.9) are a set of $(n-3)$ linear equations among the $(n+1)$ a_i's. Four a_i's are unknown, *i.e.*, are adjustable constants, so that (9.9) places constraints on the rest.

Now consider what polynomial forms for $\phi(x_1, x_2)$ give rise to associated forms for the σ_{ij}'s. For example, consider the form

$$\phi_2(x_1, x_2) = a_0 x_1^2 + a_1 x_1 x_2 + a_2 x_2^2, \qquad (9.10)$$

which yields

$$\sigma_{11} = 2a_2, \quad \sigma_{22} = 2a_0, \quad \sigma_{12} = -a_1. \qquad (9.11)$$

Consequently, (9.10) implies *a state of uniform inplane stress*.
Next, consider the form

$$\phi_3(x_1, x_2) = a_0 x_1^3 + a_1 x_1^2 x_2 + a_2 x_1 x_2^2 + a_3 x_2^3, \qquad (9.12)$$

which produces the stress components

$$\sigma_{11} = 2a_2 x_1 + 6a_3 x_2,$$
$$\sigma_{22} = 2a_1 x_2 + 6a_0 x_1, \qquad (9.13)$$
$$\sigma_{12} = -2a_1 x_1 - 2a_2 x_2.$$

Suppose the assignment $a_0 = a_1 = a_2 = 0$ is made. Then

$$\sigma_{11} = 6a_3 x_2, \qquad (9.14)$$

which corresponds to a simple state of *pure bending*. Clearly additional perspective can be developed by conjuring up other simple polynomial forms and noting that (9.2) is a linear partial differential equation for $\phi(x_1, x_2)$.

9.3 A Simple Beam Problem Continued

Consider the polynomial form

$$\phi(x_1, x_2) = c_1 x_1 x_2^3 + c_2 x_1 x_2. \qquad (9.15)$$

The fourth-order term, $c_1 x_1 x_2^3$, produces a bending stress of the form $\sigma_{11} \propto x_1 x_2$, which varies linearly with x_1. The second term, $c_2 x_1 x_2$, produces a uniform shear stress, $\sigma_{12} = -c_2$ that can be used to cancel out any otherwise unwanted shear stress arising from other terms. In fact, (9.15) yields

$$\sigma_{11} = 6c_1 x_1 x_2, \quad \sigma_{22} = 0, \quad \sigma_{12} = -3c_1 x_2^2 - c_2. \qquad (9.16)$$

The first boundary condition of (9.4) states that

$$\sigma_{12}(x_2 = \pm b) = 0 = -3c_1 b^2 - c_2 \;\Rightarrow\; c_2 = -3c_1 b^2, \qquad (9.17)$$

and thus

$$\sigma_{12} = 3c_1 (b^2 - x_2^2). \qquad (9.18)$$

The integral boundary condition of (9.5) demands that

$$\int_{-b}^{b} \sigma_{12}\, dx_2 = F \text{ on } x_1 = 0, \qquad (9.19)$$

i.e.,

$$\int_{-b}^{b} 3c_1 (b^2 - x_2^2)\, dx_2 = F \;\Rightarrow\; c_1 = F/(4b^3). \qquad (9.20)$$

Therefore, we arrive at the result

$$\phi(x_1, x_2) = \frac{F(x_1 x_2^3 - 3b^2 x_1 x_2)}{4b^3}. \qquad (9.21)$$

9.3. Simple Beam Problem

Consequently,

$$\sigma_{11} = \frac{3F}{2b^3} x_1 x_2,$$

$$\sigma_{12} = \frac{3F}{4b^3}(b^2 - x_2^2), \tag{9.22}$$

$$\sigma_{22} = 0.$$

Note that the boundary conditions for the beam at the end $x_1 = 0$ are satisfied only on an integral level; point wise satisfaction is not possible with such a simple polynomial solution. However, by Saint-Venant's principle, the solutions that correspond to statically equivalent, although pointwise different, end loads differ appreciably only within the region near the end of the beam.

9.3.1 Strains and Displacements for 2D Beams

The strains are calculated directly from the stresses as

$$e_{11} = \frac{\sigma_{11}}{E} - \nu \frac{\sigma_{22}}{E} = \frac{3F x_1 x_2}{2Eb^3},$$

$$e_{22} = \frac{\sigma_{22}}{E} - \nu \frac{\sigma_{11}}{E} = -\frac{3F\nu x_1 x_2}{2Eb^3}, \tag{9.23}$$

$$e_{12} = \frac{1+\nu}{E} \sigma_{12} = \frac{3F(1+\nu)(b^2 - x_2^2)}{4Eb^3}.$$

The strains may be integrated to obtain the displacements as follows. First, we have

$$e_{11} = \frac{\partial u_1}{\partial x_1} = \frac{3F x_1 x_2}{2Eb^3},$$

$$u_1(x_1, x_2) = \frac{3F x_1^2 x_2}{4Eb^3} + f(x_2), \tag{9.24}$$

where $f(x_2)$ is an arbitrary integration function of x_2. Likewise,

$$e_{22} = \frac{\partial u_2}{\partial x_2} = -\frac{3F\nu x_1 x_2}{2Eb^3},$$

$$u_2(x_1, x_2) = -\frac{3F\nu x_1 x_2^2}{4Eb^3} + g(x_1), \tag{9.25}$$

where $g(x_1)$ is another arbitrary integration function of x_1. Furthermore, we have

$$e_{12} = \frac{1}{2}\left(\frac{\partial u_1}{\partial x_2} + \frac{\partial u_2}{\partial x_1}\right) = \frac{3F(1+\nu)(b^2 - x_2^2)}{4Eb^3}. \tag{9.26}$$

When we substitute the expressions for u_1 and u_2 from above into (9.26) and rearrange terms, we find that

$$\frac{3F x_1^2}{8Eb^3} + \frac{1}{2} g'(x_1) = \frac{3F\nu x_2^2}{8Eb^3} - \frac{1}{2} f'(x_2) + \frac{3F(1+\nu)(b^2 - x_2^2)}{4Eb^3}. \tag{9.27}$$

Rearranged in this way, (9.27) reads

$$\mathcal{R}(x_1) = \mathcal{S}(x_2), \tag{9.28}$$

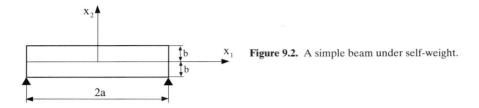

Figure 9.2. A simple beam under self-weight.

which is true only if $\mathcal{R}(x_1) = \mathcal{S}(x_2) = C$, where C is a constant. Thus,

$$g'(x_1, x_2) = -\frac{3Fx_1^2}{4Eb^3} + C,$$

$$f'(x_1, x_2) = \frac{3F\nu x_2^2}{4Eb^3} + \frac{3F(1+\nu)(b^2 - x_2^2)}{2Eb^3} - C. \quad (9.29)$$

This in turn leads to

$$g(x_1) = -\frac{Fx_1^3}{4Eb^3} + Cx_1 + B,$$

$$f(x_2) = \frac{F\nu x_2^3}{4Eb^3} + \frac{F(1+\nu)(3b^2 x_2 - x_2^3)}{2Eb^3} - Cx_2 + A. \quad (9.30)$$

By substituting equations (9.30) into (9.24) and (9.25), it is found that

$$u_1(x_1, x_2) = \frac{3Fx_1^2 x_2}{4Eb^3} + \frac{3F(1+\nu)x_2}{2Eb} - \frac{F(2+\nu)x_2^3}{4Eb^3} + A - Cx_2,$$

$$u_2(x_1, x_2) = -\frac{3F\nu x_1 x_2^2}{4Eb^3} - \frac{Fx_1^3}{4Eb^3} + B + Cx_1. \quad (9.31)$$

Physically, the constants A, B, and C describe rigid body displacements in the x_1 and x_2 directions and a counter-clockwise rotation about the x_3 axis. Possible approximate boundary conditions that may be applied to determine these constants are

$$u_1 = u_2 = \partial u_2/\partial x_1 = 0 \text{ at } x_1 = a, \ x_2 = 0, \text{ or}$$

$$u_1 = u_2 = \partial u_1/\partial x_2 = 0 \text{ at } x_1 = a, \ x_2 = 0, \text{ or} \quad (9.32)$$

$$\int_{-b}^{b} u_1 \, dx_2 = \int_{-b}^{b} u_2 \, dx_2 = \int_{-b}^{b} x_2 u_1 \, dx_2 = 0 \text{ at } x_1 = a.$$

The choice of boundary condition here is motivated by additional consideration of the actual conditions that may prevail in the physical system.

9.4 Beam Problems with Body Force Potentials

An example is now considered that involves a body force potential, in particular one that accounts for the gravitational potential associated with *self-weight*. The simple beam is illustrated in Fig. 9.2; the boundary conditions are described below. As throughout the text, ρ is the mass density, and we introduce g as the gravitational acceleration constant. Thus, the body force potential is

$$V(x_1, x_2) = \rho g x_2, \quad (9.33)$$

9.4. Body Force Potentials

where x_2 may be reckoned from any convenient reference level. This gives rise to the body force density

$$\mathbf{f} = -\frac{\partial V}{\partial \mathbf{x}} = -\rho g \mathbf{e}_2. \tag{9.34}$$

The governing equation for ϕ for the case of plane stress is, in this case,

$$\nabla^4 \phi = -(1-\nu)\nabla^2 V = 0. \tag{9.35}$$

The boundary conditions can be phrased as

$$\begin{aligned} &\sigma_{12}(x_1, \pm b) = \sigma_{22}(x_1, \pm b) = 0, \\ &\sigma_{11}(\pm a, x_2) = 0, \\ &\int_{-b}^{b} \sigma_{12}(-a, x_2)\, dx_2 = -\int_{-b}^{b} \sigma_{12}(a, x_2)\, dx_2 = -2\rho g b a. \end{aligned} \tag{9.36}$$

At this point it is fruitful to consider the symmetry that is obvious in the system. For example, for the normal stresses,

$$\begin{aligned} \sigma_{nn}(x_1, x_2) &\sim \sigma_{nn}(-x_1, x_2) \Leftrightarrow \text{an even function of } x_1, \\ \sigma_{nn}(x_1, x_2) &\sim -\sigma_{nn}(x_1, -x_2) \Leftrightarrow \text{an odd function of } x_2. \end{aligned} \tag{9.37}$$

With these considerations, a stress function of the form

$$\begin{aligned} \phi(x_1, x_2) &= c_{21} x_1^2 x_2 + c_{23} x_1^2 x_2^3 + c_{03} x_2^3 + c_{05} x_2^5 \\ &= \text{odd in } x_2, \text{ even in } x_1, \end{aligned} \tag{9.38}$$

may be tried. Note the subscripting convention used for the coefficients vis-à-vis the exponents of x_1 and x_2. Since

$$\nabla^4 (x_1^2 x_2) = \nabla^4 (x_2^3) = 0, \tag{9.39}$$

the remaining part of (9.38) must satisfy

$$\nabla^4 (c_{23} x_1^2 x_2^3 + c_{05} x_2^5) = 0, \tag{9.40}$$

which leads to

$$24 c_{23} x_2 + 120 c_{05} x_2 = 0, \quad \text{i.e.,} \quad c_{05} = -\frac{1}{5} c_{23}. \tag{9.41}$$

Consequently, the stresses are

$$\begin{aligned} \sigma_{11}(x_1, x_2) &= \rho g x_2 + 6 c_{23} x_1^2 x_2 + 6 c_{03} x_2 + 20 c_{05} x_2^3, \\ \sigma_{22}(x_1, x_2) &= \rho g x_2 + 2 c_{21} x_2 + 2 c_{23} x_2^3, \\ \sigma_{12}(x_1, x_2) &= -2 c_{21} x_1 - 6 c_{23} x_1 x_2^2. \end{aligned} \tag{9.42}$$

Next the boundary conditions are applied. For example,

$$\begin{aligned} \sigma_{12}(x_1, b) &= 0 = -2 c_{21} x_1 - 6 c_{23} x_1 b^2, \\ \sigma_{22}(x_1, b) &= 0 = \rho g b + 2 c_{21} b + 2 c_{23} b^3. \end{aligned} \tag{9.43}$$

Observe the symmetry in σ_{12} and σ_{22}, *viz.*, that σ_{12} is purely even in x_2 and σ_{22} is purely odd in x_2. Thus, the boundary conditions applied on $x_2 = b$ above may be equally well applied on $x_2 = -b$. Together they lead to

$$c_{21} = -\frac{3}{4}\rho g,$$
$$c_{23} = \frac{1}{4}\frac{\rho g}{b^2}, \qquad (9.44)$$
$$c_{05} = -\frac{1}{5}c_{23} = -\frac{1}{20}\frac{\rho g}{b^2}.$$

The boundary condition $\sigma_{11}(\pm a, x_2) = 0$ cannot be satisfied point wise with a polynomial solution of the type considered here. Global equilibrium can, however, be guaranteed by setting

$$\int_{-b}^{b} \sigma_{11}\, dx_2 = 0. \qquad (9.45)$$

Because σ_{11} is an odd function of x_2, this condition supplies no information regarding the evaluation of the remaining coefficient, c_{03}. However, equilibrium also requires that no net moment exist, and this leads to

$$\int_{-b}^{b} \sigma_{11} x_2\, dx_2 = 0 \quad \text{on} \quad x_1 = \pm a. \qquad (9.46)$$

Since σ_{11} is an even function in x_1, (9.46) can be applied at $x_1 = \pm a$. Doing so leads to

$$\frac{2}{3}\rho g b^3 + \rho g a^2 b + 4 c_{03} b^3 - \frac{2}{5}\rho g b^3 = 0,$$
$$c_{03} = -\rho g \left[\frac{1}{15} + \frac{1}{4}(a/b)^2\right]. \qquad (9.47)$$

The final results for the stresses are then

$$\sigma_{11}(x_1, x_2) = \frac{\rho g b}{I}\left[\frac{2}{5}b^2 x_2 - \frac{2}{3}x_2^3 + (x_1^2 - a^2)x_2\right],$$
$$\sigma_{22}(x_1, x_2) = \frac{\rho g b}{I}\left(-\frac{b^2 x_2}{3} + \frac{x_2^3}{3}\right), \qquad (9.48)$$
$$\sigma_{12}(x_1, x_2) = \frac{\rho g b}{I}\left(b^2 x_1 - x_2^2 x_1\right),$$

where $I = 2b^3/3$ is the cross-sectional moment of inertia per unit thickness of the beam.

9.5 Beam under Fourier Loading

The beam illustrated in Fig. 9.3 is to be subject to a general loading along its top surface ($x_2 = c$) and along its bottom surface ($x_2 = -c$). This loading is imagined to have been Fourier analyzed, as described in Chapter 3. Because of the linearity of the biharmonic

9.5. Fourier Loading

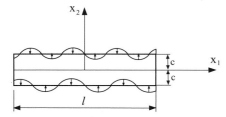

Figure 9.3. Fourier loading on a simple beam.

equation it suffices to develop a general solution for the typical Fourier component. Thus, imagine the loading to be of the form

$$\sigma_{22}(x_1, x_2 = c) = -b_m \sin(\alpha_m x_1),$$
$$\sigma_{22}(x_1, x_2 = -c) = -a_m \sin(\alpha_m x_1),$$
(9.49)

where $\alpha_m = m\pi/\ell$, m is an integer, and ℓ is the length of the beam, as indicated in Fig. 9.3. No body forces are considered and, therefore, solutions of the biharmonic equation

$$\nabla^4 \phi = 0 \tag{9.50}$$

are sought. Assume solutions of the form

$$\phi_m = \sin(\alpha_m x_1) f(x_2). \tag{9.51}$$

When this is substituted into (9.50), the result, after cancelling a redundant $\sin(\alpha_m \pi x_1)$, is

$$\alpha_m^4 f(x_2) - 2\alpha_m^2 f''(x_2) + f''''(x_2) = 0. \tag{9.52}$$

In (9.52) another common convention has been used, namely that primes denote differentiation with respect to the obvious variable, e.g., $f''(x_2) = d^2 f/dx_2^2$.

General solutions to (9.52) are of the form

$$f(x_2) \sim e^{\alpha_m x_2}, \; e^{-\alpha_m x_2}, \; x_2 e^{\pm \alpha_m x_2}. \tag{9.53}$$

It is efficient to form solutions in the form of hyperbolic functions, i.e.,

$$f(x_2) \sim (e^{\alpha_m x_2} + e^{-\alpha_m x_2}), \; (e^{\alpha_m x_2} - e^{-\alpha_m x_2}), \tag{9.54}$$

or

$$f(x_2) \sim \cosh \alpha_m x_2, \; \sinh \alpha_m x_2. \tag{9.55}$$

A suitable stress function is

$$\phi_m = [C_{1m} \cosh(\alpha_m x_2) + C_{2m} \sinh(\alpha_m x_2) \\ + C_{3m} x_2 \cosh(\alpha_m x_2) + C_{4m} x_2 \sinh(\alpha_m x_2)] \sin(\alpha_m x_1). \tag{9.56}$$

This delivers the stresses as

$$\sigma_{11} = \partial^2 \phi / \partial x_2^2 \\ = \sin(\alpha_m x_1)\{C_{1m}\alpha_m^2 \cosh(\alpha_m x_2) + C_{2m}\alpha_m^2 \sinh(\alpha_m x_2) \\ + C_{3m}\alpha_m[2 \sinh(\alpha_m x_2) + \alpha_m x_2 \cosh(\alpha_m x_2)] \\ + C_{4m}\alpha_m[2 \cosh(\alpha_m x_2) + \alpha_m x_2 \sinh(\alpha_m x_2)]\}, \tag{9.57}$$

$$\sigma_{22} = \partial^2 \phi / \partial x_1^2$$
$$= -\alpha_m^2 \sin(\alpha_m x_1)[C_{1m} \cosh(\alpha_m x_2) + C_{2m} \sinh(\alpha_m x_2) \tag{9.58}$$
$$+ C_{3m} x_2 \cosh(\alpha_m x_2) + C_{4m} x_2 \sinh(\alpha_m x_2)],$$
$$\sigma_{12} = -\partial^2 \phi / \partial x_1 \partial x_2$$
$$= -\alpha_m \cos(\alpha_m x_1)\{C_{1m}\alpha_m \sinh(\alpha_m x_2) + C_{2m}\alpha_m \cosh(\alpha_m x_2)$$
$$+ C_{3m}[\cosh(\alpha_m x_2) + \alpha_m x_2 \sinh(\alpha_m x_2)] \tag{9.59}$$
$$+ C_{4m}[\sinh(\alpha_m x_2) + \alpha_m x_2 \cosh(\alpha_m x_2)]\}.$$

The boundary condition $\sigma_{12}(x_1, x_2 = \pm c) = 0$ requires that

$$C_{1m}\alpha_m \sinh(\alpha_m c) + C_{2m}\alpha_m \cosh(\alpha_m c)$$
$$+ C_{3m}[\cosh(\alpha_m c) + \alpha_m c \sinh(\alpha_m c)] \tag{9.60}$$
$$+ C_{4m}[\sinh(\alpha_m c) + \alpha_m c \cosh(\alpha_m c)] = 0,$$

and

$$-C_{1m}\alpha_m \sinh(\alpha_m c) + C_{2m}\alpha_m \cosh(\alpha_m c)$$
$$+ C_{3m}[\cosh(\alpha_m c) + \alpha_m c \sinh(\alpha_m c)] \tag{9.61}$$
$$+ C_{4m}[\sinh(\alpha_m c) + \alpha_m c \cosh(\alpha_m c)] = 0.$$

When rearranged these provide the connections

$$C_{3m} = -C_{2m} \frac{\alpha_m \cosh(\alpha_m c)}{\cosh(\alpha_m c) + \alpha_m c \sinh(\alpha_m c)},$$
$$C_{4m} = -C_{1m} \frac{\alpha_m \sinh(\alpha_m c)}{\sinh(\alpha_m c) + \alpha_m c \cosh(\alpha_m c)}. \tag{9.62}$$

On the other hand, the boundary condition given in (9.49) requires that

$$\alpha_m^2[C_{1m} \cosh(\alpha_m c) + C_{2m} \sinh(\alpha_m c)$$
$$+ C_{3m}c \cosh(\alpha_m c) + C_{4m}c \sinh(\alpha_m c)] = b_m \tag{9.63}$$

and

$$\alpha_m^2[C_{1m} \cosh(\alpha_m c) - C_{2m} \sinh(\alpha_m c)$$
$$- C_{3m}c \cosh(\alpha_m c) + C_{4m}c \sinh(\alpha_m c)] = a_m. \tag{9.64}$$

Consequently, we obtain

$$C_{1m} = \frac{a_m + b_m}{\alpha_m^2} \frac{\sinh(\alpha_m c) + \alpha_m c \cosh(\alpha_m c)}{\sinh(2\alpha_m c) + 2\alpha_m c},$$
$$C_{2m} = -\frac{a_m - b_m}{\alpha_m^2} \frac{\cosh(\alpha_m c) + \alpha_m c \sinh(\alpha_m c)}{\sinh(2\alpha_m c) - 2\alpha_m c},$$
$$C_{3m} = \frac{a_m - b_m}{\alpha_m^2} \frac{\alpha_m \cosh(\alpha_m c)}{\sinh(2\alpha_m c) - 2\alpha_m c}, \tag{9.65}$$
$$C_{4m} = -\frac{a_m + b_m}{\alpha_m^2} \frac{\alpha_m \sinh(\alpha_m c)}{\sinh(2\alpha_m c) + 2\alpha_m c}.$$

Finally, for compactness of notation, define the functions

$$\chi_1 = [\alpha_m c \cosh(\alpha_m c) - \sinh(\alpha_m c)] \cosh(\alpha_m x_2),$$

$$\chi_2 = \alpha_m \sinh(\alpha_m c) x_2 \sinh(\alpha_m x_2),$$

$$\chi_3 = [\alpha_m c \sinh(\alpha_m c) - \cosh(\alpha_m c)] \sinh(\alpha_m x_2),$$

$$\chi_4 = \alpha_m \cosh(\alpha_m c) x_2 \cosh(\alpha_m x_2),$$

$$\chi_5 = [\alpha_m c \cosh(\alpha_m c) + \sinh(\alpha_m c)] \cosh(\alpha_m x_2),$$

$$\chi_6 = \alpha_m \sinh(\alpha_m c) x_2 \sinh(\alpha_m x_2),$$

$$\chi_7 = [\alpha_m c \sinh(\alpha_m c) + \cosh(\alpha_m c)] \sinh(\alpha_m x_2),$$

$$\chi_8 = \alpha_m \cosh(\alpha_m c) x_2 \cosh(\alpha_m x_2),$$

$$\chi_9 = \alpha_m c \cosh(\alpha_m c) \sinh(\alpha_m x_2),$$

$$\chi_{10} = \alpha_m \sinh(\alpha_m c) x_2 \cosh(\alpha_m x_2),$$

$$\chi_{11} = \alpha_m c \sinh(\alpha_m c) \cosh(\alpha_m x_2),$$

$$\chi_{12} = \alpha_m \cosh(\alpha_m c) x_2 \sinh(\alpha_m x_2).$$

With these definitions, the stresses become

$$\sigma_{11} = (a_m + b_m) \frac{\chi_1 - \chi_2}{\sinh(2\alpha_m c) + 2\alpha_m c} \sin(\alpha_m x_1)$$

$$- (a_m - b_m) \frac{\chi_3 - \chi_4}{\sinh(2\alpha_m c) - 2\alpha_m c} \sin(\alpha_m x_1), \tag{9.66}$$

$$\sigma_{22} = -(a_m + b_m) \frac{\chi_5 - \chi_6}{\sinh(2\alpha_m c) + 2\alpha_m c} \sin(\alpha_m x_1)$$

$$+ (a_m - b_m) \frac{\chi_7 - \chi_8}{\sinh(2\alpha_m c) - 2\alpha_m c} \sin(\alpha_m x_1), \tag{9.67}$$

$$\sigma_{12} = -(a_m + b_m) \frac{\chi_9 - \chi_{10}}{\sinh(2\alpha_m c) + 2\alpha_m c} \cos(\alpha_m x_1)$$

$$+ (a_m - b_m) \frac{\chi_{11} - \chi_{12}}{\sinh(2\alpha_m c) - 2\alpha_m c} \cos(\alpha_m x_1). \tag{9.68}$$

9.6 Complete Boundary Value Problems for Beams

In this section we consider a complete boundary value problem for a beam subject to a general Fourier loading, as illustrated in Fig. 9.4. The load applied to the upper surface is now

$$\sigma_{yy} = -q \sin(n\pi x/\ell), \tag{9.69}$$

where ℓ is the length of the beam. Note that in this section coordinate notation has been changed from x_1, x_2 to x, y. This is done to illustrate generality of common notations. The boundary conditions are as depicted in Fig. 9.4 and as applied explicitly below. As before,

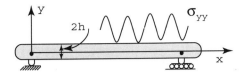

Figure 9.4. Fourier component loading on a simple beam.

we require a stress potential ϕ satisfying the biharmonic equation

$$\nabla^4 \phi = 0. \tag{9.70}$$

As discussed in Chapter 3, trial solutions of the form

$$\phi \sim e^{\alpha x} e^{\beta y}, \quad x e^{\alpha x} e^{\beta y}, \quad \text{or} \quad y e^{\alpha x} e^{\beta y} \tag{9.71}$$

are explored. When such terms are inserted into the biharmonic equation, there results the characteristic equation,

$$\alpha^4 + 2\alpha^2 \beta^2 + \beta^4 = 0 \quad \Rightarrow \quad (\alpha^2 + \beta^2)^2 = 0, \tag{9.72}$$

and thus $\alpha = \pm i\beta$, where β can be either real or imaginary. Therefore, we obtain solutions of the form

$$\begin{aligned}\phi = {} & e^{i\beta x}\left(A e^{\beta y} + B e^{-\beta y} + C y e^{\beta y} + D y e^{-\beta y}\right) \\ & + e^{-i\beta x}\left(A' e^{\beta y} + B' e^{-\beta y} + C' y e^{\beta y} + D' y e^{-\beta y}\right). \end{aligned} \tag{9.73}$$

Of course, the roles of x and y could be interchanged, i.e., $x \rightleftarrows y$, to obtain additional solutions. In addition, terms could be formed such as $\sinh(\beta y) = 1/2(e^{\beta y} - e^{-\beta y})$ or $\cosh(\beta y) = 1/2(e^{\beta y} + e^{-\beta y})$ by simply adding or subtracting exponential terms. At the end, we obtain

$$\begin{aligned}\phi = {} & \sin(\beta x)[A\sinh(\beta y) + B\cosh(\beta y) + C\beta y \sinh(\beta y) + D\beta y \cosh(\beta y)] \\ & + \cos(\beta x)[A'\sinh(\beta y) + B'\cosh(\beta y) + C'\beta y \sinh(\beta y) + D'\beta y \cosh(\beta y)] \\ & + \sin(\alpha y)[E \sinh(\alpha x) + F\cosh(\alpha x) + G\alpha x \sinh(\alpha x) + H\alpha x \cosh(\alpha x)] \\ & + \cos(\alpha y)[E'\sinh(\alpha x) + F'\cosh(\alpha x) + G'\alpha x \sinh(\alpha x) + H'\alpha x \cosh(\alpha x)] \\ & + R_0 + R_1 x + R_2 x^2 + R_3 x^3 + R_4 y + R_5 y^2 + R_6 y^3 + R_7 xy \\ & + R_8 x^2 y + R_9 xy^2. \end{aligned}$$

The boundary conditions include

$$\sigma_{yy}(x, y = h) = -q \sin(n\pi x/\ell), \quad \sigma_{xy}(x, y = h) = 0,$$

$$\sigma_{yy}(x, y = -h) = \sigma_{xy}(x, y = -h) = 0,$$

$$\sigma_{xx}(x=0, y) = 0, \quad \int_{-h}^{h} \sigma_{xy}(x=0, y)\, dy = -\frac{1}{2}\int_0^{\ell} q \sin(n\pi x/\ell)\, dx,$$

$$\sigma_{xx}(x=\ell, y) = 0, \quad \int_{-h}^{h} \sigma_{xy}(x=\ell, y)\, dy = \frac{1}{2}\int_0^{\ell} q \sin(n\pi x/\ell)\, dx.$$

9.6. Complete BVP for Beams

These suggest for ϕ the following form

$$\phi = \sin(\beta x)\left[A\sinh(\beta y) + B\cosh(\beta y) + C\beta y \sinh(\beta y) + D\beta y \cosh(\beta y)\right].$$

This form guarantees that $\sigma_{xx}(x=0, y) = 0$. It will satisfy the boundary condition at $x = \ell$, as well, if the values for β are determined as described below.

For the stresses we obtain

$$\sigma_{yy} = -\beta^2 \sin(\beta x)\left[A\sinh(\beta y) + B\cosh(\beta y) + C\beta y\sinh(\beta y) + D\beta y\cosh(\beta y)\right],$$

$$\sigma_{xy} = -\beta^2\cos(\beta x)\left\{A\cosh(\beta y) + D[\beta y\sinh(\beta y) + \cosh(\beta y)]\right\}$$
$$\quad - \beta^2\cos(\beta x)\left\{B\sinh(\beta y) + C[\beta y\cosh(\beta y) + \sinh(\beta y)]\right\}.$$

The terms have been arranged so that they are either odd or even in y. Thus, to meet the boundary condition $\sigma_{xy}(x, y=\pm h) = 0$, we require

$$A\cosh(\beta h) + D[\beta h\sinh(\beta h) + \cosh(\beta h)] = 0,$$
$$B\sinh(\beta h) + C[\beta h\cosh(\beta h) + \sinh(\beta h)] = 0. \tag{9.74}$$

This leads to

$$A = -D[\beta h\tanh(\beta h) + 1],$$
$$B = -C[\beta h\coth(\beta h) + 1], \tag{9.75}$$

and

$$\sigma_{yy} = -D\beta^2\sin(\beta x)\left\{\beta y\cosh(\beta y) - [\beta h\tanh(\beta h) + 1]\sinh(\beta y)\right\}$$
$$\quad - C\beta^2\sin(\beta x)\left\{\beta y\sinh(\beta y) - [\beta h\coth(\beta h) + 1]\cosh(\beta y)\right\}. \tag{9.76}$$

To ensure $\sigma_{yy}(x, y=-h) = 0$, we set

$$D\left\{\beta h\cosh(\beta h) - [\beta h\tanh(\beta h) + 1]\sinh(\beta h)\right\}$$
$$= C\left\{\beta h\sinh(\beta h) - [\beta h\coth(\beta h) + 1]\cosh(\beta h)\right\}. \tag{9.77}$$

Hence,

$$D\left[\frac{\beta h\cosh^2(\beta h) - \beta h\sinh^2(\beta h)}{\cosh(\beta h)} - \sinh(\beta h)\right]$$
$$= C\left[\frac{\beta h\sinh^2(\beta h) - \beta h\cosh^2(\beta h)}{\sinh(\beta h)} - \cosh(\beta h)\right]. \tag{9.78}$$

Since

$$\cosh^2(\beta h) - \sinh^2(\beta h) = 1,$$

we have

$$C = -\tanh(\beta h)\frac{\beta h - \sinh(\beta h)\cosh(\beta h)}{\beta h + \sinh(\beta h)\cosh(\beta h)}D. \tag{9.79}$$

Defining, for compactness, the parameter

$$\Upsilon \equiv \tanh(\beta h)\frac{\beta h - \sinh(\beta h)\cosh(\beta h)}{\beta h + \sinh(\beta h)\cosh(\beta h)}, \tag{9.80}$$

we can write

$$\sigma_{yy} = -D\beta^2 \sin(\beta x)[\beta y \cosh(\beta y) - (\beta h \tanh(\beta h) + 1)\sinh(\beta y)]$$
$$+ D\beta^2 \sin(\beta x)\Upsilon[\beta y \sinh(\beta y) - (\beta h \coth(\beta h) + 1)\cosh(\beta y)]. \quad (9.81)$$

Since $\sigma_{yy}(x, h) = -q\sin(n\pi x/\ell)$, there follows

$$q \sin(n\pi x/\ell) = 2\beta^2 D \sin(\beta x) \frac{\beta h - \sinh(\beta h)\cosh(\beta h)}{\cosh(\beta h)}. \quad (9.82)$$

For this to be true, we demand that

$$\beta = n\pi/\ell \quad (9.83)$$

and

$$D = \frac{q \cosh(n\pi h/\ell)}{2(n\pi/\ell)^2 [n\pi h/\ell - \sinh(n\pi h/\ell)\cosh(n\pi h/\ell)]},$$
$$C = -\frac{q \sinh(n\pi h/\ell)}{2(n\pi/\ell)^2 [n\pi h/\ell + \sinh(n\pi h/\ell)\cosh(n\pi h/\ell)]}. \quad (9.84)$$

It can be readily verified that indeed $\sigma_{xx}(0, y) = \sigma_{xx}(\ell, y) = 0$.

9.6.1 Displacement Calculations

Recall the connections between the stress potential and the stresses, *viz.*,

$$\sigma_{xx} = \frac{\partial^2 \phi}{\partial y^2}, \quad \sigma_{yy} = \frac{\partial^2 \phi}{\partial x^2}, \quad \sigma_{xy} = -\frac{\partial^2 \phi}{\partial x \partial y}. \quad (9.85)$$

For the two normal strains, and for the case of plane stress considered here, the strain-displacement relations are

$$e_{xx} = \frac{\partial u_x}{\partial x} = \frac{1}{E}\left(\frac{\partial^2 \phi}{\partial y^2} - \nu \frac{\partial^2 \phi}{\partial x^2}\right),$$
$$e_{yy} = \frac{\partial u_y}{\partial y} = \frac{1}{E}\left(\frac{\partial^2 \phi}{\partial x^2} - \nu \frac{\partial^2 \phi}{\partial y^2}\right). \quad (9.86)$$

When integrated, this yields

$$u_x = \frac{1}{E}\left[\int \frac{\partial^2 \phi}{\partial y^2} dx - \nu \frac{\partial \phi}{\partial x} + f(y)\right],$$
$$u_y = \frac{1}{E}\left[\int \frac{\partial^2 \phi}{\partial x^2} dy - \nu \frac{\partial \phi}{\partial y} + g(x)\right], \quad (9.87)$$

where $f(x)$ and $g(y)$ are integration functions. Furthermore, for the shear strain,

$$\frac{\partial u_x}{\partial y} + \frac{\partial u_y}{\partial x} = -\frac{2(1+\nu)}{E}\frac{\partial \phi}{\partial x \partial y}. \quad (9.88)$$

Now, substitute (9.87) into (9.88) and use exponential forms for ϕ of the type $\phi \sim e^{\alpha x}e^{\beta y}$, *i.e.*, ignore polynomial like terms, to obtain

$$(\beta^3/\alpha + 2\alpha\beta + \alpha^3/\beta)e^{\alpha x}e^{\beta y} + f'(y) + g'(x) = 0. \quad (9.89)$$

9.6. Complete BVP for Beams

Observe that

$$\int \frac{\partial^3 \phi}{\partial y^3} dx + 2\frac{\partial^2 \phi}{\partial x \partial y} + \int \frac{\partial^3 \phi}{\partial x^3} dy + f'(y) + g'(x) = 0. \quad (9.90)$$

But,

$$\beta^3/\alpha + 2\alpha\beta + \alpha^3/\beta = \beta^4 + 2\alpha^2\beta^2 + \alpha^4 = 0, \quad (9.91)$$

from (9.72), and thus

$$f'(y) + g'(x) = 0 \implies f'(y) = g'(x) = -\omega. \quad (9.92)$$

Therefore, we have

$$\begin{aligned} f(y) &= -\omega y + u_x^0, \\ g(x) &= -\omega x + u_y^0. \end{aligned} \quad (9.93)$$

The displacements are accordingly

$$\begin{aligned} u_x(x,y) = -\frac{1}{E}\beta \cos(\beta x)\{&A(1+\nu)\sinh(\beta y) + B(1+\nu)\cosh(\beta y) \\ &+ C[(1+\nu)\beta y \sinh(\beta y) + 2\cosh(\beta y)] \\ &+ D[(1+\nu)\beta y \cosh(\beta y) + 2\sinh(\beta y)]\} \\ &- \omega^0 y + u_x^0, \end{aligned} \quad (9.94)$$

and

$$\begin{aligned} u_y(x,y) = -\frac{1}{E}\beta \sin(\beta x)\{&A(1+\nu)\cosh(\beta y) + B(1+\nu)\sinh(\beta y) \\ &+ C[(1+\nu)\beta y \cosh(\beta y) - (1+\nu)\sinh(\beta y)] \\ &+ D[(1+\nu)\beta y \sinh(\beta y) - (1-\nu)\cosh(\beta y)]\} \\ &+ \omega^0 x + u_y^0. \end{aligned} \quad (9.95)$$

Now, referring to Fig. 9.4, set

$$u_y(0,0) = u_y(\ell,0) = u_x(0,0) = 0 \quad (9.96)$$

to obtain

$$\omega^0 = u_y^0 = 0 \quad (9.97)$$

and

$$u_x^0 = \frac{1}{E}\beta[B(1+\nu) + 2C]. \quad (9.98)$$

More general loading, specified by say $k(x)$, may be analyzed by Fourier analysis and by applying linear superposition. If $k(x)$ is an odd function, the solution just obtained will be sufficient; if not, a similar solution involving a boundary condition such as $\sigma_{yy}(x, y = h) = q\cos(n\pi/\ell)$ will be required. The generation of such a solution would follow precisely along the lines of the one obtained above.

9.7 Suggested Reading

Little, R. W. (1973), *Elasticity*, Prentice Hall, Englewood Cliffs, New Jersey.

Sokolnikoff, I. S. (1956), *Mathematical Theory of Elasticity*, 2nd ed., McGraw-Hill, New York.

Timoshenko, S. P., and Goodier, J. N. (1970), *Theory of Elasticity*, 3rd ed., McGraw-Hill, New York.

Ugural, A. C., and Fenster, S. K. (2003), *Advanced Strength and Applied Elasticity*, 4th ed., Prentice Hall, Upper Saddle River, New Jersey.

10 Solutions in Polar Coordinates

10.1 Polar Components of Stress and Strain

We introduce a polar coordinate system as illustrated in Fig. 10.1. The polar coordinates r, θ are related to the cartesian coordinates x, y by

$$r = (x^2 + y^2)^{1/2}, \quad \theta = \tan^{-1}(y/x) \tag{10.1}$$

or, through their inverse,

$$x = r\cos\theta, \quad y = r\sin\theta. \tag{10.2}$$

Thus, derivatives can be formed via (EASIEST TO REFERENCE 10.1 TO FORM $\partial \text{ER } \partial \text{VS}$)

$$\begin{aligned}
\frac{\partial}{\partial x} &= \frac{\partial r}{\partial x}\frac{\partial}{\partial r} + \frac{\partial \theta}{\partial x}\frac{\partial}{\partial \theta} = \cos\theta \frac{\partial}{\partial r} - \frac{\sin\theta}{r}\frac{\partial}{\partial \theta}, \\
\frac{\partial}{\partial y} &= \frac{\partial r}{\partial y}\frac{\partial}{\partial r} + \frac{\partial \theta}{\partial y}\frac{\partial}{\partial \theta} = \sin\theta \frac{\partial}{\partial r} + \frac{\cos\theta}{r}\frac{\partial}{\partial \theta}.
\end{aligned} \tag{10.3}$$

For second derivatives we obtain

$$\frac{\partial^2}{\partial x^2} = \left(\cos\theta\frac{\partial}{\partial r} - \frac{\sin\theta}{r}\frac{\partial}{\partial \theta}\right)^2 = \cos^2\theta\frac{\partial^2}{\partial r^2} + \sin^2\theta\left(\frac{1}{r}\frac{\partial}{\partial r} + \frac{1}{r^2}\frac{\partial^2}{\partial \theta^2}\right)$$
$$+ 2\sin\theta\cos\theta\left(\frac{1}{r^2}\frac{\partial}{\partial \theta} - \frac{1}{r}\frac{\partial^2}{\partial r\partial \theta}\right),$$

$$\frac{\partial^2}{\partial y^2} = \left(\sin\theta\frac{\partial}{\partial r} + \frac{\cos\theta}{r}\frac{\partial}{\partial \theta}\right)^2 = \sin^2\theta\frac{\partial^2}{\partial r^2} + \cos^2\theta\left(\frac{1}{r}\frac{\partial}{\partial r} + \frac{1}{r^2}\frac{\partial^2}{\partial \theta^2}\right)$$
$$- 2\sin\theta\cos\theta\left(\frac{1}{r^2}\frac{\partial}{\partial \theta} - \frac{1}{r}\frac{\partial^2}{\partial r\partial \theta}\right),$$

$$\frac{\partial^2}{\partial x\partial y} = \sin\theta\cos\theta\left(\frac{\partial^2}{\partial r^2} - \frac{1}{r}\frac{\partial}{\partial r} - \frac{1}{r^2}\frac{\partial^2}{\partial \theta^2}\right) - (\cos^2\theta - \sin^2\theta)\left(\frac{1}{r^2}\frac{\partial}{\partial \theta} - \frac{1}{r}\frac{\partial^2}{\partial r\partial \theta}\right).$$

10. Solutions in Polar Coordinates

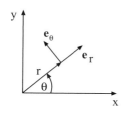
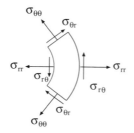

Figure 10.1. The stress components in 2D polar coordinates.

Another need is to express the components of stress, strain, and displacement in terms of the polar coordinates. This is done with the help of Fig. 10.1 that illustrates the definitions of the polar stress components. For example,

$$\sigma_{rr} = \sigma_{xx}\cos^2\theta + \sigma_{yy}\sin^2\theta + 2\sigma_{xy}\sin\theta\cos\theta$$

$$= \cos^2\theta \frac{\partial^2 \phi}{\partial y^2} + \sin^2\theta \frac{\partial^2 \phi}{\partial x^2} - 2\sin\theta\cos\theta \frac{\partial^2 \phi}{\partial x \partial y} \tag{10.4}$$

$$= \frac{1}{r}\frac{\partial \phi}{\partial r} + \frac{1}{r^2}\frac{\partial^2 \phi}{\partial \theta^2}.$$

In an entirely similar fashion it is found that

$$\sigma_{\theta\theta} = \frac{\partial^2 \phi}{\partial r^2} \tag{10.5}$$

and

$$\sigma_{r\theta} = \frac{1}{r^2}\frac{\partial \phi}{\partial \theta} - \frac{1}{r}\frac{\partial^2 \phi}{\partial r \partial \theta} = -\frac{\partial}{\partial r}\left(\frac{1}{r}\frac{\partial \phi}{\partial \theta}\right). \tag{10.6}$$

We now need to construct the field equations, in particular the biharmonic equation. The Laplacian, in this 2D (x, y) frame, is formed as

$$\frac{\partial^2}{\partial x^2} + \frac{\partial^2}{\partial y^2} = \nabla^2 = \frac{\partial^2}{\partial r^2} + \frac{1}{r}\frac{\partial}{\partial r} + \frac{1}{r^2}\frac{\partial^2}{\partial \theta^2}, \tag{10.7}$$

and, thus,

$$\nabla^4 = \nabla^2 \nabla^2 = \left(\frac{\partial^2}{\partial r^2} + \frac{1}{r}\frac{\partial}{\partial r} + \frac{1}{r^2}\frac{\partial^2}{\partial \theta^2}\right)^2. \tag{10.8}$$

As for displacements, we appeal to the definitions of unit base vectors in the polar geometry

$$\mathbf{e}_r = \cos\theta\, \mathbf{e}_x + \sin\theta\, \mathbf{e}_y,$$
$$\mathbf{e}_\theta = -\sin\theta\, \mathbf{e}_x + \cos\theta\, \mathbf{e}_y, \tag{10.9}$$

and write

$$u_x = u_r \cos\theta - u_\theta \sin\theta,$$
$$u_y = u_r \sin\theta + u_\theta \cos\theta. \tag{10.10}$$

For the strains we note, for example, that

$$e_{xx} = \frac{\partial u_x}{\partial x} = \frac{\partial u_r}{\partial r}\cos^2\theta + \left(\frac{u_\theta}{r} - \frac{\partial u_\theta}{\partial r} - \frac{1}{r}\frac{\partial u_r}{\partial \theta}\right)\sin\theta\cos\theta + \left(\frac{u_r}{r} + \frac{1}{r}\frac{\partial u_\theta}{\partial \theta}\right)\sin^2\theta. \tag{10.11}$$

10.2. Plate with Circular Hole

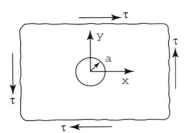

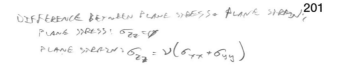

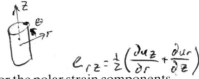

Figure 10.2. Plate with a central hole.

The use of such relations leads to the following expressions for the polar strain components

$$e_{rr} = \frac{\partial u_r}{\partial r},$$

$$e_{r\theta} = \frac{1}{2}\left(\frac{1}{r}\frac{\partial u_r}{\partial \theta} + \frac{\partial u_\theta}{\partial r} - \frac{u_\theta}{r}\right),$$

$$e_{\theta\theta} = \frac{1}{r}\frac{\partial u_\theta}{\partial \theta} + \frac{u_r}{r}.$$

(10.12)

10.2 Plate with Circular Hole

10.2.1 Far Field Shear

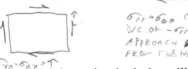

As an example, consider an infinite plate containing a circular hole, as illustrated in Fig. 10.2. The plate is assumed to be loaded by far field shear, which would represent a uniform pure shear had it not been for the hole. Thus at infinite distance from the hole the state of stress is to reduce to uniform pure shear, and the stress state is nonuniform near the hole. We seek solutions of the biharmonic equation that can account for this state of loading. Try solutions of the form

$$\phi(r,\theta) = \sum_{n=0}^{\infty} f_n(r)\cos(n\theta) + \sum_{n=1}^{\infty} g_n(r)\sin(n\theta).$$

(10.13)

Substitution into the biharmonic equation $\nabla^4 \phi = 0$ yields

$$\left(\frac{d^2}{dr^2} + \frac{1}{r}\frac{d}{dr} - \frac{n^2}{r^2}\right)^2 \{f_n(r), g_n(r)\} = 0.$$

(10.14)

Solutions are of two forms, i.e.,

$$f_n(r) = A_{n1}r^{n+2} + A_{n2}r^{-n+2} + A_{n3}r^n + A_{n4}r^{-n}, \quad n \neq 0, 1,$$

(10.15)

and

$$f_0(r) = A_{01}r^2 + A_{02}r^2 \ln r + A_{03}\ln r + A_{04}, \quad n = 0,$$

$$f_1(r) = A_{11}r^3 + A_{12}r \ln r + A_{13}r + A_{14}r^{-1}, \quad n = 1.$$

(10.16)

Similar solutions exist for $g_n(r)$. These solution forms may be used to construct the solution. Because the surface of the hole is traction free, the boundary condition there becomes

$$\sigma_{rr} = \sigma_{\theta r} = 0 \quad \text{on } r = a,$$

(10.17)

whereas far from the hole

$$\sigma_{xy} \to \tau \quad \text{as } r \to \infty, \tag{10.18}$$

where τ is the magnitude of the remotely applied shear stress. Because the far field stress state is pure shear,

$$\sigma_{xx}, \sigma_{yy} \to 0 \quad \text{as } r \to \infty. \tag{10.19}$$

The problem is most easily solved by using an elementary application of the principle of linear superposition. We imagine two problems, one of a homogeneous plate subject to pure far field shear and that of a plate with a hole whose surface is subject to traction; the traction to be imposed in this second problem will be such as to precisely annihilate the traction that would exist on a circular surface of the same radius in the homogeneous plate of the first problem. When the two solutions are added, the hole's surface becomes traction free. Moreover, as will be seen, the stresses of the second problem vanish far from the hole, so that the far field stress state for the combined problems becomes only the uniform pure shear of the first problem.

For problem (1), involving pure uniform shear, we take

$$\phi^{(1)} = -\tau xy = -\tau r^2 \sin\theta \cos\theta = -\frac{r^2}{2}\tau \sin(2\theta), \tag{10.20}$$

which clearly produces a simple state of uniform pure shear stress of magnitude τ. The corresponding stresses are

$$\sigma_{rr}^{(1)} = \tau \sin(2\theta), \quad \sigma_{r\theta}^{(1)} = \tau \cos(2\theta). \tag{10.21}$$

In problem (2) we apply the stresses $\sigma_{rr} = -\tau \sin(2\theta)$ and $\sigma_{r\theta} = -\tau \cos(2\theta)$ to the surface of the hole. To accomplish this, consider potential of the form

$$\phi^{(2)}(r,\theta) = A\sin(2\theta) + Br^{-2}\cos(2\theta), \tag{10.22}$$

which is from the inventory listed in (10.15) and (10.16). The choice of functions involving $\sin(2\theta)$ and $\cos(2\theta)$ may be taken as arrived at by *trial and error*, but is clearly motivated by similar term in the boundary conditions. The stresses associated with this potential are

$$\sigma_{rr}^{(2)} = -\left(\frac{4A}{r^2} + \frac{6B}{r^4}\right)\sin(2\theta),$$

$$\sigma_{r\theta}^{(2)} = \left(\frac{2A}{r^2} + \frac{6B}{r^4}\right)\cos(2\theta), \tag{10.23}$$

$$\sigma_{\theta\theta}^{(2)} = \frac{6B}{r^4}\sin(2\theta).$$

Next, by evaluating $\sigma_{rr}^{(2)}(r=a)$ and $\sigma_{r\theta}^{(2)}(r=a)$ and setting the results equal to $-\tau \sin(2\theta)$ and $-\tau \cos(2\theta)$, respectively, it is found that $A = \tau a^2$ and $B = -\tau a^4/2$.

The combined stresses, therefore, become

$$\sigma_{rr} = \sigma_{rr}^{(1)} + \sigma_{rr}^{(2)} = \tau\left(1 - 4a^2/r^2 + 3a^4/r^4\right)\sin(2\theta),$$

$$\sigma_{r\theta} = \sigma_{r\theta}^{(1)} + \sigma_{r\theta}^{(2)} = \tau\left(1 + 2a^2/r^2 - 3a^4/r^4\right)\cos(2\theta), \tag{10.24}$$

$$\sigma_{\theta\theta} = \sigma_{\theta\theta}^{(1)} + \sigma_{\theta\theta}^{(2)} = -\tau\left(1 + 3a^4/r^4\right)\sin(2\theta).$$

10.2. Plate with Circular Hole

It is interesting to examine the implications of such solutions *vis-à-vis* stress concentrations. Of particular interest is the normal stress that occurs at a perimeter of the hole corresponding to $\theta = 3\pi/4$, where $\sin(2\theta) = -1$. The maximum hoop stress there is $\sigma_{\theta\theta} = 4\tau$. As the state of stress at this point is uniaxial tension, because $\sigma_{rr} = \sigma_{r\theta} = 0$ on the hole's surface, the maximum shear stress at this point is simply $\tau_{\max} = 2\tau$.

10.2.2 Far Field Tension

The second problem of this type that is of interest involves an identical plate subject to far field tension (or compression). Thus the boundary conditions are the same as (10.17) on the surface of the hole, but become

$$\sigma_{xx} \to \sigma \quad \text{at } r \to \infty. \tag{10.25}$$

At $r \to \infty$ all other components of stress vanish.

The problem may be solved using an identical superposition approach. For the first problem we imagine a homogeneous plate subject to uniform far field tension; the potential corresponding to this state of stress is

$$\phi^{(1)} = \frac{1}{2}\sigma y^2 = \frac{1}{2}\sigma r^2 \sin^2\theta = \frac{1}{4}\sigma[r^2 - r^2\cos(2\theta)]. \tag{10.26}$$

$r^2 \cos(\phi \cdot \theta) \Rightarrow n = \emptyset$

Consider a second potential prescribed as

$$\phi^{(2)} = A\sigma \ln r + B\sigma \cos(2\theta) + C\sigma r^{-2}\cos(2\theta), \tag{10.27}$$

so that, when the two solutions are combined,

$$\phi = \frac{1}{4}\sigma[r^2 - r^2\cos(2\theta)] + A\sigma \ln r + B\sigma \cos(2\theta) + C\sigma r^{-2}\cos(2\theta). \tag{10.28}$$

The stresses computed from this potential are

$$\sigma_{rr} = \sigma\left[\frac{1}{2} + \frac{\cos(2\theta)}{2} + \frac{A}{r^2} - \frac{4B\cos(2\theta)}{r^2} - \frac{6C\cos(2\theta)}{r^4}\right],$$

$$\sigma_{r\theta} = -\sigma\left[\frac{\sin(2\theta)}{2} + \frac{2B\sin(2\theta)}{r^2} + \frac{6C\sin(2\theta)}{r^4}\right], \tag{10.29}$$

$$\sigma_{\theta\theta} = \sigma\left[\frac{1}{2} - \frac{\cos(2\theta)}{2} - \frac{A}{r^2} + \frac{6C\cos(2\theta)}{r^4}\right].$$

In setting the boundary condition $\sigma_{rr}(r = a) = 0$, note that two equations are generated from the first of (10.29), *viz.*,

$$\frac{1}{2} + \frac{A}{a^2} = 0,$$

$$\frac{1}{2} - \frac{4B}{a^2} - \frac{6C}{a^4} = 0. \tag{10.30}$$

These arise from the need to independently set the terms involving $\cos(2\theta)$ to zero. The second traction free boundary condition gives rise to a single equation

$$\frac{1}{2} + \frac{2B}{a^2} + \frac{6C}{a^4} = 0. \tag{10.31}$$

When the equations are solved, we find

$$A = -a^2/2, \quad B = a^2/2, \quad C = -a^4/4. \tag{10.32}$$

The resulting stresses are

$$\sigma_{rr} = \frac{\sigma}{2}\left(1 - \frac{a^2}{r^2}\right) + \frac{\sigma\cos(2\theta)}{2}\left(\frac{3a^4}{r^4} - \frac{4a^2}{r^2} + 1\right),$$

$$\sigma_{r\theta} = \frac{\sigma\sin(2\theta)}{2}\left(\frac{3a^4}{r^4} - \frac{2a^2}{r^2} - 1\right), \tag{10.33}$$

$$\sigma_{\theta\theta} = \frac{\sigma}{2}\left(1 + \frac{a^2}{r^2}\right) - \frac{\sigma\cos(2\theta)}{2}\left(\frac{3a^4}{r^4} + 1\right).$$

Of particular interest is the stress concentration at the point on the hole's surface at $\theta = \pm\pi/2$. The maximum hoop stress there is $\sigma_{\theta\theta}^{\max} = 3\sigma$.

10.3 Degenerate Cases of Solution in Polar Coordinates

We again seek solutions to the equation

$$\nabla^4 \phi = \left(\frac{\partial^2}{\partial r^2} + \frac{1}{r}\frac{\partial}{\partial r} + \frac{1}{r^2}\frac{\partial^2}{\partial \theta^2}\right)^2 \phi = 0 \tag{10.34}$$

in the form

$$\phi = \sum_{n=0}^{\infty} f_n(r)\cos(n\theta) + \sum_{n=1}^{\infty} g_n(r)\sin(n\theta). \tag{10.35}$$

We examine the terms involving $\cos(n\theta)$ for the moment. This leads to

$$\left(\frac{d^2}{dr^2} + \frac{1}{r}\frac{d}{dr} - \frac{n^2}{r^2}\right)^2 f_n(r) = 0. \tag{10.36}$$

4th order O.D.E. → need 4 solutions

In general, solutions exist of the type

$$f_n(r) = A_{n1} r^{n+2} + A_{n2} r^{-n+2} + A_{n3} r^n + A_{n4} r^{-n}, \tag{10.37}$$

as we have seen previously. But, if $n = 0$, two sets of solutions become degenerate, *viz.*, $(A_{n1} r^{n+2}, A_{n2} r^{-n+2})$ and $(A_{n3} r^n, A_{n4} r^{-n})$. If, on the other hand, $n = 1$, two other solutions become degenerate, *viz.*, $(A_{n2} r^{-n+2}, A_{n3} r^n)$. In such cases we seek to find a resolution in terms of alternative solutions to replace those that are degenerate.

Let $n = 1 + \epsilon$, that is relax the condition that n be an integer. Then the two potentially degenerate forms for $n = 1$ become

$$f(r) = Ar^{1-\epsilon} + Br^{1+\epsilon}, \tag{10.38}$$

and as $\epsilon \to 0$ the two forms become degenerate. Now, recast $f(r)$ as

$$f(r) = C(r^{1+\epsilon} + r^{1-\epsilon}) + D(r^{1+\epsilon} - r^{1-\epsilon}). \tag{10.39}$$

10.3. Degenerate Cases

In the limit as $\epsilon \to 0$, we clearly have

$$\lim_{\epsilon \to 0} C(r^{1+\epsilon} + r^{1-\epsilon}) \to 2Cr,$$

$$\lim_{\epsilon \to 0} D(r^{1+\epsilon} - r^{1-\epsilon}) \to 0. \tag{10.40}$$

Take $D = E\epsilon^{-1}$ and form again the limit to find that

$$\lim_{\epsilon \to 0} E\epsilon^{-1}(r^{1+\epsilon} - r^{1-\epsilon}) \to Er \ln r. \quad \text{(BY DEFINITION OF ln)} \tag{10.41}$$

Thus, for $n = 1$,

$$f_1(r) = A_{11} r^3 + A_{12} r \ln r + A_{13} r + A_{14} r^{-1}. \tag{10.42}$$

For $n = 0$, we find

$$\lim_{n \to 0} \frac{d(r^{n+2})}{dn} = r^2 \ln r,$$

$$\lim_{n \to 0} \frac{d(r^n)}{dn} = \ln r, \quad \text{RECALL FROM CALC:} \quad \frac{d a^{u(x)}}{dx} = a^u \left(\frac{du/dx}{\ln a} \right) \tag{10.43}$$

so that

$$f_0(r) = A_{01} r^2 + A_{02} r^2 \ln r + A_{03} \ln r + A_{04}. \tag{10.44}$$

There are, however, additional aspects of the analysis.

When $n = 0$, there are solutions of the form $\phi = A$, for which there are no stresses. We want instead solutions that give rise to stresses of the Fourier form, which would come about, say, from potentials of the type

$$\phi = Ar^\epsilon \cos(\epsilon\theta) + Br^\epsilon \sin(\epsilon\theta), \tag{10.45}$$

with the limit

$$\lim_{\epsilon \to 0} [Ar^\epsilon \cos(\epsilon\theta) + Br^\epsilon \sin(\epsilon\theta)] = A. \tag{10.46}$$

Furthermore,

$$\lim_{\epsilon \to 0} \frac{d}{d\epsilon} [Ar^\epsilon \cos(\epsilon\theta) + Br^\epsilon \sin(\epsilon\theta)] = A \ln r + B\theta. \tag{10.47}$$

The term $B\theta$ yields stresses

$$\sigma_{rr} = \sigma_{\theta\theta} = 0, \quad \sigma_{r\theta} = B/r^2. \tag{10.48}$$

When $n = 1$, the terms in $r\cos\theta$ and $r\sin\theta$ yield no stresses; a similar procedure uncovers the alternative solution [to (10.45)–(10.48)]

$$\phi = Br\theta \sin\theta + Cr\theta \cos\theta, \tag{10.49}$$

which has the associated stresses

$$\sigma_{r\theta} = \sigma_{\theta\theta} = 0, \quad \sigma_{rr} = \frac{2B\cos\theta}{r} - \frac{2C\sin\theta}{r}. \tag{10.50}$$

CAN PLAY WITH SOLNS Θ TO GET USEFUL (i.e. Θ=∅ IS SAME AS Θ=2π, ETC.)

The general solution of interest is, therefore,

$$\begin{aligned}\phi(r,\theta) = &\, A_{01}r^2 + A_{02}r^2 \ln r + A_{03} \ln r + A_{04}\theta \\
&+ \left(A_{11}r^3 + A_{12}r \ln r + A_{14}r^{-1}\right)\cos\theta + A_{13}r\theta\sin\theta \\
&+ \left(B_{11}r^3 + B_{12}r \ln r + B_{14}r^{-1}\right)\sin\theta + B_{13}r\theta\cos\theta \\
&+ \sum_{n=2}^{\infty}\left(A_{n1}r^{n+2} + A_{n2}r^{-n+2} + A_{n3}r^n + A_{n4}r^{-n}\right)\cos(n\theta) \\
&+ \sum_{n=2}^{\infty}\left(B_{n1}r^{n+2} + B_{n2}r^{-n+2} + B_{n3}r^n + B_{n4}r^{-n}\right)\sin(n\theta).
\end{aligned} \quad (10.51)$$

It should not go unnoticed that terms $r\theta\sin\theta$ and $r\theta\cos\theta$ are not single valued. This is a feature we exploit in the next section.

10.4 Curved Beams: Plane Stress

Consider the equations of equilibrium. In polar coordinates, for cases of plane stress, we have

$$\begin{aligned}\frac{\partial \sigma_{rr}}{\partial r} + \frac{\sigma_{rr} - \sigma_{\theta\theta}}{r} + \frac{1}{r}\frac{\partial \sigma_{r\theta}}{\partial \theta} + b_r &= 0, \\
\frac{\partial \sigma_{r\theta}}{\partial r} + \frac{2\sigma_{r\theta}}{r} + \frac{1}{r}\frac{\partial \sigma_{\theta\theta}}{\partial \theta} + b_\theta &= 0.\end{aligned} \quad (10.52)$$

As before, if there are no body forces, the Airy stress function satisfies

$$\nabla^4 \phi = \nabla^2 \nabla^2 \phi = \left(\frac{\partial^2}{\partial r^2} + \frac{1}{r}\frac{\partial}{\partial r} + \frac{1}{r^2}\frac{\partial^2}{\partial \theta^2}\right)^2 \phi = 0. \quad (10.53)$$

The connections to the stress components are

$$\begin{aligned}\sigma_{rr} &= \frac{1}{r}\frac{\partial \phi}{\partial r} + \frac{1}{r^2}\frac{\partial^2 \phi}{\partial \theta^2}, \\
\sigma_{\theta\theta} &= \frac{\partial^2 \phi}{\partial r^2}, \\
\sigma_{r\theta} &= -\frac{\partial}{\partial r}\left(\frac{1}{r}\frac{\partial \phi}{\partial \theta}\right).\end{aligned} \quad (10.54)$$

The strain-displacement relations are

$$\begin{aligned}e_{rr} &= \frac{\partial u_r}{\partial r}, \\
e_{r\theta} &= \frac{1}{2}\left(\frac{1}{r}\frac{\partial u_r}{\partial \theta} + \frac{\partial u_\theta}{\partial r} - \frac{u_\theta}{r}\right), \\
e_{\theta\theta} &= \frac{1}{r}\frac{\partial u_\theta}{\partial \theta} + \frac{u_r}{r}.\end{aligned} \quad (10.55)$$

10.4. Curved Beams

To complete the set of governing equations, the elastic constitutive relations for plane stress are used,

$$\epsilon_{rr} = \frac{1}{E}(\sigma_{rr} - \nu\sigma_{\theta\theta}),$$

$$\epsilon_{\theta\theta} = \frac{1}{E}(\sigma_{\theta\theta} - \nu\sigma_{rr}), \tag{10.56}$$

$$\epsilon_{r\theta} = \frac{1+\nu}{E}\sigma_{r\theta}. \quad = \frac{1}{2G}\sigma_{r\theta}$$

Consider solutions of a form that are dependent solely on r, i.e., $\phi = \phi(r)$ only. In this case, and using again the convention whereby a comma denotes differentiation, the biharmonic equation, in the absence of body forces, reduces to

$$\frac{1}{r}\left\{r\left[\frac{1}{r}(r\phi_{,r})_{,r}\right]_{,r}\right\}_{,r} = 0. \tag{10.57}$$

This equation may be integrated four times in a straightforward manner to yield

$$r\left[\frac{1}{r}(r\phi_{,r})_{,r}\right]_{,r} = C_1, \rightarrow \left[\frac{1}{r}(r\phi_{,r})_{,r}\right]_{,r} = \frac{C_1}{r}$$

$$\frac{1}{r}(r\phi_{,r})_{,r} = C_1 \ln r + C_2, \rightarrow (r\phi_{,r})_{,r} = C_1 r \ln r + C_2 r$$

$$r\phi_{,r} = C_1\left(\frac{r^2}{2}\ln r - \frac{r^2}{2}\right) + C_2\frac{r^2}{2} + C_3,$$

$$\phi_{,r} = C'_1 r \ln r + C'_2 r + C_3/r, \quad C'_1 = C_1/2, \quad C'_2 = (C_2 - C_1)/2,$$

$$\phi = C'_1\left(\frac{r^2}{2}\ln r - \frac{r^2}{2}\right) + C'_2\frac{r^2}{2} + C_3 \ln r + C_4.$$

In a slightly simpler form, the last expression becomes

$$\phi = A \ln r + B r^2 \ln r + C r^2 + D. \tag{10.58}$$

This form for ϕ gives rise to stresses (FROM 10.54)

$$\sigma_{rr} = A/r^2 + 2B \ln r + 2C,$$
$$\sigma_{\theta\theta} = -A/r^2 + 2B \ln r + 3B + 2C, \tag{10.59}$$
$$\sigma_{r\theta} = 0.$$

As for the strains, first note that

$$e_{r\theta} = \frac{1+\nu}{E}\sigma_{r\theta} = \frac{1}{2}\left(\frac{1}{r}\frac{\partial u_r}{\partial \theta} + \frac{\partial u_\theta}{\partial r} - \frac{u_\theta}{r}\right) = 0. \tag{10.60}$$

The radial strain is

$$e_{rr} = \frac{\partial u_r}{\partial r} = \frac{1}{E}\left[\frac{A(1+\nu)}{r^2} + 2C(1-\nu) + B(1-\nu)(2\ln r + 1) - 2\nu B\right]. \quad (10.61)$$

For the hoop strain we find

$$\begin{aligned}e_{\theta\theta} &= \frac{u_r}{r} + \frac{1}{r}\frac{\partial u_\theta}{\partial \theta} \\ &= \frac{1}{E}\left[-\frac{A(1+\nu)}{r^2} + 2C(1-\nu) + B(1-\nu)(2\ln r - 1) + 2B\right].\end{aligned} \quad (10.62)$$

Equations (10.61) and (10.62) provide two paths for evaluating u_r. Integrating (10.61) it is found that

$$u_r = \frac{1}{E}\left[-\frac{A(1+\nu)}{r} + 2C(1-\nu)r + B(1-\nu)(2r\ln r - r) - 2\nu Br\right] + f(\theta),$$

where $f(\theta)$ is an arbitrary function of integration. Solving for u_r in (10.62), on the other hand, gives

$$u_r = \frac{1}{E}\left[-\frac{A(1+\nu)}{r} + 2C(1-\nu)r + B(1-\nu)(2r\ln r + r) + 2Br\right] - \partial u_\theta/\partial \theta.$$

For last two expressions to be both expressions for u_r, it must be that

$$f(\theta) = -\partial u_\theta/\partial \theta + 4Br/E. \quad (10.63)$$

If we take $f(\theta) = 0$, we obtain

$$\partial u_\theta/\partial \theta = 4Br/E \quad \rightarrow \quad u_\theta = 4Br\theta/E. \quad (10.64)$$

With this choice for $f(\theta)$, (10.60) shows that $\partial u_r/\partial \theta = 0$, as expected and as is consistent with the original assumption that $u_r = u_r(r)$ only. Note also, if such nonsingle valued solutions as $u_\theta = (1/E)(4Br\theta)$ are disallowed, $B = 0$. The corresponding strains are

$$e_{rr} = \frac{du_r}{dr} = \frac{1}{E}\left[\frac{A(1+\nu)}{r^2} + 2C(1-\nu) + B(1-\nu)(2\ln r + 1) - 2\nu B\right], \quad (10.65)$$

and

$$e_{\theta\theta} = \frac{u_r}{r} = \frac{1}{E}\left[-\frac{A(1+\nu)}{r^2} + 2C(1-\nu) + B(1-\nu)(2\ln r + 1) + 2B\right]. \quad (10.66)$$

To derive the radial displacement, integrate the first of these and solve the second to obtain

$$u_r = \frac{1}{E}\left[-\frac{A(1+\nu)}{r} + 2C(1-\nu)r + B(1-\nu)(2r\ln r - r) - 2\nu Br\right] + Y,$$

where Y is a constant, and

$$u_r = \frac{1}{E}\left[-\frac{A(1+\nu)}{r} + 2C(1-\nu)r + B(1-\nu)(2r\ln r + r) + 2Br\right].$$

For these two expressions to be the same, we must have $B = Y = 0$.

10.4. Curved Beams

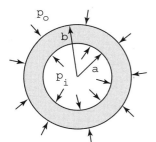

Figure 10.3. Hollow cylinder subject to imposed internal and external pressure.

10.4.1 Pressurized Cylinder

Consider the case of a hole in a thin plate subject to both external and internal pressure, as depicted in Fig. 10.3. The boundary conditions are

$$\sigma_{rr} = -p_i \quad \text{on} \quad r = a,$$
$$\sigma_{rr} = -p_o \quad \text{on} \quad r = b. \tag{10.67}$$

Take for the potential

$$\phi = A \ln r + C r^2 + D, \tag{10.68}$$

which gives the stresses

$$\sigma_{rr} = A/r^2 + 2C,$$
$$\sigma_{\theta\theta} = -A/r^2 + 2C, \tag{10.69}$$
$$\sigma_{r\theta} = 0.$$

Setting the boundary conditions prescribed above gives

$$\sigma_{rr}(r=a) = A/a^2 + 2C = -p_i,$$
$$\sigma_{rr}(r=b) = A/b^2 + 2C = -p_o. \tag{10.70}$$

When these are solved for A and C, it is found that

$$\sigma_{rr} = \frac{a^2 b^2}{b^2 - a^2} \frac{p_o - p_i}{r^2} + \frac{a^2 p_i - b^2 p_o}{b^2 - a^2},$$
$$\sigma_{\theta\theta} = -\frac{a^2 b^2}{b^2 - a^2} \frac{p_o - p_i}{r^2} + \frac{a^2 p_i - b^2 p_o}{b^2 - a^2}. \tag{10.71}$$

Some interesting special cases are obtained as follows. Let $t = (b-a)$ and $R_m = (b+a)/2$; this means that, for example,

$$b = R_m + \frac{t}{2}, \quad a = R_m - \frac{t}{2}, \quad b^2 - a^2 = 2t R_m. \tag{10.72}$$

Now, let $p_o = 0$ and assume $t/R_m \ll 1$. Then,

$$\sigma_{\theta\theta} = \frac{p_i R_m}{2t}\left(1 + \frac{R_m^2}{r^2} - \frac{t}{R_m}\right) + \mathcal{O}(t^2/R_m^2). \tag{10.73}$$

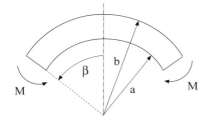

Figure 10.4. Curved beam under the bending moment M.

Recall the simple *membrane* approximation, viz., $\sigma_{\theta\theta} \approx p_i R_m/t$ for the nominal stress across the annulus wall of such a cylinder with a central hole. This may be obtained from (10.73) by neglecting the term t/R_m and evaluating the expression at $r = R_m$.

10.4.2 Bending of a Curved Beam

Consider the curved beam shown in Fig. 10.4. The beam is loaded at its ends by a bending moment M. The boundary conditions to be imposed are

$$\sigma_{rr} = \sigma_{r\theta} = 0 \quad \text{on} \quad r = a, b,$$

$$\int \sigma_{\theta\theta} \, dr = 0 \quad \text{at} \quad \theta = \pm\beta, \tag{10.74}$$

$$\int \sigma_{\theta\theta} r \, dr = -M \quad \text{at} \quad \theta = \pm\beta.$$

As the beam does not close on itself, it is possible to accept solutions that are characterized by nonsingle valued displacements, u_θ. Thus the coefficient B in (10.58) may not be zero! Consequently, by using (10.58), evaluating the stresses, and setting the above boundary conditions, it is found that

$$A/a^2 + 2B\ln a + B + 2C = 0,$$
$$A/b^2 + 2B\ln b + B + 2C = 0,$$
$$b\left(A/b^2 + 2B\ln b + B + 2C\right) - a\left(A/a^2 + 2B\ln a + B + 2C\right) = 0,$$
$$-A\ln(b/a) + B\left(b^2 \ln b - a^2 \ln a\right) + C(b^2 - a^2) = -M.$$

Solving the above, we find

$$A = -\frac{4M}{D} a^2 b^2 \ln(b/a),$$

$$B = -\frac{2M}{D}(b^2 - a^2), \tag{10.75}$$

$$C = \frac{M}{D}[(b^2 - a^2) + 2(b^2 \ln b - a^2 \ln a)],$$

$$D = (b^2 - a^2)^2 - 4a^2 b^2 \ln(b/a).$$

10.5. Axisymmetric Deformations

The corresponding stresses are

$$\sigma_{rr} = -\frac{4M}{D}\left[\frac{a^2b^2}{r^2}\ln(b/a) + b^2\ln(r/b) + a^2\ln(a/r)\right],$$

$$\sigma_{\theta\theta} = -\frac{4M}{D}\left[-\frac{a^2b^2}{r^2}\ln(b/a) + b^2\ln(r/b) + (b^2 - a^2)\right], \quad (10.76)$$

$$\sigma_{r\theta} = 0.$$

10.5 Axisymmetric Deformations

We have already encountered geometries where polar coordinates are particularly suited, namely those where the geometry is symmetric about some axis – we call that, as before, the z axis. Here we explicitly consider some cases of such axisymmetry. Suppose that the displacements are independent of the polar coordinate θ. If it happens that the stress potential is also independent of θ, then the biharmonic equation for ϕ becomes

$$\left(\frac{d^4}{dr^4} + \frac{2}{r}\frac{d^3}{dr^3} - \frac{1}{r^2}\frac{d^2}{dr^2} + \frac{1}{r^3}\frac{d}{dr}\right)\phi = 0. \quad (10.77)$$

General solutions to this reduced equation have already been developed, but an alternative approach is to introduce a change in coordinates to ξ, where ξ is defined, *via* the polar coordinate r, as

$$r = e^{\xi}. \quad (10.78)$$

In this case (10.77) is simplified to

$$\left(\frac{d^4}{d\xi^4} - 4\frac{d^3}{d\xi^3} + 4\frac{d^2}{d\xi^2}\right)\phi = 0. \quad (10.79)$$

The general solution, in terms of ξ and r, is

$$\phi = A\xi e^{2\xi} + Be^{2\xi} + C\xi + D$$
$$= Ar^2 \ln r + Br^2 + C\ln r + D. \quad (10.80)$$

We next recall the connections between the polar components of stress and the stress potential, *viz.*, relations (10.4), (10.5), and (10.6), which when specialized to the axisymmetric case considered here are

$$\sigma_{rr} = \frac{1}{r}\frac{d\phi}{dr},$$

$$\sigma_{\theta\theta} = \frac{d^2\phi}{dr^2}, \quad (10.81)$$

$$\sigma_{r\theta} = 0.$$

Substituting (10.80) into (10.81), we obtain

$$\sigma_{rr} = 2A\ln r + \frac{C}{r^2} + A + 2B,$$

$$\sigma_{\theta\theta} = 2A\ln r - \frac{C}{r^2} + 3A + 2B, \tag{10.82}$$

$$\sigma_{r\theta} = 0.$$

For solutions of this type in a simply connected body, e.g., a solid cylinder, the constants A and C must be zero so that the stresses are bounded at $r = 0$. For a multiply connected body, e.g., a hollow cylinder, this is not required. For such cases, we must examine the displacement field, as has been done earlier.

For the case of axisymmetry, relations (10.12) reduce to

$$e_{rr} = \frac{du_r}{dr},$$

$$e_{\theta\theta} = \frac{u_r}{r}, \tag{10.83}$$

$$e_{r\theta} = 0.$$

The constitutive relations become, for the case of plane stress ($\sigma_{zz} = 0$),

$$\frac{du_r}{dr} = \frac{1}{E}(\sigma_{rr} - \nu\sigma_{\theta\theta}),$$

$$\frac{u_r}{r} = \frac{1}{E}(\sigma_{\theta\theta} - \nu\sigma_{rr}). \tag{10.84}$$

For plane strain, the moduli become $1/E \leftarrow (1-\nu^2)/E$ and $\nu \leftarrow \nu/(1-\nu^2)$, while $\sigma_{zz} = \nu(\sigma_{rr} + \sigma_{\theta\theta})$. Using relations (10.82) in (10.84), we find from the first of (10.84) that

$$\frac{du_r}{dr} = \frac{1}{E}\left[2A\ln r + \frac{C}{r^2} + A + 2B - \nu\left(2A\ln r - \frac{C}{r^2} + 3A + 2B\right)\right].$$

When integrated, this becomes

$$u_r = \frac{1}{E}\left[2Ar\ln r - Ar + 2Br - \frac{C}{r} - \nu\left(2Ar\ln r + Ar + 2Br + \frac{C}{r}\right) + H\right],$$

where H is an integration constant. The second of (10.84), on the other hand, gives

$$\frac{u_r}{r} = \frac{1}{E}\left[2A\ln r - \frac{C}{r^2} + 3A + 2B - \nu\left(2A\ln r + \frac{C}{r^2} + A + 2B\right)\right].$$

When the two expressions for u_r are equated, there follows

$$4Ar - H = 0, \tag{10.85}$$

which leads to the conclusion that $A = H = 0$.

For axisymmetric deformations, there is no circumferential displacement ($u_\theta = 0$), and the equations of equilibrium, in the absence of body forces, lead directly to

$$\frac{d^2u_r}{dr^2} + \frac{1}{r}\frac{du_r}{dr} - \frac{u_r}{r^2} = 0, \tag{10.86}$$

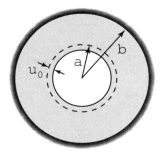

Figure 10.5. Hollow cylinder subject to an imposed displacement on its inner circular surface. The outer surface at $r = b$ is fixed.

which has the general solution

$$u_r = Kr + S\frac{1}{r}. \tag{10.87}$$

As an example, consider the case of a hollow cylinder, as in Fig. 10.5, that is subject to an imposed displacement, u_0, on its inner surface. The outer surface is fixed. Thus, the boundary conditions are $u_r = u_0$ on $r = a$, and $u_r = 0$ on $r = b$. This leads from the above to the solution

$$u_r = \frac{au_0}{b^2 - a^2}\left(\frac{b^2}{r} - r\right). \tag{10.88}$$

The strains and stresses may now be calculated directly.

10.6 Suggested Reading

Doghri, I. (2000), *Mechanics of Deformable Solids*, Springer-Verlag, Berlin.
Filonenko-Borodich, M. (1964), *Theory of Elasticity*, Noordhoff, Groningen, Netherlands.
Fung, Y.-C. (1965), *Foundations of Solid Mechanics*, Prentice Hall, Englewood Cliffs, New Jersey.
Little, R. W. (1973), *Elasticity*, Prentice Hall, Englewood Cliffs, New Jersey.
Love, A. E. H. (1944), *A Treatise on the Mathematical Theory of Elasticity*, Dover, New York.
Muskhelishvili, N. I. (1963), *Some Basic Problems of the Mathematical Theory of Elasticity*, Noordhoff, Groningen, Netherlands.
Sokolnikoff, I. S. (1956), *Mathematical Theory of Elasticity*, 2nd ed., McGraw-Hill, New York.
Timoshenko, S. P., and Goodier, J. N. (1970), *Theory of Elasticity*, 3rd ed., McGraw-Hill, New York.

11 Torsion and Bending of Prismatic Rods

In this chapter we present the Saint-Venant's theory of torsion and bending of prismatic rods. We first consider torsion of the prismatic rod by two concentrated torques at its ends, and then analyze bending of a cantilever beam by a concentrated transverse force at its end.

11.1 Torsion of Prismatic Rods

We consider here the torsional deformation of prismatic rods subject to a torque or imposed angle of *twist* per unit length along the rod. The coordinate system, as illustrated in Fig. 11.1, has the z axis along the rod and the rod's cross section in the x-y plane. The rod is deformed such that adjacent cross-sectional planes are rotated with respect to each other about the z axis; the angle of rotation ϕ is such that $d\phi = \theta dz$ where θ is the so-called *twist*, or angle of rotation per unit length along the rod's axis. The deformation is assumed to be infinitesimal in that $\theta r \ll 1$, where r represents a characteristic transverse dimension of the rod. If the rotation is small, the change in a typical radius vector $\mathbf{r}$, laying in a transverse plane, is given by

$$\delta\mathbf{r} = \delta\boldsymbol{\phi} \times \mathbf{r}, \quad = \text{DISPLACEMENT VECTOR} \tag{11.1}$$

where $\delta\boldsymbol{\phi} = \delta\phi\mathbf{e}_z$ is a vector along the z axis (measured from the mid cross section of the rod) and has a magnitude equal to the amount of rotation. The vectors $\mathbf{r}$ and $\delta\mathbf{r}$ are, respectively,

$$\mathbf{r} = x\mathbf{e}_x + y\mathbf{e}_y,$$
$$\delta\mathbf{r} = u_x\mathbf{e}_x + u_y\mathbf{e}_y. \tag{11.2}$$

Because $\delta\boldsymbol{\phi} = \theta z$, the components of the in-plane displacement vector are

$$u_x = -\theta z y, \quad u_y = \theta z x. \tag{11.3}$$

The rod will undergo deformation in the z direction, as well; this is assumed small and proportional to the twist θ. The displacement in the z direction is then given as

$$u_z = \theta \psi(x, y), \tag{11.4}$$

11.1. Torsion of Prismatic Rods

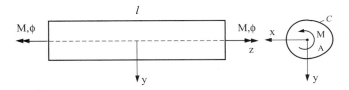

Figure 11.1. Prismatic rod in torsion. The applied torque is M, and the angle of rotation, relative to the midsection of the rod, is ϕ.

where ψ is a function of x, y, referred to as the *torsion function*. Thus each cross section of the rod becomes warped as it was originally flat. The components of strain can be calculated from

$$e_{ij} = \frac{1}{2}\left(\frac{\partial u_i}{\partial x_j} + \frac{\partial u_j}{\partial x_i}\right). \tag{11.5}$$

When this is done, the nonzero components of strain are

$$e_{xz} = \frac{1}{2}\theta\left(\frac{\partial \psi}{\partial x} - y\right), \quad e_{yz} = \frac{1}{2}\theta\left(\frac{\partial \psi}{\partial y} + x\right). \tag{11.6}$$

The application of the isotropic linear elastic constitutive relations shows that the two nonzero components of stress are

$$\sigma_{xz} = 2\mu e_{xz} = \mu\theta\left(\frac{\partial \psi}{\partial x} - y\right), \quad \sigma_{yz} = 2\mu e_{yz} = \mu\theta\left(\frac{\partial \psi}{\partial y} + x\right). \tag{11.7}$$

Given that only σ_{xz} and σ_{yz} are nonzero, the equations of equilibrium reduce to

$$\frac{\partial \sigma_{xz}}{\partial x} + \frac{\partial \sigma_{yz}}{\partial y} = 0, \tag{11.8}$$

which, after substituting (11.7), becomes

$$\nabla^2 \psi = 0, \tag{11.9}$$

where ∇^2 is the Laplacian with respect to x, y.

It is convenient at this stage to introduce two new stress potentials; the first is defined *via* its connection to the stresses as

$$\sigma_{xz} = 2\mu\theta \frac{\partial \chi}{\partial y}, \quad \sigma_{yz} = -2\mu\theta \frac{\partial \chi}{\partial x}. \tag{11.10}$$

If (11.10) is compared to (11.7), the relations between ψ and χ are found to be

$$\frac{\partial \psi}{\partial x} = y + 2\frac{\partial \chi}{\partial y}, \quad \frac{\partial \psi}{\partial y} = -x - 2\frac{\partial \chi}{\partial x}. \tag{11.11}$$

If the first of (11.11) is differentiated with respect to y, and the second with respect to x, and the resulting equations subtracted, it is found that χ must satisfy the Poisson's equation

$$\nabla^2 \chi = -1. \tag{11.12}$$

It remains to specify the boundary conditions to be applied to χ. We assume that the side surfaces of the rod are traction free, as indicated in Fig. 11.1. Thus, on the rod's surface we have

$$\boldsymbol{\sigma} \cdot \mathbf{n} = \mathbf{0} \Rightarrow \sigma_{xz} n_x + \sigma_{yz} n_y = 0, \tag{11.13}$$

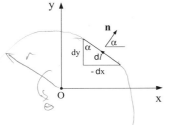

Figure 11.2. The unit normal $\mathbf{n} = \{n_x, n_y\}$ to an infinitesimal arc of length dl. By geometry, $n_x = \cos\alpha = dy/dl$ and $n_y = \sin\alpha = -dx/dl$.

where $\mathbf{n}$ is the outward pointing normal to the rod's side surface, and n_x and n_y are its x and y components. When (11.10) is substituted into the above it is found that

$$\frac{\partial \chi}{\partial y} n_x - \frac{\partial \chi}{\partial x} n_y = 0. \tag{11.14}$$

If $d\ell$ is an increment of arc length around the circumference of the rod then the normal's components are also given by the relations $n_x = dy/d\ell$ and $n_y = -dx/d\ell$ (Fig. 11.2). Using these in (11.14), it is found that

$$\frac{\partial \chi}{\partial x} dx + \frac{\partial \chi}{\partial y} dy = d\chi = 0 = \frac{d\chi}{d\ell} d\ell. \tag{11.15}$$

In other words, χ is a constant along the circumference. Since the stresses (and strains) are related to derivatives of χ, the constant in question may be taken as zero,

$$\chi = 0 \text{ on the edges of cross section.} \tag{11.16}$$

This result holds for a simply connected rod, *i.e.*, one possessing a single, continuous outer edge.

11.2 Elastic Energy of Torsion

We examine here the elastic energy of torsion. The elastic energy density is given as

$$W = \frac{1}{2} \sigma_{ij} e_{ij} = \sigma_{xz} e_{xz} + \sigma_{yz} e_{yz} = \frac{1}{2\mu}(\sigma_{xz}^2 + \sigma_{yz}^2), \tag{11.17}$$

or, in terms of χ,

$$W = 2\mu\theta^2 \left[\left(\frac{\partial \chi}{\partial x}\right)^2 + \left(\frac{\partial \chi}{\partial y}\right)^2\right] = 2\mu\theta^2 (\text{grad }\chi)^2, \tag{11.18}$$

where grad is the gradient with respect to the in-plane coordinates x, y. The total elastic energy is obtained by integration of W over the volume of the rod. The energy per unit length along the rod is

$$\mathcal{W} = 2\mu\theta^2 \int_A (\text{grad }\chi)^2 \, dA = \frac{1}{2} C\theta^2. \tag{11.19}$$

The constant C is the *torsional rigidity*, because the applied moment is linear in the twist, *i.e.*, $M \propto \theta$, and thus

$$\mathcal{W} = \int M \, d\theta = \int C\theta \, d\theta = \frac{1}{2} C\theta^2. \tag{11.20}$$

This defines C as the torsional rigidity in the relation $M = C\theta$.

BFOR ASN'T, ONLY ONE EQN CHANGES FROM ORIGINAL FORM

11.3. Rectangular Cross Section 217

But, there is a relation, *VECTOR IDENTITY:*

$$(\text{grad } \chi)^2 = \text{div}(\chi \text{grad } \chi) - \chi \nabla^2 \chi = \text{div}(\chi \text{grad } \chi) + \chi, \quad (11.21)$$

which may be verified by noting that $\nabla^2 \chi = -1$, and *FROM (11.12)*

$$\frac{\partial}{\partial x_j}\left(\chi \frac{\partial \chi}{\partial x_j}\right) = \frac{\partial \chi}{\partial x_j}\frac{\partial \chi}{\partial x_j} + \chi \nabla^2 \chi. \quad (11.22)$$

= (grad χ)·(grad χ) = (grad χ)²

Thus,

$$C = 4\mu \int_A (\text{grad } \chi)^2 \, dA = 4\mu \int_A [\text{div}(\chi \text{ grad } \chi) + \chi] \, dA. \quad (11.23)$$

The first integral contained in (11.23) vanishes, which can be be seen by transforming it using the divergence theorem,

$$\int_A \text{div}(\chi \text{ grad } \chi) \, dA = \oint_C \chi \frac{\partial \chi}{\partial n} \, d\ell = 0,$$

↳ BOUNDING SURFACE

because $\chi = 0$ on C. Thus, C is given by

$$C = 4\mu \int_A \chi \, dA. \quad (11.24)$$

Finally, we introduce the second stress potential, φ, in analogy with the function used in the previous section, viz., $\varphi \equiv 2\mu\theta\chi$. In terms of φ the above equations become

$$\begin{aligned} &\sigma_{xz} = \partial\varphi/\partial y, \quad \sigma_{yz} = -\partial\varphi/\partial x, \\ &\nabla^2\varphi = -2\mu\theta, \\ &M = 2\int_A \varphi \, dA, \\ &C = 2\frac{1}{\theta}\int_A \varphi \, dA. \end{aligned} \quad (11.25)$$

The function φ is known as the Prandtl stress function. In the next section we will use the Prandtl stress function and (11.25) as the basis for solving two classic torsional boundary value problems.

11.3 Torsion of a Rod with Rectangular Cross Section

Let $\mathcal{H} \equiv -2\mu\theta$ so that the governing equation for φ becomes

$$\begin{aligned} \nabla^2\varphi &= \mathcal{H}, \quad \text{within } C, \\ \varphi &= 0 \quad \text{on } C. \end{aligned} \quad (11.26)$$

We are concerned with a rectangular cross section, as illustrated in Fig. 11.3, so that $|x| \leq a$ and $|y| \leq b$. The boundary conditions for φ are

$$\varphi = 0 \text{ on } x = \pm a, \; |y| \leq b \quad \text{and} \quad \varphi = 0 \text{ on } y = \pm b, \; |x| \leq a. \quad (11.27)$$

Define the function $\tilde{\varphi}$ as

$$\tilde{\varphi} \equiv \frac{1}{2}\mathcal{H}(x^2 - a^2), \quad (11.28)$$

(NEED A NON ZERO BOUNDARY CONDITION → ... (11.41)

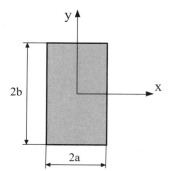

Figure 11.3. Rod of rectangular cross section.

such that

$$\nabla^2 \tilde{\varphi} = \mathcal{H} \tag{11.29}$$

and

$$\tilde{\varphi} = 0 \text{ on } x = \pm a. \tag{11.30}$$

With so defined $\tilde{\varphi}$, we express φ as

$$\varphi(x, y) = \varphi_1(x, y) + \frac{1}{2}\mathcal{H}(x^2 - a^2). \tag{11.31}$$

Thus, the boundary value problem is transformed into

$$\begin{aligned} \nabla^2 \varphi_1 &= 0 \text{ on } A, \\ \varphi_1 &= 0 \text{ on } x = \pm a, \\ \varphi_1 &= -\frac{1}{2}\mathcal{H}(x^2 - a^2) \text{ on } y = \pm b. \end{aligned} \tag{11.32}$$

We seek a solution by separation of variables as

$$\varphi_1(x, y) = f(x)g(y). \tag{11.33}$$

Upon substitution into (11.32), we find

$$\nabla^2 \varphi_1 = d^2 f(x)/dx^2 \, g(y) + f \, d^2 g(y)/dy^2 = 0, \tag{11.34}$$

which, when rearranged, yields

$$\frac{1}{f}d^2 f/dx^2 = -\frac{1}{g}d^2 g/dy^2 = -\lambda^2, \tag{11.35}$$

where λ is a constant. The two equations that follow for $f(x)$ and $g(y)$ are

$$\begin{aligned} d^2 f/dx^2 + \lambda^2 f &= 0, \\ f(x) &= A\cos(\lambda x) + B\sin(\lambda x), \end{aligned} \tag{11.36}$$

and

$$\begin{aligned} d^2 g/dy^2 - \lambda^2 g &= 0, \\ g(y) &= C\cosh(\lambda y) + D\sinh(\lambda y). \end{aligned} \tag{11.37}$$

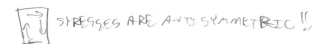

11.3. Rectangular Cross Section

To proceed, we recall the connections to stresses and note that σ_{xz} should be an odd function of y, because of symmetry, and that σ_{yz} should be an odd function of x, for the same reason. Thus, since $\sigma_{xz} = \partial\varphi/\partial y = \partial\varphi_1/\partial y$, and $\sigma_{yz} = -\partial\varphi/\partial x = \partial\varphi_1/\partial x - \mathcal{H}x$, we have $B = D = 0$ in (11.36) and (11.37). This means that

$$\varphi_1(x, y) = \bar{A}\cos(\lambda x)\cosh(\lambda y). \tag{11.38}$$

Furthermore, by imposing the boundary condition on $x = \pm a$, it is found that

$$\varphi_1 = 0 \text{ on } x = \pm a \Rightarrow$$
$$\cos(\lambda a) = 0 \Rightarrow \lambda = \frac{n\pi}{2a}, \quad n = 1, 3, 5, \ldots. \tag{11.39}$$

The general solution is, therefore,

$$\varphi_1(x, y) = \sum_{n=1,3,5,\ldots}^{\infty} A_n \cos\left(\frac{n\pi x}{2a}\right)\cosh\left(\frac{n\pi y}{2a}\right). \tag{11.40}$$

Imposing the boundary condition on $y = \pm b$ leads to

$$\sum_{n=1,3,5,\ldots}^{\infty} A_n \cos\left(\frac{n\pi x}{2a}\right)\cosh\left(\frac{n\pi b}{2a}\right) = -\frac{1}{2}\mathcal{H}(x^2 - a^2). \tag{11.41}$$

The constants A_n are easily determined by multiplying both sides of (11.41) by $\cos\left(\frac{m\pi x}{2a}\right)$ and integrating over $[-a, a]$. This yields

$$\sum_{n=1,3,5,\ldots}^{\infty} A_n \cosh\left(\frac{n\pi b}{2a}\right)\int_{-a}^{a}\cos\left(\frac{n\pi x}{2a}\right)\cos\left(\frac{m\pi x}{2a}\right)dx \quad \rightarrow \text{ORTHOGONAL TO } \cos\left(\frac{n\pi x}{2a}\right)$$
$$= -\frac{1}{2}\mathcal{H}\int_{-a}^{a}(x^2 - a^2)\cos\left(\frac{m\pi x}{2a}\right)dx. \tag{11.42}$$

Noting that

$$\int_{-a}^{a}\cos\left(\frac{n\pi x}{2a}\right)\cos\left(\frac{m\pi x}{2a}\right)dx = \begin{cases} 0, & \text{if } m \neq n, \\ a, & \text{if } m = n, \end{cases} \tag{11.43}$$

we deduce

$$A_m \cosh\left(\frac{m\pi b}{2a}\right)a = -\frac{1}{2}\mathcal{H}\int_{-a}^{a}(x^2 - a^2)\cos\left(\frac{m\pi x}{2a}\right)dx, \tag{11.44}$$

i.e.,

$$A_m = \frac{16\mathcal{H}a^2(-1)^{(m-1)/2}}{\pi^3 m^3 \cosh\left(\frac{m\pi b}{2a}\right)}. \tag{11.45}$$

Consequently, the function φ is

$$\varphi(x, y) = \frac{1}{2}\mathcal{H}(x^2 - a^2)$$
$$+ \frac{16\mathcal{H}a^2}{\pi^3}\sum_{n=1,3,5,\ldots}^{\infty}\frac{(-1)^{(n-1)/2}\cos\left(\frac{n\pi x}{2a}\right)\cosh\left(\frac{n\pi y}{2a}\right)}{n^3 \cosh\left(\frac{n\pi b}{2a}\right)}. \tag{11.46}$$

The corresponding stresses are

$$\sigma_{xz} = \frac{8\mathcal{H}a}{\pi^2} \sum_{n=1,3,5,\ldots}^{\infty} \frac{(-1)^{(n-1)/2} \cos\left(\frac{n\pi x}{2a}\right) \sinh\left(\frac{n\pi y}{2a}\right)}{n^2 \cosh\left(\frac{n\pi b}{2a}\right)}, \quad (11.47)$$

and

$$\sigma_{yz} = -\mathcal{H}x + \frac{8\mathcal{H}a}{\pi^2} \sum_{n=1,3,5,\ldots}^{\infty} \frac{(-1)^{(n-1)/2} \sin\left(\frac{n\pi x}{2a}\right) \cosh\left(\frac{n\pi y}{2a}\right)}{n^2 \cosh\left(\frac{n\pi b}{2a}\right)}. \quad (11.48)$$

Assume, with no loss in generality, that $a \leq b$. The maximum shear stress then occurs at $x = \pm a$, $y = 0$, and is equal to

$$\tau_{\max} = \sigma_{yz}(a,0) = -\mathcal{H}a \left[1 - \frac{8}{\pi^2} \sum_{n=1,3,5,\ldots}^{\infty} \frac{(-1)^{(n-1)/2} \sin\left(\frac{n\pi}{2}\right)}{n^2 \cosh\left(\frac{n\pi b}{2a}\right)}\right], \quad (11.49)$$

or, since $\sin\left(\frac{n\pi}{2}\right) = 1$ for the odd integer values of n involved,

$$\tau_{\max} = \sigma_{yz}(a,0) = -\mathcal{H}a \left[1 - \frac{8}{\pi^2} \sum_{n=1,3,5,\ldots}^{\infty} \frac{1}{n^2 \cosh\left(\frac{n\pi b}{2a}\right)}\right]. \quad (11.50)$$

Here, it is convenient to write

$$\tau_{\max} = -\mathcal{H}ak,$$

$$k = 1 - \frac{8}{\pi^2} \sum_{n=1,3,5,\ldots}^{\infty} \frac{1}{n^2 \cosh\left(\frac{n\pi b}{2a}\right)}. \quad (11.51)$$

The applied moment is

$$M = 2\int_A \varphi \, dA$$

$$= \mathcal{H}2b \int_{-a}^{a} (x^2 - a^2) \, dx \quad (11.52)$$

$$+ \frac{32\mathcal{H}a^2}{\pi^3} \sum_{n=1,3,5,\ldots}^{\infty} \frac{(-1)^{(n-1)/2} I(a,b,n)}{n^3 \cosh\left(\frac{n\pi b}{2a}\right)},$$

where

$$I(a,b,n) = \int_{-a}^{a} \cos\left(\frac{n\pi x}{2a}\right) dx \int_{-b}^{b} \cosh\left(\frac{n\pi y}{2a}\right) dy. \quad (11.53)$$

When the simple integrations are done, we find

$$M = -\frac{1}{2}\mathcal{H}\frac{(2a)^3(2b)}{3}\left[1 - \frac{192}{\pi^5}\frac{a}{b}\sum_{n=1,3,5,\ldots}^{\infty} \frac{\tanh\left(\frac{n\pi b}{2a}\right)}{n^5}\right]. \quad (11.54)$$

Once again, it is convenient to define a factor

$$k_1(b/a) = 1 - \frac{192}{\pi^5}\frac{a}{b}\sum_{n=1,3,5,\ldots}^{\infty} \frac{\tanh\left(\frac{n\pi b}{2a}\right)}{n^5}, \quad (11.55)$$

11.4. Elliptical Cross Section

so that

$$M = -\frac{1}{2}\mathcal{H}\frac{(2a)^3(2b)}{3}k_1(b/a). \tag{11.56}$$

With these, the maximum shear stress can be expressed as

$$\tau_{max} = -\mathcal{H}ak = \frac{2Mak}{(2a)^3(2b)k_1} = \frac{M}{(2a)^2(2b)}(k/k_1)$$
$$= \left(\frac{1}{k_2}\right)\frac{M}{(2a)^2(2b)}, \tag{11.57}$$

where $k_2 = k_1/k$. We list a table of computed values for the factors k, k_1 and k_2.

b/a	k	k_1	k_2
1.0	0.675	0.1406	0.208
2.0	0.930	0.229	0.246
3.0	0.985	0.263	0.267
4.0	0.997	0.281	282
5.0	0.999	0.291	0.291
∞	1.0	0.333	0.333

11.4 Torsion of a Rod with Elliptical Cross Section

Consider a rod with an elliptical cross section as shown in Fig. 11.4. The rod's periphery is given as

$= F\left(\eta(x,y)\right)$

BUT $F(\emptyset) = \emptyset$

$$\frac{x^2}{a^2} + \frac{y^2}{b^2} - 1 = 0. \tag{11.58}$$

A suitable stress function is

$$\varphi = B\left(\frac{x^2}{a^2} + \frac{y^2}{b^2} - 1\right), \tag{11.59}$$

where the constant B is readily determined by substitution of (11.59) into the governing equation developed previously. The result of this is

$$\mathcal{H} = \nabla^2\varphi = 2B\frac{a^2 + b^2}{a^2b^2},$$
$$\Rightarrow B = \mathcal{H}\frac{a^2b^2}{2(a^2 + b^2)}. \tag{11.60}$$

The torsional moment is computed from

$$M = 2\int_A \varphi \, dA, \tag{11.61}$$

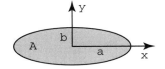

Figure 11.4. Elliptical cross section of a rod subjected to torsion.

which gives

$$\mathcal{H} = -\frac{2(a^2 + b^2)}{a^3 b^3 \pi} M. \tag{11.62}$$

For the stresses, we obtain

$$\sigma_{xz} = \frac{\partial \varphi}{\partial y} = -\frac{2y}{\pi a b^3} M,$$
$$\sigma_{yz} = -\frac{\partial \varphi}{\partial x} = \frac{2x}{\pi a b^3} M. \tag{11.63}$$

Without loss of generality, assume that $b < a$; in that case the maximum shear stress occurs at $y = b$, and is given by

$$\tau_{\max} = \frac{2M}{\pi a b^2}. \tag{11.64}$$

The evaluation of the displacement function, $u_z(x, y)$, is straightforward and leads to the result

$$u_z(x, y) = \frac{M}{G} \frac{b^2 - a^2}{\pi a^3 b^3} xy. \tag{11.65}$$

In the case of a circular cross section, *i.e.*, where $a = b$, there is no longitudinal displacement so that $u_z(x, y) = 0$. The only nonvanishing component of stress is the shear stress

$$\sigma_{z\theta} = \frac{M}{I_0} r, \quad I_0 = \frac{\pi a^4}{2}. \tag{11.66}$$

11.5 Torsion of a Rod with Multiply Connected Cross Sections

We have shown that the stress function, φ, must be constant on the boundary (or boundaries) of a rod whose side surfaces are stress free. For a simply connected shaft, *i.e.*, a solid shaft, the constant was set to 0. For a multiply connected cross section in a hollow shaft, φ is also to be constant, but the constant values on the various contours of the cross section are not the same. Such a geometry is depicted in Fig. 11.5.

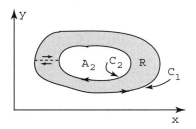

Figure 11.5. Cross section of a multiply connected rod subject to torsion. R denotes the area of the load bearing cross section between in the inner surface contour C_2, and the outer surface contour C_1. Note the continuous sense of the contour integration.

11.5. Multiply Connected Cross Sections

To establish the augmented boundary conditions on φ, consider the contour integral of the warping displacement,

$$\oint_{C_2} du_z = 0 = \oint_{C_2} \left(\frac{\partial u_z}{\partial x} dx + \frac{\partial u_z}{\partial y} dy \right). \tag{11.67}$$

The integral is equal to 0 because u_z is a single valued function. If we substitute from (11.7), we obtain

$$\oint_{C_2} du_z = \frac{1}{G} \oint_{C_2} (\sigma_{xz} dx + \sigma_{yz} dy) - \theta \oint_{C_2} (x\, dy - y\, dx)$$
$$= \frac{1}{G} \oint_{C_2} \tau\, ds - 2\theta A_2. \tag{11.68}$$

In (11.68) the third integral is easily seen, *via* Green's theorem, to be equal to $2A_2$, where A_2 is the area bounded by the inner contour C_2. To recall Green's theorem in this context, we write

$$\iint_{A_2} \left(\frac{\partial M}{\partial x} - \frac{\partial N}{\partial y} \right) dx\, dy = \oint_{C_2} (M\, dy + N\, dx)$$

and set $M = x$ and $N = -y$. The quantity τ is the magnitude of the shear stress, *i.e.*, $\tau = -\sigma_{xz} n_y + \sigma_{yz} n_x$, along the *tangent* to the contour, which becomes

$$\tau = -\frac{d\varphi}{dn}, \tag{11.69}$$

where **n** is the unit normal to the contour C_2. As the stress function must be constant on both the inner and outer contours, (11.68) supplies the additional condition to determine its constant value φ_0 on C_2. The value of φ on the outer boundary, C_1, may still be taken as 0. Thus, in addition to the boundary condition $\varphi = 0$ on C_1, φ must satisfy

$$\oint_{C_2} \tau\, ds = 2G\theta A_2 \quad \text{on } C_2. \tag{11.70}$$

The moment is readily computed from

$$M = \iint_R (\sigma_{yz} x - \sigma_{xz} y)\, dx\, dy$$
$$= -\iint_R \left(\frac{\partial \varphi}{\partial x} x + \frac{\partial \varphi}{\partial y} y \right) dx\, dy. \tag{11.71}$$

This may be rewritten as

$$M = -\iint_R \left[\frac{\partial (x\varphi)}{\partial x} + \frac{\partial (y\varphi)}{\partial y} \right] dx\, dy + 2 \iint_R \varphi\, dx\, dy, \tag{11.72}$$

and transformed, *via* Green's theorem, to

$$M = -\oint_{C_1} (x\varphi\, dy - y\varphi\, dx) - \oint_{C_2} (x\varphi\, dy - y\varphi\, dx)$$
$$+ 2 \iint_R \varphi\, dx\, dy. \tag{11.73}$$

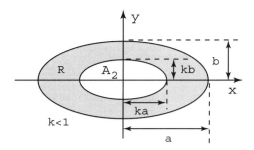

Figure 11.6. Cross section of a hollow ellipsoidal rod.

In (11.73), we take note of the senses of the contour integrals as indicated in Fig. 11.5. Since $\varphi = 0$ on C_2, (11.73) becomes

$$M = 2 \iint_R \varphi \, dx \, dy + 2\varphi_0 A_2. \tag{11.74}$$

11.5.1 Hollow Elliptical Cross Section

As an example of a multiply connected rod subject to torsion, consider the hollow rod illustrated in Fig. 11.6. The solution for the stress field in a solid rod with an ellipsoidal cross section provides a direct path to the solution for this case where the inner contour is an ellipse that is concentric with the outer surface. Examining that solution shows that all ellipsoidal contours in the solid rod are lines of shear stress, so that there is no shear stress acting on any plane section lying parallel to the axis of the rod. Therefore, it is possible to perform the heuristic procedure of "removing an elliptical section" without disturbing the stress state in the remainder of the rod. The stresses can be calculated from the potential found earlier, viz.,

$$\varphi = -\frac{a^2 b^2 G\theta}{a^2 + b^2} \left(\frac{x^2}{a^2} + \frac{y^2}{b^2} - 1 \right).$$

For a given θ, however, the moment will be less than in a solid rod with the same outer dimensions, owing to the reduced total cross section. Thus, again appealing to the solution for the solid rod,

$$\begin{aligned} M &= \frac{\pi a^3 b^3 G\theta}{a^2 + b^2} - \frac{\pi (ka)^3 (kb)^3 G\theta}{(ka)^2 + (kb)^2} \\ &= \frac{\pi G\theta}{a^2 + b^2} a^3 b^3 (1 - k^4). \end{aligned} \tag{11.75}$$

By implication, then, the stress function becomes

$$\varphi = -\frac{M}{\pi ab(1 - k^4)} \left(\frac{x^2}{a^2} + \frac{y^2}{b^2} - 1 \right). \tag{11.76}$$

The maximum shear stress is, accordingly,

$$\tau_{\max} = \frac{2M}{\pi ab^2} \frac{1}{1 - k^4}, \quad \text{for } a > b. \tag{11.77}$$

This solution may readily be verified by substitution into the governing equation and boundary conditions developed earlier in this chapter.

11.6. Bending of a Cantilever

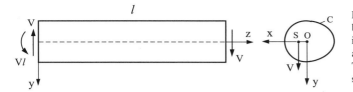

Figure 11.7. A cantilever beam under bending by a concentrated force at its end $z = l$. The principal centroidal axes of the cross section are x and y. The shear force V passes through the shear center S.

11.6 Bending of a Cantilever

Consider a cantilever beam under a transverse force V at its right end, parallel to one of the principal centroidal directions of the beam's cross section (y direction in Fig. 11.7). The force passes through the so-called shear center S of the cross section, causing no torsion of the beam. The centroid of the cross section is the point O. The reactive force and bending moment at the left end of the beam are V and Vl, where l is the length of the beam. The bending moment at an arbitrary distance z from the left end of the beam is

$$M_x = -V(l - z). \tag{11.78}$$

Adopting the Saint-Venant's semi-inverse method, assume that the normal stress σ_{zz} is distributed over the cross section in the same manner as in the case of pure bending, *i.e.*,

$$\sigma_{zz} = \frac{M_x}{I_x} y = -\frac{V(l-z)}{I_x} y, \tag{11.79}$$

where $I_x = \int_A y^2 \, dA$ is the moment of inertia of the cross-sectional area A for the x axis. Because the lateral surface of the beam is traction free, we also assume that throughout the beam

$$\sigma_{xx} = \sigma_{yy} = \sigma_{xy} = 0. \tag{11.80}$$

The objective is, thus, to determine the remaining nonvanishing stress components σ_{zx} and σ_{zy}. In the absence of body forces, the equilibrium equations reduce to

$$\frac{\partial \sigma_{zx}}{\partial z} = 0, \quad \frac{\partial \sigma_{zy}}{\partial z} = 0,$$
$$\frac{\partial \sigma_{zx}}{\partial x} + \frac{\partial \sigma_{zy}}{\partial y} = -\frac{V}{I_x} y. \tag{11.81}$$

These are satisfied by introducing the stress function

$$\Phi = \Phi(x, y), \tag{11.82}$$

such that

$$\sigma_{zx} = \frac{\partial \Phi}{\partial y},$$
$$\sigma_{zy} = -\frac{\partial \Phi}{\partial x} - \frac{V}{2I_x} y^2 + f(x), \tag{11.83}$$

where $f(x)$ is an arbitrary function of x. To derive the governing equation for Φ, we resort to the Beltrami–Michell's compatibility equations (8.86). Four of them are identically satisfied by the above assumptions, whereas the remaining two are

$$\nabla^2 \sigma_{zx} + \frac{1}{1+\nu} \frac{\partial \sigma_{zz}}{\partial x \partial z} = 0,$$
$$\nabla^2 \sigma_{zy} + \frac{1}{1+\nu} \frac{\partial \sigma_{zz}}{\partial y \partial z} = 0.$$
(11.84)

Upon the substitution of (11.83), these become

$$\frac{\partial}{\partial y}(\nabla^2 \Phi) = 0,$$
$$\frac{\partial}{\partial x}(\nabla^2 \Phi) = -\frac{\nu}{1+\nu} \frac{V}{I_x} + f''(x).$$
(11.85)

Thus,

$$\nabla^2 \Phi = -\frac{\nu}{1+\nu} \frac{V}{I_x} x + f'(x) + c.$$
(11.86)

The constant c can be given a geometric interpretation. By using Hooke's law and the above stress expressions, it can be shown that the longitudinal gradient of the material rotation $\omega_z = (u_{y,x} - u_{x,y})/2$ is

$$\frac{\partial \omega_z}{\partial z} = \frac{\partial e_{zy}}{\partial x} - \frac{\partial e_{zx}}{\partial y} = \frac{1}{2G}\left(\frac{\partial \sigma_{zy}}{\partial x} - \frac{\partial \sigma_{zx}}{\partial y}\right) = -\frac{1}{2G}\left(\frac{\nu}{1+\nu} \frac{V}{I_x} x - c\right),$$

where G is the shear modulus. We can define bending without torsion by requiring that the mean value of the relative rotation of the cross sections $\partial \omega_z/\partial z$ over the cross-sectional area is equal to zero,

$$\frac{1}{A}\int_A \frac{\partial \omega_z}{\partial z} dA = 0.$$
(11.87)

In the considered problem, this is fulfilled if we take $c = 0$ in (11.86). The local twist at the point of the cross section is then

$$\frac{\partial \omega_z}{\partial z} = -\frac{\nu}{E} \frac{V}{I_x} x.$$
(11.88)

The traction free boundary condition on the lateral surface of the beam

$$n_x \sigma_{zx} + n_y \sigma_{zy} = \frac{dy}{ds}\sigma_{zx} - \frac{dx}{ds}\sigma_{zy} = 0,$$
(11.89)

where $\mathbf{n} = \{n_x, n_y\}$ is the unit outward normal to the boundary $\mathcal{C}$ of the cross section, gives

$$\frac{d\Phi}{ds} = \left[-\frac{V}{2I_x} y^2 + f(x)\right]\frac{dx}{ds}.$$
(11.90)

An infinitesimal arc length along the boundary c is ds. In each particular problem (shape of the cross section), the function $f(x)$ will be conveniently chosen, such that along the boundary $\mathcal{C}$,

$$\frac{V}{2I_x} y^2 - f(x) = 0.$$
(11.91)

11.7. Elliptical Cross Section

This yields $\Phi = $ const. along the boundary. Since stresses depend on the spatial derivatives of Φ, the value of the constant is immaterial, and we may take $\Phi = 0$ on C. Thus, the boundary-value problem for the stress function Φ is

$$\nabla^2 \Phi = -\frac{\nu}{1+\nu} \frac{V}{I_x} x + f'(x) \quad \text{within } C, \tag{11.92}$$

$$\Phi = 0 \quad \text{on } C.$$

After solving this boundary-value problem, the shear stresses follow from (11.83). At the end cross sections of the cantilever, they must satisfy the integral boundary conditions

$$\int_A \sigma_{zx}\, dA = 0, \quad \int_A \sigma_{zy}\, dA = V,$$
$$\int_A (x\sigma_{zy} - y\sigma_{zx})\, dA = Ve, \tag{11.93}$$

where e represents the horizontal distance between the shear center S and the centroid O of the cross section.

11.7 Elliptical Cross Section

The equation of the boundary of an elliptical cross section is

$$\frac{x^2}{a^2} + \frac{y^2}{b^2} - 1 = 0, \tag{11.94}$$

where a and b are the semiaxes of the ellipse. The boundary condition $\Phi = 0$ is met by choosing

$$f(x) = \frac{Vb^2}{2I_x}\left(1 - \frac{x^2}{a^2}\right). \tag{11.95}$$

The partial differential equation (11.92) reads

$$\nabla^2 \Phi = -\left(\frac{\nu}{1+\nu} + \frac{b^2}{a^2}\right)\frac{V}{I_x} x. \tag{11.96}$$

Both, the boundary condition and this equation can be satisfied by taking

$$\Phi = Bx\left(\frac{x^2}{a^2} + \frac{y^2}{b^2} - 1\right). \tag{11.97}$$

The constant B is readily found to be

$$B = -\frac{Vb^2}{2I_x} \frac{b^2 + \dfrac{\nu}{1+\nu} a^2}{3b^2 + a^2}. \tag{11.98}$$

The corresponding shear stresses are obtained from (11.83) as

$$\sigma_{zx} = -\frac{Vxy}{I_x} \frac{b^2 + \frac{\nu}{1+\nu}a^2}{3b^2 + a^2},$$

$$\sigma_{zy} = \frac{Vb^2}{2I_x} \left[\frac{2b^2 + \frac{1}{1+\nu}a^2}{3b^2 + a^2} \left(1 - \frac{y^2}{b^2}\right) - \frac{1-2\nu}{1+\nu} \frac{x^2}{3b^2+a^2} \right]. \tag{11.99}$$

The magnitude of the maximum shear stress component σ_{zx} is

$$\sigma_{zx}^{\max} = \frac{Vab}{2I_x} \frac{b^2 + \frac{\nu}{1+\nu}a^2}{3b^2 + a^2}, \tag{11.100}$$

and it occurs at the points $x = \pm a/\sqrt{2}$, $y = \pm b/\sqrt{2}$. The maximum shear stress component σ_{zy} is

$$\sigma_{zy}^{\max} = \frac{Vb^2}{2I_x} \frac{2b^2 + \frac{1}{1+\nu}a^2}{3b^2 + a^2}, \tag{11.101}$$

occurring at the point $x = y = 0$ (center of the ellipse).

In the case of a circular cross section ($a = b = R$), we have

$$\sigma_{zx} = -\frac{V}{\pi R^4} \frac{1+2\nu}{1+\nu} xy,$$

$$\sigma_{zy} = \frac{V}{2\pi R^4} \frac{1}{1+\nu} \left[(3+2\nu)(R^2 - y^2) - (1-2\nu)x^2 \right]. \tag{11.102}$$

11.8 Suggested Reading

Cook, R. D., and Young, W. C. (1999), *Advanced Mechanics of Materials*, 2nd ed., Prentice Hall, Upper Saddle River, New Jersey.

Filonenko-Borodich, M. (1964), *Theory of Elasticity*, P. Noordhoff, Groningen, Netherlands.

Landau, L. D., and Lifshitz, E. M. (1986), *Theory of Elasticity*, 3rd English ed., Pergamon Press, New York.

Little, R. W. (1973), *Elasticity*, Prentice Hall, Englewood Cliffs, New Jersey.

Sokolnikoff, I. S. (1956), *Mathematical Theory of Elasticity*, 2nd ed., McGraw-Hill, New York.

Timoshenko, S. P., and Goodier, J. N. (1970), *Theory of Elasticity*, 3rd ed., McGraw-Hill, New York.

BOUNDARY CONDITIONS ARE WHAT y

12 Semi-Infinite Media

In this chapter solutions involving general types of loading on *half-spaces* are considered. Such solutions are developed using Fourier transforms. The media are taken to be elastically isotropic. Solutions for anisotropic elastic media are considered in subsequent chapters.

12.1 Fourier Transform of Biharmonic Equation

Consider a biharmonic function, ϕ, that satisfies the equation

$$\nabla^4 \phi = \frac{\partial^4 \phi}{\partial x^4} + 2\frac{\partial^4 \phi}{\partial x^2 \partial y^2} + \frac{\partial^4 \phi}{\partial y^4} = 0. \quad (12.1)$$

Introduce the Fourier transform in y, viz.,

$$\Phi(x, \alpha) = \int_{-\infty}^{\infty} \phi(x, y) e^{i\alpha y} \, dy, \quad (12.2)$$

and its inverse

$$\phi(x, y) = \left(\frac{1}{2\pi}\right) \int_{-\infty}^{\infty} \Phi(x, \alpha) e^{-i\alpha y} \, d\alpha. \quad (12.3)$$

It is noted here that the transform of ϕ is formed without the factor of $(2\pi)^{-1/2}$; accordingly the inverse transform contains a factor $(2\pi)^{-1}$. Apply the Fourier transform to the biharmonic equation (12.1), *i.e.*,

$$\int_{-\infty}^{\infty} \frac{\partial^4 \phi}{\partial x^4} e^{i\alpha y} \, dy + 2 \int_{-\infty}^{\infty} \frac{\partial^2}{\partial x^2} \frac{\partial^2}{\partial y^2} \phi e^{i\alpha y} \, dy + \int_{-\infty}^{\infty} \frac{\partial^4 \phi}{\partial y^4} e^{i\alpha y} \, dy = 0. \quad (12.4)$$

The differentiation with respect to x may be taken outside the integrals, and after applying the Fourier transform results from above, it is found that

$$\frac{\partial^4 \Phi}{\partial x^4} + 2(-i\alpha)^2 \frac{\partial^2 \Phi}{\partial x^2} + (-i\alpha)^4 \Phi = 0,$$

$$\frac{\partial^4 \Phi}{\partial x^4} - 2\alpha^2 \frac{\partial^2 \Phi}{\partial x^2} + \alpha^4 \Phi = 0. \quad (12.5)$$

229

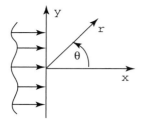

Figure 12.1. Normal loading on the surface of a half-plane.

With $\alpha > 0$, the general solution is

$$\Phi(x,\alpha) = (A+Bx)e^{-\alpha x} + (C+Dx)e^{\alpha x}. \tag{12.6}$$

When concerned with developing solutions to other problems in solid mechanics, such solutions will occasionally be written as

$$\Phi(x,\alpha) = (A+Bx)e^{-|\alpha|x} + (C+Dx)e^{|\alpha|x}. \tag{12.7}$$

The transforms of the stresses are

$$\sigma_{xx} = \frac{\partial^2 \phi}{\partial y^2} \Rightarrow \int_{-\infty}^{\infty} \sigma_{xx} e^{i\alpha y}\, dy = (-i\alpha)^2 \Phi(x,\alpha) = -\alpha^2 \Phi(x,\alpha),$$

$$\sigma_{xy} = -\frac{\partial^2 \phi}{\partial x \partial y} \Rightarrow \int_{-\infty}^{\infty} -\frac{\partial^2 \phi}{\partial x \partial y} e^{i\alpha y}\, dy = i\alpha \frac{\partial \Phi}{\partial x}, \tag{12.8}$$

$$\sigma_{yy} = \frac{\partial^2 \phi}{\partial x^2} \Rightarrow \int_{-\infty}^{\infty} \frac{\partial^2 \phi}{\partial x^2} e^{i\alpha y}\, dy = \frac{\partial^2 \Phi}{\partial x^2}.$$

The inverse transforms follow immediately as

$$\sigma_{xx} = -\frac{1}{2\pi}\int_{-\infty}^{\infty} \alpha^2 \Phi(x,\alpha) e^{-i\alpha y}\, d\alpha,$$

$$\sigma_{xy} = \frac{1}{2\pi}\int_{-\infty}^{\infty} i\alpha \frac{\partial \Phi(x,\alpha)}{\partial x} e^{-i\alpha y}\, d\alpha, \tag{12.9}$$

$$\sigma_{yy} = \frac{1}{2\pi}\int_{-\infty}^{\infty} \frac{\partial^2 \Phi(x,\alpha)}{\partial x^2} e^{-i\alpha y}\, d\alpha.$$

12.2 Loading on a Half-Plane

Consider the half space defined by $x \geq 0$, as depicted in Fig. 12.1. The loading is specified by

$$\sigma_{xx} = -p(y) \quad \text{on} \quad x = 0, \tag{12.10}$$

which describes a state of general normal loading on the external surface of the half space. There are no shear forces on this external surface, so that the other boundary condition is

$$\sigma_{xy} = 0 \quad \text{on} \quad x = 0. \tag{12.11}$$

12.2. Loading on a Half-Plane

The nature of the loading, $p(y)$, is such that it produces bounded stresses. Infinitely far into the bulk of the medium the stresses must be bounded, so that

$$\sigma_{\alpha\beta} \nrightarrow \infty \quad \text{as} \quad x \to \infty. \tag{12.12}$$

DO NOT BECOME INFINITE

As there are no body forces or temperature gradients, the Airy stress function satisfies the simple form of the biharmonic equation

$$\nabla^4 \phi = 0. \tag{12.13}$$

Introduce the Fourier transform of ϕ in the spatial variable y, viz.,

$$\Phi(x, \alpha) = \int_{-\infty}^{\infty} \phi(x, y) e^{i\alpha y} \, dy, \tag{12.14}$$

with an inverse

$$\phi(x, y) = \frac{1}{2\pi} \int_{-\infty}^{\infty} \Phi(x, \alpha) e^{-i\alpha y} \, d\alpha. \tag{12.15}$$

The acceptable solutions for the transform, $\Phi(x, \alpha)$, are of the form

$$\Phi(x, \alpha) = (A + Bx) e^{-|\alpha|x} + (C + Dx) e^{|\alpha|x}. \tag{12.16}$$

Next, invoke the boundary conditions specified above. First, form the Fourier transform of the normal loading boundary condition on $x = 0$, we have

$$\int_{-\infty}^{\infty} \sigma_{xx} e^{i\alpha y} \, dy = \int_{-\infty}^{\infty} \frac{\partial^2 \phi}{\partial y^2} e^{i\alpha y} \, dy$$

$$= -\alpha^2 \Phi(x = 0, \alpha) = -\int_{-\infty}^{\infty} p(y) e^{i\alpha y} \, dy = -P(\alpha), \tag{12.17}$$

and

$$\int_{-\infty}^{\infty} \sigma_{xy} e^{i\alpha y} \, dy = -\int_{-\infty}^{\infty} \frac{\partial^2 \phi}{\partial x \partial y} e^{i\alpha y} \, dy = i\alpha \left[\frac{\partial \Phi(x, \alpha)}{\partial x} \right]_{x=0} = 0. \tag{12.18}$$

$(-|\alpha|(A+Bx)e^{-|\alpha|x} + Be^{-|\alpha|x})$
$= -|\alpha|A + B$

Since the stresses need to be bounded at $x \to \infty$, it is clear that

$$C = D = 0, \tag{12.19}$$

whereas the transformed boundary conditions of (12.17), and (12.18) require that

$$-\alpha^2 A = -P(\alpha) \quad \text{and} \quad 0 = B - |\alpha| A. \tag{12.20}$$

REVERSE THESE BOUNDARY CONDITIONS ⊃ FOR 2ND HW PROBLEM

This leads to

$$A = P(\alpha)/\alpha^2 \quad \text{and} \quad B = P(\alpha)/|\alpha|. \tag{12.21}$$

The result for $\Phi(x, \alpha)$ is then

$$\Phi(x, \alpha) = \frac{P(\alpha)}{\alpha^2} (1 + |\alpha|x) e^{-|\alpha|x}. \tag{12.22}$$

The solution for the stresses is consequently

$$\sigma_{xx}(x, y) = -\frac{1}{2\pi}\int_{-\infty}^{\infty} P(\alpha)(1+|\alpha|x)e^{-|\alpha|x-i\alpha y}\,d\alpha,$$

$$\sigma_{xy}(x, y) = -\frac{i}{2\pi}\int_{-\infty}^{\infty} x\alpha P(\alpha)e^{-|\alpha|x-i\alpha y}\,d\alpha, \quad (12.23)$$

$$\sigma_{yy}(x, y) = -\frac{1}{2\pi}\int_{-\infty}^{\infty} P(\alpha)(1-|\alpha|x)e^{-|\alpha|x-i\alpha y}\,d\alpha.$$

There are two special cases of particular interest, viz., that arising if $P(\alpha)$ is symmetric or antisymmetric. If symmetric, $P(\alpha) = P(-\alpha)$, then

$$\sigma_{xx}(x, y) = -\frac{2}{\pi}\int_0^{\infty} P(\alpha)(1+|\alpha|x)e^{-\alpha x}\cos(\alpha y)\,d\alpha,$$

$$\sigma_{xy}(x, y) = -\frac{2x}{\pi}\int_0^{\infty} \alpha P(\alpha)e^{-\alpha x}\sin(\alpha y)\,d\alpha, \quad (12.24)$$

$$\sigma_{yy}(x, y) = -\frac{2}{\pi}\int_0^{\infty} P(\alpha)(1-|\alpha|x)e^{-\alpha x}\cos(\alpha y)\,d\alpha,$$

with

$$P(\alpha) = 2\int_0^{\infty} p(y)\cos(\alpha y)\,dy. \quad (12.25)$$

If, conversely, $P(\alpha) = -P(-\alpha)$, the integrals in (12.23) become

$$\sigma_{xx}(x, y) = -\frac{2}{\pi}\int_0^{\infty} P(\alpha)(1+|\alpha|x)e^{-\alpha x}\sin(\alpha y)\,d\alpha,$$

$$\sigma_{xy}(x, y) = \frac{2x}{\pi}\int_0^{\infty} \alpha P(\alpha)e^{-\alpha x}\cos(\alpha y)\,d\alpha, \quad (12.26)$$

$$\sigma_{yy}(x, y) = -\frac{2}{\pi}\int_0^{\infty} P(\alpha)(1-|\alpha|x)e^{-\alpha x}\sin(\alpha y)\,d\alpha,$$

with

$$P(\alpha) = \int_0^{\infty} p(y)\sin(\alpha y)\,dy. \quad (12.27)$$

12.3 Half-Plane Loading: Special Case

As a particularly interesting special case, consider loading on the external surface of a half-plane of the form

$$p(y) = \begin{cases} p_0, & \text{if } y \geq 0, \\ 0, & \text{if } y < 0. \end{cases} \quad (12.28)$$

The planar polar coordinates, to be used later to express the resulting stresses, are shown in Fig. 12.2. Consider the auxiliary function defined as

$$\delta_+(\alpha) = \frac{1}{2}\delta(\alpha) - \frac{1}{2\pi i\alpha}, \quad (12.29)$$

12.3. Half-Plane Loading

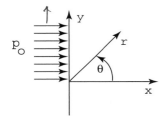

Figure 12.2. Normal loading on a half-plane surface.

where $\delta(\alpha)$ is the Dirac delta function (see Chapter 3), and note that

$$\int_{-\infty}^{\infty} \delta_+(\alpha) e^{-i\alpha y}\, d\alpha = \frac{1}{2} + \frac{1}{\pi}\int_0^{\infty} \frac{\sin(\alpha y)}{\alpha}\, d\alpha = \begin{cases} 1, & \text{if } y \geq 0 \\ 0, & \text{if } y < 0. \end{cases} \quad (12.30)$$

Thus, take

$$P(\alpha) = 2\pi p_0 \delta_+(\alpha), \quad (12.31)$$

so that

$$p(y) = \frac{1}{2\pi}\int_{-\infty}^{\infty} 2\pi p_0 \delta_+(\alpha) e^{-i\alpha y}\, d\alpha = \begin{cases} p_0, & \text{if } y \geq 0 \\ 0, & \text{if } y < 0. \end{cases} \quad (12.32)$$

From (12.23), the stresses become

$$\sigma_{xx} = -p_0 \int_{-\infty}^{\infty} \delta_+(\alpha) e^{-|\alpha|x - i\alpha y}(1 + |\alpha|x)\, d\alpha$$
$$= -\frac{p_0}{2} - \frac{p_0}{\pi}\int_0^{\infty} \frac{(1+\alpha x)e^{-\alpha x}}{\alpha}\sin(\alpha y)\, d\alpha. \quad (12.33)$$

Reducing this integral results in

$$\sigma_{xx} = -p_0 \left[1 - \frac{1}{\pi}\tan^{-1}(x/y) + \frac{xy}{\pi(x^2 + y^2)}\right]. \quad (12.34)$$

Because $\theta = \tan^{-1}(y/x)$, i.e., $x = r\cos\theta$ and $y = r\sin\theta$, and noting that $\cot\theta = \tan(\pi/2 - \theta)$, we can rewrite (12.34) as

$$\sigma_{xx} = -p_0\left[\frac{1}{2} + \frac{\theta}{\pi} + \frac{1}{2\pi}\sin(2\theta)\right]. \quad (12.35)$$

By a similar set of manipulations, it is found that

$$\sigma_{yy} = -p_0\left[\frac{1}{2} + \frac{\theta}{\pi} - \frac{1}{2\pi}\sin(2\theta)\right], \quad (12.36)$$

and

$$\sigma_{xy} = \frac{p_0}{\pi}\cos^2\theta. \quad (12.37)$$

Note that on the external surface, $x = 0$, i.e., $\theta = \pi/2$,

$$\sigma_{xx} = -p_0, \quad \sigma_{yy} = -p_0, \quad \sigma_{xy} = 0. \quad (12.38)$$

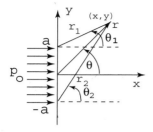

Figure 12.3. Symmetric loading on a half-plane surface.

This means that on the loading plane there exists a state of inplane hydrostatic pressure. It is also noted that rays emanating from the origin, at constant θ, are contours of equal stress.

12.4 Symmetric Half-Plane Loading

Consider the case where the loading on the half-plane is confined to a region symmetrically located about $y = 0$. By simple shifting of the position of the origin along the line $x = 0$, any situation involving a patch of loading as shown in Fig. 12.3 is embodied in this solution. Thus, if the loading is specified as

$$\sigma_{xx} = \begin{cases} -p_0, & \text{if } -a \leq y \leq a, \\ 0, & \text{otherwise,} \end{cases} \tag{12.39}$$

we have

$$P(\alpha) = \int_{-\infty}^{\infty} p(y) e^{i\alpha y} \, dy = \int_{0}^{\infty} p(y) \cos(\alpha y) \, dy = \int_{0}^{a} p_0 \cos(\alpha y) \, dy = p_0 \frac{\sin(\alpha a)}{\alpha} \tag{12.40}$$

The use of (12.24) results in

$$\sigma_{xx} = -\frac{2p_0}{\pi} \int_{0}^{\infty} \frac{1 + \alpha x}{\alpha} e^{-\alpha x} \sin(\alpha a) \cos(\alpha y) \, d\alpha,$$

$$\sigma_{xy} = -\frac{2p_0 x}{\pi} \int_{0}^{\infty} e^{-\alpha x} \sin(\alpha a) \sin(\alpha y) \, d\alpha, \tag{12.41}$$

$$\sigma_{yy} = -\frac{2p_0}{\pi} \int_{0}^{\infty} \frac{1 - \alpha x}{\alpha} e^{-\alpha x} \sin(\alpha a) \cos(\alpha y) \, d\alpha.$$

In terms of the polar angles in Fig. 12.3, it is readily shown that

$$\sigma_{xx} = -\frac{p_0}{2\pi} [\, 2(\theta_1 - \theta_2) - \sin(2\theta_1) + \sin(2\theta_2) \,],$$

$$\sigma_{xy} = \frac{p_0}{2\pi} [\, \cos(2\theta_2) - \cos(2\theta_1) \,],$$

$$\sigma_{yy} = -\frac{p_0}{2\pi} [\, 2(\theta_1 - \theta_2) + \sin(2\theta_1) - \sin(2\theta_2) \,]. \tag{12.42}$$

12.5. Alternative Approach

Figure 12.4. Normal loading on a half-plane surface.

As $r \to \infty$, $\theta_1 \to \theta_2$ and thus $\sigma_{ij} \to 0$. Note also that the maximum shear stress,

$$\tau_{max} = \left[\frac{1}{4}(\sigma_{xx} - \sigma_{yy})^2 + \sigma_{xy}^2 \right]^{1/2}, \qquad (12.43)$$

becomes

$$\tau_{max} = \frac{p_0}{\pi} \left| \sin(\theta_1 - \theta_2) \right|. \qquad (12.44)$$

Thus, contours of $\left| \sin(\theta_1 - \theta_2) \right| = $ constant are contours of equal maximum shear stress.

12.5 Half-Plane Loading: Alternative Approach

The boundary value problem specified on a half-plane by (12.10)–(12.12) is revisited here and solved *via* application of the convolution theorem discussed in Chapter 3. To emphasize the alternative approach, a change in coordinate definitions is used. In addition, the use of the factor $1/\sqrt{2\pi}$ is made both in the definition of the Fourier transform and its inverse. Recall that the state of general loading on the external surface of the half-plane $y = 0$ (Fig. 12.4) is specified as

$$\sigma_{yy} = \frac{\partial^2 \phi}{\partial x^2} = p(x),$$
$$\sigma_{xy} = 0. \qquad (12.45)$$

The Fourier transform of the Airy stress function, $\phi(x, y)$, and its inverse are

$$\Phi(\beta, y) = \frac{1}{\sqrt{2\pi}} \int_{-\infty}^{\infty} \phi(x, y) e^{i\beta x} \, dx,$$
$$\phi(x, y) = \frac{1}{\sqrt{2\pi}} \int_{-\infty}^{\infty} \Phi(\beta, y) e^{-i\beta x} \, d\beta. \qquad (12.46)$$

In the absence of body forces, the transform Φ satisfies the transformed biharmonic equation

$$\frac{\partial^4 \Phi}{\partial y^4} - 2\beta^2 \frac{\partial^2 \Phi}{\partial y^2} + \beta^4 \Phi = 0. \qquad (12.47)$$

The boundary conditions on $y = 0$ result in

$$-\beta^2 \Phi = P(\beta),$$
$$\frac{\partial \Phi}{\partial y} = 0, \qquad (12.48)$$

where $P(\beta)$ is the transform of $p(x)$, viz.,

$$P(\beta) = \frac{1}{\sqrt{2\pi}} \int_{-\infty}^{\infty} p(x) e^{i\beta x} \, dx. \qquad (12.49)$$

As in the previous section, it is noted that solutions of (12.47) that give rise to bounded stresses at $x, y \to \infty$ are of the form

$$\Phi(\beta, y) = (A + C|\beta|y)e^{-|\beta|y}. \tag{12.50}$$

Upon application of (12.48), this leads to

$$A = C = -P(\beta)/\beta^2, \tag{12.51}$$

and

$$\Phi(\beta, y) = -\frac{P(\beta)}{\beta^2}(1 + |\beta|y)e^{-|\beta|y}. \tag{12.52}$$

The transformed stresses are, therefore,

$$\Sigma_{xx} = \frac{\partial^2 \Phi}{\partial y^2} = P(\beta)(1 - |\beta|y)e^{-|\beta|y},$$

$$\Sigma_{xy} = i\beta \frac{\partial \Phi}{\partial y} = P(\beta)(i\beta y)e^{-|\beta|y}, \tag{12.53}$$

$$\Sigma_{yy} = -\beta^2 \Phi = P(\beta)(1 + |\beta|y)e^{-|\beta|y}.$$

The inverse transform of $(1 - |\beta|y)e^{-|\beta|y}$ is

$$\mathcal{F}^{-1}\left[(1 - |\beta|y)e^{-|\beta|y}\right] = 2^{3/2}\pi^{-1/2}x^2 y(x^2 + y^2)^{-2} = g(x, y), \tag{12.54}$$

that is

$$G(\beta) = \frac{1}{\sqrt{2\pi}} \int_{-\infty}^{\infty} g(x, y)e^{i\beta x}\, dx = (1 - |\beta|y)e^{-|\beta|y},$$

$$g(x, y) = \frac{1}{\sqrt{2\pi}} \int_{-\infty}^{\infty} G(\beta, y)e^{-i\beta x}\, d\beta. \tag{12.55}$$

Similarly, for σ_{yy} we obtain

$$\mathcal{F}^{-1}\left[(1 + |\beta|y)e^{-|\beta|y}\right] = 2^{3/2}\pi^{-1/2}y^3(x^2 + y^2)^{-2}, \tag{12.56}$$

whereas for σ_{xy},

$$\mathcal{F}^{-1}\left(i\beta y e^{-|\beta|y}\right) = 2^{3/2}\pi^{-1/2}xy^2(x^2 + y^2)^{-2}. \tag{12.57}$$

Next, recall the convolution theorem from Chapter 3,

$$\int_{-\infty}^{\infty} F(\beta)G(\beta)e^{-i\beta x}\, d\beta = \int_{-\infty}^{\infty} g(\eta) f(x - \eta)\, d\eta, \tag{12.58}$$

and make the association of the functions for σ_{xx}, i.e.,

$$g(\eta) \Leftrightarrow p(\eta) \quad \text{and} \quad G(\beta) \Leftrightarrow P(\beta).$$

This gives

$$\sigma_{xx} = \frac{1}{\sqrt{2\pi}} \int_{-\infty}^{\infty} \left[P(\beta)(1 - |\beta|y)e^{-|\beta|y}\right]e^{-i\beta x}\, d\beta$$

$$= \frac{2y}{\pi} \int_{-\infty}^{\infty} \frac{(x - \eta)^2 p(\eta)}{[(x - \eta)^2 + y^2]^2}\, d\eta. \tag{12.59}$$

12.6. Additional Solutions

Similar manipulations, involving the stresses σ_{xy} and σ_{yy}, result in

$$\sigma_{xy} = \frac{2y^2}{\pi} \int_{-\infty}^{\infty} \frac{(x-\eta)p(\eta)}{[(x-\eta)^2 + y^2]^2} \, d\eta, \tag{12.60}$$

and

$$\sigma_{yy} = \frac{2y^3}{\pi} \int_{-\infty}^{\infty} \frac{p(\eta)}{[(x-\eta)^2 + y^2]^2} \, d\eta. \tag{12.61}$$

A particularly interesting example that lies within this framework is a point force (actually a line of concentrated forces). Without loss of generality, place the coordinate origin at the point of the application of the force. Taking $p(x)$ to be $p(x) = p_0 \delta(x)$, we find

$$\begin{aligned} \sigma_{xx} &= -\frac{2p_0 x^2 y}{\pi (x^2+y^2)^2} = -\frac{2p_0}{\pi} \frac{\cos^2\theta \sin\theta}{r}, \\ \sigma_{xy} &= -\frac{2p_0 xy^2}{\pi (x^2+y^2)^2} = -\frac{2p_0}{\pi} \frac{\cos\theta \sin^2\theta}{r}, \\ \sigma_{yy} &= -\frac{2p_0 y^3}{\pi (x^2+y^2)^2} = -\frac{2p_0}{\pi} \frac{\sin^3\theta}{r}. \end{aligned} \tag{12.62}$$

12.6 Additional Half-Plane Solutions

This section offers additional solutions involving loading on a half-plane bounding a semi-infinite medium. The approach is different than used in the above two sections and thus the development details are provided, along with still another choice of variable names. This is done to distinguish the alternative approach employed. The general solutions are used vis-à-vis other specific forms of the loading function, $p(y)$, as indicated in Fig. 12.1.

Again, begin with the biharmonic equation

$$\nabla^4 \phi = 0, \tag{12.63}$$

and recall the definition of the Fourier transform

$$\Phi(x, \xi) = \mathcal{F}[\phi(x, y)] = \int_{-\infty}^{\infty} \phi(x, y) e^{i\xi y} \, dy, \tag{12.64}$$

with its inverse

$$\phi(x, y) = \mathcal{F}^{-1}[\Phi(x, \xi)] = \frac{1}{2\pi} \int_{-\infty}^{\infty} \Phi(x, \xi) e^{-i\xi y} \, dy. \tag{12.65}$$

Since

$$\frac{\partial^2 \Phi}{\partial y^2} = (-i\xi)^2 \Phi, \tag{12.66}$$

we have

$$\int_{-\infty}^{\infty} \nabla^2 \phi(x, y) e^{i\xi y} \, dy = \left(\frac{\partial^2}{\partial x^2} - \xi^2 \right) \int_{-\infty}^{\infty} \phi(x, y) e^{i\xi y} \, dy. \tag{12.67}$$

To form the biharmonic form, we apply the operator ∇^2 again to obtain

$$\int_{-\infty}^{\infty} \nabla^4 \phi(x,y) e^{i\xi y}\, dy = \left(\frac{\partial^2}{\partial x^2} - \xi^2\right)^2 \int_{-\infty}^{\infty} \phi(x,y) e^{i\xi y}\, dy. \tag{12.68}$$

Thus, $\nabla^4 \phi = 0$ implies

$$\left(\frac{\partial^2}{\partial x^2} - \xi^2\right)^2 \Phi(x,\xi) = 0. \tag{12.69}$$

This has been seen to have solutions of the form

$$\Phi(x,\xi) = (A + Bx)e^{-|\xi|x} + (C + Dx)e^{|\xi|x}. \tag{12.70}$$

For the transformed stresses we then obtain

$$\Sigma_{xx} = \int_{-\infty}^{\infty} \sigma_{xx} e^{i\xi y}\, dy = \int_{-\infty}^{\infty} \frac{\partial^2 \phi}{\partial y^2} e^{i\xi y}\, dy = -\xi^2 \Phi(x,\xi),$$

$$\Sigma_{xy} = \int_{-\infty}^{\infty} \sigma_{xy} e^{i\xi y}\, dy = \int_{-\infty}^{\infty} -\frac{\partial^2 \phi}{\partial x \partial y} e^{i\xi y}\, dy = i\xi \frac{\partial \Phi(x,\xi)}{\partial x}, \tag{12.71}$$

$$\Sigma_{yy} = \int_{-\infty}^{\infty} \sigma_{yy} e^{i\xi y}\, dy = \frac{\partial^2 \Phi}{\partial x^2}.$$

The inversions are

$$\sigma_{xx}(x,y) = -\frac{1}{2\pi}\int_{-\infty}^{\infty} \xi^2 \Phi(x,\xi) e^{-i\xi y}\, d\xi,$$

$$\sigma_{xy}(x,y) = \frac{1}{2\pi}\int_{-\infty}^{\infty} i\xi \frac{\partial \Phi}{\partial x} e^{-i\xi y}\, d\xi, \tag{12.72}$$

$$\sigma_{yy}(x,y) = \frac{1}{2\pi}\int_{-\infty}^{\infty} \frac{\partial^2 \Phi}{\partial x^2} e^{-i\xi y}\, d\xi.$$

12.6.1 Displacement Fields in Half-Spaces

For the case of plane strain, the isotropic elastic constitutive laws lead to

$$\frac{E}{1+\nu}\frac{\partial v}{\partial y} = \sigma_{yy} - \nu(\sigma_{xx} + \sigma_{yy}), \tag{12.73}$$

where v is the y-component of displacement. The Fourier transform of this displacement gradient is

$$\int_{-\infty}^{\infty} \frac{E}{1+\nu}\frac{\partial v}{\partial y} e^{i\xi y}\, dy = \int_{-\infty}^{\infty} \sigma_{yy} e^{i\xi y}\, dy - \nu \int_{-\infty}^{\infty} (\sigma_{xx} + \sigma_{yy}) e^{i\xi y}\, dy, \tag{12.74}$$

which yields

$$-\frac{i\xi E}{1+\nu} \int_{-\infty}^{\infty} v(x,y) e^{i\xi y}\, dy = -\frac{i\xi E}{1+\nu} v v(x,y)$$

$$= \frac{\partial^2 \Phi}{\partial x^2} - \nu\left(-\xi^2 \Phi + \frac{\partial^2 \Phi}{\partial x^2}\right). \tag{12.75}$$

12.6. Additional Solutions

By taking the inverse, we have

$$\frac{E}{1+v}v(x, y) = \frac{1}{2\pi}\int_{-\infty}^{\infty}\left[(1-v)\frac{\partial^2 \Phi}{\partial x^2} + v\xi^2\Phi\right]ie^{-i\xi y}\frac{d\xi}{\xi}. \tag{12.76}$$

In view of the definition of shear strain and the elastic constitutive relation,

$$\frac{E}{2(1+v)}\left(\frac{\partial u}{\partial y} + \frac{\partial v}{\partial x}\right) = \sigma_{xy}, \tag{12.77}$$

we thus obtain

$$\frac{E}{1+v}u(x, y) = \frac{1}{2\pi}\int_{-\infty}^{\infty}\left[(1-v)\frac{\partial^3 \Phi}{\partial x^3} + (2+v)\xi^2\frac{\partial \Phi}{\partial x}\right]e^{-i\xi y}\frac{d\xi}{\xi^2}. \tag{12.78}$$

12.6.2 Boundary Value Problem

As in the earlier sections, boundary conditions are applied that correspond to normal loading on the external surface on a semi-infinite medium. They are, in the context of the definitions used here,

$$\sigma_{xx} = -p(y), \quad \sigma_{xy} = 0 \quad \text{on} \quad x = 0,$$
$$\sigma_{ij} \quad \text{bounded as} \quad x, y \to \infty. \tag{12.79}$$

The latter of (12.79) leads to the conclusion that, with reference to (12.70), $C = D = 0$. The first two conditions of (12.79) lead to

$$-\xi^2 A = -\int_{-\infty}^{\infty} p(y)e^{i\xi y}\,dy = P(\xi),$$
$$B - |\xi|A = 0. \tag{12.80}$$

Consequently,

$$A = P(\xi)/\xi^2, \quad B = P(\xi)/|\xi|, \tag{12.81}$$

and

$$\Phi(x, \xi) = \frac{P(\xi)}{\xi^2}[1 + |\xi|x]e^{-|\xi|x}. \tag{12.82}$$

The stresses may now be calculated as

$$\sigma_{xx}(x, y) = -\frac{1}{2\pi}\int_{-\infty}^{\infty} P(\xi)e^{-|\xi|x-i\xi y}(1 + |\xi|x)\,d\xi,$$
$$\sigma_{yy}(x, y) = -\frac{1}{2\pi}\int_{-\infty}^{\infty} P(\xi)e^{-|\xi|x-i\xi y}(1 - |\xi|x)\,d\xi, \tag{12.83}$$
$$\sigma_{xy}(x, y) = -\frac{ix}{2\pi}\int_{-\infty}^{\infty} x\xi\, P(\xi)e^{-|\xi|x-i\xi y}\,d\xi.$$

Since there is an imposed state of plane strain here, $\sigma_{zz} = v(\sigma_{xx} + \sigma_{yy})$, we have

$$\sigma_{zz}(x, y) = -v\int_{-\infty}^{\infty} P(\xi)e^{-|\xi|x-i\xi y}\,d\xi. \tag{12.84}$$

For the displacements, it is found that

$$u = \frac{1+\nu}{2\pi E} \int_{-\infty}^{\infty} P(\xi) e^{-|\xi|x - i\xi y} [2(1-\nu) + |\xi|x] \frac{d\xi}{|\xi|},$$

$$v = -\frac{1+\nu}{2\pi E} \int_{-\infty}^{\infty} i P(\xi) e^{-|\xi|x - i\xi y} [(1-2\nu) - |\xi|x] \frac{d\xi}{|\xi|}. \quad (12.85)$$

The above relations take on a rather simplified form when there exists symmetry in the loading function, $p(y)$. For example, if $p(y) = p(-y)$, then

$$\sigma_{xx}(x,y) = -\frac{2}{\pi} \int_0^\infty P(\xi)(1 + \xi x) e^{-\xi x} \cos(\xi y) \, d\xi,$$

$$\sigma_{yy}(x,y) = -\frac{2}{\pi} \int_0^\infty P(\xi)(1 - \xi x) e^{-\xi x} \cos(\xi y) \, d\xi, \quad (12.86)$$

$$\sigma_{xy}(x,y) = \frac{2x}{\pi} \int_0^\infty \xi P(\xi) e^{-\xi x} \sin(\xi y) \, d\xi.$$

Correspondingly, the displacements are

$$u(x,y) = \frac{2(1+\nu)}{\pi E} \int_0^\infty P(\xi) e^{-\xi x} [2(1-\nu) + \xi x] \frac{\cos(\xi y)}{\xi} d\xi,$$

$$v(x,y) = \frac{2(1+\nu)}{\pi E} \int_0^\infty P(\xi) e^{-\xi x} [1 - 2\nu - \xi x] \frac{\sin(\xi y)}{\xi} d\xi, \quad (12.87)$$

with [MISSING FACTOR OF 2 ⇒ it was put]

$$P(\xi) = \int_0^\infty p(y) \cos(\xi y) \, dy. \quad (12.88)$$

Note that a factor of 2 has been placed on all of (12.86) and accordingly removed from (12.88).

12.6.3 Specific Example

As a specific example, consider the loading function

$$p(y) = \begin{cases} \dfrac{p_0}{\pi}(a^2 - y^2)^{-1/2}, & \text{if } 0 \le |y| \le a, \\ 0, & \text{if } |y| > a. \end{cases} \quad (12.89)$$

Note that

$$\int_{-\infty}^{\infty} p(y) \, dy = \frac{p_0}{\pi} \int_{-a}^{a} (a^2 - y^2)^{-1/2} \, dy = p_0. \quad (12.90)$$

The transform of $p(y)$ is

$$P(\xi) = \int_0^a \frac{p_0}{\pi} \frac{\cos(\xi y)}{(a^2 - y^2)^{1/2}} \, dy = 1/2 \, p_0 J_0(\xi a), \quad (12.91)$$

[BESSEL FUNCTION OF (ξa)]

$$J_0(\xi a) = \int_0^a \frac{2}{\pi} \frac{\cos(\xi y)}{(a^2 - y^2)^{1/2}} \, dy$$

12.6. Additional Solutions

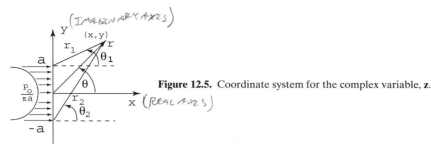

Figure 12.5. Coordinate system for the complex variable, z.

where $J_0(...)$ is a well known Bessel function. By combining terms for the various components of stress it is found that

$$\sigma_{xx} + \sigma_{yy} = -\frac{2p_0}{\pi} \int_0^\infty J_0(\xi a) e^{-\xi x} \cos(\xi y) \, d\xi,$$

$$\sigma_{yy} - \sigma_{xx} + 2i\sigma_{xy} = -\frac{2p_0 x}{\pi} \int_0^\infty \xi J_0(\xi a) e^{-\xi z} \, d\xi, \qquad (12.92)$$

where $z = x + iy$ is the complex variable, and

$$e^{-\xi z} = e^{-\xi x - i\xi y}. \qquad (12.93)$$

From standard handbooks and the properties of Bessel functions,

$$\int_0^\infty \xi J_0(\xi a) e^{-\xi z} \, d\xi = z(a^2 + z^2)^{-3/2}$$

$$= \frac{z}{(z+ia)^{3/2}(z-ia)^{3/2}}, \qquad (12.94)$$

and thus

$$\sigma_{yy} - \sigma_{xx} + 2i\sigma_{xy} = \frac{2p_0 x}{\pi} \frac{z}{(z+ia)^{3/2}(z-ia)^{3/2}}. \qquad (12.95)$$

With reference to Fig. 12.5, we have

$$z - ia = r_1 e^{i\theta_1}, \quad z + ia = r_2 e^{i\theta_2}, \qquad (12.96)$$

and

$$\sigma_{yy} - \sigma_{xx} + 2i\sigma_{xy} = \frac{2p_0 x}{\pi} \frac{re^{i\theta}}{(r_1 r_2)^{3/2} e^{i3/2\theta_1} e^{i3/2\theta_2}}$$

$$= \frac{2p_0 x}{\pi} \frac{r}{(r_1 r_2)^{3/2}} e^{i\theta - 3/2i(\theta_1 + \theta_2)}. \qquad (12.97)$$

Consequently,

$$\sigma_{xy}(x, y) = \frac{p_0 r^2 \cos\theta}{\pi (r_1 r_2)^{3/2}} \sin[\theta - 3/2(\theta_1 + \theta_2)]. \qquad (12.98)$$

It is also found that

$$\sigma_{xx} + \sigma_{yy} = -\frac{2p_0}{\pi} \Re \int_0^\infty e^{\xi z} J_0(\xi a) \, d\xi = -\frac{2p_0}{\pi} \Re \left[(a^2 - z^2)^{-1/2} \right]$$

$$= -\frac{2p_0}{\pi} \frac{\cos[1/2(\theta_1 + \theta_2)]}{(r_1 r_2)^{1/2}}, \qquad (12.99)$$

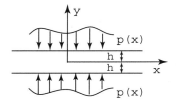

Figure 12.6. Symmetrically loaded infinite strip.

and

$$\sigma_{yy} - \sigma_{xx} = \frac{2p_0}{\pi} \frac{r^2 \cos\theta}{(r_1 r_2)^{3/2}} \cos[\theta - 3/2(\theta_1 + \theta_2)]. \quad (12.100)$$

From these relations, it readily follows that

$$\sigma_{yy}(x, y) = -\frac{p_0}{\pi (r_1 r_2)^{1/2}} \left\{ \cos[1/2(\theta_1 + \theta_2)] - \frac{r^2}{r_1 r_2} \cos\theta \cos[\theta - 3/2(\theta_1 + \theta_2)] \right\}, \quad (12.101)$$

$$\sigma_{xx}(x, y) = -\frac{p_0}{\pi (r_1 r_2)^{1/2}} \left\{ \cos[1/2(\theta_1 + \theta_2)] + \frac{r^2}{r_1 r_2} \cos\theta \cos[\theta - 3/2(\theta_1 + \theta_2)] \right\}. \quad (12.102)$$

Finally, we return to the displacements to find that

$$u(x, y) = \frac{2(1+\nu)}{\pi E} \int_0^\infty P(\xi) e^{-\xi x} [2(1-\nu) + \xi x] \frac{\cos(\xi y)}{\xi} \, d\xi, \quad (12.103)$$

which, when differentiated with respect to y, gives

$$\left(\frac{\partial u}{\partial y}\right)_{x=0} = \frac{2(1-\nu^2) p_0}{\pi E} \int_0^\infty J_0(\xi a) \sin(\xi y) \, d\xi$$

$$= \begin{cases} 0, & \text{if } |y| \leq a, \\ -\frac{2p_0(1-\nu^2)}{\pi E (y^2 - a^2)^{1/2}}, & \text{if } |y| > a. \end{cases} \quad (12.104)$$

Thus, for the assumed normal loading function, the x component of displacement is constant under the load; this is as though the surface was indented by a *rigid* punch.

12.7 Infinite Strip

In this case we consider an infinite strip loaded on both sides with symmetrical loads as shown in Fig. 12.6. We begin with the biharmonic equation

$$\nabla^4 \phi = 0, \quad (12.105)$$

and the boundary conditions

$$\sigma_{yy} = \partial^2 \phi / \partial x^2 = -p(x), \quad \text{on } -\infty \leq x \leq \infty, \ y = \pm h,$$
$$\sigma_{xy} = -\partial^2 \phi / \partial x \partial y = 0, \quad \text{on } y = \pm h. \quad (12.106)$$

12.7. Infinite Strip

Define the Fourier transform

$$\Phi(\beta, y) = \frac{1}{\sqrt{2\pi}} \int_{-\infty}^{\infty} \phi(x, y) e^{i\beta x} \, dx, \tag{12.107}$$

and its inverse

$$\phi(x, y) = \frac{1}{\sqrt{2\pi}} \int_{-\infty}^{\infty} \Phi(\beta, y) e^{-i\beta x} \, d\beta. \tag{12.108}$$

The transformed biharmonic equation becomes, as before,

$$\frac{d^4\Phi}{dy^4} - 2\beta^2 \frac{d^2\Phi}{dy^2} + \beta^4 \Phi = 0. \tag{12.109}$$

Setting the boundary condition for σ_{yy} according to the first of (12.106), yields

$$-\beta^2 \Phi = -\frac{1}{\sqrt{2\pi}} \int_{-\infty}^{\infty} p(x) e^{i\beta x} \, dx = -P(\beta). \tag{12.110}$$

To illustrate the procedure, consider loading functions, $p(x)$, that are symmetric, so that $p(x) = p(-x)$. In this case

$$P(\beta) = \frac{2}{\sqrt{2\pi}} \int_{0}^{\infty} p(x) e^{i\beta x} \, dx. \tag{12.111}$$

Similar conversions of integrals involving symmetric functions will be made in what follows. The boundary condition for σ_{xy} leads to

$$\partial \Phi / \partial y = 0, \text{ for } y = \pm h. \tag{12.112}$$

Solutions that meet these conditions are of the form

$$\Phi(\beta, y) = A \cosh(\beta y) + B \sinh(\beta y) + C\beta y \sinh(\beta y) + D\beta y \cosh(\beta y).$$

The symmetry of ϕ in y requires that $B = D = 0$. The boundary condition for σ_{xy} yields

$$\partial \Phi / \partial y |_{y=\pm h} = \beta \left\{ A \sinh(\beta h) + C [\beta h \cosh(\beta h) + \sinh(\beta h)] \right\} = 0,$$

i.e.,

$$A = -C[1 + \beta h \coth(\beta h)]. \tag{12.113}$$

The boundary condition for σ_{yy} gives the additional relation

$$-C \frac{\beta h + \sinh(\beta h) \cosh(\beta h)}{\sinh(\beta h)} = \frac{1}{\beta^2} P(\beta). \tag{12.114}$$

Thus,

$$\Phi(\beta, x) = \frac{1}{\beta^2} P(\beta) \frac{[\beta h \cosh(\beta h) + \sinh(\beta h)] \cosh(\beta y) - \beta y \sinh(\beta h) \sinh(\beta y)}{\beta h + \sinh(\beta h) \cosh(\beta h)}.$$

12.7.1 Uniform Loading: $-a \leq x \leq a$

If $p(x)$ is defined as

$$p(x) = \begin{cases} p_0, & -a \leq x \leq a, \\ 0, & |x| > a, \end{cases} \tag{12.115}$$

then
$$P(\beta) = \frac{1}{\sqrt{2\pi}} \int_{-a}^{a} p_0 \cos(\beta x) \, dx = \frac{2p_0}{\sqrt{2\pi}} \frac{\sin(\beta a)}{\beta}. \tag{12.116}$$

It is convenient to define nondimensional variables $\zeta = \beta h$, $\eta = y/h$, $\xi = x/h$, and $\alpha = a/h$. Using these definitions, it is found that

$$\phi = \frac{1}{\pi} \int_{-\infty}^{\infty} \frac{2p_0 h^2 \sin(\zeta\alpha)}{\zeta^3 [2\zeta + \sinh(2\zeta)]}$$
$$\times \{[\zeta \cosh(\zeta) + \sinh(\zeta)] \cosh(\zeta\eta) - \zeta\eta \sinh(\zeta) \sinh(\zeta\eta)\} \cos(\zeta\xi) \, d\zeta.$$

Note that here the factor h^2 is to be annihilated upon forming the stresses which involve second derivatives with respect to x and y. In fact, performing these derivatives, and noting the symmetry of the integrands involved, we obtain

$$\sigma_{xx} = -\frac{4p_0}{\pi} \int_0^{\infty} \frac{[\sinh(\zeta) + \zeta \cosh(\zeta)] \cosh(\zeta\eta) - \zeta\eta \sinh(\zeta) \sinh(\zeta\eta)}{2\zeta + \sinh(2\zeta)}$$
$$\times \frac{\sin(\zeta\alpha)}{\zeta} \cos(\zeta\xi) \, d\zeta,$$

$$\sigma_{xy} = -\frac{4p_0}{\pi} \int_0^{\infty} \frac{\zeta\eta \sinh(\zeta) \cosh(\zeta\eta) - \zeta \cosh(\zeta) \sinh(\zeta\eta)}{2\zeta + \sinh(2\zeta)} \tag{12.117}$$
$$\times \frac{\sin(\zeta\alpha)}{\zeta} \sin(\zeta\xi) \, d\zeta,$$

and

$$\sigma_{yy} = -\frac{4p_0}{\pi} \int_0^{\infty} \frac{[\sinh(\zeta) - \zeta \cosh(\zeta)] \cosh(\zeta\eta) + \zeta\eta \sinh(\zeta) \sinh(\zeta\eta)}{2\zeta + \sinh(2\zeta)}$$
$$\times \frac{\sin(\zeta\alpha)}{\zeta} \cos(\zeta\xi) \, d\zeta.$$

Strains may be calculated in a straightforward manner, and then the displacements. Because the system is fully equilibrated, the displacements will be set without the problem of undetermined rigid body displacements. Note also that the problem solved above [see (12.40) for the boundary conditions] should be recoverable by taking limits as $h \to \infty$, although taking such limits may be a formidable task.

12.7.2 Symmetrical Point Loads

If the distribution of force, $p(x)$, above is reduced to a concentrated point force, P, we let $p_0 = P/2a$, and take the limit

$$\lim_{a \to 0} \frac{P}{2a} \frac{\sin(\zeta\alpha)}{\zeta} = \frac{P}{2h}. \tag{12.118}$$

Using the previously derived results, we obtain

$$\sigma_{xx} = -\frac{2Px^2y}{\pi(x^2+y^2)^2},$$
$$\sigma_{yy} = -\frac{2Py^3}{\pi(x^2+y^2)^2}, \qquad (12.119)$$
$$\sigma_{xy} = -\frac{2Pxy^2}{\pi(x^2+y^2)^2}.$$

12.8 Suggested Reading

Little, R. W. (1973), *Elasticity*, Prentice Hall, Englewood Cliffs, New Jersey.

Muskhelishvili, N. I. (1963), *Some Basic Problems of the Mathematical Theory of Elasticity*, Noordhoff, Groningen, Netherlands.

Sokolnikoff, I. S. (1956), *Mathematical Theory of Elasticity*, 2nd ed., McGraw-Hill, New York.

Sneddon, I. N. (1951), *Fourier Transforms*, McGraw-Hill, New York.

Timoshenko, S. P., and Goodier, J. N. (1970), *Theory of Elasticity*, 3rd ed., McGraw-Hill, New York.

13 Isotropic 3D Solutions

13.1 Displacement-Based Equations of Equilibrium

Recall the Cauchy equations of equilibrium in terms of stresses

$$\frac{\partial \sigma_{ij}}{\partial x_j} + b_j = 0, \qquad (13.1)$$

along with the definition of small strains

$$e_{ij} = \frac{1}{2}\left(\frac{\partial u_i}{\partial x_j} + \frac{\partial u_j}{\partial x_i}\right), \qquad (13.2)$$

and the constitutive relations for isotropic thermoelastic media

$$\sigma_{ij} = \frac{E}{2(1+\nu)}\left(e_{ij} + \frac{\nu}{1-2\nu}e_{kk}\delta_{ij}\right) - K\alpha\Delta T \delta_{ij}. \qquad (13.3)$$

The coefficient of thermal expansion is α, and K the elastic bulk modulus, $K = E/3(1-2\nu)$. The temperature difference from the reference temperature is $\Delta T = T - T_0$. When (13.3) are used in (13.1), we obtain

$$\frac{E}{2(1+\nu)}\frac{\partial^2 u_i}{\partial x_j \partial x_j} + \frac{E}{2(1+\nu)(1-2\nu)}\frac{\partial^2 u_k}{\partial x_i \partial x_k}$$
$$= -b_i + \frac{E\alpha}{3(1-2\nu)}\frac{\partial T}{\partial x_i}. \qquad (13.4)$$

In direct tensor notation, this is written as

$$\frac{E}{2(1+\nu)}\nabla^2 \mathbf{u} + \frac{E}{2(1+\nu)(1-2\nu)}\operatorname{grad}\operatorname{div}\mathbf{u}$$
$$= -\mathbf{b} + \frac{E\alpha}{3(1-2\nu)}\operatorname{grad} T. \qquad (13.5)$$

Because the curl of a vector can be written as

$$\operatorname{curl}\mathbf{u} = \epsilon_{pqr}\frac{\partial u_r}{\partial x_q}\mathbf{e}_p, \qquad (13.6)$$

13.2. Boussinesq–Papkovitch Solutions

it is straightforward to construct the vector identity

$$\nabla^2 \mathbf{u} = \text{grad div } \mathbf{u} - \text{curl curl } \mathbf{u}. \tag{13.7}$$

When this is used in (13.5), it is found that

$$\frac{E(1-\nu)}{(1+\nu)(1-2\nu)} \text{grad div } \mathbf{u} - \frac{E}{2(1+\nu)} \text{curl curl } \mathbf{u}$$
$$= -\mathbf{b} + \frac{E\alpha}{3(1-2\nu)} \text{grad } T, \tag{13.8}$$

or

$$\frac{3(1-\nu)}{(1+\nu)} \text{grad div } \mathbf{u} - \frac{3}{2}\frac{(1-2\nu)}{(1+\nu)} \text{curl curl } \mathbf{u}$$
$$= -\frac{3(1-2\nu)}{E} \mathbf{b} + \alpha \text{ grad } T. \tag{13.9}$$

These are the displacement based equations of equilibrium for isotropic thermoelastic media.

13.2 Boussinesq–Papkovitch Solutions

There are a variety of methods for obtaining solutions to (13.9) or its alternative, (13.5). One of the methods for obtaining general solutions is as follows. In the absence of body forces and thermal gradients, (13.5) becomes

$$\text{grad div } \mathbf{u} + (1 - 2\nu)\nabla^2 \mathbf{u} = \mathbf{0}. \tag{13.10}$$

Now, if

$$v = \text{grad } \phi, \tag{13.11}$$

where ϕ is any scalar function satisfying $\nabla^2 \phi = 0$, then v is a solution of (13.10), because

$$\text{div } v = \partial v_i / \partial x_i = \nabla^2 \phi = 0, \tag{13.12}$$

and thus

$$\nabla^2 v = \text{grad } \nabla^2 \phi = \mathbf{0}. \tag{13.13}$$

Next, if

$$\omega = 4(1-\nu)\psi - \text{grad } (\mathbf{x} \cdot \psi), \tag{13.14}$$

where ψ is any vector field satisfying $\nabla^2 \psi = \mathbf{0}$, then ω is also a solution to (13.10). To see this note firstly that components of the grad $(\mathbf{x} \cdot \psi)$ are

$$(x_k \psi_k)_{,\ell} = \psi_\ell + x_k \psi_{k,\ell}, \tag{13.15}$$

because $x_{k,\ell} = \delta_{k\ell}$. Thus,

$$\omega_\ell = (3 - 4\nu)\psi_\ell - x_k \psi_{k,\ell}, \tag{13.16}$$

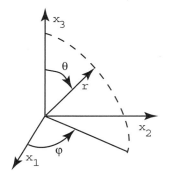

Figure 13.1. Spherical coordinate system.

and

$$\omega_{\ell,\ell} = \text{div } \boldsymbol{\omega} = (3-4\nu)\psi_{\ell,\ell} - x_k\psi_{k,\ell\ell} - \psi_{\ell,\ell} \\ = 2(1-2\nu)\psi_{\ell,\ell}, \qquad (13.17)$$

having regard to

$$\psi_{k,\ell\ell} = (\nabla^2\psi)_k = 0. \qquad (13.18)$$

Therefore,

$$\text{div } \boldsymbol{\omega} = 2(1-2\nu)\text{div } \boldsymbol{\psi}. \qquad (13.19)$$

But,

$$\omega_{\ell,kk} = (3-4\nu)\psi_{\ell,kk} - (\psi_{k,\ell} + x_m\psi_{m,\ell k})_k \\ = -2\psi_{k,\ell k}, \qquad (13.20)$$

and so

$$\nabla^2\boldsymbol{\omega} = -2\,\text{grad div }\boldsymbol{\psi}. \qquad (13.21)$$

Finally, by combining (13.21) and (13.19), we find that $\boldsymbol{\omega}$ is a solution to (13.10). Thus, general solutions to the homogeneous equation (13.10) can be constructed as

$$\mathbf{u} = \boldsymbol{v} + \boldsymbol{\omega}. \qquad (13.22)$$

Such solutions are known as Boussinesq–Papkovitch solutions.

13.3 Spherically Symmetrical Geometries

Considerable simplification, incorporating nonetheless interesting and important problems, is achieved when the problem's geometry suggests a solution in which the displacement field depends only on a radial coordinate directed from a fixed point. Such geometries are most naturally described by a spherical coordinate system, as sketched in Fig. 13.1.

13.3. Spherically Symmetrical Geometries

With respect to spherical coordinate system, the components of strain are

$$e_{rr} = \frac{\partial u_r}{\partial r}, \quad e_{\theta\theta} = \frac{1}{r}\frac{\partial u_\theta}{\partial \theta} + \frac{u_r}{r},$$

$$e_{\phi\phi} = \frac{1}{r\sin\theta}\frac{\partial u_\phi}{\partial \phi} + \frac{u_\theta}{r}\cot\theta + \frac{u_r}{r},$$

$$2e_{\theta\phi} = \frac{1}{r}\left(\frac{\partial u_\phi}{\partial \theta} - u_\phi \cot\phi\right) + \frac{1}{r\sin\theta}\frac{\partial u_\theta}{\partial \phi}, \tag{13.23}$$

$$2e_{r\theta} = \frac{\partial u_\theta}{\partial r} - \frac{u_\theta}{r} + \frac{1}{r}\frac{\partial u_r}{\partial \theta},$$

$$2e_{\phi r} = \frac{1}{r\sin\theta}\frac{\partial u_r}{\partial \phi} + \frac{\partial u_\phi}{\partial r} - \frac{u_\phi}{r}.$$

Suppose, for example, that the displacement vector and the body force vector are of the form

$$\mathbf{u} = u_r \mathbf{e}_r, \quad \mathbf{b} = b_r \mathbf{e}_r, \tag{13.24}$$

where $\mathbf{e}_r$ is the unit vector directed radially. Then,

$$\operatorname{div} \mathbf{u} = \operatorname{div}\left(u_r \frac{\mathbf{x}}{r}\right) = \left(\frac{u_r}{r}\right)\operatorname{div} \mathbf{x} + \mathbf{x}\cdot\operatorname{grad}\left(\frac{u_r}{r}\right)$$

$$= 3\frac{u_r}{r} + \frac{du_r}{dr} - \frac{u_r}{r} \tag{13.25}$$

$$= \frac{1}{r^2}\frac{d}{dr}(r^2 u_r).$$

Also,

$$\operatorname{curl} \mathbf{u} = \operatorname{curl}\left(u_r \frac{\mathbf{x}}{r}\right) = \operatorname{grad}\left(\frac{u_r}{r}\right)\times \mathbf{x} + \frac{u_r}{r}\operatorname{curl} \mathbf{x}. \tag{13.26}$$

But $\operatorname{curl} \mathbf{x} = \mathbf{0}$, and $\operatorname{grad}(u_r/r)$ is parallel to $\mathbf{e}_r \parallel \mathbf{x}$. These facts lead to the conclusion that, with such radial symmetry,

$$\operatorname{curl} \mathbf{u} = \mathbf{0}. \tag{13.27}$$

Consider the equilibrium equation (13.18) and retain just the body force term. The substitution of the results (13.25)–(13.27) gives

$$\frac{E(1-v)}{(1+v)(1-2v)}\frac{d}{dr}\left[\frac{1}{r^2}\frac{d}{dr}(r^2 u_r)\right] = -b_r. \tag{13.28}$$

A similar development goes through if a radially symmetric temperature distribution is included on the *rhs* of (13.5) or (13.9).

13.3.1 Internally Pressurized Sphere

Consider a hollow sphere with internal radius r_1, and external radius r_2. The sphere is internally pressurized at a pressure p_1 and subject to an external pressure p_2 (Fig. 13.2a). This geometry is clearly spherically symmetric and thus, with an origin at the sphere's

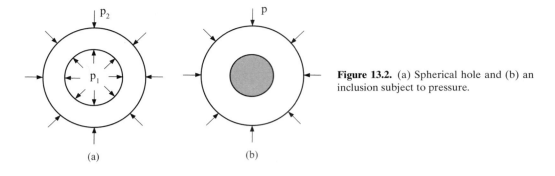

Figure 13.2. (a) Spherical hole and (b) an inclusion subject to pressure.

center, the displacement **u** is purely radial and a function of r alone. Thus, curl $\mathbf{u} = \mathbf{0}$ and (13.9) becomes

$$\text{grad div } \mathbf{u} = \mathbf{0}, \tag{13.29}$$

which leads directly to

$$\text{div } \mathbf{u} = \frac{1}{r^2} \frac{d(r^2 u_r)}{dr} = 3a, \tag{13.30}$$

where $3a$ is a constant (the 3 being incorporated for convenience). Upon integration,

$$u_r = ar + b/r^2, \tag{13.31}$$

where b is another integration constant. The components of strain are easily derived as

$$e_{rr} = a - 2b/r^3, \quad e_{\theta\theta} = e_{\phi\phi} = a + b/r^3. \tag{13.32}$$

The constants a and b are calculated by applying the stated boundary conditions to the radial stress,

$$\sigma_{rr} = \frac{E}{(1+\nu)(1-2\nu)} [(1-\nu)e_{rr} + 2\nu e_{\theta\theta}] = \frac{E}{1-2\nu} a - \frac{2E}{1+\nu} \frac{b}{r^3}. \tag{13.33}$$

The result of applying $\sigma_{rr}(r_1) = -p_1$ and $\sigma_{rr}(r_2) = -p_2$ is

$$a = \frac{p_1 r_1^3 - p_2 r_2^3}{r_2^3 - r_1^3} \frac{1-2\nu}{E}, \quad b = \frac{r_1^3 r_2^3 (p_1 - p_2)}{r_2^3 - r_1^3} \frac{1+\nu}{2E}. \tag{13.34}$$

If, for example $p_1 = p$ and $p_2 = 0$, we have

$$\sigma_{rr} = \frac{p r_1^3}{r_2^3 - r_1^3} \left(1 - \frac{r_2^3}{r^3}\right), \quad \sigma_{\phi\phi} = \sigma_{\theta\theta} = \frac{p r_1^3}{r_2^3 - r_1^3} \left(1 + \frac{r_2^3}{r^3}\right). \tag{13.35}$$

Limiting solutions for thin walled spherical shells, *i.e.*, when $r_2 = r_1 + h$ and $h/r_1 \ll 1$, are readily obtained. Another case of interest is a void in an infinite medium ($r_2 \to \infty$) subject to external pressure $p_2 = p$. In this case, with $p_1 = 0$, we obtain

$$\sigma_{rr} = -p \left(1 - \frac{r_1^3}{r^3}\right), \quad \sigma_{\phi\phi} = \sigma_{\theta\theta} = -p \left(1 + \frac{r_1^3}{2r^3}\right). \tag{13.36}$$

13.4 Pressurized Sphere: Stress-Based Solution

Another route of deriving the previous solution is as follows. In a three-dimensional problems with spherical symmetry, the only displacement component is the radial displacement $u = u(r)$. The corresponding strain components are

$$e_{rr} = \frac{du}{dr}, \quad e_{\theta\theta} = e_{\phi\phi} = \frac{u}{r}, \tag{13.37}$$

with the Saint-Venant compatibility equation

$$\frac{de_{\theta\theta}}{dr} = \frac{1}{r}(e_{rr} - e_{\theta\theta}). \tag{13.38}$$

The nonvanishing stress components are the radial stress σ_{rr} and the hoop stresses $\sigma_{\theta\theta} = \sigma_{\phi\phi}$. In the absence of body forces, the equilibrium equation is

$$\frac{d\sigma_{rr}}{dr} + \frac{2}{r}(\sigma_{rr} - \sigma_{\theta\theta}) = 0. \tag{13.39}$$

The Beltrami–Michell compatibility equation is obtained from (13.38) by incorporating the stress-strain relations

$$e_{rr} = \frac{1}{E}(\sigma_{rr} - 2\nu\sigma_{\theta\theta}), \quad e_{\theta\theta} = \frac{1}{E}[\sigma_{\theta\theta} - \nu(\sigma_{rr} + \sigma_{\theta\theta})], \tag{13.40}$$

and the equilibrium equation (13.39). This gives

$$\frac{d}{dr}(\sigma_{rr} + 2\sigma_{\theta\theta}) = 0, \tag{13.41}$$

which implies that the spherical component of stress tensor is uniform throughout the medium,

$$\frac{1}{3}(\sigma_{rr} + 2\sigma_{\theta\theta}) = A = \text{const.} \tag{13.42}$$

Combining (13.39) and (13.42) it follows that

$$\frac{d\sigma_{rr}}{dr} + \frac{3}{r}\sigma_{rr} = \frac{3}{r}A. \tag{13.43}$$

The general solution of this equation is

$$\sigma_{rr} = A + \frac{B}{r^3}. \tag{13.44}$$

The corresponding hoop stresses are

$$\sigma_{\theta\theta} = \sigma_{\phi\phi} = A - \frac{B}{2r^3}. \tag{13.45}$$

The boundary conditions for the Lamé problem of a pressurized hollow sphere are

$$\sigma_{rr}(R_1) = -p_1, \quad \sigma_{rr}(R_2) = -p_2. \tag{13.46}$$

These are satisfied provided that

$$A = \frac{p_1 R_1^3 - p_2 R_2^3}{R_2^3 - R_1^3}, \quad B = -\frac{R_1^3 R_2^3}{R_2^3 - R_1^3}(p_1 - p_2). \tag{13.47}$$

Consequently, the stress components are

$$\sigma_{rr} = \frac{R_2^3}{R_2^3 - R_1^3}\left[p_1\frac{R_1^3}{R_2^3} - p_2 - (p_1 - p_2)\frac{R_1^3}{r^3}\right], \tag{13.48}$$

$$\sigma_{\theta\theta} = \frac{R_2^3}{R_2^3 - R_1^3}\left[p_1\frac{R_1^3}{R_2^3} - p_2 + (p_1 - p_2)\frac{R_1^3}{2r^3}\right]. \tag{13.49}$$

The corresponding hoop strain is obtained by substituting (13.48) and (13.49) into the stress-strain relation (13.40). The result is

$$e_{\theta\theta} = \frac{A}{3K} - \frac{B}{4\mu}\frac{1}{r^3}, \tag{13.50}$$

where the elastic bulk modulus is $K = E/3(1 - 2\nu)$, and the elastic shear modulus is $\mu = E/2(1 + \nu)$. Thus, the radial displacement $u = r\epsilon_{\theta\theta}$ is

$$u = \frac{1}{3\kappa}\frac{p_1 R_1^3 - p_2 R_2^3}{R_2^3 - R_1^3}r + \frac{1}{4\mu}(p_1 - p_2)\frac{R_1^3 R_2^3}{R_2^3 - R_1^3}\frac{1}{r^2}. \tag{13.51}$$

For the nonpressurized hole ($p_1 = 0$) under remote pressure p_2 at infinity, the previous results give

$$\sigma_{rr} = -p_2\left(1 - \frac{R_1^3}{r^3}\right), \quad \sigma_{\theta\theta} = \sigma_{\phi\phi} = -p_2\left(1 + \frac{R_1^3}{2r^3}\right), \tag{13.52}$$

$$u = -\frac{p_2}{4\mu}\left[\frac{2(1 - 2\nu)}{1 + \nu}r + \frac{R_1^3}{r^2}\right]. \tag{13.53}$$

The displacement is here conveniently expressed by using the Poisson's ratio ν. For the pressurized cylindrical hole in an infinite medium with $p_2 = 0$, we obtain

$$\sigma_{rr} = -p_1\frac{R_1^3}{r^3}, \quad \sigma_{\theta\theta} = \sigma_{\phi\phi} = p_1\frac{R_1^3}{2r^3}, \tag{13.54}$$

$$u = \frac{p_1}{4\mu}\frac{R_1^3}{r^2}. \tag{13.55}$$

13.4.1 Pressurized Rigid Inclusion

Imagine that the cavity of the previous solution is a rigid inclusion so that the displacement at its periphery were to vanish (Fig. 13.2b). The inclusion is to be loaded by a uniform pressure applied at the outer boundary at r_2, and the inclusion's radius is still r_1. The earlier found form for the displacement field (13.31) still holds, but now we have the boundary condition that at $r = r_1$,

$$u_r(r_1) = ar_1 + b/r_1^2 = 0 \quad \Rightarrow \quad a = -b/r_1^3. \tag{13.56}$$

With this we apply the same boundary condition at $r = r_2$, viz.,

$$\sigma_{rr}(r_2) = -p, \tag{13.57}$$

13.4. Pressurized Sphere: Stress-Based Solution

to find

$$b = p\left[\frac{E}{r_1^3(1-2\nu)} + \frac{2E}{r_2^3(1+\nu)}\right]^{-1}. \tag{13.58}$$

The complete solution for the displacements, stresses and strains follows by simple calculation. Here we look at the limit as r_2 becomes very large, i.e.,

$$\lim_{r_2 \to \infty} b = \frac{r_1^3(1-2\nu)p}{E}. \tag{13.59}$$

The radial stress is then

$$\sigma_{rr} = -p - 2p\frac{1-2\nu}{1+\nu}(r_1/r)^3. \tag{13.60}$$

Note that for an incompressible material, with $\nu = 1/2$, the radial stress is everywhere constant and equal to $-p$. In addition, at the inclusion surface

$$\sigma_{rr}(r_1) = -3p\frac{1-\nu}{1+\nu}, \tag{13.61}$$

which reveals an interesting effect of Poisson's ratio on the concentration or virtual removal of pressure on the surface of the inclusion. The complete stress and strain field can be easily computed from the results listed above.

13.4.2 Disk with Circumferential Shear

It is possible to generate singular solutions of physical interest using the techniques introduced here. For example, consider the displacement field

$$\mathbf{u} = C\,\mathrm{grad}\,\theta, \tag{13.62}$$

where $\theta = \tan^{-1}(x_2/x_1)$, and C is a constant to be determined by suitable boundary conditions. Note some properties of the nonzero components of displacement gradient. Since

$$u_1 = -\frac{Cx_2}{x_1^2+x_2^2}, \quad u_2 = \frac{Cx_1}{x_1^2+x_2^2}, \tag{13.63}$$

the normal strains are

$$u_{1,1} = 2C\frac{x_2 x_1}{(x_1^2+x_2^2)^2}, \quad u_{2,2} = -2C\frac{x_2 x_1}{(x_1^2+x_2^2)^2}. \tag{13.64}$$

This shows, *inter alia*, that the deformation described by this displacement field involves no volume change, because $u_3 = 0$ and $u_{1,1} + u_{2,2} = 0$. Also, the equilibrium equations are satisfied because $\nabla^2 \mathbf{u} = \mathbf{0}$ and div $\mathbf{u} = 0$. The other relevant components of displacement gradient are

$$\begin{aligned} u_{1,2} &= -\frac{C}{x_1^2+x_2^2} + 2C\frac{x_2^2}{(x_1^2+x_2^2)^2}, \\ u_{2,1} &= \frac{C}{x_1^2+x_2^2} - 2C\frac{x_1^2}{(x_1^2+x_2^2)^2}, \end{aligned} \tag{13.65}$$

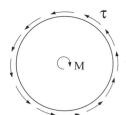

Figure 13.3. Disk-like region subject to shear.

and thus the shear strain is

$$e_{12} = \frac{1}{2}(u_{1,2} + u_{2,1}) = C\frac{x_2^2 - x_1^2}{(x_1^2 + x_2^2)^2}. \tag{13.66}$$

Consequently,

$$\sigma_{12} = 2GC\frac{x_2^2 - x_1^2}{(x_1^2 + x_2^2)^2}, \tag{13.67}$$

and

$$\sigma_{11} = -\sigma_{22} = 4GC\frac{x_1 x_2}{(x_1^2 + x_2^2)^2}. \tag{13.68}$$

The circumferential shear stress is

$$\sigma_{\theta r} = 2GC\frac{1}{x_1^2 + x_2^2} = 2GC\frac{1}{r^2}. \tag{13.69}$$

Thus, the peculiar displacement field of the problem gives rise to a singular stress field associated with uniform shear stress acting on the periphery of a disk-like region. If for example, this region were a disc of radius $r = r_1$ with applied shear stress $\sigma_{\theta r} = \tau$ on its edge, then $C = \tau r_1^2/2G$. The concentrated couple in the center of the disk, needed to equilibrate this shear stress on the periphery of the disk, is $M = 2\pi r_1^2 \tau$ (Fig. 13.3).

13.4.3 Sphere Subject to Temperature Gradients

Here we consider a sphere of radius r_2 subject to a radial temperature gradient. Again the displacement field is assumed to be purely radial. Thus, curl $\mathbf{u} = \mathbf{0}$ and (13.9) reduces to

$$\frac{d}{dr}\left[\frac{1}{r^2}\frac{d(r^2 u_r)}{dr}\right] = \alpha\frac{1+\nu}{3(1-\nu)}\frac{dT}{dr}. \tag{13.70}$$

We let $T(r_2) = 0$ such that if the sphere were uniformly cooled to that temperature no stresses would result. If (13.70) is integrated, we find

$$u_r = \alpha\frac{1+\nu}{3(1-\nu)}\left[\frac{1}{r^2}\int_0^r T(r)r^2\,dr + \frac{2(1-2\nu)}{1+\nu}\frac{r}{r_2^3}\int_0^{r_2} T(r)r^2\,dr\right]. \tag{13.71}$$

13.5 Spherical Indentation

In this section we examine the problem of indentation of an elastic half-plane by a rigid spherical indenter. The process of indentation is essentially that associated with the Brinell hardness test using a spherical indenter. We shall analyze the problem using the approach

13.5. Spherical Indentation

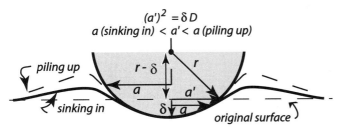

Figure 13.4. Geometry of an indentation. Note the phenomenology of either *sinking in* or *piling up* depending on the degree of nonlinearity of the indented material.

introduced by Love (1944) and later generalized in terms of the Galerkin vector potential. The geometry of indentations, in its most general form in materials that are elastic-plastic, is as sketched in Fig. 13.4. For nonlinear materials, indented material generally tends to either *sink-in* or *pile-up* beneath or around the indenter. The trends are that for materials that are linear or that display high rates of strain hardening indented material sinks away from the indenter, whereas for materials that are weakly hardening piling up is observed. The effect is that the actual contact radius will be smaller in the case of sinking in behavior and larger for the case of piling up. This, of course has an important effect on the intensity of the pressure that develops beneath the indenter and for the quantitative assessment of hardness. For the case considered here, *viz.*, that of linear elastic materials, we will find the characteristic Hertzian sinking in.

13.5.1 Displacement-Based Equilibrium

Begin with the isotropic elastic constitutive relations expressed, using Lamé constants λ and μ, as

$$\sigma_{ij} = \lambda e_{mm} \delta_{ij} + 2\mu e_{ij}. \tag{13.72}$$

Recall the connections among the isotropic elastic constants

$$E = \frac{\mu(3\lambda + 2\mu)}{\lambda + \mu}, \quad \mu = G,$$

$$\nu = \frac{\lambda}{2(\lambda + \mu)}, \quad G = \frac{E}{2(1 + \nu)}, \tag{13.73}$$

and the equations of equilibrium

$$\frac{\partial \sigma_{ij}}{\partial x_j} + b_i = 0. \tag{13.74}$$

Since $e_{kk} = \partial u_k / \partial x_k = \text{div } \mathbf{u}$, it is readily verified that

$$2 \frac{\partial e_{ij}}{\partial x_j} = \nabla^2 u_i + \frac{\partial e_{kk}}{\partial x_i}. \tag{13.75}$$

Thus,

$$(\lambda + \mu) \frac{\partial}{\partial x_i} \text{div } \mathbf{u} + \mu \nabla^2 u_i + b_i = 0. \tag{13.76}$$

Using the relation

$$\lambda + \mu = \frac{\mu}{1 - 2\nu}, \tag{13.77}$$

the equilibrium equations can be rewritten as

$$\nabla \operatorname{div} \mathbf{u} + (1 - 2\nu)\nabla^2 \mathbf{u} + \frac{1 - 2\nu}{\mu} \mathbf{b} = \mathbf{0}, \tag{13.78}$$

where $\mu = G$ is the shear modulus. In the absence of body forces, this becomes

$$\nabla \operatorname{div} \mathbf{u} + (1 - 2\nu)\nabla^2 \mathbf{u} = \mathbf{0}. \tag{13.79}$$

13.5.2 Strain Potentials

We have already seen that a suitable displacement function can be constructed from the gradient of a scalar field as

$$\mathbf{u} = \frac{1}{G}\nabla\phi. \tag{13.80}$$

To satisfy (13.79), ϕ must be such that

$$\nabla \nabla^2 \phi = \mathbf{0}, \tag{13.81}$$

which leads to the conclusion that

$$\nabla^2 \phi = \text{const.} \tag{13.82}$$

We note that (13.80) is not the most general form of displacement solution. A more general form is one in which the displacement is constructed from the second derivatives of a potential. As there are no operators of this type that transform a scalar into a vector, the potential must be a vector potential. The Galerkin vector potential, $\mathcal{G}$, is thus introduced and the most general form for the displacement $\mathbf{u}$ is given as

$$\mathbf{u} = \frac{1}{2G}\left(c\nabla^2 - \nabla \operatorname{div}\right)\mathcal{G}. \tag{13.83}$$

When (13.83) is inserted into (13.79), it is found that $c = 2(1 - \nu)$, and

$$\mathbf{u} = \frac{1}{2G}\left[2(1 - \nu)\nabla^2 \mathcal{G} - \nabla \operatorname{div} \mathcal{G}\right]. \tag{13.84}$$

The Galerkin vector potential $\mathcal{G}$ must satisfy the biharmonic equation

$$\nabla^4 \mathcal{G} = 0. \tag{13.85}$$

For the axisymmetric problem considered, we make use of one component of $\mathcal{G}$, viz., $\mathcal{G}_z$; we call $\mathcal{G}_z = \Omega$. In the (r, z) coordinate system, we then have

$$u_r = -\frac{1}{2G}\frac{\partial^2 \Omega}{\partial r \partial z}, \quad u_z = \frac{1}{2G}\left[2(1 - \nu)\nabla^2 \Omega - \frac{\partial^2 \Omega}{\partial z^2}\right], \tag{13.86}$$

$$u_\theta = 0, \quad \Omega = \Omega(r, z).$$

13.5. Spherical Indentation

The Laplacean in the (r, z) coordinates is

$$\nabla^2 = \frac{\partial^2}{\partial r^2} + \frac{1}{r}\frac{\partial}{\partial r} + \frac{\partial^2}{\partial z^2}. \tag{13.87}$$

The stresses are

$$\begin{aligned}
\sigma_{rr} &= \frac{\partial}{\partial z}\left(\nu\nabla^2\Omega - \frac{\partial^2\Omega}{\partial r^2}\right), \\
\sigma_{\theta\theta} &= \frac{\partial}{\partial z}\left(\nu\nabla^2\Omega - \frac{1}{r}\frac{\partial\Omega}{\partial r}\right), \\
\sigma_{zz} &= \frac{\partial}{\partial z}\left[(2-\nu)\nabla^2\Omega - \frac{\partial^2\Omega}{\partial z^2}\right], \\
\sigma_{rz} &= \frac{\partial}{\partial r}\left[(1-\nu)\nabla^2\Omega - \frac{\partial^2\Omega}{\partial z^2}\right],
\end{aligned} \tag{13.88}$$

where the Love's potential Ω satisfies

$$\nabla^4\Omega = 0. \tag{13.89}$$

Use has been made of the definitions of strain in a cylindrical system, *viz.*,

$$e_{rr} = \frac{\partial u_r}{\partial r}, \quad e_{\theta\theta} = \frac{u_r}{r}, \quad e_{zz} = \frac{\partial u_z}{\partial z}, \tag{13.90}$$

and

$$e_{rz} = \frac{1}{2}\left(\frac{\partial u_r}{\partial z} + \frac{\partial u_z}{\partial r}\right). \tag{13.91}$$

13.5.3 Point Force on a Half-Plane

The Love's potential for a concentrated point force of intensity p_0 is

$$\Omega = \frac{p_0}{2\pi}z\left[1 + 2\nu\sqrt{1+r^2/z^2} + (1-2\nu)\ln\left(1+\sqrt{1+r^2/z^2}\right) - \ln z\right].$$

The corresponding displacements, from (13.86), are

$$u_r = \frac{p_0(1-2\nu)}{4\pi G}\frac{1}{r}\left[\frac{1}{1-2\nu}\frac{r^2/z^2}{(1+r^2/z^2)^{3/2}} + \frac{1}{(1+r^2/z^2)^{1/2}} - 1\right], \tag{13.92}$$

and

$$u_z = \frac{p_0}{4\pi G}\frac{1}{z}\left[\frac{1}{(1+r^2/z^2)^{3/2}} + 2(1-\nu)\frac{1}{(1+r^2/z^2)^{1/2}}\right]. \tag{13.93}$$

Of particular interest is the z component of displacement, which we denote by ω when $z = 0$. In general,

$$u_z = \frac{p_0}{4\pi G}\left[\frac{z^2}{\rho^3} + 2(1-\nu)\frac{1}{\rho}\right], \tag{13.94}$$

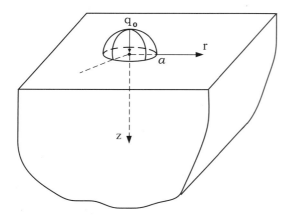

Figure 13.5. A hemispherical distributed load over the half-space.

and on the surface, $z = 0$,

$$\omega = \frac{p_0(1-\nu)}{2\pi G} \frac{1}{\rho} = \frac{p_0(1-\nu)}{2\pi G} \frac{1}{r}, \tag{13.95}$$

where $\rho = (z^2 + r^2)^{1/2}$ is the distance from the point force to the field point.

13.5.4 Hemispherical Load Distribution

Consider now the application of a distributed force whose intensity has a hemispherical form (Fig. 13.5),

$$q(r) = q_0 \left(1 - \frac{r^2}{a^2}\right)^{1/2}. \tag{13.96}$$

The displacements are found by the using the solutions given in (13.95) as a Green's function, so that

$$\omega = \frac{(1-\nu)q_0}{2aG} \int_0^{\pi/2} (a^2 - r^2 \sin^2 \varphi) \, d\varphi, \quad r \leq a, \tag{13.97}$$

$$\omega = \frac{(1-\nu)q_0}{2aG} \int_0^{\alpha} (a^2 - r^2 \sin^2 \varphi) \, d\varphi, \quad r \geq a. \tag{13.98}$$

The angles φ and α are defined in Fig. 13.6. The results of integration are

$$\omega = \omega_{\max}\left(1 - \frac{r^2}{2a^2}\right), \quad r \leq a,$$

$$\omega = \frac{\omega_{\max}}{\pi} \frac{r^2}{a^2}\left[\left(2\frac{a^2}{r^2} - 1\right)\sin^{-1}\left(\frac{a}{r}\right) + \frac{a}{r}\left(1 - \frac{a^2}{r^2}\right)^{1/2}\right], \quad r \geq a. \tag{13.99}$$

The maximum displacement at the center of the distributed load, $\omega_{\max} = \omega(0,0)$, is twice greater than the displacement at the periphery of the load, $\omega_a = \omega(a,0)$, i.e.,

$$\omega_{\max} = 2\omega_a, \quad \omega_{\max} = \frac{\pi q_0(1-\nu)}{4G} a. \tag{13.100}$$

13.5. Spherical Indentation

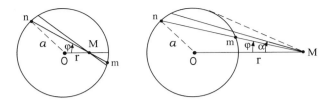

Figure 13.6. Integration angles. Note "M" signifies the typical field point, either inside or outside the circle.

The total force is

$$P = 2\pi \int_0^a q_0 \left(1 - \frac{r^2}{a^2}\right)^{1/2} r\, dr = \frac{2}{3}\pi a^2 q_0, \tag{13.101}$$

so that

$$P = \frac{2}{3}\frac{3G}{1-\nu} a\, \omega_{\max}. \tag{13.102}$$

13.5.5 Indentation by a Spherical Ball

The solution for the spherical load is an essential ingredient for constructing the Hertz solution for the elastic indentation of a half-space by a spherical indenter. Denoting by R the radius of the ball, and by δ the depth of the indentation, the equation of the sphere, on the surface of contact, is

$$r^2 + [z + (R - \delta)]^2 = R^2. \tag{13.103}$$

For shallow indentations $\delta/R \ll 1$ and $z/R \ll 1$, where in the second inequality the coordinate z is meant to be on the indented surface. Thus, when (13.103) is expanded, terms such as z^2 and δz are neglected so that, with z replaced by ω on the surface, (13.103) yields

$$\omega(r) = \delta - \frac{r^2}{2R}, \quad r \leq a. \tag{13.104}$$

Because a is a geometric mean of $\delta/2$ and $2R - \delta/2$, we have

$$a^2 = \frac{\delta}{2}\left(2R - \frac{\delta}{2}\right), \quad a^2 \approx R\delta, \tag{13.105}$$

and thus (13.104) can be rewritten as

$$\omega(r) = \delta\left(1 - \frac{r^2}{2a^2}\right), \quad r \leq a, \tag{13.106}$$

in agreement with the displacement field at the surface given in (13.99) for a hemispherical load distribution.

Alternatively, consider the displacement at the periphery of the contact circle ($r = a$). Call this $\delta_a = \omega(a, 0)$. Then,

$$\omega(r) = \delta_a + \frac{1}{2R}(a^2 - r^2), \tag{13.107}$$

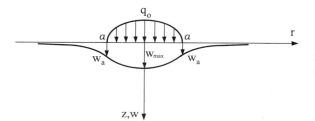

Figure 13.7. The displacement profile of the indented surface.

with a simple connection

$$\delta = \delta_a + \frac{a^2}{2R}. \tag{13.108}$$

Note that $\hat{a} \approx \sqrt{2R\delta}$ would be the "apparent radius of contact" of the sphere that was ideally pressed into the surface to a depth δ, assuming that the indented material simply did not deflect or distort. The quantity

$$c^2 = a^2/\hat{a}^2 \tag{13.109}$$

is an invariant of such indentation processes, and for this linear elastic case

$$c^2 = 1/2. \tag{13.110}$$

If (13.105) is used in (13.102), we obtain

$$P = \frac{8}{3} \frac{G}{1-\nu} R^{1/2} \delta^{3/2}, \tag{13.111}$$

or, in terms of the radius of contact,

$$P = \frac{8}{3} \frac{G}{1-\nu} \frac{1}{R} a^3. \tag{13.112}$$

An indentation hardness can be defined as

$$\mathcal{H} = \frac{P}{\pi a^2} = \frac{8}{3} \frac{G}{1-\nu} \frac{a}{R} = \frac{8}{3} \frac{G}{1-\nu} \left(\frac{\delta}{R}\right)^{1/2}. \tag{13.113}$$

The displacement of the indented surface (Fig. 13.7) is

$$\omega = \frac{a^2}{R}\left(1 - \frac{r^2}{2a^2}\right), \quad r \leq a,$$

$$\omega = \frac{r^2}{\pi R}\left[\left(2\frac{a^2}{r^2} - 1\right)\sin^{-1}\left(\frac{a}{r}\right) + \frac{a}{r}\left(1 - \frac{a^2}{r^2}\right)^{1/2}\right], \quad r \geq a. \tag{13.114}$$

Alternatively, the displacement can be expressed in terms of the depth δ of the indentation (Fig. 13.8) as

$$\omega = \delta - \frac{r^2}{2R}, \quad r \leq \sqrt{R\delta}, \tag{13.115}$$

$$\omega = \frac{r^2}{\pi R}\left[\left(2\frac{R\delta}{r^2} - 1\right)\sin^{-1}\left(\frac{\sqrt{R\delta}}{r}\right) + \frac{\sqrt{R\delta}}{r}\left(1 - \frac{R\delta}{r^2}\right)^{1/2}\right], \quad r \geq \sqrt{R\delta}.$$

13.6. Point Forces on Elastic Half-Space

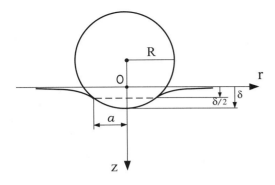

Figure 13.8. The width a and the depth δ of the indentation by a spherical ball.

The slope of the indented surface is continuous at the periphery of the contact circle and equal to

$$\frac{d\omega}{dr} = -\frac{\delta}{a}, \qquad (13.116)$$

which is unlike indentation with a rigid flat punch (where there is a slope discontinuity). On the other hand, the hemispherical loading considered here prescribes a normal stress that vanishes at the periphery of contact, in contrast to that of a flat punch (for which the normal stress is infinite at the periphery).

13.6 Point Forces on Elastic Half-Space

Here we summarize the solutions for point force loading on elastic half-spaces for 3D infinite media. The coordinate system used is shown in Fig. 13.9. We consider both normal and tangential point forces imposed on the surface of an elastically isotropic media. Relations (13.86)–(13.87) gave the radially symmetric displacement fields for a normal point force in polar coordinates. We list here the same displacement components in the Cartesian coordinate frame of Fig. 13.9. They are

$$u_x(x, y, z) = \frac{p_0}{4\pi G} \left[\frac{xz}{\rho^3} - (1-2\nu) \frac{x}{\rho(\rho+z)} \right],$$

$$u_y(x, y, z) = \frac{p_0}{4\pi G} \left[\frac{yz}{\rho^3} - (1-2\nu) \frac{y}{\rho(\rho+z)} \right], \qquad (13.117)$$

$$u_z(x, y, z) = \frac{p_0}{4\pi G} \left[\frac{z^2}{\rho^3} + \frac{2(1-\nu)}{\rho} \right],$$

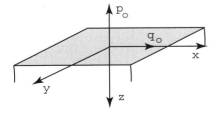

Figure 13.9. Normal and tangential point forces on elastic half-space.

where $\rho = (x^2 + y^2 + z^2)^{1/2}$. It is of specific interest to form these on the surface of the half-space, where $r^2 = x^2 + y^2$. The displacements there are

$$\bar{u}_x(x, y) = -\frac{p_0}{4\pi G} \frac{(1 - 2v)x}{r^2},$$

$$\bar{u}_y(x, y) = -\frac{p_0}{4\pi G} \frac{(1 - 2v)y}{r^2}, \quad (13.118)$$

$$\bar{u}_z(x, y) = \frac{p_0}{4\pi G} \frac{2(1 - 2v)}{r}.$$

The stresses are then readily calculated by the straightforward procedure of forming the strains and invoking the isotropic elastic constitutive relations.

The solution for the case of a tangential point load of magnitude q_0 is obtained using similar techniques as described in the previous section on spherical indentation. The results for the displacements are

$$u_x(x, y, z) = \frac{q_0}{4\pi G} \left\{ \frac{1}{\rho} + \frac{x^2}{\rho^3} + (1 - 2v) \left[\frac{1}{\rho + z} - \frac{x^2}{\rho(\rho + z)^2} \right] \right\},$$

$$u_y(x, y, z) = \frac{q_0}{4\pi G} \left[\frac{xy}{\rho^3} - (1 - 2v) \frac{xy}{\rho(\rho + z)^2} \right], \quad (13.119)$$

$$u_z(x, y, z) = \frac{q_0}{4\pi G} \left[\frac{xz}{\rho^3} + (1 - 2v) \frac{x}{\rho(\rho + z)} \right].$$

When expressed on the surface $z = 0$ of the half-space, these displacements are

$$\bar{u}_x(x, y) = \frac{q_0}{4\pi G} \left[\frac{2(1 - v)}{r} + \frac{2vx^2}{r^3} \right],$$

$$\bar{u}_y(x, y) = \frac{q_0}{4\pi G} \frac{2vxy}{r^3}, \quad (13.120)$$

$$\bar{u}_z(x, y) = \frac{q_0}{4\pi G} \frac{(1 - 2v)x}{r^2}.$$

Again, the elastic strains and then the stresses can be readily calculated from (13.119).

For later reference, we note that if a general set of point loads $\mathbf{f} = \{f_x, f_y, f_z\}$ were applied to the surface, the displacements on the surface could be expressed as

$$\bar{u}_\alpha(x, y) = \Pi_{\alpha\beta}(x, y, z) f_\beta, \quad (\alpha, \beta = x, y, z), \quad (13.121)$$

where

$$\Pi = \frac{1}{4\pi G} \begin{Bmatrix} \frac{2(1-v)}{r} + \frac{2vx^2}{r^3} & \frac{2vxy}{r^3} & -\frac{(1-2v)x}{r^2} \\ \frac{2vxy}{r^3} & \frac{2(1-v)}{r} + \frac{2vy^2}{r^3} & -\frac{(1-2v)y}{r^2} \\ \frac{(1-2v)x}{r^2} & \frac{(1-2v)y}{r^2} & \frac{2(1-2v)}{r} \end{Bmatrix}. \quad (13.122)$$

This form will be useful in solving for the elastic fields at the surfaces of half-planes subject to general loading *via* superposition techniques.

13.7 Suggested Reading

Filonenko-Borodich, M. (1964), *Theory of Elasticity*, Noordhoff, Groningen, Netherlands.

Fung, Y.-C. (1965), *Foundations of Solid Mechanics*, Prentice Hall, Englewood Cliffs, New Jersey.

Little, R. W. (1973), *Elasticity*, Prentice Hall, Englewood Cliffs, New Jersey.

Love, A. E. H. (1944), *A Treatise on the Mathematical Theory of Elasticity*, Dover, New York.

Novozhilov, V. V. (1961), *Theory of Elasticity*, Pergamon, New York.

Sokolnikoff, I. S. (1956), *Mathematical Theory of Elasticity*, 2nd ed., McGraw-Hill, New York.

Timoshenko, S. P., and Goodier, J. N. (1970), *Theory of Elasticity*, 3rd ed., McGraw-Hill, New York.

14 Anisotropic 3D Solutions

14.1 Point Force

For the time being we refer to the case of generally anisotropic elastic media, and later specialize to the case of isotropic elastic media. Recall the equilibrium equations

$$\sigma_{ij,j} + b_i = 0, \tag{14.1}$$

where b_i are the components of the body force per unit volume. The constitutive equations that connect the stresses to small elastic strains are

$$\sigma_{ij} = C_{ijkl}\, e_{kl} = C_{ijkl}\, u_{k,l}. \tag{14.2}$$

Thus, combining (14.1) and (14.2), we obtain

$$C_{ijkl}\, u_{k,lj} = b_i. \tag{14.3}$$

Suppose **b** is concentrated at a "point." Next let **b** act only in the x_m direction and, for now, let the magnitude of this concentrated force be unity. If the force acts at $\mathbf{x} = \mathbf{x}'$, we can write

$$b_i = \begin{cases} 0, & \text{if } i \neq m, \\ \delta(\mathbf{x} - \mathbf{x}'), & \text{if } i = m, \end{cases}$$

where $\delta(\mathbf{x})$ is the Dirac delta function, defined in Chapter 3. The above can be rewritten by using the Kronecker δ as

$$b_i = \delta_{im}\, \delta(\mathbf{x} - \mathbf{x}'), \quad (i = 1, 2, 3). \tag{14.4}$$

14.2 Green's Function

Using (14.4), equation (14.3) becomes

$$C_{ijkl}\, u^{(m)}_{k,lj} = -\delta_{im}\, \delta(\mathbf{x} - \mathbf{x}'). \tag{14.5}$$

14.2. Green's Function

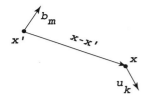

Figure 14.1. Displacement caused by a unit point force.

The direct physical interpretation of $u_k^{(m)}$ is

$u_k^{(m)}$ = *the component of displacement in the kth direction at the point* **x** *caused by a point force acting in the mth direction at* **x**′.

We note that $u_k^{(m)}$ is a tensor as it associates a vector, **u**, with another vector, **b** (Fig. 14.1). Thus, the component of tensor **G** is $G_{mk} = u_k^{(m)}$, and (14.5) becomes

$$C_{ijkl} \, G_{km,jl}(\mathbf{x}, \mathbf{x}') = -\delta_{im} \delta(\mathbf{x} - \mathbf{x}'). \tag{14.6}$$

The function G_{km} is called a Green's function. It is clearly symmetric, *i.e.*, $G_{km} = G_{mk}$.

When integrating (14.6) we demand that the influence of the point force vanish sufficiently rapid, and in particular we demand that **G** vanish at least as fast as

$$\mathbf{G} \sim \frac{1}{|\mathbf{x} - \mathbf{x}'|} = \frac{1}{r} \quad \text{as } r \to \infty. \tag{14.7}$$

Solution by Fourier transforms provides a direct route to a solution. The Fourier transform of $\mathbf{G}(\mathbf{x}, \mathbf{x}')$ is obtained from

$$\mathbf{g}(\mathbf{K}) = \int_{-\infty}^{\infty} \mathbf{G}(\mathbf{x}, \mathbf{x}') \exp(i\mathbf{K} \cdot \mathbf{x}) \, d^3\mathbf{x}, \tag{14.8}$$

where $d^3\mathbf{x} \equiv dV = dx_1 dx_2 dx_3$ and **K** is the Fourier vector in Fourier space. The inverse transform is then obtained from

$$\mathbf{G}(\mathbf{x}, \mathbf{x}') = \frac{1}{(2\pi)^3} \int_{-\infty}^{\infty} \mathbf{g}(\mathbf{K}) \exp(-i\mathbf{K} \cdot \mathbf{x}) \, d^3\mathbf{K}. \tag{14.9}$$

To proceed, multiply (14.6) by $\exp(i\mathbf{K} \cdot \mathbf{x})$ or by $(\mathbf{x} - \mathbf{x}')$, because **x**′ is a fixed position in the integrations, and integrate over all physical, *i.e.*, **x**, space to obtain

$$\int_{-\infty}^{\infty} C_{ijkl} \, G_{km,jl}(\mathbf{x}, \mathbf{x}') \exp[i\mathbf{K} \cdot (\mathbf{x} - \mathbf{x}')] \, d^3\mathbf{x}$$
$$= -\int_{-\infty}^{\infty} \delta_{im} \delta(\mathbf{x} - \mathbf{x}') \exp[i\mathbf{K} \cdot (\mathbf{x} - \mathbf{x}')] \, d^3\mathbf{x}. \tag{14.10}$$

As noted above, $d^3\mathbf{x} = d^3(\mathbf{x} - \mathbf{x}')$ because of the fixity of **x**′. Integration by parts yields, after taking into account the vanishing of **G** at afar,

$$-\int_{-\infty}^{\infty} K_j K_l C_{ijkl} G_{km}(\mathbf{x}, \mathbf{x}') d^3(\mathbf{x} - \mathbf{x}') = -\delta_{im}. \tag{14.11}$$

To reduce this expression, we define a unit vector, $\mathbf{z}$, as the unit vector along $\mathbf{K}$, i.e,

$$\mathbf{K} = K\mathbf{z}, \qquad K \equiv |\mathbf{K}|. \tag{14.12}$$

Then,

$$K_j K_l C_{ijkl} = K^2 z_j z_l C_{ijkl} = K^2 M_{ik}(\mathbf{z}). \tag{14.13}$$

This, in turn, leads to the definition of the acoustic tensor, or Christoffel stiffness tensor,

$$\mathbf{M} \equiv \mathbf{z} \cdot \mathbf{C} \cdot \mathbf{z}. \tag{14.14}$$

Thus, (14.11) becomes

$$-\int_{-\infty}^{\infty} K^2 M_{ik}(\mathbf{z}) G_{km}(\mathbf{x},\mathbf{x}') \exp[i\mathbf{K}\cdot(\mathbf{x}-\mathbf{x}')] d^3(\mathbf{x}-\mathbf{x}') = -\delta_{im}. \tag{14.15}$$

Removing items that are constant within the integral, we obtain

$$-K^2 M_{ik}(\mathbf{z}) \int_{-\infty}^{\infty} G_{km}(\mathbf{x},\mathbf{x}') \exp[i\mathbf{K}\cdot(\mathbf{x}-\mathbf{x}')] d^3(\mathbf{x}-\mathbf{x}') = -\delta_{im}. \tag{14.16}$$

Therefore,

$$-K^2 M_{ik} g_{km} = -\delta_{im}, \tag{14.17}$$

or, by inversion,

$$g_{sm} = \frac{M_{sm}^{-1}(\mathbf{z})}{K^2}. \tag{14.18}$$

The inverse transform, or $\mathbf{G}$ itself, is thus

$$G_{km}(\mathbf{x},\mathbf{x}') = \frac{1}{(2\pi)^3} \int_{-\infty}^{\infty} \frac{M_{km}^{-1}}{K^2} \exp[-i\mathbf{K}\cdot(\mathbf{x}-\mathbf{x}')] d^3\mathbf{K}. \tag{14.19}$$

Since $\mathbf{G}$ is real, we need only take the real part of (14.19), i.e.,

$$G_{km}(\mathbf{x},\mathbf{x}') = \frac{1}{(2\pi)^3} \int_{-\infty}^{\infty} \frac{M_{km}^{-1}}{K^2} \cos[\mathbf{K}\cdot(\mathbf{x}-\mathbf{x}')] d^3\mathbf{K}. \tag{14.20}$$

To perform these integrations it is convenient to introduce a unit vector $\mathbf{T}$ and a stretched vector $\boldsymbol{\lambda}$ as follows

$$\begin{aligned}\mathbf{x}-\mathbf{x}' &= |\mathbf{x}-\mathbf{x}'|\mathbf{T}, \\ \boldsymbol{\lambda} &= \mathbf{K}|\mathbf{x}-\mathbf{x}'| = K|\mathbf{x}-\mathbf{x}'|\mathbf{z}.\end{aligned} \tag{14.21}$$

With these, the inverse transform becomes

$$G_{km}(\mathbf{x},\mathbf{x}') = \frac{1}{8\pi^3 |\mathbf{x}-\mathbf{x}'|} \int_{-\infty}^{\infty} \frac{M_{km}^{-1}(\mathbf{z})}{\lambda^2} \cos(\lambda\mathbf{z}\cdot\mathbf{T}) d^3\boldsymbol{\lambda}. \tag{14.22}$$

Note, if $\mathbf{T}$ is replaced by $-\mathbf{T}$, we have $\cos(\lambda\mathbf{z}\cdot\mathbf{T}) = \cos(\lambda\mathbf{z}\cdot-\mathbf{T})$. Accordingly, we can express (14.22) as

$$G_{km}(\mathbf{x},\mathbf{x}') = G_{km}(s\mathbf{T}) = \frac{\operatorname{sgn}(s)}{s} G_{km}(\mathbf{T}), \tag{14.23}$$

where s is an algebraically signed scalar. Thus $\mathbf{G}$ scales as $1/|\mathbf{x}-\mathbf{x}'|$. In fact, $\mathbf{G}$ depends on $\mathbf{x}$ and $\mathbf{x}'$ strictly as on $(\mathbf{x}-\mathbf{x}')$.

14.2. Green's Function

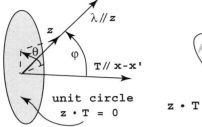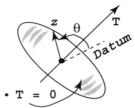

Figure 14.2. A polar coordinate system aligned with **T**.

We may now inquire about the derivatives of **G**. For example, we have

$$G_{km,p}(\mathbf{x}-\mathbf{x}') = \frac{-1}{8\pi^3|\mathbf{x}-\mathbf{x}'|^2}\int_{-\infty}^{\infty}\frac{z_p M_{km}^{-1}(\mathbf{z})}{\lambda}\sin(\lambda\mathbf{z}\cdot\mathbf{T})\,d^3\lambda. \tag{14.24}$$

Thus,

$$G_{km,p}(\mathbf{x},\mathbf{x}') = G_{km,p}(s\mathbf{T}) = \frac{\mathrm{sgn}(s)}{s^2}G_{km,p}(\mathbf{T}). \tag{14.25}$$

Similarly, we find

$$G_{km,pt}(\mathbf{x},\mathbf{x}') = G_{km,pt}(s\mathbf{T}) = \frac{\mathrm{sgn}(s)}{s^3}G_{km,pt}(\mathbf{T}), \tag{14.26}$$

and so on. In general, we obtain

$$G_{km,\alpha_1\alpha_2\ldots\alpha_n}(\mathbf{x},\mathbf{x}') = G_{km,\alpha_1\alpha_2\ldots\alpha_n}(s\mathbf{T}) = \frac{\mathrm{sgn}(s)}{s^n}G_{km,\alpha_1\alpha_2\ldots\alpha_n}(\mathbf{T}).$$

We now return to (14.22) and introduce polar coordinates aligned with **T** such as in Fig. 14.2. Then

$$d^3\lambda = \lambda^2\sin\phi\,d\lambda\,d\phi\,d\theta, \qquad \mathbf{z}\cdot\mathbf{T} = \cos\phi, \tag{14.27}$$

and

$$G_{km}(\mathbf{x}-\mathbf{x}') = \frac{1}{8\pi^3|\mathbf{x}-\mathbf{x}'|}\int_0^{2\pi}\int_0^{\pi}\int_0^{\infty}M_{km}^{-1}(\mathbf{z})\cos(\lambda\mathbf{z}\cdot\mathbf{T})\sin\phi\,d\lambda\,d\phi\,d\theta.$$

The integral over λ within the above is

$$\int_0^{\infty}\cos(\lambda\mathbf{z}\cdot\mathbf{T})\,d\lambda = \lim_{\lambda\to\infty}\frac{\sin[\lambda(\mathbf{z}\cdot\mathbf{T})]}{(\mathbf{z}\cdot\mathbf{T})}. \tag{14.28}$$

Recalling that

$$\lim_{\lambda\to\infty}\Delta_\lambda(\mathbf{z}\cdot\mathbf{T}) = \lim_{\lambda\to\infty}\frac{1}{\pi}\frac{\sin[\lambda(\mathbf{z}\cdot\mathbf{T})]}{(\mathbf{z}\cdot\mathbf{T})} = \delta(\mathbf{z}\cdot\mathbf{T}),$$

we obtain

$$\int_0^{\infty}\cos(\lambda\mathbf{z}\cdot\mathbf{T})\,d\lambda = \pi\delta(\mathbf{z}\cdot\mathbf{T}), \tag{14.29}$$

which, when substituted into (14.2), yields

$$G_{km}(\mathbf{x} - \mathbf{x}') = \frac{1}{8\pi^2 |\mathbf{x} - \mathbf{x}'|} \int_0^{2\pi} M_{km}^{-1} \, d\theta. \tag{14.30}$$

The integration over θ is about a unit circle lying on the plane $\mathbf{x} \cdot \mathbf{T} = 0$. Note also that $\delta(\mathbf{z} \cdot \mathbf{T}) = \delta(\cos\phi) = \delta(\phi - \pi/2)$.

14.3 Isotropic Green's Function

We recall that for the case of isotropic elastic media

$$C_{ijkl} = \lambda \delta_{ij} \delta_{kl} + \mu(\delta_{ik}\delta_{jl} + \delta_{il}\delta_{jk}) \tag{14.31}$$

and thus

$$M_{ik} = C_{ijkl} z_j z_l = \lambda \delta_{ij} \delta_{kl} z_j z_l + \mu(\delta_{ik}\delta_{jl} + \delta_{il}\delta_{jk}) z_j z_l$$
$$= \lambda z_i z_k + \mu \delta_{ik} z_l z_l + \mu z_i z_k. \tag{14.32}$$

But,

$$z_l z_l = z_1^2 + z_2^2 + z_3^2 = 1, \tag{14.33}$$

because $\mathbf{z}$ is a unit vector, so that

$$M_{ik} = \mu \left(\delta_{ik} + \frac{\lambda + \mu}{\mu} z_i z_k \right). \tag{14.34}$$

Now consider a matrix with components

$$Q_{li} = \frac{1}{\mu} \left(\delta_{li} + B z_l z_i \right). \tag{14.35}$$

We ask, is it possible to choose a value for B such that $Q_{li} = M_{li}^{-1}$, i.e., so that $Q_{li} M_{ik} = \delta_{lk}$? To answer the question, evaluate the product

$$Q_{li} M_{ik} = (\delta_{li} + B z_l z_i) \left(\delta_{ik} + \frac{\lambda + \mu}{\mu} z_i z_k \right)$$
$$= \delta_{li} \delta_{ik} + \frac{\lambda + \mu}{\mu} \delta_{li} z_i z_k + B \frac{\lambda + \mu}{\mu} z_l z_i z_i z_k + B z_l z_i \delta_{ik} + B z_l z_i \delta_{ik}$$
$$= \delta_{lk} + z_l z_k \left(\frac{\lambda + \mu}{\mu} + B + B \frac{\lambda + \mu}{\mu} \right). \tag{14.36}$$

Thus, if

$$\frac{\lambda + \mu}{\mu} + B + B \frac{\lambda + \mu}{\mu} = 0, \tag{14.37}$$

we have

$$B = -\frac{\lambda + \mu}{\lambda + 2\mu}, \tag{14.38}$$

and

$$M_{km}^{-1} = \frac{1}{\mu} \left(\delta_{km} + \frac{\lambda + \mu}{\lambda + 2\mu} z_k z_m \right). \tag{14.39}$$

14.3. Isotropic Green's Function

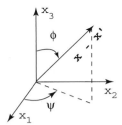

Figure 14.3. Spherical coordinate system.

Therefore, to evaluate $G_{km}(\mathbf{x} - \mathbf{x}')$ we need only express $z_k z_m$ as a function of θ noting that, for our purposes, z_k is the kth direction cosine of a unit vector which is perpendicular to $\mathbf{x} - \mathbf{x}'$. With reference to the spherical coordinate system in Fig. 14.3, the direction cosines of $\mathbf{x} - \mathbf{x}'$ are

$$\lambda_k = \frac{x_k - x'_k}{|\mathbf{x} - \mathbf{x}'|}, \tag{14.40}$$

with

$$\lambda_1 = \sin\phi \cos\psi, \quad \lambda_2 = \sin\phi \sin\psi, \quad \lambda_3 = \cos\phi, \tag{14.41}$$

and $\lambda_i \lambda_i = 1$. Then, construct $\mathbf{z}$ as

$$z_k = a_k \cos\theta + b_k \sin\theta, \tag{14.42}$$

where

$$\begin{aligned} a_1 &= \cos\phi \cos\psi, & b_1 &= -\sin\psi, \\ a_2 &= \cos\phi \sin\psi, & b_2 &= \cos\psi, \\ a_3 &= -\sin\phi, & b_3 &= 0. \end{aligned} \tag{14.43}$$

This gives

$$\begin{aligned} z_k z_m &= (a_k \cos\theta + b_k \sin\theta)(a_m \cos\theta + b_m \sin\theta) \\ &= a_k a_m \cos^2\theta + b_k b_m \sin^2\theta + (a_k b_m + a_m b_k) \sin\theta \cos\theta. \end{aligned} \tag{14.44}$$

Hence, for this isotropic case we have

$$G_{km} = \frac{1}{8\pi^2 \mu |\mathbf{x} - \mathbf{x}'|} \int_0^{2\pi} \left\{ \delta_{km} + \frac{\lambda + \mu}{\lambda + 2\mu} [a_k a_m \cos^2\theta + b_k b_m \sin^2\theta + (a_k b_m + a_m b_k) \sin\theta \cos\theta] \right\} d\theta. \tag{14.45}$$

Noting that

$$\int_0^{2\pi} \delta_{km} \, d\theta = 2\pi \delta_{km}, \quad \int_0^{2\pi} \sin\theta \cos\theta \, d\theta = 0,$$

$$\int_0^{2\pi} \sin^2\theta \, d\theta = \int_0^{2\pi} \cos^2\theta \, d\theta = \pi,$$

we thus obtain

$$G_{km} = \frac{1}{8\pi \mu |\mathbf{x} - \mathbf{x}'|} \left[2\delta_{km} - \frac{\lambda + \mu}{\lambda + 2\mu} (a_k a_m + b_k b_m) \right]. \tag{14.46}$$

It is easily verified that

$$a_k a_m + b_k b_m = \delta_{km} - \lambda_k \lambda_m = \delta_{km} - \frac{(x_k - x'_k)(x_m - x'_m)}{|\mathbf{x} - \mathbf{x}'|^2}, \qquad (14.47)$$

so that

$$G_{km}(\mathbf{x} - \mathbf{x}') = \frac{1}{8\pi\mu|\mathbf{x} - \mathbf{x}'|} \left\{ 2\delta_{km} - \frac{\lambda + \mu}{\lambda + 2\mu} \left[\delta_{km} - \frac{(x_k - x'_k)(x_m - x'_m)}{|\mathbf{x} - \mathbf{x}'|^2} \right] \right\}.$$

An alternative, useful form of this expression is obtained by observing that

$$\frac{\partial}{\partial x_m} |\mathbf{x} - \mathbf{x}'| = \frac{\partial}{\partial x_m} [(x_l - x'_l)(x_l - x'_l)]^{1/2} = \frac{x_m - x'_m}{\mathbf{x} - \mathbf{x}'}. \qquad (14.48)$$

Then,

$$\frac{\partial}{\partial x_k} \left(\frac{\partial}{\partial x_m} |\mathbf{x} - \mathbf{x}'| \right) = \frac{\partial}{\partial x_k} \left(\frac{x_k - x'_k}{|\mathbf{x} - \mathbf{x}'|} \right)$$

$$= \frac{\delta_{km}}{|\mathbf{x} - \mathbf{x}'|} - \frac{(x_k - x'_k)(x_m - x'_m)}{|\mathbf{x} - \mathbf{x}'|^2}$$

$$= \frac{1}{|\mathbf{x} - \mathbf{x}'|} \left[\delta_{km} - \frac{(x_k - x'_k)(x_m - x'_m)}{|\mathbf{x} - \mathbf{x}'|^2} \right].$$

Furthermore, we readily find

$$\nabla^2 |\mathbf{x} - \mathbf{x}'| = \frac{\partial}{\partial x_m} \frac{\partial}{\partial x_m} |\mathbf{x} - \mathbf{x}'| = \frac{1}{|\mathbf{x} - \mathbf{x}'|} \left[\delta_{mm} - \frac{(x_m - x'_m)(x_m - x'_m)}{|\mathbf{x} - \mathbf{x}'|^2} \right]. \qquad (14.49)$$

Since

$$\delta_{mm} = 3, \quad (x_m - x'_m)(x_m - x'_m) = |\mathbf{x} - \mathbf{x}'|^2, \qquad (14.50)$$

the previous expression simplifies to

$$\nabla^2 |\mathbf{x} - \mathbf{x}'| = \frac{2}{|\mathbf{x} - \mathbf{x}'|}. \qquad (14.51)$$

Thus, the alternative expression for the Green's function is

$$G_{km}(\mathbf{x} - \mathbf{x}') = \frac{1}{8\pi\mu} \left(\delta_{km} \nabla^2 - \frac{\lambda + \mu}{\lambda + 2\mu} \frac{\partial^2}{\partial x_k \partial x_m} \right) |\mathbf{x} - \mathbf{x}'|. \qquad (14.52)$$

14.4 Suggested Reading

Lekhnitskii, S. G. (1981), *Theory of Elasticity of an Anisotropic Elastic Body*, Mir Publishers, Moscow.

Love, A. E. H. (1944), *A Treatise on the Mathematical Theory of Elasticity*, Dover, New York.

Mura, T. (1987), *Micromechanics of Defects in Solids*, Martinus Nijhoff, Dordrecht, The Netherlands.

Ting, T. C. T. (1996), *Anisotropic Elasticity*, Oxford University Press, New York.

Wu, J. J., Ting, T. C. T., and Barnett, D. M., eds. (1991), *Modern Theory of Anisotropic Elasticity and Applications*, SIAM, Philadelphia.

15 Plane Contact Problems

The basic problem considered in this chapter is one in which a rigid "punch" is forced into the surface of an elastic half-space of semi-infinite extent. If the indenter is idealized as rigid, the problem is displacement driven. We examine the problem by both considering imposed forces on the surface of a half-space as well as by imposing displacements. Plane strain conditions are assumed to prevail.

15.1 Wedge Problem

Consider the problem of a "wedge" loaded at its corner by a force **f** as shown in Fig. 15.1. The geometry involved suggests that a polar coordinate system be used. We seek solutions of the form

$$\sigma = f(r)g(\theta), \tag{15.1}$$

so that all solutions are self-similar in angular form. Note that the solutions must be such that

$$\sigma \sim 1/r \Rightarrow f(r) \sim 1/r, \tag{15.2}$$

since when considering equilibrium, the traction on any arc must decrease as $1/r$, because the arc length increases in proportion to r. Thus we seek solutions to the biharmonic equation that lead to stresses corresponding to (15.2). Appealing to the inventory of solutions found earlier in Chapter 10, we try

$$\phi = C_1 r\theta \sin\theta + C_2 r\theta \cos\theta + C_3 r \ln r \cos\theta + C_4 r \ln r \sin\theta. \tag{15.3}$$

These give rise to stresses of the form

$$\sigma_{rr} = \frac{1}{r}(2C_1 \cos\theta - 2C_2 \sin\theta + C_3 \sin\theta + C_4 \sin\theta),$$

$$\sigma_{r\theta} = \frac{1}{r}(C_3 \sin\theta - C_4 \cos\theta), \tag{15.4}$$

$$\sigma_{\theta\theta} = \frac{1}{r}(C_3 \cos\theta + C_4 \sin\theta).$$

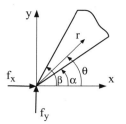

Figure 15.1. Wedge loaded by corner point forces.

The sides of the wedge are stress-free, *i.e.*,

$$\sigma_{\theta r} = \sigma_{\theta\theta} = 0 \quad \text{on} \quad \theta = \alpha, \beta, \tag{15.5}$$

which leads to $C_3 = C_4 = 0$. The constants C_1 and C_2 are determined by imposing global equilibrium

$$f_x + 2\int_\alpha^\beta \frac{C_1 \cos\theta - C_2 \sin\theta}{a} a\cos\theta \, d\theta = 0,$$

$$f_y + 2\int_\alpha^\beta \frac{C_1 \cos\theta - C_2 \sin\theta}{a} a\sin\theta \, d\theta = 0. \tag{15.6}$$

Solving (15.6) we obtain

$$(C_1 - C_2)\sin^2\theta\big|_\alpha^\beta + (C_1 + C_2)\sin\theta\cos\theta\big|_\alpha^\beta = (C_2 - C_1)(\beta - \alpha) - (f_x - f_y),$$

$$(C_1 + C_2)\sin^2\theta\big|_\alpha^\beta + (C_2 - C_1)\sin\theta\cos\theta\big|_\alpha^\beta = (C_2 + C_1)(\beta - \alpha) + (f_x - f_y).$$

Since $C_3 = C_4 = 0$, all rays are traction free, *i.e.*, $\sigma_{r\theta} = \sigma_{\theta\theta} = 0$ at each θ.

A particularly interesting case is that of a half-plane, obtained by setting $\alpha = -\pi$ and $\beta = 0$. In that case (15.6), or (15.1), yields

$$f_x + \pi C_1 = 0,$$

$$f_y - \pi C_2 = 0, \tag{15.7}$$

and thus

$$\sigma_{rr} = -\frac{2f_x}{\pi}\frac{\cos\theta}{r} - \frac{2f_y}{\pi}\frac{\sin\theta}{r}. \tag{15.8}$$

We will find convenient in the sequel to redefine the convention used to describe the applied forces. In fact, let us redefine the coordinate system as shown in Fig. 15.2. If we call $f_y = -P$, and ignore for the moment the tangential force f_x, we obtain

$$\sigma_{rr} = -\frac{2P}{\pi}\frac{\cos\hat\theta}{r}. \tag{15.9}$$

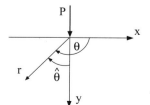

Figure 15.2. Redefined polar coordinate system.

15.1. Wedge Problem

The Cartesian components of stress for the case of an applied normal load are

$$\sigma_{xx} = \sigma_{rr} \sin^2 \hat{\theta} = -\frac{2P}{\pi} \frac{x^2 y}{(x^2 + y^2)^2},$$

$$\sigma_{yy} = \sigma_{rr} \cos^2 \hat{\theta} = -\frac{2P}{\pi} \frac{y^3}{(x^2 + y^2)^2}, \quad (15.10)$$

$$\sigma_{xy} = \sigma_{rr} \sin \hat{\theta} \cos \hat{\theta} = -\frac{2P}{\pi} \frac{xy^2}{(x^2 + y^2)^2}.$$

To proceed, recall that the strain components, for the considered two-dimensional case, are

$$e_{rr} = \frac{\partial u_r}{\partial r} = -\frac{1-\nu^2}{E} \frac{2P}{\pi} \frac{\cos \hat{\theta}}{r},$$

$$e_{\theta\theta} = \frac{u_r}{r} + \frac{1}{r}\frac{\partial u_\theta}{\partial \theta} = \frac{\nu(1+\nu)}{E} \frac{2P}{\pi} \frac{\cos \hat{\theta}}{r}, \quad (15.11)$$

$$e_{r\theta} = \frac{1}{2}\left(\frac{1}{r}\frac{\partial u_r}{\partial \theta} + \frac{\partial u_\theta}{\partial r} - \frac{u_\theta}{r}\right) = \frac{\sigma_{r\theta}}{2G} = 0.$$

The above equations can be integrated to obtain the displacement components

$$u_r = -\frac{(1-\nu^2)}{\pi E} 2P \cos \hat{\theta} \ln r - \frac{(1-2\nu)(1+\nu)}{\pi E} P\hat{\theta} \sin \hat{\theta} + a_1 \sin \hat{\theta} + a_2 \cos \hat{\theta} \quad (15.12)$$

and

$$u_\theta = \frac{(1-\nu^2)}{\pi E} 2P \sin \hat{\theta} \ln r + \frac{\nu(1+\nu)}{\pi E} 2P \sin \hat{\theta}$$

$$- \frac{(1-2\nu)(1+\nu)}{\pi E} P\hat{\theta} \cos \hat{\theta} + \frac{(1-2\nu)(1+\nu)}{\pi E} P \sin \hat{\theta} \quad (15.13)$$

$$+ a_1 \cos \hat{\theta} - a_2 \sin \hat{\theta} + a_3 r.$$

If the body does not translate or rotate, then $a_1 = a_2 = a_3 = 0$. At the surface, where $\hat{\theta} = \pm \pi/2$, we have

$$u_r|_{\hat{\theta}=\pi/2} = -u_r|_{\hat{\theta}=-\pi/2} = -\frac{(1-2\nu)(1+\nu)}{2E} P = -\frac{(1-2\nu)}{4G} P,$$

$$u_\theta|_{\hat{\theta}=\pi/2} = -u_\theta|_{\hat{\theta}=-\pi/2} = \frac{(1-\nu^2)}{\pi E} 2P \ln r + a, \quad (15.14)$$

where a is a constant combined from the remaining terms in (15.13) that do not involve r. To determine this constant, a reference point is needed from which to measure displacement. If $u_\theta = 0$ at say $r = r_0$, then

$$u_\theta|_{\hat{\theta}=\pi/2} = -u_\theta|_{\hat{\theta}=-\pi/2} = -\frac{(1-\nu^2)}{\pi E} 2P \ln(r_0/r). \quad (15.15)$$

Consider next the tangential force $f_x = F$. A similar analysis yields

$$\sigma_{rr} = -\frac{2F}{\pi} \frac{\cos \theta}{r}, \quad \sigma_{\theta\theta} = \sigma_{r\theta} = 0. \quad (15.16)$$

The corresponding Cartesian stress components are

$$\sigma_{xx} = -\frac{2F}{\pi} \frac{x^3}{(x^2+y^2)^2},$$

$$\sigma_{yy} = -\frac{2F}{\pi} \frac{xy^2}{(x^2+y^2)^2}, \tag{15.17}$$

$$\sigma_{xy} = -\frac{2F}{\pi} \frac{x^2 y}{(x^2+y^2)^2}.$$

Analysis of the displacements reveals that

$$u_r|_{\theta=0} = -u_r|_{\theta=\pi} = -\frac{1-\nu^2}{\pi E} 2F \ln r + b,$$

$$u_\theta|_{\theta=0} = u_\theta|_{\theta=\pi} = \frac{(1-2\nu)(1+\nu)}{2E} F. \tag{15.18}$$

Equation (15.18) shows an interesting effect that the entire surface at $x > 0$ is depressed by an amount proportional to F, whereas the surface behind the point force is raised by an equal amount. In a manner similar to the angular displacement due to a point normal force, the constant b in (15.18) can be determined by selection of a reference point r_t, so that

$$u_r|_{\theta=0} = -u_r|_{\theta=\pi} = -\frac{(1-\nu^2)}{\pi E} 2F \ln(r_t/r). \tag{15.19}$$

15.2 Distributed Contact Forces

The solutions listed above can be used as Green's functions for constructing the solutions to the elastic fields of distributed forces on the surface of an infinitely extended half-plane. Let the contact area be $-a \leq x \leq a$, and let the distributed normal and tangential forces be $p(x)$ and $f(x)$, respectively. The resulting stresses are obtained *via* superposition as

$$\sigma_{xx} = -\frac{2y}{\pi} \int_{-a}^{a} \frac{p(s)(x-s)^2 \, ds}{[(x-s)^2 + y^2]^2} - \frac{2}{\pi} \int_{-a}^{a} \frac{f(s)(x-s)^3 \, ds}{[(x-s)^2 + y^2]^2},$$

$$\sigma_{yy} = -\frac{2y^3}{\pi} \int_{-a}^{a} \frac{p(s) \, ds}{[(x-s)^2 + y^2]^2} - \frac{2y^2}{\pi} \int_{-a}^{a} \frac{f(s)(x-s) \, ds}{[(x-s)^2 + y^2]^2}, \tag{15.20}$$

$$\sigma_{xy} = -\frac{2y^2}{\pi} \int_{-a}^{a} \frac{p(s)(x-s) \, ds}{[(x-s)^2 + y^2]^2} - \frac{2y}{\pi} \int_{-a}^{a} \frac{f(s)(x-s)^2 \, ds}{[(x-s)^2 + y^2]^2}.$$

This generalizes the result for normal force loading given earlier to include tangential force loading. Note, however, that the sign convention for the normal load is reversed here.

When the radial displacement is given in terms of the Cartesian coordinates, the $\ln r$ terms become $\text{sgn}(x) \ln |x|$. Thus, using the means employed by Johnson (1985) to handle the sign change in u_θ at the surface, we find

15.2. Distributed Contact Forces

$$\bar{u}_x = -\frac{(1-2\nu)(1+\nu)}{2E}\left[\int_{-a}^{x} p(s)\,ds - \int_{x}^{a} p(s)\,ds\right]$$
$$-\frac{2(1-\nu^2)}{\pi E}\int_{-a}^{a} f(s)\ln|x-s|\,ds + a_1 \quad (15.21)$$

and

$$\bar{u}_y = -\frac{2(1-\nu^2)}{\pi E}\int_{-a}^{a} p(s)\ln|x-s|\,ds$$
$$+\frac{(1-2\nu)(1+\nu)}{2E}\left[\int_{-a}^{x} f(s)\,ds - \int_{x}^{a} f(s)\,ds\right] + a_2, \quad (15.22)$$

where $\bar{u}_x, \bar{u}_y$ are the displacements on the surface. Note how splitting the range of integration of the integrals handles the sign switch inherent in u_θ, as discussed above. The integration constants, here listed as a_1 and a_2, have already been discussed and do yield an arbitrary rigid body translation that is undetermined. When the distortions are computed, however, there is no ambiguity, so that

$$\frac{\partial \bar{u}_x}{\partial x} = -\frac{(1-2\nu)(1+\nu)}{E}p(x) - \frac{2(1-\nu^2)}{\pi E}\int_{-a}^{a}\frac{f(s)}{x-s}\,ds,$$
$$\frac{\partial \bar{u}_y}{\partial x} = -\frac{2(1-\nu^2)}{\pi E}\int_{-a}^{a}\frac{p(s)}{x-s}\,ds + \frac{(1-2\nu)(1+\nu)}{E}f(x). \quad (15.23)$$

The integrals in (15.23) are interpreted as Cauchy principal values where needed, that is in $-a \leq x \leq a$.

Some interesting features arise. For example, let $f(x) = 0$ and consider for the moment only normal applied forces. The normal strain on the surface, $\bar{e}_{xx}$, is in this case

$$\bar{e}_{xx} = \frac{\partial \bar{u}_x}{\partial x} = -\frac{(1-2\nu)(1+\nu)}{E}p(x). \quad (15.24)$$

On the other hand, from the elastic constitutive relation in plane strain,

$$\bar{e}_{xx} = \frac{1}{E}\left[(1-\nu^2)\bar{\sigma}_{xx} - \nu(1+\nu)\bar{\sigma}_{yy}\right]. \quad (15.25)$$

If the two expressions are to be equal, we must have

$$\bar{\sigma}_{xx} = \bar{\sigma}_{yy} = -p(x). \quad (15.26)$$

This simple result is interesting. Since in plane strain $\bar{\sigma}_{zz} = \nu(\bar{\sigma}_{xx} + \bar{\sigma}_{yy})$ and ν is, say, in a typical range $1/3 \leq \nu \leq 1/2$, the state of stress under the indenter that applies $p(x)$ is one of nearly pure hydrostatic stress. This, in turn, means that plastic deformation tends to be suppressed just under the contact surface.

15.2.1 Uniform Contact Pressure

Here we consider the problem of uniform loading over the strip $-a \leq x \leq a$ with the normal stress p_0, as depicted in Fig. 15.3. This problem has been already considered in

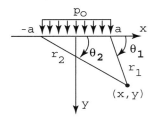

Figure 15.3. Uniform contact pressure on a half-plane.

Chapter 12 (see Fig. 12.3), where the stresses were found to be given by

$$\sigma_{xx} = -\frac{p_0}{2\pi}[2(\theta_1 - \theta_2) + \sin(2\theta_1) - \sin(2\theta_2)],$$

$$\sigma_{yy} = -\frac{p_0}{2\pi}[2(\theta_1 - \theta_2) - \sin(2\theta_1) + \sin(2\theta_2)], \quad (15.27)$$

$$\sigma_{xy} = \frac{p_0}{2\pi}[\cos(2\theta_1) - \cos(2\theta_2)].$$

The definitions of the angles θ_1 and θ_2 are indicated in Fig. 15.3. Here we compute the corresponding displacements and displacement gradients. From (15.23), as long as $-a \le x \le a$, we have

$$\frac{\partial \bar{u}_x}{\partial x} = -\frac{(1-2\nu)(1+\nu)}{E} p_0. \quad (15.28)$$

Care must be taken to note that, from (15.21), $\bar{u}_x$ no longer changes beyond $x > a$ or $x < -a$ because of the fact that $p(x) = 0$ for $|x| > a$. Then, if the origin is taken as fixed, and $|x| \le a$,

$$\bar{u}_x(x) = -\frac{(1-2\nu)(1+\nu)}{E} p_0 x, \quad (15.29)$$

and, if $|x| > a$,

$$\bar{u}_x(x) = -\frac{(1-2\nu)(1+\nu)}{E} \operatorname{sgn}(x) p_0 a. \quad (15.30)$$

As for $\bar{u}_y$, we have

$$\frac{\partial \bar{u}_y}{\partial x} = -\frac{2(1-\nu^2)}{\pi E} \int_{-a}^{a} \frac{p_0 ds}{x-s}, \quad (15.31)$$

where the principal value is to be taken if $|x| \le a$. Thus, we form

$$\int_{-a}^{a} \frac{ds}{x-s} = \lim_{\epsilon \to 0} \left(\int_{-a}^{x-\epsilon} \frac{ds}{x-s} - \int_{x+\epsilon}^{a} \frac{ds}{s-x} \right)$$

$$= \lim_{\epsilon \to 0} \left[-\ln(x-s)\big|_{-a}^{x-\epsilon} - \ln(s-x)\big|_{x+\epsilon}^{a} \right] \quad (15.32)$$

$$= \ln(x+a) - \ln(a-x).$$

Therefore,

$$\frac{\partial \bar{u}_y}{\partial x} = -\frac{2(1-\nu^2)}{\pi E} p_0 [\ln(a+x) - \ln(a-x)], \quad (15.33)$$

15.3. Displacement-Based Contact

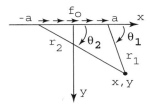

Figure 15.4. Uniform tangential force on a half-plane.

and, upon integrating,

$$\bar{u}_y(x) = -\frac{(1-\nu^2)}{\pi E} p_0 \left[(a+x)\ln\left(\frac{a+x}{a}\right)^2 + (a-x)\ln\left(\frac{a-x}{a}\right)^2 \right] + C.$$

We note that this expression holds for any range of x.

15.2.2 Uniform Tangential Force

Consider the case of a half-plane loaded on the patch $-a \le x \le a$ by a uniform tangential stress f_0. With the angles as defined in Fig. 15.4, the relations (15.20) yield the stresses

$$\sigma_{xx} = \frac{f_0}{2\pi} [4\ln(r_1/r_2) - \cos(2\theta_1) + \cos(2\theta_2)],$$

$$\sigma_{yy} = \frac{f_0}{2\pi} [\cos(2\theta_1) - \cos(2\theta_2)], \qquad (15.34)$$

$$\sigma_{xy} = -\frac{f_0}{2\pi} [2(\theta_1 - \theta_2) + \sin(2\theta_1) - \sin(2\theta_2)].$$

where $r_{1,2} = [(x \mp a)^2 + y^2]^{1/2}$. When the equations (15.22) are examined it becomes obvious that the relations obtained for the displacements produced by uniform normal pressure give the displacements for the tangential stress if we make the following transitions

$$(\bar{u}_x)_{\text{tangential}} \rightleftarrows (\bar{u}_y)_{\text{normal}},$$
$$(\bar{u}_y)_{\text{tangential}} \rightleftarrows (\bar{u}_x)_{\text{normal}}. \qquad (15.35)$$

15.3 Displacement-Based Contact: Rigid Flat Punch

Instead of imposing forces on the contact surface, suppose that displacements are imposed, say by a rigid punch of a prescribed shape. This would be equivalent to the imposition of $d\bar{u}_x/dx$ and $d\bar{u}_y/dx$ by the curved surface of the punch. The relations (15.23) now read

$$\int_{-a}^{a} \frac{f(s)}{x-s} ds = -\frac{\pi(1-2\nu)}{2(1-\nu)} p(x) - \frac{\pi E}{2(1-\nu^2)} \bar{u}'_x(x),$$

$$\int_{-a}^{a} \frac{p(s)}{x-s} ds = \frac{\pi(1-2\nu)}{2(1-\nu)} f(x) - \frac{\pi E}{2(1-\nu^2)} \bar{u}'_y(x), \qquad (15.36)$$

where $\bar{u}'_x(x) = d\bar{u}_x/dx$ and $\bar{u}'_y(x) = d\bar{u}_y/dx$.

For a frictionless flat rigid punch, with $f(x) = 0$ and $\bar{u}'_y$ prescribed, we have

$$\int_{-a}^{a} \frac{p(s)}{x-s} ds = -\frac{\pi E}{2(1-v^2)} \bar{u}'_y(x). \tag{15.37}$$

In this case, as in (15.28), we have

$$\bar{u}'_x(x) = -\frac{(1-2v)(1+v)}{E} p(x). \tag{15.38}$$

Note that the integral equation in (15.37) is singular. Its solutions of a general nature are given in Section 19.8. Here we simply note that the singular integral equation

$$PV \int_{-a}^{a} \frac{\zeta(s)}{x-s} ds = g(x) \tag{15.39}$$

has the solution, if unbounded at the end points, given by

$$\zeta(x) = \frac{1}{\pi^2(a^2-x^2)^{1/2}} \int_{-a}^{a} \frac{(a^2-s^2)^{1/2} g(s)}{x-s} ds + \frac{C}{\pi^2(a^2-x^2)^{1/2}}, \tag{15.40}$$

where

$$C = \pi \int_{-a}^{a} \zeta(x) dx. \tag{15.41}$$

For the rigid punch, $\bar{u}'_y(x) = 0$ on $-a \leq x \leq a$, and

$$\int_{-a}^{a} \frac{p(s)}{x-s} ds = 0, \tag{15.42}$$

which gives

$$p(x) = \frac{p_0}{\pi(a^2-x^2)^{1/2}}, \tag{15.43}$$

where p_0 is the total load applied by the force distribution of the punch. The derived result is, of course, the same force distribution as used earlier in Chapter 12 (see Fig. 12.5).

The displacement distribution outside the punch can be found readily from (15.21). The result is

$$\bar{u}_y(x) = -\frac{2(1-v^2)}{\pi E} \int_{-a}^{a} p(s) \ln|x-s| ds + \delta_y$$

$$= -\frac{2(1-v^2)}{\pi E} p_0 \ln\left[\frac{|x|}{a} + \left(\frac{x^2}{a^2}\right)^{1/2}\right] + \delta_y, \tag{15.44}$$

where δ_y is an integration constant determined by selecting a datum from which to measure displacement. The displacement component, $\bar{u}_x$ is found from (15.21) to be

$$\bar{u}_x(x) = -\frac{(1-2v)(1+v)}{\pi E} p_0 \sin^{-1}(x/a). \tag{15.45}$$

15.4 Suggested Reading

Fischer-Cripps, A. C. (2000), *Introduction to Contact Mechanics*, Springer-Verlag, New York.

Galin, L. A. (1961), *Contact Problems in the Theory of Elasticity*, School of Physical Sciences and Applied Mathematics, North Carolina State College.

Gladwell, G. M. L. (1980), *Contact Problems in the Classical Theory of Elasticity*, Alphen aan den Rijn, The Netherlands.

Johnson, K. L. (1985), *Contact Mechanics*, Cambridge University Press, New York.

Timoshenko, S. P., and Goodier, J. N. (1970), *Theory of Elasticity*, 3rd ed., McGraw-Hill, New York.

16 Deformation of Plates

The deformation of thin plates is considered here. Plates are assumed to be thin when their thickness is very small compared to their other two dimensions in their plane. Deformations and strains are assumed to be small and the material of the plates is isotropic linear elastic.

16.1 Stresses and Strains of Bent Plates

Plates, when bent, develop a transition from one side to the other from tension to compression and thus contain a neutral surface. If homogeneous, this neutral surface may be assumed to pass through the middle of the plate thickness (see Fig. 16.1). The coordinate system is accordingly taken so that the normal to the plane of the plate is the z axis, whereas the plane of the plate is in the x-y plane. Displacement components within the plane are assumed to be negligible compared to those normal to the plate. The displacement normal to the plate, along the z axis, is designated as w. Thus, along the neutral plane

$$u_x^{(0)} = u_y^{(0)} = 0, \quad u_z^{(0)} = w(x, y). \tag{16.1}$$

For small deflections of the plate, the displacement w is assumed to be small compared to the thickness h of the plate. Because plate is assumed to be so thin, the forces acting on the plate surface that induce deformation are typically small. Stresses within the plate, however, due to bending and shear are not. Accordingly, the plate surface is approximated as being stress free so that, if **n** is the normal to the plate,

$$\mathbf{n} \cdot \boldsymbol{\sigma} = \mathbf{0} \;\Leftrightarrow\; \sigma_{ij} n_j = 0. \tag{16.2}$$

Moreover, since the deformations are small, i.e., the plates are but slightly bent, it may be assumed that, $\mathbf{n} \approx \mathbf{e}_z$ where $\mathbf{e}_z$ is the unit base vector along the z axis. Thus we take as an approximation that $\sigma_{xz} = \sigma_{yz} = \sigma_{zz} = 0$. The linear elastic constitutive relations yield

↳ B/C NEGLIGIBLE WHEN COMPARED TO BENDING STRESSES

$$\sigma_{xz} = \frac{E}{1+\nu} e_{xz}, \quad \sigma_{yz} = \frac{E}{1+\nu} e_{yz}, \Rightarrow e_{xz} = e_{yz} \approx 0$$

$$\sigma_{zz} = \frac{E}{(1+\nu)(1-2\nu)}[(1-\nu)e_{zz} + \nu(e_{xx} + e_{yy})]. \tag{16.3}$$

$\Rightarrow e_{zz} = (16.4)$

16.2. Energy of Bent Plates

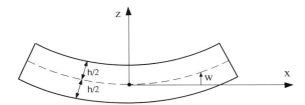

Figure 16.1. Coordinate geometry of a thin plate that undergoes deflection w.

When the definitions of the small strains are recalled, and if we take $u_z = w(x, y)$, it is found that

$$e_{xz} = \frac{\partial u_x}{\partial z} = -\frac{\partial w}{\partial x}, \quad e_{yz} = \frac{\partial u_y}{\partial z} = -\frac{\partial w}{\partial y},$$

$$e_{zz} = -\frac{\nu}{1-\nu}(e_{xx} + e_{yy}). \tag{16.4}$$

Thus, plane cross sections remain plane and orthogonal to the midsurface of the plate. The first two of (16.4) are integrated *wrt* z to find that

$$u_x = -z\frac{\partial w}{\partial x}, \quad u_y = -z\frac{\partial w}{\partial y}. \tag{16.5}$$

The constants of integration were made equal to zero, so that the displacements would be zero at $z = 0$. Now, the nonzero strain components can be calculated as

$$e_{xx} = -z\frac{\partial^2 w}{\partial x^2}, \quad e_{yy} = -z\frac{\partial^2 w}{\partial y^2}, \quad e_{xy} = -z\frac{\partial^2 w}{\partial x \partial y},$$

$$e_{xz} = e_{yz} = 0, \quad e_{zz} = \frac{\nu}{1-\nu} z\left(\frac{\partial^2 w}{\partial x^2} + \frac{\partial^2 w}{\partial y^2}\right). \tag{16.6}$$

It may be noticed, in retrospect, that from the last expression for e_{zz} and the relationship $e_{zz} = \partial u_z/\partial z$, the integration gives

$$u_z = \frac{\nu}{1-\nu}\frac{z^2}{2}\nabla^2 w + w(x, y). \tag{16.7}$$

Because the plate is thin, this is approximately equal to w, in accord with the initial assumption $u_z = w(x, y)$. The same order of approximation is common in the beam bending theory. The underlying assumptions used in the above formulation of thin plate theory are known as the Kirchhoff hypotheses.

16.2 Energy of Bent Plates

The elastic strain energy density of a plate can be expressed as

$$W = \frac{1}{2}\sigma_{ij}e_{ij} = z^2 \frac{E}{1+\nu}\left\{\frac{1}{2(1-\nu)}\left(\frac{\partial^2 w}{\partial x^2} + \frac{\partial^2 w}{\partial y^2}\right)^2 + \left[\left(\frac{\partial^2 w}{\partial x \partial y}\right)^2 - \frac{\partial^2 w}{\partial x^2}\frac{\partial^2 w}{\partial y^2}\right]\right\}.$$

The total elastic energy is obtained by integrating this expression throughout the plate. If the thickness of the plate is h, so that z runs from $z = -\frac{1}{2}h$ to $z = \frac{1}{2}h$, and because the deformations are small, it may be assumed that the area element in the plane of the plate is still just dxdy. Therefore, the total elastic energy of the plate is

$$\delta W = \delta \int_V W \, dV = \frac{Eh^3}{24(1-\nu^2)} \tag{16.8}$$

$$\times \delta \iint_A \left\{ \underbrace{\left(\frac{\partial^2 w}{\partial x^2} + \frac{\partial^2 w}{\partial y^2}\right)^2}_{\text{LAPLACIAN}} + 2(1-\nu)\left[\left(\frac{\partial^2 w}{\partial x \partial y}\right)^2 - \frac{\partial^2 w}{\partial x^2}\frac{\partial^2 w}{\partial y^2}\right] \right\} dx\, dy.$$

The plate is assumed to be thin and to deform uniformly through the thickness.

$$\Phi = \int_V w\, dv - \int_S \vec{T}\cdot\vec{u}\, ds \Rightarrow \delta \mathcal{I} = d\int_V w\, dv - d\int_S T$$

16.3 Equilibrium Equations for a Plate

↳ STRAIN ENERGY DENSITY

Imagine a variation in deflection, δw; we wish to calculate the accompanying variation in strain energy δW. To begin, examine the first term in the integral in (16.8), i.e., the integral of $\frac{1}{2}(\nabla^2 w)^2$. The factor $1/2$ is placed for later convenience. It follows that

$$\delta \frac{1}{2}\iint_A (\nabla^2 w)^2\, dx\, dy = \iint_A (\nabla^2 w)(\nabla^2 \delta w)\, dx\, dy$$

$$= \iint_A \nabla^2 w\, \mathrm{div}(\mathrm{grad}\, \delta w)\, dx\, dy \tag{16.9}$$

$$= \iint_A \mathrm{div}(\nabla^2 w\, \mathrm{grad}\, \delta w)\, dx\, dy - \iint_A \mathrm{grad}\, \delta w \cdot \mathrm{grad}\, \nabla^2 w\, dx\, dy.$$

The above transformation is verified with the help of the following identity, phrased in component form as

$$\mathrm{div}(\nabla^2 w\, \mathrm{grad}\, \delta w) = \frac{\partial}{\partial x_j}\left(\nabla^2 w\, \frac{\partial \delta w}{\partial x_j}\right) = \frac{\partial \nabla^2 w}{\partial x_j}\frac{\partial \delta w}{\partial x_j} + \nabla^2 w\, \nabla^2 \delta w.$$

The divergence theorem enables an immediate transformation of the first integral in (16.9), i.e.,

$$\iint_A \mathrm{div}(\nabla^2 w\, \mathrm{grad}\, \delta w)\, dx\, dy = \oint_C \nabla^2 w (\mathbf{n}\cdot \mathrm{grad}\, \delta w)\, d\ell = \oint_C \nabla^2 w\, \frac{\partial \delta w}{\partial n}\, d\ell, \tag{16.10}$$

where $\partial/\partial n$ denotes a derivative along the outward pointing normal to C, the curve bounding the plate area A. The second integral in (16.9) is transformed as

$$\iint_A \mathrm{grad}\, \delta w \cdot \mathrm{grad}\, \nabla^2 w\, dx\, dy = \iint_A \mathrm{div}(\delta w\, \mathrm{grad}\, \nabla^2 w)\, dx\, dy - \iint_A \delta w\, \nabla^4 w\, dx\, dy$$

$$= \oint_C \delta w(\mathbf{n}\cdot \mathrm{grad}\, \nabla^2 w)\, d\ell - \iint_A \delta w\, \nabla^4 w\, dx\, dy$$

$$= \oint_C \delta w\, \frac{\partial \nabla^2 w}{\partial n}\, d\ell - \iint_A \delta w\, \nabla^4 w\, dx\, dy.$$

16.3. Equilibrium Equations for a Plate

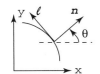

Figure 16.2. Coordinates for a plate's edge.

The above transformation can be verified by forming the following identity

$$\text{div}\,(\delta w \,\text{grad}\, \nabla^2 w) = \frac{\partial}{\partial x_j}\left(\delta w \frac{\partial \nabla^2 w}{\partial x_j}\right)$$

$$= \frac{\partial \delta w}{\partial x_j}\frac{\partial \nabla^2 w}{\partial x_j} + \delta w \frac{\partial}{\partial x_j}\frac{\partial}{\partial x_j}\nabla^2 w,$$

where the operator $\frac{\partial}{\partial x_j}\frac{\partial}{\partial x_j}\nabla^2 w = \nabla^4 w$ is the biharmonic operator. When these results are combined, there follows

$$\delta\frac{1}{2}\iint_A (\nabla^2 w)^2\, dx\, dy = \iint_A \delta w \nabla^4 w\, dx\, dy - \oint_C \delta w \frac{\partial \nabla^2 w}{\partial n}\, d\ell + \oint_C \nabla^2 w \frac{\partial \delta w}{\partial n}\, d\ell. \quad (16.11)$$

Attention is now turned to the second term in (16.8). We want to evaluate

$$\delta \iint_A \left[\left(\frac{\partial^2 w}{\partial x \partial y}\right)^2 - \frac{\partial^2 w}{\partial x^2}\frac{\partial^2 w}{\partial y^2}\right] dx\, dy \quad (16.12)$$

$$= \iint_A \left(2\frac{\partial^2 w}{\partial x \partial y}\frac{\partial^2 \delta w}{\partial x \partial y} - \frac{\partial^2 w}{\partial x^2}\frac{\partial^2 \delta w}{\partial y^2} - \frac{\partial^2 \delta w}{\partial x^2}\frac{\partial^2 w}{\partial y^2}\right) dx\, dy.$$

The integrand can be rewritten as the divergence of a certain vector, viz.,

$$I = \frac{\partial}{\partial x}\left(\frac{\partial \delta w}{\partial y}\frac{\partial^2 w}{\partial x \partial y} - \frac{\partial \delta w}{\partial x}\frac{\partial^2 w}{\partial y^2}\right) + \frac{\partial}{\partial y}\left(\frac{\partial \delta w}{\partial x}\frac{\partial^2 w}{\partial x \partial y} - \frac{\partial \delta w}{\partial y}\frac{\partial^2 w}{\partial x^2}\right).$$

Thus, after using the Stokes formula (2.3), we have

$$\delta \iint_A \left[\left(\frac{\partial^2 w}{\partial x \partial y}\right)^2 - \frac{\partial^2 w}{\partial x^2}\frac{\partial^2 w}{\partial y^2}\right] dx\, dy$$

$$= \oint_C d\ell \cos\theta \left(\frac{\partial \delta w}{\partial y}\frac{\partial^2 w}{\partial x \partial y} - \frac{\partial \delta w}{\partial x}\frac{\partial^2 w}{\partial y^2}\right) \quad (16.13)$$

$$+ \oint_C d\ell \sin\theta \left(\frac{\partial \delta w}{\partial x}\frac{\partial^2 w}{\partial x \partial y} - \frac{\partial \delta w}{\partial y}\frac{\partial^2 w}{\partial x^2}\right),$$

where θ is the angle that the outward pointing normal to the edge of the plate makes with the x axis.

The derivatives appearing in the expression for I are expressed in terms of n and ℓ (Fig. 16.2) as

$$\frac{\partial}{\partial x} = \cos\theta \frac{\partial}{\partial n} - \sin\theta \frac{\partial}{\partial \ell},$$

$$\frac{\partial}{\partial y} = \sin\theta \frac{\partial}{\partial n} + \cos\theta \frac{\partial}{\partial \ell}. \quad (16.14)$$

When these are used in (16.13), and terms rearranged, the result is assembled and combined with the contributions from (16.11). The total variation in $\mathcal{W}$ is found to be

$$\delta\mathcal{W} = \frac{Eh^3}{12(1-\nu^2)}(I_1 + I_2 + I_3), \qquad (16.15)$$

[annotation: $\frac{Eh^3}{12(1-\nu^2)} = D$ (plate bending stiffness)]

where

$$I_1 = \iint_A \nabla^4 w \, \delta w \, dx \, dy,$$

$$I_2 = -\oint_C \delta w \frac{\partial \nabla^2 w}{\partial n} d\ell \quad \text{[shear force]}$$
$$- \oint_C \delta w (1-\nu) \frac{\partial}{\partial \ell}\left[\sin\theta\cos\theta\left(\frac{\partial^2 w}{\partial y^2} - \frac{\partial^2 w}{\partial x^2}\right) + (\cos^2\theta - \sin^2\theta)\frac{\partial^2 w}{\partial x \partial y}\right] d\ell,$$

and [annotation: moment]

$$I_3 = \oint_C \frac{\partial \delta w}{\partial n}\left[\nabla^2 w + (1-\nu)\left(2\sin\theta\cos\theta\frac{\partial^2 w}{\partial x \partial y} - \sin^2\theta\frac{\partial^2 w}{\partial x^2} - \cos^2\theta\frac{\partial^2 w}{\partial y^2}\right)\right] d\ell.$$

The variation we wish to perform to determine the equilibrium configuration of the plate is that of the potential energy. For this we seek configurations that minimize Π. We thus require

$$\delta\Pi = \delta\mathcal{W} - \iint_A p\,\delta w \, dx \, dy = 0. \qquad (16.16)$$

This variation includes both surface and line integral terms, which must both vanish. For the surface integrals we obtain

$$\iint_A \left[\frac{Eh^3}{12(1-\nu^2)}\nabla^4 w - p\right] \delta w \, dx \, dy = 0. \qquad (16.17)$$

Since the variation in δw is arbitrary, this integral can only vanish if

$$\frac{Eh^3}{12(1-\nu^2)}\nabla^4 w - p = 0, \qquad (16.18)$$

i.e.,

$$\nabla^4 w = \frac{p}{D}, \quad D = \frac{Eh^3}{12(1-\nu^2)}. \qquad (16.19)$$

The coefficient D is the so-called *flexural rigidity of the plate*. The distributed load over the plate in the z direction is $p = p(x, y)$.

Boundary conditions are obtained by setting the line integrals to zero. If the edges of the plate are free, *i.e.*, no external forces act on them, δw and $\delta \partial w / \partial n$ are arbitrary and thus the integrands of the line integrals must vanish at each point on the plate's edge. This gives

$$\frac{\partial \nabla^2 w}{\partial n} + (1-\nu)\frac{\partial}{\partial \ell}\left[\sin\theta\cos\theta\left(\frac{\partial^2 w}{\partial y^2} - \frac{\partial^2 w}{\partial x^2}\right) + (\cos^2\theta - \sin^2\theta)\frac{\partial^2 w}{\partial x \partial y}\right] = 0, \qquad (16.20)$$

16.4. Shear Forces and Bending and Twisting Moments

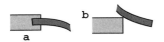

Figure 16.3. (a) Clamped vs. (b) simply supported edge conditions.

and

$$\nabla^2 w + (1-\nu)\left(2\sin\theta\cos\theta\frac{\partial^2 w}{\partial x \partial y} - \sin^2\theta\frac{\partial^2 w}{\partial x^2} - \cos^2\theta\frac{\partial^2 w}{\partial y^2}\right) = 0. \tag{16.21}$$

If, instead, the plate's edges are clamped, we have $w = 0$ and $\partial w/\partial n = 0$ (Fig. 16.3). The shear force and moment are readily calculated as follows. For example, the force Q acting on an edge point of the plate is given by $Q = \partial W/\partial w$, because $\delta W = Q\delta w$. But, looking at I_2 in (16.16), we see that its integrand is just this derivative devided by $-D$, because of the sign ahead of the integral. Likewise, the moment M at an edge point is related to energy changes as $\delta W = M\partial\delta w/\partial n$ because $\partial\delta w/\partial n$ represents, in an infinitesimal strain theory, the variation in rotation angle the moment acts on. But this is just the integrand of I_3 in (16.3) devided by D. It can be shown, with the use of (16.14) and taking $\theta = 0$ after differentiation, that these reduce to

$$Q = D\left(\frac{\partial^3 w}{\partial n^3} + \nu\frac{\partial \theta}{\partial \ell}\frac{\partial^2 w}{\partial n^2}\right),$$

$$M = D\frac{\partial^2 w}{\partial n^2}. \tag{16.22}$$

A third type of condition is where the edges lie on fixed supports so that $w = 0$, but the edges are free to rotate and thus support no moment. Also $\delta w = 0$, but $\partial\delta w/\partial n \neq 0$. Thus (16.21) remains valid, but (16.22) no longer holds, in general. The reaction force is still given by the first of (16.22) at points where the edge is supported, but at such points the reaction moment is zero. The boundary condition (16.21) can be simplified by noting that $w = 0$ when the edges are supported, and that $\partial w/\partial \ell = 0$ and $\partial^2 w/\partial \ell^2 = 0$ as well. In that case,

$$w = 0, \quad \frac{\partial^2 w}{\partial n^2} + \nu\frac{\partial \theta}{\partial \ell}\frac{\partial w}{\partial n} = 0. \tag{16.23}$$

In deriving these formulas, as in (16.22), the angle θ may, without loss of generality, be set to zero after differentiation. This simply means that we choose to measure θ from the normal at the point at which the boundary condition is invoked.

16.4 Shear Forces and Bending and Twisting Moments

The following expressions hold for the shear forces and bending and twisting moments per unit length of a rectangular plate (Fig. 16.4),

$$Q_x = D\frac{\partial}{\partial x}\left(\nabla^2 w\right), \quad Q_y = D\frac{\partial}{\partial y}\left(\nabla^2 w\right), \tag{16.24}$$

$$M_x = D\left(\frac{\partial^2 w}{\partial x^2} + \nu\frac{\partial^2 w}{\partial y^2}\right),$$

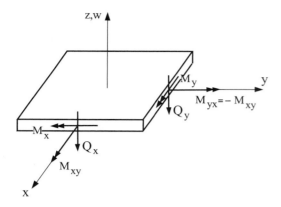

Figure 16.4. Positive directions of the shear force, bending and twisting moments in a rectangular plate element.

$$M_y = D\left(\frac{\partial^2 w}{\partial y^2} + \nu \frac{\partial^2 w}{\partial x^2}\right), \tag{16.25}$$

$$M_{xy} = D(1-\nu)\frac{\partial^2 w}{\partial x \partial y}.$$

From equilibrium conditions they are related by

$$\frac{\partial Q_x}{\partial x} + \frac{\partial Q_y}{\partial y} = p,$$

$$\frac{\partial M_x}{\partial x} + \frac{\partial M_{xy}}{\partial y} = Q_x, \tag{16.26}$$

$$\frac{\partial M_{xy}}{\partial x} + \frac{\partial M_y}{\partial y} = Q_y.$$

The effective shear forces (that need vanish at free edges) are

$$V_x = Q_x + \frac{\partial M_{xy}}{\partial y}, \quad V_y = Q_y + \frac{\partial M_{xy}}{\partial x}. \tag{16.27}$$

Finally, the stress components in the plate are

$$\sigma_{xx} = -\frac{12 M_x}{h^3} z, \quad \sigma_{yy} = -\frac{12 M_y}{h^3} z, \quad \sigma_{xy} = -\frac{12 M_{xy}}{h^3} z. \tag{16.28}$$

If we define

$$M = \frac{1}{1+\nu}(M_x + M_y), \tag{16.29}$$

we also have

$$\nabla^2 w = \frac{M}{D}, \quad \nabla^2 M = p,$$

$$Q_x = \frac{\partial M}{\partial x}, \quad Q_y = \frac{\partial M}{\partial y}. \tag{16.30}$$

In the case of circular plate, the shear forces (per unit length) can be determined from

$$Q_r = D\frac{\partial}{\partial r}(\nabla^2 w), \quad Q_\theta = D\frac{1}{r}\frac{\partial}{\partial \theta}(\nabla^2 w). \tag{16.31}$$

16.5. Examples of Plate Deformation

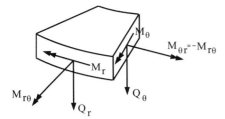

Figure 16.5. Positive directions of the shear force, bending and twisting moments in a plate element in polar coordinates.

The bending and twisting moments per unit length (Fig. 16.5) are

$$M_r = D\left[\frac{\partial^2 w}{\partial r^2} + \nu\left(\frac{1}{r}\frac{\partial w}{\partial r} + \frac{1}{r^2}\frac{\partial^2 w}{\partial \theta^2}\right)\right],$$

$$M_\theta = D\left(\frac{1}{r}\frac{\partial w}{\partial r} + \frac{1}{r^2}\frac{\partial^2 w}{\partial \theta^2} + \nu\frac{\partial^2 w}{\partial r^2}\right), \quad (16.32)$$

$$M_{r\theta} = D(1-\nu)\left(\frac{1}{r}\frac{\partial^2 w}{\partial r \partial \theta} - \frac{1}{r^2}\frac{\partial w}{\partial \theta}\right).$$

The effective shear forces are given by

$$V_r = Q_r + \frac{1}{r}\frac{\partial M_{r\theta}}{\partial \theta}, \quad V_\theta = Q_\theta + \frac{\partial M_{r\theta}}{\partial r}. \quad (16.33)$$

The corresponding stress components are

$$\sigma_{rr} = -\frac{12 M_r}{h^3} z, \quad \sigma_{\theta\theta} = -\frac{12 M_\theta}{h^3} z, \quad \sigma_{r\theta} = -\frac{12 M_{r\theta}}{h^3} z. \quad (16.34)$$

16.5 Examples of Plate Deformation

16.5.1 Clamped Circular Plate

Consider a circular plate of radius a, clamped around its edge, and loaded under the action of gravity. Let the mass density be ρ, and the plate's thickness h. We seek a solution for the deflection of the plate. In the equation of equilibrium (16.19), we identify $p = -\rho h g$, where g is the gravitational acceleration constant. Thus,

$$\nabla^4 w = 64\beta, \quad = \frac{p}{D} \quad (16.35)$$

where $\beta = -3\rho g(1-\nu)/16 h^2 E$ is a constant factor, made for convenience. Since the plate is circularly symmetric, $w = w(r)$ only, and we have

$$\nabla^2 w = \frac{1}{r}\frac{d}{dr}\left(r\frac{dw}{dr}\right), \quad (16.36)$$

and

$$\nabla^4 w = \frac{1}{r}\frac{d}{dr}\left\{r\frac{d}{dr}\left[\frac{1}{r}\frac{d}{dr}\left(r\frac{dw}{dr}\right)\right]\right\}. \quad (16.37)$$

As noted in Chapter 10, and in particular in equation (10.58), the general integral of (16.35) is

$$w(r) = \beta r^4 + Ar^2 + B + Cr^2 \ln\frac{r}{a} + F\ln\frac{r}{a}. \tag{16.38}$$

The particular solution βr^4 is included to account for the right-hand side of (16.35). We can set $F = 0$ because w is bounded everywhere, including at $r = 0$. Also, we set $C = 0$ since its term produces a singularity at $r = 0$ in $\nabla^2 w$. The boundary conditions are that at $r = a$, $w = 0$ and $dw/dr = 0$. With these, A and B are determined easily with the end result for the deflection,

$$w(r) = \beta(a^2 - r^2)^2. \tag{16.39}$$

16.5.2 Circular Plate with Simply Supported Edges

Consider a circular plate, again loaded by gravity force, whose edges are simply supported so that the boundary conditions (16.23) apply. Because for this circular plate $d\theta/d\ell = 1/r$, the conditions in (16.23) become

$$w = 0, \quad \frac{d^2w}{dr^2} + \frac{\nu}{r}\frac{dw}{dr} = 0, \quad \text{on } r = a. \tag{16.40}$$

The solution for w is similar as in previous case, and by invoking the conditions (16.40), it is found that

$$w(r) = \beta(a^2 - r^2)\left(\frac{5+\nu}{1+\nu}a^2 - r^2\right). \tag{16.41}$$

16.5.3 Circular Plate with Concentrated Force

If a clamped circular plate is loaded by a concentrated force of magnitude f at its center, we can write $p = f\delta(r)$. The integration of (16.19) then yields

$$2\pi \int_0^a r\nabla^4 w \, dr = \frac{12(1-\nu^2)}{Eh^3} \int_A f\delta(r)\,dA = \frac{12(1-\nu^2)}{Eh^3}f. \tag{16.42}$$

In this case of a concentrated point force, the term involving the coefficient C is retained so that the solution for w becomes

$$w(r) = Ar^2 + B + Cr^2 \ln\frac{r}{a}. \tag{16.43}$$

When (16.43) is used in (16.42), it is found that $C = 3(1-\nu^2)f/2\pi Eh^3$. The clamped boundary conditions, $w = 0$ and $dw/dr = 0$ at $r = a$, yield the coefficients A and B, so that

$$w(r) = \frac{3f(1-\nu^2)}{2\pi Eh^3}\left[\frac{1}{2}(a^2 - r^2) - r^2 \ln\frac{a}{r}\right]. \tag{16.44}$$

16.5.4 Peeled Surface Layer

Imagine a thin surface layer peeled off a thick elastic body, as shown in Fig. 16.6. The resistance is due to adhesive forces between the layer and the block. We seek a relation between the surface energy and a measurable feature of the process, such as the shape of the peeled layer. Note that the work done in decohering the layer is given by $\bar\gamma = 2\gamma_0 - \gamma_a$,

16.6. Rectangular Plates

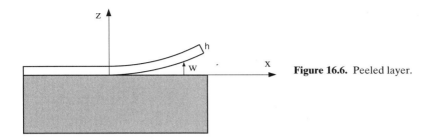

Figure 16.6. Peeled layer.

where γ_0 is the surface energy of a newly created interface (of which there are two with a surface energy of γ_0), and γ_a is the surface energy of the initial adhesive layer (which is destroyed by the peeling process).

The layer is treated as a thin plate with one edge clamped at the edge being torn off. As above, the angle made by the layer, to first order, is $\partial w/\partial x$, and its variation is $\partial \delta w/\partial x$. Thus the increment of work done by the moment M, given by the second of (16.22), and working through an increment of extension δx, is

$$M\frac{\partial \delta w}{\partial x} = M\frac{\delta x \partial^2 w}{\partial x^2}. \quad = \bar{\gamma}\delta x \tag{16.45}$$

By equating this increment of work to $\bar{\gamma}\delta x$, we find

$$\bar{\gamma} = \frac{Eh^3}{12(1-\nu^2)}\left(\frac{\partial^2 w}{\partial x^2}\right)^2. \tag{16.46}$$

↳ CRITICAL CURVATURE

16.6 Rectangular Plates

There are cases, other than those involving circular plates, where closed form solutions are possible. Notable among them are cases concerning plates with rectangular shapes such as illustrated in Fig. 16.7. We will take as a specific example cases where the edges are simply supported, so that displacements normal to the plate and moments vanish at the edges. The equation of equilibrium for a plate is

$$\nabla^4 w(x, y) = p(x, y)/D, \tag{16.47}$$

where $D = Eh^3/12(1-\nu^2)$ is the plate flexural rigidity. To meet the boundary conditions of $w = 0$ at $x = 0, a$ and $y = 0, b$, as well as the condition of vanishing moment there, w is expanded in a double Fourier sine series

$$w(x, y) = \sum_{m=1}^{\infty}\sum_{n=1}^{\infty} w_{mn} \sin\frac{m\pi x}{a}\sin\frac{n\pi y}{b}. \tag{16.48}$$

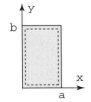

Figure 16.7. Geometry of a rectangular thin plate with simply supported edges.

The boundary conditions,

$$w(0, y) = w(a, y) = w(x, 0) = w(x, b) = 0, \tag{16.49}$$

are naturally satisfied. The moments are computed from the normal stresses, that in turn are computed from the second derivatives of $w(x, y)$, viz., $w_{,xx}$ and $w_{,yy}$, which likewise vanish along the edges. To satisfy the plate equilibrium equation, we expand the load function, $p(x, y)$, in a similar double sine series, viz.,

$$p(x, y) = \sum_{m=1}^{\infty} \sum_{n=1}^{\infty} p_{mn} \sin \frac{m\pi x}{a} \sin \frac{n\pi y}{b}. \tag{16.50}$$

The usual procedure, outlined below in a specific example, is used to compute the Fourier coefficients p_{mn}.

When (16.50) and (16.48) are substituted into (16.47), and the coefficients of like harmonics equated, it is found that

$$w_{mn} = \frac{p_{mn}}{\pi^4 D} \frac{1}{[(m/a)^2 + (n/b)^2]^2}. \tag{16.51}$$

Thus,

$$w(x, y) = \frac{1}{\pi^4 D} \sum_{m=1}^{\infty} \sum_{n=1}^{\infty} \frac{p_{mn}}{[(m/a)^2 + (n/b)^2]^2} \sin \frac{m\pi x}{a} \sin \frac{n\pi y}{b}. \tag{16.52}$$

16.6.1 Uniformly Loaded Rectangular Plate

For a uniformly loaded plate we let $p(x, y) = p_0$, and (16.50) becomes

$$p_0 = \sum_{m=1}^{\infty} \sum_{n=1}^{\infty} p_{mn} \sin \frac{m\pi x}{a} \sin \frac{n\pi y}{b}. \tag{16.53}$$

To obtain the coefficients, multiply both sides by $\sin(r\pi x/a)\sin(s\pi y/b)$, where r and s are integers, and integrate over the area of the plate. Then,

$$\int_0^a \int_0^b p_0 \sin \frac{r\pi x}{a} \sin \frac{s\pi y}{b} \, dx \, dy$$

$$= \sum_{m=1}^{\infty} \sum_{n=1}^{\infty} p_{mn} \int_0^a \sin \frac{r\pi x}{a} \sin \frac{m\pi x}{a} \, dx \int_0^b \sin \frac{s\pi y}{b} \sin \frac{n\pi y}{b} \, dy.$$

To evaluate the above integrals, recall that

$$\int_0^a \sin \frac{r\pi x}{a} \sin \frac{m\pi x}{a} \, dx = \begin{cases} a/2, & r = m, \\ 0, & r \neq m. \end{cases} \tag{16.54}$$

Since,

$$p_0 \int_0^a \sin \frac{r\pi x}{a} \, dx \int_0^b \sin \frac{s\pi y}{b} \, dy$$

$$= p_0 \frac{a}{r\pi}[\cos(r\pi) - 1]\frac{b}{s\pi}[\cos(s\pi) - 1] = \begin{cases} 0, & r \text{ or } s \text{ even}, \\ \frac{4ab}{\pi^2 rs} p_0, & r \text{ and } s \text{ odd}, \end{cases} \tag{16.55}$$

we obtain

$$p_{mn} = \frac{16 p_0}{\pi^2 mn}, \quad m, n = 1, 3, 5, \ldots, \tag{16.56}$$

and

$$w(x, y) = \frac{16}{\pi^6 D} p_0 \sum_{m=1,3,5}^{\infty} \sum_{n=1,3,5}^{\infty} \frac{1}{mn} \frac{1}{[(m/a)^2 + (n/b)^2]^2} \sin\frac{m\pi x}{a} \sin\frac{n\pi y}{b}. \tag{16.57}$$

16.7 Suggested Reading

Cook, R. D., and Young, W. C. (1999), *Advanced Mechanics of Materials*, 2nd ed., Prentice Hall, Upper Saddle River, New Jersey.

Landau, L. D., and Lifshitz, E. M. (1986), *Theory of Elasticity*, 3rd English ed., Pergamon Press, New York.

Szilard, R. (2004), *Theories and Applications of Plate Analysis: Classical, Numerical, and Engineering Methods*, Wiley, Hoboken, New Jersey.

Timoshenko, S. P., and Woinowsky-Krieger, S. (1987), *Theory of Plates and Shells*, 2nd ed., McGraw-Hill, New York.

Ugural, A. C. (1999), *Stresses in Plates and Shells*, 2nd ed., McGraw-Hill, Boston.

PART 4: MICROMECHANICS

17 Dislocations and Cracks: Elementary Treatment

17.1 Dislocations

Consider the plane strain deformation and the Airy stress function of the form

$$\phi = Cr \ln r \cos\theta, \quad C = \text{const.} \tag{17.1}$$

The corresponding stresses are obtained by using the general expressions from Chapter 10, i.e., (10.4)–(10.6). They give

$$\sigma_{rr} = \frac{1}{r}\frac{\partial \phi}{\partial r} + \frac{1}{r^2}\frac{\partial^2 \phi}{\partial \theta^2} = C\frac{\cos\theta}{r},$$

$$\sigma_{r\theta} = -\frac{\partial}{\partial r}\left(\frac{1}{r}\frac{\partial \phi}{\partial \theta}\right) = C\frac{\sin\theta}{r}, \tag{17.2}$$

$$\sigma_{\theta\theta} = \frac{\partial^2 \phi}{\partial r^2} = C\frac{\cos\theta}{r}.$$

The elastic strains may then be deduced from stresses by Hooke's law. The displacements are then obtained from strains by integration. The end result is

$$u_r = \frac{C}{G}[(1-\nu)\theta\sin\theta - 1/4\cos\theta + 1/2(1-2\nu)\ln r \cos\theta],$$
$$u_\theta = \frac{C}{G}[(1-\nu)\theta\cos\theta - 1/4\sin\theta - 1/2(1-2\nu)\ln r \sin\theta]. \tag{17.3}$$

Details of the derivation are given in the subsection below. The shear modulus is G, and ν is the Poisson's ratio. Clearly, the displacement field in this case is nonsingle valued, because

$$u_r(\theta=0) - u_r(\theta=2\pi) = 0,$$
$$u_\theta(\theta=0) - u_\theta(\theta=2\pi) = -\frac{2\pi C(1-\nu)}{G} \equiv b. \tag{17.4}$$

The physical scenario associated with this is illustrated in Fig. 17.1, which shows that the *dislocation* is created, in a heuristic manner, by first making a *cut*, and then creating a *gap* of width b, or removing a slab of material of thickness b, and then rejoining the two sides of the cut. In the case where material has been removed the *empty gap* would be filled with

17. Dislocations and Cracks

Figure 17.1. Dislocation created by a displacement discontinuity $u_\theta(0) - u_\theta(2\pi) = b$.

material of identical properties. The body is left in the state of stress

$$\sigma_{rr} = -\frac{Gb}{2\pi(1-\nu)}\frac{\cos\theta}{r},$$

$$\sigma_{r\theta} = -\frac{Gb}{2\pi(1-\nu)}\frac{\sin\theta}{r}, \quad (17.5)$$

$$\sigma_{\theta\theta} = -\frac{Gb}{2\pi(1-\nu)}\frac{\cos\theta}{r}.$$

The magnitude of the so-called Burgers vector of the dislocation is b. The stresses are singular at the center of dislocation ($r = 0$), and have both a shear and hydrostatic component. For plane strain,

$$\sigma_{zz} = \nu(\sigma_{rr} + \sigma_{\theta\theta}), \quad (17.6)$$

and thus the average normal stress is

$$\sigma_h = \frac{1}{3}(\sigma_{rr} + \sigma_{\theta\theta} + \sigma_{zz}) = -\frac{Gb(1+\nu)}{3\pi(1-\nu)}\frac{\cos\theta}{r}. \quad (17.7)$$

Note that if the dislocation were to move, say along the x axis, this motion would result in a displacement, $\mathbf{u} = b\mathbf{e}_y$ across the sections of plane so spanned by this motion. This feature allows for the construction of the stress field of a crack, as demonstrated in the next section. The vital role played by dislocations in the process of plastic deformation is dealt with in detail in later chapters of this book.

17.1.1 Derivation of the Displacement Field

Since the displacement field associated with (17.1) is nonsingle valued, the approach to derive Eqs. (17.3) is outlined here. The reader is directed to Tables 23.1–23.5 in Chapter 23 for further details on stress and displacement fields in polar coordinates.

The radial strain component is

$$e_{rr} = \frac{\partial u_r}{\partial r} = \frac{1}{E}[\sigma_{rr} - \nu(\sigma_{\theta\theta} + \sigma_{zz})], \quad (17.8)$$

which, in the present case of plane strain, becomes

$$\frac{\partial u_r}{\partial r} = \frac{C(1-2\nu)}{2G}\frac{\cos\theta}{r}. \quad (17.9)$$

When integrated this yields

$$u_r = \frac{C(1-2\nu)}{2G}\ln r \cos\theta + f(\theta), \quad (17.10)$$

where $f(\theta)$ is the integration function.

17.2. Tensile Cracks

Next consider the circumferential component of strain $e_{\theta\theta}$. From its definition in terms of the displacement components, we find

$$\frac{\partial u_\theta}{\partial \theta} = r e_{\theta\theta} - u_r. \tag{17.11}$$

When $e_{\theta\theta}$ is constructed and (17.10) used in (17.11), the integration gives

$$u_\theta = \frac{C(1-2\nu)}{2G} \sin\theta - \frac{C(1-2\nu)}{2G} \ln r \sin\theta - F(\theta) + g(r), \tag{17.12}$$

with $F(\theta) = \int f(\theta) d\theta$. Furthermore, recalling the definition of the shear strain

$$e_{r\theta} = \frac{1}{2}\left(\frac{1}{r}\frac{\partial u_r}{\partial \theta} + \frac{\partial u_\theta}{\partial r} - \frac{u_\theta}{r}\right), \tag{17.13}$$

and observing that $F''(\theta) = f'(\theta)$, there follows

$$\frac{2C(1-\nu)}{G}\frac{\sin\theta}{r} = \frac{F''(\theta)}{r} + \frac{F(\theta)}{r} + g'(r) - \frac{g(r)}{r}, \tag{17.14}$$

or

$$F''(\theta) + F(\theta) - \frac{2C(1-\nu)}{G}\sin\theta = rg'(r) - g(r) = K, \tag{17.15}$$

where K is a constant. The solution for $F(\theta)$ and $g(r)$ are, accordingly,

$$F(\theta) = -\frac{C(1-\nu)}{G}\theta\cos\theta + \frac{AC}{G}\sin\theta + \frac{BC}{G}\cos\theta + K, \tag{17.16}$$

and

$$g(r) = Kr + H. \tag{17.17}$$

The additional constants K and H correspond to rigid body motion and can be disregarded. The constants A and B, associated with the homogeneous solution to (17.15) are fixed by imposing the consistency of $f(\theta)$ with the known strains. Since

$$f(\theta) = -\frac{C(1-\nu)}{G}\cos\theta + \frac{C(1-\nu)}{G}\theta\sin\theta + \frac{AC}{G}\cos\theta - \frac{BC}{G}\sin\theta, \tag{17.18}$$

if (17.10) and (17.12) is to reproduce the known strains e_{rr}, $e_{\theta\theta}$, and $e_{r\theta}$, we must have

$$A = \frac{1}{4}(3-4\nu), \quad B = 0. \tag{17.19}$$

This specifies $F(\theta)$ and thus the displacement components u_r and u_θ, with the end results as in (17.3).

17.2 Tensile Cracks

Consider the boundary value problem associated with a *slitlike crack* of length $2a$, illustrated in Fig. 17.2. The crack lies within an infinite medium. We assume, as in the case of of the dislocation described above, a plane strain state. Thus the crack is infinitely deep in

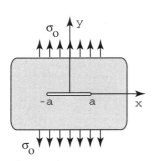

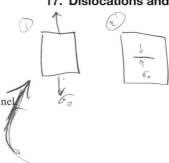

Figure 17.2. Center cracked panel

the direction normal to the x-y plane. The block is subject to a far field tensile stress of magnitude σ_0. Boundary conditions are then

$$\sigma_{xy} = \sigma_{yy} = 0 \quad \text{on} \quad -a \leq x \leq a, \; y = 0,$$
$$\sigma_{yy} \to \sigma_0 \quad \text{at} \quad r \to \infty, \tag{17.20}$$
$$\sigma_{xx}, \sigma_{xy} \to 0 \quad \text{at} \quad r \to \infty.$$

This may be solved using a simple method of linear superposition of two companion problems. For the first problem we take a simple homogeneous solution

$$\sigma_{yy}^{(1)} = \sigma_0, \quad \sigma_{xx}^{(1)}, \sigma_{xy}^{(1)} = 0. \tag{17.21}$$

For the second problem, we require

$$\sigma_{xy}^{(2)} = 0, \quad \sigma_{yy}^{(2)} = -\sigma_0 \quad \text{on} \quad -a \leq x \leq a, \; y = 0,$$
$$\sigma_{xx}^{(2)}, \sigma_{yy}^{(2)}, \sigma_{xy}^{(2)} \to 0 \quad \text{at} \quad r \to \infty. \tag{17.22}$$

Clearly, the stress state created as $\sigma = \sigma^{(1)} + \sigma^{(2)}$ will satisfy the crack boundary conditions expressed in (17.20). Problem (2) will be solved by the linear superposition of dislocation solutions obtained in the previous section. This is described next.

Let $\mathcal{B}(\xi)d\xi$ represent a continuous density of dislocations distributed between ξ and $\xi + d\xi$ (Fig. 17.3). The coordinate ξ is to lie on the x axis, between $-a \leq \xi \leq a$. This density will produce the stress

$$\sigma_{yy}(x, 0) = \sigma_{\theta\theta}(r, 0) = -\frac{G\mathcal{B}(\xi)d\xi}{2\pi(1-\nu)(x-\xi)}. \tag{17.23}$$

The stress produced by the distribution of dislocations along the entire length of the crack is

$$\sigma_{yy}(x, 0) = -\frac{G}{2\pi(1-\nu)}\int_{-a}^{a}\frac{\mathcal{B}(\xi)\,d\xi}{x-\xi} = -\sigma_0, \tag{17.24}$$

and thus

$$\int_{-a}^{a}\frac{\mathcal{B}(\xi)\,d\xi}{x-\xi} = \frac{2\pi(1-\nu)\sigma_0}{G}, \quad -a \leq x \leq a. \tag{17.25}$$

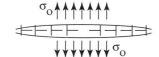

Figure 17.3. Distribution of dislocations.

17.2. Tensile Cracks

TO ENABLE THE INCORPORATION OF A FOURIER SERIES,
To solve this singular integral equation introduce the variables ϕ and θ, such that

$$x = a\cos\phi, \quad \xi = a\cos\theta. \tag{17.26}$$

The equation (17.25) then becomes

$$\int_0^\pi \frac{\mathcal{B}(\theta)\sin\theta\, d\theta}{\cos\phi - \cos\theta} = \frac{2\pi(1-\nu)\sigma_0}{G}, \quad 0 \le \phi \le \pi. \tag{17.27}$$

Call

$$A = \frac{2\pi(1-\nu)\sigma_0}{G} \tag{17.28}$$

for brevity, and note the result

$$\int_0^\pi \frac{\cos(n\theta)\, d\theta}{\cos\phi - \cos\theta} = -\frac{\pi\sin(n\phi)}{\sin\phi}. \tag{17.29}$$

Having this in mind, let $\mathcal{B}(\theta)$ be expanded in a Fourier series, viz.,

$$\mathcal{B}(\theta) = \sum_{n=0}^\infty p_n \frac{\cos(n\theta)}{\sin\theta}. \tag{17.30}$$

Then,

$$\int_0^\pi \frac{\sum_{n=0}^\infty p_n \cos(n\theta)}{\cos\phi - \cos\theta} d\theta = A, \tag{17.31}$$

i.e.,

$$\sum_{n=0}^\infty \int_0^\pi \frac{p_n \cos(n\theta)}{\cos\phi - \cos\theta} d\theta = A. \tag{17.32}$$

Using the result from (17.29), this becomes

$$-\sum_{n=0}^\infty p_n \frac{\pi\sin(n\phi)}{\sin\phi} = A. \tag{17.33}$$

In fact, with $n=1$, we have $p_1 = -A/\pi$ and

$$\mathcal{B}(\theta) = -\frac{2\sigma_0(1-\nu)}{G}\frac{\cos\theta}{\sin\theta} + \frac{C}{\sin\theta}, \tag{17.34}$$

or

$$\mathcal{B}(\xi) = -\frac{2\sigma_0(1-\nu)}{G}\frac{\xi}{\sqrt{a^2-\xi^2}} + \frac{C}{\sin\theta}. \tag{17.35}$$

But the dislocation distribution is an odd function of ξ, as there are no net dislocations on $-a \le \xi \le a$. Consequently,

$$\int_{-a}^a \mathcal{B}(\xi)\, d\xi = 0 \implies C = 0, \tag{17.36}$$

and

$$\mathcal{B}(\xi) = -\frac{2\sigma_0(1-\nu)}{G}\frac{\xi}{\sqrt{a^2-\xi^2}}. \tag{17.37}$$

This means that

$$\sigma_{yy}^{(2)} = -\frac{G}{2\pi(1-\nu)} \int_{-a}^{a} \frac{B(\xi)\,d\xi}{x-\xi} = \frac{\sigma_0}{\pi} \int_{-a}^{a} \frac{\xi\,d\xi}{(x-\xi)\sqrt{a^2-\xi^2}}$$

$$= \sigma_0\left(-1 + \frac{|x|}{\sqrt{x^2-a^2}}\right), \quad x \geq 0,\ y = 0.$$

Now, recall that the full solution is $\sigma = \sigma^{(1)} + \sigma^{(2)}$. For example, on the crack line ahead or behind the crack tip, we obtain

$$\sigma_{yy} = \frac{\sigma_0 |x|}{\sqrt{x^2-a^2}}, \quad |x| \geq a,\ y = 0. \tag{17.38}$$

Near the tip, say at $x = a + \epsilon$, where $\epsilon \ll a$, we have

$$\sigma_{yy} \to \frac{\sigma_0(a+\epsilon)}{\sqrt{2a\epsilon+\epsilon^2}} \sim \sigma_0 \sqrt{\frac{a}{2\epsilon}}, \quad \text{on } y = 0. \tag{17.39}$$

If we write that on the crack line, just ahead of the tip,

$$\sigma_{yy}(r,0) = \frac{K_I}{\sqrt{2\pi r}}, \tag{17.40}$$

where r is measured from the crack tip, then

$$K_I = \sigma_0 \sqrt{\pi a}, \tag{17.41}$$

which is the so-called ~~stress intensity factor~~ for this *center cracked panel* geometry.

Other components of stress may be generated by integration of those associated with individual dislocations.

17.3 Suggested Reading

Broek, D. (1987), *Elementary Engineering Fracture Mechanics*, 4th ed., Martinus Nijhoff, Dordrecht, The Netherlands.

Cherepanov, G. P. (1979), *Mechanics of Brittle Fracture*, McGraw-Hill, New York.

Friedel, J. (1964), *Dislocations*, Pergamon, New York.

Gdoutos, E. E. (1993), *Fracture Mechanics: An Introduction*, Kluwer Academic, Dordrecht, The Netherlands.

Hull, D., and Bacon, D. J. (1999), *Introduction to Dislocations*, 3rd ed., Butterworth-Heinemann, Boston.

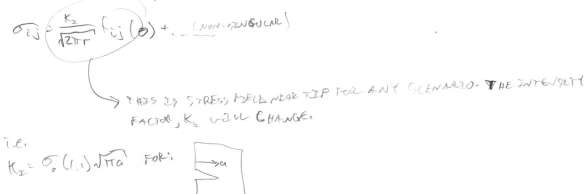

18 Dislocations in Anisotropic Media

18.1 Dislocation Character and Geometry

Imagine a cut made within an otherwise unbounded elastic medium, where the cut is over an open surface, S_{cut}, that is bounded by a line, C (see Fig. 18.1). To create the dislocation, material is displaced everywhere across the cut surface by a constant vector, $\mathbf{b}$. The displacement vector $\mathbf{b}$ is called the Burgers vector. Where such a displacement would cause a gap, extra material is imagined inserted so as to make the body continuous. Where such displacement would cause material to interpenetrate, the excess material is imagined scraped away. At all points on S_{cut}, the surface is re-bonded so that all further displacements are continuous. We note that this process causes the displacement field to be nonsingle valued across S_{cut}. In fact, if C is the line bounding the cut surface, and $\mathcal{R}$ is a closed circuit surrounding C at any point on C, then

$$\oint_{\mathcal{R}} \frac{\partial \mathbf{u}}{\partial x_m} \, \mathrm{d}x_m = \mathbf{b}, \tag{18.1}$$

where $\mathbf{u}$ is the displacement, and the integral is taken counterclockwise with $\mathbf{t}$, the unit tangent to C, taken in the positive sense. Note that the unit tangent $\mathbf{t}$ is continuous as the *dislocation line*, C, is traversed. At points on the dislocation line, where $\mathbf{t} \parallel \mathbf{b}$, $\mathbf{t} \cdot \mathbf{b} = \pm b$, whereas at points where $\mathbf{t} \perp \mathbf{b}$, $\mathbf{t} \cdot \mathbf{b} = 0$. We say that at the former type of point the dislocation line has *screw dislocation* character, whereas at the latter type of point the line has *edge dislocation* character. In general, the dislocation line has a *mixed* character and has both edge and screw character. The figure illustrates a dislocation line that has extended screw *segments* and hairpin type loop segments that, at points, have primarily edge character. The elastic field is caused by the displacement jump across S_{cut} as measured specifically by (18.1); there are no body forces or applied traction. Thus, the dislocation represents a purely internal source of stress.

To construct the elastic field, consider the identity

$$u_m(\mathbf{x}) = \int_{-\infty}^{\infty} \delta_{im}\delta(\mathbf{x} - \mathbf{x}')u_i(\mathbf{x}') \, \mathrm{d}^3\mathbf{x}' \tag{18.2}$$

and recall that

$$\delta_{im}\delta(\mathbf{x} - \mathbf{x}') = -C_{ijkl}\frac{\partial^2 G_{km}}{\partial x_l \partial x_j}, \tag{18.3}$$

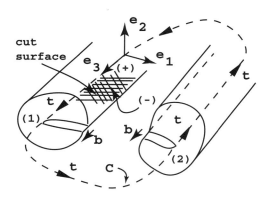

Figure 18.1. Dislocation loop and cut surface.

where **G** is the Green's function. Thus, formally, the displacement field of the dislocation can be written as

$$u_m(\mathbf{x}) = -C_{ijkl} \int_{-\infty}^{\infty} \frac{\partial^2 G_{km}}{\partial x_l \partial x_j} u_i(\mathbf{x}') \, d^3\mathbf{x}'. \tag{18.4}$$

Since **G** is a function of $\mathbf{x} - \mathbf{x}'$, we may write

$$\frac{\partial}{\partial x_j} G_{km}(\mathbf{x} - \mathbf{x}') = -\frac{\partial}{\partial x'_j} G_{km}(\mathbf{x} - \mathbf{x}'), \tag{18.5}$$

and

$$u_m(\mathbf{x}) = C_{ijkl} \int_{-\infty}^{\infty} \frac{\partial}{\partial x_l} \left[\frac{\partial G_{km}(\mathbf{x} - \mathbf{x}')}{\partial x'_j} u_i(\mathbf{x}') \right] d^3\mathbf{x}' \tag{18.6}$$

$$= C_{ijkl} \int_{-\infty}^{\infty} \frac{\partial}{\partial x_l} \left\{ \frac{\partial}{\partial x'_j} [G_{km}(\mathbf{x} - \mathbf{x}') u_i(\mathbf{x}')] - G_{km} \frac{\partial u_i(\mathbf{x}')}{\partial x'_j} \right\} d^3\mathbf{x}'.$$

We next examine the second integral in (18.6), *viz.*,

$$I = -C_{ijkl} \int_{-\infty}^{\infty} \frac{\partial}{\partial x_l} \left(G_{km} \frac{\partial u_i}{\partial x'_j} \right) d^3\mathbf{x}'. \tag{18.7}$$

First note that the stress field of the dislocation is calculated from the linear elastic constitutive relation as

$$\sigma_{kl}(\mathbf{x}') = C_{ijkl} \frac{\partial u_i(\mathbf{x}')}{\partial x'_j}, \tag{18.8}$$

because of the inherent symmetry in elastic moduli **C**. Noting (18.5) again, along with the fact that $\sigma_{kl,l} = 0$ by equilibrium, the integral I may be rewritten as

$$I = \int_{-\infty}^{\infty} \frac{\partial}{\partial x'_l} (G_{km} \sigma_{kl}) \, d^3\mathbf{x}' = \int_{S_\infty} G_{km} \sigma_{kl}(\mathbf{x}') \, dS', \tag{18.9}$$

where S_∞ is a bounding surface that retreats to infinity in the unbounded medium.

But the dislocation is actually a loop, as illustrated in Fig. 18.1, and thus possesses both positive a negative segments, *i.e.*, segments for which $\mathbf{t} \cdot \mathbf{b} > 0$ and $\mathbf{t} \cdot \mathbf{b} < 0$. At a distance, then, the elastic fields must fall off at least as fast as $1/r^2$, where r is the radial distance

18.1. Dislocation Character and Geometry

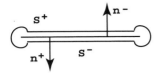

Figure 18.2. Cross section of the cut surface.

from the centroid of the loop. This is shown *via* specific example below. In fact, the fall off of the field is more like $1/r^3$, as will be verified. Moreover, the Green's function **G** falls off as $1/r$. Thus the integrand of the second integral of (18.9) falls off at least as fast as $1/r^3$, or even as fast as $1/r^4$. Thus $I \to 0$. The result for **u** is then

$$u_m(\mathbf{x}) = C_{ijkl} \int_{-\infty}^{\infty} \frac{\partial}{\partial x_l} \frac{\partial}{\partial x'_j} [G_{km} u_i(\mathbf{x}')] \, d^3\mathbf{x}', \tag{18.10}$$

or, for the distortions,

$$\frac{\partial u_m(\mathbf{x})}{\partial x_r} = u_{m,r}(\mathbf{x}) = C_{ijkl} \int_{-\infty}^{\infty} \frac{\partial^2}{\partial x_l \partial x_r} \frac{\partial}{\partial x'_j} [G_{km} u_i(\mathbf{x}')] \, d^3\mathbf{x}'$$

$$= C_{ijkl} \frac{\partial^2}{\partial x_l \partial x_r} \int_{-\infty}^{\infty} \frac{\partial}{\partial x'_j} [G_{km} u_i(\mathbf{x}')] \, d^3\mathbf{x}'. \tag{18.11}$$

An application of the divergence theorem yields

$$u_{m,r}(\mathbf{x}) = C_{ijkl} \int_{S} \frac{\partial^2}{\partial x_l \partial x_r} G_{km} u_i(\mathbf{x}') n_j \, dS', \tag{18.12}$$

where $S = S_{\text{cut}} + S_\infty$. Since $G_{km,lr} \sim 1/r^3$ and, as it happens, **u** also falls off with distance, the integral over S_∞ vanishes. What remains is the integral over S_{cut}. But across S_{cut} the displacement jump is $\mathbf{u}^- - \mathbf{u}^+ = \mathbf{b}$, on account of relation (18.1). Also, the outward pointing normal to the infinite medium is such that the unit normal to S_{cut}^-, $\mathbf{n}^-$, points in the positive direction as illustrated in Fig. 18.2. Thus we take $\mathbf{n} = \mathbf{n}^-$ as the common unit normal to S_{cut}, and write

$$u_{m,r}(\mathbf{x}) = C_{ijkl} \int_{S_{\text{cut}}} G_{km,lr}(\mathbf{x} - \mathbf{x}') b_i n_j \, dS'. \tag{18.13}$$

This is known as the Volterra's integral.

The surface integral may be converted to an integral over the bounding line $\mathcal{C}$ of the dislocation using Stoke's theorem in the form

$$\int_{S_{\text{cut}}} \epsilon_{ljk} \varphi_{r,k} n_j \, dS' = \oint_{\mathcal{C}} \varphi_r t_l \, ds, \tag{18.14}$$

where **t** is the running unit tangent vector to $\mathcal{C}$, and ds is an element of arc length. The integral in (18.13) may be rephrased as

$$u_{m,r}(\mathbf{x}) = C_{ijkl} \int_{S_{\text{cut}}} G_{km,lr}(\mathbf{x} - \mathbf{x}') b_i n_j \, dS'$$

$$= \int_{S_{\text{cut}}} \psi_j^{mr} n_j \, dS'. \tag{18.15}$$

Letting
$$\varphi_j^{mr} = \epsilon_{njp} X_n \psi_p^{mr}, \quad X_n = x_n - x'_n, \tag{18.16}$$

it follows that
$$\begin{aligned}\varphi_{j,k}^{mr} &= \epsilon_{njp} X_{n,k} \psi_p^{mr} + \epsilon_{njp} X_n \psi_{p,k}^{mr} \\ &= \epsilon_{kjp} \psi_p^{mr} + \epsilon_{njp} X_n \psi_{p,k}^{mr},\end{aligned} \tag{18.17}$$

and
$$\begin{aligned}\epsilon_{jks} \varphi_{j,k}^{mr} n_s &= \epsilon_{jks} \epsilon_{kjp} \psi_p^{mr} n_s + \epsilon_{jks} \epsilon_{njp} X_n \psi_{p,k}^{mr} n_s \\ &\quad - 2\psi_s^{mr} n_s - \psi_{p,k}^{mr} X_k n_p.\end{aligned} \tag{18.18}$$

Since ψ_p^{mr} is homogeneous of degree -3, we have
$$\left(X_i \psi_p^{mr}\right)_{,i} = 0, \quad X_k \psi_{p,k}^{mr} = -3\psi_p^{mr}. \tag{18.19}$$

Then,
$$\epsilon_{jks} \varphi_{j,k}^{mr} n_s = \psi_p^{mr} n_p, \tag{18.20}$$

and
$$\begin{aligned}\int_{S_{cut}} \epsilon_{jks} \varphi_{j,k}^{mr} n_s \, dS' &= \int_{S_{cut}} \epsilon_{jks} \left(\epsilon_{njp} X_n \psi_p^{mr}\right)_{,k} n_s \, dS' \\ &= -\oint_C \epsilon_{njp} X_n \psi_p^{mr} t_j \, ds.\end{aligned} \tag{18.21}$$

Consequently,
$$u_{m,r}(\mathbf{x}) = -\oint_C \epsilon_{njp}(x_n - x'_n) t_j b_i C_{ipsq} G_{mq,sr} \, ds. \tag{18.22}$$

Note that this conversion demonstrates an important fact that dislocations are characterized by the line as described by **t** and their Burgers vector **b**; the cut surface used to create them then becomes arbitrary, as expected.

18.2 Dislocations in Isotropic Media

In this section expressions are derived for the stress fields of infinitely long and straight dislocations in isotropic elastic media. As such the fields correspond to either states of antiplane strain or plane strain.

18.2.1 Infinitely Long Screw Dislocations

The dislocation line is taken to be $\mathbf{t} = \mathbf{e}_3$, i.e., along the unit base vector of the x_3 axis. The Burgers vector is taken such that $\mathbf{b}^s \cdot \mathbf{t} = +b^s$; the dislocation is accordingly said to be a *right-handed screw dislocation*. In other words, $\mathbf{b}^s \parallel \mathbf{t}$. The jump condition expressed in (18.1) is met with
$$u_3 = \frac{b^s}{2\pi}\theta = \frac{b^s}{2\pi}\tan^{-1}\frac{x_2}{x_1}. \tag{18.23}$$

18.2. Dislocations in Isotropic Media

This displacement field is sufficient to satisfy the equations of equilibrium. Indeed, $\sigma_{13} = G\partial u_3/\partial x_1$ and $\sigma_{23} = G\partial u_3/\partial x_2$, the single equilibrium equation reads

$$\frac{\partial \sigma_{13}}{\partial x_1} + \frac{\partial \sigma_{23}}{\partial x_2} = G\nabla^2 u_3 = 0, \tag{18.24}$$

which of course is satisfied by the harmonic function, $u_3 \sim \theta$. The stresses are, therefore,

$$\sigma_{13}(x_1, x_2) = -\frac{Gb^s}{2\pi} \frac{x_2}{x_1^2 + x_2^2}, \quad \sigma_{23}(x_1, x_2) = \frac{Gb^s}{2\pi} \frac{x_1}{x_1^2 + x_2^2}. \tag{18.25}$$

It is seen that the stress fields fall off as $1/r$ from an isolated infinitely long straight segment, but for two parallel segments that are part of the same loop, the fields fall off as $1/r^2$. Indeed, consider the two screw dislocation segments shown earlier in the figure. Let them be separated by a distance ϵ. The Burgers vector is the same for the two segments, but their unit tangents are antiparallel. Thus the stress fields are of opposite sign. Consider, for example, the stresses on the plane $x_2 = 0$. Then,

$$\sigma_{23} \sim -\frac{Gb^s}{2\pi}\left(\frac{1}{x_1} - \frac{1}{x_1 + \epsilon}\right) \sim -\frac{Gb^s}{2\pi}\frac{\epsilon}{x_1^2}. \tag{18.26}$$

18.2.2 Infinitely Long Edge Dislocations

The edge dislocation is characterized by the property that $\mathbf{b}^e = b^e \mathbf{e}_1$, whereas $\mathbf{t} = \mathbf{e}_3$. The corresponding displacement field is

$$\begin{aligned} u_1(x_1, x_2) &= \frac{b^e}{2\pi}\tan^{-1}(x_2/x_1) + \frac{b^e}{2\pi(1-\nu)}\frac{x_1 x_2}{x_1^2 + x_2^2}, \\ u_2(x_1, x_2) &= -\frac{b^e}{2\pi}\frac{1-2\nu}{2(1-\nu)}\ln r - \frac{x_1^2 - x_2^2}{4(1-\nu)(x_1^2 + x_2^2)}, \end{aligned} \tag{18.27}$$

whereas the stresses are

$$\begin{aligned} \sigma_{11} &= -\frac{Gb^e}{2\pi(1-\nu)}\frac{x_2(3x_1^2 + x_2^2)}{(x_1^2 + x_2^2)^2}, \\ \sigma_{22} &= \frac{Gb^e}{2\pi(1-\nu)}\frac{x_2(x_1^2 - x_2^2)}{(x_1^2 + x_2^2)^2}, \\ \sigma_{12} &= \frac{Gb^e}{2\pi(1-\nu)}\frac{x_1(x_1^2 - x_2^2)}{(x_1^2 + x_2^2)^2}, \\ \sigma_{33} &= -\frac{Gb^e\nu}{\pi(1-\nu)}\frac{x_2}{x_1^2 + x_2^2} = \nu(\sigma_{11} + \sigma_{22}). \end{aligned} \tag{18.28}$$

18.2.3 Infinitely Long Mixed Segments

The elastic field of an infinitely long and straight dislocation of *mixed* character is simply obtained by the linear superposition of the solutions obtained above. Here, however, we consider an infinitely long curved dislocation line lying in the plane $x_3 = 0$ with a Burgers vector as illustrated in Fig. 18.3. Our perspective is that the Burgers vector is given and fixed, but the character of the line varies because of varying orientation of the line tangent $\mathbf{t}$.

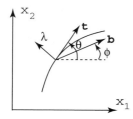

Figure 18.3. Infinitely long curved dislocation with a Burgers vector **b**.

Thus as **t** varies, *i.e.*, as θ or $(\theta - \phi)$ varies, the edge *vs.* screw dislocation character varies. Now let the Burgres vector be **b**, so that

$$\mathbf{b} = b\cos\phi \mathbf{e}_1 + b\sin\phi \mathbf{e}_2,$$
$$\mathbf{t} = \cos\theta \mathbf{e}_1 + \sin\theta \mathbf{e}_2. \tag{18.29}$$

This means that

$$b^s = b\cos(\theta - \phi) = b_t,$$
$$b^e = -b\sin(\theta - \phi) = b_\lambda, \tag{18.30}$$

where b^s and b^e are the screw and edge components of **b** respectively. We now use the solutions obtained above to construct the in-plane field of this mixed dislocation.

Let d be the normal distance from the line as reckoned by "looking right" while sighting down **t**. Then in the plane $x_3 = 0$, we find

$$\sigma_{t3} = \frac{Gb}{2\pi d}\cos(\theta - \phi),$$
$$\sigma_{\lambda 3} = -\frac{Gb}{2\pi d}\sin(\theta - \phi), \tag{18.31}$$

where t and λ are used as indices corresponding to rotated axes aligned with the dislocation line and orthogonal to it (Fig. 18.3). With respect to the fixed coordinates (x_1, x_2), we have

$$\sigma_{13} = \sigma_{t3}\cos\theta - \sigma_{\lambda 3}\sin\theta$$
$$= \frac{Gb}{2\pi d}\cos(\theta - \phi)\cos\theta + \frac{Gb}{2\pi(1-\nu)d}\sin(\theta - \phi)\sin\theta, \tag{18.32}$$

and

$$\sigma_{23} = \sigma_{t3}\sin\theta + \sigma_{\lambda 3}\cos\theta$$
$$= \frac{Gb}{2\pi d}\cos(\theta - \phi)\sin\theta - \frac{Gb}{2\pi(1-\nu)d}\sin(\theta - \phi)\cos\theta. \tag{18.33}$$

We note for later use that the in-plane stresses can be expressed as

$$\sigma_{\alpha\beta} = \frac{1}{d}\Sigma_{\alpha\beta}(\theta;\phi) = \frac{1}{d}\Sigma_{\alpha\beta}(\theta), \tag{18.34}$$

where in the second rendering of this relation the dependence on ϕ has been omitted on account of the perspective of fixed ϕ, i.e., fixed **b**. In fact,

$$\Sigma_{13}(\theta) = \frac{Gb}{2\pi}\left[\cos(\theta-\phi)\cos\theta + \frac{1}{1-\nu}\sin(\theta-\phi)\sin\theta\right], \qquad (18.35)$$
$$\Sigma_{23}(\theta) = \frac{Gb}{2\pi}\left[\cos(\theta-\phi)\sin\theta - \frac{1}{1-\nu}\sin(\theta-\phi)\cos\theta\right].$$

18.3 Planar Geometric Theorem

Here we consider the case of a dislocation loop lying entirely within a plane, and seek to construct its elastic field in that plane. Let the plane be defined as $x_3 = 0$. The dislocation line is $\mathcal{C}$ and its cut surface can be taken to lie entirely in the plane, without loss of generality. Then,

$$u_{m,r}(\mathbf{x}) = C_{ijkl}\int_{S_{\text{cut}}} G_{km,lr}(\mathbf{x}-\mathbf{x}')b_i n_j \, dS', \qquad (18.36)$$

as found earlier. The field point of interest lies in the plane $x_3 = 0$, and thus has coordinates (x_1, x_2). Let

$$X = x_1 - x_1', \quad Y = x_2 - x_2'. \qquad (18.37)$$

If we introduce a unit vector, **T**, defined to be parallel to $(\mathbf{x}-\mathbf{x}')$ then $\mathbf{x}-\mathbf{x}' = s\mathbf{T}$, where $|s| = |\mathbf{x}-\mathbf{x}'|$. Note that s is algebraically signed. Recall that the Green's function has symmetry, such that

$$G_{km}(s\mathbf{T}) = \frac{\text{sgn}(s)}{s}G_{km}(\mathbf{T}) \qquad (18.38)$$

and

$$G_{km,lr}(s\mathbf{T}) = \frac{\text{sgn}(s)}{s^3}G_{km,lr}(\mathbf{T}), \qquad (18.39)$$

i.e., both **G** and its second derivative are symmetric with respect to the sense of **T**. Thus, if θ is the angle that **T** makes with a datum drawn in the plane, we may write

$$G_{km}(\mathbf{x}-\mathbf{x}') = \frac{1}{R}G_{km}(\theta) = \frac{1}{R}G_{km}(\theta+\pi), \qquad (18.40)$$

where

$$R = |\mathbf{x}-\mathbf{x}'| = |s|, \quad \mathbf{T} = \mathbf{T}(\theta) = -\mathbf{T}(\theta+\pi). \qquad (18.41)$$

Similarly,

$$G_{km,lr}(\mathbf{x}-\mathbf{x}') = \frac{1}{R^3}G_{km,lr}(\theta) = \frac{1}{R^3}G_{km,lr}(\theta+\pi). \qquad (18.42)$$

But θ is a function of X/R and Y/R, and so the above may be expressed as a power series of the form

$$G_{km} = \frac{1}{R}\sum_{n,m}\bar{\Lambda}_{mn}\left(\frac{X}{R}\right)^m\left(\frac{Y}{R}\right)^n,$$

$$G_{km,lr} = \frac{1}{R^3}\sum_{n,m}\Lambda_{mn}\left(\frac{X}{R}\right)^m\left(\frac{Y}{R}\right)^n = I. \tag{18.43}$$

Call the second of (18.43), as noted, I, and write

$$I = \frac{1}{R^3}\Theta(\theta) = \frac{1}{R^3}\Theta(\theta+\pi),$$

$$\Theta(\theta) = \sum_{n,m}\Lambda_{mn}\left(\frac{X}{R}\right)^m\left(\frac{Y}{R}\right)^n. \tag{18.44}$$

To proceed, examine the expression

$$\mathcal{I} = \frac{\partial}{\partial X}(XI) + \frac{\partial}{\partial Y}(YI). \tag{18.45}$$

Since $R^2 = X^2 + Y^2$, we have

$$\frac{\partial R}{\partial X} = \frac{X}{R}, \quad \frac{\partial R}{\partial Y} = \frac{Y}{R},$$

$$\frac{\partial(1/R^3)}{\partial X} = -\frac{3X}{R^5}, \quad \frac{\partial(1/R^3)}{\partial Y} = -\frac{3Y}{R^5},$$

$$\frac{\partial(1/R^{m+n})}{\partial X} = -\frac{(m+n)X}{R^{m+n+2}}, \quad \frac{\partial(1/R^{m+n})}{\partial Y} = -\frac{(m+n)Y}{R^{m+n+2}}.$$

Thus, (18.45) becomes

$$\mathcal{I} = I + I - 3\frac{X^2}{R^5}\sum_{m,n}\Lambda_{mn}\left(\frac{X}{R}\right)^m\left(\frac{Y}{R}\right)^n$$

$$-3\frac{Y^2}{R^5}\sum_{m,n}\Lambda_{mn}\left(\frac{X}{R}\right)^m\left(\frac{Y}{R}\right)^n + \frac{X}{R^3}\sum_{m,n}m\Lambda_{mn}\left(\frac{X^{m-1}}{R^m}\right)\left(\frac{Y}{R}\right)^n$$

$$+\frac{Y}{R^3}\sum_{m,n}n\Lambda_{mn}\left(\frac{X}{R}\right)^m\left(\frac{Y^{n-1}}{R^n}\right) - \frac{X}{R^3}\sum_{m,n}\Lambda_{mn}\frac{(m+n)X}{r^{m+n+2}}X^m\left(\frac{Y}{R}\right)^n$$

$$-\frac{Y}{R^3}\sum_{m,n}\Lambda_{mn}\frac{(m+n)Y}{r^{m+n+2}}Y^n\left(\frac{X}{R}\right)^m = -I.$$

Therefore,

$$-I(X,Y) = \frac{\partial}{\partial X}(XI) + \frac{\partial}{\partial Y}(YI). \tag{18.46}$$

Now return to (18.36) and incorporate the terms involving b_i and C_{ijkl}, as they are constants. Further, the sums over repeated subscripts are implied (and implemented), so that (18.36) is expressed as

$$u_{m,r} = -\iint_{S_{cut}}\left[\frac{\partial}{\partial X}(XI_{mr}) + \frac{\partial}{\partial Y}(YI_{mr})\right]dX\,dY. \tag{18.47}$$

18.3. Planar Geometric Theorem

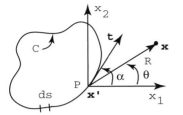

Figure 18.4. Geometry of a planar loop.

Next, recall Stokes theorem in the form

$$\oint_C (P\,dX + Q\,dY) = \iint_S \left(\frac{\partial Q}{\partial X} - \frac{\partial P}{\partial Y}\right) dX\,dY, \tag{18.48}$$

where C is the line that bounds the open surface S. Let $Q = XI_{mr}$ and $P = -YI_{mr}$ to convert the surface integral into one along the bounding line C. Also note that the loop C is characterized by the parametric equations for $x_1'(s)$ and $x_2'(s)$. Consider the typical point such as P. At this point, $dx_1'/ds = \cos\alpha$ and $dx_2'/ds = \sin\alpha$, where ds is an element of arc length and, as before, $\mathbf{t}$ is the variable unit tangent to C (see Fig. 18.4). Thus, taking the field points x_1, x_2 to be fixed in the integration, we have that $X' = dX/ds = -dx_1'/ds$, with a similar consideration for $Y' = dY/ds$. Consequently, the integral in (18.47) becomes

$$u_{m,r}(\mathbf{x}) = \oint_C (X'Y - Y'X) I_{mr}(X,Y)\,ds. \tag{18.49}$$

With reference to the figure, we can write

$$\begin{aligned} X' &= -\cos\alpha, & Y' &= -\sin\alpha, \\ X &= R\cos\theta, & Y &= R\sin\theta. \end{aligned} \tag{18.50}$$

Using these it follows that

$$(X'Y - Y'X)ds = R(-\cos\alpha\sin\theta + \sin\alpha\cos\theta)ds = -R\sin(\theta-\alpha)ds,$$

and thus

$$u_{m,r}(\mathbf{x}) = -\oint_C \frac{\Theta_{mr}(\theta)\sin(\theta-\alpha)}{R^2}\,ds. \tag{18.51}$$

Recall that $\Theta(\theta) = \Theta(\theta+\pi)$, θ, α, and R are all functions of the parameter s and therefore dependent on the size and shape of the dislocation loop. Also note that the formula expressed in (18.51) applies to any loop. We may then reinterpret (18.51) as an integral equation for the as yet unknown function, $\Theta(\theta)$, in terms of the solution, $u_{m,r}(\mathbf{x})$.

We now want to choose a somewhat convenient choice for C; we choose C to be an infinitely long and straight dislocation as depicted in Fig. 18.5. Using the geometry of the figure, we observe the following identities

$$\begin{aligned} x_1' &= s\cos\alpha, & x_2' &= s\sin\alpha, \\ s &= -d\cot(\theta-\alpha), & R &= d/\sin(\theta-\alpha), \\ ds &= \frac{d}{\sin^2(\theta-\alpha)}\,d\theta\,. \end{aligned} \tag{18.52}$$

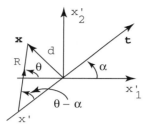

Figure 18.5. Coordinates for an infinitely long, straight dislocation.

With these, (18.51) becomes

$$u_{m,r}(d,\alpha) = -\frac{1}{d}\int_{\theta=\alpha}^{\theta=\alpha+\pi}\Theta(\theta)\sin(\theta-\alpha)\,d\theta. \tag{18.53}$$

Derivatives *wrt* to α leads to a convenient interpretation, *viz.*,

$$\frac{\partial u_{m,r}}{\partial \alpha} = \frac{1}{d}\int_{\alpha}^{\alpha+\pi}\Theta(\theta)\cos(\theta-\alpha)\,d\theta, \tag{18.54}$$

and

$$\frac{\partial^2 u_{m,r}}{\partial \alpha^2} = -\frac{1}{d}\left[\Theta(\alpha)+\Theta(\alpha+\pi)\right]+\frac{1}{d}\int_{\alpha}^{\alpha+\pi}\Theta(\theta)\sin(\theta-\alpha)\,d\theta. \tag{18.55}$$

The addition of (18.53) and (18.55) yields the desired result,

$$u_{m,r}(d,\alpha) + \frac{\partial^2 u_{m,r}(d,\alpha)}{\partial \alpha^2} = -\frac{1}{d}\left[\Theta(\alpha)+\Theta(\alpha+\pi)\right], \tag{18.56}$$

and thus

$$\Theta(\alpha) = -\frac{d}{2}\left[u_{m,r}(d,\alpha) + \frac{\partial^2 u_{m,r}(d,\alpha)}{\partial \alpha^2}\right]. \tag{18.57}$$

We have already shown that the field of an infinitely long and straight dislocation can be cast as

$$u_{m,r} = \frac{\Sigma_{mr}(\zeta)}{d}, \tag{18.58}$$

where ζ would be an angle representing the dislocation's orientation in the plane (assuming its Burgers vector is fixed). Thus, the expression for the field at an arbitrary point becomes

$$u_{m,r}(\mathbf{x}) = \frac{1}{2}\oint_C \frac{\Sigma_{mr}(\theta) + \partial^2\Sigma_{mr}(\theta)/\partial\theta^2}{R^2}\sin(\theta-\alpha)\,ds. \tag{18.59}$$

18.4 Applications of the Planar Geometric Theorem

Considerable simplification can be achieved by approximating dislocation loops as being polygonal, *i.e.*, as having faceted sides such as sketched in Fig. 18.6. The geometry sketched in the figure will be used to reduce the integral given in (18.59) above. Figure 18.7 indicates a typical segment of the loop. As we integrate about the loop,

$$R = |\mathbf{x}-\mathbf{x}'| = \frac{d}{\sin(\theta-\alpha)}, \quad s = -d\cot(\theta-\alpha), \tag{18.60}$$

18.4. Applications of Planar Geometric Theorem

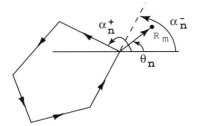

Figure 18.6. Geometry of a loop vertex.

so that

$$\frac{ds}{d\theta} = \frac{d}{\sin^2(\theta - \alpha)} = \frac{|\mathbf{x} - \mathbf{x}'|}{\sin(\theta - \alpha)} \,. \tag{18.61}$$

Thus, upon changing the integration variable from s to θ in (18.59), we obtain

$$u_{m,r}(\mathbf{x}) = 1/2 \oint_C \frac{\Sigma_{mr}(\theta) + \partial^2 \Sigma_{mr}(\theta)/\partial\theta^2}{d} \sin(\theta - \alpha)\, d\theta. \tag{18.62}$$

Integration by parts from θ_1 to θ_2, as indicated in Fig. 18.7 for the typical segment, yields

$$u_{m,r}(\mathbf{x}) = \frac{1}{2d}\left[-\Sigma_{mr}(\theta)\cos(\theta - \alpha) + \partial\Sigma_{mr}/\partial\theta\, \sin(\theta - \alpha)\right]_{\theta_1}^{\theta_2}$$
$$- \int_{\theta_1}^{\theta_2} \left[\partial\Sigma_{mr}/\partial\theta\, \cos(\theta - \alpha) + \partial\Sigma_{mr}/\partial\theta\, \cos(\theta - \alpha)\right] d\theta. \tag{18.63}$$

Thus,

$$u_{m,r}(\mathbf{x}) = \frac{1}{2d}\left[-\Sigma_{mr}(\theta)\cos(\theta - \alpha) + \partial\Sigma_{mr}/\partial\theta\, \sin(\theta - \alpha)\right]_{\theta_1}^{\theta_2}. \tag{18.64}$$

Clearly by summing over all such segments, the field of the entire loop may be constructed.
Recalling the first of (18.60), the integral in (18.63) may be recast as

$$u_{m,r}(\mathbf{x}) = \frac{1}{2}\oint_C \frac{\Sigma_{mr}(\theta) + \partial^2\Sigma_{mr}(\theta)/\partial\theta^2}{|\mathbf{x} - \mathbf{x}'|}\, d\theta. \tag{18.65}$$

Examine the second term in the integrand, viz.,

$$\oint_C \frac{\partial^2\Sigma_{mr}(\theta)/\partial\theta^2}{|\mathbf{x} - \mathbf{x}'|} d\theta = \left[\frac{\partial\Sigma_{mr}(\theta)/\partial\theta}{|\mathbf{x} - \mathbf{x}'|}\right]_1^1 - \oint_C \frac{\partial\Sigma_{mr}(\theta)/\partial\theta\, \cot(\theta - \alpha)}{|\mathbf{x} - \mathbf{x}'|} d\theta, \tag{18.66}$$

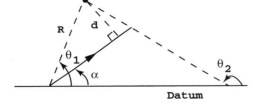

Figure 18.7. Dislocation segment.

where the limits of evaluation on the first integral indicate evaluation from a point labeled 1 to itself, and where it was noted that

$$\frac{d|\mathbf{x} - \mathbf{x}'|}{d\theta} = -|\mathbf{x} - \mathbf{x}'| \cot(\theta - \alpha). \tag{18.67}$$

The second integral on the *rhs* of (18.66) is

$$-\oint_C \frac{\partial \Sigma_{mr}(\theta)/\partial\theta \cot(\theta - \alpha)}{|\mathbf{x} - \mathbf{x}'|} d\theta = -\left[\Sigma_{mr}(\theta)\frac{\cot(\theta - \alpha)}{|\mathbf{x} - \mathbf{x}'|}\right]_1^1$$
$$+ \oint_C \frac{\Sigma_{mr}(\theta)\cot^2(\theta - \alpha)}{|\mathbf{x} - \mathbf{x}'|} d\theta$$
$$- \oint_C \frac{\Sigma_{mr}(\theta)}{|\mathbf{x} - \mathbf{x}'|} \csc^2(\theta - \alpha)(1 - d\alpha/d\theta) d\theta.$$

By recognizing that

$$\cot^2(\theta - \alpha) - \csc^2(\theta - \alpha) = -1,$$

we have

$$\frac{1}{2}\oint_C \frac{\partial^2 \Sigma_{mr}(\theta)/\partial\theta^2}{|\mathbf{x} - \mathbf{x}'|} d\theta = -\frac{1}{2}\oint_C \frac{\Sigma_{mr}(\theta)}{|\mathbf{x} - \mathbf{x}'|} d\theta + \frac{1}{2}\oint_C \frac{\Sigma_{mr}(\theta)d\alpha/d\theta}{|\mathbf{x} - \mathbf{x}'|\sin^2(\theta - \alpha)} d\theta.$$

When this result is incorporated into (18.65), the remarkably simple result follows, *viz.*,

$$\begin{aligned}u_{m,r}(\mathbf{x}) &= \frac{1}{2}\oint_C \frac{\Sigma_{mr}(\theta)d\alpha/d\theta}{|\mathbf{x} - \mathbf{x}'|\sin^2(\theta - \alpha)} d\theta \\ &= \frac{1}{2}\oint_C \frac{\Sigma_{mr}(\theta)}{|\mathbf{x} - \mathbf{x}'|\sin^2(\theta - \alpha)} d\alpha.\end{aligned} \tag{18.68}$$

Note that $d\alpha = 0$ along each segment until a corner is reached, at which point α undergoes a "jump" from, say α^- to α^+ (Fig. 18.6). But during such jumps, θ and $|\mathbf{x} - \mathbf{x}'|$ remain fixed. Thus, noting that

$$\int \frac{d\alpha}{\sin^2(\theta - \alpha)} = \cot(\theta - \alpha),$$

it is found that each corner, say the nth corner, contributes a term such as

$$\chi = \frac{1}{2}\frac{\Sigma_{mr}(\theta_n)}{|\mathbf{x} - \mathbf{x}'|_n}[\cot(\theta_n - \alpha)]_{\alpha_n^-}^{\alpha_n^+}. \tag{18.69}$$

Therefore, for the entire loop, we have

$$u_{m,r}(\mathbf{x}) = \frac{1}{2}\sum_{n=1}^N \frac{\Sigma_{mr}(\theta_n)}{|\mathbf{x} - \mathbf{x}'|_n}[\cot(\theta_n - \alpha)]_{\alpha_n^-}^{\alpha_n^+}. \tag{18.70}$$

There is a modification if the field point $\mathbf{x}$ happens to be colinear with one side, because along such a side $\theta = 0$, and

$$\int_{\theta_s}^{\theta_{s+1}} \frac{\Sigma_{mr}(\theta) + \partial^2\Sigma_{mr}/\partial\theta^2}{|\mathbf{x} - \mathbf{x}'|} d\theta = 0. \tag{18.71}$$

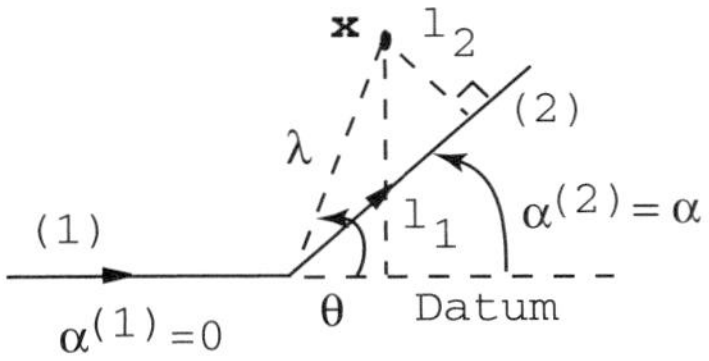

18.4. Applications of Planar Geometric Theorem

Figure 18.8. An angular dislocation.

It is assumed here that the side in question connects the s and $s+1$ corner. In this case (18.70) would have to be modified so that the segment between corners s and $s+1$ does not contribute. The result of removing this is

$$u_{m,r}(\mathbf{x}) = \frac{1}{2} \sum_{\substack{n=1 \\ n \neq s, s+1}}^{N} \frac{\Sigma_{mr}(\theta_n)}{|\mathbf{x}-\mathbf{x}'|_n} [\cot(\theta_n - \alpha)]_{\alpha_n^-}^{\alpha_n^+}$$

$$- \frac{\Sigma_{mr}(\theta_s)}{|\mathbf{x}-\mathbf{x}'|_s} \cot(\theta_s - \alpha_s^-) - \frac{\Sigma_{mr}(\theta_{s+1})\cot(\theta_{s+1} - \alpha_{s+1}^+)}{|\mathbf{x}-\mathbf{x}'|_{s+1}} \quad (18.72)$$

$$+ \frac{\partial \Sigma_{mr}(\theta_s)/\partial\theta}{|\mathbf{x}-\mathbf{x}'|_s} - \frac{\partial \Sigma_{mr}(\theta_{s+1})/\partial\theta}{|\mathbf{x}-\mathbf{x}'|_{s+1}}.$$

We can recast (18.60) or (18.65) using the fact that

$$d\alpha = \kappa \, ds, \quad ds = d\theta |\mathbf{x}-\mathbf{x}'| \csc(\theta - \alpha). \quad (18.73)$$

When this is done, we obtain

$$u_{m,r}(\mathbf{x}) = \frac{1}{2} \oint_C \kappa \Sigma_{mr}(\theta) \csc^3(\theta - \alpha) \, d\theta. \quad (18.74)$$

18.4.1 Angular Dislocations

As a further application of the previous results, consider the angular dislocation shown in Fig. 18.8. Using (18.64), it is found that, for the segment labeled 1, we have

$$u_{m,r}^{(1)} = \frac{1}{2\ell_1} \int_0^\theta \left[\Sigma_{mr}(\theta) + \partial^2 \Sigma_{mr}(\theta)/\partial\theta^2\right] \sin\theta \, d\theta$$

$$= \frac{1}{2\lambda} \left[-\Sigma_{mr}(\theta)\cot\theta + \Sigma_{mr}(0)\csc\theta + \partial\Sigma_{mr}(\theta)/\partial\theta\right]. \quad (18.75)$$

Similarly, for the segment labeled 2 it is found that

$$u_{m,r}^{(2)} = \frac{1}{2\ell_2} \int_\theta^{\alpha+\pi} \left[\Sigma_{mr}(\theta) + \partial^2 \Sigma_{mr}(\theta)/\partial\theta^2\right] \sin(\theta - \alpha) \, d\theta \quad (18.76)$$

$$= -\frac{1}{2\lambda} \left[\Sigma_{mr}(\alpha)\csc(\theta - \alpha) + \Sigma_{mr}(\theta)\cot(\theta - \alpha) - \partial\Sigma_{mr}(\theta)/\partial\theta\right].$$

When added, they yield for the full field

$$u_{m,r}(\mathbf{x}) = \frac{1}{2\lambda}\left[\Sigma_{mr}(\alpha)\csc(\theta - \alpha) + \Sigma_{mr}(0)\csc(\theta)\right]$$

$$+ \frac{1}{2\lambda}\left[\Sigma_{mr}(\theta)\frac{\sin\alpha}{\sin(\theta - \alpha)\sin\theta}\right], \quad (18.77)$$

where we have used the fact that $\Sigma_{mr}(\alpha) = \Sigma_{mr}(\alpha + \pi)$.

18.5 A 3D Geometrical Theorem

Consider again the line integral yielding the distortion field for a dislocation loop, *viz.*,

$$u_{m,r}(\mathbf{x};\mathcal{C}) = -\oint_{\mathcal{C}} b_i \epsilon_{njp} t_j (x_n - x'_n) C_{pisk} G_{sm,kr}(\mathbf{x} - \mathbf{x}') \, ds. \tag{18.78}$$

Upon using the symmetry inherent in the components C_{pisk}, let $\mathcal{P}_{pi}^m = C_{pisk} G_{sm,k}$ be the pi component of stress at $\mathbf{x}$, caused by a unit point force acting at $\mathbf{x}'$ in the m direction. Note that $\mathcal{P}_{pi}^m$ and $\mathcal{P}_{pi,r}^m$ will possess all the symmetry and scaling properties of the Green's function $\mathbf{G}$. The integral above becomes

$$u_{m,r}(\mathbf{x},\mathcal{C}) = -\oint_{\mathcal{C}} b_i \epsilon_{njp} t_j (x_n - x'_n) \mathcal{P}_{pi,r}^m(\mathbf{x} - \mathbf{x}') \, ds. \tag{18.79}$$

Let $\mathcal{C}$ be an infinitely long and straight dislocation line. Then,

$$\begin{aligned} x'_n &= t_n s, \quad \mathbf{x}' = \mathbf{t} s, \\ \epsilon_{njp} x'_n t_j &= \epsilon_{njp} t_n t_j s = 0, \end{aligned} \tag{18.80}$$

so that

$$u_{m,r}(\mathbf{x},\mathcal{C}_\infty) = -\int_{-\infty}^{\infty} b_i \epsilon_{njp} x_n t_j \mathcal{P}_{pi,r}^m(\mathbf{x} - \mathbf{s}\mathbf{t}) \, ds. \tag{18.81}$$

But, because of the scaling and symmetry properties of $\mathcal{P}_{pi,r}^m$, we have

$$\mathcal{P}_{pi,r}^m(\mathbf{x} - s\mathbf{t}) = \frac{\text{sgn}(s)}{s^3} \mathcal{P}_{pi,r}^m(\mathbf{x}/s - \mathbf{t}). \tag{18.82}$$

Define $\eta = 1/s$, $d\eta = -1/s^2 ds$, and observe that

$$\begin{aligned} u_{m,r}(\mathbf{x},\mathcal{C}_\infty) &= -\int_{0^+}^{\infty} b_i \epsilon_{njp} x_n t_j (1/s^3) \mathcal{P}_{pi,r}^m(\mathbf{x}/s - \mathbf{t}) \, ds \\ &\quad + \int_{-\infty}^{0^-} b_i \epsilon_{njp} x_n t_j (1/s^3) \mathcal{P}_{pi,r}^m(\mathbf{x}/s - \mathbf{t}) \, ds \\ &= \int_{\infty}^{0^+} b_i \epsilon_{njp} x_n t_j \eta \mathcal{P}_{pi,r}^m(\eta \mathbf{x} - \mathbf{t}) \, d\eta \\ &\quad - \int_{0^-}^{-\infty} b_i \epsilon_{njp} x_n t_j \eta \mathcal{P}_{pi,r}^m(\eta \mathbf{x} - \mathbf{t}) \, d\eta. \end{aligned}$$

Then, upon identifying this infinitely long and straight dislocation line by its constant tangent, $\mathbf{t}$, we can write

$$\begin{aligned} u_{m,r}^\infty(\mathbf{x};\mathbf{t}) &= -\int_{0^+}^{\infty} b_i \epsilon_{njp} x_n t_j \eta \mathcal{P}_{pi,r}^m(\eta \mathbf{x} - \mathbf{t}) \, d\eta \\ &\quad + \int_{-\infty}^{0^-} b_i \epsilon_{njp} x_n t_j \eta \mathcal{P}_{pi,r}^m(\eta \mathbf{x} - \mathbf{t}) \, d\eta \\ &= \int_{-\infty}^{\infty} b_i \epsilon_{njp} x_n t_j \eta \, \text{sgn}(\eta) \mathcal{P}_{pi,r}^m(\eta \mathbf{x} - \mathbf{t}) \, d\eta. \end{aligned} \tag{18.83}$$

18.5. 3D Geometrical Theorem

Operate on this equation with $(x_\alpha \partial/\partial t_\alpha)^2$, while noting that

$$x_\alpha \frac{\partial}{\partial t_\alpha} \mathcal{P}^m_{pi,r}(\eta\mathbf{x} - \mathbf{t}) = -\frac{\partial}{\partial \eta} \mathcal{P}^m_{pi,r}(\eta\mathbf{x} - \mathbf{t}). \tag{18.84}$$

It is found that

$$\begin{aligned} x_\beta x_\alpha \frac{\partial^2}{\partial t_\alpha \partial t_\beta} u^\infty_{m,r} &= 2\int_{-\infty}^{\infty} b_i \epsilon_{njp} x_n x_j \eta \, \text{sgn}(\eta) \frac{\partial \mathcal{P}^m_{pi,r}}{\partial \eta} \, d\eta \\ &- \int_{-\infty}^{\infty} b_i \epsilon_{njp} x_n t_j \eta \, \text{sgn}(\eta) \frac{\partial^2 \mathcal{P}^m_{pi,r}}{\partial \eta^2} \, d\eta. \end{aligned} \tag{18.85}$$

The first integral is clearly zero by virtue of $\epsilon_{njp} x_n x_j = 0$. The second may be integrated by parts, using the result

$$\left[\eta \, \text{sgn}(\eta) \frac{\partial \mathcal{P}^m_{pi,r}}{\partial \eta} \right]_{-\infty}^{\infty} = 0.$$

The first integration then yields

$$x_\beta x_\alpha \frac{\partial^2}{\partial t_\alpha \partial t_\beta} u^\infty_{m,r} = \int_{-\infty}^{\infty} b_i \epsilon_{njp} x_n t_j \left[\text{sgn}(\eta) + \eta \frac{d\,\text{sgn}(\eta)}{d\eta} \right] \frac{\partial \mathcal{P}^m_{pi,r}}{\partial \eta} \, d\eta.$$

But,

$$\frac{d}{d\eta} \text{sgn}(\eta) = 2\delta(\eta),$$

and the second term in the square bracket above makes zero contribution to the integral. Integrating by parts again, and noting that

$$\left[\text{sgn}(\eta) \mathcal{P}^m_{pi,r} \right]_{-\infty}^{\infty} = 0,$$

now gives

$$\begin{aligned} x_\beta x_\alpha \frac{\partial^2}{\partial t_\alpha \partial t_\beta} u^\infty_{m,r} &= -2 \int_{-\infty}^{\infty} b_i \epsilon_{njp} x_n t_j \delta(\eta) \mathcal{P}^m_{pi,r}(\eta\mathbf{x} - \mathbf{t}) \, d\eta \\ &= -2 b_i \epsilon_{njp} x_n t_j \mathcal{P}^m_{pi,r}(-\mathbf{t}) \\ &= -2 b_i \epsilon_{njp} x_n t_j \mathcal{P}^m_{pi,r}(\mathbf{t}). \end{aligned} \tag{18.86}$$

Finally, let

$$\mathbf{t} \leftarrow \mathbf{x} - \mathbf{x}' \quad \text{and} \quad \mathbf{x} \leftarrow \mathbf{t}, \tag{18.87}$$

where $\mathbf{x}'$ is a fixed point. Then, the integrand of (18.79) becomes

$$b_i \epsilon_{njp} t_n (x_j - x'_j) \mathcal{P}^m_{pi,r}(\mathbf{x} - \mathbf{x}') = -t_\alpha \frac{\partial}{\partial x_\alpha} t_\beta \frac{\partial}{\partial x_\beta} u^\infty_{m,r}(\mathbf{t}; \mathbf{x} - \mathbf{x}'), \tag{18.88}$$

and, consequently,

$$u_{m,r}(\mathbf{x}; \mathcal{C}) = -\oint_\mathcal{C} t_\alpha \frac{\partial}{\partial x_\alpha} t_\beta \frac{\partial}{\partial x_\beta} u^\infty_{m,r}(\mathbf{t}; \mathbf{x} - \mathbf{x}') \, ds. \tag{18.89}$$

18.6 Suggested Reading

Indenbom, V. L., and J. Lothe, J. (1992), *Elastic Strain Fields and Dislocation Mobility*, North-Holland, Amsterdam.

Hirth, J. P., and Lothe, J. (1982), *Theory of Dislocations*, 2nd ed., Wiley, New York.

Nabarro, F. R. N., ed. (1979), *Dislocations in Solids*, North-Holland, Amsterdam.

Nabarro, F. R. N. (1987), *Theory of Crystal Dislocations*, Dover, New York.

Mura, T. (1987), *Micromechanics of Defects in Solids*, Martinus Nijhoff, Dordrecht, The Netherlands.

Teodosiu, C. (1982), *Elastic Models of Crystal Defects*, Springer-Verlag, Berlin.

Ting, T. C. T. (1996), *Anisotropic Elasticity*, Oxford University Press, New York.

19 Cracks in Anisotropic Media

An elementary development of crack tip mechanics was given in Chapter 17. Here we provide a more advanced construction of the elastic fields at crack tips and extract from our construction some particularly important quantities that enter prominently into the physics of crack growth and crack interaction with other defects that exist within the elastic medium surrounding the crack. Of primary importance is the theoretical determination of the total mechanical energy of a cracked body and the negative of its derivative with respect to crack extension, *i.e.*, the *energy release rate*, $\mathcal{G}$, or the generalized force on the crack tip *vis-à-vis* the concept further developed in subsequent chapters. The elastic medium is taken to be arbitrarily anisotropic. The dislocation solutions and methods developed for dealing with the elastic fields of dislocations are used to construct the crack solutions.

19.1 Dislocation Mechanics: Reviewed

In Chapter 21 we show that the energy of an infinitely long and straight dislocation can be written as

$$\mathcal{E} = K_{mg} b_m b_g \ln(R/r_0), \qquad (19.1)$$

where R and r_0 were outer and inner cutoff radii respectively. The components of the dislocation's Burgers vector are b_i, and K_{mg} are the components of a positive definite second-rank tensor called the *energy factor*. **K** depends only on the direction of the dislocation line within the elastic medium and on the elastic moduli tensor **C**, *i.e.*, its components C_{ijkl}. For an isotropic medium **K** is diagonal, when phrased with respect to the basis $\{\mathbf{e}_1, \mathbf{e}_2, \mathbf{e}_3\}$ on the coordinate frame with axes (x_1, x_2, x_3), and when the dislocation line is parallel to x_3 axis, $\mathbf{t} \parallel \mathbf{e}_3$. In that case, we obtained

$$K_{11} = K_{22} = G/4\pi(1-\nu), \quad K_{33} = G/4\pi. \qquad (19.2)$$

We have already shown in Chapter 17 how a slit like crack in an elastic medium can be represented by a continuous distribution of dislocations distributed with an appropriate density along the crack line. This was done, specifically for the case of a Mode I crack. As noted in that development, the calculation of the crack tip field involves first solving a singular integral equation for the dislocation distribution and then using linear superposition to construct the field. It will be shown here that the crack extension force can

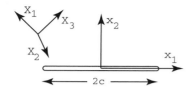

Figure 19.1. Slit-like crack and crystal frame.

also be calculated knowing the distribution of dislocations in the crack, or more particularly knowing the displacement discontinuity across the crack face that is caused by the distribution of dislocations. In fact, it is shown that the crack extension force, $\mathcal{G}$, can be calculated knowing only the inverse of the energy factor matrix, $\mathbf{K}^{-1}$, for a single straight dislocation.

Two types of cracks are considered, *viz.*, *freely slipping cracks* and cracks whose faces are not traction free but are subject to applied traction. Furthermore, the following universal result is derived: when the displacement discontinuity at the tips of the crack vanishes, the traction and stress concentration on the plane of the crack are independent of the elastic anisotropy and are, accordingly, the same as predicted by the isotropic theory. The same is true of the stress intensity factors. The same is not the case for the angular dependence of the fields around the crack or of the crack extension force, $\mathcal{G}$.

19.2 Freely Slipping Crack

Consider a slit like crack lying on the plane $x_2 = 0$, as shown in Fig. 19.1. The crack lies in the region, $|x_1| \leq c$, $x_2 = 0$, $-\infty < x_3 < \infty$ in an infinite elastic medium. For later reference, let $\mathbf{X} = \{X_1, X_2, X_3\}$ be a coordinate frame in which the elastic moduli tensor, $\mathbf{C}$, and its components, C_{ijkl}, are displayed in their simplest form, *e.g.*, for a cubic crystalline medium these would be the cube axes. As $r^2 = x_1^2 + x_2^2 \to \infty$, the state of stress is uniform and thus

$$\lim_{r \to \infty} \sigma_{ij} = \sigma_{ij}^{\mathrm{A}}. \tag{19.3}$$

We are assuming small strains so that, using the convention that commas denote differentiation, the strains are defined by

$$2e_{ij} = u_{i,j} + u_{j,i}. \tag{19.4}$$

In the absence of body forces, equilibrium requires that

$$\sigma_{ij,j} = 0. \tag{19.5}$$

Since the crack is *freely slipping* there are no tractions on the crack faces, *i.e.*,

$$\sigma_{ij} n_j = 0, \quad \text{on} \quad x_2 = 0, \ |x_1| \leq c, \tag{19.6}$$

where $\mathbf{n}$ is the unit normal to the crack plane. The convention is that on the upper crack surface ($x_2 = 0^+$), $n_1, n_3 = 0, n_2 = -1$. On the lower crack surface ($x_2 = 0^-$), $n_1, n_3 = 0, n_2 = 1$. This means that the freely slipping boundary conditions become

$$\sigma_{12} = \sigma_{22} = \sigma_{32} = 0, \quad \text{for} \quad |x_1| \leq c, \ x_2 = 0, \ |x_3| \leq \infty. \tag{19.7}$$

19.2. Freely Slipping Crack

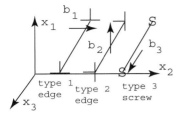

Figure 19.2. Dislocation types distributed over the crack.

This boundary value problem is solved using the method of continuously distributed dislocations as introduced earlier.

Let the solution be written in the form

$$u_i = u_i^A + u_i^D, \quad e_{ij} = e_{ij}^A + e_{ij}^D, \quad \sigma_{ij} = \sigma_{ij}^A + \sigma_{ij}^D, \tag{19.8}$$

where the D fields are constructed from the linear superposition of those of the continuously distributed dislocations of three types; these are indicated in the Fig. 19.2. The A fields refer to the uniformly stressed uncracked solid, so that, excluding arbitrary rigid body translations,

$$u_i^A = e_{ij}^A x_j, \quad e_{ij}^A = S_{ijmn} \sigma_{mn}^A, \tag{19.9}$$

where the S_{ijmn} are the components of the elastic compliance tensor (the inverse of the elastic stiffness tensor). The A fields are everywhere continuous. We note that

$$\sigma_{ij}^D \to 0 \quad \text{as} \quad r \to 0, \tag{19.10}$$

because the stress fields of individual dislocations vanish as $r^2 \to 0$. Since the A and D fields are each admissible solutions to the equilibrium and compatibility equations, we can complete the solution by choosing σ_{ij}^D so that (19.7) is satisfied, i.e.,

$$\begin{aligned}
\sigma_{12}^D &= -\sigma_{12}^A = -T_1^A, \quad \text{on} \quad |x_1| < c, \ x_2 = 0, \\
\sigma_{22}^D &= -\sigma_{22}^A = -T_2^A, \quad \text{on} \quad |x_1| < c, \ x_2 = 0, \\
\sigma_{32}^D &= -\sigma_{32}^A = -T_3^A, \quad \text{on} \quad |x_1| < c, \ x_2 = 0.
\end{aligned} \tag{19.11}$$

The D field is constructed by an integral superposition of three types of straight dislocations, the dislocations being parallel to the x_3 axis and lying in the region $|x_1| < c$, $x_2 = 0$. If we denote by $b_s f^{(s)}(t) dt$ (no sum on s) the amount of Burger's vector in the x_s direction, distributed between t and $t + dt$, then

$$\sigma_{ij}^D(x_1, x_2) = \sum_{s=1}^{3} \int_{-c}^{c} f^{(s)}(t) \sigma_{ij}^{(s)}(x_1, x_2; t, 0) \, dt, \tag{19.12}$$

where $\sigma_{ij}^{(s)}(x_1, x_2; t, 0)$ is the stress field at (x_1, x_2) due to a single straight dislocation of type s ($s = 1, 2, 3$), piercing the planes $x_3 = $ const. at the point $(t, 0)$. Thus equations (19.11) may be written as

$$\sum_{s=1}^{3} \int_{-c}^{c} f^{(s)}(t) \sigma_{i2}^{(s)}(x_1, 0; t, 0) \, dt = -T_i^A, \quad \text{for } |x_1| < c, \ i = 1, 2, 3. \tag{19.13}$$

In an infinite elastic medium,

$$\sigma_{ij}^{(s)}(x_1, 0; t, 0) = \sigma_{ij}^{(s)}(x_1 - t), \tag{19.14}$$

and, from what has been developed in Chapters 17 and 18, $\sigma_{ij}^{(s)}(x_1 - t) \sim (x_1 - t)^{-1}$.

Now consider a single dislocation parallel to the x_3 axis and piercing the planes $x_3 = $ const. at $(t, 0)$. Let its Burger's vector have components b_1, b_2, b_3. The energy per unit length can be written as

$$\mathcal{E} = K_{mq} b_m b_q \ln(R/r_0) = K_{mq} b_m b_q \int_{t+r_0}^{t+R} (x_1 - t)^{-1} \, dx_1. \tag{19.15}$$

But the formula

$$\mathcal{E} = \frac{1}{2} \sum_{s=1}^{3} \int_{t+r_0}^{t+R} \sigma_{m2}^{(s)}(x_1 - t) b_m \, dx_1 \tag{19.16}$$

is also valid. If a dislocation is of type 1 (*i.e.*, $s = 1$, $b_2 = b_3 = 0$), a comparison of (19.15) and (19.16) yields

$$\sigma_{12}^{(1)}(x_1 - t) = 2K_{11} b_1 / (x_1 - t). \tag{19.17}$$

Similar reasoning gives

$$\sigma_{22}^{(2)}(x_1 - t) = 2K_{22} b_2 / (x_1 - t), \quad \sigma_{32}^{(3)}(x_1 - t) = 2K_{33} b_3 / (x_1 - t). \tag{19.18}$$

If we consider the case $s = 1, 2$ and $b_3 = 0$, and note that

$$\sigma_{12}^{(1)}(x_1 - t) b_2 = \sigma_{12}^{(2)}(x_1 - t) b_1, \tag{19.19}$$

(because one can calculate the interaction energy between a dislocation of type 1 and type 2 by looking at the interaction of one on the other or *vice versa*), a comparison of (19.15) and (19.16) again shows that

$$\sigma_{22}^{(1)}(x_1 - t) = 2K_{21} b_1 / (x_1 - t), \quad \sigma_{12}^{(2)}(x_1 - t) = 2K_{12} b_2 / (x_1 - t). \tag{19.20}$$

Note that (19.19) essentially follows from Betti's reciprocal theorem. In general,

$$\sigma_{i2}^{(s)}(x_1 - t) = 2K_{is} b_s / (x_1 - t), \quad \text{(no sum on s)}. \tag{19.21}$$

If we use (19.21) to define

$$F_s(t) = b_s f^{(s)}(t), \quad \text{(no sum on s)}, \tag{19.22}$$

(19.13) may be concisely expressed as

$$2K_{ij} \int_{-c}^{c} (x_1 - t)^{-1} F_j(t) \, dt = -T_i^A, \quad \text{for } |x_1| < c, \tag{19.23}$$

or

$$\int_{-c}^{c} (x_1 - t)^{-1} F_j(t) \, dt = -\frac{1}{2} K_{ji}^{-1} T_i^A, \quad \text{for } |x_1| < c, \tag{19.24}$$

with the sum over the repeated index. The integrals in question, as will be the case with all singular integrals appearing in this development, are defined by their Cauchy principal values.

19.3. Crack Extension Force

We recall that the components of the inverse of K_{ji} are given by

$$K_{ji}^{-1} = K_{ij}^{-1} = \epsilon_{jmn}\epsilon_{irs} K_{mr} K_{ns}/2\epsilon_{\alpha\beta\gamma} K_{1\alpha} K_{2\beta} K_{3\lambda}, \qquad (19.25)$$

so that

$$K_{ji}^{-1} K_{is} = K_{ji} K_{is}^{-1} = \delta_{js}. \qquad (19.26)$$

Since we expect stress singularities at the crack tip, we require $F_j(\pm c)$ to be unbounded with a weak singularity. Moreover, if we demand that there be no relative displacement of the crack faces at $x_1 = \pm c$, then

$$\int_{-c}^{c} F_j(t)\, dt = 0. \qquad (19.27)$$

It is noteworthy that implicit in this analysis is the convention that the displacement discontinuity across the crack faces at $(x_1, 0)$ is given by

$$\Delta u_j(x_1) = u_j^D(x_1, x_2)\big|_{x_2=0^+}^{x_2=0^-} = \int_{-c}^{x_1} F_j(t)\, dt. \qquad (19.28)$$

The solution of (19.24) is

$$F_j(t) = K_{ji}^{-1} T_i^A \frac{t}{2\pi(c^2 - t^2)^{1/2}}. \qquad (19.29)$$

A rather important result can be extracted from the above when we consider the traction σ_{i2} on the plane of the crack ($|x_1| > c$, $x_2 = 0$). We have

$$\sigma_{i2}|_{x_2=0} = \sigma_{i2}^A + 2 \sum_{s=1}^{3} \int_{-c}^{c} \frac{F_s(t)}{b_s} \frac{K_{is} b_s}{(x_1 - t)}\, dt, \quad |x_1| > c. \qquad (19.30)$$

Since

$$T_i^A = \sigma_{i2}^A, \qquad (19.31)$$

using (19.26) and (19.29), yields

$$\sigma_{i2}|_{x_2=0} = \sigma_{i2}^A \left[1 + \frac{1}{\pi} \int_{-c}^{c} \frac{t\, dt}{(c^2 - t^2)^{1/2}(x_1 - t)} \right]$$
$$= \sigma_{i2}^A \frac{|x_1|}{(x_1^2 - c^2)^{1/2}}, \quad |x_1| > c. \qquad (19.32)$$

As $x_1 \to \pm c$, (19.32) reduces to the familiar isotropic expression for stress concentration, viz., $\sigma_{i2} \approx \sigma_{i2}^A (c/2r)^{1/2}$, where $r = |x_1| - c$. Hence, the traction and stress concentrations on the plane of the crack are independent of the elastic constants and the anisotropy of the medium, i.e., they are identical with those for a crack in an isotropic medium loaded by stresses σ_{ij}^A at infinity. It is shown below that this is also true for the case of a crack that is arbitrarily but symmetrically loaded on its faces by self-equilibrating stresses, provided that (19.27) is true.

19.3 Crack Extension Force

The energy of deformation and crack extension force may now be easily calculated. The change in total mechanical energy per unit length in the x_3 direction between the stressed cracked solid and the uncracked solid stressed homogeneously by $\sigma_{ij} = \sigma_{ij}^A$, is

$$\Delta\mathcal{E} = \frac{1}{2}\int_{-c}^{c}\sigma_{i2}^A \Delta u_i^D \, dx_1. \tag{19.33}$$

Combining (19.28), (19.29), and (19.33), we obtain

$$\Delta\mathcal{E} = -\frac{1}{8}c^2\sigma_{i2}^A \sigma_{s2}^A K_{is}^{-1}. \tag{19.34}$$

The crack extension force is defined as

$$\mathcal{G} = -\frac{\partial(\Delta\mathcal{E})}{\partial c} = \frac{1}{4}c\sigma_{i2}^A \sigma_{s2}^A K_{is}^{-1}. \tag{19.35}$$

The appropriate Griffith criterion for brittle fracture in the anisotropic medium is obtained by requiring that $\mathcal{G}$ be greater or equal to 4γ, where γ is the surface energy associated with the plane $x_2 = 0$ (the crack is assumed to extend on both ends, so that $2 \times 2\gamma = 4\gamma$). Thus, the applied stress state required to propagate the crack is determined from

$$\sigma_{i2}^A \sigma_{s2}^A K_{is}^{-1} \geq 16\gamma/c. \tag{19.36}$$

Using (19.2) allows a recovery of the isotropic Griffith criterion,

$$(1-\nu)\left[(\sigma_{12}^A)^2 + (\sigma_{22}^A)^2\right] + (\sigma_{32})^2 \geq 4G\gamma/\pi c. \tag{19.37}$$

A method is presented below for calculating K_{is}, and hence K_{is}^{-1}, using the coordinate frame denoted earlier by $\mathbf{X} = \{X_1, X_2, X_3\}$. If all quantities are referred to the $\mathbf{X}$ frame, then (19.35) becomes

$$\mathcal{G} = \frac{1}{4}c\,\sigma_{ij}^A n_j \sigma_{sm}^A n_m K_{is}^{-1}, \tag{19.38}$$

where n_m are components of the unit normal to the crack surface in the $\mathbf{X}$ frame.

As an example, imagine a cleavage crack lying in the (001) plane of a cubic crystal such as α-Fe, or an ionic crystal such as KCl or NaCl, which is stressed by far field tension p^A. The crack's normal, then, lies in the x_3 direction. The crack extension force would be given in this case by

$$\mathcal{G} = \frac{1}{4}(p^A)^2 c K_{33}^{-1}. \tag{19.39}$$

Figure 19.3 shows the variation of K_{33}^{-1} with angle in the (001) plane of three cubic materials. The effects of anisotropy appear to be modest and are appreciable only for Fe. In fact, using Voigt average isotropic elastic constants for α-Fe, $K_{33}^{-1} = 0.103 \times 10^{-10}$ cm^3/erg, which differs by between 10 and 20% of our anisotropic calculations (recall that 1 erg = 10^{-7} J).

19.4 Crack Faces Loaded by Tractions

The extension of the results for a freely slipping crack to the case of a crack whose faces are loaded by tractions $-R_i(x_1)$ is readily done. The same analysis suffices if we replace the boundary condition (19.7) by

$$\sigma_{ij}n_j = \begin{cases} -R_i(x_1), & \text{for } x_2 = 0^+, \\ R_i(x_1), & \text{for } x_2 = 0^-, \end{cases} \quad |x_1| < c, \tag{19.40}$$

19.4. Crack Faces Loaded by Tractions

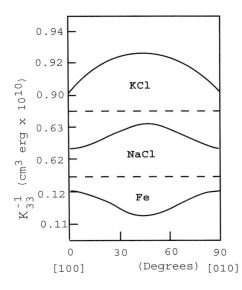

Figure 19.3. Variation of K_{33}^{-1} with angle within the (001) plane.

so that we assume that the traction on the upper and lower crack faces are equal in magnitude but opposite in direction. Clearly, the loading is symmetrical about $x_2 = 0$ and thus self-equilibrating. For example, if the loading were internal pressure of magnitude P, then $R_1 = R_3 = 0$, $R_2 = -P$. Hence the distribution function $F_j(t)$ must now satisfy

$$\int_{-c}^{c} (x_1 - t)^{-1} F_j(t) \, dt = -\frac{1}{2} K_{ij}^{-1} \left[T_i^A - R_i(x_1) \right], \quad |x_1| < c. \tag{19.41}$$

It is again required that (19.27) be satisfied when $F_j(\pm c)$ is unbounded. The integral equation may be easily solved once we specify $R_i(x_1)$. For our purposes it is sufficient to note that

$$2F_j(t) = K_{ji}^{-1} Q_i(t), \tag{19.42}$$

where $Q_i(t)$ is independent of the anisotropy of the medium and satisfies

$$\int_{-c}^{c} (x_1 - t)^{-1} Q_i(t) \, dt = -\left[T_i^A - R_i(x_1) \right], \quad |x_1| < c. \tag{19.43}$$

If $F_j(\pm c)$ is to be unbounded, then

$$Q_i(t) = \frac{T_i^A t}{\pi (c^2 - t^2)^{1/2}} + \frac{1}{\pi^2 (c^2 - t^2)^{1/2}} \int_{-c}^{c} \frac{R_i(x_1)(c^2 - x_1^2)^{1/2}}{x_1 - t} \, dx_1. \tag{19.44}$$

For cracks with no stress singularities at $x_1 = \pm c$, as in simple models of cracks relaxed by plastic deformation, $F_j(\pm c)$ must vanish, and

$$Q_i(t) = \frac{1}{\pi^2}(c^2 - t^2)^{1/2} \int_{-c}^{c} (c^2 - x_1^2)^{-1/2} (x_1 - t)^{-1} R(x_1) \, dx_1, \tag{19.45}$$

provided that the subsidiary condition

$$\int_{-c}^{c} (c^2 - s^2)^{-1/2} R_i(s) \, ds = \pi T_i^A \tag{19.46}$$

is satisfied. The traction σ_{i2} acting on the plane of the crack ($|x_1| > c$, $x_2 = 0$) is then given by (19.32) which, after using (19.26) and (19.42), reduces to

$$\sigma_{i2}|_{x_2=0} = \sigma_{i2}^A + \int_{-c}^{c} (x_1 - t)^{-1} Q_i(t)\, dt, \quad |x_1| > c. \tag{19.47}$$

Since $Q_i(t)$ depends only on the loading $T_i^A - R_i(x_1)$, the traction and also the stress concentrations, if any, on the plane of the crack are independent of the elastic anisotropy. Stress intensity factors for cracks with singular fields are computed in the next section.

If (19.27) is not satisfied, there exists a net dislocation content in the crack, i.e.,

$$\int_{-c}^{c} F_j(t)\, dt = N_j. \tag{19.48}$$

In this event, we must add to $F_j(t)$, as given by (19.29) or (19.42) and (19.44), the term

$$F_j^*(t) = N_j / \pi (c^2 - t^2)^{1/2}. \tag{19.49}$$

The traction on the crack plane is in this case no longer independent of the anisotropy because the distribution (19.49) induces an extra traction

$$\sigma_{i2}^*|_{x_2=0} = 2 K_{is} N_s (x_1^2 - c^2)^{-1/2} \operatorname{sgn}(x_1), \quad |x_1| > c. \tag{19.50}$$

19.5 Stress Intensity Factors and Crack Extension Force

The three stress intensity factors, called k_i here, are most simply defined by noting that if the stresses are singular at the crack tip, i.e., at $x_1 = c$, then

$$\sigma_{i2}|_{x_1 \to c,\, x_2 = 0} = k_i / (2\pi r)^{1/2} + \text{nonsingular terms}, \tag{19.51}$$

where $r = x_1 - c$, so that

$$k_i(c) = \lim_{x_1 \to c} (2\pi r)^{1/2} \sigma_{i2}|_{x_2=0}, \quad x_1 > c. \tag{19.52}$$

When (19.27) is satisfied, the use of (19.44) and (19.47) yields

$$k_i(c) = \sigma_{i2}^A (\pi c)^{1/2} + \lim_{x_1 \to c} \left[(2\pi r)^{1/2} \int_{-c}^{c} \frac{dt}{x_1 - t} \frac{1}{\pi^2 (c^2 - t^2)^{1/2}} \int_{-c}^{c} \frac{R_i(s)(c^2 - s^2)^{1/2}\, ds}{s - t} \right]. \tag{19.53}$$

Interchanging the order of integration and noting that

$$\frac{1}{(x_1 - t)(s - t)} = \frac{1}{s - x_1} \left(\frac{1}{x_1 - t} - \frac{1}{s - t} \right),$$

$$\int_{-c}^{c} \frac{dt}{(c^2 - t^2)^{1/2}(s - t)} = 0, \quad |s| < c \text{ (Cauchy principal value)}, \tag{19.54}$$

$$\int_{-c}^{c} \frac{dt}{(c^2 - t^2)^{1/2}(x_1 - t)} = \frac{\pi}{(x_1^2 - c^2)^{1/2}} \operatorname{sgn}(x_1), \quad |x_1| > c,$$

the limit of (19.53) as $x_1 \to c$ gives

$$k_i(c) = \sigma_{i2}^A (\pi c)^{1/2} - (\pi c)^{-1/2} \int_{-c}^{c} R_i(s) \left(\frac{c+s}{c-s} \right)^{1/2} ds. \tag{19.55}$$

19.5. Stress Intensity Factors

In a similar fashion one deduces that at the tip $x_1 \to -c$,

$$k_i(-c) = \sigma_{i2}^A (\pi c)^{1/2} - (\pi c)^{-1/2} \int_{-c}^{c} R_i(s) \left(\frac{c-s}{c+s}\right)^{1/2} ds. \tag{19.56}$$

Thus $k_i(\pm c)$ is independent of the anisotropy of the medium.

It can now be shown that the crack extension force is easily calculated from only knowing k_i and K_{ij}^{-1}. When the crack extends from $x_1 = c$ to $x_1 = c + \delta c$, the change in energy is given by

$$\delta \mathcal{E} = \frac{1}{2} \int_c^{c+\delta c} dx_1 \sigma_{i2}(x_1, 0) \int_{-c}^{x_1 - \delta c} F_i(t) \, dt. \tag{19.57}$$

Using (19.44), (19.51), and (19.55) and letting $\delta c \to 0$ (see the next subsection for details), the crack extension force is found to be

$$\mathcal{G} = -\lim_{\delta \to 0} \frac{\delta \mathcal{E}}{\delta c} = \frac{1}{8\pi} k_i k_m K_{im}^{-1}, \tag{19.58}$$

which is the desired result. Had we considered the crack tip at $x_1 = -c$ extending from $-c$ to $-(c + \delta c)$, we would have obtained (19.58) with k_i given by (19.56). Since in deriving (19.58) we considered extension of only one end of the crack, in this instance the proper Griffith criterion for brittle fracture would be $\mathcal{G} = 2\gamma$.

Equations (19.55) and (19.56) may also be used to derive formulae for the applied stress at which an equilibrium crack becomes mobile. In this instance we interpret the $R_i(x_1)$ as restraining stresses acting on the crack surfaces due to cohesive forces; usually one imagines that $R_i(x_1)$ differs appreciably from zero only in regions $c - d < |x_1| < c$, where $d \ll c$, and d is independent of c. Such a crack will propagate when $k_i(\pm c) \geq 0$, i.e., when

$$\sigma_{i2}^A \geq \frac{1}{\pi c} \int_{-c}^{c} R_i(s) \left(\frac{c \pm s}{c \mp s}\right) ds. \tag{19.59}$$

The upper and lower signs correspond to the tips $x_1 = c$ and $x_1 = -c$, respectively. The fracture criterion expressed in (19.59) depends on anisotropy only through the dependence of $R_i(s)$, the cohesive forces, on anisotropy.

The utility of (19.58) for the crack extension force is that the stress intensity factors, k_i, need be calculated only once, using either (19.55) or (19.56), for a given crack configuration, because anisotropic effects appear only through the K_{ij}^{-1}. The determination of $\mathcal{G}$ may also be executed in the **X** frame. This would be most easily done by calculating K_{ij}^{-1} in the **X** frame and computing $k_i(\mathbf{X} \text{ frame}) = A_{im} k_m(\text{crack frame})$, where A_{im} is the cosine of the angle between the X_i and x_m directions.

19.5.1 Computation of the Crack Extension Force

In this section the procedure for evaluating $\mathcal{G}$ is developed in detail. We wish to evaluate

$$\mathcal{G} = -\lim_{\delta c \to 0} \frac{\delta \mathcal{E}}{\delta c}, \tag{19.60}$$

where

$$\delta\mathcal{E} = \frac{1}{4}\int_c^{c+\delta c} \sigma_{i2}(x_1, 0)\mathrm{d}x_1 \int_{-c}^{x_1-\delta c} K_{im}^{-1} Q_m(t)\,\mathrm{d}t$$
$$= \frac{1}{2}(8\pi)^{-1/2} k_i K_{im}^{-1} \int_c^{c+\delta c} (x_1 - c)^{-1/2}\mathrm{d}x_1 \int_{-c}^{x_1-\delta c} Q_m(t)\,\mathrm{d}t,$$
(19.61)

with k_i given by (19.55). If we let $s = x_1 - c$ and integrate by parts, noting that

$$\int_{-c}^{c} Q_m(t)\,\mathrm{d}t = 0 \tag{19.62}$$

when (19.27) is satisfied, there follows

$$\delta\mathcal{E} = -(8\pi)^{-1/2} k_i K_{im}^{-1} \int_0^{\delta c} s^{1/2} Q_m(c - \delta c + s)\,\mathrm{d}s. \tag{19.63}$$

The further substitution of $s = (1 - \lambda)\delta c$ reduces (19.63) to

$$-\frac{\delta\mathcal{E}}{\delta c} = (8\pi)^{-1/2} k_i K_{im}^{-1} (\delta c)^{1/2} \int_0^1 (1 - \lambda)^{1/2} Q_m(c - \lambda\delta c)\,\mathrm{d}\lambda. \tag{19.64}$$

For the crack with singular stresses at $x_1 = \pm c$, we have

$$Q_m(t) = \frac{1}{\pi}\sigma_{m2}^A \frac{t}{(c^2 - t^2)^{1/2}} + \frac{1}{\pi^2(c^2 - t^2)^{1/2}} \int_{-c}^{c} \frac{R_m(s)(c^2 - s^2)^{1/2}}{s - t}\,\mathrm{d}s. \tag{19.65}$$

As $\delta c \to 0$, the first term in (19.65) on the *rhs* yields a contribution to $-\delta\mathcal{E}/\delta c$ given by

$$(8\pi)^{-1/2} k_i K_{im}^{-1} \sigma_{m2}^A \frac{1}{\pi}(\delta c)^{1/2} \left(\frac{c}{2\delta c}\right) \int_0^1 \left(\frac{1-\lambda}{\lambda}\right)^{1/2}\,\mathrm{d}\lambda = \frac{1}{8\pi} k_i K_{im}^{-1} \sigma_{m2}^A (\pi c)^{1/2}. \tag{19.66}$$

Interchanging the order of integration, the second term on the *rhs* of (19.65) becomes

$$\lim_{\delta c \to 0} \chi = -\frac{1}{8\pi} k_i K_{im}^{-1} \frac{1}{(\pi c)^{1/2}} \int_{-c}^{c} R_m(s)\left(\frac{c+s}{c-s}\right)^{1/2}\,\mathrm{d}s, \tag{19.67}$$

where

$$\chi = (8\pi)^{-1/2} k_i K_{im}^{-1} \frac{1}{\pi^2} (\delta c)^{1/2} \left(\frac{1}{2c\delta c}\right)^{1/2} \int_{-c}^{c} R_m(s)(c^2 - s^2)^{1/2}\,\mathrm{d}s$$
$$\times \int_0^1 \left(\frac{1-\lambda}{\lambda}\right)^{1/2} \frac{\mathrm{d}\lambda}{c - s + \lambda\delta c}.$$

In obtaining (19.67) we have noted that

$$\lim_{\delta c \to 0} \int_0^1 \left(\frac{1-\lambda}{\lambda}\right)^{1/2} \frac{\mathrm{d}\lambda}{s - c + \lambda\delta c} = -\frac{\pi}{2(c-s)}. \tag{19.68}$$

Comparing (19.66) and (19.67) with (19.55) yields the crack extension force in the form

$$\mathcal{G} = \frac{1}{8\pi} k_i K_{im}^{-1} k_m. \tag{19.69}$$

Equation (19.69) is also valid for the crack extension force at the crack tip at $x_1 = -c$, provided we use k_i as given in (19.56).

19.7. Energy Factor Matrix

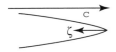

Figure 19.4. Crack tip opening.

19.6 Crack Tip Opening Displacement

Of particular interest is the result given in (19.28) for the crack tip opening displacement. Here we simply list some results for the case of isotropic elastic media. Recall that the displacement discontinuity at the crack tip is given as

$$\Delta u_i = u_i^D(x_1, x_2)|_{x_2=0+}^{x_2=0-} = \int_{-c}^{x_1} F_j(t)\, dt, \tag{19.70}$$

with

$$F_j(t) = K_{ji}^{-1} T_i^A \frac{1}{2\pi (c^2 - t^2)^{1/2}}. \tag{19.71}$$

Thus,

$$\Delta u_i = \frac{K_{ji}^{-1} \sigma_{i2}^A}{2\pi} \int_{-c}^{x_1} \frac{dt}{(c^2 - t^2)^{1/2}}. \tag{19.72}$$

Simple integration yields

$$-\Delta u_i = u_i(x_1, 0^+) - u_i(x_1, 0^-) = \frac{K_{ji}^{-1} \sigma_{i2}^A}{2\pi} (c^2 - t^2)^{1/2}. \tag{19.73}$$

Now, let $\delta_i(x_1) = u_i(x_1, 0^+) - u_i(x_1, 0^-)$ and consider the case of isotropic media. Recalling the expressions in (19.2), we have, for a Mode II crack stressed by in-plane shear,

$$\delta_1(x_1) = \frac{2(1-\nu)}{G} \sigma_{12}^A (c^2 - x_1^2)^{1/2}. \tag{19.74}$$

Consider the region just behind the crack tip, say at a distance ζ behind the tip (Fig. 19.4). Noting that $k_1 = \sigma_{12}^A \sqrt{\pi c}$, we find

$$\delta_1(\zeta) = \frac{4k_1(1-\nu)}{\sqrt{2\pi} G} \zeta^{1/2}. \tag{19.75}$$

By identical reasoning, we have

$$\delta_2(x_1) = \frac{4k_2(1-\nu)}{\sqrt{2\pi} G} \zeta^{1/2},$$

$$\delta_3(x_1) = \frac{4k_3}{\sqrt{2\pi} G} \zeta^{1/2}. \tag{19.76}$$

19.7 Dislocation Energy Factor Matrix

In this section a simple and convenient method for calculating the dislocation energy factor matrix, **K**, is developed. We begin with the line integral giving the solution for the displacement gradient of a dislocation line, i.e.,

$$u_{i,p}(\mathbf{x}) = -\epsilon_{pjw} b_m C_{wmrs} \frac{\partial}{\partial x_s} \oint_C G_{ir}(\mathbf{x} - \mathbf{x}')\, dx_j'. \tag{19.77}$$

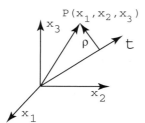

Figure 19.5. Coordinate frame for dislocation.

The integral is taken over the dislocation line, $\mathcal{C}$. We also recall that, in general, analytic solution for the Green's function does not exist, but its Fourier transform is given as

$$G_{ir}(\mathbf{x} - \mathbf{x}') = \frac{1}{8\pi^2 |\mathbf{x} - \mathbf{x}'|} \int_0^{2\pi} M_{ir}^{-1}(\mathbf{z}) \, d\theta, \qquad (19.78)$$

where the Christoffel stiffness tensor is $\mathbf{M} = \mathbf{z} \cdot \mathbf{C} \cdot \mathbf{z}$, and in this context $\mathbf{z}(\theta)$ is a unit vector lying in the plane perpendicular to $\mathbf{x} - \mathbf{x}'$. In terms of the unreduced Fourier transform of $\mathbf{G}$, we have

$$\frac{\partial}{\partial x_s} G_{ir}(\mathbf{x} - \mathbf{x}') = -\frac{i}{8\pi} \int_{-\infty}^{\infty} d^3 \mathbf{K} \, z_s \, \frac{M_{ir}^{-1}}{K} \, e^{-i\mathbf{K} \cdot (\mathbf{x} - \mathbf{x}')}. \qquad (19.79)$$

Here $\mathbf{K}$ is the Fourier vector and $\mathbf{z}$ is the unit vector along $\mathbf{K}$. Consequently, (19.77) may be rewritten as

$$u_{i,p}(\mathbf{x}) = \frac{i}{8\pi^2} \epsilon_{pjw} b_m C_{wmrs} \int_{-\infty}^{\infty} d^3 \mathbf{K} \, z_s \, \frac{M_{ir}^{-1}(\mathbf{x})}{K} \, e^{-i\mathbf{K} \cdot \mathbf{x}} \oint_{\mathcal{C}} e^{i\mathbf{K} \cdot \mathbf{x}'} \, dx_j'. \qquad (19.80)$$

For an infinitely long and straight dislocation line, laying along the unit vector $\mathbf{t}$,

$$\oint_{\mathcal{C}} e^{i\mathbf{K} \cdot \mathbf{x}'} \, dx_j' = t_j \int_{-\infty}^{\infty} e^{i\mathbf{K} \cdot \mathbf{t} s} \, ds = 2\pi t_j \delta(\mathbf{K} \cdot \mathbf{t}), \qquad (19.81)$$

where s is the distance along the line. Hence, only those Fourier vectors perpendicular to the dislocation line contribute to the displacement gradient field, and (19.81) becomes

$$u_{i,p}(\mathbf{x}) = \frac{1}{4\pi^2} \epsilon_{pjw} t_j b_m C_{wmrs} \int_{-\infty}^{\infty} d^2 \mathbf{K} \, z_s \, \frac{M_{ir}^{-1}(\mathbf{z})}{K} \sin(K\mathbf{z} \cdot \mathbf{x}). \qquad (19.82)$$

The integral in (19.82) extends over the plane defined by $\mathbf{z} \cdot \mathbf{t} = 0$, and for this plane

$$\mathbf{z} \cdot \mathbf{x} = \mathbf{z} \cdot \boldsymbol{\rho}, \qquad (19.83)$$

where $\boldsymbol{\rho}$ is the polar radius vector from the dislocation line to the point $\mathbf{x}$, as illustrated in Fig. 19.5. Accordingly, we may replace $\mathbf{x}$ with $\boldsymbol{\rho}$ in (19.82) and thereby show that the field is independent of position along the dislocation line, as it should be for an infinitely long and straight line.

In the plane $\mathbf{z} \cdot \mathbf{t} = 0$ we define a polar coordinate system centered about the dislocation line, such that

$$d^2 \mathbf{K} = K dK d\psi, \qquad (19.84)$$

where $0 \leq \psi \leq 2\pi$. Then

$$z_s = \alpha_s \cos \psi + \beta_s \sin \psi, \qquad (19.85)$$

19.7. Energy Factor Matrix

where α and β are two orthogonal unit vectors in the plane $\mathbf{z} \cdot \mathbf{t} = 0$. In particular, we may choose them such that

$$\alpha_1 = \sin\theta, \quad \alpha_2 = -\cos\theta, \quad \alpha_3 = 0,$$
$$\beta_1 = \cos\varphi\cos\theta, \quad \beta_2 = \cos\varphi\sin\theta, \quad \beta_3 = -\sin\varphi. \tag{19.86}$$

Similarly the vector ρ is written as

$$\rho_m = \rho(\alpha_m \cos\psi_0 + \beta_m \sin\psi_0), \tag{19.87}$$

so that

$$\sin(K\mathbf{z}\cdot\rho) = \sin[K\rho\cos(\psi - \psi_0)]. \tag{19.88}$$

The integral in (19.82) is now rewritten in terms of this polar coordinate system, and the integration over K yields

$$\int_0^\infty \sin[K\rho\cos(\psi - \psi_0)]\, dK = \frac{1}{\rho}\sec(\psi - \psi_0). \tag{19.89}$$

The expression for the displacements becomes

$$u_{i,p}(\rho) = \frac{1}{4\pi^2 \rho}\epsilon_{pjw} b_m t_j C_{wmrs} \int_0^{2\pi} z_s M_{ir}^{-1}(\psi)\sec(\psi - \psi_0)\, d\psi. \tag{19.90}$$

The integral in (19.90) is defined by its Cauchy principal value.

The elastic energy is given by

$$\mathcal{E} = \frac{1}{2}\int_{r_0}^R (C_{fgip} u_{i,p} b_g N_f)\, d\rho, \tag{19.91}$$

where

$$N_f = -\alpha_f \sin\psi_0 + \beta_f \cos\psi_0 = \frac{1}{\rho}\epsilon_{fsr} t_s \rho_r \tag{19.92}$$

are the components of the unit normal to the planar cut surface used (arbitrarily as usual) to create the dislocation, and across which the displacement jumps by $\mathbf{b}$. Clearly, the energy $\mathcal{E}$ is independent of ψ_0, so that

$$\mathcal{E} = \frac{1}{2\pi}\int_0^{2\pi} \mathcal{E}\, d\psi_0. \tag{19.93}$$

If we integrate (19.92) over ψ_0, interchange the order of integration, and note that

$$\sec\zeta = 2\sum_{h=0}^\infty (-1)^h \cos(2h+1)\zeta, \quad \zeta \neq -\frac{3\pi}{2}, \frac{\pi}{2}, \frac{3\pi}{2}, \ldots, \tag{19.94}$$

we obtain

$$\int_0^{2\pi} N_f \sec(\psi - \psi_0)\, d\psi_0 = 2\pi n_f, \tag{19.95}$$

with

$$n_f = -\alpha_f \sin\psi + \beta_f \cos\psi = dz_f/d\psi. \tag{19.96}$$

This yields

$$\mathcal{E} = \frac{1}{8\pi^2} \left(\ln \frac{R}{r_0}\right) b_m b_g \epsilon_{pjw} t_j C_{fgip} C_{wmrs} \int_0^{2\pi} z_s n_f M_{ir}^{-1}(\psi)\, d\psi. \qquad (19.97)$$

Next, we recognize that

$$\alpha_s \beta_f - \beta_s \alpha_f = \epsilon_{vsf} t_v, \qquad (19.98)$$

so that

$$z_s n_f = \frac{1}{2}\bigl[\epsilon_{vsf} t_v + (\alpha_s \beta_f + \beta_s \alpha_f) \cos(2\psi) + (\beta_f \beta_s - \alpha_f \alpha_s) \sin(2\psi)\bigr]. \qquad (19.99)$$

Finally, we obtain

$$\mathcal{E} = K_{mg} b_m b_g \ln(R/r_0), \qquad (19.100)$$

where

$$K_{mg} = (K_{mg}^* + K_{gm}^*) = K_{gm}, \qquad (19.101)$$

and

$$K_{mg}^* = \frac{1}{16\pi^2} \epsilon_{pjw} t_j C_{fgip} C_{wmrs} \Bigl[\epsilon_{vsf} t_v \int_0^{\pi} M_{ir}^{-1}\, d\psi + (\alpha_s \beta_f + \beta_s \alpha_f) \int_0^{\pi} M_{ir}^{-1} \cos(2\psi)\, d\psi$$
$$+ (\beta_f \beta_s - \alpha_f \alpha_s) \int_0^{\pi} M_{ir}^{-1} \sin(2\psi)\, d\psi \Bigr].$$

19.8 Inversion of a Singular Integral Equation

Certain solutions were given above for the singular integral equations involved with determining the equilibrium distributions of dislocations. Here we list a brief summary of some techniques given by Muskhelishivili (1960) for inverting integral equations of the form

$$\text{p.v.} \int_D \frac{f(t)\, dt}{t - x} = \sigma(x), \qquad (19.102)$$

where p.v. signifies the principal value of the integral. If $f(t)$ and $\sigma(x)$ are functions that are continuous in the interval D, and if D consists of p finite segments of which at q of the $2p$ ends $f(t)$ is bounded, then

$$f(x) = -\frac{1}{\pi^2}\left[\frac{R_1(x)}{R_2(x)}\right]^{1/2} \text{p.v.} \int_D \left[\frac{R_2(x)}{R_1(x)}\right]^{1/2} \frac{\sigma(t)\, dt}{t - x} + \left[\frac{R_1(x)}{R_2(x)}\right]^{1/2} P_{p-q-1}(x), \qquad (19.103)$$

provided $p - q \geq 0$. Here

$$R_1(x) = \prod_{i=1}^{q}(x - e_i), \qquad R_2(x) = \prod_{i=q+1}^{2p}(x - e_i). \qquad (19.104)$$

The polynomial $P_{p-q-1}(x)$ is an arbitrary polynomial of degree $\leq p - q - 1$, with $P_{-1} = 0$. The end points of the segments are e_i. When $p - q < 0$, the same solution is valid with the necessary and sufficient condition

$$\int_D \left[\frac{R_2(x)}{R_1(x)}\right]^{1/2} x^m \sigma(x)\, dx = 0, \qquad m = 0, 1, \ldots, q - p - 1. \qquad (19.105)$$

19.9 2D Anisotropic Elasticity – Stroh Formalism

For the sake of completeness, in this section we give a brief summary of the Stroh formalism for two-dimensional anisotropic elasticity. Consider a two-dimensional anisotropic elasticity problem in which the deformation field is independent of the x_3 coordinate, so that the displacement components are

$$u_k = u_k(x_1, x_2), \quad k = 1, 2, 3. \tag{19.106}$$

The corresponding stresses are

$$\sigma_{ij} = C_{ijkl} u_{k,l}, \tag{19.107}$$

where C_{ijkl} are the components of the anisotropic elastic moduli tensor (with respect to selected coordinate directions x_1, x_2, x_3). They are assumed to possess the usual symmetry properties with respect to the interchange of indices $i \leftrightarrows j$ and $k \leftrightarrows l$, as well as reciprocal symmetry $C_{ijkl} = C_{klij}$. In the absence of body forces, the equilibrium equations are

$$\sigma_{ij,j} = 0, \tag{19.108}$$

or, after using (19.107),

$$C_{ijkl} u_{k,jl}. \tag{19.109}$$

A general solution of (19.109) can be cast in the form

$$u_k = a_k f(z), \quad z = x_1 + p x_2, \tag{19.110}$$

where a_k and p are complex-valued constants, and f is an arbitrary function of z. Since

$$u_{k,l} = f' a_k (\delta_{1l} + p \delta_{2l}), \quad f' = df/dz,$$

$$u_{k,jl} = f'' a_k (\delta_{1j} + p \delta_{2j})(\delta_{1l} + p \delta_{2l}), \quad f'' = d^2 f/dz^2,$$

we obtain from (19.109)

$$\left[C_{i1k1} + p(C_{i1k2} + C_{i2k1}) + p^2 C_{i2k2} \right] a_k = 0, \tag{19.111}$$

whereas (19.107) gives

$$\sigma_{ij} = f'(C_{ijk1} + p C_{ijk2}) a_k. \tag{19.112}$$

In particular,

$$\sigma_{i1} = f'(C_{i1k1} + p C_{i1k2}) a_k,$$
$$\sigma_{i2} = f'(C_{i2k1} + p C_{i2k2}) a_k. \tag{19.113}$$

In view of the reciprocal symmetry of C_{ijkl}, we now define the symmetric 3×3 matrices $\mathbf{Q}$ and $\mathbf{T}$ with components

$$Q_{ik} = C_{i1k1}, \quad T_{ik} = C_{i2k2}, \tag{19.114}$$

and the nonsymmetric 3×3 matrix R with components

$$R_{ik} = C_{i1k2}. \tag{19.115}$$

Since the elastic strain energy is positive, the matrices $\mathbf{Q}$ and $\mathbf{T}$ are positive-definite. With so defined matrices $\mathbf{Q}, \mathbf{T}$, and $\mathbf{R}$, (19.111) and (19.113) can be rewritten in the matrix form as

$$\left[\mathbf{Q} + p\left(\mathbf{R} + \mathbf{R}^T\right) + p^2\mathbf{T}\right] \cdot \mathbf{a} = \mathbf{0} \tag{19.116}$$

and

$$\mathbf{t}_1 = f'(\mathbf{Q} + p\mathbf{R}) \cdot \mathbf{a}, \quad \mathbf{t}_2 = f'\left(\mathbf{R}^T + p\mathbf{T}\right) \cdot \mathbf{a}, \tag{19.117}$$

where

$$\mathbf{a} = \begin{bmatrix} a_1 \\ a_2 \\ a_3 \end{bmatrix}, \quad \mathbf{t}_1 = \begin{bmatrix} \sigma_{11} \\ \sigma_{21} \\ \sigma_{31} \end{bmatrix}, \quad \mathbf{t}_2 = \begin{bmatrix} \sigma_{12} \\ \sigma_{22} \\ \sigma_{32} \end{bmatrix}. \tag{19.118}$$

The equilibrium equations (19.108) can be cast in the vector form as

$$\mathbf{t}_{1,1} + \mathbf{t}_{2,2} = \mathbf{0}. \tag{19.119}$$

This suggests the introduction of the vector stress function φ, such that

$$\begin{aligned}\mathbf{t}_1 &= -\frac{\partial \varphi}{\partial x_2}, \quad \text{i.e.,} \quad \sigma_{i1} = -\frac{\partial \varphi_i}{\partial x_2}, \\ \mathbf{t}_2 &= \frac{\partial \varphi}{\partial x_1}, \quad \text{i.e.,} \quad \sigma_{i2} = \frac{\partial \varphi_i}{\partial x_1}.\end{aligned} \tag{19.120}$$

When these are introduced in (19.117), there follows

$$\begin{aligned}-\frac{\partial \varphi}{\partial x_2} &= f'(\mathbf{Q} + p\mathbf{R}) \cdot \mathbf{a}, \\ \frac{\partial \varphi}{\partial x_1} &= f'(\mathbf{R}^T + p\mathbf{T}) \cdot \mathbf{a}.\end{aligned} \tag{19.121}$$

Since $z = x_1 + px_2$, and

$$\frac{\partial f}{\partial x_1} = f', \quad \frac{\partial f}{\partial x_2} = pf',$$

equations (19.121) can be integrated to give

$$\varphi = -\frac{f}{p}(\mathbf{Q} + p\mathbf{R}) \cdot \mathbf{a} = f\left(\mathbf{R}^T + p\mathbf{T}\right) \cdot \mathbf{a}. \tag{19.122}$$

Thus,

$$\varphi = f(z)\mathbf{b}, \tag{19.123}$$

where

$$\mathbf{b} = \left(\mathbf{R}^T + p\mathbf{T}\right) \cdot \mathbf{a} = -\frac{1}{p}(\mathbf{Q} + p\mathbf{R}) \cdot \mathbf{a}. \tag{19.124}$$

It was shown by Stroh (1958) that the constants p, a_k, and b_k can be determined simultaneously from the six-dimensional eigenvalue problem as follows. From the first of (19.124), upon the multiplication with $\mathbf{T}^{-1}$, we have

$$-\left(\mathbf{T}^{-1} \cdot \mathbf{R}^T\right) \cdot \mathbf{a} + \mathbf{T}^{-1} \cdot \mathbf{b} = p \cdot \mathbf{a}. \tag{19.125}$$

19.9. Stroh Formalism

By multiplying this with $\mathbf{R}$, we obtain

$$p\mathbf{R} \cdot \mathbf{a} = \mathbf{R} \cdot \mathbf{T}^{-1} \cdot \mathbf{b} - (\mathbf{R} \cdot \mathbf{T}^{-1} \cdot \mathbf{R}^T) \cdot \mathbf{a}. \quad (19.126)$$

The second of (19.124) can be rearranged as

$$-\mathbf{Q} \cdot \mathbf{a} - p\mathbf{R} \cdot \mathbf{a} = p\mathbf{b}, \quad (19.127)$$

or, by using (19.126),

$$(\mathbf{R} \cdot \mathbf{T}^{-1} \cdot \mathbf{R}^T - \mathbf{Q}) \cdot \mathbf{a} - \mathbf{R} \cdot \mathbf{T}^{-1} \cdot \mathbf{b} = p \cdot \mathbf{b}. \quad (19.128)$$

Equations (19.125) and (19.128) together constitute a six-dimensional eigenvalue problem

$$\mathbb{N} \cdot v = pv, \quad (19.129)$$

where

$$\mathbb{N} = \begin{bmatrix} -\mathbf{T}^{-1} \cdot \mathbf{R}^T & \mathbf{T}^{-1} \\ \mathbf{R} \cdot \mathbf{T}^{-1} \cdot \mathbf{R}^T - \mathbf{Q} & -\mathbf{R} \cdot \mathbf{T}^{-1} \end{bmatrix} \quad (19.130)$$

is a real nonsymmetric 6×6 matrix, and

$$v = \begin{bmatrix} \mathbf{a} \\ \mathbf{b} \end{bmatrix} \quad (19.131)$$

is a six-dimensional vector with components $\{a_1, a_2, a_3, b_1, b_2, b_3\}$. The eigenvalue problem (19.129) delivers six eigenvalues $p^{(\alpha)}$ and six corresponding eigendirections $v^{(\alpha)}$ ($\alpha = 1, 2, \ldots, 6$). They depend only on the type of elastic anisotropy and the values of elastic moduli. The positive definiteness of the strain energy requires the six eigenvalues to appear as three pairs of complex conjugates. It is convenient to arrange them so that $p^{(1)}$, $p^{(2)}$, and $p^{(3)}$ have positive imaginary parts, i.e.,

$$\begin{aligned} \operatorname{Im} p^{(\alpha)} &> 0, \quad \alpha = 1, 2, 3, \\ p^{(\alpha+3)} &= \bar{p}^{(\alpha)}, \quad \alpha = 1, 2, 3, \end{aligned} \quad (19.132)$$

where overbar denotes the complex conjugation and Im the imaginary part. Correspondingly, we have

$$\begin{aligned} \mathbf{a}^{(\alpha+3)} &= \bar{\mathbf{a}}^{(\alpha)}, \quad \alpha = 1, 2, 3, \\ \mathbf{b}^{(\alpha+3)} &= \bar{\mathbf{b}}^{(\alpha)}, \quad \alpha = 1, 2, 3. \end{aligned} \quad (19.133)$$

The general solution can now be expressed by superposition as

$$\mathbf{u} = \sum_{\alpha=1}^{3} \left[\mathbf{a}^{(\alpha)} f^{(\alpha)}(z^{(\alpha)}) + \bar{\mathbf{a}}^{(\alpha)} f^{(\alpha+3)}(\bar{z}^{(\alpha)}) \right], \quad (19.134)$$

$$\varphi = \sum_{\alpha=1}^{3} \left[\mathbf{b}^{(\alpha)} f^{(\alpha)}(z^{(\alpha)}) + \bar{\mathbf{b}}^{(\alpha)} f^{(\alpha+3)}(\bar{z}^{(\alpha)}) \right], \quad (19.135)$$

in which $f^{(\alpha)}$ are arbitrary functions of their argument, and $z^{(\alpha)} = x_1 + p^{(\alpha)} x_2$, ($\alpha = 1, 2, \ldots, 6$). Equations (19.134) and (19.135) are Stroh's solutions for two-dimensional anisotropic elasticity. See also Eshelby, Read, and Shockley (1953) and, for an alternative approach, Lekhnitskii (1981).

In many applications, the functions $f^{(\alpha)}$ assume the same function form, differing only by complex scaling parameters $q^{(\alpha)}$, such that

$$f^{(\alpha)}(z^{(\alpha)}) = q^{(\alpha)} f(z^{(\alpha)}),$$
$$f^{(\alpha+3)}(\bar{z}^{(\alpha)}) = \bar{q}^{(\alpha)} \bar{f}(\bar{z}^{(\alpha)}).$$
(19.136)

Thus, by introducing the matrices

$$\mathbf{A} = [\mathbf{a}^{(1)}\ \mathbf{a}^{(2)}\ \mathbf{a}^{(3)}] = \begin{bmatrix} a_1^{(1)} & a_1^{(2)} & a_1^{(3)} \\ a_2^{(1)} & a_2^{(2)} & a_2^{(3)} \\ a_3^{(1)} & a_3^{(2)} & a_3^{(3)} \end{bmatrix},$$
(19.137)

$$\mathbf{B} = [\mathbf{b}^{(1)}\ \mathbf{b}^{(2)}\ \mathbf{b}^{(3)}] = \begin{bmatrix} b_1^{(1)} & b_1^{(2)} & b_1^{(3)} \\ b_2^{(1)} & b_2^{(2)} & b_2^{(3)} \\ b_3^{(1)} & b_3^{(2)} & b_3^{(3)} \end{bmatrix},$$
(19.138)

$$\mathbf{F} = \begin{bmatrix} f(z^{(1)}) & 0 & 0 \\ 0 & f(z^{(1)}) & 0 \\ 0 & 0 & f(z^{(1)}) \end{bmatrix},$$
(19.139)

and the vector

$$\mathbf{q} = \begin{bmatrix} q^{(1)} \\ q^{(2)} \\ q^{(3)} \end{bmatrix},$$
(19.140)

the solution (19.134) and (19.135) can be compactly expressed as

$$\mathbf{u} = 2\mathrm{Re}\,(\mathbf{A} \cdot \mathbf{F} \cdot \mathbf{q}),$$
(19.141)

$$\varphi = 2\mathrm{Re}\,(\mathbf{B} \cdot \mathbf{F} \cdot \mathbf{q}),$$
(19.142)

where Re stands for the real part. For example, for a Griffith crack of length $2c$, subject to uniform tractions $-\mathbf{t}_2^0$ over the crack faces which are parallel to x_1 direction, it can be shown that (e.g., Ting, 1991)

$$f(z) = \frac{1}{2}[(z^2 - c^2)^{1/2} - z],$$
(19.143)

$$\mathbf{q} = \mathbf{B}^{-1} \cdot \mathbf{t}_2^0, \quad \mathbf{t}_2^0 = \begin{bmatrix} \sigma_{21}^0 \\ \sigma_{22}^0 \\ \sigma_{23}^0 \end{bmatrix}.$$
(19.144)

19.9.1 Barnett–Lothe Tensors

Equations (19.141) and (19.142) are genearal solutions, provided that the state of anisotropy is such that there are indeed six independent eigenvectors $v^{(\alpha)}$. This is, for

19.9. Stroh Formalism

example, not the case with isotropic materials, for which $p = i = \sqrt{-1}$ is a triple eigenvalue with only two independent eigenvectors. Barnett and Lothe (1973) constructed a method to derive the solution which circumvent the need of solving the six-dimensional eigenvalue problem. They introduced the real matrices $\mathbf{S}$, $\mathbf{H}$, and $\mathbf{L}$, which are related to complex matrices $\mathbf{A}$ and $\mathbf{B}$ by

$$\mathbf{S} = i(2\mathbf{A} \cdot \mathbf{B}^T - \mathbf{I}), \quad \mathbf{H} = 2i\mathbf{A} \cdot \mathbf{A}^T, \quad \mathbf{L} = -2i\mathbf{B} \cdot \mathbf{B}^T, \tag{19.145}$$

and showed that

$$\mathbf{S} = -\frac{1}{\pi} \int_0^\pi \mathbf{T}^{-1}(\theta) \cdot \mathbf{R}^T(\theta) \, d\theta \,,$$

$$\mathbf{H} = \frac{1}{\pi} \int_0^\pi \mathbf{T}^{-1}(\theta) \, d\theta \,, \tag{19.146}$$

$$\mathbf{L} = \frac{1}{\pi} \int_0^\pi [\mathbf{Q}(\theta) - \mathbf{R}(\theta) \cdot \mathbf{T}^{-1}(\theta) \cdot \mathbf{R}^T(\theta)] \, d\theta \,,$$

where

$$Q_{ik}(\theta) = C_{ijkl} n_j n_l \,,$$
$$R_{ik}(\theta) = C_{ijkl} n_j m_l \,, \tag{19.147}$$
$$T_{ik}(\theta) = C_{ijkl} m_j m_l \,,$$

and $\mathbf{n} = \{\cos\theta, \sin\theta, 0\}$, $\mathbf{m} = \{-\sin\theta, \cos\theta, 0\}$. It is clear that $\mathbf{H}$ and $\mathbf{H}$ are symmetric (because $\mathbf{T}^{-1}$ and $\mathbf{Q}$ are). It can also be shown that the products $\mathbf{S} \cdot \mathbf{H}$, $\mathbf{L} \cdot \mathbf{S}$, $\mathbf{H}^{-1} \cdot \mathbf{S}$, and $\mathbf{S} \cdot \mathbf{L}^{-1}$ are antisymmetric tensors. There is furthermore a connection $\mathbf{H} \cdot \mathbf{L} - \mathbf{S} \cdot \mathbf{S} = \mathbf{I}$. The explicit representations for Barnett–Lothe tensors $\mathbf{S}$, $\mathbf{H}$, and $\mathbf{L}$ have been reported in the literature for various types of anisotropic materials, such as cubic and orthotropic (*e.g.*, Chadwick, and Smith, 1982; Ting, 1996).

For example, along the plane coinciding with the crack faces of the Griffith crack, loaded over its crack faces by uniform tractions, one has (Ting, 1991)

$$(z^2 - c^2)^{1/2} = \begin{cases} (x_1^2 - c^2)^{1/2}, & |x_1| > c, \\ \pm i(c^2 - x_1^2)^{1/2}, & |x_1| < c, \quad x_2 = \pm 0. \end{cases} \tag{19.148}$$

The displacement and traction vectors for $\pm x_1 > c$ are

$$\mathbf{u}(x_1, 0) = \pm \left[|x_1| - (x_1^2 - c^2)^{1/2}\right] \mathbf{S} \cdot \mathbf{L}^{-1} \cdot \mathbf{t_2}^0 \,,$$

$$\mathbf{t}_1(x_1, 0) = -\left[|x_1|(x_1^2 - c^2)^{-1/2} - 1\right] \mathbf{G}_2 \cdot \mathbf{t_2}^0 \,, \tag{19.149}$$

$$\mathbf{t}_2(x_1, 0) = \left[|x_1|(x_1^2 - c^2)^{-1/2} - 1\right] \mathbf{t_2}^0 \,,$$

while, for $|x_1| < c$,

$$\mathbf{u}(x_1, \pm 0) = \left[\pm (c^2 - x_1^2)^{1/2} \mathbf{I} + x_1 \mathbf{S}\right] \cdot \mathbf{L}^{-1} \cdot \mathbf{t_2}^0 \,,$$

$$\mathbf{t}_1(x_1, \pm 0) = \left[\pm x_1(c^2 - x_1^2)^{-1/2} \mathbf{G}_1 - \mathbf{G}_2\right] \cdot \mathbf{t_2}^0 \,, \tag{19.150}$$

$$\mathbf{t}_2(x_1, \pm 0) = -\mathbf{t_2}^0 \,.$$

The $\mathbf{G}$ matrices are here defined by

$$\mathbf{G}_1 = (\mathbf{R} \cdot \mathbf{T}^{-1} \cdot \mathbf{R}^T - \mathbf{Q}) \cdot \mathbf{L}^{-1}, \quad \mathbf{G}_2 = \mathbf{R} \cdot \mathbf{T}^{-1} + \mathbf{G}_1 \cdot \mathbf{L} \cdot \mathbf{S} \cdot \mathbf{L}^{-1}. \tag{19.151}$$

From the first of (19.149) or (19.150) it is observed that $\mathbf{u}(c, 0)$ is in general not equal to zero, unless $\mathbf{S} \cdot \mathbf{L}^{-1} \cdot \mathbf{t}_2^0 = \mathbf{0}$. Furthermore, the traction vector $\mathbf{t}_2(x_1, 0)$ is independent of the type of elastic anisotropy or the values of the elastic moduli. This was already demonstrated by (19.47).

19.10 Suggested Reading

Barnett, D. M., and Asaro, R. J. (1972), The Fracture Mechanics of Slit-Like Cracks in Anistropic Elastic Media, *J. Mech. Phys. Solids*, Vol. 20, pp. 353–366.

Barnett, D. M., and Lothe, J. (1973), Synthesis of the Sextic and the Integral Formalism for Dislocation, Green's Functions and Surface Waves in Anisotropic Elastic Solids, *Phys. Norv.*, Vol. 7, pp. 13–19.

Cherepanov, G. P. (1979), *Mechanics of Brittle Fracture*, McGraw-Hill, New York.

Eshelby, J. D., Read, W. T., and Shockley, W. (1953), Anisotropic Elasticity with Applications to Dislocation Theory, *Acta Metall.*, Vol. 1, pp. 251–259.

Freund, L. B. (1990), *Dynamic Fracture Mechanics*, Cambridge University Press, Cambridge.

Gdoutos, E. E. (1990), *Fracture Mechanics Criteria and Applications*, Kluwer, Dordrecht, The Netherlands.

Lekhnitskii, S. G. (1981), *Theory of Elasticity of an Anisotropic Elastic Body*, Mir Publishers, Moscow.

Kanninen, M. F., and Popelar, C. H. (1985), *Advanced Fracture Mechanics*, Oxford University Press, New York.

Mura, T. (1987), *Micromechanics of Defects in Solids*, Martinus Nijhoff, Dordrecht, The Netherlands.

Rice, J. R. (1968), Mathematical Analysis in the Mechanics of Fracture. In *Fracture – An Advanced Treatise* (H. Liebowitz, ed.), Vol. II, pp. 191–311, Academic Press, New York.

Stroh, A. N. (1958), Dislocations and Cracks in Anisotropic Elasticity, *Phil Mag.*, Vol. 3, pp. 625–646.

Ting, T. C. T. (1996), *Anisotropic Elasticity*, Oxford University Press, New York.

Wu, J. J., Ting, T. C. T., and Barnett, D. M., eds. (1991), *Modern Theory of Anisotropic Elasticity and Applications*, SIAM, Philadelphia.

20 The Inclusion Problem

The problem considered here has found application to a legion of physical applications including, *inter alia*, the theory of solid state phase transformations where the transformation (arising from second phase precipitation, allotropic transition, or uptake of solutes, or changes in chemical stoichiometry) causes a change in size and/or shape of the transformed, included, region; differences in thermal expansion of an included region and its surrounding matrix, which in turn causes incompatible thermal strains between the two; and, perhaps surprisingly, the concentrated stress and strain fields that develop around included regions that have different elastic modulus from those of their surrounding matrices. For the reason that the results of this analysis have application to such a wide variety of problem areas, and because the solution approach we adopt has heuristic value, we devote this chapter to the inclusion problem.

20.1 The Problem

In an infinitely extended elastic medium, a region – the "inclusion" – undergoes what would have been a stress free strain. Call this strain the "transformation strain," $\mathbf{e}^T$. Due to the elastic constraint of the medium, i.e., the matrix, there are internal stresses and elastic strains. What is this resulting elastic field and what are its characteristics? In particular, can an exact solution be found for this involved elastic field? The region of interest is shown in Fig. 20.1 and is denoted as V_I; the outward pointing unit normal to V_I is $\mathbf{n}$. The stress free transformation, *i.e.*, change in size and shape, can be viewed as occurring while the inclusion has been hypothetically removed from the medium, as depicted in Fig. 20.1. Thus the transformation indicated by $\mathbf{e}^T$ involves a displacement within the inclusion of the form

$$\mathbf{b}(\mathbf{x}) = \mathbf{e}^T \cdot \mathbf{x},$$
$$b_i = e^T_{ij} x_j. \tag{20.1}$$

The size and shape change associated with this transformation have caused no stresses and elastic strains if it were not for the constraint of the matrix.

We consider here the case where $\mathbf{e}^T$ is uniform, *i.e.*, does not depend on position. We denote the elastic constants as $\mathbf{C}$, so that the linear connection between stress and *elastic strain*, $\mathbf{e}^{el}$, is $\boldsymbol{\sigma} = \mathbf{C} : \mathbf{e}^{el}$ In the discussion that follows, the distinction between *elastic strain*

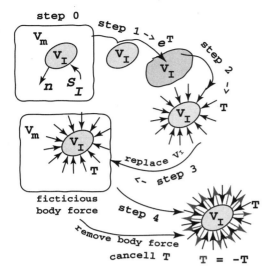

Figure 20.1. Eshelby's heuristic scheme for the inclusion setup.

and *total strain* will be made clear in specific context. For the infinitesimal strain formulation considered here the total strain is simply the sum of the elastic and the "nonelastic" strains.

20.2 Eshelby's Solution Setup

We consider the following heuristic solution approach originally devised by Eshelby (1957). The approach will involve a series of heuristic steps, each accompanied by a change in stress or strain. The concept of linear superposition is used to "build the solution."

Step 1: It is explicitly understood that the inclusion is embedded and bound to the matrix and its interface purely continuous with the matrix (medium). Thus all final displacements must be so continuous, and, as our procedure will show, they are. We imagine removing the inclusion from the medium by a purely heuristic process that produces no stress in the inclusion or in the medium. This means, of course, that no *elastic strain* has been induced in either region. Now let the inclusion transform, *i.e.*, let it undergo a homogeneous strain as prescribed by (20.1). This will induce a total strain in V_I of $\mathbf{e}^T$. At this stage, the stresses in both the inclusion, V_I, and the medium, V_m, are zero. Also, the total strain in the medium is zero at this stage. To be formal, we say

$$\mathbf{e}^{I1} = \mathbf{e}^T, \quad \sigma^{I1} = \mathbf{0},$$
$$\mathbf{e}^{m1} = \mathbf{0}, \quad \sigma^{m1} = \mathbf{0}. \tag{20.2}$$

The convention will be that the first superscript, $(..)^I$ or $(..)^m$, represents field quantities that belong to the inclusion and medium, respectively, and the second superscript indicates the contribution of that field quantity from the "step" in question.

Step 2: Now the inclusion in its hypothetically separated, yet transformed, state will no longer "fit" within the "hole" in the medium that it came from. But it has undergone a

20.2. Eshelby's Solution Setup

uniform strain, $\mathbf{e}^T$. To replace it within the medium, we may imagine applying a uniform set of boundary traction, $\mathbf{T}$, such that

$$\mathbf{T} = -\mathbf{C} : \mathbf{e}^T \cdot \mathbf{n},$$
$$T_i = -C_{ijkl}e^T_{kl}n_j. \tag{20.3}$$

This has the result of producing a uniform state of elastic strain, and therefore uniform stress, in the inclusion. No strain, or stress, has yet to be caused in the matrix. The contributions to total stress and strain from this step is

$$\mathbf{e}^{I2} = -\mathbf{e}^T, \quad \sigma^{I2} = -\mathbf{C} : \mathbf{e}^T,$$
$$\mathbf{e}^{m2} = \mathbf{0}, \quad \sigma^{m2} = \mathbf{0}. \tag{20.4}$$

Step 3: At this point the inclusion will fit perfectly into the "hole" from which it came, and therefore it may be reinserted, and the interface between it and the medium made continuous again. At this stage the state of strain and stress is

$$\mathbf{e}^{I3} = \mathbf{e}^{I1} + \mathbf{e}^{I2} = \mathbf{0}, \quad \sigma^{I3} = -\mathbf{C} : \mathbf{e}^T,$$
$$\mathbf{e}^{m3} = \mathbf{0}, \quad \sigma^{m3} = \mathbf{0}. \tag{20.5}$$

It is important to recall, however, that the elastic strain in the inclusion is at this stage, $-\mathbf{e}^T$, which of course accounts for the state of stress in the inclusion. Note, as explained above, that at this point all displacements are to be continuous across the interface between the inclusion and matrix – and at this point they already are! The final step will produce additional displacements in both the matrix and inclusion which are inherently continuous.

Although the inclusion now fits within its original "hole," there are what now appears to be an embedded layer of body force $\mathbf{T} = -\mathbf{C} : \mathbf{e}^T \cdot \mathbf{n}$ distributed around the interface S_I. To remove these fictitious body forces we apply a layer of annulling forces, $-\mathbf{T} = \mathbf{C} : \mathbf{e}^T \cdot \mathbf{n}$ around S_I, as indicated in Fig. 20.1. The application of these forces will cause displacements, elastic strains, and additional stresses, in both the medium and the inclusion; call these the "constrained field," as they occur under the constraint of the infinite elastic medium. Thus the superscript ^{4} will be replaced by ^{c} to specially designate this.

Step 4 or c: The last step involves the removal of the unwanted layer of body force, $\mathbf{T}$. Formally this is accomplished by applying a layer of body force $-\mathbf{T}$ around S_I. However, as already explained, this occurs while the inclusion is bound to the medium. Thus, all displacements that occur in this step will be continuous across the interface between the inclusion and matrix. Call these displacements, inside and outside V_I, $\mathbf{u}^c$. It is this constrained or ^{c} field we will now solve for. However, the final state for the total field quantities within the inclusion and medium will be

$$\mathbf{e}^{I4} = \mathbf{e}^{I1} + \mathbf{e}^{I2} + \mathbf{e}^c = \mathbf{e}^c, \quad \sigma^{I4} = -\mathbf{C} : \mathbf{e}^T + \sigma^c,$$
$$\mathbf{e}^{m4} = \mathbf{e}^c, \quad \sigma^{m4} = \sigma^c, \tag{20.6}$$

where the strains $\mathbf{e}^c$ are calculated from the constrained displacements, and the stresses σ^c are calculated as $\sigma^c = \mathbf{C} : \mathbf{e}^c$.

It is also important to explicitly note the total displacements in the inclusion and the medium at this final stage; they are

$$\mathbf{u} = \mathbf{u}^c \quad \text{in both } V_I \text{ and } V_m. \tag{20.7}$$

On the other hand, the *displacements that cause elastic strains, and thus stresses*, are

$$\mathbf{u} = -\mathbf{b} + \mathbf{u}^c \quad \text{in } V_I, \quad \text{and} \quad \mathbf{u} = \mathbf{u}^c \text{ in } V_m. \tag{20.8}$$

Thus, the *elastic strains* in the inclusion are

$$\mathbf{e} = \mathbf{e}^{\text{el}} = \mathbf{e}^c - \mathbf{e}^T \quad \text{in } V_I. \tag{20.9}$$

20.3 Calculation of the Constrained Fields: u^c, e^c, and σ^c

To calculate the displacements caused by the application of the layer of body force, $-\mathbf{T} = \mathbf{C} : \mathbf{e}^T \cdot \mathbf{n}$, we simply use linear superposition and the Green's function to write

$$u_p^c(\mathbf{x}) = \int_{S_I} C_{ijkl} e_{kl}^T n_j G_{ip}(\mathbf{x} - \mathbf{x}') \, dS'$$

$$= \int_{S_I} (C_{ijkl} e_{kl}^T G_{ip}) n_j \, dS'$$

$$= \int_{V_I} C_{ijkl} e_{kl}^T \frac{\partial}{\partial x_j'} G_{ip}(\mathbf{x} - \mathbf{x}') \, dV'.$$

We note that

$$G_{ip} = G_{ip}(\mathbf{x} - \mathbf{x}') \implies \frac{\partial}{\partial x_j'} G_{ip}(\mathbf{x} - \mathbf{x}') = -\frac{\partial}{\partial x_j} G_{ip}(\mathbf{x} - \mathbf{x}') \tag{20.10}$$

and thus

$$u_p^c(\mathbf{x}) = \frac{\partial}{\partial x_j} \int_{V_I} C_{ijkl} e_{kl}^T G_{ip}(\mathbf{x} - \mathbf{x}') \, dV'. \tag{20.11}$$

For the distortion, *i.e.*, the displacement gradient, we have

$$u_{p,s}^c(\mathbf{x}) = \frac{\partial u_p^c}{\partial x_s} = \frac{\partial^2}{\partial x_s \partial x_j} \int_{V_I} C_{ijkl} e_{kl}^T G_{ip}(\mathbf{x} - \mathbf{x}') \, dV'. \tag{20.12}$$

We recall that the Green's function is given by

$$G_{ip}(\mathbf{x} - \mathbf{x}') = \frac{1}{8\pi^3} \int_{-\infty}^{\infty} d^3\mathbf{k} \, \frac{M_{ip}^{-1}(\mathbf{z})}{k^2} \exp[-i\mathbf{k} \cdot (\mathbf{x} - \mathbf{x}')], \tag{20.13}$$

where $\mathbf{k}$ is the Fourier vector and $\mathbf{z}$ is a unit vector aligned along $\mathbf{k}$. Thus,

$$\frac{\partial u_p^c}{\partial x_s} = -\frac{\partial^2}{\partial x_s \partial x_j} C_{ijkl} e_{kl}^T \frac{1}{8\pi^3} \int_{-\infty}^{\infty} d^3\mathbf{k} \, \frac{M_{ip}^{-1}}{k^2} \exp(-i\mathbf{k} \cdot \mathbf{x})$$
$$\times \int_{V_I} \exp(i\mathbf{k} \cdot \mathbf{x}') \, dV'. \tag{20.14}$$

Consider the last integral first, *i.e.*,

$$\int_{V_I} \exp(i\mathbf{k} \cdot \mathbf{x}') \, dV'.$$

20.3. Calculation of the Constrained Fields

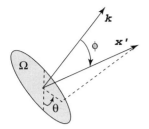

Figure 20.2. A polar coordinate system aligned with **k**.

Let V_I be an ellipsoidal volume with axes a_1, a_2, a_3, for reasons that will be clear in what follows. Define expanded variables and a polar coordinate system aligned with **k** (see Fig. 20.2), as follows

$$K_i = k_i a_i \text{ and } R_i = x'_i/a_i \quad \text{(no sum on } i),$$

$$\mathbf{k} \cdot \mathbf{x}' = \mathbf{K} \cdot \mathbf{R} = KR\cos\phi,$$

$$dV' = d^3\mathbf{x}' = a_1 a_2 a_3 \, d^3\mathbf{R},$$

$$d^3\mathbf{x}' = a_1 a_2 a_3 R^2 \sin\phi \, d\phi \, dR \, d\theta.$$

Then,

$$\int_{V_I} \exp(i\mathbf{k}\cdot\mathbf{x}')\,dV' = a_1 a_2 a_3 \int_0^1 R^2 dR \int_\Omega \exp(iKR\cos\phi)\sin\phi\,d\phi\,d\theta. \tag{20.15}$$

Note that Ω is a unit sphere such that $0 \le \theta \le 2\pi$ and $0 \le \phi \le \pi$. The integral over θ is trivial, so that

$$\int_{V_I} \exp(i\mathbf{k}\cdot\mathbf{x}')\,dV' = 2\pi a_1 a_2 a_3 \int_0^1 R^2\,dR \int_0^\pi \exp(iKR\cos\phi)\sin\phi\,d\phi. \tag{20.16}$$

The standard integral tables reveal that

$$\int_0^\pi \exp(iKR\cos\phi)\sin\phi\,d\phi = \frac{(2\pi)^{1/2}}{(KR)^{1/2}} J_{1/2}(KR), \tag{20.17}$$

where $J_{1/2}(KR)$ is a fractional Bessel function. Thus,

$$\int_0^\pi \exp(iKR\cos\phi)\sin\phi\,d\phi = \frac{2\pi a_1 a_2 a_3 (2\pi)^{1/2}}{K^{1/2}} \int_0^1 R^{3/2} J_{1/2}(KR)\,dR$$

$$= \frac{2\pi a_1 a_2 a_3 (2\pi)^{1/2}}{K^{3/2}} J_{3/2}(K),$$

with

$$K = |K| = (k_1^2 a_1^2 + k_2^2 a_2^2 + k_3^2 a_3^2)^{1/2}. \tag{20.18}$$

Consequently,

$$u^c_{p,s} = -\frac{\partial^2}{\partial x_s \partial x_j} \frac{1}{2\pi^2} a_1 a_2 a_3 (\pi/2)^{1/2} C_{ijkl} e^T_{kl} \int_{-\infty}^\infty d^3\mathbf{k}\, \frac{M^{-1}_{ip}(\mathbf{z})}{k^2} \exp(-i\mathbf{k}\cdot\mathbf{x}) \frac{J_{3/2}(K)}{K^{3/2}}. \tag{20.19}$$

If we again define

$$K_i = k_i a_i, \quad \text{(no sum on } i\text{)},$$

$$K^2 = k_i^2 a_i^2, \quad d^3\mathbf{k} = \frac{1}{a_1 a_2 a_3} d^3\mathbf{K}, \tag{20.20}$$

then

$$\frac{k_i}{K} = \frac{K_i}{K a_i} = \frac{\nu_i}{a_i}, \quad \text{(no sum on } i\text{)},$$

$$\frac{k_1^2 + k_2^2 + k_3^2}{K^2} = \frac{\nu_1^2}{a_1^2} + \frac{\nu_2^2}{a_2^2} + \frac{\nu_3^2}{a_3^2}, \tag{20.21}$$

so that

$$\mathbf{k} \cdot \mathbf{x} = K\boldsymbol{\nu} \cdot \mathbf{R}, \quad R_i = x_i/a_i, \quad \text{(no sum on } i\text{)}. \tag{20.22}$$

Now, take the real part of (20.19) to obtain

$$u_{p,s}^c = -\frac{1}{2\pi^2} \sqrt{\pi/2} \, C_{ijkl} e_{kl}^T \frac{\partial^2}{\partial x_j \partial x_s} \int_{-\infty}^{\infty} \frac{d^3\mathbf{K} M_{pi}^{-1} \cos[K(\boldsymbol{\nu} \cdot \mathbf{R})] J_{3/2}(K)}{K^2 K^{3/2}[(\nu_1/a_1)^2 + (\nu_2/a_2)^2 + (\nu_3/a_3)^2]}. \tag{20.23}$$

Next, introduce another polar coordinate system aligned with $\mathbf{R}$, so that

$$d^3\mathbf{K} = K^2 dK \sin\phi \, d\phi \, d\theta,$$

and note

$$\int_0^\infty \frac{\cos[K(\boldsymbol{\nu} \cdot \mathbf{R})] J_{3/2}(K)}{K^{3/2}} dK = 1/2\sqrt{\pi/2} \left[1 - (\boldsymbol{\nu} \cdot \mathbf{R})^2\right],$$

$$\text{if } (\boldsymbol{\nu} \cdot \mathbf{R}) \leq 1. \tag{20.24}$$

It follows that

$$u_{p,s}^c = \frac{1}{4\pi} C_{ijkl} e_{kl}^T \int_0^{2\pi} \int_0^\pi \frac{\sin\phi \, M_{pi}^{-1}(\boldsymbol{\nu})(\nu_j/a_j)(\nu_s/a_s)}{[(\nu_1/a_1)^2 + (\nu_2/a_2)^2 + (\nu_3/a_3)^2]} d\phi \, d\theta,$$

$$\text{(no sum on } j, s\text{)},$$

$$= \text{Constants}, \quad \text{if } \boldsymbol{\nu} \cdot \mathbf{R} \leq 1. \tag{20.25}$$

To form the strains we take the symmetric part of $u_{p,s}^c$, or

$$e_{p,s}^c = \frac{1}{8\pi} C_{ijkl} e_{kl}^T \int_0^{2\pi} \int_0^\pi \frac{\sin\phi \, d\phi \, d\theta \, n_j [M_{pi}^{-1}(\boldsymbol{\nu}) n_s + M_{si}^{-1}(\boldsymbol{\nu}) n_p]}{[(\nu_1/a_1)^2 + (\nu_2/a_2)^2 + (\nu_3/a_3)^2]}, \tag{20.26}$$

where

$$n_p = (\nu_p/a_p), \quad \text{(no sum on } p\text{)}.$$

We can conveniently define the components of the tensor $\mathcal{S}$ as

$$\mathcal{S}_{pskl} = \frac{1}{8\pi} C_{ijkl} \int_0^{2\pi} \int_0^\pi \frac{\sin\phi \, d\phi \, d\theta \, n_j [M_{pi}^{-1}(\boldsymbol{\nu}) n_s + M_{si}^{-1}(\boldsymbol{\nu}) n_p]}{[(\nu_1/a_1)^2 + (\nu_2/a_2)^2 + (\nu_3/a_3)^2]}, \tag{20.27}$$

such that

$$e_{ps}^c = \mathcal{S}_{pskl} e_{kl}^T. \tag{20.28}$$

20.4. Eshelby Tensor for Ellipsoidal Inclusion

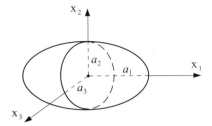

Figure 20.3. An ellipsoidal inclusion with its principal axes parallel to coordinate axes.

The fourth-order tensor with components S_{ijkl} is known as the Eshelby tensor.

In summary, it is worth recalling that:

In the Inclusion:

$$\begin{aligned}
\sigma_{ij}^I &= C_{ijps}(e_{ps}^c - e_{ps}^T), \\
e_{ps}^c &= S_{pskl}e_{kl}^T, \\
e_{ps}^{I,el} &= e_{ps}^c - e_{ps}^T, \\
u_i^{I,el} &= u_i^c - b_i, \\
e_{ps}^I &= e_{ps}^c.
\end{aligned} \tag{20.29}$$

In the medium:

$$\begin{aligned}
\sigma_{ij}^m &= C_{ijps}e_{ps}^c, \\
e_{ps}^m &= e_{ps}^c, \\
u_i^m &= u_i^{m,el} = u_i^c.
\end{aligned} \tag{20.30}$$

20.4 Components of the Eshelby Tensor for Ellipsoidal Inclusion

The components of the Eshelby tensor S_{ijkl} obey the symmetry

$$S_{ijkl} = S_{jikl} = S_{ijlk}. \tag{20.31}$$

When written with respect to coordinate axes parallel to principal axes of the ellipsoidal inclusion (Fig. 20.3), these components are

$$\begin{aligned}
S_{1111} &= \frac{3a_1^2}{8\pi(1-\nu)} I_{11} + \frac{1-2\nu}{8\pi(1-\nu)} I_1, \\
S_{1122} &= \frac{a_2^2}{8\pi(1-\nu)} I_{12} - \frac{1-2\nu}{8\pi(1-\nu)} I_1, \\
S_{1133} &= \frac{a_3^2}{8\pi(1-\nu)} I_{13} - \frac{1-2\nu}{8\pi(1-\nu)} I_1, \\
S_{1212} &= \frac{a_1^2 + a_2^2}{16\pi(1-\nu)} I_{12} + \frac{1-2\nu}{16\pi(1-\nu)} (I_1 + I_2).
\end{aligned} \tag{20.32}$$

The remaining nonzero components are obtained by cyclic permutations of $(1, 2, 3)$. All other components vanish.

The I_i integrals appearing in above expressions can be expressed in terms of elliptic integrals. Assuming that $a_1 > a_2 > a_3$, we have

$$I_1 = \frac{4\pi a_1 a_2 a_3}{(a_1^2 - a_2^2)(a_1^2 - a_3^2)^{1/2}} [F(\theta, k) - E(\theta, k)],$$

$$I_3 = \frac{4\pi a_1 a_2 a_3}{(a_2^2 - a_3^2)(a_1^2 - a_3^2)^{1/2}} \left[\frac{a_2(a_1^2 - a_3^2)^{1/2}}{a_1 a_3} - E(\theta, k) \right], \quad (20.33)$$

$$I_1 + I_2 + I_3 = 4\pi,$$

where $F(\theta, k)$ and $E(\theta, k)$ are elliptic integrals of the first and second kind, and

$$\theta = \arcsin\left(\frac{a_1^2 - a_3^2}{a_1^2}\right)^{1/2}, \quad k = \left(\frac{a_1^2 - a_2^2}{a_1^2 - a_3^2}\right)^{1/2}. \quad (20.34)$$

The I_{ij} integrals can be calculated from

$$3I_{11} + I_{12} + I_{13} = \frac{4\pi}{a_1^2},$$

$$3a_1^2 I_{11} + a_2^2 I_{12} + a_3^2 I_{13} = 3I_1, \quad (20.35)$$

$$I_{12} = \frac{1}{a_1^2 - a_2^2}(I_2 - I_1),$$

and their cyclic counterparts.

The stress components in an ellipsoidal inclusion due to uniform transformation (eigenstrain) e_{ij}^T are

$$\begin{aligned}\frac{\sigma_{11}}{2\mu} = &\left\{ \frac{a_1^2}{8\pi(1-\nu)} \left[\frac{1-\nu}{1-2\nu} 3I_{11} + \frac{\nu}{1-2\nu}(I_{21} + I_{31}) \right] \right. \\ &+ \frac{1-2\nu}{8\pi(1-\nu)} \left[\frac{1-\nu}{1-2\nu} I_1 - \frac{\nu}{1-2\nu}(I_2 + I_3) \right] - \frac{1-\nu}{1-2\nu} \right\} e_{11}^T \\ &+ \left\{ \frac{a_2^2}{8\pi(1-\nu)} \left[\frac{1-\nu}{1-2\nu} I_{12} + \frac{\nu}{1-2\nu}(3I_{22} + I_{32}) \right] \right. \\ &- \frac{1-2\nu}{8\pi(1-\nu)} \left[\frac{1-\nu}{1-2\nu} I_1 - \frac{\nu}{1-2\nu}(I_2 - I_3) \right] - \frac{\nu}{1-2\nu} \right\} e_{22}^T \\ &+ \left\{ \frac{a_3^2}{8\pi(1-\nu)} \left[\frac{1-\nu}{1-2\nu} I_{13} + \frac{\nu}{1-2\nu}(3I_{33} + I_{23}) \right] \right. \\ &- \frac{1-2\nu}{8\pi(1-\nu)} \left[\frac{1-\nu}{1-2\nu} I_1 - \frac{\nu}{1-2\nu}(I_3 - I_2) \right] - \frac{\nu}{1-2\nu} \right\} e_{33}^T,\end{aligned} \quad (20.36)$$

$$\frac{\sigma_{12}}{2\mu} = \left[\frac{a_1^2 + a_2^2}{8\pi(1-\nu)} I_{12} + \frac{1-2\nu}{8\pi(1-\nu)}(I_1 + I_2) - 1 \right] e_{12}^T, \quad (20.37)$$

with other stress components obtained from the above two expressions by cyclic permutations of $(1, 2, 3)$.

20.5 Elastic Energy of an Inclusion

The elastic energy of the entire body, *i.e.*, inclusion and medium is

$$E = \frac{1}{2}\int_V \sigma_{ij} e_{ij} \, dV = \frac{1}{2}\int_V \sigma_{ij} u_{i,j} \, dV. \tag{20.38}$$

Formally, we may divide the body into the inclusion volume and the volume of the medium, so that

$$E = \frac{1}{2}\int_{V_I} \sigma_{ij}^I u_{i,j}^{I,el} \, dV + \frac{1}{2}\int_{V_m} \sigma_{ij}^m u_{i,j}^m \, dV. \tag{20.39}$$

Since there are no body forces, *i.e.*, $\nabla \cdot \sigma = 0$, the divergence theorem shows that

$$E = \frac{1}{2}\int_{V_I} (\sigma_{ij}^I u_i^I)_{,j} \, dV + \frac{1}{2}\int_{V_m} (\sigma_{ij}^m u_i^m)_{,j} \, dV$$

$$= \frac{1}{2}\int_{S_I} \sigma_{ij}^I n_j u_i^{I,el} \, dS + \frac{1}{2}\int_{S_I} \sigma_{ij}^I \hat{n}_j u_i^{I,el} \, dS + \frac{1}{2}\int_{S_\infty} \sigma_{ij}^m n_j^\infty u_i^m \, dS.$$

Furthermore,

$$\int_{S_\infty} \sigma_{ij}^m n_j^\infty u_i^m \, dS \to 0,$$

because of the way in which the stresses and displacements fall off with distance. Thus,

$$E = \frac{1}{2}\int_{S_I} (\sigma_{ij}^I n_j u_i^{I,el} + \sigma_{ij}^I \hat{n}_j u_i^m) \, dS$$

$$= \frac{1}{2}\int_{S_I} \sigma_{ij}^I n_j (u_i^{I,el} - u_i^m) \, dS \tag{20.40}$$

$$= -\frac{1}{2}\int_{S_I} \sigma_{ij}^I n_j e_{is}^T x_s \, dS,$$

because $u_i^{I,el} - u_i^m = -b_i$. From (20.40) it then follows, with another use of the divergence theorem, that

$$E = -\frac{1}{2}\int_{V_I} \sigma_{ij}^I e_{ij}^T \, dV. \tag{20.41}$$

20.6 Inhomogeneous Inclusion: Uniform Transformation Strain

We next consider the problem of an infinite medium containing an inclusion of a material different from that of the medium. The *inhomogeneous inclusion*, which occupies the volume V_I, as before, is to undergo what would be, without the constraint of the surrounding medium, a stress free strain, $\mathbf{e}^*$. The elastic moduli tensor for this inclusion is designated as $\mathbf{C}^*$. For the case of a *homogeneous inclusion*, *i.e.*, one having the same elastic constants as the medium, after undergoing a transformation strain, $\mathbf{e}^T$, the final inclusion strain is $\mathbf{e}^c = \mathcal{S} : \mathbf{e}^T$, and the resulting inclusion stress is $\sigma = \mathbf{C} : (\mathcal{S} : \mathbf{e}^T - \mathbf{e}^T)$. We recall, if the inclusion were of ellipsoidal shape, and if $\mathbf{e}^T$ were uniform, the resulting stresses and strains in the inclusion were also uniform. We have omitted the superscript ^{I} for the

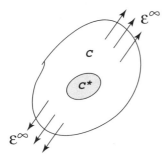

Figure 20.4. Inhomogeneous inclusion subject to far field strains.

inclusion field variables for brevity and because we will only require discussion of field quantities within the inclusions for our development. With the above in mind, imagine applying surface forces to the inhomogeneous inclusion so as to produce a strain, $\mathbf{e}^c - \mathbf{e}^*$, where we will associate $\mathbf{e}^c$ here with that of a *companion homogeneous inclusion* shortly. The resulting stresses in the inclusion are $\sigma^* = \mathbf{C} : (\mathbf{e}^c - \mathbf{e}^*)$. Note that $-\mathbf{e}^*$ strain will cancel the stress free strain, $\mathbf{e}^*$, which now leaves the total strain of the inhomogeneous inclusion to be $\mathbf{e}^c$. At this stage both the homogeneous and inhomogeneous inclusions have experienced the same total strains and thus have the same size and shape. If the stresses within the two inclusions are also identical, then we may replace the homogeneous inclusion with the inhomogeneous inclusion. This is so because the traction that would be computed from these stresses would be identical, and thus traction continuity would be maintained with the medium. For the stresses to be the same, we must be able to solve for a companion set of transformation strains for a homogeneous inclusion, $\mathbf{e}^T$, given the prescribed transformation strains, $\mathbf{e}^*$. The resulting equations are

$$\mathbf{C} : (\mathcal{S} : \mathbf{e}^T - \mathbf{e}^T) = \mathbf{C}^* : (\mathcal{S} : \mathbf{e}^T - \mathbf{e}^*),$$
$$C_{ijkl}(\mathcal{S}_{klmn}e^T_{mn} - e^T_{kl}) = C^*_{ijkl}(\mathcal{S}_{klmn}e^T_{mn} - e^*_{kl}). \tag{20.42}$$

The set (20.42) can indeed be solved for $\mathbf{e}^T$.

Consider the problem of an infinite medium containing an inclusion of a different material and subject to a far field strain $\mathbf{e}^\infty$ (Fig. 20.4). Suppose we first solve the problem of a homogeneous inclusion with the same elastic constants as the medium and the same initial size and shape, which undergoes a uniform transformation strain $\mathbf{e}^T$. We recall that in the case of this companion homogeneous inclusion, the resulting state of stress and strain would be such that the total strain would be $\mathbf{e}^c$; the elastic strain would, however, be $\mathbf{e}^c - \mathbf{e}^T$. Now, superimpose a uniform strain, $\mathbf{e}^\infty$, on this homogeneous inclusion as well as on the medium. The stress in the inclusion is

$$\sigma^{\text{hom,I}} = \mathbf{C} : (\mathbf{e}^c - \mathbf{e}^T + \mathbf{e}^\infty). \tag{20.43}$$

Next, subject the inhomogeneous inclusion, with elastic constants $\mathbf{C}$, to a uniform strain $\mathbf{e}^c + \mathbf{e}^\infty$. Note at this point that the size and shape change undergone by this inhomogeneous inclusion is identical to that of the fully transformed companion homogeneous inclusion. The stresses in the inhomogeneous inclusion are

$$\sigma^{\text{inhom,I}} = \mathbf{C}^* : (\mathbf{e}^c + \mathbf{e}^\infty). \tag{20.44}$$

20.7. Nonuniform Transformation Strain

If the stresses in both inclusions are the same, we may replace the companion homogeneous inclusion with its inhomogeneous counterpart. If the inclusions are of ellipsoidal shape this is possible because $\mathbf{e}^c = \mathcal{S} : \mathbf{e}^T$, where $\mathcal{S}$ is a constant Eshelby tensor. Thus, given $\mathbf{e}^\infty, \mathbf{C}, \mathbf{C}^*$, we may solve for the companion homogeneous $\mathbf{e}^T$ from

$$\mathbf{C} : (\mathcal{S} : \mathbf{e}^T - \mathbf{e}^T + \mathbf{e}^\infty) = \mathbf{C}^* : (\mathcal{S} : \mathbf{e}^T + \mathbf{e}^\infty),$$
$$C_{ijkl}(S_{klrs}e^T_{rs} - e^T_{kl} + e^\infty_{kl}) = C^*_{ijkl}(S_{klrs}e^T_{rs} + e^\infty_{kl}). \tag{20.45}$$

Note that the stresses and strains are everywhere the same for both the medium with the inhomogeneous and the companion homogeneous inclusions.

20.7 Nonuniform Transformation Strain Inclusion Problem

As before, we imagine a volume element – *that we will later, for clear advantage, take to be of ellipsoidal shape* – undergo what would be a stress-free change in size and shape while it is embedded within an infinite medium. The inclusion and its medium have the same elastic constants, $\mathbf{C}$. The stress-free change in size and shape is described by a *transformation strain*, $\mathbf{e}^T$, as discussed in the previous sections.

LEMMA 20.1: *If an ellipsoidal region in an infinite anisotropic linear elastic medium undergoes, in the absence of its surroundings, a stress-free transformation strain which is a polynomial of degree M in the position coordinates, x_p, then the final stress and strain state of the transformed inclusion, when constrained by its surroundings, is also a polynomial of degree M in the coordinates x_p.*

This generalization of Eshelby's original theorem for the uniform inclusion in an isotropic elastic medium has far reaching consequences - *viz.*, that all the developments of the previous section regarding the *inhomogeneous inclusion* hold true, as can be verified by direct analysis. Clearly, the above results for the uniform transformation strain are included in this generalization.

Proof: We start almost from the beginning. Imagine cutting out the volume element V_I in an undeformed infinite anisotropic linear elastic medium. As this process is heuristic, let V_I undergo a stress-free *transformation strain* $\mathbf{e}^T(\mathbf{x})$, where $\mathbf{x}$ denotes position within the inclusion. The strain $\mathbf{e}^T(\mathbf{x})$ is assumed to be continuous and differentiable, but is otherwise arbitrary. Now let $p^T_{ij}(\mathbf{x}) = C_{ijkl}e^T_{kl}$ be the stress derived from Hook's law as above, *i.e.*, if the uniform strain $\mathbf{e}^T$ were applied unconstrained to V_I. The construction of the *constrained elastic field* is accomplished by first applying $-p^T_{ij}(\mathbf{x})n_j$ to the surface S_I bounding V_I [and distributing body forces $(\partial/\partial x_j)p^T_{ij}(\mathbf{x})$ throughout V_I].

With this nomenclature, we can move to the general expression, *i.e.*, (20.11), for the components of displacement gradient,

$$\frac{\partial u^c_i}{\partial x_p} = -\frac{\partial^2}{\partial x_p \partial x_j} \int_{V_I} G_{ik}(\mathbf{x} - \mathbf{x}')p^T_{kj}(\mathbf{x}')\,dV'. \tag{20.46}$$

Take the Fourier transform of (20.46) to obtain, with $\mathbf{K}$ again being a Fourier vector,

$$\frac{\partial u_i^c}{\partial x_p} = K^2 z_p z_j g_{jk}(\mathbf{K}) \int_{V_\mathrm{I}} p_{kj}^\mathrm{T}(\mathbf{x}') \exp(i K \mathbf{z} \cdot \mathbf{x}') \, \mathrm{d}V'$$

$$= z_p z_j M_{ik}^{-1}(\mathbf{z}) \int_{V_\mathrm{I}} p_{kj}^\mathrm{T}(\mathbf{x}') \exp(i K \mathbf{z} \cdot \mathbf{x}') \, \mathrm{d}V'.$$

We recall that the Fourier transform of $\mathbf{G}$ is

$$\mathbf{g}(\mathbf{K}) = \mathbf{M}^{-1}/K^2, \tag{20.47}$$

where $\mathbf{M} \equiv \mathbf{z} \cdot \mathbf{C} \cdot \mathbf{z}$, and $\mathbf{z} = \mathbf{K}/|\mathbf{K}|$. In other words,

$$g_{ik}(\mathbf{K}) = M_{ik}^{-1}(\mathbf{K})/K^2. \tag{20.48}$$

Using the Fourier inversion described in Chapter 3, along with the spherical coordinate system, yields

$$\frac{\partial u_i^c}{\partial x_p} = (2\pi)^{-3} \int_\Omega z_p z_j M_{ik}^{-1}(\mathbf{z}) \, \mathrm{d}\Omega \int_0^\infty K^2 \exp(-i K \mathbf{z} \cdot \mathbf{x}) \, \mathrm{d}K \int_{V_\mathrm{I}} p_{kj}^\mathrm{T}(\mathbf{x}') \exp(i K \mathbf{z} \cdot \mathbf{x}') \, \mathrm{d}V'. \tag{20.49}$$

Again, $\mathrm{d}\Omega = \sin\phi \, \mathrm{d}\phi \, \mathrm{d}\theta$ is the surface area over the unit sphere, as described in the previous section. We are interested in only the real part of (20.49), and we note that

$$\nabla_x^2 \cos[K \mathbf{z} \cdot (\mathbf{x}' - \mathbf{x})] = -K^2 \cos[K \mathbf{z} \cdot (\mathbf{x}' - \mathbf{x})], \tag{20.50}$$

where

$$\nabla_x^2 \equiv \partial^2/\partial x_s \partial x_s. \tag{20.51}$$

We also recall that

$$\int_0^\infty \cos(Ks) \, \mathrm{d}K = \pi \delta(s). \tag{20.52}$$

Thus, an interchange of the order of integration yields

$$\frac{\partial u_i^c}{\partial x_p} = -\frac{1}{8\pi^2} \int_\Omega z_p z_j M_{ik}^{-1}(\mathbf{z}) \, \mathrm{d}\Omega \, \nabla_x^2 \int_{V_\mathrm{I}} p_{jk}^\mathrm{T}(\mathbf{x}') \delta(\mathbf{z} \cdot \mathbf{x}' - \mathbf{z} \cdot \mathbf{x}) \, \mathrm{d}V'. \tag{20.53}$$

We now specialize to the case of an ellipsoidal inclusion,

$$\sum_{\alpha=1}^3 (x_\alpha/a_\alpha)^2 = 1, \tag{20.54}$$

which means that $\mathbf{x}$ lies in V_I, and $\mathbf{e}^\mathrm{T}(\mathbf{x}')$ and hence $p_{jk}^\mathrm{T}(\mathbf{x}')$ are polynomials of degree M in $\mathbf{x}'$. We have, for convenience, chosen the x_i axes to coincide with the principal axes of the ellipsoid; this can be done without loss of generality because a polynomial of degree M in x_i remains a polynomial of degree M in the coordinates of any rectangular Cartesian system obtained by rotation and translation from other rectangular Cartesian system.

In general, $p_{kj}^\mathrm{T}(\mathbf{x}')$ is the sum of $M+1$ terms, with a typical r^{th} term of the form

$$A_{kjmn\ldots s} x'_m x'_n \ldots x'_s,$$

20.7. Nonuniform Transformation Strain

where $0 \leq r \leq M$ and the $A_{kjmn...s} x'_m x'_n \ldots x'_s$ are r^2 constants, whereas $x'_m x'_n \ldots x'_s$ are r products. Consider such a term substituted into (20.53), namely the integral

$$I^{(r)}_{kj} = A_{kjmn...s} \int_{V_1} x'_m x'_n \ldots x'_s \delta(\mathbf{z} \cdot \mathbf{x}' - \mathbf{z} \cdot \mathbf{x}) \, dV', \tag{20.55}$$

with

$$\sum_{\alpha=1}^{3} (x_\alpha/a_\alpha)^2 \leq 1. \tag{20.56}$$

Then,

$$\frac{\partial u^c_i}{\partial x_p} = \frac{1}{8\pi^2} \int_\Omega z_p z_j M^{-1}_{ik}(\mathbf{z}) \, d\Omega \sum_{r=0}^{M} \nabla^2_x I^{(r)}_{kj}. \tag{20.57}$$

Now, introduce the change in variables

$$t'_\alpha = x'_\alpha/a_\alpha, \quad t_\alpha = x_\alpha/a_\alpha, \tag{20.58}$$

which will convert the integral over the ellipsoid in (20.55) to an integral over the unit sphere, $|\mathbf{t}'| \leq 1$. When $\mathbf{x}$ lies within the ellipsoid, $|\mathbf{t}| \leq 1$ and (20.55) become

$$I^{(r)}_{kj} = \sum_{\alpha,\beta,\ldots\gamma=1}^{3} A_{kj\alpha\beta...\gamma} a_\alpha a_\beta \ldots a_\gamma (a_1 a_2 a_3) \int_{|\mathbf{t}'| \leq 1} t'_\alpha t'_\beta \ldots t'_\gamma \delta[\mu(\mathbf{s} \cdot \mathbf{t}' - \mathbf{s} \cdot \mathbf{t})] \, dV', \tag{20.59}$$

where

$$\mu = \left(a_1^2 z_1^2 + a_2^2 z_2^2 + a_3^2 z_3^2\right)^{1/2} > 0,$$

$$s_\alpha = z_\alpha a_\alpha/\mu, \quad \text{(no sum on } \alpha), \tag{20.60}$$

$$|\mathbf{s}| = 1.$$

The delta function in (20.59) is nonzero only for those $\mathbf{t}'$ of the form

$$\mathbf{t}' = T(\mathbf{m} \cos \psi + \mathbf{n} \sin \psi) + (\mathbf{s} \cdot \mathbf{t})\mathbf{s},$$

$$0 \leq T \leq \left[1 - (\mathbf{s} \cdot \mathbf{t})^2\right]^{1/2}, \tag{20.61}$$

and $\mathbf{m}$ and $\mathbf{n}$ are any two fixed unit vectors normal to $\mathbf{s}$. The angle ψ is a polar angle in the plane of $\mathbf{m}$ and $\mathbf{n}$. Thus, the volume integral in (20.59) is reduced to an integral over a plane circular region of radius $[1 - (\mathbf{s} \cdot \mathbf{t})^2]^{1/2}$. Since

$$\delta(\mu f) = \frac{1}{\mu} \delta(f), \tag{20.62}$$

we have

$$I^{(r)}_{kj} = \frac{a_1 a_2 a_3}{\mu} \sum_{\alpha,\beta,\ldots,\gamma=1}^{3} A_{kj\alpha\beta...\gamma} a_\alpha a_\beta \ldots a_\gamma \tag{20.63}$$

$$\times \int_0^{2\pi} d\psi \int_0^{[1-(\mathbf{s}\cdot\mathbf{t})^2]^{1/2}} [TR_\alpha + (\mathbf{s} \cdot \mathbf{t})s_\alpha] \ldots [TR_\gamma + (\mathbf{s} \cdot \mathbf{t})s_\gamma] T \, dT.$$

In here,
$$R_\alpha = m_\alpha \cos \psi + n_\alpha \sin \psi, \tag{20.64}$$

and $T dT$ is the polar element of area in the plane of integration.

The integrand in (20.63) can be expanded using the binomial theorem and is of the form
$$\sum_{w=0}^{r} T^{w+1} (\mathbf{s} \cdot \mathbf{t})^{r-w} \binom{r}{w} f_{\alpha\beta\ldots\gamma}^{(w)}(\psi), \tag{20.65}$$

where
$$\begin{cases} f_{\alpha\beta\ldots\gamma}^{(0)}(\psi) = s_\alpha s_\beta \ldots s_\gamma, \\ f_{\alpha\beta\ldots\gamma}^{(1)}(\psi) = s_\alpha s_\beta \ldots R_\gamma + s_\alpha R_\beta \ldots s_\gamma + R_\alpha s_\beta \ldots s_\alpha, \\ f_{\alpha\beta\ldots\gamma}^{(r)}(\psi) = R_\alpha R_\beta \ldots R_\gamma, \end{cases} \tag{20.66}$$

i.e., each term in $f_{\alpha\beta\ldots\gamma}^{(w)}(\psi)$ contains the product of w R's. Since
$$\int_0^{2\pi} \sin^m \psi \cos^n \psi \, d\psi = 0, \quad \text{if } m+n \text{ is odd or if both } m \text{ and } n \text{ are odd,}$$

one easily verifies that
$$\int_0^{2\pi} f_{\alpha\beta\ldots\gamma}^{(w)}(\psi) \, d\psi = 0, \quad \text{if } w \text{ is odd.} \tag{20.67}$$

Thus,
$$I_{jk}^{(r)} = \frac{a_1 a_2 a_3}{\mu} \sum_{\alpha,\beta,\ldots\gamma=1}^{3} A_{kj\alpha\beta\ldots\gamma} a_\alpha a_\beta \ldots a_\gamma \sum_{w=0}^{\langle \frac{1}{2}r \rangle} \binom{r}{2w} (\mathbf{s} \cdot \mathbf{t})^{r-2w}$$
$$\times \int_0^{2\pi} f_{\alpha\beta\ldots\gamma}^{(2w)}(\psi) \, d\psi \int_0^{[1-(\mathbf{s}\cdot\mathbf{t})^2]^{1/2}} T^{2w+1} \, dT, \tag{20.68}$$

where $\langle \frac{1}{2} r \rangle$ denotes the greatest integer not greater than $\frac{1}{2} r$. Since, from (20.58) and (20.61),
$$\mathbf{s} \cdot \mathbf{t} = \mathbf{z} \cdot \mathbf{x}/\mu, \tag{20.69}$$

and having in mind the fact that
$$\int_0^{[1-(\mathbf{s}\cdot\mathbf{t})^2]^{1/2}} T^{2w+1} \, dT = \frac{1}{2(w+1)} \left[1 - (\mathbf{s} \cdot \mathbf{t})^2\right]^{w+1}, \tag{20.70}$$

we obtain
$$I_{jk}^{(r)} = \frac{a_1 a_2 a_3}{2\mu} \sum_{\alpha,\beta,\ldots\gamma=1}^{3} A_{kj\alpha\beta\ldots\gamma} a_\alpha a_\beta \ldots a_\gamma \sum_{w=0}^{\langle \frac{1}{2}r \rangle} \binom{r}{2w} \int_0^{2\pi} f_{\alpha\beta\ldots\gamma}^{(2w)}(\psi) \, d\psi$$
$$\times \frac{1}{w+1} \left(\frac{\mathbf{z} \cdot \mathbf{w}}{\mu}\right)^{r-2w} \left[1 - \left(\frac{\mathbf{z} \cdot \mathbf{x}}{\mu}\right)^2\right]^{w+1}.$$

20.7. Nonuniform Transformation Strain

Clearly, each term in here is a polynomial in $\mathbf{z} \cdot \mathbf{x}$ of degree $r - 2w + 2(w+1) = r+2$. Thus $I_{jk}^{(r)}$ is a polynomial in x_i of degree $r+2$, and

$$\nabla_{\mathbf{x}}^2 I_{jk}^{(r)} = P_{jk}^{(r)}(\mathbf{z} \cdot \mathbf{x}), \tag{20.71}$$

where $P_{jk}^{(r)}(\mathbf{z} \cdot \mathbf{x})$ is a polynomial of degree r in x_i. Finally,

$$\sum_{r=0}^{M} \nabla_{\mathbf{x}}^2 I_{jk}^{(r)} = F_{jk}^{(M)}(\mathbf{z} \cdot \mathbf{x}), \tag{20.72}$$

where $F_{jk}^{(M)}(\mathbf{z} \cdot \mathbf{x})$ is a polynomial of degree M in x_i. Using (20.57) we immediately conclude that inside the ellipsoid the constrained elastic displacement gradients, and thus the strains and stresses, are polynomials in x_i of degree M. The total elastic strain inside the ellipsoid is given by (20.57) and $-\mathbf{e}^T(\mathbf{x})$, which is, of course, also a polynomial in x_i of degree M. Thus the proposition of the lemma is proven.

20.7.1 The Cases $M = 0, 1$

The theorem will be illustrated for the cases where $M = 0, 1$. For $M = 0$, we have $f^{(0)}(\psi) = 1$ and

$$I_{jk}^{(0)} = \pi \frac{a_1 a_2 a_3}{\mu} A_{kj} \left[1 - \left(\frac{\mathbf{z} \cdot \mathbf{x}}{\mu} \right)^2 \right]. \tag{20.73}$$

Now,

$$\nabla_{\mathbf{x}}^2 \left(\frac{\mathbf{z} \cdot \mathbf{x}}{\mu} \right)^2 = 2/\mu^2, \tag{20.74}$$

so that inside the inclusion

$$e_{ip}^c = \left[\frac{a_1 a_2 a_3}{8\pi} \int_{\Omega} \frac{(z_p z_j M_{ik}^{-1} + z_i z_j M_{pk}^{-1})}{\mu} d\Omega \right] A_{kj} = S_{ipkj} A_{kj}, \tag{20.75}$$

where the S_{ipkj} may be evaluated by simple numerical integration. For instance, using the fact that

$$M_{ik}^{-1} = \frac{1}{2} \epsilon_{isr} \epsilon_{rmn} M_{sm} M_{rn}/\Delta, \quad \Delta = \epsilon_{mns} M_{1m} M_{2n} M_{2s}, \tag{20.76}$$

we can use spherical coordinates

$$z_1 = \cos\phi \cos\theta, \quad z_2 = \sin\phi \sin\theta, \quad z_3 = \cos\phi. \tag{20.77}$$

When $M = 1$, in addition to the term in (20.75), we obtain a term corresponding to $f^{(0)}(\psi) = s_\alpha$. Then,

$$I_{jk}^{(1)} = \pi \frac{a_1 a_2 a_3}{\mu^2} \sum_{\alpha=1}^{3} A_{kj\alpha} z_\alpha a_\alpha^2 \left(\frac{\mathbf{z} \cdot \mathbf{x}}{\mu} \right) \left[1 - \left(\frac{\mathbf{z} \cdot \mathbf{x}}{\mu} \right)^2 \right]. \tag{20.78}$$

Since

$$\nabla_{\mathbf{x}}^2 \left(\frac{\mathbf{z} \cdot \mathbf{x}}{\mu} \right)^2 = 6 \frac{\mathbf{z} \cdot \mathbf{x}}{\mu^3}, \tag{20.79}$$

in addition to the term in (20.75), we obtain a contribution

$$e_{ip}^c = \left[\frac{3a_1 a_2 a_3}{8\pi} \int_\Omega \left(\frac{z_p z_j M_{ik}^{-1} + z_i z_j M_{pk}^{-1}}{\mu^5} z_s \sum_{\alpha=1}^{3} A_{kj\alpha} z_\alpha a_\alpha^2\right) d\Omega\right] x_s.$$

The integrals may be easily evaluated by numerical integration.

20.8 Inclusions in Isotropic Media

Eshelby's (1957, 1959) original development of the problem of an inclusion in a elastic medium contained insight and results of considerable interest for exploring physical phenomena. Here we present some of these. The general framework for this has already been presented and is accordingly utilized here.

20.8.1 Constrained Elastic Field

Recall the expression derived for the Green's function for an infinitely extended isotropic elastic medium, viz.,

$$G_{km}(\mathbf{x} - \mathbf{x}') = \frac{1}{8\pi\mu}\left(\delta_{km}\nabla^2 - \frac{\lambda+\mu}{\lambda+2\mu}\frac{\partial^2}{\partial x_k \partial x_m}\right)|\mathbf{x} - \mathbf{x}'|. \tag{20.80}$$

Using the result

$$\nabla^2 |\mathbf{x} - \mathbf{x}'| = \frac{2}{|\mathbf{x} - \mathbf{x}'|}, \tag{20.81}$$

(20.80) may be rewritten as follows. The displacement at the point $\mathbf{x}$ due to a unit point force with components f_i acting at the point $\mathbf{x}'$, is

$$\hat{u}_j(\mathbf{x} - \mathbf{x}') = \frac{1}{4\pi\mu}\frac{f_j}{|\mathbf{x} - \mathbf{x}'|} - \frac{1}{16\pi\mu(1-\nu)} f_\ell \frac{\partial^2}{\partial x_\ell \partial x_j}|\mathbf{x} - \mathbf{x}'|. \tag{20.82}$$

The stress associated with the uniform transformation strain, $\mathbf{e}^T$, is

$$\sigma_{ij}^T = \lambda e^T \delta_{ij} + 2\mu e_{ij}^T, \tag{20.83}$$

where $e^T \equiv \mathrm{tr}\,\mathbf{e}^T$. The *constrained elastic field* is constructed by applying traction $\sigma_{ij} n_j$ to the surface of the inclusion, i.e.,

$$u_j^c(\mathbf{x}) = \int_S \sigma_{ij}^T \hat{u}_j(\mathbf{x} - \mathbf{x}') n_k \, dS', \quad \text{(no sum on } j\text{)}. \tag{20.84}$$

The strain in the matrix and inclusion, *if measured from the original, untransformed state*, is

$$e_{ij}^c = \frac{1}{2}(u_{i,j}^c + u_{j,i}^c). \tag{20.85}$$

The stress in the matrix arises solely from this constrained field and is given by

$$\sigma_{ij}^c = \lambda e^c \delta_{ij} + 2\mu e_{ij}^c. \tag{20.86}$$

The stress in the inclusion is

$$\sigma_{ij}^I = \lambda(e^c - e^T)\delta_{ij} + 2\mu(e_{ij}^c - e_{ij}^T). \tag{20.87}$$

20.8. Inclusions in Isotropic Media

Because the σ_{ij}^T are constant, and by applying the divergence theorem to (20.84), we obtain Eshelby's result

$$u_i^c = \frac{1}{16\pi\mu(1-v)} \sigma_{jk}^T \psi_{,ijk} - \frac{1}{4\pi\mu} \sigma_{ik}^T \phi_{,k}, \qquad (20.88)$$

where

$$\phi(\mathbf{x}) = \int_V \frac{dV'}{|\mathbf{x}-\mathbf{x}'|}, \quad \text{and} \quad \psi(\mathbf{x}) = \int_V |\mathbf{x}-\mathbf{x}'| \, dV' \qquad (20.89)$$

are the Newtonian harmonic and biharmonic potentials, respectively; V is the volume of the inclusion. Note the fact that $\partial/\partial x_\ell = -\partial/\partial x'_\ell$ has been used.

Now, since $\phi(\mathbf{x})$ is the gravitational potential of a constant unit mass density in the volume V, we have

$$\nabla^2\phi(\mathbf{x}) = \begin{cases} -4\pi, & \text{inside } V, \\ 0, & \text{outside } V. \end{cases} \qquad (20.90)$$

Furthermore, it is evident that

$$\nabla^2\psi(\mathbf{x}) = 2\phi(\mathbf{x}),$$
$$\nabla^4\psi(\mathbf{x}) = 2\nabla^2\phi(\mathbf{x}). \qquad (20.91)$$

An interesting result follows immediately. Suppose we wish to know the dilatation, $u_{i,i}^c$. Since

$$\psi_{,ijki} = \psi_{,iijk} = \nabla^2(\psi_{,jk}) = 2\phi_{,jk},$$

we can construct $u_{i,i}^c$ from (20.88) as

$$u_{i,i}^c = e^c = -\frac{1-2v}{8\pi\mu(1-v)} \sigma_{ik}^T \phi_{,ik}. \qquad (20.92)$$

Suppose further, as an example, that $\mathbf{e}^T = \frac{1}{3} e^T \mathbf{I}$. Then,

$$e_{ij}^c = -\frac{1}{4\pi} \frac{1+v}{3(1-v)} e^T \phi_{,ij}, \qquad (20.93)$$

and

$$e^c = \frac{1+v}{3(1-v)} e^T = \zeta e^T. \qquad (20.94)$$

Thus, the factor ζ is a measure of the constraint of the matrix against free expansion of an inclusion.

20.8.2 Field in the Matrix

The field in the matrix is generally more difficult to determine in a simple form. The Green's function can be expressed as

$$\hat{u}_i(\mathbf{x}-\mathbf{x}') = \frac{1}{16\pi\mu(1-v)} \frac{f_j}{|\mathbf{x}-\mathbf{x}'|} \left[(3-4v)\delta_{ij} + \frac{(x_i-x'_i)(x_j-x'_j)}{|\mathbf{x}-\mathbf{x}'|^2} \right].$$

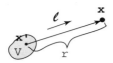

Figure 20.5. Field point in the matrix "far" from the inclusion.

Insert this into (20.84) and apply the divergence theorem to obtain

$$u_i^c(\mathbf{x}) = \frac{\sigma_{jk}^T}{16\pi\mu(1-\nu)} \int_V \frac{dV}{r^2} h_{ijk}(\boldsymbol{\ell})$$
$$= \frac{e_{jk}^T}{8\pi(1-\nu)} \int_V \frac{dV}{r^2} g_{ijk}(\boldsymbol{\ell}), \quad (20.95)$$

where r is the distance from $\mathbf{x}'$ to $\mathbf{x}$, and $\boldsymbol{\ell}$ is a unit vector parallel to $\mathbf{x} - \mathbf{x}'$, as shown in Fig. 20.5. Furthermore,

$$h_{ijk} = (1-2\nu)(\delta_{ij}\ell_k + \delta_{ik}\ell_j) - \delta_{jk}\ell_i + 3\ell_i\ell_j\ell_k,$$
$$g_{ijk} = (1-2\nu)(\delta_{ij}\ell_k + \delta_{ik}\ell_j - \delta_{jk}\ell_i) + 3\ell_i\ell_j\ell_k. \quad (20.96)$$

Suppose $\mathbf{x}$ is *far removed* from the inclusion. Then, we may remove everything from the integrand, except dV, to obtain

$$u_i^c(\mathbf{x}) \sim V\sigma_{jk}^T h_{ijk} \frac{1}{16\pi\mu(1-\nu)r^2} = V e_{jk}^T g_{ijk} \frac{1}{8\pi(1-\nu)r^2}. \quad (20.97)$$

20.8.3 Field at the Interface

Now once again the σ_{ij}^T are constant, and thus

$$u_{i,\ell}^c = \frac{1}{16\pi\mu(1-\nu)} \sigma_{ij}^T \psi_{,ijk\ell} - \frac{1}{4\pi\mu} \sigma_{ik}^T \phi_{,k\ell}. \quad (20.98)$$

Consider the jump in $u_{i,\ell}^c$ across the interface S of the inclusion. This involves

$$[[\psi_{,ijk\ell}]] \quad \text{and} \quad [[\phi_{,k\ell}]].$$

It is readily shown that

$$[[\phi_{,k\ell}]] = u_{i,\ell}^c(\text{out}) - u_{i,\ell}^c(\text{in}) = 4\pi n_k n_\ell. \quad (20.99)$$

It is left as an exercise to demonstrate this but the steps involve starting with the definition of ϕ and noting that the connection $\partial/\partial x_i = -\partial/\partial x_i'$ can be used within the integral when forming a second derivative such as $\phi_{,k\ell}$. Use of the divergence theorem, and a bit of elementary geometry, settles the proof. The same argument can be applied to $\psi_{,ij}$, which is the potential obtained by using the density $-2\phi_{,ij}/4\pi$. Thus,

$$[[\psi_{,ijk\ell}]] = 8\pi n_i n_j n_k n_\ell. \quad (20.100)$$

20.8. Inclusions in Isotropic Media

Using the results such as (20.92), there follows

$$[[e^c]] = -\frac{1}{3}\frac{1+\nu}{1-\nu}e^T - \frac{1-2\nu}{1-\nu}e'^T_{ij}n_in_j,$$

$$[[e'^c_{i\ell}]] = \frac{1}{1-\nu}e'^T_{jk}n_jn_kn_in_\ell - e'^T_{ik}n_kn_\ell - e'^T_{\ell k}n_kn_i \quad (20.101)$$

$$+ \frac{1-2\nu}{3(1-\nu)}e'^T_{jk}n_jn_k\delta_{i\ell} - \frac{1}{3}\frac{1+\nu}{1-\nu}e^T\left(n_in_\ell - \frac{1}{3}\delta_{i\ell}\right),$$

where $e'^T_{ij} = e^T_{ij} - \frac{1}{3}e^T\delta_{ij}$ is a deviatoric part of the strain tensor.

These are Eshelby's results. Several aspects of the results should be noted. First, since both e^c_{ij} and e^T_{ij} are related to σ^c_{ij} and σ^T_{ij} by the isotropic elastic constitutive law given in (20.86), the relations in (20.101) are readily expressed in terms of the stresses. Second, by evaluating the elastic field within the inclusion, the elastic field just outside is known via (20.101). This allows for the determination of stress and strain concentrations at the interface. Third, and quite obviously, the result in (20.101) is available for $[[u^c_{i,\ell}]]$ itself. This makes it useful for evaluating such quantities as *surface forces*, discussed earlier. Clearly, the strain energy density, $W(\mathbf{e})$, at points on either side of the interface can be evaluated once the field within the inclusion is known; this process is quite tractable for inclusions that are ellipsoidal in shape.

20.8.4 Isotropic Spherical Inclusion

Eshelby lists his results for an isotropic spherical inclusion and we close by providing them below. One has

$$e^c = \alpha e^T, \quad e'^c_{ij} = \beta e'^T_{ij}, \quad (20.102)$$

with

$$\alpha = \frac{1}{3}\frac{1+\nu}{1-\nu}, \quad \beta = \frac{2}{15}\frac{4-5\nu}{1-\nu}. \quad (20.103)$$

For a spherical inhomogeneous inclusion with elastic constants κ^* and μ^*, subject to a far field applied strain e^∞_{ij}, the equivalent transformation strain is given by

$$e^T = Ae^\infty, \quad e'^T_{ij} = Be'^\infty_{ij}, \quad (20.104)$$

where

$$A = \frac{\kappa^* - \kappa}{(\kappa - \kappa^*)\alpha - \kappa},$$

$$B = \frac{\mu^* - \mu}{(\mu - \mu^*)\beta - \mu}. \quad (20.105)$$

Recall that $\kappa = \lambda + \frac{2}{3}\mu$ is a *bulk modulus* governing the relation between pressure and dilatation, i.e., $\sigma = 3\kappa e$.

Of particular interest is the field that exists just outside the inclusion. The hydrostatic and deviatoric parts of the stress are readily found to be

$$\sigma = \sigma^\infty - \frac{1+\nu}{1-\nu}B\sigma'^\infty_{ij}n_in_j, \quad (20.106)$$

and

$$\sigma'_{i\ell} = (1 + \beta B)\sigma'^{\infty}_{i\ell} - B(\sigma'^{\infty}_{ik} n_k n_\ell + \sigma'^{\infty}_{\ell k} n_k n_i) + \frac{B}{1-\nu} \sigma'^{\infty}_{jk} n_j n_k n_i n_\ell$$
$$+ \frac{1-2\nu}{3(1-\nu)} B\sigma'^{\infty}_{jk} n_j n_k \delta_{i\ell} - \frac{1-2\nu}{3(1-\nu)} A\sigma^{\infty}(n_i n_\ell - \frac{1}{3}\delta_{i\ell}). \quad (20.107)$$

A particularly interesting case of an inhomogeneous inclusion in an applied remote field is that of a spherical cavity, where $\kappa^* = 0$ and $\mu^* = 0$. In that case (20.106) and (20.107) yield

$$\sigma_{i\ell} = \frac{15}{7-5\nu} \Big[(1-\nu)(\sigma^{\infty}_{i\ell} - \sigma^{\infty}_{ik} n_k n_\ell - \sigma^{\infty}_{\ell k} n_k n_i) + \sigma^{\infty}_{jk} n_j n_k n_i n_\ell$$
$$- \nu \sigma^{\infty}_{jk} n_j n_k \delta_{i\ell} + \frac{1-5\nu}{10} \sigma^{\infty}(n_i n_\ell - \delta_{i\ell})\Big]. \quad (20.108)$$

An elementary check on Eshelby's (1957) solution is to consider the case where the spherical cavity is subject to far field hydrostatic tension of magnitude σ^{∞}. Consider the hole's surface, and note that the circumferential (hoop) normal stress, tangential to the hole is indeed

$$\sigma_{tt} = \frac{3}{2}\sigma^{\infty}, \quad (20.109)$$

where the subscript tt simply indicates a tangential direction to the hole's surface. The hydrostatic tension is, of course, elevated in magnitude by the same factor of 3/2.

20.9 Suggested Reading

Asaro, R. J., and Barnett, D. M. (1974), The Non-Uniform Transformation Strain Problem for an Anisotropic Ellipsoidal Inclusion, *J. Mech. Phys. Solids*, Vol. 23, pp. 77–83.

Eshelby, J. D. (1957), The Determination of the Elastic Field of an Ellipsoidal Inclusion, and Related Problems, *Proc. Roy. Soc. Lond. A*, Vol. 241, pp. 376–396.

Eshelby, J. D. (1961), Elastic Inclusions and Inhomogeneities. In *Progress in Solid Mechanics* (I. N. Sneddon and R. Hill, eds.), Vol. 2, pp. 87–140.

Mura, T. (1987), *Micromechanics of Defects in Solids*, Martinus Nijhoff, Dordrecht, The Netherlands.

Nemat-Nasser, S., and Hori, M. (1999), *Micromechanics: Overall Properties of Heterogeneous Materials*, 2nd ed., Elsevier, Amsterdam.

Willis, J. R. (1964), Anisotropic Inclusion Problems, *Q. J. Mech. Appl. Math.*, Vol. 17, pp. 157–174.

21 Forces and Energy in Elastic Systems

21.1 Free Energy and Mechanical Potential Energy

Elasticity is concerned with the state of deformation, stress, and resulting energy in bodies subject to applied loads, displacements, or, as illustrated by the chapters on dislocations, incompatible or discontinuous displacements. For example, two such sources of stress in a body are indicated in Fig. 21.4 as S and A. Because of the presence of A and S the internal energy and the Hemholtz free energy, *i.e.*, the isothermal strain energy, are altered; in addition, the potential energy of external loading mechanisms responsible for traction, $\mathbf{T}$, on the external surfaces, would have been altered as well. In considerations of equilibrium it is not enough to simply consider energy or free energy changes of the body itself, because changes in, say, defect configuration or position cause changes in the potential energy of the external loading mechanisms. The thermodynamic potential that is appropriate to use to seek equilibrium states, or to construct forces from, therefore, must take both the energetic changes of the body and the loading mechanism into account. To begin, then, consider the so-called *total energy*,

$$\mathcal{E}_{\text{tot}} = \mathcal{E}_{\text{el}} + \mathcal{E}_{\text{ext}}, \tag{21.1}$$

where $\mathcal{E}_{\text{el}}$ is the isothermal elastic energy, or Hemholtz free energy, and $\mathcal{E}_{\text{ext}}$ is the potential energy of the loading mechanism. We now write a Gibbs equation for the change in this total energy of a body which is assumed to be at constant temperature and subject to fixed loads (*i.e.*, fixed traction). The body is thus assumed to be surrounded by a medium that serves as a *heat bath* maintaining constant temperature, and a loading mechanism exerting constant boundary traction. The body and the surrounding medium are in thermal contact and can exchange heat; together they are thermally isolated from the remainder of the universe. We also consider an external source of work, $\mathcal{W}_{\text{ext}}$, that may do work on the body, but cannot exchange heat with either the body or medium (Fig. 21.1).

The total change in internal energy of the body following an incremental process is

$$\delta U = \delta \mathcal{W}_{\text{ext}} + \int_{S_0} T_i \delta u_i \, dS - T^{(0)} \delta S^{(0)}, \tag{21.2}$$

where we have explicitly noted the dual character of the medium. First, it serves as a heat bath and thus the heat given to the body is the heat it loses in a pure thermal transfer of

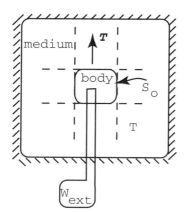

Figure 21.1. Adiabatically isolated body and medium. The traction vector is **T** and the temperature is T.

heat; this means that the heat lost to the body is just $-T^{(0)}\delta S^{(0)}$. The absolute temperature is T, and S the entropy. The superscript 0 pertains to the medium, unscripted variables in context pertain to the body, and we recall that $T = T^{(0)}$. The integral accounts for the other role of the medium as a loading mechanism capable of performing mechanical work on the body.

As the body plus medium together comprise a thermally isolated system, the second law of thermodynamics states that

$$\delta S + \delta S^{(0)} \geq 0, \tag{21.3}$$

because $\delta S_{\text{ext}} = 0$. Thus,

$$-T^{(0)}\delta S^{(0)} \leq T^{(0)}\delta S, \tag{21.4}$$

and

$$\delta U \leq \delta \mathcal{W}_{\text{ext}} + \int_{S_0} T_i \delta u_i \, dS + T\delta S, \tag{21.5}$$

or, since $S = S_0$,

$$\delta \mathcal{W}_{\text{ext}} \geq \delta U - T\delta S - \int_S T_i \delta u_i \, dS. \tag{21.6}$$

The work that must be performed externally by the external "engine" to produce this change in state is $\delta \mathcal{W}_{\text{ext}}$. When the change occurs reversibly, the equality holds and

$$\delta \mathcal{W}_{\text{ext}} \underset{\text{rev}}{=} \delta \Phi - \int_S T_i \delta u_i \, dS, \quad \delta \Phi = \delta U - T\delta S. \tag{21.7}$$

The Helmholtz free energy $\delta \Phi$ contains the "locked in" elastic strain energy within the body, whereas the surface integral describes the work effects associated with the external loading mechanism. It is common to define the potential energy of the body as

$$\Pi = \int_V W \, dV - \int_S \mathbf{T} \cdot \mathbf{u} \, dS, \tag{21.8}$$

where W is the strain energy density, and we seek the configuration, under a given set of boundary conditions, that minimizes it. This will be a solution technique we will use in later

21.2. Forces of Translation

sections. Thus, at equilibrium, Π is a minimum. Generalized forces acting on the system can be thereby defined as

$$f_\alpha = -\frac{\partial \Pi}{\partial \xi_\alpha} = -\frac{\partial \mathcal{E}_{\text{tot}}}{\partial \xi_\alpha}, \qquad (21.9)$$

where α represents a generalized displacement such as the movement of a defect like dislocation or the deflection of a plate under prescribed loading.

21.2 Forces of Translation

The objective here is to calculate the force acting on a defect, for example, a crack tip, from the formalism described above. Thus, we want the force inducing translation, which is

$$f_\ell = -\frac{\partial \mathcal{E}_{\text{tot}}}{\partial x_\ell}. \qquad (21.10)$$

We calculate the energy change in two parts. First consider the elastic energy change. As the defect translates a small distance ϵ_ℓ in the x_ℓ direction, it "drags along with it" its elastic field. After the translation, the field quantities are given as

$$\begin{aligned} u_i^\epsilon(x_j) &= u_i(x_j - \epsilon_\ell \delta_{j\ell}), \\ \sigma_{ij}^\epsilon(x_k) &= \sigma_{ij}(x_k - \epsilon_\ell \delta_{k\ell}), \end{aligned} \qquad (21.11)$$

where the superscript ϵ denotes the quantities after the translation. The strain energy density is, accordingly,

$$W^\epsilon = \frac{1}{2}\sigma_{ij}^\epsilon(x_k)e_{ij}^\epsilon(x_k) = \frac{1}{2}\sigma_{ij}(x_k - \epsilon_\ell \delta_{k\ell})e_{ij}(x_k - \epsilon_\ell \delta_{k\ell}). \qquad (21.12)$$

We can expand W^ϵ, excluding regions of singularity, as

$$\begin{aligned} W^\epsilon &= \frac{1}{2}\{[\sigma_{ij}(x_k) - \epsilon_\ell \sigma_{ij,\ell}(x_k)][e_{ij}(x_k) - \epsilon_\ell e_{ij,\ell}(x_k)]\} + \mathcal{O}(\epsilon^2) \\ &= \frac{1}{2}\sigma_{ij}e_{ij} - \frac{1}{2}\epsilon_\ell(\sigma_{ij}e_{ij})_{,\ell} + \mathcal{O}(\epsilon^2) \\ &= W - \epsilon_\ell W_{,\ell} + \mathcal{O}(\epsilon^2). \end{aligned} \qquad (21.13)$$

Therefore, the energy change in this first step is

$$\delta \mathcal{E}_{\text{el}}^{(1)} = \int_V (W^\epsilon - W)\, dV = -\epsilon_\ell \int_V W_{,\ell}\, dV + \mathcal{O}(\epsilon^2). \qquad (21.14)$$

The divergence theorem may be used to obtain, to first order in ϵ,

$$\delta \mathcal{E}_{\text{el}}^{(1)} = -\epsilon_\ell \int_S W n_\ell\, dS. \qquad (21.15)$$

Now, by simply translating the elastic field, a state of stress within the body will be created that does not satisfy the boundary conditions. That is, on a boundary with unit normal $\mathbf{n}$, we have

$$\sigma_{ij}^\epsilon n_j = (\sigma_{ij} - \epsilon_\ell \sigma_{ij,\ell})n_j + \mathcal{O}(\epsilon^2). \qquad (21.16)$$

Step 2: then involves applying a traction $\epsilon_\ell \sigma_{ij,\ell} n_j + \mathcal{O}(\epsilon^2) = \mathcal{O}(\epsilon)$ to annul the error. At the end of step 1 the surface displacements are $u_i - \epsilon_\ell u_{i,\ell} + \mathcal{O}(\epsilon^2)$, and after step 2 they attain the values $u_i(\text{final})$, which of course is unknown. Therefore, the change in elastic energy during step 2 is

$$\delta \mathcal{E}_{\text{el}}^{(2)} = \int_S \sigma_{ij}[u_i(\text{final}) - u_i + \epsilon_\ell u_{i,\ell}] n_j \, dS + \mathcal{O}(\epsilon^2). \qquad (21.17)$$

The difference $[u_i(\text{final}) - u_i]$ is of $\mathcal{O}(\epsilon)$, because the set of tractions imposed in step 2 is of $\mathcal{O}(\epsilon)$.

We now consider the change in energy associated with the loading mechanism. To first order in ϵ, this is

$$\delta \mathcal{E}_{\text{ext}} = -\int_S \sigma_{ij}[u_i(\text{final}) - u_i] n_j \, dS + \mathcal{O}(\epsilon^2), \qquad (21.18)$$

and thus

$$\begin{aligned}\delta \mathcal{E}_{\text{tot}} &= \delta \mathcal{E}_{\text{el}}^{(1)} + \delta \mathcal{E}_{\text{el}}^{(2)} + \delta \mathcal{E}_{\text{ext}} \\ &= -\epsilon_\ell \int_S (W n_\ell - \sigma_{ij} u_{i,\ell} n_j) \, dS,\end{aligned} \qquad (21.19)$$

where conveniently the unknown quantity, $[u_i(\text{final}) - u_i]$, drops out!

Therefore, appealing to the definition of force in (21.10), we find that

$$f_\ell = \int_S (W n_\ell - \sigma_{ij} u_{i,\ell} \, n_j) \, dS. \qquad (21.20)$$

This may be rephrased as

$$f_\ell = \int_S (W \delta_{j\ell} - \sigma_{ij} u_{i,\ell}) n_j \, dS. \qquad (21.21)$$

The generalized force developed here is often called the *J integral* (Rice, 1968a, 1968b). The tensor

$$\mathcal{P}_{j\ell} = W \delta_{j\ell} - \sigma_{ij} u_{i,\ell} \qquad (21.22)$$

is called the *energy momentum tensor* (Eshelby, 1956). Thus,

$$f_\ell = \int_S \mathcal{P}_{j\ell} n_j \, dS, \quad \mathbf{f} = \int_S \mathbf{n} \cdot \mathcal{P} \, dS. \qquad (21.23)$$

It is a rather straightforward exercise to show that the divergence of $\mathcal{P}$ is zero, that is

$$(W \delta_{j\ell} - \sigma_{ij} u_{i,\ell})_{,j} = 0. \qquad (21.24)$$

Thus, if S and S' are two surfaces that surround the defect s, enclosing a volume V' (Fig. 21.2), it is found that

$$\int_{V'} \operatorname{div} \mathcal{P} \, dV = 0 = \int_S \mathbf{n} \cdot \mathcal{P} \, dS + \int_{S'} \mathbf{n}' \cdot \mathcal{P} \, dS'. \qquad (21.25)$$

Therefore, since $\mathbf{n}' = -\mathbf{n}$,

$$\int_S \mathbf{n} \cdot \mathcal{P} \, dS = \int_{S'} \mathbf{n} \cdot \mathcal{P} \, dS'. \qquad (21.26)$$

Thus, the value of the generalized force on a defect does not depend on the choice of surface S surrounding the defect used to evaluate the integral in (21.21).

21.2. Forces of Translation

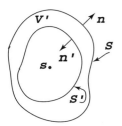

Figure 21.2. Path independence of the energy momentum tensor, or J integral. A defect labeled s is indicated as surrounded by S and S'.

21.2.1 Force on an Interface

Eshelby has extended the above development of a translational force on a defect to the consideration of a force acting on an interface. Consider the interface between two phases α and β. The surface element is to move by an incremental vector, $\delta\boldsymbol{\xi}$. For example, this may represent the result of a phase transformation $\beta \to \alpha$, where the transformation involves a change in size and shape of the material. Such problems are dealt with in detail in subsequent chapters, but for now we want to consider the possibility that due to such discontinuities in displacement, induced by a transformation, there may be net mechanical forces on the interface between the two phases. The described total energy change can be computed using the two heuristic steps, as described earlier. First, cut out the material lying between the old surface element, S, and the new element, S', and discard the material. We imagine that suitable tractions are applied to the freshly cut surface to prevent relaxation. This results in a change in elastic energy,

$$\delta\mathcal{E}^{(1)\alpha}_{\mathrm{el}} = -\int_S \delta\xi_\ell W^\alpha n_\ell \,\mathrm{d}S. \tag{21.27}$$

The direction of the unit normal **n** is taken as pointing from $\beta \to \alpha$. At this stage the displacement on S' is $u_i^\alpha + \delta\xi_\ell u_{i,\ell}^\alpha$. Now, alter the displacements to their final value, $u_i^{f\alpha}$. The work required to do this is

$$\delta\mathcal{E}^{(2)\alpha} = -\int_S \sigma_{ij}^\alpha (u_i^{f\alpha} - u_i^\alpha - \delta\xi_\ell u_{i,\ell}^\alpha) n_j \,\mathrm{d}S. \tag{21.28}$$

For phase β, similar arguments lead to

$$\delta\mathcal{E}^{(1)\beta}_{\mathrm{el}} = \int_S \delta\xi_\ell W^\beta n_\ell \,\mathrm{d}S, \tag{21.29}$$

and

$$\delta\mathcal{E}^{(2)\beta} = -\int_S \sigma_{ij}^\beta (u_i^{f\beta} - u_i^\beta - \delta\xi_\ell u_{i,\ell}^\beta) n_j \,\mathrm{d}S. \tag{21.30}$$

But during this process of interface motion, $u_i^{f\alpha} - u_i^\alpha = u_i^{f\beta} - u_i^\beta$, so that

$$\delta\mathcal{E}_{\mathrm{tot}} = -\int_S n_j (\mathcal{P}_{j\ell}^\alpha - \mathcal{P}_{j\ell}^\beta) \delta\xi_\ell \,\mathrm{d}S. \tag{21.31}$$

Thus, the force on the interface becomes

$$f_\ell^{\alpha\beta} = \int_S n_j [[\mathcal{P}_{j\ell}]] \,\mathrm{d}S. \tag{21.32}$$

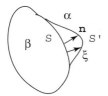

Figure 21.3. Moving interface.

If we form $f_\ell^{\alpha\beta} n_\ell$, we can define

$$\mathcal{F} = [[W]] - \mathbf{T} \cdot [[\partial \mathbf{u}/\partial n]], \tag{21.33}$$

as the normal force per unit area acting on the interface.

This result should be extended in two obvious ways. In the first place, the energy considered so far has only included the elastic energy of deformation. If, however, the interface is one between two phases undergoing the transformation $\alpha \to \beta$, then the chemical part of the free energy change needs to be included. This would add a contribution to the free energy change

$$\int_S \delta\xi_\ell (G^\beta - G^\alpha) n_\ell \, dS = -\int_S \delta\xi_\ell [[G]] n_\ell \, dS. \tag{21.34}$$

The second contribution to the total energy change is that due to the surface energy, γ, per unit area. This contributes a term

$$\int_S \delta\xi_\ell \, \kappa n_\ell \gamma n_\ell \, dS, \tag{21.35}$$

where κ is the *double mean curvature* of the surface. Note that this curvature is measured positive for a surface as seen from the α side in Fig. 21.3, *i.e.*, where the radii of curvature are constructed on the α side.

Including these two effects yields for the total normal force on the interface

$$\mathcal{F} = [[\mathcal{W}]] - \mathbf{T} \cdot [[\partial \mathbf{u}/\partial n]] - \kappa\gamma, \tag{21.36}$$

where $\mathcal{W} = W + G$. With the sense of the curvature as taken here, the surface energy term induces a retarding effect on interface motion if $\kappa > 0$, *i.e.*, if additional surface area is created by forward motion.

21.2.2 Finite Deformation Energy Momentum Tensor

The energy momentum tensor also exists for finite elastic deformations, and has similar form to that derived for the infinitesimal strain case. The development was originally also due to Eshelby (1970, 1975). Recall that the position of a material particle in the deformed configuration, $\mathbf{x}$, is given in terms of its position in the reference configuration by

$$\mathbf{x}(\mathbf{X}) = \mathbf{X} + \mathbf{u}(\mathbf{X}). \tag{21.37}$$

In component form this gives for the displacement

$$u_i(X_m) = x_i(X_m) - X_i. \tag{21.38}$$

21.2. Forces of Translation

The equilibrium equations with respect to the nominal stress, **P**, and without body forces, are

$$\frac{\partial P_{ij}}{\partial X_j} = 0. \tag{21.39}$$

Denoting the strain energy density per unit reference volume by W, we have $W = W(\text{Grad } \mathbf{u}, \mathbf{X})$, and

$$P_{ij} = \frac{\partial W}{\partial u_{i,j}}. \tag{21.40}$$

Recall that **P** is not symmetric, but satisfies

$$P_{\ell i} - P_{j\ell} = P_{ji}u_{\ell,i} - P_{\ell i}u_{j,i}. \tag{21.41}$$

The energy momentum tensor is derived by considering the set of field equations that result from causing the ratio of the current volume and reference volume to be stationary (extreme). This gives

$$\frac{\partial}{\partial X_j} \frac{\partial W}{\partial u_{i,j}} = \frac{\partial W}{\partial u_i}, \tag{21.42}$$

with $W = W(u_i, u_{i,j}, X_i)$. Consider the gradient of W with components $\partial W/\partial X_\ell$, and then the partial derivative of W holding the displacements and displacement gradients fixed, i.e.,

$$\left(\frac{\partial W}{\partial X_\ell}\right)_{\text{exp}} = \left(\frac{\partial W}{\partial X_\ell}\right)_{u_i, u_{i,j}, X_{m \neq \ell}}. \tag{21.43}$$

If the material were homogeneous and free of singularities or dislocations, then

$$\left(\frac{\partial W}{\partial X_\ell}\right)_{\text{exp}} = 0. \tag{21.44}$$

But now we have

$$\frac{\partial W}{\partial X_\ell} = \frac{\partial W}{\partial u_i} u_{i,\ell} + \frac{\partial W}{\partial u_{i,j}} u_{i,j\ell} + \left(\frac{\partial W}{\partial X_\ell}\right)_{\text{exp}}. \tag{21.45}$$

When (21.42) is used to replace $\partial W/\partial u_i$ with $\partial(\partial W/\partial u_{i,j})/\partial X_j$, the first two terms on the *rhs* of (21.45) become

$$\frac{\partial}{\partial X_j}\left(\frac{\partial W}{\partial u_{i,j}} u_{i,\ell}\right).$$

Thus, since formally $\partial W/\partial X_i = \partial(W\delta_{ij})/\partial X_j$, we can write

$$\frac{\partial \mathcal{P}_{j\ell}}{\partial X_j} = \left(\frac{\partial W}{\partial X_\ell}\right)_{\text{exp}}, \tag{21.46}$$

where

$$\mathcal{P}_{j\ell} = W\delta_{j\ell} - \frac{\partial W}{\partial u_{i,j}} u_{i,\ell} \tag{21.47}$$

is the energy momentum tensor. Also, because of (21.40), we have

$$\mathcal{P}_{j\ell} = W\delta_{j\ell} - P_{ij}u_{i,\ell}. \tag{21.48}$$

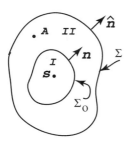

Figure 21.4. Two sources of stress, labeled A and S, in an elastic medium.

The integral defined by

$$F_\ell = \int_S \mathcal{P}_{j\ell} n_j \, dS, \tag{21.49}$$

is equal to zero if S encloses a volume in which $\frac{\partial W}{\partial u_{i,j}} u_{i,\ell} = 0$, i.e., one that is homogeneous and free of defects. The argument shows that, in the case that S does enclose a defect, i.e., a singularity or displacement discontinuity, the integral given in (21.49) represents the force of translation acting on the defect. The proof follows the same route as detailed in the previous section.

21.3 Interaction Between Defects and Loading Mechanisms

We return to the general situation of two sources of stress that are called A and S in Fig. 21.4. As noted earlier, for internal sources of stress such as dislocations or inclusions, the elastic fields are caused by discontinuous displacement fields, but even in those cases the stresses and strains are continuous. Since linear superposition applies, the total field is the sum of the fields due to all sources, i.e.,

$$\sigma_{ij} = \sigma_{ij}^A + \sigma_{ij}^S, \quad e_{ij} = e_{ij}^A + e_{ij}^S. \tag{21.50}$$

Equilibrium applies to each field separately, so that

$$\sigma_{ij,j}^A = \sigma_{ij,j}^S = 0. \tag{21.51}$$

The total elastic energy is then

$$\begin{aligned}
\mathcal{E}_{el} &= \frac{1}{2} \int_V (\sigma_{ij}^A + \sigma_{ij}^S)(e_{ij}^A + e_{ij}^S) \, dV \\
&= \frac{1}{2} \int_V \sigma_{ij}^A e_{ij}^A \, dV + \frac{1}{2} \int_V \sigma_{ij}^S e_{ij}^S \, dV + \frac{1}{2} \int_V (\sigma_{ij}^A e_{ij}^S + \sigma_{ij}^S e_{ij}^A) \, dV \\
&= \mathcal{E}_s^A + \mathcal{E}_s^S + \mathcal{E}_{int}^{A-S}.
\end{aligned} \tag{21.52}$$

Here, $\mathcal{E}_s^A$ and $\mathcal{E}_s^S$ are "self-energies" of the A and S systems, as if they were isolated, and $\mathcal{E}_{int}^{A-S}$ is the *interaction energy* of the two sources. Clearly, all forces that these two systems are to exert on each other derive from $\mathcal{E}_{int}^{A-S}$, which we examine now. The interaction energy is

$$\mathcal{E}_{int}^{A-S} = \frac{1}{2} \int_V (\sigma_{ij}^A e_{ij}^S + \sigma_{ij}^S e_{ij}^A) \, dV. \tag{21.53}$$

21.3. Interaction Between Defects and Loading Mechanisms

In general, we expect that $\mathbf{u}^A$ and/or $\mathbf{u}^S$ may be discontinuous. Since $\sigma_{ij} = C_{ijkl}e_{kl}$, we can write

$$C_{ijkl}e_{kl}^A e_{ij}^S = \sigma_{ij}^A e_{ij}^S = \sigma_{kl}^S e_{kl}^A, \tag{21.54}$$

because of the symmetry $C_{ijkl} = C_{klij}$. Thus, the two terms in the integrand of (21.53) are equal. Now construct a surface, Σ_0, surrounding S such that outside this surface $\mathbf{u}^S$ is continuous. Then, we may rewrite (21.53) as

$$\mathcal{E}_{\text{int}}^{A-S} = \int_I \sigma_{ij}^S e_{ij}^A \, dV + \int_{II} \sigma_{ij}^A e_{ij}^S \, dV, \tag{21.55}$$

where, as seen in Fig. 21.4, I and II are the two volume elements so formed by the separator Σ_0. Recognizing that the integrands of (21.55) can be put into the form $\sigma_{ij}u_{i,j} = (\sigma_{ij}u_i)_{,j}$, because of the equilibrium stated in (21.51), the application of the divergence theorem leads to

$$\int_I \sigma_{ij}^S e_{ij}^A \, dV = \int_{\Sigma_0} \sigma_{ij}^S n_j u_i^A \, dS,$$

$$\int_{II} \sigma_{ij}^A e_{ij}^S \, dV = \int_{\Sigma} \sigma_{ij}^A \hat{n}_j u_i^S \, dS - \int_{\Sigma_0} \sigma_{ij}^A n_j u_i^S \, dS. \tag{21.56}$$

The outward pointing normal to the surface Σ is $\hat{n}$, and the minus sign of the last integral appears because the outward pointing normal on Σ_0 from volume I has been used in the integrand. The nature of Σ will be defined by application to specific examples below. For now, we find that

$$\mathcal{E}_{\text{int}}^{A-S} = \int_{\Sigma_0} (\sigma_{ij}^S n_j u_i^A - \sigma_{ij}^A n_j u_i^S) \, dS + \int_{\Sigma} \sigma_{ij}^A \hat{n}_j u_i^S \, dS. \tag{21.57}$$

Suppose that A is an internal source of stress, so that on Σ, which is now taken as the external surface of the body containing A and S, $\sigma_{ij}^A \hat{n}_j = 0$. Then, not only is the last integral in (21.57) zero, but the integral that would be associated with the potential energy of the external loading mechanism, as well. Thus, we obtain that

$$\mathcal{E}_{\text{int}}^{A-S} = \mathcal{E}_{\text{tot}} = \int_{\Sigma_0} (\sigma_{ij}^S u_i^A - \sigma_{ij}^A u_i^S) n_j \, dS \tag{21.58}$$

is the total energy of interaction between A and S.

Now let A be an external source of stress, so that $\sigma_{ij}^A \hat{n}_j \neq 0$. Then, to $\mathcal{E}_{\text{int}}^{A-S}$ we add the potential energy associated with the loading mechanism, i.e.,

$$-\int_{\Sigma} T_i^A u_i^S \, dS = -\int_{\Sigma} \sigma_{ij}^A \hat{n}_j u_i^S \, dS, \tag{21.59}$$

and arrive again at the conclusion that the total interaction energy is given by

$$\mathcal{E}_{\text{int}}^{A-S} = \mathcal{E}_{\text{tot}} = \int_{\Sigma_0} (\sigma_{ij}^S u_i^A - \sigma_{ij}^A u_i^S) n_j \, dS. \tag{21.60}$$

Furthermore, in this case we can let $\Sigma_0 \to \Sigma$ and, since S is an internal source of stress so that $\sigma_{ij}^S \hat{n}_j = 0$, we have

$$\mathcal{E}_{\text{int}}^{A-S} = \mathcal{E}_{\text{tot}} = -\int_{\Sigma} \sigma_{ij}^A \hat{n}_j u_i^S \, dS. \tag{21.61}$$

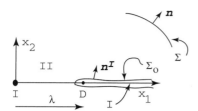

Figure 21.5. Dislocation and inclusion.

21.3.1 Interaction Between Dislocations and Inclusions

Consider the interaction of an infinitely long and straight dislocation line and an inclusion. The stress field σ_{ij}^I of an inclusion vanishes with distance from it as $1/r^3$. Let superscripts I and D denote field quantities associated with the inclusion and dislocation, respectively. The interaction energy is then

$$\mathcal{E}_{\text{int}}^{\text{I-D}} = \frac{1}{2} \int_V (\sigma_{ij}^I u_{i,j}^D + \sigma_{ij}^D u_{i,j}^I) \, dV. \tag{21.62}$$

The reciprocity $\sigma_{ij}^I u_{i,j}^D = \sigma_{ij}^D u_{i,j}^I$ within the above integrand is observed, as before. The total volume is divided *via* the construction of a surface Σ_0 made so that in the two volume elements so created one or the other of $\mathbf{u}^I$ or $\mathbf{u}^D$ is continuous (Fig. 21.5). Then,

$$\begin{aligned}
\mathcal{E}_{\text{int}}^{\text{I-D}} &= \int_I \sigma_{ij}^D u_{i,j}^I \, dV + \int_{II} \sigma_{ij}^I u_{i,j}^D \, dV \\
&= \int_I (\sigma_{ij}^D u_i^I)_{,j} \, dV + \int_{II} (\sigma_{ij}^I u_i^D)_{,j} \, dV \\
&= \int_{\Sigma_0} \sigma_{ij}^D n_j u_i^I \, dS - \int_{\Sigma_0} \sigma_{ij}^I n_j u_i^D \, dS + \int_{\Sigma} \sigma_{ij}^I u_i^D n_j \, dS.
\end{aligned} \tag{21.63}$$

We note that the outer surface of region II may be taken as the free surface of the body (at least as far as the inclusion's stress field is concerned). Moreover, we may first choose the half-plane defined by $x_2 = 0$ and $x_1 \geq 0$ as the cut surface of the dislocation, S_{cut}, and let $\Sigma_0 \to S_{\text{cut}}$. The displacement field of the inclusion is continuous across Σ_0 and so is the stress field of the dislocation. We note that on Σ_0, $\sigma_{ij}^D n_j^+ = -\sigma_{ij}^D n_j^-$, where the normals are defined in the figure; thus along those parts of Σ_0 called Σ_0^+ and Σ_0^- the combined integrals cancel. Now, around the loop of vanishing radius ϵ, the displacement of the inclusion is uniquely defined, and

$$\oint_\epsilon \sigma_{ij}^D n_j \, dS = 0, \tag{21.64}$$

because this would be, in fact, just the total content of body force at that point, which is zero for the dislocation. Thus, the total contribution to the interaction energy is

$$\begin{aligned}
\mathcal{E}_{\text{int}}^{\text{I-D}} &= -\int_{\Sigma_0} \sigma_{ij}^I u_i^D n_j \, dS = -\int_{\Sigma_0} \sigma_{ij}^I n_j^+ \left[u_i^{D(+)} - u_i^{D(-)} \right] dS \\
&= \int_{\Sigma_0} \sigma_{ij}^I n_j^+ b_i \, dS.
\end{aligned} \tag{21.65}$$

21.3. Interaction Between Defects and Loading Mechanisms

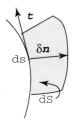

Figure 21.6. Dislocation segment movement.

To calculate a force from this interaction energy we first consider this integral per unit length normal to the plane of the figure, and write

$$\mathcal{E}_{\text{int}}^{I-D} = -\int_\lambda^\infty \sigma_{ij}^I n_j^+ \left[u_i^{D(+)} - u_i^{D(-)} \right] dS = \int_\lambda^\infty \sigma_{ij}^I n_j^+ b_i \, dS. \quad (21.66)$$

It follows that

$$f_\lambda = -\frac{\partial \mathcal{E}_{\text{int}}^{I-D}}{\partial \lambda} = \sigma_{ij}^I(\lambda) n_j^+ b_i. \quad (21.67)$$

The normal $\mathbf{n}^+$ is $\mathbf{n}^I$ oriented upward. This result can be generalized in terms of the force acting on an arbitrary segment of curved dislocation, which is discussed next.

21.3.2 Force on a Dislocation Segment

The above result can be used to readily generate an expression for the force on an arbitrary segment of dislocation. Consider a segment of dislocation depicted in Fig. 21.6. The dislocation is acted on by a stress field labeled A. Let the arc length of this segment be initially ds and let it move by the normal vector distance $\delta \mathbf{n}$, thus traversing the planar element dS. In terms of the force acting on this segment, $\hat{\mathbf{f}}$, we have

$$-\delta \mathcal{E}_{\text{tot}} = \hat{\mathbf{f}} \cdot \delta \mathbf{n} = -\sigma_{ij}^A b_i \delta S_j, \quad (21.68)$$

where $dS_j = \eta_j dS$, and η would be the unit normal to the area element traversed. But,

$$\delta S_j = (\delta \mathbf{n} \times d\mathbf{s})_j + \mathcal{O}(\kappa \delta n^2), \quad d\mathbf{s} = \mathbf{t} ds, \quad (21.69)$$

where κ is the curvature at the point in question on the segment, and $\mathbf{t}$ is the unit tangent to the loop at the point of the segment. Thus,

$$dS_j = \epsilon_{jkl} ds_l \delta n_k = \epsilon_{jkl} t_l \delta n_k ds. \quad (21.70)$$

Consequently,

$$f_k \delta n_k = -\sigma_{ij}^A b_i \epsilon_{jkl} t_l \delta n_k, \quad (21.71)$$

where $f_k = \hat{f}_k / ds$. Therefore, the force *per unit length* of dislocation segment is

$$f_k = \epsilon_{jlk} \sigma_{ij}^A b_i t_l. \quad (21.72)$$

This is called the Peach–Koehler force.

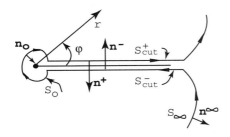

Figure 21.7. Dislocation cut surface and integration path.

21.4 Elastic Energy of a Dislocation

Consider an infinitely long and straight dislocation indicated in Fig. 21.7. Per unit length, the elastic energy can be written as

$$\mathcal{E} = \frac{1}{2} \int_V \sigma_{ij} e_{ij} \, dV$$

$$= \frac{1}{2} \int_V \sigma_{ij} u_{i,j} \, dV = \frac{1}{2} \int_V (\sigma_{ij} u_i)_{,j} \, dV \qquad (21.73)$$

$$= \frac{1}{2} \int_S \sigma_{ij} u_i n_j \, dS,$$

where S is the surface that bounds the total volume V and is constructed so that the dislocation's displacement field is single valued within V. This is illustrated in the figure, where the outward pointing normal, $\mathbf{n}$, to S is also shown. Now, the integrals are divided as

$$\mathcal{E} = \frac{1}{2} \int_{S^{\pm}_{\text{cut}}} \sigma_{ij} u_i n_j \, dS + \frac{1}{2} \int_{S_\infty} \sigma_{ij} u_i n_j \, dS + \frac{1}{2} \int_{S_0} \sigma_{ij} u_i n_j \, dS. \qquad (21.74)$$

Considering the angle φ, we note that $\sigma_{ij} = \Sigma_{ij}(\varphi)/r$, $n_j = n_j(\varphi)$ and thus $T_i = \sigma_{ij} n_j = h_i(\varphi)/r$. As there are no body forces, we have

$$\int_0^{2\pi} h_i(\varphi) \, d\varphi = 0. \qquad (21.75)$$

However,

$$\int_0^{2\pi} u_i(\varphi) \, d\varphi = b_i. \qquad (21.76)$$

The second two integrals cancel, because $dS = r\,d\varphi$, and $h_i^\infty = -h_i^0$ (the directions of the outward pointing normals are opposite). The integrals left on S_{cut} become

$$\mathcal{E} = \frac{1}{2} \int_{S^{\pm}_{\text{cut}}} \sigma_{ij} n_j^{\pm} u_i^{\pm} \, dS = \frac{1}{2} \int_S \sigma_{ij} n_j^{-} (u_i^{-} - u_i^{+}) \, dS$$

$$= \frac{1}{2} \int_S \sigma_{ij} n_j^{-} b_i \, dS, \qquad (21.77)$$

where we have used $\mathbf{n}^{-}$ to describe the normal and $\mathbf{b} = \mathbf{u}^{-} - \mathbf{u}^{+}$. Since again $\sigma_{ij} = \Sigma_{ij}(\varphi)/r$, this can be rewritten as

$$\mathcal{E} = \frac{1}{2} \int_0^\infty \frac{\Sigma_{i2}(0) b_i}{x_1} \, dx_1. \qquad (21.78)$$

21.5. In-Plane Stresses

Note from the geometry that $\mathbf{n} \parallel x_2$. The integral is logarithmicaly singular and we introduce the so-called *cutoff radii*, r_0 and R, to write

$$\mathcal{E} = \frac{1}{2} \int_{r_0}^{R} \frac{\Sigma_{i2}(0)b_i}{x_1} dx_1 = \frac{1}{2} \Sigma_{i2}(0) b_i \ln(R/r_0). \tag{21.79}$$

Recalling that $\Sigma_{i2} \propto K_{ij} b_j$, (21.79) may be rewritten as

$$\mathcal{E} = K_{ij} b_i b_j \ln(R/r_0). \tag{21.80}$$

For the infinitely long and straight screw dislocation in an isotropic medium, our previous results show that

$$\mathcal{E}_{\text{screw}} = \frac{Gb_s^2}{4\pi} \ln(R/r_0), \tag{21.81}$$

whereas for the edge dislocation

$$\mathcal{E}_{\text{edge}} = \frac{Gb_e^2}{4\pi(1-\nu)} \ln(R/r_0). \tag{21.82}$$

For an infinitely long and straight dislocation with $\mathbf{t} \parallel x_3$ and $\mathbf{b} = b_e \mathbf{e}_1 + b_e \mathbf{e}_2 + b_s \mathbf{e}_3$, we find

$$\mathbf{K} = \frac{G}{4\pi} \begin{pmatrix} b_e/(1-\nu) & 0 & 0 \\ 0 & b_e/(1-\nu) & 0 \\ 0 & 0 & b_s \end{pmatrix} \tag{21.83}$$

21.5 In-Plane Stresses of Straight Dislocation Lines

The energy factor matrix introduced in the previous section, $\mathbf{K}$, has an interesting and useful connection to the traction caused by a dislocation in planes that contain the dislocation line. We explore that connection herein.

Recall two expressions for the energy of a straight dislocation line, i.e.,

$$\mathcal{E} = K_{ij} b_i b_j \ln(R/r_0), \tag{21.84}$$

and

$$\mathcal{E} = \frac{1}{2} \int_V \sigma_{ij} u_{i,j} \, dV = \frac{1}{2} \int_S \sigma_{ij} u_i \, dS$$
$$= \frac{1}{2} \int_{r_0}^{R} \sigma_{ij} n_j b_i \, dS = \frac{1}{2} \int_{r_0}^{R} \sigma_{ij} n_j b_i \, dx_1. \tag{21.85}$$

In (21.85) the cut has been taken along the x_1 axis, that is on the plane where $x_2 = 0$; the dislocation is lying along the x_3 axis.

Consider a dislocation with a Burgers vector with a nonzero component b_1 ($b_2 = b_3 = 0$). From (21.84) then

$$\mathcal{E} = K_{11} b_1 b_1 \ln(R/r_0) = \int_{r_0}^{R} \frac{K_{11} b_1 b_1}{x_1} dx_1. \tag{21.86}$$

But, from (21.85), the energy is also given as

$$\mathcal{E} = \frac{1}{2}\int_{r_0}^{R} \sigma_{ij}^{(1)} n_j b_i \, dx_1 = \frac{1}{2}\int_{r_0}^{R} \sigma_{12}^{(1)} b_1 \, dx_1, \tag{21.87}$$

where $\sigma_{12}^{(1)}$ is the stress component in the plane $x_2 = 0$ caused by the dislocation with Burgers vector b_1. When (21.87) is compared to (21.86), we find that

$$\sigma_{12}^{(1)} = \frac{2K_{11}b_1}{x_1}. \tag{21.88}$$

Similarly, we find that for a dislocation with Burgers vector b_2 ($b_1 = b_3 = 0$),

$$\sigma_{22}^{(2)} = \frac{2K_{22}b_2}{x_1}, \tag{21.89}$$

where $\sigma_{22}^{(2)}$ is the stress component on the plane $x_2 = 0$ caused by the dislocation with Burgers vector b_2. Finally, if the Burgers vector had only a nonzero component b_3, we would find that

$$\sigma_{23}^{(3)} = \frac{2K_{33}b_3}{x_1}. \tag{21.90}$$

Note that the above results were arrived at by considering the *self-energy* of individual dislocations with single nonzero components of Burgers vector. Now consider interaction effects between two dislocations or between the stresses arising from different components of Burgers vector of a single dislocation. Let a dislocation line be again along the x_3 axis, but with a Burgers vector $\mathbf{b} = b_1\mathbf{e}_1 + b_2\mathbf{e}_2$. From (21.84), the energy is

$$\mathcal{E} = K_{ij}b_i b_j \ln(R/r_0) = K_{11}b_1 b_1 \ln(R/r_0) + K_{22}b_2 b_2 \ln(R/r_0) \\ + (K_{12}b_1 b_2 + K_{21}b_2 b_1)\ln(R/r_0). \tag{21.91}$$

Since $K_{12} = K_{21}$, this becomes

$$\mathcal{E} = K_{ij}b_i b_j \ln(R/r_0) = K_{11}b_1 b_1 \ln(R/r_0) + K_{22}b_2 b_2 \ln(R/r_0) + 2K_{12}b_1 b_2 \ln(R/r_0). \tag{21.92}$$

Alternatively, we could write that this energy is

$$\begin{aligned}\mathcal{E} &= \frac{1}{2}\int_{r_0}^{R} \sigma_{ij} n_j b_i \, dx_1 \\ &= \frac{1}{2}\int_{r_0}^{R} \left[\sigma_{ij}^{(1)} n_j + \sigma_{ij}^{(2)} n_j\right] b_i \, dx_1 \\ &= \frac{1}{2}\int_{r_0}^{R} \left[\sigma_{12}^{(1)} b_1 + \sigma_{22}^{(1)} b_2 + \sigma_{12}^{(2)} b_1 + \sigma_{22}^{(2)} b_2\right] dx_1.\end{aligned} \tag{21.93}$$

Comparing (21.93) with (21.92), having in mind that $\sigma_{12}^{(1)}$ and $\sigma_{22}^{(2)}$ are already known, we find that

$$\frac{1}{2}\int_{r_0}^{R}\left[\sigma_{22}^{(1)}b_2 + \sigma_{12}^{(2)}b_1\right] dx_1 = \int_{r_0}^{R} \frac{2K_{12}b_1 b_2}{x_1} \, dx_1. \tag{21.94}$$

21.6. Chemical Potential

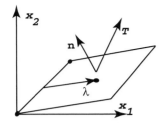

Figure 21.8. Arbitrary cut surface.

Furthermore, the reciprocity $\sigma_{22}^{(1)} b_2 = \sigma_{12}^{(2)} b_1$ holds. To see this, write

$$\frac{1}{2} \int_V \sigma_{ij}^{(1)} e_{ij}^{(2)} \, dV = \frac{1}{2} \int_V \sigma_{ij}^{(2)} e_{ij}^{(1)} \, dV$$

$$= \frac{1}{2} \int_V \sigma_{ij}^{(1)} u_{i,j}^{(2)} \, dV = \frac{1}{2} \int_S \sigma_{ij}^{(1)} n_j b_i^{(2)} \, dS \qquad (21.95)$$

$$= \frac{1}{2} \int_{r_0}^{R} \sigma_{22}^{(1)} b_2^{(2)} \, dx_1 = \frac{1}{2} \int_V \sigma_{ij}^{(2)} u_{i,j}^{(1)} \, dV$$

$$= \frac{1}{2} \int_{r_0}^{R} \sigma_{12}^{(2)} b_1^{(1)} \, dx_1.$$

Hence, we find that

$$\sigma_{12}^{(2)} = \frac{2 K_{12} b_2}{x_1},$$
$$\sigma_{22}^{(1)} = \frac{2 K_{21} b_1}{x_1}. \qquad (21.96)$$

In general, we obtain by using this kind of analysis that

$$\sigma_{i2} = \frac{2 K_{is} b_s}{x_1}. \qquad (21.97)$$

It should be noted at this point that our arguments and procedures would be true for any cut plane that contains the dislocation line, and that could have been used to calculate the energy. Thus, for any plane containing the dislocation line, and with λ as the distance from that line,

$$\sigma_{ij} n_j = T_i = \frac{K_{is} b_s}{\lambda}, \qquad (21.98)$$

i.e., the traction is the same on any plane containing the dislocation line (Fig. 21.8).

21.6 Chemical Potential

The chemical potential is defined as the differential change in Gibbs free energy with respect to the addition of a unit amount of a chemical specie, or in the case considered here, of matter. We wish to specifically evaluate the influence of stress and elastic strain state on that quantity. Thus we consider the addition of a small amount of material to the surface of a stressed body.

Consider the potential of two states, 1 and 2, before and after the addition of matter. We have

$$\Pi_1 = \int_{V_1} W(e_{ij}^{(1)}) dV - \int_{\Sigma_1} T_i^{(1)} u_i^{(1)} dS,$$
$$\Pi_2 = \int_{V_2} W(e_{ij}^{(2)}) dV - \int_{\Sigma_1} T_i^{(2)} u_i^{(2)} dS.$$
(21.99)

The system is to act at constant temperature and state of applied traction on its external surface. The latter constraint sets $\mathbf{T}^{(1)} = \mathbf{T}^{(2)}$. Thus, considering the change in potential, $\Pi_2 - \Pi_1$, the surface integrals of (21.99) contribute

$$\int_{\Sigma_1} T_i^{(1)} u_i^{(1)} dS - \int_{\Sigma_2} T_i^{(2)} u_i^{(2)} dS = -\int_{\Sigma_1} T_i^{(1)}[u_i^{(2)} - u_i^{(1)}] dS - \int_{\Delta\Sigma} T_i^{(1)}[u_i^{(2)} - u_i^{(1)}] dS,$$

where $\Delta\Sigma$ is the increment of surface area to which matter was added. Naturally, if $\Delta\Sigma$ should lie on a traction free part of the surface, the last integral would vanish.

Next, focus attention on an infinitesimal volume element at the site of $\Delta\Sigma$; in state 1 it is v_1 and in state 2 it is v_2. The goal is to let $v_2 - v_1 \to 0$. Then,

$$\Pi_2 - \Pi_1 = \bar{W}(e_{ij}^{(2)}) \int_{v_2} dV + \int_{V_2 - v_2} W(e_{ij}^{(2)}) dV$$
$$- \bar{W}(e_{ij}^{(1)}) \int_{v_1} dV - \int_{V_1 - v_1} W(e_{ij}^{(1)}) dV \qquad (21.100)$$
$$+ \int_{\Sigma_1} T_i^{(1)}[u_i^{(1)} - u_i^{(2)}] dS - \int_{\Delta\Sigma} T_i^{(1)}[u_i^{(2)} - u_i^{(1)}] dS,$$

where $\bar{W}(e_{ij}^{(1,2)})$ is the mean value of strain energy density in these vanishingly small volume elements. But, after combining terms, (21.100) is equivalent to

$$\Pi_2 - \Pi_1 = \bar{W}(e_{ij}^{(1)})(v_2 - v_1) + \int_{V_2} \left\{ \int_{e_{ij}^{(2)}}^{e_{ij}^{(1)}} [\sigma_{ij}^{(1)} - \sigma_{ij}] de_{ij} \right\} dV - \int_{\Delta\Sigma} T_i^{(1)}[u_i^{(2)} - u_i^{(1)}] dS,$$
(21.101)

or, by using the divergence theorem, to

$$\Pi_2 - \Pi_1 = \bar{W}(e_{ij}^{(1)})(v_2 - v_1) + \int_{\Sigma} \left\{ \int_{u_i^{(2)}}^{u_i^{(1)}} [T_i^{(1)} - T_i] du_i \right\} dS - \int_{\Delta\Sigma} T_i^{(1)}[u_i^{(2)} - u_i^{(1)}] dS.$$
(21.102)

The result may be rewritten as

$$\Pi_2 - \Pi_1 = \bar{W}(e_{ij}^{(1)})(v_2 - v_1) + \int_{v_2 - v_1} \left\{ \int_{e_{ij}^{(2)}}^{e_{ij}^{(1)}} [\sigma_{ij}^{(1)} - \sigma_{ij}] de_{ij} \right\} dV - \int_{\Delta\Sigma} T_i^{(1)}[u_i^{(2)} - u_i^{(1)}] dS$$
(21.103)
$$= \bar{W}(e_{ij}^{(1)})(v_2 - v_1) + \int_{e_{ij}^{(2)}}^{e_{ij}^{(1)}} [\sigma_{ij}^{(1)} - \sigma_{ij}] de_{ij} - \int_{\Delta\Sigma} T_i^{(1)}[u_i^{(2)} - u_i^{(1)}] dS.$$

But

$$[u_i^{(2)} - u_i^{(1)}] dS = (v_2 - v_1) n_i dS,$$

21.6. Chemical Potential

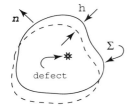

Figure 21.9. Movement of an internal source of stress by the removal and addition of matter to the external surface.

because the change in volume on the surface is

$$v_2 - v_1 = [u_i^{(2)} - u_i^{(1)}] n_i \, dS, \quad (21.104)$$

where **n** is the unit normal on $\Delta \Sigma$. Now, divide by $v_2 - v_1$ and let $v_2 - v_1 \to 0$. The integral over strain vanishes as the two states merge, and

$$\frac{\partial \Pi}{\partial v} = W(e_{ij}^{(1)}) - T_n^{(1)}, \quad (21.105)$$

where $T_n^{(1)}$ is the normal traction at the point on the surface where mass is added, *i.e.*, $T_n^{(1)} = T_i^{(1)} n_i$. Therefore, we have deduced that the chemical potential is

$$\mu = W(\mathbf{e}) - T_n. \quad (21.106)$$

21.6.1 Force on a Defect due to a Free Surface

Consider a purely internal source of stress, such as a defect within a body whose external surface is traction free, as depicted in Fig. 21.9. If the surface did have applied traction, then by ignoring it we would be calculating only that contribution to the total force arising from the free surface. This is the so-called *image force*. From the energy momentum tensor, this force would be calculated from the integral

$$J_\ell = \int_C (W n_\ell - \sigma_{ij} n_j \partial u_i / \partial x_\ell) \, dS. \quad (21.107)$$

Now take the contour C to be Σ, the external surface of the body. On Σ we have $\sigma_{ij} n_j = 0$, and hence

$$J_\ell = \int_\Sigma W(\mathbf{e}) n_\ell \, dS. \quad (21.108)$$

The movement of such an internal source of stress could be accomplished by imagining a process of mass removal and addition to the external surface as illustrated in the figure. In particular, if mass is removed in the direction x_ℓ from the surface element whose outward pointing normal is **n**, and if the distance along x_ℓ where mass is removed is h, then the change in free energy is, by using (21.106),

$$\Delta \Pi = -\int_0^h \int_\Sigma W(e_{ij}) n_\ell \, dS \, dh. \quad (21.109)$$

Thus, the "force" on the defect can be constructed as

$$f_\ell = -\partial \Delta \Pi / \partial h = \int_\Sigma W(e_{ij}) n_\ell \, dS, \quad (21.110)$$

which reproduces the result obtained from the energy momentum tensor.

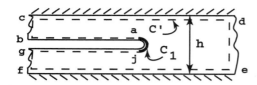

Figure 21.10. Crack in a strip clamped at its upper and lower edges and subject to extension normal to the crack line.

21.7 Applications of the J Integral

Here we present some applications of the J integral to calculate the force on particular defects in solid bodies.

21.7.1 Force on a Clamped Crack

Consider a crack in a strip, shown in Fig. 21.10, that is loaded *via* the action of applying a displacement to the upper and lower surfaces, *i.e.*, the surfaces along the lines c-d and e-f are displaced away from each other by a distance δ. As the original height of the strip is h, this action produces a strain $e = \delta/h$. The force acting on the crack tip would be computed by evaluating J_1 along the contour C_1 (axis x_1 being parallel to crack faces). Path independence, however, allows the alternate evaluation of J_1 along the contour C'. This is rather straightforward in this case. First, note that along the paths b-c and f-g, the stresses vanish, as does the strain energy density and the traction; thus, there is no contribution from these segments. Next, along the paths c-d and e-f, the normal is orthogonal to the direction of propagation of the crack and the displacement is confined by the rigid clamp such that $u_{i,1}$ vanishes; thus, there is no contribution from these segments, either. Finally, along the segment d-e, the strain energy density is given, assuming a state of plane stress, as

$$W = \frac{E}{2(1-\nu^2)} \left(\frac{\delta}{h}\right)^2. \tag{21.111}$$

Integration of this along the segment d-e gives

$$\int W \, dy = \frac{Eh}{2(1-\nu^2)} \left(\frac{\delta}{h}\right)^2, \tag{21.112}$$

which is the only contribution to the total contour integral. Thus, we obtain

$$J_1 = \frac{E}{2(1-\nu^2)} \frac{\delta^2}{h}. \tag{21.113}$$

21.7.2 Application of the Interface Force to Precipitation

As an application of the concept of a force on an interface, consider the case of the precipitation of a second phase particle within an elastic matrix. We idealize the situation by ignoring any inelastic deformation, *i.e.*, dislocation activity. Further, let this process occur in an infinite matrix (the parent phase) under the influence of a remotely applied hydrostatic stress state of intensity σ_∞. The process is driven (that is, if it is spontaneous) by the combination of a negative change in the chemical part of the Gibbs free energy, $\Delta G < 0$, and the interaction energy between the remotely applied stress field and the transformation

21.7. J Integral

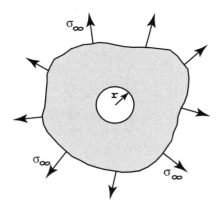

Figure 21.11. Formation of a critical nucleus and growing precipitate in an elastic matrix.

strain associated with the phase change. Here we assume that the transformation strain is purely dilatational, *i.e.*,

$$e_{ij}^T = \frac{1}{3} e^T \delta_{ij}. \tag{21.114}$$

Of course, it is possible that, even if $\Delta G > 0$, the transformation may be driven forward by an intense enough remote stress, and this is conventionally mapped on "pressure *vs.* temperature" phase diagram for single component systems.

As the phase forms, an elastic field is created along with its energy. This problem was dealt with in great detail in previous chapter and will not be included here. The results are only used. Specifically, we consider the precipitate to be spherical and of radius r (Fig. 21.11). The solution for the elastic field can be determined by the methods developed in Chapter 20. The matrix and precipitate are considered to be isotropic in their elastic properties and of the same modulus. With this in mind, the normal interface force from (21.36) becomes

$$\mathcal{F} = -\Delta G - \frac{E}{9(1-\nu)} (\epsilon^T)^2 + \epsilon^T \sigma_\infty - \frac{2}{r} \gamma, \tag{21.115}$$

where γ is the surface energy of the α/β interface. Of course, for growth we must have $\mathcal{F} > 0$, or, more specifically,

$$-\Delta G + \epsilon^T \sigma_\infty > \frac{E}{9(1-\nu)} (\epsilon^T)^2. \tag{21.116}$$

We note that, even if (21.116) were satisfied, there is the nucleation problem to face, because if γ is significant in magnitude, the surface energy term dominates (in a negative way) for small values of r; that is, we must acquire a nucleus of *critical size*, r^*. This is readily attained by setting $\mathcal{F} = 0$ in (21.115), which yields

$$r^*(T) = \frac{2\gamma}{-\Delta G - E/[9(1-\nu)](\epsilon^T)^2 + \epsilon^T \sigma_\infty}. \tag{21.117}$$

We have explicitly noted the very important temperature dependence of the process, coming most strongly through the term involving ΔG, and more weakly through the temperature dependence of E, ν, ϵ^T, and γ. Now, if $r < r^*$, the nucleus is unstable and the work done on the system to attain the nucleus of radius $r > r^*$ is

$$\Delta G^* = -\int_0^{r^*} 4\pi r^2 \mathcal{F}\, dr. \tag{21.118}$$

In (21.118) we have employed the often used nomenclature ΔG^* to denote the magnitude of the *activation energy of nucleation*. This energy arises through statistical fluctuations and is of paramount importance with respect to the kinetics of the nucleation process. The evaluation of (21.118) for particular cases is left to the reader's interest.

21.8 Suggested Reading

Eshelby, J. D. (1951), The Force on an Elastic Singularity, *Phil. Trans. R. Soc. A*, Vol. 244, pp. 87–112.

Eshelby, J. D. (1956), The Continuum Theory of Lattice Defects. In *Solid State Physics* (F. Seitz and D. Turnbull, eds.), Vol. 3, Academic Press, New York, pp. 79–144.

Eshelby, J. D. (1970), Energy Relations and the Energy-Momentum Tensor in Continuum Mechanics. In *Inelastic Behavior of Solids* (M. F. Kanninen, W. F. Adier, A. R. Rosenfield, and R. I. Janee, eds.), pp. 77–115, McGraw-Hill, New York.

Eshelby, J. D. (1975), The Elastic Energy-Momentum Tensor, *J. of Elasticity*, Vol. 5, pp. 321–335.

Kienzler, R., and Maugin, G. A., eds. (2001), *Configurational Mechanics of Materials*, Springer-Verlag, Wien.

Maugin, G. A. (1993), *Material Inhomogeneities in Elasticity*, Chapman & Hall, New York.

Mura, T. (1987), *Micromechanics of Defects in Solids*, Martinus Nijhoff Publishers, Dordrecht, The Netherlands.

Rice, J. R. (1968), Mathematical Analysis in the Mechanics of Fracture. In *Fracture – An Advanced Treatise* (H. Liebowitz, ed.), Vol. II, pp. 191–311, Academic Press, New York.

Rice, J. R. (1985), Conserved Integrals and Energetic Forces. In *Fundamentals of Deformation and Fracture* (B. A. Bilby, K. J. Miller, and J. R. Willis, eds.), Cambridge University Press, Cambridge.

22 Micropolar Elasticity

22.1 Introduction

In a micropolar continuum the deformation is described by the displacement vector and an independent rotation vector. The rotation vector specifies the orientation of a triad of director vectors attached to each material particle. A particle (material element) can experience a microrotation without undergoing a macrodisplacement. An infinitesimal surface element transmits a force and a couple vector, which give rise to nonsymmetric stress and couple-stress tensors. The former is related to a nonsymmetric strain tensor and the latter to a nonsymmetric curvature tensor, defined as the gradient of the rotation vector. This type of the continuum mechanics was originally introduced by Voigt (1887) and the brothers Cosserat (1909). In a simplified micropolar theory, the so-called couple-stress theory, the rotation vector is not independent of the displacement vector, but related to it in the same way as in classical continuum mechanics.

The physical rationale for the extension of the classical to micropolar and couple-stress theory was that the classical theory was not able to predict the size effect experimentally observed in problems which had a geometric length scale comparable to material's microstructural length, such as the grain size in a polycrystalline or granular aggregate. For example, the apparent strength of some materials with stress concentrators such as holes and notches is higher for smaller grain size; for a given volume fraction of dispersed hard particles, the strengthening of metals is greater for smaller particles; the bending and torsional strengths are higher for very thin beams and wires. The classical theory was also in disagreement with experiments for high-frequency ultrashort wave propagation problems, if the wavelength becomes comparable to the material's microstructural length. In the presence of couple-stresses, shear waves propagate dispersively (with a frequency dependent wave speed). Couple-stresses are also expected to affect the singular nature of the crack tip fields. The couple-stress and related nonlocal and strain-gradient theories of elastic and inelastic material response are also of interest to describe the deformation mechanisms and manufacturing of micro and nanostructured materials and devices, as well as inelastic localization and instability phenomena.

22.2 Basic Equations of Couple-Stress Elasticity

In a micropolar continuum the deformation is described by the displacement vector and an independent rotation vector. In the couple-stress theory, the rotation vector φ_i is not independent of the displacement vector u_i but is subject to the constraint

$$\varphi_i = \frac{1}{2}\epsilon_{ijk}\omega_{jk} = \frac{1}{2}\epsilon_{ijk}u_{k,j}, \quad \omega_{ij} = \epsilon_{ijk}\varphi_k, \tag{22.1}$$

as in classical continuum mechanics. The skew-symmetric alternating tensor is ϵ_{ijk}, and ω_{ij} are the rectangular components of the infinitesimal rotation tensor. The latter is related to the displacement gradient and the symmetric strain tensor by $u_{j,i} = e_{ij} + \omega_{ij}$, where

$$e_{ij} = \frac{1}{2}(u_{j,i} + u_{i,j}), \quad \omega_{ij} = \frac{1}{2}(u_{j,i} - u_{i,j}). \tag{22.2}$$

The comma designates the partial differentiation with respect to Cartesian coordinates x_i.

A surface element dS transmits a force vector $T_i\,dS$ and a couple vector $M_i\,dS$. The surface forces are in equilibrium with the nonsymmetric Cauchy stress t_{ij}, and the surface couples are in equilibrium with the nonsymmetric couple-stress m_{ij}, such that

$$T_i = n_j t_{ji}, \quad M_i = n_j m_{ji}, \tag{22.3}$$

where n_j are the components of the unit vector orthogonal to the surface element under consideration. In the absence of body couples, the differential equations of equilibrium are

$$t_{ji,j} + f_i = 0, \quad m_{ji,j} + \epsilon_{ijk}t_{jk} = 0. \tag{22.4}$$

The body forces are denoted by f_i. By decomposing the stress tensor into its symmetric and antisymmetric part,

$$t_{ij} = \sigma_{ij} + \tau_{ij}, \quad (\sigma_{ij} = \sigma_{ji}, \quad \tau_{ij} = -\tau_{ji}), \tag{22.5}$$

from the moment equilibrium equation it readily follows that the antisymmetric part can be determined as

$$\tau_{ij} = -\frac{1}{2}\epsilon_{ijk}m_{lk,l}. \tag{22.6}$$

If the gradient of the couple-stress vanishes at some point, the stress tensor is symmetric at that point.

The rate of strain energy per unit volume is

$$\dot{W} = \sigma_{ij}\dot{e}_{ij} + m_{ij}\dot{\kappa}_{ij}, \tag{22.7}$$

where

$$\kappa_{ij} = \varphi_{j,i} \tag{22.8}$$

is a nonsymmetric curvature tensor. In view of the identity

$$\omega_{ij,k} = e_{kj,i} - e_{ki,j}, \tag{22.9}$$

the curvature tensor can also be expressed as

$$\kappa_{ij} = -\epsilon_{jkl}e_{ik,l}. \tag{22.10}$$

22.3. Displacement Equations of Equilibrium

These are the compatibility equations for curvature and strain fields. In addition, there is an identity

$$\kappa_{ij,k} = \kappa_{kj,i} \, (= \varphi_{j,ik}), \tag{22.11}$$

which represents the compatibility equations for curvature components. The compatibility equations for strain components are the usual Saint-Venant's compatibility equations. Since e_{ij} is symmetric and ϵ_{ijk} skew-symmetric, from (22.10) it follows that the curvature tensor in couple-stress theory is a deviatoric tensor ($\kappa_{kk} = 0$).

Assuming that the elastic strain energy is a function of the strain and curvature tensors, $W = W(e_{ij}, \kappa_{ij})$, the differentiation and the comparison with (22.7) establishes the constitutive relations of couple-stress elasticity,

$$\sigma_{ij} = \frac{\partial W}{\partial e_{ij}}, \quad m_{ij} = \frac{\partial W}{\partial \kappa_{ij}}. \tag{22.12}$$

In the case of isotropic material, with the quadratic strain energy,

$$W = \frac{1}{2}\lambda \, e_{kk}e_{ll} + \mu \, e_{kl}e_{kl} + 2\alpha \, \kappa_{kl}\kappa_{kl} + 2\beta\kappa_{kl}\kappa_{lk}, \tag{22.13}$$

where μ, λ, α, and β are the Lamé-type constants of isotropic couple-stress elasticity. The stress and couple-stress tensors are in this case

$$\sigma_{ij} = 2\mu \, e_{ij} + \lambda \, e_{kk}\delta_{ij}, \quad m_{ij} = 4\alpha \, \kappa_{ij} + 4\beta \, \kappa_{ji}. \tag{22.14}$$

By the positive-definiteness of the strain energy, it follows that $\alpha + \beta > 0$, and $\alpha - \beta > 0$. Thus, α is positive, but not necessarily β. Since the curvature tensor is deviatoric, from the second (22.14) it follows that the couple-stress is also a deviatoric tensor ($m_{kk} = 0$). In some problems the curvature tensor may be symmetric, and then the couple-stress is also symmetric, regardless of the ratio α/β.

If the displacement components are prescribed at a point of the bounding surface of the body, the normal component of the rotation vector at that point cannot be prescribed independently. This implies (*e.g.*, Mindlin and Tiersten, 1962; Koiter, 1964) that at any point of a smooth boundary we can specify three reduced stress tractions

$$\bar{T}_i = n_j t_{ji} - \frac{1}{2} \epsilon_{ijk} n_j (n_p m_{pq} n_q)_{,k}, \tag{22.15}$$

and two tangential couple-stress tractions

$$\bar{M}_i = n_j m_{ji} - (n_j m_{jk} n_k) n_i. \tag{22.16}$$

22.3 Displacement Equations of Equilibrium

The couple-stress gradient can be expressed, from (22.10) and (22.14), as

$$m_{lk,l} = -2\alpha \, \epsilon_{kpq} u_{p,qll}, \tag{22.17}$$

independently of the material parameter β. The substitution into (22.6) gives an expression for the antisymmetric part of the stress tensor

$$\tau_{ij} = -2\alpha \, \omega_{ij,kk} = -2\alpha \, \nabla^2 \omega_{ij}, \tag{22.18}$$

which is also independent of β. The Laplacian operator is $\nabla^2 = \partial^2/\partial x_k \partial x_k$. Consequently, by adding (22.14) and (22.18), the total stress tensor is

$$t_{ij} = 2\mu\, e_{ij} + \lambda\, e_{kk}\, \delta_{ij} - 2\alpha\, \nabla^2 \omega_{ij}\,. \tag{22.19}$$

Incorporating this into the force equilibrium equations (22.4), we obtain the equilibrium equations in terms of the displacement components

$$\nabla^2 u_i - l^2 \nabla^4 u_i + \frac{\partial}{\partial x_i}\left[\frac{1}{1-2\nu}(\nabla \cdot \mathbf{u}) + l^2 \nabla^2 (\nabla \cdot \mathbf{u})\right] = 0, \tag{22.20}$$

where $\nabla \cdot \mathbf{u} = u_{k,k}$. The biharmonic operator is $\nabla^4 = \nabla^2 \nabla^2$, and

$$l^2 = \frac{\alpha}{\mu}, \quad 1 + \frac{\lambda}{\mu} = \frac{1}{1-2\nu}\,. \tag{22.21}$$

The Poisson coefficient is denoted by ν. Upon applying to (22.20) the partial derivative $\partial/\partial x_i$, there follows $\nabla^2 e_{kk} = 0$. Thus, the volumetric strain is governed by the same equation as in classical elasticity, without couple-stresses. The substitution into (22.20) yields the final form of the displacement equations of equilibrium

$$\nabla^2 u_i - l^2 \nabla^4 u_i + \frac{1}{1-2\nu}\frac{\partial}{\partial x_i}(\nabla \cdot \mathbf{u}) = 0\,. \tag{22.22}$$

Three components of displacement and only two tangential components of rotation may be specified on the boundary. Alternatively, three reduced stress tractions and two tangential couple-stress tractions may be specified on a smooth boundary.

The general solution of (22.22) can be cast in the form

$$u_i = U_i - l^2 \frac{\partial}{\partial x_i}(\nabla \cdot \mathbf{U}) - \frac{1}{4(1-\nu)}\frac{\partial}{\partial x_i}\left[\varphi + \mathbf{x} \cdot (1 - l^2 \nabla^2)\mathbf{U}\right], \tag{22.23}$$

where the scalar potential φ and the vector potential U_i are solutions of the Laplace's and Helmholtz partial differential equations

$$\nabla^2 \varphi = 0, \quad \nabla^2\left(U_i - l^2 \nabla^2 U_i\right) = 0\,. \tag{22.24}$$

The general solution of the latter equation can be obtained by observing that

$$U_i - l^2 \nabla^2 U_i = U_i^0 \tag{22.25}$$

must be a harmonic function, satisfying the Laplacian equation $\nabla^2 U_i^0 = 0$. Thus, the general solution can be expressed as $U_i = U_i^0 + U_i^*$, where

$$U_i^* - l^2 \nabla^2 U_i^* = 0\,. \tag{22.26}$$

22.4 Correspondence Theorem of Couple-Stress Elasticity

For equilibrium problems of couple-stress elasticity with prescribed displacement boundary conditions, and with no body forces or body couples present, we state the following theorem.

22.5. Plane Strain Problems

Theorem: If $u_i = \hat{u}_i$ is a solution of the Navier equations of elasticity without couple-stresses,

$$\nabla^2 \hat{u}_i + \frac{1}{1-2\nu} \frac{\partial}{\partial x_i} (\boldsymbol{\nabla} \cdot \hat{\mathbf{u}}) = 0, \tag{22.27}$$

then $\hat{u}_i$ is also a solution of differential equations (22.22) for couple-stress elasticity.

Proof: It suffices to prove that $\hat{u}_i$ is a biharmonic function. By applying the Laplacian operator to (22.27), we obtain

$$\nabla^4 \hat{u}_i + \frac{1}{1-2\nu} \frac{\partial}{\partial x_i} \nabla^2(\boldsymbol{\nabla} \cdot \hat{\mathbf{u}}) = 0. \tag{22.28}$$

Since $\boldsymbol{\nabla} \cdot \hat{\mathbf{u}}$ is a harmonic function, as can be verified from (22.27) by applying the partial derivatives $\partial/\partial x_i$, equation (22.28) reduces to

$$\nabla^4 \hat{u}_i = 0. \tag{22.29}$$

This shows that $\hat{u}_i$ is a biharmonic function, which completes the proof.

We now prove that the stress tensor in couple-stress elasticity with prescribed displacement boundary conditions, and without body forces or body couples, is a symmetric tensor. From (22.27) it readily follows by partial differentiation that the rotation components are harmonic functions ($\nabla^2 \omega_{ij} = 0$, $\nabla^2 \varphi_i = 0$), and substitution into (22.18) gives $\tau_{ij} = 0$. In general, the couple-stress tensor is still nonsymmetric, although in the case of antiplane strain with prescribed displacement boundary conditions it becomes a symmetric tensor.

22.5 Plane Strain Problems of Couple-Stress Elasticity

In plane strain elasticity the displacement components are

$$u_1 = u_1(x_1, x_2), \quad u_2 = u_2(x_1, x_2), \quad u_3 = 0. \tag{22.30}$$

The nonvanishing strain, rotation, and curvature components are

$$e_{11} = \frac{\partial u_1}{\partial x_1}, \quad e_{22} = \frac{\partial u_2}{\partial x_2}, \quad e_{12} = \frac{1}{2}\left(\frac{\partial u_2}{\partial x_1} + \frac{\partial u_1}{\partial x_2}\right),$$

$$\varphi_3 = \omega_{12} = \frac{1}{2}\left(\frac{\partial u_2}{\partial x_1} - \frac{\partial u_1}{\partial x_2}\right), \tag{22.31}$$

$$\kappa_{13} = \frac{\partial \varphi_3}{\partial x_1}, \quad \kappa_{23} = \frac{\partial \varphi_3}{\partial x_2}.$$

The stress-strain relations are

$$\sigma_{11} = (2\mu + \lambda)e_{11} + \lambda e_{22}, \quad \sigma_{22} = (2\mu + \lambda)e_{22} + \lambda e_{11},$$
$$\sigma_{12} = 2\mu e_{12}, \quad \tau_{12} = -2\alpha \nabla^2 \varphi_3. \tag{22.32}$$

The normal stress $\sigma_{33} = \lambda(e_{11} + e_{22})$. The couple-stress–curvature relations are

$$m_{13} = 4\alpha \kappa_{13}, \quad m_{31} = 4\beta \kappa_{13}, \quad m_{23} = 4\alpha \kappa_{23}, \quad m_{32} = 4\beta \kappa_{23}. \tag{22.33}$$

The elastic strain energy per unit volume is

$$W = \frac{1}{2\mu}\left[\sigma_{12}^2 + \frac{1}{2(1+\nu)}\left(\sigma_{11}^2 + \sigma_{22}^2 - 2\nu\sigma_{11}\sigma_{22} - \sigma_{33}^2\right)\right] + \frac{1}{8\alpha}(m_{13}^2 + m_{23}^2). \quad (22.34)$$

Equations (22.31)–(22.34) can be easily rewritten in terms of polar coordinate components. For example, we have

$$e_{rr} = \frac{\partial u_r}{\partial r}, \quad e_{\theta\theta} = \frac{1}{r}\left(\frac{\partial u_\theta}{\partial \theta} + u_r\right), \quad e_{r\theta} = \frac{1}{2r}\left(\frac{\partial u_r}{\partial \theta} + r\frac{\partial u_\theta}{\partial r} - u_\theta\right),$$

$$\varphi_3 = \omega_{r\theta} = \frac{1}{2r}\left[\frac{\partial(ru_\theta)}{\partial r} - \frac{\partial u_r}{\partial \theta}\right], \quad \kappa_{r3} = \frac{\partial \varphi_3}{\partial r}, \quad \kappa_{\theta 3} = \frac{1}{r}\frac{\partial \varphi_3}{\partial \theta}.$$

22.5.1 Mindlin's Stress Functions

The rectangular components of stress and couple-stress tensors can be expressed in terms of the functions Φ and Ψ as

$$t_{11} = \frac{\partial^2 \Phi}{\partial x_2^2} - \frac{\partial^2 \Psi}{\partial x_1 \partial x_2}, \quad t_{22} = \frac{\partial^2 \Phi}{\partial x_1^2} + \frac{\partial^2 \Psi}{\partial x_1 \partial x_2},$$

$$t_{12} = -\frac{\partial^2 \Phi}{\partial x_1 \partial x_2} - \frac{\partial^2 \Psi}{\partial x_2^2}, \quad t_{21} = -\frac{\partial^2 \Phi}{\partial x_1 \partial x_2} + \frac{\partial^2 \Psi}{\partial x_1^2}, \quad (22.35)$$

$$m_{13} = \frac{\partial \Psi}{\partial x_1}, \quad m_{23} = \frac{\partial \Psi}{\partial x_2},$$

where the functions Φ and Ψ satisfy the partial differential equations

$$\nabla^4 \Phi = 0, \quad \nabla^2 \Psi - l^2 \nabla^4 \Psi = 0. \quad (22.36)$$

The curvature-strain compatibility equations require that the functions Φ and Ψ be related by

$$\frac{\partial}{\partial x_1}\left(\Psi - l^2 \nabla^2 \Psi\right) = -2(1-\nu)l^2 \frac{\partial}{\partial x_2}(\nabla^2 \Phi),$$

$$\frac{\partial}{\partial x_2}\left(\Psi - l^2 \nabla^2 \Psi\right) = 2(1-\nu)l^2 \frac{\partial}{\partial x_1}(\nabla^2 \Phi). \quad (22.37)$$

The solution of the equation for Ψ in (22.36) can be expressed as $\Psi = \Psi^0 + \Psi^*$, where

$$\nabla^2 \Psi^0 = 0, \quad \Psi^* - l^2 \nabla^2 \Psi^* = 0. \quad (22.38)$$

Thus, (22.37) can be rewritten as

$$\frac{\partial \Psi_0}{\partial x_1} = -2(1-\nu)l^2 \frac{\partial}{\partial x_2}(\nabla^2 \Phi),$$

$$\frac{\partial \Psi_0}{\partial x_2} = 2(1-\nu)l^2 \frac{\partial}{\partial x_1}(\nabla^2 \Phi). \quad (22.39)$$

The counterparts of (22.35) and (22.39) in polar coordinates are

$$t_{rr} = \frac{1}{r}\frac{\partial \Phi}{\partial r} + \frac{1}{r^2}\frac{\partial^2 \Phi}{\partial \theta^2} - \frac{1}{r}\frac{\partial^2 \Psi}{\partial r \partial \theta} + \frac{1}{r^2}\frac{\partial \Psi}{\partial \theta},$$

22.6. Edge Dislocation in Couple-Stress Elasticity

$$t_{\theta\theta} = \frac{\partial^2 \Phi}{\partial r^2} + \frac{1}{r}\frac{\partial^2 \Psi}{\partial r \partial \theta} - \frac{1}{r^2}\frac{\partial \Psi}{\partial \theta},$$

$$t_{r\theta} = -\frac{1}{r}\frac{\partial^2 \Phi}{\partial r \partial \theta} + \frac{1}{r^2}\frac{\partial \Phi}{\partial \theta} - \frac{1}{r}\frac{\partial \Psi}{\partial r} - \frac{1}{r^2}\frac{\partial^2 \Psi}{\partial \theta^2}, \quad (22.40)$$

$$t_{\theta r} = -\frac{1}{r}\frac{\partial^2 \Phi}{\partial r \partial \theta} + \frac{1}{r^2}\frac{\partial \Phi}{\partial \theta} + \frac{\partial^2 \Psi}{\partial r^2},$$

$$m_{r3} = \frac{\partial \Psi}{\partial r}, \quad m_{\theta 3} = \frac{1}{r}\frac{\partial \Psi}{\partial \theta},$$

and

$$\frac{\partial}{\partial r}(\Psi - l^2 \nabla^2 \Psi) = -2(1-\nu)l^2 \frac{1}{r}\frac{\partial}{\partial \theta}(\nabla^2 \Phi),$$
$$\frac{1}{r}\frac{\partial}{\partial \theta}(\Psi - l^2 \nabla^2 \Psi) = 2(1-\nu)l^2 \frac{\partial}{\partial r}(\nabla^2 \Phi). \quad (22.41)$$

22.6 Edge Dislocation in Couple-Stress Elasticity

For an edge dislocation in an infinite medium, the only boundary condition is the displacement discontinuity b, imposed for example along the plane $x_1 > 0$, $x_2 = 0$. Thus, by the correspondence theorem, the displacement field is as in classical elasticity, i.e.,

$$u_1 = \frac{b}{2\pi}\left[\tan^{-1}\frac{x_2}{x_1} + \frac{1}{2(1-\nu)}\frac{x_1 x_2}{x_1^2 + x_2^2}\right],$$
$$u_2 = -\frac{b}{2\pi}\frac{1}{4(1-\nu)}\left[(1-2\nu)\ln\frac{x_1^2 + x_2^2}{b^2} + \frac{x_1^2 - x_2^2}{x_1^2 + x_2^2}\right]. \quad (22.42)$$

The stresses are

$$\sigma_{11} = -\frac{\mu b}{2\pi(1-\nu)}\frac{x_2(3x_1^2 + x_2^2)}{(x_1^2 + x_2^2)^2},$$

$$\sigma_{22} = \frac{\mu b}{2\pi(1-\nu)}\frac{x_2(x_1^2 - x_2^2)}{(x_1^2 + x_2^2)^2},$$

$$\sigma_{12} = \frac{\mu b}{2\pi(1-\nu)}\frac{x_1(x_1^2 - x_2^2)}{(x_1^2 + x_2^2)^2}, \quad (22.43)$$

$$\sigma_{33} = -\frac{\nu \mu b}{\pi(1-\nu)}\frac{x_2}{x_1^2 + x_2^2}.$$

The rotation and curvature components are

$$\varphi_3 = -\frac{b}{2\pi}\frac{x_1}{x_1^2 + x_2^2},$$

$$\kappa_{13} = \frac{b}{2\pi}\frac{x_1^2 - x_2^2}{(x_1^2 + x_2^2)^2}, \quad \kappa_{23} = \frac{b}{2\pi}\frac{2x_1 x_2}{(x_1^2 + x_2^2)^2}. \quad (22.44)$$

The corresponding couple-stresses are

$$m_{13} = \frac{2\alpha b}{\pi} \frac{x_1^2 - x_2^2}{(x_1^2 + x_2^2)^2}, \quad m_{31} = \frac{2\beta b}{\pi} \frac{x_1^2 - x_2^2}{(x_1^2 + x_2^2)^2},$$
$$m_{23} = \frac{2\alpha b}{\pi} \frac{2x_1 x_2}{(x_1^2 + x_2^2)^2}, \quad m_{31} = \frac{2\beta b}{\pi} \frac{2x_1 x_2}{(x_1^2 + x_2^2)^2}.$$
(22.45)

In polar coordinates, the displacements are

$$u_r = \frac{b}{2\pi}\left\{\theta\cos\theta + \frac{1}{4(1-\nu)}\left[1 - (1-2\nu)\ln\frac{r^2}{b^2}\right]\sin\theta\right\},$$
$$u_\theta = -\frac{b}{2\pi}\left\{\theta\sin\theta + \frac{1}{4(1-\nu)}\left[1 + (1-2\nu)\ln\frac{r^2}{b^2}\right]\cos\theta\right\},$$
(22.46)

and the stresses

$$\sigma_{rr} = \sigma_{\theta\theta} = -\frac{\mu b}{2\pi(1-\nu)}\frac{\sin\theta}{r},$$
$$\sigma_{r\theta} = \frac{\mu b}{2\pi(1-\nu)}\frac{\cos\theta}{r},$$
$$\sigma_{33} = -\frac{\nu\mu b}{\pi(1-\nu)}\frac{\sin\theta}{r}.$$
(22.47)

The rotation and curvature components are

$$\varphi_3 = -\frac{b}{2\pi}\frac{\cos\theta}{r}, \quad \kappa_{r3} = \frac{b}{2\pi}\frac{\cos\theta}{r^2}, \quad \kappa_{\theta 3} = \frac{b}{2\pi}\frac{\sin\theta}{r^2},$$
(22.48)

with the corresponding couple-stresses

$$m_{r3} = \frac{2\alpha b}{\pi}\frac{\cos\theta}{r^2}, \quad m_{3r} = \frac{2\beta b}{\pi}\frac{\cos\theta}{r^2},$$
$$m_{\theta 3} = \frac{2\alpha b}{\pi}\frac{\sin\theta}{r^2}, \quad m_{3\theta} = \frac{2\beta b}{\pi}\frac{\sin\theta}{r^2}.$$
(22.49)

The stress components decay with a distance from the center of dislocation as r^{-1}, whereas the couple-stresses decay as r^{-2}. These also specify the orders of the singularities at the dislocation core when $r \to 0$. The displacement and rotation fields for an edge dislocation in polar elasticity, without the constraint (22.1), can be found in Nowacki (1986).

22.6.1 Strain Energy

The strain energy (per unit length in x_3 direction) stored within a cylinder bounded by the radii r_0 and R (Fig. 22.1) is

$$E = \int_{r_0}^{R}\int_0^{2\pi} W r \, dr \, d\theta,$$
(22.50)

22.6. Edge Dislocation in Couple-Stress Elasticity

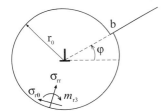

Figure 22.1. A material of the dislocation core is removed and its effect on the remaining material represented by the indicated stress and couple-stress tractions over the surface $r = r_0$. The slip discontinuity of amount b is imposed along the cut at an angle φ.

where the specific strain energy (per unit volume) is

$$W = \frac{1}{2\mu}\left[\sigma_{r\theta}^2 + (1-2\nu)\sigma_{rr}^2\right] + \frac{1}{8\alpha}\left(m_{r3}^2 + m_{\theta 3}^2\right)$$
$$= \frac{\mu b^2}{8\pi^2(1-\nu)^2}\frac{1}{r^2}\left(1 - 2\nu\sin^2\theta\right) + \frac{\alpha b^2}{2\pi^2}\frac{1}{r^4}. \tag{22.51}$$

Upon the substitution into (22.50) and integration, there follows

$$E = \frac{\mu b^2}{4\pi(1-\nu)}\ln\frac{R}{r_0} + \frac{\alpha b^2}{2\pi}\left(\frac{1}{r_o^2} - \frac{1}{R^2}\right). \tag{22.52}$$

The second term on the right-hand side is the strain energy contribution from the couple-stresses. The presence of this term is associated with the work done by the couple-stresses on the surfaces $r = r_0$ and $r = R$. This can be seen by writing an alternative expression for the strain energy,

$$E = \frac{1}{2}\int_{r_0}^{R}\sigma_{r\theta}(r,0)\,b\,dr + \frac{1}{2}\int_0^{2\pi}M_3\,\varphi_3\,R\,d\theta - \frac{1}{2}\int_0^{2\pi}M_3\,\varphi_3\,r_0\,d\theta, \tag{22.53}$$

with $M_3 = m_{r3}$ given by (22.49), and φ_3 given in (22.48). The work of the tractions σ_{rr} and $\sigma_{r\theta}$ on the displacements u_r and u_θ over the surface $r = R$ cancels the work of the tractions σ_{rr} and $\sigma_{r\theta}$ over the surface $r = r_0$. These terms are thus not explicitly included in (22.53). The second term in (22.52) is the strain energy contribution due to last two work terms in (22.53). For example, in a metallic crystal with the dislocation density $\rho = 10^{10}$ cm^{-2}, the radius of influence of each dislocation (defined as the average distance between dislocations) is of the order of $\rho^{-1/2} = 100$ nm. For an fcc crystal with the lattice parameter $a = 0.4$ nm, and the Burgers vector along the closed packed direction $b = a/\sqrt{2}$, the radius R can be approximately taken as $R = 200b$. By choosing the material length l to be the lattice parameter ($l = \sqrt{2}b$), the couple-stress modulus is $\alpha = 2\mu b^2$, and by selecting $r_0 = 2b$ and $\nu = 1/3$, the strain energy contribution from couple-stresses in (22.52) is 14.5% of the strain energy without couple stresses. The calculations are sensitive to selected value of the dislocation core radius; larger the value of r_0 smaller the effect of couple-stresses in the region beyond r_0 (Lubarda, 2003a). The strain energy contribution from couple-stresses is likely to be lowered by inclusion of the micropolar effects. It is known that the effect of couple-stresses on stress concentration is less pronounced if the microrotations are assumed to be independent of the displacement field (Eringen, 1999).

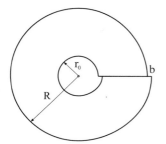

Figure 22.2. A slip discontinuity of amount b is imposed along the horizontal cut from the inner radius r_0 to the outer radius R. The inner and outer surface of the cylinder are free from stresses and couple-stresses.

22.7 Edge Dislocation in a Hollow Cylinder

The so-called hollow dislocation along the axis of a circular cylinder with inner radius r_0 and the outer radius R is shown in Fig. 22.2. Both surfaces of the cylinder are required to be stress and couple stress free. The displacement discontinuity of amount b is imposed along the horizontal cut from r_0 to R. The solution is derived from the infinite body solution by superposing an additional solution that cancels the stresses and couple-stresses over the inner and outer surface associated with the solution for an edge dislocation in an infinite medium. Thus, we require that the superposed solution satisfies the boundary conditions

$$t_{rr}(R,\theta) = \frac{\mu b}{2\pi(1-\nu)} \frac{\sin\theta}{R}, \quad t_{rr}(r_0,\theta) = \frac{\mu b}{2\pi(1-\nu)} \frac{\sin\theta}{r_0},$$

$$t_{r\theta}(R,\theta) = -\frac{\mu b}{2\pi(1-\nu)} \frac{\cos\theta}{R}, \quad t_{r\theta}(r_0,\theta) = -\frac{\mu b}{2\pi(1-\nu)} \frac{\cos\theta}{r_0}, \quad (22.54)$$

$$m_{r3}(R,\theta) = -\frac{2\alpha b}{\pi} \frac{\cos\theta}{R^2}, \quad m_{r3}(r_0,\theta) = -\frac{2\alpha b}{\pi} \frac{\cos\theta}{r_0^2}.$$

This can be accomplished by using the following structure of the Mindlin stress functions

$$\Phi = \left(A_0 r^3 + B_0 \frac{1}{r}\right) \sin\theta, \quad (22.55)$$

$$\Psi = \left[Ar + B\frac{1}{r} + CI_1\left(\frac{r}{l}\right) + DK_1\left(\frac{r}{l}\right)\right] \cos\theta. \quad (22.56)$$

From the conditions (22.41) it readily follows that

$$A = -16(1-\nu)l^2 A_0, \quad B = 0. \quad (22.57)$$

The stress and couple-stress components of the superposed solution are accordingly

$$t_{rr} = \left[2A_0 r - 2B_0 \frac{1}{r^3} + C\frac{1}{rl} I_2\left(\frac{r}{l}\right) - D\frac{1}{rl} K_2\left(\frac{r}{l}\right)\right] \sin\theta,$$

$$t_{r\theta} = -\left[2A_0 r - 2B_0 \frac{1}{r^3} + C\frac{1}{rl} I_2\left(\frac{r}{l}\right) - D\frac{1}{rl} K_2\left(\frac{r}{l}\right)\right] \cos\theta, \quad (22.58)$$

$$m_{r3} = -\left\{16(1-\nu)l^2 A_0 - C\frac{1}{2l}\left[I_0\left(\frac{r}{l}\right) + I_2\left(\frac{r}{l}\right)\right] + D\frac{1}{2l}\left[K_0\left(\frac{r}{l}\right) + K_2\left(\frac{r}{l}\right)\right]\right\} \cos\theta.$$

22.7. Edge Dislocation in a Hollow Cylinder

The expressions for the derivatives of the modified Bessel functions with respect to r/l are used (Watson 1995).

After a lengthy but straightforward derivation, it follows that

$$A_0 = \frac{\mu b}{2\pi(1-\nu)} \frac{R^2 - r_0^2}{Ra_1 - r_0 a_2} - \frac{2\alpha b}{\pi} \frac{l}{R^2} \frac{Rb_1 - r_0 b_2}{Ra_1 - r_0 a_2}, \quad (22.59)$$

$$B_0 = \frac{1}{2} Rr_0 \left[\frac{\mu b}{2\pi(1-\nu)} \frac{Ra_2 - r_0 a_1}{Ra_1 - r_0 a_2} - \frac{2\alpha b}{\pi} \frac{l}{R^2} \frac{a_2 b_1 - a_1 b_2}{Ra_1 - r_0 a_2} \right],$$

and

$$C = 16(1-\nu)l^3 \frac{c_1}{c} A_0 - \frac{2\alpha b}{\pi} \frac{l}{R^2} \frac{c_2}{c},$$

$$D = -16(1-\nu)l^3 \frac{d_1}{c} A_0 + \frac{2\alpha b}{\pi} \frac{l}{R^2} \frac{d_2}{c}. \quad (22.60)$$

The introduced parameters are

$$a_1 = 2R^3 - 16(1-\nu)\frac{l^3}{c}(c_1 d_3 - d_1 c_3),$$

$$a_2 = 2r_0^3 - 16(1-\nu)\frac{l^3}{c}(c_1 d_4 - d_1 c_4), \quad (22.61)$$

$$b_1 = \frac{1}{c}(c_2 d_3 - d_2 c_3), \quad b_2 = \frac{1}{c}(c_2 d_4 - d_2 c_4),$$

where

$$c_1 = \frac{1}{2}\left[K_0\left(\frac{R}{l}\right) + K_2\left(\frac{R}{l}\right)\right] - \frac{1}{2}\left[K_0\left(\frac{r_0}{l}\right) + K_2\left(\frac{r_0}{l}\right)\right],$$

$$c_2 = -\frac{1}{2}\left[K_0\left(\frac{r_0}{l}\right) + K_2\left(\frac{r_0}{l}\right)\right] + \frac{R^2}{2r_0^2}\left[K_0\left(\frac{R}{l}\right) + K_2\left(\frac{R}{l}\right)\right], \quad (22.62)$$

$$c_3 = \frac{R}{l} K_2\left(\frac{R}{l}\right), \quad c_4 = \frac{r_0}{l} K_2\left(\frac{r_0}{l}\right),$$

and

$$d_1 = -\frac{1}{2}\left[I_0\left(\frac{R}{l}\right) + I_2\left(\frac{R}{l}\right)\right] + \frac{1}{2}\left[I_0\left(\frac{r_0}{l}\right) + I_2\left(\frac{r_0}{l}\right)\right],$$

$$d_2 = \frac{1}{2}\left[I_0\left(\frac{r_0}{l}\right) + I_2\left(\frac{r_0}{l}\right)\right] - \frac{R^2}{2r_0^2}\left[I_0\left(\frac{R}{l}\right) + I_2\left(\frac{R}{l}\right)\right], \quad (22.63)$$

$$d_3 = -\frac{R}{l} I_2\left(\frac{R}{l}\right), \quad d_4 = -\frac{r_0}{l} I_2\left(\frac{r_0}{l}\right).$$

The parameter c is defined by

$$c = \frac{1}{4}\left[K_0\left(\frac{R}{l}\right) + K_2\left(\frac{R}{l}\right)\right]\left[I_0\left(\frac{r_0}{l}\right) + I_2\left(\frac{r_0}{l}\right)\right] \\ - \frac{1}{4}\left[K_0\left(\frac{r_0}{l}\right) + K_2\left(\frac{r_0}{l}\right)\right]\left[I_0\left(\frac{R}{l}\right) + I_2\left(\frac{R}{l}\right)\right]. \tag{22.64}$$

If $R \to \infty$, we obtain the solution for an edge dislocation with a stress free hollow core in an infinite medium. In this case $A_0 = A = B = D = 0$, and

$$B_0 = -\frac{\mu b r_0^2}{4\pi(1-\nu)} - \frac{2\alpha b}{\pi}\frac{K_2(r_0/l)}{K_0(r_0/l) + K_2(r_0/l)},$$

$$D = \frac{4\alpha b l}{\pi r_0^2}\frac{1}{K_0(r_0/l) + K_2(r_0/l)}. \tag{22.65}$$

If couple-stresses are neglected, (22.59) and (22.60) yield

$$A_0 = \frac{\mu b}{4\pi(1-\nu)}\frac{1}{R^2+r_0^2}, \quad B_0 = -R^2 r_0^2 A_0 \quad A = C = D = 0. \tag{22.66}$$

The corresponding stresses are

$$\sigma_{rr} = -\frac{\mu b}{2\pi(1-\nu)}\left[\frac{1}{r} - \frac{1}{R^2+r_0^2}\left(r + \frac{R^2 r_0^2}{r^3}\right)\right]\sin\theta,$$

$$\sigma_{r\theta} = \frac{\mu b}{2\pi(1-\nu)}\left[\frac{1}{r} - \frac{1}{R^2+r_0^2}\left(r + \frac{R^2 r_0^2}{r^3}\right)\right]\cos\theta, \tag{22.67}$$

$$\sigma_{\theta\theta} = -\frac{\mu b}{2\pi(1-\nu)}\left[\frac{1}{r} - \frac{1}{R^2+r_0^2}\left(3r - \frac{R^2 r_0^2}{r^3}\right)\right]\sin\theta,$$

in agreement with the solution for the Volterra edge dislocation from classical elasticity (Love, 1944).

22.8 Governing Equations for Antiplane Strain

For the antiplane strain problems, the displacements are

$$u_1 = u_2 = 0, \quad u_3 = w(x_1, x_2). \tag{22.68}$$

The nonvanishing strain, rotation, and curvature components are

$$e_{13} = e_{31} = \frac{1}{2}\frac{\partial w}{\partial x_1}, \quad e_{23} = e_{32} = \frac{1}{2}\frac{\partial w}{\partial x_2},$$

$$\varphi_1 = \omega_{23} = \frac{1}{2}\frac{\partial w}{\partial x_2}, \quad \varphi_2 = \omega_{31} = -\frac{1}{2}\frac{\partial w}{\partial x_1}, \tag{22.69}$$

$$\kappa_{11} = -\kappa_{22} = \frac{1}{2}\frac{\partial^2 w}{\partial x_1 \partial x_2}, \quad \kappa_{12} = -\frac{1}{2}\frac{\partial^2 w}{\partial x_1^2}, \quad \kappa_{21} = \frac{1}{2}\frac{\partial^2 w}{\partial x_2^2}.$$

22.8. Governing Equations for Antiplane Strain

It readily follows that

$$\nabla^2 \omega_{13} = \frac{1}{2} \frac{\partial}{\partial x_1}\left(\nabla^2 w\right), \quad \nabla^2 \omega_{23} = \frac{1}{2} \frac{\partial}{\partial x_2}\left(\nabla^2 w\right), \tag{22.70}$$

so that, from (22.18),

$$\tau_{31} = -\tau_{13} = \alpha \frac{\partial}{\partial x_1}\left(\nabla^2 w\right), \quad \tau_{32} = -\tau_{23} = \alpha \frac{\partial}{\partial x_2}\left(\nabla^2 w\right). \tag{22.71}$$

Consequently, from (22.19),

$$\begin{aligned} t_{13} &= \mu \frac{\partial}{\partial x_1}\left(w - l^2 \nabla^2 w\right), & t_{31} &= \mu \frac{\partial}{\partial x_1}\left(w + l^2 \nabla^2 w\right), \\ t_{23} &= \mu \frac{\partial}{\partial x_2}\left(w - l^2 \nabla^2 w\right), & t_{32} &= \mu \frac{\partial}{\partial x_2}\left(w + l^2 \nabla^2 w\right). \end{aligned} \tag{22.72}$$

The couple-stresses are related to the curvature components by

$$\begin{aligned} m_{11} &= 4(\alpha+\beta)\kappa_{11} = 2(\alpha+\beta)\frac{\partial^2 w}{\partial x_1 \partial x_2}, \\ m_{22} &= 4(\alpha+\beta)\kappa_{22} = -2(\alpha+\beta)\frac{\partial^2 w}{\partial x_1 \partial x_2} = -m_{11}, \\ m_{12} &= 4\alpha\kappa_{12} + 4\beta\kappa_{21} = -2\alpha\frac{\partial^2 w}{\partial x_1^2} + 2\beta\frac{\partial^2 w}{\partial x_2^2}, \\ m_{21} &= 4\alpha\kappa_{21} + 4\beta\kappa_{12} = 2\alpha\frac{\partial^2 w}{\partial x_2^2} - 2\beta\frac{\partial^2 w}{\partial x_1^2}. \end{aligned} \tag{22.73}$$

It is noted that

$$m_{12} - m_{21} = 2(\beta - \alpha)\nabla^2 w. \tag{22.74}$$

Since displacement field is isochoric ($\nabla \cdot \mathbf{u} = 0$), the displacement equations of equilibrium (22.22) reduce to a single equation

$$\nabla^2 w - l^2 \nabla^4 w = 0. \tag{22.75}$$

The general solution can be expressed as

$$w = w^0 + w^*, \tag{22.76}$$

where w^0 and w^* are the solutions of the partial differential equations

$$\begin{aligned} \nabla^2 w^0 &= 0, \\ w^* - l^2 \nabla^2 w^* &= 0. \end{aligned} \tag{22.77}$$

In view of (22.76) and (22.77), the following identities hold

$$\nabla^2 w = \frac{1}{l^2} w^*, \tag{22.78}$$

and

$$w - l^2 \nabla^2 w = w^0, \quad w + l^2 \nabla^2 w = w^0 + 2w^*, \tag{22.79}$$

which can be conveniently used to simplify the stress expressions (22.72).

22.8.1 Expressions in Polar Coordinates

When expressed in terms of polar coordinates, the general solutions of (22.77), obtained by separation of variables, are

$$w^0 = (A_0 + B_0 \ln r)(C_0 + \theta) + \sum_{n=1}^{\infty} (A_n r^n + B_n r^{-n})(C_n \cos n\theta + \sin n\theta), \qquad (22.80)$$

$$w^* = \left[A_0^* I_0\left(\frac{r}{l}\right) + B_0^* K_0\left(\frac{r}{l}\right)\right](C_0^* + \theta)$$
$$+ \sum_{n=1}^{\infty} \left[A_n^* I_n\left(\frac{r}{l}\right) + B_n^* K_n\left(\frac{r}{l}\right)\right](C_n^* \cos n\theta + \sin n\theta). \qquad (22.81)$$

The functions $I_n(\rho)$ and $K_n(\rho)$ (with $\rho = r/l$) in (22.81) are the modified Bessel functions of the first and second kind (of the order n).

The nonvanishing strain, rotation, and curvature components in polar coordinates are

$$e_{\theta 3} = e_{3\theta} = \frac{1}{2r} \frac{\partial w}{\partial \theta}, \qquad e_{r3} = e_{3r} = \frac{1}{2} \frac{\partial w}{\partial r},$$
$$\varphi_r = \omega_{\theta 3} = \frac{1}{2r} \frac{\partial w}{\partial \theta}, \qquad \varphi_\theta = \omega_{3r} = -\frac{1}{2} \frac{\partial w}{\partial r}, \qquad (22.82)$$

and

$$\kappa_{rr} = \frac{\partial \varphi_r}{\partial r} = \frac{1}{2} \frac{\partial}{\partial r}\left(\frac{1}{r} \frac{\partial w}{\partial \theta}\right), \qquad \kappa_{r\theta} = \frac{\partial \varphi_\theta}{\partial r} = -\frac{1}{2} \frac{\partial^2 w}{\partial r^2},$$

$$\kappa_{\theta r} = \frac{1}{r} \frac{\partial \varphi_r}{\partial \theta} - \frac{\varphi_\theta}{r} = \frac{1}{2r^2} \frac{\partial^2 w}{\partial \theta^2} + \frac{1}{2r} \frac{\partial w}{\partial r}, \qquad (22.83)$$

$$\kappa_{\theta\theta} = \frac{1}{r} \frac{\partial \varphi_\theta}{\partial \theta} + \frac{\varphi_r}{r} = -\frac{1}{2} \frac{\partial}{\partial r}\left(\frac{1}{r} \frac{\partial w}{\partial \theta}\right) = -\kappa_{rr}.$$

Note that

$$\kappa_{r\theta} - \kappa_{\theta r} = \kappa_{12} - \kappa_{21} = -\frac{1}{2} \nabla^2 w. \qquad (22.84)$$

The corresponding couple-stress components are (Fig. 22.3)

$$m_{rr} = -m_{\theta\theta} = 2(\alpha + \beta) \frac{\partial}{\partial r}\left(\frac{1}{r} \frac{\partial w}{\partial \theta}\right),$$

$$m_{r\theta} = -2(\alpha + \beta) \frac{\partial^2 w}{\partial r^2} + 2\beta \nabla^2 w, \qquad (22.85)$$

$$m_{\theta r} = -2(\alpha + \beta) \frac{\partial^2 w}{\partial r^2} + 2\alpha \nabla^2 w.$$

The shear stresses are

$$t_{r3} = \mu \frac{\partial}{\partial r}(w - l^2 \nabla^2 w), \qquad t_{\theta 3} = \mu \frac{1}{r} \frac{\partial}{\partial \theta}(w - l^2 \nabla^2 w),$$
$$t_{3r} = \mu \frac{\partial}{\partial r}(w + l^2 \nabla^2 w), \qquad t_{3\theta} = \mu \frac{1}{r} \frac{\partial}{\partial \theta}(w + l^2 \nabla^2 w). \qquad (22.86)$$

22.8. Governing Equations for Antiplane Strain

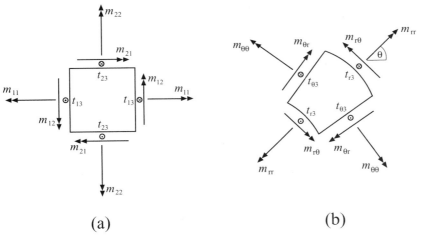

Figure 22.3. (a) A material element with sides parallel to coordinates directions x_1 and x_2 under conditions of antiplane strain; (b) The corresponding element with sides parallel to polar directions r and θ.

In view of (22.79), the stress components t_{r3} and $t_{\theta 3}$ (or t_{13} and t_{23}) do not depend explicitly on w^*, i.e.,

$$t_{r3} = \mu \frac{\partial w^0}{\partial r}, \quad t_{\theta 3} = \mu \frac{1}{r}\frac{\partial w^0}{\partial \theta},$$

$$t_{3r} = t_{r3} + 2\mu \frac{\partial w^*}{\partial r}, \quad t_{3\theta} = t_{\theta 3} + 2\mu \frac{1}{r}\frac{\partial w^*}{\partial \theta}. \tag{22.87}$$

The couple-stresses, however, affect the values of t_{r3} and $t_{\theta 3}$ through the imposed boundary conditions. For example, along an unstressed circular boundary $r = R$ around the origin, the reduced tractions must vanish,

$$\bar{t}_{r3} = t_{r3} - \frac{1}{2R}\frac{\partial m_{rr}}{\partial \theta} = 0, \quad m_{r\theta} = 0. \tag{22.88}$$

Also note that

$$m_{r\theta} = -2(\alpha + \beta)\frac{\partial^2 w}{\partial r^2} + \frac{2\beta}{l^2}w^*, \quad m_{\theta r} = -2(\alpha + \beta)\frac{\partial^2 w}{\partial r^2} + \frac{2\alpha}{l^2}w^*. \tag{22.89}$$

The elastic strain energy per unit volume is

$$W = \frac{1}{2\mu}\left(\sigma_{r3}^2 + \sigma_{\theta 3}^2\right) + \frac{1}{4(\alpha + \beta)}\left\{m_{rr}^2 + \frac{1}{2(\alpha - \beta)}\left[\alpha(m_{r\theta}^2 + m_{\theta r}^2) - 2\beta m_{r\theta}m_{\theta r}\right]\right\}. \tag{22.90}$$

22.8.2 Correspondence Theorem for Antiplane Strain

For antiplane strain problems with prescribed displacement boundary conditions, the correspondence theorem of couple-stress elasticity reads: If $w = w^0$ is a solution of differential equation of elasticity without couple-stresses $\nabla^2 w^0 = 0$, then w^0 is also a solution of differential equations (22.75) for couple-stress elasticity.

The proof is simple. Since w^0 is a harmonic function, it is also a biharmonic function, satisfying (22.75). For prescribed displacement boundary conditions, the function w^0 specifies the displacement field in both nonpolar and couple-stress elasticity.

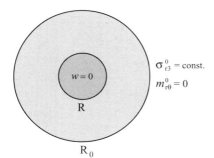

Figure 22.4. A circular annulus of inner radius R and outer radius R_0 under shear stress σ_{r3}^0 over the outer surface. The inner surface of the cylinder is fixed ($w = 0$).

The stress and couple-stress tensors in antiplane strain problems of couple-stress elasticity, in the case of prescribed displacement boundary conditions, are symmetric tensors. Indeed, since the displacement field is a harmonic function, the antisymmetric stress components in (22.72) vanish, *i.e.*, $\tau_{13} = \tau_{23} = 0$. Thus, the total stress tensor is a symmetric tensor. From (22.8) it further follows that the curvature tensor is a symmetric tensor ($\kappa_{12} = \kappa_{21}$). This implies from (22.73) that the couple-stress tensor is also symmetric ($m_{12} = m_{21}$), regardless of the ratio α/β.

22.9 Antiplane Shear of Circular Annulus

A simple but illustrative problem of couple-stress elasticity is the antiplane shearing of a circular annulus. Suppose that the inner surface $r = R$ is fixed, whereas the constant shearing stress σ_{r3}^0 is applied on the outer surface $r = R_0$ (Fig. 22.4). The corresponding displacement field is independent of θ and given by

$$w(r) = R_0 \left[A \ln \frac{r}{R} + B + C I_0 \left(\frac{r}{l} \right) + D K_0 \left(\frac{r}{l} \right) \right]. \tag{22.91}$$

The integration constants are specified from the boundary conditions

$$w(R) = 0, \quad t_{r3}(R_0) = \sigma_{r3}^0, \quad m_{r\theta}(R_0) = 0. \tag{22.92}$$

The fourth boundary condition is obtained by specifying an additional information about the bonded interface, such as the magnitude of slope dw/dr or the couple-stress $m_{r\theta}$ at $r = R$. We will proceed by adopting the first choice, *i.e.*, by assuming that the rotation

$$\varphi_\theta(R) = -\frac{1}{2} \left(\frac{dw}{dr} \right)_{r=R} = \hat{\varphi}_\theta \tag{22.93}$$

is known at the interface. It readily follows from (22.87)–(22.89) that

$$A = \frac{\sigma_{r3}^0}{\mu}, \quad B = -C I_0 \left(\frac{R}{l} \right) - D K_0 \left(\frac{R}{l} \right), \tag{22.94}$$

and

$$C = \frac{l}{R_0} \frac{\left[K_1 \left(\frac{R}{l} \right) - b \frac{R_0}{R} \right] \frac{\sigma_{r3}^0}{\mu} - 2b \hat{\varphi}_\theta}{b I_1 \left(\frac{R}{l} \right) + a K_1 \left(\frac{R}{l} \right)},$$

$$D = \frac{l}{R_0} \frac{\left[I_1 \left(\frac{R}{l} \right) + a \frac{R_0}{R} \right] \frac{\sigma_{r3}^0}{\mu} + 2a \hat{\varphi}_\theta}{b I_1 \left(\frac{R}{l} \right) + a K_1 \left(\frac{R}{l} \right)}, \tag{22.95}$$

where

$$a = \frac{\alpha}{\alpha+\beta}\frac{R_0}{l}\left[I_0\left(\frac{R_0}{l}\right) - I_1\left(\frac{R_0}{l}\right)\right], \quad b = \frac{\alpha}{\alpha+\beta}\frac{R_0}{l}\left[K_0\left(\frac{R_0}{l}\right) + K_1\left(\frac{R_0}{l}\right)\right].$$

Consequently,

$$t_{r3}(r) = \frac{R_0}{r}\sigma_{r3}^0,$$

$$t_{3r}(r) = \frac{R_0}{r}\left\{\sigma_{r3}^0 + 2\mu\frac{r}{l}\left[CI_1\left(\frac{r}{l}\right) - DK_1\left(\frac{r}{l}\right)\right]\right\}, \tag{22.96}$$

$$m_{r\theta}(r) = 2(\alpha+\beta)\left\{\frac{R_0}{r^2}\frac{\sigma_{r3}^0}{\mu} - C\frac{R_0}{l^2}\left[\frac{\alpha}{\alpha+\beta}I_0\left(\frac{r}{l}\right) - \frac{l}{r}I_1\left(\frac{r}{l}\right)\right]\right.$$
$$\left. - D\frac{R_0}{l^2}\left[\frac{\alpha}{\alpha+\beta}K_0\left(\frac{r}{l}\right) + \frac{l}{r}K_1\left(\frac{r}{l}\right)\right]\right\},$$

$$m_{\theta r}(r) = 2(\alpha+\beta)\left\{\frac{R_0}{r^2}\frac{\sigma_{r3}^0}{\mu} - C\frac{R_0}{l^2}\left[\frac{\beta}{\alpha+\beta}I_0\left(\frac{r}{l}\right) - \frac{l}{r}I_1\left(\frac{r}{l}\right)\right]\right. \tag{22.97}$$
$$\left. - D\frac{R_0}{l^2}\left[\frac{\beta}{\alpha+\beta}K_0\left(\frac{r}{l}\right) + \frac{l}{r}K_1\left(\frac{r}{l}\right)\right]\right\}.$$

If the bonded interface cannot support the couple-stress $m_{r\theta}$, we set the right-hand side of the above expression for $m_{r\theta}$ equal to zero and calculate the corresponding rotation $\hat{\varphi}_\theta$.

22.10 Screw Dislocation in Couple-Stress Elasticity

The displacement field for a screw dislocation with imposed displacement discontinuity b along the plane $x_1 > 0$, $x_2 = 0$ is as in classical elasticity, i.e.,

$$w = \frac{b}{2\pi}\tan^{-1}\frac{x_2}{x_1} = \frac{b}{2\pi}\theta. \tag{22.98}$$

This follows from the correspondence theorem, since only the displacement boundary conditions are prescribed. The stresses and couple-stresses associated with (22.98) are

$$\sigma_{13} = -\frac{\mu b}{2\pi}\frac{x_2}{x_1^2 + x_2^2}, \quad \sigma_{23} = \frac{\mu b}{2\pi}\frac{x_1}{x_1^2 + x_2^2},$$

$$m_{11} = -m_{22} = -\frac{(\alpha+\beta)b}{\pi}\frac{x_1^2 - x_2^2}{(x_1^2 + x_2^2)^2}, \tag{22.99}$$

$$m_{12} = m_{21} = -\frac{(\alpha+\beta)b}{\pi}\frac{2x_1 x_2}{(x_1^2 + x_2^2)^2}.$$

The components of the stress and couple-stress tensors along the polar directions are

$$\sigma_{\theta 3} = \frac{\mu b}{2\pi}\frac{1}{r}, \quad \sigma_{r3} = 0,$$
$$m_{rr} = -m_{\theta\theta} = -\frac{(\alpha+\beta)b}{\pi}\frac{1}{r^2}, \quad m_{r\theta} = m_{\theta r} = 0. \tag{22.100}$$

The stress components decay with a distance from the center of dislocation as r^{-1}, whereas the couple-stresses decay as r^{-2}. These also specify the orders of the singularities at the dislocation core when $r \to 0$.

22.10.1 Strain Energy

The strain energy (per unit length in x_3 direction) stored within a cylinder bounded by the radii r_0 and R is

$$E = \int_{r_0}^{R} W 2\pi r \, dr, \tag{22.101}$$

where the specific strain energy (per unit volume) is

$$W = \frac{1}{2\mu} \sigma_{\theta 3}^2 + \frac{1}{4(\alpha+\beta)} m_{rr}^2 = \frac{\mu b^2}{8\pi^2} \frac{1}{r^2} + \frac{(\alpha+\beta)b^2}{4\pi^2} \frac{1}{r^4}. \tag{22.102}$$

Upon the substitution into (22.101) and integration, there follows

$$E = \frac{\mu b^2}{4\pi} \ln \frac{R}{r_0} + \frac{(\alpha+\beta)b^2}{4\pi} \left(\frac{1}{r_0^2} - \frac{1}{R^2} \right). \tag{22.103}$$

The second term on the right-hand side is the strain energy contribution from the couple-stresses. The presence of this term is associated with the work done by the couple-stresses on the surfaces $r = r_0$ and $r = R$. This can be seen by writing an alternative expression for the strain energy,

$$E = \frac{1}{2} \int_{r_0}^{R} \sigma_{\theta 3}(r,0) b \, dr + \frac{1}{2} \int_{0}^{2\pi} M_r \, \varphi_r \, R \, d\theta - \frac{1}{2} \int_{0}^{2\pi} M_r \, \varphi_r \, r_0 \, d\theta, \tag{22.104}$$

with $M_r = m_{rr}$ given by (22.100), and with

$$\varphi_r = \omega_{\theta z} = \frac{1}{2r} \frac{\partial w}{\partial \theta} = \frac{1}{4\pi} \frac{b}{r} \tag{22.105}$$

being the r component of the rotation vector. Since $\sigma_{r3} = 0$, there is no work of stress traction on the displacement w over the surfaces $r = r_0$ and $r = R$. The second term in (22.103) is the strain energy contribution due to last two work terms in (22.104). For example, if we set $R = 200b$, $r_0 = 2b$, and $\alpha + \beta = 2\mu b^2$, the energy contribution from couple-stresses in (22.103) is 10.9% of the strain energy without couple-stresses. In the classical elasticity a cylindrical surface around the screw dislocation at its center is stress free. On the other hand, the solution derived in this section is characterized by the presence of the constant couple stress m_{rr} along that surface. However, since m_{rr} in (22.100) does not depend on θ, the reduced traction $\bar{t}_{r3}$ vanishes on the cylindrical surface $r = \text{const}$.

22.11 Configurational Forces in Couple-Stress Elasticity

Returning to general theory of couple-stress elasticity, it is appealing to extend the energy considerations from Chapter 21 to the framework of couple-stress theory. Thus, we now derive expressions for the interaction energy between the stress systems due to internal and external sources of stress, the configurational force on a singularity or the source of internal stress, and the energy-momentum tensor of elastic couple-stress field.

22.11. Configurational Forces

22.11.1 Reciprocal Properties

Consider two equilibrium elastic fields, designated by the superscripts A and B. Clearly, in analogy with (21.54),

$$\sigma_{ji}^A e_{ji}^B = \sigma_{ji}^B e_{ji}^A, \quad m_{ji}^A \kappa_{ji}^B = m_{ji}^B \kappa_{ji}^A, \tag{22.106}$$

and

$$(t_{ji}^A - \tau_{ji}^A) u_{i,j}^B = (t_{ji}^B - \tau_{ji}^B) u_{i,j}^A. \tag{22.107}$$

The equilibrium equations (22.4) provide the relationships

$$\begin{aligned} t_{ji}^A u_{i,j}^B - t_{ji}^B u_{i,j}^A &= (t_{ji}^A u_i^B - t_{ji}^B u_i^A)_{,j} + f_i^A u_i^B - f_i^B u_i^A, \\ \tau_{ji}^A u_{i,j}^B - \tau_{ji}^B u_{i,j}^A &= -(m_{ji}^A \varphi_i^B - m_{ji}^B \varphi_i^A)_{,j}, \end{aligned} \tag{22.108}$$

so that the substitution into (22.107) gives

$$(t_{ji}^A u_i^B + m_{ji}^A \varphi_i^B - t_{ji}^B u_i^A - m_{ji}^B \varphi_i^A)_{,j} = f_i^B u_i^A - f_i^A u_i^B. \tag{22.109}$$

Integrating over the volume of the body and using the Gauss divergence theorem, we thus obtain

$$\begin{aligned} &\int_S (t_{ji}^A u_i^B + m_{ji}^A \varphi_i^B) n_j \mathrm{d}S + \int_V f_i^A u_i^B \mathrm{d}V = \\ &\int_S (t_{ji}^B u_i^A + m_{ji}^B \varphi_i^A) n_j \mathrm{d}S + \int_S f_i^B u_i^A \mathrm{d}V, \end{aligned} \tag{22.110}$$

which is Betti's reciprocal theorem of couple-stress elasticity.

In the absence of body forces, from (22.109) it follows that the vector

$$v_j(A, B) = t_{ji}^A u_i^B + m_{ji}^A \varphi_i^B - t_{ji}^B u_i^A - m_{ji}^B \varphi_i^A \tag{22.111}$$

has zero divergence. Thus,

$$\int_{S_1} v_j(A, B) n_j \mathrm{d}S = \int_{S_2} v_j(A, B) n_j \mathrm{d}S, \tag{22.112}$$

for any two surfaces S_1 and S_2 which do not embrace any singularity of $v_j(A, B)$. In particular, if there are no singularities of $v_j(A, B)$ within a surface S,

$$\int_S v_j(A, B) n_j \mathrm{d}S = 0. \tag{22.113}$$

If the material is homogeneous, in addition to (22.106) we have

$$\sigma_{ji,k}^A e_{ji}^B = \sigma_{ji}^B e_{ji,k}^A, \quad m_{ji,k}^A \kappa_{ji}^B = m_{ji}^B \kappa_{ji,k}^A. \tag{22.114}$$

Consequently, for a homogeneous material (22.113) also applies when $v_j(A, B)$ is replaced by

$$\vartheta_{jk}(A, B) = t_{ji,k}^A u_i^B + m_{ji,k}^A \varphi_i^B - t_{ji}^B u_{i,k}^A - m_{ji}^B \varphi_{i,k}^A. \tag{22.115}$$

22.11.2 Energy due to Internal Sources of Stress

Suppose that there are two systems of internal stresses in the body of volume V bounded by the surface S. The sources of internal stress system A lie entirely within the surface S_0, and those of the system B lie entirely outside of S_0. If E_A and E_B are the values of the elastic strain energy when A and B alone exists in the body, we may write the total strain energy as $E = E_A + E_B + E_{AB}$ when the sources coexist in the body. Here,

$$E_{AB} = \frac{1}{2}\int_V (\sigma_{ji}^A e_{ji}^B + m_{ji}^A \kappa_{ji}^B + \sigma_{ji}^B e_{ji}^A + m_{ji}^B \kappa_{ji}^A)dV \tag{22.116}$$

is the interaction energy between A and B. Noting that $e_{ij}^B = \frac{1}{2}(u_{i,j}^B + u_{j,i}^B)$ and $\kappa_{ij}^B = \varphi_{j,i}^B$ in the volume V_0 within the surface S_0, and $e_{ij}^A = \frac{1}{2}(u_{i,j}^A + u_{j,i}^A)$ and $\kappa_{ij}^A = \varphi_{j,i}^A$ in the volume $V - V_0$ (but not conversely), and in view of the reciprocity properties, (22.116) can be rewritten as

$$E_{AB} = \int_{V_0}(\sigma_{ji}^A e_{ji}^B + m_{ji}^A \kappa_{ji}^B)dV + \int_{V-V_0}(\sigma_{ji}^B e_{ji}^A + m_{ji}^B \kappa_{ji}^A)dV. \tag{22.117}$$

Furthermore, $\sigma_{ji}^A e_{ji}^B = \sigma_{ji}^A u_{i,j}^B = (t_{ji}^A - \tau_{ji}^A)u_{i,j}^B$, and since from equilibrium equations without body forces

$$t_{ji}^A u_{i,j}^B = (t_{ji}^A u_i^B)_{,j}, \quad \tau_{ji}^A u_{i,j}^B = -m_{ji,j}^A \varphi_i^B, \tag{22.118}$$

we obtain

$$\sigma_{ji}^A e_{ji}^B + m_{ji}^A \kappa_{ji}^B = (t_{ji}^A u_i^B + m_{ji}^A \varphi_i^B)_{,j}. \tag{22.119}$$

An analogous equation holds when the superscripts A and B are interchanged in (22.119). Substituting these two equations into (22.117) and applying the Gauss divergence theorem gives

$$E_{AB} = \int_{S_0}(t_{ji}^A u_i^B + m_{ji}^A \varphi_i^B - t_{ji}^B u_i^A - m_{ji}^B \varphi_i^A)n_j dS, \tag{22.120}$$

where n_j is the outward normal to S_0. The surface integral over S vanishes, because no load is there applied. Thus, the interaction energy between two internal stress systems can be expressed by the integral over any surface S_0 that separates the two stress sources. Note that the bracketed terms of the integrand in (22.120) are equal to $v_j(A, B)$ by (22.111).

22.11.3 Energy due to Internal and External Sources of Stress

Denote by A the stress system due to internal sources within the volume V, and by B the stress system due to external surface load T_i^B and M_i^B applied over S. The total potential energy of the system is $\Pi = \Pi_A + \Pi_B + \Pi_{AB}$, where

$$\begin{aligned}\Pi_A &= E_A = \frac{1}{2}\int_V(\sigma_{ji}^A e_{ji}^A + m_{ji}^A \kappa_{ji}^A)dV, \\ \Pi_B &= E_B - \int_S (T_i^B u_i^B + M_i^B \varphi_i^B)dS\end{aligned} \tag{22.121}$$

22.11. Configurational Forces

represent the potential energies of two systems when they act alone, and

$$\Pi_{AB} = E_{AB} - \int_S (T_i^B u_i^A + M_i^B \varphi_i^A) \mathrm{d}S \tag{22.122}$$

is the interaction potential energy between the two systems. The interaction strain energy E_{AB} between stress systems due to internal and external stress sources is equal to zero. Indeed, in this case u_i^B and φ_i^B exist throughout the volume V, and from (22.116) and the reciprocity property we have

$$E_{AB} = \int_V (\sigma_{ji}^A e_{ji}^B + m_{ji}^A \kappa_{ji}^B) \, \mathrm{d}V = \int_V (t_{ji}^A u_i^B + m_{ji}^A \varphi_i^B)_{,j} \, \mathrm{d}V. \tag{22.123}$$

Thus, upon application of the Gauss theorem, we obtain $E_{AB} = 0$, because $T_i^A = 0$ and $M_i^A = 0$ on S. The response of the body to external loading in couple-stress theory is, therefore, the same whether the body is self-stressed or not, as in classical elasticity.

The interaction potential energy between external and internal stress systems is consequently

$$\Pi_{AB} = -\int_S (t_{ji}^B u_i^A + m_{ji}^B \varphi_i^A) n_j \, \mathrm{d}S. \tag{22.124}$$

The right-hand side of (22.124) can be rewritten in the same form as the right-hand side of (22.120), provided that S_0 is taken to be any surface that entirely encompasses the sources of internal stress A. This follows from the Gauss divergence theorem applied to the region $V - V_0$, noting that $T_i^A = M_i^A = 0$ on S, and that $v_j(A, B)$ is divergence free in the region between S and S_0.

In fact, if $(u_i^C, \varphi_i^C, t_{ij}^C, m_{ij}^C)$ is any elastic field free of singularities within S_0 (thus $v_j(B, C)$ is divergence free in S_0), we can write

$$\Pi_{AB} = \int_{S_0} [(t_{ji}^A - t_{ji}^C) u_i^B + (m_{ji}^A - m_{ji}^C) \varphi_i^B \\ - t_{ji}^B (u_i^A - u_i^C) - m_{ji}^B (\varphi_i^A - \varphi_i^C)] n_j \mathrm{d}S, \tag{22.125}$$

Since the field C is arbitrary field without singularities within S_0, $(A - C)$ field in (22.125) can be any field which has the same singularities within S_0 as does the field A. In applications it is convenient to take $(A - C)$ to be the elastic field of internal sources A, considered to be emerged in an infinite body. Thus, the interaction potential energy between the internal system A and external system B can be written as

$$\Pi_{AB} = \int_{S_0} (t_{ji}^\infty u_i^B + m_{ji}^\infty \varphi_i^B - t_{ji}^B u_i^\infty - m_{ji}^B \varphi_i^\infty) n_j \mathrm{d}S. \tag{22.126}$$

This is a generalization of the corresponding Eshelby's result from classical elasticity.

22.11.4 The Force on an Elastic Singularity

The force on a singularity A due to external system B can be defined as the negative gradient of interaction energy with respect to the location of the singularity, i.e.,

$$J_k = -\frac{\partial \Pi_{AB}}{\partial x_k^A}. \tag{22.127}$$

To elaborate on this expression, it is convenient to use (22.126) for the interaction energy Π_{AB}. The stress field in an infinite body due to singularity at A evidently satisfies the property

$$t_{ij}^{\infty}(x_k, x_k^A + \mathrm{d}x_k^A) = t_{ij}^{\infty}(x_k - \mathrm{d}x_k^A, x_k^A), \tag{22.128}$$

because moving the singularity for $\mathrm{d}x_k^A$ toward the point of observation x_k, or approaching the point of observation toward the singularity by $\mathrm{d}x_k^A$, equally effects the stress at the point of observation. Thus,

$$\frac{\partial t_{ij}^{\infty}}{\partial x_k^A} = -\frac{\partial t_{ij}^{\infty}}{\partial x_k}. \tag{22.129}$$

Analogous expressions apply to couple-stress, displacement and rotation gradients. Substitution into (22.127), therefore, gives

$$J_k = \int_{S_0} \left(t_{ji,k}^{\infty} u_i^B + m_{ji,k}^{\infty} \varphi_i^B - t_{ji}^B u_{i,k}^{\infty} - m_{ji}^B \varphi_{i,k}^{\infty}\right) n_j \mathrm{d}S. \tag{22.130}$$

Since $t_{ij}^A = t_{ij}^{\infty} + t_{ij}^I$ (and similarly for other fields), where the superscript I denotes the image field, free of singularities within S_0, (22.130) can be rewritten as

$$J_k = \int_{S_0} \left(t_{ji,k}^A u_i^B + m_{ji,k}^A \varphi_i^B - t_{ji}^B u_{i,k}^A - m_{ji}^B \varphi_{i,k}^A\right) n_j \mathrm{d}S. \tag{22.131}$$

The terms within the brackets in the integrand are equal to $\vartheta_{jk}(A, B)$, defined by (22.115). Recall that $\vartheta_{jk}(I, B)$ is divergence free within S_0.

The right-hand side of (22.131) is symmetric with respect to superscripts A and B. Thus, the force on a singularity can also be expressed as

$$J_k = \int_{S_0} \left(t_{ji,k}^B u_i^A + m_{ji,k}^B \varphi_i^A - t_{ji}^A u_{i,k}^B - m_{ji}^A \varphi_{i,k}^B\right) n_j \mathrm{d}S. \tag{22.132}$$

To see that the right-hand sides of (22.131) and (22.132) are equal, we can form their difference, which is

$$\int_{S_0} [\vartheta_{jk}(A, B) - \vartheta_{jk}(B, A)] n_j \mathrm{d}S = \int_{S_0} v_{j,k}(A, B) n_j \mathrm{d}S = \int_{S_0} v_{j,j}(A, B) n_k \mathrm{d}S. \tag{22.133}$$

The last integral is equal to zero, because from (22.111) we obtain $v_{j,j}(A, B) = 0$, in view of the reciprocity properties and (22.118). In (22.133) we also assumed that u_i^A is single-valued on S_0.

22.12 Energy-Momentum Tensor of a Couple-Stress Field

As in classical elasticity, it is possible to develop a general expression for the force on an elastic singularity. To that goal, consider a body of volume V, loaded by the force traction T_k and couple traction M_k over its external surface S. The body contains a singularity which is a source of internal stress (*e.g.*, a dislocation, an interstitial atom, or other lattice defect). The total potential energy is

$$\Pi = \int_V W \mathrm{d}V - \int_S (T_j u_j + M_j \varphi_j) \mathrm{d}S, \tag{22.134}$$

22.12. Energy-Momentum Tensor

where W is the elastic strain energy density, defined in linear couple-stress theory by (22.13). If the singularity is moved a small distance ϵ in the direction n_k, the potential energy alters due to the change of the elastic strain energy and the load potential. The change of the elastic strain energy is

$$\delta_k \int_V W \mathrm{d}V = -\epsilon \int_S W n_k \mathrm{d}S + \int_S [T_j(\delta_k u_j + \epsilon u_{j,k}) + M_j(\delta_k \varphi_j + \epsilon \varphi_{j,k})] \mathrm{d}S, \quad (22.135)$$

to within linear terms in ϵ. This can be established by the same consideration as in the original Eshelby's derivation of classical elasticity. The symbol δ_k indicates a variation associated with an infinitesimal displacement of the singularity in the direction n_k. The change of the load potential due to displacement of the singularity is

$$\delta_k \int_S (T_j u_j + M_j \varphi_j) \mathrm{d}S = \int_S (T_j \delta_k u_j + M_j \delta_k \varphi_j) \mathrm{d}S + O(\epsilon^2). \quad (22.136)$$

Thus, the total change of the potential energy becomes

$$\delta_k \Pi = -\epsilon \int_S (W n_k - T_j u_{j,k} - M_j \varphi_{j,k}) \, \mathrm{d}S + O(\epsilon^2). \quad (22.137)$$

Since the force on the singularity can be defined as

$$J_k = -\lim_{\epsilon \to 0} \frac{\delta_k \Pi}{\epsilon}, \quad (22.138)$$

we obtain, from (22.137),

$$J_k = \int_S (W n_k - T_j u_{j,k} - M_j \varphi_{j,k}) \mathrm{d}S. \quad (22.139)$$

This can be rewritten as

$$J_k = \int_S \mathcal{P}_{ik} n_i \, \mathrm{d}S, \quad (22.140)$$

where

$$\mathcal{P}_{ik} = W \delta_{ik} - t_{ij} u_{j,k} - m_{ij} \varphi_{j,k} \quad (22.141)$$

is the energy-momentum tensor of the couple-stress elastic field.

Pursuing Eshelby's (1956) analysis further, assume that in addition to a considered singularity A and external load B, there are other sources of internal stress, collectively denoted by C. Each of the elastic fields in the body, such as the displacement field u_i, can be represented as

$$u_i = u_i^\infty + u_i^I + u_i^B + u_i^C, \quad (22.142)$$

where the superscript I designates the image field of the singularity A, considered to be in an infinite medium. The total force on the singularity A is then

$$J_k = J_k^I + J_k^B + J_k^C, \quad (22.143)$$

where

$$J_k^X = \int_S (t_{ji,k}^\infty u_i^X + m_{ji,k}^\infty \varphi_i^X - t_{ji}^X u_{i,k}^\infty - m_{ji}^X \varphi_{i,k}^\infty) n_j \mathrm{d}S, \quad (22.144)$$

for each of the superscripts $X = I, B, C$. Indeed, a typical cross term in (22.139) is

$$(X, Y) = \int_S \left[\frac{1}{2}(\sigma^X_{lk} e^Y_{kl} + m^X_{lk}\varphi^Y_{k,l})n_i - (t^X_{jk} u^Y_{k,i} + m^X_{jk}\varphi^Y_{k,i})n_j\right] dS. \tag{22.145}$$

This can be rewritten, by using (22.119), as

$$(X, Y) = \int_S \left[\frac{1}{2}(t^X_{lk} u^Y_k + m^X_{lk}\varphi^Y_k)_{,i} n_l - (t^X_{jk} u^Y_{k,i} + m^X_{jk}\varphi^Y_{k,i})n_j\right] dS. \tag{22.146}$$

The property was employed that the surface integral of $\Phi_{,l} n_i$ is equal to that of $\Phi_{,i} n_l$ for any tensor field Φ defined within S. Thus, upon differentiation, it follows that

$$(X, Y) = \frac{1}{2} \int_S \vartheta_{ji}(X, Y) n_j dS. \tag{22.147}$$

This is equal to zero because $\vartheta_{ji}(X, Y)$ is a divergence free field, unless one or other of the labels X and Y stands for ∞. The term (∞, ∞) also vanishes because the products of the elastic fields in an infinite medium rapidly approach zero at large distances from the singularity.

Returning to (22.139), if S does not embrace a singularity, a conservation law $J_k = 0$ of couple-stress elasticity is obtained. There is also a conservation law $L_k = 0$, where

$$L_k = \int_S \epsilon_{kij}(W x_j n_i + T_i u_j + M_i \varphi_j - T_l u_{l,i} x_j - M_l \varphi_{l,i} x_j) dS, \tag{22.148}$$

i.e., written more compactly by using the energy-momentum tensor,

$$L_k = \int_S \epsilon_{kij}(\mathcal{P}_{li} x_j + t_{li} u_j + m_{li}\varphi_j) n_l dS. \tag{22.149}$$

22.13 Basic Equations of Micropolar Elasticity

In the infinitesimal micropolar elasticity the deformation is described by the displacement vector u_i and an independent rotation vector φ_i, which are both functions of the position vector x_i. A surface element dS transmits a force vector $T_i\, dS$ and a couple vector $M_i\, dS$. The surface forces are in equilibrium with the nonsymmetric Cauchy stress t_{ij}, and the surface couples are in equilibrium with the nonsymmetric couple-stress m_{ij}, such that

$$T_i = n_j t_{ji}, \quad M_i = n_j m_{ji}, \tag{22.150}$$

where n_j are the components of the unit vector orthogonal to the surface element under consideration. In the absence of body forces and body couples, the integral forms of the force and moment equilibrium conditions are

$$\int_S T_i\, dS = 0, \quad \int_S (\epsilon_{ijk} x_j T_k + M_i) dS = 0. \tag{22.151}$$

Upon using (22.150) and the Gauss divergence theorem, equation (22.151) yields the differential equations of equilibrium

$$t_{ji,j} = 0, \quad m_{ji,j} + \epsilon_{ijk} t_{jk} = 0. \tag{22.152}$$

22.13. Basic Equations of Micropolar Elasticity

For elastic deformations of micropolar continuum, the increase of the strain energy is due to external work done by the surface forces and couples, *i.e.*,

$$\int_V \dot{W} \, dV = \int_S (T_i \dot{u}_i + M_i \dot{\varphi}_i) \, dS. \tag{22.153}$$

The strain energy per unit volume is W, and the superposed dot denotes the time derivative. Incorporating (22.150) and (22.152) and using the divergence theorem gives

$$\dot{W} = t_{ij} \dot{\gamma}_{ij} + m_{ij} \dot{\kappa}_{ij}, \tag{22.154}$$

where

$$\gamma_{ij} = u_{j,i} - \epsilon_{ijk}\varphi_k, \quad \kappa_{ij} = \varphi_{j,i} \tag{22.155}$$

are the nonsymmetric strain and curvature tensors, respectively. The symmetric and antisymmetric parts of γ_{ij} are

$$\gamma_{(ij)} = e_{ij} = \frac{1}{2}(u_{j,i} + u_{i,j}),$$

$$\gamma_{<ij>} = \omega_{ij} - \epsilon_{ijk}\varphi_k, \quad \omega_{ij} = \frac{1}{2}(u_{j,i} - u_{i,j}). \tag{22.156}$$

In general, both e_{kk} and κ_{kk} are different from zero. In addition, there is an identity $\kappa_{ij,k} = \kappa_{kj,i} = \varphi_{j,ik}$. Assuming that the strain energy is a function of the strain and curvature tensors, $W = W(\gamma_{ij}, \kappa_{ij})$, the differentiation and the comparison with (22.154) establishes the constitutive relations of micropolar elasticity

$$t_{ij} = \frac{\partial W}{\partial \gamma_{ij}}, \quad m_{ij} = \frac{\partial W}{\partial \kappa_{ij}}. \tag{22.157}$$

In the case of material linearity, the strain energy is a quadratic function of the strain and curvature components

$$W = \frac{1}{2} C_{ijkl} \gamma_{ij} \gamma_{kl} + \frac{1}{2} K_{ijkl} \kappa_{ij} \kappa_{kl}. \tag{22.158}$$

The fourth-order tensors of micropolar elastic moduli are C_{ijkl} and K_{ijkl}. Since the strain and curvature tensors are not symmetric, only reciprocal symmetries hold $C_{ijkl} = C_{klij}$ and $K_{ijkl} = K_{klij}$. The stresses associated with (22.158) are $t_{ij} = C_{ijkl} \gamma_{kl}$ and $m_{ij} = K_{ijkl} \kappa_{kl}$. In the case of isotropic micropolar elasticity, we have

$$C_{ijkl} = (\mu + \bar{\mu}) \delta_{ik} \delta_{jl} + (\mu - \bar{\mu}) \delta_{il} \delta_{jk} + \lambda \delta_{ij} \delta_{kl},$$

$$K_{ijkl} = (\alpha + \bar{\alpha}) \delta_{ik} \delta_{jl} + (\alpha - \bar{\alpha}) \delta_{il} \delta_{jk} + \beta \delta_{ij} \delta_{kl}, \tag{22.159}$$

where $\mu, \bar{\mu}, \lambda$ and $\alpha, \bar{\alpha}, \beta$ are the Lamé-type constants of isotropic micropolar elasticity. The symmetric and antisymmetric parts of the stress tensors are then

$$t_{(ij)} = 2\mu \, e_{ij} + \lambda \, e_{kk} \delta_{ij}, \quad t_{<ij>} = 2\bar{\mu} (\omega_{ij} - \epsilon_{ijk}\varphi_k),$$

$$m_{(ij)} = 2\alpha \, \kappa_{(ij)} + \beta \, \kappa_{kk} \delta_{ij}, \quad m_{<ij>} = 2\bar{\alpha} \, \kappa_{<ij>}. \tag{22.160}$$

More generally, suppose that the elastic strain energy of a nonlinear isotropic material is given by

$$W = W\left(I_\gamma, II_\gamma, \bar{II}_\gamma, III_\gamma, I_\kappa, II_\kappa, \bar{II}_\kappa, III_\kappa\right), \tag{22.161}$$

where

$$I_\gamma = \gamma_{kk}, \quad II_\gamma = \gamma_{ij}\gamma_{ij}, \quad \bar{II}_\gamma = \gamma_{ij}\gamma_{ji}, \quad III_\gamma = \frac{1}{6}\epsilon_{ijk}\epsilon_{lmn}\gamma_{il}\gamma_{jm}\gamma_{kn},$$

and similarly for the first-, second-, and third-order invariants of the curvature tensor κ_{ij}. It follows that

$$\begin{aligned} t_{ij} &= c_1\delta_{ij} + c_2\gamma_{ij} + \bar{c}_2\gamma_{ji} + c_3\epsilon_{ikl}\epsilon_{jmn}\gamma_{km}\gamma_{ln}, \\ m_{ij} &= k_1\delta_{ij} + k_2\kappa_{ij} + \bar{k}_2\kappa_{ji} + k_3\epsilon_{ikl}\epsilon_{jmn}\kappa_{km}\kappa_{ln}, \end{aligned} \quad (22.162)$$

with

$$\begin{aligned} c_1 &= \frac{\partial W}{\partial I_\gamma}, & c_2 &= 2\frac{\partial W}{\partial II_\gamma}, & \bar{c}_2 &= 2\frac{\partial W}{\partial \bar{II}_\gamma}, & c_3 &= \frac{1}{2}\frac{\partial W}{\partial III_\gamma}, \\ k_1 &= \frac{\partial W}{\partial I_\kappa}, & k_2 &= 2\frac{\partial W}{\partial II_\kappa}, & \bar{k}_2 &= 2\frac{\partial W}{\partial \bar{II}_\kappa}, & k_3 &= \frac{1}{2}\frac{\partial W}{\partial III_\kappa}. \end{aligned} \quad (22.163)$$

22.14 Noether's Theorem of Micropolar Elasticity

The original Noether's (1918) theorem on invariant variational principles states that there is a conservation law for the Euler–Lagrange differential equations associated with each infinitesimal symmetry group of the Lagrangian functional. A comprehensive treatment of the general and various restricted forms of Noether's theorem, with a historical outline, can be found in the book by Olver (1986). Noether's theorem was applied by Günther (1962), and Knowles and Sternberg (1972) to derive the conservation integrals of infinitesimal nonpolar elasticity. When evaluated over a closed surface that does not embrace any singularity, these integrals give rise to conservation laws $J_k = 0$, $L_k = 0$, and $M = 0$. The law $J_k = 0$ applies to anisotropic linear or nonlinear material, the law $L_k = 0$ to isotropic linear or nonlinear material, and $M = 0$ to anisotropic linear material. If the surface embraces a singularity or inhomogeneity (defect), Eshelby (1951, 1956) has shown that the value of J_k is not equal to zero but represents a configurational or energetic force on the embraced defect (vacancy, inclusion, dislocation). As discussed in Chapter 21, the path-independent J integral of plane fracture mechanics has proved to be of great practical importance in modern fracture mechanics, allowing the prediction of the behavior at the crack tip from the values of the remote field quantities (Rice, 1968a, 1968b). Budiansky and Rice (1973) interpreted the L_k and M integrals as the energetic forces (potential energy release rates) conjugate to rotation (by erosion/addition of material) and self-similar expansion (erosion) of the traction-free void. Freund (1978) used the M conservation law for certain plane elastic crack problems to calculate the elastic stress intensity factor without solving the corresponding boundary value problem.

Consider a family of coordinate mappings defined by a vector-valued function

$$\hat{\mathbf{x}} = \mathbf{f}(\mathbf{x}, \eta), \quad \eta \in (-\eta_*, \eta_*), \quad (22.164)$$

such that $\mathbf{f}(\mathbf{x}, 0) = \mathbf{x}$ for all position vectors $\mathbf{x}$. Consider also the families of the displacement and rotation mappings

$$\hat{\mathbf{u}} = \mathbf{g}(\mathbf{u}, \eta), \quad \hat{\varphi} = \mathbf{h}(\varphi, \eta), \quad (22.165)$$

22.14. Noether's Theorem of Micropolar Elasticity

such that $\mathbf{g}(\mathbf{u}, 0) = \mathbf{u}$ and $\mathbf{h}(\boldsymbol{\varphi}, 0) = \boldsymbol{\varphi}$ for all displacement and rotation vectors $\mathbf{u} = \mathbf{u}(\mathbf{x})$ and $\boldsymbol{\varphi} = \boldsymbol{\varphi}(\mathbf{x})$. Finally, introduce a one-parameter family of functionals

$$E_\eta = \int_{\hat{V}} W(\hat{\gamma}_{ij}, \hat{\kappa}_{ij}) \, d\hat{V}, \tag{22.166}$$

where

$$\hat{\gamma}_{ij} = \frac{\partial \hat{u}_j}{\partial \hat{x}_i} - \epsilon_{ijk} \hat{\varphi}_k, \quad \hat{\kappa}_{ij} = \frac{\partial \hat{\varphi}_j}{\partial \hat{x}_i}, \tag{22.167}$$

and

$$d\hat{V} = \det\left(\frac{\partial \hat{x}_j}{\partial x_i}\right) dV = \det(f_{j,i}) \, dV. \tag{22.168}$$

When the parameter η is equal to zero, we have

$$E_0 = E = \int_V W(\gamma_{ij}, \kappa_{ij}) \, dV, \tag{22.169}$$

which is the total strain energy within the volume V. The family E_η is, therefore, the family of functionals induced from the functional E by the families of mappings $\mathbf{f}$, $\mathbf{g}$, and $\mathbf{h}$.

Definition: The functional E is considered to be invariant at $(\mathbf{u}, \boldsymbol{\varphi})$ with respect to $\mathbf{f}$, $\mathbf{g}$, and $\mathbf{h}$, if

$$E_\eta = E, \quad \eta \in (-\eta_*, \eta_*), \tag{22.170}$$

and infinitesimally invariant if

$$\left(\frac{\partial E_\eta}{\partial \eta}\right)_{\eta=0} = 0. \tag{22.171}$$

Theorem: If $\mathbf{u}$ and $\boldsymbol{\varphi}$ satisfy the equilibrium equations

$$\frac{\partial}{\partial x_j}\left(\frac{\partial W}{\partial \gamma_{ji}}\right) = 0, \quad \frac{\partial}{\partial x_j}\left(\frac{\partial W}{\partial \kappa_{ji}}\right) + \epsilon_{ijk} t_{jk} = 0, \tag{22.172}$$

for all $\mathbf{x}$ in V, then the total strain energy E is infinitesimally invariant at $(\mathbf{u}, \boldsymbol{\varphi})$ with respect to mappings $\mathbf{f}$, $\mathbf{g}$, and $\mathbf{h}$, if and only if

$$\frac{\partial}{\partial x_i}(a_i W + b_j t_{ij} + c_j m_{ij}) = 0, \tag{22.173}$$

where

$$a_i = f_i'(\mathbf{x}, 0),$$

$$b_i = g_i'(\mathbf{u}, 0) - f_k'(\mathbf{x}, 0) u_{i,k}, \tag{22.174}$$

$$c_i = h_i'(\boldsymbol{\varphi}, 0) - f_k'(\mathbf{x}, 0) \varphi_{i,k}.$$

The prime designates the derivative with respect to the parameter η, such that

$$f'_i(\mathbf{x}, 0) = \left[\frac{\partial}{\partial \eta} f_i(\mathbf{x}, \eta)\right]_0. \qquad (22.175)$$

For brevity, the subscript 0 is used to indicate that the quantity within the brackets is evaluated at $\eta = 0$. The condition (22.173) implies the conservation law in the integral form

$$\int_S (a_i n_i W + T_i b_i + M_i c_i) \, dS = 0, \qquad (22.176)$$

for every surface S bounding a regular subregion of V.

Proof: By differentiating (22.166) with respect to η and then setting $\eta = 0$, there follows

$$\left(\frac{\partial E_\eta}{\partial \eta}\right)_0 = \int_V \left[W f'_{k,k}(\mathbf{x}, 0) + \frac{\partial W}{\partial \gamma_{ij}}\left(\frac{\partial \hat{\gamma}_{ij}}{\partial \eta}\right)_0 + \frac{\partial W}{\partial \kappa_{ij}}\left(\frac{\partial \hat{\kappa}_{ij}}{\partial \eta}\right)_0\right] dV. \qquad (22.177)$$

The partial derivatives with respect to η appearing in (22.177) can be evaluated by using (22.167) and (22.168). This gives

$$\left(\frac{\partial \hat{\gamma}_{ij}}{\partial \eta}\right)_0 = \frac{\partial g'_j(\mathbf{u}, 0)}{\partial u_k} u_{k,i} - f'_{k,i}(\mathbf{x}, 0) u_{j,k} - \epsilon_{ijk} h'_k(\varphi, 0),$$

$$\left(\frac{\partial \hat{\kappa}_{ij}}{\partial \eta}\right)_0 = \frac{\partial h'_j(\varphi, 0)}{\partial \varphi_k} \kappa_{ik} - f'_{k,i}(\mathbf{x}, 0) \kappa_{kj}, \qquad (22.178)$$

$$\left[\frac{\partial (d\hat{V})}{\partial \eta}\right]_0 = f'_{k,k}(\mathbf{x}, 0) \, dV.$$

The integrand in (22.177) is continuous on V, so that the integral vanishes if and only if its integrand vanishes at each $\mathbf{x}$. The leading term of the integrand can be eliminated by using the identity

$$\frac{\partial}{\partial x_k}[W f'_k(\mathbf{x}, 0)] = W f'_{k,k}(\mathbf{x}, 0) + f'_k(\mathbf{x}, 0)\left(\frac{\partial W}{\partial \gamma_{ij}} \gamma_{ij,k} + \frac{\partial W}{\partial \kappa_{ij}} \kappa_{ij,k}\right).$$

Accordingly, the integrand in (22.177) becomes

$$\frac{\partial}{\partial x_k}[W f'_k(\mathbf{x}, 0)] + D_{ij} \frac{\partial W}{\partial \gamma_{ij}} + d_{ij} \frac{\partial W}{\partial \kappa_{ij}} = 0, \qquad (22.179)$$

where

$$D_{ij} = \left(\frac{\partial \hat{\gamma}_{ij}}{\partial \eta}\right)_0 - f'_k(\mathbf{x}, 0) \gamma_{ij,k},$$

$$d_{ij} = \left(\frac{\partial \hat{\kappa}_{ij}}{\partial \eta}\right)_0 - f'_k(\mathbf{x}, 0) \kappa_{ij,k}. \qquad (22.180)$$

Introducing the vectors b_i and c_i, defined by (22.174), it can be readily verified that

$$b_{j,i} = D_{ij} + \epsilon_{ijk} c_k, \quad c_{j,i} = d_{ij}. \qquad (22.181)$$

22.15. Conservation Integrals

Thus, there is an identity

$$\frac{\partial}{\partial x_i}\left(\frac{\partial W}{\partial \gamma_{ij}} b_j + \frac{\partial W}{\partial \kappa_{ij}} c_j\right) = \frac{\partial W}{\partial \gamma_{ij}} D_{ij} + \frac{\partial W}{\partial \kappa_{ij}} d_{ij} \qquad (22.182)$$
$$+ \frac{\partial W}{\partial \gamma_{ij}} \epsilon_{ijk} c_k + \frac{\partial}{\partial x_i}\left(\frac{\partial W}{\partial \gamma_{ij}}\right) b_j + \frac{\partial}{\partial x_i}\left(\frac{\partial W}{\partial \kappa_{ij}}\right) c_j.$$

In view of the equilibrium equations (22.172), the last two terms on the right-hand side of (22.182) are together equal to $-c_j \epsilon_{jkl} t_{kl}$, so that

$$\frac{\partial W}{\partial \gamma_{ij}} \epsilon_{ijk} c_k - c_j \epsilon_{jkl} t_{kl} = 0. \qquad (22.183)$$

Consequently, (22.182) reduces to

$$\frac{\partial}{\partial x_i}(t_{ij} b_j + m_{ij} c_j) = \frac{\partial W}{\partial \gamma_{ij}} D_{ij} + \frac{\partial W}{\partial \kappa_{ij}} d_{ij}. \qquad (22.184)$$

Substituting (22.184) into (22.179) gives the desired result of (22.173). The conservation law (22.176) follows by applying the Gauss divergence theorem. The Knowles and Sternberg's (1972) proof for infinitesimal nonpolar elasticity follows by taking $W = W(e_{ij})$ and by setting $m_{ij} = 0$ and $t_{<ij>} = 0$.

22.15 Conservation Integrals in Micropolar Elasticity

The strain energy in micropolar elasticity is invariant under the mappings

$$\hat{x}_i = x_i^0 \eta + Q_{ij}(\eta) x_j, \quad \hat{u}_i = Q_{ij}(\eta) u_j, \quad \hat{\varphi}_i = Q_{ij}(\eta) \varphi_j, \qquad (22.185)$$

where x_i^0 is a constant vector, $Q_{ij}(\eta)$ is an orthogonal tensor in the case of an isotropic material, and $Q_{ij} = \delta_{ij}$ (Kronecker delta) in the case of a fully anisotropic material. Thus, since the invariance necessarily implies an infinitesimal invariance, the corresponding conservation laws follow from (22.176). Indeed, we have

$$f_i'(\mathbf{x}, 0) = x_i^0 + q_{ij} x_j, \quad g_i'(\mathbf{u}, 0) = q_{ij} u_j, \quad h_i'(\varphi, 0) = q_{ij} \varphi_j, \qquad (22.186)$$

and

$$a_i = x_i^0 + q_{ij} x_j,$$
$$b_i = q_{ij} u_j - (x_m^0 + q_{mn} x_n) u_{i,m}, \qquad (22.187)$$
$$c_i = q_{ij} \varphi_j - (x_m^0 + q_{mn} x_n) \varphi_{i,m},$$

where $q_{ij} = Q_{ij}(0)$. When this is substituted into (22.176), we obtain

$$x_i^0 \int_S (W n_i - T_i u_{i,j} - M_i \varphi_{i,j}) \, dS$$
$$+ q_{ij} \int_S (W n_i x_j + T_i u_j + M_i \varphi_j - T_l u_{l,i} x_j - M_l \varphi_{l,i} x_j) \, dS = 0.$$

For a fully anisotropic material $q_{ij} = 0$, and by choosing the vector x_i^0 to be a unit vector in the direction k ($x_i^0 = \delta_{ik}$, for each value of $k = 1, 2, 3$), the above equation gives

$$J_k = \int_S (W n_k - T_k u_{k,j} - M_k \varphi_{k,j}) \, dS = 0. \qquad (22.188)$$

For an isotropic material q_{ij} are the components of an orthogonal tensor, and we can take $q_{ij} = \epsilon_{ijk}$ for $k = 1, 2, 3$, so that, in addition to (22.188), there is a conservation law

$$L_k = \epsilon_{ijk} \int_S (W n_i x_j + T_i u_j + M_i \varphi_j - T_l u_{l,i} x_j - M_l \varphi_{l,i} x_j) \, dS = 0. \tag{22.189}$$

If the micropolar terms are omitted, the above conservation laws reduce to those of the classical nonpolar elasticity (Knowles and Sternberg, 1972; Budiansky and Rice, 1973).

Using the energy-momentum tensor of micropolar elastic field,

$$\mathcal{P}_{ij} = W \delta_{ij} - t_{ik} u_{k,j} - m_{ik} \varphi_{k,j}, \tag{22.190}$$

the derived conservation integrals can be recast as

$$J_k = \int_S \mathcal{P}_{jk} n_j \, dS, \tag{22.191}$$

$$L_k = \epsilon_{ijk} \int_S (\mathcal{P}_{li} x_j + t_{li} u_j + m_{li} \varphi_j) n_l \, dS. \tag{22.192}$$

22.16 Conservation Laws for Plane Strain Micropolar Elasticity

In two-dimensional plane strain problems within (x_1, x_2) plane, the components φ_3, M_3, m_{13}, and m_{23} are generally different from zero, whereas other rotation, moment, and couple-stress components are equal to zero. By taking S to be a cylindrical surface with its generatrix parallel to x_3 axis and with its two flat bases bounded by a curve C, integration in (22.191) and (22.192) gives (per unit length in x_3 direction)

$$J_\alpha = \int_C \mathcal{P}_{\beta\alpha} n_\beta \, dC, \tag{22.193}$$

$$L = \epsilon_{\alpha\beta 3} \int_S (\mathcal{P}_{\gamma\alpha} x_\beta + t_{\gamma\alpha} u_\beta) n_\gamma \, dC. \tag{22.194}$$

The energy-momentum tensor of the plane strain micropolar elasticity is

$$\mathcal{P}_{\alpha\beta} = W \delta_{\alpha\beta} - t_{\alpha\gamma} u_{\gamma,\beta} - m_{\alpha 3} \varphi_{3,\beta}. \tag{22.195}$$

The summation in repeated Greek indices is over 1 and 2. The J_1 integral from (22.193) was used by Atkinson and Leppington (1974) to calculate the energy release rate for a semi-infinite crack within a strip of thickness h. Xia and Hutchinson (1996) also used the J_1 integral to study the elastoplastic crack tip field in a strain-gradient dependent material described by the deformation-type theory of plasticity (see Chapter 26).

22.17 *M* Integral of Micropolar Elasticity

In contrast to classical elasticity, there is no M conservation law of micropolar elasticity. In two-dimensional case this was originally observed by Atkinson and Leppington (1977), and elaborated in the three-dimensional context by Lubarda and Markenscoff (2003). Consider a family of scale-changes

$$\hat{x}_i = (1 + \eta) x_i, \quad \hat{u}_i = \left(1 - \frac{\eta}{2}\right) u_i. \tag{22.196}$$

22.17. *M* Integral of Micropolar Elasticity

It is easily verified that the total strain energy of a nonpolar elastic material with a quadratic strain energy representation is infinitesimally invariant under (22.196). This is, however, not so in the case of micropolar elasticity, because the material length parameter, whose square is the ratio of the representative micropolar elastic moduli $l^2 = [K]/[C]$, remains unaltered by the transformation (22.196). Indeed, since the angle change corresponding to (22.196) is

$$\hat{\varphi}_i = \frac{1 - \eta/2}{1 + \eta} \varphi_i , \qquad (22.197)$$

which follows from a simple dimensional argument ($u \sim x\varphi$, $\hat{u} \sim \hat{x}\hat{\varphi}$), we have

$$\hat{\gamma}_{ij} = \frac{1 - \eta/2}{1 + \eta} \gamma_{ij}, \quad \hat{\kappa}_{ij} = \frac{1 - \eta/2}{(1 + \eta)^2} \kappa_{ij} . \qquad (22.198)$$

We assume here that $\eta_* < 1$. Thus, for a linear micropolar material with a quadratic strain energy representation,

$$E_\eta = \frac{1}{2} \left(1 - \frac{\eta}{2}\right)^2 \int_V \left[(1 + \eta) t_{ij} \gamma_{ij} + \frac{1}{1 + \eta} m_{ij} \kappa_{ij}\right] dV , \qquad (22.199)$$

so that $E_\eta \neq E_0 = E$, and

$$\left(\frac{\partial E_\eta}{\partial \eta}\right)_0 = -\int_V m_{ij} \kappa_{ij}\, dV \neq 0 . \qquad (22.200)$$

This shows that the total strain energy E is not infinitesimally invariant with respect to the considered family of scale changes. Consequently, there is no *M* conservation law in micropolar elasticity. Actually, the value of the *M* integral is equal to the expression in (22.200). This follows from

$$f_i'(\mathbf{x}, 0) = x_i, \quad g_i'(\mathbf{u}, 0) = -\frac{1}{2} u_i, \quad h_i'(\varphi, 0) = -\frac{3}{2} \varphi_i , \qquad (22.201)$$

and

$$M = \int_S \left(W x_i n_i - \frac{1}{2} T_i u_i - \frac{3}{2} M_i \varphi_i - T_i x_j u_{i,j} - M_i x_j \varphi_{i,j}\right) dS . \qquad (22.202)$$

Upon using the Gauss divergence theorem, the evaluation of the last integral gives

$$M = -\int_V m_{ij} \kappa_{ij}\, dV . \qquad (22.203)$$

In the derivation it is noted that

$$W_{,k} = t_{ij} \gamma_{ij,k} + m_{ij} \kappa_{ij,k} , \qquad (22.204)$$

and that, for the quadratic strain energy representation, the identities hold

$$t_{ij} \gamma_{ij,k} = t_{ij,k} \gamma_{ij}, \quad m_{ij} \kappa_{ij,k} = m_{ij,k} \kappa_{ij} . \qquad (22.205)$$

In terms of the energy-momentum tensor (22.190), the *M* integral of (22.202) can be rewritten as

$$M = \int_S \left(P_{ij} x_j - \frac{1}{2} t_{ij} u_j - \frac{3}{2} m_{ij} \varphi_j\right) n_i\, dS . \qquad (22.206)$$

The plane strain counterpart is

$$M = \int_C (\mathcal{P}_{\alpha\beta} x_\beta - m_{\alpha 3} \varphi_3) n_\alpha \, dC. \quad (22.207)$$

If the polar effects are neglected, the couple-stress vanishes and the conservation law $M = 0$ of the classical linear isotropic elasticity is obtained. This is

$$M = \int_S \left(W x_i n_i - \frac{1}{2} T_i u_i - T_i x_j u_{i,j} \right) dS = 0 \quad (22.208)$$

for three-dimensional nonpolar elasticity, and

$$M = \int_C (W x_\alpha n_\alpha - T_\alpha x_\beta u_{\alpha,\beta}) \, dC = 0 \quad (22.209)$$

for two-dimensional nonpolar elasticity.

22.18 Suggested Reading

Brulin, O., and Hsieh, R. K. T., eds. (1982), *Mechanics of Micropolar Media*, World Scientific, Singapore.

Dhaliwal, R. S., and Singh, A. (1987), Micropolar Thermoelasticity. In *Thermal Stresses II* (R. B. Hetnarski, ed.), pp. 269–328, Elsevier, Amsterdam.

Eringen, A. C. (1999), *Microcontinuum Field Theories*, Springer-Verlag, New York.

Jasiuk, I., and Ostoja-Starzewski, M. (1995), Planar Cosserat Elasticity of Materials with Holes and Intrusions, *Appl. Mech. Rev.*, Vol. 48, pp. S11–S18.

Koiter, W. T. (1964), Couple-Stresses in the Theory of Elasticity, *Proc. Ned. Akad. Wet. (B)*, Vol. 67, I: pp. 17–29, II: pp. 30–44.

Lubarda, V. A. (2003), Circular Inclusions in Anti-Plane Strain Couple Stress Elasticity, *Int. J. Solids Struct.*, Vol. 40, pp. 3827–3851.

Lubarda, V. A., and Markenscoff, X. (2003), On Conservation Integrals in Micropolar Elasticity, *Phil. Mag. A*, Vol. 83, pp. 1365–1377.

Mindlin, R. D., (1964), Micro-Structure in Linear Elasticity, *Arch. Ration. Mech. Anal.*, Vol. 16, pp. 51–78.

Mindlin, R. D., and Tiersten, H. F. (1962), Effects of Couple-Stresses in Linear Elasticity, *Arch. Ration. Mech. Anal.*, Vol. 11, pp. 415–448.

Nowacki, W. (1986), *Theory of Asymmetric Elasticity*, Pergamon, Oxford, UK.

Stojanović, R. (1970), *Recent Developments in the Theory of Polar Continua*, CISM Lecture Notes, Springer-Verlag, Wien.

Toupin, R. A. (1962), Perfectly Elastic Materials with Couple Stresses, *Arch. Ration. Mech. Anal.*, Vol. 11, pp. 385–414.

PART 5: THIN FILMS AND INTERFACES

23 Dislocations in Bimaterials

23.1 Introduction

This chapter contains a detailed analysis of straight (edge and screw) dislocations near the bimaterial interface or near the free surface of a semi-infinite medium. A general elastic study of straight dislocations and dislocation loops was presented earlier in Chapter 18. Here, we focus on those aspects of the dislocation analysis that are of importance to study the strain relaxation in thin films. We first derive the stress fields for straight dislocations near bimaterial interfaces and then specialize these results to dislocations near the free surfaces or rigid boundaries. An energy analysis of dislocations beneath the free surface of a semi-infinite body is also presented.

23.2 Screw Dislocation Near a Bimaterial Interface

Consider a screw dislocation near a bimaterial interface at a distance h from it (Fig. 23.1a). The Burgers vector of dislocation is in the z direction and has the magnitude b_z. The shear modulus and Poisson's ratio of the material (1) are μ_1 and ν_1; those of material (2) are μ_2 and ν_2. A complete adherence between the two materials is assumed, so that the displacement and traction components are continuous across the interface. The only nonvanishing displacement component is the out-of-plane displacement

$$u_z^{(1)} = \frac{b_z}{2\pi}(\theta_1 - c\,\theta_2), \qquad (23.1)$$

$$u_z^{(2)} = \frac{b_z}{2\pi}[(1+c)\theta_1 - c\pi], \qquad (23.2)$$

where the constant c is a nondimensional shear modulus mismatch parameter, defined by

$$c = \frac{\mu_1 - \mu_2}{\mu_1 + \mu_2}. \qquad (23.3)$$

The above displacement field is continuous across the interface, and the slip discontinuity of amount b_z is imposed along the cut from $x = h$ to $x \to \infty$. The two sets of polar coordinates

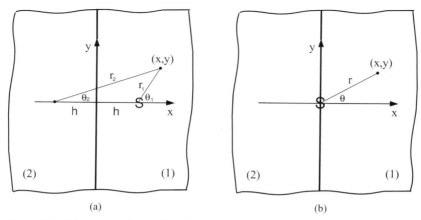

Figure 23.1. (a) A screw dislocation with the Burgers vector b_z at a distance h from the bimaterial interface. (b) An interface screw dislocation.

shown in Fig. 23.1a are related to the Cartesian coordinates (x, y) by
$$r_1^2 = (x - h)^2 + y^2, \quad r_2^2 = (x + h)^2 + y^2,$$
$$\tan \theta_1 = \frac{y}{x - h}, \quad \tan \theta_2 = \frac{y}{x + h}. \tag{23.4}$$

The utilized angles have the range
$$0 \le \theta_1 \le 2\pi, \quad -\pi \le \theta_2 \le \pi. \tag{23.5}$$

The stress components associated with the displacement field in (23.1) and (23.2) are obtained from Hooke's law as
$$\sigma_{zx}^{(1)} = \mu_1 \frac{\partial u_z^{(1)}}{\partial x} = -\frac{\mu_1 b_z}{2\pi} \left(\frac{y}{r_1^2} - c \frac{y}{r_2^2} \right), \tag{23.6}$$

$$\sigma_{zy}^{(1)} = \mu_1 \frac{\partial u_z^{(1)}}{\partial y} = \frac{\mu_1 b_z}{2\pi} \left(\frac{x - h}{r_1^2} - c \frac{x + h}{r_2^2} \right), \tag{23.7}$$

and
$$\sigma_{zx}^{(2)} = \mu_2 \frac{\partial u_z^{(2)}}{\partial x} = -\frac{\mu_2 b_z}{2\pi} (1 + c) \frac{y}{r_1^2}, \tag{23.8}$$

$$\sigma_{zy}^{(2)} = \mu_2 \frac{\partial u_z^{(2)}}{\partial y} = \frac{\mu_2 b_z}{2\pi} (1 + c) \frac{x - h}{r_1^2}. \tag{23.9}$$

The force exerted by the interface on the dislocation is
$$F_x = b_z \sigma_{zy}^{(1)}(h, 0). \tag{23.10}$$

Excluding the divergent part of the stress at the core center of the dislocation, this gives
$$F_x = -c \frac{\mu_1 b_z^2}{4\pi h}. \tag{23.11}$$

Depending on the sign of c, the dislocation may be attracted or repelled by the interface.

23.2. Screw Dislocation Near an Interface

23.2.1 Interface Screw Dislocation

For an interface screw dislocation (Fig. 23.1b) we take $h = 0$, $r_1 = r_2 = r$, $\theta_1 = \theta$, and

$$\theta_2 = \begin{cases} \theta, & y \geq 0^+, \\ \theta - 2\pi, & y \leq 0^-, \end{cases} \qquad (23.12)$$

with $0 \leq \theta \leq 2\pi$. The displacement field is

$$u_z^{(1)} = \frac{b_z}{2\pi}(1-c)\theta, \qquad y \geq 0^+, \qquad (23.13)$$

$$u_z^{(1)} = \frac{b_z}{2\pi}[(1-c)\theta + 2\pi c], \qquad y \leq 0^-, \qquad (23.14)$$

$$u_z^{(2)} = \frac{b_z}{2\pi}[(1+c)\theta - c\pi]. \qquad (23.15)$$

The corresponding stresses are

$$\sigma_{zx}^{(1)} = -\frac{\mu_1 b_z}{2\pi}\frac{(1-c)y}{r^2}, \qquad (23.16)$$

$$\sigma_{zy}^{(1)} = \frac{\mu_1 b_z}{2\pi}\frac{(1-c)x}{r^2}, \qquad (23.17)$$

and

$$\sigma_{zx}^{(2)} = -\frac{\mu_2 b_z}{2\pi}\frac{(1+c)y}{r^2}, \qquad (23.18)$$

$$\sigma_{zy}^{(2)} = \frac{\mu_2 b_z}{2\pi}\frac{(1+c)x}{r^2}. \qquad (23.19)$$

23.2.2 Screw Dislocation in a Homogeneous Medium

If two materials have the same shear modulus (*e.g.*, homogeneous medium), the parameter $c = 0$ and, with $h = 0$, the results reduce to

$$u_z = \frac{b_z}{2\pi}\theta, \qquad (23.20)$$

$$\sigma_{zx} = -\frac{\mu b_z}{2\pi}\frac{y}{r^2}, \qquad (23.21)$$

$$\sigma_{zy} = \frac{\mu b_z}{2\pi}\frac{x}{r^2}. \qquad (23.22)$$

23.2.3 Screw Dislocation Near a Free Surface

If a dislocation is near the free surface of a semi-infinite homogeneous medium, we take $\mu_2 = 0$ and $c = 1$. The displacement and stress fields in this case are

$$u_z = \frac{b_z}{2\pi}(\theta_1 - \theta_2), \qquad (23.23)$$

$$\sigma_{zx} = -\frac{\mu b_z}{2\pi}\left(\frac{y}{r_1^2} - \frac{y}{r_2^2}\right), \qquad (23.24)$$

$$\sigma_{zy} = \frac{\mu b_z}{2\pi}\left(\frac{x-h}{r_1^2} - \frac{x+h}{r_2^2}\right). \qquad (23.25)$$

The displacement discontinuity is imposed along the cut from $x = 0$ to $x = h$. The fields are recognized to be the superposition of two infinite medium fields, one for a dislocation at the point $x = h$, $y = 0$ and the other for an opposite (image) dislocation at the point $x = -h$, $y = 0$.

The force exerted by the free surface on the dislocation is

$$F_x = -\frac{\mu_1 b_z^2}{4\pi h}. \qquad (23.26)$$

The dislocation is attracted by the free surface.

23.2.4 Screw Dislocation Near a Rigid Boundary

If the second material is rigid, we take $\mu_2 \to \infty$ and $c = -1$. The displacement and stress fields are

$$u_z = \frac{b_z}{2\pi}(\theta_1 + \theta_2) - \frac{b_z}{2}, \qquad (23.27)$$

$$\sigma_{zx} = -\frac{\mu b_z}{2\pi}\left(\frac{y}{r_1^2} + \frac{y}{r_2^2}\right), \qquad (23.28)$$

$$\sigma_{zy} = \frac{\mu b_z}{2\pi}\left(\frac{x-h}{r_1^2} + \frac{x+h}{r_2^2}\right). \qquad (23.29)$$

The displacement vanishes along the rigid boundary, and the discontinuity is imposed along the cut from $x = h$ to $x \to \infty$. The fields are recognized to be the superposition of two infinite medium fields, one for a dislocation at the point $x = h$, $y = 0$ and the other for an alike dislocation at the point $x = -h$, $y = 0$.

The force exerted by the rigid boundary on the dislocation is

$$F_x = \frac{\mu_1 b_z^2}{4\pi h}. \qquad (23.30)$$

The dislocation is repelled by the rigid boundary.

23.3 Edge Dislocation (b_x) Near a Bimaterial Interface

Two edge dislocations near the bimaterial interface, one with the Burgers vector b_x and the other with the Burgers vector b_y, are shown in Figs. 23.2a and b. Consider first the stress field for an edge dislocation with the Burgers vector b_x. The corresponding Airy stress functions are (Dundurs, 1969)

$$\Phi^{(1)} = -k_1 b_x \left[r_1 \ln r_1 \sin \theta_1 + (a-1) r_2 \ln r_2 \sin \theta_2 \right.$$
$$\left. - qh\left(\sin 2\theta_2 - 2h\frac{\sin \theta_2}{r_2}\right) + a\beta r_2 \theta_2 \cos \theta_2 \right], \qquad (23.31)$$

$$\Phi^{(2)} = -ak_1 b_x [r_1 \ln r_1 \sin \theta_1 - \beta(r_1 \theta_1 \cos \theta_1 + 2h\theta_1)]. \qquad (23.32)$$

23.3. Edge Dislocation (b_x) Near an Interface

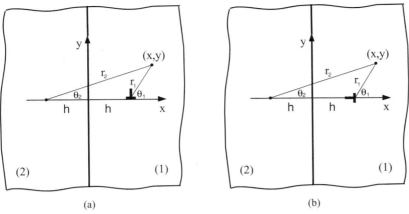

Figure 23.2. (a) An edge dislocation with the Burgers vector b_x at a distance h from the bimaterial interface. (b) The same for an edge dislocation with the Burgers vector b_y.

The parameters are introduced

$$k_1 = \frac{\mu_1}{2\pi(1-\nu_1)}, \quad a = \frac{1+\alpha}{1-\beta^2}, \quad q = \frac{\alpha-\beta}{1+\beta}, \tag{23.33}$$

and

$$\alpha = \frac{(1-\nu_1)\mu_2 - (1-\nu_2)\mu_1}{(1-\nu_1)\mu_2 + (1-\nu_2)\mu_1}, \quad 2\beta = \frac{(1-2\nu_1)\mu_2 - (1-2\nu_2)\mu_1}{(1-\nu_1)\mu_2 + (1-\nu_2)\mu_1}. \tag{23.34}$$

The Airy stress function yields the stress components according to

$$\sigma_{xx} = \frac{\partial^2 \Phi}{\partial y^2}, \quad \sigma_{yy} = \frac{\partial^2 \Phi}{\partial x^2}, \quad \sigma_{xy} = -\frac{\partial^2 \Phi}{\partial x \partial y}. \tag{23.35}$$

The polar coordinates counterparts are

$$\sigma_{rr} = \frac{1}{r}\frac{\partial \Phi}{\partial r} + \frac{1}{r^2}\frac{\partial^2 \Phi}{\partial \theta^2}, \quad \sigma_{\theta\theta} = \frac{\partial^2 \Phi}{\partial r^2}, \quad \sigma_{r\theta} = -\frac{\partial}{\partial r}\left(\frac{1}{r}\frac{\partial \Phi}{\partial \theta}\right). \tag{23.36}$$

The Tables 23.1–23.5 (reproduced from the results listed by Dundurs and Mura, 1964) give the displacement and stress components corresponding to some typical forms of the Airy stress function. In these tables, ϑ is the Kolosov constant defined, as in (8.136), by

$$\vartheta = \begin{cases} 3 - 4\nu, & \text{plane strain}, \\ \dfrac{3-\nu}{1+\nu}, & \text{plane stress}. \end{cases} \tag{23.37}$$

By using the results from Table 23.4, we obtain

$$\sigma_{xx}^{(1)} = -k_1 b_x y \left[\frac{3(x-h)^2 + y^2}{r_1^4} + q\frac{3(x+h)^2 + y^2}{r_2^4} \right. \\ \left. + 4qhx\frac{3(x+h)^2 - y^2}{r_2^6} + a\beta\frac{1}{r_2^2} \right], \tag{23.38}$$

Table 23.1. *Rectangular displacement components*

Φ	$2Gu_x$	$2Gu_y$
$r \ln r \cos\theta$	$\frac{1}{2}(\vartheta - 1)\ln r - \frac{x^2}{r^2}$	$\frac{1}{2}(\vartheta + 1)\theta - \frac{xy}{r^2}$
$r \ln r \sin\theta$	$-\frac{1}{2}(\vartheta + 1)\theta - \frac{xy}{r^2}$	$\frac{1}{2}(\vartheta - 1)\ln r + \frac{x^2}{r^2}$
$r\theta \cos\theta$	$\frac{1}{2}(\vartheta - 1)\theta + \frac{xy}{r^2}$	$-\frac{1}{2}(\vartheta + 1)\ln r - \frac{x^2}{r^2}$
$r\theta \sin\theta$	$\frac{1}{2}(\vartheta + 1)\ln r - \frac{x^2}{r^2}$	$\frac{1}{2}(\vartheta - 1)\theta - \frac{xy}{r^2}$
$\frac{\cos\theta}{r}$	$\frac{1}{r^2} + \frac{2x^2}{r^4}$	$\frac{2xy}{r^4}$
$\frac{\sin\theta}{r}$	$\frac{2xy}{r^4}$	$\frac{1}{r^2} - \frac{2x^2}{r^4}$
$\cos 2\theta$	$(\vartheta - 3)\frac{x}{r^2} + \frac{4x^3}{r^4}$	$-(\vartheta + 1)\frac{y}{r^2} + \frac{4x^2 y}{r^4}$
$\sin 2\theta$	$(\vartheta - 1)\frac{y}{r^2} + \frac{4x^2 y}{r^4}$	$(\vartheta + 3)\frac{x}{r^2} - \frac{4x^3}{r^4}$
θ	$\frac{y}{r^2}$	$-\frac{x}{r^2}$
$\ln r$	$-\frac{x}{r^2}$	$-\frac{y}{r^2}$

Table 23.2. *Radial displacement component*

Φ	$2\mu u_r$
$r \ln r \cos\theta$	$\frac{1}{2}[(\vartheta + 1)\theta \sin\theta + (\vartheta - 1)\ln r \cos\theta - \cos\theta]$
$r \ln r \sin\theta$	$\frac{1}{2}[-(\vartheta + 1)\theta \cos\theta + (\vartheta - 1)\ln r \sin\theta - \sin\theta]$
$r\theta \cos\theta$	$\frac{1}{2}[(\vartheta - 1)\theta \cos\theta - (\vartheta + 1)\ln r \sin\theta + \sin\theta]$
$r\theta \sin\theta$	$\frac{1}{2}[(\vartheta - 1)\theta \sin\theta + (\vartheta + 1)\ln r \cos\theta - \cos\theta]$
$\frac{\cos\theta}{r}$	$\frac{\cos\theta}{r^2}$
$\frac{\sin\theta}{r}$	$\frac{\sin\theta}{r^2}$
$\cos 2\theta$	$(\vartheta + 1)\frac{\cos 2\theta}{r}$
$\sin 2\theta$	$(\vartheta + 1)\frac{\sin 2\theta}{r}$
θ	0
$\ln r$	$-\frac{1}{r}$

23.3. Edge Dislocation (b_x) Near an Interface

Table 23.3. *Circumferential displacement component*

Φ	$2\mu u_\theta$
$r \ln r \cos\theta$	$\frac{1}{2}[(\vartheta+1)\theta\cos\theta - (\vartheta-1)\ln r \sin\theta - \sin\theta]$
$r \ln r \sin\theta$	$\frac{1}{2}[(\vartheta+1)\theta\sin\theta + (\vartheta-1)\ln r \cos\theta + \cos\theta]$
$r\theta \cos\theta$	$\frac{1}{2}[-(\vartheta-1)\theta\sin\theta - (\vartheta+1)\ln r \cos\theta - \cos\theta]$
$r\theta \sin\theta$	$\frac{1}{2}[(\vartheta-1)\theta\cos\theta - (\vartheta+1)\ln r \sin\theta - \sin\theta]$
$\dfrac{\cos\theta}{r}$	$\dfrac{\sin\theta}{r^2}$
$\dfrac{\sin\theta}{r}$	$-\dfrac{\cos\theta}{r^2}$
$\cos 2\theta$	$-(\vartheta-1)\dfrac{\sin 2\theta}{r}$
$\sin 2\theta$	$(\vartheta-1)\dfrac{\cos 2\theta}{r}$
θ	$-\dfrac{1}{r}$
$\ln r$	0

Table 23.4. *Rectangular stress components*

Φ	σ_{xx}	σ_{yy}	σ_{xy}
$r \ln r \cos\theta$	$-\dfrac{x}{r^2} + \dfrac{2x^3}{r^4}$	$\dfrac{3x}{r^2} - \dfrac{2x^3}{r^4}$	$-\dfrac{y}{r^2} + \dfrac{2x^2 y}{r^4}$
$r \ln r \sin\theta$	$\dfrac{y}{r^2} + \dfrac{2x^2 y}{r^4}$	$\dfrac{y}{r^2} - \dfrac{2x^2 y}{r^4}$	$\dfrac{x}{r^2} - \dfrac{2x^3}{r^4}$
$r\theta \cos\theta$	$-\dfrac{2x^2 y}{r^4}$	$-\dfrac{2y}{r^2} + \dfrac{2x^2 y}{r^4}$	$-\dfrac{2x}{r^2} + \dfrac{2x^3}{r^4}$
$r\theta \sin\theta$	$\dfrac{2x^3}{r^4}$	$\dfrac{2x}{r^2} - \dfrac{2x^3}{r^4}$	$\dfrac{2x^2 y}{r^4}$
$\dfrac{\cos\theta}{r}$	$\dfrac{6x}{r^4} - \dfrac{8x^3}{r^6}$	$-\dfrac{6x}{r^4} + \dfrac{8x^3}{r^6}$	$\dfrac{2y}{r^4} - \dfrac{8x^2 y}{r^6}$
$\dfrac{\sin\theta}{r}$	$\dfrac{2y}{r^4} - \dfrac{8x^2 y}{r^6}$	$-\dfrac{2y}{r^4} + \dfrac{8x^2 y}{r^6}$	$-\dfrac{6x}{r^4} + \dfrac{8x^3}{r^6}$
$\cos 2\theta$	$\dfrac{12x^2}{r^4} - \dfrac{16x^4}{r^6}$	$\dfrac{4}{r^2} - \dfrac{20x^2}{r^4} + \dfrac{16x^4}{r^6}$	$\dfrac{8xy}{r^4} - \dfrac{16x^3 y}{r^6}$
$\sin 2\theta$	$\dfrac{4xy}{r^4} - \dfrac{16x^3 y}{r^6}$	$-\dfrac{12xy}{r^4} + \dfrac{16x^3 y}{r^6}$	$\dfrac{2}{r^2} - \dfrac{16x^2}{r^4} + \dfrac{16x^4}{r^6}$
θ	$-\dfrac{2xy}{r^4}$	$\dfrac{2xy}{r^4}$	$-\dfrac{1}{r^2} + \dfrac{2x^2}{r^4}$
$\ln r$	$-\dfrac{1}{r^2} + \dfrac{2x^2}{r^4}$	$\dfrac{1}{r^2} - \dfrac{2x^2}{r^4}$	$\dfrac{2xy}{r^4}$

Table 23.5. *Polar stress components*

Φ	σ_{rr}	$\sigma_{\theta\theta}$	$\sigma_{r\theta}$
$r \ln r \cos\theta$	$\dfrac{\cos\theta}{r}$	$\dfrac{\cos\theta}{r}$	$\dfrac{\sin\theta}{r}$
$r \ln r \sin\theta$	$\dfrac{\sin\theta}{r}$	$\dfrac{\sin\theta}{r}$	$-\dfrac{\cos\theta}{r}$
$r\theta \cos\theta$	$-2\dfrac{\sin\theta}{r}$	0	0
$r\theta \sin\theta$	$2\dfrac{\cos\theta}{r}$	0	0
$\dfrac{\cos\theta}{r}$	$-2\dfrac{\cos\theta}{r^3}$	$2\dfrac{\cos\theta}{r^3}$	$-2\dfrac{\sin\theta}{r^3}$
$\dfrac{\sin\theta}{r}$	$-2\dfrac{\sin\theta}{r^3}$	$2\dfrac{\sin\theta}{r^3}$	$2\dfrac{\cos\theta}{r^3}$
$\cos 2\theta$	$-4\dfrac{\cos 2\theta}{r^2}$	0	$-2\dfrac{\sin 2\theta}{r^2}$
$\sin 2\theta$	$-4\dfrac{\sin 2\theta}{r^2}$	0	$2\dfrac{\cos 2\theta}{r^2}$
θ	0	0	$\dfrac{1}{r^2}$
$\ln r$	$\dfrac{1}{r^2}$	$-\dfrac{1}{r^2}$	0

$$\sigma_{yy}^{(1)} = k_1 b_x y \left[\frac{(x-h)^2 - y^2}{r_1^4} + q \frac{(x+h)^2 - y^2}{r_2^4} + a\beta \frac{1}{r_2^2} \right. \tag{23.39}$$
$$\left. - 4qh \frac{2(x+h)^3 - 3x(x+h)^2 + 2(x+h)y^2 + xy^2}{r_2^6} \right],$$

$$\sigma_{xy}^{(1)} = k_1 b_x \left\{ \frac{(x-h)[(x-h)^2 - y^2]}{r_1^4} + q \frac{(x+h)[(x+h)^2 - y^2]}{r_2^4} \right. \tag{23.40}$$
$$\left. - 2qh \frac{(x+h)^4 - 2x(x+h)^3 + 6x(x+h)y^2 - y^4}{r_2^6} + a\beta \frac{x+h}{r_2^2} \right\},$$

and

$$\sigma_{xx}^{(2)} = -a k_1 b_x y \left[\frac{3(x-h)^2 + y^2}{r_1^4} + 2\beta \frac{x^2 - h^2}{r_1^4} \right], \tag{23.41}$$

$$\sigma_{yy}^{(2)} = a k_1 b_x y \left[\frac{(x-h)^2 - y^2}{r_1^4} + 2\beta \frac{2h(x-h) - y^2}{r_1^4} \right], \tag{23.42}$$

$$\sigma_{xy}^{(2)} = a k_1 b_x \left\{ \frac{(x-h)[(x-h)^2 - y^2]}{r_1^4} + 2\beta \frac{h(x-h)^2 - xy^2}{r_1^4} \right\}. \tag{23.43}$$

The force exerted by the interface on the dislocation is obtained from

$$F_x = b_x \sigma_{xy}^{(1)}(h, 0), \tag{23.44}$$

23.3. Edge Dislocation (b_x) Near an Interface

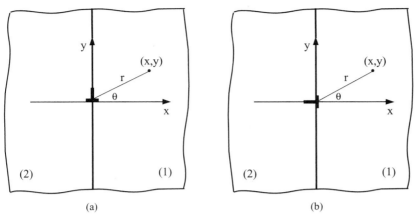

Figure 23.3. (a) An interface dislocation with the Burgers vector b_x, and (b) with the Burgers vector b_y.

by excluding the divergent part of the stress at the core center of the dislocation. This gives

$$F_x = -(1-a)\frac{k_1 b_x^2}{2h}. \tag{23.45}$$

Depending on the value of a, the dislocation may be attracted or repelled by the interface.

23.3.1 Interface Edge Dislocation

The Airy stress functions for an interface dislocation with the Burgers vector b_x (Fig. 23.3a) are

$$\Phi^{(1)} = -ak_1 b_x [r \ln r \sin\theta + \beta r\theta \cos\theta], \tag{23.46}$$

$$\Phi^{(2)} = -ak_1 b_x [r \ln r \sin\theta - \beta r\theta \cos\theta]. \tag{23.47}$$

The rectangular stress components in the material (1) are

$$\sigma_{xx}^{(1)} = -ak_1 b_x y \frac{(3-2\beta)x^2 + y^2}{(x^2+y^2)^2}, \tag{23.48}$$

$$\sigma_{yy}^{(1)} = ak_1 b_x y \frac{x^2 - (1-2\beta)y^2}{(x^2+y^2)^2}, \tag{23.49}$$

$$\sigma_{xy}^{(1)} = ak_1 b_x x \frac{x^2 - (1-2\beta)y^2}{(x^2+y^2)^2}. \tag{23.50}$$

The same expressions apply for the material (2), provided that β is replaced with $-\beta$. The expressions have simple counterparts in polar coordinates. The radial stresses are

$$\sigma_{rr}^{(1)} = -ak_1 b_x (1-2\beta)\frac{\sin\theta}{r}, \tag{23.51}$$

$$\sigma_{rr}^{(2)} = -ak_1 b_x (1+2\beta)\frac{\sin\theta}{r}. \tag{23.52}$$

The hoop and shear stresses in both media are

$$\sigma_{\theta\theta} = -ak_1b_x\frac{\sin\theta}{r}, \quad \sigma_{r\theta} = ak_1b_x\frac{\cos\theta}{r}. \tag{23.53}$$

These expressions can be deduced by using the results listed in Tables 23.4 and 23.5.
The radial displacements are

$$u_r^{(1)} = -\frac{ak_1b_x}{2\mu_1}\{[1 - 2\nu_1 - 2\beta(1-\nu_1)]\ln\frac{r}{b_x}\sin\theta \\ - [2(1-\nu_1) - \beta(1-2\nu_1)]\theta\cos\theta - \frac{\mu_1}{2\mu_2}(1+\beta)\sin\theta\}, \quad y \geq 0^+, \tag{23.54}$$

$$u_r^{(1)} = b_x\cos\theta - \frac{ak_1b_x}{2\mu_1}\{[1 - 2\nu_1 - 2\beta(1-\nu_1)]\ln\frac{r}{b_x}\sin\theta \\ - [2(1-\nu_1) - \beta(1-2\nu_1)](\theta - 2\pi)\cos\theta - \frac{\mu_1}{2\mu_2}(1+\beta)\sin\theta\}, \quad y \leq 0^-, \tag{23.55}$$

$$u_r^{(2)} = -\frac{ak_1b_x}{2\mu_2}\{[1 - 2\nu_2 + 2\beta(1-\nu_2)]\ln\frac{r}{b_x}\sin\theta \\ - [2(1-\nu_2) + \beta(1-2\nu_2)]\theta\cos\theta - \frac{1}{2}(1+\beta)\sin\theta\}. \tag{23.56}$$

These expressions satisfy the continuity across the interface and give a slip discontinuity across $y = 0$, $x > 0$. The circumferential displacements are

$$u_\theta^{(1)} = -\frac{ak_1b_x}{2\mu_1}\{[1 - 2\nu_1 - 2\beta(1-\nu_1)]\ln\frac{r}{b_x}\cos\theta \\ + [2(1-\nu_1) - \beta(1-2\nu_1)]\theta\sin\theta \\ + [1 - \beta - \frac{\mu_1}{2\mu_2}(1+\beta)]\cos\theta\}, \quad y \geq 0^+, \tag{23.57}$$

$$u_\theta^{(1)} = -\frac{ak_1b_x}{2\mu_1}\{[1 - 2\nu_1 - 2\beta(1-\nu_1)]\ln\frac{r}{b_x}\cos\theta \\ + [2(1-\nu_1) - \beta(1-2\nu_1)]\theta\sin\theta - C_2'\sin\theta \\ + [1 - \beta - \frac{\mu_1}{2\mu_2}(1+\beta)]\cos\theta\}, \quad y \leq 0^-, \tag{23.58}$$

$$u_\theta^{(2)} = -\frac{ak_1b_x}{2\mu_2}\{[1 - 2\nu_2 + 2\beta(1-\nu_2)]\ln\frac{r}{b_x}\cos\theta \\ + [2(1-\nu_2) + \beta(1-2\nu_2)]\theta\sin\theta - C_2''\sin\theta + \frac{1}{2}(1+\beta)\cos\theta\}. \tag{23.59}$$

The continuity conditions across the interface specify the constants C_2' and C_2'' as

$$C_2' = \pi\{2(1-\nu_1) - \beta(1-2\nu_1) - \frac{\mu_1}{\mu_2}[2(1-\nu_2) + \beta(1-2\nu_2)]\}, \\ C_2'' = \frac{\pi}{2}\{2(1-\nu_2) + \beta(1-2\nu_2) - \frac{\mu_2}{\mu_1}[2(1-\nu_1) - \beta(1-2\nu_1)]\}. \tag{23.60}$$

23.3. Edge Dislocation (b_x) Near an Interface

It is recalled that the rigid body displacements in polar coordinates are

$$u_r = C_1 \cos\theta + C_2 \sin\theta, \quad u_\theta = -C_1 \sin\theta + C_2 \cos\theta + C_3 r, \tag{23.61}$$

with the rectangular counterparts $u_x = C_1 + C_3 y$ and $u_y = C_2 - C_3 x$.

23.3.2 Edge Dislocation in an Infinite Medium

If two materials have the same elastic properties, the introduced parameters are $\alpha = \beta = q = 0$ and $a = 1$. Taking $h = 0$, the Airy stress function becomes

$$\Phi = -k b_x r \ln r \sin\theta, \quad k = \frac{\mu}{2\pi(1-\nu)}, \tag{23.62}$$

with the corresponding stresses

$$\sigma_{xx} = -k b_x y \frac{3x^2 + y^2}{(x^2 + y^2)^2}, \tag{23.63}$$

$$\sigma_{yy} = k b_x y \frac{x^2 - y^2}{(x^2 + y^2)^2}, \tag{23.64}$$

$$\sigma_{xy} = k b_x x \frac{x^2 - y^2}{(x^2 + y^2)^2}. \tag{23.65}$$

The displacements are

$$u_x = \frac{b_x}{2\pi}\left[\tan^{-1}\frac{y}{x} + \frac{1}{2(1-\nu)}\frac{xy}{x^2+y^2}\right], \tag{23.66}$$

$$u_y = -\frac{b_x}{2\pi}\left[\frac{1-2\nu}{4(1-\nu)}\ln\frac{x^2+y^2}{b_x^2} + \frac{1}{4(1-\nu)}\frac{x^2-y^2}{x^2+y^2}\right]. \tag{23.67}$$

The counterparts in polar coordinates are

$$\sigma_{rr} = \sigma_{\theta\theta} = -k b_x \frac{\sin\theta}{r}, \quad \sigma_{r\theta} = k b_x \frac{\cos\theta}{r}, \tag{23.68}$$

with the displacements

$$u_r = -\frac{k b_x}{2\mu}\left[(1-2\nu)\ln\frac{r}{b_x}\sin\theta - 2(1-\nu)\theta\cos\theta - \frac{1}{2}\sin\theta\right], \tag{23.69}$$

$$u_\theta = -\frac{k b_x}{2\mu}\left[(1-2\nu)\ln\frac{r}{b_x}\cos\theta + 2(1-\nu)\theta\sin\theta + \frac{1}{2}\cos\theta\right]. \tag{23.70}$$

23.3.3 Edge Dislocation Near a Free Surface

For an edge dislocation near a free surface (Fig. 23.4a) we take $\mu_2 = 0$, so that $\alpha = q = -1$ and $a = 0$. The Airy stress function becomes

$$\Phi = -k b_x \left[r_1 \ln r_1 \sin\theta_1 - r_2 \ln r_2 \sin\theta_2 \right.$$
$$\left. + h\left(\sin 2\theta_2 - 2h\frac{\sin\theta_2}{r_2}\right)\right]. \tag{23.71}$$

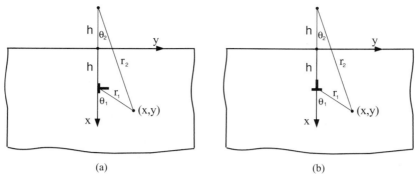

Figure 23.4. (a) A dislocation with the Burgers vector b_x at a distance h from the free surface. (b) Same for a dislocation with the Burgers vector b_y.

The corresponding stresses are

$$\sigma_{xx} = -kb_x y \left[\frac{3(x-h)^2 + y^2}{r_1^4} - \frac{3(x+h)^2 + y^2}{r_2^4} - 4hx \frac{3(x+h)^2 - y^2}{r_2^6} \right], \quad (23.72)$$

$$\sigma_{yy} = kb_x y \left[\frac{(x-h)^2 - y^2}{r_1^4} - \frac{(x+h)^2 - y^2}{r_2^4} \right.$$
$$\left. + 4h \frac{2(x+h)^3 - 3x(x+h)^2 + 2(x+h)y^2 + xy^2}{r_2^6} \right], \quad (23.73)$$

$$\sigma_{xy} = kb_x \left\{ \frac{(x-h)[(x-h)^2 - y^2]}{r_1^4} - \frac{(x+h)[(x+h)^2 - y^2]}{r_2^4} \right.$$
$$\left. + 2h \frac{(x+h)^4 - 2x(x+h)^3 + 6x(x+h)y^2 - y^4}{r_2^6} \right\}. \quad (23.74)$$

These expressions were originally derived by Head (1953a, 1953b).

The force exerted by the free surface on the dislocation is

$$F_x = -\frac{k_1 b_x^2}{2h}, \quad (23.75)$$

which is an attraction toward the free surface.

23.3.4 Edge Dislocation Near a Rigid Boundary

In this case we take $\mu_2 \to \infty$, and with $\mu_1 = \mu$ and $\nu_1 = \nu$, we have

$$\alpha = -1, \quad 2\beta = \frac{1-2\nu}{1-\nu}, \quad q = \frac{1}{3-4\nu},$$
$$a = \frac{8(1-\nu)^2}{3-4\nu}, \quad a\beta = \frac{4(1-\nu)(1-2\nu)}{3-4\nu}. \quad (23.76)$$

The Airy stress function is

$$\Phi = -kb_x \left[r_1 \ln r_1 \sin\theta_1 + (a-1) r_2 \ln r_2 \sin\theta_2 \right.$$
$$\left. - qh \left(\sin 2\theta_2 - 2h \frac{\sin\theta_2}{r_2} \right) + a\beta r_2 \theta_2 \cos\theta_2 \right]. \quad (23.77)$$

23.4. Edge Dislocation (b_y) Near an Interface

The corresponding stresses are

$$\sigma_{xx} = -kb_x y \left[\frac{3(x-h)^2 + y^2}{r_1^4} + q \frac{3(x+h)^2 + y^2}{r_2^4} \right. \\ \left. + 4qhx \frac{3(x+h)^2 - y^2}{r_2^6} + a\beta \frac{1}{r_2^2} \right], \tag{23.78}$$

$$\sigma_{yy} = kb_x y \left[\frac{(x-h)^2 - y^2}{r_1^4} + q \frac{(x+h)^2 - y^2}{r_2^4} + a\beta \frac{1}{r_2^2} \right. \\ \left. - 4qh \frac{2(x+h)^3 - 3x(x+h)^2 + 2(x+h)y^2 + xy^2}{r_2^6} \right], \tag{23.79}$$

$$\sigma_{xy} = kb_x \left\{ \frac{(x-h)[(x-h)^2 - y^2]}{r_1^4} + q \frac{(x+h)[(x+h)^2 - y^2]}{r_2^4} \right. \\ \left. - 2qh \frac{(x+h)^4 - 2x(x+h)^3 + 6x(x+h)y^2 - y^4}{r_2^6} + a\beta \frac{x+h}{r_2^2} \right\}. \tag{23.80}$$

23.4 Edge Dislocation (b_y) Near a Bimaterial Interface

Consider next an edge dislocation with the Burgers vector b_y (Fig. 23.2b). The corresponding Airy stress functions are

$$\Phi^{(1)} = k_1 b_y \left[r_1 \ln r_1 \cos\theta_1 + (a-1) r_2 \ln r_2 \cos\theta_2 \right. \\ \left. - qh \left(2 \ln r_2 - \cos 2\theta_2 + 2h \frac{\cos\theta_2}{r_2} \right) - a\beta r_2 \theta_2 \sin\theta_2 \right], \tag{23.81}$$

$$\Phi^{(2)} = ak_1 b_y [r_1 \ln r_1 \cos\theta_1 + \beta(2h \ln r_1 + r_1 \theta_1 \sin\theta_1)]. \tag{23.82}$$

This gives

$$\sigma_{xx}^{(1)} = k_1 b_y \left\{ \frac{(x-h)[(x-h)^2 - y^2]}{r_1^4} + q \frac{(x+h)[(x+h)^2 - y^2]}{r_2^4} \right. \\ \left. - 2qh \frac{(x+h)^4 + 2x(x+h)^3 - 6x(x+h)y^2 - y^4}{r_2^6} - a\beta \frac{x+h}{r_2^2} \right\}, \tag{23.83}$$

$$\sigma_{yy}^{(1)} = k_1 b_y \left\{ \frac{(x-h)[(x-h)^2 + 3y^2]}{r_1^4} + q \frac{(x+h)[(x+h)^2 + 3y^2]}{r_2^4} \right. \\ \left. - 2qh \frac{(x+h)^4 - 2x(x+h)^3 + 6x(x+h)y^2 - y^4}{r_2^6} + a\beta \frac{x+h}{r_2^2} \right\}, \tag{23.84}$$

$$\sigma_{xy}^{(1)} = k_1 b_y \left\{ \frac{y[(x-h)^2 - y^2]}{r_1^4} + q \frac{y[(x+h)^2 - y^2]}{r_2^4} \right. \\ \left. - 4qhxy \frac{3(x+h)^2 - y^2}{r_2^6} - a\beta \frac{y}{r_2^2} \right\}, \tag{23.85}$$

and

$$\sigma_{xx}^{(2)} = ak_1 b_y \left[\frac{(x-h)[(x-h)^2 - y^2]}{r_1^4} + 2\beta \frac{x(x-h)^2 - hy^2}{r_1^4} \right], \quad (23.86)$$

$$\sigma_{yy}^{(2)} = ak_1 b_y \left[\frac{(x-h)[(x-h)^2 + 3y^2]}{r_1^4} - 2\beta \frac{h(x-h)^2 - xy^2}{r_1^4} \right], \quad (23.87)$$

$$\sigma_{xy}^{(2)} = ak_1 b_y \left[\frac{y[(x-h)^2 - y^2]}{r_1^4} + 2\beta y \frac{x^2 - h^2}{r_1^4} \right]. \quad (23.88)$$

23.4.1 Interface Edge Dislocation

The Airy stress functions in this case (Fig. 23.3b) are

$$\Phi^{(1)} = ak_1 b_y [r \ln r \cos\theta - \beta r\theta \sin\theta], \quad (23.89)$$

$$\Phi^{(2)} = ak_1 b_y [r \ln r \cos\theta + \beta r\theta \sin\theta]. \quad (23.90)$$

The stress components in the material (1) are

$$\sigma_{xx}^{(1)} = ak_1 b_y x \frac{(1-2\beta)x^2 - y^2}{(x^2 + y^2)^2}, \quad (23.91)$$

$$\sigma_{yy}^{(1)} = ak_1 b_y x \frac{x^2 + (3-2\beta)y^2}{(x^2 + y^2)^2}, \quad (23.92)$$

$$\sigma_{xy}^{(1)} = ak_1 b_y y \frac{(1-2\beta)x^2 - y^2}{(x^2 + y^2)^2}. \quad (23.93)$$

The same expressions hold in material (2), provided that β is replaced with $-\beta$. The expressions have simple counterparts in polar coordinates. The radial stresses are

$$\sigma_{rr}^{(1)} = ak_1 b_y (1 - 2\beta) \frac{\cos\theta}{r}, \quad (23.94)$$

$$\sigma_{rr}^{(2)} = ak_1 b_y (1 + 2\beta) \frac{\cos\theta}{r}. \quad (23.95)$$

The hoop and shear stresses in both media are given by

$$\sigma_{\theta\theta} = ak_1 b_y \frac{\cos\theta}{r}, \quad \sigma_{r\theta} = ak_1 b_y \frac{\sin\theta}{r}. \quad (23.96)$$

The circumferential displacements are

$$u_\theta^{(1)} = -\frac{ak_1 b_y}{2\mu_1} \{[1 - 2\nu_1 - 2\beta(1-\nu_1)] \ln\frac{r}{b_y} \sin\theta$$
$$- [2(1-\nu_1) - \beta(1-2\nu_1)]\theta \cos\theta$$
$$+ \frac{\mu_1}{2\mu_2}(1+\beta) \sin\theta\}, \quad y \geq 0^+, \quad (23.97)$$

23.4. Edge Dislocation (b_y) Near an Interface

$$u_\theta^{(1)} = b_y \cos\theta - \frac{ak_1 b_y}{2\mu_1}\{[1 - 2\nu_1 - 2\beta(1-\nu_1)]\ln\frac{r}{b_y}\sin\theta$$
$$- [2(1-\nu_1) - \beta(1-2\nu_1)](\theta - 2\pi)\cos\theta \qquad (23.98)$$
$$+ \frac{\mu_1}{2\mu_2}(1+\beta)\sin\theta\}, \quad y \le 0^-,$$

$$u_\theta^{(2)} = -\frac{ak_1 b_y}{2\mu_2}\{[1 - 2\nu_2 + 2\beta(1-\nu_2)]\ln\frac{r}{b_y}\sin\theta$$
$$- [2(1-\nu_2) + \beta(1-2\nu_2)]\theta\cos\theta \qquad (23.99)$$
$$+ \frac{1}{2}(1+\beta)\sin\theta\}.$$

These expressions satisfy the continuity across the interface and give a displacement discontinuity of amount b_y across $y = 0, x > 0$.

The radial displacements are

$$u_r^{(1)} = \frac{ak_1 b_y}{2\mu_1}\{[1 - 2\nu_1 - 2\beta(1-\nu_1)]\ln\frac{r}{b_y}\cos\theta$$
$$+ [2(1-\nu_1) - \beta(1-2\nu_1)]\theta\sin\theta \qquad (23.100)$$
$$- [1 - \beta - \frac{\mu_1}{2\mu_2}(1+\beta)]\cos\theta\}, \quad y \ge 0^+,$$

$$u_r^{(1)} = \frac{ak_1 b_y}{2\mu_1}\{[1 - 2\nu_1 - 2\beta(1-\nu_1)]\ln\frac{r}{b_y}\cos\theta$$
$$+ [2(1-\nu_1) - \beta(1-2\nu_1)]\theta\sin\theta - C_2'\sin\theta \qquad (23.101)$$
$$- [1 - \beta - \frac{\mu_1}{2\mu_2}(1+\beta)]\cos\theta\}, \quad y \le 0^-,$$

$$u_r^{(2)} = \frac{ak_1 b_y}{2\mu_2}\{[1 - 2\nu_2 + 2\beta(1-\nu_2)]\ln\frac{r}{b_y}\cos\theta$$
$$+ [2(1-\nu_2) + \beta(1-2\nu_2)]\theta\sin\theta - C_2''\sin\theta \qquad (23.102)$$
$$- \frac{1}{2}(1+\beta)\cos\theta\}.$$

The continuity conditions across the interface specify the constants C_2' and C_2'', as given by (23.60).

The force exerted by the interface on the dislocation is obtained from

$$F_x = b_y \sigma_{yy}^{(1)}(h, 0), \qquad (23.103)$$

by excluding the divergent part of the stress at the core center of the dislocation. This gives

$$F_x = -(1-a)\frac{k_1 b_y^2}{2h}. \qquad (23.104)$$

Depending on the value of a, the dislocation may be attracted or repelled by the interface.

23.4.2 Edge Dislocation in an Infinite Medium

The Airy stress function is

$$\Phi = kb_y r \ln r \cos\theta, \qquad (23.105)$$

with the corresponding stress components

$$\sigma_{xx} = kb_y x \frac{x^2 - y^2}{(x^2 + y^2)^2}, \qquad (23.106)$$

$$\sigma_{yy} = kb_y x \frac{x^2 + 3y^2}{(x^2 + y^2)^2}, \qquad (23.107)$$

$$\sigma_{xy} = kb_y y \frac{x^2 - y^2}{(x^2 + y^2)^2}. \qquad (23.108)$$

The displacements are

$$u_x = \frac{b_y}{2\pi} \left[\frac{1-2\nu}{4(1-\nu)} \ln \frac{x^2+y^2}{b_y^2} - \frac{1}{4(1-\nu)} \frac{x^2-y^2}{x^2+y^2} \right], \qquad (23.109)$$

$$u_y = \frac{b_y}{2\pi} \left[\tan^{-1} \frac{y}{x} - \frac{1}{2(1-\nu)} \frac{xy}{x^2+y^2} \right]. \qquad (23.110)$$

The counterparts in polar coordinates are

$$\sigma_{rr} = \sigma_{\theta\theta} = kb_y \frac{\cos\theta}{r}, \quad \sigma_{r\theta} = kb_y \frac{\sin\theta}{r}, \qquad (23.111)$$

with the displacements

$$u_r = \frac{b_y}{2\pi} \left[\frac{1-2\nu}{2(1-\nu)} \ln \frac{r}{b_y} \cos\theta + \theta \sin\theta - \frac{1}{4(1-\nu)} \cos\theta \right], \qquad (23.112)$$

$$u_\theta = -\frac{b_y}{2\pi} \left[\frac{1-2\nu}{2(1-\nu)} \ln \frac{r}{b_y} \sin\theta - \theta \cos\theta + \frac{1}{4(1-\nu)} \sin\theta \right]. \qquad (23.113)$$

23.4.3 Edge Dislocation Near a Free Surface

In this case (Fig. 23.4b) we take $\mu_2 = 0$, so that $\alpha = q = -1$ and $a = 0$. The Airy stress function becomes

$$\Phi = kb_y \left[r_1 \ln r_1 \cos\theta_1 - r_2 \ln r_2 \cos\theta_2 \right.$$
$$\left. + h \left(2 \ln r_2 - \cos 2\theta_2 + 2h \frac{\cos\theta_2}{r_2} \right) \right]. \qquad (23.114)$$

The stresses are

$$\sigma_{xx} = kb_y \left\{ \frac{(x-h)[(x-h)^2 - y^2]}{r_1^4} - \frac{(x+h)[(x+h)^2 - y^2]}{r_2^4} \right.$$
$$\left. + 2h \frac{(x+h)^4 + 2x(x+h)^3 - 6x(x+h)y^2 - y^4}{r_2^6} \right], \qquad (23.115)$$

23.5. Strain Energy of Dislocation Near an Interface

$$\sigma_{yy} = kb_y \left\{ \frac{(x-h)[(x-h)^2 + 3y^2]}{r_1^4} - \frac{(x+h)[(x+h)^2 + 3y^2]}{r_2^4} \right. $$
$$\left. + 2h \frac{(x+h)^4 - 2x(x+h)^3 + 6x(x+h)y^2 - y^4}{r_2^6} \right\}, \quad (23.116)$$

$$\sigma_{xy} = kb_y \left\{ \frac{y[(x-h)^2 - y^2]}{r_1^4} - \frac{y[(x+h)^2 - y^2]}{r_2^4} \right. $$
$$\left. + 4hxy \frac{3(x+h)^2 - y^2}{r_2^6} \right\}. \quad (23.117)$$

The force exerted by the free surface on the dislocation is

$$F_x = -\frac{k_1 b_y^2}{2h}. \quad (23.118)$$

The dislocation is attracted by the free surface.

23.4.4 Edge Dislocation Near a Rigid Boundary

In this case we take $\mu_2 \to \infty$, which implies (23.76). The Airy stress function is

$$\Phi = kb_y \left[r_1 \ln r_1 \cos\theta_1 + (a-1) r_2 \ln r_2 \cos\theta_2 \right. $$
$$\left. - qh \left(2 \ln r_2 - \cos 2\theta_2 + 2h \frac{\cos\theta_2}{r_2} \right) - a\beta r_2 \theta_2 \sin\theta_2 \right]. \quad (23.119)$$

The corresponding stresses are

$$\sigma_{xx} = kb_y \left\{ \frac{(x-h)[(x-h)^2 - y^2]}{r_1^4} + q \frac{(x+h)[(x+h)^2 - y^2]}{r_2^4} \right. $$
$$\left. - 2qh \frac{(x+h)^4 + 2x(x+h)^3 - 6x(x+h)y^2 - y^4}{r_2^6} - a\beta \frac{x+h}{r_2^2} \right\}, \quad (23.120)$$

$$\sigma_{yy} = kb_y \left\{ \frac{(x-h)[(x-h)^2 + 3y^2]}{r_1^4} + q \frac{(x+h)[(x+h)^2 + 3y^2]}{r_2^4} \right. $$
$$\left. - 2qh \frac{(x+h)^4 - 2x(x+h)^3 + 6x(x+h)y^2 - y^4}{r_2^6} + a\beta \frac{x+h}{r_2^2} \right\}, \quad (23.121)$$

$$\sigma_{xy} = kb_y \left\{ \frac{y[(x-h)^2 - y^2]}{r_1^4} + q \frac{y[(x+h)^2 - y^2]}{r_2^4} \right. $$
$$\left. - 4qhxy \frac{3(x+h)^2 - y^2}{r_2^6} - a\beta \frac{y}{r_2^2} \right\}. \quad (23.122)$$

23.5 Strain Energy of a Dislocation Near a Bimaterial Interface

The elastic strain energy per unit dislocation length within a large cylinder of radius $R \gg h$ around the core of the dislocation, excluding the core itself, can be calculated by using the

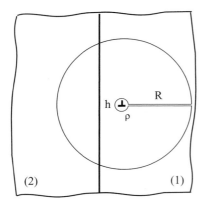

Figure 23.5. The dislocation core surface of radius ρ, the remote contour of radius R, and the cut surface from ρ to R used to create a dislocation near the bimaterial interface.

divergence theorem. The energy consists of the work done by the traction on the slip discontinuity along the surface of the cut used to create a dislocation, and the work done by tractions on the corresponding displacements along the surface of the dislocation core and the surface of the remote contour (Fig. 23.5). Thus, for a dislocation with the Burgers vector b_x,

$$E = \frac{1}{2} b_x \int_{h+\rho}^{R} \sigma_{xy}^{(1)}(x,0) dx + E_R + E_\rho , \qquad (23.123)$$

where ρ is a selected small radius of the dislocation core. The shear stress $\sigma_{xy}^{(1)}(x,0)$ is obtained from (23.40) as

$$\sigma_{xy}^{(1)}(x,0) = k_1 b_x \left[\frac{1}{x-h} + (q+a\beta)\frac{1}{x+h} + 2qh\frac{x-h}{(x+h)^3} \right]. \qquad (23.124)$$

The substitution into (23.123) and integration gives

$$E = \frac{1}{2} k_1 b_x^2 \left(\ln \frac{2h}{\rho} + a \ln \frac{R}{2h} + \frac{q}{2} \right) + E_R + E_\rho . \qquad (23.125)$$

For a sufficiently small radius ($\rho \ll h$), the core energy E_ρ can be calculated by replacing the dislocation core with a cylindrical hole, whose surface is subjected to tractions of an isolated dislocation in an infinite homogeneous medium, along with the corresponding displacements. The work of these is

$$E_\rho = -\frac{1}{2} \int_0^{2\pi} (\sigma_{rr} u_r + \sigma_{r\theta} u_\theta) \rho d\theta . \qquad (23.126)$$

With the displacement discontinuity imposed along the x axis, we have, from (23.69) and (23.70),

$$u_r = -\frac{k_1 b_x}{2\mu_1} \left[(1-2\nu_1) \ln \frac{r}{b_x} \sin\theta - 2(1-\nu_1)\theta \cos\theta - \frac{1}{2}\sin\theta \right], \qquad (23.127)$$

$$u_\theta = -\frac{k_1 b_x}{2\mu_1} \left[(1-2\nu_1) \ln \frac{r}{b_x} \cos\theta + 2(1-\nu_1)\theta \sin\theta + \frac{1}{2}\cos\theta \right]. \qquad (23.128)$$

23.5. Strain Energy of Dislocation Near an Interface

The stress components (independently of the cut) are

$$\sigma_{rr} = \sigma_{\theta\theta} = -k_1 b_x \frac{\sin\theta}{r}, \quad \sigma_{r\theta} = k_1 b_x \frac{\cos\theta}{r}. \tag{23.129}$$

The substitution into (23.126) and integration gives

$$E_\rho = -\frac{1}{8} k_1 b_x^2 \frac{1 - 2\nu_1}{1 - \nu_1}. \tag{23.130}$$

The energy contribution E_R can be calculated by using the stress and displacement fields of an interfacial dislocation, because the distance h between the dislocation and the interface is not observed at a far remote contour R. These fields are given by (23.51)–(23.53) for the stresses, and (23.54)–(23.59) for the displacements. The substitution into

$$E_R = \frac{1}{2} \int_0^{2\pi} (\sigma_{rr} u_r + \sigma_{r\theta} u_\theta) R \, d\theta \tag{23.131}$$

gives, upon integration,

$$E_R = \frac{1}{4} a k_1 b_x^2 (1 - 2\beta) + \frac{1}{8} (a k_1 b_x)^2 \left[\beta \frac{C_2'}{\mu_1} - \pi \left(\frac{1-\beta}{\mu_1} + \frac{1+\beta}{\mu_2} \right) \right]. \tag{23.132}$$

If $\beta = 0$ (homogeneous medium), the above reduces to

$$E_R = \frac{1}{8} k b_x^2 \frac{1 - 2\nu}{1 - \nu} = -E_\rho. \tag{23.133}$$

The strain energy for a dislocation with the Burgers vector b_y can be calculated from

$$E = \frac{1}{2} b_y \int_{h+\rho}^R \sigma_{yy}^{(1)}(x, 0) \, dx + E_R + E_\rho, \tag{23.134}$$

with appropriate values of E_R and E_ρ. The normal stress $\sigma_{yy}^{(1)}(x, 0)$ is obtained from (23.84) as

$$\sigma_{yy}^{(1)}(x, 0) = k_1 b_y \left[\frac{1}{x - h} + (q + a\beta) \frac{1}{x + h} + 2qh \frac{x - h}{(x + h)^3} \right]. \tag{23.135}$$

The substitution into (23.134) and integration gives

$$E = \frac{1}{2} k_1 b_y^2 \left(\ln \frac{2h}{\rho} + a \ln \frac{R}{2h} + \frac{q}{2} \right) + E_R + E_\rho. \tag{23.136}$$

By an analogous analysis, as for the misfit dislocation with the Burgers vector b_x, it follows that

$$E_R = -\frac{1}{4} a k_1 b_y^2 - \frac{\pi}{4} (a k_1 b_y)^2 \left[\left(\frac{1}{\mu_1} + \frac{1}{\mu_2} \right) (1 + 2\beta^2) \right. \\
\left. - \beta \left(\frac{3 + A_1}{\mu_1} - \frac{3 - 3A_2}{\mu_2} \right) \right], \tag{23.137}$$

where

$$A_1 = 2(1 - \nu_1) - \beta(1 - 2\nu_1), \quad A_2 = 2(1 - \nu_2) + \beta(1 - 2\nu_2). \tag{23.138}$$

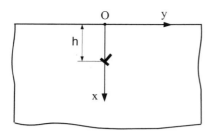

Figure 23.6. A general straight dislocation with the edge components b_x and b_y, and the screw component b_z, at a distance h from the free surface.

If $\beta = 0$ (homogeneous medium), the above reduces to

$$E_R = -\frac{1}{8} k b_y^2 \frac{3-2v}{1-v}. \tag{23.139}$$

This is just the negative of the core energy

$$E_\rho = \frac{1}{8} k b_y^2 \frac{3-2v}{1-v}, \tag{23.140}$$

so that $E_R + E_\rho = 0$ in a homogeneous medium (see also Hirth and Lothe, 1982).

The elastic strain energy for a screw dislocation within a large cylinder around the dislocation is

$$E = \frac{1}{2} b_z \int_{h+\rho}^{R} \sigma_{zy}^{(1)}(x,0) dx. \tag{23.141}$$

There is no contribution from the tractions at the remote contour of radius R, since $\sigma_{zr} = 0$ for an interface screw dislocation, and no work is done on the u_z displacement. Similarly, there is no contribution from the tractions over the core surface for a dislocation in an infinite homogeneous medium. Since the shear stress $\sigma_{zy}^{(1)}(x,0)$ is obtained from (23.7) as

$$\sigma_{zy}^{(1)}(x,0) = \frac{\mu_1 b_z}{2\pi} \left(\frac{1}{x-h} - c \frac{1}{x+h} \right), \tag{23.142}$$

substitution into (23.141) and integration gives

$$E = \frac{\mu_1 b_z^2}{4\pi} \left[\ln \frac{2h}{\rho} + (1-c) \ln \frac{R}{2h} \right]. \tag{23.143}$$

23.5.1 Strain Energy of a Dislocation Near a Free Surface

For the analysis of thin film dislocations, a particularly important special case of the previous results is the case of a dislocation near the free surface in a semi-infinite body. In this case the parameters $a = 0$, $c = 1$, and from (23.132) and (23.137) it follows that $E_R = 0$. This was expected on physical grounds because for a dislocation in a semi-infinite medium the stresses decay with the distance r from the dislocation as $1/r^2$. The energy contribution from the core surface for a general straight dislocation with the Burgers vector $\mathbf{b} = \{b_x, b_y, b_z\}$ (Fig. 23.6) is the sum of (23.130) and (23.140), which is

$$E_\rho = \frac{k}{4} \left[b_x^2 - b_y^2 - \frac{1}{2(1-v)} (b_x^2 + b_y^2) \right]. \tag{23.144}$$

23.6. Suggested Reading

There is no interaction energy between the individual components of the dislocation, as can be checked by inspection. The total elastic strain energy, outside the core region of the dislocation, is

$$E = -\frac{1}{2} \int_0^{h-\rho} [b_x \sigma_{xy}(x,0) + b_y \sigma_{yy}(x,0) + b_z \sigma_{zy}(x,0)] \, dx + E_\rho. \tag{23.145}$$

This implies the slip discontinuity along the x axis is from 0 to h. This is also equal to

$$E = \frac{1}{2} \int_{h+\rho}^{\infty} [b_x \sigma_{xy}(x,0) + b_y \sigma_{yy}(x,0) + b_z \sigma_{zy}(x,0)] \, dx + E_\rho, \tag{23.146}$$

which corresponds to slip discontinuity from h to infinity. Upon integration (or by specializing the earlier results for the dislocation near a bimaterial interface), there follows

$$E = \frac{k}{2} \left\{ [b_x^2 + b_y^2 + (1-\nu)b_z^2] \ln \frac{2h}{\rho} - \frac{1}{4(1-\nu)} [(3-4\nu)b_x^2 - b_y^2] \right\}. \tag{23.147}$$

The result is originally due to Freund (1987, 1990).

The force exerted by the free surface on the dislocation is the negative gradient of E with respect to h

$$F_x = -\frac{\partial E}{\partial h}. \tag{23.148}$$

This gives

$$F_x = -\frac{k}{2} [(1-a)(b_x^2 + b_y^2) + c(1-\nu)b_z^2] \frac{1}{h}. \tag{23.149}$$

23.6 Suggested Reading

Dundurs, J. (1969), Elastic Interactions of Dislocations with Inhomogeneities. In *Mathematical Theory of Dislocations* (T. Mura, ed.), ASME, New York, pp. 70–115.

Dundurs, J., and Mura, T. (1964), Interaction Between Edge Dislocation and a Circular Inclusion, *J. Mech. Phys. Solids*, Vol. 12, pp. 177–189.

Head, A. K. (1953a), The Interaction of Dislocations and Boundaries, *Phil. Mag.*, Vol. 44, pp. 92–64.

Head, A. K. (1953b), Edge Dislocations in Inhomogeneous Media, *Proc. Phys. Soc. Lond.*, Vol. B66, pp. 793–801.

Hirth, J. P., and Lothe, J. (1982), *Theory of Dislocations*, 2nd ed., Wiley, New York.

Lubarda, V. A. (1997), Energy Analysis of Dislocation Arrays Near Bimaterial Interfaces, *Int. J. Solids Struct.*, Vol 34, pp. 1053–1073.

Mura, T. (1987), *Micromechanics of Defects in Solids*, Martinus Nijhoff, Dordrecht, The Netherlands.

24 Strain Relaxation in Thin Films

24.1 Dislocation Array Beneath the Free Surface

Consider an infinite array of dislocations with uniform spacing p at a distance h below the free surface of a semi-infinite body (Fig. 24.1). The stresses due to this array at an arbitrary point can be obtained by superposition of stresses due to each dislocation alone. Of particular importance for the subsequent energy analysis are the stress components σ_{xy}, σ_{yy}, and σ_{zy}. These are derived by Lubarda (1998) as

$$\sigma_{xy} = \frac{\pi k b_x}{p} T_x + \frac{\pi k b_y}{p} T_y \sin \psi, \tag{24.1}$$

$$\sigma_{yy} = \frac{\pi k b_y}{p} Y_y + \frac{\pi k b_x}{p} Y_x \sin \psi, \tag{24.2}$$

$$\sigma_{zy} = \frac{\pi k (1-\nu) b_z}{p} \left(\frac{\sinh \vartheta}{C} - \frac{\sinh \varphi}{A} \right). \tag{24.3}$$

The parameter $k = \mu / 2\pi(1-\nu)$, and

$$\begin{aligned} T_x &= \frac{D\vartheta}{C^2} - \frac{B\vartheta}{A^2} - \frac{2\varphi_0(\varphi - \varphi_0)}{A^3}(B - \sin^2 \psi) \sinh \varphi, \\ T_y &= \frac{1}{A} - \frac{1}{C} + \frac{\vartheta \sinh \vartheta}{C^2} - \frac{\varphi \sinh \varphi}{A^2} + \frac{2\varphi_0(\varphi - \varphi_0)}{A^3}(B + \sinh^2 \varphi), \end{aligned} \tag{24.4}$$

$$Y_x = \frac{1}{A} - \frac{1}{C} + \frac{\vartheta \sinh \vartheta}{C^2} - \frac{(\varphi - 4\varphi_0)\sinh \varphi}{A^2} - \frac{2\varphi_0(\varphi - \varphi_0)}{A^3}(B + \sinh^2 \varphi), \tag{24.5}$$

$$Y_y = \frac{2\sinh \vartheta}{C} - \frac{2\sinh \varphi}{A} - \frac{D\vartheta}{C^2} + \frac{B(\varphi + 2\varphi_0)}{A^2} - \frac{2\varphi_0(\varphi - \varphi_0)}{A^3}(B - \sin^2 \psi)\sinh \varphi.$$

The following abbreviations have been used

$$\begin{aligned} A &= \cosh \varphi - \cos \psi, & B &= \cosh \varphi \cos \psi - 1, \\ C &= \cosh \vartheta - \cos \psi, & D &= \cosh \vartheta \cos \psi - 1. \end{aligned} \tag{24.6}$$

24.1. Dislocation Array Beneath the Free Surface

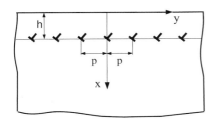

Figure 24.1. An infinite array of dislocations with uniform spacing p at a distance h bellow the free surface of a semi-infinite body.

The nondimensional variables in (24.6) are

$$\vartheta = 2\pi \frac{x-h}{p}, \quad \varphi = 2\pi \frac{x+h}{p}, \quad \varphi_0 = 2\pi \frac{h}{p}, \quad \psi = 2\pi \frac{y}{p}. \tag{24.7}$$

In particular, the stress components along the vertical plane containing a dislocation from the array can be obtained from the above formulas by substituting $y = 0$. The expressions can be conveniently written as

$$\sigma_{xy}(x,0) = \frac{\pi k b_x}{2p} \left[\frac{\vartheta}{\sinh^2(\vartheta/2)} - \frac{\varphi}{\sinh^2(\varphi/2)} + \frac{2\varphi_0}{\sinh^2(\varphi/2)} A(\varphi) \right], \tag{24.8}$$

$$\sigma_{yy}(x,0) = \frac{\pi k b_y}{2p} \left[4 \left(\coth \frac{\vartheta}{2} - \coth \frac{\varphi}{2} \right) - \frac{\vartheta}{\sinh^2(\vartheta/2)} \right. $$
$$\left. + \frac{\varphi}{\sinh^2(\varphi/2)} + \frac{2\varphi_0}{\sinh^2(\varphi/2)} A(\varphi) \right], \tag{24.9}$$

$$\sigma_{zy}(x,0) = \frac{\mu b_z}{2p} \left(\coth \frac{\vartheta}{2} - \coth \frac{\varphi}{2} \right), \tag{24.10}$$

where

$$A(\varphi) = 1 - (\varphi - \varphi_0) \coth \frac{\varphi}{2}. \tag{24.11}$$

Along the vertical planes midway between the two dislocations from the array, the stresses are

$$\sigma_{xy}\left(x, \frac{p}{2}\right) = \frac{\pi k b_x}{2p} \left[-\frac{\vartheta}{\cosh^2(\vartheta/2)} + \frac{\varphi}{\cosh^2(\varphi/2)} - \frac{2\varphi_0}{\cosh^2(\varphi/2)} B(\varphi) \right], \tag{24.12}$$

$$\sigma_{yy}\left(x, \frac{p}{2}\right) = \frac{\pi k b_y}{2p} \left[4 \left(\tanh \frac{\vartheta}{2} - \tanh \frac{\varphi}{2} \right) + \frac{\vartheta}{\cosh^2(\vartheta/2)} \right.$$
$$\left. - \frac{\varphi}{\cosh^2(\varphi/2)} - \frac{2\varphi_0}{\cosh^2(\varphi/2)} B(\varphi) \right], \tag{24.13}$$

$$\sigma_{zy}\left(x, \frac{p}{2}\right) = \frac{\mu b_z}{2p} \left(\tanh \frac{\vartheta}{2} - \tanh \frac{\varphi}{2} \right), \tag{24.14}$$

where

$$B(\varphi) = 1 - (\varphi - \varphi_0) \tanh \frac{\varphi}{2}. \tag{24.15}$$

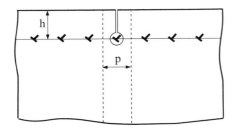

Figure 24.2. A strip of width p around a dislocation from the periodic array of spacing p. Indicated are the surface of the dislocation core, and the vertical cut from the free surface to the core used to create the dislocation.

24.2 Energy of a Dislocation Array

The strain energy per unit length of a dislocation in the strip of width p (Fig. 24.2), outside the dislocation core region, is

$$E^d = -\frac{1}{2}\int_0^{h-\rho}[b_x\sigma_{xy}(x,0) + b_y\sigma_{yy}(x,0) + b_z\sigma_{zy}(x,0)]\,dx + E_\rho. \tag{24.16}$$

This implies the displacement discontinuity is from 0 to h. The energy E_ρ is the contribution from the tractions on the dislocation core surface of radius ρ. For a sufficiently small core radius ($\rho \ll h$), E_ρ can be calculated by replacing the dislocation core with a cylindrical hole, whose surface is subjected to tractions of an isolated dislocation in an infinite homogeneous medium, along with the corresponding displacements. With a displacement discontinuity imposed along the cut along the x axis from 0 to h, this is

$$E_\rho = \frac{1}{4}k\left[b_x^2 - b_y^2 - \frac{1}{2(1-\nu)}(b_x^2 + b_y^2)\right]. \tag{24.17}$$

The stresses $\sigma_{xy}(x,0)$, $\sigma_{yy}(x,0)$, and $\sigma_{zy}(x,0)$ are given by (24.8)–(24.10). Upon substitution into (24.16) and integration, it follows that

$$E^d = \frac{k}{2}\left\{(b_x^2 + b_y^2)\left[\ln\frac{\sinh\varphi_0}{\sinh\rho_0} - \frac{\varphi_0^2}{2\sinh^2\varphi_0} + \frac{1}{4(1-\nu)}\right]\right.$$
$$\left. + (b_x^2 - b_y^2)\left(\frac{1}{2} - \varphi_0\coth\varphi_0\right) + (1-\nu)b_z^2\ln\frac{\sinh\varphi_0}{\sinh\rho_0}\right\}. \tag{24.18}$$

The nondimensional variables $\varphi_0 = 2\pi h/p$ and $\rho_0 = \pi\rho/p$ are used. For a sufficiently small core radius, $\sinh\rho_0$ can be replaced by ρ_0. The result in this form was derived by Lubarda (1997). The original result given in terms of exponential functions is due to Willis et al. (1990, 1991).

If $h \gg p$, (24.18) gives

$$E^d = [2b_y^2 + (1-\nu)b_z^2]\frac{\pi k h}{p}, \tag{24.19}$$

which is proportional to h. This is so because for $h \gg p$ the stress field in the layer above the dislocation array becomes essentially uniform

$$\sigma_{yy} = -\frac{4\pi k b_y}{p}, \quad \sigma_{zy} = -\frac{\mu b_z}{p}, \tag{24.20}$$

as if the array consists of continuously distributed infinitesimal dislocations.

24.3. Strained-Layer Epitaxy

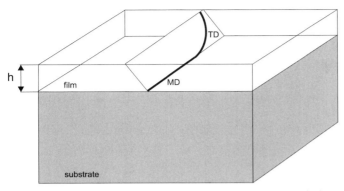

Figure 24.3. A threading dislocation segment across the layer and a long misfit dislocation left behind at the interface between the layer and its substrate.

The energy for an isolated dislocation near the free surface is obtained when $h \ll p$, which reproduces (23.147), i.e.,

$$E_0^d = \frac{k}{2} \left\{ [b_x^2 + b_y^2 + (1-\nu)b_z^2] \ln \frac{2h}{\rho} - \frac{1}{4(1-\nu)} [(3-4\nu)b_x^2 - b_y^2] \right\}. \tag{24.21}$$

The force on an individual dislocation from the array in Fig. 24.1 can be calculated from

$$F_x = -\frac{\partial E^d}{\partial h}, \tag{24.22}$$

which gives

$$F_x = -\frac{\pi k}{p} \coth \varphi_0 \left\{ (b_x^2 + b_y^2) \left[1 - \frac{2\varphi_0}{\sinh 2\varphi_0} (1 - \varphi_0 \coth \varphi_0) \right] \right.$$
$$\left. - (b_x^2 - b_y^2) \left(1 - \frac{2\varphi_0}{\sinh 2\varphi_0} \right) + (1-\nu)b_z^2 \right\}. \tag{24.23}$$

24.3 Strained-Layer Epitaxy

Thin films comprise important parts of many electronic, optoelectronic, and magnetic devices. When the lattice parameters of the film and a substrate match, the film grows without a mismatch strain. If the lattice parameters differ, strain is needed to achieve perfect atomic registry across the interface (strained-layer epitaxy). The elastic energy stored in the film can cause the onset and propagation of structural defects in the layer. These defects are generally undesirable, because they can degrade electrical and optical performance of the layer and heterostructural device. If a dislocation is nucleated, for example, as a half loop from irregularities at the free surface, or if it extends from the substrate to the free surface of the layer, it is desirable that the dislocation expand into the configuration with a threading segment across the layer and a long misfit dislocation left behind at the interface between the layer and its substrate (Fig. 24.3). The driving force provided by the misfit energy in the layer pushes the threading segment until it exits at the edges of the film. Only the misfit dislocation is left, which relaxes the strain in the layer and causes nonalignment between the layer and substrate lattices. The smallest layer thickness at which the first misfit dislocation forms during epitaxial growth is known as the critical layer thickness. Comprehensive reviews of the subject are given by Matthews (1979), Nix (1989), Fitzgerald (1991), van der Merwe (1991), Freund (1993), and in the book by Freund and Suresh (2004). If the layer is grown beyond its critical thickness and more dislocations enter, it is of interest to determine the relationship between the dislocation spacing in the interface array and the layer thickness, for any given amount of initial mismatch strain, crystalline orientation and material properties.

Let the layer be bonded to the substrate, with dislocations at the interface, and let the initial uniform misfit strains be e_{yy}^m, e_{zz}^m, e_{zy}^m, with the corresponding stresses

$$\sigma_{yy}^m = \frac{2\mu}{1-\nu}(e_{yy}^m + \nu e_{zz}^m), \quad \sigma_{zz}^m = \frac{2\mu}{1-\nu}(e_{zz}^m + \nu e_{yy}^m), \quad \sigma_{zy}^m = 2\mu e_{zy}^m. \tag{24.24}$$

The total elastic strain energy per unit length of a dislocation within the strip of width p is

$$E = E^d + E^m + E^{d,m}. \tag{24.25}$$

Here, E^d is the energy associated with dislocations alone, given by (24.18), while E^m is the energy associated with the misfit strain alone, which is

$$E^m = \frac{1}{2}\left(\sigma_{yy}^m e_{yy}^m + \sigma_{zz}^m e_{zz}^m + 2\sigma_{zy}^m e_{zy}^m\right) hp. \tag{24.26}$$

The interaction energy $E^{d,m}$ is the work of uniform misfit stresses on dislocation jump displacements along the cut from the free surface to $x = h$, i.e.,

$$E^{d,m} = -(\sigma_{yy}^m b_y + \sigma_{zy}^m b_z)h. \tag{24.27}$$

During the layer deposition, all elastic accommodation is assumed to take place in the layer, with the substrate, being much thicker than the layer, essentially behaving as a rigid elastic half-space.

24.4 Conditions for Dislocation Array Formation

The dislocation array will not form at the interface if the process is not energetically favored. We can thus require as a necessary, but not sufficient condition, that $E \leq E^m$ in order for an array to form. The difference $F = E^m - E$ can be interpreted as the total driving force on each threading dislocation in the array, when all dislocations are imagined to simultaneously form. In view of (24.25), we therefore have

$$F = -(E^d + E^{d,m}), \tag{24.28}$$

where E^d is given by (24.18), and $E^{d,m}$ by (24.27). If the array is at the interface, $F \geq 0$. For arbitrary $\varphi_0 = 2\pi h/p$, the limiting condition $F = 0$ gives the relationship between the layer thickness h and dislocation spacing p. Actually, it gives the smallest dislocation spacing for which the array could exist in the film of a given thickness. The arrays with larger spacing could also exist at this film thickness, but they would be associated with the condition $F > 0$. Alternatively, for a given dislocation spacing of the array, the condition $F = 0$ specifies the smallest film thickness required to support the array at the interface. Thicker film could also support the considered array, but they would correspond to $F > 0$. Since E^d is positive, in order that $F \geq 0$, the interaction energy $E^{d,m}$ must be negative, and its magnitude greater or equal to E^d.

24.4. Conditions for Dislocation Array Formation

In the limit $\varphi_0 \to 0$, the condition $F = 0$ reduces to the Matthews–Blakeslee (1974) equation for the critical film thickness, associated with the introduction of an isolated dislocation. In this case $F = -(E_0^d + E^{d,m})$, where E_0^d is given by (24.21). If $\varphi_0 \to \infty$, i.e., $h >> p$, from (24.28) it follows that the smallest dislocation spacing of the interface array for a very thick layer, is independent of h and is given by

$$p = k\pi[2b_y^2 + (1-\nu)b_z^2]\left(\sigma_{yy}^m b_y + \sigma_{zy}^m b_z\right)^{-1}. \quad (24.29)$$

Physically, the independence of h is a consequence of the fact that for $h >> p$ the stress field in the layer becomes essentially constant, and both energies, far ahead and far behind the threading dislocation segments, are proportional to the layer thickness.

The misfit dislocation, far behind the threading dislocation segment, must be under a force directed away from the free surface, so that misfit dislocations are not pulled out to the free surface. This force is given by the negative gradient of the energy difference $E - E^m$ with respect to the film thickness. Although this must be positive, the misfit dislocation will not advance into the substrate, because the external stress in a thick substrate is zero, whereas the free surface exerts only an attractive force on the misfit dislocation.

As an illustration, consider the layer and substrate that share the same cubic lattice and orientation, with the interface parallel to their (001) crystallographic planes. If the lattice parameters of the layer and substrate are a_l and a_s, the fractional mismatch of the lattice parameter is

$$e^m = \frac{a_s - a_l}{a_l}. \quad (24.30)$$

The associated misfit strain components are

$$e_{yy}^m = e_{zz}^m = e^m, \quad e_{zy}^m = 0, \quad (24.31)$$

and the biaxial stress state is

$$\sigma_{yy}^m = \sigma_{zz}^m = 4\pi k(1+\nu)e^m, \quad \sigma_{zy}^m = 0. \quad (24.32)$$

If the layer/substrate system is Ge_xSi_{1-x}/Si, where x is the fraction of lattice sites in the layer occupied by Ge atoms, the lattice parameter of the layer is (approximately, by Vegard's rule)

$$a_l = xa_{Ge} + (1-x)a_{Si}, \quad (24.33)$$

and of substrate $a_s = a_{Si}$. Since $a_{Si} = 0.54305$ nm and $a_{Ge} = 0.56576$ nm, the misfit strain is $e^m \approx -0.042x$. For $x = 0.25$, this gives $e^m \approx -0.01$. The dislocation array consists of dislocations along $[1\bar{1}0]$ crystallographic direction on the (111) glide planes (Fig. 24.4). The dislocation Burgers vector is along the $[0\bar{1}1]$, so that relative to the (xyz) coordinate system $b_x = -b/\sqrt{2}$, $b_y = -b/2$, and $b_z = b/2$, where b is equal to $a_l/\sqrt{2}$. Consequently, (24.18) becomes

$$E^d = \frac{kb^2}{8}\left[(4-\nu)\ln\frac{\sinh\varphi_0}{\rho_0} - \frac{3\varphi_0^2}{2\sinh^2\varphi_0} - \varphi_0\coth\varphi_0 + \frac{5-2\nu}{4(1-\nu)}\right], \quad (24.34)$$

whereas (24.27) gives

$$E^{d,m} = 2\pi k(1+\nu)e^m bh. \quad (24.35)$$

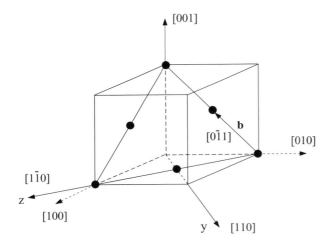

Figure 24.4. A cubic lattice and orientation, with the interface parallel to (001) crystallographic plane. The Burgers vector of dislocation is $\mathbf{b} = \frac{1}{2}[0\bar{1}1]$. The coordinate axes y and z are along the diagonals of the lattice cube. The x axis is parallel to the crystallographic direction [001].

The relaxation occurs, because $E^{d,m}$ is negative. Relaxation actually proceeds with the formation of two orthogonal dislocation arrays; the other array consists of dislocations along [110] direction on the $(1\bar{1}1)$ glide planes. Since the incorporation of the contribution from the second array is straightforward, we proceed with the consideration of one array only (the relaxation via the formation of two orthogonal dislocations arrays is considered is Problems 24.2 and 24.3 of Chapter 34). With $\nu = 0.3$, and with the core radius ρ equal to the length of the Burgers vector b, the critical layer thickness is $h_{cr} = 19.25\,b$, where $b = 0.388$ nm. This satisfies the condition $F_x > 0$, which requires $h > 5.66\,b$. The relationship between the dislocation spacing and the layer thickness, resulting from the condition $F = 0$, is shown in Fig. 24.5. The dislocation spacing p tends to infinity when $h \to h_{cr}$. The thickness $h_0 = 29.25\,b$ corresponds to a dislocation spacing $p_0 = -(b/2)e^m = 50\,b$ at which the array completely relaxes the initial mismatch strain.

24.5 Frank and van der Merwe Energy Criterion

Frank and van der Merwe proposed that, for a given layer thickness h, dislocations in the array will arrange themselves by choosing the periodicity p which minimizes the energy per unit area of the free surface. This energy is E/p, and the criterion requires

$$\frac{d}{dp}\left(\frac{E}{p}\right) = 0, \quad i.e., \quad p\frac{dE}{dp} = E. \tag{24.36}$$

Since E is given by (24.25), from the above condition it follows that

$$p\frac{dE^d}{dp} = E^d + E^{d,m}. \tag{24.37}$$

The right-hand side of (24.37) is equal to $-F$. Defining the force f by

$$f = -p\frac{dE^d}{dp}, \tag{24.38}$$

the Frank and van der Merwe energy condition can be expressed as

$$F = f, \tag{24.39}$$

24.5. Frank and van der Merwe Criterion

where

$$f = \frac{k}{2}\left\{\left[(b_x^2 + b_y^2)\varphi_0 \coth \varphi_0 - 2b_y^2\right]\frac{\varphi_0^2}{\sinh^2 \varphi_0} \right. \quad (24.40)$$
$$\left. + \left[2b_y^2 + (1-\nu)b_z^2\right]\varphi_0 \coth \varphi_0 - \left[b_x^2 + b_y^2 + (1-\nu)b_z^2\right]\right\}.$$

It can be verified that f is always positive (or equal to zero in the limit $\varphi_0 \to 0$), as expected, because from the criterion $E \leq E^m$, we reason that F cannot be negative. A physical interpretation of f can be given as follows. The force on a single threading dislocation entering alone into an epitaxial layer which already contains an array of spacing p is the negative gradient of the specific energy with respect to dislocation density, i.e., (Gosling et al. 1992),

$$G = -\frac{d(E/p)}{d(1/p)}, \quad (24.41)$$

Indeed, if n is the large number of dislocations in the array before one additional dislocation is inserted, the force can be written as

$$G = nE(p) - (n+1)E(p + dp). \quad (24.42)$$

Since dislocations distribute within the same domain, $np = (n+1)(p + dp)$. Thus, upon expanding,

$$G = p\frac{dE}{dp} - E = F - f. \quad (24.43)$$

The force f is, therefore, the difference between the force F on a threading dislocation associated with a simultaneous formation of all dislocations in the array of spacing p, and the force G on a single dislocation when it alone enters an epitaxial layer already containing an array of spacing p. In the latter case, it is assumed that the array remains periodic upon the introduction of new dislocation by appropriate adjustment of its spacing (if necessary, by dislocation climb). Since f is never negative, it follows that $F \geq G$ (equality sign applying only at infinitely large dislocation spacing). From this we can again conclude that the equilibrium spacing predicted by the Frank and van der Merwe criterion ($G = 0$) must be greater than that obtained from the condition $F = 0$. Indeed, if for an array $F = 0$, then $G < 0$, and one dislocation would tend to leave the array, which would increase the spacing among the remaining dislocations. The process would gradually continue until the condition $G = 0$ is reached, which gives the equilibrium array configuration according to Frank and van der Merwe criterion.

For the layer/substrate system under consideration, the predictions based on the condition $F = f$, and the condition $F = 0$, are shown in Fig. 24.5. As previously discussed, for a given film thickness, a larger spacing is predicted by the Frank and van der Merwe criterion. If the array with dislocation spacing according to Frank and van der Merwe would be formed by a simultaneous threading of all dislocations, the corresponding force F on the threading segments would not be zero, but large and positive. Arrays are observed that do not correspond to the minimum energy, or the most relaxed configuration, so that the actual spacing may indeed be greater or smaller than that predicted by the Frank and van der Merwe criterion. One reason for this is that during the process of their gradual introduction, dislocations cannot easily readjust their positions to minimize the total energy.

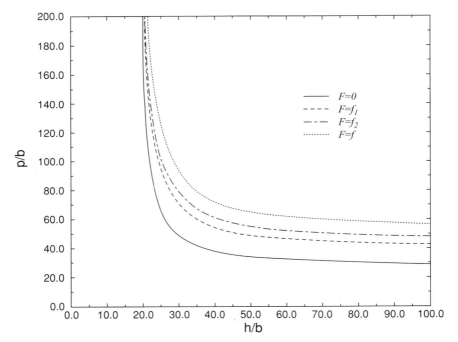

Figure 24.5. Dislocation spacing p vs. layer thickness h (scaled by the length of the Burgers vector) according to the condition based on simultaneous array formation, $F = 0$; Frank and van der Merwe energy minimum condition, $F = f$; and the conditions based on two different processes of gradual array formation, $F = f_1$, and $F = f_2$ ($0 < f_1 < f_2 < f$). From Lubarda (1999).

Experiments, however, indicate that for a film thickness that exceeds the critical thickness by a factor of 2 or 3, the dislocation spacing is substantially greater than that predicted by the condition $F = 0$ (Freund, 1993).

The critical layer thicknesses according to the Matthews and Frank and van der Merwe criteria are identical. The critical thickness according to the Frank and van der Merwe criterion follows from (24.39) in the limit $p \to \infty$, which implies that very few dislocations are introduced in the film. Because f goes to zero as p goes to infinity, the condition (24.39) reduces to $F = 0$, which is the Matthews condition for the critical layer thickness. It is interesting to note that at this value of the layer thickness, the stationary value of the specific energy E/p, reached asymptotically in the limit $p \to \infty$, is actually a local energy maximum. There is also an energy minimum (being very slightly lower than the local energy maximum), which occurs at a large but finite value of the dislocation spacing. This was originally observed by Jain *et al.* (1992). However, by taking the critical film thickness to be just slightly smaller than that associated with the condition $F = 0$, the value of the energy in the limit of infinite spacing becomes an energy minimum, and a single dislocation can be deposited at the interface in a stable manner.

24.6 Gradual Strain Relaxation

During the film growth beyond its critical thickness, dislocations gradually enter to form the misfit dislocation array at the interface between the film and its substrate. To uniformly

24.6. Gradual Strain Relaxation

relax the film, dislocations tend to form periodic arrays. A misfit dislocation already deposited at the interface on a particular glide plane relaxes the elastic strain on adjacent glide planes, reducing a tendency for another misfit dislocation there. The gradual relaxation is a difficult process, which involves time effects and the kinetics of dislocation nucleation and motion for any given temperature of the film growth. A simplified model of gradual relaxation was suggested by Freund (1993). Imagine that in the process of the formation of an array of spacing p, at some instant the array of uniform spacing $2p$ is first formed. Denote the corresponding energy within the width $2p$ by $E(2p)$. The film is assumed to be thick enough for the energy difference

$$F(2p) = 2E^m - E(2p) = -E^d(2p) - E^{d,m} \qquad (24.44)$$

to be positive. This is required to make the configuration energetically preferred relative to the film configuration without dislocations. The actual order by which dislocations entered the film to form the considered array is irrelevant for the present discussion, because the energy $E(2p)$ does not depend on that order. Thus, the value of $F(2p)$ is also independent of the order, although $F(2p)$ can be interpreted as a driving force on each threading dislocation in the array, if they would simultaneously form. For $F(2p)$ to be positive, the film thickness must be greater than the thickness associated with the condition $F(2p) = 0$.

The second set of dislocations is next introduced by the glide of their threading segments along the planes midway between the glide planes of the first set. After this set is introduced, an array of dislocation spacing p is formed. The corresponding elastic strain energy per unit length of dislocation, stored within the width $2p$, can be written as

$$2E(p) = E(2p) + E^d(2p) + E^{d,m} + E^{d,d}(2p). \qquad (24.45)$$

Here, $E(p)$ is the energy within the width p, given by (24.18), and $E^{d,d}(2p)$ is the interaction energy between the two sets of dislocations (two arrays of spacing $2p$). The film of a given thickness will prefer the array of spacing p rather than the array of spacing $2p$, if $2E(p) < E(2p)$, regardless of the order by which the second set is introduced. The difference

$$F_1 = E(2p) - 2E(p) \qquad (24.46)$$

is a driving force on each threading dislocation from the second set, if they all were introduced simultaneously. Substituting (24.44) and (24.45) into (24.46), we obtain

$$F_1 = F(p) - f_1, \quad f_1 = \frac{1}{2} E^{d,d}(2p). \qquad (24.47)$$

The force $F(p)$ is given by (24.28). An expression for the interaction energy $E^{d,d}(2p)$ can be conveniently obtained by using the fact that the elastic strain energy does not depend on the sequence by which dislocations are introduced in the array. Thus, the energy within the width $2p$, associated with sequential formation of the array, given by (24.45), must be equal to the energy associated with simultaneous formation of the whole array, which is $2E(p) = 2E^d(p) + E^m + 2E^{d,m}$. By equating this to the right-hand side of (24.45), we find

$$E^{d,d}(2p) = 2[E^d(p) - E^d(2p)]. \qquad (24.48)$$

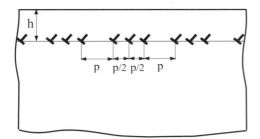

Figure 24.6. An array of nonuniform dislocation spacing, which alters between p and $p/2$.

Substituting (24.18), this becomes explicitly

$$E^{d,d}(2p) = k\left\{[b_x^2 + b_y^2 + (1-\nu)b_z^2]\ln\left(\cosh\frac{\varphi_0}{2}\right) + \frac{1}{8}(b_x^2 + b_y^2)\frac{\varphi_0^2}{\cosh^2(\varphi_0/2)} - \frac{1}{2}(b_x^2 - b_y^2)\varphi_0\tanh\frac{\varphi_0}{2}\right\}.$$

For the previously considered Ge-Si layer/substrate system, (24.6) simplifies to

$$E^{d,d}(2p) = \frac{kb^2}{4}\left[(4-\nu)\ln\left(\cosh\frac{\varphi_0}{2}\right) + \frac{3\varphi_0^2}{8\cosh^2(\varphi_0/2)} - \frac{1}{2}\varphi_0\tanh\frac{\varphi_0}{2}\right]. \quad (24.49)$$

It can be verified that this energy is always positive. It is also observed that $E^{d,d}$ is a monotonically increasing function of φ_0, so that $E^{d,d}(p) > E^{d,d}(2p)$. The plot of the relationship between the dislocation spacing p and the layer thickness h, associated with the condition $F_1 = 0$, i.e., $F(p) = f_1$, is shown in Fig. 24.5. For a given film thickness, the predicted dislocation spacing is greater in the case of sequential, rather than simultaneous, array formation, associated with the condition $F(p) = 0$.

An additional increase in predicted dislocation spacing is obtained if the relaxation process is more gradual. For example, imagine that, in the transition from the array of spacing $2p$ to the array of spacing p, an intermediate configuration is first reached which contains a periodic array of period $4p$. The dislocation spacing in this array is nonuniform, and it varies from p to $2p$. The array is shown in Fig. 24.6, if p there is replaced by $2p$. This configuration can be obtained by the introduction of a new dislocation between every second pair of dislocations of the array of spacing $2p$, i.e., by an appropriate introduction of an additional array of spacing $4p$. The corresponding energy, within the width $4p$, is

$$E_2 = 2E(2p) + E^d(4p) + E^{d,m} + E^{d,d}(2p). \quad (24.50)$$

The driving force for the transition from the configuration with the array of spacing $2p$ to the considered intermediate configuration is $2E(2p) - E_2$, which is assumed to be positive. The driving force from the intermediate configuration to the configuration with the array of spacing p is $F_2 = E_2 - 4E(p)$, which gives

$$F_2 = F(p) - f_2, \quad f_2 = 3E^d(p) - 2E^d(2p) - E^d(4p) - E^{d,d}(2p). \quad (24.51)$$

The plot of the relationship between the film thickness and the dislocation spacing, resulting from the condition $F_2 = 0$, i.e., $F(p) = f_2$ is shown in Fig. 24.5. The results demonstrate that, for a given film thickness, the predicted spacing is greater than that associated with

24.8. Stronger Stability Criteria

the condition $F_1 = 0$. Thus, the more gradual the relaxation process, the less dense is the array deposited at the interface.

24.7 Stability of Array Configurations

Consider a periodic array of spacing p at the interface between the film and its substrate. If the film is sufficiently thick, additional dislocations will enter to relax the film. If the film is too thin, some dislocations will recede. For example, if enough dislocations enter so that, from the array of spacing p, an array of spacing $p/2$ is formed, the driving force for the transition is $E(p) - 2E(p/2) = F(p) - E^{d,d}(p)$. If the film resists the transition, this force must be negative, hence

$$F(p) < E^{d,d}(p). \tag{24.52}$$

On the other hand, if the film is too thin, it may not support the array of dislocation density as high as $1/p$, and some dislocations will recede. Imagine that every second dislocation from the array leaves the film. The driving force for this recession is $-F_1 = -F(p) + E^{d,d}(2p)/2$, as given by (24.47). If the recession is not preferred, the force must be negative, and

$$F(p) > \frac{1}{2} E^{d,d}(2p). \tag{24.53}$$

Combining the inequalities (24.52) and (24.53), we obtain the bounds that define the range of the (p, h) values for which the array can stably exist at the interface, at least regarding the considered perturbations of its structure. The bounds are shown in Fig. 24.7, where they are designated by B1. Recall that the stable range was previously bounded from below by the condition $F > 0$, which defines the lowest lower bound.

A higher lower bound can be obtained from the condition $F_2 > 0$, associated with the recession in which every fourth dislocation leaves the array of uniform spacing p. From (24.51), we have

$$F(p) > 3E^d(p) - 2E^d(2p) - E^d(4p) - E^{d,d}(2p). \tag{24.54}$$

On the other hand, a lower upper bound can be obtained by considering a possible transition of the array of spacing p into the array of nonuniform spacing, which is obtained by the entrance of additional dislocation midway between every second pair of dislocations in the array of uniform spacing p (Fig. 24.6). The energy of this configuration, within the width $2p$, is $2E(p) + E^d(2p) + E^{d,m} + E^{d,d}(p)$. Thus, if the transition should not occur,

$$F(p) < E^{d,d}(p) + E^d(2p) - E^d(p). \tag{24.55}$$

The bounds defined by (24.54) and (24.55) are also shown in Fig. 24.7, designated there by B2, and they are within the previously defined bounds B1.

24.8 Stronger Stability Criteria

Stronger stability conditions and restrictions on possible bounds can be obtained by using the analysis of Gosling *et al.* (1992), who introduced the criterion based on the conditions for the entrance, or recession, of a single dislocation from the periodic array. If the dislocation

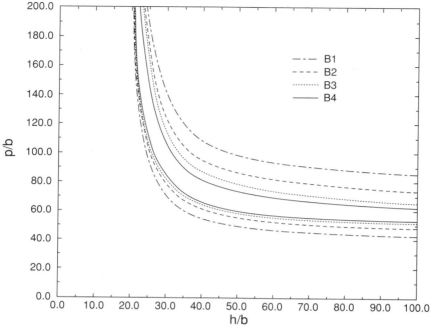

Figure 24.7. Four sets of bounds (B1 through B4) which define range of stable array configurations with respect to assumed perturbations in the periodic array structure, as described in the text (from Lubarda, 1999).

spacing can adjust so that the array maintains its periodicity upon the entrance of a new dislocation, the force on a threading dislocation entering the array would be $G = F - f$. Because the adjustment of spacing generally requires dislocation climb (which may not be operative at low temperatures), it may be assumed that dislocations in the array remain fixed as the new misfit dislocation deposits at the interface. The array is then considered to be stable if this deposition is resisted. The driving force on a threading dislocation, midway between the two dislocations of the periodic array of spacing p, is the difference between the energies of the two configurations: the energy of the configuration with a periodic array, and the energy of the same configuration with an inserted new dislocation. The latter is equal to the former, plus the energy of the added dislocation E_0^d, given by (24.21), plus the work done to introduce the new dislocation against the stress of the existing array and the misfit stress, which is $E^{d,d}(p) + E^{d,m}$. Thus, the force to drive a dislocation into the array is

$$F_{(+)} = -E^{d,d}(p) - E^{d,m} - E_0^d = F(p) - E^{d,d}(p) + E^d(p) - E_0^d. \qquad (24.56)$$

If the introduction of new dislocation is resisted, then $F_{(+)} < 0$, and

$$F(p) < E^{d,d}(p) - [E^d(p) - E_0^d]. \qquad (24.57)$$

It was additionally proposed by Gosling *et al.* (1992) that the recession of a single dislocation from the array (creation of a vacancy in its periodic structure), should also be resisted. A simple way to calculate the force that tends to drive a dislocation out of the array, while all other dislocations remain fixed, is to imagine that a negative dislocation (whose Burgers vector is opposite to that of dislocations in the array) is introduced to

24.9. Further Stability Bounds

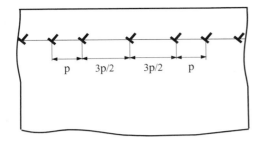

Figure 24.8. A perturbed array configuration used to derive a lower bound of the stable array configurations (designated by B4 in Fig. 24.7).

annihilate one of dislocations from the array. The energy difference between the two configurations is then equal to the energy of the added negative dislocation, E_0^d, plus the work done to introduce the negative dislocation against the stress of the existing periodic array and the misfit stress. This work is

$$\int_0^{h-\rho}[b_x\sigma_{xy}(x,0)+b_y\sigma_{yy}(x,0)+b_z\sigma_{zy}(x,0)]dx+2E_\rho-E^{d,m}=-2E^d(p)-E^{d,m}. \tag{24.58}$$

Equation (24.16) was used to express the integral in (24.58) in terms of other introduced energy contributions. Thus, the force to drive dislocation out of the array is

$$F_{(-)}=2E^d(p)+E^{d,m}-E_0^d=E^d(p)-E_0^d-F(p). \tag{24.59}$$

A dislocation will not recede from the array if $F_{(-)} < 0$, i.e.,

$$F(p) > E^d(p) - E_0^d. \tag{24.60}$$

For the considered layer/substrate system, the bounds defined by (24.57) and (24.60) are shown in Fig. 24.7, where they are designated by B3. These bounds are within the bounds B1. This was clear for the upper bounds, because $E^d(p) - E_0^d$ is positive. It was also expected for the lower bounds on physical grounds, because the array may be in a stable configuration regarding the recession of every second dislocation from the array, but in an unstable configuration regarding a slighter disturbance, due to recession of a single dislocation. For example, suppose the film is in the state corresponding to a point on the lower bound curve. With further film growth, dislocation spacing remains constant until the point on the upper bound curve is reached, at which instant a new dislocation can enter the film.

24.9 Further Stability Bounds

A new set of bounds which define possible range of stable array configurations can be constructed by comparing the array configuration with two perturbed neighboring configurations, as follows.

24.9.1 Lower Bound

Consider a perturbed array configuration, which contains one dislocation at a distance $3p/2$ from the two neighboring dislocations, whereas the rest of the array has uniform spacing p (Fig. 24.8). This configuration can be obtained from a perfectly periodic array of spacing p

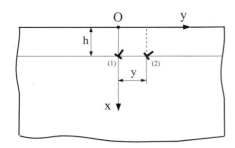

Figure 24.9. Two dislocations with different Burgers vectors, beneath the free surface at the horizontal distance y between each other.

by recession of two neighboring dislocations and by injection of one new dislocation along the slip plane midway between the two receding dislocations. One may also think that one dislocation has receded, whereas one dislocation, ahead or behind the receding dislocation, has subsequently positioned itself in the middle between the two dislocations of the perturbed array. The change of energy between the perturbed and unperturbed configurations can be calculated as follows. First, introduce a new dislocation midway between the two neighboring dislocations of the perfect array. As previously shown, the energy increases by $E_0^d + E^{d,d}(p) + E^{d,m}$. Introduce next two negative dislocations to annihilate two dislocations, ahead and behind the inserted dislocation. This further increases the energy by $2[E_0^d - 2E^d(p) - E^{d,m}]$, plus the interaction energy among the three dislocations, which is $E^{in} = E^{(-,-)} + 2E^{(+,-)}$.

The interaction energy between two dislocations, both at depth h below the free surface of a semi-infinite body, and at the horizontal distance y from each other (Fig. 24.9), can be calculated from

$$E^{(1,2)} = -\int_0^h \left[b_x^{(2)} \sigma_{xy}^{(1)}(x,y) + b_y^{(2)} \sigma_{yy}^{(1)}(x,y) + b_z^{(2)} \sigma_{zy}^{(1)}(x,y) \right] dx. \qquad (24.61)$$

One dislocation has the Burgers vector $\mathbf{b}^{(1)}$ and the other $\mathbf{b}^{(2)}$. The substitution of the Head's stress expressions for a single dislocation beneath the free surface (see Chapter 23) into (24.61), and integration gives

$$E^{(1,2)} = \frac{k}{2} \Big\{ \left[b_x^{(1)} b_x^{(2)} + b_y^{(1)} b_y^{(2)} + (1-\nu) b_z^{(1)} b_z^{(2)} \right] \ln(1 + \eta^2)$$
$$- \left(b_x^{(1)} b_y^{(2)} - b_y^{(1)} b_x^{(2)} \right) \frac{2\eta^3}{(1+\eta^2)^2} \qquad (24.62)$$
$$- \left[(1 + 3\eta^2) b_x^{(1)} b_x^{(2)} - (3+\eta^2) b_y^{(1)} b_y^{(2)} \right] \frac{\eta^2}{(1+\eta^2)^2} \Big\},$$

where $\eta = 2h/y$. The energy $E^{(-,-)}$ is obtained from (24.62) by taking $\mathbf{b}^{(1)} = \mathbf{b}^{(2)} = -\mathbf{b}$, and $y = p$. The energy $E^{(+,-)}$ is obtained if $\mathbf{b}^{(1)} = -\mathbf{b}^{(2)} = \mathbf{b}$, and $y = p/2$. Therefore, we obtain

$$E^{(-,-)} = \frac{k}{2} \Big\{ \left[b_x^2 + b_y^2 + (1-\nu) b_z^2 \right] \ln(1 + \eta^2)$$
$$- \left[(1+3\eta^2) b_x^2 - (3+\eta^2) b_y^2 \right] \frac{\eta^2}{(1+\eta^2)^2} \Big\}, \qquad (24.63)$$

where now $\eta = 2h/p$. The interaction energy between the positive and negative dislocation $E^{(+,-)}$ is given by the same expression, with η replaced by 2η, and k by $-k$. The force which

24.9. Further Stability Bounds

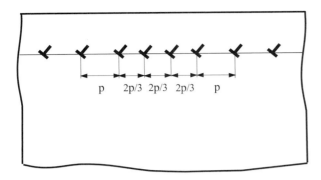

Figure 24.10. A perturbed array configuration used to derive an upper bound of the stable array configurations (designated by B4 in Fig. 24.7).

drives the perfect array into the perturbed array is the negative of the corresponding energy change, which gives

$$F_* = -F(p) + 3[E^d(p) - E_0^d] - E^{d,d}(p) - E^{in}. \tag{24.64}$$

If the perturbation is resisted, $F_* < 0$, i.e.,

$$F(p) > 3[E^d(p) - E_0^d] - E^{d,d}(p) - E^{in}. \tag{24.65}$$

This defines a lower bound for the stable configuration of the perfect array (B4 in Fig. 24.7), which is slightly higher than the lower bound B3 defined by (24.60). This was expected to be the case, because the symmetric configuration in Fig. 24.8 is more relaxed than the configuration with the array containing the vacancy in its periodic structure, with the surrounding dislocations being fixed.

Since dislocation adjustment may require climb, it is supportive to the above consideration to give an additional or alternative interpretation of the condition $F_* < 0$. Imagine that the periodic array of uniform spacing p is completed, except for two missing dislocations next to each other. Denote the corresponding energy by E. (Total energies of infinite arrays are infinitely large, but their differences are finite, and will only be needed in this discussion). If two dislocations enter and complete the perfect array, the energy becomes E_a, and the driving force for this to occur would be $F_a = E - E_a$. If, instead of two, only one dislocation enters, midway between the two missing dislocations of the perfect array, the energy is E_b, and the corresponding force would be $F_b = E - E_b$. If $F_a > F_b$, the case (a) is preferred, because then $E_a < E_b$. On the other hand, the force that would drive the configuration (a) into (b) is $F_* = E_a - E_b = F_b - F_a$. Thus, the condition $F_* < 0$ again gives $F_a > F_b$, which means that the perfect array (a) would be preferred to the perturbed array (b).

24.9.2 Upper Bound

A lower upper bound can be obtained by considering a perturbed array configuration shown in Fig. 24.10. This configuration can be obtained from the perfectly periodic array of spacing p by recession of one dislocation, and by symmetric injection of two new dislocations at the distance $p/3$ from the receding dislocation. Alternatively, one may consider that a perfect array was created, except for one missing dislocation, and the competition is taking place whether one more dislocation will enter and complete the perfect array or

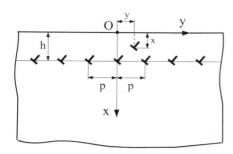

Figure 24.11. A dislocation at the position (x, y) approaching a dislocation array of spacing p, beneath the free surface.

whether two dislocations will symmetrically enter to form the perturbed array in Fig. 24.10. Imagine that a negative dislocation is introduced to cancel one dislocation from the perfect array and two additional dislocations are then injected. The energy change relative to the perfect configuration is equal to the energy of the added negative dislocation, E_0^d, plus the work done to introduce the negative dislocation against the stress of the perfect array and the misfit stress, which is given by (24.58), plus the energy associated with the introduction of the two dislocations. This is $2E_0^d + 2E^{d,m}$, plus the interaction energy of the two dislocations with the perfect array $2E^{int}(p/3)$, plus the interaction energy among the three added dislocations, $E^{in} = E^{(+,+)} + 2E^{(+,-)}$. Thus, the force that would drive the perfect array into the perturbed array is

$$F^* = F(p) + 3[E^d(p) - E_0^d] - 2E^{int}(\frac{p}{3}) - E^{in}. \quad (24.66)$$

The interaction energy $E^{(+,+)}$ is given by (24.63), with $\eta = 3h/p$. The interaction energy $E^{(+,-)}$ is given by the same expression, with η replaced by 2η, and k by $-k$.

The general expression for the interaction energy associated with the introduction of an additional dislocation into the periodic array, anywhere between two dislocations in the array (Fig. 24.11), can be calculated from

$$E^{int}(x, y) = -\int_0^x [b_x \sigma_{xy}(x, y) + b_y \sigma_{yy}(x, y) + b_z \sigma_{zy}(x, y)] dx. \quad (24.67)$$

The substitution of the stress expressions (24.1)–(24.3) gives upon integration the following expression for the interaction energy

$$E^{int}(x, y) = k[b_x^2 I_x + b_y^2 I_y + b_x b_y I_{xy} + (1 - \nu)b_z^2 I_z], \quad (24.68)$$

where

$$I_x = I_z - \frac{1}{2}\frac{\varphi \sinh \varphi}{\cosh \varphi - \cos \psi} + \frac{1}{2}\frac{\vartheta \sinh \vartheta}{\cosh \vartheta - \cos \psi} + \varphi_0(\varphi - \varphi_0)\frac{1 - \cosh \varphi \cos \psi}{(\cosh \varphi - \cos \psi)^2},$$

$$I_y = I_x + \frac{\varphi \sinh \varphi}{\cosh \varphi - \cos \psi} - \frac{\vartheta \sinh \vartheta}{\cosh \vartheta - \cos \psi},$$

$$I_{xy} = \vartheta \sin \psi \left(\frac{1}{\cosh \vartheta - \cos \psi} - \frac{1}{\cosh \varphi - \cos \psi}\right), \quad (24.69)$$

$$I_z = \frac{1}{2}\ln\frac{\cosh \varphi - \cos \psi}{\cosh \vartheta - \cos \psi}.$$

24.9. Further Stability Bounds

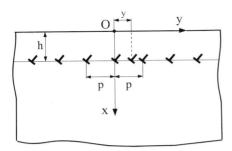

Figure 24.12. A dislocation injected between two neighboring dislocations from a dislocation array of spacing p.

The interaction energy for $x = h$ (Fig. 24.12) is consequently

$$E^{\text{int}}(y) = k\,[b_x^2 I_x + b_y^2 I_y + (1-\nu)b_z^2 I_z], \tag{24.70}$$

where

$$I_x = I_z - \frac{\varphi_0 \sinh 2\varphi_0}{\cosh 2\varphi_0 - \cos\psi} + \frac{\varphi_0^2(1 - \cosh 2\varphi_0 \cos\psi)}{(\cosh 2\varphi_0 - \cos\psi)^2},$$

$$I_y = I_x + \frac{2\varphi_0 \sinh 2\varphi_0}{\cosh 2\varphi_0 - \cos\psi}, \tag{24.71}$$

$$I_z = \frac{1}{2}\ln\frac{\cosh 2\varphi_0 - \cos\psi}{1 - \cos\psi}.$$

In (24.9.2)–(24.9.2), $\varphi_0 = 2\pi h/p$, and $\psi = 2\pi y/p$. The interaction energy $E^{\text{int}}(p/3)$ is obtained by taking $\psi = 2\pi/3$.

Having $E^{\text{int}}(p/3)$ calculated, the perturbation is resisted if $F^* < 0$, i.e.,

$$F(p) < 2E^{\text{int}}\left(\frac{p}{3}\right) + E^{\text{in}} - 3[E^{\text{d}}(p) - E_0^{\text{d}}]. \tag{24.72}$$

This defines an upper bound for the stable configuration of the perfect array (B4 in Fig. 24.7), which is lower than the upper bound B3 defined by (24.57), because the symmetric configuration in Fig. 24.10 is more relaxed than the perfect array configuration with an additional dislocation exactly midway between the two dislocations of the array. It is interesting to note that the four lower bounds shown in Fig. 24.7 are closer to each other than the four upper bounds, so that bounds are more sensitive to perturbation modes involving the entrance of new dislocations, than the recession of some of the dislocations from the array.

From a purely energetic point of view, which does not take into account possible mechanisms by which configurations can alter, the stable configuration would be unique and given by the Frank and van der Merwe criterion. The corresponding curve is always between the upper and lower bound of any considered set of bounds, because f in (24.39) is always between these bounds. However, during film growth dislocations may be entering in such a way that the minimum energy spacing cannot actually be attained at a given film thickness, because complete readjustment of already taken dislocation positions would be required (either by climb or recession of some and entrance of other dislocations). In view of this, then, any spacing between derived upper and lower bounds could in principle correspond to a given film thickness, depending on the sequence or the order by which dislocations entered in the process of film growth. Inevitably, the dislocation spacing will be more or less nonuniform, although dislocations will try their best to form, as nearly as possible, periodic arrays and minimize the total energy of the system. Furthermore, the order by which dislocations enter depends on the location and strength of available dislocation sources. The rate of film growth has also an obvious effect on dislocation spacing that is eventually taken by the array at the final film thickness.

24.10 Suggested Reading

Fitzgerald, E. A. (1991), Dislocations in Strained-Layer Epitaxy: Theory, Experiment, and Applications, *Mater. Sci. Reports*, Vol. 7, pp. 87–142.

Freund, L. B. (1993), The Mechanics of Dislocations in Strained-Layer Semiconductor Materials, *Adv. Appl. Mech.*, Vol. 30, pp. 1–66.

Freund, L. B., and Suresh, S. (2003), *Thin Film Materials: Stress, Defect Formation and Surface Evolution*, Cambridge University Press, New York.

Gosling, T. J., Jain, S. C., Willis, J. R., Atkinson, A., and Bullough, R. (1992), Stable Configurations in Strained Epitaxial Layers, *Phil. Mag. A*, Vol. 66, pp. 119–132.

Lubarda, V. A. (1999), Dislocations Arrarys at the Interface between an Epitaxial Layer and Its Substrate, *Math. Mech. Solids*, Vol. 4, pp. 411–431.

Matthews, J. W. (1979), Misfit Dislocations. In *Dislocations in Solids* (F. R. N. Nabarro, ed.), Vol. 2, pp. 461–545, North-Holland, Amsterdam.

Nix, W. D. (1969), Mechanical Properties of Thin Films, *Metall. Trans. A*, Vol. 20A, pp. 2217–2245.

Van der Merwe, J. H. (1991), Strain Relaxation in Epitaxial Overlayers, *J. Electr. Mat.*, Vol. 20, pp. 793–803.

25 Stability of Planar Interfaces

The breakdown of an initially flat, or smooth, surface into one characterized by surface roughness is an important type of phenomena occurring, *inter alia*, during the growth of thin films or at surfaces of solids subject to remotely applied stress in environments that induce mass removal or transport. In the case of thin films, stresses arise due to lattice mismatch and/or differences in coefficients of thermal expansion. The sources of stress are, indeed, legion but the effect can be to induce roughness, and surface restructuring, that may be either deleterious, or in some cases desirable, if the patterning can be controlled. The phenomena was first studied by Asaro and Tiller (1972) and has since been pursued by others. Our purpose is to develop some of the guiding principles, but we note that the topic is far from being thoroughly worked out. In particular, we make many simplifying assumptions, one being the assumption of surface isotropy. We also ignore some important physical attributes of surfaces, such as *surface stress*, which have recently been added to the description of surface patterning (Freund and Suresh, 2003).

25.1 Stressed Surface Problem

We consider here the phenomena of the breakdown of planar interfaces subject to stress into interfaces characterized by undulated topology. The phenomena is governed by those same driving forces that lead to crack growth and the growth of defects, such as inclusions, that cause internal stresses. Figure 25.1 illustrates such a process occurring under the action of a remotely applied tensile stress, although the same process can be driven by remotely applied shear stress. Moreover, attention will be confined to cases where mass transport is effected by surface diffusion and dissolution followed by diffusion through a bulk medium. An exactly parallel development can be made where mass transport is *via* bulk solid diffusion.

The process involves the development of stress concentrations at the *troughs* of the (slightly) rough surface; this, in turn, causes higher levels of strain energy density and chemical potential (defined in the next section). The gradients of chemical potential cause mass flow, *via* diffusion, from the troughs to the *hills*, thus leading to deepening of the troughs and an increase in undulation. The impediment is the the energy required to create additional surface area. This leads to a stability problem as described below. We begin by considering a half-plane whose free surface has become slightly undulated; the

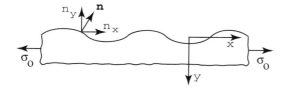

Figure 25.1. Undulated stressed surface.

surface profile is assumed to be sinusoidal with a form $y = a\cos(\omega x)$. The slope of the surface is assumed small and the intent is to create an asymptotic solution of order $\mathcal{O}(a\omega)$. The normal to the surface is **n**, where

$$n_x = \sin[\tan^{-1}(-a\omega \sin(\omega x))] \sim -a\omega \sin(\omega x) + \mathcal{O}(a^2\omega^2),$$
$$n_y = \cos[\tan^{-1}(-a\omega \sin(\omega x))] \sim 1 + \mathcal{O}(a^2\omega^2).$$
(25.1)

This surface is traction free.

Once again the total problem will be solved by superposing two companion problems. The first problem is simply a half-plane subject to far field tension of magnitude σ_0 – this will create tractions on the sinusoidal surface, **T**, that we shall evaluate. The second problem is one in which the opposite tractions $-\mathbf{T}$ are applied to the surface of the half-plane; when these are added the effect is to annihilate those existing in the first problem, thereby creating a sinusoidal surface that is traction free, as desired. Clearly, the potential for the first problem is $\phi^{(1)} = \frac{1}{2}\sigma_0 y^2$, with the stress $\sigma_{xx}^{(1)} = \sigma_0$. For problem (2), we proceed as follows. Consider the potential

$$\phi^{(2)} = a\sigma_0 y e^{-\omega y} \cos(\omega x).$$
(25.2)

The corresponding stresses are

$$\sigma_{xx}^{(2)} = \frac{\partial^2 \phi^{(2)}}{\partial y^2} = -2a\omega\sigma_0 e^{-\omega y} \cos(\omega x) + a\omega^2 \sigma_0 y e^{-\omega y} \cos(\omega x),$$

$$\sigma_{yy}^{(2)} = \frac{\partial^2 \phi^{(2)}}{\partial x^2} = -a\omega^2 \sigma_0 y e^{-\omega y} \cos(\omega x),$$
(25.3)

$$\sigma_{xy}^{(2)} = -\frac{\partial^2 \phi^{(2)}}{\partial x \partial y} = a\omega\sigma_0 e^{-\omega y} \sin(\omega x) - a\omega^2 \sigma_0 y e^{-\omega y} \sin(\omega x).$$

On the surface, which is taken asymptotically to be at $y = a\cos(\omega x)$, the traction components are, to first order in $a\omega$,

$$T_x^{(2)} = \sigma_{xx}^{(2)} n_x + \sigma_{xy}^{(2)} n_y = a\omega\sigma_0 \sin(\omega x) + \mathcal{O}(a^2\omega^2),$$
$$T_y^{(2)} = \sigma_{yy}^{(2)} n_y + \sigma_{xy}^{(2)} n_x = 0 + \mathcal{O}(a^2\omega^2).$$
(25.4)

Note from problem (1) that these traction components are, again to first order in $a\omega$,

$$T_x^{(1)} = -a\omega\sigma_0 \sin(\omega x) + \mathcal{O}(a^2\omega^2),$$
$$T_y^{(1)} = 0 + \mathcal{O}(a^2\omega^2).$$
(25.5)

Thus, to first order in $a\omega$, the full solution for the stress potential is

$$\phi(x, y) = \frac{1}{2}\sigma_0 y^2 + a\sigma_0 y e^{-\omega y} \cos(\omega x).$$
(25.6)

25.2. Chemical Potential

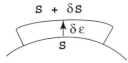

Figure 25.2. Surface area and curvature.

The corresponding stresses are

$$\sigma_{xx} = \sigma_0 - a\sigma_0(\omega^2 y - 2\omega)e^{-\omega y}\cos(\omega x),$$
$$\sigma_{yy} = -\omega^2 a\sigma_0 y e^{-\omega y}\cos(\omega x), \quad (25.7)$$
$$\sigma_{xy} = -\omega a\sigma_0(1 - \omega y)e^{-\omega y}\sin(\omega x).$$

The strain energy density on $y = a\cos(\omega x)$ is given, to first order in $a\omega$, by

$$W(x) = \sigma_0^2[1 + 4a\omega\cos(\omega x)]/2E + \mathcal{O}(a^2\omega^2). \quad (25.8)$$

The above will be used to construct the chemical potential for matter on the undulated surface and to set up the equations of diffusion that govern the evolution of the surface profile.

25.2 Chemical Potential

The Gibbs free energy was introduced in previous chapters. Here we introduce the chemical potential μ as the partial molar derivative of G. Accordingly, μ is defined as the *work required to add a unit amount of material at constant temperature to the body*. Moreover, we are specifically concerned with the chemical potential on the surface of the body. When material is added to a surface it is necessarily added normal to the surface, and if the surface has curvature this will change the surface area and thus require additional free energy associated with the surface energy, γ. This energy is to be added to that associated with the elastic strain energy.

Consider the addition of mass to a surface as depicted in Fig. 25.2. The increment in area is related to curvature by

$$\delta S = S\kappa\delta\epsilon, \quad (25.9)$$

where $\delta\epsilon$ is the normal growth of the surface, S is the area to which material is being added, and κ is the curvature at the point where material is added. In two dimensions, the curvature is given as

$$\kappa = \frac{d^2 y/dx^2}{[1 + (dy/dx)^2]^{3/2}} \approx -a\omega\cos(\omega x) + \mathcal{O}(a^2\omega^2). \quad (25.10)$$

Now, the surface area change can also be written as

$$\delta S = \kappa(S\delta\epsilon) = \kappa\delta v, \quad (25.11)$$

where δv is the increment of volume of added material. Thus, the extra amount of work required to add material associated with surface energy is $\gamma\kappa\delta v$. Let the unit amount be per atom, and define v_0 as the *atomic volume*. Then, the amount of work required to create (or destroy) surface area is, per atom, $\gamma\kappa v_0$.

Recall that the body is elastically strained. To add a unit amount of material at a point, it is necessary to bring it to the same state of elastic strain – this adds an additional amount of required work, $W(x)v_0$. The total chemical potential is, therefore,

$$\mu = \mu_0 + W(x)v_0 + \gamma \kappa v_0, \qquad (25.12)$$

where μ_0 represents all other contributions to μ, not affected by position on the surface. It is gradients in μ that cause mass flow, which we assume occurs by surface diffusion.

25.3 Surface Diffusion and Interface Stability

Suppose that the body contains β atoms per unit area. Gradients in the chemical potential give rise to a diffusive flux, J, calculated from

$$J = -\frac{D}{kT} \nabla \mu, \qquad (25.13)$$

where k is Boltzman's constant, T is the absolute temperature, and D is the surface diffusion constant. For the case considered,

$$\nabla \mu = \frac{\partial \mu}{\partial x} + \mathcal{O}(a^2 \omega^2). \qquad (25.14)$$

Define r_n as the *rate of normal motion of the surface*. Then,

$$r_n = v_0 \,\mathrm{div}\, J = -\frac{D\beta v_0}{kT} \frac{\partial^2 \mu}{\partial x^2}. \qquad (25.15)$$

The rate of normal advance is related to the time change in the curve $y(x)$ defining the surface as

$$r_n = \left(1 + \frac{dy}{dx}\right)^{-1/2} \frac{dy}{dt}, \qquad (25.16)$$

and so, to first order,

$$\frac{\partial y}{\partial t} = -\frac{D\beta v_0}{KT} \frac{\partial^2 \mu}{\partial x^2}. \qquad (25.17)$$

The differential equation in (25.17) may be solved using the definitions in (25.10) and (25.12). At $t = 0$ and at $x = 0$, *i.e.*, at the depth of a trough at time $t = 0$, we find that

$$V = \frac{da}{dt} = B\left(-\gamma^2 \omega^4 + \frac{2\sigma_0^2}{E} \omega^3\right), \qquad (25.18)$$

where $B = av_0^2 \beta D/kT$. For $V \geq 0$, that is for the interface to become *unstable*, in the sense that its undulated amplitude begins to grow, we must have

$$\sqrt{2}\,\sigma_0 \geq (E\gamma \omega)^{1/2} = \left(\frac{2\pi E\gamma}{\lambda}\right)^{1/2}, \qquad (25.19)$$

where λ is the wavelength of the sinusoidal surface.

Assume that all wavelengths are possible and that the one that dominates is that which grows fastest, *i.e.*, set

$$\frac{\partial V}{\partial \omega} = 0 \quad \text{and} \quad \frac{\partial^2 V}{\partial \omega^2} < 0, \qquad (25.20)$$

25.4. Interface Stability

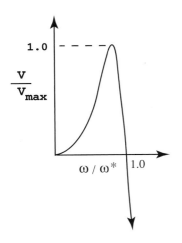

Figure 25.3. Velocity of surface amplitude growth vs. frequency.

to obtain

$$\omega^{\star} = \frac{3}{2} \frac{\sigma_0}{E\gamma}, \qquad (25.21)$$

and

$$V = V_{\max} = 0.1 B \frac{\sigma_0^8}{\gamma^3 E^4}. \qquad (25.22)$$

Relations such as (25.21) and (25.22) certainly suggest a very strong sensitivity of interface stability to the material properties involved in the process, such as interface energy and elastic modulus (Fig. 25.3).

25.4 Volume Diffusion and Interface Stability

The chemical potential on the surface of the stressed solid was given in (25.12), and in full it is

$$\mu - \mu_0 = \frac{v_0 \sigma_0^2}{2E} [1 + 4a\omega e^{-\omega y} \cos(\omega x)] - \gamma v_0 a \omega^2 \cos(\omega x), \qquad (25.23)$$

where the expression for strain energy density was evaluated on $y = 0$. Now we assume that the stressed solid is in contact with a fluid that is capable of dissolving the solid and forming with it a perfect solution. With that assumption, we take the chemical potential of the solid constituent in the fluid to be

$$\mu - \mu_0 = kT \ln c, \qquad (25.24)$$

where c is the concentration. If c_0 is the equilibrium concentration between an unstressed flat solid and the liquid, we obtain the relation

$$kT \ln \frac{c}{c_0} = \frac{v_0 \sigma_0^2}{2E} [1 + 4a\omega e^{-\omega y} \cos(\omega x)] - \gamma v_0 a \omega^2 \cos(\omega x), \qquad (25.25)$$

which can be expressed as

$$c = c_0 e^{f(\sigma_0, \omega)}. \qquad (25.26)$$

Since $f(\sigma_0, \omega)$ is typically much less that unity, we may expand (25.26) to first order and obtain

$$\frac{c}{c_0} = 1 - \frac{v_0}{kT} \frac{\sigma_0^2}{2E} \left[1 + 4a\omega e^{-\omega y} \cos(\omega x)\right] \tag{25.27}$$
$$+ \gamma \left(\frac{v_0}{kT}\right) \omega^2 \cos(\omega x).$$

This relation is the interface boundary condition for a fluid diffusion problem wherein matter will dissolve at the valleys and deposit at the peaks, or *vice versa*, depending on the sign of the chemical potential gradient. Assuming that equilibrium is maintained between the solid and the fluid, and that the interface moves sufficiently slowly that the transport is nearly steady state at all times, the solution to the transport problem, subject to (25.27) as the boundary condition, is given by

$$c - c_0 = -c_0 \left\{\frac{v_0}{kT} \frac{\sigma_0^2}{2E} \left[1 + 4a\omega e^{-\omega y_0} \cos(\omega x)\right]\right\} e^{-\omega y} \tag{25.28}$$
$$+ c_0 \left[\gamma \left(\frac{v_0}{kT}\right) a\omega^2 \cos(\omega x)\right] e^{-\omega y}.$$

The rate of change of the amplitude, a, is

$$\dot{a} = \left(\frac{\partial y}{\partial t}\right)_{x=0} = v_0 D_L \left(\frac{\partial c}{\partial y}\right)_{\substack{y=0 \\ x=0}}$$
$$= D_L c_0 \frac{v_0^2}{kT} a \left[\frac{\sigma_0^2}{2E} \left(\frac{\omega}{a} + 4\omega^2\right) - \gamma \omega^3\right], \tag{25.29}$$

with $D_L c_0 v_0^2 a/kT = B^*$ for later reference.

For growth of the wave, we must have

$$\frac{\sigma_0^2}{2E} \left(\frac{\omega}{a} + 4\omega^2\right) > \gamma \omega^3, \tag{25.30}$$

i.e.,

$$\sigma_0 > (2E\gamma\omega)^{1/2} \left(\frac{1}{a\omega} + 4\right)^{-1/2}. \tag{25.31}$$

The frequency, ω^*, yielding a maximum in $\dot{a}$ is obtained from

$$\left(\frac{\partial \dot{a}}{\partial \omega}\right)_{\omega^*} = 0 = \frac{\sigma_0^2}{2E} \left(\frac{1}{a} + 8\omega^*\right) - 3\gamma\omega^2, \tag{25.32}$$

as

$$\omega^* = \frac{2}{3} \frac{\sigma_0^2}{2E} \pm \frac{1}{2} \sqrt{\left(\frac{4}{3} \frac{\sigma_0^2}{E\gamma}\right)^2 + 4\left(\frac{\sigma_0^2}{6aE\gamma}\right)}. \tag{25.33}$$

For very small values of a, the second term within the radical will be comparable to the first and no real simplification is possible. It is best to leave the expression as it is. Thus, (25.33) provides us with ω^* for Mode I type loading and, with (25.29), we have

$$\dot{a}(\omega^*) = V = B^* \left[\frac{\sigma_0^2}{2E} \left(\frac{\omega^*}{a} + 4\omega^{*2}\right) - \gamma\omega^{*3}\right]. \tag{25.34}$$

25.4. Interface Stability

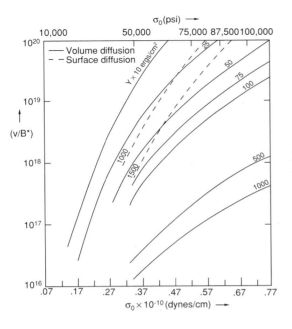

Figure 25.4. Velocity normalized with B (from Asaro and Tiller, 1972).

In Figs. 25.4 and 25.5, the results for V/B^* have been plotted as a function of σ_0 for a range of values of γ and compared with the similar predictions for growth via surface diffusion as described earlier; see (25.23). We note that the dependence on applied stress is not as strong for this case of volume diffusion as was the case for surface diffusion

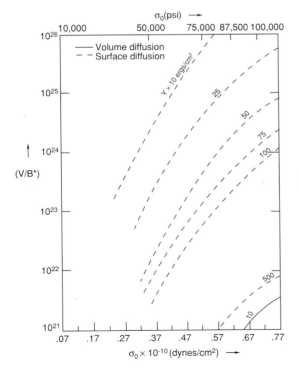

Figure 25.5. Velocity normalized with B continued (from Asaro and Tiller, 1972).

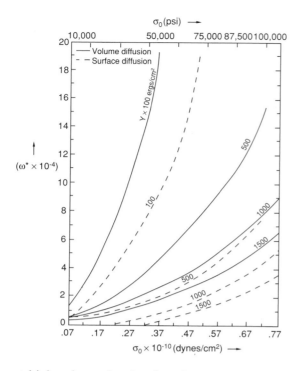

Figure 25.6. Optimal wave frequency (from Asaro and Tiller, 1972).

at high values of γ, but is quite comparable in this dependence at low values of γ. The strong dependence on γ should, of course, also be noted. For these calculations, the value of $E = 10^{12}$ dynes/cm^2 and $a = 10^{-5}$ cm were used (1 dyne = 10^{-5} N). For a complete comparison between the volume diffusion (*i.e.*, dissolution) case and the surface diffusion case we must also consider the relative magnitude of B^* and the coefficient B of (25.22); we recall that for surface diffusion

$$V = 0.1 B_{sd} \sigma_0^8 / \gamma^3 E^4,$$
$$B_{sd} = a v_0^2 \beta D_s / kT, \tag{25.35}$$

where β is the surface density of atoms. We have renamed the surface diffusion coefficient D_s for clarity. Thus, if we similarly rename $B^* = B_{vd}$, we must consider the ratio $B_{sd}/B_{vd} = \beta D_s / c_0 D_L$. If this ratio were small, the two mechanisms would give comparable rates of growth of the rough surface. If this ratio were near unity, the surface diffusion mechanism is strongly dominant over the volume/dissolution mechanism. Of course, in reality both mechanisms operate concurrently.

The presence of surface adsorption leading to a reduction in γ will produce a strong increase in V unless there is a compensatory reduction in B. For the surface diffusion case, β may be reduced, so that B_{sd} may also be reduced. For the case of dissolution/volume diffusion, a significant affect on B_{vd} is not anticipated, but there may be an interference with the assumption of surface equilibrium that would have a negative effect on the overall kinetics. Such effects of surface adsorption may, of course, be marked at a liquid/metal interface.

In Fig. 25.6 the optimum frequency ω^* has been plotted as a function of σ_0 and it is noted that the dissolution mechanism produces a comparable, but smaller optimum wavelengths,

25.5. 2D Surface Profiles and Surface Stability

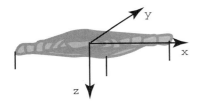

Figure 25.7. 2D surface roughness.

λ^*, than the surface diffusion mechanism for similar stress levels. Depending on the surface energy, there seems to be a critical stress range beyond which ω^* increases very sharply. For example, for $\gamma \sim 100$ erg/cm^2, and $\sigma_0 \leq 50{,}000$ to $75{,}000$ psi, λ^* decreases rapidly and the roughness sharpens drastically (1 erg = 10^{-7} J and 1 psi =6.895 kPa). Finally, it should be noted that the analysis of volume diffusion as considered here should also apply to the case of diffusion *via* vacancy diffusion in the volume of the solid beneath the undulated surface. Equations (25.27) and (25.28) are still applicable, but with c_0 being the equilibrium vacancy concentration in the solid in the unstressed state. Equations (25.29) to (25.34) also hold, with D_{bd} replacing D_{vd} in the expression for B. Thus, the figures also hold, with B changing *via* D and c_0.

25.5 2D Surface Profiles and Surface Stability

As seen in the above discussion of 1D surface profiles, modeled as sinusoidal perturbations, the tendency for an initially nearly smooth surface in a stressed body to break down into one characterized by growing roughness, is a competition between the release of strain energy and the creation of surface energy. The driving force for the process is properly described by the gradient in chemical potential, as originally introduced in this context by Asaro and Tiller (1972). Thus we want to evaluate

$$\mu = \mu_0 + W(x, y)v^0 + \kappa\gamma. \tag{25.36}$$

The coordinate geometry for the 2D surface profiles is as illustrated in Fig. 25.7. The surface energy γ is so far taken to be isotropic and independent of orientation and state of surface stress; $W(x, y)$ is the surface strain energy density. Alternatively, we could calculate the rate of change of the total free energy, G_{tot}, due to growth of the surface,

$$\dot{G}_{\text{tot}} = \int_S \mu v_n \, dS, \tag{25.37}$$

where v_n is the normal growth rate described earlier and inquire as to its algebraic sign. This in itself would be sufficient to assess stability, but an evaluation of μ would be desirable to carry out detailed kinetic simulations, as dictated by the appropriate mechanisms of mass transport such as surface diffusion, bulk diffusion, and dissolution and bulk transport. Since our purposes here are illustrative only, we will examine stability only. To calculate μ it is necessary to evaluate the elastic field at the surface and evaluate the strain energy density $W(x, y)$. A general procedure by Asaro and Tiller (1972) to accomplish this is based on a linear superposition. The method was described previously for 1D sinusoidal surface profiles; here we extend the analysis to a 2D sinusoidal profile.

Let the surface take on the perturbed shape given by

$$z = a\cos(\omega_x x)\cos(\omega_y y), \tag{25.38}$$

where $\omega_x = 2\pi/\lambda_x$ and $\omega_y = 2\pi/\lambda_y$ are the respective frequencies. We assume that the surface is *shallow*, i.e., $a\omega_x, a\omega_y \ll 1$. The body is loaded by a biaxially applied uniform stress, σ_0, rather than by the uniaxial stress as in the case of 1D surface profiles. A uniform state of stress of this type produces a traction on a so perturbed surface of the form

$$T_x = -\sigma_0 \frac{\partial z}{\partial x}, \quad T_y = -\sigma_0 \frac{\partial z}{\partial y}, \quad T_z = 0, \qquad (25.39)$$

to first order in the $a\omega$'s. Thus, if we write the total stress field as

$$\boldsymbol{\sigma} = \boldsymbol{\sigma}^{(1)} + \boldsymbol{\sigma}^{(2)}(x, y, z), \qquad (25.40)$$

where $\boldsymbol{\sigma}^{(1)} = \sigma_0\,\mathbf{e}_x\mathbf{e}_x + \sigma_0\,\mathbf{e}_y\mathbf{e}_y$ is the uniform biaxial stress state, then the perturbed field, $\boldsymbol{\sigma}^{(2)}$, is obtained by solving the problem of applying the negative traction, given by (25.39) above, to the surface; this has the effect of producing a traction free perturbed surface, as desired. The solution for the surface displacements is

$$\bar{u}_\alpha^{(2)}(x, y) = \sigma_0 \iint_{-\infty}^{\infty} \left[\Pi_{\alpha x}(x-\xi, y-\eta) \frac{\partial z}{\partial x} + \Pi_{\alpha y}(x-\xi, y-\eta) \frac{\partial z}{\partial y} \right] d\xi\, d\eta,$$

where the surface is taken as $z = z(x, y)$, and the matrix $\boldsymbol{\Pi}$ has been given earlier in (13.122) as

$$\boldsymbol{\Pi} = \frac{1}{4\pi G} \begin{Bmatrix} \dfrac{2(1-\nu)}{r} + \dfrac{2\nu x^2}{r^3} & \dfrac{2\nu xy}{r^3} & -\dfrac{(1-2\nu)x}{r^2} \\ \dfrac{2\nu xy}{r^3} & \dfrac{2(1-\nu)}{r} + \dfrac{2\nu y^2}{r^3} & -\dfrac{(1-2\nu)y}{r^2} \\ \dfrac{(1-2\nu)x}{r^2} & \dfrac{(1-2\nu)y}{r^2} & \dfrac{2(1-\nu)}{r} \end{Bmatrix}. \qquad (25.41)$$

Freund and Suresh (2003) have carried out this analysis and found by straightforward integration that

$$\bar{u}_x^{(2)} = -\frac{2\sigma_0 a \lambda_y}{\bar{E}\sqrt{\lambda_x^2 + \lambda_y^2}} \sin(\omega_x x)\cos(\omega_y y),$$

$$\bar{u}_y^{(2)} = -\frac{2\sigma_0 a \lambda_x}{\bar{E}\sqrt{\lambda_x^2 + \lambda_y^2}} \cos(\omega_x x)\sin(\omega_y y), \qquad (25.42)$$

$$\bar{u}_z^{(2)} = 0.$$

The stresses follow immediately by calculating the strains and using the isotropic constitutive relations.

The change in energy is the energy involved with creating the perturbation and the perturbed field. This was also calculated by Freund and Suresh (2003). Per period over the perturbation it is found that

$$\delta\mathcal{E}_{el}^{(2)} = \int_0^{\lambda_x} \int_0^{\lambda_y} \frac{1}{2} [\bar{u}_x^{(2)}\sigma_{xz} + \bar{u}_y^{(2)}\sigma_{yz}]\, dx\, dy$$

$$= -\frac{\pi \sigma_0^2 a^2}{2\bar{E}} \sqrt{\lambda_x^2 + \lambda_y^2}. \qquad (25.43)$$

25.6. Asymptotic Stresses

The change in energy due to the creation of surface is calculated by noting that, to first order,

$$d(\delta S) = \frac{1}{2}\left[\left(\frac{\partial z}{\partial x}\right)^2 + \left(\frac{\partial z}{\partial y}\right)^2\right] dx\, dy. \tag{25.44}$$

Thus, this contribution to the energy change is

$$\delta\mathcal{E}_{\text{surface}} = \gamma \int_0^{\lambda_x}\int_0^{\lambda_y} \frac{1}{2}\left[\left(\frac{\partial z}{\partial x}\right)^2 + \left(\frac{\partial z}{\partial y}\right)^2\right] dx\, dy$$
$$= \frac{\pi^2 \gamma a^2}{2}\frac{\lambda_x^2 + \lambda_y^2}{\lambda_x \lambda_y}. \tag{25.45}$$

Therefore, if

$$\delta\mathcal{E} = \delta\mathcal{E}_{\text{el}}^{(2)} + \delta\mathcal{E}_{\text{surface}} = 0 \tag{25.46}$$

defines the condition for stability, i.e., for growth *versus* decay of the perturbation, the critical stress is obtained as

$$\left(\frac{\sigma_0^2}{\bar{E}}\right)_{\text{cr}} = \gamma\pi\,\frac{\sqrt{\lambda_x^2 + \lambda_y^2}}{\lambda_x \lambda_y} = \gamma\pi\sqrt{\frac{1}{\lambda_x^2} + \frac{1}{\lambda_y^2}}. \tag{25.47}$$

Defining

$$\frac{1}{\lambda^*} = \sqrt{\frac{1}{\lambda_x^2} + \frac{1}{\lambda_y^2}}, \tag{25.48}$$

the stability condition (25.47) becomes

$$\left(\frac{\sigma_0^2}{\bar{E}}\right)_{\text{cr}} = \frac{\gamma\pi}{\lambda^*} \quad\Rightarrow\quad \left(\frac{2\sigma_0^2}{\bar{E}}\right)_{\text{cr}} = \gamma\omega^*. \tag{25.49}$$

This may be compared to the original criterion of this type developed by Asaro and Tiller (1972), which is $\omega^* = 2\pi/\lambda^* = 2\sigma_0^2/(\gamma E)$. The difference is simply the change from a uniaxial stress state to one of uniform biaxial stress.

An interesting observation is that these results suggest that initially surface profiles may tend to take on a 1D shape. For example, look at the stability criterion by forming the ratio of the magnitude of the released elastic energy to the positive surface energy, viz.,

$$\frac{\delta\mathcal{E}_{\text{el}}^{(2)}}{\delta\mathcal{E}_{\text{surface}}} = \frac{\lambda^*}{\lambda_x}\sqrt{1 + \frac{\lambda_x^2}{\lambda_y^2}}. \tag{25.50}$$

When this ratio becomes greater than unity, the perturbation may grow. The smallest value of λ_x for which this happens is when $\lambda_x = \lambda^*$ and $\lambda_y \to \infty$. Of course, after initial growth the patterning may, and probably will, become multidimensional.

25.6 Asymptotic Stresses for 1D Surface Profiles

In the previous section a general procedure was developed for calculating the asymptotic stresses on 2D surface profiles. Here we revisit the 1D case given out at the outset of the

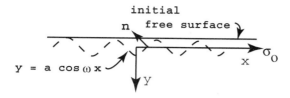

Figure 25.8. Creation of a perturbed free surface.

chapter *via* the original analysis of Asaro and Tiller (1972). We simply seek to establish the similar procedure for constructing the surface stress field and strain energy density required to calculate the surface chemical potential. The procedure is developed *via* the general sinusoidal profile $y = a\cos(\omega x)$.

Consider the surface to be in a state of uniform biaxial stress of magnitude σ_0. If the surface was perturbed in shape with the profile $y = a\cos(\omega x)$, $(a\omega \ll 1)$, then we would write the stress state as

$$\sigma = \sigma_0 \mathbf{e}_x \mathbf{e}_x + \sigma^{(2)}, \tag{25.51}$$

where, as before, $\sigma^{(2)}$ is the perturbed field. The goal is to determine $\sigma^{(2)}$ to $\mathcal{O}(a\omega)$. The traction induced over the sketched surface in Fig. 25.8 by the biaxial stress is, to first order,

$$T_x = \sigma_0 n_x \sim \sigma_0 \partial y/\partial x \sim -a\omega\sigma_0 \sin(\omega x) + \mathcal{O}(a^2\omega^2),$$
$$T_y = 0 + \mathcal{O}(a^2\omega^2). \tag{25.52}$$

To render the perturbed surface traction free we need to impose a shear traction

$$-T_x = a\omega\sigma_0 \cos(\omega x) = -\sigma_0 \frac{\partial y}{\partial x}. \tag{25.53}$$

This defines the perturbed field, called (2) above. Recalling the point force solution given by (15.17), yields

$$\sigma^{(2)}_{xx}(x) = \frac{2\sigma_0}{\pi} \int_{-\infty}^{\infty} \frac{\partial y(\xi)/\partial \xi}{x - \xi} \, d\xi + \mathcal{O}(a^2\omega^2),$$
$$\sigma^{(2)}_{yy}(x) = 0. \tag{25.54}$$

Of course, $\sigma^{(2)}_{xy}$ is already specified as

$$\sigma^{(2)}_{xy} = a\omega\sigma_0 \sin(\omega x) = -\sigma_0 \frac{\partial y}{\partial x}. \tag{25.55}$$

This procedure holds for general profiles so long as $|\partial y/\partial x| \ll 1$.

When applied to the case where $y = a\cos(\omega x)$, we find

$$\sigma^{(2)}_{xx} = -\frac{2\sigma_0 a\omega}{\pi} \int_{-\infty}^{\infty} \frac{\sin(\omega \xi)}{x - \xi} \, d\xi. \tag{25.56}$$

The (real) Cauchy principal value is needed, which is

$$PV \int_{-\infty}^{\infty} \frac{\sin(\omega \xi)}{x - \xi} \, d\xi = -\pi \cos(\omega x).$$

This yields the Asaro–Tiller field, viz.,

$$\sigma_{xx}^{(2)} \sim 2a\omega\sigma_0 \cos(\omega x) + \mathcal{O}(a^2\omega^2),$$
$$\sigma_{yy}^{(2)} \sim 0 + \mathcal{O}(a^2\omega^2), \qquad (25.57)$$
$$\sigma_{xy}^{(2)} \sim a\omega\sigma_0 \sin(\omega x) + \mathcal{O}(a^2\omega^2).$$

The surface strain energy density is readily computed as

$$W(x) = \frac{\sigma_0^2}{E_b} + \frac{\sigma_{xx}^{(2)}}{\bar{E}}\left[\sigma_0 + \frac{1}{2}\sigma_{xx}^{(2)}\right] + \mathcal{O}(a^2\omega^2), \qquad (25.58)$$

where $E_b = E/(1-\nu)$ is the biaxial modulus and $\bar{E} = E/(1-\nu^2)$ is the plane strain modulus. Note that the $\mathcal{O}(a^2\omega^2)$ term embedded in (25.58) would be omitted from consideration in an asymptotic analysis, but is nonetheless retained for completeness. Indeed, if one had the general field $\sigma_{xx} = \sigma_0 + \sigma_{xx}^{(2)}$, then (25.58) would be the energy density.

25.7 Suggested Reading

Asaro, R. J., and Tiller, W. A. (1972), Interface Morphology Development during Stress Corrosion Cracking. Part I: Via Surface Diffusion, *Metall. Trans.*, Vol. 3, pp. 1789–1796.

Cammarata, R. C. (1994), Surface and Interface Stress Effects in Thin Films, *Progress in Surface Science*, Vol. 46, pp. 1–38.

Freund, L. B., and Suresh, S. (2003), *Thin Film Materials: Stress, Defect Formation and Surface Evolution*, Cambridge University Press, New York.

Gao, H., and Nix, W. D. (1999), Surface Roughening of Heteroepitaxial Thin Films, *Ann. Rev. Mater. Sci.*, Vol. 29, pp. 173–209.

Suo, Z. (1997), Motions of Microscopic Surfaces in Materials, *Adv. Appl. Mech.*, Vol. 33, pp. 193–294.

PART 6: PLASTICITY AND VISCOPLASTICITY

26 Phenomenological Plasticity

When stressed beyond a critical stress, ductile materials such as metals and alloys display a nonlinear *plastic response*. This is sketched in Fig. 26.1 for a uniaxial tensile test of a smooth specimen, where some relevant terms are defined. In general, *plastic yielding* is gradual when resolved at typical strain levels (*e.g.*, $\sim 10^{-4}$). A critical stress, called the *yield stress*, is defined, which is the stress in uniaxial tension, or compression, required to cause a small, yet finite, permanent strain that is not recovered after unloading. It is common to take this onset *yield strain* as $e_y = 0.002 = 0.2\%$. Some common general features of plastic flow, with reference to stress *vs.* strain curves, are:

- The σ *vs.* e response is nonlinear and characterized by a decreasing intensity of *strain hardening*, measured by the slope $d\sigma/de$, as the strain increases. Generally, $d\sigma/de \geq 0$.
- Unloading is nearly elastic.
- Plastic deformation of nonporous metals is essentially incompressible, *i.e.*, volume preserving. A discussion of the physical basis for plastic deformation in subsequent chapters will explain why this is so.

As noted, a schematic *stress-strain curve* during uniaxial loading and unloading of an elastoplastic material is shown in Fig. 26.1. The initial yield stress is Y (later terms such as σ_y will be used to denote yield stress). Note that the yield stress is now ideally represented as a stress level at which an abrupt transition from linear, purely elastic, to nonlinear, elastic-plastic deformation occurs. Below this stress level the deformation is elastic, *i.e.*, reversible upon the removal of stress. For linear elasticity, Hooke's law relates the stress and strain by $\sigma = Ee$, where E is Young's modulus of elasticity (the slope of the linear portion of the curve in Fig. 26.1). When the stress is increased beyond the initial yield limit, the material deforms plastically. The nonlinear portion of the stress-strain curve is associated with plastic mechanisms of deformation due to dislocations (in metals) and is referred to as the work or strain hardening of the material. At an arbitrary state of stress $\sigma > Y$, the total strain e consists of the elastic part $e^e = \sigma/E$, which is recoverable upon removal of the stress, and the plastic part $e^p = e - \sigma/E$, which is the permanent (residual) strain left upon removal of the stress. It is generally assumed that plastic deformation does not affect elastic properties of the material, so that the unloading slope (Young's modulus E) remains the same as before the plastic deformation took place. The physical basis for

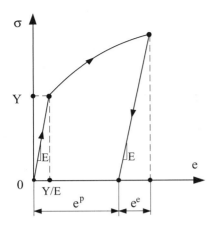

Figure 26.1. The stress strain curve during tensile loading and unloading of an elastoplastic material. The initial yield stress is Y, and the Young's modulus of elasticity is E. At an arbitrary stage of elastoplastic deformation, the total strain is the sum of elastic and plastic parts ($e = e^e + e^p$).

this assumption is made clear in the discussion of crystalline plasticity given in subsequent chapters.

The concern of this chapter is the formulation of constitutive equations that describe elastoplastic material response under multiaxial states of stress. While purely elastic deformation is a history independent process, in which the current strain depends only on the current stress, elastic-plastic states of strain depend on the entire history of loading and deformation. Consequently, elastoplastic constitutive equations are more appropriately expressed in an incremental or rate-type form by relating the rate of deformation to the rate of stress. Derivation of a variety of these equations is presented in this chapter. Various constitutive models are considered, applicable to either metals or geomaterials. Time-independent response is mainly considered, although an introduction to time-dependent constitutive behavior is also given. A more detailed development of strain rate and time dependent elastic-plastic deformation is given in the following chapters that are concerned with physically based constitutive theory.

26.1 Yield Criteria for Multiaxial Stress States

A central concept in the classical theory of rate independent plasticity is the *yield surface*, which defines the multiaxial states of stress at the threshold of plastic deformation. If the stress states are within the yield surface, the corresponding changes of deformation are purely elastic. Plastic deformation is possible only when the stress state is on the current yield surface. The yield surface evolves with the progression of plastic deformation due to strain hardening of the material. We analyze the shape of the initial yield surface first, leaving the hardening models for later sections. The yield surface can be expressed as a hypersurface in the six-dimensional stress space

$$f(\sigma_{ij}) = 0. \tag{26.1}$$

For isotropic material, the onset of plastic deformation does not depend on the directions of principal stresses, but only on their magnitudes. Thus, f can be expressed in terms of the stress invariants as

$$f(I_1, I_2, I_3) = 0, \tag{26.2}$$

26.2. Von Mises Yield Criterion

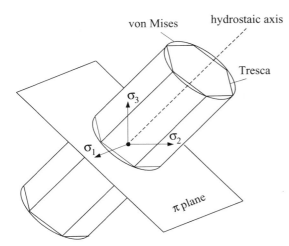

Figure 26.2. Von Mises and Tresca yield surfaces in principal stress space. The yield cylinder and the yield prism have their axis parallel to the hydrostatic axis, which is perpendicular to the π plane ($\sigma_1 + \sigma_2 + \sigma_3 = 0$).

where

$$I_1 = \sigma_{ii}, \quad I_2 = \frac{1}{2}(\sigma_{ij}\sigma_{ij} - \sigma_{ii}\sigma_{jj}), \quad I_3 = \det(\sigma_{ij}). \tag{26.3}$$

For nonporous metals at moderate pressures, the yield is not influenced by hydrostatic pressure, and the yield surface can be expressed in terms of the deviatoric part of stress only, i.e.,

$$f(J_2, J_3) = 0, \tag{26.4}$$

where

$$J_1 = 0, \quad J_2 = \frac{1}{2}\sigma'_{ij}\sigma'_{ij}, \quad J_3 = \det(\sigma'_{ij}). \tag{26.5}$$

26.2 Von Mises Yield Criterion

The most frequently utilized yield criterion is the von Mises criterion

$$J_2 = k^2, \tag{26.6}$$

where k is the material parameter (yield stress in pure shear). In an expanded form,

$$J_2 = \frac{1}{6}\left[(\sigma_{11} - \sigma_{22})^2 + (\sigma_{22} - \sigma_{33})^2 + (\sigma_{33} - \sigma_{11})^2 + 6(\sigma_{12}^2 + \sigma_{23}^2 + \sigma_{31}^2)\right]. \tag{26.7}$$

If the yield stress in uniaxial tension is Y, then according to the von Mises criterion $Y^2/3 = k^2$, so that $k = Y/\sqrt{3}$. The von Mises yield locus in the principal stress space is shown in Fig. 26.2.

Two physical interpretations of the von Mises criterion are possible. The elastic strain energy (per unit volume) is

$$W = \frac{1}{2}\sigma_{ij}e_{ij} = \frac{1}{2}\left(\sigma'_{ij} + \frac{1}{3}\sigma_{kk}\delta_{ij}\right)\left(e'_{ij} + \frac{1}{3}e_{kk}\delta_{ij}\right), \tag{26.8}$$

i.e.,

$$W = \frac{1}{2}\sigma'_{ij}e'_{ij} + \frac{1}{6}\sigma_{ii}e_{jj}. \tag{26.9}$$

Since $\sigma'_{ij} = 2\mu e'_{ij}$, where μ denotes the elastic shear modulus, the portion of the elastic strain energy associated with the deviatoric part of stress and the resulting shape change of the material, is

$$W' = \frac{1}{2}\sigma'_{ij}e'_{ij} = \frac{1}{4\mu}\sigma'_{ij}\sigma'_{ij} = \frac{1}{2\mu}J_2. \tag{26.10}$$

Thus, the von Mises yield criterion can be interpreted as the criterion according to which the yield begins when the deviatoric work W' reaches the critical value $(k^2/2\mu = Y^2/6\mu)$.

The second interpretation of the von Mises criterion is based on the octahedral shear stress. An octahedral plane is the material plane that makes equal angles with the principal stress directions at the considered point of the material. Thus, with respect to principal stress axes $\mathbf{e}_i$, the unit vector normal to an octahedral plane is $\mathbf{n} = \pm(\mathbf{e}_1 + \mathbf{e}_2 + \mathbf{e}_3)/\sqrt{3}$. Since the stress tensor can be expressed relative to its principal directions via the spectral decomposition

$$\boldsymbol{\sigma} = \sigma_1\mathbf{e}_1\mathbf{e}_1 + \sigma_2\mathbf{e}_2\mathbf{e}_2 + \sigma_3\mathbf{e}_3\mathbf{e}_3, \tag{26.11}$$

where the product such as $\mathbf{e}_1\mathbf{e}_1$ represents a dyadic product, the traction vector over the octahedral plane is

$$\mathbf{t}_n = \mathbf{n} \cdot \boldsymbol{\sigma} = \pm\frac{1}{\sqrt{3}}(\sigma_1\mathbf{e}_1 + \sigma_2\mathbf{e}_2 + \sigma_3\mathbf{e}_3). \tag{26.12}$$

The magnitude of the octahedral traction is obtained from

$$t_n^2 = \frac{1}{3}(\sigma_1^2 + \sigma_2^2 + \sigma_3^3), \tag{26.13}$$

whereas the associated normal stress is

$$\sigma_n = \mathbf{t}_n \cdot \mathbf{n} = \mathbf{n} \cdot \boldsymbol{\sigma} \cdot \mathbf{n} = \frac{1}{3}(\sigma_1 + \sigma_2 + \sigma_3). \tag{26.14}$$

Therefore, the square of the octahedral shear stress is

$$\tau_n^2 = t_n^2 - \sigma_n^2 = \frac{1}{9}\left[(\sigma_1 - \sigma_2)^2 + (\sigma_2 - \sigma_3)^2 + (\sigma_3 - \sigma_1)^3\right]. \tag{26.15}$$

It can be shown that the octahedral shear stress is the average shear stress over all planes passing through the considered material point. Since the second invariant of the deviatoric part of stress tensor, expressed in terms of principal stresses, is

$$J_2 = \frac{1}{6}\left[(\sigma_1 - \sigma_2)^2 + (\sigma_2 - \sigma_3)^2 + (\sigma_3 - \sigma_1)^2\right], \tag{26.16}$$

we have the connection

$$\tau_n^2 = \frac{2}{3}J_2. \tag{26.17}$$

Consequently, the von Mises yield criterion can be interpreted as the criterion according to which the yield begins when the octahedral shear stress reaches the critical value $(\sqrt{2/3}\,k = \sqrt{2}\,Y/3)$.

26.3. Tresca Yield Criterion

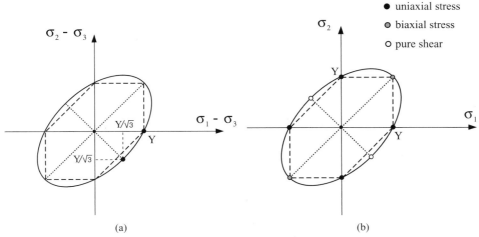

Figure 26.3. (a) The von Mises ellipse and the Tresca hexagon in the plane of principal stress differences; (b) the same in the case of plane stress ($\sigma_3 = 0$).

We note that the von Mises yield criterion can be rewritten as

$$(\sigma_1 - \sigma_3)^2 - (\sigma_1 - \sigma_3)(\sigma_2 - \sigma_3) + (\sigma_2 - \sigma_3)^2 = Y^2. \tag{26.18}$$

This is an ellipse at 45° relative to the coordinates $\sigma_1 - \sigma_3$ and $\sigma_2 - \sigma_3$ (Fig. 26.3a). In the case of plane stress ($\sigma_3 = 0$), we have

$$\sigma_1^2 - \sigma_1\sigma_2 + \sigma_2^2 = Y^2, \tag{26.19}$$

which is an ellipse at 45° relative to the coordinates σ_1 and σ_2 (Fig. 26.3b). Clearly, the yield stress in pure shear ($\sigma_2 = -\sigma_1$) is equal to $Y/\sqrt{3}$, whereas the yield stress in biaxial tension ($\sigma_2 = \sigma_1$) is the same as in uniaxial tension (Y).

In the combined tension-torsion (σ, τ) test (as it occurs in a thin-walled tube under an axial force and a twisting moment), the principal stresses are

$$\sigma_{1,2} = \frac{\sigma}{2} \pm \frac{1}{2}(\sigma^2 + 4\tau^2)^{1/2}, \tag{26.20}$$

and the yield criterion (26.19) gives

$$\sigma^2 + 3\tau^2 = Y^2. \tag{26.21}$$

This is an ellipse with the semiaxes of length Y and $Y/\sqrt{3}$ along the σ and τ axes, respectively (Fig. 26.4).

26.3 Tresca Yield Criterion

The Tresca criterion is a maximum shear stress criterion: plastic deformation begins when the maximum shear stress reaches the critical value $k_T = Y/2$, which is the maximum shear stress at yield in uniaxial tension. Analytically, the Tresca yield criterion is

$$\tau_{\max} = \frac{1}{2}(\sigma_{\max} - \sigma_{\min}) = k_T. \tag{26.22}$$

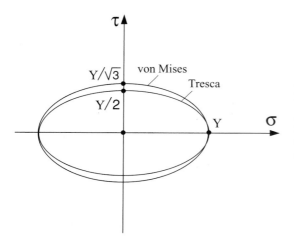

Figure 26.4. The von Mises and Tresca ellipses in the case of combined tension–torsion (σ, τ) test. According to von Mises criterion the yield stress in pure shear is $Y/\sqrt{3}$, whereas according to Tresca criterion it is $Y/2$, where Y is the yield stress in simple tension.

Thus, the intermediate principal stress according to Tresca criterion does not have any effect on the yield.

The maximum shear stress criterion can be recast in terms of principal stresses by any of the following forms

$$\frac{1}{2} \max\left(|\sigma_1 - \sigma_2|, |\sigma_2 - \sigma_3|, |\sigma_3 - \sigma_1|\right) = k_T,$$

$$\frac{1}{4}\left(|\sigma_1 - \sigma_2| + |\sigma_2 - \sigma_3| + |\sigma_3 - \sigma_1|\right) = k_T, \tag{26.23}$$

$$\left[\frac{1}{4}(\sigma_1 - \sigma_2)^2 - k_T^2\right]\left[\frac{1}{4}(\sigma_2 - \sigma_3)^2 - k_T^2\right]\left[\frac{1}{4}(\sigma_3 - \sigma_1)^2 - k_T^2\right] = 0.$$

Since $\sigma_1 - \sigma_2 = \sigma_1' - \sigma_2'$, and similarly for other principal stress differences, and since the invariants of the deviatoric part of stress are

$$J_1 = (\sigma_1' - \sigma_2') + (\sigma_2' - \sigma_3') + (\sigma_3' - \sigma_1') = 0,$$

$$J_2 = \frac{1}{6}\left[(\sigma_1' - \sigma_2')^2 + (\sigma_2' - \sigma_3')^2 + (\sigma_3' - \sigma_1')^2\right], \tag{26.24}$$

$$J_3 = \sigma_1' \sigma_2' \sigma_3',$$

the analytical representation of the maximum shear stress criterion (26.3) in terms of the invariants J_2 and J_3 is

$$4J_2^3 - 27J_3^2 - 36k_T^2 J_2^2 + 96k_T^4 J_2 - 64k_T^6 = 0. \tag{26.25}$$

Because of its complexity, this representation of the yield criterion is rarely used.

The Tresca yield locus for plane stress $(\sigma_3 = 0)$ is depicted in Fig. 26.3b. The same plot applies to three-dimensional states of stress, but with respect to the stress differences as the coordinates (Fig. 26.3a). In the combined tension–torsion (σ, τ) test, the Tresca yield criterion gives

$$\sigma^2 + 4\tau^2 = Y^2. \tag{26.26}$$

26.4. Mohr–Coulomb Yield Criterion

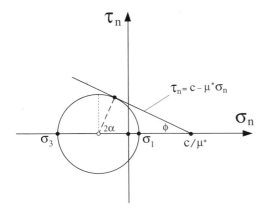

Figure 26.5. The Mohr's circle corresponding to three-dimensional state of stress with maximum stress σ_1 and minimum stress σ_3, and the Mohr–Coulomb envelope $\tau_n = c - \mu^* \sigma_n$.

This is an ellipse with the semiaxes of length Y and $Y/2$ along the σ and τ axes, respectively (Fig. 26.4).

The von Mises yield criterion is in better agreement with experimental data than the Tresca criterion. To optimize the fit with experimental data, a generalization of the von Mises criterion in the form

$$\left(1 - c \frac{J_3^2}{J_2^3}\right)^\alpha J_2 = k^2 \tag{26.27}$$

has been suggested. Usually, α is taken to be equal to one, and c is an appropriate parameter. Because of uncertainties in the description of the subsequent work hardening, the benefits of using the representation (26.27) over the simple von Mises criterion $J_2 = k^2$ are hardly worth. Further discussion can be found in the book by Hill (1950), and review articles by Drucker (1960) and Naghdi (1960).

26.4 Mohr–Coulomb Yield Criterion

For geomaterials such as rocks and soils and for concrete, the inelastic deformation occurs by frictional sliding over the plane of shearing, and thus the normal stress over that plane affects the yield. A simple criterion that accounts for this is a Mohr–Coulomb yield criterion $\tau_n = \tau_f + c$, where $\tau_f = -\mu^* \sigma_n$ is the friction stress ($\mu^* = \tan\phi$ is the coefficient of internal friction, ϕ being the angle of friction), and c is the cohesive strength of the material. Failure occurs if the Mohr's circle corresponding to the stress state $(\sigma_1, \sigma_2, \sigma_3)$ touches the Mohr's envelope $\tau_n = c - \mu^* \sigma_n$ (Fig. 26.5). The corresponding plane of failure has a normal at an angle α from the axis of the least tensile stress (σ_3). From Fig. 26.5, $2\alpha = \pi/2 - \phi$ and

$$\sigma_n = \frac{\sigma_1 + \sigma_3}{2} + \frac{\sigma_1 - \sigma_3}{2} \cos 2\alpha,$$

$$\tau_n = \frac{\sigma_1 - \sigma_3}{2} \sin 2\alpha. \tag{26.28}$$

Thus, the Mohr–Coulomb yield criterion is

$$\sigma_1 - \sigma_3 + (\sigma_1 + \sigma_3) \sin\phi = 2c \cos\phi, \tag{26.29}$$

or, in all six sextants,

$$\sigma_{\max} - \sigma_{\min} + (\sigma_{\max} + \sigma_{\min}) \sin\phi = 2c \cos\phi. \tag{26.30}$$

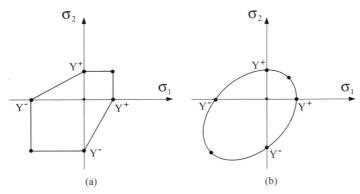

Figure 26.6. (a) The Mohr–Coulomb hexagon for plane stress ($\sigma_3 = 0$). The yield stress in simple tension is Y^+ and in compression Y^-. (b) The Drucker–Prager ellipse for plane stress.

It can be easily verified that, according to this criterion, the yield stress in pure shear is $\tau_Y = c\cos\phi$, whereas the yield stresses in uniaxial tension and compression are $Y^+ = 2c\tan\alpha$ and $Y^- = 2c\cot\alpha$, respectively.

We note that the Mohr–Coulomb yield criterion is of the general form

$$f(J_2, J_3) = a - bI_1 \,. \tag{26.31}$$

Indeed, the criterion (26.29) can be rewritten as

$$\sigma_1 - \sigma_3 + \frac{1}{3}[(\sigma_1 - \sigma_2) - (\sigma_2 - \sigma_3)]\sin\phi = 2c\cos\phi - \frac{2}{3}I_1\sin\phi, \tag{26.32}$$

where $I_1 = \sigma_1 + \sigma_2 + \sigma_3$. The left hand side of the above equation is an isotropic function of deviatoric stress, which can thus be expressed in terms of J_2 and J_3.

The Mohr–Coulomb yield locus for plane stress ($\sigma_3 = 0$) is depicted in Fig. 26.6a.

26.4.1 Drucker–Prager Yield Criterion

For porous metals, concrete and geomaterials like soils and rocks, plastic deformation has its origin in pressure dependent microscopic processes. The corresponding yield condition depends on both deviatoric and hydrostatic parts of the stress tensor. Drucker and Prager (1952) suggested that yielding in soils occurs when the shear stress on octahedral planes overcomes cohesive and frictional resistance to sliding on these planes, *i.e.*, when

$$\tau_{\text{oct}} = \tau_f + \sqrt{\frac{2}{3}}\,k, \tag{26.33}$$

where

$$\tau_{\text{oct}} = \left(\frac{2}{3}J_2\right)^{1/2}, \quad \tau_f = -\mu^*\sigma_{\text{oct}} = -\frac{1}{3}\mu^*I_1. \tag{26.34}$$

The yield stress in simple shear is k, the coefficient of internal friction is μ^*, the first invariant of the Cauchy stress tensor is I_1, and J_2 is the second invariant of the deviatoric part of the Cauchy stress. The yield condition is consequently

$$f = J_2^{1/2} + \frac{1}{3}\mu_* I_1 - k = 0 \,. \tag{26.35}$$

26.4. Mohr–Coulomb Yield Criterion

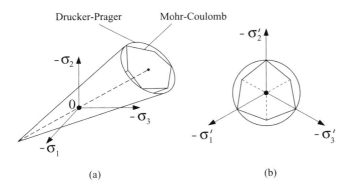

Figure 26.7. The Drucker–Prager cone and the Mohr–Coulomb pyramid matched along the compressive meridian shown in (a) principal stress space and (b) deviatoric plane.

A modified friction parameter $\mu_* = \sqrt{3/2}\,\mu^*$ is conveniently introduced. The criterion geometrically represents a cone in the principal stress space with its axis parallel to the hydrostatic axis (Fig. 26.7). The radius of the circle in the deviatoric (π) plane is $\sqrt{2}\,k$, where k is the yield stress in simple shear. The angle of the cone is $\tan^{-1}(\sqrt{2}\,\mu_*/3)$. The yield stresses in uniaxial tension and compression are, according to (26.35),

$$Y^+ = \frac{3k}{\sqrt{3}+\mu_*}, \quad Y^- = \frac{3k}{\sqrt{3}-\mu_*}. \tag{26.36}$$

For the yield condition to be physically meaningful, the restriction holds $\mu_* < \sqrt{3}$.

If the compressive states of stress are considered positive (as commonly done in geomechanics, e.g., Jaeger and Cook, 1976), a minus sign appears in front of the second term of f in (26.35). In the case of plane stress ($\sigma_3 = 0$), the Drucker–Prager criterion reduces to an ellipse in the (σ_1, σ_2) plane, with the center at the point $\sigma_1 = \sigma_2 = -2\mu_* k/(1-4\mu_*^2/3)$, Fig. 26.6b. The magnitudes of the yield stresses in simple tension and compression are as in (26.36), whereas the yield stresses in equal biaxial tension and compression are, respectively,

$$\sigma^+ = \frac{3k}{\sqrt{3}+2\mu_*}, \quad \sigma^- = \frac{3k}{\sqrt{3}-2\mu_*}. \tag{26.37}$$

When the Drucker–Prager cone is applied to porous rocks, it overestimates the yield stress at higher pressures, and inadequately predicts inelastic volume changes. To circumvent the former, DiMaggio and Sandler (1971) introduced an ellipsoidal cap to close the cone at the certain level of pressure. This cap is described by the equation

$$\frac{J_2}{K^2} + \frac{\left(\frac{1}{3}I_1+b\right)^2}{(a-b)^2} = 1, \tag{26.38}$$

where a is the magnitude of the compressive stress that would alone cause the crushing of the material (apex of the cap in Fig. 26.8). Other shapes of the cap were also used. Details can be found in Chen and Han (1988). See also Ortiz (1985) and Lubarda *et al.* (1996).

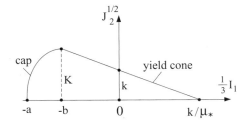

Figure 26.8. The Drucker–Prager yield criterion shown in the coordinates of stress invariants I_1 and J_2. The yield stress in pure shear is k, and the frictional parameter is μ_*. At high pressure a cap is used to close the cone.

26.5 Gurson Yield Criterion for Porous Metals

Based on a rigid-perfectly plastic analysis of spherically symmetric deformation around a spherical cavity, Gurson (1977) suggested a yield criterion for porous metals in the form

$$f = J_2 + \frac{2}{3} v Y^2 \cosh\left(\frac{I_1}{2Y}\right) - (1 + v^2)\frac{Y^2}{3} = 0, \tag{26.39}$$

where v is the porosity (void/volume fraction) and Y is the tensile yield stress of the matrix material (Fig. 26.9). To improve the agreement with experimental data on ductile void growth, Tvergaard (1982) introduced two additional material parameters (q_1, q_2) in the structure of the Gurson yield criterion, such that

$$f = J_2 + \frac{2}{3} v Y^2 q_1 \cosh\left(\frac{q_2 I_1}{2Y}\right) - (1 + q_1^2 v^2)\frac{Y^2}{3} = 0. \tag{26.40}$$

Further refinements of the model were suggested by Tvergaard and Needleman (1984) and Mear and Hutchinson (1985). If there is no porosity $(v = 0)$, both (26.39) and (26.40) reduce to von Mises yield criterion $J_2 = Y^2/3$.

26.6 Anisotropic Yield Criteria

An anisotropic generalization of the von Mises yield criterion, due to Hill (1950), replaces J_2 with a general quadratic function of the stress components

$$\frac{1}{2} A_{ijkl} \sigma_{ij} \sigma_{kl} = k^2, \tag{26.41}$$

where the anisotropy tensor A_{ijkl} has the same symmetries as the elastic stiffness tensor ($A_{ijkl} = A_{jikl} = A_{klij}$). Decomposing the stress tensor into its deviatoric and hydrostatic parts

$$\sigma_{ij} = \sigma'_{ij} - p \delta_{ij}, \quad p = -\frac{1}{3}\sigma_{kk}, \tag{26.42}$$

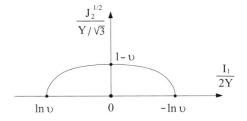

Figure 26.9. Gurson yield condition for porous metals with the void/volume fraction v. The tensile yield stress of the matrix material is Y.

26.7. Elastic-Plastic Constitutive Equations

the yield criterion (26.41) can be rewritten as

$$\frac{1}{2} A_{ijkl} \sigma'_{ij} \sigma'_{kl} - p A_{ijkk} \sigma'_{ij} + \frac{1}{2} p^2 A_{iijj} = k^2. \tag{26.43}$$

Consequently, if the plastic yield is independent of p, we must have $A_{ijkk} = 0$, and the yield criterion reduces to

$$\frac{1}{2} A_{ijkl} \sigma'_{ij} \sigma'_{kl} = k^2. \tag{26.44}$$

The von Mises isotropic yield condition is recovered if

$$A_{ijkl} = \frac{1}{2}(\delta_{ik}\delta_{jl} + \delta_{il}\delta_{jk}) - \frac{1}{3}\delta_{ij}\delta_{kl}. \tag{26.45}$$

The most well-known anisotropic yield criterion of the type considered in this section is the Hill's orthotropic yield criterion. Hill (1948) introduced the yield condition, expressed relative to the principal axes of orthotropy, in the form

$$F(\sigma_{22} - \sigma_{33})^2 + G(\sigma_{22} - \sigma_{11})^2 + H(\sigma_{11} - \sigma_{22})^2 + 2L\sigma_{23}^2 + 2M\sigma_{31}^2 + 2N\sigma_{12}^2 = 1,$$

where F, G, H, L, M, N are the material parameters. If the tensile yield stresses in the directions of the principal axes of orthotropy are Y_1, Y_2, and Y_3, then

$$\frac{1}{Y_1^2} = G + H, \quad \frac{1}{Y_2^2} = H + F, \quad \frac{1}{Y_3^2} = F + G. \tag{26.46}$$

Consequently, the material parameters F, G, and H are specified by

$$F = \frac{1}{2}\left(\frac{1}{Y_2^2} + \frac{1}{Y_3^2} - \frac{1}{Y_1^2}\right), \tag{26.47}$$

$$G = \frac{1}{2}\left(\frac{1}{Y_3^2} + \frac{1}{Y_1^2} - \frac{1}{Y_2^2}\right), \tag{26.48}$$

$$H = \frac{1}{2}\left(\frac{1}{Y_1^2} + \frac{1}{Y_2^2} - \frac{1}{Y_3^2}\right). \tag{26.49}$$

Furthermore, if T_{12}, T_{23}, and T_{31} are the yield stresses under pure shear in the corresponding planes of orthotropy, we have

$$L = \frac{1}{2T_{23}^2}, \quad M = \frac{1}{2T_{31}^2}, \quad N = \frac{1}{2T_{12}^2}. \tag{26.50}$$

In the case of transverse isotropy (with the axis 3 as the axis of transverse isotropy), the relationships hold

$$N = F + 2H = G + 2H, \quad L = M. \tag{26.51}$$

For a fully isotropic material, $L = M = N = 3F = 3G = 3H$.

26.7 Elastic-Plastic Constitutive Equations

In the rate-type theory of plasticity, the rate of deformation tensor **D** (symmetric part of the velocity gradient $\mathbf{L} = \dot{\mathbf{F}} \cdot \mathbf{F}^{-1}$) is decomposed into the elastic and plastic parts, such that

$$\mathbf{D} = \mathbf{D}^e + \mathbf{D}^p. \tag{26.52}$$

The elastic part gives a reversible part of the strain increment $\mathbf{D}^e dt$ in an infinitesimal loading–unloading cycle of the stress increment $\overset{\triangledown}{\tau} dt$. Thus, if $\mathbf{M}^e$ is the fourth-order tensor of elastic compliances, we have

$$\mathbf{D}^e = \mathbf{M}^e : \overset{\triangledown}{\tau}, \tag{26.53}$$

where

$$\overset{\triangledown}{\tau} = \dot{\sigma} + \sigma \operatorname{tr} \mathbf{D} - \mathbf{W} \cdot \sigma + \sigma \cdot \mathbf{W} \tag{26.54}$$

is the Jaumann rate of the Kirchhoff stress $\tau = (\det \mathbf{F}) \sigma$, with the current state as the reference. For infinitesimally small elastic deformation of isotropic materials, the rectangular components of the elastic compliance tensor (inverse of the elastic moduli tensor) are

$$M^e_{ijkl} = \frac{1}{2\mu} \left[\frac{1}{2} (\delta_{ik}\delta_{jl} + \delta_{il}\delta_{jk}) - \frac{\lambda}{2\mu + 3\lambda} \delta_{ij}\delta_{kl} \right]. \tag{26.55}$$

The Lamé elastic constants are λ and μ, and δ_{ij} denotes the Kronecker delta.

In the so-called associative theory of plasticity, the plastic part of the rate of deformation tensor is codirectional with the outward normal to the yield surface in stress space, *i.e.*,

$$\mathbf{D}^p = \dot{\gamma} \frac{\partial f}{\partial \sigma}, \quad \dot{\gamma} > 0. \tag{26.56}$$

It can be shown that this normality rule follows from the plasticity postulate, which states that the net work in an isothermal cycle of strain must be positive if a cycle involves plastic deformation at some stage (Ilyushin, 1961; Hill and Rice, 1973). In the context of infinitesimal strain, Drucker (1951,1959) introduced a plasticity postulate based on stress cycles, which also leads to normality rule and the convexity of the yield surface. Assuming that the response is incrementally linear, the loading index $\dot{\gamma}$ can be expressed as

$$\dot{\gamma} = \frac{1}{H} \left(\frac{\partial f}{\partial \sigma} : \overset{\triangledown}{\tau} \right), \tag{26.57}$$

where H is a scalar parameter dependent on the history of deformation. For a given representation of the yield function and its evolution, this parameter can be determined from the consistency condition for continuing plastic deformation $\dot{f} = 0$.

In the material is in the hardening range ($H > 0$), three types of response can be identified, depending on the direction of stress increment. These are

$$\frac{\partial f}{\partial \sigma} : \overset{\triangledown}{\tau} \begin{cases} > 0, & \text{plastic loading}, \\ = 0, & \text{neutral loading}, \\ < 0, & \text{elastic unloading}. \end{cases} \tag{26.58}$$

Thus, in the case of plastic loading

$$\mathbf{D}^p = \frac{1}{H} \left(\frac{\partial f}{\partial \sigma} : \overset{\triangledown}{\tau} \right) \frac{\partial f}{\partial \sigma}. \tag{26.59}$$

Physically, the plastic increment of the rate of deformation tensor $\mathbf{D}^p dt$ is the residual strain increment left upon a cycle of application and removal of the stress increment $\overset{\triangledown}{\tau} dt$.

26.8. Isotropic Hardening

The overall elastoplastic constitutive structure is obtained by combining (26.53) and (26.59). This gives

$$\mathbf{D} = \left(\mathbf{M}^e + \frac{1}{H}\frac{\partial f}{\partial \boldsymbol{\sigma}}\frac{\partial f}{\partial \boldsymbol{\sigma}}\right) : \overset{\triangledown}{\boldsymbol{\tau}}. \qquad (26.60)$$

The scalar parameter H can be positive, negative or equal to zero. Three types of response are thus possible within this constitutive framework. They are

$$H > 0, \quad \frac{\partial f}{\partial \boldsymbol{\sigma}} : \overset{\triangledown}{\boldsymbol{\tau}} > 0, \quad \text{hardening,}$$

$$H < 0, \quad \frac{\partial f}{\partial \boldsymbol{\sigma}} : \overset{\triangledown}{\boldsymbol{\tau}} < 0, \quad \text{softening,} \qquad (26.61)$$

$$H = 0, \quad \frac{\partial f}{\partial \boldsymbol{\sigma}} : \overset{\triangledown}{\boldsymbol{\tau}} = 0, \quad \text{ideally plastic.}$$

Starting from the current yield surface in stress space, the stress point moves outward in the case of hardening, inward in the case of softening, and tangentially to the yield surface in the case of ideally plastic response. In the case of softening, $\mathbf{D}$ is not uniquely determined by the prescribed stress rate $\overset{\triangledown}{\boldsymbol{\tau}}$, because either (26.60) or the elastic unloading expression applies,

$$\mathbf{D} = \mathbf{M}^e : \overset{\triangledown}{\boldsymbol{\tau}}. \qquad (26.62)$$

In the case of ideally plastic response, the plastic part of the strain rate is indeterminate to the extent of an arbitrary positive multiple, since $\dot{\gamma}$ in (26.57) is indeterminate.

Clearly, from (26.59),

$$\overset{\triangledown}{\boldsymbol{\tau}} : \mathbf{D}^p = \frac{1}{H}\left(\frac{\partial f}{\partial \boldsymbol{\sigma}} : \overset{\triangledown}{\boldsymbol{\tau}}\right)^2, \qquad (26.63)$$

which is positive in the hardening range and negative in the softening range of the material response. The inverted form of (26.60) is

$$\overset{\triangledown}{\boldsymbol{\tau}} = \left[\boldsymbol{\Lambda}^e - \frac{1}{h}\left(\boldsymbol{\Lambda}^e : \frac{\partial f}{\partial \boldsymbol{\sigma}}\right)\left(\frac{\partial f}{\partial \boldsymbol{\sigma}} : \boldsymbol{\Lambda}^e\right)\right] : \mathbf{D}, \qquad (26.64)$$

where $\boldsymbol{\Lambda}^e$ is the elastic stiffness tensor (inverse of $\mathbf{M}^e$), and

$$h = H + \frac{\partial f}{\partial \boldsymbol{\sigma}} : \boldsymbol{\Lambda}^e : \frac{\partial f}{\partial \boldsymbol{\sigma}}. \qquad (26.65)$$

26.8 Isotropic Hardening

Suppose that the yield function is an isotropic function of stress throughout the deformation process, i.e.,

$$f(\boldsymbol{\sigma}, K) = 0, \qquad (26.66)$$

where

$$K = K(\vartheta) \qquad (26.67)$$

is a scalar function of deformation history which defines the size of the current yield surface. The hardening model in which the yield surface expands during plastic deformation, preserving its shape, is known as the isotropic hardening model. Because f is taken to be an isotropic function of stress, the material is assumed to be isotropic. The history parameter ϑ can be taken to be the generalized plastic strain, defined by

$$\vartheta = \int_0^t (2\mathbf{D}^\mathrm{p} : \mathbf{D}^\mathrm{p})^{1/2} \, dt. \tag{26.68}$$

The consistency condition for continuing plastic deformation, which requires that the stress state remains on the yield surface during plastic deformation, is

$$\dot{f} = \frac{\partial f}{\partial \boldsymbol{\sigma}} : \overset{\triangledown}{\boldsymbol{\tau}} + \frac{\partial f}{\partial K} \frac{\partial K}{\partial \vartheta} \dot{\vartheta} = 0. \tag{26.69}$$

Since, from (26.56), (26.57), and (26.68),

$$\dot{\vartheta} = (2\mathbf{D}^\mathrm{p} : \mathbf{D}^\mathrm{p})^{1/2} = \frac{1}{H}\left(2\frac{\partial f}{\partial \boldsymbol{\sigma}} : \frac{\partial f}{\partial \boldsymbol{\sigma}}\right)^{1/2}\left(\frac{\partial f}{\partial \boldsymbol{\sigma}} : \overset{\triangledown}{\boldsymbol{\tau}}\right), \tag{26.70}$$

the substitution into (26.69) gives

$$H = -\frac{\partial f}{\partial K}\frac{\mathrm{d}K}{\mathrm{d}\vartheta}\left(2\frac{\partial f}{\partial \boldsymbol{\sigma}} : \frac{\partial f}{\partial \boldsymbol{\sigma}}\right)^{1/2}. \tag{26.71}$$

26.8.1 J_2 Flow Theory of Plasticity

Suppose that the yield criterion is of the von Mises type

$$f = J_2 - k^2(\vartheta) = 0, \quad J_2 = \frac{1}{2}\boldsymbol{\sigma}' : \boldsymbol{\sigma}', \tag{26.72}$$

where $\boldsymbol{\sigma}'$ is a deviatoric part of $\boldsymbol{\sigma}$. It readily follows that

$$\frac{\partial f}{\partial \boldsymbol{\sigma}} = \boldsymbol{\sigma}' \tag{26.73}$$

and

$$H = 4k^2 h_\mathrm{t}^\mathrm{p}. \tag{26.74}$$

The plastic tangent modulus in the pure shear test is

$$h_\mathrm{t}^\mathrm{p} = \frac{\mathrm{d}k}{\mathrm{d}\vartheta}. \tag{26.75}$$

Equation (26.73) implies that plastic deformation is isochoric,

$$\mathrm{tr}\,\mathbf{D}^\mathrm{p} = 0. \tag{26.76}$$

The total rate rate of deformation is

$$\mathbf{D} = \left(\mathbf{M}^\mathrm{e} + \frac{1}{2h_\mathrm{t}^\mathrm{p}}\frac{\boldsymbol{\sigma}'\boldsymbol{\sigma}'}{\boldsymbol{\sigma}':\boldsymbol{\sigma}'}\right) : \overset{\triangledown}{\boldsymbol{\tau}}. \tag{26.77}$$

The plastic loading condition in the hardening range is

$$\boldsymbol{\sigma}' : \overset{\triangledown}{\boldsymbol{\tau}} > 0. \tag{26.78}$$

26.9. Kinematic Hardening

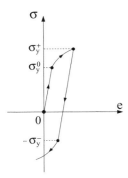

Figure 26.10. Illustration of the Bauschinger effect ($|\sigma_y^-| < \sigma_y^+$) in uniaxial tension.

The inverse equation is

$$\overset{\triangledown}{\tau} = \left(\Lambda^e - \frac{2\mu}{1 + h_t^p/\mu} \frac{\sigma' \sigma'}{\sigma' : \sigma'} \right) : \mathbf{D}, \tag{26.79}$$

which applies for

$$\sigma' : \mathbf{D} > 0. \tag{26.80}$$

In retrospect, the plastic rate of deformation can be expressed either in terms of the stress rate or the total rate of deformation as

$$\mathbf{D}^p = \frac{1}{2h_t^p} \frac{\sigma' : \overset{\triangledown}{\tau}}{\sigma' : \sigma'} \sigma' = \frac{1}{1 + h_t^p/\mu} \frac{\sigma' : \mathbf{D}}{\sigma' : \sigma'} \sigma'. \tag{26.81}$$

The derivation of elastoplastic constitutive equations within the framework of infinitesimal strain, with a historical perspective, can be found in the books by Hill (1950), Kachanov (1971), and Johnson and Mellor (1973).

26.9 Kinematic Hardening

To account for the Bauschinger effect (Fig. 26.10) and anisotropic hardening, and thus provide better description of material response under cyclic loading, a simple model of kinematic hardening was introduced by Melan (1938) and Prager (1955, 1956). According to this model, the initial yield surface does not change its size and shape during plastic deformation, but translates in the stress space according to some prescribed rule (Fig. 26.11). If the yield criterion is pressure-independent, it is assumed that

$$f(\sigma' - \alpha, K_0) = 0, \quad K_0 = \text{const.}, \tag{26.82}$$

where α represents the current center of the yield locus in the deviatoric plane (back stress), and f is an isotropic function of the stress difference $\sigma' - \alpha$ (Fig. 26.12). The size of the yield locus is specified by the constant K_0. The consistency condition for continuing plastic deformation is

$$\frac{\partial f}{\partial \sigma} : \left(\overset{\triangledown}{\tau} - \overset{\triangledown}{\alpha} \right) = 0. \tag{26.83}$$

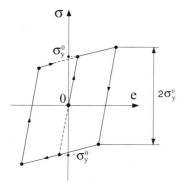

Figure 26.11. One-dimensional stress-strain response according to linear kinematic hardening model.

Suppose that the yield surface instantaneously translates so that the evolution of back stress is governed by

$$\overset{\triangledown}{\boldsymbol{\alpha}} = c(\boldsymbol{\alpha}, \vartheta)\mathbf{D}^{\text{p}} + \mathbf{C}(\boldsymbol{\alpha}, \vartheta)(\mathbf{D}^{\text{p}} : \mathbf{D}^{\text{p}})^{1/2}, \qquad (26.84)$$

where c and $\mathbf{C}$ are the appropriate scalar and tensor functions of $\boldsymbol{\alpha}$ and ϑ. This representation is in accord with assumed time independence of plastic deformation, which requires (26.84) to be homogeneous function of degree one in the components of plastic rate of deformation. Since the plastic rate of deformation is

$$\mathbf{D}^{\text{p}} = \dot{\gamma}\frac{\partial f}{\partial \boldsymbol{\sigma}}, \qquad (26.85)$$

the substitution of (26.84) into (26.83) gives the loading index

$$\dot{\gamma} = \frac{1}{H}\left(\frac{\partial f}{\partial \boldsymbol{\sigma}} : \overset{\triangledown}{\boldsymbol{\tau}}\right). \qquad (26.86)$$

The hardening parameter H is defined by

$$H = c\left(\frac{\partial f}{\partial \boldsymbol{\sigma}} : \frac{\partial f}{\partial \boldsymbol{\sigma}}\right) + \left(\frac{\partial f}{\partial \boldsymbol{\sigma}} : \frac{\partial f}{\partial \boldsymbol{\sigma}}\right)^{1/2}\left(\mathbf{C} : \frac{\partial f}{\partial \boldsymbol{\sigma}}\right). \qquad (26.87)$$

If the yield condition is specified by

$$f = \frac{1}{2}(\boldsymbol{\sigma}' - \boldsymbol{\alpha}) : (\boldsymbol{\sigma}' - \boldsymbol{\alpha}) - k_0^2 = 0, \qquad (26.88)$$

then

$$\frac{\partial f}{\partial \boldsymbol{\sigma}} = \boldsymbol{\sigma}' - \boldsymbol{\alpha}. \qquad (26.89)$$

The loading index is

$$\dot{\gamma} = \frac{1}{H}(\boldsymbol{\sigma}' - \boldsymbol{\alpha}) : \overset{\triangledown}{\boldsymbol{\tau}}, \qquad (26.90)$$

with

$$H = 2ck_0^2 + \sqrt{2}k_0\,\mathbf{C} : (\boldsymbol{\sigma}' - \boldsymbol{\alpha}). \qquad (26.91)$$

26.9. Kinematic Hardening

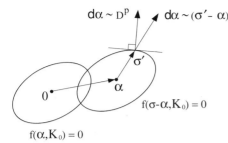

Figure 26.12. Translation of the yield surface according to kinematic hardening model. The center of the yield surface is the back stress α. Its evolution is governed by $d\alpha \sim \mathbf{D}^p$ according to Prager's model, and by $d\alpha \sim (\sigma' - \alpha)$ according to Ziegler's model.

26.9.1 Linear and Nonlinear Kinematic Hardening

When $\mathbf{C} = \mathbf{0}$ and $c = 2h_t^p$, the model with evolution equation (26.84) reduces to Prager's linear kinematic hardening (Fig. 26.12). The corresponding elastoplastic constitutive equation is

$$\mathbf{D} = \left[\mathbf{M}^e + \frac{1}{2h_t^p} \frac{(\sigma' - \alpha)(\sigma' - \alpha)}{(\sigma' - \alpha):(\sigma' - \alpha)} \right] : \overset{\triangledown}{\tau}, \qquad (26.92)$$

with plastic loading condition in the hardening range

$$(\sigma' - \alpha) : \overset{\triangledown}{\tau} > 0. \qquad (26.93)$$

The inverse equation is

$$\overset{\triangledown}{\tau} = \left[\mathbf{\Lambda}^e - \frac{2\mu}{1 + h_t^p/\mu} \frac{(\sigma' - \alpha)(\sigma' - \alpha)}{(\sigma' - \alpha):(\sigma' - \alpha)} \right] : \mathbf{D}, \qquad (26.94)$$

provided that

$$(\sigma' - \alpha) : \mathbf{D} > 0. \qquad (26.95)$$

Alternatively, by using the suggestion by Ziegler (1959), the evolution equation for back stress can be taken in the form

$$\overset{\triangledown}{\alpha} = \dot{\beta} (\sigma' - \alpha). \qquad (26.96)$$

The proportionality factor $\dot{\beta}$ is determined from the consistency condition in terms of σ and α (Fig. 26.12). Detailed analysis is available in the book by Chakrabarty (1987). Duszek and Perzyna (1991) used an evolution equation that is a linear combination of the Prager and Ziegler hardening rules.

A nonlinear kinematic hardening model of Armstrong and Frederick (1966) is obtained if $\mathbf{C}$ in (26.84) is taken to be proportional to back stress, $\mathbf{C} = -c_0\,\alpha$, where c_0 is a constant material parameter. In this case

$$\overset{\triangledown}{\alpha} = 2h\,\mathbf{D}^p - c_0\,\alpha\,(\mathbf{D}^p : \mathbf{D}^p)^{1/2}, \qquad (26.97)$$

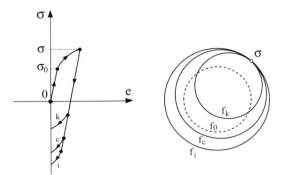

Figure 26.13. Geometric illustration of isotropic, kinematic, and combined hardening. The initial yield surface (f_0) expands in the case of isotropic, translates in the case of kinematic (f_k), and expands and translates in the case of combined hardening (f_c).

with h as another material parameter. The added nonlinear term in (26.97), referred to as a recall term, gives rise to hardening moduli for reversed plastic loading that are in better agreement with experimental data. It follows that

$$\mathbf{D}^p = \frac{1}{2h(1-m)} \frac{(\boldsymbol{\sigma}'-\boldsymbol{\alpha}):\overset{\triangledown}{\boldsymbol{\tau}}}{(\boldsymbol{\sigma}'-\boldsymbol{\alpha}):(\boldsymbol{\sigma}'-\boldsymbol{\alpha})} (\boldsymbol{\sigma}'-\boldsymbol{\alpha}), \qquad (26.98)$$

where

$$m = \frac{c_0}{2h} \frac{(\boldsymbol{\sigma}'-\boldsymbol{\alpha}):\boldsymbol{\alpha}}{[(\boldsymbol{\sigma}'-\boldsymbol{\alpha}):(\boldsymbol{\sigma}'-\boldsymbol{\alpha})]^{1/2}}. \qquad (26.99)$$

If the yield surface expands and translates during plastic deformation, we have a combined isotropic–kinematic hardening (Fig. 26.13). For example, consider

$$\frac{1}{2}(\boldsymbol{\sigma}'-\boldsymbol{\alpha}):(\boldsymbol{\sigma}'-\boldsymbol{\alpha}) = k_\alpha^2(\vartheta), \qquad (26.100)$$

where $\boldsymbol{\alpha}$ represents the current center of the yield surface and $k_\alpha(\vartheta)$ is its current radius. If the evolution equation for the back stress $\boldsymbol{\alpha}$ is given by (26.97), we obtain

$$\mathbf{D}^p = \frac{1}{2h_\alpha^p + 2h(1-m)} \frac{(\boldsymbol{\sigma}'-\boldsymbol{\alpha}):\overset{\triangledown}{\boldsymbol{\tau}}}{(\boldsymbol{\sigma}'-\boldsymbol{\alpha}):(\boldsymbol{\sigma}'-\boldsymbol{\alpha})} (\boldsymbol{\sigma}'-\boldsymbol{\alpha}). \qquad (26.101)$$

The rate of the yield surface expansion is $h_\alpha^p = dk_\alpha/d\vartheta$, and the parameter m is specified by (26.99). For purely kinematic hardening $h_\alpha^p = 0$, whereas for a purely isotropic hardening $h_\alpha^p = h_t^p$ (plastic tangent modulus in simple shear).

More involved, multisurface hardening models were proposed by Mróz (1967, 1976), Dafalias and Popov (1975, 1976), and Krieg (1975). A review of these models is given in the book by Khan and Huang (1995). See also the books by Lubliner (1990) and Lubarda (2002).

26.10 Constitutive Equations for Pressure-Dependent Plasticity

Suppose that the pressure-dependent yield criterion is of the type

$$f(J_2, I_1, K) = 0, \qquad (26.102)$$

where J_2 and I_1 stand for the second invariant of the deviatoric stress and the first invariant of the stress, whereas K designates an appropriate history dependent parameter. If it is

26.10. Pressure-Dependent Plasticity

assumed that plastic part of the rate of deformation tensor is normal to the yield surface, we have

$$\mathbf{D}^{\mathrm{p}} = \dot{\gamma} \frac{\partial f}{\partial \boldsymbol{\sigma}}, \quad \frac{\partial f}{\partial \boldsymbol{\sigma}} = \frac{\partial f}{\partial J_2} \boldsymbol{\sigma}' + \frac{\partial f}{\partial I_1} \mathbf{I}. \tag{26.103}$$

The unit second-order tensor whose components are δ_{ij} is denoted by $\mathbf{I}$. The loading index can be expressed from the consistency condition $\dot{f} = 0$ as

$$\dot{\gamma} = \frac{1}{H} \left(\frac{\partial f}{\partial J_2} \boldsymbol{\sigma}' + \frac{\partial f}{\partial I_1} \mathbf{I} \right) : \stackrel{\triangledown}{\boldsymbol{\tau}}, \tag{26.104}$$

where H is a hardening modulus. The substitution of (26.104) into (26.103) gives

$$\mathbf{D}^{\mathrm{p}} = \frac{1}{H} \left[\left(\frac{\partial f}{\partial J_2} \boldsymbol{\sigma}' + \frac{\partial f}{\partial I_1} \mathbf{I} \right) \left(\frac{\partial f}{\partial J_2} \boldsymbol{\sigma}' + \frac{\partial f}{\partial I_1} \mathbf{I} \right) \right] : \stackrel{\triangledown}{\boldsymbol{\tau}}. \tag{26.105}$$

The volumetric part of the plastic rate of deformation is

$$\mathrm{tr}\,\mathbf{D}^{\mathrm{p}} = \frac{3}{H} \frac{\partial f}{\partial I_1} \left(\frac{\partial f}{\partial J_2} \boldsymbol{\sigma}' + \frac{\partial f}{\partial I_1} \mathbf{I} \right) : \stackrel{\triangledown}{\boldsymbol{\tau}}. \tag{26.106}$$

In the case of the Drucker–Prager yield criterion for geomaterials, we have

$$\frac{\partial f}{\partial J_2} = \frac{1}{2} J_2^{-1/2}, \quad \frac{\partial f}{\partial I_1} = \frac{1}{3} \mu_* \tag{26.107}$$

and

$$H = h_\mathrm{t}^\mathrm{p} = \frac{\mathrm{d}k}{\mathrm{d}\vartheta}, \quad \vartheta = \int_0^t (2\,\mathbf{D}^{\mathrm{p}\prime} : \mathbf{D}^{\mathrm{p}\prime})^{1/2}\,\mathrm{d}t. \tag{26.108}$$

The relationship $k = k(\vartheta)$ between the shear yield stress k under a given superimposed pressure, and the generalized shear plastic strain ϑ, is assumed to be known. Note that $\dot{\vartheta} = \dot{\gamma}$.

In the case of the Gurson yield criterion for porous metals, we have

$$\frac{\partial f}{\partial J_2} = 1, \quad \frac{\partial f}{\partial I_1} = \frac{1}{3} \upsilon Y \sinh\left(\frac{I_1}{2Y}\right) \tag{26.109}$$

and

$$H = \frac{2}{3} \upsilon (1 - \upsilon) Y^3 \sinh\left(\frac{I_1}{2Y}\right) \left[\upsilon - \cosh\left(\frac{I_1}{2Y}\right) \right]. \tag{26.110}$$

The change in porosity during plastic deformation is specified by the evolution equation, such as

$$\dot{\upsilon} = (1 - \upsilon)\,\mathrm{tr}\,\mathbf{D}^{\mathrm{p}}. \tag{26.111}$$

Other evolution equations, which take into account nucleation and growth of voids, have also been considered (*e.g.*, Tvergaard and Needleman, 1984). From (26.111) and (26.103) it follows that the porosity evolves according to

$$\dot{\upsilon} = \dot{\gamma}\,\upsilon(1 - \upsilon) Y \sinh\left(\frac{I_1}{2Y}\right). \tag{26.112}$$

26.11 Nonassociative Plasticity

Constitutive equations in which plastic part of the rate of deformation is normal to the yield surface $f = 0$, i.e.,

$$\mathbf{D}^{\mathrm{p}} = \dot{\gamma}\frac{\partial f}{\partial \boldsymbol{\sigma}}, \qquad (26.113)$$

are referred to as the associative flow rules. A sufficient condition for this constitutive structure is that the material obeys the Ilyushin's work postulate (Ilyushin, 1961). However, many pressure-dependent dilatant materials, with internal frictional effects, are not well described by the associative flow rules. For example, the associative flow rules largely overestimate inelastic volume changes in geomaterials such as rocks and soils (Rudnicki and Rice, 1975; Rice, 1977), and in certain high-strength steels exhibiting the strength-differential effect by which the yield strength is higher in compression than in tension (Spitzig, Sober, and Richmond, 1975). For such materials, plastic part of the rate of strain is taken to be normal to the plastic potential surface

$$g = 0, \qquad (26.114)$$

which is distinct from the yield surface $f = 0$. The resulting constitutive structure,

$$\mathbf{D}^{\mathrm{p}} = \dot{\gamma}\frac{\partial g}{\partial \boldsymbol{\sigma}}, \qquad (26.115)$$

is referred to as a nonassociative flow rule.

The consistency condition $\dot{f} = 0$ gives the loading index

$$\dot{\gamma} = \frac{1}{H}\left(\frac{\partial f}{\partial \boldsymbol{\sigma}} : \overset{\nabla}{\boldsymbol{\tau}}\right), \qquad (26.116)$$

where H is a hardening modulus. Thus,

$$\mathbf{D}^{\mathrm{p}} = \frac{1}{H}\left(\frac{\partial f}{\partial \boldsymbol{\sigma}} : \overset{\nabla}{\boldsymbol{\tau}}\right)\frac{\partial g}{\partial \boldsymbol{\sigma}}. \qquad (26.117)$$

The overall elastic-plastic constitutive structure is

$$\mathbf{D} = \left[\mathbf{M}^{\mathrm{e}} + \frac{1}{H}\left(\frac{\partial g}{\partial \boldsymbol{\sigma}}\frac{\partial f}{\partial \boldsymbol{\sigma}}\right)\right] : \overset{\nabla}{\boldsymbol{\tau}}. \qquad (26.118)$$

Because $g \neq f$, the elastoplastic compliance tensor in (26.118) does not possess a reciprocal symmetry.

In an inverted form, the constitutive equation (26.118) becomes

$$\overset{\nabla}{\boldsymbol{\tau}} = \left[\boldsymbol{\Lambda}^{\mathrm{e}} - \frac{1}{h}\left(\boldsymbol{\Lambda}^{\mathrm{e}} : \frac{\partial g}{\partial \boldsymbol{\sigma}}\right)\left(\frac{\partial f}{\partial \boldsymbol{\sigma}} : \boldsymbol{\Lambda}^{\mathrm{e}}\right)\right] : \mathbf{D}, \qquad (26.119)$$

where

$$h = H + \frac{\partial f}{\partial \boldsymbol{\sigma}} : \boldsymbol{\Lambda}^{\mathrm{e}} : \frac{\partial g}{\partial \boldsymbol{\sigma}}. \qquad (26.120)$$

26.12 Plastic Potential for Geomaterials

To better describe inelastic behavior of geomaterials whose yield is governed by the Drucker–Prager yield criterion of (26.35), a nonassociative flow rule is used with the plastic potential (Fig. 26.14) given by

$$g = J_2^{1/2} + \frac{1}{3}\beta I_1 - k = 0. \qquad (26.121)$$

26.12. Plastic Potential for Geomaterials

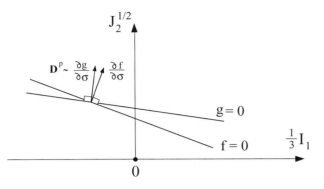

Figure 26.14. Illustration of a nonassociative flow rule. The plastic rate of deformation $\mathbf{D}^p$ is normal to the flow potential $g = 0$, which is distinct from the yield surface $f = 0$.

The material parameter β is in general different from the friction parameter μ_* of (26.35). Thus,

$$\mathbf{D}^p = \dot{\gamma}\frac{\partial g}{\partial \boldsymbol{\sigma}} = \dot{\gamma}\left(\frac{1}{2}J_2^{-1/2}\boldsymbol{\sigma}' + \frac{1}{3}\beta\mathbf{I}\right). \tag{26.122}$$

The loading index $\dot{\gamma}$ is determined from the consistency condition. Assuming known the relationship $k = k(\vartheta)$ between the shear yield stress and the generalized shear plastic strain ϑ, defined by (26.108), the consistency condition $\dot{f} = 0$ gives

$$\dot{\gamma} = \frac{1}{H}\left(\frac{1}{2}J_2^{-1/2}\boldsymbol{\sigma}' + \frac{1}{3}\mu_*\mathbf{I}\right):\overset{\triangledown}{\boldsymbol{\tau}}, \quad H = h_t^p = \frac{dk}{d\vartheta}. \tag{26.123}$$

The substitution of (26.123) into (26.122) yields

$$\mathbf{D}^p = \frac{1}{H}\left[\left(\frac{1}{2}J_2^{-1/2}\boldsymbol{\sigma}' + \frac{1}{3}\beta\mathbf{I}\right)\left(\frac{1}{2}J_2^{-1/2}\boldsymbol{\sigma}' + \frac{1}{3}\mu_*\mathbf{I}\right)\right]:\overset{\triangledown}{\boldsymbol{\tau}}. \tag{26.124}$$

The deviatoric and spherical parts of the plastic rate of deformation are

$$\mathbf{D}^{p\prime} = \frac{1}{2H}\frac{\boldsymbol{\sigma}'}{J_2^{1/2}}\left(\frac{\boldsymbol{\sigma}':\overset{\triangledown}{\boldsymbol{\tau}}}{2J_2^{1/2}} + \frac{1}{3}\mu_*\,\mathrm{tr}\,\overset{\triangledown}{\boldsymbol{\tau}}\right), \tag{26.125}$$

$$\mathrm{tr}\,\mathbf{D}^p = \frac{\beta}{H}\left(\frac{\boldsymbol{\sigma}':\overset{\triangledown}{\boldsymbol{\tau}}}{2J_2^{1/2}} + \frac{1}{3}\mu_*\,\mathrm{tr}\,\overset{\triangledown}{\boldsymbol{\tau}}\right). \tag{26.126}$$

To physically interpret the parameter β, we observe from (26.122) that

$$(2\,\mathbf{D}^{p\prime}:\mathbf{D}^{p\prime})^{1/2} = \dot{\gamma}, \quad \mathrm{tr}\,\mathbf{D}^p = \beta\dot{\gamma}, \tag{26.127}$$

i.e.,

$$\beta = \frac{\mathrm{tr}\,\mathbf{D}^p}{(2\,\mathbf{D}^{p\prime}:\mathbf{D}^{p\prime})^{1/2}}. \tag{26.128}$$

Thus, β is the ratio of the volumetric and shear part of the plastic rate of deformation, which is often called the dilatancy factor (Rudnicki and Rice, 1975). The frictional parameter and inelastic dilatancy of the material actually change with the progression of inelastic deformation, but are here treated as constants. This is further analyzed by Nemat-Nasser and Shokooh (1980) and Nemat-Nasser (1983).

The deviatoric and spherical parts of the total rate of deformation are, respectively,

$$\mathbf{D}' = \frac{\overset{\triangledown}{\boldsymbol{\tau}}'}{2\mu} + \frac{1}{2H}\frac{\boldsymbol{\sigma}'}{J_2^{1/2}}\left(\frac{\boldsymbol{\sigma}':\overset{\triangledown}{\boldsymbol{\tau}}}{2J_2^{1/2}} + \frac{1}{3}\mu_* \operatorname{tr}\overset{\triangledown}{\boldsymbol{\tau}}\right), \qquad (26.129)$$

$$\operatorname{tr}\mathbf{D} = \frac{1}{3\kappa}\operatorname{tr}\overset{\triangledown}{\boldsymbol{\tau}} + \frac{\beta}{H}\left(\frac{\boldsymbol{\sigma}':\overset{\triangledown}{\boldsymbol{\tau}}}{2J_2^{1/2}} + \frac{1}{3}\mu_* \operatorname{tr}\overset{\triangledown}{\boldsymbol{\tau}}\right). \qquad (26.130)$$

These can be inverted to give the deviatoric and spherical parts of the stress rate. The results are

$$\overset{\triangledown}{\boldsymbol{\tau}}' = 2\mu\left[\mathbf{D}' - \frac{1}{c}\frac{\boldsymbol{\sigma}'}{J_2^{1/2}}\left(\frac{\boldsymbol{\sigma}':\mathbf{D}}{2J_2^{1/2}} + \mu_*\frac{\kappa}{2\mu}\operatorname{tr}\mathbf{D}\right)\right], \qquad (26.131)$$

$$\operatorname{tr}\overset{\triangledown}{\boldsymbol{\tau}} = \frac{3\kappa}{c}\left[\left(1 + \frac{H}{\mu}\right)\operatorname{tr}\mathbf{D} - \beta\frac{\boldsymbol{\sigma}':\mathbf{D}}{J_2^{1/2}}\right], \qquad (26.132)$$

where

$$c = 1 + \frac{H}{\mu} + \mu_*\beta\frac{\kappa}{\mu}. \qquad (26.133)$$

If the friction coefficient μ_* is equal to zero, (26.131) and (26.132) reduce to

$$\overset{\triangledown}{\boldsymbol{\tau}}' = 2\mu\left[\mathbf{D}' - \frac{1}{1+H/\mu}\frac{\boldsymbol{\sigma}':\mathbf{D}}{2J_2}\boldsymbol{\sigma}'\right], \qquad (26.134)$$

$$\operatorname{tr}\overset{\triangledown}{\boldsymbol{\tau}} = 3\kappa\left(\operatorname{tr}\mathbf{D} - \frac{\beta}{1+H/\mu}\frac{\boldsymbol{\sigma}':\mathbf{D}}{J_2^{1/2}}\right). \qquad (26.135)$$

With a vanishing dilatancy factor ($\beta = 0$), (26.134) and (26.135) coincide with the constitutive equations of isotropic hardening pressure-independent metal plasticity.

26.13 Rate-Dependent Plasticity

In the modeling of rate-dependent plastic response of metals and alloys, there is no yield surface in the model and plastic deformation commences from the onset of loading, although it may be exceedingly small below certain levels of applied stress. In his analysis of rate-dependent behavior of metals, Rice (1970, 1971) showed that the plastic rate of deformation can be derived from a scalar flow potential Ω, as its gradient

$$\mathbf{D}^p = \frac{\partial \Omega(\boldsymbol{\sigma}, T, \mathcal{H})}{\partial \boldsymbol{\sigma}}, \qquad (26.136)$$

provided that the rate of shearing on any given slip system within a crystalline grain depends on local stresses only through the resolved shear stress. The history of deformation is represented by the pattern of internal rearrangements $\mathcal{H}$, and the absolute temperature is T. Geometrically, the plastic part of the strain rate is normal to surfaces of constant

26.13. Rate-Dependent Plasticity

flow potential in stress space. Time-independent behavior can be recovered, under certain idealizations – neglecting creep and rate effects, as an appropriate limit. In this limit, at each instant of deformation there is a range in stress space over which the flow potential is constant. The current yield surface is then a boundary of this range, a singular clustering of all surfaces of constant flow potential.

The power-law representation of the flow potential is

$$\Omega = \frac{2\dot{\gamma}^0}{m+1} \left(\frac{J_2^{1/2}}{k} \right)^m J_2^{1/2}, \quad J_2 = \frac{1}{2} \sigma' : \sigma', \tag{26.137}$$

where $k = k(T, \mathcal{H})$ is the reference shear stress, $\dot{\gamma}^0$ is the reference shear strain rate to be selected for each material, and m is the material parameter. The corresponding plastic part of the rate of deformation is

$$\mathbf{D}^p = \dot{\gamma}^0 \left(\frac{J_2^{1/2}}{k} \right)^m \frac{\sigma'}{J_2^{1/2}}. \tag{26.138}$$

The equivalent plastic strain ϑ, defined by (26.69), is usually used as the only history parameter, and the reference shear stress depends on ϑ and T according to

$$k = k^0 \left(1 + \frac{\vartheta}{\vartheta^0} \right)^\alpha \exp\left(-\beta \frac{T - T_0}{T_m - T_0} \right). \tag{26.139}$$

Here, k^0 and ϑ^0 are the normalizing stress and strain, T_0 and T_m are the room and melting temperatures, and α and β are the material parameters. The total rate of deformation is

$$\mathbf{D} = \mathbf{M}^e : \stackrel{\triangledown}{\tau} + \dot{\gamma}^0 \left(\frac{J_2^{1/2}}{k} \right)^m \frac{\sigma'}{J_2^{1/2}}. \tag{26.140}$$

The instantaneous elastic compliance tensor is $\mathbf{M}^e$. From the onset of loading the deformation rate consists of elastic and plastic constituents, although for large m the plastic contribution may be small if J_2 is less than k. The inverted form of (26.140), expressing $\stackrel{\triangledown}{\tau}$ in terms of $\mathbf{D}$, is

$$\stackrel{\triangledown}{\tau} = \Lambda^e : \mathbf{D} - 2\mu\dot{\gamma}^0 \left(\frac{J_2^{1/2}}{k} \right)^m \frac{\sigma'}{J_2^{1/2}}. \tag{26.141}$$

Another representation of the flow potential, according to the Johnson–Cook (1983) model, is

$$\Omega = \frac{2\dot{\gamma}^0}{a} k \exp\left[a \left(\frac{J_2^{1/2}}{k} - 1 \right) \right]. \tag{26.142}$$

The reference shear stress is

$$k = k^0 \left[1 + b \left(\frac{\vartheta}{\vartheta^0} \right)^c \right] \left[1 - \left(\frac{T - T_0}{T_m - T_0} \right)^d \right], \tag{26.143}$$

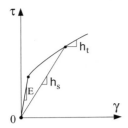

Figure 26.15. Nonlinear stress-strain response in pure shear. Indicated are the initial elastic modulus E, the secant modulus h_s, and the tangent modulus h_t.

where a, b, c, and d are the material parameters. The corresponding plastic part of the rate of deformation becomes

$$\mathbf{D}^p = \dot{\gamma}^0 \exp\left[a\left(\frac{J_2^{1/2}}{k} - 1\right)\right] \frac{\sigma'}{J_2^{1/2}}. \tag{26.144}$$

26.14 Deformation Theory of Plasticity

Simple plasticity theory has been suggested for proportional loading and small deformation by Hencky (1924) and Ilyushin (1947, 1963). Assume that the loading is such that all stress components increase proportionally, *i.e.*,

$$\sigma = c(t)\,\sigma^*, \tag{26.145}$$

where σ^* is the stress tensor at an instant t^* and $c(t)$ is a monotonically increasing function of t, with $c(t^*) = 1$. This implies that the principal directions of σ remain fixed during the deformation process and parallel to those of σ^*.

Because the stress components proportionally increase, and no elastic unloading takes place, it is assumed that elastoplastic response can be described macroscopically by the constitutive structure of nonlinear elasticity, in which the total strain is a function of the total stress. Thus, we decompose the strain tensor into elastic and plastic parts,

$$\mathbf{e} = \mathbf{e}^e + \mathbf{e}^p, \tag{26.146}$$

and assume that

$$\mathbf{e}^e = \mathbf{M}^e : \sigma, \quad \mathbf{e}^p = \varphi\,\sigma', \tag{26.147}$$

where $\mathbf{M}^e$ is the elastic compliance tensor, and φ is an appropriate scalar function to be determined in accord with experimental data. The prime designates a deviatoric part, so that plastic strain tensor is assumed to be traceless.

Suppose that a nonlinear relationship

$$\tau = \tau(\gamma) \tag{26.148}$$

between the stress and strain is available from the elastoplastic shear test. Let the secant and tangent moduli be defined by (Fig. 26.15)

$$h_s = \frac{\tau}{\gamma}, \quad h_t = \frac{d\tau}{d\gamma}, \tag{26.149}$$

26.14. Deformation Theory of Plasticity

and let

$$\tau = \left(\frac{1}{2}\boldsymbol{\sigma}' : \boldsymbol{\sigma}'\right)^{1/2}, \quad \gamma = (2\,\mathbf{e}' : \mathbf{e}')^{1/2}. \tag{26.150}$$

Since

$$\mathbf{e}' = \left(\frac{1}{2\mu} + \varphi\right)\boldsymbol{\sigma}', \tag{26.151}$$

the substitution into (26.150) gives

$$\varphi = \frac{1}{2h_s} - \frac{1}{2\mu}. \tag{26.152}$$

26.14.1 Rate-Type Formulation of Deformation Theory

Although the deformation theory of plasticity is a total strain theory, deformation theory can be cast in the rate-type form. This is important for the comparison with the flow theory of plasticity and for extending the application of the resulting constitutive equations beyond the proportional loading. The rate-type formulation is also needed whenever the considered boundary value problem is being solved in an incremental manner.

In place of (26.146) and (26.147), we take

$$\mathbf{D} = \mathbf{D}^e + \mathbf{D}^p, \tag{26.153}$$

$$\mathbf{D}^e = \mathbf{M}^e : \overset{\nabla}{\boldsymbol{\tau}}, \tag{26.154}$$

$$\mathbf{D}^p = \dot{\varphi}\,\boldsymbol{\sigma}' + \varphi\,\overset{\nabla}{\boldsymbol{\tau}}'. \tag{26.155}$$

The deviatoric and spherical parts of the total rate of deformation tensor are accordingly

$$\mathbf{D}' = \dot{\varphi}\,\boldsymbol{\sigma}' + \left(\frac{1}{2\mu} + \varphi\right)\overset{\nabla}{\boldsymbol{\tau}}', \tag{26.156}$$

$$\operatorname{tr}\mathbf{D} = \frac{1}{3\kappa}\operatorname{tr}\overset{\nabla}{\boldsymbol{\tau}}. \tag{26.157}$$

To derive an expression for the rate $\dot{\varphi}$, we use

$$\tau\dot{\tau} = \frac{1}{2}\boldsymbol{\sigma}' : \overset{\nabla}{\boldsymbol{\tau}}, \quad \gamma\dot{\gamma} = 2\,\mathbf{e}' : \mathbf{D}. \tag{26.158}$$

In view of (26.149), (26.151), and (26.152), this gives

$$\frac{1}{2}\boldsymbol{\sigma}' : \overset{\nabla}{\boldsymbol{\tau}} = h_t\,\boldsymbol{\sigma}' : \mathbf{D}'. \tag{26.159}$$

When (26.156) is incorporated into (26.159), there follows

$$\dot{\varphi} = \frac{1}{2}\left(\frac{1}{h_t} - \frac{1}{h_s}\right)\frac{\boldsymbol{\sigma}' : \overset{\nabla}{\boldsymbol{\tau}}}{\boldsymbol{\sigma}' : \boldsymbol{\sigma}'}. \tag{26.160}$$

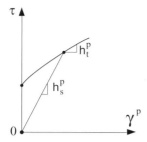

Figure 26.16. Shear stress *vs.* plastic shear strain. The plastic secant modulus is h_s^p, and the plastic tangent modulus is h_t^p.

Substituting (26.160) into (26.156), the deviatoric part of the total rate of deformation becomes

$$\mathbf{D}' = \frac{1}{2h_s}\left[\overset{\triangledown}{\boldsymbol{\tau}}' + \left(\frac{h_s}{h_t}-1\right)\frac{\boldsymbol{\sigma}':\overset{\triangledown}{\boldsymbol{\tau}}}{\boldsymbol{\sigma}':\boldsymbol{\sigma}'}\boldsymbol{\sigma}'\right]. \quad (26.161)$$

Equation (26.161) can be inverted to express the deviatoric part of $\overset{\triangledown}{\boldsymbol{\tau}}$ as

$$\overset{\triangledown}{\boldsymbol{\tau}}' = 2h_s\left[\mathbf{D}' - \left(1-\frac{h_t}{h_s}\right)\frac{\boldsymbol{\sigma}':\mathbf{D}}{\boldsymbol{\sigma}':\boldsymbol{\sigma}'}\boldsymbol{\sigma}'\right]. \quad (26.162)$$

During the initial, purely elastic, stage of deformation $h_t = h_s = \mu$. The onset of plasticity, beyond which (26.161) and (26.162) apply, occurs when τ, defined by the second invariant of the deviatoric stress in (26.150), reaches the initial yield stress in shear. The resulting theory is often referred to as the J_2 deformation theory of plasticity.

If plastic secant and tangent moduli are used (Fig. 26.16), related to secant and tangent moduli with respect to total strain by

$$\frac{1}{h_t} - \frac{1}{h_t^p} = \frac{1}{h_s} - \frac{1}{h_s^p} = \frac{1}{\mu}, \quad (26.163)$$

the plastic part of the rate of deformation can be rewritten as

$$\mathbf{D}^p = \frac{1}{2h_s^p}\overset{\triangledown}{\boldsymbol{\tau}}' + \left(\frac{1}{2h_t^p} - \frac{1}{2h_s^p}\right)\frac{\boldsymbol{\sigma}':\overset{\triangledown}{\boldsymbol{\tau}}}{\boldsymbol{\sigma}':\boldsymbol{\sigma}'}\boldsymbol{\sigma}'. \quad (26.164)$$

26.14.2 Application beyond Proportional Loading

Deformation theory agrees with flow theory of plasticity only under proportional loading, because then specification of the final state of stress also specifies the stress history. For general (nonproportional) loading, more accurate and physically appropriate is the flow theory of plasticity, particularly with an accurate modeling of the yield surface and the hardening characteristics. Budiansky (1959), however, indicated that deformation theory can be successfully used for certain nearly proportional loading paths, as well. The stress rate $\overset{\triangledown}{\boldsymbol{\tau}}'$ in (26.164) then does not have to be codirectional with $\boldsymbol{\sigma}'$. The first and third term (both proportional to $1/2h_s^p$) in (26.164) do not cancel each other in this case (as they do for proportional loading), and the plastic part of the rate of deformation depends on both components of the stress rate $\overset{\triangledown}{\boldsymbol{\tau}}'$, one in the direction of $\boldsymbol{\sigma}'$ and the other normal to it. In contrast, according to the flow theory with the von Mises smooth yield surface,

26.15. J_2 Corner Theory

the component of the stress rate $\stackrel{\nabla}{\tau}{}'$ normal to σ' (thus tangential to the yield surface) does not affect the plastic part of the rate of deformation. Physical theories of plasticity (Batdorf and Budiansky, 1954; Sanders, 1955; Hill, 1967) indicate that the yield surface of a polycrystalline aggregate develops a vertex at its loading stress point, so that infinitesimal increments of stress in the direction normal to σ' indeed cause further plastic flow ("vertex softening"). Because the structure of the deformation theory of plasticity under proportional loading does not use any notion of the yield surface, Budiansky suggested that (26.164) can be adopted to describe the response when the yield surface develops a vertex. If (26.164) is rewritten in the form

$$\mathbf{D}^{\mathrm{p}} = \frac{1}{2h_{\mathrm{s}}^{\mathrm{p}}} \left(\stackrel{\nabla}{\tau}{}' - \frac{\sigma' : \stackrel{\nabla}{\tau}}{\sigma' : \sigma'} \sigma' \right) + \frac{1}{2h_{\mathrm{t}}^{\mathrm{p}}} \frac{\sigma' : \stackrel{\nabla}{\tau}}{\sigma' : \sigma'} \sigma', \qquad (26.165)$$

the first term on the right-hand side gives the response to component of the stress increment normal to σ'. The associated plastic modulus is $h_{\mathrm{s}}^{\mathrm{p}}$. The plastic modulus associated with the component of the stress increment in the direction of σ' is $h_{\mathrm{t}}^{\mathrm{p}}$. Therefore, for continued plastic flow with small deviations from proportional loading (so that all yield segments which intersect at the vertex are active – fully active loading), (26.165) can be used as a model of a pointed vertex (Stören and Rice, 1975). The idea was used by Rudnicki and Rice (1975) to model the inelastic behavior of fissured rocks.

For the full range of directions of the stress increment, the relationship between the rates of stress and plastic deformation is not necessarily linear, although it is homogeneous in these rates, in the absence of time-dependent (creep) effects. A corner theory that predicts continuous variation of the stiffness and allows increasingly nonproportional increments of stress was formulated by Christoffersen and Hutchinson (1979). This is discussed in the next subsection. When applied to the analysis of necking in thin sheets under biaxial stretching, the results were in better agreement with experiments than those obtained from the theory with a smooth yield surface. Further discussion can be found in Section 30.5 of this book.

26.15 J_2 Corner Theory

In phenomenological J_2 corner theory of plasticity, proposed by Christoffersen and Hutchinson (1979), the instantaneous elastoplastic moduli for nearly proportional loading are chosen equal to the J_2 deformation theory moduli, whereas for increasing deviation from proportional loading the moduli increase smoothly until they coincide with elastic moduli for stress increments directed along or within the corner of the yield surface. The yield surface in the neighborhood of the loading point in deviatoric stress space (Fig. 26.17) is a cone around the axis

$$l = \frac{\sigma'}{(\sigma' : \mathbf{M}_{\mathrm{def}}^{\mathrm{p}} : \sigma')^{1/2}}, \qquad (26.166)$$

where $\mathbf{M}_{\mathrm{def}}^{\mathrm{p}}$ is the plastic compliance tensor of the deformation theory. The angular measure θ of the stress rate direction, relative to the cone axis, is defined by

$$\cos\theta = \frac{l : \mathbf{M}_{\mathrm{def}}^{\mathrm{p}} : \stackrel{\nabla}{\tau}}{(\stackrel{\nabla}{\tau} : \mathbf{M}_{\mathrm{def}}^{\mathrm{p}} : \stackrel{\nabla}{\tau})^{1/2}}. \qquad (26.167)$$

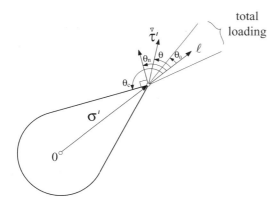

Figure 26.17. Near proportional or total loading range at the yield vertex of J_2 corner theory is a cone with the angle θ_0 around the axis $l \sim \boldsymbol{\sigma}'$. The vertex cone is defined by the angle θ_c, and $\theta_n = \theta_c - \pi/2$.

The conical surface separating elastic unloading and plastic loading is $\theta = \theta_c$, so that plastic rate of deformation falls within the range $0 \leq \theta \leq \theta_n$, where $\theta_n = \theta_c - \pi/2$. The range of near proportional loading is $0 \leq \theta \leq \theta_0$. The angle θ_0 is a suitable fraction of θ_n. The range of near proportional loading is the range of stress-rate directions for which no elastic unloading takes place on any of the yield vertex segments. This range is also called the fully active or total loading range.

The stress-rate potential at the corner is defined by

$$\Pi = \Pi^e + \Pi^p, \quad \Pi^p = f(\theta)\Pi^p_{\text{def}}. \tag{26.168}$$

The elastic contribution to the stress-rate potential is

$$\Pi^e = \frac{1}{2}\overset{\triangledown}{\boldsymbol{\tau}} : \mathbf{M}^e : \overset{\triangledown}{\boldsymbol{\tau}}. \tag{26.169}$$

The plastic stress-rate potential of the J_2 deformation theory can be written, from (26.161), as

$$\Pi^p_{\text{def}} = \frac{1}{2}\overset{\triangledown}{\boldsymbol{\tau}} : \mathbf{M}^p_{\text{def}} : \overset{\triangledown}{\boldsymbol{\tau}}, \quad \mathbf{M}^p_{\text{def}} = \frac{1}{2h_s}\left[\left(1 - \frac{h_s}{\mu}\right)\mathbf{A} + \left(\frac{h_s}{h_t} - 1\right)\frac{\boldsymbol{\sigma}'\boldsymbol{\sigma}'}{\boldsymbol{\sigma}':\boldsymbol{\sigma}'}\right],$$

where

$$A_{ijkl} = \frac{1}{2}(\delta_{ik}\delta_{jl} + \delta_{il}\delta_{jk}) - \frac{1}{3}\delta_{ij}\delta_{kl}. \tag{26.170}$$

The plastic stress-rate potential Π^p_{def} is weighted by the cone transition function $f(\theta)$ to obtain the plastic stress-rate potential Π^p of the J_2 corner theory.

In the range of near proportional loading

$$0 \leq \theta \leq \theta_0, \quad f(\theta) = 1, \tag{26.171}$$

whereas in the elastic unloading range

$$\theta_c \leq \theta \leq \pi, \quad f(\theta) = 0. \tag{26.172}$$

26.16. Rate-Dependent Flow Theory

In the transition region $\theta_0 \leq \theta \leq \theta_c$, the function $f(\theta)$ decreases monotonically and smoothly from one to zero in a way which ensures convexity of the plastic-rate potential,

$$\Pi^p(\overset{\triangledown}{\tau}_2) - \Pi^p(\overset{\triangledown}{\tau}_1) \geq \frac{\partial \Pi^p}{\partial \overset{\triangledown}{\tau}_1} : (\overset{\triangledown}{\tau}_2 - \overset{\triangledown}{\tau}_1). \tag{26.173}$$

A simple choice of $f(\theta)$ meeting these requirements is

$$f(\theta) = \cos^2\left(\frac{\pi}{2} \frac{\theta - \theta_0}{\theta_c - \theta_0}\right), \quad \theta_0 \leq \theta \leq \theta_c. \tag{26.174}$$

The specification of the angles θ_c and θ_0 in terms of the current stress measure is discussed by Christoffersen and Hutchinson (1979).

The rate independence of the material response requires

$$\mathbf{D}^p = \frac{\partial \Pi^p}{\partial \overset{\triangledown}{\tau}} = \frac{\partial^2 \Pi^p}{\partial \overset{\triangledown}{\tau} \partial \overset{\triangledown}{\tau}} : \overset{\triangledown}{\tau} = \mathbf{M}^p : \overset{\triangledown}{\tau} \tag{26.175}$$

to be a homogeneous function of degree one, and Π^p to be a homogeneous function of degree two in the stress rate $\overset{\triangledown}{\tau}$. The function $\Pi^p(\overset{\triangledown}{\tau})$ is quadratic in the region of nearly proportional loading, but highly nonlinear in the transition region, because of nonlinearity associated with $f(\theta)$. The plastic rate of deformation is accordingly a linear function of $\overset{\triangledown}{\tau}$ in the region of nearly proportional loading, but a nonlinear function in the transition region.

26.16 Rate-Dependent Flow Theory

In this section we introduce a multiplicative decomposition of the total deformation gradient into elastic and plastic parts to provide an additional kinematical framework for dealing with finite elastic and plastic deformation. We introduce this in the specific context of a strain rate, and thus time dependent, J_2 flow theory of plasticity. We will introduce a simple description of kinetic behavior for the strain rates and for the strain hardening rates for the purpose of making the discussion explicit. We note, however, that these concepts are developed in far more detail in the chapters to follow concerned with crystalline plasticity.

26.16.1 Multiplicative Decomposition $\mathbf{F} = \mathbf{F}^e \cdot \mathbf{F}^p$

Consider the current elastoplastically deformed configuration of the material sample. Let $\mathbf{F}$ be the deformation gradient that maps an infinitesimal material element $d\mathbf{X}$ from the initial to $d\mathbf{x}$ in the current configuration, such that $d\mathbf{x} = \mathbf{F} \cdot d\mathbf{X}$. Introduce an intermediate configuration by elastically destressing the current configuration to zero stress (Fig. 26.18). Such configuration differs from the initial configuration by residual (plastic) deformation and from the current configuration by reversible (elastic) deformation. If $d\mathbf{x}^p$ is the material element in the intermediate configuration, corresponding to $d\mathbf{x}$ in the current configuration, then $d\mathbf{x} = \mathbf{F}^e \cdot d\mathbf{x}^p$, where $\mathbf{F}^e$ represents the deformation gradient associated with the elastic loading from the intermediate to current configuration. If the deformation gradient

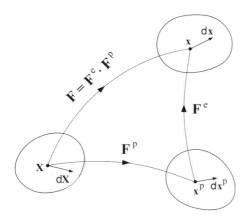

Figure 26.18. Schematic representation of the multiplicative decomposition of deformation gradient into its elastic and plastic parts. The intermediate configuration is obtained from the current configuration by elastic destressing to zero stress.

of the plastic transformation is $\mathbf{F}^p$, such that $d\mathbf{x}^p = \mathbf{F}^p \cdot d\mathbf{X}$, the multiplicative decomposition of the total deformation gradient into its elastic and plastic parts follows as

$$\mathbf{F} = \mathbf{F}^e \cdot \mathbf{F}^p. \tag{26.176}$$

This decomposition was introduced in the phenomenological rate independent theory of plasticity by Lee (1969). In the case when elastic destressing to zero stress is not physically achievable because of possible onset of reverse inelastic deformation before the state of zero stress is reached, the intermediate configuration can be conceptually introduced by virtual destressing to zero stress, locking all inelastic structural changes that would take place during the actual destressing.

The deformation gradients $\mathbf{F}^e$ and $\mathbf{F}^p$ are not uniquely defined because the intermediate unstressed configuration is not unique. Arbitrary local material rotations can be superposed to the intermediate configuration, preserving it unstressed. In applications, however, the decomposition (26.176) can be made unique by additional specifications, dictated by the nature of the considered material model. For example, for elastically isotropic materials the elastic stress response depends only on the elastic stretch $\mathbf{V}^e$ and not on the rotation $\mathbf{R}^e$ from the polar decomposition $\mathbf{F}^e = \mathbf{V}^e \cdot \mathbf{R}^e$. Consequently, the intermediate configuration can be specified uniquely by requiring that the elastic unloading takes place without rotation ($\mathbf{F}^e = \mathbf{V}^e$). An alternative choice will be pursued below.

The velocity gradient in the current configuration at time t is defined by

$$\mathbf{L} = \dot{\mathbf{F}} \cdot \mathbf{F}^{-1}. \tag{26.177}$$

The superposed dot designates the material time derivative. By introducing the multiplicative decomposition of deformation gradient (26.176), the velocity gradient becomes

$$\mathbf{L} = \dot{\mathbf{F}}^e \cdot \mathbf{F}^{e-1} + \mathbf{F}^e \cdot \left(\dot{\mathbf{F}}^p \cdot \mathbf{F}^{p-1}\right) \cdot \mathbf{F}^{e-1}. \tag{26.178}$$

The rate of deformation $\mathbf{D}$ and the spin $\mathbf{W}$ are, respectively, the symmetric and antisymmetric part of $\mathbf{L}$,

$$\mathbf{D} = \left(\dot{\mathbf{F}}^e \cdot \mathbf{F}^{e-1}\right)_s + \left[\mathbf{F}^e \cdot \left(\dot{\mathbf{F}}^p \cdot \mathbf{F}^{p-1}\right) \cdot \mathbf{F}^{e-1}\right]_s, \tag{26.179}$$

$$\mathbf{W} = \left(\dot{\mathbf{F}}^e \cdot \mathbf{F}^{e-1}\right)_a + \left[\mathbf{F}^e \cdot \left(\dot{\mathbf{F}}^p \cdot \mathbf{F}^{p-1}\right) \cdot \mathbf{F}^{e-1}\right]_a. \tag{26.180}$$

26.17. Elastic and Plastic Contributions

Since $\mathbf{F}^e$ is specified up to an arbitrary rotation, and since the stress response of elastically isotropic materials does not depend on the rotation, we shall choose the unloading program such that

$$\left[\mathbf{F}^e \cdot \left(\dot{\mathbf{F}}^p \cdot \mathbf{F}^{p-1}\right) \cdot \mathbf{F}^{e-1}\right]_a = \mathbf{0}. \tag{26.181}$$

With this choice, the rate of deformation and the spin tensors become

$$\mathbf{D} = \left(\dot{\mathbf{F}}^e \cdot \mathbf{F}^{e-1}\right)_s + \mathbf{F}^e \cdot \left(\dot{\mathbf{F}}^p \cdot \mathbf{F}^{p-1}\right) \cdot \mathbf{F}^{e-1}, \tag{26.182}$$

$$\mathbf{W} = \left(\dot{\mathbf{F}}^e \cdot \mathbf{F}^{e-1}\right)_a. \tag{26.183}$$

26.17 Elastic and Plastic Constitutive Contributions

It is assumed that the material is elastically isotropic in its initial undeformed state and that plastic deformation does not affect its elastic properties. The elastic response is then given by

$$\boldsymbol{\tau} = \mathbf{F}^e \cdot \frac{\partial \Phi^e(\mathbf{E}^e)}{\partial \mathbf{E}^e} \cdot \mathbf{F}^{eT}. \tag{26.184}$$

The elastic strain energy per unit unstressed volume, Φ^e, is an isotropic function of the Lagrangian strain $\mathbf{E}^e = \left(\mathbf{F}^{eT} \cdot \mathbf{F}^e - \mathbf{I}\right)/2$. Plastic deformation is assumed to be incompressible ($\det \mathbf{F}^e = \det \mathbf{F}$), so that $\boldsymbol{\tau} = (\det \mathbf{F})\boldsymbol{\sigma}$ is the Kirchhoff stress. By differentiating (26.184), we obtain

$$\dot{\boldsymbol{\tau}} - \left(\dot{\mathbf{F}}^e \cdot \mathbf{F}^{e-1}\right) \cdot \boldsymbol{\tau} - \boldsymbol{\tau} \cdot \left(\dot{\mathbf{F}}^e \cdot \mathbf{F}^{e-1}\right)^T = \hat{\boldsymbol{\Lambda}}^e : \left(\dot{\mathbf{F}}^e \cdot \mathbf{F}^{e-1}\right)_s. \tag{26.185}$$

The rectangular components of $\hat{\boldsymbol{\Lambda}}^e$ are

$$\hat{\Lambda}^e_{ijkl} = F^e_{im} F^e_{jn} \frac{\partial^2 \Phi^e}{\partial E^e_{mn} \partial E^e_{pq}} F^e_{kp} F^e_{lq}. \tag{26.186}$$

Equation (26.185) can be equivalently written as

$$\dot{\boldsymbol{\tau}} - \left(\dot{\mathbf{F}}^e \cdot \mathbf{F}^{e-1}\right)_a \cdot \boldsymbol{\tau} + \boldsymbol{\tau} \cdot \left(\dot{\mathbf{F}}^e \cdot \mathbf{F}^{e-1}\right)_a = \boldsymbol{\Lambda}^e : \left(\dot{\mathbf{F}}^e \cdot \mathbf{F}^{e-1}\right)_s. \tag{26.187}$$

The modified elastic moduli tensor $\boldsymbol{\Lambda}^e$ has the components

$$\Lambda^e_{ijkl} = \hat{\Lambda}^e_{ijkl} + \frac{1}{2}\left(\tau_{ik}\delta_{jl} + \tau_{jk}\delta_{il} + \tau_{il}\delta_{jk} + \tau_{jl}\delta_{ik}\right). \tag{26.188}$$

In view of (26.183), we can rewrite (26.187) as

$$\stackrel{\nabla}{\boldsymbol{\tau}} = \boldsymbol{\Lambda}^e : \left(\dot{\mathbf{F}}^e \cdot \mathbf{F}^{e-1}\right)_s, \tag{26.189}$$

where

$$\stackrel{\nabla}{\boldsymbol{\tau}} = \dot{\boldsymbol{\tau}} - \mathbf{W} \cdot \boldsymbol{\tau} + \boldsymbol{\tau} \cdot \mathbf{W} \tag{26.190}$$

is the Jaumann rate of the Kirchhoff stress with respect to total spin. By inversion, (26.189) gives the elastic rate of deformation as

$$\mathbf{D}^e = \left(\dot{\mathbf{F}}^e \cdot \mathbf{F}^{e-1}\right)_s = \boldsymbol{\Lambda}^{e-1} : \stackrel{\nabla}{\boldsymbol{\tau}}. \tag{26.191}$$

Physically, the strain increment $\mathbf{D}^e \, dt$ is a reversible part the total strain increment $\mathbf{D} \, dt$, which is recovered upon loading–unloading cycle of the stress increment $\overset{\triangledown}{\boldsymbol{\tau}} \, dt$. The remaining part of the total rate of deformation,

$$\mathbf{D}^p = \mathbf{D} - \mathbf{D}^e, \tag{26.192}$$

is the plastic part, which gives a residual strain increment left after the considered infinitesimal cycle of stress.

26.17.1 Rate-Dependent J_2 Flow Theory

Classical J_2 flow theory uses the yield surface as generated earlier as a flow potential. Thus the current yield criteria $\bar{\sigma} = \kappa$ defines a series of yield surfaces in stress space, where κ serves the role of a scaling parameter. Here we rephrase the yield criterion of (26.6) in terms of the *effective stress*, $\bar{\sigma} = (3/2 \sigma'_{ij} \sigma'_{ij})^{1/2}$; κ is then the uniaxial yield stress Y. J_2 flow theory assumes that $\mathbf{D}^p \parallel \boldsymbol{\sigma}'$. This amounts to taking

$$\mathbf{D}^p \parallel \frac{\partial \bar{\sigma}}{\partial \boldsymbol{\sigma}'}, \tag{26.193}$$

or

$$D^p_{ij} \parallel \frac{\partial \bar{\sigma}}{\partial \sigma'_{ij}} = \frac{3}{2} \frac{\sigma'_{ij}}{\bar{\sigma}}, \tag{26.194}$$

Thus, we can write

$$\mathbf{D}^p = \dot{\bar{e}}^p \frac{3}{2} \frac{\boldsymbol{\sigma}'}{\bar{\sigma}}, \tag{26.195}$$

where $\dot{\bar{e}}^p$ is an effective plastic strain rate whose specification requires an additional model statement. By incorporating (26.195) we can write, from (26.189),

$$\overset{\triangledown}{\boldsymbol{\tau}} = \boldsymbol{\Lambda}^e : \mathbf{D}^e = \boldsymbol{\Lambda}^e : (\mathbf{D} - \mathbf{D}^p) = \boldsymbol{\Lambda}^e : \left(\mathbf{D} - \dot{\bar{e}}^p \frac{3}{2} \frac{\boldsymbol{\sigma}'}{\bar{\sigma}} \right). \tag{26.196}$$

For the present we use a simple power law expression of the form

$$\dot{\bar{e}}^p = \dot{e}_0 \left(\frac{\bar{\sigma}}{g} \right)^{1/m}, \tag{26.197}$$

where $\dot{e}_0$ is a reference strain rate and $1/m$ represents a strain rate sensitivity coefficient. We note that for common metals, $50 < 1/m < 200$. For values of $1/m \sim 100$, or larger, the materials will display a very nearly rate independent response in the sense that $\bar{\sigma}$ will track g, the *hardness*, at nearly any value of strain rate.

Strain hardening is described as an evolution of the hardness function g. As an example, we adopt the model with

$$g(\bar{e}^p) = \sigma_0 \left(1 + \frac{\bar{e}^p}{e_y} \right)^n, \tag{26.198}$$

where

$$\dot{\bar{e}}^p = \left(\frac{2}{3} \mathbf{D}^p : \mathbf{D}^p \right)^{1/2}, \quad \bar{e}^p = \int_0^t \left(\frac{2}{3} \mathbf{D}^p : \mathbf{D}^p \right)^{1/2} dt \tag{26.199}$$

are the *effective plastic strain rate* and *effective plastic strain*, respectively. The remaining parameters are the material parameters; the initial yield stress is σ_0, and n is the hardening exponent.

26.18 A Rate Tangent Integration

In this section we introduce an integration method, originally proposed by Peirce et al. (1982) for phenomenological constitutive theories discussed above and by Peirce et al. (1983) for physically based theories that are discussed in the following chapters. The main concern is the estimate of the plastic strain increments over finite time steps during the integration of relations such as (26.196). Consider, for example, the expression for the plastic part of the rate of deformation in the phenomenological J_2 flow theory, viz.,

$$\mathbf{D}^\mathrm{p} = \dot{\bar{e}}(\bar{e}, \beta_i)\mathbf{p}, \qquad (26.200)$$

where $\mathbf{p}$ is the *direction* of viscoplastic straining, and the *effective plastic strain*, $\bar{e}$, is defined as

$$\bar{e} = \int_0^t \left(\frac{\mathbf{D}^\mathrm{p} : \mathbf{D}^\mathrm{p}}{\mathbf{p} : \mathbf{p}}\right)^{1/2} dt. \qquad (26.201)$$

A set of scalar variables β_i define the "strength of the material." Of course, the effective plastic strain rate depends on the effective stress, $\bar{\sigma}$, as well. For the J_2 flow theory, we have

$$\mathbf{p} = \frac{3}{2} \frac{\boldsymbol{\tau}'}{\bar{\tau}}, \qquad (26.202)$$

where

$$\boldsymbol{\tau}' = \boldsymbol{\tau} - \frac{1}{3}(\boldsymbol{\tau} : \mathbf{I})\mathbf{I}, \quad \bar{\tau}^2 = \frac{3}{2}\boldsymbol{\tau} : \boldsymbol{\tau}. \qquad (26.203)$$

The total rate of deformation is consequently

$$\mathbf{D} = \mathbf{D}^* + \mathbf{D}^\mathrm{p} = \boldsymbol{\Lambda}^{\mathrm{e}\,-1} : \stackrel{\triangledown}{\boldsymbol{\tau}} + \dot{\bar{e}}(\bar{e}, \beta_i)\mathbf{p}, \qquad (26.204)$$

which can be inverted to give

$$\stackrel{\triangledown}{\boldsymbol{\tau}} = \boldsymbol{\Lambda}^\mathrm{e} : \mathbf{D} - \dot{\bar{e}}\,\mathbf{P}, \qquad (26.205)$$

where

$$\mathbf{P} = \boldsymbol{\Lambda}^\mathrm{e} : \mathbf{p}. \qquad (26.206)$$

As a model evolutionary law for the β_i, we assume

$$\dot{\beta}_i = \mathbf{A}_i : \mathbf{D}^\mathrm{p} + \mathbf{B}_i : \stackrel{\triangledown}{\boldsymbol{\tau}}. \qquad (26.207)$$

The $\mathbf{A}_i$ and $\mathbf{B}_i$ can, in general, depend on the current values of $\boldsymbol{\tau}, \bar{e}$, and the β_i. The objective is to estimate an effective $\dot{\bar{e}}\Delta t$ over a small, but finite time increment Δt. To that goal, we write for an increment in $\Delta\bar{e}$, over the time step Δt,

$$\Delta\bar{e} = \Delta t\left[(1-\theta)\dot{\bar{e}}_t + \theta\dot{\bar{e}}_{t+\Delta t}\right]. \qquad (26.208)$$

We then expand $\dot{\bar{e}}_{t+\Delta t}$ in a Taylor series about the time t to obtain

$$\dot{\bar{e}}_{t+\Delta t} \approx \dot{\bar{e}}_t + \left(\frac{\partial \dot{\bar{e}}}{\partial \bar{e}}\right) \dot{\bar{e}}_t \Delta t + \sum_i \left(\frac{\partial \dot{\bar{e}}}{\partial \beta_i}\right) \dot{\beta}_{i,t} \Delta t. \quad (26.209)$$

The use of (26.207) in (26.209) gives

$$\Delta \bar{e} \approx \Delta t \, \dot{\bar{e}}_t + \theta (\Delta t)^2 \left[\frac{\partial \dot{\bar{e}}}{\partial \bar{e}} \dot{\bar{e}} + \sum_i \frac{\partial \dot{\bar{e}}}{\partial \beta_i} (\mathbf{A}_i : \mathbf{D}^p + \mathbf{B}_i : \overset{\triangledown}{\tau})\right]. \quad (26.210)$$

Upon the substitution of (26.200) and (26.205) into (26.210), we obtain

$$\Delta \bar{e} \approx \Delta t \, \dot{\bar{e}}_t + (\theta \Delta t)$$

$$\times \left[\frac{\partial \dot{\bar{e}}}{\partial \bar{e}} \Delta \bar{e} + \sum_i \frac{\partial \dot{\bar{e}}}{\partial \beta_i} \mathbf{B}_i : \mathbf{\Lambda}^e : \mathbf{D} \Delta t + \sum_i \frac{\partial \dot{\bar{e}}}{\partial \beta_i} (\mathbf{A}_i : \mathbf{p} - \mathbf{B}_i : \mathbf{P}) \Delta \bar{e}\right].$$

When this relation is solved for $\Delta \bar{e}$, and then divided by Δt, there follows

$$\dot{\bar{e}} = \frac{\dot{\bar{e}}_t}{1+\xi} + \frac{1}{\bar{H}} \frac{\xi}{1+\xi} \mathbf{Q} : \mathbf{D}, \quad (26.211)$$

where

$$\mathbf{Q} = \sum_i \frac{\partial \dot{\bar{e}}}{\partial \beta_i} \mathbf{B}_i : \mathbf{\Lambda}^e,$$

$$\bar{H} = -\frac{\partial \dot{\bar{e}}}{\partial \bar{e}} - \sum_i \frac{\partial \dot{\bar{e}}}{\partial \beta_i} (\mathbf{A}_i : \mathbf{p} - \mathbf{B}_i : \mathbf{P}), \quad (26.212)$$

$$\xi = (\theta \Delta t) \bar{H}.$$

For the J_2 flow theory itself, take as a specific model for evolution,

$$\dot{\bar{e}} = \dot{\bar{e}}(\bar{e}, \bar{\sigma}), \quad (26.213)$$

and note that

$$\dot{\bar{\sigma}} = \mathbf{p} : \overset{\triangledown}{\tau}. \quad (26.214)$$

Then,

$$\mathbf{Q} = \frac{\partial \dot{\bar{\sigma}}}{\partial \bar{\sigma}} \mathbf{P}. \quad (26.215)$$

It follows that

$$\bar{H} = -\frac{\partial \dot{\bar{e}}}{\partial \bar{e}} + \frac{\partial \dot{\bar{e}}}{\partial \bar{\sigma}} (\mathbf{p} : \mathbf{\Lambda}^e : \mathbf{p}), \quad (26.216)$$

and

$$\mathbf{\Lambda}^{\tan} = \mathbf{\Lambda}^e - \frac{\xi}{1+\xi} \frac{1}{h} \mathbf{P}\mathbf{P}. \quad (26.217)$$

Here,

$$h = \frac{\bar{H}}{\partial \dot{\bar{e}}/\partial \bar{e}} = -\left(\frac{\partial \dot{\bar{e}}}{\partial \bar{e}}\right)\left(\frac{\partial \dot{\bar{e}}}{\partial \bar{\sigma}}\right)^{-1} + \mathbf{p} : \mathbf{\Lambda}^e : \mathbf{p}, \quad (26.218)$$

with
$$\xi = \theta \Delta t \left(\frac{\partial \dot{e}}{\partial \bar{\sigma}}\right) h. \tag{26.219}$$

We note that $\boldsymbol{\Lambda}^{\text{tan}}$ is symmetric, and the constitutive relation takes the final form

$$\overset{\triangledown}{\boldsymbol{\tau}} = \boldsymbol{\Lambda}^{\text{tan}} : \mathbf{D} - \frac{\dot{e}}{1+\xi} \mathbf{P}. \tag{26.220}$$

26.19 Plastic Void Growth

As an illustration of the application of some of the derived equations in this chapter, we present in this section an analysis of plastic void growth under triaxial tension. More involved studies of plastic void growth under different states of stress, and based on the more involved material models, can be found in Rice and Tracey (1969), Needleman (1972), Budiansky, Hutchinson, and Slutsky (1982), Huang, Hutchinson, and Tvergaard (1991), Needleman, Tvergaard, and Hutchinson (1992), and others.

Consider a spherical void of initial radius R_0 in an isotropic infinite medium under remote triaxial tension σ. The stress state at an arbitrary point consists of the radial stress σ_{rr} and the hoop stress $\sigma_{\theta\theta} = \sigma_{\phi\phi}$. Because of spherical symmetry, the stress components depend only on the radial distance r and not on the spherical angles θ and ϕ. In the absence of body forces the equilibrium equation is

$$\frac{d\sigma_{rr}}{dr} - \frac{2}{r}(\sigma_{\theta\theta} - \sigma_{rr}) = 0. \tag{26.221}$$

If the material response is within the infinitesimal elastic range, the compatibility equation is

$$\frac{d}{dr}(\sigma_{rr} + 2\sigma_{\theta\theta}) = 0, \tag{26.222}$$

which implies that the spherical component of stress tensor is uniform throughout the medium, and thus equal to σ, i.e.,

$$\sigma_{rr} + 2\sigma_{\theta\theta} = 3\sigma. \tag{26.223}$$

Combining (26.221) and (26.223) it follows that

$$\sigma_{rr} = \sigma\left(1 - \frac{R_0^3}{r^3}\right), \quad \sigma_{\theta\theta} = \sigma\left(1 + \frac{1}{2}\frac{R_0^3}{r^3}\right). \tag{26.224}$$

The corresponding radial displacement is

$$u = \frac{\sigma}{E}\left[(1-2\nu)r + \frac{1}{2}(1+\nu)\frac{R_0^3}{r^2}\right], \tag{26.225}$$

where E is Young's modulus and ν is Poisson's ratio of the material.

According to either von Mises or Tresca yield criterion, the plastic deformation commences when

$$\sigma_{\theta\theta} - \sigma_{rr} = Y. \tag{26.226}$$

The initial yield stress of the material in uniaxial tension is denoted by Y. Thus, the threshold value of applied stress σ for the onset of plastic deformation at the surface of the void

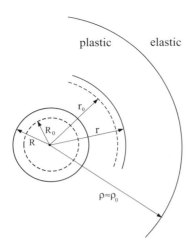

Figure 26.19. The elastic-plastic interface around the spherical void of initial radius R_0. The radius of the expanded void is R. The initial and current radii of material elements within the plastic zone are r_0 and r. Since deformation is infinitesimal beyond the elastic-plastic interface, its deformed and undeformed radii are nearly equal to each other ($\rho \approx \rho_0$).

($r = R_0$) is $\sigma_y = 2Y/3$. With further increase of stress beyond this value, the plastic zone expands outward and at an arbitrary instant of deformation the elastic-plastic boundary has reached the radius ρ (Fig. 26.19). The corresponding radius of the expanded void is R. A material element within the current plastic zone, which was initially at the radius r_0, is currently at the radius r. We assume that ρ is sufficiently large, so that large strains take place in the vicinity of the void, whereas infinitesimal strains characterize the elastic zone beyond the radius ρ. This means that $\rho \approx \rho_0$ and similarly for any radius beyond the elastic-plastic boundary. The deformation gradient tensor associated with the spherical expansion of the void, expressed in spherical coordinates, is

$$\mathbf{F} = \begin{bmatrix} \dfrac{dr}{dr_0} & 0 & 0 \\ 0 & \dfrac{r}{r_0} & 0 \\ 0 & 0 & \dfrac{r}{r_0} \end{bmatrix}. \tag{26.227}$$

The corresponding volume change of an infinitesimal material element is

$$\frac{dV}{dV_0} = \det \mathbf{F} = \left(\frac{r}{r_0}\right)^2 \frac{dr}{dr_0}. \tag{26.228}$$

If plastic deformation is assumed to be isochoric, the volume change is entirely due to elastic deformation, and

$$\frac{dV - dV_0}{dV} = \frac{1 - 2\nu}{E}(\sigma_{rr} + 2\sigma_{\theta\theta}). \tag{26.229}$$

Combining (26.228) and (26.229), there follows

$$\left(\frac{r_0}{r}\right)^2 \frac{dr_0}{dr} = 1 - \frac{1 - 2\nu}{E}(\sigma_{rr} + 2\sigma_{\theta\theta}). \tag{26.230}$$

Incorporating the equilibrium equation (26.221), rewritten as

$$\sigma_{rr} + 2\sigma_{\theta\theta} = \frac{1}{r^2} \frac{d}{dr}\left(r^3 \sigma_{rr}\right), \tag{26.231}$$

26.19. Plastic Void Growth

equation (26.230) becomes

$$r_0^2\, dr_0 = r^2 dr - \frac{1-2v}{E}\, d\left(r^3 \sigma_{rr}\right). \quad (26.232)$$

Upon integration, using the boundary condition $\sigma_{rr}(R) = 0$, we obtain

$$\sigma_{rr} = \frac{E}{3(1-2v)}\left(\frac{r^3 - r_0^3}{r^3} - \frac{R^3 - R_0^3}{r^3}\right). \quad (26.233)$$

This expression holds everywhere in the medium, regardless of the constitutive description in the plastic zone $r \leq \rho$.

On the other hand, in the elastic zone $r \geq \rho$, we have

$$\sigma_{rr} = \sigma - \frac{2}{3} Y \frac{\rho^3}{r^3}, \quad \sigma_{\theta\theta} = \sigma + \frac{1}{3} Y \frac{\rho^3}{r^3}, \quad (26.234)$$

which ensures that $\sigma_{\theta\theta}(\rho) - \sigma_{rr}(\rho) = Y$, and $\sigma_{rr} + 2\sigma_{\theta\theta} = 3\sigma$ everywhere in the elastic region. Thus, $\sigma_{rr}(\rho) = \sigma - 2Y/3$, and from (26.233) we obtain

$$\sigma = \frac{2}{3} Y + \frac{E}{3(1-2v)}\left(\frac{\rho^3 - \rho_0^3}{\rho^3} - \frac{R^3 - R_0^3}{\rho^3}\right). \quad (26.235)$$

Since the strain is elastic and infinitesimal at the elastic-plastic interface, there is an approximation

$$\frac{\rho^3 - \rho_0^3}{\rho^3} \approx 3e_{\theta\theta}(\rho) = \frac{3(1-2v)}{E}\sigma + \frac{1+v}{E} Y, \quad (26.236)$$

and the substitution into (26.235) gives

$$R^3 = R_0^3 + 3(1-v)\frac{Y}{E}\rho^3. \quad (26.237)$$

At the advanced stages of plastic deformation, when R becomes much greater than R_0, the above result implies that the ratio ρ/R approaches a constant value

$$\frac{\rho}{R} = \left[\frac{E}{3(1-v)Y}\right]^{1/3}. \quad (26.238)$$

Equations (26.237) and (26.238) hold for any type of hardening in the plastic zone, but the analysis cannot be pursued further without using the constitutive description of plastic deformation. For an elastoplastic material with a general nonlinear hardening the relationship between the applied stress σ and the size of the plastic zone ρ can be determined only numerically. The closed form solutions are attainable for two particular material models: an elastic ideally plastic material, and an incompressible elastic linearly hardening material.

26.19.1 Ideally Plastic Material

In the case of a nonhardening material, we have $\sigma_{\theta\theta} - \sigma_{rr} = Y$ throughout the plastic zone, and from the equilibrium equation the stress state in the plastic zone ($r \leq \rho$) is

$$\sigma_{rr} = Y \ln \frac{r^2}{R^2}, \quad \sigma_{\theta\theta} = Y\left(1 + \ln \frac{r^2}{R^2}\right). \quad (26.239)$$

Since the stress field in the surrounding elastic region ($r \geq \rho$) is specified by (26.234), the continuity condition for the radial stress component at the elastic-plastic interface gives

$$\sigma = \frac{2}{3} Y \left(1 + 3 \ln \frac{\rho}{R}\right), \quad \rho = R \exp\left(\frac{\sigma}{2Y} - \frac{1}{3}\right). \tag{26.240}$$

The substitution of (26.240) into (26.237) furnishes an expression for the current radius of the void in terms of the applied stress

$$R = R_0 \left[1 - 3(1-\nu)\frac{Y}{E} \exp\left(\frac{3\sigma}{2Y} - 1\right)\right]^{-1/3}. \tag{26.241}$$

At this stress level, the plastic zone has spread to

$$\rho = R_0 \frac{\exp\left(\frac{\sigma}{2Y} - \frac{1}{3}\right)}{\left[1 - 3(1-\nu)\frac{Y}{E} \exp\left(\frac{3\sigma}{2Y} - 1\right)\right]^{1/3}}. \tag{26.242}$$

The inverted form, giving σ in terms of ρ, is

$$\sigma = \frac{2Y}{3} \left\{1 - \ln\left[\frac{R_0^3}{\rho^3} + 3(1-\nu)\frac{Y}{E}\right]\right\}. \tag{26.243}$$

The limiting stress value, obtained as R or ρ increases indefinitely, is

$$\sigma_{\mathrm{cr}} = \frac{2Y}{3} \left\{1 - \ln\left[3(1-\nu)\frac{Y}{E}\right]\right\}. \tag{26.244}$$

The unlimited void growth under this level of stress is referred to as an unstable cavitation. This stress level is also the critical (bifurcation) stress at which a nonhomogeneous deformation bifurcates from the homogeneous by the sudden void formation.

26.19.2 Incompressible Linearly Hardening Material

If the material is both elastically and plastically incompressible, we have $\det \mathbf{F} = 1$ in (26.228), and

$$r^3 - r_0^3 = R^3 - R_0^3 = \rho^3 - \rho_0^3 = \frac{3Y}{2E} \rho^3. \tag{26.245}$$

The deformation gradient tensor is in this case

$$\mathbf{F} = \begin{bmatrix} \dfrac{r_0^2}{r^2} & 0 & 0 \\ 0 & \dfrac{r}{r_0} & 0 \\ 0 & 0 & \dfrac{r}{r_0} \end{bmatrix}. \tag{26.246}$$

26.19. Plastic Void Growth

By using the multiplicative decomposition of the deformation gradient into its elastic and plastic parts, we can write

$$\mathbf{F} = \mathbf{F}^e \cdot \mathbf{F}^p = \begin{bmatrix} \frac{r_p^2}{r^2} & 0 & 0 \\ 0 & \frac{r}{r_p} & 0 \\ 0 & 0 & \frac{r}{r_p} \end{bmatrix} \begin{bmatrix} \frac{r_0^2}{r_p^2} & 0 & 0 \\ 0 & \frac{r_p}{r_0} & 0 \\ 0 & 0 & \frac{r_p}{r_0} \end{bmatrix}. \tag{26.247}$$

The logarithmic strain

$$\mathbf{E} = \begin{bmatrix} -2\ln\frac{r}{r_0} & 0 & 0 \\ 0 & \ln\frac{r}{r_0} & 0 \\ 0 & 0 & \ln\frac{r}{r_0} \end{bmatrix} \tag{26.248}$$

can be additively decomposed into its elastic and plastic parts ($\mathbf{E} = \mathbf{E}^e + \mathbf{E}^p$) as

$$\mathbf{E} = \begin{bmatrix} -2\ln\frac{r}{r_p} & 0 & 0 \\ 0 & \ln\frac{r}{r_p} & 0 \\ 0 & 0 & \ln\frac{r}{r_p} \end{bmatrix} + \begin{bmatrix} -2\ln\frac{r_p}{r_0} & 0 & 0 \\ 0 & \ln\frac{r_p}{r_0} & 0 \\ 0 & 0 & \ln\frac{r_p}{r_0} \end{bmatrix}. \tag{26.249}$$

For the proportional loading, the generalized plastic strain is

$$\bar{E}^p = \left(\frac{2}{3}\mathbf{E}^p : \mathbf{E}^p\right)^{1/2} = 2\ln\frac{r_p}{r_0}. \tag{26.250}$$

If the material is linearly hardening with respect to this strain measure, the equivalent yield stress at an arbitrary stage of deformation can be expressed as

$$\sigma_{\theta\theta} - \sigma_{rr} = Y + 2k\ln\frac{r_p}{r_0} = Y + 2k\left(\ln\frac{r}{r_0} - \ln\frac{r}{r_p}\right), \tag{26.251}$$

where k is the hardening modulus. By using Hooke's law for the elastic component of strain,

$$\mathbf{E}^e = \frac{3}{2E}\left[\sigma - \frac{1}{3}(\operatorname{tr}\sigma)\mathbf{I}\right], \tag{26.252}$$

we obtain

$$\ln\frac{r}{r_p} = \frac{1}{2E}(\sigma_{\theta\theta} - \sigma_{rr}). \tag{26.253}$$

The substitution of (26.253) into (26.251) yields

$$\sigma_{\theta\theta} - \sigma_{rr} = \hat{Y} + 2\hat{k}\ln\frac{r}{r_0}, \tag{26.254}$$

with the abbreviations

$$\hat{Y} = \frac{Y}{1 + k/E}, \quad \hat{k} = \frac{k}{1 + k/E}. \tag{26.255}$$

The distinction between the barred and nonbarred quantities is important for strong hardening rates and high values of the ratio k/E. Physically, the stress $\hat{Y}$ and the hardening modulus $\hat{k}$ appear in the relationship between the uniaxial stress and the total strain ($\sigma = \hat{Y} + \hat{k}E$), whereas Y and k appear in the relationship between uniaxial stress and plastic strain ($\sigma = Y + kE_p$).

When (26.254) is introduced into (26.221), the integration gives

$$\sigma_{rr} = 2\hat{Y} \ln \frac{r}{R} + \frac{4\hat{k}}{3} \int_R^r \ln \frac{r^3}{r_0^3} \frac{dr}{r}, \tag{26.256}$$

or, in view of (26.245),

$$\sigma_{rr} = 2\hat{Y} \ln \frac{r}{R} + \frac{4\hat{k}}{9} \sum_{n=1}^{\infty} \frac{1}{n^2} \left[\left(\frac{R^3 - R_0^3}{R^3} \right)^n - \left(\frac{R^3 - R_0^3}{r^3} \right)^n \right]. \tag{26.257}$$

Evaluating at $r = \rho$, and equating the result with $\sigma_{rr}(\rho) = \sigma - 2Y/3$, following from (26.234), provides an expression for the applied stress σ in terms of the plastic zone radius ρ. This is

$$\sigma = \frac{2}{3} Y + 2\hat{Y} \ln \frac{\rho}{R} + \frac{4\hat{k}}{9} \sum_{n=1}^{\infty} \frac{1}{n^2} \left[\left(\frac{R^3 - R_0^3}{R^3} \right)^n - \left(\frac{3Y}{2E} \right)^n \right]. \tag{26.258}$$

The radius R of the expanded void is given in terms of ρ by (26.245). If the plastic zone surrounding the void has spread to a large radius ρ, we have $R^3 \gg R_0^3$, and by neglecting higher than linear terms in $3Y/2E$, we obtain

$$\sigma_{cr} = \frac{2Y/3}{1 + k/E} \left(1 + \ln \frac{2E}{3Y} + \frac{\pi^2}{9} \frac{k}{Y} \right). \tag{26.259}$$

This is the critical stress for cavitation instability in an incompressible elastic linearly hardening material. If the ratio $k/E \ll 1$, the above result coincides with the result of Bishop, Hill, and Mott (1945) for a pressurized spherical cavity in an infinite solid. This is so because the solutions for a pressurized spherical void and a spherical void under remote triaxial tension differ by a state of uniform spherical stress throughout the medium, which for an incompressible material has no effect on the deformation field. If $k = 0$ is substituted in (26.259), and $\nu = 1/2$ in (26.244), the two expressions give the critical stress for cavitation instability in an incompressible elastic-ideally plastic material.

A similar analysis for the cylindrical void (see Problem 26.8 of Chapter 34) yields an expression for the applied biaxial stress σ in terms of the plastic zone radius ρ,

$$\sigma = \frac{Y}{\sqrt{3}} + \hat{Y} \ln \frac{\rho}{R} + \frac{\hat{k}}{4} \sum_{n=1}^{\infty} \frac{1}{n^2} \left[\left(\frac{R^2 - R_0^2}{R^2} \right)^n - \left(\frac{\sqrt{3}Y}{E} \right)^n \right]. \tag{26.260}$$

The radius of the expanded void R is given in terms of ρ by

$$R^2 = R_0^2 + \frac{\sqrt{3}Y}{E} \rho^2. \tag{26.261}$$

If the plastic zone around the void has spread to a large extent, we have

$$\sigma_{\mathrm{cr}} = \frac{Y/\sqrt{3}}{1+k/E}\left(1 + \ln\frac{E}{\sqrt{3}Y} + \frac{\pi^2}{18}\frac{\sqrt{3}k}{Y}\right), \qquad (26.262)$$

which is the critical stress for a cylindrical cavitation instability in an incompressible elastic linearly hardening material.

26.20 Suggested Reading

Chakrabarty, J. (1987), *Theory of Plasticity*, McGraw-Hill, New York.

Chen, W. F., and Han, D. J. (1988), *Plasticity for Structural Engineers*, Springer-Verlag, New York.

Cristescu, N. (1967), *Dynamic Plasticity*, North-Holland, Amsterdam.

Cristescu, N., and Suliciu, I. (1982), *Viscoplasticity*, Martinus Nijhoff, The Hague.

Desai, C. S., and Siriwardane, H. J. (1984), *Constitutive Laws for Engineering Materials – With Emphasis on Geological Materials*, Prentice Hall, Englewood Cliffs, New Jersey.

Han, W., and Reddy, B. D. (1999), *Plasticity – Mathematical Theory and Numerical Analysis*, Springer-Verlag, New York.

Hill, R. (1950), *The Mathematical Theory of Plasticity*, Oxford University Press, London.

Jaeger, J. C., and Cook, N. G. W. (1976), *Fundamentals of Rock Mechanics*, Chapman and Hall, London.

Johnson, W., and Mellor, P. B. (1973), *Engineering Plasticity*, Van Nostrand Reinhold, London.

Kachanov, L. M. (1971), *Foundations of Theory of Plasticity*, North-Holland, Amsterdam.

Khan, A. S., and Huang, S. (1995), *Continuum Theory of Plasticity*, Wiley, New York.

Lubarda, V. A. (2002), *Elastoplasticity Theory*, CRC Press, Boca Raton, Florida.

Lubliner, J. (1990), *Plasticity Theory*, Macmillan, New York.

Martin, J. B. (1975), *Plasticity: Fundamentals and General Results*, MIT Press, Cambridge, Massachusetts.

Maugin, G. A. (1992), *The Thermomechanics of Plasticity and Fracture*, Cambridge University Press, Cambridge.

Nemat-Nasser, S. (2004), *Plasticity: A Treatise on Finite Deformation of Heterogeneous Inelastic Materials*, Cambridge University Press, New York.

Salençon, J. (1977), *Application of the Theory of Plasticity in Soil Mechanics*, Wiley, Chichester, UK.

Simo, J. C., and Hughes, T. J. R. (1998), *Computational Inelasticity*, Springer-Verlag, New York.

Temam, R. (1985), *Mathematical Problems in Plasticity*, Gauthier-Villars, Paris.

Yang, W., and Lee, W. B. (1993), *Mesoplasticity and Its Applications*, Springer-Verlag, Berlin.

Zyczkowski, M. (1981), *Combined Loadings in the Theory of Plasticity*, PWN, Polish Scientific Publishers, Warszawa.

27 Micromechanics of Crystallographic Slip

Fundamental concepts concerning the micromechanics of crystalline plasticity are reviewed in this chapter. An overview of deformation mechanisms is given for crystalline materials that possess grain sizes that are said to be "traditional," *i.e.*, larger than about $2\,\mu$m in diameter. Some brief comments are made about the trends in deformation mechanisms when the grain sizes are much below this range (nanograins).

27.1 Early Observations

In a series of articles published between 1898 and 1900 Ewing and Rosenhain summarized their metallographic studies of deformed polycrystalline metals. The conclusion they reached concerning the mechanisms of plastic deformation provided a remarkably accurate picture of crystalline plasticity. Figure 27.1 is a schematic diagram, including some surrounding text, taken from their 1900 overview article. Figure 27.2 is one of their many excellent optical micrographs of deformed polycrystalline metals; the particular micrograph in Fig. 27.2 is of polycrystalline lead. They identified the steps *a-e* in Fig. 27.1 as "slip-steps" caused by the emergence of "slip bands," which formed along crystallographic planes, at the specimen surfaces (thereby coining these two well-known phrases).

Traces of the crystalline slip planes were indicated by the dashed lines. The line labeled C was indicated by them to be a grain boundary separating two grains; the grains, they concluded, were *crystals* with a more or less homogeneous crystallographic orientation. Slip steps corresponding to the diagram of Fig. 27.1 are clearly visible in the micrograph of Fig. 27.2. Ewing and Rosenhain had not only concluded from their slip-line studies that metals and alloys were crystalline and composed of aggregates of crystallites (*i.e.*, grains), but also that plastic deformation took place by simple shearing caused by the sliding of only certain families of crystal planes over each other in certain crystallographic directions lying in the planes. They also noted that certain metals deformed by "twinning" in addition to slip, which is also a crystallographic phenomenon.

Ewing and Rosenhain's (1900) early slip-line studies further indicated that the particular crystalline structure of the metals were preserved during plastic straining. This view was consistent with their conclusion that the simple shearing process of plastic flow involved only *crystallographic planes* sliding in *crystallographic directions*, but included the further assumption that slip progressed in whole multiples of lattice spacings so that during slip

27.1. Early Observations

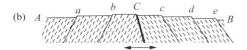

Figure 27.1. Schematic diagram taken from Ewing and Rosenhain (1900) indicating slip steps on the surface of plastically deformed metal: (a) before straining; (b) after straining. Slip was concluded to occur along crystallographic plane. Ewing and Rosenhain say, "The diagram, Fig. 27.1, is intended to represent a section through the upper part of two contiguous surface grains, having cleavage or gliding places as indicated by the dotted lines, AB being a portion of the polished surface, C being the junction between the two grains." "When the metal is strained beyond its elastic limit, as say by a pull in the direction of the arrows, yielding takes place by finite amounts of slips at a limited number of places, in the manner shown at a, b, c, d, e. This exposes short portions of inclined cleavage or gliding surfaces, and when viewed in the microscope under normally incident light these surfaces appear black because they return no light to the microscope. They consequently show as dark lines or narrow bands extending over the polished surface in directions which depend on the intersection of the polished surface with the surfaces of slip."

atoms were transported to equivalent lattice sites within the crystal structure. They recognized, however, that grains deformed inhomogeneously, *i.e.*, only certain planes in the family of possible slip planes underwent slip, and this they attributed to a random distribution of microscopic imperfections of a nonspecified type that triggered the slipping process. They further argued that slip would disrupt these imperfections, which in turn would make it more difficult to activate further slip and thus cause strain hardening.

A remarkable aspect of these early studies of metal plasticity is that it was not until 1912 that Von Laue first diffracted X rays from copper sulfate crystals and not until 1913 that W. H. Bragg and W. L. Bragg made the first crystal structure determination for ionic crystals (Bragg and Bragg, 1933). In 1919 Hull published his results on the crystal structures of

Figure 27.2. Optical micrograph of deformed polycrystalline lead taken from Ewing and Rosenhain (1900).

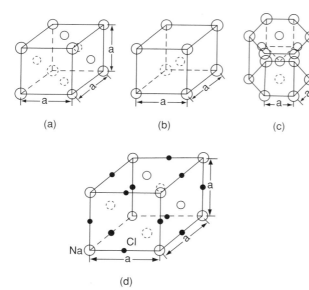

Figure 27.3. Unit cells for four common crystal structures: (a) fcc, (b) bcc, (c) hcp, and (d) NaCl type (from Asaro, 1983b).

various common metals that provided proof and documentation of the crystalline structure of metals. As examples, Figs. 27.3 show the "unit cells" of the face-centered-cubic (fcc), body-centered-cubic (bcc), and hexagonal-closed-packed (hcp) crystal structures. Aluminum, copper, nickel, and gold, along with γ-iron (austenite), are fcc crystals; niobium, molybdenum, and α-iron (ferrite) are bcc; whereas zinc, magnesium, and cadmium are hcp. As a fourth example, Fig. 27.3d shows the crystal structure of a typical ionic crystal such as NaCl or LiF. It is easy to confirm that this "NaCl crystal structure" is based on the fcc point lattice but that each lattice site such as shown in Fig. 27.3a has four atoms positions $(0, 1/2a, 0)$, $(1/2a, 0, 0)$, and $(0, 0, 1/2a)$ with respect to it.

A good deal of the present quantitative understanding of plastic deformation in crystalline materials is due to Taylor and coworkers, in particular to Taylor and Elam (1923, 1925). Their pioneering experiments carried out in the 1920s again firmly established the crystallographic nature of slip, but then with the aid of X-ray diffraction. They made a detailed study of aluminum single crystals and interpreted the kinematics of deformation in terms of the crystallography. For fcc aluminum single crystals they identified the slip planes as the family of octahedral {111} planes and the slip directions as the particular <110> type directions lying in the {111} planes. Figure 27.4 illustrates the (111) and the [10$\bar{1}$] direction, one of the three crystallographically identical <110> type directions lying

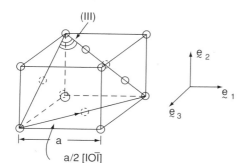

Figure 27.4. Unit cell and one of the 12 crystallographically equivalent slip systems for fcc crystals (from Asaro, 1983b).

27.1. Early Observations

in the plane. Note that the Ewing and Rosenhain (1900) conclusion that crystal structure is preserved by shearing requires that slips occur in units of "perfect lattice vectors" along the 110 directions. In Fig. 27.4 the shortest of these along $[10\bar{1}]$ is a $a/2[10\bar{1}]$, as indicated. Furthermore, as there are 4 independent (but crystallographically identical) {111} planes and 3 <110> type directions in each plane, there are 12 distinct "slip systems" of the type {111} <110> in fcc crystals.

Taylor and Elam (1923, 1925) had also established for their aluminum crystals what is now commonly referred to as the *Schmid law* (1924) of a critical resolved shear stress for plastic yielding. They phrased it as follows: "Of the twelve crystallographically similar possible modes of shearing, the one for which the component of shear-stress in the direction was greatest was the one which actually occurred" (1925, p. 28). Taylor and Elam had therefore identified the component of shear stress resolved in the slip plane and in the direction of slip, the "resolved shear stress," defined, as above, as the combination of slip plane and slip direction. We shall adopt this nomenclature and refer to a particular set of *active* slip systems as the "slip mode."

During this same period (1920–1930) research in Germany on crystalline plasticity was conducted by Polanyi, Masing, Schmid, and Orowan, among others. They had reached many of the conclusions during their studies of zinc (with a hcp crystal structure) that Taylor and Elam had with fcc aluminum. In particular, in 1922 Polanyi had not only observed crystallographic slip in zinc but had also noted in polycrystalline zinc that grains with initially random crystallographic orientations tended to assume preferred orientations after finite straining, *i.e.*, they developed texture. In 1924 Schmid, using data on zinc single crystals subject to tension, suggested that plastic yield would begin on a slip system when the resolved shear stress reached a critical value, independent of the orientation of the tensile axis and thus of other components of stress resolved on the lattice. This was a clear statement of the Schmid law. It should be pointed out, however, that although experiments conducted in uniaxial tension or compression often yield an approximate confirmation of Schmid's law, deviations from it are likely and have been found, as later discussion will allow.

The early work of the 1920s included some important observations of strain-hardening behavior in crystalline plasticity. Taylor and Elam (1923, 1925), in particular, noted that the individual slip systems hardened with strain and that slip on one slip system hardens other slip systems, even if the latter are not active. This is known as "latent hardening," and because it plays an important role in crystal mechanics the Taylor and Elam (1925) results will now be briefly discussed.

Figure 27.5a illustrates a single crystal in tension, oriented for single slip, *i.e.*, the resolved shear stress is highest on the slip system $(\bar{1}11)[110]$. The vectors **s** and **m** are unit vectors lying in the [110] and $[\bar{1}11]$ directions, respectively, so that (**s**, **m**) now defines the slip system that is denoted the "primary" slip system. For the present the crystal is modeled as rigid-plastic, because the points being illustrated are not substantially affected by elasticity. Figure 27.5d is Taylor and Elam's stereographic plot of the orientation of the tensile axis, with respect to the crystal axes, as a function of the extensional strain; the point marked 0 represents the initial orientation. As the crystal slips, the orientation of the tensile axis AB changes with respect to the crystal axes and therefore to the slip direction **s** and the slip plane normal **m**, as shown in Fig. 27.5b. In fact, since most tensile machines constrain the loading axis to remain fixed in orientation relative to the laboratory (*e.g.*, vertical in Fig. 27.5), the crystal undergoes a rigid rotation, shown in Fig. 27.5e. The result is that the

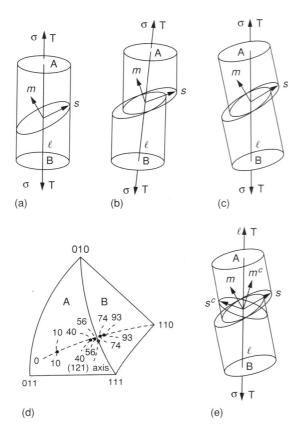

Figure 27.5. (a) Schematic diagrams for a crystal undergoing single slip in tension. The material fiber l and AB are considered to remain vertical, which causes a rigid lattice rotation (b) to (c). (d) Taylor and Elam's (1925) result for the change in crystallographic orientation of the tensile axis during deformation. (e) The relative orientation of the so-called conjugate slip system (from Asaro, 1983b).

slip direction rotates toward the tensile axis about an axis **r** orthogonal to both, where the unit vector **r** is given by

$$\mathbf{r} = \mathbf{s} \times \mathbf{T}/[1 - (\mathbf{s} \cdot \mathbf{T})^2]^{1/2}. \tag{27.1}$$

On the stereographic projection the pole of the tensile axis appears to rotate along the great circle common to **s** and **T**. Now consider increments of extension, shear, and rotation starting at some current stage taken as reference, as indicated in Fig. 27.5e. The slip system that is said to be "conjugate" to the primary system, $(11\bar{1})[011]$, is also indicated. Conjugacy refers to the fact that single slip on the system $(11\bar{1})[011]$ will induce rotations during tensile straining that lead to large resolved shear stresses and inevitable slip on the system $(11\bar{1})[011]$. The rate of rotation β about **r** is easily shown to be

$$\dot{\beta} = \dot{\gamma}(\mathbf{T} \cdot \mathbf{m})[1 - (\mathbf{s} \cdot \mathbf{T})^2]^{1/2}, \tag{27.2}$$

where $\dot{\gamma}$ is the plastic shearing rate on the primary system. The resolved shear stress τ is given by

$$\tau = \sigma(\mathbf{m} \cdot \mathbf{T})(\mathbf{s} \cdot \mathbf{T}) = S\sigma, \tag{27.3}$$

with a similar relation holding for the conjugate system. In (27.3), S is defined as the "Schmid factor." Taylor and Elam (1925) noted that if the strain-hardening rate on the active slip system (call this the "self-hardening rate") were equal to the latent-hardening rate

27.1. Early Observations

of the conjugate slip system, conjugate slip would become active when the two resolved shear stresses were equal. This occurs when the tensile axis rotates to any point on the [111]-[010] symmetry boundary in Fig. 27.5d. The dotted arc in Fig. 27.5d is the great circle on the sterographic projection whose pole is **r** and is the path of rotation for the tensile axis. The "tick" marks indicate the crystallographic orientations of the tensile axis expected from pure single slip, and the dots the orientations measured by Taylor and Elam (1925) at the specific extensional strains (in percentages). Until the tensile axis approaches the symmetry boundary, the predictions of single slip are closely matched by experiment. The X-ray measurements indicate that the tensile axis "overshoots" the symmetry boundary, which leads to a larger resolved shear stress on the conjugate, *i.e.*, latent system. The measurements also indicate that the rotations are nonetheless much less than expected from single slip, which means that conjugate slip does occur, but at a lower rate than primary slip. If the double mode of slip were symmetric once the symmetry boundary were reached, the tensile axis would rotate to the [121] pole and remain there. In this case the rotations caused by primary and conjugate slip exactly cancel. Such behavior was observed in some of their other tests. Taylor and Elam (1925) had thus demonstrated two very important aspects of crystal strain hardening: (1) slip systems hardened on other systems (whether they themselves are active or not), and (2) this latent hardening is at least comparable in magnitude to self-hardening. The observations of "overshooting" indicated that latent-hardening rates are often somewhat larger than self-hardening rates, so that a slightly larger shear stress is required on the previously latent system to activate it. Taylor and Elam summarized the hardening behavior as follows: "It is found that though the double slipping does, in fact, begin when the two planes get to the position in which they make equal angles with the axis, the rate of slipping on the original slip-plane is sometimes greater than it is on the new one" (1925, p. 29). They went on to note that "the process cannot be followed very far, however, because the specimen usually breaks when only a comparatively small amount of double slipping has occurred" (1925, p. 29). As discussed later, crystals typically undergo necking and intense localized shearing after double slip begins.

In passing, we note an interesting consequence of the kinematics of crystalline slip, again using the rigid plastic crystal model of Fig. 27.5 for a crystal in tension. According to the Schmid rule, the resolved shear stress τ must remain at the critical yield value τ_c for slip to continue; τ_c increases with shear strain at a rate given by $\dot{\tau}_c = h\dot{\gamma}$, where h is the current strain-hardening (self-hardening) rate of the active slip system. Differentiating (27.3) with respect to time, evaluating $\dot{\mathbf{s}}$ and $\dot{\mathbf{m}}$ by noting that

$$\dot{\mathbf{s}} = \Omega^* \cdot \mathbf{s}, \quad \dot{\mathbf{m}} = \Omega^* \cdot \mathbf{m}, \tag{27.4}$$

where

$$\Omega^* = -(\mathbf{sT} - \mathbf{Ts})\dot{\gamma}\mathbf{T} \cdot \mathbf{m} \tag{27.5}$$

is the rigid lattice spin rate, and setting the result equal to $h\dot{\gamma}$ yields

$$\dot{\sigma} = \left[\frac{h}{\cos^2\phi \cos^2\theta} + \frac{\sigma \cos(2\phi)}{\cos^2\phi} \right] \dot{e}. \tag{27.6}$$

Here $\dot{e}$ is the current rate of extension along the tensile direction. This rate is taken from the current state as the reference state of strain. The first term, due to strain hardening, leads to an increase in true tensile stress with extension (if $h > 0$), whereas the second

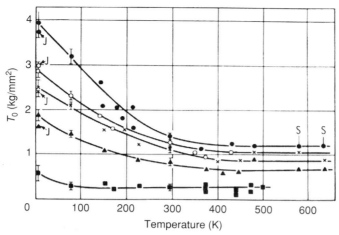

Figure 27.6. Data for the critical resolved shear stress for initial yield *vs.* temperature for various materials. ● 70:30 brass; × 90:10; ○ 80:20; ▲ 95:5; ■ Cu. Reproduced with permission from Mitchell (1964).

term is due to the change in τ, at fixed σ, caused by rotation of the lattice with respect to the fixed tensile axis. If $\cos 2\phi < 0$ and $h/\sigma < |\cos 2\phi| \cos^2 \theta$, the instantaneous modulus governing extension is negative, and the crystal softens. This softening is associated with purely geometrical effects and not material strain softening, and for this reason is called *geometrical softening*. Evidently, if lattice rotations that cause geometrical softening occurred locally within the crystal's gauge section, they would promote nonuniform and perhaps localized deformation. This is true for single crystals, as well as for individual grains of polycrystals. We show later that geometrical softening does indeed play a vital role in the phenomenology of crystalline slip. In fact, this effect will lead to the formation of zones, later called *patches*, that are the precursors to the formation of substructures.

Thus, by 1930 much of the macroscopic phenomenology of crystalline plasticity, aside from strain rate effects, had been documented. The fundamental connection was observed between the resolved shear stress and notions of self-hardening and latent hardening of slip systems. Temperature had long been known to have an important influence on strength, as shown in Fig. 27.6 which is taken from Mitchell's (1964) review. A tenable micromechanistic theory, however, was missing. For example, in 1926 Frenkel made a simple but plausible calculation of the theoretical shear strength of a perfect crystal. The result indicated that the shear stress required to slide an entire unit area of crystal plane over the overlaying plane by one lattice spacing is not less than about 1/10 the elastic shear modulus, a value several orders of magnitude higher than observed yield strengths (see Fig. 27.6). As Orowan (1963) pointed out much later, it was also realized that Becker's thermal fluctuation theory could not resolve this large discrepancy. The satisfactory resolution of this problem led to the theoretical discovery of crystal dislocation in 1934.

27.2 Dislocations

Figures 27.7–27.9 were all published in 1934 and they describe essentially the same crystal defect known as an "edge" dislocation. In all three cases the dislocation is drawn in a

27.2. Dislocations

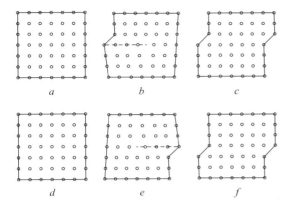

Figure 27.7. Taylor's (1934) model picture for a crystal dislocation. Glide of the dislocation causes material on either side of the slip plane to move by one lattice spacing, a, b, c, positive dislocation; d, e, f, negative dislocation.

simple cubic crystal and was used by Taylor (1934), Orowan (1934), and Polanyi (1934) to explain the micromechanics of slip. As is easily appreciated from the Taylor model in Fig. 27.7, slip is caused by the glide of the dislocation across the slip plane one lattice spacing at a time. The result is to displace the material on either side of the plane by the unit lattice spacing. Taylor (1934) argued that the shear stress required to cause incremental dislocation motion would be very low and thus propagation of such defects would result in shear strengths consistent with those observed. Crystal symmetry is preserved after glide and, in fact, even if the dislocation is trapped within the crystal, as in Fig. 27.7b or 27.7e, because the disregistry in the crystal structure is highly localized in the dislocation "core" region.

The precise specification of the dislocation involves both the displacement vector of the material and the dislocation line, including its sense; in Fig. 27.7 the dislocation line is directed into the plane and the displacement vector is orthogonal to the dislocation line. A second type of dislocation involves a displacement vector that is collinear with the dislocation line. These two dislocations are known as "edge" and "screw" dislocations, respectively, and actually coexist as shown in Fig. 27.10. Figure 27.10 illustrates that a glide dislocation can be formed within a continuum by making a cut over the surface, S, bounded by the line C, and displacing material on either side of the cut by the vector $\mathbf{b}$. The cut is then rejoined so that the material across the cut surface has been permanently slipped by $\mathbf{b}$; $\mathbf{t}$ is the unit tangent vector to C and thus the segments of C for which $\mathbf{t} \cdot \mathbf{b} = 0$ are "edge" segments and those for which $\mathbf{t} \cdot \mathbf{b}/b = \pm 1$ are "screw" segments. The vector $\mathbf{b}$ is known as the Burgers vector, and for crystal dislocations it must be one of the perfect lattice vectors, if slip is to preserve crystal symmetry. The Schmid rule can now be reinterpreted in terms of a critical force required either to move or to generate dislocations. This force if defined

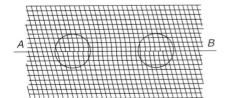

Figure 27.8. Orowan's (1934) model for a crystal dislocation.

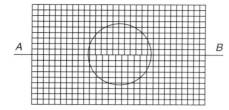

Figure 27.9. Polanyi's (1934) model for a crystal dislocation.

to be work conjugate to the gliding motion and hence is equivalent to the resolved shear stress multiplied by the magnitude of **b**.

Taylor's 1934 analysis also provided basis for a quantitative understanding of strain hardening. He recognized that dislocations were sources of internal elastic strain and stress and that dislocations would interact with each other through their mutual stress fields. Figure 27.11 shows two of Taylor's (1934) interaction problems involving dislocations of opposite and similar signs. The stress fields for these dislocations had already been calculated in linear isotropic elastic media by Timpe (1905), Volterra (1907), and Love (1944). The interactions in Fig. 27.11 can be either attractive or repulsive, depending on the relative sign of dislocations, but in either case the maximum resolved shear stress acting on the dislocations required to slip them by each other occurs at $|x| = h$ and is equal to $Gb/2\pi h$, where G is the elastic shear modulus. Thus the shear stress required to move dislocations within a "substructure" of other dislocations is considerably larger than that required to move an isolated dislocation. Furthermore, once dislocations and slip have been generated on a plane, it becomes easier to generate slip on planes further removed from those currently active. However, this leads to an increase in dislocation density, a decrease in the average spacing between dislocations, and hence to smaller values of h. This in turn leads to an increase in the value of the "passing stress."

One model picture envisioned by Taylor for the dislocation arrangement is shown in Fig. 27.12. To see how this leads to a specific strain-hardening relation we consider the following dimensional arguments. Taylor assumed that dislocations would be generated at one surface of the crystal and would move a distance x along the slip plane; the maximum value x may have is L, which is either the dimension of the crystal or the distance to a boundary. The average dislocation moves by $L/2$. The average spacing between active slip planes in Fig. 27.12 is $d/2$. The spacing between dislocations is a, so the average "dislocation density" ρ is $2/ad$; the shear strain accumulated in this substructure, γ, is $L\frac{2}{d}\frac{1}{2}b/L = \rho Lb/2 = Lb/ad$. Dislocation density is often measured as the number of dislocation lines penetrating a unit area of plane, as here, or as the total length of dislocation lines per unit volume. Now since the passing stress is proportional to h^{-1}, and h scales with $\rho^{-1/2}$ if the dislocation distribution is regular, it follows from dimensional considerations alone that

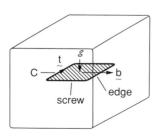

Figure 27.10. A dislocation loop in continuum is produced by first making a cut on surface S and then displacing material across it by **b** (from Asaro, 1983b).

27.2. Dislocations

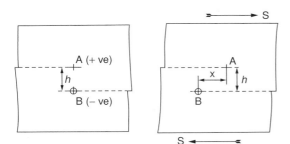

Figure 27.11. Taylor's (1934) model for the elastic interaction of two edge dislocations gliding by each other.

the passing stress is proportional to $\rho^{1/2}$, or actually to $Gb\rho^{1/2}$, and hence to $G(\gamma b/L)^{1/2}$. Finally, if τ_i represents the shear stress required to move an isolated dislocation, the stress-strain law derived above has the form

$$\tau_c = \tau_i + \kappa G(b\gamma/L)^{1/2} = \tau_i + \eta G b \rho^{1/2}, \quad (27.7)$$

where κ and η are constants determined by the geometry of the dislocation distribution.

Although Taylor's model arrangement for the distribution of dislocations is oversimplified, it serves to focus on an important cause of strain hardening – the elastic interaction between dislocations. The dimensional form of his strain-hardening law reflects the experimentally observed fact that, in metals at least, strain-hardening rates generally decrease with strain. Furthermore, the proportionality between the flow stress and the square root of dislocation density is also well documented, especially in fcc metals, although, as we shall see later, this dimensional rule follows from other micromechanical models for dislocation interaction, as well. Finally, we note that the indicated dependence of τ_c on the microstructural dimension L reflects the well-established fact that yield and flow strength increase with decreases in grain (or subgrain) size, *i.e.*, with refinement of the microstructure in general.

27.2.1 Some Basic Properties of Dislocations in Crystals

A linear elastic theory of dislocations, notable due to Volterra (1905), was available at the time Taylor published his 1934 analysis of dislocation interactions. This theory has since been extended to include elastic anisotropy and a variety of techniques for solving for the elastic fields of complex arrays of dislocations (see, for example, Hirth and Lothe 1982, Asaro *et al.*, 1973, and Asaro and Barnett, 1976, and Chapter 18 of this book). For infinitely long and straight dislocations, for example, it follows from this linear theory that the elastic energy per unit length of dislocation line can be expressed as (see Section 21.5)

$$E = \mathcal{K}_{ij} b_i b_j \ln(R/r_0), \quad (27.8)$$

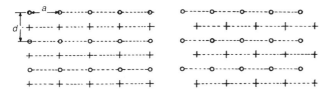

Figure 27.12. Taylor's (1934) idealized arrangement for the dislocation substructure in a finitely deformed crystal.

Table 27.1. *Slip systems for common crystal types*

Crystal structure	Burgers vector	Slip plane (at 20° C)
fcc	$a/2$ <110>	{111}
bcc	$a/2$ <111>	{110} and {112}
NaCl	$a/2$ <110>	{110}

where $\mathcal{K}$ is an energy factor that depends on the direction of the dislocation line with respect to the crystal axes, and R and r_0 are outer and inner cutoff radii (not specified in the linear theory). The quadratic form in the components of the Burgers vector indicated that stable dislocations take on the shortest Burgers vector possible – the energies of crystal bonding, along with the evidence that slip does not affect crystal structure, indicate that **b** is a perfect lattice vector. Thus, with reference to Fig. 27.4, $\mathbf{b} = a/2$ <110> for fcc crystals. The slip systems for fcc metals crystals are then of the type $\{111\} a/2$ <110>, where a is the lattice parameter shown in Fig. 27.3a. Table 27.1 indicates some common slip systems for the cubic crystal types illustrated in Fig. 27.3.

Although the Burgers vector of a crystal dislocation is a perfect lattice vector, it is possible for dislocations to "dissociate" and form "partial dislocations," as illustrated in Fig. 27.13 for a glide dislocation in a fcc crystal. The extent of dissociation depends not only on the crystal geometry, but also on the chemistry. A simple analysis, using linear isotropic elasticity, follows.

The glide force acting on a segment of dislocation line is by definition the force that is work conjugate to the gliding motion. Then, if τ is the shear stress resolved in the plane of glide at the dislocation line and in the direction of the Burgers vector, the force is τb per unit length of dislocation line. A more precise definition of τ, accounting for finite lattice elasticity, is given in the chapter to follow. In general, the force per unit length, due to stress σ, acting on a segment of dislocation with unit tangent **t** is (Peach and Koehler, 1950)

$$\mathbf{f} = \mathbf{t} \times (\mathbf{b} \cdot \boldsymbol{\sigma}). \tag{27.9}$$

For a general discussion of forces on elastic singularities the reader is referred to Eshelby (1951, 1958) and Chapter 21 of this book. It can be shown that the in-plane traction acting on any plane containing an infinitely long straight dislocation line is given by $\mathbf{T} = 2\mathcal{K} \cdot \mathbf{b}/d$, where d is the distance from the line that is taken to be positive if it falls to the right of the dislocation line (see, for example, Barnett and Asaro, 1972). With reference to Fig. 27.14, x_1, x_2, x_3 are mutually orthogonal directions with x_3 along the dislocation line. The point in question lies along x_1, which is to the right. In this coordinate frame, the energy factor $\mathcal{K}$ is diagonal, with $\mathcal{K}_{11} = \mathcal{K}_{22} = G/[4\pi(1-\nu)]$ and $\mathcal{K}_{33} = G/4\pi$; ν is Poisson's ratio. It is now easy to show that any two parallel dislocations *repel* each other on their common plane

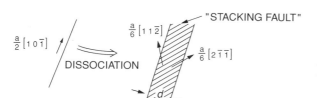

Figure 27.13. Screw dislocation dissociating in an fcc crystal. Partial dislocations are produced by the dissociation of a perfect dislocation (from Asaro, 1983b).

27.2. Dislocations

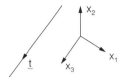

Figure 27.14. Coordinate frame used to reference the elements of the **K** matrix (from Asaro, 1983b).

with a force given by $2\mathbf{b}^{(1)} \cdot \mathcal{K} \cdot \mathbf{b}^{(2)}/d$; if this quantity is negative, they attract each other. However, as the partial dislocations form according to the reaction

$$a/2[10\bar{1}] = a/6[11\bar{2}] + a/6[2\bar{1}\bar{1}], \tag{27.10}$$

a "faulted" region is created across which the equilibrium atomic stacking sequence is disrupted. It is convenient to express the energy associated with this "stacking fault" as a surface tension, *i.e.*, the reversible isothermal work required to create a unit area of fault, Γ. Hirth and Lothe (1982) quoted some values for this as follows: aluminum, 20×10^{-6} J/cm²; silver, 1.7×10^{-6} J/cm²; and copper, 7.3×10^{-6} J/cm². The total force, including the fault tension acting between two partial dislocations, per unit length, is

$$2\mathbf{b}^{(1)} \cdot \mathcal{K} \cdot \mathbf{b}^{(2)}/\delta - \Gamma. \tag{27.11}$$

At equilibrium, this force vanishes, which yields

$$\delta_{eq} = [2\mathbf{b}^{(1)} \cdot \mathcal{K} \cdot \mathbf{b}^{(2)}]/\Gamma. \tag{27.12}$$

If the undissociated dislocation in Fig. 27.13 were a perfect screw dislocation, the estimated equilibrium spacings δ_{eq} would be on the order of 0.2 nm, for aluminum and 2 nm, for copper. In practical terms this suggests that the partial dislocations are not "extended" in aluminum, but may well be in copper and almost certainly would be in silver.

An important feature of dissociated, or extended, dislocation is that they may not readily undergo such micromechanical processes as *cross-slip*. Figure 27.15 is an illustration of such a process in a fcc crystal adopted from Asaro and Rice (1977). Dislocations in fcc crystals may glide in any {111} plane containing the Burgers vector and the dislocation line. However, the individual partial dislocations do not lie in the cross-slip plane and so the extended dislocation on the primary slip system (**s**, **m**) must first develop a constricted segment that can bow out on the cross-slip plane. Once the dislocation has bowed by a critical amount on the cross-slip plane, it may then continue to glide on an adjacent primary plane. Micromechanical processes such as cross-slip lead to deviations from the Schmid law, because stress components other than the resolved shear stress τ_{ms} affect the constriction and bowing process.

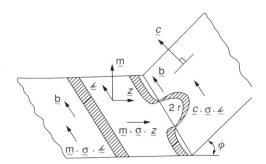

Figure 27.15. Idealized model for the cross-slip of an extended screw dislocation (from Asaro, 1983b).

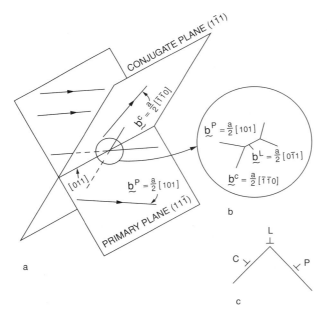

Figure 27.16. Interaction between dislocations in a primary-conjugate relationship (from Asaro, 1983b).

27.2.2 Strain Hardening, Dislocation Interactions, and Dislocation Multiplication

Taylor's (1934) original analysis of strain hardening focused on the elastic interaction of essentially straight nonintersecting dislocations belonging to one slip system. An entire different and revealing picture of dislocation substructure emerges when interactions between dislocations belonging to two or more systems are accounted for. Figure 27.16 illustrates an interaction for fcc crystals proposed by Lomer (1951) between dislocations having a primary-conjugate relationship. Suppose that the conjugate slip system, $(1\bar{1}1)a/2[110]$, is not active but that a certain initial "grown-in" density of its dislocations exists in the crystal. When the primary system, taken to be $(11\bar{1})a/2[101]$, yields, its dislocations necessarily intersect conjugate dislocations along the line common to both slip planes, [011]. At the juncture the two dislocations merge according to the reaction

$$a/2[101] + a/2[\bar{1}\bar{1}0] = a/2[0\bar{1}1], \tag{27.13}$$

where $\mathbf{b}^L = a/2[0\bar{1}1]$ is the Burgers vector of the product Lomer dislocation. The basis for the reaction lies in the fact that the elastic energies of all three dislocations involved in reaction (27.13) are equal and so the formation of the Lomer dislocation leads to a new decrease in the total elastic energy. A more rigorous method of analyzing dislocation reactions of this type has been given by Asaro and Hirth (1974). The Lomer dislocation is an edge dislocation, because $\mathbf{t}^L = 1/\sqrt{2}[011]$ and $\mathbf{t}^L \cdot \mathbf{b}^L = 0$, and if it were to glide it would have to do so on the (100) plane containing $\mathbf{t}^L$ and $\mathbf{b}^L$. Since glide on {100} type planes is very difficult in fcc crystals, the Lomer dislocation is immobile and thus impedes further glide of the original primary dislocation. What is further evident is that glide of conjugate dislocations would also be more difficult should the conjugate system become highly stressed. Figure 27.16c shows a view of the two slip planes along the [011] direction. Subsequent primary *and* conjugate dislocations are impeded in their motion, since they are both *repelled* from the Lomer dislocation with a force equal to $Ga^2/[8\pi(1-\nu)d]$. Both

27.2. Dislocations

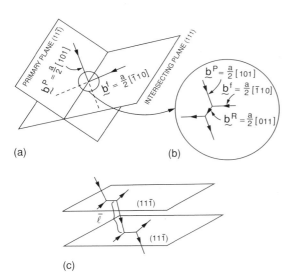

Figure 27.17. (a, b) Interaction between primary dislocations and "forest" dislocations. (c) Formation of "nodes" and discrete dislocation segments (from Asaro, 1983b).

the primary and conjugate slip systems are then hardened in a roughly symmetrical way, which helps explain Taylor and Elam's (1923, 1925) observations of latent hardening.

Another type of dislocation reaction leading to strain hardening and latent hardening is illustrated in Fig. 27.17. The primary slip system and primary dislocations are again taken to be $(11\bar{1})a/2[101]$ and the other system, whose dislocations are referred to as "forest" dislocations, is $(111)a/2[\bar{1}10]$. When the two dislocations meet, they undergo the reaction

$$a/2[101] + a/2[\bar{1}10] = a/2[011], \qquad (27.14)$$

i.e., they react to form a product segment lying in both planes along the [110] direction with the Burgers vector along [011]. The dislocation with Burgers vector $a/2[011]$ lies in the $(11\bar{1})$ plane and so it may glide in the primary plane. However, since the "nodes" shown in Fig. 27.17 are energetically stable, they essentially pin and thus impede the motion of the original primary *and* forest dislocations. If adjacent primary planes are considered, the effect is to produce a three-dimensional network, as suggested in Fig. 27.17c. As a consequence, dislocations are not arranged in the ideal fashion envisioned in the Taylor model of Fig. 27.12, but in a segmented distribution such as suggested by the idealized model in Fig. 27.17c.

The determination of the crystal's strength depends on the resistance of the network to the motion of dislocation *segments*. If we assume for a moment that the nodes are rigid and act as pinning sites, the segments move by bowing out in the manner illustrated in

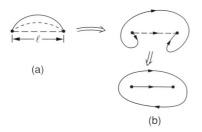

Figure 27.18. (a) Idealized model for the movement of dislocation segment pinned at nodes. (b) Bowing out of the segment to form a dislocation loop (from Asaro, 1983b).

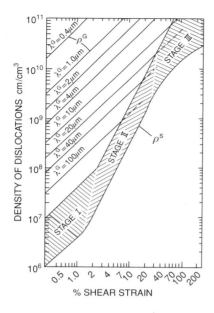

Figure 27.19. Plot showing how the dislocation density increases with shear strain (from Ashby, 1971).

Fig. 27.18a. The simplest analysis of this follows if the dislocation is taken to be a line with average tension $\mathcal{L}$ and if the bowing is assumed to proceed in circular arcs. Then at any stage that is stable, $\tau b = \mathcal{L}/r$, where τ is the resolved shear stress for the dislocation and r is the radius of curvature of the bowing segment. If the shear stress exceeds a critical value $\tau_c > 2\mathcal{L}/b\ell$ corresponding to $r = \ell/2$, the bowing becomes unstable and the loop closes on itself, as shown in Fig. 27.18b, The result is that a new dislocation loop is produced and the original segment restored, whereupon it may bow again; the critical stress required for this is given by $\tau_c = 2\mathcal{L}/b\ell$.

The mechanism just described constitutes one possible source of dislocations and represents a type of source introduced by Frank and Read (1950). In well-annealed crystals dislocation densities can be on the order of 10^6 cm^{-2} and even less in certain cases. However, after modest plastic strains of, say, 10%, this may increase to 10^{10} cm^{-2}, as indicated in Fig. 27.19 taken from Ashby (1971). A dislocation source involving double cross-slip was suggested by Koehler (1952). As shown in Fig. 27.20 (see also Fig. 27.15), once a dislocation segment has cross-slipped onto the cross-slip plane, it may then cross-slip onto a primary slip plane again. This segment may then bow out and operate as a dislocation source. Figure 27.19 indicates how dislocation density ρ depends on shear strain. Figure 27.21 shows the corresponding relationships for shear stress and $\rho^{1/2}$. As mentioned earlier, most models for strain hardening result in such relationships. For example, if the

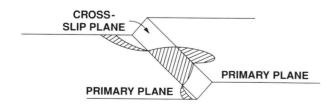

Figure 27.20. Koehler's model of the so-called double cross-slip mechanism for dislocation multiplication (from Asaro, 1983b).

27.3. Other Strengthening Mechanisms

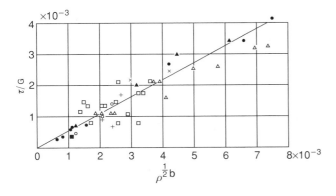

Figure 27.21. Collection of data for single crystals for the flow stress on a slip system vs. the average dislocation density (from Ashby, 1971).

flow stress is determined by the stress τ_c, which is proportional to ℓ^{-1}, then, by dimensional arguments, ℓ^{-1} should be proportional to $\rho^{1/2}$.

27.3 Other Strengthening Mechanisms

Most crystalline materials derive their strength from a combination of the intrinsic mechanisms described previously and the interaction of dislocations with other microstructural features. The yield strength and strain-hardening characteristics of polycrystals depend in an important way on grain size, as shown by the two examples in Fig. 27.22. In both these cases a close correspondence with the well-known Hall–Petch relation (Hall, 1951; Petch, 1953) between yield strength σ_y and average grain diameter d is found, $\sigma_y = \sigma_i + k_y d^{-1/2}$, at least for the range of grain sizes included by the data. Although the Hall–Petch relation bears a superficial resemblance to Taylor's flow stress of (27.7), the physical underpinning of it is different. The yield strength σ_y is determined in large part by the process by which slip is transferred from grains that yield at the smallest plastic strain to the surrounding grains and not solely by the interaction of dislocations and strain-hardening processes occurring in the grain interiors. As pointed out by Embury (1971), the local stress necessary to propagate slip may be determined by the critical conditions for (1) the unpinning of existing dislocations or the operation of dislocation sources in unyielded grains or (2) the creation of dislocations in the grain boundaries that glide into the unyielded grain. Furthermore, as is evident form the Thompson and Baskes (1973) data, the Hall–Petch relation holds, even in the range of grain sizes shown, only for proof stresses defined at plastic strains less than 0.01. The relation breaks down at high strains, which indicates that grain size has a significant influence on strain-hardening behavior, especially at the finer grain sizes. Thompson (1975) correctly pointed out the implications of such grain size dependence regarding the development of continuum polycrystalline models, all of which have to date ignored grain size *per se*. In short, the data indicate that grains of polycrystals, especially grains less than 10^{-2} mm in diameter, do not strain harden as if they were single crystals subjected to comparable strains.

Finally, we note that the development of a segmented network of dislocations, as described in the previous section, typically evolves into the formation of dislocation cells after finite strain. An example of cell structure in single-crystal aluminum, taken from the work of Chiem and Duffy (1981), is shown in Fig. 27.23a. The cell walls are characterized

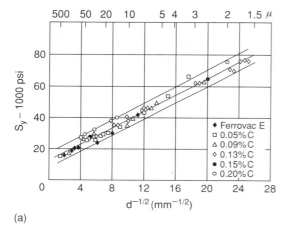

(a)

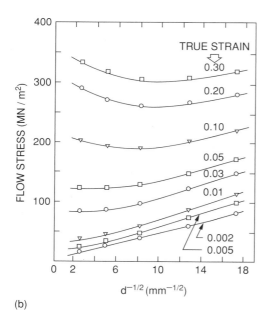

(b)

Figure 27.22. Morrison's data for the dependence of yield strength on grain size. Reproduced with permission from Morrison (1966).

by high dislocation density, as described, for example, by Embury (1971) and Kuhlman-Wilsdorf (1975), and by the existence of dislocations with two or three different Burgers vectors in them. Cell size in turn plays an important role in determining strength, as shown in Fig. 27.23b, which is also taken from Chiem and Duffy (1981).

The present discussion does not cover the wide range of strengthening and strain-hardening phenomena in crystalline materials and the importance of chemistry and microstructure. For example, topics such as solute strengthening and precipitation hardening have not been discussed and the reader is encouraged, in this regard, to consult Kelly and Nicholson (1971). The intent was rather to present a very basic picture that serves both to identify important phenomena and construct continuum constitutive laws which allow for a more precise analysis of them. As it happens, one of the most important of these phenomena is latent hardening, and some of the available data for this are discussed next.

27.4. Measurements of Latent Hardening

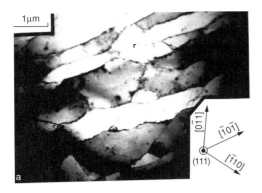

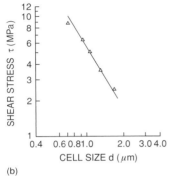

Figure 27.23. (a) Dislocation cell structure in aluminum single crystals and (b) the relationship between flow strength and cell diameter (from Chiem and Duffy, 1981).

27.4 Measurements of Latent Hardening

Latent hardening has been measured in essentially two ways. One method involves deforming large crystals in single slip and the machining specimens aligned for slip on another slip system from them. Yield stresses on previously latent systems are then measured and compared to the flow stress reached on the original, primary, system. The results are typically described by a variation of the "latent-hardening ratio," the ratio of these two flow stresses. A second method involves measurements of lattice rotations in single crystals that occur during tensile straining in single slip, as described previously in connection with Fig. 27.5. The rate of rotation with strain can be measured, and discrepancies with the single-slip predictions can be used to detect yielding on latent systems. The extent of overshoot can also be measured and used as an indication of the magnitude of latent hardening. These two techniques are not equivalent, because they subject the specimen to different strain histories; in the first only one system is active at any stage, whereas in the second at least two systems are simultaneously active at the point where yield stresses are measured for the initially latent system. To describe the experimental results quantitatively, it is helpful to assume for the moment a rate-independent hardening rule of the form

$$d\tau_c^{(\alpha)} = \sum_\beta h_{\alpha\beta} \, d\gamma^{(\beta)}, \tag{27.15}$$

where $\tau_c^{(\alpha)}$ is the current yield strength on the α slip system, and $h_{\alpha\beta}$ are the hardening rates. The off-diagonal terms in the matrix **h** represent latent hardening.

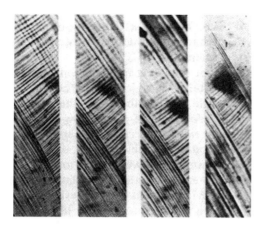

Figure 27.24. Optical micrographs of the gauge section of α-Brass crystals deformed in tension. The crystals were deformed in single slip and the onset of conjugate slip was determined by the appearance of slip lines belonging to the conjugate system. Reproduced with permission from Piercy *et al.* (1955).

In his original analysis of polycrystals, Taylor (1938a, 1938b) assumed that the latent-hardening rates were equal to self-hardening rates, *i.e.*, $h_{\alpha\beta} = h$ for all α and β. He was undoubtedly motivated to assume such isotropic hardening by his experiments with Elam (Taylor and Elam, 1923, 1925) presented above. It is important to note, however, that overshoots of the symmetry positions, where equal shear stresses existed on the conjugate system, of 2-3° were common in those experiments, indicating that slip on the initially latent conjugate system required a slightly larger resolved shear stress than on the active primary system. The Taylor isotropic rule was evidently meant to be approximate, because its background experiments actually indicated that the latent-hardening rate was slightly larger than the self-hardening rate.

A brief review of the measurements made by 1970 was given by Kocks (1970), who concluded that the average ratio of latent-hardening to self-hardening rates is nearly unity for coplanar systems (*i.e.*, systems sharing the same plane) and between 1 and 1.4 for noncoplaner systems. This appears to be a very reasonable approximate range, especially for pure metals with intermediate or high stacking fault energies (*e.g.*, Al and Cu) at finite strains, although the authors reported to variation of latent-hardening ratios with strain to be more complicated than described by Jackson and Basinski (1967) for copper and by Kocks and Brown (1966) for aluminum. Franciosi *et al.* (1980) found that the latent-strengthening ratios first increase from unity to peak values of 1.6-2.2 for aluminum and to peak values nearly twice as high for copper. It is important to note, though, that these peak values occur at prestrains of only 0.2% or so. The ratios rapidly decrease at larger strains and level off at values near 1.3 and 1.5 for aluminum and copper, respectively.

The experimental study of Piercy *et al.* (1955) on α-brass crystals is an interesting example of earlier work on latent hardening. Their crystals were initially oriented for single slip in tension, and lattice rotations as described in connection with Fig. 27.5 were measured by X-ray diffraction. The point where conjugate slip began was determined "by continual microscopical examination (at 15 x) of the specimen during the period of linear hardening, after easy glide" (Piercy *et al.*, 1955, p. 332). Figure 27.24 shows a micrograph of the gauge section of one of their crystals. The crystallographic orientation of the tensile axis was then determined at the point where activity on the conjugate system was detected and the resolved shear stress on the conjugate and primary systems calculated. Piercy *et al.* (1955) found the latent-hardening ratio to be essentially constant and equal to $\tau_c^{(c)}/\tau_c^{(p)} = 1.28$,

27.4. Measurements of Latent Hardening

thus indicating that the average rate of latent hardening on the conjugate system was approximately 28% greater than the self-hardening rate on the primary system. Note that their experiments are of the second type described previously and involve simultaneous activity on the primary and conjugate systems. The amount of overshoot they observed, 4°-7°, was rather large but consistent with the measurements made earlier on copper-aluminum by Elam (1927a,b). She also found overshoot of several degrees. α-Brass and copper-5 at% aluminum alloys have low stacking fault energies, and this appears to contribute to strong latent hardening. Mitchell and Thornton (1964) have also reported large amounts of overshoot in α-brass of up to 7°.

The large amounts of overshoot reported in α-brass and Cu-Al alloys and the high latent hardening they imply do not appear to be typical of other fcc crystals. Taylor was able to interpret his experiments conducted with Elam on aluminum in terms of isotropic hardening, although it is important to note that up to 2°-3° of overshoot were reported (Taylor and Elam, 1925). Conjugate slip was definitely observed through, even as the symmetry line was approached. Mitchell and Thornton (1964) also reported that conjugate slip began in copper crystals before the symmetry boundary was reached. They nonetheless found that the tensile axis rotated past the symmetry position and followed a path that fell between the predictions of single and symmetric double slip. Ramaswami et al. (1965) reported very little or no overshoot in pure silver crystals, and between 2° and 3° in alloy crystals of silver-10% gold. They estimated the ratio of latent-hardening to self-hardening rates to be 0.95 for silver and 1.05 for silver-gold alloy. Their results for silver are interesting, considering its low stacking fault energy, but the trend toward larger overshoots with alloying, and therefore lower stacking fault energy, is consistent with the results on α-brass Cu-Al alloys. Chang and Asaro (1981) found that in age-hardened aluminum-copper alloys overshoot ranged from 0° to 4° at most. As in many other reported cases, conjugate slip was often observed to begin before the tensile axis reached the symmetry line.

Linear relationships between flow (or yield) stresses on latent systems, $\tau_c^{(L)}$, and flow stresses on active primary systems, $\tau_c^{(p)}$, of the form $\tau_c^{(L)} = A \tau_c^{(p)} + B$ have been reported in several of these studies *after a strain of several percentages*. Piercy et al. (1955) found $A = 1.28$ and $B = 0$ for α-brass; Jackson and Basinski (1967) found $A = 1.36$ and $B = 1.49$ MPa for copper; for silver and silver-10% gold Ramaswami et al. (1965) found (where the latent system was what they called the "half-critical" system) $A = 1.4$ and 1.2, respectively, with $B = 0$ in both cases. Thus Kocks' suggestion that the average ratio of latent-hardening to self-hardening rates lies between 1 and 1.4 is consistent with available measurements.

It must be realized that the direct measurements of latent hardening are limited to only a few studies in which the hardening of latent systems is measured following single slip. Furthermore, the measurements described above have all been interpreted as if plastic flow were rate insensitive, and so the influence of strain rate is uncertain. In a study on aluminum crystals, Joshi and Green (1980) found for a few crystals that larger strain rates were accompanied by larger amounts of overshoot – for other crystals no systematic correlation between strain rate and overshoot was detected. In other studies, such as those of Mitchell and Thornton (1964) on copper, as well as those of Joshi and Green (1980), the observed lattice rotations during the transitions from single to double slip are not readily described in detail by the simple relation suggested above with A constant. Figure 27.25 shows data from these two studies of the observed lattice rotation, compared in the one case to the predicted rotation assuming single slip and in the other to the extensional strain;

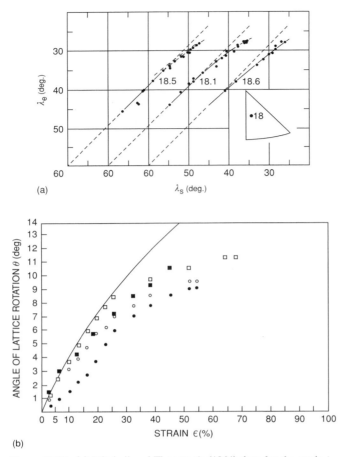

Figure 27.25. (a) Mitchell and Thornton's (1964) data for the angle λ_e between the tensile axis and primary slip direction and λ_s, the angle expected from single slip. Note that deviations from the single-slip predictions occur before the symmetry position is reached, but that the tensile axis overshoots the symmetry position. The dotted lines indicate the values for λ_e, assuming equal slipping on the primary and conjugate systems once symmetry position is reached. (b) Joshi and Green's (1980) data for the angle of lattice rotation versus tensile strain. The solid line is based on the prediction of single slip. Reproduced with permission from Springer-Verlag and from Taylor & Francis Ltd.

Fig. 27.25 also shows, by dotted lines, the rotation expected if the hardening were isotropic and symmetric double slip began when the tensile axis reached the symmetry line. This indicates that, although a constant A (larger than 1) accounts for the observed average ratio of hardening rates, it cannot account for both the apparent premature conjugate slip *and* subsequent overshoot. In a later section an alternative description of overshoot is suggested based on a rate-dependent constitutive law, which may help to explain these observations.

Finally, we note that the limited experimental data available suggests that rate of latent hardening may also be influenced by the ratio of strain rates on the active systems. For example, the high latent-hardening ratios of 1.36 and 1.5 in copper reported by Jackson and

27.5. Observations of Slip

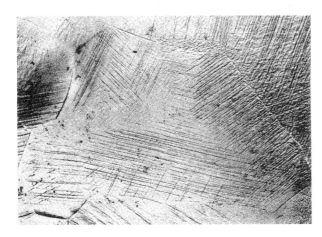

Figure 27.26. Optical micrograph of the slip-line pattern in polycrystalline aluminum. Note the discontinuous slip mode leading to a pattern of *patchy slip* (Boas and Ogilvie, 1954). Reproduced with permission from Elsevier.

Basinski (1967) and Franciosi *et al.* (1980), respectively, are not consistent with the lattice rotations and conjugate slip reported by Mitchell and Thornton (1964). The latter found that conjugate slip began even before the symmetry line corresponding to equal shear stresses on the primary and conjugate systems was reached. These high latent-hardening ratios would have required large amounts of overshoot to activate the conjugate system. The two measurements differ in that the former maintains the crystal in a single-slip mode, first on the primary system, then on one of the latent systems, whereas the latter measurement allows the crystal to deform in multiple slip. The inclusion of rate sensitivity may account for some of these differences, but there may still be influences having to do more with the ratio of shearing rates that will only be sorted out by further experiment.

27.5 Observations of Slip in Single Crystals and Polycrystals at Modest Strains

When slip is confined to one system, *i.e.*, single slip, macroscopic deformation often occurs uniformly over the gauge section of uniformly stressed crystals. When more than one system is active, however, observations suggest that this is generally not the case. For example, Fig. 27.24 shows the *patchy* slip-line pattern that develops in α-brass crystals when they are stretched in tension so that double primary-conjugate slip eventually occurs after primary single slip. The patches consist of regions of double slip. Piercy *et al.* (1955) attributed this nonuniform slip mode to latent hardening. They argued that "these results prove the reality of latent-hardening, in the sense that the slip lines of the one system experience difficulty in breaking through the active slip lines of the other one" (p. 337). A very similar kind of patchy slip is observed within the grains of polycrystals, as shown, for example, in Fig. 27.26 taken from the work of Boas and Ogilvie (1954) on aluminum. In both these examples, the loading was uniform, and in the polycrystalline case the microscopic deformation field that included the grain in question was also uniform. Nonetheless, the slip mode was highly nonuniform. One significance of patchy slip is that the slip mode, and therefore the lattice rotation with respect to the material, becomes difficult to specify for an entire grain. Thus grain, or crystal, reorientation becomes ambiguous. At small strains this may not be important, but at finite strains, where crystallographic texture develops owing to finite amounts of grain reorientation, specification of the slip mode and thus

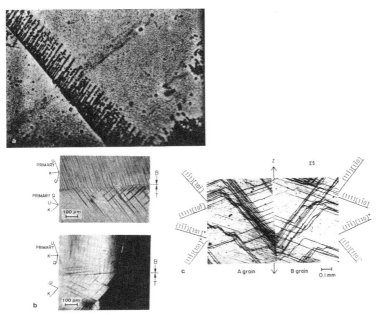

Figure 27.27. Slip bands initiated at (a) a grain boundary of iron-silicon polycrystals (Worthington and Smith, 1964), (b) a grain boundary in a Cu-4 wt% Al alloy (Swearengen and Taggart, 1971) and (c) multiple slip induced at the bicrystal boundary due to the impingement of slip bands. The asterisk indicates an induced slip system. Strain was 4.2% (Miura and Saeki, 1978). Reproduced with permission from Elsevier.

the relative rotation of the lattice with respect to the material is vital to the evolution of constructive behavior. Initiation of the nonuniform slip mode occurs owing to the initiation of slip at grain boundaries, as shown for an iron-silicon alloy in Fig 27.27a. On the other hand, a nonuniform slip mode may be triggered by the impingement of slip bands at grain boundaries, resulting in stress concentrations, as shown in Fig. 27.27b for copper-aluminum alloy. A third example of a nonuniform slip mode is shown in Fig. 27.27c for an aluminum bicrystal. This latter example is taken from the work of Miura and Saeki (1978), who discussed the stress concentrations at the bicrystal (grain) boundary, caused by the impinging slip bands belonging to the slip systems highly stressed by the applied "induced" slip activity on other systems. In this way full compatibility between the crystals was maintained. On the other hand, Peirce *et al.* (1982) have shown that small spatial variations in stress coupled with relatively strong latent hardening ($A \approx 1.4$) would bring about patchy slip in single crystals. Whatever the mechanism, the result is to produce a highly nonuniform mode of slip throughout the grain, as well as a nonuniform dislocation density. It is common that the dislocation density becomes much larger in regions adjacent to the grain boundaries, where complex multiple slip modes are generated, than in the grain interiors. Remarkably, Miura and Saeki (1978) found that, provided that the slip mode within the grains was multiple slip, there was very little difference in the strain-hardening properties of their bicrystals and the component single crystals comprising them. When the grain slip mode was single slip, however, they reported that the grain boundaries caused multiple slip, even at very small strains, and this effect on the overall strain hardening

27.6. Nanocrystalline Grains

of the grain will be negligible at extremely large grain sizes. When the grains are fine and the more highly dislocated boundary zones occupy a proportionately larger volume, the influence of the boundary-induced slip modes should be more significant. This may well be an important factor contributing to the grain size dependence of polycrystalline strain hardening evidenced in Fig. 27.22b. As yet these effects have not been included in continuum polycrystalline models.

27.6 Deformation Mechanisms in Nanocrystalline Grains

It has been recognized for several decades that metals and alloys with grain dimensions smaller than 100 nm (so-called *nanostructured* or *nanocrystalline* metals and alloys) generally exhibit substantially higher strength than their *microstructured* or *microcrystalline* counterparts with grain dimensions typically larger than, say, 1 μm (*e.g.*, Hall, 1951; Petch, 1953; Gleiter, 1989; Kumar *et al.*, 2003). Moreover, there is mounting experimental evidence from recent investigations that points to the following additional mechanical characteristics of nanostructured metals:

- The plastic deformation characteristics on nanocrystalline metals are much more sensitive to the rate of loading than those of microcrystalline metals; the strain-rate sensitivity index, defined below, is an order of magnitude higher for metals with nanocrystalline structures (Lu *et al.*, 2001; Wang and Ma, 2003; Dalla Torre *et al.*, 2002; Schwaiger *et al.*, 2003; Wei *et al.*, 2004).
- The activation volume, which is broadly defined as the rate of decrease of activation enthalpy with respect to flow stress at fixed temperature and which influences the rate-controlling mechanisms in the plastic deformation of engineering metals and alloys, is some two orders of magnitude smaller for nanocrystalline metals than for microcrystalline metals (Wang and Ma, 2004; Lu *et al.*, 2004).
- Whereas the abundance of grain boundaries providing obstacles to dislocation motion during plastic deformation generally leads to enhanced strength in nanocrystalline metals, twin boundaries (which are special kinds of coherent internal interfaces) are also known to obstruct dislocation motion. Recent experiments show that the introduction of nanoscale twins within *ultrafine crystalline* metals, with average grain size within the 100 nm to 1 μm range, leads to significant increases in flow stress and hardness. The extent of such strengthening is comparable to that achievable by nanocrystalline grain refinement (Lu *et al.*, 2005).
- The incorporation of nanoscale twins during the processing of metals with ultrafine grains is also known to increase the loading rate sensitivity by almost an order of magnitude and decrease the activation volume by two orders of magnitude as compared to the values observed in microcrystalline metals (Lu *et al.*, 2005).

Thermally activated mechanisms contributing to plastic deformation processes in metals and alloys are often quantitatively interpreted by examining the rate sensitivity index, m, and activation volume, υ. The nondimensional strain rate sensitivity index is defined as (*e.g.*, Conrad, 1965)

$$m = \frac{\sqrt{3}kT}{\upsilon\sigma} = \frac{3\sqrt{3}kT}{\upsilon H}, \qquad (27.16)$$

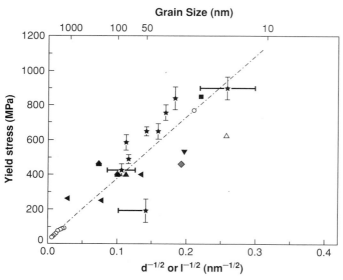

Figure 27.28. A plot of the yield stress as a function of inverse square root grain size from experimental data obtained on nanocrystalline, ultrafine and microcrystalline pure Cu. Also indicated are data (denoted by the symbols with a scatter-range) for Cu with controlled growth twins for which the twin width l is plotted in place of the grain size d. The different shaded symbols denote data from different sources.

where k is the Boltzman constant, T the absolute temperature, σ the uniaxial flow stress, H is the hardness (which is generally assumed to be three times the flow stress), and

$$v = \sqrt{3}kT \left(\frac{\partial \ln \dot{e}}{\partial \sigma} \right). \tag{27.17}$$

The strain rate $\dot{e}$ is measured in uniaxial tension, for instance. For a general discussion of thermally activated rate theory as applied to deformation kinetics, see Krausz and Eyring (1975). The strain rate sensitivity index, m, in (27.16) is the exponent used in common phenomenological power law relations for the rate dependence of plastic flow, viz.,

$$\sigma \propto \dot{e}^m \quad \text{or} \quad \dot{e} \propto \sigma^{1/m}. \tag{27.18}$$

Such simple power law rate forms are used in other chapters of this book concerned with strain rate dependent phenomenological and physically based theories of plasticity.

The implications of grain refinement on strengthening and hardening in face-centered cubic metals is summarized in Figs. 27.28 and 27.29 for Cu and Ni. Available data from Gertsman et al. (1994), Sanders et al. (1997), Legros et al. (2000), Wang et al. (2002), Valiev et al. (2002), Champion et al. (2003), Youssef et al. (2004), and Lu et al. (2004) on the dependence of flow stress on grain size for polycrystalline Cu, over a range from nanometers to micrometers, are plotted in Fig. 27.28. The flow stress varies more or less linearly with $d^{-1/2}$, where d is the grain size. The shaded symbols in Fig. 27.28 pertain to ultrafine-grained Cu with growth twins, where the twin width l is plotted in place of the grain size. Note that the twin width exhibits the same connection to flow stress as the grain size in nanotwinned Cu.

27.6. Nanocrystalline Grains

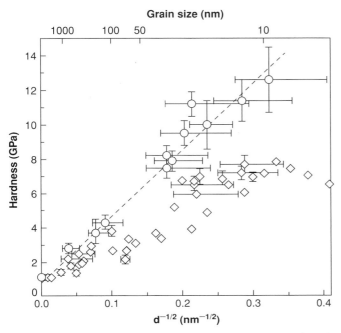

Figure 27.29. Variation of hardness as a function of grain size (top abscissa axis), or inverse square root grain size (bottom abscissa axis), for polycrystalline Ni with grain size from nanocrystalline, ultrafine, and microcrystalline regimes from several investigators, indicated by the data points. The points with a scatter-range denote data extracted from instrumented nanoindentation experiments on thin Ni foils produced by pulsed laser deposition on different hard substrates (Knapp and Follstaedt, 2004).

Figure 27.29 is a plot of indentation hardness versus decreasing grain size for polycrystalline Ni spanning the nanostructured to the microstructured grain dimensions. The data points indicated in blue denote experimental results obtained on nanocrystalline and ultrafine crystalline Ni produced by such techniques as electrodeposition, powder consolidation and inert gas condensation followed by compaction (Hughes *et al.*, 1986; Mitra *et al.*, 2001; Dalla Torre *et al.*, 2002; Legros *et al.*, 2003; Schwaiger *et al.*, 2003; Wei *et al.*, 2004). These results indicate that the hardness of Ni departs significantly from the classical Hall–Petch-type behavior when the grain size reduced typically below about 100 nm. In addition, a lowering of hardness with grain refinement has been reported for Ni below a grain size of about 8 nm (Schuh *et al.* 2002). Such a transition in strengthening to weakening with grain refinement has been postulated from computational simulations (Schiotz *et al.*, 1998; Van Swygenhoven *et al.* 2001; Yamakov *et al.*, 2002; Kumar *et al.*, 2003) and by invoking concepts of grain boundary sliding associated with room temperature creep (Chokshi *et al.*, 1989). Experimental simulations of indentation of two-dimensional nanocrystalline structures employing the polycrystalline bubble raft analogs also revealed a transition from primarily dislocation nucleation at grain boundary triple junctions to a greater propensity for grain boundary sliding when the average grain size of the crystals in the draft was reduced below about 7 nm (Van Vliet *et al.*, 2003). By contrast, hardness values (showed by the red data points in Fig. 27.29) extracted from instrumented nanoindentation for nanocrystalline

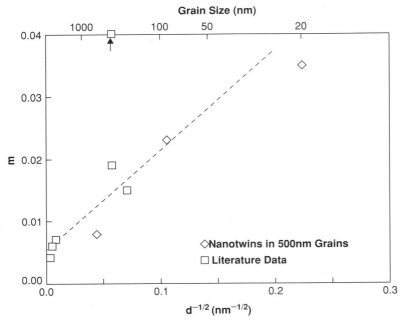

Figure 27.30. A plot of the effect of grain size on the loading rate sensitivity index, m, of pure Cu and Ni at room temperature from available literature data. The data point indicated by the vertical arrow represents estimated value for $m \sim 0.1$ and is not plotted to scale. Also indicated are two points, taken from Lu *et al.* (2004) and denoted by open diamonds, for pure Cu where twins with a width of 20 nm or 90 nm were introduced by pulsed electrodeposition inside grains with an average size of approximately 500 nm. For these cases, the twin width is plotted instead of the grain size.

Ni foils pulse-laser-deposited on different hard substrates (Knapp and Follstaedt, 2004) appear to show strengthening with grain refinement down to about 8 nm or so in a manner consistent with the expectations predicated on the classical Hall–Petch behavior. It is evident from the results displayed in Fig. 27.29 that, despite the growing body of experimental results on the deformation characteristics of nanostructured metals, considerable uncertainty exists about the mechanisms responsible for deformation, especially at very small grain sizes.

Figure 27.30 provides a summary of the experimental results available to date in the literature on the variation of m as a function of grain size for microcrystalline, ultrafine, and nanocrystalline metals and alloys. It is evident here that a reduction in grain size from the microcrystalline to the nanocrystalline regime causes an order of magnitude increase in the strain rate sensitivity of plastic deformation. Results are shown for Cu with a fixed grain size of 500 μm that was pulse-electrodeposited in such a way that the grains contained nanoscale twins with widths of approximately 20 or 90 nm. The twin width is indicated in Fig. 27.30 for these cases in place of the grain size. Available data indeed indicate a trend of increased rate sensitivity at higher strain rates. This was also observed for ultrafine-grained Cu of a grain size of approximately 300 nm (Wang and Ma, 2004).

A summary of literature data (Carreker and Hibbard, 1953; Gray *et al.*, 1997; Follansbee and Kocks, 1988; Conrad and Narayan, 2000; Conrad, 2003; Schwaiger *et al.*, 2003; Wei

27.6. Nanocrystalline Grains

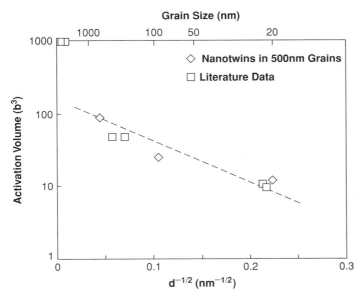

Figure 27.31. A plot of the effect of grain size on the activation volume, measured in units of b^3, for pure Cu and Ni from the literature data. Also indicated are two data points, denoted by open diamonds (corresponding to the same set of experiments for which m values were shown in Fig. 27.30), for pure Cu where twins with a width of 20 or 90 nm were introduced via pulsed electrodeposition inside grains with an average size of 500 nm (from Lu *et al.*, 2004). For these cases, the twin width is plotted instead of the grain size.

et al., 2004; Ma, 2004; Wang and Ma, 2004; Lu *et al.*, 2004) on the effect of grain size on the activation volume of Cu and Ni is shown in Fig. 27.31. Note the decrease in activation volume with grain refinement. The activation volume for the ultrafine-grained Cu specimens with nanotwins is also indicated in this figure, with the twin width replacing the grain size as the characteristic structural length scale. There is the hundredfold increase in activation volume as the spacing of the internal interface is varied from about 20 to about 100 nm.

Although the observations clearly illustrate the effects of nanocrystalline grains and nanoscale twins on strength, hardness, rate sensitivity of deformation, and activation volume, the mechanisms underlying such trends are not well understood at the present time. In addition, no quantitative analysis is available at the time of this writing for rationalizing the observed variations in parameters as the characteristic structural length scales are altered by several orders of magnitude. The subject remains a research area. Some brief comments are made that outline some of the basic issues involved.

In addition to the above phenomenology, when the grain sizes of metals or alloys transits through the micrometer down to the nanometre, there are accompanying transitions in the mechanisms of inelastic deformation as well as significant changes in constitutive properties including, *inter alia*, levels of strength and strain hardening. There is direct experimental evidence for these transitions, theoretical evidence *vis-à-vis* molecular dynamics simulations of nanocrystalline deformation (Van Swygenhoven, 2003, and Van Swygenhoven and co-workers 2001, 2002), as well as suspicions that arise from what is known about the mechanisms of plastic deformation in crystalline metals and alloys. For example, in fcc metals with grain sizes in the micron and larger size range, plastic deformation occurs *via* the generation and motion of intragranular slip, *i.e.*, dislocation motion. This process is evidently shut off at grain sizes somewhat below a micron. This is natural to understand simply by noting that the crystallographic shear stresses required to generate and move dislocation segments that exist within the well-characterized networks that evolve during plastic flow are on the order of $\tau \approx Gb/\ell$, where G is the shear modulus, b is the Burgers vector, and ℓ is the segment length. This has been discussed above. But if dislocations are to be confined to the intragranular space, which might be taken as the definition of a grain, ℓ must be less than the grain diameter, d, and in fact simulations of the operation of *Frank–Read sources* would suggest $\ell \leq d/4 - d/3$. This would lead to the conclusion that $\tau \geq (3-4)G(b/d)$, or $\tau/G \geq (3-4)(b/d)$. For pure Ni, the shear modulus is $G \approx 82$ GPa, and if $d \approx 1\,\mu$m, we obtain $\tau \geq 82$ MPa, which is reasonable. If, however, $d \approx 30$ nm, then $\tau \geq 3280$ MPa, which is too large by at least a factor of nearly 3! Not surprisingly, experimental evidence shows that at grain sizes in this nanocrystalline range, grains seem free of dislocations in their interior. Likewise molecular dynamic simulations of deformation of such nanocrystalline polycrystals do not reveal dislocation activity as part of the deformation process.

Recently, Asaro, Krysl, and Kad (2003) developed a simple yet compelling mechanistic model for the likely scenario for the anticipated transitions in deformation mechanisms that appears to be quite consistent with experimental observations that exist to date and that leads to a complete constitutive theory amenable to computational analysis. The model is based on the notion, shown to follow naturally from dislocation and partial dislocation mechanics, that at grain sizes below, say, 400 nm deformation occurs *via* grain boundary dislocation emission; at grain sizes below, say, 50 nm deformation occurs instead by the emission of partial dislocations, and at even finer grain sizes deformation occurs by grain boundary sliding. These mechanisms and characteristic grain sizes are influenced by temperature, lattice spacing, and stacking fault energy. In fact, stacking fault energy is revealed to be a particularly important factor in determining deformation mechanism and strength level. Moreover, it is shown that the proposed mechanisms lead to strain rate sensitivities that are inherent to the various mechanisms and that are also part of deciding which mechanism is dominant at particular imposed strain rates.

27.6.1 Background: AKK Model

Figure 27.32 can well serve as a point of departure in reviewing the model of Asaro–Krysl–Kad. The figure illustrates a process of emission of a dislocation from a grain boundary into the interior of a grain. The figure describes the details associated with the emission of a partial dislocation, but it can also be used to explain the result of the emission of a perfect dislocation. As explained by AKK, as the segment is emitted into the grain, it creates two trailing segments in the "side grain boundaries," and, in the case of a partial dislocation, a stacking fault within the grain. The energy (per unit length) of these segments is taken as $1/2 Gb^2$ for the perfect dislocation, and $1/6 Gb^2$ for the partial dislocation. The explanation for the latter value lies simply in the fact that for a Schockley partial dislocation in a fcc crystal, the magnitude of the Burgers vector is $b_{\text{partial}} = 1/3 b_{\text{perfect}}$. As evident, $b \equiv b_{\text{perfect}}$ is the magnitude of the perfect Burgers vector. Now in the case of the emission of a perfect

27.6. Nanocrystalline Grains

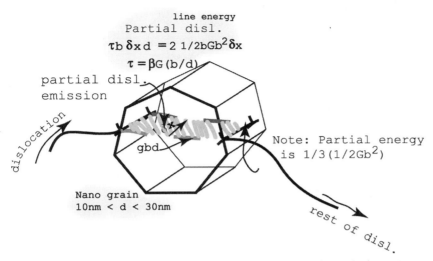

Figure 27.32. Emission of a partial dislocation into a nanocrystalline grain (from Asaro et al., 2003).

dislocation, the minimum required resolved shear stress is that required to perform the work of creating the two residual segments in the side boundaries, or $\tau b d \delta x = 2(\frac{1}{2} G b^2) \delta x$. This leads to the remarkably simple result, $\tau/G = b/d$. The d^{-1} scaling of stress level derives simply from the fact that the area over which work can be performed by the applied shear stress itself scales with d. As noted by AKK, this leads to forecasted shear stresses that are too high for grain sizes less than, say, 50 nm for typical fcc metals for which data exist.

For the case of the emission of a partial dislocation, there is a reduced requirement for work associated with the residual segments in the side boundaries (owing to their lesser energy per unit length), but the additional requirement of creating a stacking fault with an energy Γ per unit area. Consider the primary slip system to be $b = a/2[10\bar{1}]$; its two partials are thereby $b^{(1)} = a/6[2\bar{1}\bar{1}]$ and $b^{(2)} = a/6[11\bar{2}]$. Recall that for the perfect lattice dislocation the energy per unit length is, within the limits of linear elasticity,

$$E = \mathcal{K}_{ij} b_i b_j \ln(R/r_0). \tag{27.19}$$

To complete the geometry, let the slip plane normal be $m = 1/\sqrt{3}[111]$ and the slip direction be $s = 1/\sqrt{2}[10\bar{1}]$; this leaves the unit vector in the slip plane orthogonal to s as $z = 1/\sqrt{6}[\bar{1}2\bar{1}]$. $\mathcal{K}$ is the *energy factor matrix* defined earlier, and for convenience we repeat that for an elastically isotropic material, $\mathcal{K}_{mm} = \mathcal{K}_{zz} = G/4\pi(1-\nu)$ and $\mathcal{K}_{ss} = G/4\pi$. Now consider the extended dislocation, extended through the distance δ and with the stacking fault energy, Γ. The energy of this dislocation is

$$E = \mathcal{K}_{ij} b_i^{(1)} b_j^{(1)} \ln(R/r_0) + \mathcal{K}_{ij} b_i^{(2)} b_j^{(2)} \ln(R/r_0) + 2\mathcal{K}_{ij} b_i^{(1)} b_j^{(2)} \ln(R/r_0) + \Gamma(\delta - r_0).$$

Minimizing E by choice of δ yields

$$\delta_{eq} = 2E_{12}/\Gamma, \tag{27.20}$$

where $E_{12} \equiv \mathcal{K}_{ij} b_i^{(1)} b_j^{(2)}$. Assume the partial (1) to be fixed and let partial (2) move, *i.e.*, extend relative to (1); if it is *vice versa* the signs of the driving stresses would change. We then consider the total energy change when the dislocation extends, including the work done by the applied stresses τ_{ms} and τ_{mz}, *viz.*,

$$\Delta E = -2E_{12} \ln(\delta/r_0) + [\tau_{ms} b_s^{(2)} + \tau_{mz} b_z^{(2)}](\delta - r_0) + \Gamma(\delta - r_0). \tag{27.21}$$

If ΔE is minimized relative to δ, the equilibrium spacing becomes

$$\delta_{eq} = 2E_{12}/\bar{\Gamma}, \tag{27.22}$$

where $\bar{\Gamma} \equiv \Gamma + \tau_{ms} b_s^{(2)} + \tau_{mz} b_z^{(2)}$. It should be apparent that deviations from the Schmid rule of a critical resolved shear stress are inherent in any such analysis, but at this juncture we consider only the influence of the Schmid stress τ_{ms}. We leave the exploration of such deviations to a later study and keep in mind the findings of Asaro and Rice (1977) of the importance of such deviations from what amounts to a normality rule for flow *vis-à-vis* the promotion of localized plastic flow. Specifically, Asaro and Rice (1977) found that such deviations would promote the onset of intense localized plastic flow, either in the form of intense shear bands or "kinking-type" bands. Equations (27.20)–(27.22) allow us to calculate the shear stress required to drive a partial dislocation across a grain of dimension d, *i.e.*, δ is now set to $\delta = d$. We do this by assuming that the terms involving τ_{mz} are of the same order in magnitude as those involving τ_{ms}, because in the average grain this is as likely to be as true as not. Note how the mechanics at this scale lead to a much less accurate picture for the notion of the Schmid rule.

Consider, again, Fig. 27.32, which illustrates the extension of a partial dislocation from the boundary into the intragranular region; we imagine that it traverses the entire grain. As it does, it produces two segments whose energy per unit length in an fcc crystal is, in fact, $(1/3)(Gb^2/2)$. This leads, as before, to a contribution to the required shear stress of $(1/3)G(b/d)$. We next define $\alpha \equiv d/\delta_{eq}$ and if we set $\delta = d$, (27.22) yields, when combined with the stress required to create the additional two segments, the following result

$$\frac{\tau_{ms}}{G} \frac{b_s^{(1)}}{|b|} + \frac{\tau_{mz}}{G} \frac{b_z^{(1)}}{|b|} = \frac{\alpha - 1}{\alpha} \tilde{\Gamma} + \frac{1}{3} \frac{b}{d}. \tag{27.23}$$

Here, $\alpha \equiv d/\delta_{eq}$ and $\tilde{\Gamma} \equiv \Gamma/Gb$. Thus, if we define

$$\tau^{(\alpha)} = \tau_{ms} \frac{b_s^{(1)}}{|b|} + \tau_{mz} \frac{b_z^{(1)}}{|b|}, \tag{27.24}$$

we obtain

$$\frac{\tau^{(\alpha)}}{G} \approx \frac{1}{3} \frac{b}{d} + \frac{\alpha - 1}{\alpha} \tilde{\Gamma}. \tag{27.25}$$

We also recall that δ_{eq} is defined as the equilibrium spacing of Schockley partials in the absence of applied stress and is given as

$$\delta_{eq} = \frac{1}{12\pi} \frac{Gb}{\Gamma}. \tag{27.26}$$

The coordinate system of the dislocation's slip system is such that m is the unit normal to the slip plane, s is unit vector along the direction of the perfect Burgers vector, $\tilde{b}$, and z is the third direction of a right-handed triad and lies in the slip plane as well.

27.6. Nanocrystalline Grains

Table 27.2. *Predicted strength levels for Cu, Ni, Ag, and Pd*

d (nm)	50	30	20	10	δ_{eq}	$\Gamma/(Gb)$
Cu	211 MPa	248 MPa	291 MPa	421 MPa	1.6 nm	1/250
Ni	960 MPa	1027 MPa	1115 MPa	1381 MPa	0.5 nm	1/100
Ag	125 MPa	158 MPa	198 MPa	321 MPa	3.4 nm	1/447
Pd	719 MPa	763 MPa	820 MPa	988 MPa	0.54 nm	1/75

Note that for Cu this leads to required stresses considerably lower than would be the case if only a perfect dislocation were emitted into the interior of the grain. For example, if $d = 25.6$ nm, $b/d \approx 1/100$ and the required shear stress suggested by the result Gb/d, coming from a picture involving the emission of a perfect dislocation, would be $G/100$ vs. the $G/300$, coming from the first term in the above result. As $\Gamma/(Gb) \approx 1/250$ for Cu, the second term yields a maximum contribution of $G/250$ which, in combination with the contribution of $G/300$, yields a total result approximately 33% less than Gb/d would. Of course, as the grain size decreases even further, the differences between the predictions increase. The other obvious feature of this result is the effect of stacking fault energy that appears as a primary influence on the required stress. We note that for an fcc crystal, (27.20) yields $\delta_{eq} = (1/12\pi)Gb/\Gamma$. It is useful to note what this yields regarding equilibrium partial dislocation extensions without stress. Some numerical predictions are listed in Table 27.2. Note that the combination of a relatively high stacking fault energy and modulus for Ni results in a predicted high strength level. On the other hand, the strength level for Pd, which has a high stacking fault energy, is similarly predicted to be relatively high, especially as compared to Cu, which has a comparable shear modulus to that of Pd. Ag is clearly predicted to show the lowest strength, which, although not surprising, should be appreciated to suggest that nanostructuring will not lead to exceedingly high strengths. Note that for Ag, $\delta_{eq} \approx 3.4$ nm, and thus when $d \sim 10$ nm little is to be gained from further reductions in grin size!

Two examples of experimental data are now shown to illustrate the consistency of these predictions and what is known experimentally. Figure 27.33 is taken from Dalla Torre *et al.*

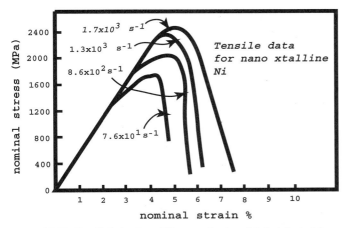

Figure 27.33. Tensile behavior of 20 nm grain-size nickel. Adapted from Dalla Torre *et al.* (2002).

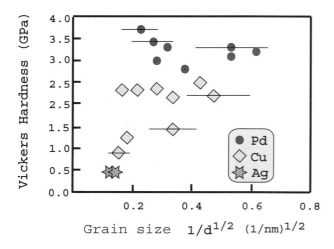

Figure 27.34. Hardness data for nanocrystalline Cu, Ag, and Pd. Adapted from Nieman et al. (1992).

(2002) and shows stress vs. strain data in tension for Ni with a grain size of approximately 20 nm. In the first place, we note that for this Ni the ultimate strength at the highest imposed strain rate ($\dot{e} = 1.7 \times 10^3\,\text{s}^{-1}$) is approximately $\sigma_{\text{ult}} \approx 2400$ MPa. If we take $\sigma \approx 2\tau$ at yield and just after, Table 27.2 suggests that $\sigma_{\text{ult}} \approx 2 \times 1115$ MPa $= 2230$ MPa, which is in reasonably good agreement with the high strain rate measurement of Dalla Torre et al. (2002). We also note the essential absence of strain hardening, which may be also forecast from the AKK model as discussed below. There is, however, much more that requires comment concerning this data.

Between the highest strain rate imposed and the lowest, viz., $\dot{e} = 8.6 \times 10^2\,\text{s}^{-1}$, there is an approximately 800 MPa reduction in flow stress. As discussed below, it may be no coincidence that the first term in 27.25 is for Ni with $d \approx 20$ nm, $1/3(b/d) \approx 4.58 \times 10^{-3}$, which, when multiplied by the shear modulus of Ni, is 376 MPa. When multiplied by 2 (to link shear stress to tensile stress), a tensile stress increment of $\delta\tau \approx 752$ MPa is obtained. A speculation then may be that residual segments of partial dislocation in the side boundaries may be capable of "annealing out" or dissipating if sufficient time is allowed, i.e., if the imposed strain rate is sufficiently low. A further forecast, and one which follows for this first speculation, is that at lower imposed strain rates no further reduction in strength will be observed, until still another transition to grain boundary sliding sets in.

At the highest imposed strain rate, we can estimate the grain size, say for Ni, at which there would be a transition from perfect dislocation emission to partial dislocation emission by simply equating (b/d) to the *rhs* of (27.25) as

$$\frac{b}{d} = \frac{1}{3}\frac{b}{d} + \tilde{\alpha}\tilde{\Gamma}, \tag{27.27}$$

where $\tilde{\alpha} \equiv 1 - 1/\alpha$. Since for Ni, $\tilde{\alpha} \approx 1$, the estimate is that the transition would occur at a grain size of approximately $d \approx 70\,b$ or $d \approx 20$ nm! This result is somewhat surprising and may be in need for further refinement.

Figure 27.34 illustrates additional data obtained *via* hardness measurements on three other fcc metals, viz., Cu, Ag, and Pd. The shear modulii of Cu and Pd are similar; that for Ag is approximately 15% less. With this in mind, and with the understanding that the grain sizes for these metals in this set of data are in the range $3\,\text{nm} \leq d \leq 25\,\text{nm}$, it is

27.6. Nanocrystalline Grains

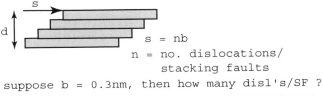

Figure 27.35. How many partial dislocations or stacking faults are emitted into a nanocrystalline grain?

clear that the ordering of strength appears to be Pd ≥ Cu ≥ Ag. This is, of course, what is predicted form the AKK model, that is without the appearance of grain boundary sliding. Grain boundary sliding *per se* is not discussed here, but it should be noted that it occurs at sufficiently high temperatures and in cases where the grain sizes fall below, say, 5 nm or so. It then appears as a transition in deformation mode.

27.6.2 Perspective on Discreteness

It is revealing to inquire as to the actual number of dislocations or stacking faults that would need to be emitted into a typical grain whose dimensions are on the order of 20 or 30 nm. Figure 27.35 presents a simple analysis of this for perspective. The important point to be illustrated here is that to induce strains that are less than, say 10%, the number of defects involved is itself quite modest and typically less than 10. This, in turn, illustrates the rather discrete nature of the deformation process that occurs in grains of this size. We will, nonetheless, view the process in what follows as occurring in a continuous manner, despite its obvious discrete nature. For analyzing polycrystalline regions, this is viewed as representing suitable averaging. In other words, we take the view that the discrete slip increments that given grains undergo over a short time scale on the order of 10^{-6}s can be *time stretched* over much longer times, provided the average ensemble's strain increment is the same in both cases. This means that in the analysis, and at any given moment, the number of defects emitted will be noninteger, yet the average plastic strain in a representative polycrystalline aggregate will be equal to what would have been produced by discrete slip events in a discrete group of grains.

27.6.3 Dislocation and Partial Dislocation Slip Systems

The slip systems associated with perfect dislocations are the usual 12 (or 24) fcc systems involving the octahedral {111} planes, and the face diagonal <110> directions lying within those planes. The choice of 12 or 24 is simply one of allowing the slipping rate to possess algebraic sign or to be strictly positive, respectively. In the latter case slip systems describing slip in *positive* and *negative* directions are required. Figure 27.36 illustrates the kinematics associated with partial dislocations, which are themselves associated with *parent* perfect

Table 27.3. *List of fault systems*

Slip Plane	Burgers Vector	Leading partial, b^+	Trailing partial, b^-
(111)	$[10\bar{1}]$	$[11\bar{2}]$	$[2\bar{1}\bar{1}]$
(111)	$[0\bar{1}1]$	$[1\bar{2}1]$	$[\bar{1}\bar{1}2]$
(111)	$[\bar{1}10]$	$[\bar{2}11]$	$[\bar{1}2\bar{1}]$
$(\bar{1}11)$	$[110]$	$[211]$	$[12\bar{1}]$
$(\bar{1}11)$	$[0\bar{1}1]$	$[\bar{1}\bar{2}1]$	$[1\bar{1}2]$
$(\bar{1}11)$	$[\bar{1}0\bar{1}]$	$[\bar{1}1\bar{2}]$	$[\bar{2}1\bar{1}]$
$(1\bar{1}1)$	$[110]$	$[121]$	$[21\bar{1}]$
$(1\bar{1}1)$	$[0\bar{1}\bar{1}]$	$[1\bar{1}\bar{2}]$	$[\bar{1}\bar{2}\bar{1}]$
$(1\bar{1}1)$	$[\bar{1}01]$	$[\bar{2}11]$	$[\bar{1}12]$
$(11\bar{1})$	$[101]$	$[2\bar{1}1]$	$[112]$
$(11\bar{1})$	$[1\bar{1}\bar{1}]$	$[1\bar{2}\bar{1}]$	$[\bar{1}\bar{1}\bar{2}]$
$(11\bar{1})$	$[\bar{1}10]$	$[\bar{1}21]$	$[\bar{2}11]$

dislocations. The parent, *i.e.*, perfect dislocation Burgers vector is denoted as $\tilde{b}$. Partial systems are associated with each parent fcc slip system, and for our purposes we may restrict attention to slipping in a single sense along each <110> direction. This is because each perfect slip system naturally defines two partial systems, one for slip along the parent <110> direction and the second for slipping in the opposite direction. Thus slip in what may be referred to as the + and − directions, in any {111} plane, can be described by 24 partial systems. Note that partial slip events occur in particular sequences, *e.g.*, slipping over the perfect Burgers vector, $\tilde{b}$, occurs in the sequence, $\tilde{b}^+$ followed by $\tilde{b}^-$. Slipping in the direction $-\tilde{b}$, on the other hand, occurs in the sequence $-\tilde{b}^-$ followed by $-\tilde{b}^+$. Now we recall that slip events involving partial dislocations, *i.e.*, the emission of stacking faults into the grains, involves emitting only the leading partial dislocation. Therefore, from the perfect slip system (m, b), we obtain the two partial slip systems, (m, b^+) and $(m, -b^-)$. Table 27.3 lists the 24 such partial slip systems, each associated with a perfect slip system.

The restriction of allowing only the "leading" partial dislocation to be emitted derives from several rationales. First, we recall from the dislocation analysis above that on energetic

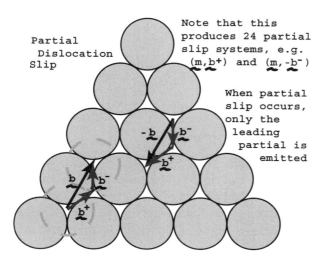

Figure 27.36. Partial dislocation slip systems.

27.7. Suggested Reading

grounds the emission of a single partial dislocation is favored over the emission of a perfect dislocation at sufficiently small grain sizes. Second, and entirely consistent with this view, are the results of a series of molecular dynamics simulations performed by Van Swygenhoven and co-workers (2001–2003).

We close this discussion with the complete list of partial dislocation, *i.e.*, stacking fault based, slip systems in fcc crystals. The slip, or actually fault, systems listed herein would provide the basis for the kinematics of fault induced shear deformation, as perfect dislocations do for slip in larger grain materials.

27.7 Suggested Reading

Asaro, R. J., Krysl, P., and Kad, B. (2003), Deformation Mechanism Transitions in Nanoscale FCC Metals, *Phil. Mag. Letters*, Vol. 83, pp. 733–743.

Chang, Y. W., and Asaro, R. J. (1981), An Experimental Study of Shear Localization in Aluminum-Copper Single Crystals., *Acta Metall.*, Vol. 29, pp. 241–257.

Cottrell, A. H. (1961), *Dislocations and Plastic Flow in Crystals*, Oxford University Press, London.

Cottrell, A. H. (1964), *The Mechanical Properties of Matter*, Wiley, New York.

Kocks, U. F., Tomé, C. N., and Wenk, H.-R. (1998), *Texture and Anisotropy: Preferred Orientations in Polycrystals and Their Effect on Materials Properties*, Cambridge University Press, Cambridge.

Kelly, A., Groves, G. W., and Kidd, P. (2000), *Crystallography and Crystal Defects*, Wiley, New York.

Kumar, K. S., Van Swygenhoven, H., and Suresh, S. (2003), Mechanical Behavior of Nanocrystalline Metals and Alloys, *Acta Mater.*, Vol. 51, pp. 5743–5774.

McClintock, F. A., and Argon, A. S. (1966), *Mechanical Behavior of Materials*, Addison-Wesley, Reading, Mass.

Schmid, E., and Boas, W. (1968), *Plasticity of Crystals*, Chapman and Hall, London.

Taylor, G. I. (1934), The Mechanism of Plastic Deformation of Crystals. Part I: Theoretical, *Proc. Roy. Soc. London, Sec. A*, Vol. 145, pp. 362–387.

28 Crystal Plasticity

This chapter develops the mathematical framework for the physical theory of plasticity outlined in the previous chapter. Examples of the use of the theory are given in subsequent chapters, including the chapter devoted to the development of a special form of the theory developed herein as applied to laminates. In the derivations that follows, the summation convention is used but augmented by the stipulation that a subscript enclosed in $(\cdot)$ is not to be included in a summation; for instance, $\mathbf{s}_{(\alpha)}\mathbf{m}_\alpha$ is simply the dyadic product of the two vectors, $\mathbf{s}_\alpha$ and $\mathbf{m}_\alpha$, not summed over the index α.

28.1 Basic Kinematics

In a crystalline solid an increment of deformation is imagined to occur in two steps. The first, starting from the reference state, occurs by a process of simple shears on slip systems, as described in the previous chapter. Following this is a process of lattice deformation; the basic kinematic scheme is as shown in Fig. 28.1. Thus, the deformation gradient is decomposed as

$$\mathbf{F} = \mathbf{F}^* \cdot \mathbf{F}^p. \tag{28.1}$$

The velocity gradient due to slip is

$$\mathbf{L}^p = \dot{\mathbf{F}}^p \cdot \mathbf{F}^{p-1} = \dot{\gamma}_{(\alpha)} \mathbf{s}_\alpha \mathbf{m}_\alpha. \tag{28.2}$$

In the summation α goes from 1 to the number of active slip systems, n. For the symmetric and skew parts of $\mathbf{L}^p$ we have

$$\mathbf{D}^p = \dot{\gamma}_{(\alpha)} \mathrm{sym}(\mathbf{s}_\alpha \mathbf{m}_\alpha), \quad \mathbf{W}^p = \dot{\gamma}_{(\alpha)} \mathrm{skew}(\mathbf{s}_\alpha \mathbf{m}_\alpha), \tag{28.3}$$

where $\dot{\gamma}_\alpha$ is the rate of simple shearing in the direction of $\mathbf{s}_\alpha$ across planes whose normal is $\mathbf{m}_\alpha$; $\mathbf{s}_\alpha$ and $\mathbf{m}_\alpha$ are unit vectors, so that $\mathbf{s}_{(\alpha)} \cdot \mathbf{m}_\alpha = 0$. Note that for an fcc crystal there are 12 distinct systems of the type ([110], [111]), and for the rate-independent case, it is taken that slip along the ±[110] directions represent two different systems. This leads to a total of 24 systems. In later sections, when dealing with the rate-dependent case, we will revert back to a specification of 12 systems, where slipping can occur in the positive and negative sense along each slip direction. For example, when phrased *wrt* components

28.1. Basic Kinematics

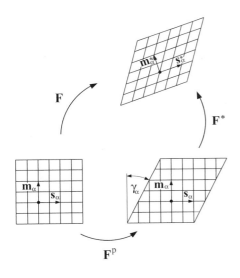

Figure 28.1. Kinematic model of elastoplastic deformation of single crystal. The material flows through the crystalline lattice by crystallographic slip, which gives rise to deformation gradient $\mathbf{F}^p$. The material with embedded lattice is then deformed elastically from the intermediate to the current configuration. The lattice vectors in the two configurations are related by $\mathbf{s}_\alpha^* = \mathbf{F}^* \cdot \mathbf{s}_\alpha$ and $\mathbf{m}_\alpha^* = \mathbf{s}^\alpha \cdot \mathbf{F}^{*-1}$.

formed on crystal axes $\{\mathbf{a}_i\}$, i.e., $\mathbf{s}_\alpha = s_\alpha^i \mathbf{a}_i$, then $s_\alpha^1 = -1/\sqrt{2}, s_\alpha^2 = 1/\sqrt{2}, s_\alpha^3 = 0; m_\alpha^1 = m_\alpha^2 = m_\alpha^3 = 1/\sqrt{3}$ would be the components of the vectors defining one of the typical slip systems. We use covariant and contravariant base vectors and components throughout this chapter.

The deformation $\mathbf{F}^*$ is the deformation of the lattice plus rigid body motions, all imagined to occur after the plastic shears. In the plastic part, material is imagined to flow through the lattice *via* the shears; the lattice is sensibly unaffected by this. In the lattice deformation, the lattice with the material embedded on it is deformed and translated/rotated; thus the material fully participates in this lattice motion. The total velocity gradient,

$$\mathbf{L} = \dot{\mathbf{F}} \cdot \mathbf{F}^{-1} = \mathbf{D} + \mathbf{W}, \tag{28.4}$$

consequently becomes

$$\begin{aligned}\mathbf{L} &= (\mathbf{F}^* \cdot \mathbf{F}^p)^\cdot \cdot (\mathbf{F}^* \cdot \mathbf{F}^p)^{-1} \\ &= \dot{\mathbf{F}}^* \cdot \mathbf{F}^{*-1} + \mathbf{F}^* \cdot \dot{\mathbf{F}}^p \cdot \mathbf{F}^{p-1} \cdot \mathbf{F}^{*-1}.\end{aligned} \tag{28.5}$$

We define the lattice velocity gradient, and the shear induced plastic velocity gradient in the intermediate configuration, as

$$\mathbf{L}^* = \dot{\mathbf{F}}^* \cdot \mathbf{F}^{*-1}, \quad \mathbf{L}^p = \dot{\mathbf{F}}^p \cdot \mathbf{F}^{p-1}. \tag{28.6}$$

What is defined the plastic part of $\mathbf{L}$ in the deformed configuration contains more than the effect of simple shears occurring in the intermediate reference state. Accordingly, we define the plastic velocity gradient in the current configuration as

$$\ell^p = \mathbf{F}^* \cdot \mathbf{L}^p \cdot \mathbf{F}^{*-1} \tag{28.7}$$

and thus

$$\mathbf{L} = \mathbf{L}^* + \ell^p. \tag{28.8}$$

Consequently, we arrive at the following decompositions

$$\mathbf{L} = \mathbf{D} + \mathbf{W}, \quad \mathbf{L}^* = \mathbf{D}^* + \mathbf{W}^*,$$
$$\mathbf{L}^p = \mathbf{D}^p + \mathbf{W}^p, \quad \ell^p = \mathbf{d}^p + \mathbf{w}^p. \tag{28.9}$$

To proceed, consider a typical slip direction vector, $\mathbf{s}_\alpha$. This vector convects with the lattice deformation, because it is embedded in the lattice. Consequently, we can write

$$\mathbf{s}_\alpha^* = \mathbf{F}^* \cdot \mathbf{s}_\alpha, \qquad \mathbf{s}_\alpha^* = \mathbf{s}_\alpha \cdot \mathbf{F}^{*T},$$
$$\mathbf{s}_\alpha = \mathbf{F}^{*-1} \cdot \mathbf{s}_\alpha^*, \qquad \mathbf{s}_\alpha = \mathbf{s}_\alpha^* \cdot \mathbf{F}^{*-T}, \tag{28.10}$$
$$\dot{\mathbf{s}}_\alpha^* = \dot{\mathbf{F}}^* \cdot \mathbf{s}_\alpha = \dot{\mathbf{F}}^* \cdot \mathbf{F}^{*-1} \cdot \mathbf{s}_\alpha^* \quad \Rightarrow \quad \dot{\mathbf{s}}_\alpha^* = \mathbf{L}^* \cdot \mathbf{s}_\alpha^*.$$

Recalling that in the reference configuration $\mathbf{m}_\alpha \cdot \mathbf{s}_{(\alpha)} = 0$, we will choose to define $\mathbf{m}_\alpha^*$ as follows

$$\mathbf{m}_\alpha \cdot \mathbf{F}^{*-1} \cdot \mathbf{F}^* \cdot \mathbf{s}_{(\alpha)} = 0, \qquad \mathbf{m}_\alpha^* \cdot \mathbf{s}_{(\alpha)}^* = 0, \tag{28.11}$$

where

$$\mathbf{m}_\alpha^* = \mathbf{m}_\alpha \cdot \mathbf{F}^{*-1}, \qquad \mathbf{m}_\alpha^* = \mathbf{F}^{*-T} \cdot \mathbf{m}_\alpha,$$
$$\mathbf{m}_\alpha = \mathbf{m}_\alpha^* \cdot \mathbf{F}^*, \qquad \mathbf{m}_\alpha = \mathbf{F}^{*T} \cdot \mathbf{m}_\alpha^*. \tag{28.12}$$

To demonstrate that $\mathbf{m}_\alpha^*$, so defined, is orthogonal to the deformed slip plane, note that an area element in the reference state in the α slip plane is, by the Nanson's relation of Chapter 4, transformed into

$$\hat{\mathbf{m}}_\alpha^* dS^* = J^* \mathbf{m}_\alpha \cdot \mathbf{F}^{*-1} dS. \tag{28.13}$$

The unit normal to the deformed area is $\hat{\mathbf{m}}_\alpha^*$, and $J = \det \mathbf{F}^*$. Thus,

$$\frac{1}{J^*} \frac{dS^*}{dS} \hat{\mathbf{m}}_\alpha^* = \mathbf{m}_\alpha \cdot \mathbf{F}^{*-1} = \mathbf{m}_\alpha^*, \tag{28.14}$$

which is clearly normal to the deformed slip plane area element ($\mathbf{m}_\alpha^* \cdot \mathbf{s}_{(\alpha)}^* = 0$).

The rates $\dot{\mathbf{s}}_\alpha$ and $\dot{\mathbf{m}}_\alpha$ are inherently equal to zero, as the vectors $\mathbf{s}_\alpha$ and $\mathbf{m}_\alpha$ are fixed in the reference configuration. The rate of change of $\mathbf{s}_\alpha^*$ is obtained by differentiating $\mathbf{s}_\alpha^* = \mathbf{F}^* \cdot \mathbf{s}_\alpha$, which gives

$$\dot{\mathbf{s}}_\alpha^* = \dot{\mathbf{F}}^* \cdot \mathbf{s}_\alpha = \dot{\mathbf{F}}^* \cdot \mathbf{F}^{*-1} \cdot \mathbf{F}^* \cdot \mathbf{s}_\alpha = \mathbf{L}^* \cdot \mathbf{s}_\alpha^*. \tag{28.15}$$

Similarly, the rate of change of $\mathbf{m}_\alpha^*$ is obtained by differentiating $\mathbf{m}_\alpha = \mathbf{m}_\alpha^* \cdot \mathbf{F}^*$, as follows

$$0 = \dot{\mathbf{m}}_\alpha^* \cdot \mathbf{F}^* + \mathbf{m}_\alpha^* \cdot \dot{\mathbf{F}}^* \quad \Rightarrow \quad \dot{\mathbf{m}}_\alpha^* = -\mathbf{m}_\alpha^* \cdot \left(\dot{\mathbf{F}}^* \cdot \mathbf{F}^{*-1} \right) = -\mathbf{m}_\alpha^* \cdot \mathbf{L}^*. \tag{28.16}$$

Introduce reference covariant and contravariant base vectors, $\mathbf{B}_i$ and $\mathbf{B}^i$. If the coordinate systems are Cartesian, $\mathbf{B}_i = \mathbf{B}^i$. Introduce also a convected set of base vectors, $\mathbf{b}_i$ and $\mathbf{b}^i$, such that

$$\mathbf{b}_i = \mathbf{F}^* \cdot \mathbf{B}_i, \quad \mathbf{b}_i = \mathbf{B}_i \cdot \mathbf{F}^{*T}, \quad \mathbf{B}_i = \mathbf{F}^{*-1} \cdot \mathbf{b}_i, \quad \mathbf{B}_i = \mathbf{b}_i \cdot \mathbf{F}^{*-T},$$
$$\mathbf{b}^i = \mathbf{B}^i \cdot \mathbf{F}^{*-1}, \quad \mathbf{b}^i = \mathbf{F}^{*T} \cdot \mathbf{B}^i, \quad \mathbf{B}^i = \mathbf{b}^i \cdot \mathbf{F}^*, \quad \mathbf{B}^i = \mathbf{F}^{*T} \cdot \mathbf{b}^i, \tag{28.17}$$
$$\dot{\mathbf{b}}_i = \dot{\mathbf{F}}^* \cdot \mathbf{B}_i = \dot{\mathbf{F}}^* \cdot \mathbf{F}^{*-1} \cdot \mathbf{b}_i = \mathbf{L}^* \cdot \mathbf{b}_i.$$

28.2. Stress and Stress Rates

Similarly,

$$\dot{\mathbf{B}}^i = 0 = \dot{\mathbf{b}}^i \cdot \mathbf{F}^* + \mathbf{b}^i \cdot \dot{\mathbf{F}}^* \quad \Rightarrow \quad \dot{\mathbf{b}}^i = -\mathbf{b}^i \cdot \mathbf{L}^*. \tag{28.18}$$

Of course, $\mathbf{b}_i$ and $\mathbf{b}^j$ are a set of reciprocal base vectors, as are $\mathbf{B}_i$ and $\mathbf{B}^j$, and so

$$\mathbf{b}^i \cdot \mathbf{b}_j = \mathbf{B}^i \cdot \mathbf{F}^{*-1} \cdot \mathbf{F}^* \cdot \mathbf{B}_j = \mathbf{B}^i \cdot \mathbf{B}_j = \delta^i_{.j}. \tag{28.19}$$

Now, reconsider the velocity gradient

$$\ell^p = \mathbf{F}^* \cdot \mathbf{L}^p \cdot \mathbf{F}^{*-1}, \quad \mathbf{L}^p = \dot{\gamma}_{(\alpha)} \mathbf{s}_\alpha \mathbf{m}_\alpha. \tag{28.20}$$

In a more expanded form, this is

$$\ell^p = \dot{\gamma}_{(\alpha)} \mathbf{F}^* \cdot \mathbf{s}_\alpha \mathbf{m}_\alpha \cdot \mathbf{F}^{*-1} = \dot{\gamma}_{(\alpha)} \mathbf{s}^*_\alpha \mathbf{m}^*_\alpha. \tag{28.21}$$

Its symmetric and antisymmetric parts are

$$\mathbf{d}^p = \dot{\gamma}_{(\alpha)} \operatorname{sym}(\mathbf{s}^*_\alpha \mathbf{m}^*_\alpha), \quad \mathbf{w}^p = \dot{\gamma}_{(\alpha)} \operatorname{skew}(\mathbf{s}^*_\alpha \mathbf{m}^*_\alpha). \tag{28.22}$$

If we define

$$\mathbf{P}_\alpha = \operatorname{sym}(\mathbf{s}^*_{(\alpha)} \mathbf{m}^*_\alpha), \quad \mathbf{Q}_\alpha = \operatorname{skew}(\mathbf{s}^*_{(\alpha)} \mathbf{m}^*_\alpha), \tag{28.23}$$

the above can be rewritten as

$$\begin{aligned}\mathbf{d}^p &= \dot{\gamma}_\alpha \mathbf{P}_\alpha, \quad \mathbf{w}^p = \dot{\gamma}_\alpha \mathbf{Q}_\alpha, \\ \ell^p &= \dot{\gamma}_\alpha (\mathbf{P}_\alpha + \mathbf{Q}_\alpha).\end{aligned} \tag{28.24}$$

28.2 Stress and Stress Rates

Consider the two Jaumann rates of Kirchhoff stress, *viz.*,

$$\begin{aligned}\stackrel{\triangledown}{\boldsymbol{\tau}} &= \dot{\boldsymbol{\tau}} - \mathbf{W} \cdot \boldsymbol{\tau} - \boldsymbol{\tau} \cdot \mathbf{W}^T, \\ \stackrel{\triangledown}{\boldsymbol{\tau}}^* &= \dot{\boldsymbol{\tau}} - \mathbf{W}^* \cdot \boldsymbol{\tau} - \boldsymbol{\tau} \cdot \mathbf{W}^{*T}.\end{aligned} \tag{28.25}$$

The difference in these is

$$\begin{aligned}\stackrel{\triangledown}{\boldsymbol{\tau}}^* - \stackrel{\triangledown}{\boldsymbol{\tau}} &= (\mathbf{W} - \mathbf{W}^*) \cdot \boldsymbol{\tau} + \boldsymbol{\tau} \cdot (\mathbf{W} - \mathbf{W}^*)^T \\ &= \mathbf{w}^p \cdot \boldsymbol{\tau} + \boldsymbol{\tau} \cdot \mathbf{w}^{pT},\end{aligned} \tag{28.26}$$

or, alternatively, using the second of (28.24),

$$\stackrel{\triangledown}{\boldsymbol{\tau}}^* - \stackrel{\triangledown}{\boldsymbol{\tau}} = \dot{\gamma}_\alpha (\mathbf{Q}_\alpha \cdot \boldsymbol{\tau} - \boldsymbol{\tau} \cdot \mathbf{Q}_\alpha). \tag{28.27}$$

Thus,

$$\stackrel{\triangledown}{\boldsymbol{\tau}}^* - \stackrel{\triangledown}{\boldsymbol{\tau}} = \dot{\gamma}_\alpha \boldsymbol{\beta}_\alpha, \tag{28.28}$$

where

$$\boldsymbol{\beta}_\alpha = \mathbf{Q}_\alpha \cdot \boldsymbol{\tau} - \boldsymbol{\tau} \cdot \mathbf{Q}_\alpha. \tag{28.29}$$

If we take the elastic constitutive law to be

$$\stackrel{\triangledown}{\boldsymbol{\tau}}^* = \boldsymbol{\Lambda} : \mathbf{D}^* = \boldsymbol{\Lambda} : (\mathbf{D} - \mathbf{d}^p), \tag{28.30}$$

as will be later justified, we obtain

$$\overset{\triangledown}{\boldsymbol{\tau}} + \dot{\gamma}_\alpha \boldsymbol{\beta}_\alpha = \boldsymbol{\Lambda} : (\mathbf{D} - \dot{\gamma}_\alpha \mathbf{P}_\alpha), \qquad (28.31)$$

or

$$\begin{aligned}\overset{\triangledown}{\boldsymbol{\tau}} &= \boldsymbol{\Lambda} : (\mathbf{D} - \dot{\gamma}_\alpha \mathbf{P}_\alpha) - \dot{\gamma}_\alpha \boldsymbol{\beta}_\alpha \\ &= \boldsymbol{\Lambda} : \left[\mathbf{D} - \dot{\gamma}_\alpha \left(\mathbf{P}_\alpha + \boldsymbol{\Lambda}^{-1} : \boldsymbol{\beta}_\alpha \right) \right]. \end{aligned} \qquad (28.32)$$

The elastic compliance tensor is $\boldsymbol{\Lambda}^{-1}$ is the inverse to the elastic moduli tensor $\boldsymbol{\Lambda}$. The sums in (28.32) over repeated α are from 1 to n.

28.2.1 Resolved Shear Stress

The resolved shear stress on the α slip system may be defined as

$$\tau_\alpha = \hat{\mathbf{m}}^*_{(\alpha)} \cdot \boldsymbol{\sigma} \cdot \hat{\mathbf{s}}^*_\alpha, \qquad (28.33)$$

where the $\hat{}$ denotes a unit vector. Asaro and Rice (1977) used, instead, the definition

$$\tau_\alpha = \mathbf{m}^*_{(\alpha)} \cdot \boldsymbol{\sigma} \cdot \mathbf{s}^*_\alpha. \qquad (28.34)$$

It is thus prudent to explore the implications and meaning of these deferring choices. Toward this end let the slip system vectors be expressed as

$$\begin{aligned} \mathbf{s}_\alpha &= (s_\alpha)^i \mathbf{B}_i, \quad \mathbf{m}_\alpha = (m_\alpha)_i \mathbf{B}^i, \\ \mathbf{s}^*_\alpha &= \mathbf{F}^* \cdot \mathbf{s}_\alpha = (s_\alpha)^i \mathbf{F}^* \cdot \mathbf{B}_i = (s_\alpha)^i \mathbf{b}_i, \\ \mathbf{m}^*_\alpha &= \mathbf{m}_\alpha \cdot \mathbf{F}^{*-1} = (m_\alpha)_i \mathbf{B}^i \cdot \mathbf{F}^{*-1} = (m_\alpha)_i \mathbf{b}^i. \end{aligned} \qquad (28.35)$$

Then,

$$\begin{aligned} \tau_\alpha &= \mathbf{m}^* \cdot \boldsymbol{\tau} \cdot \mathbf{s}^* = \{m_{(\alpha)}\}_i \{s_\alpha\}^j \mathbf{b}^i \cdot \boldsymbol{\tau} \cdot \mathbf{b}_j, \\ \tau_\alpha &= \{m_{(\alpha)}\}_i \{s_\alpha\}^j \tau^i_{\cdot j}, \end{aligned} \qquad (28.36)$$

where $\tau^i_{\cdot j}$ are the mixed components of Kirchhoff stress on the convected and reciprocal $\mathbf{b}_j$ and $\mathbf{b}^i$ base vectors. Recall that $\mathbf{b}_i = \mathbf{F}^* \cdot \mathbf{B}_i$, $\mathbf{b}^i = \mathbf{B}^i \cdot \mathbf{F}^{*-1}$, and

$$\begin{aligned} \dot{\mathbf{b}}_i &= \mathbf{L}^* \cdot \mathbf{b}_i, & \dot{\mathbf{b}}^i &= -\mathbf{b}^i \cdot \mathbf{L}^*, \\ \dot{\mathbf{s}}^*_\alpha &= \mathbf{L}^* \cdot \mathbf{s}^*_\alpha, & \dot{\mathbf{m}}^*_\alpha &= -\mathbf{m}^*_\alpha \cdot \mathbf{L}^*. \end{aligned} \qquad (28.37)$$

The rate of τ_α, defined in (28.34), is

$$\begin{aligned} \dot{\tau}_\alpha &= \dot{\mathbf{m}}^*_\alpha \cdot \boldsymbol{\tau} \cdot \mathbf{s}^*_{(\alpha)} + \mathbf{m}^*_\alpha \cdot \dot{\boldsymbol{\tau}} \cdot \mathbf{s}^*_{(\alpha)} + \mathbf{m}^*_\alpha \cdot \boldsymbol{\tau} \cdot \dot{\mathbf{s}}^*_{(\alpha)} \\ &= \mathbf{m}^*_\alpha \cdot (\dot{\boldsymbol{\tau}} - \mathbf{L}^* \cdot \boldsymbol{\tau} + \boldsymbol{\tau} \cdot \mathbf{L}^*) \cdot \mathbf{s}^*_{(\alpha)}. \end{aligned} \qquad (28.38)$$

Since, from (28.25),

$$\dot{\boldsymbol{\tau}} = \overset{\triangledown}{\boldsymbol{\tau}}^* + \mathbf{W}^* \cdot \boldsymbol{\tau} - \boldsymbol{\tau} \cdot \mathbf{W}^*,$$

28.2. Stress and Stress Rates

we obtain

$$\begin{aligned}\dot{t}_\alpha &= \mathbf{m}_\alpha^* \cdot \left(\overset{\triangledown}{\boldsymbol{\tau}}{}^* + \mathbf{W}^* \cdot \boldsymbol{\tau} - \mathbf{L}^* \cdot \boldsymbol{\tau} + \boldsymbol{\tau} \cdot \mathbf{L}^* - \boldsymbol{\tau} \cdot \mathbf{W}^*\right) \cdot \mathbf{s}_{(\alpha)}^* \\ &= \mathbf{m}_\alpha^* \cdot \left(\overset{\triangledown}{\boldsymbol{\tau}}{}^* + \boldsymbol{\tau} \cdot \mathbf{D}^* - \mathbf{D}^* \cdot \boldsymbol{\tau}\right) \cdot \mathbf{s}_{(\alpha)}^* \\ &= \mathbf{m}_\alpha^* \cdot (\boldsymbol{\Lambda} : \mathbf{D}^* + \boldsymbol{\tau} \cdot \mathbf{D}^* - \mathbf{D}^* \cdot \boldsymbol{\tau}) \cdot \mathbf{s}_{(\alpha)}^* \\ &= \mathbf{m}_\alpha^* \cdot (\boldsymbol{\Lambda} : \mathbf{D}^*) \cdot \mathbf{s}_{(\alpha)}^* + \mathbf{m}_\alpha^* \cdot (\boldsymbol{\tau} \cdot \mathbf{D}^* - \mathbf{D}^* \cdot \boldsymbol{\tau}) \cdot \mathbf{s}_{(\alpha)}^*.\end{aligned} \qquad (28.39)$$

To proceed consider the quantity

$$\boldsymbol{\beta}_\alpha : \mathbf{D}^* = (\mathbf{Q}_\alpha \cdot \boldsymbol{\tau} - \boldsymbol{\tau} \cdot \mathbf{Q}_\alpha) \cdot \mathbf{D}^*. \qquad (28.40)$$

Recall that $\boldsymbol{\beta}_\alpha$ was introduced in (28.29) and $\mathbf{Q}_\alpha$ in (28.23), so that

$$\boldsymbol{\beta}_\alpha = \frac{1}{2}\left(\mathbf{s}_\alpha^* \mathbf{m}_{(\alpha)}^* \cdot \boldsymbol{\tau} - \mathbf{m}_\alpha^* \mathbf{s}_{(\alpha)}^* \cdot \boldsymbol{\tau}\right) - \frac{1}{2}\left(\boldsymbol{\tau} \cdot \mathbf{s}_\alpha^* \mathbf{m}_{(\alpha)}^* - \boldsymbol{\tau} \cdot \mathbf{m}_\alpha^* \mathbf{s}_{(\alpha)}^*\right). \qquad (28.41)$$

Upon expanding (28.40), we have

$$\begin{aligned}\boldsymbol{\beta}_\alpha : \mathbf{D}^* &= \frac{1}{2}\left(\mathbf{s}_\alpha^* \mathbf{m}_{(\alpha)}^* \cdot \boldsymbol{\tau}\right) : \mathbf{D}^* - \frac{1}{2}\left(\mathbf{m}_\alpha^* \mathbf{s}_{(\alpha)}^* \cdot \boldsymbol{\tau}\right) : \mathbf{D}^* \\ &\quad - \frac{1}{2}\left(\boldsymbol{\tau} \cdot \mathbf{s}_\alpha^* \mathbf{m}_{(\alpha)}^*\right) : \mathbf{D}^* + \frac{1}{2}\left(\boldsymbol{\tau} \cdot \mathbf{m}_\alpha^* \mathbf{s}_{(\alpha)}^*\right) : \mathbf{D}^*.\end{aligned} \qquad (28.42)$$

If this is expressed in component form, temporarily suppressing the Greek subscripts for clarity, we obtain

$$\begin{aligned}(\boldsymbol{\beta}_\alpha)_{ij} D_{ij}^* &= \frac{1}{2} s_i^* m_\ell^* \tau_{\ell j} D_{ij}^* - \frac{1}{2} m_i^* s_\ell^* \tau_{\ell j} D_{ij}^* \\ &\quad - \frac{1}{2} \tau_{i\ell} s_\ell^* m_j^* D_{ij}^* + \frac{1}{2} \tau_{i\ell} m_\ell^* s_j^* D_{ij}^* \\ &= \frac{1}{2} s_i^* m_\ell^* \tau_{\ell j} D_{ij}^* + \frac{1}{2} \tau_{j\ell} m_\ell^* s_i^* D_{ij}^* \\ &\quad - \frac{1}{2} m_i^* s_\ell^* \tau_{\ell j} D_{ij}^* - \frac{1}{2} \tau_{j\ell} s_\ell^* m_i^* D_{ij}^* \\ &= m_\ell^* \left(\tau_{\ell j} D_{ji}^* - D_{\ell j}^* \tau_{ji}\right) s_i^*.\end{aligned} \qquad (28.43)$$

This establishes the identity

$$\boldsymbol{\beta}_\alpha : \mathbf{D}^* = \mathbf{m}_\alpha^* \cdot (\boldsymbol{\tau} \cdot \mathbf{D}^* - \mathbf{D}^* \cdot \boldsymbol{\tau}) \cdot \mathbf{s}_{(\alpha)}^*. \qquad (28.44)$$

Since this appears in (28.39), that equation can be written as

$$\dot{t}_\alpha = \mathbf{m}_\alpha^* \cdot (\boldsymbol{\Lambda} : \mathbf{D}^*) \cdot \mathbf{s}_{(\alpha)}^* + \boldsymbol{\beta}_\alpha : \mathbf{D}^*. \qquad (28.45)$$

But,

$$\mathbf{m}_\alpha^* \cdot (\boldsymbol{\Lambda} : \mathbf{D}^*) \cdot \mathbf{s}_{(\alpha)}^* = \mathbf{m}_\alpha^* \mathbf{s}_{(\alpha)}^* : \boldsymbol{\Lambda} : \mathbf{D}^*, \qquad (28.46)$$

and, because of the inherent symmetry in $\boldsymbol{\Lambda}$, we have

$$\mathbf{m}_\alpha^* \mathbf{s}_{(\alpha)}^* : \boldsymbol{\Lambda} : \mathbf{D}^* = \operatorname{sym}\left(\mathbf{m}_\alpha^* \mathbf{s}_{(\alpha)}^*\right) : \boldsymbol{\Lambda} : \mathbf{D}^* = \mathbf{P}_\alpha : \boldsymbol{\Lambda} : \mathbf{D}^*. \qquad (28.47)$$

Thus, (28.45) becomes

$$\dot{\tau}_\alpha = \mathbf{P}_\alpha : \mathbf{\Lambda} : \mathbf{D}^* + \boldsymbol{\beta}_\alpha : \mathbf{D}^* = (\mathbf{P}_\alpha : \mathbf{\Lambda} + \boldsymbol{\beta}_\alpha) : \mathbf{D}^*. \tag{28.48}$$

This can be further expanded as

$$\begin{aligned}
\dot{\tau}_\alpha &= (\mathbf{P}_\alpha : \mathbf{\Lambda} + \boldsymbol{\beta}_\alpha) : (\mathbf{D} - \mathbf{d}^{\mathrm{p}}) \\
&= (\mathbf{P}_\alpha : \mathbf{\Lambda} + \boldsymbol{\beta}_\alpha) : \mathbf{D} - (\mathbf{P}_\alpha : \mathbf{\Lambda} + \boldsymbol{\beta}_\alpha) : \dot{\gamma}_\delta \mathbf{P}_\delta \\
&= (\mathbf{P}_\alpha : \mathbf{\Lambda} + \boldsymbol{\beta}_\alpha) : \mathbf{D} - \dot{\gamma}_\delta (\mathbf{P}_\alpha : \mathbf{\Lambda} : \mathbf{P}_\delta + \boldsymbol{\beta}_\alpha : \mathbf{P}_\delta).
\end{aligned} \tag{28.49}$$

28.2.2 Rate-Independent Strain Hardening

In the rate-independent case, the strain hardening on each slip system is taken as

$$\dot{\tau}_\alpha = h_{\alpha\delta} \dot{\gamma}_\delta. \tag{28.50}$$

Thus, by equating with (28.49),

$$h_{\alpha\delta} \dot{\gamma}_\delta = (\mathbf{P}_\alpha : \mathbf{\Lambda} + \boldsymbol{\beta}_\alpha) : \mathbf{D} - \dot{\gamma}_\delta (\mathbf{P}_\alpha : \mathbf{\Lambda} : \mathbf{P}_\delta + \boldsymbol{\beta}_\alpha : \mathbf{P}_\delta). \tag{28.51}$$

This leads to

$$(h_{\alpha\delta} + \mathbf{P}_\alpha : \mathbf{\Lambda} : \mathbf{P}_\delta + \boldsymbol{\beta}_\alpha : \mathbf{P}_\delta) \dot{\gamma}_\delta = (\mathbf{P}_\alpha : \mathbf{\Lambda} + \boldsymbol{\beta}_\alpha) : \mathbf{D}. \tag{28.52}$$

By defining

$$h_{\alpha\delta} + \mathbf{P}_\alpha : \mathbf{\Lambda} : \mathbf{P}_\delta + \boldsymbol{\beta}_\alpha : \mathbf{P}_\delta = N_{\alpha\delta}, \tag{28.53}$$

the more compact form of (28.52) is

$$N_{\alpha\delta} \dot{\gamma}_\delta = (\mathbf{P}_\alpha : \mathbf{\Lambda} + \boldsymbol{\beta}_\alpha) : \mathbf{D}. \tag{28.54}$$

If we introduce $M_{\alpha\delta} = N_{\alpha\delta}^{-1}$ such that

$$M_{\theta\alpha} N_{\alpha\delta} = \delta_{\theta\delta}, \tag{28.55}$$

the solution of (28.54) can be expressed as

$$\dot{\gamma}_\alpha = M_{\alpha\delta} (\mathbf{P}_\delta : \mathbf{\Lambda} + \boldsymbol{\beta}_\delta) : \mathbf{D}. \tag{28.56}$$

When (28.56) is substituted into (28.32), we obtain

$$\overset{\nabla}{\boldsymbol{\tau}} = \mathbf{\Lambda} : \left\{ \mathbf{D} - [M_{\alpha\delta} (\mathbf{P}_\delta : \mathbf{\Lambda} + \boldsymbol{\beta}_\delta) : \mathbf{D}] \left(\mathbf{P}_\alpha + \mathbf{\Lambda}^{-1} : \boldsymbol{\beta}_\alpha \right) \right\}. \tag{28.57}$$

Letting

$$\begin{aligned}
\psi^\alpha &= M_{\alpha\delta} (\mathbf{P}_\delta : \mathbf{\Lambda} + \boldsymbol{\beta}_\delta), \\
\chi^\alpha &= \mathbf{P}_\alpha + \mathbf{\Lambda}^{-1} : \boldsymbol{\beta}_\alpha,
\end{aligned} \tag{28.58}$$

the more compact form of (28.57) is

$$\begin{aligned}
\overset{\nabla}{\boldsymbol{\tau}} &= \mathbf{\Lambda} : [\mathbf{D} - (\psi^\alpha : \mathbf{D}) \chi^\alpha] \\
&= \mathbf{\Lambda} : \mathbf{D} - (\psi^\alpha : \mathbf{D}) (\mathbf{\Lambda} : \chi^\alpha).
\end{aligned} \tag{28.59}$$

28.3. Convected Elasticity

This, however, can be simplified further. Regressing to index notation, we have

$$\overset{\triangledown}{\tau}_{ij} = \Lambda_{ijkl} D_{kl} - \psi^\alpha_{kl} D_{kl} \Lambda_{ijmn} \chi^\alpha_{mn} = \left(\Lambda_{ijkl} - \Lambda_{ijmn} \chi^\alpha_{mn} \psi^\alpha_{kl} \right) D_{kl}. \tag{28.60}$$

Thus, by using (28.58),

$$\overset{\triangledown}{\tau} = [\Lambda - (\mathbf{P}_\alpha : \Lambda + \boldsymbol{\beta}_\alpha) M_{\alpha\delta} (\mathbf{P}_\delta : \Lambda + \boldsymbol{\beta}_\delta)] : \mathbf{D}. \tag{28.61}$$

Consequently, the rate-independent constitutive law for elastoplastic deformation of single crystals reads

$$\overset{\triangledown}{\tau} = \Lambda^{\text{e-p}} : \mathbf{D},$$
$$\Lambda^{\text{e-p}} = \Lambda - (\mathbf{P}_\alpha : \Lambda + \boldsymbol{\beta}_\alpha) M_{\alpha\delta} (\mathbf{P}_\delta : \Lambda + \boldsymbol{\beta}_\delta). \tag{28.62}$$

However, there are a number of strong proviso's that exist in above derivation. First, even when the inverse $M_{\alpha\delta}$ exists, it is not symmetric, i.e.,

$$M_{\alpha\delta} \neq M_{\delta\alpha}, \tag{28.63}$$

which implies that

$$\Lambda^{\text{e-p}}_{ijkl} \neq \Lambda^{\text{e-p}}_{klij}. \tag{28.64}$$

Second, it should be noted that the inverse $M_{\alpha\delta}$ may not always exist and thus additional stipulation is required to determine the $\dot{\gamma}_\alpha$ for a given $\mathbf{D}$. For instance, if all slip systems are of equal strength and with equal strain hardening properties, in fcc crystals with more than five linearly independent slip systems a unique set required to construct a given $\mathbf{D}$ is impossible to find without auxiliary requirements. Details can be found in the books by Havner (1992) and Lubarda (2002).

28.3 Convected Elasticity

Consider the rate of work per unit reference volume, i.e.,

$$\mathcal{R} = \boldsymbol{\tau} : \mathbf{D} = \boldsymbol{\tau} : \mathbf{D}^* + \boldsymbol{\tau} : \mathbf{d}^{\text{p}} = \boldsymbol{\tau} : \mathbf{D}^* + \dot{\gamma}_\alpha \boldsymbol{\tau} : \mathbf{P}_\alpha. \tag{28.65}$$

Because of the symmetry of $\boldsymbol{\tau}$, we may write

$$\boldsymbol{\tau} : \mathbf{P}_\alpha = \boldsymbol{\tau} : \text{sym}\,(\mathbf{s}^*_\alpha \mathbf{m}^*_\alpha) = \tau_\alpha = \mathbf{s}^*_\alpha \cdot \boldsymbol{\tau} \cdot \mathbf{m}^*_\alpha. \tag{28.66}$$

Thus,

$$\boldsymbol{\tau} : \mathbf{D} = \boldsymbol{\tau} : \mathbf{D}^* + \tau_\alpha \dot{\gamma}_\alpha. \tag{28.67}$$

Since the rate of lattice Lagrangian strain is

$$\dot{\mathbf{E}}^* = \mathbf{F}^{*T} \cdot \mathbf{D}^* \cdot \mathbf{F}^*, \quad \mathbf{D}^* = \mathbf{F}^{*-T} \cdot \dot{\mathbf{E}}^* \cdot \mathbf{F}^{*-1}, \tag{28.68}$$

we have

$$\boldsymbol{\tau} : \mathbf{D}^* = \boldsymbol{\tau} : \left(\mathbf{F}^{*-T} \cdot \dot{\mathbf{E}}^* \cdot \mathbf{F}^{*-1} \right) = \left(\mathbf{F}^{*-1} \cdot \boldsymbol{\tau} \cdot \mathbf{F}^{*-T} \right) : \dot{\mathbf{E}}^*. \tag{28.69}$$

Hence,

$$\boldsymbol{\tau} : \mathbf{D} = \left(\mathbf{F}^{*-1} \cdot \boldsymbol{\tau} \cdot \mathbf{F}^{*-T} \right) : \dot{\mathbf{E}}^* + \tau_\alpha \dot{\gamma}_\alpha. \tag{28.70}$$

The stress measure

$$\mathbf{S}^* = \mathbf{F}^{*-1} \cdot \boldsymbol{\tau} \cdot \mathbf{F}^{*-T} \tag{28.71}$$

is the second Piola–Kirchhoff stress relative to the lattice. If plasticity does not affect the elasticity, and if we assume the material is hyperelastic, then a reference volume based work increment is given as

$$\delta \Phi = S^{*ij} \delta E^*_{ij}. \tag{28.72}$$

This gives

$$S^{*ij} = \frac{\partial \Phi}{\partial E^*_{ij}}, \quad \dot{S}^{*ij} = \frac{\partial^2 \Phi}{\partial E^*_{ij} \partial E^*_{k\ell}} \dot{E}^*_{k\ell}. \tag{28.73}$$

We recall that slip induced plasticity considered here is volume preserving, so that $\det \mathbf{F} = \det \mathbf{F}^*$. The components of the elastic moduli tensor are

$$\Lambda^{ijkl} = \frac{\partial^2 \Phi}{\partial E^*_{ij} \partial E^*_{k\ell}}, \tag{28.74}$$

such that

$$\boldsymbol{\Lambda}^* = \Lambda^{ijk\ell} \mathbf{B}_i \mathbf{B}_j \mathbf{B}_k \mathbf{B}_\ell. \tag{28.75}$$

For later reference, we also define

$$\boldsymbol{\Lambda} = \Lambda^{ijk\ell} \mathbf{b}_i \mathbf{b}_j \mathbf{b}_k \mathbf{b}_\ell. \tag{28.76}$$

The vectors $\mathbf{b}_i$ are convected with the lattice deformation, as given in (28.17). Hence, the rate-type hyperelastic relation is

$$\dot{\mathbf{S}}^* = \boldsymbol{\Lambda}^* : \dot{\mathbf{E}}^*. \tag{28.77}$$

Taking the derivative of (28.71) yields, after some manipulation (see Problem 28.3 in Chapter 34),

$$\overset{\triangledown}{\boldsymbol{\tau}}{}^* = \mathbf{F}^* \cdot \left(\boldsymbol{\Lambda}^* : \dot{\mathbf{E}}^* \right) \cdot \mathbf{F}^{*T} + \mathbf{D}^* \cdot \boldsymbol{\tau} + \boldsymbol{\tau} \cdot \mathbf{D}^*, \tag{28.78}$$

or, in terms of $\mathbf{D}^*$,

$$\overset{\triangledown}{\boldsymbol{\tau}}{}^* = \mathbf{F}^* \cdot \left[\boldsymbol{\Lambda}^* : (\mathbf{F}^{*T} \cdot \mathbf{D}^* \cdot \mathbf{F}^*) \right] \cdot \mathbf{F}^{*T} + \mathbf{D}^* \cdot \boldsymbol{\tau} + \boldsymbol{\tau} \cdot \mathbf{D}^*. \tag{28.79}$$

We will examine the first term in (28.79). Toward this end, because $\mathbf{I} = \mathbf{B}_k \mathbf{B}^k$, we have

$$\mathbf{F}^* = \mathbf{b}_k \mathbf{B}^k, \quad \mathbf{F}^{*T} = \mathbf{B}^k \mathbf{b}_k. \tag{28.80}$$

Then, the inner term of the entire expression in question becomes

$$\begin{aligned}
\left[\boldsymbol{\Lambda}^* : (\mathbf{F}^{*T} \cdot \mathbf{D}^* \cdot \mathbf{F}^*) \right] &= \Lambda^{ijk\ell} \mathbf{B}_i \mathbf{B}_j \mathbf{B}_k \mathbf{B}_\ell : (\mathbf{B}^m \mathbf{b}_m \cdot \mathbf{D}^* \cdot \mathbf{b}_n \mathbf{B}^n) \\
&= \Lambda^{ijk\ell} D^*_{mn} \mathbf{B}_i \mathbf{B}_j \delta^m_{.k} \delta^n_{.\ell} \\
&= \Lambda^{ijmn} D^*_{mn} \mathbf{B}_i \mathbf{B}_j.
\end{aligned} \tag{28.81}$$

28.4. Rate-Dependent Slip

The full expression for the first term in (28.79) is accordingly

$$\begin{aligned}
\mathbf{F}^* \cdot \left(\Lambda^{ijmn} D^*_{mn} \mathbf{B}_i \mathbf{B}_j\right) \cdot \mathbf{F}^{*T} &= \Lambda^{ijmn} D^*_{mn} \mathbf{b}_k \mathbf{B}^k \cdot \mathbf{B}_i \mathbf{B}_j \cdot \mathbf{B}^\ell \mathbf{b}_\ell \\
&= \Lambda^{ijmn} D^*_{mn} \mathbf{b}_i \mathbf{b}_j = \Lambda^{ijmn} D^*_{k\ell} \delta^\ell_{.m} \delta^\ell_{.n} \mathbf{b}_i \mathbf{b}_j \\
&= \Lambda^{ijmn} D^*_{k\ell} (\mathbf{b}^k \cdot \mathbf{b}_m)(\mathbf{b}^\ell \cdot \mathbf{b}_n) \mathbf{b}_i \mathbf{b}_j = \Lambda^{ijmn} D^*_{k\ell} (\mathbf{b}_i \mathbf{b}_j \mathbf{b}_m \mathbf{b}_n) : (\mathbf{b}^k \mathbf{b}^\ell) \\
&= \Lambda : \mathbf{D}^*.
\end{aligned}$$

Thus, (28.79) becomes

$$\overset{\nabla}{\tau}{}^* = \Lambda : \mathbf{D}^* + \mathbf{D}^* \cdot \tau + \tau \cdot \mathbf{D}^*, \qquad (28.82)$$

$$\Lambda = \Lambda^{ijmn} \mathbf{b}_i \mathbf{b}_j \mathbf{b}_m \mathbf{b}_n.$$

If, as is typical for metals and alloys, $\|\Lambda\| \gg \|\tau\|$, in some suitable norm or component by component, then (28.82) may be approximated as

$$\overset{\nabla}{\tau}{}^* \approx \Lambda : \mathbf{D}^*. \qquad (28.83)$$

This approximation amounts to assuming that the magnitude of a typical component of stress is small as compared to a typical component of elastic modulus tensor. Note also that the tensor, Λ, used in (28.30) has now been properly defined and qualified.

28.4 Rate-Dependent Slip

As discussed in the previous chapter, slip in crystalline materials is rate-dependent, typically governed by thermally activated processes. Here we adopt a simple, isothermal, phenomenological description of the rate dependence, and write

$$\dot{\gamma}_\alpha = \dot{\gamma}_0 \frac{\tau_{(\alpha)}}{g_\alpha} \left|\frac{\tau_{(\alpha)}}{g_\alpha}\right|^{1/m-1}, \qquad (28.84)$$

where m is the so-called strain rate sensitivity parameter and $\dot{\gamma}_0$ is a reference strain rate. As before, $g_\alpha(\gamma)$ is a *hardness* parameter that depends on the accumulated shear strain, γ, defined as

$$\gamma = \sum_\alpha |\gamma_\alpha|. \qquad (28.85)$$

The hardness increases with on-going shear strain, and $g_\alpha(0) = \tau_0$. As $m \to 0$, the material displays sensibly rate-independent behavior. Strain hardening is prescribed by an evolutionary law akin to (28.50), *i.e.*,

$$\dot{g}_\alpha = h_{\alpha\beta} |\dot{\gamma}_\beta|. \qquad (28.86)$$

Various forms for the hardening matrix, $\mathbf{h}$, were discussed in Chapter 27. A simple, yet flexible, form for $\mathbf{h}$ is

$$h_{\alpha\beta} = h(\gamma)[q + (1-q)\delta_{\alpha\beta}], \qquad (28.87)$$

where q is the ratio of the latent and self-hardening rates.

28.4.1 A Rate Tangent Modulus

When relations such as (28.84) are used in constitutive relations such as (28.32), there are issues of stability and accuracy of numerical integration algorithms that arise because of the inherent severe nonlinearity of the kinetic laws for slip. Peirce, Asaro, and Needleman (1983), and later Harren and Asaro (1988), developed a *reduced modulus* scheme, or *rate tangent modulus* scheme, for dealing with such severe nonlinearity. The basic goal is to estimate an *effective slip rate* to use over a small yet finite time step in such relations as (28.32); that is, we seek $\dot{\gamma}_{\text{eff}} = \Delta\gamma/\Delta t$ for a given time step Δt.

With the convention that a subscript $,t$ means evaluate at time t, the kinetic law for slip is written as

$$\dot{\gamma}_{\alpha,t} = \dot{\gamma}_\alpha\left[\tau_{(\alpha),t}; g_{(\alpha),t}\right] \tag{28.88}$$

and accordingly

$$\dot{\gamma}_{\alpha,t+\Delta t} = \dot{\gamma}_\alpha\left[\tau_{(\alpha),t+\Delta t}; g_{(\alpha),t+\Delta t}\right]$$
$$= \dot{\gamma}_\alpha\left[\tau_{(\alpha),t} + \Delta\tau_{(\alpha)}; g_{(\alpha),t} + \Delta g_{(\alpha)}\right]. \tag{28.89}$$

Imagine that $\dot{\gamma}_{\alpha,t+\Delta t}$ is expanded in a Taylor series about the current time t, i.e.,

$$\dot{\gamma}_{\alpha,t+\Delta t} = \dot{\gamma}_\alpha\left[\tau_{(\alpha),t}; g_{(\alpha),t}\right] + \left.\frac{\partial \dot{\gamma}_\alpha}{\partial \tau_{(\alpha)}}\right|_t \Delta\tau_{(\alpha)} + \left.\frac{\partial \dot{\gamma}_\alpha}{\partial g_{(\alpha)}}\right|_t \Delta g_{(\alpha)}. \tag{28.90}$$

Thus, given the form (28.84), we have

$$\left.\frac{\partial \dot{\gamma}_\alpha}{\partial \tau_{(\alpha)}}\right|_t = \frac{\dot{\gamma}_0}{m\tau_{(\alpha),t}}\left[\frac{\tau_{(\alpha),t}}{g_{(\alpha),t}}\right]^{1/m} = \frac{\dot{\gamma}_{\alpha,t}}{m\tau_{(\alpha),t}},$$

$$\left.\frac{\partial \dot{\gamma}_\alpha}{\partial g_{(\alpha)}}\right|_t = -\frac{\dot{\gamma}_0}{m g_{(\alpha),t}}\left[\frac{\tau_{(\alpha),t}}{g_{(\alpha),t}}\right]^{1/m} = -\frac{\dot{\gamma}_{\alpha,t}}{m g_{(\alpha),t}}. \tag{28.91}$$

When (28.91) is substituted into (28.90), we obtain

$$\dot{\gamma}_{\alpha,t+\Delta t} = \dot{\gamma}_{\alpha,t}\left\{1 + \frac{1}{m}\left[\frac{\Delta\tau_{(\alpha)}}{\tau_{(\alpha),t}} - \frac{\Delta g_{(\alpha)}}{g_{(\alpha),t}}\right]\right\}. \tag{28.92}$$

Now, write the difference relation

$$\frac{1}{\Delta t}(\gamma_{\alpha,t+\Delta t} - \gamma_{\alpha,t}) = \frac{\Delta\gamma_\alpha}{\Delta t} = (1-\theta)\dot{\gamma}_{\alpha,t} + \theta\dot{\gamma}_{\alpha,t+\Delta t}, \tag{28.93}$$

where $\theta = 0$ leads to a forward Euler integration scheme and $\theta = 1$ to an Euler integration scheme. Substituting (28.92) into (28.93) yields

$$\frac{\Delta\gamma_\alpha}{\Delta t} = \dot{\gamma}_{(\alpha),t}\left\{1 + \frac{\theta}{m}\left[\frac{\Delta\tau_{(\alpha)}}{\tau_{(\alpha),t}} - \frac{\Delta g_{(\alpha)}}{g_{(\alpha),t}}\right]\right\}. \tag{28.94}$$

Recalling the relation for $\dot{\tau}_\alpha$ from (28.49), and defining

$$\mathbf{R}_\alpha = \mathbf{P}_\alpha : \mathbf{\Lambda} + \boldsymbol{\beta}_\alpha, \tag{28.95}$$

we obtain

$$\dot{\tau}_\alpha = \mathbf{R}_\alpha : \mathbf{D} - \mathbf{R}_\alpha : \mathbf{P}_\beta \dot{\gamma}_\beta. \tag{28.96}$$

28.4. Rate-Dependent Slip

The hardening relation gives $\dot{g}_\alpha$ as

$$\dot{g}_\alpha = \text{sgn}[\dot{\gamma}_{(\beta)}]h_{\alpha\beta}\dot{\gamma}_\beta. \tag{28.97}$$

Thus, if

$$\dot{\tau}_{\alpha,t}\Delta t = \Delta\tau_\alpha, \quad \dot{g}_{\alpha,t}\Delta t = \Delta g_\alpha, \tag{28.98}$$

we obtain

$$\Delta\tau_\alpha = \mathbf{R}_{\alpha,t}:\mathbf{D}_t\Delta t - \mathbf{R}_{\alpha,t}:\mathbf{P}_{\beta,t}\dot{\gamma}_{\beta,t}\Delta t,$$
$$\Delta g_\alpha = \text{sgn}[\dot{\gamma}_{(\beta),t}]h_{\alpha\beta,t}\dot{\gamma}_{\beta,t}\Delta t. \tag{28.99}$$

When the relations (28.99) are substituted into (28.94), there follows

$$\Delta\gamma_\alpha + \frac{\dot{\gamma}_{(\alpha),t}\theta\Delta t}{m\tau_{(\alpha),t}}\mathbf{R}_{\alpha,t}:\mathbf{P}_{\beta,t}\dot{\gamma}_{\beta,t}\Delta t + \frac{\dot{\gamma}_{(\alpha),t}\theta\Delta t}{m\tau_{(\alpha),t}}\frac{\tau_{(\alpha),t}}{g_{(\alpha),t}}\text{sgn}[\dot{\gamma}_{(\beta),t}]h_{\alpha\beta,t}\dot{\gamma}_{\beta,t}\Delta t$$
$$= \left[\dot{\gamma}_{\alpha,t} + \frac{\dot{\gamma}_{(\alpha),t}\theta\Delta t}{m\tau_{(\alpha),t}}\mathbf{R}_{\alpha,t}:\mathbf{D}_t\right]\Delta t. \tag{28.100}$$

Using again the Euler estimate

$$\dot{\gamma}_{\beta,t}\Delta t = \Delta\gamma_\beta \tag{28.101}$$

in (28.100), there follows

$$\left\{\delta_{\alpha\beta} + \frac{\dot{\gamma}_{(\alpha)}\theta\Delta t}{m\tau_{(\alpha)}}\left[\mathbf{R}_\alpha:\mathbf{P}_\beta + \text{sgn}[\dot{\gamma}_{(\beta)}]\frac{\tau_{(\alpha)}}{g_{(\alpha)}}h_{\alpha\beta}\right]\right\}\Delta\gamma_\beta$$
$$= \left[\dot{\gamma}_{\alpha,t} + \frac{\dot{\gamma}_{(\alpha)}\theta\Delta t}{m\tau_{(\alpha)}}\mathbf{R}_\alpha:\mathbf{D}\right]\Delta t. \tag{28.102}$$

If we define

$$N_{\alpha\beta} = \delta_{\alpha\beta} + \frac{\dot{\gamma}_{(\alpha)t}\theta\Delta t}{m\tau_{(\alpha)}}\left[\mathbf{R}_\alpha:\mathbf{P}_\beta + \text{sgn}[\dot{\gamma}_{(\beta)}]\frac{\tau_{(\alpha)}}{g_{(\alpha)}}h_{\alpha\beta}\right],$$
$$\mathbf{Q}_\alpha = \frac{\dot{\gamma}_{(\alpha)}\theta\Delta t}{m\tau_{(\alpha)}}\mathbf{R}_\alpha, \tag{28.103}$$

the equation (28.102) becomes

$$N_{\alpha\beta}\Delta\gamma_\beta = (\dot{\gamma}_\alpha + \mathbf{Q}_\alpha:\mathbf{D})\Delta t. \tag{28.104}$$

Note that $\text{sgn}(\tau_\beta)$ may be used instead of $\text{sgn}(\dot{\gamma}_\beta)$ if convenient. If the inversion of $N_{\alpha\beta}$ is defined by

$$M_{\delta\alpha}N_{\alpha\beta} = \delta_{\delta\beta}, \tag{28.105}$$

we can rewrite (28.104) as

$$\Delta\gamma_\alpha = (M_{\alpha\beta}\dot{\gamma}_\beta + M_{\alpha\beta}\mathbf{Q}_\beta:\mathbf{D})\Delta t. \tag{28.106}$$

The $N_{\alpha\beta}$ and $M_{\delta\alpha}$ may be compared to their rate-independent analogs in (28.53)–(28.56), but we now note that the inverse $M_{\delta\alpha}$ always exists. This is evident from the fact that with Δt small enough, $N_{\alpha\beta}$ is but a perturbation away from unity.

Defining, for compactness,

$$M_{\alpha\beta}\dot{\gamma}_\beta = \dot{f}_\alpha, \quad M_{\alpha\beta}\mathbf{Q}_\beta = \mathbf{F}_\alpha, \tag{28.107}$$

(28.106) can be recast in the form

$$\Delta\gamma_\alpha = \left(\dot{f}_\alpha + \mathbf{F}_\alpha : \mathbf{D}\right)\Delta t. \tag{28.108}$$

Thus, we arrive at the effective slip rate

$$\dot{\gamma}_{\alpha,t}^{\diamond} = \dot{f}_{\alpha,t} + \mathbf{F}_{\alpha,t} : \mathbf{D}_t. \tag{28.109}$$

When (28.109) is used in (28.31), *i.e.*, in either of

$$\overset{\nabla}{\boldsymbol{\tau}} = \boldsymbol{\Lambda} : \mathbf{D} - \dot{\gamma}_\alpha \left(\boldsymbol{\Lambda} : \mathbf{P}_\alpha + \boldsymbol{\beta}_\alpha\right)$$
$$= \boldsymbol{\Lambda} : \mathbf{D} - \dot{\gamma}_\alpha \mathbf{R}_\alpha, \tag{28.110}$$

in place of $\dot{\gamma}_\alpha$, we obtain

$$\overset{\nabla}{\boldsymbol{\tau}}_t = \boldsymbol{\Lambda} : \mathbf{D}_t - \dot{\gamma}_{\alpha,t}^{\diamond} \mathbf{R}_{\alpha,t}$$
$$= \boldsymbol{\Lambda} : \mathbf{D}_t - \left(\dot{f}_{\alpha,t} + \mathbf{F}_{\alpha,t} : \mathbf{D}_t\right)\mathbf{R}_{\alpha,t} \tag{28.111}$$
$$= (\boldsymbol{\Lambda} - \mathbf{R}_{\alpha,t}\mathbf{F}_{\alpha,t}) : \mathbf{D}_t - \dot{f}_{\alpha,t}\mathbf{R}_{\alpha,t}.$$

Of course, from (28.110), we also have

$$\mathbf{R}_\alpha = \boldsymbol{\Lambda} : \mathbf{P}_\alpha + \boldsymbol{\beta}_\alpha. \tag{28.112}$$

By defining

$$\boldsymbol{\Lambda}_t^R = \boldsymbol{\Lambda} - \mathbf{R}_{\alpha,t}\mathbf{F}_{\alpha,t}, \tag{28.113}$$

so that

$$\overset{\nabla}{\boldsymbol{\tau}}_t = \boldsymbol{\Lambda}_t^R : \mathbf{D}_t - \dot{f}_{\alpha,t}\mathbf{R}_{\alpha,t}, \tag{28.114}$$

we finally obtain

$$\overset{\nabla}{\boldsymbol{\tau}} = \mathbf{C}_t^R : \mathbf{D} - \dot{f}_\alpha \mathbf{R}_{\alpha,t}, \tag{28.115}$$

where

$$\boldsymbol{\Lambda}^R = \boldsymbol{\Lambda} - \frac{\dot{\gamma}_{(\beta)}\theta\Delta t}{m\tau_{(\beta)}}\left(\mathbf{P}_\alpha : \boldsymbol{\Lambda} + \boldsymbol{\beta}_\alpha\right)M_{\alpha\beta}\left(\mathbf{P}_\beta : \boldsymbol{\Lambda} + \boldsymbol{\beta}_\beta\right). \tag{28.116}$$

Note that, in general, the moduli $\Lambda^R_{ijk\ell}$ do not possess reciprocal symmetry, *i.e.*,

$$\Lambda^R_{ijk\ell} \neq \Lambda^R_{k\ell ij}. \tag{28.117}$$

28.5 Crystalline Component Forms

Consider a typical crystalline grain, and let the local covariant and contravariant base vectors of the reference configuration be $\{\mathbf{a}_i\}$ and $\{\mathbf{a}^i\}$, respectively. Note that for cubic

28.5. Crystalline Component Forms

crystals, it is most natural to take these as Cartesian, in which case $\{\mathbf{a}_i\} = \{\mathbf{a}^i\}$. Let $\{\mathbf{a}_i^*\}$ and $\{\mathbf{a}^{*i}\}$ be the convected base vectors with respect to $\mathbf{F}^*$, and let $\{\bar{\mathbf{a}}_i\}$ and $\{\bar{\mathbf{a}}^i\}$ be those convected with the total $\mathbf{F}$. Thus, we have

$$\mathbf{a}_i^* = \mathbf{F}^* \cdot \mathbf{a}_i = \mathbf{a}_i \cdot \mathbf{F}^{*T}, \quad \mathbf{a}_i = \mathbf{F}^{*-1} \cdot \mathbf{a}_i^* = \mathbf{a}_i^* \cdot \mathbf{F}^{*-T},$$
$$\mathbf{a}^{*i} \cdot \mathbf{F}^* = \mathbf{a}^i, \quad \mathbf{F}^{*T} \cdot \mathbf{a}^{*i} = \mathbf{a}^i, \quad \mathbf{a}^{*i} = \mathbf{a}^i \cdot \mathbf{F}^{*-1}, \quad \mathbf{a}^{*i} = \mathbf{F}^{*-T} \cdot \mathbf{a}^i. \tag{28.118}$$

Similarly, we have

$$\bar{\mathbf{a}}_i = \mathbf{F} \cdot \mathbf{a}_i, \quad \bar{\mathbf{a}}_i = \mathbf{a}_i \cdot \mathbf{F}^T, \quad \mathbf{a}_i = \mathbf{F}^{-1} \cdot \bar{\mathbf{a}}_i, \quad \mathbf{a}_i = \bar{\mathbf{a}}_i \cdot \mathbf{F}^{-T},$$
$$\mathbf{a}^i = \bar{\mathbf{a}}^i \cdot \mathbf{F}, \quad \mathbf{a}^i = \mathbf{F}^T \cdot \bar{\mathbf{a}}^i, \quad \bar{\mathbf{a}}^i = \mathbf{a}^i \cdot \mathbf{F}^{-1}, \quad \bar{\mathbf{a}}^i = \mathbf{F}^{-T} \cdot \mathbf{a}^i. \tag{28.119}$$

For later reference, we record the metrics

$$\mathbf{a}_i \cdot \mathbf{a}_j = a_{ij}, \quad \mathbf{a}^i \cdot \mathbf{a}^j = a^{ij}, \quad a = \det(a_{ij}),$$
$$\mathbf{a}_i^* \cdot \mathbf{a}_j^* = a_{ij}^*, \quad \mathbf{a}^{*i} \cdot \mathbf{a}^{*j} = a^{*ij}, \quad a^* = \det(a_{ij}^*), \tag{28.120}$$
$$\bar{\mathbf{a}}_i \cdot \bar{\mathbf{a}}_j = \bar{a}_{ij}, \quad \bar{\mathbf{a}}^i \cdot \bar{\mathbf{a}}^j = \bar{a}^{ij}, \quad \bar{a} = \det(\bar{a}_{ij}).$$

We have already shown in (28.82) that

$$\mathbf{\Lambda} = \Lambda^{ijk\ell} \mathbf{a}_i^* \mathbf{a}_j^* \mathbf{a}_k^* \mathbf{a}_\ell^*, \tag{28.121}$$

and

$$\overset{\nabla}{\boldsymbol{\tau}}{}^* = \mathbf{\Lambda} : \mathbf{D}^*. \tag{28.122}$$

Now, consider the reduced stiffness tensor $\mathbf{C}^R$ of (28.116), and the constitutive relation developed in (28.110), viz.,

$$\overset{\nabla}{\boldsymbol{\tau}} = \mathbf{\Lambda}^R : \mathbf{D} - \dot{f}_\alpha \left(\mathbf{P}_\alpha : \mathbf{\Lambda} + \boldsymbol{\beta}_\alpha \right). \tag{28.123}$$

Letting

$$\mathbf{P}_\alpha = P_{ij}^\alpha \mathbf{a}^{*i} \mathbf{a}^{*j}, \quad \boldsymbol{\beta}_\alpha = \beta_\alpha^{ij} \mathbf{a}_i^* \mathbf{a}_j^*, \tag{28.124}$$

yields

$$\mathbf{\Lambda}^R = \Lambda^{Rijk\ell} \mathbf{a}_i^* \mathbf{a}_j^* \mathbf{a}_k^* \mathbf{a}_\ell^*, \tag{28.125}$$

where

$$\Lambda^{Rijk\ell} = \Lambda^{ijk\ell} - \frac{\dot{\gamma}_{(\beta)} \theta \Delta t}{m \tau_{(\beta)}} \left(P_{mn}^\alpha \Lambda^{mnij} + \beta_\alpha^{ij} \right) M_{\alpha\beta} \left(P_{rs}^\beta \Lambda^{rsk\ell} + \beta_\beta^{k\ell} \right). \tag{28.126}$$

But, the component forms

$$\overset{\nabla}{\boldsymbol{\tau}} = \overset{\nabla}{\tau}{}^{ij} \bar{\mathbf{a}}_i \bar{\mathbf{a}}_j, \quad \mathbf{D} = D_{k\ell} \bar{\mathbf{a}}^k \bar{\mathbf{a}}^\ell, \quad \boldsymbol{\tau} = \tau^{ij} \bar{\mathbf{a}}_i \bar{\mathbf{a}}_j, \tag{28.127}$$

are used and hence the components of $\mathbf{\Lambda}^R$ on the $\{\bar{\mathbf{a}}_i\}$ bases are needed. To this end, we write

$$\mathbf{\Lambda}^R = \Lambda^{Rijk\ell} \mathbf{a}_i^* \mathbf{a}_j^* \mathbf{a}_k^* \mathbf{a}_\ell^* = \bar{\Lambda}^{Rijk\ell} \bar{\mathbf{a}}_i \bar{\mathbf{a}}_j \bar{\mathbf{a}}_k \bar{\mathbf{a}}_\ell. \tag{28.128}$$

Then
$$\bar{\Lambda}^{Rijk\ell} = \bar{\mathbf{a}}^j \cdot (\bar{\mathbf{a}}^i \cdot \mathbf{\Lambda}^R \cdot \bar{\mathbf{a}}^\ell) \cdot \bar{\mathbf{a}}^k$$
$$= \Lambda^{Rmnrs}(\bar{\mathbf{a}}^i \cdot \mathbf{a}_m^*)(\bar{\mathbf{a}}^j \cdot \mathbf{a}_n^*)(\bar{\mathbf{a}}^\ell \cdot \mathbf{a}_s^*)(\bar{\mathbf{a}}^k \cdot \mathbf{a}_r^*). \tag{28.129}$$

Since
$$\bar{\mathbf{a}}^i \cdot \mathbf{a}_m^* = (\mathbf{a}^i \cdot \mathbf{F}^{-1}) \cdot (\mathbf{F}^* \cdot \mathbf{a}_m) = \mathbf{a}^i \cdot \mathbf{F}^{p-1} \cdot \mathbf{a}_m = \{F^{p-1}\}^i_{.m}, \tag{28.130}$$

and similarly for other dot products, we obtain
$$\bar{\Lambda}^{Rijk\ell} = \Lambda^{Rmnrs}\{F^{p-1}\}^i_{.m}\{F^{p-1}\}^j_{.n}\{F^{p-1}\}^\ell_{.s}\{F^{p-1}\}^k_{.r}. \tag{28.131}$$

The same is required for $\mathbf{R}_\alpha$, see (28.112), and so write
$$\mathbf{R}_\alpha = R^{ij}_\alpha \mathbf{a}_i^* \mathbf{a}_j^* = \bar{R}^{ij}_\alpha \bar{\mathbf{a}}_i \bar{\mathbf{a}}_j. \tag{28.132}$$

By analogous manipulations, we arrive at
$$\bar{R}^{ij}_\alpha = R^{mn}_\alpha \{F^{p-1}\}^i_{.m}\{F^{p-1}\}^j_{.n}. \tag{28.133}$$

Consequently,
$$\overset{\triangledown}{\tau}{}^{ij} = \bar{\Lambda}^{Rijk\ell} D_{k\ell} - \dot{f}_\alpha \bar{R}^{ij}_\alpha. \tag{28.134}$$

The explicit forms for $\mathbf{P}_\alpha$, β_α and $\mathbf{Q}_\alpha$ are needed on the lattice bases to compute (28.134). Recall first that
$$\mathbf{P}_\alpha = \mathrm{sym}\,(\mathbf{s}_\alpha^* \mathbf{m}_\alpha^*) = \frac{1}{2}(\mathbf{s}_\alpha^* \mathbf{m}_\alpha^* + \mathbf{m}_\alpha^* \mathbf{s}_\alpha^*),$$
$$\mathbf{Q}_\alpha = \mathrm{skew}\,(\mathbf{s}_\alpha^* \mathbf{m}_\alpha^*) = \frac{1}{2}(\mathbf{s}_\alpha^* \mathbf{m}_\alpha^* - \mathbf{m}_\alpha^* \mathbf{s}_\alpha^*). \tag{28.135}$$

Then, since
$$\mathbf{s}_\alpha = s_\alpha^i \mathbf{a}_i, \quad \mathbf{m}_{(\alpha)} = m_{(\alpha)i} \mathbf{a}^i,$$
$$\mathbf{s}_\alpha^* = \mathbf{F}^* \cdot s_\alpha^i \mathbf{a}_i = s_\alpha^i (\mathbf{F}^* \cdot \mathbf{a}_i) = s_\alpha^i \mathbf{a}_i^*, \tag{28.136}$$
$$\mathbf{m}_{(\alpha)}^* = \mathbf{F}^{*-T} \cdot m_{(\alpha)i} \mathbf{a}^i = m_{(\alpha)i} \mathbf{a}^{*i},$$

the tensors
$$\mathbf{P}_\alpha = P^\alpha_{ij} \mathbf{a}^{*i} \mathbf{a}^{*j}, \quad \mathbf{Q}_\alpha = Q^\alpha_{ij} \mathbf{a}^{*i} \mathbf{a}^{*j} \tag{28.137}$$

have the components
$$P^\alpha_{ij} = \frac{1}{2}\left[s_\alpha^k a^*_{ki} m_{(\alpha)j} + m_{(\alpha)i} a^*_{jk} s_\alpha^k\right],$$
$$Q^\alpha_{ij} = \frac{1}{2}\left[s_\alpha^k a^*_{ki} m_{(\alpha)j} - m_{(\alpha)i} a^*_{jk} s_\alpha^k\right]. \tag{28.138}$$

Furthermore,
$$\beta_\alpha = \mathbf{Q}_\alpha \cdot \tau - \tau \cdot \mathbf{Q}_\alpha = \beta^{ij}_\alpha \mathbf{a}_i^* \mathbf{a}_j^*, \tag{28.139}$$

with
$$\tau = \tau^{k\ell} \bar{\mathbf{a}}_k \bar{\mathbf{a}}_\ell = \tau^{*k\ell} \mathbf{a}_k^* \mathbf{a}_\ell^*. \tag{28.140}$$

28.5. Crystalline Component Forms

It is easy enough to show that

$$\tau^{*k\ell} = \mathbf{a}^{*k} \cdot \tau \cdot \mathbf{a}^{*\ell} = \tau^{mn} \{F^p\}^k_{.m} \{F^p\}^\ell_{.n}, \tag{28.141}$$

and thus

$$\beta^{ij}_\alpha = Q^\alpha_{mn} \left(a^{*im} \tau^{*nj} + a^{*mj} \tau^{*in} \right). \tag{28.142}$$

Consequently,

$$\beta^{ij}_\alpha = Q^\alpha_{mn} \tau^{k\ell} \left[a^{*im} (F^p)^n_{.k} (F^p)^j_{.\ell} + a^{*mj} (F^p)^i_{.k} (F^p)^n_{.\ell} \right]. \tag{28.143}$$

Note that the symmetry of τ leads immediately to

$$\beta^{ij}_\alpha = \beta^{ji}_\alpha. \tag{28.144}$$

From the definition of $\mathbf{R}_\alpha$ this, in turn, means that

$$R^{ij}_\alpha = R^{ji}_\alpha, \tag{28.145}$$

which yields the result

$$\bar{R}^{ji}_\alpha = R^{mn}_\alpha \{F^{p-1}\}^j_{.n} \{F^{p-1}\}^i_{.m}. \tag{28.146}$$

28.5.1 Additional Crystalline Forms

Examine the rate of Lagrangian strain, *viz.*,

$$\dot{\mathbf{E}} = \mathbf{F}^T \cdot \mathbf{D} \cdot \mathbf{F}, \quad \mathbf{L} = \dot{\mathbf{F}} \cdot \mathbf{F}^{-1}. \tag{28.147}$$

Since

$$\mathbf{D} = \frac{1}{2} \left(\dot{\mathbf{F}} \cdot \mathbf{F}^{-1} + \mathbf{F}^{-T} \cdot \dot{\mathbf{F}}^T \right), \tag{28.148}$$

and thus

$$\mathbf{F}^T \cdot \mathbf{D} \cdot \mathbf{F} = \text{sym} \left(\mathbf{F}^T \cdot \dot{\mathbf{F}} \right), \tag{28.149}$$

we have

$$\dot{\mathbf{E}} = \text{sym} \left(\mathbf{F}^T \cdot \dot{\mathbf{F}} \right). \tag{28.150}$$

Of course, (28.150) is most naturally expressed in components on the $\{\mathbf{a}_i\}$ basis, *i.e.*,

$$\dot{E}_{mn} = \text{sym} \left(F^k_{.m} \dot{F}_{kn} \right). \tag{28.151}$$

Thus, if $\mathbf{D}$ is expressed most naturally on the $\{\bar{\mathbf{a}}^m\}$ basis as

$$\mathbf{D} = D_{mn} \bar{\mathbf{a}}^m \bar{\mathbf{a}}^n, \tag{28.152}$$

we have

$$D_{mn} = \text{sym} \left(F^k_{.m} \dot{F}_{kn} \right) = \dot{E}_{mn}. \tag{28.153}$$

Next, consider the nominal stress, $\mathbf{P}$, defined *via*

$$\mathbf{P} = \mathbf{F}^{-1} \cdot \tau. \tag{28.154}$$

Note that this definition of the nominal stress **P** is the transpose of the tensor definition used in Chapter 5. If **P** is expressed on the reference basis $\{\mathbf{a}_i\}$, then

$$\mathbf{P} = P^{ij}\mathbf{a}_i\mathbf{a}_j, \quad P_{ij} = \mathbf{a}^i \cdot \mathbf{P} \cdot \mathbf{a}^j. \tag{28.155}$$

Thus,

$$P^{ij} = \mathbf{a}^i \cdot \mathbf{F}^{-1} \cdot \boldsymbol{\tau} \cdot \mathbf{a}^j = \bar{\mathbf{a}}^i \cdot \boldsymbol{\tau} \cdot \mathbf{a}^j. \tag{28.156}$$

But,

$$\mathbf{a}^j = \mathbf{a}^j \cdot \mathbf{I} = \mathbf{a}^j \cdot (\bar{\mathbf{a}}_k \bar{\mathbf{a}}^k) = (\mathbf{a}^j \cdot \bar{\mathbf{a}}_k)\bar{\mathbf{a}}^k = F^j_{.k}\bar{\mathbf{a}}^k. \tag{28.157}$$

Using (28.157) in (28.156), we then obtain

$$P^{ij} = \tau^{iq} F^j_{.q}. \tag{28.158}$$

Now, reconsider the constitutive relation (28.134), *viz.*,

$$\overset{\nabla}{\tau}{}^{ij} = \bar{\Lambda}^{Rijk\ell} D_{k\ell} - \dot{f}_\alpha \bar{R}^{ij}_\alpha. \tag{28.159}$$

Since

$$\begin{aligned}\overset{\nabla}{\tau}{}^{ij} &= \dot{\tau}^{ij} + D^i_{.k}\tau^{kj} + \tau^{ik} D^j_{.k} \\ &= \dot{\tau}^{ij} + \bar{a}^{ip}\tau^{jq} D_{pq} + \bar{a}^{jp}\tau^{iq} D_{pq}, \end{aligned} \tag{28.160}$$

(28.159) can be rewritten as

$$\dot{\tau}^{ij} = \bar{\Lambda}^{Rijk\ell} D_{k\ell} - \bar{\bar{\Lambda}}^{ijk\ell} D_{k\ell} - \dot{f}_\alpha \bar{R}^{ij}_\alpha, \tag{28.161}$$

where

$$\bar{\bar{\Lambda}}^{ijk\ell} = \bar{a}^{ik}\tau^{j\ell} + \bar{a}^{jk}\tau^{i\ell}, \tag{28.162}$$

which has the symmetry

$$\bar{\bar{\Lambda}}^{ijk\ell} = \bar{\bar{\Lambda}}^{ij\ell k}. \tag{28.163}$$

Defining

$$\hat{\Lambda}^{ijk\ell} = \bar{\Lambda}^{Rijk\ell} - \bar{\bar{\Lambda}}^{ijk\ell}, \tag{28.164}$$

which also possesses the symmetry in $k\ell \rightleftharpoons \ell k$, we have

$$\dot{\tau}^{ij} = \hat{\Lambda}^{ijk\ell} D_{k\ell} - \dot{f}_\alpha \bar{R}^{ij}_\alpha. \tag{28.165}$$

Now, use (28.153) in (28.165), noting the $k\ell$ symmetry, to obtain

$$\dot{\tau}^{ik} = \hat{\Lambda}^{ikmn} F^s_{.m} \dot{F}_{sn} - \dot{f}_\alpha \bar{R}^{ik}_\alpha. \tag{28.166}$$

Hence, by differentiating (28.158), we find

$$\dot{P}^{ij} = \dot{\tau}^{ik} F^j_{.k} + \tau^{ik} \dot{F}^j_{.k}, \tag{28.167}$$

and, after incorporating (28.166),

$$\dot{P}^{ij} = \hat{\Lambda}^{iprk} F^\ell_{.r} F^j_{.p} \dot{F}_{\ell k} + \tau^{ik} a^{j\ell} \dot{F}_{\ell k} - \dot{f}_\alpha \bar{R}^{ik}_\alpha F^j_{.k}. \tag{28.168}$$

28.6. Suggested Reading

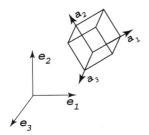

Figure 28.2. Laboratory *vs.* local crystal coordinates.

Finally, if we define

$$K^{ijk\ell} = \hat{\Lambda}^{iprk} F^{\ell}_{.r} F^{j}_{.p} + \tau^{ik} a^{j\ell},$$

$$\dot{B}^{ij} = \dot{f}_{\alpha} \bar{R}^{ik}_{\alpha} F^{j}_{.k},$$
(28.169)

we obtain

$$\dot{P}^{ij} = K^{ijk\ell} \dot{F}_{\ell k} - \dot{B}^{ij}.$$
(28.170)

28.5.2 Component Forms on Laboratory Axes

Suppose the given grain (crystallite) in question is referred to a set of fixed laboratory axes, *i.e.*, the $\{\mathbf{e}_i\}$ basis. Define the transformation

$$\Phi_{ij}\mathbf{e}_j = \mathbf{a}_i,$$
(28.171)

as shown in Fig. 28.2. Assuming that the bases $\{\mathbf{a}_i\}$ and $\{\mathbf{a}^i\}$ are indeed Cartesian, (28.170) can equally well be written as

$$\dot{P}_{ij} = K_{ijk\ell} \dot{F}_{\ell k} - \dot{B}_{ij}.$$
(28.172)

Then, if

$$\mathbf{K} = K_{ijk\ell}\, \mathbf{a}_i \mathbf{a}_j \mathbf{a}_k \mathbf{a}_\ell = \mathcal{K}_{ijk\ell}\, \mathbf{e}_i \mathbf{e}_j \mathbf{e}_k \mathbf{e}_\ell,$$
(28.173)

we have

$$\mathcal{K}_{mnrs} = \Phi_{im}\Phi_{jn} K_{ijk\ell}\, \Phi_{kr}\Phi_{\ell s}.$$
(28.174)

Similarly, if $\dot{\mathcal{P}}_{ij}$ are the rates of nominal stress on the laboratory basis,

$$\dot{\mathcal{P}}_{mn} = \Phi_{im}\dot{P}_{ij}\Phi_{jn},$$
(28.175)

it follows that

$$\dot{\mathcal{B}}_{mn} = \Phi_{im}\dot{B}_{ij}\Phi_{jn}, \quad \dot{\mathcal{F}}_{rs} = \Phi_{\ell r}\dot{F}_{\ell k}\Phi_{ks}.$$
(28.176)

On the $\{\mathbf{e}_i\}$ basis, we accordingly have

$$\dot{\mathcal{P}}_{mn} = \mathcal{K}_{mnrs}\dot{\mathcal{F}}_{rs} - \dot{\mathcal{B}}_{mn}.$$
(28.177)

28.6 Suggested Reading

Asaro, R. J. (1983a), Crystal Plasticity, *J. Appl. Mech.*, Vol. 50, pp. 921–934.

Asaro, R. J. (1983b), Micromechanics of Crystals and Polycrystals, *Adv. Appl. Mech.*, Vol. 23, pp. 1–115.

Bassani, J. L. (1990), Single Crystal Hardening, *Appl. Mech. Rev.*, Vol. 43, pp. S320–S327.
Bassani, J. L. (1993), Plastic Flow of Crystals, *Adv. Appl. Mech.*, Vol. 30, pp. 191–258.
Havner, K. S. (1992), *Finite Plastic Deformation of Crystalline Solids*, Cambridge University Press, Cambridge, UK.
Hill, R. (1978), Aspects of Invariance in Solid Mechanics, *Adv. Appl. Mech.*, Vol. 18, pp. 1–75.
Hill, R., and Rice, J. R. (1972), Constitutive Analysis of Elastic-Plastic Crystals at Arbitrary Strains, *J. Mech. Phys. Solids*, Vol. 20, pp. 401–413.
Lubarda, V. A. (2002), *Elastoplasticity Theory*, CRC Press, Boca Raton, Florida.
Nemat-Nasser, S. (2004), *Plasticity: A Treatise on Finite Deformation of Heterogeneous Inelastic Materials*, Cambridge University Press, New York.

29 The Nature of Crystalline Deformation: Localized Plastic Deformation

The previous three chapters have laid out some of the basic phenomenological features of plastic deformation in crystals and have developed a mathematical constitutive framework for analyzing crystalline deformation. It is not the purpose herein to provide an exhaustive treatment of particular case studies, in particular through the review of various numerical studies that have been performed, as this is the subject of a rather different volume. We do, however, explore some of the phenomenological implications of the mechanisms and theory developed above *vis-à-vis* the nature of crystalline deformation. In particular, we will explore the natural tendency of plastic deformation to become highly nonuniform and in fact localized into patterns that can, *inter alia*, evolve into bands of intensely localized slip, kinking patterns, and the sort of heterogeneous patterns of slip on different systems that were referred to as "patchy slip" in Chapter 27. These examples of localized deformation are important because they often lead to material failure, as well as to the evolution of internal substructure that, in turn, directly influences evolving material response. On the other hand, the analysis of these deformation patterns serves to highlight some rather fundamental aspects of the process of crystalline deformation *via* the process of slip. This serves to reveal and, in part explain, some of the basic implications of the type of theory we have outlined herein.

Specifically, we examine two aspects of the theory regarding the stability of uniform deformation, *viz.*, the implications of deviations from the Schmid rule of a critical resolved shear stress, as well as the effects of lattice kinematics, on the formation of localized modes of deformation. As will become clear from the development, deviations from the Schmid rule, in fact, represent deviations from the rule of plastic normality in the flow law. In addition, there are geometrical effects that involve discontinuous rates of lattice rotation that can destabilize uniform plastic flow and lead to the formation of modes such as kinks and shear bands. Some empirical perspective is needed first.

29.1 Perspectives on Nonuniform and Localized Plastic Flow

The origins and phenomenology of localized deformation in crystals has been reviewed by Asaro (1983a, 1983b) and Dao and Asaro (1996); here we provide a summary of those findings.

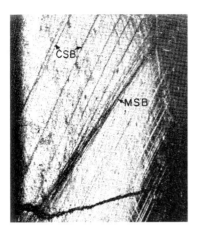

Figure 29.1. CSB and subsequent MSB formation in a GP-II containing single crystal of Al-2.8 wt% Cu deformed in uniaxial tension. The CSB's are very nearly aligned with the most active slip system, whereas the MSB's are characteristically misoriented by several degrees to the active slip planes (from Chang and Asaro, 1981).

29.1.1 Coarse Slip Bands and Macroscopic Shear Bands in Simple Crystals

Most ductile crystals display the formation of intense shear bands, if deformed sufficiently far into the plastic range prior to, but often leading to, fracture. For example, precipitation hardened alloys containing GP-I and GP-II zones display intense shear bands leading to dramatic shear fractures as shown earlier. Elam (1927) performed early experiments on Al-Zn and Cu-Be single crystals, Beevers and Honeycombe (1962) on Al-Cu, and Price and Kelly (1964) on Al-Cu, Al-Zn, Al-Ag, and Cu-Be. They observed the same trends with respect to the natural tendency toward the localization of deformation. Chang and Asaro (1981) have performed systematic experiments on Al-2.8 wt% Cu single crystals in tension and compression and Harren, Deve, and Asaro (1988) performed tests on the same Al-Cu crystals Chang and Asaro (1981) tested, but this time in plane strain compression as well as uniaxial tension and compression. The phenomena studied by Chang and Asaro (1981) and by Harren *et al.* (1988) revealed that the process of localization occurred by the formation of what they termed *coarse slip bands* (CSB) that were followed after further deformation by *macroscopic shear bands* (MSB). The same phenomena occur in nominally pure, single-phase crystals and thus appear to be an entirely natural part of the finite strain deformation process.

Figure 29.1 shows CSB and MSB formation in an Al-2.8 wt% Cu crystal containing GP-II zones. The crystal was oriented so that slip began on the *primary* slip plane, whereas the lattice rotations described earlier eventually induced slip on a second system (*i.e.*, the conjugate system). The CSB's formed very nearly aligned with the primary slip system, and subsequently clustered, and finally within a cluster of CSB's a MSB formed. The MSB's are characteristically misoriented from the CSB's (and the most active slip system) as is evident in the figure and this misorientation had been shown by Asaro (1979) to be because of an abrupt lattice misorientation that, in turn, causes a *geometrical softening* as described below. Chang and Asaro (1981) showed that MSB's sometimes formed on the conjugate slip system after CSB's formed on the primary slip system. Very similar phenomenology was observed in compression. For example, Fig. 29.2 shows the formation of CSB's on both the primary and conjugate systems in plane strain compression. Eventually, MSB's formed in this state of compression as described by Harren *et al.* (1988).

29.1. Localized Flow

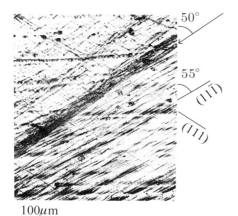

Figure 29.2. CSB formation on both the primary and conjugate systems followed by MSB formation in a GP-II containing single crystal of Al-2.8 wt% Cu deformed in plane strain compression. As in the case of uniaxial tension, the MSB's are characteristically misoriented with respect to the active slip planes, again associated with geometrical softening (from Harren et al., 1988).

29.1.2 Coarse Slip Bands and Macroscopic Shear Bands in Ordered Crystals

CSB patterns are observed in ordered intermetallic compounds, including Ni_3Ga ($L1_2$) (Takeuchi and Kuramoto, 1973), TiAl ($L1_0$) (Kawabata et al., 1985), and Ti_3AL (DO_3) (Minonishi, 1991). Figure 29.3 shows examples of compression tests along a fixed orientation at three different temperatures for Ni_3Ga. CSB's are clearly observed at the low temperature of 77 K, as well at the higher temperature of 993 K. At the intermediate temperature of 458 K the deformation appears rather uniform. The analysis performed by Dao and Asaro (1996), using methods described below, indicated that the strain hardening rates were too high to meet the critical conditions required for localization, even though the magnitude of the deviation from the Schmid rule was largest at this intermediate temperature. Heredia and Pope (1988) studied Ni_3Al ($L1_1$) single crystals in tension. They found that the crystals often fractured on {111}-type shear planes. Such abruptly occurring localized plastic flow has been reported in crystals of other ordered intermetallic compounds such as Ni-Al (Wasilewski et al., 1967) as well as in Ti-Al alloys containing duplex γ/α microstructures (Inui et al., 1992). What appears to be common to these materials

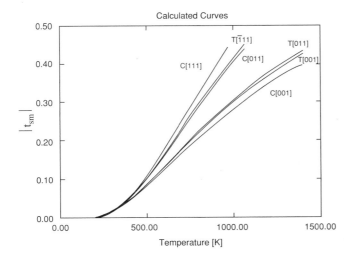

Figure 29.3. Compression tests along a fixed orientation at three temperatures for Ni_3Ga, (a) at 77 K, (b) at 458 K, and (c) at 993 K (from Dao and Asaro, 1996).

is that they display stable, and often high, rates of strain hardening, yet fail *via* intensely concentrated shearing.

As noted above, the theory of crystal plasticity can describe much of the observed behavior. Indeed localized plastic deformation of the type described here has been analyzed, and we explore some of that analysis as an application of the theory in what follows.

29.2 Localized Deformation in Single Slip

29.2.1 Constitutive Law for the Single Slip Crystal

For the sake of compactness we develop a suitable constitutive law for a crystal undergoing single slip based on the framework described in the previous chapter. Let **s** and **m** be the slip direction and slip plane normal, respectively. Let them be unit vectors. Then, the rate of deformation and spin rates are, respectively,

$$\mathbf{D} = \mathbf{D}^* + \mathbf{P}\dot{\gamma}, \quad 2\mathbf{P} = \mathbf{sm} + \mathbf{ms},$$
$$\mathbf{W} = \mathbf{W}^* + \mathbf{Q}\dot{\gamma}, \quad 2\mathbf{Q} = \mathbf{sm} - \mathbf{ms}. \qquad (29.1)$$

The elastic response is of the form

$$\overset{\triangledown}{\sigma}{}^* + \sigma \operatorname{tr} \mathbf{D}^* = \mathcal{L} : \mathbf{D}^*, \qquad (29.2)$$

where the Jaumann rate of stress on axes that spin with the lattice, $\overset{\triangledown}{\sigma}{}^*$, is given as

$$\overset{\triangledown}{\sigma}{}^* = \dot{\sigma} - \mathbf{W}^* \cdot \sigma + \sigma \cdot \mathbf{W}^*. \qquad (29.3)$$

The elastic moduli have the reciprocal symmetry, $\mathcal{L}_{ijk\ell} = \mathcal{L}_{k\ell ij}$. Of course, the *lhs* of (29.2) is the lattice based Jaumann rate of Kirchhoff stress, $\overset{\triangledown}{\tau}{}^*$, when the reference state is chosen to instantaneously coincide with the current state. In this regard, also note that Λ from (28.30) is related to $\mathcal{L}$ by $\Lambda = (\det \mathbf{F})\mathcal{L}$, because $\overset{\triangledown}{\tau}{}^* = (\det \mathbf{F})(\overset{\triangledown}{\sigma}{}^* + \sigma \operatorname{tr} \mathbf{D}^*)$.

Expressing $\mathbf{D}^*$ and $\mathbf{W}^*$ in terms of $\mathbf{D}$, $\mathbf{W}$, and $\dot{\gamma}$, the constitutive law becomes

$$\overset{\triangledown}{\sigma} + \sigma \operatorname{tr}(\mathbf{D}) = \mathcal{L} : (\mathbf{D} - \mathbf{P}'\dot{\gamma}), \qquad (29.4)$$

where

$$\mathbf{P}' = \mathbf{P} + \mathcal{L}^{-1} : (\mathbf{Q} \cdot \sigma - \sigma \cdot \mathbf{Q}), \qquad (29.5)$$

and

$$\overset{\triangledown}{\sigma} = \dot{\sigma} - \mathbf{W} \cdot \sigma + \sigma \cdot \mathbf{W}. \qquad (29.6)$$

29.2.2 Plastic Shearing with Non-Schmid Effects

Begin by writing an expression for an increment of slip, in accordance with the Schmid rule,

$$d\gamma = d\tau_{ms}/h, \qquad (29.7)$$

where $d\tau_{ms}$ is the increment of resolved shear stress following an increment of slip strain, and h is the strain hardening modulus on the slip system. For compactness of notation

29.2. Localized Deformation

here, γ is used to represent the plastic shear strain γ_{ms}, the shear strain on the $(\mathbf{s}, \mathbf{m})$ slip system. An expression for $d\gamma$ such as this must be refined and generalized in several ways. First, it is necessary to state precisely how $d\tau_{ms}$ is to be related to a stress increment such as $\stackrel{\triangledown}{\boldsymbol{\sigma}}{}^*$. We recall that a particular way of computing an increment of resolved shear stress, $d\tau_{ms}$, was used in the previous chapter. Here we explore the implications of more general definitions of such increments. Next, as noted in Chapter 27 on the micromechanics of slip, there are good reasons for suspecting that *small* departures from the Schmid rule are to be expected, in the sense that increments of components of stress, other than the resolved shear stress, will influence slip on a particular slip system. To describe such effects, let $\mathbf{z}$ be the unit vector that, together with $\mathbf{s}$ and $\mathbf{m}$, form a right handed triad; $\mathbf{z}$ is in the slip plane along with $\mathbf{s}$. Thus, in place of (29.7), we write the flow law as

$$d\gamma = \frac{1}{h}(d\tau_{ms} + \alpha_{ss}d\tau_{ss} + \alpha_{mm}d\tau_{mm} + \alpha_{zz}d\tau_{zz} + 2\alpha_{sz}d\tau_{sz} + 2\alpha_{mz}d\tau_{mz}). \quad (29.8)$$

Here each component of α gives the decrement on the Schmid stress, τ_{ms}, required for flow per unit increase of the corresponding non-Schmid stress. Examples of such effects, and estimates of their values, are given below.

We now turn attention to defining the increments in the resolved shear stress $d\tau_{ms}$ (and accordingly increments in the other $d\tau_{ij}$'s). In *all* such definitions, we take

$$\tau_{ms} = \mathbf{m} \cdot \boldsymbol{\sigma} \cdot \mathbf{s}, \quad (29.9)$$

where $\mathbf{s}$ and $\mathbf{m}$ are unit vectors in the current state, which is taken as the reference state. The choice is, however, how to form the increments $\dot{\mathbf{s}}dt$ and $\dot{\mathbf{m}}dt$. The most direct way is to require that $\mathbf{s}$ and $\mathbf{m}$ remain orthogonal unit vectors, with $\mathbf{s}$ remaining in the slip plane as the plane rotates. Then,

$$\begin{aligned}\dot{\mathbf{s}} &= (\mathbf{D}^* + \mathbf{W}^*) \cdot \mathbf{s} - \mathbf{s}(\mathbf{s} \cdot \mathbf{D}^* \cdot \mathbf{s}), \\ \dot{\mathbf{m}} &= -\mathbf{m} \cdot (\mathbf{D}^* + \mathbf{W}^*) + \mathbf{m}(\mathbf{m} \cdot \mathbf{D}^* \cdot \mathbf{m}).\end{aligned} \quad (29.10)$$

The rate of resolved shear stress becomes

$$\begin{aligned}\dot{\tau}_{ms} &= \dot{\mathbf{m}} \cdot \boldsymbol{\sigma} \cdot \mathbf{s} + \mathbf{m} \cdot \dot{\boldsymbol{\sigma}} \cdot \mathbf{s} + \mathbf{m} \cdot \boldsymbol{\sigma} \cdot \dot{\mathbf{s}} \\ &= \mathbf{m} \cdot [\stackrel{\triangledown}{\boldsymbol{\sigma}}{}^* - \mathbf{D}^* \cdot \boldsymbol{\sigma} + \boldsymbol{\sigma} \cdot \mathbf{D}^* + \boldsymbol{\sigma}(\mathbf{m} \cdot \mathbf{D}^* \cdot \mathbf{m} - \mathbf{s} \cdot \mathbf{D}^* \cdot \mathbf{s})] \cdot \mathbf{s}.\end{aligned}$$

An alternative definition follows from choosing to convect $\mathbf{s}$ with the lattice slip plane, so that its length changes, and to choose $\mathbf{m}$ as the reciprocal base vector normal to the deformed slip plane. In this case

$$\dot{\mathbf{s}} = (\mathbf{D}^* + \mathbf{W}^*) \cdot \mathbf{s}, \quad \dot{\mathbf{m}} = -\mathbf{m} \cdot (\mathbf{D}^* + \mathbf{W}^*), \quad (29.11)$$

which gives for the rate of Schmid stress,

$$\dot{\tau}_{ms} = \mathbf{m} \cdot (\stackrel{\triangledown}{\boldsymbol{\sigma}}{}^* - \mathbf{D}^* \cdot \boldsymbol{\sigma} + \boldsymbol{\sigma} \cdot \mathbf{D}^*) \cdot \mathbf{s}. \quad (29.12)$$

A *third* choice follows if we convect $\mathbf{s}$ with the lattice and choose $\mathbf{m}$ so that it is normal to the deformed slip plane, but of a length that decreases in proportion to slip plane area. Then,

$$\dot{\mathbf{s}} = (\mathbf{D}^* + \mathbf{W}^*) \cdot \mathbf{s}, \quad \dot{\mathbf{m}} = -\mathbf{m} \cdot (\mathbf{D}^* + \mathbf{W}^*) + \mathbf{m}\,\mathrm{tr}\,(\mathbf{D}^*), \quad (29.13)$$

which gives

$$\dot{t}_{ms} = \mathbf{m} \cdot [\overset{\triangledown}{\boldsymbol{\sigma}}{}^* + \boldsymbol{\sigma}\operatorname{tr}(\mathbf{D}^*) - \mathbf{D}^* \cdot \boldsymbol{\sigma} + \boldsymbol{\sigma} \cdot \mathbf{D}^*] \cdot \mathbf{s}. \tag{29.14}$$

This particular version of $d\tau_{ms}$ is of some special interest because it is the definition of the resolved shear stress that is precisely the work conjugate to $\dot{\gamma}$, and thus leads to a precise statement of normality.

There are unlimited alternative definitions for the rate of the Schmid stress, and as a *fourth* and final example we again require **s** and **m** to be orthogonal unit vectors, but simply rotate them at the lattice spin rate, so that

$$\dot{\mathbf{s}} = \mathbf{W}^* \cdot \mathbf{s}, \quad \dot{\mathbf{m}} = \mathbf{W}^* \cdot \mathbf{m}. \tag{29.15}$$

In this case,

$$\dot{t}_{ms} = \mathbf{m} \cdot \overset{\triangledown}{\boldsymbol{\sigma}}{}^* \cdot \mathbf{s}. \tag{29.16}$$

All the above expressions for the Schmid rate have in common the feature that

$$\dot{t}_{ms} = \mathbf{m} \cdot [\overset{\triangledown}{\boldsymbol{\sigma}}{}^* + \boldsymbol{\sigma}\operatorname{tr}(\mathbf{D}^*)] \cdot \mathbf{s} + \boldsymbol{\sigma} : \mathcal{H} : \mathbf{D}^*, \tag{29.17}$$

where $\mathcal{H}$ is a fourth-rank tensor that depends on the precise way in which the base vectors **s** and **m** deform with the lattice and that has components that are of $\mathcal{O}(1)$. By inverting (29.2), all such rates can be written in the form

$$\begin{aligned}\dot{t}_{ms} &= (\mathbf{sm} + \boldsymbol{\sigma} : \mathcal{H} : \mathcal{L}^{-1}) : [\overset{\triangledown}{\boldsymbol{\sigma}}{}^* + \boldsymbol{\sigma}\operatorname{tr}(\mathbf{D}^*)] \\ &= (\mathbf{P} + \boldsymbol{\sigma} : \mathcal{H} : \mathcal{L}^{-1}) : [\overset{\triangledown}{\boldsymbol{\sigma}}{}^* + \boldsymbol{\sigma}\operatorname{tr}(\mathbf{D}^*)].\end{aligned} \tag{29.18}$$

In the second version of the above expression, we have noted that the term in [...] is symmetric.

We examine now the specification of the various non-Schmid stress increments appearing in (29.8). Each of these increments is multiplied by an α. If the α's are small fractions of unity, there will be contributions of negligibly small size, of $\mathcal{O}(\alpha\sigma/\mathcal{L})$ in comparison to terms of $\mathcal{O}(1)$ or $\mathcal{O}(\alpha)$ or $\mathcal{O}(\sigma/\mathcal{L})$, made to the bracketed term in (29.8) (*e.g.*, at the level of choosing a specific form for $\mathcal{H}$ in (29.18)). On the other hand, looking ahead to the final result for h_{cr} at localization, if the α's are not small then the correction due to the retention of the $\mathcal{O}(\sigma/\mathcal{L})$-terms in the non-Schmid stress rates would in any case be negligible. This is a fortunate circumstance, because the current understanding of crystalline slip barely enables a definite choice of the α's, much less a precise specification of the non-Schmid stress rates. Even though the precise form of the Schmid rate $\dot{t}_{ms}$ cannot be specified according to our present understanding of the physics of crystalline slip, in this case we are helped by the remarkable fact that our result for h_{cr}, to the order of accuracy that we determine it, turns out to be independent of how we choose $\mathcal{H}$.

In view of the above remarks it suffices to write expressions of the type $\dot{t}_{mz} = \mathbf{m} \cdot [\overset{\triangledown}{\boldsymbol{\sigma}}{}^* + \boldsymbol{\sigma}\operatorname{tr}(\mathbf{D}^*)] \cdot \mathbf{z}, \dot{t}_{ss} = \mathbf{s} \cdot [\overset{\triangledown}{\boldsymbol{\sigma}}{}^* + \boldsymbol{\sigma}\operatorname{tr}(\mathbf{D}^*)] \cdot \mathbf{s}$, and so on for the non-Schmid stress rates, because precision of the order $\mathcal{O}(\sigma\mathbf{D}^*)$ is evidently unnecessary. *This, by the way, justifies the*

29.2. Localized Deformation

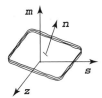

Figure 29.4. Cartesian coordinates aligned with unit vectors of the crystal's slip system. The surface of the band of localized deformation has unit normal **n**.

definitions of resolved shear stress used in previous chapters. Thus, using (29.18), the plastic shear rate of (29.8) can be written as

$$\dot{\gamma} = \frac{1}{h} \mathbf{Q}' : [\overset{\triangledown}{\sigma}^* + \sigma \operatorname{tr}(\mathbf{D}^*)], \quad (29.19)$$

where

$$\mathbf{Q}' = \mathbf{P} + \sigma : \mathcal{H} : \mathcal{L}^{-1} + \alpha. \quad (29.20)$$

The tensor α of non-Schmid effects, chosen without loss of generality to be symmetric, has the matrix components, taken on axes aligned with the triad **s**, **m**, **z** and ordered in the same sense,

$$\alpha = \begin{bmatrix} \alpha_{ss} & 0 & \alpha_{sz} \\ 0 & \alpha_{mm} & \alpha_{mz} \\ \alpha_{sz} & \alpha_{mz} & \alpha_{zz} \end{bmatrix}. \quad (29.21)$$

The elastic relation (29.2) can be used to rewrite (29.19) as

$$\dot{\gamma} = \frac{1}{h} \mathbf{Q}' : \mathcal{L} : (\mathbf{D} - \mathbf{P}\dot{\gamma}), \quad (29.22)$$

which can be solved for $\dot{\gamma}$ to obtain

$$\dot{\gamma} = \frac{1}{h + \mathbf{Q}' : \mathcal{L} : \mathbf{P}} \mathbf{Q}' : \mathcal{L} : \mathbf{D}. \quad (29.23)$$

By substituting this into (29.4), we obtain the form of the constitutive rate equation needed for the localization analysis, *viz.*,

$$\overset{\triangledown}{\sigma} + \sigma \operatorname{tr}(\mathbf{D}) = \left[\mathcal{L} - \frac{(\mathcal{L} : \mathbf{P}')(\mathbf{Q}' : \mathcal{L})}{h + \mathbf{Q}' : \mathcal{L} : \mathbf{P}} \right] : \mathbf{D}. \quad (29.24)$$

When the tensor α is zero and when $\mathcal{H}$ is chosen to yield (29.14), it is readily verified that $\mathbf{Q}' = \mathbf{P}'$; the bracketed tensor of constitutive moduli in (29.24) is then said to exhibit "normality."

29.2.3 Conditions for Localization

Hill (1962) has presented the general theory of bifurcation of a homogeneous elastic-plastic flow field into a band of localized deformation [or in Hadamard's (1903) terminology, into a "stationary discontinuity"]. There is first the kinematical restriction that for localization in a thin planar band of unit normal **n** (see Fig. 29.4) the velocity gradient field, $v_{i,j}$, inside the band can differ from that outside, namely $v_{i,j}^0$, by only an expression of the form

$$v_{i,j} - v_{i,j}^0 = g_i n_j. \quad (29.25)$$

In addition, there is the requirement of continuing equilibrium that

$$n_i \dot{\sigma}_{ij} - n_i \dot{\sigma}_{ij}^0 = 0 \tag{29.26}$$

at incipient localization, where $\dot{\sigma}$ is the stress rate within the band and $\dot{\sigma}^0$ that outside of it. The latter requirement is merely a statement of an ongoing traction continuity.

If the constitutive rate relation is assumed to have the form

$$\dot{\sigma}_{ij} = C_{ijk\ell} v_{k,\ell}, \tag{29.27}$$

and if the same set of constitutive coefficients, $\mathbf{C}$, apply both inside and outside the band at incipient localization, then (29.25) and (29.26) will be satisfied simultaneously if

$$(n_i C_{ijk\ell} n_\ell) g_k = 0, \tag{29.28}$$

where $\mathbf{n} \cdot \mathbf{C} \cdot \mathbf{n}$ is considered to be a second-rank matrix. Of course, once the critical value of some constitutive parameter, say h, for localization is known as a function of $\mathbf{n}$ from (29.28), it is then necessary to determine the orientation $\mathbf{n}$ at which the critical state is first attained.

It is necessary to first identify the modulus $\mathbf{C}$ used in (29.28), and so we write

$$\dot{\sigma} = \left[\mathcal{L} - \frac{(\mathcal{L} : \mathbf{P}')(\mathbf{Q}' : \mathcal{L})}{h + \mathbf{Q}' : \mathcal{L} : \mathbf{P}} \right] : \mathbf{D} + \mathbf{W} \cdot \sigma - \sigma \cdot \mathbf{W} - \sigma \operatorname{tr}(\mathbf{D}), \tag{29.29}$$

recognizing that $\mathbf{D}$ and $\mathbf{W}$ can be expressed in terms of the velocity gradient. To form the expression analogous to (29.28), we multiply (29.29) from the left with $\mathbf{n}$, and write

$$\frac{1}{2} (\mathbf{gn} + \mathbf{ng}) \quad \text{and} \quad \frac{1}{2} (\mathbf{gn} - \mathbf{ng})$$

in place of $\mathbf{D}$ and $\mathbf{W}$, respectively. This gives

$$0 = \left[(\mathbf{n} \cdot \mathcal{L} \cdot \mathbf{n}) - \frac{(\mathbf{n} \cdot \mathcal{L} : \mathbf{P}')(\mathbf{Q}' : \mathcal{L} \cdot \mathbf{n})}{h + \mathbf{Q}' : \mathcal{L} : \mathbf{P}} \right] \cdot \mathbf{g} + \mathbf{A} \cdot \mathbf{g}, \tag{29.30}$$

where

$$\mathbf{A} = \frac{1}{2} [(\mathbf{n} \cdot \sigma \cdot \mathbf{n}) \mathbf{I} - \sigma - (\mathbf{n} \cdot \sigma) \mathbf{n} - \mathbf{n} (\sigma \cdot \mathbf{n})]. \tag{29.31}$$

Here, $\mathbf{I}$ has the usual meaning as the second-rank identity tensor.

Now consider the acoustic tensor, $\mathbf{n} \cdot \mathcal{L} \cdot \mathbf{n}$, and form its inverse, $(\mathbf{n} \cdot \mathcal{L} \cdot \mathbf{n})^{-1}$, which we may assume to exist as long as the stresses are not significant fractions of the elastic moduli themselves. Multiplying (29.30) with this inverse, we obtain

$$0 = \left\{ [\mathbf{I} + (\mathbf{n} \cdot \mathcal{L} \cdot \mathbf{n})^{-1} \cdot \mathbf{A}] - \frac{[(\mathbf{n} \cdot \mathcal{L} \cdot \mathbf{n})^{-1} \cdot (\mathbf{n} \cdot \mathcal{L} : \mathbf{P}')] (\mathbf{Q}' : \mathcal{L} \cdot \mathbf{n})}{h + \mathbf{Q}' : \mathcal{L} : \mathbf{P}} \right\} \cdot \mathbf{g}. \tag{29.32}$$

Since $\mathbf{A}$ in (29.32) has elements which are of order σ, the tensor

$$\mathbf{I} + (\mathbf{n} \cdot \mathcal{L} \cdot \mathbf{n})^{-1} \cdot \mathbf{A}$$

29.2. Localized Deformation

differs from unity by terms of $\mathcal{O}(\sigma/\mathcal{L})$, which, in representative cases, are small fractions of unity. Thus, we may assume that this tensor has an inverse and we express it as an expansion series

$$[\mathbf{I} + (\mathbf{n} \cdot \mathcal{L} \cdot \mathbf{n})^{-1} \cdot \mathbf{A}]^{-1} = \mathbf{I} - (\mathbf{n} \cdot \mathcal{L} \cdot \mathbf{n})^{-1} \cdot \mathbf{A}$$
$$+ [(\mathbf{n} \cdot \mathcal{L} \cdot \mathbf{n})^{-1} \cdot \mathbf{A}] \cdot [(\mathbf{n} \cdot \mathcal{L} \cdot \mathbf{n})^{-1} \cdot \mathbf{A}] - \cdots.$$

When (29.32) is multiplied by this inverse, we arrive at an expression in the form

$$\left(\mathbf{I} - \frac{\mathbf{ab}}{h + \mathbf{Q}' : \mathcal{L} : \mathbf{P}}\right) \cdot \mathbf{g} = \mathbf{0}, \tag{29.33}$$

where the vectors $\mathbf{a}$ and $\mathbf{b}$ are given by

$$\mathbf{a} = [\mathbf{I} + (\mathbf{n} \cdot \mathcal{L} \cdot \mathbf{n})^{-1} \cdot \mathbf{A}]^{-1} \cdot (\mathbf{n} \cdot \mathcal{L} \cdot \mathbf{n})^{-1} \cdot (\mathbf{n} \cdot \mathcal{L} : \mathbf{P}'),$$
$$\mathbf{b} = \mathbf{Q}' : \mathcal{L} \cdot \mathbf{n}. \tag{29.34}$$

Upon multiplying (29.34) from the left by $\mathbf{b}$, we then obtain

$$\left(1 - \frac{\mathbf{b} \cdot \mathbf{a}}{h + \mathbf{Q}' : \mathcal{L} : \mathbf{P}}\right) \cdot (\mathbf{b} \cdot \mathbf{g}) = 0. \tag{29.35}$$

In view of (29.23), the term $\mathbf{b} \cdot \mathbf{g}$ cannot vanish for nonzero $\mathbf{g}$ unless the bifurcation mode involves no plastic strain. Thus, the only relevant condition allowing a nonzero $\mathbf{g}$ is that the coefficient of $\mathbf{b} \cdot \mathbf{g}$ vanish, which yields the condition for the critical value of h at localization,

$$h + \mathbf{Q}' : \mathcal{L} : \mathbf{P} = \mathbf{b} \cdot \mathbf{a}. \tag{29.36}$$

It is easy to verify that the localization mode has the form $\mathbf{g} \propto \mathbf{a}$. Therefore, using the expressions for $\mathbf{a}$ and $\mathbf{b}$, we obtain

$$h_{\text{cr}} = -\mathbf{Q}' : \mathcal{L} : \mathbf{P} + (\mathbf{Q}' : \mathcal{L} \cdot \mathbf{n})[\mathbf{I} + (\mathbf{n} \cdot \mathcal{L} \cdot \mathbf{n})^{-1} \cdot \mathbf{A}]^{-1} \cdot (\mathbf{n} \cdot \mathcal{L} \cdot \mathbf{n})^{-1} \cdot (\mathbf{n} \cdot \mathcal{L} : \mathbf{P}'). \tag{29.37}$$

Of course, (29.37) may be readily evaluated for particular cases numerically, but there is reason for insight to consider expansions of it in terms of the order of magnitudes of the various parameters within it. Asaro and Rice (1977) have done that as described below.

29.2.4 Expansion to the Order of σ

To review the origin of the terms in (29.37), we first note that $\mathbf{P}$ is of $\mathcal{O}(1)$ and it represents the plastic flow direction tensor. $\mathbf{P}'$ is defined by (29.5) and differs from $\mathbf{P}$ by terms of $\mathcal{O}(\sigma/\mathcal{L})$. $\mathbf{A}$ is defined by (29.31) and is of $\mathcal{O}(\sigma)$. Finally, $\mathbf{Q}'$ is defined by (29.20) and involves the term $\mathbf{P}$. This term is of $\mathcal{O}(\sigma/\mathcal{L})$ involving the fourth-rank tensor $\mathcal{H}$ introduced to account for lattice deformation effects on the Schmid stress rate, and the term α, which accounts for the non-Schmid effects on yielding and subsequent plastic flow. We recall that the specification of $\mathbf{Q}'$ neglected terms of $\mathcal{O}(\alpha\sigma/\mathcal{L})$, which in any event we would be *hard-pressed* to specify anyway. For this reason we delete from the *rhs* of (29.37) all terms of the orders

$$\alpha\sigma, \ \alpha\sigma^2/\mathcal{L}, \ \alpha\sigma^3/\mathcal{L}^2, \ldots \quad \text{and} \quad \sigma^2/\mathcal{L}, \ \sigma^3/\mathcal{L}^2, \ldots.$$

When this is done, we obtain

$$h_{\text{cr}} = h_{\text{cr}}^{\mathcal{L}} + h_{\text{cr}}^{\alpha\mathcal{L}} + h_{\text{cr}}^{\sigma}, \tag{29.38}$$

where the three terms have the order of their superscripts and are defined by

$$\begin{aligned} h_{\text{cr}}^{\mathcal{L}} &= -\mathbf{P}:\mathcal{L}:\mathbf{P} + (\mathbf{P}:\mathcal{L}\cdot\mathbf{n})\cdot(\mathbf{n}\cdot\mathcal{L}\cdot\mathbf{n})^{-1}\cdot(\mathbf{n}\cdot\mathcal{L}:\mathbf{P}), \\ h_{\text{cr}}^{\alpha\mathcal{L}} &= -\alpha:\mathcal{L}:\mathbf{P} + (\alpha:\mathcal{L}\cdot\mathbf{n})\cdot(\mathbf{n}\cdot\mathcal{L}\cdot\mathbf{n})^{-1}\cdot(\mathbf{n}\cdot\mathcal{L}:\mathbf{P}), \end{aligned} \tag{29.39}$$

$$\begin{aligned} h_{\text{cr}}^{\sigma} &= -\sigma:\mathcal{H}:\mathbf{P} + (\sigma:\mathcal{H}\cdot\mathbf{n})\cdot(\mathbf{n}\cdot\mathcal{L}\cdot\mathbf{n})^{-1}\cdot(\mathbf{n}\cdot\mathcal{L}:\mathbf{P}) \\ &\quad + (\mathbf{P}:\mathcal{L}\cdot\mathbf{n})\cdot(\mathbf{n}\cdot\mathcal{L}\cdot\mathbf{n})^{-1}\cdot[\mathbf{n}\cdot(\mathbf{Q}\cdot\sigma - \sigma\cdot\mathbf{Q})] \\ &\quad - (\mathbf{P}:\mathcal{L}\cdot\mathbf{n})\cdot(\mathbf{n}\cdot\mathcal{L}\cdot\mathbf{n})^{-1}\cdot\mathbf{A}\cdot(\mathbf{n}\cdot\mathcal{L}\cdot\mathbf{n})^{-1}\cdot(\mathbf{n}\cdot\mathcal{L}:\mathbf{P}). \end{aligned} \tag{29.40}$$

When expressed in this order, it is convenient to represent these terms as

$$h_{\text{cr}} = \mathcal{L} F_0(\mathbf{n}) + \alpha F_1(\mathbf{n}) + \sigma F_2(\mathbf{n}). \tag{29.41}$$

In (29.41) the functions F are all of $\mathcal{O}(1)$, and $\mathcal{L}, \alpha, \sigma$ are meant to be representative members of their corresponding tensors.

We begin the exploration of the most critical orientation for $\mathbf{n}$ by examining the case where both α and $\sigma/\mathcal{L}$ are sufficiently small that we can approximate (29.41) by

$$h_{\text{cr}} \approx \mathcal{L} F_0(\mathbf{n}). \tag{29.42}$$

Consider the first of (29.39), which is rewritten as

$$\mathcal{L} F_0(\mathbf{n}) = -\mathbf{P}:\mathcal{N}:\mathbf{P}, \tag{29.43}$$

where the fourth-rank tensor $\mathcal{N}$ is defined by

$$\mathcal{N} = \mathcal{L} - (\mathcal{L}\cdot\mathbf{n})\cdot(\mathbf{n}\cdot\mathcal{L}\cdot\mathbf{n})^{-1}\cdot(\mathbf{n}\cdot\mathcal{L}). \tag{29.44}$$

We observe that $\mathcal{N}$ has the same symmetry as does $\mathcal{L}$, and that

$$\mathbf{n}\cdot\mathcal{N} = \mathcal{N}\cdot\mathbf{n} = \mathbf{0}. \tag{29.45}$$

We now show that the tensor $\mathcal{N}$ is, in fact, the tensor of incremental elastic moduli governing *plane stress* states in a plane perpendicular to $\mathbf{n}$. Specifically, we let the unit vectors $\mathbf{u}, \mathbf{v}$, and $\mathbf{n}$ form a right-handed triad and we form

$$\mathbf{d} = 2D_{nu}\mathbf{u} + 2D_{nv}\mathbf{v} + D_{nn}\mathbf{n}, \quad \text{(no sum on } u, v, n\text{)} \tag{29.46}$$

and

$$\mathbf{D}' = D_{uu}\mathbf{u}\mathbf{u} + D_{uv}(\mathbf{u}\mathbf{v} + \mathbf{v}\mathbf{u}) + D_{vv}\mathbf{v}\mathbf{v}, \tag{29.47}$$

where the D_{ij}'s ($i, j = u, v, n$) are components of an arbitrary rate of stretching tensor $\mathbf{D}$ on the directions of the corresponding unit vectors. Furthermore, we observe that

$$\mathbf{D} = \mathbf{D}' + \frac{1}{2}(\mathbf{n}\mathbf{d} + \mathbf{d}\mathbf{n}). \tag{29.48}$$

29.2. Localized Deformation

For an increment of elastic plane stress deformation in the $\mathbf{u} - \mathbf{v}$ plane, in the sense that the corotational increment of Kirchhoff stress has no components associated with the normal $\mathbf{n}$ to that plane, it is evidently necessary that

$$\mathbf{0} = \mathbf{n} \cdot \mathcal{L} : \mathbf{D} = \mathbf{n} \cdot \mathcal{L} : \mathbf{D}' + (\mathbf{n} \cdot \mathcal{L} \cdot \mathbf{n}) \cdot \mathbf{d} \qquad (29.49)$$

and hence that

$$\mathbf{d} = -(\mathbf{n} \cdot \mathcal{L} \cdot \mathbf{n})^{-1} \cdot (\mathbf{n} \cdot \mathcal{L} : \mathbf{D}'). \qquad (29.50)$$

Thus the corotational Kirchhoff rate can be written in terms of the rate of deformation $\mathbf{D}'$ in the plane of stretching as

$$\overset{\triangledown}{\sigma} + \sigma \mathrm{tr}(\mathbf{D}) = \mathcal{L} : \mathbf{D} = \mathcal{L} : \mathbf{D}' + (\mathcal{L} \cdot \mathbf{n}) \cdot \mathbf{d}$$
$$= [\mathcal{L} - (\mathcal{L} \cdot \mathbf{n}) \cdot (\mathbf{n} \cdot \mathcal{L} \cdot \mathbf{n})^{-1} \cdot (\mathbf{n} \cdot \mathcal{L})] : \mathbf{D}' = \mathcal{N} : \mathbf{D}', \qquad (29.51)$$

where (29.44) has been used for $\mathcal{N}$. This establishes the interpretation of $\mathcal{N}$ as the plane stress elastic modulus tensor. Because lattice distortions are small in all cases that we consider, the quadratic form

$$V = \mathbf{D}' : \mathcal{N} : \mathbf{D}' \qquad (29.52)$$

may be assumed to be *positive definite* in $\mathbf{D}'$. On the other hand,

$$\mathcal{N} : \mathbf{D} = \mathcal{N} : \mathbf{D}' + (\mathcal{N} \cdot \mathbf{n}) \cdot \mathbf{d} = \mathcal{N} : \mathbf{D}', \qquad (29.53)$$

because of (29.45), and thus the quadratic form V can also be written

$$V = \mathbf{D} : \mathcal{N} : \mathbf{D}. \qquad (29.54)$$

We see, therefore, that V is a positive definite function of $\mathbf{D}'$ but a positive *semidefinite* function of $\mathbf{D}$. In particular,

$$V = \mathbf{D} : \mathcal{N} : \mathbf{D} = 0, \quad \text{if and only if} \quad \mathbf{D}' = \mathbf{0}. \qquad (29.55)$$

Making use of these results in (29.42) and (29.43), we see that the critical value, h_{cr}, for localization on a plane with normal $\mathbf{n}$ must be *either negative or zero*, the latter occurring when n is chosen so that $\mathbf{P}$ has no components in the plane perpendicular to $\mathbf{n}$. It is straightforward to show that there are two, and only two, orientations $\mathbf{n}$ that allow this condition of $h_{\mathrm{cr}} = 0$, and these are given by

$$\text{case (i):} \ \mathbf{n} = \mathbf{m}, \quad \text{and} \quad \text{case (ii):} \ \mathbf{n} = \mathbf{s}. \qquad (29.56)$$

Thus, to summarize, when we approximate (29.41) by (28.40), we find that the critical plastic-hardening modulus at the inception of localization is

$$h_{\mathrm{cr}} = 0, \qquad (29.57)$$

and the plane of localization is either the slip plane, where $\mathbf{n} = \mathbf{m}$, or a plane that we shall refer to as the *kink* plane, where $\mathbf{n} = \mathbf{s}$.

29.2.5 Perturbations about the Slip and Kink Plane Orientations

To study the influence of the terms of $\mathcal{O}(\alpha\mathcal{L})$ and $\mathcal{O}(\sigma)$ in (29.38), we begin by expanding (29.38) in a series in $\mathbf{n}$, first about $\mathbf{n} = \mathbf{m}$ and then about $\mathbf{n} = \mathbf{s}$. In this we wish to be mindful of the fact that $\sigma/\mathcal{L}$ is of order 10^{-2} or smaller in representative cases, but that the α's, which we estimate later based on models for dislocation processes such as cross-slip, and from experimental data, may be appreciably larger, say on the order of 10^{-1}.

For the perturbation about $\mathbf{n} = \mathbf{m}$, we write

$$\mathbf{n} = \mathbf{m} + \boldsymbol{\epsilon}, \tag{29.58}$$

where $\boldsymbol{\epsilon}$ is understood to be small and to be chosen so that $\mathbf{n}$, like $\mathbf{m}$, is a *unit* vector. Also, we define the second-rank tensors

$$\mathbf{M} = \mathbf{m} \cdot \mathcal{L} \cdot \mathbf{m}, \quad \mathbf{E} = \mathbf{m} \cdot \mathcal{L} \cdot \boldsymbol{\epsilon}, \quad \mathbf{H} = \boldsymbol{\epsilon} \cdot \mathcal{L} \cdot \boldsymbol{\epsilon}, \tag{29.59}$$

and we observe that, by the definition of $\mathbf{P}$, we have

$$\mathbf{P} : \mathcal{L} \cdot \mathbf{n} = \mathbf{s} \cdot (\mathbf{M} + \mathbf{E}), \quad \mathbf{P} : \mathcal{L} : \mathbf{P} = \mathbf{s} \cdot \mathbf{M} \cdot \mathbf{s}. \tag{29.60}$$

Further,

$$\begin{aligned}(\mathbf{n} \cdot \mathcal{L} \cdot \mathbf{n})^{-1} &= \left(\mathbf{M} + \mathbf{E} + \mathbf{E}^T + \mathbf{H}\right)^{-1} \\ &= \left[\mathbf{I} + \mathbf{M}^{-1} \cdot (\mathbf{E} + \mathbf{E}^T + \mathbf{H})\right]^{-1} \cdot \mathbf{M}^{-1},\end{aligned} \tag{29.61}$$

where $\mathbf{E}^T$ is the transpose of $\mathbf{E}$. Since $\boldsymbol{\epsilon}$ is small, so also are $\mathbf{E}$ and $\mathbf{H}$, and the inverse of the bracketed matrix can be expanded in a series to give

$$\begin{aligned}(\mathbf{n} \cdot \mathcal{L} \cdot \mathbf{n})^{-1} = &\mathbf{M}^{-1} - \mathbf{M}^{-1} \cdot (\mathbf{E} + \mathbf{E}^T) \cdot \mathbf{M}^{-1} - \mathbf{M}^{-1} \cdot \mathbf{H} \cdot \mathbf{M}^{-1} \\ &+ \mathbf{M}^{-1} \cdot (\mathbf{E} + \mathbf{E}^T) \cdot \mathbf{M}^{-1} \cdot (\mathbf{E} + \mathbf{E}^T) \cdot \mathbf{M}^{-1} + \cdots \mathcal{O}(\epsilon^3).\end{aligned}$$

By using these results and reading off the various terms of the first set of brackets in (29.39), and comprising $\mathcal{L}F_0(\mathbf{n})$, we find, after some algebra, that

$$\mathcal{L}F_0(\mathbf{n}) = -\mathbf{s} \cdot (\mathbf{H} - \mathbf{E}^T \cdot \mathbf{M}^{-1} \cdot \mathbf{E}) \cdot \mathbf{s} + \mathcal{O}(\mathcal{L}\epsilon^3). \tag{29.62}$$

Next, by using the definitions of $\mathbf{M}$, $\mathbf{E}$, $\mathbf{H}$, and by writing $\mathbf{n} - \mathbf{m}$ for $\boldsymbol{\epsilon}$, there follows

$$\mathcal{L}F_0(\mathbf{n}) = -(\mathbf{n} - \mathbf{m}) \cdot (\mathbf{s} \cdot \mathcal{M} \cdot \mathbf{s}) \cdot (\mathbf{n} - \mathbf{m}) + \mathcal{O}(\mathcal{L}|\mathbf{n} - \mathbf{m}|^3). \tag{29.63}$$

The fourth-rank tensor $\mathcal{M}$ is given by

$$\mathcal{M} = \mathcal{L} - (\mathcal{L} \cdot \mathbf{m}) \cdot (\mathbf{m} \cdot \mathcal{L} \cdot \mathbf{m})^{-1} \cdot (\mathbf{m} \cdot \mathcal{L}), \tag{29.64}$$

and corresponds to the tensor $\mathcal{N}$ when $\mathbf{n}$ is set to $\mathbf{m}$. Since the vector $\mathbf{n} - \mathbf{m}$ can have no component in the direction of $\mathbf{m}$ to the order considered ($\mathbf{n}$ and $\mathbf{m}$ are unit vectors), we observe from the properties discussed earlier for $\mathcal{N}$ that (29.63) is a *negative definite* quadratic form.

Consider the term $\alpha \mathcal{L}F_1(\mathbf{n})$, which is defined by the second set of brackets in (29.39). We can write, using the usual definitions,

$$\alpha \mathcal{L}F_1(\mathbf{n}) = -\boldsymbol{\alpha} : \mathcal{N} : \mathbf{P} = -(\boldsymbol{\alpha} : \mathcal{N} \cdot \mathbf{m}) \cdot \mathbf{s}. \tag{29.65}$$

Observe that $F_1(\mathbf{m}) = 0$, because $\mathcal{M} \cdot \mathbf{m} = 0$. By expanding the expression for $\alpha \mathcal{L}F_1(\mathbf{n})$, using various results from above, we obtain

$$\alpha \mathcal{L}F_1(\mathbf{n}) = (\boldsymbol{\alpha} : \mathcal{M} \cdot \mathbf{s}) \cdot (\mathbf{n} - \mathbf{m}) + \mathcal{O}(\alpha \mathcal{L}|\mathbf{n} - \mathbf{m}|^2). \tag{29.66}$$

29.2. Localized Deformation

Finally, $\sigma F_2(\mathbf{n})$ of (29.40) is evaluated. We first calculate $F_2(\mathbf{m})$, noting that

$$(\mathbf{m} \cdot \mathcal{L} \cdot \mathbf{m})^{-1} \cdot (\mathbf{m} \cdot \mathcal{L} : \mathbf{P}) = (\mathbf{m} \cdot \mathcal{L} \cdot \mathbf{m})^{-1} \cdot (\mathbf{m} \cdot \mathcal{L} \cdot \mathbf{m}) \cdot \mathbf{s} = \mathbf{s}, \tag{29.67}$$

so that

$$\alpha \mathcal{L} F_2(\mathbf{m}) = -\sigma : \mathcal{H} : \left[\frac{1}{2}(\mathbf{ms} + \mathbf{sm})\right] + (\sigma : \mathcal{H} \cdot \mathbf{m}) \cdot \mathbf{s}$$
$$+ \mathbf{s} \cdot \left\{\mathbf{m} \cdot \left[\frac{1}{2}(\mathbf{sm} - \mathbf{ms}) \cdot \sigma - \frac{1}{2}\sigma \cdot (\mathbf{sm} - \mathbf{ms})\right]\right\}$$
$$- \mathbf{s} \cdot \left\{\frac{1}{2}\left[(\mathbf{m} \cdot \sigma \cdot \mathbf{m})\mathbf{I} - \sigma - (\mathbf{m} \cdot \sigma)\mathbf{m} - \mathbf{m}(\sigma \cdot \mathbf{m})\right]\right\} \cdot \mathbf{s}.$$

Observing that the two terms involving $\mathcal{H}$ cancel one another, because of the symmetry of $\mathcal{H}$, and simplifying the remaining terms, we come to the remarkable conclusion that

$$\sigma F_2(\mathbf{m}) = 0, \tag{29.68}$$

which applies *irrespective* of the several different generalizations of the Schmid stress considered above. Therefore, for any choice for $\mathcal{H}$, we have

$$\sigma F_2(\mathbf{n}) = \mathcal{O}(\sigma|\mathbf{n} - \mathbf{m}|). \tag{29.69}$$

By combining all the terms for h_{cr}, we obtain

$$h_{\text{cr}} = -(\mathbf{n} - \mathbf{m}) \cdot (\mathbf{s} \cdot \mathcal{M} \cdot \mathbf{s}) \cdot (\mathbf{n} - \mathbf{m}) + (\alpha : \mathcal{M} \cdot \mathbf{s}) \cdot (\mathbf{n} - \mathbf{m})$$
$$+ \mathcal{O}(\mathcal{L}|\mathbf{n} - \mathbf{m}|^3, \alpha\mathcal{L}|\mathbf{n} - \mathbf{m}|^2, \sigma|\mathbf{n} - \mathbf{m}|, \alpha\sigma, \sigma^2/\mathcal{L}). \tag{29.70}$$

Now, since $\mathbf{n} - \mathbf{m}$ has no component in the direction of $\mathbf{m}$ to the order considered, and since the (mm, sm, zm) components of $\mathbf{s} \cdot \mathcal{M} \cdot \mathbf{s}$ vanish, because

$$\mathbf{m} \cdot (\mathbf{s} \cdot \mathcal{M} \cdot \mathbf{s}) = (\mathbf{s} \cdot \mathcal{M} \cdot \mathbf{s}) \cdot \mathbf{m} = 0, \tag{29.71}$$

by (29.64), we shall henceforth understand the notation $\mathbf{s} \cdot \mathcal{M} \cdot \mathbf{s}$ to represent a *plane* second-rank tensor. Specifically, this is a tensor that has only components with the indices ss, sz, zs, and zz and whose components agree with those of its three-dimensional counterpart. The inverse operation to $(\mathbf{s} \cdot \mathcal{M} \cdot \mathbf{s}) \cdot \mathbf{q} = \mathbf{p}$ is defined only for vectors $\mathbf{p}$ and $\mathbf{q}$ lying in the $\mathbf{s}$-$\mathbf{z}$ plane. In this sense, the inverse $(\mathbf{s} \cdot \mathcal{M} \cdot \mathbf{s})^{-1}$ exists as a *plane* tensor, and

$$(\mathbf{s} \cdot \mathcal{M} \cdot \mathbf{s}) \cdot \mathbf{q} = \mathbf{p} \quad \text{implies} \quad \mathbf{q} = (\mathbf{s} \cdot \mathcal{M} \cdot \mathbf{s})^{-1} \cdot \mathbf{p}, \tag{29.72}$$

for associated vectors $\mathbf{p}$ and $\mathbf{q}$ lying in the $\mathbf{s}$-$\mathbf{z}$ plane.

In terms of this inverse, the orientation $\mathbf{n}$ which maximizes the *rhs* of (29.70) is

$$\mathbf{n} = \mathbf{m} + \frac{1}{2}(\mathbf{s} \cdot \mathcal{M} \cdot \mathbf{s})^{-1} \cdot (\mathbf{s} \cdot \mathcal{M} : \alpha) + \mathcal{O}(\alpha^2, \sigma/\mathcal{L}). \tag{29.73}$$

When this expression is inserted into (29.70), we find that the critical hardening rate at the onset of localization is

$$h_{\text{cr}} = \frac{1}{4}(\alpha : \mathcal{M} \cdot \mathbf{s}) \cdot (\mathbf{s} \cdot \mathcal{M} \cdot \mathbf{s})^{-1} \cdot (\mathbf{s} \cdot \mathcal{M} : \alpha) + \mathcal{O}(\alpha\sigma, \sigma^2/\mathcal{L}, \alpha^3\mathcal{L}). \tag{29.74}$$

A parallel calculation can be carried out for case (ii), in which we perturb about the kink-plane $\mathbf{n} = \mathbf{s}$. The results are given in terms of a tensor

$$\zeta = \mathcal{L} - (\mathcal{L} \cdot \mathbf{s}) \cdot (\mathbf{s} \cdot \mathcal{L} \cdot \mathbf{s})^{-1} \cdot (\mathbf{s} \cdot \mathcal{L}). \tag{29.75}$$

The critical orientation is given by

$$\mathbf{n} = \mathbf{s} + \frac{1}{2}(\mathbf{m} \cdot \boldsymbol{\zeta} \cdot \mathbf{m})^{-1} \cdot (\mathbf{m} \cdot \boldsymbol{\zeta} : \boldsymbol{\alpha}) + \mathcal{O}(\alpha^2, \sigma/\mathcal{L}), \tag{29.76}$$

and the corresponding hardening rate is

$$h_{\text{cr}} = \frac{1}{4}(\boldsymbol{\alpha} : \boldsymbol{\zeta} \cdot \mathbf{m}) \cdot (\mathbf{m} \cdot \boldsymbol{\zeta} \cdot \mathbf{m})^{-1} \cdot (\mathbf{m} \cdot \boldsymbol{\zeta} : \boldsymbol{\alpha}) + \mathcal{O}(\alpha\sigma, \sigma^2/\mathcal{L}, \alpha^3\mathcal{L}). \tag{29.77}$$

It is interesting to note from (29.74) and from the result $\mathbf{m} \cdot \mathcal{M} = 0$, that only the components of $\boldsymbol{\alpha}$ in the slip-plane (specifically $\alpha_{ss}, \alpha_{sz} = \alpha_{zs}$, and α_{zz}) affect the value of h_{cr} for localization, at least to quadratic order in the α's. Similarly, from (29.77), only components of $\boldsymbol{\alpha}$ in the kink-plane affect the result for h_{cr}. Also, we see that the critical value of h for the onset of localization is indeed *positive* when the non-Schmid effects, represented by the α's, are present.

We note that the terms represented explicitly in (29.74) and (29.77) have the order of $\alpha^2/\mathcal{L}$. So long as α is much larger than $\sigma/\mathcal{L}$, the terms of order $\alpha\sigma$ and $\sigma^2/\mathcal{L}$, represented by $\mathcal{O}(\ldots)$ in (29.74) and (29.77), will be negligible in comparison to what is retained. On the other hand, the neglected and retained terms are of the *same order* when the α's are of order $\sigma/\mathcal{L}$. But in this case, h_{cr} will be such a small fraction of σ (say, $10^{-2}\sigma$ or less) that it is to be expected that local necking, setting in when h is of the order of σ, will have long preceded the attainment of conditions for localizations of the types considered here. In any event, as we have emphasized earlier, it does not seem possible at present to specify the constitutive relation with enough precision to determine suitably the terms of order $\alpha\sigma$ and $\sigma^2/\mathcal{L}$ that we neglected in the above expressions for h_{cr}.

29.2.6 Isotropic Elastic Moduli

Suppose that $\mathcal{L}$ has the isotropic form

$$\mathcal{L}_{ijk\ell} = G(\delta_{ik}\delta_{j\ell} + \delta_{i\ell}\delta_{jk}) + \lambda\delta_{ij}\delta_{k\ell}, \tag{29.78}$$

where G and λ have their usual meanings as the Lamé moduli. Then we note that

$$\boldsymbol{\alpha} : \mathcal{L} \cdot \mathbf{s} = 2G\boldsymbol{\alpha} \cdot \mathbf{s} + \lambda \mathbf{s}\,\text{tr}(\boldsymbol{\alpha}),$$
$$\mathbf{m} \cdot \mathcal{L} \cdot \mathbf{m} = G(\mathbf{I} + \mathbf{mm}) + \lambda \mathbf{mm}, \tag{29.79}$$
$$\mathbf{m} \cdot \mathcal{L} \cdot \mathbf{s} = G\mathbf{sm} + \lambda \mathbf{ms},$$

as well as

$$(\mathbf{m} \cdot \mathcal{L} \cdot \mathbf{m})^{-1} = G^{-1}(\mathbf{I} - \xi \mathbf{mm}), \quad \xi = (\lambda + G)/(\lambda + 2G). \tag{29.80}$$

With these expressions, and with (29.64), we find

$$\mathbf{s} \cdot \mathcal{M} \cdot \mathbf{s} = G(\mathbf{zz} + 4\xi \mathbf{ss}), \tag{29.81}$$

so that its inverse, in the sense discussed earlier, is

$$(\mathbf{s} \cdot \mathcal{M} \cdot \mathbf{s})^{-1} = G^{-1}(\mathbf{zz} + \mathbf{ss}/4\xi). \tag{29.82}$$

Furthermore,

$$\boldsymbol{\alpha} : \mathcal{M} \cdot \mathbf{s} = 2G\{\alpha_{zs}\mathbf{z} + [(2\xi - 1)\alpha_{zz} + 2\xi\alpha_{ss}]\mathbf{s}\}. \tag{29.83}$$

29.2. Localized Deformation

These last two expressions enable us to carry out the perturbation about $\mathbf{n} = \mathbf{m}$, in which case we find from (29.73) that the orientation of the plane of localization is

$$\mathbf{n} = \mathbf{m} + \alpha_{sz}\mathbf{z} + \frac{1}{4\xi}[(2\xi - 1)\alpha_{zz} + 2\xi\alpha_{ss}]\mathbf{s} + \mathcal{O}(\alpha^2, \sigma/\mathcal{L}), \qquad (29.84)$$

with for the corresponding critical hardening rate

$$h_{\mathrm{cr}} = G\left\{\alpha_{sz}^2 + \frac{1}{4\xi}[(2\xi - 1)\alpha_{zz} + 2\xi\alpha_{ss}]^2\right\} + \mathcal{O}(\alpha\sigma, \sigma^2/G, \alpha^3 G). \qquad (29.85)$$

In a similar manner, the perturbation about $\mathbf{n} = \mathbf{s}$ leads to the orientation of the localization plane

$$\mathbf{n} = \mathbf{s} + \alpha_{mz}\mathbf{z} + \frac{1}{4\xi}[(2\xi - 1)\alpha_{zz} + 2\xi\alpha_{ss}]\mathbf{m} + \mathcal{O}(\alpha^2, \sigma/G), \qquad (29.86)$$

with the corresponding critical hardening rate

$$h_{\mathrm{cr}} = G\left\{\alpha_{mz}^2 + \frac{1}{4\xi}[(2\xi - 1)\alpha_{zz} + 2\xi\alpha_{mm}]^2\right\} + \mathcal{O}(\alpha\sigma, \sigma^2/G, \alpha^3 G. \qquad (29.87)$$

The results given so far are for *small* α. It is possible, however, to write out explicitly the entire expressions for $\mathcal{L}F_0(\mathbf{n})$ and $\alpha\mathcal{L}F_1(\mathbf{n})$ in the isotropic case. Since such expressions are wanted only when the α's are large, there is no need to retain the terms of order σ in (29.41), and it suffices to write

$$h_{\mathrm{cr}} = \mathcal{L}F_0(\mathbf{n}) + \alpha\mathcal{L}F_1(\mathbf{n}). \qquad (29.88)$$

Because α enters linearly in (29.39), there is no truncation in this expression. By using (29.78) in (29.41), and noting that

$$(\mathbf{n}\cdot\mathcal{L}\cdot\mathbf{n})^{-1} = G^{-1}(\mathbf{I} - \xi\mathbf{nn}), \qquad (29.89)$$

equation (29.88) can be written as

$$h/G = -1 + (\mathbf{n}\cdot\mathbf{m})^2 + (\mathbf{n}\cdot\mathbf{s})^2 - 4\xi(\mathbf{n}\cdot\mathbf{m})^2(\mathbf{n}\cdot\mathbf{s})^2 + 2[\mathbf{n}\cdot\alpha\cdot(\mathbf{sm}+\mathbf{ms})\cdot\mathbf{n}$$
$$- 2\xi(\mathbf{n}\cdot\mathbf{m})(\mathbf{n}\cdot\mathbf{s})(\mathbf{n}\cdot\alpha\cdot\mathbf{n}) + (2\xi - 1)(\mathbf{n}\cdot\mathbf{m})(\mathbf{n}\cdot\mathbf{s})\,\mathrm{tr}(\alpha)].$$

This latter result may be used to numerically search for the most critical set of conditions when maximum accuracy is desired.

29.2.7 Particular Cases for Localization

In what follows we examine some particular cases of non-Schmid effects, *i.e.*, the pressure sensitivity of plastic flow, the models for cross slip, and an analysis of non-Schmid effects in intermetallic compound Ni_3Al.

Consider first the possibility of pressure sensitivity of the Schmid stress for plastic flow. Suppose, for example, that

$$(\tau_{ms})_{\mathrm{onset}} = \tau_0 + \kappa p, \qquad (29.90)$$

where τ_0 is the flow stress at zero pressure, $p = -\frac{1}{3}\sigma_{kk}$ is the mean pressure and κ is a dimensionless parameter. Then, the Schmid stress increment $d\tau_{ms}$ is replaced by

$$d\tau_{ms} + \frac{1}{3}(d\tau_{mm} + d\tau_{ss} + d\tau_{zz}). \qquad (29.91)$$

Comparing this to (29.8), we find that

$$\alpha_{ij} = \frac{1}{3} \kappa \delta_{ij}. \tag{29.92}$$

This yields the same value for h_{cr} from (29.85) or (29.87); by taking $\xi = 2/3$, corresponding to taking $\lambda = G$ for instance, we obtain

$$h_{\mathrm{cr}} \approx 0.12 \kappa^2 G. \tag{29.93}$$

The parameter κ is readily interpreted in terms of the difference between yield strengths in uniaxial tension, σ_t, and compression, σ_c. If the slip system were oriented for a maximum resolved shear stress, i.e., at 45° to the tensile axis so that $\tau_{ms} = \frac{1}{2}\sigma$, then (29.90) gives

$$\frac{1}{2}(\sigma_c - \sigma_t) = \kappa \frac{1}{3}(\sigma_c + \sigma_t), \tag{29.94}$$

or

$$\sigma_c - \sigma_t = \frac{4\kappa}{3} \frac{\sigma_c + \sigma_t}{2}. \tag{29.95}$$

Now the *strength differential* is $4\kappa/3$ times the mean strength. Hence, making the definition $SD = 4\kappa/3$,

$$h_{\mathrm{cr}} \approx 0.07 (SD)^2 G. \tag{29.96}$$

Of course, this would also yield the connection

$$\alpha_{mm} = \alpha_{ss} = \alpha_{zz} = \frac{1}{4} SD. \tag{29.97}$$

Localization in crystals is generally observed to take place in the range of hardening between $5 \times 10^{-4} G$ and $5 \times 10^{-3} G$. If this is to be explained by a pressure sensitivity of yielding, it would be necessary to have values of SD in the range of 0.085 to 0.27. Such values seem, according to the survey of Dao and Asaro (1996), larger than are observed, except for martensitic high-strength steels and low symmetry crystals such as Zn. Spitzig, Sober, and Richmond (1975), for example, found values for the strength differential effect in the range 0.07-0.10, whereas in Zn crystals Barendareght and Sharpe (1973) report values for a normal stress dependence, as deduced by Dao and Asaro (1996), that would suggest $\alpha_{mm} \approx 0.1$.

Next we consider the models for cross slip. Asaro and Rice (1977) suggested that cross slip, which tends to occur in the later stages of deformation in single crystals, also provides a fundamental mechanism that leads to deviations from the Schmid rule. The process is illustrated in Fig. 29.5. Details of the cross slip process are given in the original work of Asaro and Rice (1977); here we list only some of the results from their analysis of the cross-slip process. We note that in the process of cross-slip an extended dislocation gliding on the primary plane will first have to constrict along a segment of critical length, so that it may successfully "bow out onto" the cross slip plane. The process of constriction is influenced by the component of shear stress τ_{mz} resolved in the primary slip plane, but acting in the direction orthogonal to the primary slip direction, **s**. Following this constriction, the now perfect dislocation must bow onto the cross-slip plane, which involves a process influenced by the component of shear stress, τ_{cs}, resolved on the cross-slip system. Of course, τ_{cs} is built up in significant measure by the shear stress component τ_{zs}. In this way we see the

29.2. Localized Deformation

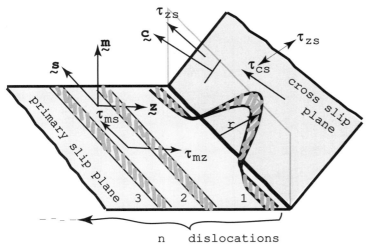

Figure 29.5. Cross-slip model of Asaro and Rice. Local obstacles are bypassed by partial dislocation segments of screw dislocations on the primary plane through a process of constriction and cross-slip onto the cross-slip plane. The cross-slipped dislocations, after bypassing their obstacles, cross-slip back to an adjacent primary plane. Adapted from Asaro and Rice (1977).

roles that the non-Schmid components τ_{mz} and τ_{zs} play in the total process of ongoing primary slip. We describe these effects by writing

$$d\gamma = \frac{1}{h}(d\tau_{ms} + \alpha d\tau_{zs} + \beta d\tau_{mz}), \tag{29.98}$$

so that α describes the effect of the stress component τ_{mz} in the primary slip-plane, and β the effect of the shear stress on the plane whose normal is along $\mathbf{z}$. In terms of the coefficients of α, we have the connections

$$\alpha_{sz} = \alpha_{zs} = \frac{1}{2}\alpha, \quad \alpha_{mz} = \alpha_{zm} = \frac{1}{2}\beta. \tag{29.99}$$

With no other effects in the model, all other components of α vanish. Thus, for localization on a plane near the slip plane, we have

$$\mathbf{n} = \mathbf{m} + \frac{1}{2}\alpha\mathbf{z}, \tag{29.100}$$

and

$$h_{cr} = \frac{1}{4}\alpha^2 G. \tag{29.101}$$

For localization near the kink plane orientation, we have

$$\mathbf{n} = \mathbf{s} + \frac{1}{2}\beta\mathbf{z}, \tag{29.102}$$

and

$$h_{cr} = \frac{1}{4}\beta^2 G. \tag{29.103}$$

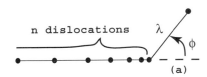
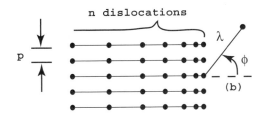

Figure 29.6. Arrays of dislocations piled up at a barrier. ϕ is the angle of inclination of the cross-slip plane. The dots represent dislocations in the pileups. (a) An isolated array which may have a single or a double end. (b) A group of parallel arrays with spacing p. Adapted from Barnett and Asaro (1972).

Asaro and Rice (1977) have provided a rather detailed analysis of the cross-slip process shown in Fig. 29.5. In fact, their models describe the effects of having the primary dislocations *piled up* at their obstacles in various forms of arrays, as sketched in Fig. 29.6. Several types, or geometries, of arrays were considered in an attempt to understand the effect of slip geometry on the potential magnitudes of the parameters α and β introduced above. The arrays were characterized by their length, ℓ, and in the case of parallel arrays, by their average spacing, p. As examples of the results, let b_s be the magnitude of the "perfect dislocation" of the primary slip system and let ℓ be the average (or typical) length of a slip line (*i.e.*, a pileup). Then, standard results indicate that the number of dislocations in each type of pileup would be

$$n = \pi \tau_{ms} \ell/(Gb_s), \quad \text{for a single ended pileup,}$$
$$n = 2\tau_{ms} \ell/(Gb_s), \quad \text{for a double ended pileup,} \quad (29.104)$$
$$n \approx \tau_{ms} p/(Gb_s), \quad \text{for a parallel array,}$$

and where the last of (29.104) is most accurate if $p/\ell \ll 1$. With these scenarios, Asaro and Rice (1977) found estimates for α to be

$$\alpha \approx 1.1(b_s/\ell)^{1/2}, \quad \text{for a single ended pileup,}$$
$$\alpha \approx 1.6(b_s/\ell)^{1/2}, \quad \text{for a double ended pileup,} \quad (29.105)$$
$$\alpha \approx 2.8(b_s/p)^{1/2}, \quad \text{for a parallel array.}$$

Now, ℓ typically ranges from 10^{-7} to 10^{-6} m in the later stages of deformation (stages II and III); Burgers vectors are in the range $b_s \approx 2.7 \times 10^{-10}$ m. With $\ell = 10^{-7}$ m, we find $\alpha = 0.06$ and 0.08 for the model cases of isolated single and double ended pileups, whereas for sequential pileups, taking $p = 0.5 \ell$, we find $\alpha = 0.21$. Of course, for $\ell = 10^{-6}$ m, all the numbers are smaller by approximately $\frac{1}{3} \ell$. We note that ℓ typically decreases with ongoing strain, with the implication that the relative importance of τ_{zs} with respect to τ_{ms}, and hence the size of α increases. Thus, since $h_{cr} \approx \frac{1}{4} \alpha^4 G$, the critical hardening modulus, below which localization occurs, increases with strain in stage II and early stage III hardening. This pattern continues until the localization condition is met.

29.2. Localized Deformation

Table 29.1. *Non-Schmid coefficients for* Ni_3Al *at* 298 K

Stress state	α_{ss}	α_{mm}	α_{zz}	α_{mz}	α_{sz}
$\tau_{sm} > 0, \tau_{cb} > 0$	0	0.008	−0.008	0.008	−0.015
$\tau_{sm} < 0, \tau_{cb} < 0$	0	−0.008	0.008	−0.008	−0.015
$\tau_{sm} > 0, \tau_{cb} < 0$	0	0.008	−0.008	0.008	0.014
$\tau_{sm} < 0, \tau_{cb} > 0$	0	−0.008	0.008	−0.008	0.014

To compare the predicted critical hardening rate, $\frac{1}{4}\alpha^2 G$, with experimental values, we need only remember that near the stage II-III transition, h/G is of the order $1/300 \approx 3 \times 10^{-3}$ and, of course, generally falling with increasing strain. However, if $\alpha \approx 0.1$, which is representative of the estimates of Asaro and Rice (1977), then $\frac{1}{4}\alpha^2 G$ is of the order $2.5 \times 10^{-3} G$. Thus, the predicted magnitudes of these non-Schmid effects lead to critical hardening rates that are consistent with experiment. On the other hand, Chang and Asaro (1981) conducted detailed experimental studies on single crystals of Al-Cu and observed the sequential formation of what they described as CSB's and MSB's (see Fig. 29.1). The CSB's were interpreted as being localized modes of deformation of the general type considered here; indeed, these modes formed when the crystals were still deforming by a sensibly single slip mode. They estimated the critical hardening rates at which such modes first appeared, and extracted values of α in the range $\alpha \approx 0.08\text{-}0.09$. These are, again, consistent with the range of values forecast by the Asaro and Rice (1977) modeling.

There are at least two scenarios for the value of the parameter β. First, if the only barrier to cross-slip were the constriction of the partial dislocations comprising the extended primary dislocations, then

$$\beta = b_z^{(2)}/nb_s^{(2)} \approx 0.58/n, \tag{29.106}$$

where the superscript on the b_z and b_s signifies the Burgers vector of the trailing partial dislocation (called 2 in Asaro and Rice model). Of course, the number of dislocations in the pileup can be related to ℓ and the stress level *via* (29.103). However, if both the processes of constriction of partial dislocations and bowing onto the cross-slip plane are simultaneously important for cross-slip, then β is computed as

$$\beta = \frac{0.3\varsigma(\delta - r_0)(b_z^{(2)}/b_s)\sec(\frac{1}{2}\phi)}{(2\pi\ell r_0)^{1/2}},$$
$$\beta = \frac{0.3(\delta - r_0)(b_z^{(2)}/b_s)\sec(\frac{1}{2}\phi)}{(pr_0)^{1/2}}. \tag{29.107}$$

Here $\varsigma = 1$ or $\varsigma = \sqrt{2}$ for the isolated single or double ended pileup, respectively.

We end this subsection with an analysis of non-Schmid effects in intermetallic compound Ni_3Al. Dao and Asaro (1996) reviewed the data of Paidar *et al.* (1984) for ordered Ni_3Al and interpreted their data on yield strength *vis-à-vis* the existence of non-Schmid effects. Table 29.1 shows some of their results for Ni_3Al deformed at a temperature of $T = 298$ K.

The values of the non-Schmid coefficients extracted this way are noticeably dependent on stress state. Note also that Table 29.1 defines the stress component τ_{cb} as the shear stress

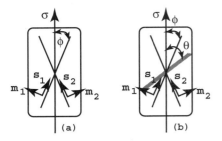

Figure 29.7. Geometry for an idealized crystal subject to nominally uniaxial tension. The crystal is presumed to deform by either slip on two slip systems (*i.e.*, the primary and conjugate systems) or more than two. This system is idealized as being two-dimensional. A potential band of localized deformation is also shown, inclined at the angle θ to the crystal's axis.

resolved on the cross-slip plane and in the direction of the primary Burgers vector. The values of the non-Schmid coefficients are also very much dependent upon the temperature of testing. For example, at temperatures of 600-800 K the values of the nonzero coefficients, although similar in sign, have magnitudes that are some 3-4 times as large.

29.3 Localization in Multiple Slip

Although modes of deformation involving primarily single slip are often nearly achieved in single crystals, crystalline deformation invariably occurs *via* activity on more than one slip system. In polycrystals, compatibility constraints among the differently oriented grains impose multiple slip system activity in all grains even under the simplest states of globally imposed strains. In single crystals, oriented initially for slip on a single system, lattice rotations (of a type already discussed) lead to states of multiple slip often after what turns out to be quite modest globally imposed strains. Aside from the global multiple slip states just mentioned, slip on other than the primary system can occur locally through the action of concentrated stresses developed at the tips of slip bands or at substructural sites such as dislocation boundaries. As shown by Asaro and co-workers (Asaro, 1979; Chang and Asaro, 1981; Peirce *et al.*, 1982; Harren *et al.*, 1988; Dao and Asaro, 1993, 1994) multiple slip states give rise to several critical differences in phenomenology that have important implications *vis-à-vis* inhomogeneous, and indeed localized, deformation. Two of these are the appearance of local vertexes on what would be an idealized strain rate independent yield surface, and the phenomena of *geometrical softening* within the deformation patterns that, in turn, induce localized flow. Of course, the effects of latent hardening have already been discussed with respect to its strong influence on promoting (indeed causing) inhomogeneous plastic flow. We examine some of the fundamental aspects of these geometrical and energetic effects on deformation patterning *via* analysis of the original model of Asaro (1979) as generalized by Dao and Asaro (1996)

29.3.1 Double Slip Model

Consider the crystal subject to tension as depicted in Fig. 29.7. In this two-dimensional geometry, both slip systems are symmetrically oriented about the tensile axis by the angle ϕ. We first develop a suitable constitutive model for this double slip crystal model and perform a bifurcation analysis akin to that performed for the crystal undergoing only single slip.

29.3. Localization in Multiple Slip

29.3.2 Constitutive Law for the Double Slip Crystal

We begin by writing a suitable hardening relation for the ξ-slip system ($\xi = 1, 2$) by making a suitable choice for $\mathcal{H}$, as we have seen that its precise specification is of minor importance. With that, we write the generalization of (29.19) for single slip as

$$\left(\mathbf{P}_\xi + \mathbf{\Phi}^\xi \cdot \boldsymbol{\alpha} \cdot \mathbf{\Phi}^{\xi T}\right) : \overset{\nabla}{\boldsymbol{\sigma}}{}^* = h_{\xi\beta}\dot{\gamma}_\beta, \quad \text{(no sum on } \xi\text{)}. \tag{29.108}$$

Here $\mathbf{\Phi}^\xi$ is a transformation matrix that transforms the coefficients of $\boldsymbol{\alpha}$ from their local base vectors formed on the slip system ξ to those based on a set of fixed laboratory base vectors, $\{\mathbf{e}_j\}$; for example, we may do this via the connection $\Phi^\xi_{ij} = \mathbf{e}_i \cdot \mathbf{a}^\xi_j$, where the $\{\mathbf{a}^\xi_i\}$ are the slip system based basis. For this 2-dimensional crystal model we need to keep the in-plane components of $\boldsymbol{\alpha}$, viz., α_{ss} and α_{mm}. To account for incompressibility, it is convenient to define α^+ and α^- as

$$\alpha^+ = \alpha_{ss} + \alpha_{mm}, \qquad \alpha^- = \alpha_{ss} - \alpha_{mm} \tag{29.109}$$

and

$$\boldsymbol{\alpha}' = \begin{bmatrix} \alpha_{ss} - \frac{1}{2}\alpha^+ & 0 \\ 0 & \alpha_{mm} - \frac{1}{2}\alpha^+ \end{bmatrix} = \begin{bmatrix} \frac{1}{2}\alpha^- & 0 \\ 0 & -\frac{1}{2}\alpha^- \end{bmatrix}. \tag{29.110}$$

In terms of these so defined quantities, (29.108) becomes

$$\mathbf{Q}'_\xi : \overset{\nabla}{\boldsymbol{\sigma}}{}^* + \frac{1}{2}\alpha^+ \dot{p} = h_{\xi\beta}\dot{\gamma}_\beta, \tag{29.111}$$

with

$$\mathbf{Q}'_\xi = \mathbf{P}_\xi + \mathbf{\Phi}^\xi \cdot \boldsymbol{\alpha}' \cdot \mathbf{\Phi}^{\xi T}, \quad \text{(no sum on } \xi\text{)}, \tag{29.112}$$

and where the pressure rate can be expressed as

$$\dot{p} = \frac{1}{2}\left(\overset{\nabla}{\sigma}_{11} + \overset{\nabla}{\sigma}_{22}\right) = \frac{1}{2}(\dot{\sigma}_{11} + \dot{\sigma}_{22}). \tag{29.113}$$

The development of the constitutive relations that govern this crystal follows directly, and straightforwardly, from the procedure developed by Asaro (1979). For example, Asaro's (1979) relation (Eq. 3.15) becomes

$$\mathbf{D} = \mathcal{L}^{-1} : \left(\overset{\nabla}{\boldsymbol{\sigma}} + \boldsymbol{\sigma}\,\mathrm{tr}\,\mathbf{D}\right) + \hat{\mathbf{P}}_\xi N^{-1}_{\xi\beta}(\mathbf{Q}'_\beta : \mathcal{L} : \mathbf{D}) + \frac{1}{2}\hat{\mathbf{P}}_\xi \sum_\beta N^{-1}_{\xi\beta}\alpha^+ \dot{p}, \tag{29.114}$$

where

$$\begin{aligned} N_{\xi\beta} &= h_{\xi\beta} + \mathbf{Q}'_\xi : \mathcal{L} : \mathbf{P}_\beta, \\ \hat{\mathbf{P}}_\xi &= \mathbf{P}_\xi + \mathcal{L}^{-1} : (\mathbf{Q}_\xi \cdot \boldsymbol{\sigma} - \boldsymbol{\sigma} \cdot \mathbf{Q}_\xi). \end{aligned} \tag{29.115}$$

To complete the constitutive analysis we next assume that the crystal's elasticity is isotropic and incompressible; this has the effect of simplifying the relations and yet preserving the phenomena. Then $\mathrm{tr}\,\mathbf{D} = 0$, and the first term in (29.114) becomes

$$\frac{1}{2G}\overset{\nabla}{\boldsymbol{\sigma}}.$$

Next we take a simple, yet acceptably general, form for the hardening matrix $h_{11} = h_{22} = h$ and $h_{12} = h_{21} = h_1$; we also write the tensile stress as simply $\sigma_{22} = \sigma$. Together, these yield

$$N_{\xi\beta}^{-1} = \frac{1}{\det(\mathbf{N})} \begin{bmatrix} h + G & -h_1 + G\cos(4\phi) - G\alpha^- \sin(4\phi) \\ -h_1 + G\cos(4\phi) - G\alpha^- \sin(4\phi) & h + G \end{bmatrix},$$

and finally

$$D_{11} = \frac{1}{2G}\overset{\triangledown}{\sigma}_{11} - \frac{G\sin(2\phi)[\sin(2\phi) + \alpha^- \cos(2\phi)]}{h + h_1 + 2G\sin^2(2\phi) + G\alpha^- \sin(4\phi)}(D_{22} - D_{11}) - \frac{\delta \dot{p}}{2G}, \quad (29.116)$$

$$D_{22} = \frac{1}{2G}\overset{\triangledown}{\sigma}_{22} + \frac{G\sin(2\phi)[\sin(2\phi) + \alpha^- \cos(2\phi)]}{h + h_1 + 2G\sin^2(2\phi) + G\alpha^- \sin(4\phi)}(D_{22} - D_{11}) + \frac{\delta \dot{p}}{2G}, \quad (29.117)$$

$$D_{12} = \frac{1}{2G}\overset{\triangledown}{\sigma}_{12} + \frac{G[2\cos(2\phi) - \sigma/G][\cos(2\phi) - \alpha^- \sin(2\phi)]}{h - h_1 + 2G\cos^2(2\phi) - G\alpha^- \sin(4\phi)} D_{12}, \quad (29.118)$$

along with the incompressibility constraint

$$D_{11} + D_{22} = 0. \quad (29.119)$$

The parameter δ is defined as

$$\delta = \frac{2G\alpha^+ \sin(2\phi)}{h + h_1 + 2G\sin^2(2\phi) + G\alpha^- \sin(4\phi)}. \quad (29.120)$$

Alternatively, the above relations may be rephrased in a form used to study the influence of pressure sensitivity in bifurcation analysis, in the context of plane strain tension, as studied by Biot (1973), Hill and Hutchinson (1975), and Needleman (1979), viz.,

$$\overset{\triangledown}{\sigma}_{11} = 2\mu^* D_{11} + (1 + \delta)\dot{p},$$
$$\overset{\triangledown}{\sigma}_{22} = 2\mu^* D_{22} + (1 - \delta)\dot{p}, \quad (29.121)$$
$$\overset{\triangledown}{\sigma}_{12} = 2\mu D_{12},$$

and $D_{11} + D_{22} = 0$. The moduli in (29.121) will be identified after noting that the constitutive relation can be rewritten, by combining the first two expressions, as

$$\overset{\triangledown}{\sigma}_{22} - \overset{\triangledown}{\sigma}_{11} = 2\mu^*(D_{22} - D_{11}) - 2\delta \dot{p}. \quad (29.122)$$

In the above, the moduli are identified as

$$2\mu^* = \frac{2G(h + h_1)}{h + h_1 + 2G\sin^2(2\phi) + G\alpha^- \sin(4\phi)},$$
$$2\mu = \frac{2G[h - h_1 + \sigma \cos(2\phi) - \sigma\alpha^- \sin(2\phi)]}{h - h_1 + 2G\cos^2(2\phi) - G\alpha^- \sin(4\phi)}. \quad (29.123)$$

When $G \gg \sigma$ and $G \gg h, h_1$, we have

$$2\mu^* = \frac{h + h_1}{\sin^2(2\phi) + \alpha^- \sin(2\phi)\cos(2\phi)},$$
$$2\mu = \frac{h - h_1 + \sigma \cos(2\phi) - \sigma\alpha^- \sin(2\phi)}{\cos^2(2\phi) - \alpha^- \sin(2\phi)\cos(2\phi)}, \quad (29.124)$$

29.3. Localization in Multiple Slip

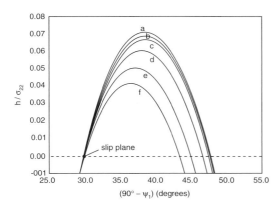

Figure 29.8. The critical ratio, $(h/\sigma)_{cr}$, for a band of localized shear inclined by the angle θ. The figure is for the case where $\phi = 30°$. The cases included are for: (a) $\alpha_{ss} = 0.08$, $\alpha_{mm} = 0$; (b) $\alpha_{ss} = 0.04$, $\alpha_{mm} = 0.04$; (c) $\alpha_{ss} = 0$, $\alpha_{mm} = 0.08$; (d) $\alpha_{ss} = 0$, $\alpha_{mm} = 0.05$; (e) $\alpha_{ss} = 0$, $\alpha_{mm} = 0$; and (f) $\alpha_{ss} = 0$, $\alpha_{mm} = -0.05$, respectively (from Dao and Asaro, 1996).

and in this case

$$\delta = \frac{\alpha^+}{\sin(2\phi) + \alpha^- \cos(2\phi)}. \tag{29.125}$$

When the compatibility and equilibrium conditions listed in (29.25) and (29.26) are applied, the criterion for localization becomes

$$(1+\delta)(\mu - \frac{1}{2}\sigma)n_1^4 + [2(2\mu^* - \mu) - \delta\sigma]n_1^2 n_2^2 + (1-\delta)(\mu + \frac{1}{2}\sigma)n_2^4 = 0. \tag{29.126}$$

As before, **n** is the unit normal to the plane of localization. With the simple representations, $n_1^2 = \cos^2\theta$ and $n_2^2 = \sin^2\theta$, equation (29.126) can be rephrased as

$$\left(\frac{h}{\sigma}\right)_{cr} = \frac{\cos(2\theta) - \cos^2(2\theta)/\cos(2\phi) + \delta(1 - \cos(2\theta)/\cos(2\phi))}{(1-q)\cos^2(2\theta)/B + (1+q)\sin^2(2\theta)/A + (1-q)\delta\cos^2(2\theta)/B},$$

where

$$A = \sin^2(2\phi) + \alpha^- \sin(2\phi)\cos(2\phi),$$
$$B = \cos^2(2\phi) - \alpha^- \sin(2\phi)\cos(2\phi), \tag{29.127}$$
$$q = h_1/h,$$

and δ has been given earlier in (29.120). When the Schmid rule holds, *i.e.*, when $\alpha^- = \alpha^+ = 0$, we obtain

$$\left(\frac{h}{\sigma}\right)_{cr} = \frac{\cos(2\theta) - \cos^2(2\theta)/\cos(2\phi)}{(1-q)\cos^2(2\theta)/\cos^2(2\phi) + (1+q)\sin^2(2\theta)/\sin^2(2\phi)}. \tag{29.128}$$

We note that for crystal's undergoing multiple slip, the conditions for localization take the form of critical ratios of hardening rate to the prevailing stress. Examples are shown in Fig. 29.8.

To interpret these results note first that for localization on either slip system, *i.e.*, with $\theta = \phi$ in the context of this model, the critical conditions become $(h/\sigma)_{cr} = 0$, which is consistent with the analysis given earlier for single slip. On the other hand, with deviations from the Schmid rule, for example if $\alpha_{ss} + \alpha_{mm} \approx 0.08$, the critical hardening to stress ratios are seen to be some 30-40% larger than when Schmid's rule applies. Furthermore, in this particular orientation the optimal orientation of the band of localization is only

Figure 29.9. X-ray (Berg–Barrett) topograph of the gage section of an Al-2.8 wt% Cu single crystal subject to tensile deformation. The contrast seen is indicative an abrupt transition in lattice orientation in the shear band relative to the surrounding matrix (from Chang and Asaro, 1981).

slightly affected by the existence of non-Schmid effects, although the range of orientations where localization is possible with positive h is increased. Now, as pointed out by Asaro (1979), the optimal orientations for localized bands of deformation are characteristically misoriented from the slip systems, and this is a direct reflection of some of the more important mechanistic underpinnings of the phenomena of localization itself. It is readily shown, for example, that with a band oriented at an angle $\theta \approx 36°\text{-}39°$, *i.e.*, near slip system 2, the strain is concentrated on slip system 2; this indicates that this band is indeed a shear band. But for slip to concentrate in this way, it is necessary that the lattice itself undergoes a reorientation in the same sense as the misorientation of the band. The geometry of this simple crystal then shows that such a rotation of the lattice causes a *geometrical softening* within the band, which promotes strain concentration. These effects of lattice rotations are examined in the next section by way of a brief survey of numerical results based on the kinds of models introduced here. In particular, we will be interested in exploring the trends with respect to the development of deformation patterns.

Before closing this section we do mention that Dao and Asaro (1996) have performed the same type of rate independent bifurcation analysis for fully three-dimensional geometries. The results are substantially consistent with the phenomena revealed by the two-dimensional analysis described here. The reader is referred to their work for further details.

29.4 Numerical Results for Crystalline Deformation

29.4.1 Additional Experimental Observations

Experimental observations have been presented in this chapter, as well as in previous chapters, concerned with an overview of the micromechanics of plastic deformation in crystalline materials. In this section we present additional observations that bear directly upon the analysis described above. Some results of numerical analysis using the constitutive theory laid out in this and, more specifically, the previous chapter are then described.

Figure 29.9 shows an X-ray topograph of the gage section of an Al-Cu single crystal deformed in tension. The rather obvious contrast is caused by what turns out to be an abrupt

29.4. Numerical Results

transition in the orientation of the crystal's lattice from inside a shear band with respect to the surrounding matrix. As noted previously, such lattice misorientations are characteristic of shear bands that eventually develop in ductile crystals. Analysis of such misorientations, in fact, shows that they are due to the formation of dislocation tilt boundaries that are, in turn, formed by arrays of Lomer dislocations (discussed earlier in Chapter 27). This makes sense because fcc crystals like these, when deformed in tension, will invariably develop a double slip mode involving the primary slip system and the system that is *conjugate* to it. The slip mode may, in fact, involve more than these two systems, but *primary-conjugate* slip is clearly expected. Thus the dislocation interactions that can explain the observed lattice rotations are naturally expected.

To understand some of the salient mechanistic features of the localization process, and in particular the role that geometrical softening plays, we recall the compatibility condition imposed on the jump in velocity gradient across the band, *i.e.*, relation (29.25), which is here rewritten as

$$\Delta(v_{i,j}) = v_{i,j}|_{\text{band}} - v_{i,j}|_{\text{matrix}} = g_i n_j = \lambda \bar{g}_i n_j. \tag{29.129}$$

A unit vector along the direction of velocity jump is $\bar{\mathbf{g}}$. Given the geometry defined in Fig. 29.7, we have $(\bar{g}_1, \bar{g}_2) = (\sin\theta, \cos\theta)$ and $(n_1, n_2) = (-\cos\theta, \sin\theta)$. If we recognize that this jump involves essentially plastic deformation, and after setting $\Delta \mathbf{D} = \lambda/2(\bar{\mathbf{g}}\mathbf{n} + \mathbf{n}\bar{\mathbf{g}})$, we find that

$$\Delta(\dot{\gamma}^{(1)} - \dot{\gamma}^{(2)}) = \lambda \cos(2\theta)/\cos(2\phi). \tag{29.130}$$

Similarly, from the jump in material spin rate, $\Delta \mathbf{W} = \lambda/2(\bar{\mathbf{g}}\mathbf{n} - \mathbf{n}\bar{\mathbf{g}})$, we find by forming both sides of the first of (29.1) (extended in an obvious way to include two slip systems) that

$$\Delta W_{12} = \Delta W_{12}^* + \frac{\lambda}{2}\cos(2\theta)/\cos(2\phi) = \frac{\lambda}{2},$$

$$\Delta W_{12}^* = \frac{\lambda}{2}[1 - \cos(2\theta)/\cos(2\phi)], \tag{29.131}$$

$$\Delta W_{12}^* = \frac{\Delta D_{22}}{\sin(2\theta)}[1 - \cos(2\theta)/\cos(2\phi)].$$

Since elastic contributions to this difference are negligible, ΔW_{12}^* is interpreted as a finite spin of the lattice in the band relative to that outside. If we take, as another representative example, $\phi = 35°$ and $\theta = 40°$, we find that $\Delta W_{12}^* \approx 0.246\lambda$ (radians) or 14.1λ (degrees). To gain some appreciation for the magnitude of rotation implied by this, let $\lambda dt = 0.01$, where dt is a *time increment*, *i.e.*, let an approximately 10% excess shear strain increment develop in the band. Then $\Delta W_{12}^* dt \approx 1.41°$. In other words if, say, some 30% strain were to develop within the band, there would result a roughly 4° lattice rotation jump across the band/matrix interface. It is important to note that the sense of this lattice spin is to *increase* the Schmid factor within the band relative to the surrounding matrix. Thus the kinematics of localized shearing imply a *local geometrical softening* of the slip plane with which the shear band is most closely aligned. These are precisely the sort of lattice misorientations given evidence in Fig. 29.9; other examples are shown next.

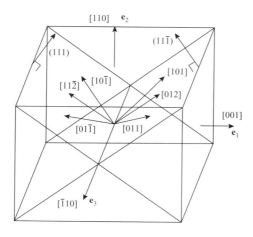

Figure 29.10. Sketch of the crystallographic geometry of a plane strain compression test (from Harren et al., 1988).

Harren, Deve, and Asaro (1988) performed experimental studies on single crystals of Al-2.8 wt% Cu crystals in states of plane strain compression. The geometry is sketched in Fig. 29.10; a photomicrograph of a shear band that formed after a compressive strain of approximately $\epsilon \approx 0.44$ was shown in Fig. 29.2. The symmetry in this orientation suggests that the deformation mode would be symmetric slip on four slip systems leading to a near state of plane strain. Of course, the small elastic strains that form include *out-of-plane* strains. The sense of the lattice rotations is to bring the most active slip planes within the shear band toward the orientation of maximum resolved shear stress, *i.e.*, the 45° inclination to the compression axis (which is vertical). This again leads to a geometrical softening within the shear band.

29.4.2 Numerical Observations

The experimental studies briefly reviewed have been analyzed in detail by Asaro and co-workers (*e.g.*, by Harren, Deve, and Asaro, 1988). As discussed in the previous chapter, material response is characteristically rate dependent and the fully strain rate dependent constitutive theory outlined there was used. In particular, the kinetic-hardening of the power law form

$$\dot{\gamma}_\alpha = \dot{\gamma}_0 \, \text{sgn}\,(\tau_\alpha) \left| \frac{\tau_{(\alpha)}}{g_{(\alpha)}} \right|^{1/m} \tag{29.132}$$

was incorporated. For the crystals studied, the hardening function, g_α, was fitted to tensile and compression tests, and was found to be

$$g(\gamma_a) = g_0 + h_\infty \gamma_\alpha + (g_\infty - g_0) \tanh\left[\left(\frac{h_0 - h_\infty}{g_\infty - g_0}\right) \gamma_\alpha\right], \tag{29.133}$$

where $g_0 = g(\gamma_\alpha = 0) = 123$ MPa, $h_0 = 120$ MPa, $g_\infty = 133$ MPa, and $h_\infty = 11$ MPa. The latent hardening ratios were taken as unity, *i.e.*, isotropic hardening among the slip systems was assumed. Aged hardened alloys such as these are weakly rate sensitive and a value of $m = 0.005$ was selected. The contours clearly indicate the intensely concentrated nature

29.4. Numerical Results

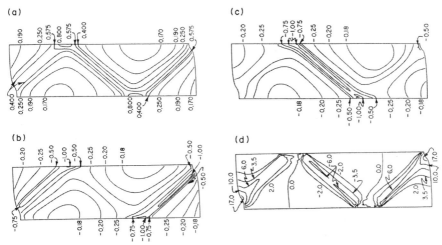

Figure 29.11. Contours of (a) maximum principal logarithmic strain; (b) accumulated glide strain γ_1 on system 1; (c) accumulated glide strain γ_2 on system 2; (d) lattice rotation in degrees (from Harren et al., 1988).

of the plastic deformation. Figure 29.11 demonstrates the geometrical softening described above.

The reader is referred to Harren, Deve, and Asaro (1988) for the legion of details, but here we close with one last account of some numerically and experimentally obtained observations for inhomogeneous flow in polycrystals. This is to be added to the already mentioned observations of inhomogeneous flow in single crystals (*e.g., patchy slip*). Figure 29.12 shows a result, using the same constitutive data as given above, of a simulation of plane strain compression of polycrystals of the same Al-Cu material. The figure is for one grain in an assembly of 27 grains all with hexagonal shape. The short lines are traces of the most active slip system in the grain as per location within the intragrain space. Note that although each grain, including the one shown, initially had a uniform lattice orientation –

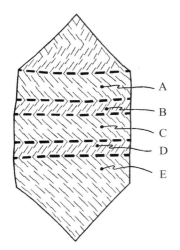

Figure 29.12. Traces of the most active slip planes within a given grain in polycrystal. This is the result of a finite element simulation of an Al-Cu polycrystal (from Harren and Asaro, 1989).

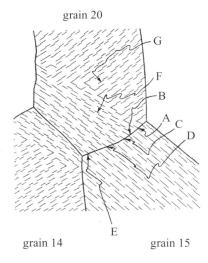

Figure 29.13. Traces of the most active slip planes at a grain boundary *triple point*. Note how the traces tend to become aligned within a shear band (from Harren and Asaro, 1989).

but different than that of its neighboring grains – the lattice has *broken up* into subregions of distinctly different orientation. The grains become *patchy* and the boundaries between the misoriented regions are quite distinct, *i.e.*, the boundaries are sharp. This was suggested to be the continuum analog to the formation of sub-boundaries within the grain. It should also be noted that the hardening was again taken to be isotropic. If the latent hardening were to be biased, as experimental evidence indicates (see Chapter 27), the "patchyness" would only be increased. Thus we again see how the fundamental trends in deformation are to produce substructure which, in turn, directly influences the subsequent response of the material.

Figure 29.13 shows the same sort of trace pattern near a *triple point*, where a shear band has formed and propagated through the polycrystal. There are several features to note here. First, the grains shown have all broken up into patches, but there is clear organization there. The lattices in each grain have rotated so that the most active slip planes supporting the concentrated shear have become closely aligned – again a kind of geometrical softening. This pattern of lattice rotation has been documented by Asaro and co-workers (see Harren, Deve, and Asaro, 1988 for details).

29.5 Suggested Reading

Asaro, R. J., and Rice, J. R. (1977), Strain Localization in Ductile Single Crystals, *J. Mech. Phys. Solids*, Vol. 25, pp. 309–338.

Dao, M., and Asaro, R. J. (1996), Localized Deformation Modes and non-Schmid Effects in Crystalline Solids. Part I: Critical Conditions for Localization, *Mech. Mater.*, Vol. 23, pp. 71–102.

De Borst, R., and Van der Giessen, E., eds. (1998), *Material Instabilities in Solids*, Wiley, Chichester, UK.

Harren, S. V., Deve, H. E., and Asaro, R. J. (1988), Shear Band Formation in Plane Strain Compression, *Acta Metall.*, Vol. 36, pp. 2435–2480.

Havner, K. S. (1992), *Finite Plastic Deformation of Crystalline Solids*, Cambridge University Press, Cambridge, UK.

29.5. Suggested Reading

Hill, R. (1978), Aspects of Invariance in Solid Mechanics, *Adv. Appl. Mech.*, Vol. 18, pp. 1–75.

Lubarda, V. A. (2002), *Elastoplasticity Theory*, CRC Press, Boca Raton, Florida.

Needleman, A., and Tvergaard, V. (1992), Analyses of Plastic Flow Localization in Metals, *Appl. Mech. Rev.*, Vol. 47, pp. S3–S18.

Nemat-Nasser, S. (2004), *Plasticity: A Treatise on Finite Deformation of Heterogeneous Inelastic Materials*, Cambridge University Press, New York.

Peirce, D., Asaro, R. J., and Needleman, A. (1982), An Analysis of Nonuniform and Localized Deformation in Ductile Single Crystals, *Acta Metall.*, Vol. 30, pp. 1087–1119.

Peirce, D., Asaro, R. J., and Needleman, A. (1983), Material Rate Dependency and Localized Deformation in Crystalline Solids, *Acta Metall.*, Vol. 31, pp. 1951–1976.

Rice, J. R. (1977), The Localization of Plastic Deformation. In *Theoretical and Applied Mechanics* (W. T. Koiter, ed.), pp. 207–220, North-Holland, Amsterdam.

30 Polycrystal Plasticity

In this chapter we explore the transition from the plastic response of single crystals to that of polycrystalline aggregates. The treatment given here is not meant to be exhaustive but rather to reveal some of the more fundamental issues involved. Suggested reading provides the link to the rather large volume of research conducted during the past two decades on the subject. The basic issues to be explored include the link between the micromechanical mechanisms of deformation on the scale of individual grains and macroscopic elastic-plastic response. One particular aggregate model is developed in detail and used to examine several physical phenomena. Among these are the development of crystallographic texture and anisotropic macroscopic response. We use the model to perform "numerical experiments" to define yield surfaces as they might be measured experimentally. We note how such surfaces naturally develop structure that is described as *corners* and explore the significance of this *vis-à-vis* the plastic strain response to sudden changes in strain path. We study this path dependent behavior further by appealing to simple rate-independent flow and deformation theories thus completing the link between microscopic and macroscopic behavior. The development of anisotropic plastic behavior is shown to occur after only modest deformation of initially isotropic aggregates.

30.1 Perspectives on Polycrystalline Modeling and Texture Development

Polycrystals are continuous 3D collections of grains (crystallites), which, as assumed herein, can deform by cyrstallographic slip. As such, the actual solution to a problem of a deforming polycrystal is that of a highly complex elastic-plastic boundary value problem for a large collection of anisotropic, continuous, and fully contiguous crystals. (If we assume that the process of deformation is, indeed, intragranular slip then, as noted in previous chapter, we also assume that the grain sizes are in the $2\,\mu$m range and larger.) Specific solutions will depend on the shape of the grains and their orientations, both of which change with plastic strain. These changes are what produce material and crystallographic texture. Rigorous solutions of this type are possible and could, in principle, be pursued using the constitutive framework presented in the forgoing chapters. Specific calculations have, indeed, been performed for the study of specific phenomena such as the localization of plastic strain. What we seek here, however, are models for aggregates of small to modest numbers of grains that describe, in some approximate way, the behavior of truly large aggregates. Many

30.1. Modeling and Texture Development

such models have been proposed and one of the first was that of Taylor (1938a, 1938b). Here we review, in detail, the more recent *extended Taylor model*, developed by Asaro and Needleman (1985), in which the deformation is modeled as finite and rate-dependent. This extends Taylor's original small strain, rate-independent, model. Some perspective follows first.

One of the earliest polycrystal models, proposed by Sachs (1928), was based on the assumption that each grain was subject to the same stress state. This was taken to be a state of uniaxial stressing. The grains were taken to respond as though they were isolated single crystals, which meant they initially deformed by single slip on their most highly stressed slip system. The model was refined by Kochendörfer (1941), who further stipulated that each grain was subjected to the same stretch. Then, as pointed out by Bishop and Hill (1951a, 1951b), with the assumption of identical strain hardening in all grains, common to the Sachs (1928) and Kochendörfer (1941) analyses, each grain fits the relation

$$\sigma/\tau = d\gamma/d\epsilon = M, \quad (30.1)$$

where σ and $d\epsilon$ are the axial stress in a grain and the macroscopic aggregate strain increment, respectively, and τ and $d\gamma$ are the shear strength and slip shear increment, respectively. The factor M depends only on geometry and in particular on the relationship between the loading axis and the crystal slip systems. Then, as also noted by Bishop and Hill (1951a, 1951b), if each grain is taken to be at the same stage of strain hardening and if M is taken as a constant, independent of strain, an aggregate stress-strain relation can be defined as an average over all orientations, *viz.*,

$$\sigma = \bar{M}\tau(\gamma) = \bar{M}\tau(\bar{M}\epsilon). \quad (30.2)$$

The average value for $\bar{M}$ determined by Sachs (1928) assuming an isotropic aggregate, *i.e.*, uniform coverage of all grain orientations, is 2.2. It might be noted, though, that the value of $\bar{M}$ found by Taylor (1938a, 1938b), which produced a close correlation between measured polycrystal tensile stress strain curves and those computed from experimentally measured fcc single crystal shear stress-shear strain curves, was approximately 3.06. (It should be indicated, however, that this correlation was done for polycrystals of relatively large grain sizes, where grain size effects are not dominant.) For later reference we also note that, within the context of Taylor's model described below, if all slip systems have equal strength, τ, then M in the relation (30.1) can be defined as $\sum_\alpha d\gamma_\alpha/d\epsilon$, where the summation is over all shear increments on the active slip systems; $\bar{M}$ would then be interpreted as the average value of M over all grains in the aggregate.

Aside from numerical inconsistencies with experimentally measured str-ess *vs.* strain behavior, there are at least two other objections to the Sach's model. In the first place, equilibrium of the stresses cannot be established across grain boundaries if each grain is subject to only simple tension of different amounts. Second, there is no way to maintain compatibility among the grains when each is assumed to deform by single slip, even if the deformation is extension of a uniform amount. To overcome the latter objection, Taylor (1938a, 1938b) proposed a model that strictly enforces compatibility by imposing the same set of strains – identified with the aggregate strain – on each grain. This is also a basis of Asaro and Needleman's (1985) aggregate model.

The basic idea underlying the Taylor model rests on the experimental observations, quoted by Taylor (1938a), that most of the grains of a polycrystal undergo about the same strain. His idea then was to *a priori* subject the all grains – treated as isolated single crystals – to the presupposed aggregate strain. Specifically, he analyzed states of uniaxial tension and compression of single phase fcc aggregates. The polycrystals were made initially isotropic by choosing the grains to have a uniform coverage of all crystallographic orientations – this provided further justification for Taylor's additional assumption of axisymmetric uniaxial deformation of initially isotropic fcc polycrystals. The procedure was to then determine the combination of slip systems and corresponding shear strains and stress states in each grain required to produce the specified strain. The aggregate stress was taken to be the average of the stresses generated in each grain. The selection of slip systems required to produce an arbitrary strain increment is, however, not unique within the strain rate independent framework he was using – a fact we have already encountered in previous chapters. Since the lattice rotations depend on the choice of active slip systems, and therefore so will the textures that develop, the predicted textures are also nonunique. Taylor made the physically intuitive assumption that, of all possible choices for combinations of active slip systems, the appropriate selection was that for which the cumulative shears would be minimized. His criterion was actually that the net internal work be minimized, and if it is further assumed that all slip systems are hardened equally so that they have the same shear strength, this amounts to a minimization of the cumulative shears.

Bishop and Hill (1951a, 1951b) later recast this theory for polycrystals and based it on the principle of maximum work, a version of which they derived for a single crystal. In particular, from the principle of maximum work, they derived inequalities between external work, computed as the product of macroscopic stress and strain increments, and internal work computed as the integral over the volumes of grains of the products of crystallographic shear strength and assumed slip increments, and used these to set bounds on the critical stress state required to induce yield. Indeed, the primary aim of the Bishop and Hill theory was the computation of yield surfaces. This is done using the extended Taylor model of Asaro and Needleman (1985) below.

The Taylor–Bishop–Hill theories have been used extensively and successfully to predict textures, and more recently, polycrystal stress-strain response. This literature is not reviewed here, but suggested reading provides an overview of the more recent work. Here we develop the finite strain, strain rate dependent theory of Asaro and Needleman (1985), which provides insight into some of the basic issues involved in constructing aggregate models. This model, above all others, has been used by them and others (see, for example, Bronkhorst *et al.*, 1992; Anand and Kothari, 1996) to explore a wide range of phenomenology.

30.2 Polycrystal Model

We consider an aggregate of grains, with the constitutive response of each grain characterized by a constitutive law of the form (28.110), *i.e.*,

$$\overset{\triangledown}{\tau} = \Lambda : \mathbf{D} - \sum_\alpha \mathbf{R}^{(\alpha)} \dot{\gamma}_\alpha, \tag{30.3}$$

where all quantities have been defined in Chapter 28. The aggregate occupies a region of volume V with external surface S_{ext}. We confine attention to quasistatic deformation processes, neglect body forces and presume that the aggregate is subject to all around

30.2. Polycrystal Model

displacement boundary conditions. These displacement boundary conditions are taken to be such as to give rise to homogeneous deformations in a homogeneous sample. In this model the transition from the microscopic response of the individual grains to the macroscopic response of the aggregate is provided *via* the averaging theorems of Hill (1972). However, these averaging theorems pertain to fields that satisfy both equilibrium and compatibility throughout the volume. In the approximate model to be used here, compatibility is satisfied and equilibrium holds within each grain, but equilibrium may be violated between grains.

A convenient choice of variables for expressing the aggregate response is the deformation gradient $\mathbf{F}$ and the unsymmetrical nominal stress, $\mathbf{P}$. We recall that the nominal stress is related to the Kirchhoff stress τ by

$$\mathbf{P} = \mathbf{F}^{-1} \cdot \tau, \tag{30.4}$$

and that it is related to the force transmitted across a material element by $\mathbf{N} \cdot \mathbf{P} \, dS$, where $\mathbf{N}$ is the unit normal to the area element and dS is the magnitude of the area in the reference configuration. We take $\mathbf{F}$ to be a deformation gradient that satisfies compatibility within each grain and between grains, and $\mathbf{P}$ to be a nominal stress field that satisfies equilibrium within each grain. Then, from the divergence theorem,

$$\int_V \mathbf{P} \cdot \cdot \mathbf{F} \, dV = \sum_{\text{grains}} \int_{V_g} \mathbf{P} \cdot \cdot \mathbf{F} \, dV \\ = \int_{S_{\text{int}}} \Delta \mathbf{T} \cdot \mathbf{u} \, dS + \oint_{S_{\text{ext}}} \mathbf{T} \cdot \mathbf{u} \, dS. \tag{30.5}$$

Here, $\mathbf{P} \cdot \cdot \mathbf{F} = P_{ij} F_{ji}$, the grain volume is V_g, the internal grain boundaries are denoted by S_{int}, and $\Delta \mathbf{T}$ is the traction difference across grain boundaries. For an equilibrium stress field $\Delta \mathbf{T} = \mathbf{0}$. We note that an identity of the form (30.5) also holds with $\mathbf{F}$ or $\mathbf{P}$ both replaced by the corresponding rate quantity, $\dot{\mathbf{F}}$ and $\dot{\mathbf{P}}$, this being so because the equilibrium equations expressed in terms of $\mathbf{P}$ are $P_{ji,j} = 0$ or in rate form $\dot{P}_{ji,j} = 0$. We have assumed here that the base vectors to which the components of $\mathbf{P}$ are phrased are Cartesian.

The homogeneous displacement boundary condition mentioned above takes the form $\mathbf{u} = \bar{\mathbf{F}} \cdot \mathbf{x}$ on S_{ext}. Substituting the prescribed value of $\mathbf{F}$ into the last integral on the *rhs* of (30.5) gives

$$\sum_{\text{grains}} \int_{V_g} \mathbf{P} \cdot \cdot \mathbf{F} \, dV - \int_{S_{\text{int}}} \Delta \mathbf{T} \cdot \mathbf{u} \, dS = \left(\oint_{S_{\text{ext}}} \mathbf{x} \mathbf{T} \, dS \right) \cdot \cdot \bar{\mathbf{F}}. \tag{30.6}$$

The basic assumption underlying the averaging procedure for an approximate stress field is that the contribution from the lack of equilibrium across grain boundaries is small compared to the volume integral contribution. The error is difficult to quantify, but this approximation is inherent in any derivation of aggregate properties based on an approach that does not enforce equilibrium between grains. Invoking this approximation, (30.6) is replaced by

$$\sum_{\text{grains}} \int_{V_g} \mathbf{P} \cdot \cdot \mathbf{F} \, dV = \left(\oint_{S_{\text{ext}}} \mathbf{x} \mathbf{T} \, dS \right) \cdot \cdot \bar{\mathbf{F}}. \tag{30.7}$$

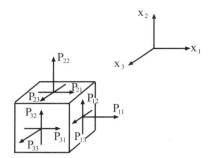

Figure 30.1. Coordinate geometry used to describe material directions and stress state in homogeneously deformed polycrystals. The x_2 axis will be the axis of tension or compression. Components of nominal stress are illustrated.

In particular, we specify $\mathbf{F}$ to have the uniform value $\bar{\mathbf{F}}$ throughout the aggregate, which permits (30.7) to be written as

$$\sum_{\text{grains}} \int_{V_g} \mathbf{P} \cdot \cdot \mathbf{F} \, dV = V \bar{\mathbf{P}} \cdot \cdot \bar{\mathbf{F}}, \tag{30.8}$$

where

$$\bar{\mathbf{P}} = \frac{1}{V} \sum_{\text{grains}} \int_{V_g} \mathbf{P} \, dV, \quad V = \sum_{\text{grains}} V_g. \tag{30.9}$$

Since $\bar{\mathbf{F}}$ is, trivially, the average value of $\mathbf{F}$ in the aggregate, the average Kirchhoff stress is given by

$$\bar{\tau} = \bar{\mathbf{F}} \cdot \bar{\mathbf{P}}. \tag{30.10}$$

The relations (30.8) and (30.10), or their rate forms, serve as the basis for calculating aggregate response.

We have confined attention to displacement boundary conditions. Other boundary conditions such as prescribed traction could be also considered. In any case, we presume the sample is large compared to any heterogeneities so that it can be argued that constitutive properties are independent of the boundary conditions, as long as these are consistent with macroscopically uniform deformation.

30.3 Extended Taylor Model

The essential kinematical assumption in Taylor's original model was that each grain in the aggregate is subject to the same homogeneous deformation field as the aggregate. We make the same assumption here, but now in the context of finite strains and rate-dependent material response. Specific constitutive relations have already been derived and will be recalled as in (30.3) above. Here we illustrate the procedure by considering two basic deformation histories, one corresponding to monotonic tension, the other to monotonic compression; the specification of other histories would be a straightforward extension of the methods laid out.

Consider the "sample" as depicted in Fig. 30.1. Take the x_2 axis to be the tensile or compression direction and imagine prescribing $\dot{F}_{22}$. The basic requirements for tensile

30.3. Extended Taylor Model

loading is that a line of material particles initially along the x_2 axis remains along the tensile axis, so that

$$\bar{\mathbf{F}} \cdot \mathbf{e}_2 = \bar{F}_{22} \mathbf{e}_2, \qquad (30.11)$$

and thus

$$\bar{F}_{12} = \bar{F}_{13} = 0. \qquad (30.12)$$

This is also true for the rates, *i.e.*, $\dot{\bar{F}}_{12} = \dot{\bar{F}}_{13} = 0$.

Compressive loading is assumed to be applied by rigid plattens that require the surfaces perpendicular to the loading axis to remain perpendicular to that axis. Referring to Fig. 30.1, this means that any line of material particles in the $x_1 - x_3$ plane must remain in that plane. We also constrain the body not to spin about the loading axis. Therefore, we demand

$$\bar{\mathbf{F}} \cdot \mathbf{e}_1 = \bar{F}_{11} \mathbf{e}_1, \quad \bar{\mathbf{F}} \cdot \mathbf{e}_2 = \bar{F}_{33} \mathbf{e}_3, \qquad (30.13)$$

so that

$$\bar{F}_{21} = \bar{F}_{31} = \bar{F}_{13} = \bar{F}_{23} = 0. \qquad (30.14)$$

For tensile loading the remaining boundary conditions are

$$\begin{aligned}
\bar{T}_1 = \bar{P}_{11} = 0, \quad \bar{T}_2 = \bar{P}_{12} = 0, \quad &\text{on } x_1 = \text{const.}, \\
\bar{T}_1 = \bar{P}_{31} = 0, \quad \bar{T}_2 = \bar{P}_{32} = 0, \quad &\text{on } x_3 = \text{const.}, \\
\bar{T}_3 = \bar{P}_{33} = 0, \quad &\text{on } x_3 = \text{const., for axisymmetric loading}, \\
\text{or} \quad \bar{F}_{33} = 1, \quad &\text{for plane strain loading}, \\
\bar{F}_{21} = 0.
\end{aligned} \qquad (30.15)$$

For compressive loading,

$$\begin{aligned}
\bar{T}_1 = \bar{P}_{11} = 0, \quad &\text{on } x_1 = \text{const.}, \\
\bar{T}_1 = \bar{P}_{21} = 0, \quad &\text{on } x_2 = \text{const.}, \\
\bar{T}_3 = \bar{P}_{23} = 0, \quad &\text{on } x_2 = \text{const.}, \\
\bar{T}_3 = \bar{P}_{33} = 0, \quad &\text{on } x_3 = \text{const., for axisymmetric loading}, \\
\text{or} \quad \bar{F}_{33} = 1, \quad &\text{for plane strain loading.}
\end{aligned} \qquad (30.16)$$

These boundary conditions are chosen to meet the kinematic and traction constraints of homogeneous tensile and compressive loading and to fix overall rigid body rotations. The particular combinations of nominal stress and deformation gradient components are consistent with the virtual work expression $\bar{\mathbf{P}} \cdot \cdot \, d\mathbf{F}$ in that one component of each pair is prescribed. The boundary conditions are also such that for an orthotropic specimen the shearing deformations and rotations vanish, that is for orthotropic material response $\bar{\mathbf{F}}$ is diagonal. The boundary conditions do not demand such orthotropic response so that, when it emerges, it is a consequence of the material behavior, and in particular of the material's lack of such symmetry!

30.4 Model Calculational Procedure

We recall here the development of the single crystal constitutive relations laid out in equations (28.171)–(28.177). These were concerned with expressing the constitutive relation for a single crystal with respect to laboratory coordinates. When this is done, the nominal stress rate within, say the Kth grain, is from (28.172) given by

$$\dot{\mathbf{P}}^{(K)} = \mathbf{K}^{(K)} \cdot \cdot \dot{\mathbf{F}}^{(K)} - \dot{\mathbf{B}}^{(K)}. \tag{30.17}$$

The "laboratory axes" are taken to be aligned with the $\{\mathbf{e}_i\}$ basis, and the transformation between the local crystal axes of the Kth grain and the laboratory axes is given by

$$\Phi_{ij}^{(K)} \mathbf{e}_j = \mathbf{a}_i. \tag{30.18}$$

Thus, in creating a "polycrystal" it is necessary to create a collection of $\Phi^{(K)}$'s. Following the development from above, the aggregate's nominal stress rate is

$$\dot{\bar{\mathbf{P}}} = \bar{\mathbf{K}} \cdot \cdot \dot{\bar{\mathbf{F}}} - \dot{\bar{\mathbf{B}}}, \tag{30.19}$$

where, for an aggregate of N grains,

$$\dot{\bar{\mathbf{P}}} = \frac{1}{N} \sum_{K=1}^{N} \dot{\mathbf{P}}^{(K)},$$

$$\bar{\mathbf{K}} = \frac{1}{N} \sum_{K=1}^{N} \mathbf{K}^{(K)}, \quad \dot{\bar{\mathbf{B}}} = \frac{1}{N} \sum_{K=1}^{N} \dot{\mathbf{B}}^{(K)}. \tag{30.20}$$

The homogeneous boundary value problems outlined in (30.11)–(30.16) are then posed by first constructing the aggregate relation (30.19), at a particular time, and invoking the conditions on either components of nominal stress rate, or a conjugate component of deformation rate; integration follows in discrete time steps. Several examples are presented below.

The kinetic and hardening laws used in the examples follow from those presented earlier. Specifically, the shearing rate, according to (28.84), is given by

$$\dot{\gamma}_\alpha = \dot{\gamma}_0 \frac{\tau_{(\alpha)}}{g_\alpha} \left| \frac{\tau_{(\alpha)}}{g_\alpha} \right|^{1/m-1}. \tag{30.21}$$

The hardening rule is

$$\dot{g}^{(\alpha)} = \sum_\beta h_{\alpha\beta} |\dot{\gamma}^{(\beta)}|, \tag{30.22}$$

where

$$h_{\alpha\beta} = qh + (1-q)h\delta_{\alpha\beta}. \tag{30.23}$$

The parameter q thus serves to describe the level of latent hardening and two values are used, viz., $q = 1$ and $q = 1.4$. The hardening function was selected to be

$$h(\gamma) = h_0 \operatorname{sech}^2 \left(\frac{h_0 \gamma}{\tau_s - \tau_0} \right). \tag{30.24}$$

Values used included $\tau_s/\tau_0 = 1.8$ and $h_0/\tau_0 = 8.9$. These values were in fact taken from experimental data from single crystal tests performed on Al-2.8 wt% Cu crystals. This

30.4. Model Calculational Procedure

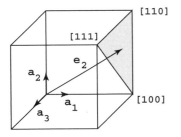

Figure 30.2. Unit cube used to describe crystallographic directions. The standard triangle, with corners at [100], [110], [111] is used to represent the independent directions of a cubic crystal. An *inverse pole figure* is a plot of a particular laboratory axis within this standard triangle.

numerology leads to a nearly complete saturation of strain hardening after strains on the order of $\gamma \approx 0.1$. For our cubic crystal, the elastic constants were $C_{11} = 842\,\tau_0$, $C_{12} = 607\,\tau_0$, $C_{44} = 377\,\tau_0$, which are also typical for these crystals. The utilized strain rate sensitivity exponent for these crystals was, as measured, $m = 0.005$.

30.4.1 Texture Determinations

Initially isotropic fcc aggregates were subjected to axisymmetric tension and compression and to plane strain tension and compression. The results for texture will be presented in the simplest form, *viz.*, in terms of *inverse pole figures* for the $\mathbf{e}_2$ laboratory axis. This construction is explained in Fig. 30.2. Figure 30.3 shows four results for the cases mentioned above of axisymmetric and plane strain tension and compression. The development of strong crystallographic textures is evident and it should be noted that this occurs after only modest amounts of strain. It needs to be mentioned, however, that this is so if the material's strain rate sensitivity is not too high. In axisymmetric tension, the specimen's axis tends to become aligned with high symmetry directions, including [111] and [100], which promote multiple slip. In axisymmetric compression, the situation is more complex, and

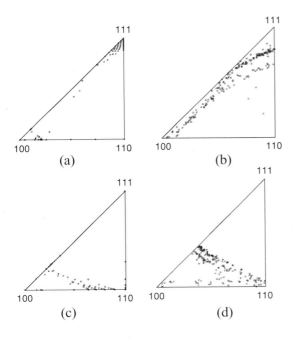

Figure 30.3. Four inverse pole figures: (a) for axisymmetric tension following a strain of 0.89 and for the case of $q = 1.4$; (b) for axisymmetric compression following a strain of 0.89 and for the case of $q = 1.0$; (c) for plane strain tension following a strain of 1.34 and for the case of $q = 1.4$; (d) for axisymmetric compression following a strain of 1.53 and for the case of $q = 1.0$ (from Asaro and Needleman, 1985).

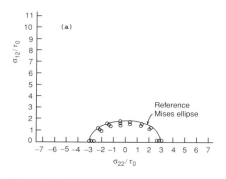
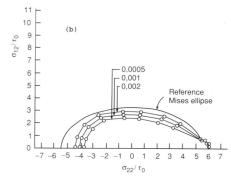

Figure 30.4. (a) Initial yield surface determined from an initially isotropic polycrystal model. Note the apparent consistency between the calculated initial yield surface and a von Mises ellipse. (b) Subsequent yield surface determined following an axisymmetric tensile strain of 0.20. The plastic strain offset values used to produce the data points are 0.0005, 0.001, and 0.002 (from Asaro and Needleman, 1985).

the compression axis lies in a "band" of orientations as seen. In plane strain tension and compression, we again see a band of orientations develop, but the trends toward anisotropy following only modest strains are nonetheless clear. Asaro and Needleman (1985) carried out extensive analysis of the effects of strain (stress) state, as well as of material properties, on texture development and the reader is encouraged to explore this topic in more detail with reference to their original work.

30.4.2 Yield Surface Determination

To provide further insight into the implications of texture development and the path dependence of strain hardening on constitutive response, polycrystal models of this type can be used to construct "yield surfaces" by a procedure that closely matches their experimental determination. At various stages in each loading history, the polycrystals are unloaded to a state of zero average stress. They are then reloaded along various prescribed stress paths until specified "offset plastic strains" are accumulated; the stress states at these point are marked out to form a yield surface. This was done in the simulations of Asaro and Needleman (1985), as it is experimentally. Initial yield surfaces for the initially isotropic polycrystal models are determined first as shown in Fig. 30.4. The definitions of effective stress and strain used here were

$$\bar{\sigma}^2 = \frac{3}{2}\sigma'_{ij}\sigma'_{ij}, \quad \bar{\epsilon} = \int_0^t \left(\frac{2}{3}D'_{ij}D'_{ij}\right)^{1/2} dt. \quad (30.25)$$

As an example of the procedure, consider the calculation of a section of a yield surface corresponding to normal tension or compression along the x_2 direction combined with shearing in the x_1-x_2 plane. Define **N** as the nominal stress based on the configuration at unloading after the imposed prestrain. This is the stress measure of interest, because during the reloading process it is desirable to measure stress and strain from the unloaded state. We also note, since strains remain small during the "probing" for the yield surface,

30.4. Model Calculational Procedure

there is little difference between Cauchy stress and nominal stress values. Then, for this reloading process, the boundary conditions imposed are

$$\dot{N}_{11} = 0, \quad \dot{N}_{23} = 0, \quad \dot{N}_{31} = 0, \quad \dot{N}_{32} = 0, \quad \dot{N}_{33} = 0,$$

$$\frac{1}{2}(\dot{N}_{21} + \dot{N}_{12}) = c\dot{N}_{22}, \quad (30.26)$$

$$\dot{f}_{12} - \dot{f}_{21} = 0, \quad \dot{f}_{31} = 0,$$

where $\dot{f}_{22}$ and the stress ratio c are prescribed, and

$$\mathbf{f} = \mathbf{F} \cdot \mathbf{F}_i^{-1},$$
$$\mathbf{N} = \mathbf{F}_i \cdot \mathbf{P} / \det \mathbf{F}_i. \quad (30.27)$$

In the above $\mathbf{F}_i$ is the deformation gradient from the initial state to the state where unloading is complete. We note that the imposed boundary conditions (30.26) give, to a very close approximation, σ_{22} and σ_{12} as the only nonvanishing stresses. As a final note on procedure, we mention the slope of the effective stress vs. effective strain during the first stages of reloading is calculated so that "elastic strains" can be estimated. These are then subtracted from the total strains to give the plastic strain tensor.

Figure 30.4 illustrates that an initially isotropic polycrystalline aggregate does, indeed, produce a yield surface that is consistent with a von Mises ellipse. The ellipse drawn is constructed by using the average Taylor factor for an fcc polycrystal of $M = 3.06$, so that the initial yield strength is given as $3.06\,\tau_0$. The subsequent yield surfaces take on a pronounced distorted shape indicating several effects. The surfaces display anisotropy, caused by the development of texture, and a kind of Bauschinger effect in that the magnitude of the subsequent flow strength in compression is clearly less than that for continued tensile deformation. The shape of the surface also displays a "corner," the significance of which lies more in the subsequent flow response than yielding *per se*. Cornerlike structure typically develops at the current stress point on the yield surface.

We close this section with an explanation of this effect. In the next section we delve more deeply into this effect on flow in the context of simple rate-independent, phenomenological plasticity models. An initially isotropic polycrystal is subject to a preplastic strain in plane strain tension of amount $\bar{\epsilon} = 0.23$ (although the precise amount of this prestrain is unimportant for the phenomena to be described). At this stage, the material is subject to a strain path consisting of a fixed ratio of shear to ongoing tension. In particular, a fixed ratio of $\dot{F}_{12}/\dot{F}_{22}$ is imposed. Figure 30.5 shows the result as a function of this ratio. Note that this type of straining represents a *deviation from proportional straining* which was, as noted, plane strain tension. If the material were rate-independent, and a J_2 flow theory used to describe plastic response, the instantaneous response would have been purely elastic, i.e., quite stiff. A rate-dependent material can only respond instantaneously in an elastic manner, but plastic response will follow after what amounts to a quite modest, or small, amount of straining. This is seen in the figure, again as a function of the nonproportional strain ratio. The response to nonproportional strain paths as seen here is quite characteristic of so-called corner theories of plasticity. We complete this chapter with a discussion of this phenomenology in the context of deformation theories of plasticity.

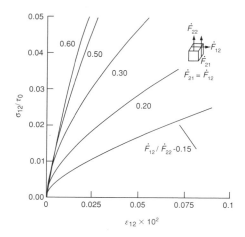

Figure 30.5. Shear stress vs. shear strain response following a plane strain tensile prestrain of 0.23. Note that the shear stiffness is reduced if the loading is nearly proportional (from Asaro and Needleman, 1985).

30.5 Deformation Theories and Path-Dependent Response

As previously discussed in Section 26.14, deformation theories, or *total strain* theories, can be developed by considering that the strain, or stress, can be calculated from potential functions. Consider, therefore, the dual potentials φ and ψ, defined *via*

$$e_{ij} = \partial\varphi/\partial\sigma_{ij}, \quad \sigma_{ij} = \partial\psi/\partial e_{ij}, \quad \varphi + \psi = \sigma_{ij}e_{ij}. \tag{30.28}$$

Assume that φ and ψ are symmetric functions in the sense that, when stress and strain are calculated as indicated, $\sigma_{ij} = \sigma_{ji}$ and $e_{ij} = e_{ji}$. Also, to model so-called *power law materials*, we assume that

$$\begin{aligned}\varphi \text{ is homogeneous of degree } n+1,\\ \psi \text{ is homogeneous of degree } m+1.\end{aligned} \tag{30.29}$$

As a consequence of such homogeneity, we have

$$\sigma_{ij}\partial\varphi/\partial\sigma_{ij} = \sigma_{ij}e_{ij} = (n+1)\varphi = e_{ij}\partial\psi/\partial e_{ij} = (m+1)\psi. \tag{30.30}$$

It also follows that

$$\varphi = m\psi, \quad \psi = n\varphi, \quad mn = 1, \quad (m+1) = (n+1)/n, \quad \text{etc.} \tag{30.31}$$

As a specific model for power law materials, *i.e.*, those for which stress is a power of strain, take

$$\varphi(\sigma_{ij}) = \kappa \left[\tau(\sigma_{ij})/\kappa\right]^{n+1}/(n+1), \quad \psi = \kappa \left[\gamma(e_{ij})\right]^{m+1}/(m+1). \tag{30.32}$$

Here, κ is a material constant with the dimension of stress, and the functions $\tau(\sigma_{ij})$ and $\gamma(e_{ij})$ are to be homogeneous of degree 1, to preserve the stated homogeneity of (30.29). Thus, for instance, $\tau =$ const. serves to parameterize surfaces of constant potential φ, and similarly for γ and ψ. Considering (30.30), we then have

$$\tau = \kappa\gamma^{1/n} = \kappa\gamma^m, \tag{30.33}$$

and

$$(n+1)\varphi = (m+1)\psi = \tau\gamma = \sigma_{ij}e_{ij}, \tag{30.34}$$

30.5. Deformation Theories

so that

$$e_{ij} = \gamma \partial \tau / \partial \sigma_{ij}, \quad \sigma_{ij} = \tau \partial \gamma / \partial e_{ij}. \tag{30.35}$$

Note that τ and γ are work conjugate because

$$\tau d\gamma = \sigma_{ij} d e_{ij}. \tag{30.36}$$

30.5.1 Specific Model Forms

For a general form, we may write

$$\tau = (M_{ijk\ell} \sigma_{ij} \sigma_{k\ell})^{1/2}, \quad \gamma = (L_{ijk\ell} e_{ij} e_{k\ell})^{1/2},$$
$$M_{ijk\ell} = M_{jik\ell} = M_{ij\ell k} = M_{k\ell ij}. \tag{30.37}$$

Then, using (30.35), we obtain

$$e_{ij}/\gamma = M_{ijk\ell} \sigma_{k\ell}/\tau, \quad \sigma_{ij}/\tau = L_{ijk\ell} e_{k\ell}/\gamma, \tag{30.38}$$

and thus $M_{ijk\ell}$ and $L_{ijk\ell}$ are the inverses of each other. If the material were isotropic, (30.38) take on the form

$$\tau = \sqrt{\frac{3}{2}} \{[(1-2\nu)/3(1+\nu)](\sigma_{kk})^2 + \sigma'_{ij}\sigma'_{ij}\}^{1/2},$$
$$\gamma = \sqrt{\frac{2}{3}} \{[(1+\nu)/3(1-2\nu)](e_{kk})^2 + e'_{ij}e'_{ij}\}^{1/2}, \tag{30.39}$$

Here, ν is a kind of Poisson's ratio *wrt* a uniaxial type loading. Thus, using (30.35) and (30.39), we obtain

$$\frac{1}{3} e'_{ij}/\gamma = \frac{1}{2} \sigma'_{ij}/\tau, \quad \frac{1}{3} e_{kk}/\gamma = \frac{1}{2} [(1-2\nu)/(1+\nu)] \sigma_{kk}/\tau. \tag{30.40}$$

Suppose we take $\nu = 1/2$, so that the material is incompressible, or do this for the purpose of using the formalism to describe only the *plastic part* of the deformation. Then, $e_{kk} = 0$ and

$$\tau = \left(\frac{3}{2} \sigma'_{ij}\sigma'_{ij}\right)^{1/2}, \quad \gamma = \left(\frac{2}{3} e'_{ij}e'_{ij}\right)^{1/2}, \quad \frac{1}{3} e'_{ij}/\gamma = \frac{1}{2} \sigma'_{ij}/\tau. \tag{30.41}$$

These can be expressed as

$$\sigma'_{ij} = \frac{2}{3}(\tau/\gamma)e'_{ij},$$
$$\sigma'_{ij} = \frac{2}{3} h^p_s e'_{ij}, \tag{30.42}$$

where h^p_s is a *secant modulus* for a universal curve of effective stress, τ, vs. effective strain, γ.

Before using these results to discuss the "corner effects" introduced above, we take an alternative and illustrative approach to such a constitutive theory. We do this *via* a rate form of the constitutive theory.

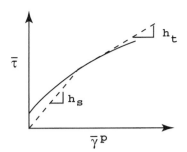

Figure 30.6. Effective stress vs. effective strain relation. Note the definition of a *tangent modulus* and a *secant modulus*.

30.5.2 Alternative Approach to a Deformation Theory

Define two *effective stress* measures, *viz.*,

$$\text{effective shear stress:} \quad \bar{\tau}^2 = \frac{1}{2}\sigma'_{ij}\sigma'_{ij},$$
$$\text{effective tensile stress:} \quad \bar{\sigma}^2 = \frac{3}{2}\sigma'_{ij}\sigma'_{ij}. \tag{30.43}$$

They are clearly related by

$$\bar{\sigma} = \sqrt{3}\,\bar{\tau}. \tag{30.44}$$

Correspondingly, define conjugate effective (plastic) strains as

$$\bar{\gamma}^p = (2e^p_{ij}e^p_{ij})^{1/2}, \quad \bar{e}^p = \left(\frac{2}{3}e^p_{ij}e^p_{ij}\right)^{1/2}. \tag{30.45}$$

Note again the assumption of a universal *hardening relation* in terms of $\bar{\tau}$ vs. $\bar{\gamma}^p$ (or $\bar{\sigma}$ vs. $\bar{e}^p$); two moduli may be defined, as in Fig. 30.6, *viz.*,

$$h_t = d\bar{\tau}/d\bar{\gamma}^p = \frac{1}{3}d\bar{\sigma}/d\bar{e}^p = \text{tangent modulus},$$
$$h_s = \bar{\tau}/\bar{\gamma}^p = \frac{1}{3}\bar{\sigma}/\bar{e}^p = \text{secant modulus}. \tag{30.46}$$

30.5.3 Nonproportional Loading

Constitutive equation for deformation theories of the type just presented can be cast in the form

$$e^p_{ij} = \lambda \sigma'_{ij}. \tag{30.47}$$

To determine λ, square both sides of (30.47) to obtain

$$e^p_{ij}e^p_{ij} = \lambda^2 \sigma'_{ij}\sigma'_{ij}, \tag{30.48}$$

which means that the effective plastic shear strain is

$$(\bar{\gamma}^p)^2 = 4\lambda^2 \bar{\tau}^2 \quad \Rightarrow \quad \lambda = 1/2h_s. \tag{30.49}$$

Now, differentiate (30.47) with respect to time to obtain

$$\dot{e}^p_{ij} = \dot{\lambda}\sigma'_{ij} + \lambda\dot{\sigma}'_{ij}. \tag{30.50}$$

30.5. Deformation Theories

Since $\lambda = \bar{\gamma}^p/2\bar{\tau}$, we find

$$\dot{\lambda} = \frac{\dot{\bar{\tau}}}{2\bar{\tau}}\left(\frac{1}{h_t} - \frac{1}{h_s}\right). \tag{30.51}$$

Thus, the relation for the plastic strain rate becomes

$$\dot{e}_{ij}^p = \frac{1}{2h_s}\dot{\sigma}_{ij}' + \frac{1}{4}\left(\frac{1}{h_t} - \frac{1}{h_s}\right)\frac{\sigma_{ij}'\sigma_{k\ell}'}{\bar{\tau}^2}\dot{\sigma}_{k\ell}'. \tag{30.52}$$

Note that

$$\frac{\dot{\bar{\tau}}}{2\bar{\tau}} = \frac{1}{4}\frac{\sigma_{k\ell}'\dot{\sigma}_{k\ell}'}{\bar{\tau}^2}. \tag{30.53}$$

Imagine a program of loading whereby a specimen is subject to purely axial loading so that only $\sigma_{22} \equiv \sigma \neq 0$; in particular, $\sigma_{12} \equiv \tau = 0$. After some stage of deformation, impose a stress path, wherein *both* σ and τ are applied, say in a fixed ratio, $d\sigma/d\tau = $ const. Then for the strain rates, $\dot{e}_{22}^p = \dot{e}^p$ and $\dot{e}_{12}^p = \dot{\gamma}^p$, we have

$$\dot{e}^p = \frac{1}{3h_s}\dot{\sigma} + \left(\frac{1}{h_t} - \frac{1}{h_s}\right)\sigma\frac{(d\sigma/d\tau) + 3\tau}{3(\sigma^2 + 3\tau^2)}\dot{\tau},$$

$$\dot{\gamma}^p = \frac{1}{h_s}\dot{\tau} + \left(\frac{1}{h_t} - \frac{1}{h_s}\right)\tau\frac{(d\sigma/d\tau) + 3\tau}{2(\sigma^2 + 3\tau^2)}\dot{\tau}. \tag{30.54}$$

Since at the moment when this multiaxial loading is imposed, $\tau = 0$, the *instantaneous response* is

$$\dot{e}^p = \frac{1}{3h_s}\dot{\sigma} + \left(\frac{1}{h_t} - \frac{1}{h_s}\right)\frac{d\sigma/d\tau}{3}\dot{\tau},$$

$$\dot{\gamma}^p = \frac{1}{h_s}\dot{\tau}. \tag{30.55}$$

What is important to note here is that the instantaneous shear response is finite; in fact from (30.54) we may note that for a given level of stress, the plastic shear strain rate increases with $d\sigma/d\tau$; in other words the smaller the departure from nonproportional stressing, the *less stiff* is the apparent shear response. This may be compared with the strain rate-dependent simulations shown in Fig. 30.5. The behavior is very different from what would be expected from a material described by say J_2 flow theory, as we show next.

We recall that within a simple version of a small strain flow theory, the rate of plastic strain can be written as

$$\dot{e}_{ij}^p = \frac{1}{4h_t}\frac{\sigma_{ij}'}{\bar{\tau}^2}\sigma_{k\ell}'\dot{\sigma}_{k\ell}', \tag{30.56}$$

which applies for the rate of plastic straining so long as there is active loading, i.e.,

$$\sigma_{ij}'\dot{\sigma}_{ij}' > 0. \tag{30.57}$$

Imagine the same program of loading as above, then

$$\dot{e}^p = \frac{\sigma(\sigma\dot{\sigma} + 3\tau\dot{\tau})}{3h_t(\sigma^2 + 3\tau^2)},$$

$$\frac{1}{2}\dot{\gamma} = \dot{e}_{12}^p = \frac{\tau(\sigma\dot{\sigma} + 3\tau\dot{\tau})}{2h_t(\sigma^2 + 3\tau^2)}. \tag{30.58}$$

When the program is such that $d\sigma/d\tau =$ const., we have

$$\dot{e}^p = \frac{\sigma[\sigma(d\sigma/d\tau) + 3\tau]}{3h_t(\sigma^2 + 3\tau^2)}\dot{\tau},$$

$$\dot{\gamma}^p = \frac{\tau[\sigma(d\sigma/d\tau) + 3\tau]}{h_t(\sigma^2 + 3\tau^2)}\dot{\tau},$$

(30.59)

and, if *instantaneously* $\tau = 0$, we obtain

$$\dot{e}^p = \frac{1}{3h_t}(d\sigma/d\tau)\dot{\tau},$$

$$\dot{\gamma}^p = 0.$$

(30.60)

Thus, we observe the effect alluded to earlier concerning the response of materials to nonproportional stress or strain paths. An important aspect of the corner structure is its implication regarding this kind of strain path dependence to material response. Experimental evidence suggests behavior very much consistent with what was revealed above *vis-à-vis* the simulations performed using the rate-dependent aggregate model. The path dependence of the stress-strain response just revealed has significance in such processes as localized deformation, plastic buckling, as well as other deformation modes that involve sudden departures from proportional straining. It is easy enough to see that, as such response is less stiff, modes that incur nonproportional strain increments are facilitated. As shown by Asaro (1985) and Pan and Rice (1985), this path dependent response is directly influenced by rate sensitivity. Rate sensitivity has the expected effect of increasing stiffness in response to such modes, thereby providing another type of stabilizing influence on plastic flow.

30.6 Suggested Reading

Asaro, R. J. (1983), Micromechanics of Crystals and Polycrystals, *Adv. Appl. Mech.*, Vol. 23, pp. 1–115.

Asaro, R. J., and Needleman, A. (1985), Texture Development and Strain Hardening in Rate Dependent Polycrystals, *Acta Metall.*, Vol. 33, pp. 923–953.

Bunge, H.-J. (1982), *Texture Analysis in Materials Science – Mathematical Methods*, Butterworths, London.

Harren, S. V., Lowe, T. C., Asaro, R. J., and Needlemann, A. (1989), Analysis of Large-Strain Shear in Rate-Dependent Face-Centered Cubic Polycrystals: Correlation of Micro- and Macromechanics, *Phil. Trans. Roy. Soc. London, Ser. A*, Vol. 328, pp. 442–500.

Havner, K. S. (1992), *Finite Plastic Deformation of Crystalline Solids*, Cambridge University Press, Cambridge, UK.

Kocks, U. F., Tomé, C. N., and Wenk, H.-R. (1998), *Texture and Anisotropy: Preferred Orientations in Polycrystals and Their Effect on Materials Properties*, Cambridge University Press, Cambridge, UK.

Lubarda, V. A. (2002), *Elastoplasticity Theory*, CRC Press, Boca Raton, Florida.

Yang, W., and Lee, W. B. (1993), *Mesoplasticity and Its Applications*, Springer-Verlag, Berlin.

31 Laminate Plasticity

Crystalline materials deform by a process of crystalline slip, whereby material is transported *via* shear across distinct crystal planes and only in certain distinct crystallographic directions in those planes. This process imparts a strong directionality to the plastic flow process and specifies a clear kinematic definition to the plastic spin. In what follows the theory is developed around a model for a *laminated material*; this is done to demonstrate the generality of the approach to a broader range of materials where slip is kinematically mediated by fixed directions.

31.1 Laminate Model

We consider the fiber reinforced plastic (FRP) material to be composed of an essentially orthotropic laminate, which contains a sufficient number of plies so that homogenization is a reasonable way to describe the material behavior. The principal directions of the fibers are described by a set of mutually orthogonal unit base vectors, $\mathbf{a}_i$, as depicted in Fig. 31.1. The resulting orthotropic elastic response of the laminated composite will thus be fixed on and described by these vectors. The material can also deform *via* slipping in the plane of the laminate, *i.e., via interlaminar shear*, and this slipping is confined to the interlaminar plane. Slipping is possible in all directions in the plane, but not necessarily with equal ease. We thus introduce two slip systems, aligned with the *slip directions* $\mathbf{s}_1$ and $\mathbf{s}_2$. The normal to the laminate plane is $\mathbf{m}$, so that $\mathbf{s}_1 \cdot \mathbf{m} = 0$ and $\mathbf{s}_2 \cdot \mathbf{m} = 0$. It may well be natural, but not necessary, to take $\mathbf{s}_1$ and $\mathbf{s}_2$ to be orthogonal, *i.e.*, $\mathbf{s}_1 \cdot \mathbf{s}_2 = 0$, but note that due to elastic distortions they may not remain so during deformation. These vectors will be called $\mathbf{s}_1^*$, $\mathbf{s}_2^*$, and $\mathbf{m}^*$ in the deformed state. Since $\mathbf{m}^*$ is to be the normal to the slip plane, *i.e.*, the plane of the laminate, it will always be the case that $\mathbf{s}_1^* \cdot \mathbf{m}^* = 0$ and $\mathbf{s}_2^* \cdot \mathbf{m}^* = 0$, as naturally described by our expressions for the kinematics of laminate deformation. In fact, it is possible to take the slip system vectors to be coincident with the laminate base vectors, $\mathbf{a}_i$, and insure that they are convected, so that the above stated orthogonality is preserved; there is no need to do this, however. Even though both slip systems have the same slip plane normal, *i.e.*, $\mathbf{m}$, it will be convenient for symmetry of expression to refer to $\mathbf{m}_1$ and $\mathbf{m}_2$ in the expressions below. This makes it easier to establish correlations with the body of theory for crystal plasticity. Figure 31.1 illustrates the total deformation from the *reference configuration*, $\mathcal{B}_0$, to occur in two stages, producing a viscoplastic and an elastic

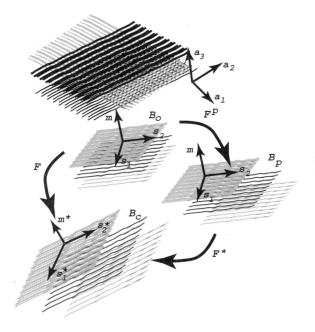

Figure 31.1. Kinematics of elastic and plastic deformation.

deformation, respectively. The viscoplastic deformation occurs by the *flow* of the material through the framework, *i.e.*, the lattice, of the laminate *via* the interlaminar shears. The spatial velocity gradient of this plastic flow is thus written as

$$\dot{\mathbf{F}}^p \cdot \mathbf{F}^{p-1} = \sum_{1}^{2} \dot{\gamma}_\alpha \mathbf{s}_\alpha \mathbf{m}_\alpha, \tag{31.1}$$

where $\mathbf{F}^p$ is the plastic deformation gradient and $\dot{\gamma}_\alpha$ is the rate of shearing on the αth slip system. The value of $\mathbf{F}^p$ is obtained by the path dependent integration of (31.1). This process produces an intermediate configuration, $\mathcal{B}_p$. Next the current, *i.e.*, deformed, configuration $\mathcal{B}_c$ is reached from $\mathcal{B}_p$ by elastically distorting and rigidly rotating the laminate (*i.e.*, lattice) along with material, *i.e.*, fabric and matrix, embedded on it. This second step of deformation is described by the elastic deformation gradient $\mathbf{F}^*$. Hence, the decomposition

$$\mathbf{F} = \mathbf{F}^* \cdot \mathbf{F}^p \tag{31.2}$$

is obtained, where $\mathbf{F}$ is the total elastic-plastic deformation. This decomposition was introduced in this context for crystal plasticity by Asaro and Rice (1977), and for phenomenological plasticity by Lee (1969); see Chapters 26 and 28.

The driving force for slip on the αth system is taken to be primarily caused by the *resolved shear stress*, τ_α, on that system. This is written as

$$\tau_\alpha = \mathbf{m}_\alpha^* \cdot \boldsymbol{\tau} \cdot \mathbf{s}_\alpha^*, \quad \mathbf{s}_\alpha^* = \mathbf{F}^* \cdot \mathbf{s}_\alpha, \quad \mathbf{m}_\alpha^* = \mathbf{m}_\alpha \cdot \mathbf{F}^{*-1}, \tag{31.3}$$

where $J = \det \mathbf{F}$ is the Jacobian, $\boldsymbol{\tau} = J\boldsymbol{\sigma}$ is the Kirchhoff stress, and $\boldsymbol{\sigma}$ is the Cauchy (true) stress. Here, in the current (deformed) state, $\mathbf{s}_\alpha^*$ and $\mathbf{m}_\alpha^*$ are respectively along the αth slip direction and the slip plane normal. Note that $\mathbf{s}_1$ and $\mathbf{s}_2$ are made to convect with the lattice framework, whereas $\mathbf{m}$ is the reciprocal base vector that remains normal to the plane that contains both $\mathbf{s}_1$ and $\mathbf{s}_2$. This definition of τ_α is used because it makes τ_α conjugate to $\dot{\gamma}_\alpha$,

31.1. Laminate Model

i.e., $\sum_{\alpha=1}^{2} \tau_\alpha \dot{\gamma}_\alpha$ is precisely the plastic dissipation rate per unit reference volume. On the other hand, it is entirely possible that the shear resistance of interlaminar zones is sensitive to the normal stress that acts on the plane. Then, the shear rate would depend on the stress component $\tau_{mm} = \mathbf{m}^* \cdot \boldsymbol{\tau} \cdot \mathbf{m}^*$. In this case a more general load parameter might be prescribed for the αth slip system, viz., $\hat{\tau}_\alpha = \|\tau_\alpha\| + \beta \tau_{mm}$, where $\hat{\tau} \geq 0$. The parameter β represents are frictional resistance because of the compaction of fabric by normal compressive stresses. As shown by Asaro and Rice (1977), and then again by Dao and Asaro (1996), the appearance of such terms in the loading parameter represents a deviation from the law of normality which, in turn, has implications for localized deformation.

The constitutive description of the plasticity on each slip system is cast in terms of the resolved shear stress on that system and the current slip rate on that system, as

$$\dot{\gamma}_\alpha = \dot{\gamma}_0 \, \mathrm{sgn}(\tau_\alpha) \left| \frac{\tau_\alpha}{g_\alpha} \right|^{1/m}. \tag{31.4}$$

The current value of the resolved shear stress is τ_α, and g_α is the current value of the slip system hardness. When frictional effects are important, τ_α should be replaced by $\hat{\tau}_\alpha$. In (31.4), m is the *strain rate sensitivity* exponent, $\dot{\gamma}_0$ is a reference shear strain rate. As $m \to 0$, the rate independent behavior is achieved, and in that limit g_α corresponds to the slip system strength, τ_α, at least in absolute value. This limit is unlikely for the polymer resin systems of interest here. Also, for creeplike behavior, we expect $0.15 \leq m \leq 1$, whereas for viscoplastic response $m \leq 0.1$.

In general, we anticipate the possibility that there may be hardening of the interlaminar layers following slip, and this is described by a path dependent evolution equation of the form

$$\dot{g}_\alpha = \sum_{\beta=1}^{2} h_{\alpha\beta}(\gamma_a) |\dot{\gamma}_\beta|, \quad \gamma_a = \int_0^t \sum_\alpha |\dot{\gamma}_\alpha| dt, \tag{31.5}$$

where $h_{\alpha\beta}$ is a hardening matrix of (nonnegative) hardening moduli, and γ_a is the accumulated sum of the slips. The initial conditions for this evolution are specified as $g_\alpha(\gamma_\alpha = 0) = g_0$, and the form of $h_{\alpha\beta}$ is

$$h_{\alpha\beta}(\gamma_a) = g'(\gamma_a) q_{\alpha\beta}. \tag{31.6}$$

The prime denotes differentiation with respect to γ_a, and $q_{\alpha\beta}$ is a matrix that describes the cross hardening between the two slipping directions. A possible and quite general form for $g(\gamma_a)$ is

$$g(\gamma_a) = g_0 + h_\infty \gamma_a + (g_\infty - g_0) \tanh\left(\frac{h_0 - h_\infty}{g_\infty - g_0}\right). \tag{31.7}$$

If the laminate is subjected to a monotonically increasing shear strain $\gamma > 0$ along one direction, then $\gamma = \gamma_a$ and the relation $g = g(\gamma_a)$ can be interpreted as being the relation of hardness vs. shear strain for that slip system. Also, in (31.7), $g_0 = g(\gamma_a = 0)$, $h_0 = g'(\gamma_a = 0)$, and $h_\infty = g'(\gamma_a \to \infty)$. If $h_\infty \equiv 0$, then $g_\infty = g(\gamma_a \to \infty)$. At present there is little data to guide the calibration of hardening laws such as (31.7), but its flexibility should provide the ability to do so.

The description of the laminate's constitutive response is completed with a specification of its elasticity, which is expressed in terms of the Green strain of the fabric framework

$\mathbf{E}^* = (1/2)(\mathbf{F}^{*T} \cdot \mathbf{F}^* - \mathbf{I})$, and the laminate framework-based second Piola–Kirchhoff stress $\mathbf{S}^* = \mathbf{F}^{*-1} \cdot \boldsymbol{\tau} \cdot \mathbf{F}^{*-T}$. For anisotropic elastic solids the elastic response may be written as

$$S^*_{ij} = \frac{\partial \Phi}{\partial E^*_{ij}}, \tag{31.8}$$

where

$$\mathbf{S}^* = S^*_{ij} \mathbf{a}_i \mathbf{a}_j, \quad \mathbf{E}^* = E^*_{ij} \mathbf{a}_i \mathbf{a}_j. \tag{31.9}$$

The strain energy of the fabric framework per unit reference volume is $\Phi = \Phi(E^*_{ij})$. Alternatively, in rate form, we have

$$\dot{\mathbf{S}}^* = \boldsymbol{\Lambda}^* : \dot{\mathbf{E}}^*, \quad \boldsymbol{\Lambda}^* = \Lambda^*_{ijkl} \mathbf{a}_i \mathbf{a}_j \mathbf{a}_k \mathbf{a}_l, \quad \Lambda^*_{ijkl} = \frac{\partial^2 \Phi}{\partial E^*_{ij} \partial E^*_{kl}}. \tag{31.10}$$

On the other hand, (31.10) can, and typically will, be constructed from a laminate theory based on discrete analysis of the layering of thin plies.

31.2 Additional Kinematical Perspective

The decomposition of (31.2) results in the total velocity gradient taking the form

$$\dot{\mathbf{F}} \cdot \mathbf{F}^{-1} = \dot{\mathbf{F}}^* \cdot \mathbf{F}^{*-1} + \mathbf{F}^* \cdot \sum_{\alpha=1}^{2} \dot{\gamma}_\alpha \mathbf{s}_\alpha \mathbf{m}_\alpha \cdot \mathbf{F}^{*-1}. \tag{31.11}$$

By forming symmetric and antisymmetric parts of this velocity gradient, we obtain the decompositions of the rate of deformation and the spin rates, *viz.*,

$$\mathbf{D} = \mathbf{D}^* + \mathbf{D}^p, \quad \mathbf{W} = \mathbf{W}^* + \mathbf{W}^p, \tag{31.12}$$

where

$$\mathbf{D}^p = \sum_{\alpha=1}^{2} \mathbf{P}_\alpha \dot{\gamma}_\alpha, \quad \mathbf{W}^p = \sum_{\alpha=1}^{2} \mathbf{Q}_\alpha \dot{\gamma}_\alpha. \tag{31.13}$$

In (31.13), the symmetric and antisymmetric tensors $\mathbf{P}_\alpha$ and $\mathbf{Q}_\alpha$ are defined as

$$\mathbf{P}_\alpha = \frac{1}{2}(\mathbf{s}^*_\alpha \mathbf{m}^*_\alpha + \mathbf{m}^*_\alpha \mathbf{s}^*_\alpha), \quad \mathbf{Q}_\alpha = \frac{1}{2}(\mathbf{s}^*_\alpha \mathbf{m}^*_\alpha - \mathbf{m}^*_\alpha \mathbf{s}^*_\alpha). \tag{31.14}$$

The elastic parts of $\mathbf{D}$ and $\mathbf{W}$, *viz.*, $\mathbf{D}^*$ and $\mathbf{W}^*$, are formed by taking the symmetric and antisymmetric parts of $\mathbf{L}^*$ in (31.11), *i.e.*,

$$\mathbf{D}^* = \left(\dot{\mathbf{F}}^* \cdot \mathbf{F}^{*-1}\right)_{\text{sym}}, \quad \mathbf{W}^* = \left(\dot{\mathbf{F}}^* \cdot \mathbf{F}^{*-1}\right)_{\text{asym}}. \tag{31.15}$$

Note that the interpretation of how $\mathbf{s}$ and $\mathbf{m}$ convect with the laminate framework, as introduced in (31.3), is a natural outcome of the kinematical scheme shown in Fig. 31.1.

31.3 Final Constitutive Forms

The description of the laminate constitutive response is completed with a specification of its elasticity, which is expressed in terms of the Green strain of the framework $\mathbf{E}^*$, as given in (31.10). Having the description of this response, the entire constitutive theory

31.3. Final Constitutive Forms

can be expressed in terms of the Green strain $\mathbf{E} = 1/2(\mathbf{F}^T \cdot \mathbf{F} - \mathbf{I})$, and the second Piola–Kirchhoff stress, $\mathbf{S} = \mathbf{F}^{-1} \cdot \tau \cdot \mathbf{F}^{-T}$. Straightforward tensor manipulations of the above relations yield the governing constitutive rate form

$$\dot{\mathbf{S}} = \mathbf{\Lambda} : \dot{\mathbf{E}} - \sum_{\alpha=1}^{2} \dot{\gamma}_\alpha \mathbf{X}_\alpha, \tag{31.16}$$

where

$$\Lambda_{ijrn} = F^{p-1}_{ik} F^{p-1}_{jl} \Lambda^*_{klpq} F^{p-1}_{rp} F^{p-1}_{nq}, \tag{31.17}$$

and

$$\mathbf{\Lambda} = \Lambda_{ijkl} \mathbf{a}_i \mathbf{a}_j \mathbf{a}_k \mathbf{a}_l, \qquad \mathbf{F}^{p-1} = F^{p-1}_{ij} \mathbf{a}_i \mathbf{a}_j. \tag{31.18}$$

In (31.16), the fourth-order tensor $\mathbf{X}_\alpha$ is defined by

$$\mathbf{X}_\alpha = \mathbf{F}^{p-1} \cdot (\mathbf{\Lambda}^* : \mathbf{A}_\alpha + 2\mathbf{B}_\alpha) \cdot \mathbf{F}^{p-1}, \tag{31.19}$$

with

$$\mathbf{A}_\alpha = \text{sym}\left(\mathbf{F}^{*T} \cdot \mathbf{F}^* \cdot \mathbf{s}_\alpha \mathbf{m}_\alpha\right), \qquad \mathbf{B}_\alpha = \text{sym}\left(\mathbf{s}_\alpha \mathbf{m}_\alpha \cdot \mathbf{S}^*\right). \tag{31.20}$$

It is useful to express the constitutive relations in terms of the nominal stress $\mathbf{P}$, and its conjugate $\mathbf{F}$ itself. The transformation is straightforward since $\mathbf{P} = \mathbf{F}^{-1} \cdot \tau = \mathbf{S} \cdot \mathbf{F}^T$ and, consequently, $\dot{\mathbf{P}} = \dot{\mathbf{S}} \cdot \mathbf{F}^T + \mathbf{S} \cdot \dot{\mathbf{F}}^T$. This then yields

$$\dot{\mathbf{P}} = \mathbf{M} : \dot{\mathbf{F}} - \sum_{\alpha=1}^{2} \dot{\gamma}_\alpha \mathbf{Y}_\alpha, \tag{31.21}$$

with

$$M_{ijkl} = F^{p-1}_{ia} F^*_{jb} \Lambda^*_{abcd} F^{p-1}_{kc} F^*_{id} + S_{ik}\delta_{jl},$$

$$\mathbf{F}^* = F^*_{ij} \mathbf{a}_i \mathbf{a}_j, \qquad \mathbf{S} = S_{ij} \mathbf{a}_i \mathbf{a}_j, \tag{31.22}$$

$$\mathbf{I} = \delta_{ij} \mathbf{a}_i \mathbf{a}_j, \qquad \mathbf{M} = M_{ijkl} \mathbf{a}_i \mathbf{a}_j \mathbf{a}_k \mathbf{a}_l,$$

and

$$\mathbf{Y}_\alpha = \mathbf{X}_\alpha \cdot \mathbf{F}^T. \tag{31.23}$$

31.3.1 Rigid-Plastic Laminate in Single Slip

A particularly illuminating, yet simple application of the theory presented above is the example of a *rigid-plastic* laminate undergoing a process of single slip, as depicted in Fig. 31.2. The case being considered is that of uniaxial tension. For this case, the decomposition of the deformation gradient becomes straightforward, *viz.*,

$$\mathbf{F} = \mathbf{F}^* \cdot \mathbf{F}^p = \mathbf{R}^* \cdot \mathbf{F}^p, \tag{31.24}$$

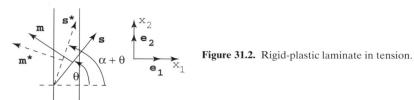

Figure 31.2. Rigid-plastic laminate in tension.

where $\mathbf{R}^*$ is the orthogonal tensor representing the laminate (lattice) rotation. As Fig. 31.2 implies, this is

$$\mathbf{R}^* = \begin{Bmatrix} \cos\alpha & -\sin\alpha & 0 \\ \sin\alpha & \cos\alpha & 0 \\ 0 & 0 & 1 \end{Bmatrix}. \tag{31.25}$$

Let $\mathbf{s}$ and $\mathbf{m}$ lie in the $x_1 - x_2$ plane and the tensile axis be parallel to the x_2 axis. Further, let θ be the angle between the initial slip direction, $\mathbf{s}$, and the x_1 axis. Also, let $\mathbf{e}_1$ and $\mathbf{e}_2$ be unit vectors along the x_1 and x_2 axes, respectively. Since α is the angle by which the lattice rotates during the actual deformation, the slip direction takes on components

$$\mathbf{s}^* = \mathbf{R}^* \cdot \mathbf{s} = \begin{Bmatrix} \cos(\theta+\alpha) \\ \sin(\theta+\alpha) \\ 0 \end{Bmatrix}. \tag{31.26}$$

On the other hand, the slip plane normal, also shown in Fig. 31.2, becomes

$$\mathbf{m}^* = \mathbf{m} \cdot \mathbf{F}^{*-1} = \mathbf{R}^* \cdot \mathbf{m} = \begin{Bmatrix} -\sin(\theta+\alpha) \\ \cos(\theta+\alpha) \\ 0 \end{Bmatrix}. \tag{31.27}$$

The resolved shear stress in this single slip system is, *via* simple geometry,

$$\tau = \sigma \sin(\theta+\alpha)\cos(\theta+\alpha), \tag{31.28}$$

where σ is the value of the tensile stress.

The plastic part of the deformation gradient is

$$\mathbf{F}^p = \mathbf{I} + \gamma \mathbf{sm}, \tag{31.29}$$

and the total deformation gradient becomes

$$\begin{aligned}\mathbf{F} = \mathbf{R}^* \cdot \mathbf{F}^p =\ & \begin{Bmatrix} \cos\alpha & -\sin\alpha & 0 \\ \sin\alpha & \cos\alpha & 0 \\ 0 & 0 & 1 \end{Bmatrix} \\ & + \gamma \begin{Bmatrix} -\cos(\theta+\alpha)\sin\theta & \cos(\theta+\alpha)\cos\theta & 0 \\ -\sin(\theta+\alpha)\sin\theta & \sin(\theta+\alpha)\cos\theta & 0 \\ 0 & 0 & 1 \end{Bmatrix}.\end{aligned} \tag{31.30}$$

But,

$$\mathbf{F} \cdot \mathbf{e}_2 = F_{22}\mathbf{e}_2 = \lambda \mathbf{e}_2, \tag{31.31}$$

31.4. Suggested Reading

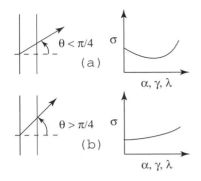

Figure 31.3. (a) Geometrical softening vs. (b) geometrical hardening.

where λ is the stretch along the tensile axis. Thus, from (31.30) and (31.31), it is found that

$$-\sin\alpha + \gamma \cos(\theta + \alpha)\cos\theta = 0, \quad \cos\alpha + \gamma \sin(\theta + \alpha)\cos\theta = \lambda. \tag{31.32}$$

Let the laminated material be *ideally plastic* with a constant hardness g_0, and further let it be rate independent (this will not affect the points to be made using this example). Then,

$$\sigma = \frac{g_0}{\sin(\theta + \alpha)\cos(\theta + \alpha)},$$

$$\gamma = \frac{\sin\alpha}{\cos(\theta + \alpha)\cos\theta}, \tag{31.33}$$

$$\lambda = \cos\alpha + \frac{\sin\alpha \sin(\theta + \alpha)}{\cos(\theta + \alpha)}.$$

Also, from (31.32), we have

$$\frac{\gamma \cos(\theta + \alpha)\cos\theta}{\gamma \sin(\theta + \alpha)\cos\theta} = \frac{\sin\alpha}{\lambda - \cos\alpha}, \tag{31.34}$$

which, in turn, leads to

$$\cot(\theta + \alpha) = \frac{\sin\alpha}{\lambda - \cos\alpha}. \tag{31.35}$$

It is instructive to use these results to explore the relationship of the magnitude of the tensile stress to any parameter such as λ, α, or γ that marks out ongoing deformation; two generic cases are sketched in Fig. 31.3. The phenomenon illustrated is called *geometrical softening*, as it relates to the decrease in the magnitude of σ with ongoing deformation due to lattice rotations that position the slip system in more favorable orientations for slip.

31.4 Suggested Reading

Dao, M., and Asaro, R. J. (1996), Localized Deformation Modes and non-Schmid Effects in Crystalline Solids. Part I: Critical Conditions for Localization, *Mech. Mater.*, Vol. 23, pp. 71–102.

PART 7: BIOMECHANICS

32 Mechanics of a Growing Mass

32.1 Introduction

A general constitutive theory of the stress-modulated growth of biomaterials is presented in this chapter with a particular accent given to pseudoelastic living tissues. The governing equations of the mechanics of solids with a growing mass are derived within the framework of finite deformation continuum thermodynamics. The analysis of stress-modulated growth of living soft tissues, bones, and other biomaterials has been an important research topic in biomechanics during past several decades. Early work includes a study of the relationship between the mechanical loads and uniform growth by Hsu (1968) and a study of the mass deposition and resorption processes in a living bone by Cowin and Hegedus (1976a, 1976b). The latter work provided a set of governing equations for the so-called adaptive elasticity theory, in which an elastic material adopts its structure to applied loading. In contrast to hard tissues which undergo only small deformations, soft tissues such as blood vessels, tendons, or ligaments can experience large deformations. Fundamental contributions were made by Fung and his co-workers (*e.g.*, Fung 1993, 1995) in the analytical description of the volumetrically distributed mass growth and by Skalak *et al.* (1982) for the mass growth by deposition or resorption on a surface. Hard tissues, such as bones and teeth, grow by deposition on a surface (apposition). Changes in porosity, mineral content and mass density are because of internal remodeling. Soft tissues grow by volumetric, also referred to as interstitial, growth. In general a tissue consists of cells and extracellular matrix. The growth and remodeling of a tissue take place during normal developmental growth, healing processes and pathological conditions. The growth and remodeling of tissues are affected, usually enhanced, by the use of tissues. Tissues remodel their structure to adjust to loading conditions they experience. For example, compressive stress stimulates the formation of new bone and is used in fracture healing. Weightless condition in space, or immobilization, conversely, may cause skeletal muscle atrophy. Hypertension causes the arterial wall to thicken, with little change in the outer diameter of the artery. Adaptive growth and remodeling of the heart because of high blood pressure is known as cardiac hypertrophy. This may be characterized by an increased ventricular wall thickness with little change in cavity size (concentric hypertrophy), or by an enlargement of the cavity accompanied by the wall thickness increase which keeps the ratio of radius to thickness approximately the same (eccentric hypertrophy). The latter growth may not be necessarily pathological, because it also takes place during normal hearth growth and exercise, when the ventricular cavity enlarges to accommodate the greater blood volume. The review articles by Taber (1995) and Humphrey (1995, 2003) offer further discussion and analysis of the biomechanical aspects of growth (mass change), remodeling (property change), and morphogenesis (shape change), with the reference to original work.

32.2 Continuity Equation

Let r_g be the time rate of mass growth per unit current volume. Then

$$\frac{d}{dt}(dm) = r^g \, dV, \tag{32.1}$$

where d/dt stands for the material time derivative. For $r^g > 0$ the mass growth occurs, and for $r^g < 0$ the mass resorption takes place. If $\rho = dm/dV$ is the mass density, we obtain from (32.1)

$$\frac{d\rho}{dt} dV + \rho \frac{d}{dt}(dV) = r^g \, dV. \tag{32.2}$$

Because the volume rate is proportional to the divergence of velocity field,

$$\frac{d}{dt}(dV) = (\nabla \cdot \mathbf{v}) \, dV, \tag{32.3}$$

the substitution into (32.2) gives the continuity equation for the continuum with a growing mass

$$\frac{d\rho}{dt} + \rho \nabla \cdot \mathbf{v} = r^g. \tag{32.4}$$

The integral form of the continuity equation follows from the identity

$$\frac{d}{dt} \int_V dm = \int_V \frac{d}{dt}(dm), \tag{32.5}$$

where V is the current volume of the considered material sample. This gives

$$\frac{d}{dt} \int_V \rho \, dV = \int_V r^g \, dV. \tag{32.6}$$

Initially, before the deformation and mass growth, $\rho = \rho_0$ (initial mass density).

For isochoric (volume preserving) deformation and growth

$$\nabla \cdot \mathbf{v} = 0, \quad \frac{d\rho}{dt} = r^g, \tag{32.7}$$

whereas for incompressible materials

$$\frac{d\rho}{dt} = 0, \quad \nabla \cdot \mathbf{v} = \frac{1}{\rho} r^g. \tag{32.8}$$

32.2.1 Material Form of Continuity Equation

If $\mathbf{F}$ is the deformation gradient and $J = \det \mathbf{F}$, then

$$\frac{d}{dt}(\rho \, dV) = r^g \, dV \quad \Rightarrow \quad \frac{d}{dt}(\rho J \, dV_0) = r^g J \, dV_0, \tag{32.9}$$

32.2. Continuity Equation

where dV_0 is the initial volume element, and $dV = J\,dV_0$. We assume that material points are everywhere dense during the mass growth, so that in any small neighborhood around the particle there are always points that existed before the growth. Thus,

$$\frac{d}{dt}(\rho J) = r^g J. \tag{32.10}$$

Furthermore, introduce the time rate of mass growth per unit initial volume r_0^g, such that

$$\frac{d}{dt}(dm) = r^g\,dV = r_0^g\,dV_0. \tag{32.11}$$

It follows that the rate of mass growth per unit initial and current volume are related by

$$r_0^g = r^g J. \tag{32.12}$$

Consequently, we can rewrite (32.10) as

$$\frac{d}{dt}(\rho J) = r_0^g. \tag{32.13}$$

Upon the time integration this gives

$$\rho J = \rho_0 + \int_0^t r_0^g\,d\tau. \tag{32.14}$$

The physical interpretation of the integral on the right-hand side is available from

$$\int_0^t r_0^g\,d\tau = \frac{(dm)_t - (dm)_0}{dV_0}. \tag{32.15}$$

In (32.14) we have

$$\rho = \rho\left[\mathbf{x}(\mathbf{X}, t)\right], \quad J = J\left[\mathbf{x}(\mathbf{X}, t)\right], \quad r_0^g = r_0^g\left[\mathbf{x}(\mathbf{X}, \tau)\right], \tag{32.16}$$

where $\mathbf{x}$ is the current position of the material point initially at $\mathbf{X}$, and $0 \leq \tau \leq t$. Equation (32.14) represents a material form of the continuity equation for the continuum with a growing mass. It will be convenient in the sequel to designate the quantity ρJ by ρ_0^g, i.e.,

$$\rho_0^g = \frac{(dm)_t}{dV_0} = \rho J, \quad \frac{d\rho_0^g}{dt} = r_0^g. \tag{32.17}$$

For a continuum without the mass growth ($dm = $ const.), $\rho_0^g = \rho_0$.

If the volume remains preserved throughout the course of deformation and growth (densification), $J = 1$ and

$$\rho_0^g = \rho, \quad r^g = r_0^g,$$
$$\rho = \rho_0 + \int_0^t r_0^g\,d\tau. \tag{32.18}$$

For an incompressible material that remains incompressible during the mass growth ($\rho = \rho_0$), equation (32.14) gives

$$J = 1 + \frac{1}{\rho_0}\int_0^t r_0^g\,d\tau. \tag{32.19}$$

32.2.2 Quantities per Unit Initial and Current Mass

Consider a quantity A per unit current mass and the corresponding quantity A_0 per unit initial mass, defined such that $A (\mathrm{d}m)_t = A_0 (\mathrm{d}m)_0$. It follows that

$$\rho A \mathrm{d}V = \rho_0 A_0 \mathrm{d}V_0, \tag{32.20}$$

i.e.,

$$\rho_0 A_0 = \rho_0^g A. \tag{32.21}$$

Thus,

$$A_0 = \left(1 + \frac{1}{\rho_0} \int_0^t r_0^g \, \mathrm{d}\tau \right) A. \tag{32.22}$$

By differentiating (32.21), we also have

$$\rho_0 \frac{\mathrm{d}A_0}{\mathrm{d}t} = J \left(\rho \frac{\mathrm{d}A}{\mathrm{d}t} + r^g A \right), \tag{32.23}$$

$$\rho_0^g \frac{\mathrm{d}A}{\mathrm{d}t} = \rho_0 \left(\frac{\mathrm{d}A_0}{\mathrm{d}t} - \frac{r_0^g}{\rho_0^g} A_0 \right). \tag{32.24}$$

It is noted that

$$\frac{\mathrm{d}A_0}{\mathrm{d}t} \neq \left(\frac{\mathrm{d}A}{\mathrm{d}t} \right)_0, \tag{32.25}$$

where the latter quantity is defined by

$$\rho \frac{\mathrm{d}A}{\mathrm{d}t} \mathrm{d}V = \rho_0 \left(\frac{\mathrm{d}A}{\mathrm{d}t} \right)_0 \mathrm{d}V_0, \tag{32.26}$$

which implies

$$\rho_0 \left(\frac{\mathrm{d}A}{\mathrm{d}t} \right)_0 = \rho_0^g \frac{\mathrm{d}A}{\mathrm{d}t}. \tag{32.27}$$

Evidently, by comparing (32.24) and (32.27), there is a connection

$$\left(\frac{\mathrm{d}A}{\mathrm{d}t} \right)_0 = \frac{\mathrm{d}A_0}{\mathrm{d}t} - \frac{r_0^g}{\rho_0^g} A_0. \tag{32.28}$$

32.3 Reynolds Transport Theorem

The integration of (32.20) gives

$$\frac{\mathrm{d}}{\mathrm{d}t} \int_V \rho A \mathrm{d}V = \int_{V_0} \rho_0 \frac{\mathrm{d}A_0}{\mathrm{d}t} \mathrm{d}V_0. \tag{32.29}$$

The substitution of (32.23) into (32.29) yields

$$\frac{\mathrm{d}}{\mathrm{d}t} \int_V \rho A \mathrm{d}V = \int_V \left(\rho \frac{\mathrm{d}A}{\mathrm{d}t} + r^g A \right) \mathrm{d}V. \tag{32.30}$$

32.3. Reynolds Transport Theorem

This, of course, also follows directly from

$$\frac{d}{dt}\int_V \rho \mathcal{A} dV = \int_V \frac{d}{dt}(\rho \mathcal{A} dV) = \int_V \rho \frac{d\mathcal{A}}{dt} dV + \int_V \mathcal{A} \frac{d}{dt}(\rho dV), \quad (32.31)$$

because $d(\rho dV)/dt = r^g dV$.

It is instructive to provide an alternative derivation. Consider the quantity $\mathcal{A}$ per unit current volume. Since

$$\frac{d}{dt}\int_V \mathcal{A} dV = \int_V \frac{d}{dt}(\mathcal{A} dV), \quad (32.32)$$

upon the differentiation under the integral sign on the right-hand side, we deduce an important formula of continuum mechanics

$$\frac{d}{dt}\int_V \mathcal{A} dV = \int_V \left(\frac{d\mathcal{A}}{dt} + \mathcal{A}\nabla \cdot \mathbf{v}\right) dV. \quad (32.33)$$

This result also follows from

$$\frac{d}{dt}\int_V \mathcal{A} dV = \frac{d}{dt}\int_{V_0} \mathcal{A}_0 dV_0 = \int_{V_0} \frac{d\mathcal{A}_0}{dt} dV_0, \quad (32.34)$$

where $\mathcal{A}_0$ is the quantity per unit initial volume ($\mathcal{A}_0 = \mathcal{A} J$). Upon differentiation,

$$\frac{d\mathcal{A}_0}{dt} = J\left(\frac{d\mathcal{A}}{dt} + \mathcal{A}\nabla \cdot \mathbf{v}\right), \quad (32.35)$$

and substitution into (32.34) yields (32.33). The formula (32.30) follows from (32.33) by writing $\mathcal{A} = \rho A$, and by using the continuity equation (32.4). See also a related analysis by Kelly (1964) and Green and Naghdi (1965).

Returning to (32.29), we can write

$$\frac{d}{dt}\int_V \rho A dV = \int_V \frac{d}{dt}(\rho A dV) = \int_V \frac{d}{dt}(\rho A) dV + \int_V \rho A \frac{d}{dt}(dV). \quad (32.36)$$

But,

$$\frac{d}{dt}(dV) = \frac{\partial v_i}{\partial x_i} dV, \quad \frac{d}{dt}(\rho A) = \frac{\partial}{\partial t}(\rho A) + \frac{\partial}{\partial x_i}(\rho A) v_i. \quad (32.37)$$

The substitution into (32.36) then gives

$$\frac{d}{dt}\int_V \rho A dV = \int_V \frac{\partial(\rho A)}{\partial t} dV + \int_V \frac{\partial}{\partial x_i}(\rho A v_i) dV. \quad (32.38)$$

Finally, upon applying the Gauss divergence theorem to the second integral on the *rhs*, there follows

$$\frac{d}{dt}\int_V \rho A dV = \int_V \frac{\partial(\rho A)}{\partial t} dV + \int_S \rho A v_i n_i dS, \quad (32.39)$$

which is the Reynolds transport formula of the continuum mechanics. Combining this with (32.30), we also have

$$\int_V \left(\rho \frac{dA}{dt} + r^g A\right) dV = \int_V \frac{\partial(\rho A)}{\partial t} dV + \int_S \rho A v_i n_i dS. \quad (32.40)$$

If there is no mass growth, $r^g = 0$.

32.4 Momentum Principles

The first Euler's law of motion for the continuum with a growing mass can be expressed as

$$\frac{d}{dt}\int_V \rho\mathbf{v}\,dV = \int_S \mathbf{t}\,dS + \int_V \rho\mathbf{b}\,dV + \int_V r^g\mathbf{v}\,dV. \tag{32.41}$$

In addition to applied surface $\mathbf{t}$ and body $\mathbf{b}$ forces, the time rate of change of the momentum is affected by the momentum rate associated with a growing mass. This is given by the last integral on the right-hand side of (32.41). Since by the Reynolds transport theorem

$$\frac{d}{dt}\int_V \rho\mathbf{v}\,dV = \int_V \left(\rho\frac{d\mathbf{v}}{dt} + r^g\mathbf{v}\right)dV, \tag{32.42}$$

the substitution into (32.41) yields the usual differential equations of motion

$$\nabla\cdot\boldsymbol{\sigma} + \rho\mathbf{b} = \rho\frac{d\mathbf{v}}{dt}. \tag{32.43}$$

The Cauchy stress $\boldsymbol{\sigma}$ is related to the traction vector $\mathbf{t}$ by $\mathbf{t} = \mathbf{n}\cdot\boldsymbol{\sigma}$, where $\mathbf{n}$ is the unit outward normal to the surface S bounding the volume V.

An alternative derivation proceeds by applying the momentum principle to an infinitesimal parallelepiped of volume dV, whose sides are parallel to the coordinate directions $\mathbf{e}_i$ ($i = 1, 2, 3$), i.e.,

$$\frac{d}{dt}(\mathbf{v}\rho\,dV) = \mathbf{b}\rho\,dV + \frac{\partial\mathbf{t}_i}{\partial x_i}dV + \mathbf{v}r^g\,dV. \tag{32.44}$$

The traction vector over the side with a normal $\mathbf{e}_i$ is denoted by $\mathbf{t}_i$, so that $\partial\mathbf{t}_i/\partial x_i$ multiplied by dV is the net force from the surface tractions on all sides of the element. Incorporating

$$\frac{d}{dt}(\mathbf{v}\rho\,dV) = \frac{d\mathbf{v}}{dt}\rho\,dV + \mathbf{v}r^g\,dV \tag{32.45}$$

into (32.44), we obtain

$$\rho\frac{d\mathbf{v}}{dt} = \rho\mathbf{b} + \frac{\partial\mathbf{t}_i}{\partial x_i}. \tag{32.46}$$

Having regard to $\mathbf{t}_i = \mathbf{e}_i\cdot\boldsymbol{\sigma}$, and

$$\frac{\partial\mathbf{t}_i}{\partial x_i} = \nabla\cdot\boldsymbol{\sigma}, \tag{32.47}$$

the substitution into (32.46) yields the equations of motion (32.43).

The material form (relative to the initial configuration) of the first Euler's law of motion is

$$\frac{d}{dt}\int_{V_0} \rho_0^g\mathbf{v}\,dV_0 = \int_{S_0} \mathbf{t}_0\,ds_0 + \int_{V_0} \rho_0\mathbf{b}_0\,dV_0 + \int_{V_0} r_0^g\mathbf{v}\,dV_0, \tag{32.48}$$

where

$$\rho_0\mathbf{b}_0 = \rho_0^g\mathbf{b}. \tag{32.49}$$

Since

$$\frac{d}{dt}\int_{V_0} \rho_0^g\mathbf{v}\,dV_0 = \int_{V_0}\left(\rho_0^g\frac{d\mathbf{v}}{dt} + r_0^g\mathbf{v}\right)dV_0, \tag{32.50}$$

32.5. Energy Equation

the substitution into (32.48) yields the material form of the differential equations of motion,

$$\nabla_0 \cdot \mathbf{P} + \rho_0 \mathbf{b}_0 = \rho_0^g \frac{d\mathbf{v}}{dt}. \tag{32.51}$$

The nominal stress $\mathbf{P}$ is related to the nominal traction $\mathbf{t}_0$ by $\mathbf{t}_0 = \mathbf{n}_0 \cdot \mathbf{P}$, where $\mathbf{n}_0$ is the unit normal to the surface S_0 bounding the initial volume V_0. The well-known connections $\mathbf{t}\,dS = \mathbf{t}_0\,dS_0$ and $\mathbf{P} = \mathbf{F}^{-1} \cdot \boldsymbol{\tau}$ are recalled, where $\boldsymbol{\tau} = J\boldsymbol{\sigma}$ is the Kirchhoff stress. The accompanying traction boundary condition is $\mathbf{n}_0 \cdot \mathbf{P} = \mathbf{p}_n$, over the part of the bounding surface where the traction $\mathbf{p}_n$ is prescribed.

The integral form of the second Euler's law of motion for the continuum with a growing mass is

$$\frac{d}{dt}\int_V (\mathbf{x} \times \rho\mathbf{v})\,dV = \int_S (\mathbf{x} \times \mathbf{t})\,dS + \int_V (\mathbf{x} \times \rho\mathbf{b})\,dV + \int_V r^g (\mathbf{x} \times \mathbf{v})\,dV. \tag{32.52}$$

Since by the Reynolds transport theorem

$$\frac{d}{dt}\int_V (\mathbf{x} \times \rho\mathbf{v})\,dV = \int_V \left[\rho \frac{d}{dt}(\mathbf{x} \times \mathbf{v}) + r^g (\mathbf{x} \times \mathbf{v})\right] dV, \tag{32.53}$$

the substitution into (32.52) gives

$$\int_V \rho \frac{d}{dt}(\mathbf{x} \times \mathbf{v})\,dV = \int_S (\mathbf{x} \times \mathbf{t})\,dS + \int_V \rho (\mathbf{x} \times \mathbf{b})\,dV. \tag{32.54}$$

This is the same expression as in the mass-conserving continuum, which therefore implies the symmetry of the Cauchy stress tensor ($\boldsymbol{\sigma} = \boldsymbol{\sigma}^T$).

32.4.1 Rate-Type Equations of Motion

By differentiating (32.51) we obtain the rate-type equations of motion

$$\nabla_0 \cdot \frac{d\mathbf{P}}{dt} + \rho_0 \frac{d\mathbf{b}_0}{dt} = \rho_0^g \frac{d^2\mathbf{v}}{dt^2} + r_0^g \frac{d\mathbf{v}}{dt}. \tag{32.55}$$

The rate of the body force is

$$\frac{d\mathbf{b}_0}{dt} = \frac{1}{\rho_0}\left(\rho_0^g \frac{d\mathbf{b}}{dt} + r_0^g \mathbf{b}\right) = \left(\frac{d\mathbf{b}}{dt}\right)_0 + \frac{r_0^g}{\rho_0^g}\mathbf{b}_0, \tag{32.56}$$

in accordance with the general recipe (32.28). The accompanying rate-type boundary condition is

$$\mathbf{n}_0 \cdot \frac{d\mathbf{P}}{dt} = \frac{d\mathbf{p}_n}{dt}, \tag{32.57}$$

for the part of the bounding surface where the rate of traction is prescribed.

32.5 Energy Equation

The rate at which the external surface and body forces are doing work on the current mass is given by the standard expression

$$\mathcal{P} = \int_S \mathbf{t} \cdot \mathbf{v}\,dS + \int_V \rho \mathbf{b} \cdot \mathbf{v}\,dV = \int_V \left[\rho \frac{d}{dt}\left(\frac{1}{2}\mathbf{v} \cdot \mathbf{v}\right) + \boldsymbol{\sigma} : \mathbf{D}\right] dV, \tag{32.58}$$

where **D** is the rate of deformation tensor, the symmetric part of the velocity gradient $\mathbf{L} = \mathbf{v}\nabla$. If **q** is the rate of heat flow by conduction across the surface element $\mathbf{n}\,dS$, and w is the rate of heat input per unit current mass because of distributed internal or external heat sources, the total heat input rate is

$$Q = -\int_S \mathbf{q} \cdot \mathbf{n}\,dS + \int_V \rho w\,dV = \int_V (-\nabla \cdot \mathbf{q} + \rho w)\,dV. \tag{32.59}$$

The first law of thermodynamics (conservation of energy) for the continuum with a growing mass can be expressed as

$$\frac{d}{dt}\int_V \rho \left(\frac{1}{2}\mathbf{v}\cdot\mathbf{v} + u\right)dV = \mathcal{P} + \mathcal{Q} + \int_V r^g \left(\frac{1}{2}\mathbf{v}\cdot\mathbf{v} + u\right)dV + \int_V \rho \mathcal{R}^g r^g\,dV. \tag{32.60}$$

The third term on the right-hand side is the rate of kinetic and internal (u) energy associated with the current mass growth. The last term represents an average rate of (chemical) energy associated with the mass growth. We introduced the affinity $\mathcal{R}^g$, conjugate to the flux r^g, such that $\mathcal{R}^g r^g$ represents the rate of energy per unit current mass (see, for example, Fung, 1990). Since by the Reynolds transport theorem

$$\frac{d}{dt}\int_V \rho \left(\frac{1}{2}\mathbf{v}\cdot\mathbf{v} + u\right)dV = \int_V \rho \frac{d}{dt}\left(\frac{1}{2}\mathbf{v}\cdot\mathbf{v} + u\right)dV \\ + \int_V r^g \left(\frac{1}{2}\mathbf{v}\cdot\mathbf{v} + u\right)dV, \tag{32.61}$$

the substitution of (32.58), (32.59), and (32.61) into (32.60) yields the local form of the energy equation

$$\frac{du}{dt} = \frac{1}{\rho}\boldsymbol{\sigma}:\mathbf{D} - \frac{1}{\rho}\nabla\cdot\mathbf{q} + w + \mathcal{R}^g r^g. \tag{32.62}$$

This can also be deduced directly by applying the energy balance to an infinitesimal parallelepiped with the sides along the coordinate directions $\mathbf{e}_i$. The net rate of work of the traction vectors over all sides of the element is

$$\frac{\partial}{\partial x_i}(\mathbf{t}_i \cdot \mathbf{v}) = \nabla \cdot (\boldsymbol{\sigma} \cdot \mathbf{v}), \tag{32.63}$$

multiplied by the volume dV, and from

$$\frac{d}{dt}\left[\left(\frac{1}{2}\mathbf{v}\cdot\mathbf{v} + u\right)\rho\,dV\right] = \left[\frac{\partial}{\partial x_i}(\mathbf{t}_i\cdot\mathbf{v}) + \rho\mathbf{b}\cdot\mathbf{v} - \nabla\cdot\mathbf{q} + \rho w \right. \\ \left. + r^g\left(\frac{1}{2}\mathbf{v}\cdot\mathbf{v} + u\right) + \rho\mathcal{R}^g r^g\right]dV \tag{32.64}$$

we deduce by differentiation the energy equation (32.62).

32.5.1 Material Form of Energy Equation

If the initial configuration is used to cast the expressions, we have

$$\mathcal{P} = \int_{V_0}\left[\rho_0^g \frac{d}{dt}\left(\frac{1}{2}\mathbf{v}\cdot\mathbf{v}\right) + \mathbf{P}\cdot\cdot\frac{d\mathbf{F}}{dt}\right]dV_0, \tag{32.65}$$

$$\mathcal{Q} = \int_{V_0}(-\nabla_0\cdot\mathbf{q}_0 + \rho_0 w_0)\,dV_0, \tag{32.66}$$

32.6. Entropy Equation

where ∇_0 is the gradient operator with respect to initial coordinates, and $\mathbf{q}_0 = J\mathbf{F}^{-1} \cdot \mathbf{q}$ is the nominal heat flux vector. The energy equation for the whole continuum is then

$$\frac{d}{dt}\int_{V_0}\left[\rho_0^g\left(\frac{1}{2}\mathbf{v}\cdot\mathbf{v}\right) + \rho_0 u_0\right]dV_0 = \int_{V_0} r_0^g\left(\frac{1}{2}\mathbf{v}\cdot\mathbf{v} + \frac{\rho_0}{\rho_0^g}u_0\right)dV_0 \qquad (32.67)$$
$$+ \int_{V_0}\rho_0\mathcal{R}_0^g r_0^g\, dV_0 + \mathcal{P} + \mathcal{Q}.$$

The internal energy and the internal heat rate per unit initial mass are denoted by u_0 and w_0, such that

$$\rho_0 u_0 = \rho_0^g u, \qquad \rho_0 w_0 = \rho_0^g w. \qquad (32.68)$$

Performing the differentiation on the left-hand side of (32.67), and substituting (32.65) and (32.66), yields

$$\frac{du_0}{dt} - \frac{r_0^g}{\rho_0^g}u_0 = \frac{1}{\rho_0}\mathbf{P}\cdot\cdot\frac{d\mathbf{F}}{dt} - \frac{1}{\rho_0}\nabla_0\cdot\mathbf{q}_0 + w_0 + \mathcal{R}_0^g r_0^g, \qquad (32.69)$$

which is a dual equation to energy equation (32.62). If there is no mass growth, $r_0^g = 0$, and (32.69) reduces to the classical expression for the material form of the energy equation (e.g., Truesdell and Toupin, 1960).

The affinities $\mathcal{R}^g$ and $\mathcal{R}_0^g$ are related by

$$\mathcal{M}\frac{d}{dt}(dm) = \rho\,\mathcal{R}^g\,(r^g\,dV) = \rho_0\,\mathcal{R}_0^g\,(r_0^g\,dV_0), \qquad (32.70)$$

where $\mathcal{M}$ is the affinity conjugate to the mass flux, so that

$$\rho\,\mathcal{R}^g = \rho_0\,\mathcal{R}_0^g. \qquad (32.71)$$

Also, there is a connection between the rates of internal energy

$$\rho_0^g\frac{du}{dt} = \rho_0\left(\frac{du_0}{dt} - \frac{r_0^g}{\rho_0^g}u_0\right) = \rho_0\left(\frac{du}{dt}\right)_0. \qquad (32.72)$$

This follows by differentiation from the first of (32.68), or from the general results (32.24) and (32.28).

32.6 Entropy Equation

Let the rate of dissipation due to the rate of mass growth be

$$\mathcal{T}^g\frac{d}{dt}(dm) = \rho\,\Gamma^g r^g\,dV, \qquad \mathcal{T}^g = \rho\,\Gamma^g. \qquad (32.73)$$

Suppose that ξ^ν ($\nu = 1, 2, \ldots, n$) are the internal variables that describe in some average sense the microstructural changes that occurred at the considered material particle during the deformation process. These, for example, can be used to describe the local structural remodeling caused by deformation and growth. Conceptually similar variables are used in the thermodynamic analysis of inelastic deformation processes of metals and other materials (see Chapter 6). The rate of dissipation due to structural changes, per unit current mass, can be expressed as

$$f^\nu\frac{d\xi^\nu}{dt} \quad (\text{sum on } \nu), \qquad (32.74)$$

where f^ν are the thermodynamic forces (affinities) conjugate to the fluxes $d\xi^\nu/dt$, similarly as Γ^g is conjugate to r^g. The total rate of dissipation, which is the product of the absolute temperature T and the entropy production rate γ (per unit current mass) is then

$$T\gamma = \Gamma^g r^g + f^\nu \frac{d\xi^\nu}{dt}. \tag{32.75}$$

The second law of thermodynamics requires that $\gamma > 0$.

The integral form of the entropy equation for the continuum with a growing mass is

$$\frac{d}{dt}\int_V \rho s\, dV = -\int_S \frac{1}{T}\mathbf{q}\cdot\mathbf{n}\, dS + \int_V \frac{w}{T}\rho\, dV \\ + \int_V r^g s\, dV + \int_V \rho\gamma\, dV, \tag{32.76}$$

where s stands for the entropy per unit current mass. Upon applying the Reynolds transport theorem to the left-hand side of (32.76), there follows

$$\int_V \rho \frac{ds}{dt} dV = -\int_S \frac{1}{T}\mathbf{q}\cdot\mathbf{n}\, dS + \int_V \frac{w}{T}\rho\, dV + \int_V \rho\gamma\, dV. \tag{32.77}$$

This leads to a local form of the entropy equation

$$\frac{ds}{dt} = -\frac{1}{\rho}\nabla\cdot\left(\frac{1}{T}\mathbf{q}\right) + \frac{1}{T}\left(w + \Gamma^g r^g + f^\nu \frac{d\xi^\nu}{dt}\right). \tag{32.78}$$

If the temperature gradients are negligible, this reduces to

$$T\frac{ds}{dt} = -\frac{1}{\rho}\nabla\cdot\mathbf{q} + w + \Gamma^g r^g + f^\nu \frac{d\xi^\nu}{dt}. \tag{32.79}$$

32.6.1 Material Form of Entropy Equation

The rate of dissipation per unit initial mass is

$$T\gamma_0 = \Gamma_0^g r_0^g + f_0^\nu \frac{d\xi_0^\nu}{dt}. \tag{32.80}$$

The relationships with the quantities per unit current mass are

$$\rho_0 \Gamma_0^g = \rho \Gamma^g, \quad \rho_0 f_0^\nu = \rho f^\nu, \quad \frac{d\xi_0^\nu}{dt} = J\frac{d\xi^\nu}{dt}. \tag{32.81}$$

If s_0 is the entropy per unit initial mass, the integral form of the entropy equation becomes

$$\frac{d}{dt}\int_{V_0} \rho_0 s_0\, dV_0 = -\int_{S_0} \frac{1}{T}\mathbf{q}_0\cdot\mathbf{n}_0\, ds_0 + \int_{V_0} \frac{w_0}{T}\rho_0\, dV_0 \\ + \int_{V_0} r_0^g \frac{\rho_0}{\rho_0^g} s_0\, dV_0 + \int_{V_0} \rho_0\gamma_0\, dV_0. \tag{32.82}$$

The corresponding local equation is

$$\frac{ds_0}{dt} - \frac{r_0^g}{\rho_0^g}s_0 = -\frac{1}{\rho_0}\nabla_0\cdot\left(\frac{1}{T}\mathbf{q}_0\right) + \frac{1}{T}\left(w_0 + \Gamma_0^g r_0^g + f_0^\nu \frac{d\xi_0^\nu}{dt}\right). \tag{32.83}$$

32.7. General Constitutive Framework

If the temperature gradients are negligible, this reduces to

$$T\left(\frac{ds_0}{dt} - \frac{r_0^g}{\rho_0^g} s_0\right) = -\frac{1}{\rho_0} \nabla_0 \cdot \mathbf{q}_0 + w_0 + \Gamma_0^g r_0^g + f_0^\nu \frac{d\xi_0^\nu}{dt}. \tag{32.84}$$

It is observed, from (32.28), that

$$\frac{ds_0}{dt} - \frac{r_0^g}{\rho_0^g} s_0 = \left(\frac{ds}{dt}\right)_0 = \frac{\rho_0^g}{\rho_0} \frac{ds}{dt}. \tag{32.85}$$

32.6.2 Combined Energy and Entropy Equations

When (32.79) is combined with (32.62), there follows

$$\frac{du}{dt} = \frac{1}{\rho} \boldsymbol{\sigma} : \mathbf{D} + T\frac{ds}{dt} + (\mathcal{R}^g - \Gamma^g) r^g - f^\nu \frac{d\xi^\nu}{dt}. \tag{32.86}$$

Dually, when (32.85) is combined with (32.69), we obtain an expression for the rate of internal energy per unit initial mass. This is

$$\frac{du_0}{dt} - \frac{r_0^g}{\rho_0^g} u_0 = \frac{1}{\rho_0} \mathbf{P} \cdot \cdot \frac{d\mathbf{F}}{dt} + T\left(\frac{ds_0}{dt} - \frac{r_0^g}{\rho_0^g} s_0\right) + (\mathcal{R}_0^g - \Gamma_0^g) r_0^g - f_0^\nu \frac{d\xi_0^\nu}{dt}.$$

It is noted that

$$\Pi = \frac{1}{\rho} \boldsymbol{\sigma} : \mathbf{D} = \frac{1}{\rho_0^g} \boldsymbol{\tau} : \mathbf{D}, \quad \boldsymbol{\tau} = J\boldsymbol{\sigma}. \tag{32.87}$$

Here, Π is the stress power per unit current mass, whereas

$$\Pi_0 = \frac{1}{\rho_0} \mathbf{P} \cdot \cdot \frac{d\mathbf{F}}{dt} \tag{32.88}$$

is the stress power per unit initial mass. These quantities are, of course, different and related by $\rho_0 \Pi_0 = \rho_0^g \Pi$. It is also recalled that the stress $\mathbf{S}$ conjugate to the material strain $\mathbf{E}$ is defined by

$$\mathbf{P} \cdot \cdot \frac{d\mathbf{F}}{dt} = \boldsymbol{\tau} : \mathbf{D} = \mathbf{S} : \frac{d\mathbf{E}}{dt}. \tag{32.89}$$

32.7 General Constitutive Framework

Suppose that the internal energy per unit current mass is given by

$$u = u\left(\mathbf{E}, s, \rho_0^g, \xi^\nu\right). \tag{32.90}$$

Its time rate is

$$\frac{du}{dt} = \frac{\partial u}{\partial \mathbf{E}} : \frac{d\mathbf{E}}{dt} + \frac{\partial u}{\partial s}\frac{ds}{dt} + \frac{\partial u}{\partial \rho_0^g} r_0^g + \frac{\partial u}{\partial \xi^\nu}\frac{d\xi^\nu}{dt}. \tag{32.91}$$

When this is compared to (32.86), there follows

$$\mathbf{S} = \rho_0^g \frac{\partial u}{\partial \mathbf{E}}, \qquad (32.92)$$

$$T = \frac{\partial u}{\partial s}, \qquad (32.93)$$

$$\mathcal{R}^g - \Gamma^g = J \frac{\partial u}{\partial \rho_0^g}, \qquad (32.94)$$

$$f^\nu = -\frac{\partial u}{\partial \xi^\nu}. \qquad (32.95)$$

On the other hand, by introducing the Helmholtz free energy

$$\phi\left(\mathbf{E}, T, \rho_0^g, \xi^\nu\right) = u\left(\mathbf{E}, s, \rho_0^g, \xi^\nu\right) - Ts, \qquad (32.96)$$

we have by, differentiation and incorporation of (32.86),

$$\frac{d\phi}{dt} = \frac{1}{\rho_0^g} \mathbf{S} : \frac{d\mathbf{E}}{dt} - s \frac{dT}{dt} + (\mathcal{R}^g - \Gamma^g) r^g - f^\nu \frac{d\xi^\nu}{dt}. \qquad (32.97)$$

Thus, ϕ is a thermodynamic potential such that

$$\mathbf{S} = \rho_0^g \frac{\partial \phi}{\partial \mathbf{E}}, \qquad (32.98)$$

$$s = -\frac{\partial \phi}{\partial T}, \qquad (32.99)$$

$$\mathcal{R}^g - \Gamma^g = J \frac{\partial \phi}{\partial \rho_0^g}, \qquad (32.100)$$

$$f^\nu = -\frac{\partial \phi}{\partial \xi^\nu}. \qquad (32.101)$$

The Maxwell-type relationships hold

$$\frac{\partial \mathbf{S}}{\partial T} = -\rho_0^g \frac{\partial s}{\partial \mathbf{E}}, \quad \frac{\partial \mathbf{S}}{\partial s} = \rho_0^g \frac{\partial T}{\partial \mathbf{E}}, \qquad (32.102)$$

and

$$\frac{\partial \mathbf{S}}{\partial \xi^\nu} = -\rho_0^g \frac{\partial f^\nu}{\partial \mathbf{E}}, \quad \frac{\partial}{\partial \mathbf{E}} \left[\rho \left(\mathcal{R}^g - \Gamma^g\right)\right] = \frac{\partial \mathbf{S}}{\partial \rho_0^g} - \frac{\mathbf{S}}{\rho_0^g}. \qquad (32.103)$$

32.7.1 Thermodynamic Potentials per Unit Initial Mass

In an alternative formulation we can introduce the internal energy per unit initial mass as

$$u_0 = u_0\left(\mathbf{F}, s_0, \rho_0^g, \xi_0^\nu\right). \qquad (32.104)$$

32.7. General Constitutive Framework

The function u_0 is an objective function of $\mathbf{F}$, *e.g.*, dependent on the right Cauchy–Green deformation tensor $\mathbf{F}^T \cdot \mathbf{F}$. Its time rate is

$$\frac{du_0}{dt} = \frac{\partial u_0}{\partial \mathbf{F}} \cdot \cdot \frac{d\mathbf{F}}{dt} + \frac{\partial u_0}{\partial s_0} \frac{ds_0}{dt} + \frac{\partial u_0}{\partial \rho_0^g} r_0^g + \frac{\partial u_0}{\partial \xi_0^\nu} \frac{d\xi_0^\nu}{dt}. \tag{32.105}$$

The comparison with (32.6.2) establishes the constitutive structures

$$\mathbf{P} = \rho_0 \frac{\partial u_0}{\partial \mathbf{F}}, \tag{32.106}$$

$$T = \frac{\partial u_0}{\partial s_0}, \tag{32.107}$$

$$\mathcal{R}_0^g - \Gamma_0^g = \frac{\partial u_0}{\partial \rho_0^g} - \frac{1}{\rho_0^g}(u_0 - T s_0), \tag{32.108}$$

$$f_0^\nu = -\frac{\partial u_0}{\partial \xi_0^\nu}. \tag{32.109}$$

If the Helmholtz free energy per unit initial mass is selected as a thermodynamic potential, *i.e.*,

$$\phi_0\left(\mathbf{F}, T, \rho_0^g, \xi_0^\nu\right) = u_0\left(\mathbf{F}, s_0, \rho_0^g, \xi_0^\nu\right) - T s_0, \tag{32.110}$$

we obtain, by differentiation and incorporation of (32.6.2),

$$\frac{d\phi_0}{dt} - \frac{r_0^g}{\rho_0^g}\phi_0 = \frac{1}{\rho_0}\mathbf{P} \cdot \cdot \frac{d\mathbf{F}}{dt} - s_0 \frac{dT}{dt} + \left(\mathcal{R}_0^g - \Gamma_0^g\right)r_0^g - f_0^\nu \frac{d\xi_0^\nu}{dt}. \tag{32.111}$$

Since

$$\frac{d\phi_0}{dt} = \frac{\partial \phi_0}{\partial \mathbf{F}} \cdot \cdot \frac{d\mathbf{F}}{dt} + \frac{\partial \phi_0}{\partial T}\frac{dT}{dt} + \frac{\partial \phi_0}{\partial \rho_0^g}r_0^g + \frac{\partial \phi_0}{\partial \xi_0^\nu}\frac{d\xi_0^\nu}{dt}, \tag{32.112}$$

there follows

$$\mathbf{P} = \rho_0 \frac{\partial \phi_0}{\partial \mathbf{F}}, \tag{32.113}$$

$$s_0 = -\frac{\partial \phi_0}{\partial T}, \tag{32.114}$$

$$\mathcal{R}_0^g - \Gamma_0^g = \frac{\partial \phi_0}{\partial \rho_0^g} - \frac{1}{\rho_0^g}\phi_0, \tag{32.115}$$

$$f^\nu = -\frac{\partial \phi_0}{\partial \xi_0^\nu}. \tag{32.116}$$

32.7.2 Equivalence of the Constitutive Structures

The equivalence of the constitutive structures, such as (32.92) and (32.106), or (32.93) and (32.107), is easily verified. The equivalence of (32.94) and (32.108) is less transparent and merits an explicit demonstration. To that goal, write the internal energy per unit initial mass as

$$u_0 = \frac{\rho_0^g}{\rho_0} u\left(\mathbf{E}, s, \rho_0^g, \xi^\nu\right) = \frac{\rho_0^g}{\rho_0} u\left(\mathbf{E}, \frac{\rho_0}{\rho_0^g} s_0, \rho_0^g, \xi^\nu\right). \tag{32.117}$$

By taking the gradient with respect to ρ_0^g, it follows that

$$\frac{\partial u_0}{\partial \rho_0^g} = \frac{1}{\rho_0} u - \frac{1}{\rho_0^g} s_0 \frac{\partial u}{\partial s} + \frac{\rho_0^g}{\rho_0} \frac{\partial u}{\partial \rho_0^g}. \tag{32.118}$$

In view of (32.68), (32.93) and (32.94), this can be rewritten as

$$\frac{\partial u_0}{\partial \rho_0^g} = \frac{1}{\rho_0^g} (u_0 - T s_0) + \frac{\rho_0^g}{\rho_0} \frac{1}{J} (\mathcal{R}^g - \Gamma^g). \tag{32.119}$$

Recalling that

$$\rho_0 (\mathcal{R}_0^g - \Gamma_0^g) = \rho (\mathcal{R}^g - \Gamma^g), \tag{32.120}$$

equation (32.119) becomes

$$\frac{\partial u_0}{\partial \rho_0^g} = \frac{1}{\rho_0^g} (u_0 - T s_0) + \mathcal{R}_0^g - \Gamma_0^g, \tag{32.121}$$

which is equivalent to (32.108). Similar derivation proceeds to establish the equivalence of (32.100) and (32.115).

32.8 Multiplicative Decomposition of Deformation Gradient

Let $\mathcal{B}_0$ be the initial configuration of the material sample, which is assumed to be stress free. If the original material sample supported a residual distribution of internal stress, we can imagine that the sample is dissected into small pieces to relieve the residual stress. In this case $\mathcal{B}_0$ is an incompatible configuration. For the constitutive analysis, however, this does not pose a problem, because it is sufficient to analyze any one of the stress free pieces. Let $\mathbf{F}$ be a local deformation gradient that relates an infinitesimal material element $d\mathbf{X}$ from $\mathcal{B}_0$ to $d\mathbf{x}$ in the deformed configuration $\mathcal{B}$ at time t, i.e.,

$$d\mathbf{x} = \mathbf{F} \cdot d\mathbf{X}. \tag{32.122}$$

The deformation gradient $\mathbf{F}$ is produced by the mass growth and deformation because of externally applied and growth induced stresses. Introduce an intermediate configuration $\mathcal{B}^g$ by instantaneous elastic distressing of the current configuration $\mathcal{B}$ to zero stress (Fig. 32.1). Define a local elastic deformation gradient $\mathbf{F}^e$ that maps an infinitesimal element $d\mathbf{x}^g$ from $\mathcal{B}^g$ to $d\mathbf{x}$ in $\mathcal{B}$, such that

$$d\mathbf{x} = \mathbf{F}^e \cdot d\mathbf{x}^g. \tag{32.123}$$

Similarly, define a local growth deformation gradient $\mathbf{F}^g$ that maps an infinitesimal element $d\mathbf{X}$ from $\mathcal{B}_0$ to $d\mathbf{x}^g$ in $\mathcal{B}^g$, such that

$$d\mathbf{x}^g = \mathbf{F}^g \cdot d\mathbf{X}. \tag{32.124}$$

32.8. Multiplicative Decomposition

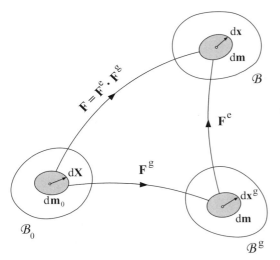

Figure 32.1. The intermediate configuration $\mathcal{B}^g$ is obtained from the current configuration $\mathcal{B}$ by instantaneous elastic destressing to zero stress. The mass of an infinitesimal volume element in the initial configuration $\mathcal{B}_0$ is dm_0. The mass of the corresponding elements in $\mathcal{B}^g$ and $\mathcal{B}$ is dm.

Substituting (32.124) into (32.123) and comparing with (32.122) establishes the multiplicative decomposition of the deformation gradient into its elastic and growth parts

$$\mathbf{F} = \mathbf{F}^e \cdot \mathbf{F}^g . \tag{32.125}$$

This decomposition is formally analogous to the decomposition of elastoplastic deformation gradient into its elastic and plastic parts, and was introduced in biomechanics by Rodrigez *et al.* (1994).

32.8.1 Strain and Strain-Rate Measures

A Lagrangian type strain measures associated with the deformation gradients $\mathbf{F}^e$ and $\mathbf{F}^g$ are

$$\mathbf{E}^e = \frac{1}{2}\left(\mathbf{F}^{eT} \cdot \mathbf{F}^e - \mathbf{I}\right), \quad \mathbf{E}^g = \frac{1}{2}\left(\mathbf{F}^{gT} \cdot \mathbf{F}^g - \mathbf{I}\right). \tag{32.126}$$

The total strain can be expressed in terms of these measures as

$$\mathbf{E} = \frac{1}{2}\left(\mathbf{F}^T \cdot \mathbf{F} - \mathbf{I}\right) = \mathbf{E}^g + \mathbf{F}^{gT} \cdot \mathbf{E}^e \cdot \mathbf{F}^g . \tag{32.127}$$

Because $\mathbf{E}^e$ and $\mathbf{E}^g$ are defined with respect to different reference configurations, clearly $\mathbf{E} \neq \mathbf{E}^e + \mathbf{E}^g$. On the other hand, the velocity gradient becomes

$$\mathbf{L} = \dot{\mathbf{F}} \cdot \mathbf{F}^{-1} = \dot{\mathbf{F}}^e \cdot \mathbf{F}^{e-1} + \mathbf{F}^e \cdot \left(\dot{\mathbf{F}}^g \cdot \mathbf{F}^{g-1}\right) \cdot \mathbf{F}^{e-1} . \tag{32.128}$$

The symmetric and antisymmetric parts of the second term on the far right-hand side will be conveniently denoted by

$$\mathbf{d}^g = \left[\mathbf{F}^e \cdot \left(\dot{\mathbf{F}}^g \cdot \mathbf{F}^{g-1}\right) \cdot \mathbf{F}^{e-1}\right]_s , \quad \boldsymbol{\omega}^g = \left[\mathbf{F}^e \cdot \left(\dot{\mathbf{F}}^g \cdot \mathbf{F}^{g-1}\right) \cdot \mathbf{F}^{e-1}\right]_a .$$

The rates of elastic and growth strains can now be expressed in terms of the rate of total strain $\dot{\mathbf{E}}$ and the velocity gradient of intermediate configuration $\dot{\mathbf{F}}^g \cdot \mathbf{F}^{g-1}$ as

$$\dot{\mathbf{E}}^e = \mathbf{F}^{g-T} \cdot \dot{\mathbf{E}} \cdot \mathbf{F}^{g-1} - \mathbf{F}^{eT} \cdot \mathbf{d}^g \cdot \mathbf{F}^e , \tag{32.129}$$

$$\dot{\mathbf{E}}^g = \mathbf{F}^{gT} \cdot \left(\dot{\mathbf{F}}^g \cdot \mathbf{F}^{g-1}\right)_s \cdot \mathbf{F}^g . \tag{32.130}$$

Being defined relative to different reference configurations, these naturally do not sum up to give the rate of total strain ($\dot{\mathbf{E}} \neq \dot{\mathbf{E}}^e + \dot{\mathbf{E}}^g$).

32.9 Density Expressions

If the mass of an infinitesimal volume element in the initial configuration is $dm_0 = \rho_0 \, dV_0$, the mass of the corresponding element in the configurations $\mathcal{B}_g$ and $\mathcal{B}$ is

$$dm = \rho^g \, dV^g = \rho \, dV. \tag{32.131}$$

Since

$$dm = dm_0 + \int_0^t r_0^g \, d\tau \, dV_0, \tag{32.132}$$

and

$$dV^g = J^g \, dV_0, \quad J^g = \det \mathbf{F}^g, \tag{32.133}$$

we have

$$\rho^g J^g = \rho_0 + \int_0^t r_0^g \, d\tau. \tag{32.134}$$

In addition, from (32.131) there follows

$$\rho^g J^g = \rho J, \quad \rho^g = \rho J^e, \tag{32.135}$$

because $dV = J^e \, dV^g$ and $J = J^e J^g$, where $J^e = \det \mathbf{F}^e$.

If $\rho^g = \rho_0$ throughout the mass growth, equation (32.134) reduces to

$$J^g = 1 + \frac{1}{\rho_0} \int_0^t r_0^g \, d\tau. \tag{32.136}$$

For elastically incompressible material, $J^e = 1$ and $\rho = \rho^g$.

From (32.134) and (32.135), we further have

$$\frac{d}{dt}(\rho J) = \frac{d}{dt}(\rho^g J^g) = r_0^g. \tag{32.137}$$

This yields the continuity equations for the densities ρ and ρ^g, i.e.,

$$\frac{d\rho}{dt} + \rho \, \mathrm{tr}\left(\dot{\mathbf{F}} \cdot \mathbf{F}^{-1}\right) = r^g, \tag{32.138}$$

$$\frac{d\rho^g}{dt} + \rho^g \, \mathrm{tr}\left(\dot{\mathbf{F}}^g \cdot \mathbf{F}^{g-1}\right) = r^g J^e. \tag{32.139}$$

In view of the additive decomposition

$$\mathrm{tr}\left(\dot{\mathbf{F}} \cdot \mathbf{F}^{-1}\right) = \mathrm{tr}\left(\dot{\mathbf{F}}^e \cdot \mathbf{F}^{e-1}\right) + \mathrm{tr}\left(\dot{\mathbf{F}}^g \cdot \mathbf{F}^{g-1}\right), \tag{32.140}$$

which results from (32.128) and the cyclic property of the trace of a matrix product, from (32.138) and (32.139) it is readily found that

$$J^e \frac{d\rho}{dt} + \rho^g \, \mathrm{tr}\left(\dot{\mathbf{F}}^e \cdot \mathbf{F}^{e-1}\right) = \frac{d\rho^g}{dt}. \tag{32.141}$$

32.10. Elastic Stress Response

If the growth takes place in a density preserving manner ($\rho^g = \rho_0 =$ const.), we have $\rho J = \rho_0 J^g$, $\rho_0 = \rho J^e$, and

$$\frac{d\rho}{dt} + \rho \operatorname{tr}\left(\dot{\mathbf{F}}^e \cdot \mathbf{F}^{e-1}\right) = 0, \tag{32.142}$$

$$\rho \operatorname{tr}\left(\dot{\mathbf{F}}^g \cdot \mathbf{F}^{g-1}\right) = r^g. \tag{32.143}$$

The last expression also follows directly from (32.1) by using $dm = \rho_0 \, dV^g$, and

$$\frac{d}{dt}(dV^g) = \operatorname{tr}\left(\dot{\mathbf{F}}^g \cdot \mathbf{F}^{g-1}\right) dV^g. \tag{32.144}$$

If material is, in addition, elastically incompressible, then $\operatorname{tr}\left(\dot{\mathbf{F}}^e \cdot \mathbf{F}^{e-1}\right) = 0$ and $d\rho/dt = 0$ (*i.e.*, $\rho = \rho_0 =$ const.).

In the case when the mass growth occurs by densification only, *i.e.*, when $dV^g = dV_0$ (volume preserving mass growth), we have $J^g = 1$ and

$$\operatorname{tr}\left(\dot{\mathbf{F}}^g \cdot \mathbf{F}^{g-1}\right) = 0, \quad \frac{d\rho^g}{dt} = r^g J. \tag{32.145}$$

If, in addition, the material is elastically incompressible, then $dV = dV^g = dV_0$, $J = 1$, $\rho = \rho^g$, and

$$\operatorname{tr}\left(\dot{\mathbf{F}} \cdot \mathbf{F}^{-1}\right) = 0, \quad \frac{d\rho}{dt} = r^g. \tag{32.146}$$

This, for example, could occur in an incompressible tissue which increases its mass by an increasing concentration of collagen molecules. In some cases, on the other hand, it may be unlikely that the material can increase its mass by densification only, while being elastically incompressible. More often, the mass growth by densification occurs in porous materials, which are commonly characterized by elastic compressibility.

32.10 Elastic Stress Response

Consider an isothermal deformation and growth process of an isotropic tissue, which remains isotropic during the growth and deformation (for anisotropic tissues, see Lubarda and Hoger, 2002). The elastic strain energy per unit current mass is then given by an isotropic function of the elastic strain $\mathbf{E}^e$, *viz.*,

$$\phi\left(\mathbf{E}^e, \rho_0^g\right) = \phi\left[\mathbf{F}^{g-T} \cdot (\mathbf{E} - \mathbf{E}^g) \cdot \mathbf{F}^{g-1}, \rho_0^g\right]. \tag{32.147}$$

Introduce the stress tensor $\mathbf{S}^e$ such that $\mathbf{S}^e : d\mathbf{E}^e$ is the increment of elastic work per unit unstressed volume in the configuration $\mathcal{B}^g$. Since $\rho^g \phi$ is the elastic strain energy per unit unstressed volume, we can write

$$\mathbf{S}^e : d\mathbf{E}^e = \frac{\partial (\rho^g \phi)}{\partial \mathbf{E}^e} : d\mathbf{E}^e, \quad \mathbf{S}^e = \frac{\partial (\rho^g \phi)}{\partial \mathbf{E}^e}. \tag{32.148}$$

On the other hand, the strain energy per unit initial volume in the configuration $\mathcal{B}_0$ is $\rho_0^g \phi$, and

$$\mathbf{S} : d\mathbf{E} = \frac{\partial (\rho_0^g \phi)}{\partial \mathbf{E}} : d\mathbf{E}, \quad \mathbf{S} = \frac{\partial (\rho_0^g \phi)}{\partial \mathbf{E}}, \tag{32.149}$$

in accord with (32.98). In view of (32.147), the partial differentiation gives

$$\mathbf{S} = \frac{\partial(\rho_0^g \phi)}{\partial \mathbf{E}^e} : \frac{\partial \mathbf{E}^e}{\partial \mathbf{E}} = \mathbf{F}^{g-1} \cdot \frac{\partial(\rho_0^g \phi)}{\partial \mathbf{E}^e} \cdot \mathbf{F}^{g-T}. \tag{32.150}$$

Since

$$\mathbf{S} = \mathbf{F}^{-1} \cdot \boldsymbol{\tau} \cdot \mathbf{F}^{-T}, \quad \boldsymbol{\tau} = J\boldsymbol{\sigma}, \tag{32.151}$$

combining (32.150) and (32.151) and using the multiplicative decomposition $\mathbf{F} = \mathbf{F}^e \cdot \mathbf{F}^g$, there follows

$$\mathbf{S}^e = \mathbf{F}^{e-1} \cdot \boldsymbol{\tau}^e \cdot \mathbf{F}^{e-T}, \quad \boldsymbol{\tau}^e = J^e \boldsymbol{\sigma}. \tag{32.152}$$

Thus, we have

$$\boldsymbol{\tau}^e = \mathbf{F}^e \cdot \frac{\partial(\rho_0^g \phi)}{\partial \mathbf{E}^e} \cdot \mathbf{F}^{e T}, \tag{32.153}$$

and

$$\boldsymbol{\tau} = \mathbf{F} \cdot \frac{\partial(\rho_0^g \phi)}{\partial \mathbf{E}} \cdot \mathbf{F}^T = \mathbf{F}^e \cdot \frac{\partial(\rho_0^g \phi)}{\partial \mathbf{E}^e} \cdot \mathbf{F}^{e T}. \tag{32.154}$$

In terms of the right Cauchy–Green deformation tenors $\mathbf{C} = \mathbf{F}^T \cdot \mathbf{F}$ and $\mathbf{C}^e = \mathbf{F}^{e T} \cdot \mathbf{F}^e$, this can be rewritten as

$$\boldsymbol{\tau} = 2\mathbf{F} \cdot \frac{\partial(\rho_0^g \phi)}{\partial \mathbf{C}} \cdot \mathbf{F}^T = 2\mathbf{F}^e \cdot \frac{\partial(\rho_0^g \phi)}{\partial \mathbf{C}^e} \cdot \mathbf{F}^{e T}. \tag{32.155}$$

32.11 Partition of the Rate of Deformation

The elastic part of the rate of deformation tensor will be defined by a kinetic relation

$$\mathbf{D}^e = \mathcal{L}^{e-1} : \overset{\triangledown}{\boldsymbol{\tau}}, \tag{32.156}$$

where

$$\overset{\triangledown}{\boldsymbol{\tau}} = \dot{\boldsymbol{\tau}} - \mathbf{W} \cdot \boldsymbol{\tau} + \boldsymbol{\tau} \cdot \mathbf{W} \tag{32.157}$$

is the Jaumann derivative of the Kirchhoff stress relative to material spin $\mathbf{W}$, and $\mathcal{L}^e$ is the corresponding fourth-order elastic moduli tensor. The remaining part of the rate of deformation will be referred to as the growth part of the rate of deformation tensor, such that

$$\mathbf{D} = \mathbf{D}^e + \mathbf{D}^g. \tag{32.158}$$

To derive an expression for $\mathbf{D}^g$, we differentiate the second of (32.154) and obtain

$$\dot{\boldsymbol{\tau}} = \left(\dot{\mathbf{F}}^e \cdot \mathbf{F}^{e-1}\right) \cdot \boldsymbol{\tau} + \boldsymbol{\tau} \cdot \left(\dot{\mathbf{F}}^e \cdot \mathbf{F}^{e-1}\right)^T + \mathbf{F}^e \cdot \left(\mathbf{\Lambda}^e : \dot{\mathbf{E}}^e\right) \cdot \mathbf{F}^{e T} + \frac{\partial \boldsymbol{\tau}}{\partial \rho_0^g} r_0^g, \tag{32.159}$$

where

$$\mathbf{\Lambda}^e = \frac{\partial^2(\rho_0^g \phi)}{\partial \mathbf{E}^e \partial \mathbf{E}^e} = 4 \frac{\partial^2(\rho_0^g \phi)}{\partial \mathbf{C}^e \partial \mathbf{C}^e}, \tag{32.160}$$

32.12. Elastic Moduli Tensor

and

$$\frac{\partial \boldsymbol{\tau}}{\partial \rho_0^g} = \mathbf{F}^e \cdot \frac{\partial^2(\rho_0^g \phi)}{\partial \mathbf{E}^e \, \partial \rho_0^g} \cdot \mathbf{F}^{eT} = 2\, \mathbf{F}^e \cdot \frac{\partial^2(\rho_0^g \phi)}{\partial \mathbf{C}^e \, \partial \rho_0^g} \cdot \mathbf{F}^{eT} . \tag{32.161}$$

Since

$$\dot{\mathbf{E}}^e = \mathbf{F}^{eT} \cdot \left(\dot{\mathbf{F}}^e \cdot \mathbf{F}^{e-1} \right)_s \cdot \mathbf{F}^e , \tag{32.162}$$

the substitution into (32.159) gives

$$\dot{\boldsymbol{\tau}} - \left(\dot{\mathbf{F}}^e \cdot \mathbf{F}^{e-1} \right)_a \cdot \boldsymbol{\tau} + \boldsymbol{\tau} \cdot \left(\dot{\mathbf{F}}^e \cdot \mathbf{F}^{e-1} \right)_a = \mathcal{L}^e : \left(\dot{\mathbf{F}}^e \cdot \mathbf{F}^{e-1} \right)_s + \frac{\partial \boldsymbol{\tau}}{\partial \rho_0^g} r_0^g .$$

The rectangular components of the elastic moduli tensor $\mathcal{L}^e$ are

$$\mathcal{L}^e_{ijkl} = F^e_{im} F^e_{jn} \Lambda^e_{mnpq} F^e_{kp} F^e_{lq} + \frac{1}{2} \left(\tau_{ik} \delta_{jl} + \tau_{jk} \delta_{il} + \tau_{il} \delta_{jk} + \tau_{jl} \delta_{ik} \right) . \tag{32.163}$$

When the antisymmetric part of (32.128) is inserted into (32.163), there follows

$$\overset{\triangledown}{\boldsymbol{\tau}} = \mathcal{L}^e : \left(\dot{\mathbf{F}}^e \cdot \mathbf{F}^{e-1} \right)_s - \boldsymbol{\omega}^g \cdot \boldsymbol{\tau} + \boldsymbol{\tau} \cdot \boldsymbol{\omega}^g + \frac{\partial \boldsymbol{\tau}}{\partial \rho_0^g} r_0^g . \tag{32.164}$$

By taking the symmetric part of (32.128) we have

$$\left(\dot{\mathbf{F}}^e \cdot \mathbf{F}^{e-1} \right)_s = \mathbf{D} - \mathbf{d}^g , \tag{32.165}$$

so that (32.164) can be rewritten as

$$\mathcal{L}^{e-1} : \overset{\triangledown}{\boldsymbol{\tau}} = \mathbf{D} - \mathbf{d}^g - \mathcal{L}^{e-1} : \left(\boldsymbol{\omega}^g \cdot \boldsymbol{\tau} - \boldsymbol{\tau} \cdot \boldsymbol{\omega}^g - \frac{\partial \boldsymbol{\tau}}{\partial \rho_0^g} r_0^g \right) . \tag{32.166}$$

According to (32.156), the left-hand side is the elastic part of the rate of deformation tensor, so that the growth part is given by

$$\mathbf{D}^g = \mathbf{d}^g + \mathcal{L}^{e-1} : \left(\boldsymbol{\omega}^g \cdot \boldsymbol{\tau} - \boldsymbol{\tau} \cdot \boldsymbol{\omega}^g - \frac{\partial \boldsymbol{\tau}}{\partial \rho_0^g} r_0^g \right) . \tag{32.167}$$

32.12 Elastic Moduli Tensor

For isotropic materials the elastic strain energy is an isotropic function of elastic deformation tensor, *i.e.*, the function of its principal invariants

$$\phi = \phi \left(\mathbf{C}^e, \rho_0^g \right) = \phi \left(I_C, II_C, III_C, \rho_0^g \right) . \tag{32.168}$$

The principal invariants are

$$I_C = \text{tr}\, \mathbf{C}^e , \quad II_C = \frac{1}{2} \left[\text{tr}\left(\mathbf{C}^{e\,2} \right) - \left(\text{tr}\, \mathbf{C}^e \right)^2 \right] , \quad III_C = \det \mathbf{C}^e . \tag{32.169}$$

The Kirchhoff stress is

$$\boldsymbol{\tau} = 2 \left(c_2 \mathbf{I} + c_0 \mathbf{B}_e + c_1 \mathbf{B}^{e\,2} \right) , \tag{32.170}$$

where $\mathbf{B}^e = \mathbf{F}^e \cdot \mathbf{F}^{eT}$ is the left Cauchy–Green deformation tensor, and

$$c_0 = \frac{\partial(\rho_0^g \phi)}{\partial I_C} - I_C \frac{\partial(\rho_0^g \phi)}{\partial II_C} , \quad c_1 = \frac{\partial(\rho_0^g \phi)}{\partial II_C} , \quad c_2 = III_C \frac{\partial(\rho_0^g \phi)}{\partial III_C} . \tag{32.171}$$

If the mass growth occurs isotropically, the growth part of the deformation gradient is

$$\mathbf{F}^g = \vartheta^g \mathbf{I}, \tag{32.172}$$

where ϑ^g is the isotropic stretch ratio because of volumetric mass growth. It readily follows that the velocity gradient in the intermediate configuration $\mathcal{B}^g$ is

$$\dot{\mathbf{F}}^g \cdot \mathbf{F}^{g-1} = \frac{\dot{\vartheta}^g}{\vartheta^g} \mathbf{I}, \tag{32.173}$$

whereas in the configuration $\mathcal{B}$,

$$\mathbf{L} = \dot{\mathbf{F}}^e \cdot \mathbf{F}^{e-1} + \frac{\dot{\vartheta}^g}{\vartheta^g} \mathbf{I}. \tag{32.174}$$

Since $\boldsymbol{\omega}^g = \mathbf{0}$, the growth part of the rate of deformation tensor is

$$\mathbf{D}^g = \frac{\dot{\vartheta}^g}{\vartheta^g} \mathbf{I} - \mathcal{L}^{e-1} : \left(\frac{\partial \boldsymbol{\tau}}{\partial \rho_0^g} r_0^g \right), \tag{32.175}$$

which follows from (32.167). The elastic part of the deformation gradient is

$$\mathbf{F}^e = \frac{1}{\vartheta^g} \mathbf{F}. \tag{32.176}$$

The rectangular components of the elastic moduli tensor $\boldsymbol{\Lambda}^e$ appearing in (32.160) and (32.163), are

$$\begin{aligned}\Lambda^e_{ijkl} = 4\Big[&a_1 \delta_{ij}\delta_{kl} + a_2 \delta_{(ik}\delta_{jl)} + a_3 \delta_{(ij} C^e_{kl)} + a_4 C^e_{ij} C^e_{kl} + a_5 \delta_{(ij} C^{e-1}_{kl)} \\ &+ a_6 C^e_{(ij} C^{e-1}_{kl)} + a_7 C^{e-1}_{ij} C^{e-1}_{kl} + a_8 C^{e-1}_{(ik} C^{e-1}_{jl)} \Big].\end{aligned} \tag{32.177}$$

The coefficients a_i ($i = 1, 2, \ldots, 8$) are defined in the subsection below. The symmetrization with respect to i and j, k and l, and ij and kl is used in (32.177), such that

$$\delta_{(ij} C^e_{kl)} = \frac{1}{2} \left(\delta_{ij} C^e_{kl} + C^e_{ij} \delta_{kl} \right), \tag{32.178}$$

and similarly for other terms.

In the case of elastically incompressible material, there is a geometric constraint $III_C = 1$, so that the Cauchy stress becomes

$$\boldsymbol{\sigma} = -p\mathbf{I} + \frac{2}{J} \left(c_0 \mathbf{B}^e + c_1 \mathbf{B}^{e\,2} \right), \tag{32.179}$$

where p is an arbitrary pressure, indeterminate by the constitutive analysis. (In the unstressed configuration we take p to be equal to p_0 such that the overall stress is there equal to zero). The rectangular components of the elastic moduli tensor $\boldsymbol{\Lambda}^e$ are given by (32.177) with the coefficients $a_5 = a_6 = a_7 = a_8 = 0$, i.e.,

$$\Lambda^e_{ijkl} = 4 \Big[a_1 \delta_{ij}\delta_{kl} + a_2 \delta_{(ik}\delta_{jl)} + a_3 \delta_{(ij} C^e_{kl)} + a_4 C^e_{ij} C^e_{kl} \Big]. \tag{32.180}$$

32.12.1 Elastic Moduli Coefficients

The coefficients appearing in the expressions for the elastic moduli components Λ^e_{ijkl} of (32.177) are defined in terms of the gradients of elastic strain energy with respect to the invariants of the elastic deformation tensor as follows:

$$a_1 = \frac{\partial^2(\rho_0^g\phi)}{\partial I_C^2} - 2I_C \frac{\partial^2(\rho_0^g\phi)}{\partial I_C \partial II_C} + I_C^2 \frac{\partial^2(\rho_0^g\phi)}{\partial II_C^2} - \frac{\partial(\rho_0^g\phi)}{\partial II_C}, \tag{32.181}$$

$$a_2 = \frac{\partial(\rho_0^g\phi)}{\partial II_C}, \tag{32.182}$$

$$a_3 = 2\left[\frac{\partial^2(\rho_0^g\phi)}{\partial I_C \partial II_C} - I_C \frac{\partial^2(\rho_0^g\phi)}{\partial II_C^2}\right], \tag{32.183}$$

$$a_4 = \frac{\partial^2(\rho_0^g\phi)}{\partial II_C^2}, \tag{32.184}$$

$$a_5 = 2\left[III_C \frac{\partial^2(\rho_0^g\phi)}{\partial III_C \partial I_C} - III_C I_C \frac{\partial^2(\rho_0^g\phi)}{\partial II_C \partial III_C}\right], \tag{32.185}$$

$$a_6 = 2 III_C \frac{\partial^2(\rho_0^g\phi)}{\partial II_C \partial III_C}, \tag{32.186}$$

$$a_7 = III_C^2 \frac{\partial^2(\rho_0^g\phi)}{\partial III_C^2} + III_C \frac{\partial(\rho_0^g\phi)}{\partial III_C}, \tag{32.187}$$

$$a_8 = -III_C \frac{\partial(\rho_0^g\phi)}{\partial III_C}. \tag{32.188}$$

32.13 Elastic Strain Energy Representation

Various forms of the strain energy function were proposed in the literature for different biological materials. The articles by Holzapfel *et al.* (2000) and Sacks (2000) and the book by Taber (2004) contain a number of pertinent references. Following Fung's (1973, 1995) proposal for a vascular soft tissue modeled as an incompressible elastic material, we consider the following structure of the elastic strain energy per unit initial volume

$$\rho_0^g \phi = \frac{1}{2}\alpha_0\left[\exp(Q) - Q - 1\right] + \frac{1}{2}q - \frac{1}{2}p(III_C - 1), \tag{32.189}$$

where Q and q are the polynomials in the invariants of $\mathbf{C}^e$, which include terms up to the fourth order in elastic stretch ratios, *i.e.*,

$$Q = \alpha_1 (I_C - 3) + \alpha_2 (II_C - 3) + \alpha_3 (I_C - 3)^2,$$
$$q = \beta_1 (I_C - 3) + \beta_2 (II_C - 3) + \beta_3 (I_C - 3)^2. \tag{32.190}$$

The incompressibility constraint in (32.189) is $III_C - 1 = 0$, and the pressure p plays the role of the Lagrangian multiplier. The α's and β's are the material parameters. In order that the intermediate configuration is unstressed, we require that $\beta_1 - 2\beta_2 = Jp_0$. If the material constants are such that $\beta_1 = 2\beta_2$, then $p_0 = 0$.

32.14 Evolution Equation for Stretch Ratio

The constitutive formulation is completed by specifying an appropriate evolution equation for the stretch ratio ϑ^g. In a particular, but for the tissue mechanics important special case when the growth takes place in a density preserving manner ($\rho^g = \rho_0$), from (32.143) and (32.173) we have

$$\mathrm{tr}\left(\dot{\mathbf{F}}^g \cdot \mathbf{F}^{g-1}\right) = 3\frac{\dot{\vartheta}^g}{\vartheta^g} = \frac{r^g}{\rho}. \tag{32.191}$$

Thus, recalling that $r^g/\rho = r_0^g/\rho_0^g$, the rate of mass growth $r_0^g = d\rho_0^g/dt$ can be expressed in terms of the rate of stretch $\dot{\vartheta}^g$ as

$$r_0^g = 3\rho_0^g \frac{\dot{\vartheta}^g}{\vartheta^g}. \tag{32.192}$$

Upon integration of (32.192) using the initial conditions $\vartheta_0^g = 1$ and $\rho_0^g = \rho_0$, we obtain

$$\rho_0^g = \rho_0 \left(\vartheta^g\right)^3. \tag{32.193}$$

A plausible evolution equation for the stretch ratio ϑ^g is

$$\dot{\vartheta}^g = f_\vartheta\left(\vartheta^g, \mathrm{tr}\,\mathbf{S}^e\right). \tag{32.194}$$

The tensor $\mathbf{S}^e$ is the symmetric Piola–Kirchhoff stress with respect to intermediate configuration $\mathcal{B}^g$ where the stretch ratio ϑ^g is defined. For isotropic mass growth, only spherical part of this tensor is assumed to affect the change of the stretch ratio. In view of (32.152), this can be expressed in terms of the Cauchy stress σ and the elastic deformation as

$$\mathrm{tr}\,\mathbf{S}^e = J^e\,\mathbf{B}^{e-1} : \sigma. \tag{32.195}$$

The simplest evolution of the stretch ratio incorporates a linear dependence on stress, such that

$$\dot{\vartheta}^g = k_\vartheta\left(\vartheta^g\right)\mathrm{tr}\,\mathbf{S}^e. \tag{32.196}$$

This implies that the growth-equilibrium stress is equal to zero (i.e., $\dot{\vartheta}^g = 0$ when $\mathrm{tr}\,\mathbf{S}^e = 0$). The coefficient k_ϑ may be constant or dependent on ϑ^g. For example, k_ϑ may take one value during the development of the tissue and another value during the normal maturity. Yet another value may be characteristic for abnormal conditions, such as occur in thickening of blood vessels under hypertension. To prevent an unlimited growth at nonzero stress, the following expression for the function k_ϑ in (32.196) may be adopted

$$k_\vartheta\left(\vartheta^g\right) = k_+ \left(\frac{\vartheta_+^g - \vartheta^g}{\vartheta_+^g - 1}\right)^{m_+}, \quad \mathrm{tr}\,\mathbf{S}^e > 0, \tag{32.197}$$

where $\vartheta_+^g > 1$ is the limiting value of the stretch ratio that can be reached by mass growth and k_+ and m_+ are the appropriate constants (material parameters). If the mass growth is homogeneous throughout the body, ϑ_+^g is constant, but for nonuniform mass growth caused by nonuniform biochemical properties ϑ_+^g may be different at different points (for example, inner and outer layers of aorta may have different growth potentials, in addition to stress-modulated growth effects). We assume that the stress-modulated growth occurs under tension, whereas resorption takes place under compression. In the latter case

$$k_\vartheta(\vartheta^g) = k_- \left(\frac{\vartheta^g - \vartheta_-^g}{1 - \vartheta_-^g} \right)^{m_-}, \quad \operatorname{tr} \mathbf{S}^e < 0, \qquad (32.198)$$

where $\vartheta_-^g < 1$ is the limiting value of the stretch ratio that can be reached by mass resorption. For generality, we assume that the resorption parameters k_- and m_- are different than those in growth. Other evolution equations were suggested in the literature motivated by the possibilities of growth and resorption. The most well-known is the evolution equation for the mass growth in terms of a nonlinear function of stress, which includes three growth-equilibrium states of stress (Fung, 1990). The material parameters that appear in these expressions are specified in accordance with the experimental data obtained for the particular tissue. Appealing tests include those with a transmural radial cut through the blood vessel, which relieves the residual stresses due to differential growth of its inner and outer layers. The opening angle then provides a convenient measure of the circumferential residual strain. Detailed analysis can be found in the original work and reviews by Liu and Fung (1988, 1989), Fung (1993), Humphrey (1995), Taber and Eggers (1996), and other researchers in the field of biomechanics.

32.15 Suggested Reading

Beatty, M. F. (1987), Topics in Finite Elasticity: Hyperelasticity of Rubber, Elastomers, and Biological Tissues – with Examples, *Appl. Mech. Rev.*, Vol. 40, pp. 1699–1734.

Cowin, S. C. (1990), Structural Adaptation of Bones, *Appl. Mech. Rev.*, Vol. 43, pp. S126–S133.

Cowin, S. C. (2004), Tissue Growth and Remodeling, *Annu. Rev. Biomed. Eng.*, Vol. 6, pp. 77–107.

D'Arcy Thomson, W. (1942), *On Growth and Form*, Cambridge University Press, Cambridge.

Datta, A. K. (2002), *Biological and Bioenvironmental Heat and Mass Transfer*, Marcel Dekker, New York.

Fung, Y.-C. (1990), *Biomechanics: Motion, Flow, Stress, and Growth*, Springer-Verlag, New York.

Fung, Y.-C. (1993), *Biomechanics: Mechanical Properties of Living Tissues*, 2nd ed., Springer-Verlag, New York.

Humphrey, J. D. (1995), Mechanics of the Arterial Wall: Review and Directions, *Critical Reviews in Biomed. Engrg.*, Vol. 23, pp. 1–162.

Humphrey, J. D. (2002), *Cardiovascular Solid Mechanics: Cells, Tissues, and Organs*, Springer-Verlag, New York.

Lubarda, V. A., and Hoger, A. (2002), On the Mechanics of Solids with A Growing Mass, *Int. J. Solids Struct.*, Vol. 39, pp. 4627–4664.

Palsson, B. Ø., and Bhatia, S. N. (2004), *Tissue Engineering*, Pearson – Prentice Hall, Upper Saddle River, New Jersey.

Ratner, B. D., Hoffman, A. S., Schoen, F. J., and Lemons, J. E., eds. (2004), *Biomaterials Science: An Introduction to Materials in Medicine*, 2nd ed., Elsevier, San Diego.

Sacks, M. S. (2000), Biaxial Mechanical Evaluation of Planar Biological Materials, *J. of Elasticity*, Vol. 61, pp. 199–246.

Taber, L. A. (1995), Biomechanics of Growth, Remodeling, and Morphogenesis, *Appl. Mech. Rev.*, Vol. 48, pp. 487–545.

Taber, L. A. (2004), *Nonlinear Theory of Elasticity – Application to Biomechanics*, World Scientific, New Jersey.

33 Constitutive Relations for Membranes

In this chapter we consider the behavior of essentially 2D membranes. The membranes may be linear or nonlinear and are generally considered to undergo arbitrarily large deformations. More specifically, the membranes considered here are modeled after biological membranes such as those that comprise cell walls or the layers that exist within biomineralized structures, *e.g.*, shells or teeth. The discussion is preliminary and meant to provide a brief introduction to the basic concepts involved in the constitutive modeling of such structures.

33.1 Biological Membranes

Figure 33.1a illustrates an idealized view of the red blood cell that shows its hybrid structure consisting of an outer bilipid membrane and an attached cytoskeleton; Fig. 33.1b shows a micrograph of a section of the cytoskeleton that is illustrated schematically in Fig. 33.1a. The cell membrane is a hybrid, *i.e.*, composite structure consisting of an outer bilipid layer that is supported (*i.e.*, reinforced) by a network attached to it on the cytoplasmic side, which is on the inside of the cell. The cytoskeleton is built up from mostly tetramers, and higher order polypeptides, of the protein *spectrin* attached at actin nodes. We note that the spectrin network has close to a sixfold nodal coordination. It has been known that the nonlinear elastic properties of the membrane depend sensitively on the details of the topology that includes, *inter alia*, nodal coordination, spectrin segment length, and the statistical distribution of such topological parameters. We will not attempt here a detailed description of the nodal connections within the cytoskeleton, except to note that the nodes are comprised of proteins such as actin; the attachment to the bilipid membrane involves, *inter alia*, the membrane protein ankyrin.

The bilipid membrane is held together by essentially hydrophobic forces. As such it contributes to a lesser degree to the shear stiffness of the membrane than to the stiffness resisting changes in membrane area. The shear stiffness is contributed to by the cytoskeleton, which also contributes to the resistance to area change. In what follows we will be concerned with the behavior of the cytoskeleton, although the constitutive framework that is developed is general enough to provide a reasonable starting point for modeling the entire composite membrane.

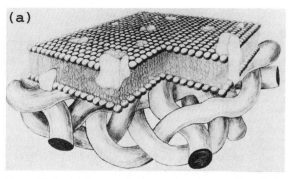
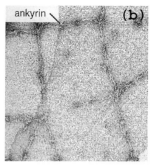

Figure 33.1. (a) Schematic illustration of the cell wall of a red blood cell. Note the hybrid structure consisting of an outer bilipid membrane and an attached inner cytoskeleton described in the text. (b) A micrograph of a section of the cytoskeleton consisting of a network of biopolymer filaments of the protein spectrin connected *via* actrin nodes, or *via* hexamer or octamer junctions. Reproduced with permission from Evans and Skalak (1980).

We consider the membrane to be a 2D continuum and view all deformation to be uniform through the thickness. Moreover, we assume the membranes to be isotropic in the plane, an assumption that will become less valid as the strains in the membrane increase. Thus, there are two fundamental ways in which the membrane can deform: (1) *via* area changes that are governed by a *bulk* or *area modulus* and (2) *via* shear governed by a *shear modulus*. Both moduli are defined below with respect to their associated modes of deformation.

33.2 Membrane Kinematics

We imagine the surface of the membrane as being a collage of vanishing small tiles such as the tile located at the point p in Fig. 33.2. During an increment of deformation a reference element of material lying in the surface, d**X**, is transformed to d**x** in the deformed state. The lengths of these elements are

$$dS = (d\mathbf{X} \cdot d\mathbf{X})^{1/2}, \quad ds = (d\mathbf{x} \cdot d\mathbf{x})^{1/2}. \tag{33.1}$$

Furthermore, we describe the positions of points **x**, that were at **X** in the reference state, *via* the mapping

$$\mathbf{x} = \mathbf{x}(\mathbf{X}), \tag{33.2}$$

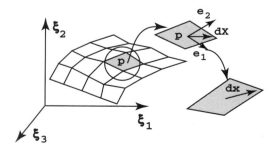

Figure 33.2. Geometry of a membrane viewed as built from a collage of small tiles. Note also that the deformation maps material elements lying in the membrane from the reference to the deformed state.

33.2. Membrane Kinematics

as has been done before for general 3D deformations. Thus,

$$d\mathbf{x} = \frac{\partial \mathbf{x}}{\partial \mathbf{X}} \cdot d\mathbf{X} = \mathcal{F} \cdot d\mathbf{X}, \tag{33.3}$$

where the 2D deformation gradient is, accordingly, defined as

$$\mathcal{F} = \frac{\partial \mathbf{x}}{\partial \mathbf{X}}. \tag{33.4}$$

Unsurprisingly, when we consider such constructions as the change in the square of the length of such material fibers, we find

$$ds^2 - dS^2 = \frac{\partial x_k}{\partial X_i} \frac{\partial x_k}{\partial X_j} dX_i dX_j - \delta_{ij} dX_i dX_j, \quad (i, j, k = 1, 2). \tag{33.5}$$

Thus, we arrive at the definition of a 2D Green strain tensor, *viz.*,

$$\mathcal{E} = \frac{1}{2} \left(\mathcal{F}^T \cdot \mathcal{F} - \mathbf{I} \right). \tag{33.6}$$

This means, *inter alia*, that

$$ds^2 - dS^2 = 2 d\mathbf{X} \cdot \mathcal{E} \cdot d\mathbf{X}. \tag{33.7}$$

Consider the principal values of $\mathcal{E}$, the associated principal directions, and the corresponding principal stretches, λ_1 and λ_2. If $\mathcal{E}_1$ and $\mathcal{E}_2$ are the two principal components of Green strain, then

$$\mathcal{E}_1 = \frac{1}{2}(\lambda_1^2 - 1), \quad \mathcal{E}_2 = \frac{1}{2}(\lambda_2^2 - 1). \tag{33.8}$$

In terms of principal values and axes, the quantity $\lambda_1 \lambda_2$ represents the local area per unit reference area. Later we use the quantity α, defined as

$$\alpha = \lambda_1 \lambda_2 - 1, \tag{33.9}$$

which represents the fractional change in reference area, as a primary measure of area strain of the membrane.

Invariants follow directly as an extension of our already established 3D results. For example, the 2D trace of $\mathcal{E}$,

$$\operatorname{tr} \mathcal{E} = \mathcal{E}_{11} + \mathcal{E}_{22} = \mathcal{E}_1 + \mathcal{E}_2 = \frac{1}{2}(\lambda_1^2 + \lambda_2^2 - 2), \tag{33.10}$$

is an invariant. Another invariant is

$$(\mathcal{E}_1 - \mathcal{E}_2)^2 = \frac{1}{4}(\lambda_1^4 - 2\lambda_1^2 \lambda_2^2 + \lambda_2^4), \tag{33.11}$$

which can be expressed as

$$(\mathcal{E}_1 - \mathcal{E}_2)^2 = \operatorname{tr}^2 \mathcal{E} - 4\mathcal{E}_1 \mathcal{E}_2. \tag{33.12}$$

This, in turn, establishes $\mathcal{E}_1 \mathcal{E}_2$ as still another invariant because $(\mathcal{E}_1 - \mathcal{E}_2)$ and $\operatorname{tr} \mathcal{E}$ are both invariants. But,

$$-\mathcal{E}_1 \mathcal{E}_2 = \frac{1}{4}(\lambda_1^2 + \lambda_2^2 - \lambda_1^2 \lambda_2^2 - 1), \tag{33.13}$$

which leads to a useful set of deformation parameters introduced below.

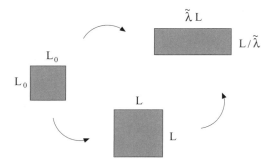

Figure 33.3. Total deformation decomposed into a change in area, plus a change in shape with fixed final area. The length $L = (1+\alpha)^{1/2} L_0$.

Consider the pair of functions of the stretches

$$\alpha = \lambda_1 \lambda_2 - 1,$$
$$\beta = \frac{1}{2\lambda_1 \lambda_2} (\lambda_1^2 + \lambda_2^2) - 1. \tag{33.14}$$

We have already noted the physical meaning of α. If $\lambda_1 = \lambda_2$, the deformation is one of pure 2D dilatation (*i.e.*, a pure area change at constant shape) and thus $\beta = 0$. Now consider two fibers along the principal directions and call them dX_1 and dX_2; in the deformed state they become dx_1 and dx_2, respectively. The change in eccentricity of the parallelepiped they form is

$$\frac{dx_1}{dx_2} - \frac{dX_1}{dX_2} = \frac{dX_1}{dX_2}\left(\frac{\lambda_1}{\lambda_2} - 1\right). \tag{33.15}$$

For a reference and initially square element, the change is simply $\lambda_1/\lambda_2 - 1$.

Next, take the symmetric part of this, with respect to λ_1 and λ_2, to define a parameter that is independent of basing the eccentricity measure on the x_1 or x_2 coordinates, *i.e.*, define

$$\beta = \frac{1}{2}\left(\frac{\lambda_1}{\lambda_2} - 1\right) + \frac{1}{2}\left(\frac{\lambda_2}{\lambda_1} - 1\right)$$
$$= \frac{1}{2\lambda_1 \lambda_2} (\lambda_1^2 + \lambda_2^2) - 1. \tag{33.16}$$

To provide even more interpretation to this measure, consider the stretch ratio

$$\tilde{\lambda} = \frac{\lambda_1}{\sqrt{\lambda_1 \lambda_2}}, \quad \tilde{\lambda}^{-1} = \frac{\lambda_2}{\sqrt{\lambda_1 \lambda_2}}. \tag{33.17}$$

In terms of $\tilde{\lambda}$, we have

$$\beta = \frac{1}{2}(\tilde{\lambda}^2 + \tilde{\lambda}^{-2}) - 1. \tag{33.18}$$

For a shape preserving deformation $\tilde{\lambda} = 1$ and, of course, $\beta = 0$.

The above amounts to decomposing the deformation into the sequential processes of a pure change in area, followed by a shape change at constant (final) area. Thus, aside from a possible rigid body rotation, the process is as depicted in Fig. 33.3. By their nature, α and β (or α and $\tilde{\lambda}$) are linearly independent. It is easily verified that the Jacobian of α and $\tilde{\lambda}$ is equal to $\sqrt{\lambda_1/\lambda_2}$ which cannot be equal to zero, thus demonstrating the independence. Having

33.3 Constitutive Laws for Membranes

established α and β as independent and geometrically relevant deformation measures, we use them in the construction of a hyperelastic constitutive framework.

33.3 Constitutive Laws for Membranes

As in previous chapters, we define the total free energy as Φ, but here we use the measures of deformation α and β. Thus,

$$\Phi = \Phi(\alpha, \beta). \tag{33.19}$$

During an incremental process, the change of Φ is, formally,

$$d\Phi = \left(\frac{\partial \Phi}{\partial \alpha}\right)_\beta d\alpha + \left(\frac{\partial \Phi}{\partial \beta}\right)_\alpha d\beta. \tag{33.20}$$

Let σ_1 and σ_2 be the principal values of Cauchy stress in the membrane's surface. Then $\lambda_1\lambda_2\sigma_1$ and $\lambda_1\lambda_2\sigma_2$ are the principal values of surface Kirchhoff stress. The increment of work done per unit reference area is, accordingly,

$$\begin{aligned} d\Phi &= \lambda_1\lambda_2\sigma_1 d\ln\lambda_1 + \lambda_1\lambda_2\sigma_2 d\ln\lambda_2 \\ &= \lambda_2\sigma_1 d\lambda_1 + \lambda_1\sigma_2 d\lambda_2. \end{aligned} \tag{33.21}$$

Since λ_1 and λ_2 are independent variables, we have the connections

$$\frac{\partial \alpha}{\partial \lambda_1} = \lambda_2, \quad \frac{\partial \alpha}{\partial \lambda_2} = \lambda_1,$$

$$\frac{\partial \beta}{\partial \lambda_1} = \frac{1}{2\lambda_1^2\lambda_2}(\lambda_1^2 - \lambda_2^2), \quad \frac{\partial \beta}{\partial \lambda_2} = \frac{1}{2\lambda_1\lambda_2^2}(\lambda_1^2 - \lambda_2^2). \tag{33.22}$$

By equating (33.21) and (33.20), and using (33.22), we find

$$\sigma_1 = \left(\frac{\partial \Phi}{\partial \alpha}\right)_\beta + \frac{\lambda_1^2 - \lambda_2^2}{2\lambda_1^2\lambda_2^2}\left(\frac{\partial \Phi}{\partial \beta}\right)_\alpha,$$

$$\sigma_2 = \left(\frac{\partial \Phi}{\partial \alpha}\right)_\beta + \frac{\lambda_2^2 - \lambda_1^2}{2\lambda_1^2\lambda_2^2}\left(\frac{\partial \Phi}{\partial \beta}\right)_\alpha. \tag{33.23}$$

The 2D pressure is

$$\bar{p} = \frac{1}{2}(\sigma_1 + \sigma_2) = \left(\frac{\partial \Phi}{\partial \alpha}\right)_\beta, \tag{33.24}$$

whereas the maximum shear stress is

$$\bar{\tau} = \frac{1}{2}|\sigma_1 - \sigma_2| = \frac{|\lambda_1^2 - \lambda_2^2|}{2\lambda_1^2\lambda_2^2}\left(\frac{\partial \Phi}{\partial \beta}\right)_\alpha. \tag{33.25}$$

Evans and Skalak (1980) have shown that in terms of $\tilde{\lambda}$, the maximum shear stress becomes

$$\bar{\tau} = \frac{|\tilde{\lambda}^2 - \tilde{\lambda}^{-2}|}{2(1+\alpha)}\left(\frac{\partial \Phi}{\partial \beta}\right)_\alpha, \tag{33.26}$$

so that the change in Φ can be expressed as

$$d\Phi = \bar{p}d\alpha + 2\bar{\tau}(1+\alpha)d\ln\tilde{\lambda}. \tag{33.27}$$

Suppose that Φ has the phenomenological form

$$\Phi = \bar{p}_0\alpha + \kappa\alpha^2/2 + \cdots. \tag{33.28}$$

Then,

$$\frac{\partial\Phi}{\partial\alpha} \approx \bar{p}_0 + \kappa\alpha + \cdots, \tag{33.29}$$

where $\bar{p}_0$ is a reference pressure (tension) of the surface. Relative to this state of reference pressure, we may calculate $\bar{p}$ as

$$\bar{p} = \kappa\alpha, \quad \kappa = \left(\frac{\partial^2\Phi}{\partial\alpha^2}\right)_{\beta,\alpha=0}. \tag{33.30}$$

As for the shear modulus, we note that the maximum Lagrangian shear strain is

$$\mathcal{E}_s = \frac{1}{4}|\lambda_1^2 - \lambda_2^2|. \tag{33.31}$$

Thus,

$$\bar{\tau} = \frac{2}{\lambda_1^2\lambda_2^2}\left(\frac{\partial\Phi}{\partial\beta}\right)_\alpha \mathcal{E}_s. \tag{33.32}$$

If we set

$$\mu = \left(\frac{\partial\Phi}{\partial\beta}\right)_\alpha, \tag{33.33}$$

we obtain

$$\bar{\tau} = \frac{2\mu}{\lambda_1^2\lambda_2^2}\mathcal{E}_s. \tag{33.34}$$

In terms of the maximum Eulerian shear strain, which is

$$e_s = \frac{1}{4}|\lambda_1^{-2} - \lambda_2^{-2}|, \tag{33.35}$$

equation (33.34) becomes

$$\bar{\tau} = 2\mu e_s, \tag{33.36}$$

consistent with typical linear theories. Note, however, that μ defined *via* (33.33) above, need not be independent of e_s.

33.4 Limited Area Compressibility

In the limit when membranes display relatively strong resistance to area expansion *vis-à-vis* shear extension, the free energy may be approximated by

$$\phi \approx \frac{1}{2}\kappa\alpha^2 + \mu\beta, \quad \frac{\partial^2\Phi}{\partial^2\alpha} = \kappa, \quad \frac{\partial\Phi}{\partial\beta} = \mu. \tag{33.37}$$

33.5. Simple Triangular Networks

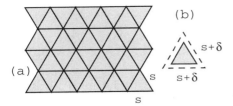

Figure 33.4. (a) A simple triangular network composed of isosceles triangles. In such a simple network each node is co-ordinated with six triangular elements. (b) An element which has undergone a pure area change without a change in shape.

Within this level of approximation, we obtain

$$\sigma_1 = \kappa\alpha + \mu\,\frac{\lambda_1^2 - \lambda_2^2}{2\lambda_1^2\lambda_2^2},$$

$$\sigma_2 = \kappa\alpha + \mu\,\frac{\lambda_2^2 - \lambda_1^2}{2\lambda_1^2\lambda_2^2}.$$
(33.38)

There is no reason, however, to view these as linear.

33.5 Simple Triangular Networks

As an example, consider the case of simple triangular network such as shown in Fig. 33.4. The network elements are assumed to have sides of equal length, s_0, and to have undergone a stretch to length $s_0 + \delta$. The elements edges are assumed to be simple linear springs, with spring constant k. This means, when stretched to a length s, from the rest length of s_0, the energy stored is

$$\phi = \frac{1}{2}k(s - s_0)^2.$$
(33.39)

Per node, however, this becomes

$$\phi_n = 3\phi = \frac{3}{2}k\delta^2,$$
(33.40)

since there are three edges per node for this geometry. If we now note that the reference area of an element is $A_e = \sqrt{3}/4\, s_0^2$, then we may calculate the energy per unit reference area as

$$\delta\Phi = \phi_n/A_n = \sqrt{3}k(\delta/s_0)^2,$$
(33.41)

where $A_n = 2A_e$ is the area per node.

An alternative expression for this follows from (33.28). When the geometry is worked out, we obtain

$$\delta\Phi \approx 2\kappa(\delta/s_0)^2.$$
(33.42)

Comparing this with (33.41), it is found that

$$\kappa = \frac{\sqrt{3}}{2}k.$$
(33.43)

A similar analysis considering pure shear leads to the estimate for the shear modulus

$$\mu = \frac{\sqrt{3}}{4}k.$$
(33.44)

33.6 Suggested Reading

Boal, D. (2002), *Mechanics of the Cell*, Cambridge University Press, New York.

Evans, E. A., and Skalak, R. (1980), *Mechanics and Thermodynamics of Biomembranes*, CRC Press, Boca Raton, Florida.

Green, A. E., and Zerna, W. (1954), *Theoretical Elasticity*, Oxford University Press, London.

PART 8: SOLVED PROBLEMS

34 Solved Problems for Chapters 1–33

CHAPTER 1

Problem 1.1. Prove that

$$\mathbf{a} \cdot (\mathbf{b} \times \mathbf{c})\mathbf{r} = (\mathbf{a} \cdot \mathbf{r})\mathbf{b} \times \mathbf{c} + (\mathbf{b} \cdot \mathbf{r})\mathbf{c} \times \mathbf{a} + (\mathbf{c} \cdot \mathbf{r})\mathbf{a} \times \mathbf{b}.$$

Solution: Begin by considering the product $\mathbf{a} \times [(\mathbf{b} \times \mathbf{c}) \times \mathbf{r}]$. This yields

$$\mathbf{a} \times [(\mathbf{b} \times \mathbf{c}) \times \mathbf{r}] = \mathbf{a} \times [(\mathbf{b} \cdot \mathbf{r})\mathbf{c} - (\mathbf{c} \cdot \mathbf{r})\mathbf{b}] = -(\mathbf{b} \cdot \mathbf{r})\mathbf{c} \times \mathbf{a} - (\mathbf{c} \cdot \mathbf{r})\mathbf{a} \times \mathbf{b}.$$

By setting $\mathbf{b} \times \mathbf{c} = \mathbf{v}$, we obtain

$$\mathbf{a} \times [(\mathbf{b} \times \mathbf{c}) \times \mathbf{r}] = \mathbf{a} \times (\mathbf{v} \times \mathbf{r}) = (\mathbf{a} \cdot \mathbf{r})\mathbf{b} \times \mathbf{c} - \mathbf{a} \cdot (\mathbf{b} \times \mathbf{c})\mathbf{r}.$$

Thus,

$$-(\mathbf{b} \cdot \mathbf{r})\mathbf{c} \times \mathbf{a} - (\mathbf{c} \cdot \mathbf{r})\mathbf{a} \times \mathbf{b} = (\mathbf{a} \cdot \mathbf{r})\mathbf{b} \times \mathbf{c} - \mathbf{a} \cdot (\mathbf{b} \times \mathbf{c})\mathbf{r},$$

and so

$$\mathbf{a} \cdot (\mathbf{b} \times \mathbf{c})\mathbf{r} = (\mathbf{a} \cdot \mathbf{r})\mathbf{b} \times \mathbf{c} + (\mathbf{b} \cdot \mathbf{r})\mathbf{c} \times \mathbf{a} + (\mathbf{c} \cdot \mathbf{r})\mathbf{a} \times \mathbf{b}.$$

Problem 1.2. Prove that

$$\epsilon_{pqs}\epsilon_{mnr} = \begin{vmatrix} \delta_{mp} & \delta_{mq} & \delta_{ms} \\ \delta_{np} & \delta_{nq} & \delta_{ns} \\ \delta_{rp} & \delta_{rq} & \delta_{rs} \end{vmatrix}.$$

Solution: Let the determinant of $\mathbf{A}$ be

$$\det \mathbf{A} = \begin{vmatrix} A_{11} & A_{12} & A_{13} \\ A_{21} & A_{22} & A_{23} \\ A_{31} & A_{32} & A_{33} \end{vmatrix}.$$

An interchange of rows or columns causes a change in sign, *i.e.*,

$$\begin{vmatrix} A_{21} & A_{22} & A_{23} \\ A_{11} & A_{12} & A_{13} \\ A_{31} & A_{32} & A_{33} \end{vmatrix} = \begin{vmatrix} A_{12} & A_{11} & A_{13} \\ A_{22} & A_{21} & A_{23} \\ A_{32} & A_{31} & A_{33} \end{vmatrix} = -\det \mathbf{A}.$$

For an arbitrary number of row interchanges, we have

$$\begin{vmatrix} A_{m1} & A_{m2} & A_{m3} \\ A_{n1} & A_{n2} & A_{n3} \\ A_{r1} & A_{r2} & A_{r3} \end{vmatrix} = \epsilon_{mnr} \det \mathbf{A},$$

and for an arbitrary number of column interchanges,

$$\begin{vmatrix} A_{1p} & A_{1q} & A_{1s} \\ A_{2p} & A_{2q} & A_{2s} \\ A_{3p} & A_{3q} & A_{3s} \end{vmatrix} = \epsilon_{pqs} \det \mathbf{A}.$$

Thus, for an arbitrary number of row and column interchanges,

$$\begin{vmatrix} A_{mp} & A_{mq} & A_{ms} \\ A_{np} & A_{nq} & A_{ns} \\ A_{rp} & A_{rq} & A_{rs} \end{vmatrix} = \epsilon_{pqs}\epsilon_{mnr} \det \mathbf{A}.$$

The proof is completed by setting $A_{ij} = \delta_{ij}$, in which case $\det \mathbf{A} = 1$.

Problem 1.3. Prove that

$$\epsilon_{pqs}\epsilon_{snr} = \delta_{pn}\delta_{qr} - \delta_{pr}\delta_{qn},$$

$$\epsilon_{pqs}\epsilon_{sqr} = -2\delta_{pr}.$$

Solution: Expand the determinant in Problem 1.2, *i.e.*,

$$\epsilon_{pqs}\epsilon_{mnr} = \delta_{mp}(\delta_{nq}\delta_{rs} - \delta_{ns}\delta_{rq}) + \delta_{mq}(\delta_{ns}\delta_{rp} - \delta_{np}\delta_{rs}) + \delta_{ms}(\delta_{np}\delta_{rq} - \delta_{nq}\delta_{rp}).$$

Setting $m = s$,

$$\epsilon_{pqs}\epsilon_{snr} = \delta_{sp}(\delta_{nq}\delta_{rs} - \delta_{ns}\delta_{rq}) + \delta_{sq}(\delta_{ns}\delta_{rp} - \delta_{np}\delta_{rs}) + \delta_{ss}(\delta_{np}\delta_{rq} - \delta_{nq}\delta_{rp})$$

$$= \delta_{rp}\delta_{nq} - \delta_{pn}\delta_{rq} + \delta_{qn}\delta_{rp} - \delta_{np}\delta_{qr} + 3\delta_{np}\delta_{rq} - 3\delta_{nq}\delta_{rp}$$

$$= \delta_{np}\delta_{rq} - \delta_{nq}\delta_{rp}.$$

Then setting $q = n$, we obtain

$$\epsilon_{pqs}\epsilon_{sqr} = \delta_{qp}\delta_{rq} - \delta_{qq}\delta_{rp} = \delta_{pr} - 3\delta_{pr} = -2\delta_{pr}.$$

Problem 1.4. Prove that

$$\det \mathbf{A} = \epsilon_{ijk} A_{i1} A_{j2} A_{k3} = \frac{1}{6} \epsilon_{ijk}\epsilon_{\alpha\beta\gamma} A_{i\alpha} A_{j\beta} A_{k\gamma}.$$

Solution: We first note that

$$\mathbf{A} \cdot \mathbf{e}_k = (A_{ij}\mathbf{e}_i\mathbf{e}_j) \cdot \mathbf{e}_k = A_{ij}\mathbf{e}_i\delta_{jk} = A_{ik}\mathbf{e}_i.$$

Thus, since

$$(\mathbf{A} \cdot \mathbf{e}_1) \cdot [(\mathbf{A} \cdot \mathbf{e}_2) \times (\mathbf{A} \cdot \mathbf{e}_3)] = \begin{vmatrix} A_{11} & A_{12} & A_{13} \\ A_{21} & A_{22} & A_{23} \\ A_{31} & A_{32} & A_{33} \end{vmatrix} = \det \mathbf{A},$$

Problems 1.4–1.5

we have
$$\det \mathbf{A} = A_{i1}\mathbf{e}_i \cdot (A_{j2}\mathbf{e}_j \times A_{k3}\mathbf{e}_k) = A_{i1}A_{j2}A_{k3}\mathbf{e}_i \cdot (\mathbf{e}_j \times \mathbf{e}_k) = \epsilon_{ijk}A_{i1}A_{j2}A_{k3},$$

because
$$\epsilon_{ijk} = \mathbf{e}_i \cdot (\mathbf{e}_j \times \mathbf{e}_k).$$

Since a noncyclic interchange of two columns changes the sign of the determinant, we can write
$$\epsilon_{\alpha\beta\gamma} \det \mathbf{A} = \begin{vmatrix} A_{1\alpha} & A_{1\beta} & A_{1\gamma} \\ A_{2\alpha} & A_{2\beta} & A_{2\gamma} \\ A_{3\alpha} & A_{3\beta} & A_{3\gamma} \end{vmatrix}.$$

Recalling that
$$\mathbf{a} \cdot (\mathbf{b} \times \mathbf{c}) = \begin{vmatrix} a_1 & b_1 & c_1 \\ a_2 & b_2 & c_2 \\ a_3 & b_3 & c_3 \end{vmatrix} = \epsilon_{ijk}a_ib_jc_k,$$

the previous determinant can be written as
$$\begin{vmatrix} A_{1\alpha} & A_{1\beta} & A_{1\gamma} \\ A_{2\alpha} & A_{2\beta} & A_{2\gamma} \\ A_{3\alpha} & A_{3\beta} & A_{3\gamma} \end{vmatrix} = \epsilon_{ijk}A_{i\alpha}A_{j\beta}A_{k\gamma}.$$

Thus,
$$\epsilon_{\alpha\beta\gamma} \det \mathbf{A} = \epsilon_{ijk}A_{i\alpha}A_{j\beta}A_{k\gamma}.$$

Multiplying above with $\epsilon_{\alpha\beta\gamma}$ and recalling that $\epsilon_{\alpha\beta\gamma}\epsilon_{\alpha\beta\gamma} = 6$, we finally obtain
$$\det \mathbf{A} = \frac{1}{6}\epsilon_{ijk}\epsilon_{\alpha\beta\gamma}A_{i\alpha}A_{j\beta}A_{k\gamma}.$$

Problem 1.5. Prove the invariance of
$$\chi_2 = ([\mathbf{A} \cdot \mathbf{f}, \mathbf{A} \cdot \mathbf{g}, \mathbf{h}] + [\mathbf{f}, \mathbf{A} \cdot \mathbf{g}, \mathbf{A} \cdot \mathbf{h}] + [\mathbf{A} \cdot \mathbf{f}, \mathbf{g}, \mathbf{A} \cdot \mathbf{h}])/[\mathbf{f}, \mathbf{g}, \mathbf{h}].$$

Solution: Represent the base vectors $\{\mathbf{f}, \mathbf{g}, \mathbf{h}\}$ in terms of the unit orthogonal basis $\{\mathbf{e}_i\}$, i.e.,
$$\mathbf{f} = f_p\mathbf{e}_p, \quad \mathbf{g} = g_q\mathbf{e}_q, \quad \mathbf{h} = h_r\mathbf{e}_r.$$

Then, the first term in χ_2 becomes
$$[\mathbf{A} \cdot \mathbf{f}, \mathbf{A} \cdot \mathbf{g}, \mathbf{h}] = [\mathbf{A} \cdot (f_p\mathbf{e}_p), \mathbf{A} \cdot (g_q\mathbf{e}_q), h_r\mathbf{e}_r]$$
$$= [f_p(\mathbf{A} \cdot \mathbf{e}_p), g_q(\mathbf{A} \cdot \mathbf{e}_q), h_r\mathbf{e}_r]$$
$$= f_pg_qh_r[\mathbf{A} \cdot \mathbf{e}_p, \mathbf{A} \cdot \mathbf{e}_q, \mathbf{e}_r].$$

Performing similar manipulations on the second two terms of χ_2, we find
$$\chi_2 = \frac{f_pg_qh_r}{[\mathbf{f}, \mathbf{g}, \mathbf{h}]}([\mathbf{A} \cdot \mathbf{e}_p, \mathbf{A} \cdot \mathbf{e}_q, \mathbf{e}_r] + [\mathbf{e}_p, \mathbf{A} \cdot \mathbf{e}_q, \mathbf{A} \cdot \mathbf{e}_r] + [\mathbf{A} \cdot \mathbf{e}_p, \mathbf{e}_q, \mathbf{A} \cdot \mathbf{e}_r]).$$

Recalling that $\mathbf{e}_i \times \mathbf{e}_j = \epsilon_{ijs}\mathbf{e}_s$, the above becomes, with $p=1, q=2, r=3$,

$$\chi_2 = \frac{f_p g_q h_r \epsilon_{pqr}}{[\mathbf{f}, \mathbf{g}, \mathbf{h}]} \left([\mathbf{A} \cdot \mathbf{e}_1, \mathbf{A} \cdot \mathbf{e}_2, \mathbf{e}_3] + [\mathbf{e}_1, \mathbf{A} \cdot \mathbf{e}_2, \mathbf{A} \cdot \mathbf{e}_3] + [\mathbf{A} \cdot \mathbf{e}_1, \mathbf{e}_2, \mathbf{A} \cdot \mathbf{e}_3] \right).$$

Note that

$$\begin{aligned}
[\mathbf{f}, \mathbf{g}, \mathbf{h}] &= \mathbf{f} \cdot (\mathbf{g} \times \mathbf{h}) = \mathbf{f} \cdot (g_q \mathbf{e}_q \times h_r \mathbf{e}_r) \\
&= \mathbf{f} \cdot [g_q h_r (\mathbf{e}_q \times \mathbf{e}_r)] = \mathbf{f} \cdot (g_q h_r \epsilon_{qri} \mathbf{e}_i) \\
&= \mathbf{f} \cdot (\epsilon_{iqr} g_q h_r \mathbf{e}_i) = f_p \mathbf{e}_p \cdot (\epsilon_{iqr} g_q h_r \mathbf{e}_i) \\
&= \epsilon_{iqr} f_p g_q h_r (\mathbf{e}_p \cdot \mathbf{e}_i) = \epsilon_{pqr} (f_p g_q h_r).
\end{aligned}$$

Thus,

$$\chi_2 = \left([\mathbf{A} \cdot \mathbf{f}, \mathbf{A} \cdot \mathbf{g}, \mathbf{h}] + [\mathbf{f}, \mathbf{A} \cdot \mathbf{g}, \mathbf{A} \cdot \mathbf{h}] + [\mathbf{A} \cdot \mathbf{f}, \mathbf{g}, \mathbf{A} \cdot \mathbf{h}] \right) / [\mathbf{f}, \mathbf{g}, \mathbf{h}]$$

is indeed invariant to the basis $\{\mathbf{f}, \mathbf{g}, \mathbf{h}\}$.

Problem 1.6. If $\mathbf{B}$ is a skew symmetric tensor for which $b_i = \frac{1}{2}\epsilon_{ijk} B_{jk}$, prove that

$$B_{pq} = \epsilon_{pqi} b_i.$$

Solution: Multiply the starting equation by ϵ_{pqi} and use the $\epsilon - \delta$ identity to show that

$$\epsilon_{pqi} b_i = \frac{1}{2} \epsilon_{pqi} \epsilon_{ijk} B_{jk} = \frac{1}{2} (\delta_{pj}\delta_{qk} - \delta_{pk}\delta_{qj}) B_{jk}$$

$$= \frac{1}{2}(B_{pq} - B_{qp}) = \frac{1}{2}(B_{pq} + B_{pq}) = B_{pq}.$$

Problem 1.7. Prove that

$$[(\mathbf{A} \cdot \mathbf{a}) \times (\mathbf{A} \cdot \mathbf{b})] \cdot (\mathbf{A} \cdot \mathbf{c}) = (\det \mathbf{A})(\mathbf{a} \times \mathbf{b}) \cdot \mathbf{c}.$$

Solution: Expressed in rectangular coordinates, the left-hand side is

$$\epsilon_{ijk} A_{i\alpha} a_\alpha A_{j\beta} a_\beta A_{k\gamma} a_\gamma.$$

Since, from Problem 1.4,

$$\epsilon_{\alpha\beta\gamma}(\det \mathbf{A}) = \epsilon_{ijk} A_{i\alpha} A_{j\beta} A_{k\gamma},$$

we obtain

$$\epsilon_{ijk} A_{i\alpha} a_\alpha A_{j\beta} a_\beta A_{k\gamma} a_\gamma = (\det \mathbf{A}) \epsilon_{\alpha\beta\gamma} a_\alpha a_\beta a_\gamma = (\det \mathbf{A})(\mathbf{a} \times \mathbf{b}) \cdot \mathbf{c}.$$

Problem 1.8. If $f = f(\mathbf{A})$ is a scalar function of the second-order tensor $\mathbf{A}$, then

$$df = \frac{\partial f}{\partial \mathbf{A}} \cdot \cdot \, d\mathbf{A} = \frac{\partial f}{\partial A_{ij}} dA_{ij}.$$

Prove that

$$\frac{\partial f}{\partial \mathbf{A}} = \frac{\partial f}{\partial A_{ji}} \mathbf{e}_i \mathbf{e}_j.$$

Problems 1.8–1.10

Solution: Since $d\mathbf{A} = dA_{ij}\mathbf{e}_i\mathbf{e}_j$, we have

$$\frac{\partial f}{\partial \mathbf{A}} \cdot \cdot d\mathbf{A} = \frac{\partial f}{\partial A_{kl}}\mathbf{e}_l\mathbf{e}_k \cdot \cdot dA_{ij}\mathbf{e}_i\mathbf{e}_j = \frac{\partial f}{\partial A_{kl}}dA_{ij}\delta_{ki}\delta_{lj} = \frac{\partial f}{\partial A_{ij}}dA_{ij}.$$

Problem 1.9. The principal invariants of the second-order tensor $\mathbf{A}$ are

$$I_A = \text{tr}\,\mathbf{A}, \quad II_A = \frac{1}{2}\left[(\text{tr}\,\mathbf{A})^2 - \text{tr}(\mathbf{A}^2)\right], \quad III_A = \det \mathbf{A}.$$

Show that

$$\det \mathbf{A} = \frac{1}{6}\left[2\text{tr}(\mathbf{A}^3) - 3(\text{tr}\mathbf{A})\text{tr}(\mathbf{A}^2) + (\text{tr}\,\mathbf{A})^3\right].$$

Also, if $\mathbf{A}^{-1}$ exists, show that

$$\mathbf{A}^{-1} = \frac{1}{III_A}\left(\mathbf{A}^2 - I_A\mathbf{A} + II_A\mathbf{I}\right).$$

Solution: The Cayley–Hamilton theorem states

$$\mathbf{A}^3 - I_A\mathbf{A}^2 + II_A\mathbf{A} - III_A\mathbf{I} = \mathbf{0},$$

so that, upon taking the trace,

$$3(\det \mathbf{A}) = \text{tr}\left(\mathbf{A}^3\right) - I_A\text{tr}\left(\mathbf{A}^2\right) + II_A\text{tr}\,\mathbf{A}.$$

This can be easily rewritten in terms of the traces of $\mathbf{A}$, $\mathbf{A}^2$ and $\mathbf{A}^3$.

An expression for the inverse $\mathbf{A}^{-1}$ can be obtained by multiplying the above Cayley–Hamilton matrix equation with $\mathbf{A}^{-1}$. It follows that

$$(\det \mathbf{A})\mathbf{A}^{-1} = \mathbf{A}^2 - I_A\mathbf{A} + II_A\mathbf{I}.$$

Problem 1.10. Derive the expressions for the gradients of the three invariants I_A, II_A, and III_A with respect to $\mathbf{A}$.

Solution: For the first invariant we have

$$\frac{\partial I_A}{\partial \mathbf{A}} = \frac{\partial I_A}{\partial A_{ji}}\mathbf{e}_i\mathbf{e}_j = \frac{\partial A_{kk}}{\partial A_{ji}}\mathbf{e}_i\mathbf{e}_j = \delta_{kj}\delta_{ki}\mathbf{e}_i\mathbf{e}_j = \mathbf{e}_k\mathbf{e}_k.$$

Thus,

$$\frac{\partial I_A}{\partial \mathbf{A}} = \mathbf{I}.$$

Next, we have

$$\frac{\partial(\text{tr}\,\mathbf{A}^2)}{\partial \mathbf{A}} = \frac{\partial(A_{mn}A_{nm})}{\partial A_{ji}}\mathbf{e}_i\mathbf{e}_j = \left(\frac{\partial A_{mn}}{\partial A_{ji}}A_{nm} + A_{mn}\frac{\partial A_{nm}}{\partial A_{ji}}\right)\mathbf{e}_i\mathbf{e}_j$$

$$= (\delta_{jm}\delta_{ni}A_{nm} + A_{mn}\delta_{nj}\delta_{mi})\mathbf{e}_i\mathbf{e}_j = 2A_{ij}\mathbf{e}_i\mathbf{e}_j = 2\mathbf{A}.$$

Furthermore,

$$\frac{\partial(\text{tr}\,\mathbf{A})^2}{\partial \mathbf{A}} = 2(\text{tr}\,\mathbf{A})\frac{\partial(\text{tr}\,\mathbf{A})}{\partial \mathbf{A}} = 2I_A\mathbf{I}.$$

Thus,
$$\frac{\partial II_A}{\partial \mathbf{A}} = I_A \mathbf{I} - \mathbf{A}.$$

Finally, to derive an expression for the gradient of det $\mathbf{A}$, we take the partial derivative $\partial/\partial A_{\alpha\beta}$ of

$$3(\det \mathbf{A}) = A_{ij} A_{jk} A_{ki} - I_A A_{ij} A_{ji} - II_A A_{ii}.$$

After a somewhat lengthy but straightforward derivation, is follows that

$$\frac{\partial(\det \mathbf{A})}{\partial A_{\alpha\beta}} = A_{\beta\gamma} A_{\gamma\alpha} - I_A A_{\beta\alpha} + II_A \delta_{\beta\alpha}.$$

Thus,
$$\frac{\partial III_A}{\partial \mathbf{A}} = \mathbf{A}^2 - I_A \mathbf{A} + II_A \mathbf{I} = (\det \mathbf{A}) \mathbf{A}^{-1}.$$

Problem 1.11. Consider the tensors

$$K_{ijkl} = \frac{1}{3} \delta_{ij} \delta_{kl}, \quad J_{ijkl} = I_{ijkl} - K_{ijkl},$$

where δ_{ij} is the second-order unit tensor (Kronecker delta), and $I_{ijkl} = (\delta_{ik}\delta_{jl} + \delta_{il}\delta_{jk})/2$ is the fourth-order unit tensor. Show that in the trace operation with any second-order tensor $\mathbf{A}$, the tensor $\mathbf{J}$ extracts its deviatoric part, while the tensor $\mathbf{K}$ extracts its spherical part.

Solution: We first observe that

$$\mathbf{K} : \mathbf{K} = \mathbf{K}, \quad \mathbf{J} : \mathbf{J} = \mathbf{J}, \quad \mathbf{K} : \mathbf{J} = \mathbf{J} : \mathbf{K} = \mathbf{0}.$$

For example,

$$K_{ijmn} K_{mnkl} = \frac{1}{9} \delta_{ij} \delta_{mn} \delta_{mn} \delta_{kl} = \frac{1}{3} \delta_{ij} \delta_{kl} = K_{ijkl},$$

$$J_{ijmn} J_{mnkl} = (I_{ijmn} - K_{ijmn})(I_{mnkl} - K_{mnkl})$$
$$= I_{ijkl} - K_{ijkl} - K_{ijkl} + K_{ijkl} = J_{ijkl},$$

and similarly for other identities. Clearly, then,

$$K_{ijkl} A_{kl} = \frac{1}{3} \delta_{ij} \delta_{kl} A_{kl} = \frac{1}{3} A_{kk} \delta_{ij},$$

which are the spherical components of A_{ij}. Finally,

$$J_{ijkl} A_{kl} = (I_{ijkl} - K_{ijkl}) A_{kl} = A_{ij} - \frac{1}{3} A_{kk} \delta_{ij} = A'_{ij},$$

which are the deviatoric components of A_{ij}.

Problem 1.12. Consider the rotation $\mathbf{Q}$ of an orthogonal basis defined by the unit vectors $\{\mathbf{e}_i\}$. The new triad of orthogonal unit vectors is

$$\mathbf{e}_i^* = \mathbf{Q} \cdot \mathbf{e}_i = Q_{\alpha i} \mathbf{e}_\alpha.$$

Problems 1.12–1.13

Show that **Q** has the same components in both bases, *i.e.*,

$$\mathbf{Q} = Q_{ij}\mathbf{e}_i\mathbf{e}_j = Q^*_{ij}\mathbf{e}^*_i\mathbf{e}^*_j, \quad Q^*_{ij} = Q_{ij}.$$

Solution: It readily follows that

$$\mathbf{Q} = Q_{\alpha\beta}\mathbf{e}_\alpha\mathbf{e}_\beta = Q^*_{ij} Q_{\alpha i} Q_{\beta j}\mathbf{e}_\alpha\mathbf{e}_\beta,$$

i.e.,

$$Q_{\alpha\beta} = Q^*_{ij} Q_{\alpha i} Q_{\beta j}.$$

Multiplying this by $Q^T_{m\alpha} Q^T_{n\beta}$, and recalling that for an orthogonal tensor $Q^T_{m\alpha} Q_{\alpha n} = \delta_{mn}$, we obtain

$$Q^*_{mn} = Q_{mn}.$$

This is also obvious from the relationship between any two induced tensors, say, **A** and **A**∗. These are related by

$$\mathbf{A} = A_{ij}\mathbf{e}_i\mathbf{e}_j = A^*_{ij}\mathbf{e}^*_i\mathbf{e}^*_j, \quad \mathbf{A}^* = A^*_{ij}\mathbf{e}_i\mathbf{e}_j.$$

Thus,

$$\mathbf{A}^* = \mathbf{Q}^T \cdot \mathbf{A} \cdot \mathbf{Q}, \quad A^*_{ij} = Q^T_{ik} A_{kl} Q_{lj}.$$

If $A_{kl} = Q_{kl}$, this gives $Q^*_{ij} = Q_{ij}$.

Yet another prove follows directly from the relationships between the two sets of unit vectors. Indeed,

$$\mathbf{e}^*_j = \mathbf{Q} \cdot \mathbf{e}_j = Q_{\alpha j}\mathbf{e}_\alpha \quad \Rightarrow \quad Q_{ij} = \mathbf{e}_i \cdot \mathbf{e}^*_j,$$

and, dually,

$$\mathbf{e}_i = \mathbf{Q}^T \cdot \mathbf{e}^*_i = Q^*_{i\alpha}\mathbf{e}^*_\alpha \quad \Rightarrow \quad Q^*_{ij} = \mathbf{e}_i \cdot \mathbf{e}^*_j.$$

Thus, again, $Q^*_{ij} = Q_{ij}$.

Problem 1.13. Prove that the following tensors are isotropic tensors:

(a) $\mathbf{I} = \delta_{ij}\mathbf{e}_i\mathbf{e}_j$;
(b) $\boldsymbol{\epsilon} = \epsilon_{ijk}\mathbf{e}_i\mathbf{e}_j\mathbf{e}_k$;
(c) $\mathbf{A} = \delta_{ij}\delta_{kl}\mathbf{e}_i\mathbf{e}_j\mathbf{e}_k\mathbf{e}_l$, $\mathbf{B} = \delta_{ik}\delta_{jl}\mathbf{e}_i\mathbf{e}_j\mathbf{e}_k\mathbf{e}_l$, $\mathbf{C} = \delta_{il}\delta_{jk}\mathbf{e}_i\mathbf{e}_j\mathbf{e}_k\mathbf{e}_l$.

Solution: (a) If the tensor is an isotropic tensor, it has the same rectangular components in any two coordinate systems that differ by rigid body rotation. Thus, we need to prove that

$$\mathbf{I} = \delta_{ij}\mathbf{e}_i\mathbf{e}_j = I^*_{ij}\mathbf{e}^*_i\mathbf{e}^*_j, \quad I^*_{ij} = \delta_{ij},$$

where the triad of orthogonal vectors $\{\mathbf{e}_i\}$ is obtained from $\{\mathbf{e}_i\}$ an arbitrary rigid body rotation **Q**. Since

$$\mathbf{e}^*_i = \mathbf{Q} \cdot \mathbf{e}_i = Q_{\alpha i}\mathbf{e}_\alpha,$$

we have

$$\delta_{\alpha\beta}\mathbf{e}_\alpha\mathbf{e}_\beta = I^*_{ij} Q_{\alpha i} Q_{\beta j}\mathbf{e}_\alpha\mathbf{e}_\beta,$$

i.e.,
$$\delta_{\alpha\beta} = I^*_{ij} Q_{\alpha i} Q_{\beta j}.$$

Multiplying this by $Q^T_{m\alpha} Q^T_{n\beta}$, and recalling that for an orthogonal tensor $Q^T_{m\alpha} Q_{\alpha n} = \delta_{mn}$, we obtain

$$I^*_{mn} = \delta_{mn}.$$

(b) We need to prove that

$$\epsilon = \epsilon_{ijk} \mathbf{e}_i \mathbf{e}_j \mathbf{e}_k = \epsilon^*_{ijk} \mathbf{e}^*_i \mathbf{e}^*_j \mathbf{e}^*_k, \quad \epsilon^*_{ijk} = \epsilon_{ijk}.$$

By using $\mathbf{e}^*_i = \mathbf{Q} \cdot \mathbf{e}_i = Q_{\alpha i} \mathbf{e}_\alpha$, we have

$$\epsilon_{\alpha\beta\gamma} \mathbf{e}_\alpha \mathbf{e}_\beta \mathbf{e}_\gamma = \epsilon^*_{ijk} Q_{\alpha i} Q_{\beta j} Q_{\gamma k} \mathbf{e}_\alpha \mathbf{e}_\beta \mathbf{e}_\gamma,$$

i.e.,

$$\epsilon_{\alpha\beta\gamma} = \epsilon^*_{ijk} Q_{\alpha i} Q_{\beta j} Q_{\gamma k}.$$

Multiplying this by $Q^T_{m\alpha} Q^T_{n\beta} Q^T_{p\gamma}$, we obtain

$$\epsilon_{\alpha\beta\gamma} Q_{\alpha m} Q_{\beta n} Q_{\gamma p} = \epsilon^*_{mnp}.$$

But, from Problem 1.4, the left-hand side is

$$\epsilon_{\alpha\beta\gamma} Q_{\alpha m} Q_{\beta n} Q_{\gamma p} = (\det \mathbf{Q}) \epsilon_{mnp},$$

and because $\det \mathbf{Q} = 1$, we obtain

$$\epsilon^*_{mnp} = \epsilon_{mnp}.$$

(c) The proof of isotropic nature of tensors **A**, **B**, and **C** is similar. We thus prove the isotropy of only one of them, say **B**. We need to show that

$$\mathbf{B} = \delta_{ik}\delta_{jl} \mathbf{e}_i \mathbf{e}_j \mathbf{e}_k \mathbf{e}_l = B^*_{ijkl} \mathbf{e}^*_i \mathbf{e}^*_j \mathbf{e}^*_k \mathbf{e}^*_l, \quad B^*_{ijkl} = \delta_{ik}\delta_{jl}.$$

As before, by using $\mathbf{e}^*_i = \mathbf{Q} \cdot \mathbf{e}_i = Q_{\alpha i} \mathbf{e}_\alpha$, we have

$$\delta_{\alpha\gamma}\delta_{\beta\delta} \mathbf{e}_\alpha \mathbf{e}_\beta \mathbf{e}_\gamma \mathbf{e}_\delta = B^*_{ijkl} Q_{\alpha i} Q_{\beta j} Q_{\gamma k} Q_{\delta l} \mathbf{e}_\alpha \mathbf{e}_\beta \mathbf{e}_\gamma \mathbf{e}_\delta,$$

i.e.,

$$\delta_{\alpha\gamma}\delta_{\beta\delta} = B^*_{ijkl} Q_{\alpha i} Q_{\beta j} Q_{\gamma k} Q_{\delta l}.$$

Multiplying this by $Q^T_{m\alpha} Q^T_{n\beta} Q^T_{p\gamma} Q_{q\delta}$, we obtain

$$B^*_{mnpq} = \delta_{mp}\delta_{nq}.$$

Problem 1.14. The following matrix often arises in two dimensional problems of solid mechanics

$$[A_{ij}] = \begin{bmatrix} A_{11} & 0 & 0 \\ 0 & A_{22} & A_{23} \\ 0 & A_{32} & A_{33} \end{bmatrix}.$$

Find its inverse.

Problems 1.14–2.1

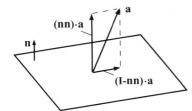

Figure 1-1. Projecting the vector **a** onto the plane with unit normal **n**.

Solution: Denoting by

$$\Delta = A_{22}A_{33} - A_{23}A_{32},$$

which is assumed to be different from zero, it readily follows that

$$[A_{ij}^{-1}] = \frac{1}{\Delta}\begin{bmatrix} \Delta/A_{11} & 0 & 0 \\ 0 & A_{33} & -A_{23} \\ 0 & -A_{32} & A_{22} \end{bmatrix}.$$

Problem 1.15. Derive an expression for the second-order tensor that projects the vector **a** onto the plane with unit normal **n**.

Solution: The vector **a** can be decomposed into the component parallel to **n** and the component orthogonal to **n** (Fig. 1-1), as

$$\mathbf{a} = (\mathbf{n}\cdot\mathbf{a})\mathbf{n} + [\mathbf{a} - (\mathbf{n}\cdot\mathbf{a})\mathbf{n}].$$

Thus,

$$\mathbf{a} = (\mathbf{n}\,\mathbf{n})\cdot\mathbf{a} + (\mathbf{I} - \mathbf{n}\,\mathbf{n})\cdot\mathbf{a},$$

where **I** is the unit second-order tensor. The required projection tensor is clearly

$$\mathbf{I} - \mathbf{n}\,\mathbf{n}.$$

CHAPTER 2

Problem 2.1. Prove that

$$\text{div}\,\mathbf{u}(\mathbf{x}) = \nabla\cdot\mathbf{u}(\mathbf{x}) = \partial u_p/\partial x_p.$$

Solution: By definition,

$$\nabla\cdot\mathbf{u}(\mathbf{x}) = \text{tr}\,[\text{grad}\,\mathbf{u}(\mathbf{x})],$$

where

$$\text{grad}\,\mathbf{u}(\mathbf{x}) = \partial u_p/\partial x_q\,\mathbf{e}_p\mathbf{e}_q.$$

Thus,

$$\text{tr}[\text{grad } \mathbf{u}(\mathbf{x})] = [\partial u_p/\partial x_q \mathbf{e}_p \mathbf{e}_q \cdot \mathbf{e}_1, \mathbf{e}_2, \mathbf{e}_3] + [\mathbf{e}_1, \partial u_p/\partial x_q \mathbf{e}_p \mathbf{e}_q \cdot \mathbf{e}_2, \mathbf{e}_3]$$
$$+ [\mathbf{e}_1, \mathbf{e}_2, \partial u_p/\partial x_q \mathbf{e}_p \mathbf{e}_q \cdot \mathbf{e}_3]$$
$$= \partial u_p/\partial x_1 [\mathbf{e}_p, \mathbf{e}_2, \mathbf{e}_3] + \partial u_p/\partial x_2 [\mathbf{e}_1, \mathbf{e}_p, \mathbf{e}_3] + \partial u_p/\partial x_3 [\mathbf{e}_1, \mathbf{e}_2, \mathbf{e}_p]$$
$$= \partial u_1/\partial x_1 + \partial u_2/\partial x_2 + \partial u_3/\partial x_3 = \partial u_p/\partial x_p,$$

i.e.,

$$\nabla \cdot \mathbf{u}(\mathbf{x}) = \partial u_p/\partial x_p.$$

Problem 2.2. Prove that

$$\int_S \mathbf{n} \times (\mathbf{a} \times \mathbf{x}) \, dS = 2\mathbf{a} V,$$

where V is the volume of the region bounded by the closed surface S, and $\mathbf{n}$ is the outward pointing normal to S; $\mathbf{x}$ is the position vector within V.

Solution: Consider the qth component of the integral, viz.,

$$\int_S \epsilon_{qpi} n_p \epsilon_{ijk} a_j x_k \, dS.$$

The divergence theorem leads to

$$\int_V (\epsilon_{qpi} \epsilon_{ijk} a_j x_k)_{,p} \, dV,$$

and, because $\mathbf{a}$ is a constant vector, we have

$$\int_V \epsilon_{qpi} \epsilon_{ijk} a_j x_{k,p} \, dV = \int_V (\delta_{qj} \delta_{pk} - \delta_{qk} \delta_{pj}) a_j x_{k,p} \, dV$$
$$= \int_V (a_q x_{p,p} - a_p x_{q,p}) \, dV$$
$$= \int_V (a_q \delta_{pp} - a_p \delta_{qp}) \, dV$$
$$= \int_V (3a_q - a_q) \, dV = 2a_q V.$$

Problem 2.3. Prove that

$$\mathbf{A} \times \nabla = -\left(\nabla \times \mathbf{A}^T\right)^T.$$

Solution: The expanded form of the left-hand side is

$$\mathbf{A} \times \nabla = A_{ij} \mathbf{e}_i \mathbf{e}_j \times \frac{\partial}{\partial x_k} \mathbf{e}_k = \frac{\partial A_{ij}}{\partial x_k} \mathbf{e}_i \epsilon_{jkl} \mathbf{e}_l = \epsilon_{jkl} \frac{\partial A_{ij}}{\partial x_k} \mathbf{e}_i \mathbf{e}_l.$$

The expanded form of the cross product $\nabla \times \mathbf{A}^T$ is

$$\frac{\partial}{\partial x_k} \mathbf{e}_k \times A_{ij} \mathbf{e}_j \mathbf{e}_i = \frac{\partial A_{ij}}{\partial x_k} \epsilon_{kjl} \mathbf{e}_l \mathbf{e}_i = -\epsilon_{jkl} \frac{\partial A_{ij}}{\partial x_k} \mathbf{e}_l \mathbf{e}_i.$$

Problems 2.3–2.6

Consequently,

$$-\left(\nabla \times \mathbf{A}^T\right)^T = \epsilon_{jkl} \frac{\partial A_{ij}}{\partial x_k} \mathbf{e}_i \mathbf{e}_l,$$

which completes the proof.

Problem 2.4. Prove that
(a) $\nabla \cdot (\mathbf{A} \cdot \mathbf{a}) = (\nabla \cdot \mathbf{A}) \cdot \mathbf{a} + \mathbf{A} : (\nabla \mathbf{a})$,
(b) $\nabla \cdot (\mathbf{A} \cdot \mathbf{B}) = (\nabla \cdot \mathbf{A}) \cdot \mathbf{B} + (\mathbf{A}^T \cdot \nabla) \cdot \mathbf{B}$.

Solution: (a) The expanded form of the left-hand side is

$$\frac{\partial}{\partial x_i} \mathbf{e}_i \cdot A_{jk} \mathbf{e}_j \mathbf{e}_k \cdot a_l \mathbf{e}_l = \frac{\partial}{\partial x_i}(A_{jk} a_l) \delta_{ij} \delta_{kl} = \frac{\partial}{\partial x_i}(A_{ik} a_k).$$

Thus, by partial differentiation

$$\frac{\partial}{\partial x_i}(A_{ik} a_k) = \frac{\partial A_{ik}}{\partial x_i} a_k + A_{ik} \frac{\partial a_k}{\partial x_i} = (\nabla \cdot \mathbf{A}) \cdot \mathbf{a} + \mathbf{A} : (\nabla \mathbf{a}).$$

(b) In an expanded form, the left-hand side is

$$\nabla \cdot (\mathbf{A} \cdot \mathbf{B}) = \frac{\partial}{\partial x_i} \mathbf{e}_i \cdot (A_{kl} B_{lj} \mathbf{e}_k \mathbf{e}_j) = \frac{\partial}{\partial x_i}(A_{il} B_{lj}) \mathbf{e}_j$$

$$= \left(\frac{\partial}{\partial x_i} A_{il}\right) B_{lj} \mathbf{e}_j + \left(A_{li}^T \frac{\partial}{\partial x_i}\right) B_{lj} \mathbf{e}_j$$

$$= (\nabla \cdot \mathbf{A}) \cdot \mathbf{B} + (\mathbf{A}^T \cdot \nabla) \cdot \mathbf{B}.$$

Problem 2.5. If $\mathbf{n}$ is the outward unit vector normal to the closed surface S bounding the volume V, prove that

$$\int_S \mathbf{n} \, dS = \mathbf{0}.$$

Solution: By the Gauss divergence theorem, applied to a scalar field f, we have

$$\int_S n_i f \, dS = \int_V \frac{\partial f}{\partial x_i} \, dV.$$

By taking $f = 1$, we obtain

$$\int_S n_i \, dS = 0.$$

Problem 2.6. Derive the expression for the gradient operator ∇ in the cylindrical coordinates (r, θ, z).

Solution: The position vector $\mathbf{x}$ of an arbitrary point, expressed in Cartesian and cylindrical coordinates (Fig. 2-1), is

$$\mathbf{x} = x_1 \mathbf{e}_1 + x_2 \mathbf{e}_2 + x_3 \mathbf{e}_3 = r \mathbf{e}_r + z \mathbf{e}_z,$$

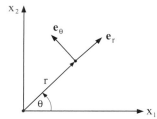

Figure 2-1. The unit vectors of a polar coordinate system.

where
$$x_1 = r\cos\theta, \quad x_2 = r\sin\theta, \quad x_3 = z,$$

$$\mathbf{e}_r = \cos\theta\,\mathbf{e}_1 + \sin\theta\,\mathbf{e}_2, \quad \mathbf{e}_\theta = -\sin\theta\,\mathbf{e}_1 + \cos\theta\,\mathbf{e}_2, \quad \mathbf{e}_z = \mathbf{e}_3.$$

The gradient operator in Cartesian coordinates is
$$\nabla = \frac{\partial}{\partial x_1}\mathbf{e}_1 + \frac{\partial}{\partial x_2}\mathbf{e}_2 + \frac{\partial}{\partial x_3}\mathbf{e}_3.$$

Since
$$\frac{\partial}{\partial x_i} = \frac{\partial}{\partial r}\frac{\partial r}{\partial x_i} + \frac{\partial}{\partial \theta}\frac{\partial \theta}{\partial x_i} \quad (i=1,2), \quad \frac{\partial}{\partial x_3} = \frac{\partial}{\partial z},$$

and
$$\frac{\partial r}{\partial x_1} = \cos\theta, \quad \frac{\partial r}{\partial x_2} = \sin\theta, \quad \frac{\partial \theta}{\partial x_1} = -\frac{\sin\theta}{r}, \quad \frac{\partial \theta}{\partial x_2} = \frac{\cos\theta}{r},$$

we obtain
$$\nabla = \frac{\partial}{\partial r}\mathbf{e}_r + \frac{1}{r}\frac{\partial}{\partial \theta}\mathbf{e}_\theta + \frac{\partial}{\partial z}\mathbf{e}_z.$$

CHAPTER 3

We are here interested in the Fourier series expansion of functions of a single variable, $f(x)$, in terms of the orthogonal functions sin and cos, *viz.*,

$$f(x) = \frac{a_0}{2} + \sum_{k=1}^{\infty}[a_k\cos(kx) + b_k\sin(kx)],$$

where $f(x)$ is presumed to be periodically extended from the interval $-\pi \leq x \leq \pi$. The coefficients are then computed from

$$a_n = \frac{1}{\pi}\int_{-\pi}^{\pi} f(x)\cos(nx)\,dx, \quad (n=0,1,2,\ldots),$$

$$b_n = \frac{1}{\pi}\int_{-\pi}^{\pi} f(x)\sin(nx)\,dx, \quad (n=1,2,\ldots).$$

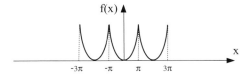

Figure 3-1. A periodic even extension of the function $f(x) = x^2$.

Problem 3.1. Expand $f(x) = x^2$ ($-\pi \le x \le \pi$) in a Fourier series. The function is even and its extension, as shown in the Fig. 3-1, is periodic.

Solution: The Fourier coefficients are

$$a_0 = \frac{2}{\pi} \int_0^\pi x^2 \, dx = \frac{2\pi^2}{3},$$

and

$$a_n = \frac{2}{\pi} \int_0^\pi x^2 \cos(nx) \, dx = (-1)^n \frac{4}{n^2},$$

whereas

$$b_n = 0,$$

because of the symmetry of x^2. Thus,

$$x^2 = \frac{\pi^2}{3} + 4\left[\cos x - \frac{\cos(2x)}{2^2} + \frac{\cos(3x)}{3^2} - \cdots\right].$$

Problem 3.2. Expand $f(x) = |x|$ ($-\pi \le x \le \pi$) in a Fourier series (see Fig. 3-2).

Solution: Here again, the function is seen to be even with an extension as shown in the figure. The expanded function will converge to the function $|x|$ within $-\pi \le x \le \pi$ and to its periodic extension outside that interval. The calculation of the coefficients gives

$$a_0 = \frac{2}{\pi} \int_0^\pi x \, dx = \pi,$$

and

$$a_n = \frac{2}{\pi} \int_0^\pi x \cos(nx) \, dx = \frac{2}{\pi n^2}[\cos(n\pi) - 1] = \frac{2}{\pi n^2}[(-1)^n - 1].$$

Therefore,

$$|x| = \frac{\pi}{2} - \frac{4}{\pi}\left[\cos x + \frac{\cos(3x)}{3^2} + \frac{\cos(5x)}{5^2} + \cdots\right].$$

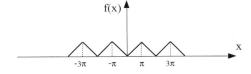

Figure 3-2. A periodic even extension of the function $f(x) = |x|$.

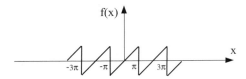

Figure 3-3. The periodic odd function $f(x) = x$.

Problem 3.3. Expand $f(x) = x$ $(-\pi \leq x \leq \pi)$ in a Fourier series (see Fig. 3-3).

Solution: This function is clearly odd and is discontinuous at the points $x = (2k+1)\pi$ ($k = 0, \pm 1, \pm 2, \ldots$). Since $f(x)$ is odd, we obtain

$$a_n = 0,$$

$$b_n = \frac{2}{\pi} \int_0^\pi x \sin(nx)\,dx = -\frac{2}{n} \cos(n\pi) = \frac{2}{n}(-1)^{n+1}.$$

Therefore,

$$x = 2\left[\sin x - \frac{\sin(2x)}{2} + \frac{\sin(3x)}{3} - \cdots\right].$$

Problem 3.4. Find the Fourier series for the function

$$f(x) = |\sin(\omega x)|, \quad |x| \leq \frac{\pi}{2\omega}.$$

Solution: Since $f(x)$ is an even function (Fig. 3-4), we have

$$f(x) = \frac{a_0}{2} + \sum_{n=1}^\infty a_n \cos\frac{n\pi x}{L}, \quad L = \frac{\pi}{2\omega},$$

$$a_n = \frac{2}{L} \int_0^L f(x) \cos\frac{n\pi x}{L}\,dx, \quad n = 0, 1, 2, \ldots.$$

Thus,

$$a_n = \frac{4\omega}{\pi} \int_0^{\pi/2\omega} \sin(\omega x) \cos(2\omega n x)\,dx.$$

In view of the trigonometric identity

$$2 \sin\alpha \cos\beta = \sin(\alpha + \beta) + \sin(\alpha - \beta),$$

the above can be rewritten as

$$a_n = \frac{2\omega}{\pi} \int_0^{\pi/2\omega} [\sin\omega(1+2n)x + \sin\omega(1-2n)x]\,dx.$$

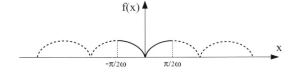

Figure 3-4. The function $f(x) = |\sin(\omega x)|$.

Upon integration, this is

$$a_n = \frac{2\omega}{\pi}\left[-\frac{\cos\omega(1+2n)x}{\omega(1+2n)} - \frac{\cos\omega(1-2n)x}{\omega(1-2n)}\right]_0^{\pi/2\omega},$$

i.e.,

$$a_n = -\frac{4}{\pi(4n^2-1)}.$$

Thus,

$$f(x) = \frac{2}{\pi} - \frac{4}{\pi}\sum_{n=1}^{\infty}\frac{\cos(2\omega nx)}{4n^2-1}.$$

CHAPTER 4

Problem 4.1. If $\mathbf{A}$ is an invertible tensor, it can be decomposed as

$$\mathbf{A} = \mathbf{Q}\cdot\mathbf{U} = \mathbf{V}\cdot\mathbf{Q}.$$

If $\mathbf{U}$ has eigenvectors $\mathbf{p}_i$ and associated eigenvalues λ_i, and $\mathbf{V}$ has eigenvectors $\mathbf{q}_i$ with associated eigenvalues μ_i, prove that

$$\lambda_i = \mu_i \quad \text{and} \quad \mathbf{q}_i = \mathbf{Q}\cdot\mathbf{p}_i.$$

Solution: First note that

$$\mathbf{A} = \mathbf{Q}\cdot\mathbf{U} = \mathbf{V}\cdot\mathbf{Q},$$
$$\mathbf{Q}^{-1}\cdot\mathbf{Q}\cdot\mathbf{U} = \mathbf{Q}^{-1}\cdot\mathbf{V}\cdot\mathbf{Q},$$

and, thus,

$$\mathbf{U} = \mathbf{Q}^{-1}\cdot\mathbf{V}\cdot\mathbf{Q}.$$

For $\mathbf{U}$ to have eigenvalues λ_i and eigenvectors $\mathbf{p}_i$,

$$(\mathbf{U} - \lambda_i\mathbf{I})\cdot\mathbf{p}_i = \mathbf{0} \quad \text{and} \quad \det(\mathbf{U} - \lambda_i\mathbf{I}) = 0.$$

For $\mathbf{V}$ to have eigenvalues μ_i and eigenvectors $\mathbf{q}_i$,

$$(\mathbf{V} - \mu_i\mathbf{I})\cdot\mathbf{q}_i = \mathbf{0} \quad \text{and} \quad \det(\mathbf{V} - \mu_i\mathbf{I}) = 0.$$

Then,

$$\det(\mathbf{U} - \lambda\mathbf{I}) = \det(\mathbf{Q}^{-1}\cdot\mathbf{V}\cdot\mathbf{Q} - \lambda\mathbf{I})$$
$$= \det[\mathbf{Q}^{-1}\cdot(\mathbf{V} - \lambda\mathbf{I})\cdot\mathbf{Q}]$$
$$= \det(\mathbf{Q}^{-1})\det(\mathbf{V} - \lambda\mathbf{I})\det(\mathbf{Q})$$
$$= \det(\mathbf{V} - \lambda\mathbf{I}),$$

because $\det\mathbf{Q} = \det\mathbf{Q}^{-1} = 1$. This demonstrates that $\lambda_i = \mu_i$.

Next note that

$$\mathbf{U} \cdot \mathbf{p}_i = \lambda_i \mathbf{p}_i = \mathbf{Q}^{-1} \cdot \mathbf{V} \cdot (\mathbf{Q} \cdot \mathbf{p}_i),$$

$$\mathbf{Q} \cdot \lambda_i \mathbf{p}_i = \mathbf{V} \cdot (\mathbf{Q} \cdot \mathbf{p}_i),$$

$$\lambda_i (\mathbf{Q} \cdot \mathbf{p}_i) = \mathbf{V} \cdot (\mathbf{Q} \cdot \mathbf{p}_i).$$

Since $\lambda_i = \mu_i$, the above becomes

$$\mu_i (\mathbf{Q} \cdot \mathbf{p}_i) = \mathbf{V} \cdot (\mathbf{Q} \cdot \mathbf{p}_i),$$

and thus

$$\mathbf{q}_i = \mathbf{Q} \cdot \mathbf{p}_i.$$

Problem 4.2. For a simple shearing deformation

$$x_1 = X_1 + k_1 X_2, \quad x_2 = X_2, \quad x_3 = X_3, \quad (k = \text{const.}),$$

find the tensors $\mathbf{F}$, $\mathbf{C}$, $\mathbf{B}$, $\mathbf{E}$, $\mathbf{U}$, $\mathbf{V}$ and $\mathbf{R}$, where $\mathbf{F} = \mathbf{V} \cdot \mathbf{R} = \mathbf{R} \cdot \mathbf{U}$, $\mathbf{C} = \mathbf{F}^T \cdot \mathbf{F}$, $\mathbf{B} = \mathbf{F} \cdot \mathbf{F}^T$, and $2\mathbf{E} = \mathbf{C} - \mathbf{I}$.

Solution: We clearly have

$$\mathbf{F} = \begin{bmatrix} 1 & k & 0 \\ 0 & 1 & 0 \\ 0 & 0 & 1 \end{bmatrix}, \quad \mathbf{B} = \mathbf{F} \cdot \mathbf{F}^T = \begin{bmatrix} 1+k^2 & k & 0 \\ k & 1 & 0 \\ 0 & 0 & 1 \end{bmatrix},$$

$$\mathbf{C} = \mathbf{F}^T \cdot \mathbf{F} = \begin{bmatrix} 1 & k & 0 \\ k & 1+k^2 & 0 \\ 0 & 0 & 1 \end{bmatrix}, \quad \mathbf{E} = \frac{1}{2}(\mathbf{C} - \mathbf{I}) = \begin{bmatrix} 0 & k/2 & 0 \\ k/2 & k^2/2 & 0 \\ 0 & 0 & 0 \end{bmatrix}.$$

Since $\mathbf{C} = \mathbf{R}^T \cdot \mathbf{B} \cdot \mathbf{R}$, and restricting for convenience to nontrivial two-by-two submatrices, we can write

$$\begin{bmatrix} 1 & k \\ k & 1+k^2 \end{bmatrix} = \begin{bmatrix} c & -s \\ s & c \end{bmatrix}^T \cdot \begin{bmatrix} 1+k^2 & k \\ k & 1 \end{bmatrix} \cdot \begin{bmatrix} c & -s \\ s & c \end{bmatrix},$$

where

$$\mathbf{R} = \begin{bmatrix} c & -s & 0 \\ s & c & 0 \\ 0 & 0 & 1 \end{bmatrix}, \quad c = \cos \varphi, \quad s = \sin \varphi,$$

is the rotation matrix. From the previous matrix equation it follows that $kc + 2s = 0$, i.e., $\tan \varphi = -k/2$, so that

$$c = \frac{1}{(1+k^2/4)^{1/2}}, \quad s = -\frac{k/2}{(1+k^2/4)^{1/2}}.$$

Problems 4.2–4.4

Consequently, the rotation matrix of the polar decomposition is

$$\mathbf{R} = \begin{bmatrix} \dfrac{1}{(1+k^2/4)^{1/2}} & \dfrac{k/2}{(1+k^2/4)^{1/2}} & 0 \\ -\dfrac{k/2}{(1+k^2/4)^{1/2}} & \dfrac{1}{(1+k^2/4)^{1/2}} & 0 \\ 0 & 0 & 1 \end{bmatrix}.$$

Having this determined, the right and left deformation tensors follow by simple matrix multiplication as $\mathbf{U} = \mathbf{R}^T \cdot \mathbf{F}$ and $\mathbf{V} = \mathbf{F} \cdot \mathbf{R}^T$.

Problem 4.3. The Rivlin–Ericksen tensors $\mathbf{A}_n$ are defined by

$$\frac{d^n}{dt^n}(ds)^2 = d\mathbf{x} \cdot \mathbf{A}_n \cdot d\mathbf{x}, \quad (n = 1, 2, 3, \ldots).$$

The square of an infinitesimal material length is $(ds)^2 = d\mathbf{x} \cdot d\mathbf{x}$. It can be shown that

$$\mathbf{A}_{n+1} = \dot{\mathbf{A}}_n + \mathbf{A}_n \cdot \mathbf{L} + \mathbf{L}^T \cdot \mathbf{A}_n,$$

where the superposed dot designates a time differentiation, and $\mathbf{L}$ is the velocity gradient. Since $\mathbf{A}_1 = 2\mathbf{D}$ ($\mathbf{D}$ being the rate of deformation tensor, symmetric part of $\mathbf{L}$), the above formula gives for $n = 1$

$$\mathbf{A}_2 = \dot{\mathbf{D}} + \mathbf{D} \cdot \mathbf{L} + \mathbf{L}^T \cdot \mathbf{D}.$$

Show that the tensor $\mathbf{A}_2$ is objective, *i.e.*, under the rigid body rotation $\mathbf{Q}(t)$, superposed to the current configuration ($\mathbf{F}^* = \mathbf{Q} \cdot \mathbf{F}$), it transforms as $\mathbf{A}_2^* = \mathbf{Q} \cdot \mathbf{A}_2 \cdot \mathbf{Q}^T$.

Solution: We have

$$\frac{d^2}{dt^2}(ds)^2 = d\mathbf{x} \cdot \mathbf{A}_2 \cdot d\mathbf{x}.$$

Under rigid body rotation of the current configuration, $d\mathbf{x}^* = \mathbf{Q} \cdot d\mathbf{x} = d\mathbf{x} \cdot \mathbf{Q}^T$ and $ds^* = ds$. Consequently,

$$\frac{d^2}{dt^2}(ds^*)^2 = d\mathbf{x}^* \cdot \mathbf{A}_2^* \cdot d\mathbf{x}^* = d\mathbf{x} \cdot \mathbf{Q}^T \cdot \mathbf{A}_2^* \cdot \mathbf{Q} \cdot d\mathbf{x}.$$

Therefore,

$$\mathbf{A}_2^* = \mathbf{Q} \cdot \mathbf{A}_2 \cdot \mathbf{Q}^T.$$

Alternatively, the above result can be deduced directly from

$$\mathbf{A}_2 = 2(\dot{\mathbf{D}} + \mathbf{D} \cdot \mathbf{L} + \mathbf{L}^T \cdot \mathbf{D}), \quad \mathbf{A}_2^* = 2(\dot{\mathbf{D}}^* + \mathbf{D}^* \cdot \mathbf{L}^* + \mathbf{L}^{*T} \cdot \mathbf{D}^*),$$

by using the transformation rules

$$\mathbf{L}^* = \dot{\mathbf{Q}} \cdot \mathbf{Q}^{-1} + \mathbf{Q} \cdot \mathbf{L} \cdot \mathbf{Q}^T, \quad \mathbf{D}^* = \mathbf{Q} \cdot \mathbf{D} \cdot \mathbf{Q}^T.$$

Problem 4.4. Examine the nature of the deformation given by

$$x_1 = X_1, \quad x_2 = X_2 + \alpha X_3, \quad x_3 = X_3 + \alpha X_2, \quad \alpha = \text{const.}$$

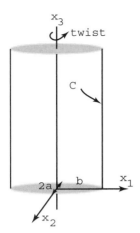

Figure 4-1. A "ruler" subject to pure torsion.

Solution: Consider the base vectors in both the reference and current configurations to be coincident. We want to consider the implications of this deformation as regards the deformation of line elements and surface elements. For example, the displacements associated with this deformation are

$$u_1 = x_1 - X_1 = 0, \quad u_2 = x_2 - X_2 = \alpha X_3, \quad u_3 = x_3 - X_3 = \alpha X_2.$$

Inverting these, we obtain

$$X_1 = x_1, \quad X_2 = (x_2 - \alpha x_3)/(1 - \alpha^2), \quad X_3 = (x_3 - \alpha x_2)/(1 - \alpha^2),$$

and

$$u_1 = 0, \quad u_2 = \alpha(x_3 - \alpha x_2)/(1 - \alpha^2), \quad u_3 = \alpha(x_2 - \alpha x_3)/(1 - \alpha^2).$$

From the above, for example, we can determine that the linear material fiber given by $X_1 = 0$, $X_2 + X_3 = 1/(1+\alpha)$ occupies the position $x_1 = 0$, $x_2 = x_3$ after the deformation. Also the circular region $x_1 = 0$, $X_2^2 + X_3^2 = 1/(1 - \alpha^2)$ is deformed into the elliptical region $(1 + \alpha^2)x_2^2 - 4\alpha x_2 x_3 + (1 + \alpha^2)x_3^2 = (1 - \alpha^2)$. Other deformation features can be found similarly by direct substitution.

Problem 4.5. Consider the slender object (possibly a "ruler") subject to pure torsion about its long axis as sketched in Fig. 4-1. The mapping function for this deformation is given as

$$x_1 = X_1 \cos(\alpha X_3) - X_2 \sin(\alpha X_3),$$
$$x_2 = X_1 \sin(\alpha X_3) + X_2 \cos(\alpha X_3),$$
$$X_3 = X_3.$$

Calculate the stretch along the fiber marked C (*i.e.*, the outer most edge), which lies along the unit vector $\mathbf{e}_3 \| x_3$.

Problems 4.5–4.6

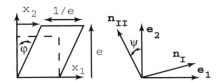

Figure 4-2. A two-dimensional deformation with combined extension and shear.

Solution: A straightforward calculation of the right Cauchy–Green deformation tensor yields

$$\mathbf{C} = \begin{bmatrix} 1 & 0 & -\alpha X_2 \\ 0 & 1 & \alpha X_1 \\ -\alpha X_2 & \alpha X_1 & \alpha^2(X_1^2 + X_2^2) + 1 \end{bmatrix}.$$

Then, from

$$\lambda^2(\mathbf{e}_3) = \mathbf{e}_3 \cdot \mathbf{C} \cdot \mathbf{e}_3,$$

we obtain

$$\lambda(C) = \sqrt{\alpha^2 b^2 + 1},$$

where the edge C is positioned at $x_1 = b$, $x_2 = 0$.

Problem 4.6. Consider the two-dimensional deformation described in Fig. 4-2, for which the deformation gradient is

$$\mathbf{F} = \begin{bmatrix} 1/e & \gamma \\ 0 & e \end{bmatrix}, \quad \det \mathbf{F} = 1.$$

The polar decomposition theorem states that $\mathbf{F} = \mathbf{V} \cdot \mathbf{R}$, where $\mathbf{V}$ is the symmetric left stretch tensor and $\mathbf{R}$ is an orthogonal tensor. The two eigenvalues of $\mathbf{V}$ are related by $\lambda_{II} = 1/\lambda_I$, because the deformation is volume preserving. The principal directions associated with these eigenvalues are $\mathbf{n}_I$ and $\mathbf{n}_{II} = d\mathbf{n}_I/d\psi$, where ψ is the angle depicted in the figure. With

$$\mu \equiv \frac{1}{2}(e + 1/e),$$

show that

$$\lambda_I = 1/\lambda_{II} = \sqrt{\mu^2 + \gamma^2/4} + \sqrt{\mu^2 + \gamma^2/4 - 1},$$

and

$$\psi = \tan^{-1}(2\chi/e\gamma),$$

where

$$\chi = \mu^2 - \mu/e - \gamma^2/4 + \sqrt{(\mu^2 + \gamma^2/4)(\mu^2 + \gamma^2/4 - 1)}.$$

Also show that

$$\sin(2\psi) = \frac{\chi e \gamma}{\chi^2 + (1/2e\gamma)^2},$$

$$\cos(2\psi) = \frac{(1/2e\gamma)^2 - \chi^2}{\chi^2 + (1/2e\gamma)^2},$$

and

$$\mathbf{R} = \begin{bmatrix} \cos\alpha & \sin\alpha \\ -\sin\alpha & \cos\alpha \end{bmatrix},$$

where

$$\alpha = \tan^{-1}\left(\frac{\gamma}{2\mu}\right).$$

Solution: Since $\mathbf{V}$ is symmetric, let it have the form

$$\mathbf{V} = \begin{bmatrix} u & v \\ v & q \end{bmatrix},$$

and let $\mathbf{R}$ have the form as listed above. Then, the product $\mathbf{V} \cdot \mathbf{R}$ leads to the following four equations, $viz.$,

$$u\cos\alpha - v\sin\alpha = 1/e,$$

$$u\sin\alpha + v\cos\alpha = \gamma,$$

$$v\cos\alpha - q\sin\alpha = 0,$$

$$v\sin\alpha + q\cos\alpha = e.$$

Combining the second and third of these, and then again the first and fourth, yields

$$(u+q)\sin\alpha = \gamma,$$

$$(u+q)\cos\alpha = 2\mu.$$

Thus,

$$\tan\alpha = \frac{\gamma}{2\mu},$$

and

$$\sin\alpha = \frac{\gamma}{\sqrt{4\mu^2 + \gamma^2}}, \quad \cos\alpha = \frac{2\mu}{\sqrt{4\mu^2 + \gamma^2}}.$$

Substitution of these relations into the original four relations yields

$$\frac{v\gamma}{\sqrt{4\mu^2 + \gamma^2}} + \frac{q(2\mu)}{\sqrt{4\mu^2 + \gamma^2}} = e,$$

or

$$v\gamma + 2\mu q = e\sqrt{4\mu^2 + \gamma^2}.$$

The third of the original four relations gives

$$q\tan\alpha = v,$$

Problems 4.6–4.7

and, therefore,

$$v = \frac{\gamma}{2\mu}q.$$

Thus,

$$\frac{\gamma}{2\mu}q + 2\mu q = e\sqrt{4\mu^2 + \gamma^2},$$

which leads to

$$q = \frac{\mu e}{\sqrt{\mu^2 + \gamma^2/4}},$$

$$v = \frac{e\gamma/2}{\sqrt{\mu^2 + \gamma^2/4}},$$

$$u = \frac{\mu/e + \gamma^2/2}{\sqrt{\mu^2 + \gamma^2/4}}.$$

The eigenvalues are found by solving for the roots of

$$\begin{vmatrix} u - \lambda & v \\ v & q - \lambda \end{vmatrix} = 0,$$

which yields the desired solutions

$$\lambda_I = 1/\lambda_{II} = \sqrt{\mu^2 + \gamma^2/4} + \sqrt{\mu^2 + \gamma^2/4 - 1}.$$

The eigenvector $\mathbf{n}^I$ is found from

$$\begin{bmatrix} u & v \\ v & q \end{bmatrix} \cdot \begin{bmatrix} n_1^I \\ n_2^I \end{bmatrix} = \lambda_I \begin{bmatrix} n_1^I \\ n_2^I \end{bmatrix}.$$

In general, it is found that

$$\frac{n_2^I}{n_1^I} = \tan \psi,$$

with ψ as given above. The remaining relations follow from elementary geometric manipulations.

Problem 4.7. Construct a simple scheme for calculating the components of Green strain, and stretch, within a triangular region by knowing, perhaps *via* measurement, the elongations of its sides.

Solution: Consider the schematic drawing shown in Fig. 4-3. Figure 4-3a shows a single side of the triangular region shown in Fig. 4-3b. In Fig. 4-3a, the initial length of the side p is dS and after the deformation it is ds. The two lengths are connected *via* the Green strain by

$$ds^2 - dS^2 = 2d\mathbf{X} \cdot \mathbf{E} \cdot d\mathbf{X} = 2E_{ij}dX_i dX_j$$

$$= 2E_{11}\,dX_1^2 + 4E_{12}\,dX_1 dX_2 + 2E_{22}\,dX_2^2.$$

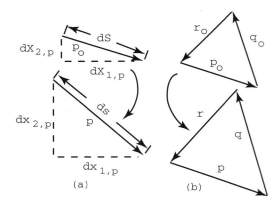

Figure 4-3. (a) Initial and deformed infinitesimal length (single side). (b) An infinitesimal triangular region before and after the deformation.

Now, construct similar formulae for sides q and r, shown in Fig. 4-3b, and combine the three such formulae in the expression

$$\mathbf{A} \cdot \mathbf{x} = \mathbf{b},$$

where

$$\mathbf{A} = \begin{bmatrix} 2\mathrm{d}X_{1,p}^2 & 4\mathrm{d}X_{1,p}\mathrm{d}X_{2,p} & 2\mathrm{d}X_{2,p}^2 \\ 2\mathrm{d}X_{1,q}^2 & 4\mathrm{d}X_{1,q}\mathrm{d}X_{2,q} & 2\mathrm{d}X_{2,q}^2 \\ 2\mathrm{d}X_{1,r}^2 & 4\mathrm{d}X_{1,r}\mathrm{d}X_{2,r} & 2\mathrm{d}X_{2,r}^2 \end{bmatrix},$$

$$\mathbf{b} = \begin{bmatrix} \mathrm{d}s_p^2 - \mathrm{d}S_p^2 \mathrm{d}s_q^2 - \mathrm{d}S_q^2 \\ \mathrm{d}s_r^2 - \mathrm{d}S_r^2 \end{bmatrix},$$

$$\mathbf{x} = \{E_{11}, E_{12}, E_{22}\}.$$

The above are readily solved for the three components of Green strain, and from these the principal values are obtained. Calling them E_1 and E_2, the principal stretches are

$$\lambda_1 = \sqrt{2E_1 + 1}, \quad \lambda_2 = \sqrt{2E_2 + 1}.$$

Problem 4.8. Consider the deformation mapping

$$x_1 = X_1 + X_2(e^t - 1), \quad x_2 = X_1(e^{-t} - 1) + X_2, \quad x_3 = X_3,$$

which has the inverse

$$X_1 = \frac{-x_1 + x_2(e^t - 1)}{1 - e^t - e^{-t}}, \quad X_2 = \frac{x_1(e^{-t} - 1) - x_2}{1 - e^t - e^{-t}}, \quad X_3 = x_3.$$

Compute the deformation gradient, and the velocity gradient for this mapping.

Solution: We first note that

$$\mathbf{F} = \partial \mathbf{x}/\partial \mathbf{X} = \begin{bmatrix} 1 & e^t - 1 & 0 \\ e^{-t} - 1 & 1 & 0 \\ 0 & 0 & 1 \end{bmatrix},$$

and

$$\dot{\mathbf{F}} = \partial \dot{\mathbf{x}}/\partial \mathbf{X} = \begin{bmatrix} 0 & e^t & 0 \\ -e^{-t} & 0 & 0 \\ 0 & 0 & 0 \end{bmatrix}.$$

The inverse can be computed as

$$\mathbf{F}^{-1} = \partial \mathbf{X}/\partial \mathbf{x} = \begin{bmatrix} \dfrac{1}{e^t + e^{-t} - 1} & \dfrac{-(e^t - 1)}{e^t + e^{-t} - 1} & 0 \\ \dfrac{-(e^{-t} - 1)}{e^t + e^{-t} - 1} & 0 & 0 \\ 0 & 0 & 1 \end{bmatrix}.$$

Since $\mathbf{L} = \dot{\mathbf{F}} \cdot \mathbf{F}^{-1}$, the components of the velocity gradient are

$$L_{11} = \frac{-e^t(e^{-t} - 1)}{e^t + e^{-t} - 1}, \quad L_{12} = \frac{e^t}{e^t + e^{-t} - 1},$$

$$L_{21} = \frac{-e^{-t}}{e^t + e^{-t} - 1}, \quad L_{22} = \frac{e^{-t}(e^t - 1)}{e^t + e^{-t} - 1},$$

$$L_{ij} = 0, \text{ otherwise.}$$

Problem 4.9. Consider the following deformation field

$$x_1 = X_1 + kX_2^2 t^2,$$
$$x_2 = X_2 + kX_2 t,$$
$$x_3 = X_3.$$

At $t = 0$ the corners of a unit square are at $A(0, 0, 0)$, $B(0, 1, 0)$, $C(1, 1, 0)$, and $D(1, 0, 0)$. Sketch the deformed shape of this square at $t = 2$. Calculate the spatial velocity and the acceleration fields.

Solution: The deformed shape is as shown in the Fig. 4-4. The velocity field is obtained as

$$v_1 = \frac{\partial x_1}{\partial t} = 2kX_2^2 t = 2k\frac{x_2 t}{(1 + kt)^2},$$

$$v_2 = \frac{\partial x_2}{\partial t} = kX_2 = k\frac{x_2}{(1 + kt)},$$

$$v_3 = 0.$$

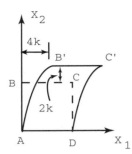

Figure 4-4. Deformed shape.

The acceleration field is

$$a_1 = \frac{\partial^2 x_1}{\partial t^2} = 2kX_2^2 = 2k\frac{x_2^2}{(1+kt)^2}, \quad a_2 = a_3 = 0.$$

Problem 4.10. Prove that

$$d(\ln J)/dt = \text{div } \mathbf{v},$$

where $\mathbf{v}$ is the material velocity.

Solution: First recall that

$$J = \epsilon_{PQR} x_{1,P} x_{2,Q} x_{3,R},$$

where the notation $\partial x_i / \partial X_p \equiv x_{i,P}$ has been used. For the time rate of J, we have

$$\dot{J} = \epsilon_{PQR}(\dot{x}_{1,P} x_{2,Q} x_{3,R} + x_{1,P} \dot{x}_{2,Q} x_{3,R} + x_{1,P} x_{2,Q} \dot{x}_{3,R}).$$

Since $\dot{x}_{i,H} = v_{i,s} x_{s,H}$, the above becomes

$$\dot{J} = \epsilon_{PQR}(v_{1,s} x_{s,P} x_{2,Q} x_{3,R} + x_{1,P} v_{2,s} x_{s,Q} x_{3,R} + x_{1,P} x_{2,Q} v_{3,s} x_{s,R}).$$

Note that there are nine terms in the above, but only three are nonvanishing. They yield

$$\dot{J} = v_{1,1} J + v_{2,2} J + v_{3,3} J,$$

and thus

$$\dot{J} = J \nabla \cdot \mathbf{v}, \quad \text{i.e.,} \quad d(\ln J)/dt = \text{div } \mathbf{v}.$$

Problem 4.11. Prove that the time rate of the surface element, dS_p, is

$$\frac{dS_p}{dt} = \frac{\partial v_q}{\partial x_q} dS_p - \frac{\partial v_q}{\partial x_p} dS_q.$$

Solution: Without loss of generality the surface element may be taken to have edge generators initially along the X_2 and X_3 axes in the reference state. Then, for the edge that was along the X_s axis, $d\mathbf{x}^{(s)} = \mathbf{F} \cdot d\mathbf{X}^{(s)}$, and

$$dS_i = \epsilon_{ijk} \frac{\partial x_j}{\partial X_2} dX_2 \frac{\partial x_k}{\partial X_3} dX_3.$$

Problems 4.11–4.12

Since,

$$\frac{\partial x_i}{\partial X_i}\mathrm{d}S_i = \epsilon_{ijk}\frac{\partial x_i}{\partial X_1}\frac{\partial x_j}{\partial X_2}\frac{\partial x_k}{\partial X_3}\mathrm{d}X_2\mathrm{d}X_3 = J\mathrm{d}X_2\mathrm{d}X_3,$$

we find

$$\frac{\partial X_1}{\partial x_p}\frac{\partial x_i}{\partial X_1}\mathrm{d}S_i = \delta_{ip}\mathrm{d}S_i = \mathrm{d}S_p = \frac{\partial X_1}{\partial x_p}J\mathrm{d}X_2\mathrm{d}X_3.$$

Thus,

$$\frac{\mathrm{d}S_p}{\mathrm{d}t} = \left(\frac{\partial X_1}{\partial x_p}J\frac{\partial v_q}{\partial x_q} - J\frac{\partial X_1}{\partial x_q}\frac{\partial v_q}{\partial x_p}\right)\mathrm{d}X_2\mathrm{d}X_3$$

$$= \frac{\partial v_q}{\partial x_q}\mathrm{d}S_p - \frac{\partial v_q}{\partial x_p}\mathrm{d}S_q.$$

Problem 4.12. For the motion

$$x_1 = X_1, \quad x_2 = X_2 + X_1(e^{-2t} - 1), \quad x_3 = X_3 + X_1(e^{-3t} - 1),$$

compute the rate of deformation tensor and compare it to the rate of the small strain tensor, $e_{ij} = (\partial u_i/\partial x_j + \partial u_j/\partial x_i)/2$.

Solution: First note that the displacement components for this motion are

$$u_1 = 0, \quad u_2 = x_1(e^{-2t} - 1), \quad u_3 = x_1(e^{-3t} - 1),$$

with the velocity components

$$v_1 = 0, \quad v_2 = -2x_1 e^{-2t}, \quad v_3 = -3x_1 e^{-3t}.$$

Now, decompose the velocity gradient into its symmetric and antisymmetric parts, as

$$(L_{ij}) = \left(\frac{\partial v_i}{\partial x_j}\right) = \begin{pmatrix} 0 & 0 & 0 \\ -2e^{-2t} & 0 & 0 \\ -3e^{-3t} & 0 & 0 \end{pmatrix} = \begin{pmatrix} 0 & -e^{-2t} & -3/2e^{-3t} \\ -e^{-2t} & 0 & 0 \\ -3/2e^{-3t} & 0 & 0 \end{pmatrix}$$

$$+ \begin{pmatrix} 0 & e^{-2t} & 3/2e^{-3t} \\ -e^{-2t} & 0 & 0 \\ -3/2e^{-3t} & 0 & 0 \end{pmatrix} = (D_{ij}) + (W_{ij}).$$

Likewise,

$$\left(\frac{\partial u_i}{\partial x_j}\right) = \begin{pmatrix} 0 & 0 & 0 \\ e^{-2t} & 0 & 0 \\ e^{-3t} & 0 & 0 \end{pmatrix} = \frac{1}{2}\begin{pmatrix} 0 & e^{-2t} & e^{-3t} \\ e^{-2t} & 0 & 0 \\ e^{-3t} & 0 & 0 \end{pmatrix} + \frac{1}{2}\begin{pmatrix} 0 & -e^{-2t} & -e^{-3t} \\ e^{-2t} & 0 & 0 \\ e^{-3t} & 0 & 0 \end{pmatrix}.$$

Consequently,

$$\frac{\mathrm{d}}{\mathrm{d}t}\left(\frac{\partial u_i}{\partial x_j}\right) = \begin{pmatrix} 0 & -e^{-2t} & -3/2e^{-3t} \\ -e^{-2t} & 0 & 0 \\ -3/2e^{-3t} & 0 & 0 \end{pmatrix} + \begin{pmatrix} 0 & e^{-2t} & 3/2e^{-3t} \\ -e^{-2t} & 0 & 0 \\ -3/2e^{-3t} & 0 & 0 \end{pmatrix}.$$

Therefore, in this case

$$\frac{d}{dt}(e_{ij}) = (D_{ij}).$$

Problem 4.13. The deformation gradient at a point in a body is

$$\mathbf{F} = 0.2\mathbf{e}_1\mathbf{e}_1 - 0.1\mathbf{e}_1\mathbf{e}_2 + 0.3\mathbf{e}_2\mathbf{e}_1 + 0.4\mathbf{e}_2\mathbf{e}_2 + 0.1\mathbf{e}_3\mathbf{e}_3,$$

where $\mathbf{e}_i$ ($i = 1, 2, 3$) are the Cartesian unit base vectors.
(a) Determine the Cauchy–Green deformation tensors $\mathbf{B}$ and $\mathbf{C}$.
(b) Determine the eigenvalues $\lambda^{(i)}$ and eigenvectors $\mathbf{n}^{(i)}$ and $\mathbf{N}^{(i)}$ of $\mathbf{B}$ and $\mathbf{C}$, respectively.
(c) Verify that

$$\mathbf{F} = \sum_{i=1}^{3} \sqrt{\lambda^{(i)}}\, \mathbf{n}^{(i)} \mathbf{N}^{(i)}.$$

(d) Calculate $\mathbf{R}$, $\mathbf{U}$, and $\mathbf{V}$ by using the relationships

$$\mathbf{R} = \sum_{i=1}^{3} \mathbf{n}^{(i)} \mathbf{N}^{(i)}, \quad \mathbf{U} = \sum_{i=1}^{3} \sqrt{\lambda^{(i)}}\, \mathbf{N}^{(i)} \mathbf{N}^{(i)}, \quad \mathbf{V} = \sum_{i=1}^{3} \sqrt{\lambda^{(i)}}\, \mathbf{n}^{(i)} \mathbf{n}^{(i)}.$$

Solution: (a) The deformation gradient matrix is

$$\mathbf{F} = \begin{bmatrix} 0.2 & -0.1 & 0 \\ 0.3 & 0.4 & 0 \\ 0 & 0 & 0.1 \end{bmatrix},$$

whereas the right and left Cauchy–Green deformation tensors are

$$\mathbf{C} = \mathbf{F}^T \cdot \mathbf{F} = \begin{bmatrix} 0.13 & 0.1 & 0 \\ 0.1 & 0.17 & 0 \\ 0 & 0 & 0.01 \end{bmatrix}, \quad \mathbf{B} = \mathbf{F} \cdot \mathbf{F}^T = \begin{bmatrix} 0.05 & 0.02 & 0 \\ 0.02 & 0.25 & 0 \\ 0 & 0 & 0.01 \end{bmatrix}.$$

(b) Upon solving the eigenvalue problem

$$\mathbf{C} \cdot \mathbf{N}^{(i)} = \lambda^{(i)} \mathbf{N}^{(i)},$$

there follows

$$\lambda^{(1)} = 0.252, \quad \mathbf{N}^{(1)} = \begin{bmatrix} 0.634 \\ 0.773 \\ 0 \end{bmatrix}; \quad \lambda^{(2)} = 0.048, \quad \mathbf{N}^{(2)} = \begin{bmatrix} -0.773 \\ 0.634 \\ 0 \end{bmatrix}.$$

Similarly, by solving the eigenvalue problem

$$\mathbf{B} \cdot \mathbf{n}^{(i)} = \lambda^{(i)} \mathbf{n}^{(i)},$$

there follows

$$\lambda^{(1)} = 0.252, \quad \mathbf{n}^{(1)} = \begin{bmatrix} 0.0985 \\ 0.9953 \\ 0 \end{bmatrix}; \quad \lambda^{(2)} = 0.048, \quad \mathbf{n}^{(2)} = \begin{bmatrix} -0.9953 \\ 0.0985 \\ 0 \end{bmatrix}.$$

The third eigenvalue and the corresponding eigenvector are in each case

$$\lambda^{(3)} = 0.01, \quad \mathbf{N}^{(3)} = \mathbf{n}^{(3)} = \begin{bmatrix} 0 \\ 0 \\ 1 \end{bmatrix}.$$

(c) It can be readily verified that

$$\mathbf{F} = \sum_{i=1}^{3} \sqrt{\lambda^{(i)}} \, \mathbf{n}^{(i)} \mathbf{N}^{(i)} = \begin{bmatrix} 0.2 & -0.1 & 0 \\ 0.3 & 0.4 & 0 \\ 0 & 0 & 0.1 \end{bmatrix}.$$

(d) The rotation matrix is

$$\mathbf{R} = \sum_{i=1}^{3} \mathbf{n}^{(i)} \mathbf{N}^{(i)} = \begin{bmatrix} 0.832 & -0.554 & 0 \\ 0.554 & 0.832 & 0 \\ 0 & 0 & 1 \end{bmatrix},$$

whereas the stretch tensors are

$$\mathbf{U} = \sum_{i=1}^{3} \sqrt{\lambda^{(i)}} \, \mathbf{N}^{(i)} \mathbf{N}^{(i)} = \begin{bmatrix} 0.333 & 0.139 & 0 \\ 0.139 & 0.388 & 0 \\ 0 & 0 & 0.1 \end{bmatrix},$$

$$\mathbf{V} = \sum_{i=1}^{3} \sqrt{\lambda^{(i)}} \, \mathbf{n}^{(i)} \mathbf{n}^{(i)} = \begin{bmatrix} 0.222 & 0.028 & 0 \\ 0.028 & 0.5 & 0 \\ 0 & 0 & 0.1 \end{bmatrix}.$$

Problem 4.14. Let ∇^0 be the gradient operator with respect to material coordinates X_i. Prove that

$$\nabla^0 \cdot \left[(\det \mathbf{F}) \mathbf{F}^{-1} \right] = \mathbf{0},$$

where $\mathbf{F}$ is the deformation gradient.

Solution: By the Nanson's relation we have

$$n_i \, dS = (\det \mathbf{F}) F_{ik}^{-T} n_k^0 \, dS^0.$$

Upon integration, and recalling the result from Problem 2.5, we obtain

$$0 = \int_S n_i \, dS = \int_{S^0} (\det \mathbf{F}) F_{ik}^{-T} n_k^0 \, dS^0.$$

Thus, by applying the Gauss divergence theorem

$$\int_{S^0} (\det \mathbf{F}) F_{ik}^{-T} n_k^0 \, dS^0 = \int_{V^0} \frac{\partial}{\partial X_k} \left[(\det \mathbf{F}) F_{ki}^{-1} \right] dV^0 = 0.$$

This holds for the whole volume V^0, or any part of it, so that we must have locally, at any point of the deformed body,

$$\frac{\partial}{\partial X_k} \left[(\det \mathbf{F}) F_{ki}^{-1} \right] = 0.$$

Problem 4.15. Consider cylindrical coordinates (r, θ, z), and the corresponding displacement vector $\mathbf{u} = \{u_r, u_\theta, u_z\}$. The ∇ operator in the cylindrical coordinates (see Problem 2.6) is

$$\nabla = \frac{\partial}{\partial r} \mathbf{e}_r + \frac{1}{r} \frac{\partial}{\partial \theta} \mathbf{e}_\theta + \frac{\partial}{\partial z} \mathbf{e}_z.$$

(a) Derive the cylindrical components of the strain tensor $\mathbf{e} = \frac{1}{2}(\nabla \mathbf{u} + \mathbf{u} \nabla)$.
(b) Write the cylindrical components of the acceleration vector $\mathbf{a}$ in spatial cylindrical coordinates.
(c) Derive the cylindrical components of the velocity gradient tensor $\mathbf{L} = \mathbf{v} \nabla$, and its symmetric and antisymmetric parts $\mathbf{D}$ and $\mathbf{W}$ (rate of deformation and spin tensors).

Solution: (a) The outer (tensor) product of the vectors ∇ and $\mathbf{u}$ is

$$\nabla \mathbf{u} = \left(\mathbf{e}_r \frac{\partial}{\partial r} + \mathbf{e}_\theta \frac{1}{r} \frac{\partial}{\partial \theta} + \mathbf{e}_z \frac{\partial}{\partial z} \right)(u_r \mathbf{e}_r + u_\theta \mathbf{e}_\theta + u_z \mathbf{e}_z).$$

Recalling that

$$\frac{\partial \mathbf{e}_r}{\partial \theta} = \mathbf{e}_\theta, \quad \frac{\partial \mathbf{e}_\theta}{\partial \theta} = -\mathbf{e}_r,$$

we have, for example,

$$\mathbf{e}_\theta \frac{1}{r} \frac{\partial}{\partial \theta}(u_\theta \mathbf{e}_\theta) = \mathbf{e}_\theta \frac{1}{r} \frac{\partial}{\partial \theta}(u_\theta \mathbf{e}_\theta) = \mathbf{e}_\theta \frac{1}{r} \left(\frac{\partial u_\theta}{\partial \theta} \mathbf{e}_\theta + u_\theta \frac{\partial \mathbf{e}_\theta}{\partial \theta} \right)$$

$$= \frac{1}{r} \frac{\partial u_\theta}{\partial \theta} \mathbf{e}_\theta \mathbf{e}_\theta - \frac{1}{r} u_\theta \mathbf{e}_\theta \mathbf{e}_r.$$

Continuing like this with other tensor products, we obtain

$$\nabla \mathbf{u} = \begin{bmatrix} \dfrac{\partial u_r}{\partial r} & \dfrac{\partial u_\theta}{\partial r} & \dfrac{\partial u_z}{\partial r} \\ \dfrac{1}{r} \dfrac{\partial u_r}{\partial \theta} - \dfrac{u_\theta}{r} & \dfrac{1}{r} \dfrac{\partial u_\theta}{\partial \theta} + \dfrac{u_r}{r} & \dfrac{1}{r} \dfrac{\partial u_z}{\partial \theta} \\ \dfrac{\partial u_r}{\partial z} & \dfrac{\partial u_\theta}{\partial z} & \dfrac{\partial u_z}{\partial z} \end{bmatrix}.$$

The matrix $\mathbf{u} \nabla$ is the transpose of this, *i.e.*,

$$\mathbf{u} \nabla = (\nabla \mathbf{u})^T.$$

The strain tensor can then be easily calculated as

$$\mathbf{e} = \frac{1}{2}(\mathbf{u} \nabla + \nabla \mathbf{u}).$$

Problem 4.15

This gives the following strain-displacement relations in cylindrical coordinates

$$e_{rr} = \frac{\partial u_r}{\partial r},$$

$$e_{\theta\theta} = \frac{1}{r}\frac{\partial u_\theta}{\partial \theta} + \frac{u_r}{r},$$

$$e_{zz} = \frac{\partial u_z}{\partial z},$$

$$e_{r\theta} = \frac{1}{2}\left(\frac{1}{r}\frac{\partial u_r}{\partial \theta} + \frac{\partial u_\theta}{\partial r} - \frac{u_\theta}{r}\right),$$

$$e_{rz} = \frac{1}{2}\left(\frac{\partial u_r}{\partial z} + \frac{\partial u_z}{\partial r}\right),$$

$$e_{\theta z} = \frac{1}{2}\left(\frac{\partial u_\theta}{\partial z} + \frac{1}{r}\frac{\partial u_z}{\partial \theta}\right).$$

(b) Consider the velocity vector in spatial coordinates $\mathbf{v} = \mathbf{v}(r, \theta, z)$. Applying the ∇ operator with respect to spatial coordinates, we obtain the velocity gradient components

$$[\mathbf{L}] = [\mathbf{v}\nabla] = \begin{bmatrix} \dfrac{\partial v_r}{\partial r} & \dfrac{1}{r}\dfrac{\partial v_r}{\partial \theta} - \dfrac{v_\theta}{r} & \dfrac{\partial v_r}{\partial z} \\ \dfrac{\partial v_\theta}{\partial r} & \dfrac{1}{r}\dfrac{\partial v_\theta}{\partial \theta} + \dfrac{v_r}{r} & \dfrac{\partial v_\theta}{\partial z} \\ \dfrac{\partial v_z}{\partial r} & \dfrac{1}{r}\dfrac{\partial v_z}{\partial \theta} & \dfrac{\partial v_z}{\partial z} \end{bmatrix}.$$

Since

$$\mathbf{D} = \mathbf{L} + \mathbf{L}^T, \quad \mathbf{W} = \mathbf{L} - \mathbf{L}^T,$$

we obtain for the rate of deformation and spin tensors

$$[\mathbf{D}] = \frac{1}{2}\begin{bmatrix} 2\dfrac{\partial v_r}{\partial r} & \dfrac{1}{r}\dfrac{\partial v_r}{\partial \theta} + \dfrac{\partial v_\theta}{\partial r} - \dfrac{v_\theta}{r} & \dfrac{\partial v_r}{\partial z} + \dfrac{\partial v_z}{\partial r} \\ \dfrac{1}{r}\dfrac{\partial v_r}{\partial \theta} + \dfrac{\partial v_\theta}{\partial r} - \dfrac{v_\theta}{r} & 2\left(\dfrac{1}{r}\dfrac{\partial v_\theta}{\partial \theta} + \dfrac{v_r}{r}\right) & \dfrac{\partial v_\theta}{\partial z} + \dfrac{1}{r}\dfrac{\partial v_z}{\partial \theta} \\ \dfrac{\partial v_r}{\partial z} + \dfrac{\partial v_z}{\partial r} & \dfrac{\partial v_\theta}{\partial z} + \dfrac{1}{r}\dfrac{\partial v_z}{\partial \theta} & 2\dfrac{\partial v_z}{\partial z} \end{bmatrix},$$

$$[\mathbf{W}] = \frac{1}{2}\begin{bmatrix} 0 & \dfrac{1}{r}\dfrac{\partial v_r}{\partial \theta} - \dfrac{\partial v_\theta}{\partial r} - \dfrac{v_\theta}{r} & \dfrac{\partial v_r}{\partial z} - \dfrac{\partial v_z}{\partial r} \\ -\dfrac{1}{r}\dfrac{\partial v_r}{\partial \theta} + \dfrac{\partial v_\theta}{\partial r} + \dfrac{v_\theta}{r} & 0 & \dfrac{\partial v_\theta}{\partial z} - \dfrac{1}{r}\dfrac{\partial v_z}{\partial \theta} \\ -\dfrac{\partial v_r}{\partial z} + \dfrac{\partial v_z}{\partial r} & -\dfrac{\partial v_\theta}{\partial z} + \dfrac{1}{r}\dfrac{\partial v_z}{\partial \theta} & 0 \end{bmatrix}.$$

(c) The time rate of the velocity vector

$$\mathbf{v} = v_r \mathbf{e}_r + v_\theta \mathbf{e}_\theta + v_z \mathbf{e}_z$$

is

$$\mathbf{a} = \dot{\mathbf{v}} = \dot{v}_r \mathbf{e}_r + \dot{v}_\theta \mathbf{e}_\theta + \dot{v}_z \mathbf{e}_z + v_r \dot{\mathbf{e}}_r + v_\theta \dot{\mathbf{e}}_\theta + v_z \dot{\mathbf{e}}_z.$$

Since the time rates of the unit vectors parallel to cylindrical coordinates are

$$\dot{\mathbf{e}}_r = \frac{v_\theta}{r} \mathbf{e}_\theta \quad \dot{\mathbf{e}}_\theta = -\frac{v_\theta}{r} \mathbf{e}_r, \quad \dot{\mathbf{e}}_z = \mathbf{0},$$

we obtain

$$\mathbf{a} = \dot{\mathbf{v}} = \left(\dot{v}_r - \frac{v_\theta^2}{r} \right) \mathbf{e}_r + \left(\dot{v}_\theta + \frac{v_r v_\theta}{r} \right) \mathbf{e}_\theta + \dot{v}_z \mathbf{e}_z.$$

This defines the physical components of the acceleration vector in cylindrical coordinates,

$$a_r = \dot{v}_r - \frac{v_\theta^2}{r}, \quad a_\theta = \dot{v}_\theta + \frac{v_r v_\theta}{r}, \quad a_z = \dot{v}_z.$$

The physical components of a vector in the case of orthogonal curvilinear coordinates are the components in a local rectangular Cartesian system with axes parallel to the coordinate curves.

Alternatively, the acceleration in spatial coordinates is

$$\mathbf{a} = \frac{\partial \mathbf{v}}{\partial t} + \mathbf{v} \cdot (\nabla \mathbf{v}).$$

Using the expression for $\nabla \mathbf{v}$ and performing its dot product with $\mathbf{v}$, we obtain

$$a_r = \frac{\partial v_r}{\partial t} + v_r \frac{\partial v_r}{\partial r} + \frac{v_\theta}{r} \frac{\partial v_r}{\partial \theta} + v_z \frac{\partial v_r}{\partial z} - \frac{1}{r} v_\theta^2,$$

$$a_\theta = \frac{\partial v_\theta}{\partial t} + v_r \frac{\partial v_\theta}{\partial r} + \frac{v_\theta}{r} \frac{\partial v_\theta}{\partial \theta} + v_z \frac{\partial v_\theta}{\partial z} + \frac{1}{r} v_r v_\theta,$$

$$a_z = \frac{\partial v_z}{\partial t} + v_r \frac{\partial v_z}{\partial r} + \frac{v_\theta}{r} \frac{\partial v_z}{\partial \theta} + v_z \frac{\partial v_z}{\partial z}.$$

Problem 4.16. Derive the components of the deformation gradient in cylindrical coordinates (Fig. 4-5).

Solution: The position of the material point in the initial undeformed configuration of the body is specified by the position vector, relative to the reference cylindrical coordinate basis, as

$$\mathbf{X} = R \mathbf{e}_R + Z \mathbf{e}_Z.$$

Its position in the deformed configuration, relative to the cylindrical basis in that configuration, is

$$\mathbf{x} = r \mathbf{e}_r + z \mathbf{e}_z,$$

with $\mathbf{e}_z = \mathbf{e}_Z$. The material line element in two configurations are, respectively,

$$d\mathbf{X} = dR \mathbf{e}_R + R d\mathbf{e}_R + dZ \mathbf{e}_Z = dR \mathbf{e}_R + R d\Theta \mathbf{e}_\Theta + dZ \mathbf{e}_Z,$$

$$d\mathbf{x} = dr \mathbf{e}_r + r d\mathbf{e}_r + dz \mathbf{e}_z = dr \mathbf{e}_r + r d\theta \mathbf{e}_\theta + dz \mathbf{e}_z.$$

Problems 4.16–4.17

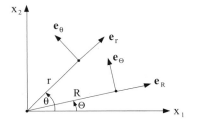

Figure 4-5. The unit vectors of a polar coordinate system in undeformed and deformed configuration.

Let the motion (deformation) be described by
$$r = r(R, \Theta, Z), \quad \theta = \theta(R, \Theta, Z), \quad z = z(R, \Theta, Z).$$

The physical components of the vector d**x** are then
$$dr = \frac{\partial r}{\partial R} dR + \frac{1}{R}\frac{\partial r}{\partial \Theta} R d\Theta + \frac{\partial r}{\partial Z} dZ,$$

$$r d\theta = r\frac{\partial \theta}{\partial R} dR + \frac{r}{R}\frac{\partial \theta}{\partial \Theta} R d\Theta + r\frac{\partial \theta}{\partial Z} dZ,$$

$$dz = \frac{\partial z}{\partial R} dR + \frac{1}{R}\frac{\partial z}{\partial \Theta} R d\Theta + \frac{\partial z}{\partial Z} dZ.$$

Consequently, by writing the deformation gradient (a two-point tensor) in terms of its physical components (relative to bases in undeformed and deformed configurations) as
$$\mathbf{F} = F_{iJ}\, \mathbf{e}_i\, \mathbf{e}_J, \quad (i = r, \theta, z; J = R, \Theta, Z),$$

we have
$$d\mathbf{x} = \mathbf{F} \cdot d\mathbf{X}, \quad dx_i = F_{iJ}\, dX_J.$$

When written in matrix form, this is
$$\begin{bmatrix} dr \\ r d\theta \\ dz \end{bmatrix} = \begin{bmatrix} \dfrac{\partial r}{\partial R} & \dfrac{1}{R}\dfrac{\partial r}{\partial \Theta} & \dfrac{\partial r}{\partial Z} \\ r\dfrac{\partial \theta}{\partial R} & \dfrac{r}{R}\dfrac{\partial \theta}{\partial \Theta} & r\dfrac{\partial \theta}{\partial Z} \\ \dfrac{\partial z}{\partial R} & \dfrac{1}{R}\dfrac{\partial z}{\partial \Theta} & \dfrac{\partial z}{\partial Z} \end{bmatrix} \cdot \begin{bmatrix} dR \\ R d\Theta \\ dZ \end{bmatrix}.$$

Problem 4.17. Evaluate the time rate of the deformation gradient $\dot{\mathbf{F}}$ in cylindrical coordinates, and use the result to obtain an expression for the tensor
$$\dot{F}_{iK} F^{-1}_{Kj}\, \mathbf{e}_i\, \mathbf{e}_j.$$

Solution: Since
$$\mathbf{F} = F_{iJ}\, \mathbf{e}_i\, \mathbf{e}_J, \quad (i = r, \theta, z; J = R, \Theta, Z),$$

the time differentiation gives
$$\dot{\mathbf{F}} = \dot{F}_{iJ}\, \mathbf{e}_i\, \mathbf{e}_J + F_{iJ}\, \dot{\mathbf{e}}_i\, \mathbf{e}_J.$$

The velocity gradient is then

$$\mathbf{L} = \dot{\mathbf{F}} \cdot \mathbf{F}^{-1} = \dot{F}_{iK}\,\mathbf{e}_i\,\mathbf{e}_K \cdot F^{-1}_{Mj}\,\mathbf{e}_M\,\mathbf{e}_j + F_{iK}\,\dot{\mathbf{e}}_i\,\mathbf{e}_K \cdot F^{-1}_{Mj}\,\mathbf{e}_M\,\mathbf{e}_j =$$
$$= \dot{F}_{iK}\,F^{-1}_{Kj}\,\mathbf{e}_i\,\mathbf{e}_j + \dot{\mathbf{e}}_i\,\mathbf{e}_i\,.$$

Since the time rates of the unit vectors parallel to cylindrical coordinates are

$$\dot{\mathbf{e}}_r = \dot{\theta}\,\mathbf{e}_\theta = \frac{v_\theta}{r}\,\mathbf{e}_\theta\,,\qquad \dot{\mathbf{e}}_\theta = -\dot{\theta}\,\mathbf{e}_r = -\frac{v_\theta}{r}\,\mathbf{e}_r\,,\qquad \dot{\mathbf{e}}_z = \mathbf{0}\,,$$

we have

$$\dot{\mathbf{e}}_i\,\mathbf{e}_i = \dot{\theta}(\mathbf{e}_\theta\,\mathbf{e}_r - \mathbf{e}_r\,\mathbf{e}_\theta) = \frac{v_\theta}{r}(\mathbf{e}_\theta\,\mathbf{e}_r - \mathbf{e}_r\,\mathbf{e}_\theta)\,.$$

Thus,

$$\dot{F}_{iK}\,F^{-1}_{Kj}\,\mathbf{e}_i\,\mathbf{e}_j = \mathbf{L} - \mathbf{\Omega} = \mathbf{D} + \hat{\mathbf{W}}\,,$$

where, expressed in the $\{\mathbf{e}_r,\mathbf{e}_\theta,\mathbf{e}_z\}$ basis,

$$[\mathbf{\Omega}] = \frac{v_\theta}{r}\begin{bmatrix} 0 & -1 & 0 \\ 1 & 0 & 0 \\ 0 & 0 & 0 \end{bmatrix}.$$

Consequently, by using the results for $[\mathbf{L}]$ from Problem 4.15,

$$\left[\dot{F}_{iK}\,F^{-1}_{Kj}\right] = [\mathbf{L} - \mathbf{\Omega}] = \begin{bmatrix} \dfrac{\partial v_r}{\partial r} & \dfrac{1}{r}\dfrac{\partial v_r}{\partial \theta} & \dfrac{\partial v_r}{\partial z} \\[6pt] \dfrac{\partial v_\theta}{\partial r} - \dfrac{v_\theta}{r} & \dfrac{1}{r}\dfrac{\partial v_\theta}{\partial \theta} + \dfrac{v_r}{r} & \dfrac{\partial v_\theta}{\partial z} \\[6pt] \dfrac{\partial v_z}{\partial r} & \dfrac{1}{r}\dfrac{\partial v_z}{\partial \theta} & \dfrac{\partial v_z}{\partial z} \end{bmatrix}.$$

Note that

$$\left[\dot{F}_{iK}\,F^{-1}_{Kj}\right]_{\text{sym}} = [\mathbf{D}] =$$

$$= \frac{1}{2}\begin{bmatrix} 2\dfrac{\partial v_r}{\partial r} & \dfrac{1}{r}\dfrac{\partial v_r}{\partial \theta} + \dfrac{\partial v_\theta}{\partial r} - \dfrac{v_\theta}{r} & \dfrac{\partial v_r}{\partial z} + \dfrac{\partial v_z}{\partial r} \\[6pt] \dfrac{1}{r}\dfrac{\partial v_r}{\partial \theta} + \dfrac{\partial v_\theta}{\partial r} - \dfrac{v_\theta}{r} & 2\left(\dfrac{1}{r}\dfrac{\partial v_\theta}{\partial \theta} + \dfrac{v_r}{r}\right) & \dfrac{\partial v_\theta}{\partial z} + \dfrac{1}{r}\dfrac{\partial v_z}{\partial \theta} \\[6pt] \dfrac{\partial v_r}{\partial z} + \dfrac{\partial v_z}{\partial r} & \dfrac{\partial v_\theta}{\partial z} + \dfrac{1}{r}\dfrac{\partial v_z}{\partial \theta} & 2\dfrac{\partial v_z}{\partial z} \end{bmatrix},$$

and

$$\left[\dot{F}_{iK}F_{Kj}^{-1}\right]_{\text{asym}} = [\hat{\mathbf{W}}] = [\mathbf{W}] - [\mathbf{\Omega}] =$$

$$= \frac{1}{2}\begin{bmatrix} 0 & \dfrac{1}{r}\dfrac{\partial v_r}{\partial \theta} - \dfrac{\partial v_\theta}{\partial r} + \dfrac{v_\theta}{r} & \dfrac{\partial v_r}{\partial z} - \dfrac{\partial v_z}{\partial r} \\ -\dfrac{1}{r}\dfrac{\partial v_r}{\partial \theta} + \dfrac{\partial v_\theta}{\partial r} - \dfrac{v_\theta}{r} & 0 & \dfrac{\partial v_\theta}{\partial z} - \dfrac{1}{r}\dfrac{\partial v_z}{\partial \theta} \\ -\dfrac{\partial v_r}{\partial z} + \dfrac{\partial v_z}{\partial r} & -\dfrac{\partial v_\theta}{\partial z} + \dfrac{1}{r}\dfrac{\partial v_z}{\partial \theta} & 0 \end{bmatrix}.$$

CHAPTER 5

Problem 5.1. Consider the stress tensor whose components are

$$\boldsymbol{\sigma} = \begin{bmatrix} \sigma & \sigma & \sigma \\ \sigma & \sigma & \sigma \\ \sigma & \sigma & \sigma \end{bmatrix}.$$

(a) Determine the principal values and associated principal directions of $\boldsymbol{\sigma}$.
(b) Determine the maximum shear stress.
(c) Determine the spherical and deviatoric parts of $\boldsymbol{\sigma}$.

Solution: (a) Begin with the equation

$$\boldsymbol{\sigma} \cdot \mathbf{n} = \lambda \mathbf{n}.$$

The resulting determinant equation for the eigenvalues is

$$\det(\boldsymbol{\sigma} - \lambda \mathbf{I}) = 0,$$

or

$$\begin{vmatrix} \sigma - \lambda & \sigma & \sigma \\ \sigma & \sigma - \lambda & \sigma \\ \sigma & \sigma & \sigma - \lambda \end{vmatrix} = 0.$$

After evaluating the determinant,

$$(\sigma - \lambda)^3 - 3\sigma^2(\sigma - \lambda) + 2\sigma^3 = 0,$$

or

$$\lambda^2(3\sigma - \lambda) = 0.$$

Thus the eigenvalues are $\lambda_1 = 3\sigma$, $\lambda_2 = \lambda_3 = 0$.

Next, for the eigenvector $\mathbf{n}^{(1)}$ associated with $\lambda_1 = 3\sigma$, we return to the eigenvalue equation and find

$$\sigma n_1^{(1)} + \sigma n_2^{(1)} + \sigma n_3^{(1)} = 3\sigma n_1^{(1)},$$
$$\sigma n_1^{(1)} + \sigma n_2^{(1)} + \sigma n_3^{(1)} = 3\sigma n_2^{(1)},$$
$$\sigma n_1^{(1)} + \sigma n_2^{(1)} + \sigma n_3^{(1)} = 3\sigma n_3^{(1)}.$$

Thus, $n_1^{(1)} = n_2^{(1)} = n_3^{(1)} = 1/\sqrt{3}$. Note that the equations for $\mathbf{n}^{(2)}$ and $\mathbf{n}^{(3)}$ do not produce unique values for them. Instead, they may be any two unit vectors in the plane normal to $\mathbf{n}^{(1)}$.

(b) The maximum shear stress is

$$\tau_{max} = \frac{\sigma_{max} - \sigma_{min}}{2} = \frac{3\sigma - 0}{2} = \frac{3\sigma}{2}.$$

(c) The spherical part of stress state is $\sigma_{ij}^{sph} = (\operatorname{tr} \boldsymbol{\sigma}/3)\delta_{ij} = \sigma\delta_{ij}$. Thus, the deviatoric part of stress tensor is

$$\boldsymbol{\sigma}' = \boldsymbol{\sigma} - \boldsymbol{\sigma}^{sph} = \begin{bmatrix} 0 & \sigma & \sigma \\ \sigma & 0 & \sigma \\ \sigma & \sigma & 0 \end{bmatrix}.$$

Problem 5.2. Derive the expressions for the stress gradients of the two invariants of deviatoric part of the Cauchy stress

$$J_2 = \frac{1}{2}\operatorname{tr}(\boldsymbol{\sigma}'^2), \quad J_3 = \frac{1}{3}\operatorname{tr}(\boldsymbol{\sigma}'^3).$$

Solution: The deviatoric part of the Cauchy stress has the rectangular components

$$\sigma'_{ij} = \sigma_{ij} - \frac{1}{3}\sigma_{kk}\delta_{ij}.$$

Since

$$\frac{\partial \sigma'_{mn}}{\partial \sigma_{pq}} = \frac{\partial}{\partial \sigma_{pq}}\left(\sigma_{mn} - \frac{1}{3}\sigma_{rr}\delta_{mn}\right) = \frac{1}{2}(\delta_{mp}\delta_{nq} + \delta_{mq}\delta_{np}) - \frac{1}{3}\delta_{mn}\delta_{pq},$$

there follows

$$\frac{\partial J_2}{\partial \sigma_{ij}} = \frac{\partial}{\partial \sigma_{ij}}\left(\frac{1}{2}\sigma'_{kl}\sigma'_{kl}\right) = \cdots = \sigma'_{ij},$$

$$\frac{\partial J_3}{\partial \sigma_{ij}} = \frac{\partial}{\partial \sigma_{ij}}\left(\frac{1}{3}\sigma'_{kl}\sigma'_{lq}\sigma'_{qk}\right) = \cdots = \sigma'_{ik}\sigma'_{kj} - \frac{2}{3}J_2\delta_{ij}.$$

Problem 5.3. If $\boldsymbol{\tau} = (\det \mathbf{F})\boldsymbol{\sigma}$ is the Kirchhoff stress, prove that

$$\nabla^0 \cdot (\mathbf{F}^{-1} \cdot \boldsymbol{\tau}) = (\det \mathbf{F})\nabla \cdot \boldsymbol{\sigma},$$

where ∇^0 and ∇ are the gradient operators with respect to material and spatial coordinates, respectively ($\nabla = \mathbf{F}^{-T} \cdot \nabla^0$).

Problems 5.3–5.4   675

Solution: (a) We first recall the identity proven in Problem 2.4b, rewritten here by using the gradient operator ∇^0 as

$$\nabla^0 \cdot (\mathbf{A} \cdot \mathbf{B}) = (\nabla^0 \cdot \mathbf{A}) \cdot \mathbf{B} + (\mathbf{A}^T \cdot \nabla^0) \cdot \mathbf{B}.$$

By choosing

$$\mathbf{A} = (\det \mathbf{F})\mathbf{F}^{-1}, \quad \mathbf{B} = \boldsymbol{\sigma},$$

it follows that

$$\nabla^0 \cdot [(\det \mathbf{F})\mathbf{F}^{-1} \cdot \boldsymbol{\sigma}] = \{\nabla^0 \cdot [(\det \mathbf{F})\mathbf{F}^{-1}]\} \cdot \boldsymbol{\sigma} + [(\det \mathbf{F})\mathbf{F}^{-T} \cdot \nabla^0] \cdot \boldsymbol{\sigma}.$$

The first term on the right-hand side is equal to zero by the result from Problem 4.14, and since $\mathbf{F}^{-T} \cdot \nabla^0 = \nabla$, we have the proof.

Problem 5.4. The deformed equilibrium configuration of the body is defined by the deformation mapping

$$x_1 = \frac{2}{3} X_1, \quad x_2 = X_2 - \frac{\sqrt{3}}{3} X_3, \quad x_3 = \sqrt{3}\, X_2 + \frac{1}{3} X_3.$$

The corresponding Cauchy stress within the body, relative to unit vectors $\mathbf{e}_1$, $\mathbf{e}_2$, and $\mathbf{e}_3$, is

$$\boldsymbol{\sigma} = 90 \begin{bmatrix} 0 & 0 & 0 \\ 0 & 1 & \sqrt{3} \\ 0 & \sqrt{3} & 3 \end{bmatrix} \text{ (MPa)}.$$

(a) Determine the tensors **C**, **E**, **B**, **e**, **U**, **V**, and **R**.
(b) Determine the principal directions of **U** and **V**.
(c) Determine the corresponding first and second Piola–Kirchhoff stress tensors.
(d) Determine the true, nominal, and pseudo traction vectors associated with the planes $\mathbf{m} = \{0, 1/2, \sqrt{3}/2\}$ and $\mathbf{n} = \{0, -\sqrt{3}/2, 1/2\}$ in the deformed state. Sketch the undeformed element twice – once under the first Piola–Kirchoff stress components, and once under the second Piola–Kirchhoff stress components;
(e) Determine the components of the rotated tensor $\boldsymbol{\sigma}^* = \mathbf{Q} \cdot \boldsymbol{\sigma} \cdot \mathbf{Q}^T$, where $\mathbf{Q} = \mathbf{R}^T$. Verify that these components are the same as the components of the tensor $\boldsymbol{\sigma}$ on the axes $\mathbf{e}_1$, **m**, and **n**. Sketch the element under these stress components.

Solution: (a) The deformation gradient and its transpose are

$$\mathbf{F} = \begin{bmatrix} \frac{2}{3} & 0 & 0 \\ 0 & 1 & -\frac{\sqrt{3}}{3} \\ 0 & \sqrt{3} & \frac{1}{3} \end{bmatrix}, \quad \mathbf{F}^T = \begin{bmatrix} \frac{2}{3} & 0 & 0 \\ 0 & 1 & \sqrt{3} \\ 0 & -\frac{\sqrt{3}}{3} & \frac{1}{3} \end{bmatrix}.$$

The left Cauchy–Green deformation tensor and the Lagrangian strain are

$$\mathbf{C} = \mathbf{F}^T \cdot \mathbf{F} = \begin{bmatrix} \frac{4}{9} & 0 & 0 \\ 0 & 4 & 0 \\ 0 & 0 & \frac{4}{9} \end{bmatrix}, \quad \mathbf{E} = \frac{1}{2}(\mathbf{C} - \mathbf{I}) = \begin{bmatrix} -\frac{5}{18} & 0 & 0 \\ 0 & \frac{3}{2} & 0 \\ 0 & 0 & -\frac{5}{18} \end{bmatrix}.$$

Since **C** is a diagonal matrix, we can calculate the right stretch tensor and its inverse immediately as

$$\mathbf{U} = \mathbf{C}^{1/2} = \begin{bmatrix} \frac{2}{3} & 0 & 0 \\ 0 & 2 & 0 \\ 0 & 0 & \frac{2}{3} \end{bmatrix}, \quad \mathbf{U}^{-1} = \begin{bmatrix} \frac{3}{2} & 0 & 0 \\ 0 & \frac{1}{2} & 0 \\ 0 & 0 & \frac{3}{2} \end{bmatrix}.$$

From the polar decomposition theorem $\mathbf{F} = \mathbf{R} \cdot \mathbf{U}$, it then follows that

$$\mathbf{R} = \mathbf{F} \cdot \mathbf{U}^{-1} = \begin{bmatrix} 1 & 0 & 0 \\ 0 & \frac{1}{2} & -\frac{\sqrt{3}}{2} \\ 0 & \frac{\sqrt{3}}{2} & \frac{1}{2} \end{bmatrix}, \quad \mathbf{V} = \mathbf{F} \cdot \mathbf{R}^T = \begin{bmatrix} \frac{2}{3} & 0 & 0 \\ 0 & 1 & \frac{\sqrt{3}}{3} \\ 0 & \frac{\sqrt{3}}{3} & \frac{5}{3} \end{bmatrix}.$$

The left Cauchy–Green deformation tensor can be calculated directly from its definition. Since it is not diagonal, its inverse can be calculated more conveniently from its relationship with **C** (which happened to be diagonal). Thus,

$$\mathbf{B} = \mathbf{F} \cdot \mathbf{F}^T = \mathbf{V}^2 = \begin{bmatrix} \frac{4}{9} & 0 & 0 \\ 0 & \frac{4}{3} & \frac{8\sqrt{3}}{9} \\ 0 & \frac{8\sqrt{3}}{9} & \frac{28}{9} \end{bmatrix},$$

$$\mathbf{B}^{-1} = \mathbf{R} \cdot \mathbf{C}^{-1} \cdot \mathbf{R}^T = \begin{bmatrix} \frac{9}{4} & 0 & 0 \\ 0 & \frac{7}{4} & -\frac{\sqrt{3}}{2} \\ 0 & -\frac{\sqrt{3}}{2} & \frac{3}{4} \end{bmatrix}.$$

The Eulerian strain is

$$\mathbf{e} = \frac{1}{2}(\mathbf{I} - \mathbf{B}^{-1}) = \begin{bmatrix} -\frac{5}{8} & 0 & 0 \\ 0 & -\frac{3}{8} & -\frac{\sqrt{3}}{4} \\ 0 & -\frac{\sqrt{3}}{4} & \frac{1}{2} \end{bmatrix}.$$

It can be easily verified that $\mathbf{E} = \mathbf{F}^T \cdot \mathbf{e} \cdot \mathbf{F}$.

(b) Since **U** is diagonal in the reference frame $\{\mathbf{e}_i\}$, we have

$$\mathbf{U} = \frac{2}{3}\mathbf{e}_1\,\mathbf{e}_1 + 2\mathbf{e}_2\,\mathbf{e}_2 + \frac{2}{3}\mathbf{e}_3\,\mathbf{e}_3.$$

Thus, its principal values and principal directions are

$$U_1 = \frac{2}{3}, \quad \mathbf{n}_1^U = \mathbf{e}_1; \quad U_2 = 2, \quad \mathbf{n}_2^U = \mathbf{e}_2; \quad U_3 = \frac{2}{3}, \quad \mathbf{n}_3^U = \mathbf{e}_3.$$

Actually, since $U_1 = U_3 = 2/3$, any direction in the plane normal to $\mathbf{n}_2^U$ is the principal direction of **U**, associated with the eigenvalue $2/3$ (Fig. 5-1).

The eigenvalues of **V** are equal to those of **U**, while the corresponding eigendirections are related by $\mathbf{n}_i^V = \mathbf{R} \cdot \mathbf{n}_i^U$. Thus, we obtain

$$V_1 = \frac{2}{3}, \quad \mathbf{n}_1^V = \mathbf{e}_1; \quad V_2 = 2, \quad \mathbf{n}_2^V = \left\{0, \frac{1}{2}, \frac{\sqrt{3}}{2}\right\};$$

$$V_3 = \frac{2}{3}, \quad \mathbf{n}_3^U = \left\{0, -\frac{\sqrt{3}}{2}, \frac{1}{2}\right\}.$$

Actually, since $V_1 = V_3 = 2/3$, any direction in the plane normal to $\mathbf{n}_2^V$ is the principal direction of **V**, associated with the eigenvalue $2/3$.

Problem 5.4

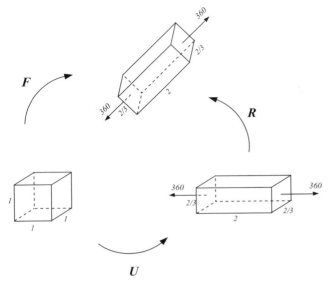

Figure 5-1. A unit cube in the undeformed configuration, imagined to be first streched along principal directions of **U** and then rotated by **R** to its final configuration in the deformed state.

(c) The first (nonsymmetric) Piola–Kirchhoff stress tensor is

$$\mathbf{P} = (\det \mathbf{F})\boldsymbol{\sigma} \cdot \mathbf{F}^{-T} = 80 \begin{bmatrix} 0 & 0 & 0 \\ 0 & 1 & 0 \\ 0 & \sqrt{3} & 0 \end{bmatrix} \text{ (MPa)}.$$

The inverse of the deformation gradient matrix was conveniently determined from

$$\mathbf{F}^{-1} = \mathbf{U}^{-1} \cdot \mathbf{R}^T = \begin{bmatrix} \frac{3}{2} & 0 & 0 \\ 0 & \frac{1}{4} & \frac{\sqrt{3}}{4} \\ 0 & -\frac{3\sqrt{3}}{4} & \frac{3}{4} \end{bmatrix},$$

and $\det \mathbf{F} = 8/9$. The second (symmetric) Piola–Kirchhoff stress tensor is

$$\mathbf{S} = \mathbf{F}^{-1} \cdot \mathbf{P} = 80 \begin{bmatrix} 0 & 0 & 0 \\ 0 & 1 & 0 \\ 0 & 0 & 0 \end{bmatrix} \text{ (MPa)}.$$

The undeformed material elements under these stress components are shown in Figs. 5-2b and 5-2c. Note, if $\hat{\mathbf{P}} = \mathbf{P}^T$ was used as a definition for the nominal stress, there would be shear stress $\hat{P}_{23} = \sqrt{3}$, whereas $\hat{P}_{32} = 0$. The symmetric Piola–Kirchhoff stress is unaffected by these two definitions of the nonsymmetric Piola–Kirchhoff stress, because

$$\mathbf{S} = \mathbf{F}^{-1} \cdot \mathbf{P} = \hat{\mathbf{P}} \cdot \mathbf{F}^{-T}.$$

(d) The true traction vectors on the planes with unit normal vectors **m** and **n** are

$$\mathbf{t}_m = \boldsymbol{\sigma} \cdot \mathbf{m} = 360 \begin{bmatrix} 0 \\ \frac{1}{2} \\ \frac{\sqrt{3}}{2} \end{bmatrix}, \quad \mathbf{t}_n = \boldsymbol{\sigma} \cdot \mathbf{n} = 360 \begin{bmatrix} 0 \\ 0 \\ 0 \end{bmatrix} \text{ (MPa)}.$$

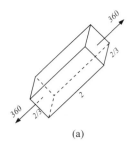

Figure 5-2. (a) Cauchy stress components on the deformed element with its sides parallel to $\mathbf{e}_1$, $\mathbf{m}$, and $\mathbf{n}$ axes. (b) Nonsymmetric Piola–Kirchhoff stress components on the unit cube in the undeformed configuration. (c) Symmetric Piola–Kirchhoff stress components.

The nominal traction $\mathbf{t}_m^0$ is

$$\mathbf{t}_m^0 = \mathbf{t}_m \frac{dS}{dS^0},$$

where

$$\frac{dS}{dS^0} = \frac{\det \mathbf{F}}{(\mathbf{m} \cdot \mathbf{B} \cdot \mathbf{m})^{1/2}}.$$

Since $\mathbf{m} \cdot \mathbf{B} \cdot \mathbf{m} = 4$ and $\det \mathbf{F} = 8/9$, we obtain $dS/dS^0 = 4/9$, and thus

$$\mathbf{t}_m^0 = 160 \begin{bmatrix} 0 \\ \frac{1}{2} \\ \frac{\sqrt{3}}{2} \end{bmatrix} \text{ (MPa)}.$$

The pseudotraction $\hat{\mathbf{t}}_m$ is defined such that

$$\mathbf{F} \cdot \hat{\mathbf{t}}_m \, dS^0 = \mathbf{t}_m \, dS.$$

This gives

$$\hat{\mathbf{t}}_m = \mathbf{F}^{-1} \cdot \mathbf{t}_m^0 = 80 \begin{bmatrix} 0 \\ 1 \\ 0 \end{bmatrix} \text{ (MPa)}.$$

The unit normal to the undeformed area dS^0, corresponding to deformed area dS with the unit normal $\mathbf{m}$, is

$$\mathbf{m}^0 = \frac{\mathbf{F}^T \cdot \mathbf{m}}{(\mathbf{m} \cdot \mathbf{B} \cdot \mathbf{m})^{1/2}} = \begin{bmatrix} 0 \\ 1 \\ 0 \end{bmatrix}.$$

Problems 5.4–5.5

The unit normal to the undeformed area dS^0, corresponding to deformed area dS with the unit normal $\mathbf{n}$, is

$$\mathbf{n}^0 = \frac{\mathbf{F}^T \cdot \mathbf{n}}{(\mathbf{n} \cdot \mathbf{B} \cdot \mathbf{n})^{1/2}} = \begin{bmatrix} 0 \\ 0 \\ 1 \end{bmatrix},$$

since $\mathbf{n} \cdot \mathbf{B} \cdot \mathbf{n} = 4/9$ (with the corresponding areas ratio $dS/dS^0 = 4/3$). Since $\mathbf{t}_n = \mathbf{0}$, the corresponding nominal and pseudo tractions also vanish, $\mathbf{t}_n^0 = \hat{\mathbf{t}}_n = \mathbf{0}$.

(e) The rotated stress tensor $\boldsymbol{\sigma}^*$ is

$$\boldsymbol{\sigma}^* = \mathbf{R}^T \cdot \boldsymbol{\sigma} \cdot \mathbf{R} = 360 \begin{bmatrix} 0 & 0 & 0 \\ 0 & 1 & 0 \\ 0 & 0 & 0 \end{bmatrix} \text{ (MPa)}.$$

The components of $\boldsymbol{\sigma}^*$ on the axes $\{\mathbf{e}_i\}$ are the same as the components of the stress tensor $\boldsymbol{\sigma}$ on the axes $\mathbf{e}_1$, $\mathbf{M}$ and $\mathbf{n}$. Indeed, because $\mathbf{t}_1 = \mathbf{t}_n = \mathbf{0}$ and $\mathbf{t}_m = 360\mathbf{m}$, we obtain

$$\boldsymbol{\sigma} = (\mathbf{e}_1 \mathbf{t}_1) + (\mathbf{m}\, \mathbf{t}_m) + (\mathbf{n}\, \mathbf{t}_n) = 360\,(\mathbf{m}\,\mathbf{m}) = 360 \begin{bmatrix} 0 & 0 & 0 \\ 0 & 1 & 0 \\ 0 & 0 & 0 \end{bmatrix}_{\mathbf{e}_1, \mathbf{m}, \mathbf{n}} \text{ (MPa)}.$$

The stress state is shown in Fig. 5-2a.

Problem 5.5. Consider the deformation process with the velocity gradient $\mathbf{L}$.

(a) Derive the expression for the Oldroyd convected rate of the Cauchy stress, observed in the coordinate frame $\{\mathbf{e}_i\}$ that is embedded in the material ($\dot{\mathbf{e}}_i = \mathbf{L} \cdot \mathbf{e}_i$), i.e., prove that

$$\overset{\diamond}{\boldsymbol{\sigma}} = \dot{\boldsymbol{\sigma}} - \mathbf{L} \cdot \boldsymbol{\sigma} - \boldsymbol{\sigma} \cdot \mathbf{L}^T.$$

(b) Prove that $\overset{\diamond}{\boldsymbol{\sigma}}$ is an objective stress rate.
(c) Derive the relationships between the Jaumann, Oldroyd, and Cotter–Rivlin stress rates.

Solution: (a) The stress tensor $\boldsymbol{\sigma}$ can be expressed in the basis $\{\mathbf{e}_i\}$ as

$$\boldsymbol{\sigma} = \sigma_{ij} \mathbf{e}_i \mathbf{e}_j.$$

If the basis $\{\mathbf{e}_i\}$ is convected with the material, then

$$\dot{\mathbf{e}}_i = \mathbf{L} \cdot \mathbf{e}_i = L_{ki} \mathbf{e}_k.$$

Thus,

$$\dot{\boldsymbol{\sigma}} = \dot{\sigma}_{ij} \mathbf{e}_i \mathbf{e}_j + \sigma_{ij} \dot{\mathbf{e}}_i \mathbf{e}_j + \sigma_{ij} \mathbf{e}_i \dot{\mathbf{e}}_j,$$

becomes

$$\dot{\boldsymbol{\sigma}} = \dot{\sigma}_{ij} \mathbf{e}_i \mathbf{e}_j + L_{ik} \sigma_{kj} \mathbf{e}_i \mathbf{e}_j + \sigma_{ik} L_{jk} \mathbf{e}_i \mathbf{e}_j,$$

i.e.,

$$\dot{\boldsymbol{\sigma}} = \dot{\sigma}_{ij} \mathbf{e}_i \mathbf{e}_j - \mathbf{L} \cdot \boldsymbol{\sigma} - \boldsymbol{\sigma} \cdot \mathbf{L}^T.$$

This leads to definition of corotational stress rate, observed in the frame that instantaneously convects with the material,

$$\dot{\sigma}_{ij} \mathbf{e}_i \mathbf{e}_j = \dot{\boldsymbol{\sigma}} - \mathbf{L} \cdot \boldsymbol{\sigma} - \boldsymbol{\sigma} \cdot \mathbf{L}^T.$$

The stress rate

$$\overset{\diamond}{\boldsymbol{\sigma}} = \dot{\boldsymbol{\sigma}} - \mathbf{L} \cdot \boldsymbol{\sigma} - \boldsymbol{\sigma} \cdot \mathbf{L}^T$$

is known as the Oldroyd convected rate of the Cauchy stress.

(b) We first recall that under a time dependent rigid-body rotation $\mathbf{Q} = \mathbf{Q}(t)$, the deformation gradient changes to $\mathbf{F}^* = \mathbf{Q} \cdot \mathbf{F}$, while the Cauchy stress and the velocity gradient change to

$$\boldsymbol{\sigma}^* = \mathbf{Q} \cdot \boldsymbol{\sigma} \cdot \mathbf{Q}^T, \quad \mathbf{L}^* = \dot{\mathbf{Q}} \cdot \mathbf{Q}^{-1} + \mathbf{Q} \cdot \mathbf{L} \cdot \mathbf{Q}^T.$$

By combining these, we find

$$\mathbf{L}^* \cdot \boldsymbol{\sigma}^* + \boldsymbol{\sigma}^* \cdot \mathbf{L}^{*T} = \mathbf{Q} \cdot (\mathbf{L} \cdot \boldsymbol{\sigma} + \boldsymbol{\sigma} \cdot \mathbf{L}^T + \mathbf{Q}^T \cdot \dot{\mathbf{Q}} \cdot \boldsymbol{\sigma} + \boldsymbol{\sigma} \cdot \dot{\mathbf{Q}}^T \cdot \mathbf{Q}) \cdot \mathbf{Q}^T.$$

On the other hand, by differentiating $\boldsymbol{\sigma}^* = \mathbf{Q} \cdot \boldsymbol{\sigma} \cdot \mathbf{Q}^T$, we obtain

$$\dot{\boldsymbol{\sigma}}^* = \mathbf{Q} \cdot (\dot{\boldsymbol{\sigma}} + \mathbf{Q}^T \cdot \dot{\mathbf{Q}} \cdot \boldsymbol{\sigma} + \boldsymbol{\sigma} \cdot \dot{\mathbf{Q}}^T \cdot \mathbf{Q}) \cdot \mathbf{Q}^T.$$

Consequently,

$$\overset{\diamond}{\boldsymbol{\sigma}}^* = \dot{\boldsymbol{\sigma}}^* - \mathbf{L}^* \cdot \boldsymbol{\sigma}^* - \boldsymbol{\sigma}^* \cdot \mathbf{L}^{*T}$$
$$= \mathbf{Q} \cdot (\dot{\boldsymbol{\sigma}} - \mathbf{L} \cdot \boldsymbol{\sigma} - \boldsymbol{\sigma} \cdot \mathbf{L}^T) \cdot \mathbf{Q}^T = \mathbf{Q} \cdot \overset{\diamond}{\boldsymbol{\sigma}} \cdot \mathbf{Q}^T,$$

which demonstrates that the Oldroyd convected rate $\overset{\diamond}{\boldsymbol{\sigma}}$ of the Cauchy stress is indeed objective.

(c) The Jaumann corotational stress rate is defined by

$$\overset{\triangledown}{\boldsymbol{\sigma}} = \dot{\boldsymbol{\sigma}} - \mathbf{W} \cdot \boldsymbol{\sigma} - \boldsymbol{\sigma} \cdot \mathbf{W}.$$

Since $\mathbf{L} = \mathbf{D} + \mathbf{W}$, we obtain

$$\overset{\diamond}{\boldsymbol{\sigma}} = \dot{\boldsymbol{\sigma}} - (\mathbf{D} + \mathbf{W}) \cdot \boldsymbol{\sigma} - \boldsymbol{\sigma} \cdot (\mathbf{D} - \mathbf{W}) = \overset{\triangledown}{\boldsymbol{\sigma}} - \mathbf{D} \cdot \boldsymbol{\sigma} - \boldsymbol{\sigma} \cdot \mathbf{D}.$$

Furthermore, since the two convected rates are

$$\overset{\diamond}{\boldsymbol{\sigma}} = \dot{\boldsymbol{\sigma}} - \mathbf{L} \cdot \boldsymbol{\sigma} - \boldsymbol{\sigma} \cdot \mathbf{L}^T, \quad \overset{\triangledown}{\boldsymbol{\sigma}} = \dot{\boldsymbol{\sigma}} + \mathbf{L}^T \cdot \boldsymbol{\sigma} + \boldsymbol{\sigma} \cdot \mathbf{L},$$

we obtain

$$\overset{\diamond}{\boldsymbol{\sigma}} = \overset{\triangledown}{\boldsymbol{\sigma}} + 2(\mathbf{D} \cdot \boldsymbol{\sigma} + \boldsymbol{\sigma} \cdot \mathbf{D}),$$

$$\overset{\triangledown}{\boldsymbol{\sigma}} = \overset{\triangledown}{\boldsymbol{\sigma}} + \mathbf{D} \cdot \boldsymbol{\sigma} + \boldsymbol{\sigma} \cdot \mathbf{D}.$$

Problem 5.6. The Jaumann and the Cotter–Rivlin convected rate of the second-order tensor $\mathbf{A}$ are defined by

$$\overset{\triangledown}{\mathbf{A}} = \dot{\mathbf{A}} - \mathbf{W} \cdot \mathbf{A} + \mathbf{A} \cdot \mathbf{W}, \quad \overset{\triangle}{\mathbf{A}} = \dot{\mathbf{A}} + \mathbf{L}^T \cdot \mathbf{A} + \mathbf{A} \cdot \mathbf{L},$$

Problems 5.6–5.7

where $\mathbf{L}$ is the velocity gradient and $\mathbf{W}$ is its antisymmetric part ($\mathbf{L} = \mathbf{D} + \mathbf{W}$).

(a) Show that for the rate of deformation $\mathbf{D}$,

$$\overset{\triangle}{\mathbf{D}} = \overset{\triangledown}{\mathbf{D}} + 2\mathbf{D}^2,$$

$$\overset{\triangle\triangle}{\mathbf{D}} = \overset{\triangledown\triangledown}{\mathbf{D}} + 3(\overset{\triangledown}{\mathbf{D}} \cdot \mathbf{D} + \mathbf{D} \cdot \overset{\triangledown}{\mathbf{D}}) + 4\mathbf{D}^3.$$

(b) Show that for the Cauchy stress tensor $\boldsymbol{\sigma}$,

$$\overset{\triangle}{\boldsymbol{\sigma}} = \overset{\triangledown}{\boldsymbol{\sigma}} + \boldsymbol{\sigma} \cdot \mathbf{D} + \mathbf{D} \cdot \boldsymbol{\sigma},$$

$$\overset{\triangle\triangle}{\boldsymbol{\sigma}} = \overset{\triangledown\triangledown}{\boldsymbol{\sigma}} + 2(\overset{\triangledown}{\boldsymbol{\sigma}} \cdot \mathbf{D} + \mathbf{D} \cdot \overset{\triangledown}{\boldsymbol{\sigma}}) + (\boldsymbol{\sigma} \cdot \overset{\triangledown}{\mathbf{D}} + \overset{\triangledown}{\mathbf{D}} \cdot \boldsymbol{\sigma})$$
$$+ 2\mathbf{D} \cdot \boldsymbol{\sigma} \cdot \mathbf{D} + \mathbf{D}^2 \cdot \boldsymbol{\sigma} + \boldsymbol{\sigma} \cdot \mathbf{D}^2.$$

Solution: (a) Substitute $\mathbf{L} = \mathbf{D} + \mathbf{W}$ into

$$\overset{\triangle}{\mathbf{D}} = \dot{\mathbf{D}} + \mathbf{L}^T \cdot \mathbf{D} + \mathbf{D} \cdot \mathbf{L},$$

and use

$$\overset{\triangledown}{\mathbf{D}} = \dot{\mathbf{D}} - \mathbf{W} \cdot \mathbf{D} + \mathbf{D} \cdot \mathbf{W}.$$

It follows that

$$\overset{\triangle}{\mathbf{D}} = \overset{\triangledown}{\mathbf{D}} + 2\mathbf{D}^2.$$

Similarly, by starting from

$$\overset{\triangle\triangle}{\mathbf{D}} = \overset{\dot{\triangle}}{\mathbf{D}} + \mathbf{L}^T \cdot \overset{\triangle}{\mathbf{D}} + \overset{\triangle}{\mathbf{D}} \cdot \mathbf{L},$$

and using

$$\overset{\triangledown\triangledown}{\mathbf{D}} = \overset{\dot{\triangledown}}{\mathbf{D}} - \mathbf{W} \cdot \overset{\triangledown}{\mathbf{D}} + \overset{\triangledown}{\mathbf{D}} \cdot \mathbf{W},$$

it follows that

$$\overset{\triangle\triangle}{\mathbf{D}} = \overset{\triangledown\triangledown}{\mathbf{D}} + 3(\overset{\triangledown}{\mathbf{D}} \cdot \mathbf{D} + \mathbf{D} \cdot \overset{\triangledown}{\mathbf{D}}) + 4\mathbf{D}^3.$$

(b) Use

$$\overset{\triangledown}{\boldsymbol{\sigma}} = \dot{\boldsymbol{\sigma}} + \mathbf{L}^T \cdot \boldsymbol{\sigma} + \boldsymbol{\sigma} \cdot \mathbf{L},$$

$$\overset{\triangle\triangle}{\boldsymbol{\sigma}} = \overset{\dot{\triangledown}}{\boldsymbol{\sigma}} + \mathbf{L}^T \cdot \overset{\triangledown}{\boldsymbol{\sigma}} + \overset{\triangledown}{\boldsymbol{\sigma}} \cdot \mathbf{L},$$

and follow the procedure from part (a).

Problem 5.7. Prove that

$$\boldsymbol{\sigma}' : \overset{\triangledown}{\boldsymbol{\sigma}} = \boldsymbol{\sigma}' : \dot{\boldsymbol{\sigma}}.$$

Solution: Since
$$\overset{\triangledown}{\sigma}_{ij} = \dot{\sigma}_{ij} - W_{ik}\sigma_{kj} + \sigma_{ik}W_{kj},$$

we have
$$\sigma'_{ij}\overset{\triangledown}{\sigma}_{ij} = \sigma'_{ij}\dot{\sigma}_{ij} - \sigma'_{ij}W_{ik}\sigma_{kj} + \sigma'_{ij}\sigma_{ik}W_{kj}.$$

But, in view of the symmetry of σ'_{ij},
$$\sigma'_{ij}\sigma_{ik}W_{kj} - \sigma'_{ij}W_{ik}\sigma_{kj} = \sigma'_{ij}\sigma_{ik}W_{kj} - \sigma'_{ji}W_{ik}\sigma_{kj}$$
$$= \sigma'_{ij}\sigma_{ik}W_{kj} - \sigma'_{ij}W_{jk}\sigma_{ki}$$
$$= 2\sigma'_{ij}\sigma_{ik}W_{kj} = 0,$$

because $\sigma'_{ij}\sigma_{ik} = \sigma'_{ij}\sigma'_{ik}$ is symmetric, whereas W_{kj} is antisymmetric in jk.

The proof also follows directly by observing that $J = \sigma' : \sigma'$ is a scalar invariant of σ', so that $\overset{\triangledown}{J} = \dot{J}$.

Problem 5.8. Consider cylindrical coordinates (r, θ, z). The ∇ operator in the cylindrical coordinates (see Problem 2.6) is
$$\nabla = \frac{\partial}{\partial r}\mathbf{e}_r + \frac{1}{r}\frac{\partial}{\partial \theta}\mathbf{e}_\theta + \frac{\partial}{\partial z}\mathbf{e}_z.$$

(a) Write $\nabla \cdot \sigma$ in an expanded form.
(b) Write three scalar equations of motion in cylindrical coordinates.

Solution: (a) Decomposing the stress tensor on the components along the unit vectors of the cylindrical coordinate system, we have
$$\sigma = \sigma_{rr}\mathbf{e}_r\mathbf{e}_r + \sigma_{\theta\theta}\mathbf{e}_\theta\mathbf{e}_\theta + \sigma_{zz}\mathbf{e}_z\mathbf{e}_z$$
$$+ \sigma_{r\theta}\mathbf{e}_r\mathbf{e}_\theta + \sigma_{\theta r}\mathbf{e}_\theta\mathbf{e}_r + \sigma_{\theta z}\mathbf{e}_\theta\mathbf{e}_z$$
$$+ \sigma_{z\theta}\mathbf{e}_z\mathbf{e}_\theta + \sigma_{zr}\mathbf{e}_z\mathbf{e}_r + \sigma_{rz}\mathbf{e}_r\mathbf{e}_z.$$

Recalling from Problem 4.15 that
$$\frac{\partial \mathbf{e}_r}{\partial \theta} = \mathbf{e}_\theta, \quad \frac{\partial \mathbf{e}_\theta}{\partial \theta} = -\mathbf{e}_r,$$

and applying the divergence operator to σ, there follows
$$\nabla \cdot \sigma = \left[\frac{1}{r}\frac{\partial(r\sigma_{rr})}{\partial r} + \frac{1}{r}\frac{\partial \sigma_{r\theta}}{\partial \theta} + \frac{\partial \sigma_{rz}}{\partial z} - \frac{1}{r}\sigma_{\theta\theta}\right]\mathbf{e}_r$$
$$+ \left[\frac{1}{r}\frac{\partial(r\sigma_{\theta r})}{\partial r} + \frac{1}{r}\frac{\partial \sigma_{\theta\theta}}{\partial \theta} + \frac{\partial \sigma_{\theta z}}{\partial z} + \frac{1}{r}\sigma_{\theta r}\right]\mathbf{e}_\theta$$
$$+ \left[\frac{1}{r}\frac{\partial(r\sigma_{zr})}{\partial r} + \frac{1}{r}\frac{\partial \sigma_{z\theta}}{\partial \theta} + \frac{\partial \sigma_{zz}}{\partial z}\right]\mathbf{e}_z.$$

(b) Equations of motion are
$$\nabla \cdot \sigma + \mathbf{b} = \rho \mathbf{a},$$

Problems 5.8–5.10

where **b** is the body force per unit volume. By using the derived expression for $\nabla \cdot \boldsymbol{\sigma}$, the three scalar equations of motion in cylindrical coordinates are

$$\frac{\partial \sigma_{rr}}{\partial r} + \frac{1}{r}\frac{\partial \sigma_{r\theta}}{\partial \theta} + \frac{\partial \sigma_{rz}}{\partial z} + \frac{1}{r}(\sigma_{rr} - \sigma_{\theta\theta}) + b_r = \rho a_r,$$

$$\frac{\partial \sigma_{\theta r}}{\partial r} + \frac{1}{r}\frac{\partial \sigma_{\theta\theta}}{\partial \theta} + \frac{\partial \sigma_{\theta z}}{\partial z} + \frac{2}{r}\sigma_{\theta r} + b_\theta = \rho a_\theta,$$

$$\frac{\partial \sigma_{zr}}{\partial r} + \frac{1}{r}\frac{\partial \sigma_{z\theta}}{\partial \theta} + \frac{\partial \sigma_{zz}}{\partial z} + \frac{1}{r}\sigma_{zr} + b_z = \rho a_z.$$

The cylindrical components of the acceleration vector **a** are listed in Problem 4.15.

Problem 5.9. Consider the Cauchy stress tensor $\boldsymbol{\sigma}$ in cylindrical coordinates. If $\{\mathbf{e}_r, \mathbf{e}_\theta, \mathbf{e}_z\}$ is a local basis with respect to which the stress tensor is expressed, derive the expressions for $\dot{\boldsymbol{\sigma}}$.

Solution: By differentiating the stress tensor

$$\boldsymbol{\sigma} = \sigma_{rr}\mathbf{e}_r\mathbf{e}_r + \sigma_{\theta\theta}\mathbf{e}_\theta\mathbf{e}_\theta + \sigma_{zz}\mathbf{e}_z\mathbf{e}_z$$
$$+ \sigma_{r\theta}\mathbf{e}_r\mathbf{e}_\theta + \sigma_{\theta r}\mathbf{e}_\theta\mathbf{e}_r + \sigma_{\theta z}\mathbf{e}_\theta\mathbf{e}_z$$
$$+ \sigma_{z\theta}\mathbf{e}_z\mathbf{e}_\theta + \sigma_{zr}\mathbf{e}_z\mathbf{e}_r + \sigma_{rz}\mathbf{e}_r\mathbf{e}_z,$$

having in mind that

$$\dot{\mathbf{e}}_r = \frac{v_\theta}{r}\mathbf{e}_\theta \quad \dot{\mathbf{e}}_\theta = -\frac{v_\theta}{r}\mathbf{e}_r, \quad \dot{\mathbf{e}}_z = \mathbf{0},$$

we obtain

$$\dot{\boldsymbol{\sigma}} = \left(\dot{\sigma}_{rr} - 2\sigma_{r\theta}\frac{v_\theta}{r}\right)\mathbf{e}_r\mathbf{e}_r + \left(\dot{\sigma}_{\theta\theta} + 2\sigma_{r\theta}\frac{v_\theta}{r}\right)\mathbf{e}_\theta\mathbf{e}_\theta + \dot{\sigma}_{zz}\mathbf{e}_z\mathbf{e}_z$$
$$+ \left[\dot{\sigma}_{r\theta} + (\sigma_{rr} - \sigma_{\theta\theta})\frac{v_\theta}{r}\right](\mathbf{e}_r\mathbf{e}_\theta + \mathbf{e}_\theta\mathbf{e}_r)$$
$$+ \left(\dot{\sigma}_{rz} - \sigma_{\theta z}\frac{v_\theta}{r}\right)(\mathbf{e}_r\mathbf{e}_z + \mathbf{e}_z\mathbf{e}_r)$$
$$+ \left(\dot{\sigma}_{\theta z} + \sigma_{rz}\frac{v_\theta}{r}\right)(\mathbf{e}_\theta\mathbf{e}_z + \mathbf{e}_z\mathbf{e}_\theta).$$

The right-hand side can be conveniently rewritten in the matrix form, relative to the instantaneous basis $\{\mathbf{e}_r, \mathbf{e}_\theta, \mathbf{e}_z\}$, as

$$\dot{\boldsymbol{\sigma}} = \begin{bmatrix} \dot{\sigma}_{rr} & \dot{\sigma}_{r\theta} & \dot{\sigma}_{rz} \\ \dot{\sigma}_{\theta r} & \dot{\sigma}_{\theta\theta} & \dot{\sigma}_{\theta z} \\ \dot{\sigma}_{zr} & \dot{\sigma}_{z\theta} & \dot{\sigma}_{zz} \end{bmatrix}_{\mathbf{e}\,\text{basis}} + [\Omega]\cdot[\sigma] - [\sigma]\cdot[\Omega],$$

where

$$[\sigma] = \begin{bmatrix} \sigma_{rr} & \sigma_{r\theta} & \sigma_{rz} \\ \sigma_{\theta r} & \sigma_{\theta\theta} & \sigma_{\theta z} \\ \sigma_{zr} & \sigma_{z\theta} & \sigma_{zz} \end{bmatrix}, \quad [\Omega] = \begin{bmatrix} 0 & -v_\theta/r & 0 \\ v_\theta/r & 0 & 0 \\ 0 & 0 & 0 \end{bmatrix}.$$

Problem 5.10. Derive the expressions for the Jaumann rate of the cylindrical stress components σ_{zz} and $\sigma_{z\theta}$.

Solution: The material spin is the antisymmetric part of the velocity gradient $\mathbf{L} = \mathbf{v}\nabla$, i.e.,

$$\mathbf{W} = \frac{1}{2}(\mathbf{v}\nabla - \nabla\mathbf{v}).$$

When expressed in the instantaneous basis $\{\mathbf{e}_r, \mathbf{e}_\theta, \mathbf{e}_z\}$, we have (see Problem 4.15)

$$[\nabla \mathbf{v}] = \begin{bmatrix} \dfrac{\partial v_r}{\partial r} & \dfrac{\partial v_\theta}{\partial r} & \dfrac{\partial v_z}{\partial r} \\[6pt] \dfrac{1}{r}\dfrac{\partial v_r}{\partial \theta} - \dfrac{v_\theta}{r} & \dfrac{1}{r}\dfrac{\partial v_\theta}{\partial \theta} + \dfrac{v_r}{r} & \dfrac{1}{r}\dfrac{\partial v_z}{\partial \theta} \\[6pt] \dfrac{\partial v_r}{\partial z} & \dfrac{\partial v_\theta}{\partial z} & \dfrac{\partial v_z}{\partial z} \end{bmatrix}.$$

Thus,

$$[\mathbf{W}] = \frac{1}{2}\begin{bmatrix} 0 & \dfrac{1}{r}\dfrac{\partial v_r}{\partial \theta} - \dfrac{\partial v_\theta}{\partial r} - \dfrac{v_\theta}{r} & \dfrac{\partial v_r}{\partial z} - \dfrac{\partial v_z}{\partial r} \\[6pt] -\dfrac{1}{r}\dfrac{\partial v_r}{\partial \theta} + \dfrac{\partial v_\theta}{\partial r} + \dfrac{v_\theta}{r} & 0 & \dfrac{\partial v_\theta}{\partial z} - \dfrac{1}{r}\dfrac{\partial v_z}{\partial \theta} \\[6pt] -\dfrac{\partial v_r}{\partial z} + \dfrac{\partial v_z}{\partial r} & -\dfrac{\partial v_\theta}{\partial z} + \dfrac{1}{r}\dfrac{\partial v_z}{\partial \theta} & 0 \end{bmatrix}.$$

The Jaumann rate of $\boldsymbol{\sigma}$ is

$$\overset{\triangledown}{\boldsymbol{\sigma}} = \dot{\boldsymbol{\sigma}} - \mathbf{W}\cdot\boldsymbol{\sigma} + \boldsymbol{\sigma}\cdot\mathbf{W}.$$

By using the expression for $\dot{\boldsymbol{\sigma}}$ from Problem 5.9, there follows

$$\begin{bmatrix} \overset{\triangledown}{\sigma}_{rr} & \overset{\triangledown}{\sigma}_{r\theta} & \overset{\triangledown}{\sigma}_{rz} \\ \overset{\triangledown}{\sigma}_{\theta r} & \overset{\triangledown}{\sigma}_{\theta\theta} & \overset{\triangledown}{\sigma}_{\theta z} \\ \overset{\triangledown}{\sigma}_{zr} & \overset{\triangledown}{\sigma}_{z\theta} & \overset{\triangledown}{\sigma}_{zz} \end{bmatrix} = \begin{bmatrix} \dot{\sigma}_{rr} & \dot{\sigma}_{r\theta} & \dot{\sigma}_{rz} \\ \dot{\sigma}_{\theta r} & \dot{\sigma}_{\theta\theta} & \dot{\sigma}_{\theta z} \\ \dot{\sigma}_{zr} & \dot{\sigma}_{z\theta} & \dot{\sigma}_{zz} \end{bmatrix} - [\hat{\mathbf{W}}]\cdot[\boldsymbol{\sigma}] + [\boldsymbol{\sigma}]\cdot[\hat{\mathbf{W}}].$$

The components of the spin matrix

$$[\hat{\mathbf{W}}] = [\mathbf{W}] - [\boldsymbol{\Omega}]$$

are

$$[\hat{\mathbf{W}}] = \frac{1}{2}\begin{bmatrix} 0 & \dfrac{1}{r}\dfrac{\partial v_r}{\partial \theta} - \dfrac{\partial v_\theta}{\partial r} + \dfrac{v_\theta}{r} & \dfrac{\partial v_r}{\partial z} - \dfrac{\partial v_z}{\partial r} \\[6pt] -\dfrac{1}{r}\dfrac{\partial v_r}{\partial \theta} + \dfrac{\partial v_\theta}{\partial r} - \dfrac{v_\theta}{r} & 0 & \dfrac{\partial v_\theta}{\partial z} - \dfrac{1}{r}\dfrac{\partial v_z}{\partial \theta} \\[6pt] -\dfrac{\partial v_r}{\partial z} + \dfrac{\partial v_z}{\partial r} & -\dfrac{\partial v_\theta}{\partial z} + \dfrac{1}{r}\dfrac{\partial v_z}{\partial \theta} & 0 \end{bmatrix}.$$

To kinematically interpret the Jaumann stress rate components in cylindrical coordinates, suppose that we want to calculate the rate of stress in a basis $\{\hat{\mathbf{e}}_r, \hat{\mathbf{e}}_\theta, \hat{\mathbf{e}}_z\}$ that momentarily

Problem 5.10

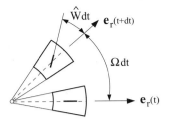

Figure 5-3. A schematic representation of the spin $\hat{\mathbf{W}}$. A material line element (parallel to a principal direction of $\mathbf{D}$) is rotated by $\hat{\mathbf{W}} = \mathbf{W} - \boldsymbol{\Omega}$ relative to the rotated $\mathbf{e}_r$ coordinate direction.

coincides with the basis $\{\mathbf{e}_r, \mathbf{e}_\theta, \mathbf{e}_z\}$, but rotates instantaneously with the material spin $\mathbf{W}$ (Fig. 5-3). The time rate of the unit vectors of such basis are

$$\dot{\hat{\mathbf{e}}}_r = \mathbf{W} \cdot \mathbf{e}_r, \quad \dot{\hat{\mathbf{e}}}_\theta = \mathbf{W} \cdot \mathbf{e}_\theta, \quad \dot{\hat{\mathbf{e}}}_z = \mathbf{W} \cdot \mathbf{e}_z.$$

It readily follows that

$$\dot{\boldsymbol{\sigma}} = \begin{bmatrix} \dot{\sigma}_{rr} & \dot{\sigma}_{r\theta} & \dot{\sigma}_{rz} \\ \dot{\sigma}_{\theta r} & \dot{\sigma}_{\theta\theta} & \dot{\sigma}_{\theta z} \\ \dot{\sigma}_{zr} & \dot{\sigma}_{z\theta} & \dot{\sigma}_{zz} \end{bmatrix}_{\hat{\mathbf{e}}\text{ basis}} + [\mathbf{W}] \cdot [\boldsymbol{\sigma}] - [\boldsymbol{\sigma}] \cdot [\mathbf{W}].$$

If this is equated to the corresponding expression for $\dot{\boldsymbol{\sigma}}$ from the Problem 5.9, there follows

$$\begin{bmatrix} \dot{\sigma}_{rr} & \dot{\sigma}_{r\theta} & \dot{\sigma}_{rz} \\ \dot{\sigma}_{\theta r} & \dot{\sigma}_{\theta\theta} & \dot{\sigma}_{\theta z} \\ \dot{\sigma}_{zr} & \dot{\sigma}_{z\theta} & \dot{\sigma}_{zz} \end{bmatrix}_{\hat{\mathbf{e}}\text{ basis}} = \begin{bmatrix} \dot{\sigma}_{rr} & \dot{\sigma}_{r\theta} & \dot{\sigma}_{rz} \\ \dot{\sigma}_{\theta r} & \dot{\sigma}_{\theta\theta} & \dot{\sigma}_{\theta z} \\ \dot{\sigma}_{zr} & \dot{\sigma}_{z\theta} & \dot{\sigma}_{zz} \end{bmatrix}_{\mathbf{e}\text{ basis}} - [\hat{\mathbf{W}}] \cdot [\boldsymbol{\sigma}] + [\boldsymbol{\sigma}] \cdot [\hat{\mathbf{W}}].$$

These are the components of the Jaumann rate of stress in cylindrical coordinates. For example,

$$\overset{\triangledown}{\sigma}_{zz} = \dot{\sigma}_{zz} - \hat{W}_{zk}\sigma_{kz} + \sigma_{zk}\hat{W}_{kz},$$

with sum on k over r, θ, z. This gives

$$\overset{\triangledown}{\sigma}_{zz} = \dot{\sigma}_{zz} + 2\sigma_{rz}\hat{W}_{rz} - 2\sigma_{z\theta}\hat{W}_{z\theta},$$

i.e.,

$$\overset{\triangledown}{\sigma}_{zz} = \dot{\sigma}_{zz} + \sigma_{rz}\left(\frac{\partial v_r}{\partial z} - \frac{\partial v_z}{\partial r}\right) + \sigma_{z\theta}\left(\frac{\partial v_\theta}{\partial z} - \frac{1}{r}\frac{\partial v_z}{\partial \theta}\right).$$

Similarly,

$$\overset{\triangledown}{\sigma}_{z\theta} = \dot{\sigma}_{z\theta} - \hat{W}_{zk}\sigma_{k\theta} + \sigma_{zk}\hat{W}_{k\theta},$$

which gives

$$\overset{\triangledown}{\sigma}_{z\theta} = \dot{\sigma}_{z\theta} + \sigma_{r\theta}\hat{W}_{rz} - \sigma_{zr}\hat{W}_{\theta r} - (\sigma_{\theta\theta} - \sigma_{zz})\hat{W}_{z\theta},$$

i.e.,

$$\overset{\triangledown}{\sigma}_{z\theta} = \dot{\sigma}_{z\theta} + \frac{1}{2}\sigma_{r\theta}\left(\frac{\partial v_r}{\partial z} - \frac{\partial v_z}{\partial r}\right) + \frac{1}{2}\sigma_{zr}\left(\frac{1}{r}\frac{\partial v_r}{\partial \theta} - \frac{\partial v_\theta}{\partial r}\right)$$
$$+ \frac{1}{2}(\sigma_{\theta\theta} - \sigma_{zz})\left(\frac{\partial v_\theta}{\partial z} - \frac{1}{r}\frac{\partial v_z}{\partial \theta}\right).$$

CHAPTER 6

Problem 6.1. The equation of state for n moles of an ideal gas is

$$pV = nRT,$$

where $R = 8.314$ J/K is the universal gas constant, p is the pressure, V is the volume, and T is the temperature. Assume that the specific heat at constant volume c_v is constant.

(a) Derive the expression for the corresponding Helmholtz free energy $\phi = \phi(V, T)$, and the entropy $s = s(V, T)$. Assume that $\phi(V_0, T_0) = -T_0 s_0$, so that internal energy u vanishes in the reference state.

(b) Use the obtained results for $\phi(V, T)$ and $s(V, T)$, and the relationship $\phi = u - TS$, to derive an expression for the internal energy $u = u(T)$. Confirm the result by an independent derivation starting from the energy equation.

(c) Show that

$$c_p - c_v = nR.$$

(d) Derive the expression for the enthalpy by using the connection $h = u + pV$. Confirm the result by starting from the differential expression

$$dh = V dp + T ds.$$

(e) Show that

$$\frac{dT}{T} = \frac{dp}{p} + \frac{dV}{V}.$$

(f) Derive the expressions for the coefficient of thermal expansion α and the compressibility coefficient β of an ideal gas.

Solution: (a) We start from

$$p = \frac{nRT}{V} = -\frac{\partial \phi}{\partial V},$$

where ϕ is the free energy within n moles. The integration gives

$$\phi = -nRT \ln \frac{V}{V_0} + \tilde{\phi}(T),$$

where $\tilde{\phi}(T)$ is the integration function. The corresponding entropy of n moles is

$$s = -\frac{\partial \phi}{\partial T} = nR \ln \frac{V}{V_0} - \frac{d\tilde{\phi}}{dT}.$$

The specific heat for n moles is

$$c_v = T \frac{\partial s}{\partial T} = -T \frac{d^2 \tilde{\phi}}{dT^2},$$

which is assumed to be a given constant c_v. Upon integration, we find

$$\frac{d^2 \tilde{\phi}}{dT^2} = -\frac{c_v}{T} \quad \Rightarrow \quad \frac{d\tilde{\phi}}{dT} = -c_v \ln \frac{T}{T_0} - s_0,$$

Problem 6.1

i.e.,
$$\tilde{\phi}(T) = -c_v T \ln \frac{T}{T_0} + c_v(T - T_0) - s_0 T.$$

Consequently,
$$\phi = -nRT \ln \frac{V}{V_0} + c_v \left(T - T_0 - T \ln \frac{T}{T_0}\right) - s_0 T,$$

$$s = nR \ln \frac{V}{V_0} + c_v \ln \frac{T}{T_0} + s_0.$$

(b) By substituting the above expressions for ϕ and s into $u = \phi + Ts$, we readily find that
$$u = c_v(T - T_0).$$

Remarkably, this is a function of temperature alone. An independent derivation of this result begins with the energy equation
$$du = -pdV + Tds.$$

From the entropy expression from part (a), we have
$$ds = nR\frac{dV}{V} + c_v \frac{dT}{T},$$

i.e.,
$$Tds = nRT\frac{dV}{V} + c_v dT = pV\frac{dV}{V} + c_v dT.$$

By substituting this into the energy equation, we obtain
$$du = c_v dT \quad \Rightarrow \quad u = c_v(T - T_0).$$

(c) Since $V = nRT/p$, the entropy expression can be written as
$$s = nR \ln \frac{V}{V_0} + c_v \ln \frac{T}{T_0} + s_0 = nR \ln \frac{nRT}{pV_0} + c_v \ln \frac{T}{T_0} + s_0.$$

This gives
$$\left(\frac{\partial s}{\partial T}\right)_p = \frac{nR}{T} + c_v \frac{1}{T},$$

and, therefore,
$$c_p = T\left(\frac{\partial s}{\partial T}\right)_p = nR + c_v \quad \Rightarrow \quad c_p - c_v = nR.$$

(d) Since $h = u + pV$, by substituting $u = C_v(T - T_0)$ and $pV = nRT$ we obtain
$$h = c_v(T - T_0) + nRT = c_p(T - T_0) + (c_p - c_v)T_0,$$

i.e.,
$$h = c_p(T - T_0) + nRT_0.$$

An independent derivation proceeds from

$$dh = Vdp + Tds,$$

and

$$Tds = pdV + c_v dT.$$

It follows that

$$dh = d(pV) + c_v dT = d(nRT) + c_v dT = (nR + c_v)dT,$$

i.e.,

$$dh = c_p dT \quad \Rightarrow \quad h = c_p(T - T_0) + nRT_0.$$

(e) By taking a differential of $pV = nRT$, we have

$$pdV + Vdp = nRdT.$$

Therefore, upon the division with $pV = nRT$,

$$\frac{dp}{p} + \frac{dV}{V} = \frac{dT}{T}.$$

(f) From $pV = nRT$, we have $V = nRT/p$, and

$$\alpha = \frac{1}{V}\left(\frac{\partial V}{\partial T}\right)_p = \frac{1}{T}, \quad \beta = -\frac{1}{V}\left(\frac{\partial V}{\partial p}\right)_T = \frac{1}{p}.$$

Problem 6.2. A total of n moles of an ideal gas occupying the volume V_0 at initial temperature T_0 is isentropically (reversibly and adiabatically) compressed to a final volume V.
(a) Derive an expression for the corresponding temperature T.
(b) Derive an expression for the corresponding pressure p.
(c) Show that along an isentropic path

$$pV^\gamma = \text{const.}, \quad \gamma = \frac{c_p}{c_v}.$$

Solution: (a) For an isentropic process the entropy remains constant, i.e.,

$$s = nR \ln \frac{V}{V_0} + c_v \ln \frac{T}{T_0} + s_0 = s_0.$$

Thus,

$$\frac{nR}{c_v} \ln \frac{V}{V_0} + \ln \frac{T}{T_0} = 0,$$

which gives

$$T = T_0 \left(\frac{V_0}{V}\right)^{nR/c_v}.$$

(b) Since

$$T = \frac{pV}{nR}, \quad T_0 = \frac{p_0 V_0}{nR},$$

Problems 6.2–6.3

the substitution into the above result for T gives

$$\frac{p}{p_0} = \left(\frac{V_0}{V}\right)^{1+nR/c_v} = \left(\frac{V_0}{V}\right)^{c_p/c_v}.$$

(c) From the result in part (c), we recognize that

$$pV^{c_p/c_v} = p_0 V_0^{c_p/c_v},$$

i.e.,

$$pV^\gamma = \text{const.}, \quad \gamma = \frac{c_p}{c_v}.$$

Problem 6.3. One mole of Ni initially at $T_0 = 298$ K and $p_0 = 1$ atm is taken through an isobaric process to temperature $T_1 = 800$ K. Calculate the corresponding change in:
(a) molar volume $V_1 - V_0$, by using a linear approximation $V(p_0, T) = V(p_0, T_0)[1 + \alpha(T - T_0)]$, $\alpha = \text{const}$;
(b) enthalpy $h_1 - h_0$;
(c) entropy $s_1 - s_0$;
(d) Gibbs energy $g_1 - g_0$.

The process is continued with one mole of Ni at $T_1 = 800$ K by isothermal compression from $p_1 = 1$ atm to $p_2 = 1000$ atm. Calculate the corresponding change in:
(e) molar volume $V_2 - V_1$, by using a linear approximation $V(p, T_1) = V(p_1, T_1)[1 - \beta(p - p_1)]$, $\beta = \text{const}$;
(f) Gibbs energy $g_2 - g_1$;
(g) entropy $s_2 - s_1$;
(h) enthalpy $h_2 - h_1$.

THERMODYNAMIC DATA FOR ONE MOLE OF NICKEL:

$$c_p = a + bT, \quad a = 16.99 \text{ J/K}, \quad b = 2.95 \times 10^{-2} \text{ J/K}^2, \quad T \in [298 - 1726] \text{ K},$$

$$\Delta s_f^0 = 2.42 \text{ cal/K} \quad (\text{not needed}) \quad \text{at} \quad T = 1726 \text{ K},$$

$$\alpha = 4 \times 10^{-5} \, K^{-1}, \quad \beta = 1.5 \times 10^{-6} \, \text{atm}^{-1},$$

$$s_{298}^0 = s(1 \text{ atm}, 298 \, K) = 7.14 \text{ cal/K},$$

$$V_{298}^0 = V(1 \text{ atm}, 298 \, K) = 6.57 \text{ cm}^3.$$

Solution: (a) From the given volume-temperature expression, we calculate

$$V_1 = V(p_0, T_1) = V(p_0, T_0)[1 + \alpha(T_1 - T_0)] = \cdots = 6.7 \text{ cm}^3,$$

so that $V_1 - V_0 = 0.13 \text{ cm}^3$.

(b) Along an isobaric path, $p = $ const., and $dh = c_p dT$. Thus,

$$h_1 - h_0 = \int_{T_0}^{T_1} c_p(T) dT = \int_{T_0}^{T_1} (a + bT) dT.$$

This gives

$$h_1 - h_0 = a(T_1 - T_0) + \frac{1}{2}b(T_1^2 - T_0^2) = \cdots = 16,659 \, \text{J}.$$

(c) Along an isobaric path $T ds = c_p dT$, so that

$$s_1 - s_0 = \int_{T_0}^{T_1} \frac{c_p(T)}{T} dT = \int_{T_0}^{T_1} \left(\frac{a}{T} + b\right) dT.$$

This gives

$$s_1 - s_0 = a \ln \frac{T_1}{T_0} + b(T_1 - T_0) = \cdots = 31.59 \, \text{J/K}.$$

(d) Along an isobaric path $dg = -s dT$, so that

$$g_1 - g_0 = -\int_{T_0}^{T_1} s(T) dT = -\int_{T_0}^{T_1} \left[a \ln \frac{T}{T_0} + b(T - T_0)\right] dT - s_0(T_1 - T_0).$$

This gives

$$g_1 - g_0 = -\left[a\left(T_1 \ln \frac{T_1}{T_0} - T_1\right) + aT_0 + \frac{1}{2}b(T_1 - T_0)^2\right] - s_0(T_1 - T_0),$$

which is equal to $g_1 - g_0 = -23,605 \, \text{J}$.

(e) The volume change is $V_2 - V_1 = V(p_2, T_1) - V(p_1, T_1)$, where

$$V(p_2, T_1) = V(p_1, T_1)[1 - \beta(p_2 - p_1)] = \cdots = 6.69 \, \text{cm}^3.$$

Thus, $V_2 - V_1 = -0.01 \, \text{cm}^3$.

(f) During an isothermal path $dg = V dp$, which gives

$$g_2 - g_1 = \int_{p_1}^{p_2} V(p) dp = \int_{p_1}^{p_2} V_1[1 - \beta(p - p_1)] dp,$$

i.e.,

$$g_2 - g_1 = V_1 \left[(p_2 - p_1) - \frac{1}{2}\beta(p_2 - p_1)^2\right] = \cdots = 677.52 \, \text{J}.$$

(f) During an isothermal path, $T_1 ds = l_p dp = -\alpha T_1 V dp$, which gives

$$ds = -\alpha V dp = -\alpha dg.$$

Thus,

$$s_2 - s_1 = -\alpha(g_2 - g_1) = \cdots = -0.0271 \, \text{J/K}.$$

Problems 6.3–6.4

(h) Since

$$dh = (1 - \alpha T_1)Vdp = (1 - \alpha T_1)dg = dg + T_1 ds,$$

we obtain

$$h_2 - h_1 = g_2 - g_1 + T_1(s_2 - s_1) = \cdots = 655.84 \, \text{J}.$$

Problem 6.4. Derive thermoelastic potentials for uniaxial stress state.

Solution: For the uniaxial state of stress

$$\sigma_{ij} = \sigma \, \delta_{i1}\delta_{j1}, \quad \sigma_{ij}\sigma_{ij} = \sigma^2, \quad \sigma_{kk} = \sigma.$$

The corresponding thermodynamic potentials are obtained from the general expressions listed in the text. The following are the results, expressed in terms of different sets of independent variables.

$$u(\sigma, s) = \frac{1}{2E_S}\sigma^2 - \frac{T_0}{2c_p^0}(s - s_0)^2 + T_0(s - s_0),$$

$$u(\sigma, T) = \frac{1}{2E_T}\sigma^2 + \frac{\alpha_0}{3}T\sigma + \frac{c_p^0}{2T_0}(T^2 - T_0^2),$$

$$\phi(\sigma, s) = \frac{1}{2E_S}\sigma^2 + \frac{\alpha_0 T_0}{3c_p^0}s\sigma - \frac{T_0}{2c_p^0}(s^2 - s_0^2) - T_0 s_0,$$

$$\phi(\sigma, T) = \frac{1}{2E_T}\sigma^2 - \frac{c_p^0}{2T_0}(T - T_0)^2 - s_0 T,$$

$$g(\sigma, s) = -\frac{1}{2E_S}\sigma^2 + \frac{\alpha_0 T_0}{3c_p^0}s_0\sigma - \frac{T_0}{2c_p^0}(s^2 - s_0^2) - T_0 s_0,$$

$$g(\sigma, T) = -\frac{1}{2E_T}\sigma^2 - \frac{\alpha_0}{3}(T - T_0)\sigma - \frac{c_p^0}{2T_0}(T - T_0)^2 - s_0 T,$$

$$h(\sigma, s) = -\frac{1}{2E_S}\sigma^2 - \frac{\alpha_0 T_0}{3c_p^0}(s - s_0)\sigma + \frac{T_0}{2c_p^0}(s - s_0)^2 + T_0(s - s_0),$$

$$h(\sigma, T) = -\frac{1}{2E_T}\sigma^2 + \frac{\alpha_0}{3}T_0\sigma + \frac{c_p^0}{2T_0}(T^2 - T_0^2).$$

The corresponding thermoelastic constitutive equations are

$$e_{ij} = \frac{\sigma}{2\mu}\left(\delta_{i1}\delta_{j1} - \frac{\nu_T}{1 + \nu_T}\delta_{ij}\right) + \frac{\alpha_0}{3}(T - T_0)\delta_{ij},$$

or
$$e_{ij} = \frac{\sigma}{2\mu}\left(\delta_{i1}\delta_{j1} - \frac{\nu_S}{1+\nu_S}\delta_{ij}\right) + \frac{\alpha_0 T_0}{3c_p^0}(s-s_0)\delta_{ij},$$

and
$$s - s_0 = \frac{\alpha_0}{3}\sigma + \frac{c_p^0}{T_0}(T - T_0).$$

Problem 6.5. Derive the thermoelastic potentials for the state of spherical stress.

Solution: For the spherical state of stress
$$\sigma_{ij} = \sigma\,\delta_{ij}, \qquad \sigma_{ij}\sigma_{ij} = 3\sigma^2, \qquad \sigma_{kk} = 3\sigma.$$

The thermodynamic potentials are found to be
$$u(\sigma, s) = \frac{1}{2\kappa_S}\sigma^2 - \frac{T_0}{2c_p^0}(s-s_0)^2 + T_0(s-s_0),$$

$$u(\sigma, T) = \frac{1}{2\kappa_T}\sigma^2 + \alpha_0 T\sigma + \frac{c_p^0}{2T_0}(T^2 - T_0^2),$$

$$\phi(\sigma, s) = \frac{1}{2\kappa_S}\sigma^2 + \frac{\alpha_0 T_0}{c_p^0}s\sigma - \frac{T_0}{2c_p^0}(s^2 - s_0^2) - T_0 s_0,$$

$$\phi(\sigma, T) = \frac{1}{2\kappa_T}\sigma^2 - \frac{c_p^0}{2T_0}(T-T_0)^2 - s_0 T$$

$$g(\sigma, s) = -\frac{1}{2\kappa_S}\sigma^2 + \frac{\alpha_0 T_0}{c_p^0}s_0\sigma - \frac{T_0}{2c_p^0}(s^2 - s_0^2) - T_0 s_0,$$

$$g(\sigma, T) = -\frac{1}{2\kappa_T}\sigma^2 - \alpha_0(T-T_0)\sigma - \frac{c_p^0}{2T_0}(T-T_0)^2 - s_0 T,$$

$$h(\sigma, s) = -\frac{1}{2\kappa_S}\sigma^2 - \frac{\alpha_0 T_0}{c_p^0}(s-s_0)\sigma + \frac{T_0}{2c_p^0}(s-s_0)^2 + T_0(s-s_0),$$

$$h(\sigma, T) = -\frac{1}{2\kappa_T}\sigma^2 + \alpha_0 T_0\sigma + \frac{c_p^0}{2T_0}(T^2 - T_0^2).$$

The corresponding constitutive equations are
$$e = \frac{\sigma}{3\kappa_T} + \frac{\alpha_0}{3}(T-T_0), \quad \text{or} \quad e = \frac{\sigma}{3\kappa_S} + \frac{\alpha_0 T_0}{3c_p^0}(s-s_0),$$

and
$$s - s_0 = \alpha_0\sigma + \frac{c_p^0}{T_0}(T - T_0).$$

Problem 6.6

Problem 6.6. For a binary solution (components A and B dissolved in one another at a given pressure), the molar enthalpy of solution is

$$H_m = X_A H_m^A + X_B H_m^B + \omega X_A X_B,$$

where H_m^A and H_m^B are the (arbitrary) molar enthalpies of pure A and B in their reference states, and X_A and X_B are the molar concentrations of A and B. In the above equation,

$$H_m^{\text{mix}} = \omega X_A X_B,$$

is the enthalpy of mixing. The parameter ω characterizes the interaction of components (atoms) A and B. If $\omega = 0$, the solution is referred to as an ideal solution. If $\omega = \text{const.} \neq 0$, the solution is referred to as a regular solution. If $\omega = \omega(T)$, the solution is a nonregular solution.

(a) Consider a regular solution. Using an expression for the molar entropy

$$S_m = X_A S_m^A + X_B S_m^B - R(X_A \ln X_A + X_B \ln X_B),$$

derive the corresponding expression for the molar Gibbs energy G_m. Denote the molar Gibbs energies of pure A and B by G_m^A and G_m^B.

(b) Derive the corresponding chemical potentials μ_A and μ_B.
(c) Verify that $G_m = X_A \mu_A + X_B \mu_B$.

Solution: (a) The molar Gibbs energy is

$$G_m = H_m - T S_m.$$

The substitution of the expressions for H_m and S_m gives

$$G_m = X_A G_m^A + X_B G_m^B + RT(X_A \ln X_A + X_B \ln X_B) + \omega X_A X_B,$$

where $G_m^A = H_m^A - T S_m^A$ and $G_m^B = H_m^B - T S_m^B$ are the molar Gibbs energies of pure components A and B.

(b) It was derived in the text, see (6.250) and (6.251), that the chemical potentials of a binary system are

$$\mu_A = G_m - X_B \frac{dG_m}{dX_B},$$

$$\mu_B = G_m + (1 - X_B) \frac{dG_m}{dX_B}.$$

For the regular solution with the molar Gibbs energy as derived in part (a), we have

$$\frac{dG_m}{dX_B} = G_m^B - G_m^A + \omega(1 - 2X_B) + RT \ln \frac{X_B}{1 - X_B}.$$

Thus, we obtain

$$\mu_A = G_m^A + RT \ln X_A + \omega X_B^2,$$

$$\mu_B = G_m^B + RT \ln X_B + \omega X_A^2.$$

(c) The relationship

$$G_m = X_A \mu_A + X_B \mu_B$$

follows by inspection, having in mind that

$$X_A X_B^2 + X_B X_A^2 = X_A X_B (X_A + X_B) = X_A X_B.$$

CHAPTER 7

Problem 7.1. Show that the general constitutive expression for isotropic Cauchy elastic material is $\boldsymbol{\sigma} = \mathbf{h}(\mathbf{V})$, where $\mathbf{h}$ is an isotropic function of its argument.

Solution: For an arbitrary anisotropic Cauchy elastic material, the stress response is given by (7.14), *i.e.*,

$$\boldsymbol{\sigma} = \mathbf{R} \cdot \mathbf{h}(\mathbf{U}) \cdot \mathbf{R}^T.$$

Since the right and left stretch tensors are related by $\mathbf{U} = \mathbf{R}^T \cdot \mathbf{V} \cdot \mathbf{R}$, the above becomes

$$\boldsymbol{\sigma} = \mathbf{R} \cdot \mathbf{h}(\mathbf{R}^T \cdot \mathbf{V} \cdot \mathbf{R}) \cdot \mathbf{R}^T.$$

If the material is elastically isotropic, the tensor function $\mathbf{h}$ is an isotropic function of its argument, so that

$$\mathbf{h}(\mathbf{R}^T \cdot \mathbf{V} \cdot \mathbf{R}) = \mathbf{R}^T \cdot \mathbf{h}(\mathbf{V}) \cdot \mathbf{R}.$$

By substituting this into the previous equation, we obtain the desired result

$$\boldsymbol{\sigma} = \mathbf{h}(\mathbf{V}).$$

Problem 7.2. If $\Phi = \Phi(\mathbf{B})$ is an isotropic function of $\mathbf{B}$, show that $\mathbf{B}$ commutes with $\partial \Phi / \partial \mathbf{B}$, and that principal directions of $\boldsymbol{\sigma}$ are parallel to those of $\mathbf{B}$.

Solution: Since $\Phi = \Phi(\mathbf{B})$ is an isotropic function, it can be expressed as

$$\Phi = \Phi(I_B, II_B, III_B),$$

where

$$I_B = \operatorname{tr} \mathbf{B}, \quad II_B = \frac{1}{2}\left[\operatorname{tr}(\mathbf{B}^2) - (\operatorname{tr} \mathbf{B})^2\right], \quad III_B = \det \mathbf{B}.$$

Recalling the results from Problem 1.10, we have

$$\frac{\partial I_B}{\partial \mathbf{B}} = \mathbf{I}, \quad \frac{\partial II_B}{\partial \mathbf{B}} = \mathbf{B} - I_B \mathbf{I}, \quad \frac{\partial III_B}{\partial \mathbf{B}} = \mathbf{B}^2 - I_B \mathbf{B} - II_B \mathbf{I}.$$

Thus,

$$\frac{\partial \Phi}{\partial \mathbf{B}} = \frac{\partial \Phi}{\partial I_B} \mathbf{I} + \frac{\partial \Phi}{\partial II_B}(\mathbf{B} - I_B \mathbf{I}) + \frac{\partial \Phi}{\partial III_B}(\mathbf{B}^2 - I_B \mathbf{B} - II_B \mathbf{I}),$$

i.e.,

$$\frac{\partial \Phi}{\partial \mathbf{B}} = b_0 \mathbf{I} + b_1 \mathbf{B} + b_2 \mathbf{B}^2,$$

Problems 7.2–7.3

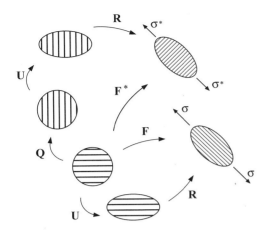

Figure 7-1. Schematics of the deformation process which involves the application of $\mathbf{R} \cdot \mathbf{U}$ to unrotated and rotated initial configuration. Because of elastic anisotropy the required stresses to produce the deformation in two cases are different.

where b_i are scalar functions of the invariants of $\mathbf{B}$. Since $\mathbf{B}^2$ has parallel principal directions to those of $\mathbf{B}$, the above equation shows that the gradient $\partial\Phi/\partial\mathbf{B}$ also has its principal directions parallel to those of $\mathbf{B}$. Consequently, $\partial\Phi/\partial\mathbf{B}$ commutes with $\mathbf{B}$, so that

$$\mathbf{B} \cdot \frac{\partial \Phi}{\partial \mathbf{B}} = \frac{\partial \Phi}{\partial \mathbf{B}} \cdot \mathbf{B} .$$

Since, from (7.28),

$$\sigma = \frac{1}{(\det \mathbf{B})^{1/2}} \left(\mathbf{B} \cdot \frac{\partial \Phi}{\partial \mathbf{B}} + \frac{\partial \Phi}{\partial \mathbf{B}} \cdot \mathbf{B} \right),$$

we conclude that the principal directions of the Cauchy stress σ are parallel to those of $\mathbf{B}$ (and thus also to those of the stretch tensor $\mathbf{V} = \mathbf{B}^{1/2}$).

Problem 7.3. Consider an anisotropic Cauchy-elastic material in its undeformed configuration. If this is subjected to deformation gradient $\mathbf{F} = \mathbf{R} \cdot \mathbf{U}$, the stress response is

$$\sigma = \mathbf{R} \cdot \mathbf{f}(\mathbf{U}) \cdot \mathbf{R}^T ,$$

where $\mathbf{f}$ is the second-order tensor function, whose representation depends on the type of elastic anisotropy of the material.
(a) Find the stress response if the material is rotated by $\mathbf{Q}$ prior to application of $\mathbf{R} \cdot \mathbf{U}$, *i.e.*, find the stress response σ^* corresponding to the deformation gradient $\mathbf{F}^* = \mathbf{R} \cdot \mathbf{U} \cdot \mathbf{Q}$ (Fig. 7-1).
(b) If material is isotropic, the rotation $\mathbf{Q}$ prior to $\mathbf{U}$ does not affect the stress response. What condition does this impose on the function $\mathbf{f}$?

Solution: (a) Since

$$\mathbf{F}^* = \mathbf{R} \cdot \mathbf{U} \cdot \mathbf{Q} = \mathbf{R} \cdot \mathbf{Q} \cdot \mathbf{Q}^T \cdot \mathbf{U} \cdot \mathbf{Q} ,$$

and $\mathbf{F}^* = \mathbf{R}^* \cdot \mathbf{U}^*$, we conclude that

$$\mathbf{U}^* = \mathbf{Q}^T \cdot \mathbf{U} \cdot \mathbf{Q} , \quad \mathbf{R}^* = \mathbf{R} \cdot \mathbf{Q} .$$

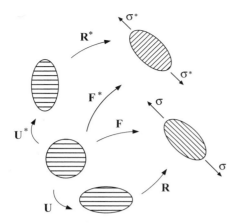

Figure 7-2. Schematics of the deformation process which illustrates the stretch tensors $\mathbf{U}$ and $\mathbf{U}^* = \mathbf{Q}^T \cdot \mathbf{U} \cdot \mathbf{Q}$. Relative to the fixed coordinate frame, the principal directions of $\mathbf{U}^*$ are obtained from those of $\mathbf{U}$ by the rotation $\mathbf{Q}$. Their principal values are the same.

This ensures that $\mathbf{U}^*$ is symmetric. The schematics of deformation process involving $\mathbf{U}^*$ and $\mathbf{R}^*$ is shown in Fig. 7-2. Thus, the stress response corresponding to $\mathbf{F}^*$ is

$$\boldsymbol{\sigma}^* = \mathbf{R}^* \cdot \mathbf{f}(\mathbf{U}^*) \cdot \mathbf{R}^{*T} = \mathbf{R} \cdot \mathbf{Q} \cdot \mathbf{f}(\mathbf{Q}^T \cdot \mathbf{U} \cdot \mathbf{Q}) \cdot \mathbf{Q}^T \cdot \mathbf{R}^T.$$

(b) If the stress response is unaffected by rotation $\mathbf{Q}$ prior to the application of $\mathbf{R} \cdot \mathbf{U}$ (isotropic materials), then the condition $\boldsymbol{\sigma}^* = \boldsymbol{\sigma}$ gives

$$\mathbf{R} \cdot \mathbf{Q} \cdot \mathbf{f}(\mathbf{Q}^T \cdot \mathbf{U} \cdot \mathbf{Q}) \cdot \mathbf{Q}^T \cdot \mathbf{R}^T = \mathbf{R} \cdot \mathbf{f}(\mathbf{U}) \cdot \mathbf{R}^T,$$

i.e.,

$$\mathbf{Q} \cdot \mathbf{f}(\mathbf{Q}^T \cdot \mathbf{U} \cdot \mathbf{Q}) \cdot \mathbf{Q}^T = \mathbf{f}(\mathbf{U}).$$

Consequently,

$$\mathbf{f}(\mathbf{Q}^T \cdot \mathbf{U} \cdot \mathbf{Q}) = \mathbf{Q}^T \cdot \mathbf{f}(\mathbf{U}) \cdot \mathbf{Q}.$$

A tensor function $\mathbf{f}$ that satisfies this condition for any orthogonal $\mathbf{Q}$ is said to be an isotropic tensor function of its argument $\mathbf{U}$. It can be shown that such a function can be expressed as

$$\mathbf{f}(\mathbf{U}) = a_0 \mathbf{I} + a_1 \mathbf{U} + a_2 \mathbf{U}^2,$$

where a_i are scalar functions of the principal invariants of $\mathbf{U}$.

Problem 7.4. In the theory of Saint-Venant and Kirchhoff for large deformation of compressible elastic material it is assumed that the strain energy is

$$\Phi = \frac{1}{2}(\lambda + 2\mu) I_E^2 + 2\mu I I_E,$$

where I_E and II_E are the invariants of the Green (Lagrangian) strain $\mathbf{E}$. Derive the corresponding expressions for the second Piola–Kirchhoff and the Cauchy stress tensors.

In the derivation, recall that $I_E = (I_B - 3)/2$ and that by the Cayley–Hamilton theorem

$$\mathbf{B}^2 = I_B \mathbf{B} + II_B \mathbf{I} + III_B \mathbf{B}^{-1}.$$

Problems 7.4–7.5

Solution: We start with
$$\mathbf{S} = \frac{\partial \Phi}{\partial \mathbf{E}} = (\lambda + 2\mu) I_E \frac{\partial I_E}{\partial \mathbf{E}} + 2\mu \frac{\partial II_E}{\partial \mathbf{E}}.$$

Since
$$\frac{\partial I_E}{\partial \mathbf{E}} = \mathbf{I}, \quad \frac{\partial II_E}{\partial \mathbf{E}} = \mathbf{E} - I_E \mathbf{I},$$

we obtain
$$\mathbf{S} = \lambda I_E \mathbf{I} + 2\mu \mathbf{E}.$$

To derive an expression for the Cauchy stress we substitute above into the expression
$$\sigma = \frac{1}{\det \mathbf{F}} \mathbf{F} \cdot \mathbf{S} \cdot \mathbf{F}^T.$$

By observing the connection $2\mathbf{F} \cdot \mathbf{E} \cdot \mathbf{F}^T = \mathbf{B}^2 - \mathbf{B}$, and by using the Cayley–Hamilton theorem to eliminate $\mathbf{B}^2$, we obtain
$$\sigma = III_B^{-1/2} \left\{ \mu II_B \mathbf{I} + \left[\frac{1}{2}\lambda(I_B - 3) + \mu(I_B - 1) \right] \mathbf{B} + \mu III_B \mathbf{B}^{-1} \right\}.$$

Problem 7.5. An initially rectangular parallelepiped ($0 \leq X_1 \leq a, 0 \leq X_2 \leq b, 0 \leq X_3 \leq c$) of a compressible finite elastic material is subject to the homogeneous deformation
$$x_1 = X_1 + k_1 X_2, \quad x_2 = X_2 + k_2 X_1, \quad x_3 = X_3,$$
where k_1 and k_2 are constants.

(a) Show that $k_1 k_2 < 1$.
(b) If the constitutive law is $\sigma = b_0 \mathbf{I} + b_1 \mathbf{B} + b_2 \mathbf{B}^2$, find all stress components and show that $\sigma_{11} - \sigma_{22} = (k_1 - k_2)\sigma_{12}$.

Solution: (a) The deformation gradient is
$$\mathbf{F} = \begin{bmatrix} 1 & k_1 & 0 \\ k_2 & 1 & 0 \\ 0 & 0 & 1 \end{bmatrix}, \quad \det \mathbf{F} = 1 - k_1 k_2.$$

Because $\det \mathbf{F} > 0$, we conclude that $k_1 k_2 < 1$.
(b) The left Cauchy–Green deformation tensor is
$$\mathbf{B} = \mathbf{F} \cdot \mathbf{F}^T = \begin{bmatrix} 1 + k_1^2 & k_1 + k_2 & 0 \\ k_1 + k_2 & 1 + k_2^2 & 0 \\ 0 & 0 & 1 \end{bmatrix},$$

By substituting into $\sigma = b_0 \mathbf{I} + b_1 \mathbf{B} + b_2 \mathbf{B}^2$, it readily follows that
$$\sigma_{11} = b_0 + b_1(1 + k_1^2) + b_2[(1 + k_1^2)^2 + (k_1 + k_2)^2],$$
$$\sigma_{22} = b_0 + b_1(1 + k_2^2) + b_2[(1 + k_2^2)^2 + (k_1 + k_2)^2],$$
$$\sigma_{12} = b_1(k_1 + k_2) + b_2[(k_1 + k_2)(2 + k_1^2 + k_2^2)],$$
$$\sigma_{33} = b_0 + b_1 + b_2, \quad \sigma_{13} = \sigma_{23} = 0.$$

Thus, by inspection, $\sigma_{11} - \sigma_{22} = (k_1 - k_2)\sigma_{12}$.

Problem 7.6. Consider a rectangular block under tensile stress in X_1 direction (simple extension test), which causes the stretch in that direction of amount λ_1. If the material of the block is an incompressible Mooney–Rivlin rubber with

$$\Phi = \frac{1}{2}\mu\left[\left(\frac{1}{2}+\beta\right)(I_B - 3) - \left(\frac{1}{2}-\beta\right)(II_B + 3)\right],$$

where $\mu > 0$ and $-1/2 \leq \beta \leq 1/2$ are material constants, find the stress required to produce this deformation.

Solution: The deformation is described by

$$x_1 = \lambda_1 X_1, \quad x_2 = \lambda_2 X_2, \quad x_3 = \lambda_2 X_3,$$

with the corresponding deformation gradient

$$\mathbf{F} = \begin{bmatrix} \lambda_1 & 0 & 0 \\ 0 & \lambda_2 & 0 \\ 0 & 0 & \lambda_2 \end{bmatrix}.$$

Note that because of elastic isotropy $\lambda_3 = \lambda_2$. Since material is also incompressible, we have $\det \mathbf{F} = \lambda_1 \lambda_2^2 = 1$, so that $\lambda_2 = 1/\sqrt{\lambda_1}$. The left Cauchy–Green deformation tensor and its inverse are thus

$$\mathbf{B} = \begin{bmatrix} \lambda_1^2 & 0 & 0 \\ 0 & 1/\lambda_1 & 0 \\ 0 & 0 & 1/\lambda_1 \end{bmatrix}, \quad \mathbf{B}^{-1} = \begin{bmatrix} 1/\lambda_1^2 & 0 & 0 \\ 0 & \lambda_1 & 0 \\ 0 & 0 & \lambda_1 \end{bmatrix}.$$

The finite elasticity constitutive law for an incompressible material is

$$\sigma = -p_0 \mathbf{I} + 2\left[\left(\frac{\partial \Phi}{\partial I_B}\right)\mathbf{B} + \left(\frac{\partial \Phi}{\partial II_B}\right)\mathbf{B}^{-1}\right],$$

where p_0 is the hydrostatic pressure. In view of the given strain energy representation, this gives

$$\sigma = -p_0 \mathbf{I} + \mu\left[\left(\frac{1}{2}+\beta\right)\mathbf{B} - \left(\frac{1}{2}-\beta\right)\mathbf{B}^{-1}\right].$$

The normal stress components are

$$\sigma_{11} = -p_0 + \mu\left[\left(\frac{1}{2}+\beta\right)\lambda_1^2 - \left(\frac{1}{2}-\beta\right)\frac{1}{\lambda_1^2}\right],$$

$$\sigma_{22} = \sigma_{33} = -p_0 + \mu\left[\left(\frac{1}{2}+\beta\right)\frac{1}{\lambda_1} - \left(\frac{1}{2}-\beta\right)\lambda_1\right] = 0.$$

The last expression specifies the pressure

$$p_0 = \mu\left[\left(\frac{1}{2}+\beta\right)\frac{1}{\lambda_1} - \left(\frac{1}{2}-\beta\right)\lambda_1\right].$$

Problems 7.6–7.7

Consequently, the required longitudinal stress is

$$\sigma_{11} = \mu \frac{\lambda_1^3 - 1}{\lambda_1} \left[\frac{1}{2} + \beta + \left(\frac{1}{2} - \beta \right) \frac{1}{\lambda_1} \right].$$

Problem 7.7. Determine the stress and strain state in a rectangular block made from a Mooney–Rivlin material under simple shear of amount φ in the direction X_1.

Solution: The deformation is described by

$$x_1 = X_1 + k X_2, \quad x_2 = X_2, \quad x_3 = X_3,$$

where $k = \tan \varphi$. The corresponding deformation gradient is

$$\mathbf{F} = \begin{bmatrix} 1 & k & 0 \\ 0 & 1 & 0 \\ 0 & 0 & 1 \end{bmatrix}.$$

The right Cauchy–Green deformation tensor and the Lagrangian strain are accordingly

$$\mathbf{C} = \mathbf{F}^T \cdot \mathbf{F} = \begin{bmatrix} 1 & k & 0 \\ k & 1+k^2 & 0 \\ 0 & 0 & 1 \end{bmatrix}, \quad \mathbf{E} = \frac{1}{2}(\mathbf{C} - \mathbf{I}) = \frac{1}{2} \begin{bmatrix} 0 & k & 0 \\ k & k^2 & 0 \\ 0 & 0 & 0 \end{bmatrix}.$$

The left Cauchy–Green deformation tensor and its inverse are

$$\mathbf{B} = \mathbf{F} \cdot \mathbf{F}^T = \begin{bmatrix} 1+k^2 & k & 0 \\ k & 1 & 0 \\ 0 & 0 & 1 \end{bmatrix}, \quad \mathbf{B}^{-1} = \begin{bmatrix} 1 & -k & 0 \\ -k & 1+k^2 & 0 \\ 0 & 0 & 1 \end{bmatrix}.$$

The finite elasticity constitutive law for an incompressible material is

$$\sigma = -p_0 \mathbf{I} + 2 \left[\left(\frac{\partial \Phi}{\partial I_B} \right) \mathbf{B} + \left(\frac{\partial \Phi}{\partial II_B} \right) \mathbf{B}^{-1} \right],$$

where p_0 is the pressure. Since for the Mooney–Rivlin material

$$\Phi = \frac{1}{2} \mu \left[\left(\frac{1}{2} + \beta \right) (I_B - 3) - \left(\frac{1}{2} - \beta \right) (II_B + 3) \right],$$

we obtain

$$\sigma = -p_0 \mathbf{I} + \mu \left[\left(\frac{1}{2} + \beta \right) \mathbf{B} - \left(\frac{1}{2} - \beta \right) \mathbf{B}^{-1} \right].$$

The nonvanishing stress components are thus

$$\sigma_{11} = -p_0 + 2\beta\mu + \left(\frac{1}{2} + \beta \right) \mu k^2, \quad \sigma_{22} = -p_0 + 2\beta\mu - \left(\frac{1}{2} - \beta \right) \mu k^2,$$

$$\sigma_{33} = -p_0 + 2\beta\mu, \quad \sigma_{12} = \mu k.$$

Note that $\sigma_{11} - \sigma_{22} = k\sigma_{12}$, regardless of the material parameters μ and β. If the planes $x_3 = \text{const.}$ are stress free, we have $\sigma_{33} = 0$, which specifies the pressure

$$p_0 = 2\beta\mu.$$

The nonvanishing stress components in this case are consequently

$$\sigma_{11} = \left(\frac{1}{2} + \beta\right)\mu k^2, \quad \sigma_{22} = -\left(\frac{1}{2} - \beta\right)\mu k^2, \quad \sigma_{12} = \mu k.$$

Because $E_{12} = k$, we have $\sigma_{12} = \mu E_{12}$, so that μ can be interpreted as the generalized shear modulus. The presence of normal stress due to simple shear of a nonlinear elastic solid is known as the Poynting effect.

Problem 7.8. An isotropic compressible elastic material is characterized by the Blatz–Ko strain energy function

$$\Phi = \frac{1}{2}\mu \left(2III_B^{1/2} - II_B III_B^{-1} - 5\right),$$

where μ is the material parameter, and

$$II_B = \frac{1}{2}\left[\text{tr}(\mathbf{B}^2) - (\text{tr }\mathbf{B})^2\right], \quad III_B = \det \mathbf{B}.$$

Derive the corresponding constitutive equation for the Cauchy stress tensor in terms of μ and $\mathbf{B}$.

Solution: From (7.31) of Chapter 7 we have

$$(\det \mathbf{F})\sigma = 2\left[\left(III_B \frac{\partial \Phi}{\partial III_B} + II_B \frac{\partial \Phi}{\partial II_B}\right)\mathbf{I} + \left(\frac{\partial \Phi}{\partial I_B}\right)\mathbf{B} + \left(III_B \frac{\partial \Phi}{\partial II_B}\right)\mathbf{B}^{-1}\right].$$

The strain energy gradients with respect to its invariants for the Blatz–Ko energy function are

$$\frac{\partial \Phi}{\partial I_B} = 0, \quad \frac{\partial \Phi}{\partial II_B} = -\frac{1}{2}\mu III_B^{-1}, \quad \frac{\partial \Phi}{\partial III_B} = \frac{1}{2}\mu \left(III_B^{-1/2} + II_B III_B^{-2}\right).$$

Thus

$$(\det \mathbf{F})\sigma = \mu \left(III_B^{1/2} \mathbf{I} - \mathbf{B}^{-1}\right).$$

Since $\det \mathbf{F} = III_B^{1/2}$, the above reduces to

$$\sigma = \mu \left(\mathbf{I} - III_B^{-1/2} \mathbf{B}^{-1}\right).$$

Problem 7.9. The incompressible Reiner–Rivlin fluid is defined by the constitutive equation

$$\sigma = -p\mathbf{I} + a_1(II_D, III_D)\mathbf{D} + a_2(II_D, III_D)\mathbf{D}^2,$$

where p is an arbitrary pressure, and a_1, a_2 are given scalar functions of the indicated invariants of $\mathbf{D}$. Derive the expressions for the stress components if the fluid is undergoing a simple shearing flow, with velocity

$$v_1 = kx_2, \quad v_2 = v_3 = 0, \quad (k = \text{const.}).$$

Solution: The velocity gradient, the rate of deformation tensor and its square are, respectively,

$$\mathbf{L} = \begin{bmatrix} 0 & k & 0 \\ 0 & 0 & 0 \\ 0 & 0 & 0 \end{bmatrix}, \quad \mathbf{D} = \begin{bmatrix} 0 & k/2 & 0 \\ k/2 & 0 & 0 \\ 0 & 0 & 0 \end{bmatrix}, \quad \mathbf{D}^2 = \begin{bmatrix} k^2/4 & 0 & 0 \\ 0 & k^2/4 & 0 \\ 0 & 0 & 0 \end{bmatrix}.$$

The corresponding stress is

$$\sigma = \begin{bmatrix} -p + a_2 k^2/4 & a_1 k/2 & 0 \\ a_1 k/2 & -p + a_2 k^2/4 & 0 \\ 0 & 0 & -p \end{bmatrix}.$$

Note that

$$I_D = \operatorname{tr} \mathbf{D} = 0, \quad II_D = \frac{1}{2}\left[\operatorname{tr}(\mathbf{D}^2) - (\operatorname{tr} \mathbf{D})^2\right] = \frac{k^2}{4}, \quad III_D = \det \mathbf{D} = 0.$$

If $a_2 = 0$, we obtain the stress response for an incompressible Newtonian fluid, $\sigma = -p\mathbf{I} + a_1 \mathbf{D}$.

Problem 7.10. If $\Phi = \Phi(E_{ij})$ is the strain energy function, the complementary energy $\Phi^* = \Phi^*(S_{ij})$ is defined such that

$$\Phi(E_{ij}) + \Phi^*(S_{ij}) = S_{ij} E_{ij}.$$

Thus, from

$$\frac{\partial \Phi}{\partial E_{ij}} dE_{ij} + \frac{\partial \Phi^*}{\partial S_{ij}} dS_{ij} = S_{ij} dE_{ij} + E_{ij} dS_{ij},$$

we obtain

$$S_{ij} = \frac{\partial \Phi}{\partial E_{ij}}, \quad E_{ij} = \frac{\partial \Phi^*}{\partial S_{ij}}.$$

If E_{ij} is a homogeneous function of stress of degree n, prove that

$$\Phi = \frac{n}{n+1} S_{ij} E_{ij}.$$

Solution: Since E_{ij} is a homogeneous function of stress of degree n, and since $E_{ij} = \partial \Phi^*/\partial S_{ij}$, we conclude that $\Phi^*(S_{ij})$ must be a homogeneous function of degree $n + 1$. Consequently,

$$S_{ij} \frac{\partial \Phi^*}{\partial S_{ij}} = (n+1)\Phi^*,$$

i.e.,

$$\Phi^* = \frac{1}{n+1} S_{ij} \frac{\partial \Phi^*}{\partial S_{ij}} = \frac{1}{n+1} S_{ij} E_{ij}.$$

Since

$$\Phi = S_{ij} E_{ij} - \Phi^*,$$

we obtain

$$\Phi = \frac{n}{n+1} S_{ij} E_{ij}.$$

If $n = 1$, we recover the result of linear theory

$$\Phi = \Phi^* = \frac{1}{2} S_{ij} E_{ij}.$$

Problem 7.11. Consider a pressurized spherical balloon of the initial (zero pressure) radius R_0 and initial thickness $t_0 \ll R_0$. The elastic strain energy function of the incompressible material of the balloon is

$$\Phi = \sum_{k=1}^{3} \frac{\mu_k}{\alpha_k} \left(\lambda_1^{\alpha_k} + \lambda_2^{\alpha_k} + \lambda_3^{\alpha_k} - 3 \right),$$

where μ_k and α_k are material parameters. Derive the relationship between the internal (inflation) pressure p and the circumferential stretch ratio λ, assuming a spherical mode of the inflation process.

Solution: By spherical symmetry, the stretch ratios are $\lambda_1 = \lambda_2 = \lambda = R/R_0$ and $\lambda_3 = 1/\lambda^2 = t/t_0$, the latter being the stretch ratio in the direction perpendicular to the surface of the balloon. The radius and the thickness of the balloon in the deformed configuration are R and t, respectively. The circumferential stress components are thus

$$\sigma_i = -p_0 + \lambda_i \frac{\partial \Phi}{\partial \lambda_i}, \quad (i = 1, 2),$$

whereas in the direction perpendicular to the surface of the balloon,

$$0 = -p_0 + \lambda_3 \frac{\partial \Phi}{\partial \lambda_3} \quad \Rightarrow \quad p_0 = \sum_{k=1}^{3} \mu_k \lambda^{-2\alpha_k}.$$

Consequently,

$$\sigma_1 = \sigma_2 = \sigma = \sum_{k=1}^{3} \mu_k \left(\lambda^{\alpha_k} - \lambda^{-2\alpha_k} \right).$$

Imagining balloon to be cut into two halves, the equilibrium condition gives

$$2R\pi t \sigma = R^2 \pi p \quad \Rightarrow \quad p = 2 \frac{t}{R} \sigma.$$

Since by the incompressibility constraint $t/t_0 = (R_0/R)^2 = \lambda^{-2}$, we have $t/R = (t_0/R_0)\lambda^{-3}$ and, therefore,

$$p = 2 \frac{t_0}{R_0} \sum_{k=1}^{3} \mu_k \left(\lambda^{\alpha_k - 3} - \lambda^{-2\alpha_k - 3} \right).$$

Problem 7.12. Consider a tension–torsion–inflation test of an elastic hollow cylinder. The initial length of the tube is L, and its initial inner and outer radii are R_1 and R_2. The tube is deformed such that its planar cross sections remain planar, rotating around the tube's longitudinal axis. The angle of rotation of the end $Z = L$ of the tube, relative to the end

Problem 7.12

$Z = 0$, is ϕ. The deformed length of the tube is l, and its deformed radii become r_1 and r_2. The stretch ratio in the longitudinal direction is l/L.

(a) Write down the deformation mapping in cylindrical coordinates, and determine the corresponding components of the deformation gradient, the Cauchy–Green deformation tensors, and the Lagrangian strain.
(b) Specialize the results in the case of an incompressible isotropic material.
(c) If the strain energy is of the Mooney–Rivlin type, determine the corresponding stress components within the tube.

Solution: (a) The deformation mapping is described by

$$r = r(R),$$

$$\theta = \Theta + \frac{Z}{L}\phi,$$

$$z = \frac{l}{L}Z,$$

where r, θ, z are the cylindrical coordinates in the deformed, and R, Θ, Z in the undeformed state. Since, from Problem 4.16, the components of the deformation gradient in cylindrical coordinates are

$$[F_{iJ}] = \begin{bmatrix} \dfrac{\partial r}{\partial R} & \dfrac{1}{R}\dfrac{\partial r}{\partial \Theta} & \dfrac{\partial r}{\partial Z} \\[6pt] r\dfrac{\partial \theta}{\partial R} & \dfrac{r}{R}\dfrac{\partial \theta}{\partial \Theta} & r\dfrac{\partial \theta}{\partial Z} \\[6pt] \dfrac{\partial z}{\partial R} & \dfrac{1}{R}\dfrac{\partial z}{\partial \Theta} & \dfrac{\partial z}{\partial Z} \end{bmatrix},$$

we obtain

$$[F_{iJ}] = \begin{bmatrix} \dfrac{dr}{dR} & 0 & 0 \\[6pt] 0 & \dfrac{r}{R} & \dfrac{r\phi}{L} \\[6pt] 0 & 0 & \dfrac{l}{L} \end{bmatrix}.$$

The right Cauchy–Green deformation tensor $\mathbf{C} = \mathbf{F}^T \cdot \mathbf{F}$ has the components

$$[C_{IJ}] = [F^T_{Ik}]\cdot[F_{kJ}] = \begin{bmatrix} \dfrac{dr}{dR} & 0 & 0 \\[6pt] 0 & \dfrac{r}{R} & 0 \\[6pt] 0 & \dfrac{r\phi}{L} & \dfrac{l}{L} \end{bmatrix} \begin{bmatrix} \dfrac{dr}{dR} & 0 & 0 \\[6pt] 0 & \dfrac{r}{R} & \dfrac{r\phi}{L} \\[6pt] 0 & 0 & \dfrac{l}{L} \end{bmatrix}.$$

This gives

$$[C_{IJ}] = \begin{bmatrix} \left(\dfrac{dr}{dR}\right)^2 & 0 & 0 \\ 0 & \dfrac{r^2}{R^2} & \dfrac{r^2\phi}{RL} \\ 0 & \dfrac{r^2\phi}{RL} & \dfrac{r^2\phi^2 + l^2}{L^2} \end{bmatrix}.$$

Thus, the physical components of the Lagrangian strain are

$$[E_{IJ}] = \dfrac{1}{2}[C_{IJ} - \delta_{IJ}] = \dfrac{1}{2}\begin{bmatrix} \left(\dfrac{dr}{dR}\right)^2 - 1 & 0 & 0 \\ 0 & \dfrac{r^2}{R^2} - 1 & \dfrac{r^2\phi}{RL} \\ 0 & \dfrac{r^2\phi}{RL} & \dfrac{r^2\phi^2 + l^2}{L^2} - 1 \end{bmatrix}.$$

The left Cauchy–Green deformation tensor $\mathbf{B} = \mathbf{F} \cdot \mathbf{F}^T$ has the components

$$[B_{ij}] = [F_{iK}] \cdot [F^T_{Kj}] = \begin{bmatrix} \left(\dfrac{dr}{dR}\right)^2 & 0 & 0 \\ 0 & \dfrac{r^2}{R^2} + \dfrac{r^2\phi^2}{l^2} & \dfrac{r\phi l}{l^2} \\ 0 & \dfrac{r\phi l}{l^2} & \dfrac{l^2}{l^2} \end{bmatrix}.$$

This has an inverse (see Problem 1.14)

$$[B_{ij}^{-1}] = \begin{bmatrix} \left(\dfrac{dR}{dr}\right)^2 & 0 & 0 \\ 0 & \dfrac{R^2}{r^2} & -\dfrac{R^2\phi}{rl} \\ 0 & -\dfrac{R^2\phi}{rl} & \dfrac{L^2}{l^2} + \dfrac{R^2\phi^2}{l^2} \end{bmatrix}.$$

(b) If material is incompressible, $\det \mathbf{F} = 1$. This gives

$$\dfrac{dr}{dR}\dfrac{r}{R}\dfrac{l}{L} = 1 \quad \Rightarrow \quad r^2(R) = r^2(R_1) + \dfrac{L}{l}(R^2 - R_1^2).$$

If material is isotropic, the stretch ratios in r and θ directions due to stretching in the z direction must be the same, i.e.,

$$\dfrac{dr}{dR} = \dfrac{r}{R} \quad \Rightarrow \quad r(R) = \dfrac{r(R_1)}{R_1}R.$$

Problem 7.12

If material is both incompressible and isotropic, then

$$\frac{r}{R} = \sqrt{\frac{L}{l}}.$$

In this case,

$$[F_{iJ}] = \begin{bmatrix} \sqrt{\dfrac{L}{l}} & 0 & 0 \\ 0 & \sqrt{\dfrac{L}{l}} & \dfrac{R\phi}{\sqrt{Ll}} \\ 0 & 0 & \dfrac{l}{L} \end{bmatrix},$$

$$[E_{IJ}] = \frac{1}{2}\begin{bmatrix} \dfrac{L}{l} - 1 & 0 & 0 \\ 0 & \dfrac{L}{l} - 1 & \dfrac{R\phi}{l} \\ 0 & \dfrac{R\phi}{l} & \dfrac{R^2\phi^2}{Ll} + \dfrac{l^2}{L^2} - 1 \end{bmatrix}.$$

The components of the left Cauchy–Green deformation tensor and its inverse are

$$[B_{ij}] = \begin{bmatrix} \dfrac{L}{l} & 0 & 0 \\ 0 & \dfrac{L}{l} + \dfrac{R^2\phi^2}{Ll} & \sqrt{\dfrac{l}{L}}\dfrac{R\phi}{L} \\ 0 & \sqrt{\dfrac{l}{L}}\dfrac{R\phi}{L} & \dfrac{l^2}{L^2} \end{bmatrix},$$

$$[B_{ij}^{-1}] = \begin{bmatrix} \dfrac{l}{L} & 0 & 0 \\ 0 & \dfrac{l}{L} & -\dfrac{R\phi}{\sqrt{Ll}} \\ 0 & -\dfrac{R\phi}{\sqrt{Ll}} & \dfrac{L^2}{l^2} + \dfrac{R^2\phi^2}{l^2} \end{bmatrix}.$$

(c) The Cauchy stress corresponding to the Mooney–Rivlin type elasticity is

$$\sigma = -p_0(r)\mathbf{I} + \mu\left[\left(\frac{1}{2} + \beta\right)\mathbf{B} - \left(\frac{1}{2} - \beta\right)\mathbf{B}^{-1}\right].$$

In view of the physical components of **B** and **B**$^{-1}$ from part (b), this gives for σ_{rr} and $\sigma_{\theta\theta}$ stress components

$$\sigma_{rr} = -p_0(r) + \mu\left(\frac{1}{2} + \beta\right)\frac{L}{l} - \mu\left(\frac{1}{2} - \beta\right)\frac{l}{L},$$

$$\sigma_{\theta\theta} = -p_0(r) + \mu\left(\frac{1}{2} + \beta\right)\left(\frac{L}{l} + \frac{R^2\phi^2}{Ll}\right) - \mu\left(\frac{1}{2} - \beta\right)\frac{l}{L}.$$

Then,

$$\frac{d\sigma_{rr}}{dr} = -\frac{dp_0}{dr},$$

$$\frac{\sigma_{rr} - \sigma_{\theta\theta}}{r} = -\mu\left(\frac{1}{2} + \beta\right)\frac{R}{L}\frac{\phi^2}{\sqrt{Ll}}.$$

The only nontrivial equilibrium equation is

$$\frac{d\sigma_{rr}}{dr} + \frac{\sigma_{rr} - \sigma_{\theta\theta}}{r} = 0.$$

Since

$$\frac{dp_0}{dr} = \frac{dp_0}{dR}\frac{dR}{dr} = \frac{dp_0}{dR}\sqrt{\frac{l}{L}},$$

the substitution in the above equation and integration gives

$$p_0 = -\mu\left(\frac{1}{2} + \beta\right)\frac{R^2\phi^2}{2Ll} - C = -\mu\left(\frac{1}{2} + \beta\right)\frac{r^2\phi^2}{2L^2} - C.$$

The integration constant C is arbitrary, but can be specified if one wants, for example, to match the applied pressure on either inner or outer surface of the tube. The pressure on other surface is then as given by the solution. The complete stress distribution within the tube is

$$\sigma_{rr} = \mu\left[\left(\frac{1}{2} + \beta\right)\left(\frac{L}{l} + \frac{r^2\phi^2}{2L^2}\right) - \left(\frac{1}{2} - \beta\right)\frac{l}{L}\right] + C,$$

$$\sigma_{\theta\theta} = \mu\left[\left(\frac{1}{2} + \beta\right)\left(\frac{L}{l} + \frac{3r^2\phi^2}{2L^2}\right) - \left(\frac{1}{2} - \beta\right)\frac{l}{L}\right] + C,$$

$$\sigma_{zz} = \mu\left[\left(\frac{1}{2} + \beta\right)\left(\frac{l^2}{L^2} + \frac{r^2\phi^2}{2L^2}\right) - \left(\frac{1}{2} - \beta\right)\left(\frac{L^2}{l^2} + \frac{r^2\phi^2}{Ll}\right)\right] + C,$$

$$\sigma_{z\theta} = \mu\left[\left(\frac{1}{2} + \beta\right)\frac{l}{L} - \left(\frac{1}{2} - \beta\right)\right]\frac{r}{L} + C.$$

Note the variation of σ_{zz} with radius r required to keep the cross sections planar. Also, recall that the radial deformation is specified by $r = \sqrt{L/l}\,R$, so that inner and outer radii become after deformation

$$r_1 = \sqrt{\frac{L}{l}}\,R_1, \quad r_2 = \sqrt{\frac{L}{l}}\,R_2.$$

Problem 7.13. Consider again a tension–torsion–inflation test of a cylindrical tube from the previous problem. Derive the cylindrical components of the velocity gradient by using:

(a) the relationship, from Problem 4.17,

$$[\mathbf{L}] = \left[\dot{F}_{iK} F^{-1}_{Kj}\right] + [\Omega],$$

(b) the velocity field and the relationship for $[\mathbf{L}]$ from Problem 4.15.

Solution: (a) The deformation gradient matrix and its inverse are

$$[F_{iJ}] = \begin{bmatrix} \dfrac{dr}{dR} & 0 & 0 \\ 0 & \dfrac{r}{R} & \dfrac{r\phi}{L} \\ 0 & 0 & \dfrac{l}{L} \end{bmatrix}, \quad [F^{-1}_{iJ}] = \begin{bmatrix} \dfrac{dR}{dr} & 0 & 0 \\ 0 & \dfrac{R}{r} & -\dfrac{R\phi}{l} \\ 0 & 0 & \dfrac{L}{l} \end{bmatrix}.$$

Furthermore,

$$[\dot{F}_{iJ}] = \begin{bmatrix} \dfrac{d\dot{r}}{dR} & 0 & 0 \\ 0 & \dfrac{\dot{r}}{R} & \dfrac{\dot{r}\phi + r\dot\phi}{L} \\ 0 & 0 & \dfrac{\dot{l}}{L} \end{bmatrix}.$$

Consequently, upon the multiplication, we obtain

$$[\dot{F}_{iK}][F^{-1}_{Kj}] = \begin{bmatrix} \dfrac{d\dot{r}}{dr} & 0 & 0 \\ 0 & \dfrac{\dot{r}}{r} & \dfrac{r\dot\phi}{l} \\ 0 & 0 & \dfrac{\dot{l}}{l} \end{bmatrix}.$$

Since, from Problem 4.17,

$$[\Omega] = \dot\theta \begin{bmatrix} 0 & -1 & 0 \\ 1 & 0 & 0 \\ 0 & 0 & 0 \end{bmatrix}, \quad \dot\theta = \dfrac{Z}{L}\dot\phi = \dfrac{z}{l}\dot\phi,$$

there follows

$$[\mathbf{L}] = \left[\dot{F}_{iK} F_{Kj}^{-1}\right] + [\Omega] = \begin{bmatrix} \dfrac{d\dot{r}}{dr} & -\dfrac{z\dot{\phi}}{l} & 0 \\ \dfrac{z\dot{\phi}}{l} & \dfrac{\dot{r}}{r} & \dfrac{r\dot{\phi}}{l} \\ 0 & 0 & \dfrac{\dot{l}}{l} \end{bmatrix}.$$

Note that for an incompressible material,

$$\frac{d\dot{r}}{dr} = -\left(\frac{\dot{r}}{r} + \frac{\dot{l}}{l}\right).$$

For an incompressible material that is also isotropic, $r = \sqrt{L/l}\, R$ and

$$\frac{d\dot{r}}{dr} = \frac{\dot{r}}{r} = -\frac{1}{2}\frac{\dot{l}}{l}.$$

(b) The physical components of the velocity vector in the considered problem, expressed in spatial coordinates, are

$$v_r = \dot{r},$$

$$v_\theta = r\dot{\theta} = r\frac{z}{l}\dot{\phi},$$

$$v_z = \frac{\dot{l}}{l} z.$$

If this is substituted into the expression for the velocity gradient from Problem 4.15, i.e.,

$$[\mathbf{L}] = \begin{bmatrix} \dfrac{\partial v_r}{\partial r} & \dfrac{1}{r}\dfrac{\partial v_r}{\partial \theta} - \dfrac{v_\theta}{r} & \dfrac{\partial v_r}{\partial z} \\ \dfrac{\partial v_\theta}{\partial r} & \dfrac{1}{r}\dfrac{\partial v_\theta}{\partial \theta} + \dfrac{v_r}{r} & \dfrac{\partial v_\theta}{\partial z} \\ \dfrac{\partial v_z}{\partial r} & \dfrac{1}{r}\dfrac{\partial v_z}{\partial \theta} & \dfrac{\partial v_z}{\partial z} \end{bmatrix},$$

we recover the results from part (a).

Problem 7.14. Derive the components of the Jaumann rate of the Cauchy stress in cylindrical coordinates at a point of an isotropic cylindrical tube in a tension–torsion–inflation test.

Solution: The nonvanishing physical components of the Cauchy stress tensor in the considered problem are

$$\begin{bmatrix} \sigma_{rr} & 0 & 0 \\ 0 & \sigma_{\theta\theta} & \sigma_{\theta z} \\ 0 & \sigma_{z\theta} & \sigma_{zz} \end{bmatrix}.$$

Problem 7.14

The Jaumann components are, by using the results from Problem 5.10,

$$[\overset{\triangledown}{\sigma}] = [\dot{\sigma}] - [\hat{\mathbf{W}}] \cdot [\sigma] + [\sigma] \cdot [\hat{\mathbf{W}}].$$

By using the results from the previous problem,

$$[\hat{\mathbf{W}}] = [\mathbf{W}] - [\Omega] = \frac{1}{2}\begin{bmatrix} 0 & 0 & 0 \\ 0 & 0 & \dfrac{r\dot{\phi}}{l} \\ 0 & -\dfrac{r\dot{\phi}}{l} & 0 \end{bmatrix},$$

we obtain

$$[\sigma] \cdot [\hat{\mathbf{W}}] - [\hat{\mathbf{W}}] \cdot [\sigma] = \begin{bmatrix} 0 & 0 & 0 \\ 0 & -\sigma_{z\theta} & \dfrac{1}{2}(\sigma_{\theta\theta} - \sigma_{zz}) \\ 0 & \dfrac{1}{2}(\sigma_{\theta\theta} - \sigma_{zz}) & \sigma_{z\theta} \end{bmatrix}\dfrac{r\dot{\phi}}{l},$$

which is, as expected, a symmetric traceless matrix. Thus, the nonvanishing Jaumann rates of the Cauchy stress in cylindrical coordinates are

$$\overset{\triangledown}{\sigma}_{rr} = \dot{\sigma}_{rr},$$

$$\overset{\triangledown}{\sigma}_{\theta\theta} = \dot{\sigma}_{\theta\theta} - \sigma_{z\theta}\frac{r}{l}\dot{\phi},$$

$$\overset{\triangledown}{\sigma}_{zz} = \dot{\sigma}_{zz} + \sigma_{z\theta}\frac{r}{l}\dot{\phi},$$

$$\overset{\triangledown}{\sigma}_{z\theta} = \dot{\sigma}_{z\theta} + \frac{1}{2}(\sigma_{\theta\theta} - \sigma_{zz})\frac{r}{l}\dot{\phi}.$$

This, of course, also follows directly from the general results of Problem 5.10 by substituting there the velocity field $v_r = \dot{r}$, $v_\theta = rz\dot{\phi}/l$, $v_z = z\dot{l}/l$.

In a tension–torsion test of a thin-wall tube, we may assume that $\sigma_{rr} = \sigma_{\theta\theta} = 0$, and the nonvanishing Jaumann rates of stress are

$$\overset{\triangledown}{\sigma}_{\theta\theta} = -\sigma_{z\theta}\frac{r}{l}\dot{\phi},$$

$$\overset{\triangledown}{\sigma}_{zz} = \dot{\sigma}_{zz} + \sigma_{z\theta}\frac{r}{l}\dot{\phi},$$

$$\overset{\triangledown}{\sigma}_{z\theta} = \dot{\sigma}_{z\theta} - \frac{1}{2}\sigma_{zz}\frac{r}{l}\dot{\phi}.$$

CHAPTER 8

Problem 8.1. The components of elastic moduli tensor of an isotropic material are

$$C_{ijkl} = \lambda \delta_{ij}\delta_{kl} + 2\mu I_{ijkl} .$$

This can be rewritten as

$$C_{ijkl} = 2\mu J_{ijkl} + 3\kappa K_{ijkl} ,$$

where $\kappa = \lambda + 2\mu/3$, and

$$K_{ijkl} = \frac{1}{3}\delta_{ij}\delta_{kl} , \quad J_{ijkl} = I_{ijkl} - K_{ijkl} .$$

The fourth-order unit tensor is $I_{ijkl} = (\delta_{ik}\delta_{jl} + \delta_{il}\delta_{jk})/2$. Show that the components of the corresponding elastic compliance tensor are

$$S_{ijkl} = \frac{1}{2\mu} J_{ijkl} + \frac{1}{3\kappa} K_{ijkl} .$$

Solution: Evidently

$$S_{ijmn} C_{mnkl} = C_{ijmn} S_{mnkl} = I_{ijkl} ,$$

because $\mathbf{K} : \mathbf{K} = \mathbf{K}, \mathbf{J} : \mathbf{J} = \mathbf{J}, \mathbf{K} : \mathbf{J} = \mathbf{J} : \mathbf{K} = \mathbf{0}$ (see Problem 1.11). Thus, the components S_{ijkl}, as given above, are the components of the inverse tensor of the elastic moduli tensor, i.e., the components of the elastic compliance tensor.

Problem 8.2. The stress-strain relationships for a cubic crystal are

$$\sigma_{11} = c_{11}e_{11} + c_{12}(e_{22} + e_{33}) ,$$
$$\sigma_{22} = c_{11}e_{22} + c_{12}(e_{33} + e_{11}) ,$$
$$\sigma_{33} = c_{11}e_{33} + c_{12}(e_{11} + e_{22}) ,$$

and

$$\sigma_{12} = 2c_{44}e_{12}, \quad \sigma_{23} = 2c_{44}e_{23}, \quad \sigma_{31} = 2c_{44}e_{31} .$$

(a) Derive the inverse relationships expressing the strain components e_{ij} in terms of stress components σ_{ij} and the elastic constants c_{ij}.
(b) Derive the relationships between c_{11}, c_{12}, c_{44} and E, ν, G (Young's modulus, Poisson's ratio, shear modulus).

Solution: (a) First observe that

$$\sigma_{kk} = (c_{11} + 2c_{12})e_{kk}, \quad e_{kk} = \frac{1}{c_{11} + 2c_{12}} \sigma_{kk} .$$

Thus,

$$\sigma_{11} = (c_{11} - c_{12})e_{11} + c_{12}e_{kk} ,$$

and

$$e_{11} = \frac{1}{c_{11} - c_{12}} \left(\sigma_{11} - \frac{c_{12}}{c_{11} + 2c_{12}} \sigma_{kk} \right).$$

Similar derivation proceeds to find the expressions for e_{22} and e_{33}. For the shear strain we have trivially $e_{12} = \sigma_{12}/2c_{44}$ and likewise for e_{23} and e_{31}.

(b) Comparing

$$e_{11} = \frac{1}{E} \sigma_{11} - \frac{\nu}{E} (\sigma_{22} + \sigma_{33}),$$

with

$$e_{11} = \frac{1}{c_{11} - c_{12}} \left[\left(1 - \frac{c_{12}}{c_{11} + 2c_{12}} \right) \sigma_{11} - \frac{c_{12}}{c_{11} + 2c_{12}} (\sigma_{22} + \sigma_{33}) \right],$$

there follows

$$E = \frac{(c_{11} - c_{12})(c_{11} + 2c_{12})}{c_{11} + c_{12}}, \quad \nu = \frac{c_{12}}{c_{11} + c_{12}}.$$

For the shear modulus we have $G = c_{44}$.

Problem 8.3. For a thermally anisotropic elastic solid, the relationship between the infinitesimal thermal strain e_{ij} and the temperature difference $\theta - \theta_0$ is $e_{ij} = \alpha_{ij}(\theta - \theta_0)$, where α_{ij} are the components of the second-order tensor of thermal coefficients.

(a) If (x_1, x_2) is a plane of thermal symmetry, derive the restrictions on α_{ij} and write the resulting matrix $[\alpha_{ij}]$.
(b) If (x_2, x_3) is also a plane of thermal symmetry, derive the additional restrictions on α_{ij} and write the resulting matrix $[\alpha_{ij}]$.
(c) If x_3 is the axis of thermal transverse isotropy, derive the restrictions on α_{ij} and write the resulting matrix $[\alpha_{ij}]$.
(d) Write the matrix $[\alpha_{ij}]$ for thermally isotropic material.

Solution: (a) The symmetry transformation is

$$\mathbf{Q} = \begin{bmatrix} 1 & 0 & 0 \\ 0 & 1 & 0 \\ 0 & 0 & -1 \end{bmatrix}.$$

The components are thus $Q_{ij} = \delta_{ij} - 2\delta_{i3}\delta_{j3}$. Imposing the symmetry condition $\mathbf{Q}^T \cdot \boldsymbol{\alpha} \cdot \mathbf{Q} = \boldsymbol{\alpha}$, there follows $\alpha_{13} = \alpha_{23} = 0$, so that

$$\boldsymbol{\alpha} = \begin{bmatrix} \alpha_{11} & \alpha_{12} & 0 \\ \alpha_{21} & \alpha_{22} & 0 \\ 0 & 0 & \alpha_{33} \end{bmatrix}.$$

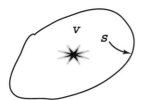

Figure 8-1. An internal source of stress.

(b) In addition to symmetry transformation from part (a), we also have the symmetry transformation $Q_{ij} = \delta_{ij} - 2\delta_{i1}\delta_{j1}$. This leads to $\alpha_{21} = 0$, and thus

$$\alpha = \begin{bmatrix} \alpha_{11} & 0 & 0 \\ 0 & \alpha_{22} & 0 \\ 0 & 0 & \alpha_{33} \end{bmatrix}.$$

(c) In this case

$$\mathbf{Q} = \begin{bmatrix} \cos\varphi & -\sin\varphi & 0 \\ \sin\varphi & \cos\varphi & 0 \\ 0 & 0 & 1 \end{bmatrix}.$$

Because transversely isotropic material is also orthotropic, begin with the matrix

$$\alpha = \begin{bmatrix} \alpha_{11} & 0 & 0 \\ 0 & \alpha_{22} & 0 \\ 0 & 0 & \alpha_{33} \end{bmatrix},$$

and show that $\alpha_{11} = \alpha_{22}$. Thus,

$$\alpha = \begin{bmatrix} \alpha_{11} & 0 & 0 \\ 0 & \alpha_{11} & 0 \\ 0 & 0 & \alpha_{33} \end{bmatrix}.$$

(d) In the case of thermally isotropic material $\alpha_{11} = \alpha_{22} = \alpha_{33} = \alpha$, so that

$$\alpha = \begin{bmatrix} \alpha & 0 & 0 \\ 0 & \alpha & 0 \\ 0 & 0 & \alpha \end{bmatrix}.$$

Problem 8.4. For a purely *internal source of stress*, that has no associated body forces or applied surface traction (Fig. 8-1), the volume average of each component of stress is zero, i.e.,

$$\int_V \sigma \, dV = \mathbf{0}.$$

This, in turn, means that the net dilatation is zero. Prove the above.

Solution: Write $\sigma_{ij} = \partial(\sigma_{i\ell}x_j)/\partial x_\ell$. This is true because the equilibrium equations for this case are $\partial\sigma_{i\ell}/\partial x_\ell = 0$, and $\partial x_\ell/\partial x_j = \delta_{j\ell}$. Thus,

$$\int_V \partial(\sigma_{i\ell}x_j)/\partial x_\ell \, dV = \int_S \sigma_{i\ell}n_\ell x_j \, dS = 0,$$

Problems 8.4–8.5

because the surface S is traction free. Noting that for a linear elastic body, the connection between stress and strain is itself linear, $\sigma_{ij} = C_{ijkl}e_{kl}$, we conclude that all components of strain have vanishing average values, as well.

Problem 8.5. Given the stress distribution

$$\sigma = \begin{bmatrix} x_1^2 x_2 - \frac{2}{3}x_2^3 & \sigma_{12}(x_1, x_2) & 0 \\ \sigma_{12}(x_1, x_2) & \frac{1}{3}x_2^3 & 0 \\ 0 & 0 & \frac{1}{3}(x_1^2 x_2 - \frac{1}{3}x_2^3) \end{bmatrix},$$

find σ_{12} so that the stress distribution is in equilibrium with zero body force, and that the traction vector on the plane $x_1 = x_2$, with the unit normal $\mathbf{n} = \{-1, 1, 0\}/\sqrt{2}$, is given by

$$\mathbf{T}_n = \frac{2\sqrt{2}}{3}\left[(-x_1^3 + a^3)\mathbf{e}_1 + (x_1^3 - a^3)\mathbf{e}_2\right],$$

where a is a given constant.

Solution: With $b_i = 0$, the equations of equilibrium are

$$\frac{\partial \sigma_{ij}}{\partial x_j} = 0.$$

Thus,

$$\frac{\partial \sigma_{11}}{\partial x_1} + \frac{\partial \sigma_{12}}{\partial x_2} + \frac{\partial \sigma_{13}}{\partial x_3} = 2x_1 x_2 + \frac{\partial \sigma_{12}}{\partial x_2} = 0,$$

$$\frac{\partial \sigma_{12}}{\partial x_1} + \frac{\partial \sigma_{22}}{\partial x_2} + \frac{\partial \sigma_{23}}{\partial x_3} = \frac{\partial \sigma_{12}}{\partial x_1} + x_2^2 = 0,$$

and the third of the equilibrium equations vanishes identically. Now,

$$2x_1 x_2 + \frac{\partial \sigma_{12}}{\partial x_2} = 0 \rightarrow \sigma_{12} = -x_1 x_2^2 + f(x_1),$$

$$\frac{\partial \sigma_{21}}{\partial x_1} + x_2^2 = 0 \rightarrow \sigma_{21} = -x_1 x_2^2 + g(x_2).$$

Consequently, $f(x_1) = g(x_2) = C = $ const., and

$$\sigma = \begin{bmatrix} x_1^2 x_2 - \frac{2}{3}x_2^3 & -x_1 x_2^2 + C & 0 \\ -x_1 x_2^2 + C & \frac{1}{3}x_2^3 & 0 \\ 0 & 0 & \frac{1}{3}(x_1^2 x_2 - \frac{1}{3}x_2^3) \end{bmatrix}.$$

The unit normal of the plane $x_1 = x_2$ is $\mathbf{n} = \pm\{-1, 1, 0\}/\sqrt{2}$. Taking the plus sign, we obtain

$$\mathbf{T}_n = \sigma \cdot \mathbf{n} = \frac{1}{\sqrt{2}} \begin{bmatrix} -\frac{4}{3}x_1^3 + C \\ \frac{4}{3}x_1^3 - C \\ 0 \end{bmatrix}.$$

Therefore, by comparing with a given expression for $\mathbf{T}_n$, we find $C = 4a^3/3$, and

$$\sigma_{12} = -x_1 x_2^2 + \frac{4}{3}a^3.$$

Problem 8.6. Show that the isotropic Hooke's law may be written as

$$\sigma_{ij} = 2\mu\left(e_{ij} + \frac{\nu}{1-2\nu}e_{kk}\delta_{ij}\right),$$

and hence derive the following form of the equilibrium equations for an elastic body in terms of the displacements

$$\mu\left(\nabla^2 u_i + \frac{1}{1-2\nu}\frac{\partial^2 u_j}{\partial x_i \partial x_j}\right) + b_i = 0.$$

Solution: By taking a trace of (8.13), i.e.,

$$e_{ij} = \frac{1}{E}[(1+\nu)\sigma_{ij} - \nu\sigma_{kk}\delta_{ij}],$$

we obtain

$$e_{kk} = \frac{1-2\nu}{E}\sigma_{kk}.$$

Substituting this back in the original expression, there follows

$$\sigma_{ij} = 2\mu\left(e_{ij} + \frac{\nu}{1-2\nu}e_{kk}\delta_{ij}\right).$$

Since $e_{ij} = (u_{i,j} + u_{j,i})/2$, the above becomes

$$\sigma_{ij} = 2\mu\left[\frac{1}{2}(u_{i,j} + u_{j,i}) + \frac{\nu}{1-2\nu}u_{k,k}\delta_{ij}\right].$$

When this is inserted in the equations of equilibrium,

$$\sigma_{ij,j} + b_i = 0,$$

we obtain

$$\mu\left(u_{i,jj} + \frac{1}{1-2\nu}u_{j,ij}\right) + b_i = 0,$$

which is the desired expression.

Problem 8.7. Assume that the body force **b** per unit volume is derivable from a harmonic potential function φ, so that

$$\mathbf{b} = -\nabla\varphi, \quad \nabla^2\varphi = 0.$$

(a) Write the corresponding Beltrami–Michell compatibility equations.
(b) Show that the first invariant of stress tensor, σ_{kk}, is a harmonic function, i.e., $\nabla^2\sigma_{kk} = 0$.
(c) Show that each stress component σ_{ij} is a biharmonic function, i.e., $\nabla^4\sigma_{ij} = 0$.

Solution: The Beltrami–Michell equations of (8.84) are

$$\sigma_{ij,kk} + \frac{1}{1+\nu}\sigma_{kk,ij} = -\frac{\nu}{1-\nu}b_{k,k}\delta_{ij} - b_{i,j} - b_{j,i}.$$

Since $\mathbf{b} = -\nabla\varphi$, we have

$$b_i = -\varphi_{,i} \quad \Rightarrow \quad b_{i,j} = -\varphi_{,ij}, \quad b_{j,i} = -\varphi_{,ji}.$$

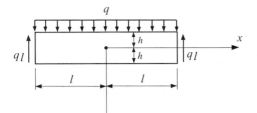

Figure 9-1. A simply supported beam under uniform loading of intensity q.

Thus, the Beltrami–Michell compatibility equations become

$$\sigma_{ij,kk} + \frac{1}{1+\nu}\sigma_{kk,ij} = 2\varphi_{,ij}.$$

(b) The contraction $i = j$ of the above equation give

$$\sigma_{ii,kk} + \frac{1}{1+\nu}\sigma_{kk,ii} = 2\varphi_{,ii}.$$

Since

$$\varphi_{,ii} = \nabla^2\varphi = 0,$$

we conclude that

$$\sigma_{kk,ii} = \nabla^2\sigma_{kk} = 0,$$

i.e., the first invariant of the stress tensor, σ_{kk}, is a harmonic function.

(c) By applying the Laplacian $\nabla^2 = \partial^2/\partial x_l \partial x_l$ derivative to the previously derived Beltrami-Michell compatibility equations, there follows

$$\sigma_{ij,kkll} + \frac{1}{1+\nu}\sigma_{kk,ijll} = 2\varphi_{,ijll}.$$

But,

$$\sigma_{kk,ijll} = (\sigma_{kk,ll})_{,ij} = 0, \quad \varphi_{,ijll} = (\varphi_{,ll})_{,ij} = 0,$$

and we conclude that

$$\sigma_{ij,kkll} = 0, \quad \nabla^4\sigma_{ij} = 0.$$

Thus each stress component σ_{ij} is a biharmonic function.

CHAPTER 9

Problem 9.1. For a simply supported beam of unit thickness, loaded with uniformly distributed load q over its top side, the airy stress function is

$$\phi = -\frac{3q}{4h^3}\left(\frac{1}{6}x^2y^3 - \frac{1}{2}h^2x^2y + \frac{1}{3}h^3x^2 - \frac{1}{30}y^5\right) + \frac{q}{8h^3}\left(l^2 - \frac{2}{5}h^2\right)y^3.$$

Derive the stress components and verify the point-wise boundary conditions at $y = \pm h$, and the integral boundary conditions at $x = \pm l$ (see Fig. 9-1).

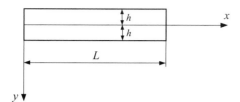

Figure 9-2. A rectangular beam of height $2h$ and length L with the utilized coordinate axes to express the stress function Φ.

Solution: The stresses are

$$\sigma_{xx} = \frac{\partial^2 \phi}{\partial y^2} = \frac{3q}{4h^3}\left(l^2 - x^2 + \frac{2}{3}y^3 - \frac{2}{5}h^2 y\right),$$

$$\sigma_{yy} = \frac{\partial^2 \phi}{\partial x^2} = -\frac{3q}{4h^3}\left(\frac{1}{3}y^3 - h^2 y + \frac{2}{3}h^3\right),$$

$$\sigma_{xy} = -\frac{\partial^2 \phi}{\partial x \partial y} = -\frac{3q}{4h^3}\left(h^2 - y^2\right)x.$$

It can be easily verified that

$$\sigma_{xy}(x, \pm h) = 0, \quad \sigma_{yy}(x, h) = 0, \quad \sigma_{yy}(x, -h) = -q,$$

and

$$\int_{-h}^{h} \sigma_{xy}(\pm l, y)\,dy = \mp ql, \quad \int_{-h}^{h} \sigma_{xx}(\pm l, y)\,dy = 0, \quad \int_{-h}^{h} y\sigma_{xx}(\pm l, y)\,dy = 0.$$

The first of these integral boundary conditions is the condition for the total shear force at the ends of the beam, and the last two are the conditions of the vanishing axial force and bending moment at the ends of the beam.

Problem 9.2. The Airy stress function for a rectangular beam of height $2h$ and length L is

$$\Phi = \frac{\tau_0}{4}\left(xy - \frac{xy^2}{h} - \frac{xy^3}{h^2} + \frac{Ly^2}{h} + \frac{Ly^3}{h^2}\right).$$

The coordinate axes are as shown in Fig. 9-2. Calculate the corresponding traction components on the four sides of the beam. In particular, show that the lower side of the beam is traction free and that there is no normal stress on the side $x = L$.

Solution: The stress components are

$$\sigma_{xx} = \frac{\partial^2 \Phi}{\partial y^2} = \frac{\tau_0}{4}\left(-2\frac{x}{h} - 6\frac{xy}{h^2} + \frac{2L}{h} + \frac{6Ly}{h^2}\right),$$

$$\sigma_{yy} = \frac{\partial^2 \Phi}{\partial x^2} = 0,$$

$$\sigma_{xy} = -\frac{\partial^2 \Phi}{\partial x \partial y} = -\frac{\tau_0}{4}\left(1 - 2\frac{y}{h} - 3\frac{y^2}{h^2}\right).$$

Consequently, the stress components over the four sides of the beam are

$$\sigma_{xy}(x, h) = \tau_0, \quad \sigma_{xy}(x, -h) = 0,$$

$$\sigma_{xy}(0, y) = \sigma_{xy}(L, y) = -\frac{\tau_0}{4}\left(1 - 2\frac{y}{h} - 3\frac{y^2}{h^2}\right),$$

$$\sigma_{xx}(0, y) = \frac{\tau_0}{4}\left(2\frac{L}{h} + 6\frac{Ly}{h^2}\right), \quad \sigma_{xx}(L, y) = 0.$$

CHAPTER 10

Problem 10.1. Derive the solution of a two-dimensional Lamé problem (pressurized hollow cylinder under conditions of plain-strain) by using a stress-based formulation.

Solution: The only displacement component is the radial displacement $u = u(r)$. The corresponding strain components are

$$e_{rr} = \frac{du}{dr}, \quad e_{\theta\theta} = \frac{u}{r},$$

with the Saint-Venant compatibility equation

$$\frac{de_{\theta\theta}}{dr} = \frac{1}{r}(e_{rr} - e_{\theta\theta}).$$

The nonvanishing stress components are the radial stress σ_{rr}, the hoop stress $\sigma_{\theta\theta}$, and the longitudinal stress σ_{zz}. Since for plain strain $e_{zz} = 0$, the Hooke's law gives $\sigma_{zz} = \nu(\sigma_{rr} + \sigma_{\theta\theta})$. In the absence of body forces, the equilibrium equation is

$$\frac{d\sigma_{rr}}{dr} + \frac{1}{r}(\sigma_{rr} - \sigma_{\theta\theta}) = 0.$$

The Beltrami–Michell compatibility equation is obtained from the Saint-Venant compatibility equation by incorporating the stress-strain relations

$$e_{rr} = \frac{1}{E}[\sigma_{rr} - \nu(\sigma_{\theta\theta} + \sigma_{zz})], \quad e_{\theta\theta} = \frac{1}{E}[\sigma_{\theta\theta} - \nu(\sigma_{rr} + \sigma_{zz})],$$

and the equilibrium equation. This gives

$$\frac{d}{dr}(\sigma_{rr} + \sigma_{\theta\theta}) = 0,$$

which implies that the spherical component of stress tensor is uniform throughout the medium,

$$\frac{1}{3}(\sigma_{rr} + \sigma_{\theta\theta} + \sigma_{zz}) = \frac{1+\nu}{3}(\sigma_{rr} + \sigma_{\theta\theta}) = \frac{2A(1+\nu)}{3} = \text{const.}$$

Combining this with the equilibrium equation gives

$$\frac{d\sigma_{rr}}{dr} + \frac{2}{r}\sigma_{rr} = \frac{2}{r}A.$$

The general solution of this equation is
$$\sigma_{rr} = A + \frac{B}{r^2}.$$

The corresponding hoop stress is
$$\sigma_{\theta\theta} = A - \frac{B}{r^2}.$$

The boundary conditions for the Lamé problem of a pressurized hollow cylinder are
$$\sigma_{rr}(R_1) = -p_1, \quad \sigma_{rr}(R_2) = -p_2.$$

These are satisfied provided that
$$A = \frac{p_1 R_1^2 - p_2 R_2^2}{R_2^2 - R_1^2}, \quad B = -\frac{R_1^2 R_2^2}{R_2^2 - R_1^2}(p_1 - p_2).$$

Consequently, the stress components are
$$\sigma_{rr} = \frac{R_2^2}{R_2^2 - R_1^2}\left[p_1 \frac{R_1^2}{R_2^2} - p_2 - (p_1 - p_2)\frac{R_1^2}{r^2}\right],$$

$$\sigma_{\theta\theta} = \frac{R_2^2}{R_2^2 - R_1^2}\left[p_1 \frac{R_1^2}{R_2^2} - p_2 + (p_1 - p_2)\frac{R_1^2}{r^2}\right].$$

The corresponding hoop strain is obtained by substituting last two equations into the stress-relation. The result is
$$e_{\theta\theta} = \frac{1}{2\mu}\left[(1 - 2\nu)A - \frac{B}{r^2}\right].$$

Thus, the radial displacement $u = r e_{\theta\theta}$ is
$$u = \frac{1}{2\mu} \frac{R_2^2 r}{R_2^2 - R_1^2}\left[(1 - 2\nu)\left(p_1 \frac{R_1^2}{R_2^2} - p_2\right) + (p_1 - p_2)\frac{R_1^2}{r^2}\right].$$

For the nonpressurized hole ($p_1 = 0$) under remote pressure p_2 at infinity, the previous results give
$$\sigma_{rr} = -p_2\left(1 - \frac{R_1^2}{r^2}\right), \quad \sigma_{\theta\theta} = -p_2\left(1 + \frac{R_1^2}{r^2}\right),$$

$$u = -\frac{p_2}{2\mu}\left[(1 - 2\nu)r + \frac{R_1^2}{r}\right].$$

For the pressurized cylindrical hole in an infinite medium with $p_2 = 0$, we have
$$\sigma_{rr} = -p_1 \frac{R_1^2}{r^2}, \quad \sigma_{\theta\theta} = p_1 \frac{R_1^2}{r^2},$$

$$u = \frac{p_1}{2\mu} \frac{R_1^2}{r}.$$

Problem 10.2. Derive the solution of the previous problem (pressurized hollow cylinder under conditions of plain-strain) by using a displacement-based formulation.

Problems 10.2–10.3

Solution: In the displacement-based approach, the substitution of the strain-displacement expressions

$$e_{rr} = \frac{du}{dr}, \quad e_{\theta\theta} = \frac{u}{r},$$

into the stress-strain relations give

$$\sigma_{rr} = 2\mu e_{rr} + \lambda(e_{rr} + e_{\theta\theta}) = (\lambda + 2\mu)\frac{du}{dr} + \lambda\frac{u}{r},$$

$$\sigma_{\theta\theta} = 2\mu e_{\theta\theta} + \lambda(e_{rr} + e_{\theta\theta}) = \lambda\frac{du}{dr} + (\lambda + 2\mu)\frac{u}{r}.$$

When this is introduced into the equilibrium equation

$$\frac{d\sigma_{rr}}{dr} + \frac{1}{r}(\sigma_{rr} - \sigma_{\theta\theta}) = 0,$$

there follows

$$\frac{d^2u}{dr^2} + \frac{1}{r}\frac{du}{dr} - \frac{u}{r^2} = 0.$$

The solution of this equation is

$$u = Cr + \frac{D}{r}.$$

Its substitution back into above expressions for the stress components gives

$$\sigma_{rr} = 2(\lambda + \mu)C - 2\mu\frac{D}{r^2}, \quad \sigma_{\theta\theta} = 2(\lambda + \mu)C + 2\mu\frac{D}{r^2}.$$

The boundary conditions

$$\sigma_{rr}(R_1) = -p_1, \quad \sigma_{rr}(R_2) = -p_2.$$

specify the integration constants

$$C = \frac{1}{2(\lambda + \mu)} \frac{p_1 R_1^2 - p_2 R_2^2}{R_2^2 - R_1^2}, \quad D = \frac{1}{2\mu} \frac{R_1^2 R_2^2}{R_2^2 - R_1^2} (p_1 - p_2).$$

Evidently, these are related to the integration constants of the stress-based solution by $A = 2(\lambda + \mu)C$ and $B = -2\mu D$. The Poisson's ratio can be incorporated in the results by recalling that $\lambda + \mu = \mu/(1 - 2\nu)$.

Problem 10.3. Derive the stress distribution in a rotating circular disk whose thickness is small relative to its radius R. The density of the disk is ρ, and the angular speed of the disk is ω.

Solution: We can treat the inertial force due to rotation as the body force $\rho\omega^2 r$. Because of symmetry, the only nonvanishing stress components are σ_{rr} and $\sigma_{\theta\theta}$, and the (dynamic) equilibrium equation becomes

$$\frac{d\sigma_{rr}}{dr} + \frac{1}{r}(\sigma_{rr} - \sigma_{\theta\theta}) + \rho\omega^2 r = 0.$$

If $u = u(r)$ is the radial displacement, the strain-displacement expressions are

$$e_{rr} = \frac{du}{dr}, \quad e_{\theta\theta} = \frac{u}{r}.$$

The plane stress Hooke's law then gives

$$\sigma_{rr} = \frac{E}{1-\nu^2}\left(\frac{du}{dr} + \nu\frac{u}{r}\right),$$

$$\sigma_{\theta\theta} = \frac{E}{1-\nu^2}\left(\frac{u}{r} + \nu\frac{du}{dr}\right).$$

When this is substituted into the equilibrium equation, we obtain

$$\frac{d^2u}{dr^2} + \frac{1}{r}\frac{du}{dr} - \frac{u}{r^2} = -\frac{1-\nu^2}{E}\rho\omega^2 r.$$

The solution of this equation is

$$u = Cr + \frac{D}{r} - \frac{1-\nu^2}{8E}\rho\omega^2 r^3.$$

Since $u(0) = 0$, we have $D = 0$. The resulting stresses are

$$\sigma_{rr} = \frac{E}{1-\nu}C - \frac{3+\nu}{8}\rho\omega^2 r^2,$$

$$\sigma_{\theta\theta} = \frac{E}{1-\nu}C - \frac{1+3\nu}{8}\rho\omega^2 r^2.$$

If the outer surface $r = R$ of the disk is stress free, $\sigma_{rr}(R) = 0$, which is satisfied when

$$C = \frac{(1-\nu)(3+\nu)}{8E}\rho\omega^2 r^2.$$

Thus, the stress distribution within the disk is

$$\sigma_{rr} = \frac{3+\nu}{8}\rho\omega^2(R^2 - r^2),$$

$$\sigma_{\theta\theta} = \frac{3+\nu}{8}\rho\omega^2 R^2 - \frac{1+3\nu}{8}\rho\omega^2 r^2.$$

The similar analysis can be performed for a disk with a circular hole of radius R_0 at its center. The boundary conditions are in this case $\sigma_{rr}(R_0) = 0$ and $\sigma_{rr}(R) = 0$.

Problem 10.4. Revisit the problem of a wedge loaded by a point load at its apex (Fig. 10-1), and solve it by using a different approach than used in the text. In particular, calculate the stresses in the solid region occupied by the wedge.

Solution: We seek solutions of the biharmonic equation of the form

$$\phi = rf(\theta),$$

Problem 10.4

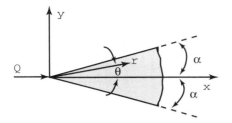

Figure 10-1. Semi-infinite wedge loaded with a point load at its apex.

based on the following consideration. We know that the shear stress vanishes so that $\sigma_{r\theta} = 0$. To ensure static equilibrium against the applied force Q, it will be necessary for the radial stress, σ_{rr}, to be of the form

$$\sigma_{rr} \sim \frac{1}{r} g(\theta),$$

because the length of any arc, *i.e.*, at any value of θ, scales proportionately with radius r. A quick perusal of the relations between the derivatives of ϕ and the components of stress will reveal the logic. Then if the above form of ϕ is substituted into the biharmonic equation, there results the ordinary differential equation

$$\frac{d^4 f(\theta)}{d\theta^4} + 2 \frac{d^2 f(\theta)}{d\theta^2} + f(\theta) = 0.$$

The boundary conditions are

$$\int_{-\alpha}^{\alpha} (\sigma_{rr} \cos\theta - \sigma_{r\theta} \sin\theta) r \, d\theta + Q = 0,$$

$$\int_{-\alpha}^{\alpha} (\sigma_{rr} \sin\theta + \sigma_{r\theta} \cos\theta) r \, d\theta = 0,$$

$$\int_{-\alpha}^{\alpha} \sigma_{r\theta} r^2 \, d\theta = 0.$$

We note that the above three conditions hold for any value of r. In addition, there are also the conditions

$$\sigma_{\theta\theta} = \sigma_{r\theta} = 0, \quad \text{at } \theta = \pm\alpha,$$

$$\sigma_{rr} = \sigma_{\theta\theta} = \sigma_{r\theta} = 0, \quad \text{at } r \to \infty.$$

The general solution for $f(\theta)$ is

$$f(\theta) = A \sin\theta + B \cos\theta + C\theta \sin\theta + D\theta \cos\theta.$$

The constants may be calculated by applying the boundary conditions listed above. When this is done, the results for the stresses are

$$\sigma_{rr} = -\frac{2Q}{2\alpha + \sin(2\alpha)} \frac{\cos\theta}{r},$$

$$\sigma_{\theta\theta} = \sigma_{r\theta} = 0,$$

as we have seen in the text using a somewhat different approach.

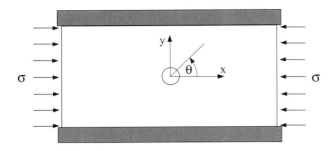

Figure 10-2. Small hole in a large plate constrained in the y-direction.

Problem 10.5. A large rectangular plate with a small central hole is compressed in the longitudinal direction, while its lateral expansion is prevented by rigid boundaries, as shown in Fig. 10-2. Assuming the plane stress conditions, determine the hoop stress along the circumference of the hole.

Solution: The remote stress σ_{yy} along the fixed boundaries is obtained from the condition there

$$e_{yy} = \frac{1}{E}(\sigma_{xx} - \nu\sigma_{yy}) = 0.$$

Since $\sigma_{xx} = -\sigma$, we obtain $\sigma_{yy} = -\nu\sigma$ along the fixed boundaries. The hoop stress along the circumference of the hole due to longitudinal compression of magnitude σ in a plate with unconstrained boundaries is, from the results in the text,

$$\sigma_{\theta\theta} = -\sigma(1 - 2\cos 2\theta).$$

Similarly, the hoop stress along the circumference of the hole due to lateral compression of magnitude $\nu\sigma$ alone is

$$\sigma_{\theta\theta} = -\nu\sigma(1 + 2\cos 2\theta).$$

Thus, by the superposition, the total hoop stress is

$$\sigma_{\theta\theta} = -\sigma[1 + \nu - 2(1 - \nu)\cos 2\theta].$$

The hoop stress at $\theta = \pi/2$ is equal to $\sigma_{\theta\theta} = -(3 - \nu)\sigma$, whereas at $\theta = 0$ it is $\sigma_{\theta\theta} = (1 - 3\nu)\sigma$. The former is always compressive, while the latter is tensile for $\nu < 1/3$ and compressive for $\nu > 1/3$.

Problem 10.6. Derive the compatibility equation in polar coordinates for an axisymmetric plane strain problem.

Solution: In the case of axial symmetry, the only nonvanishing displacement is the radial displacement $u_r = u_r(r)$, with the corresponding strains

$$e_{rr} = \frac{du_r}{dr}, \quad e_{\theta\theta} = \frac{u_r}{r}.$$

Problems 10.6–10.7

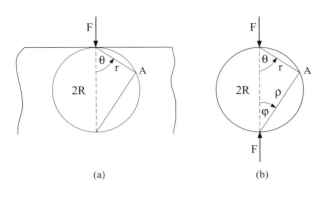

Figure 10-3. (a) A concentrated force F at the boundary of a half-space. Indicated is an arbitrary point A on the circle of diameter $2R$ within a half space. (b) A circular disk under two equal but opposite forces F along its diameter $2R$. The indicated directions r and rho through the point A are orthogonal.

Thus, by differentiating $u_r = r e_{\theta\theta}$, we obtain

$$e_{\theta\theta} + r \frac{d e_{\theta\theta}}{dr} = e_{rr},$$

which is a compatibility equation for the involved strain components.

More generally, for plane strain problems without axial symmetry, the polar components of the strain tensor are

$$e_{rr} = \frac{\partial u_r}{\partial r}, \quad e_{\theta\theta} = \frac{1}{r}\frac{\partial u_\theta}{\partial \theta} + \frac{u_r}{r}, \quad e_{r\theta} = \frac{1}{2}\left(\frac{1}{r}\frac{\partial u_r}{\partial \theta} + \frac{\partial u_\theta}{\partial r} - \frac{u_\theta}{r}\right).$$

The corresponding compatibility equation is

$$\frac{1}{r^2}\frac{\partial^2 e_{rr}}{\partial \theta^2} + \frac{\partial^2 e_{\theta\theta}}{\partial r^2} - \frac{2}{r}\frac{\partial^2 e_{r\theta}}{\partial r \partial \theta} - \frac{1}{r}\frac{\partial e_{rr}}{\partial r} + \frac{2}{r}\frac{\partial e_{\theta\theta}}{\partial r} - \frac{2}{r^2}\frac{\partial e_{r\theta}}{\partial \theta} = 0,$$

as can be verified by inspection.

Problem 10.7. The Airy stress function for the stress field in a half space due to a concentrated force F on its boundary (Flamant solution) is

$$\Phi = -\frac{F}{\pi} r\theta \sin\theta,$$

with the resulting stresses

$$\sigma_{rr} = -\frac{2F}{\pi}\frac{\cos\theta}{r}, \quad \sigma_{\theta\theta} = \sigma_{r\theta} = 0.$$

(a) Evaluate the stress state along the circle of an arbitrary diameter $d = 2R$ (Fig. 10-3a).
(b) Using the result from part (a), deduce the Airy stress function for the stress field in a circular disk due to two equal and opposite forces, shown in Fig 10-3b.
(c) Evaluate the normal stress along the horizontal and vertical diameter of the disk.

Solution: By simple geometry, at any point A of the circle shown in Fig. 10-3a,

$$2R \cos\theta = r.$$

Thus, the radial stress is constant and equal to

$$\sigma_{rr} = -\frac{F}{\pi R},$$

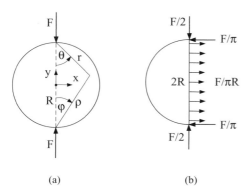

Figure 10-4. (a) A circular disk under two equal but opposite forces F along its diameter $2R$. The two set of coordinates (r, θ) and (ρ, φ) used to express the corresponding Airy stress function. (b) The stress state along the vertical diameter of the disk.

at any point of this circle (except the point of the application of the force, due to stress singularity there).

(b) The solution for the circular disk under a pair of equal but opposite forces along its diameter (Michell's problem) can be obtained by an appropriate superposition. For any point A on the boundary of the disk, the directions r and ρ are orthogonal. The sum of two Flamant's solutions from part (a), one for each of two forces, gives

$$\sigma_{rr} = -\frac{2F}{\pi}\frac{\cos\theta}{r}, \quad \sigma_{\rho\rho} = -\frac{2F}{\pi}\frac{\cos\varphi}{\rho}.$$

But,

$$\frac{r}{\cos\theta} = \frac{\rho}{\cos\varphi} = 2R,$$

so that

$$\sigma_{rr} = \sigma_{\rho\rho} = -\frac{F}{\pi R}$$

at any point A of the considered circle of diameter $2R$. Therefore, there is a state of equal biaxial compression at A, so that A feels the same stress over any plane. To achieve the traction free condition over the boundary of the disk, we then superpose a third solution – an equal biaxial tension of amount $F/\pi R$ within the plane of the disk (its Airy stress function being $Fr^2/\pi d$). The Airy stress function for the stress field in the disk is consequently

$$\Phi = \frac{F}{\pi}\left(\frac{r^2}{2R} - r\theta\sin\theta - \rho\varphi\sin\varphi\right).$$

Since, from Fig. 10-4a,

$$r\sin\theta = \rho\sin\varphi = x,$$

the above simplifies to

$$\Phi = \frac{F}{\pi}\left[\frac{r^2}{2R} - x(\theta + \varphi)\right].$$

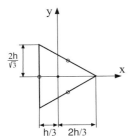

Figure 11-1. Cross section of a triangular shaft subject to torsion. The open circles indicate the symmetrically located positions of maximum shear stress as determined by the given solution.

(c) The rectangular stress components within the disk can be evaluated from

$$\sigma_{xx} = \frac{\partial^2 \Phi}{\partial y^2}, \quad \sigma_{yy} = \frac{\partial^2 \Phi}{\partial x^2}, \quad \sigma_{xy} = -\frac{\partial^2 \Phi}{\partial x \partial y},$$

by using the derived expression for the Airy stress function Φ, and the geometric relationships

$$r^2 = x^2 + (R-y)^2, \quad \rho^2 = x^2 + (R+y)^2,$$

$$\tan\theta = \frac{x}{R-y}, \quad \tan\varphi = \frac{x}{R+y}.$$

It readily follows that

$$\sigma_{xx} = \frac{2F}{\pi}\left\{\frac{1}{2R} - \frac{x^2(R-y)}{[x^2+(R-y)^2]^2} - \frac{x^2(R+y)}{[x^2+(R+y)^2]^2}\right\},$$

$$\sigma_{yy} = \frac{2F}{\pi}\left\{\frac{1}{2R} - \frac{(R-y)^3}{[x^2+(R-y)^2]^2} - \frac{(R+y)^3}{[x^2+(R+y)^2]^2}\right\},$$

$$\sigma_{xy} = \frac{2F}{\pi}\left\{\frac{x(R-y)^2}{[x^2+(R-y)^2]^2} - \frac{x(R+y)^2}{[x^2+(R+y)^2]^2}\right\}.$$

The normal stress along the horizontal diameter is obtained for $y = 0$, which gives

$$\sigma_{yy}(x,0) = \frac{F}{\pi R}\left[1 - \frac{4R^4}{(x^2+R^2)^2}\right],$$

The normal stress along the vertical diameter is obtained for $x = 0$, which gives

$$\sigma_{xx}(0,y) = \frac{F}{\pi R}.$$

The net horizontal force due to this stress is $2F/\pi$, which is balanced by two concentrated horizontal forces due to stress concentration, as indicated in Fig. 10-4b.

CHAPTER 11

Problem 11.1. Consider the torsion of a rod with the cross section sketched in Fig. 11-1. Determine the stress function and use it to determine the stresses in the cross section and the maximum stresses in the cross section.

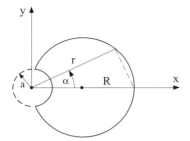

Figure 11-2. A circular cross section weekend by a circular groove of radius a.

Solution: Consider the stress function made up the product of three functions that are annihilated on each of the three sides of the triangular cross section, viz.,

$$\varphi = c(x - \sqrt{3}y - 2/3h)(x + \sqrt{3}y - 2/3h)(x + 1/3h).$$

When this is substituted into the governing equation, $\nabla^2 \varphi = -2G\theta = \mathcal{H}$, it is found that

$$c = -\frac{\mathcal{H}}{4h}.$$

Thus, for the stresses, we have

$$\sigma_{xz} = \partial\varphi/\partial y = \frac{15\sqrt{3}M}{h^5} y(3x + h),$$

$$\sigma_{yz} = -\partial\varphi/\partial x = \frac{15\sqrt{3}M}{2h^5}(-3x^2 + 2hx + 3y^2),$$

where the moment M is given by

$$M = 2\int \varphi\, dA = \frac{\sqrt{3}G\theta h^4}{45}.$$

From the above it is readily found that the maximum stresses occur at the three locations, $(h/6, h/3\sqrt{3})$, $(h/6, -h/3\sqrt{3})$, and $(-1/3h, 0)$, and are equal to

$$\sigma_{max} = \frac{15\sqrt{3}M}{2h^3}.$$

Problem 11.2. Determine the maximum shear stress in the rod of the grooved cross section shown in Fig. 11-2, if the rod is subject to the angle of twist θ.

Solution: The Prandtl's stress function is

$$\varphi = -\frac{G\theta}{2}(r^2 - a^2)\left(1 - 2\frac{R}{r}\cos\alpha\right),$$

because this satisfies the Poisson's differential equation $\nabla^2\varphi = -2G\theta$ in the interior of the cross section, and the boundary condition $\varphi = 0$ at the circles $r = a$ and $r = 2R\cos\alpha$. Rewritten in terms of the Cartesian coordinates, the above becomes

$$\varphi = -\frac{G\theta}{2}\left(x^2 + y^2 - 2Rx + 2Ra^2\frac{x}{x^2 + y^2} - \frac{1}{4}a^2\right).$$

Problems 11.2–11.3

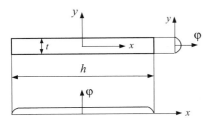

Figure 11-3. A narrow rectangular section under torsion, with an anticipated shape of the stress function φ.

The stress concentration occurs at the bottom of the groove ($x = a$, $y = 0$). Thus, the maximum shear stress is the magnitude of σ_{zy} at this point. Since

$$\sigma_{zy} = -\frac{\partial \varphi}{\partial x} = G\theta \left[x - R + Ra^2 \frac{y^2 - x^2}{(x^2 + y^2)^2} \right],$$

we obtain

$$\tau_{max} = G\theta(2R - a).$$

The angle of twist θ is related by the applied torsional moment through

$$M = 2 \int_A \varphi \, dA.$$

Problem 11.3. Derive the relationship between the applied torque T and the angle of twist θ for: (a) a thin-walled open tube, (b) a thin-walled closed tube, and (c) a closed tube with fins.

Solution: (a) For a thin rectangle the stress function is independent of x, except near the ends $x = \pm h/2$ (Fig. 11-3). Thus, by taking

$$\varphi = c \left(\frac{t^2}{4} - y^2 \right),$$

we have

$$\nabla^2 \varphi = 0 \quad \Rightarrow \quad c = G\theta \quad \Rightarrow \quad \varphi = G\theta \left(\frac{t^2}{4} - y^2 \right).$$

Since

$$T = 2 \int_A \varphi \, dA = 2G\theta h \int_{-t/2}^{t/2} \left(\frac{t^2}{4} - y^2 \right) dy = G\theta \frac{ht^3}{3},$$

there follows

$$\theta = \frac{T}{GI_t}, \quad I_t = \frac{1}{3} ht^3.$$

The corresponding shear stress is $\tau = \tau_{zx} = -2Ty/I_t$ (except near the ends $x = \pm h/2$, where both components of shear stress can be present, albeit of lesser magnitude than $\tau_{max} = Tt/I_t$). The same expressions can be used for a curved thin-walled open section,

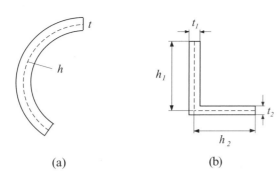

(a)   (b)

Figure 11-4. Thin-walled open sections with the indicated geometric characteristics used to calculate I_t.

such as shown in Fig. 11-4. If the thickness changes along the section, we can use

$$I_t = \frac{1}{3}\int t(s)^3 ds, \quad \text{or} \quad I_t = \frac{1}{3}\sum_{i=1}^{n} h_i t_i^3,$$

depending whether the change is continuous or not.

(b) A thin-walled closed section is a doubly connected region. Thus, by taking $\varphi = 0$ on the outer and $\varphi = f = \text{const.}$ on the inner contour of the cross section, and in view of the small thickness t of the section, the shear stress is parallel to the midline of the section and equal to the slope of φ across the thickness. This is nearly constant across the thickness and equal to

$$\tau(s) = \frac{f}{t(s)}.$$

Thus the shear flow $\tau(s)t(s) = f$ is constant around the section. The applied torque is related to the shear flow by the equilibrium requirement (Fig. 11-5b)

$$T = \oint_C hq\, ds = 2A_0 q,$$

where A_0 is the area within the midline C of the cross section. This gives

$$\tau(s) = \frac{T}{2A_0 t(s)}.$$

The relationship between the applied torque and the angle o twist can be conveniently derived by equating the work done by the applied torque to the strain energy in the tube, *i.e.*,

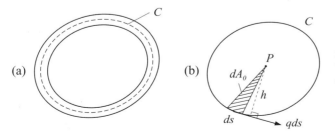

Figure 11-5. A thin-walled closed section under torsion. The midline of the section is C and q is the shear flow.

Problems 11.3–11.4

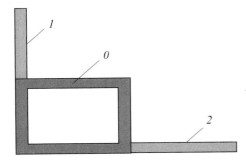

Figure 11-6. The cross section of a thin-walled closed tube with two attached fins.

$$\frac{1}{2}T(L\theta) = L\oint_C \frac{\tau^2}{2G} t\,ds\,.$$

Upon the substitution of the shear stress expression $\tau = T/2A_0 t$, there follows

$$\theta = \frac{T}{GI_t}\,, \quad I_t = \frac{4A_0^2}{\oint_C \dfrac{ds}{t(s)}}\,.$$

(c) For the thin-walled closed tube with two attached fins (Fig. 11-6), the applied torque is distributed between the tube and fins, whereas the angle of twist is same for the whole section, i.e.,

$$T = T_0 + T_1 + T_2\,, \quad \theta = \theta_0 = \theta_1 = \theta_2\,.$$

Consequently, by using the torque-twist relationships from parts (a) and (b), we obtain

$$T = G\theta I_t\,, \quad I_t = \frac{4A_0^2}{\oint_C \dfrac{ds}{t(s)}} + \frac{1}{3}(h_1 t_1^3 + h_2 t_2^3)\,.$$

If there are n fins, we have

$$I_t = \frac{4A_0^2}{\oint_C \dfrac{ds}{t(s)}} + \sum_{i=1}^n \frac{1}{3}h_i t_i^3\,.$$

Problem 11.4. By using the results for the bending of the cantilever beam of circular cross section, determine the location of the shear center of a semicircular cross section (Fig. 11-7a). The torsional moment of inertia for the semi-circle of radius R is $I_t = 0.298\,R^4$.

Solution: For the bending of a cantilever with circular cross section, there is a symmetry with respect to vertical diameter, so that there is no shear stress σ_{zx} along this diameter. Thus, shear stress distribution from this problem also satisfies the boundary conditions for a cantilever of a semicircular cross section. The shear stresses are given by (11.102), which can be rewritten as

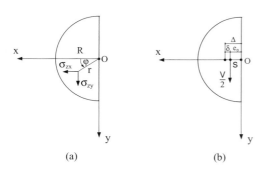

Figure 11-7. (a) A semicircular cross section of radius R. (b) The shear center S of the cross section.

$$\sigma_{zx} = -\frac{V}{\pi R^4} \frac{1+2\nu}{1+\nu} r^2 \sin\varphi \cos\varphi,$$

$$\sigma_{zy} = \frac{V}{2\pi R^4} \frac{1}{1+\nu} \left[(3+2\nu)(R^2 - r^2 \sin^2\varphi) - (1-2\nu) r^2 \cos^2\varphi\right].$$

The moment of these stresses for the point O is

$$M_O = \int_0^R \int_{-\pi/2}^{\pi/2} r\sigma_{z\varphi} r \, d\varphi \, dr, \quad \sigma_{z\varphi} = \sigma_{zy} \cos\varphi - \sigma_{zx} \sin\varphi.$$

It readily follows that

$$M_O = \frac{4}{15\pi} \frac{3+4\nu}{1+\nu} VR.$$

Since half of the force V is carried by each half of the circular section, the listed shear stresses are also present in a semicircular cross section under force $V/2$, provided that this force passes through the point at the distance Δ from the point O (Fig. 11-7b), such that

$$\frac{V}{2}\Delta = M_O \quad \Rightarrow \quad \Delta = \frac{8}{15\pi} \frac{3+4\nu}{1+\nu} R.$$

However, there is a net twist in each half of the circular cross section during the bending by the force V. This is

$$\bar{\theta} = \frac{2}{\pi R^2} \int_{A/2} \frac{\partial \omega_z}{\partial z} dA, \quad \frac{\partial \omega_z}{\partial z} = -\frac{\nu}{E} \frac{4V}{\pi R^4} x.$$

Since

$$\int_{A/2} x \, dA = \frac{4R}{3\pi} \frac{A}{2}, \quad A = \pi R^2,$$

we obtain

$$\bar{\theta} = -\frac{\nu}{E} \frac{16}{3\pi^2} \frac{V}{R^3}.$$

The minus sign indicates the clockwise rotation. If we don't want this net twist during the bending of a cantilever with a semicircular cross section, the force $V/2$ should act through the shear center S (Fig. 11-7b), such that the torque $T = (V/2)\delta$ cancels $\bar{\theta}$, i.e.,

Problems 11.4–11.5

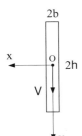

Figure 11-8. A narrow rectangular cross section of width 2b and height 2h.

$$\frac{(V/2)\delta}{GI_t} = \frac{\nu}{E} \frac{16}{3\pi^2} \frac{V}{R^3}.$$

Since $I_t = 0.298 R^4$ (see, for example, the book by Timoshenko and Goodier, who derive this result), the above equation gives

$$\delta = 0.298 \frac{16}{3\pi^2} \frac{\nu}{1+\nu} R.$$

The shear center S is at the distance $e_0 = \Delta - \delta$ from the poit O. For example, if $\nu = 0.3$, we obtain $\Delta = 0.548\,R$, $\delta = 0.037\,R$, and $e_0 = 0.511\,R$.

Problem 11.5. Derive the shear stress and displacement fields for the cantilever beam of a narrow rectangular cross section (Fig. 11-8).

Solution: As shown in the text, the stress function satisfies

$$\nabla^2 \Phi = -\frac{\nu}{1+\nu} \frac{V}{I_x} x + f'(x) \quad \text{within the rectangle},$$

and

$$\frac{d\Phi}{ds} = \left[-\frac{V}{2I_x} y^2 + f(x) \right] \frac{dx}{ds} \quad \text{on the boundary.}$$

By taking $f = Vh^2/2I_x$, where $I_x = 4bh^3/3$, we have $d\Phi/ds = 0$ along the sides $y = \pm h$. Since $dx/ds = 0$ along the sides $x = \pm b$, we achieve the boundary condition $\Phi = 0$ along all four sides of the rectangle. The differential equation becomes

$$\nabla^2 \Phi = -\frac{\nu}{1+\nu} \frac{V}{I_x} x.$$

Since $h \gg b$, we may assume that approximately $\Phi = \Phi(x)$ (except near the ends $y = \pm h$). Thus,

$$\frac{d^2\Phi}{dx^2} = -\frac{\nu}{1+\nu} \frac{V}{I_x} x \quad \Rightarrow \quad \Phi(x) = \frac{\nu}{1+\nu} \frac{V}{6I_x} (b^2 x - x^3).$$

The corresponding stresses are

$$\sigma_{zx} = \frac{\partial \Phi}{\partial y} = 0,$$

$$\sigma_{zy} = -\frac{\partial \Phi}{\partial x} - \frac{V}{2I_x} y^2 + f(x) = \frac{V}{2I_x}\left[h^2 - y^2 + \frac{\nu}{1+\nu}\left(x^2 - \frac{b^2}{3}\right)\right].$$

If the small variation of σ_{zy} with x is ignored, we recover the shear stress expression from the elementary beam theory,

$$\sigma_{zy} = \frac{V}{2I_x}(h^2 - y^2).$$

To determine the corresponding inplane displacements u_y and u_z, we use the strain-displacement relations and the Hooke's law. This gives

$$e_{yy} = \frac{\partial u_y}{\partial y} = -\nu \frac{\sigma_{zz}}{E} = \frac{\nu V(l-z)}{EI_x} y,$$

$$e_{zz} = \frac{\partial u_z}{\partial z} = \frac{\sigma_{zz}}{E} = -\frac{V(l-z)}{EI_x} y,$$

$$e_{zy} = \frac{1}{2}\left(\frac{\partial u_z}{\partial y} + \frac{\partial u_y}{\partial z}\right) = \frac{1+\nu}{E}\sigma_{zy} = \frac{1+\nu}{E}\frac{V}{2I_x}(h^2 - y^2).$$

Upon integration, there follows

$$u_y = \frac{V}{2EI_x}\left[lz^2 - \frac{z^3}{3} + 2(1+\nu)h^2 z + \nu(l-z)y^2\right],$$

$$u_z = -\frac{V}{2EI_x}\left[(2lz - z^2)y + (2+\nu)\frac{y^3}{3}\right].$$

The integration constants were determined from the conditions $u_y = u_z = 0$ and $u_{z,y} = 0$ for $y = z = 0$.

The deflection of the axis $y = 0$ is

$$u_y(0, z) = \frac{V}{2EI_x}\left[lz^2 - \frac{z^3}{3} + 2(1+\nu)h^2 z\right].$$

The first part of this expression represents the deflection due to bending, and the second term due to shearing deformation.

CHAPTER 12

Problem 12.1. Consider the loading over the portion $|y| \leq a$ on the surface of a half-plane, given by

$$\sigma_{xx} = -p_0\left[1 - (y/a)^2\right]^{1/2}, \quad \text{on } x = 0,$$

$$\sigma_{xy} = 0, \quad \text{on } x = 0.$$

Problem 12.1

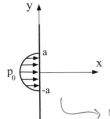
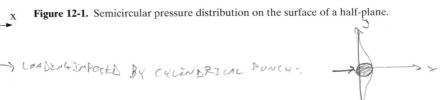

Figure 12-1. Semicircular pressure distribution on the surface of a half-plane.

Evaluate the stresses and discuss the displacement field (see Fig. 12-1).

Solution: The total load is

$$P = p_0 \int_{-a}^{a} \left[1 - (y/a)^2\right]^{1/2} dy = p_0 a.$$

Since the loading is symmetric, the stresses can be calculated from

$$\sigma_{xx} = -\frac{2}{\pi} \int_0^\infty p(\xi)(1 + \xi x)e^{-\xi x} \cos(\xi y)\, d\xi,$$

$$\sigma_{yy} = -\frac{2}{\pi} \int_0^\infty p(\xi)(1 - \xi x)e^{-\xi x} \cos(\xi y)\, d\xi,$$

$$\sigma_{xy} = -\frac{2x}{\pi} \int_0^\infty \xi p(\xi) e^{-\xi x} \sin(\xi y)\, d\xi,$$

with

$$p(\xi) = \int_0^a \left[1 - (y/a)^2\right]^{1/2} \cos(\xi y)\, dy = \frac{p_0 \pi}{2\xi} J_1(\xi a).$$

Thus,

$$\sigma_{xx} + \sigma_{yy} = -\frac{2}{\pi} \int_0^\infty 2p(\xi) e^{-\xi x} \cos(\xi y)\, d\xi,$$

and

$$\sigma_{yy} - \sigma_{xx} + 2i\sigma_{xy} = \frac{4x}{\pi} \int_0^\infty p(\xi) \xi e^{-\xi y}[\cos(\xi y) - i \sin(\xi y)]\, d\xi.$$

The latter equation can be written as

$$\sigma_{yy} - \sigma_{xx} + 2i\sigma_{xy} = \frac{4x}{\pi} \int_0^\infty p(\xi) \xi e^{-\xi x - i\xi y}\, d\xi.$$

If now we define $z = x + iy$, we obtain $e^{-\xi x - i\xi y} = e^{-\xi z}$ and

$$\sigma_{yy} - \sigma_{xx} + 2i\sigma_{xy} = 2p_0 x \int_0^\infty J_1(\xi a) e^{-\xi z}\, d\xi$$

$$= \frac{2p_0 x}{a}\left[1 - z(z^2 + a^2)^{-1/2}\right].$$

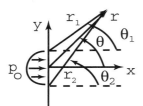

Figure 12-2. Coordinate system based on loading along a strip on a half-plane.

At this stage it is useful and informative to examine the limit of the above relation as $z \to \infty$. For example, let $y = 0$, $z = x$ and let $x \to \infty$ to obtain

$$\sigma_{yy} - \sigma_{xx} + 2i\sigma_{xy} \to \frac{2P}{\pi}\frac{1}{x}.$$

This limit may be compared with that obtained earlier in connection with the problem of uniform loading on a strip of half-plane between $-a \le x \le a$, for which the total load was also P. The two limits are identical which is, of course, to be expected because at great distances from the loading source both load appear as concentrated point forces located at the origin.

Now consider the geometry shown in Fig. 12-2, and let $z = re^{i\theta}$, $z + ia = r_2 e^{i\theta_2}$, $z - ia = r_1 e^{-i\theta_1}$. Then,

$$\sigma_{yy} - \sigma_{xx} + 2i\sigma_{xy} = \frac{2p_0 x}{a}\left[1 - \frac{r}{(r_1 r_2)^{1/2}} e^{i\theta - 1/2i(\theta_1 + \theta_2)}\right].$$

Thus, by taking the imaginary and real parts of the above expression, we find that

$$\sigma_{xy} = -\frac{p_0 x}{a}\frac{r}{(r_1 r_2)^{1/2}} \sin\left[\theta - \frac{1}{2}(\theta_1 + \theta_2)\right]$$

$$= -\frac{p_0 r^2 \cos\theta}{a(r_1 r_2)^{1/2}} \sin\left[\theta - \frac{1}{2}(\theta_1 + \theta_2)\right],$$

and

$$\sigma_{yy} - \sigma_{xx} = \frac{2p_0 r \cos\theta}{a}\left\{1 - \frac{r}{(r_1 r_2)^{1/2}} \cos\left[\theta - \frac{1}{2}(\theta_1 + \theta_2)\right]\right\}.$$

Furthermore,

$$\sigma_{yy} + \sigma_{xx} = -\frac{4}{\pi}\int_0^\infty p(\xi)e^{-\xi x}\cos(\xi y)\,d\xi$$

$$= -2p_0 \Re \int_0^\infty J_1(\xi a)\frac{e^{-\xi z}}{\xi}\,d\xi$$

$$= -2p_0 \Re\left[\frac{(z-ia)^{1/2}(z+ia)^{1/2} - z}{a}\right]$$

$$= -\frac{2p_0}{a}\left\{(r_1 r_2)^{1/2}\cos\left[\frac{1}{2}(\theta_1 + \theta_2)\right] - r\cos\theta\right\}.$$

Combining these results, we obtain

$$\sigma_{xx} = \frac{p_0}{a(r_1 r_2)^{1/2}}\left\{-(r_1 r_2)\cos\left[\frac{1}{2}(\theta_1 + \theta_2)\right]\right.$$

$$\left. + r^2 \cos\theta \cos\left[\theta - \frac{1}{2}(\theta_1 + \theta_2)\right]\right\},$$

Problems 12.1–12.2

$$\sigma_{yy} = \frac{p_0}{a(r_1 r_2)^{1/2}} \{2r(r_1 r_2)^{1/2} \cos\theta$$
$$- (r_1 r_2) \cos\left[\frac{1}{2}(\theta_1 + \theta_2)\right] - r^2 \cos\theta \cos\left[\theta - \frac{1}{2}(\theta_1 + \theta_2)\right]\},$$

$$\sigma_{xy} = -\frac{p_0 r^2 \cos\theta}{a(r_1 r_2)^{1/2}} \sin\left[\theta - \frac{1}{2}(\theta_1 + \theta_2)\right].$$

Finally, we may explore the general shape of the displacement caused by this pressure distribution. We recall that, in general, the x component of displacement is given by the integral

$$u = \frac{2(1+\nu)}{\pi E} \int_0^\infty p(\xi) e^{-\xi x} [2(1-\nu) + \xi x] \frac{\cos(\xi y)}{\xi} d\xi.$$

To explore the shape of the displacement at the surface, consider

$$\left.\frac{\partial u}{\partial y}\right|_{x=0} = -\frac{4(1-\nu^2)}{\pi E} \int_0^\infty p(\xi) \sin(\xi y) d\xi$$
$$= -\frac{2(1-\nu^2)}{\pi E} \int_0^\infty \frac{J_1(\xi a)}{\xi} \sin(\xi y) d\xi$$
$$= -\frac{2(1-\nu^2)}{\pi E} \sin[\sin^{-1}(y/a)], \quad \text{at } y \leq a$$
$$= -\frac{a}{(y + \sqrt{y^2 - a^2})}, \quad \text{at } y \geq a.$$

At $y \approx 0$, the form of the displacement gradient is

$$\frac{\partial u}{\partial y} \sim y/a,$$

which suggests that the trough of the displacement is circular.

In the context of contact mechanics, this problem is revisited as Problem 15.2.

Problem 12.2. Consider the application of a normal stress on the surface of a half-plane of the form

$$\sigma_{yy} \sim -A\cos(\alpha x).$$

Determine the stress state that is caused by this loading and the general shape of the resulting surface indentation.

Solution: Consider the stress function

$$\phi(x, y) = \frac{p_\alpha}{\alpha^2}(1 + \alpha y)e^{-\alpha y} \cos(\alpha x).$$

The corresponding stresses are

$$\sigma_{yy} = \frac{\partial^2 \phi}{\partial x^2} = -p_\alpha(1+\alpha y)e^{-\alpha y}\cos(\alpha x),$$

$$\sigma_{xx} = \frac{\partial^2 \phi}{\partial y^2} = -p_\alpha(1-\alpha y)e^{-\alpha y}\cos(\alpha x),$$

$$\sigma_{xy} = -\frac{\partial^2 \phi}{\partial x \partial y} = -p_\alpha \alpha y e^{-\alpha y}\sin(\alpha x).$$

On $y=0$ we have $\sigma_{yy} = -p_\alpha \cos(\alpha x)$ and $\sigma_{xy} = 0$.

To examine the displacement field, under conditions of plane strain, consider

$$e_{yy} = \frac{\partial u_y}{\partial y} = \frac{1}{E}\left[(1-\nu^2)\frac{\partial^2 \phi}{\partial x^2} - \nu(1+\nu)\frac{\partial^2 \phi}{\partial y^2}\right],$$

which, upon integration, yields

$$u_y = \frac{1}{E}\left[(1-\nu^2)\int \frac{\partial^2 \phi}{\partial x^2}\,dy - \nu(1+\nu)\frac{\partial \phi}{\partial y}\right] + \text{const.}$$

But, we note that

$$\frac{\partial \phi}{\partial y} = -p_\alpha y e^{-\alpha y}\cos(\alpha x) = 0, \quad \text{on } y=0,$$

and so

$$u_y = \frac{2p_\alpha(1-\nu^2)}{\alpha E}\cos(\alpha x) + \text{const.}$$

Thus, it appears that to create a surface displacement of the form $y = a(\alpha)\cos(\alpha x)$ requires an applied surface traction to be

$$\sigma_{yy} = \frac{\alpha E a(\alpha)}{2(1-\nu^2)}\cos(\alpha x).$$

CHAPTER 13

Problem 13.1. Derive an expression for the change of volume caused by elastic deformation, without solving the particular boundary value problem, provided that the traction field is known over the bounding surface of the body, and the body force field inside the body.

Solution: Let σ_{ij} be the stress field within the body of volume V, and let $\hat{u}_i$ be any displacement field (not necessarily related to σ_{ij} and not necessarily infinitesimal). Define the tensor

$$\hat{e}_{ij} = \frac{1}{2}(\hat{u}_{i,j} + \hat{u}_{j,i}).$$

Then,

$$\int_V \sigma_{ij}\hat{e}_{ij}\,dV = \int_V \sigma_{ij}\hat{u}_{i,j}\,dV = \int_V\left[(\sigma_{ij}\hat{u}_i)_{,j} + b_i\hat{u}_i\right]dV,$$

Problems 13.1–13.2

because by the equilibrium equations $\sigma_{ij,j} + b_i = 0$. Applying the Gauss divergence theorem to the first integral on the right-hand side, we have

$$\int_V (\sigma_{ij}\hat{u}_i)_{,j} \, dV = \int_S \sigma_{ij}\hat{u}_i n_j \, dS,$$

where n_j are the components of the unit vector normal to the bounding surface S. Therefore, recalling that $T_i = \sigma_{ij} n_j$, one has

$$\int_V \sigma_{ij}\hat{e}_{ij} \, dV = \int_S T_i \hat{u}_i \, dS + \int_V b_i \hat{u}_i \, dV.$$

By choosing $\hat{u}_i = x_i$ (and, thus, $\hat{e}_{ij} = \delta_{ij}$), the previous equation becomes

$$\int_V \sigma_{kk} \, dV = \int_S T_i x_i \, dS + \int_V b_i x_i \, dV.$$

This yields an expression for the average normal stress within the body, regardless of the material properties and entirely in terms of the specified surface tractions and body forces. If the volume change is because of infinitesimal elastic deformation only, then $\sigma_{kk} = 3\kappa e_{kk}$ by Hooke's law, and the resulting volume change is

$$\Delta V = \int_V e_{kk} \, dV = \frac{1}{3\kappa} \left(\int_S T_i x_i \, dS + \int_V b_i x_i \, dV \right).$$

In particular, if the stress distribution within the body is self-equilibrated ($\sigma_{ij} \neq 0$, but $T_i = 0$ and $b_i = 0$), as occurs around an inclusion or dislocation within an externally unloaded body, the corresponding volume change is equal to zero ($\Delta V = 0$).

Problem 13.2. Derive the solution of a pressurized sphere by the displacement method.

Solution: In a three-dimensional problems with spherical symmetry, the only displacement component is the radial displacement $u = u(r)$. The corresponding strain components are

$$e_{rr} = \frac{du}{dr}, \quad e_{\theta\theta} = e_{\phi\phi} = \frac{u}{r}.$$

The nonvanishing stress components are the radial stress σ_{rr} and the hoop stresses $\sigma_{\theta\theta} = \sigma_{\phi\phi}$. In the absence of body forces, the equilibrium equation is

$$\frac{d\sigma_{rr}}{dr} + \frac{2}{r}(\sigma_{rr} - \sigma_{\theta\theta}) = 0.$$

In the displacement-based approach, the compatibility equation is not needed. The substitution of the strain-displacement expressions into the stress-strain relations gives

$$\sigma_{rr} = 2\mu e_{rr} + \lambda(e_{rr} + 2e_{\theta\theta}) = (\lambda + 2\mu)\frac{du}{dr} + 2\lambda\frac{u}{r},$$

$$\sigma_{\theta\theta} = 2\mu e_{\theta\theta} + \lambda(e_{rr} + 2e_{\theta\theta}) = \lambda\frac{du}{dr} + 2(\lambda + \mu)\frac{u}{r},$$

where λ and μ are the Lamé elastic constants. When this is introduced into the equilibrium equation, there follows

$$\frac{d^2 u}{dr^2} + \frac{2}{r}\frac{du}{dr} - 2\frac{u}{r^2} = 0.$$

The solution of this differential equation is

$$u = Cr + \frac{D}{r^2}.$$

Its substitution back into the stress-strain relations gives

$$\sigma_{rr} = 3\kappa C - 4\mu \frac{D}{r^3}, \quad \sigma_{\theta\theta} = 3\kappa C + 2\mu \frac{D}{r^3},$$

where $3\kappa = 2\mu + 3\lambda$. The boundary conditions,

$$\sigma_{rr}(R_1) = -p_1, \quad \sigma_{rr}(R_2) = -p_2,$$

specify the integration constants as

$$C = \frac{1}{3\kappa} \frac{p_1 R_1^3 - p_2 R_2^3}{R_2^3 - R_1^3}, \quad D = \frac{1}{4\mu} \frac{R_1^3 R_2^3}{R_2^3 - R_1^3} (p_1 - p_2).$$

Evidently, these are related to the integration constants of the stress-based solution discussed in the text by $A = 3\kappa C$ and $B = -4\mu D$.

Problem 13.3. Consider a hollow sphere whose inner surface $r = R_1$ is held at constant temperature T_1 and the outer surface $r = R_2$ at constant temperature $T_2 < T_1$ (steady heat flow from inside out). Determine the stress distribution in the sphere.

Solution: The only displacement component is the radial displacement $u = u(r)$. The corresponding strain components are

$$e_{rr} = \frac{du}{dr}, \quad e_{\theta\theta} = e_{\phi\phi} = \frac{u}{r}.$$

The nonvanishing stress components are the radial stress σ_{rr} and the hoop stresses $\sigma_{\theta\theta} = \sigma_{\phi\phi}$. The equilibrium equation is

$$\frac{d\sigma_{rr}}{dr} + \frac{2}{r}(\sigma_{rr} - \sigma_{\theta\theta}) = 0.$$

The thermoelastic stress-strain relations are

$$e_{rr} = \frac{1}{E}(\sigma_{rr} - 2\nu\sigma_{\theta\theta}) + \alpha(T - T_0),$$

$$e_{\theta\theta} = \frac{1}{E}[\sigma_{\theta\theta} - \nu(\sigma_{rr} + \sigma_{\theta\theta})] + \alpha(T - T_0),$$

where T_0 is the reference temperature, and α is the coefficient of linear thermal expansion. A straightforward inversion of the above expressions gives

$$\sigma_{rr} = \frac{E}{(1+\nu)(1-2\nu)}[(1-\nu)e_{rr} + 2\nu e_{\theta\theta} - (1+\nu)\alpha(T - T_0)],$$

$$\sigma_{\theta\theta} = \frac{E}{(1+\nu)(1-2\nu)}[e_{\theta\theta} + \nu e_{rr} - (1+\nu)\alpha(T - T_0)].$$

Problems 13.3–13.4

The substitution into the equilibrium equation then yields

$$\frac{d^2 u}{dr^2} + \frac{2}{r}\frac{du}{dr} - 2\frac{u}{r^2} = \frac{1+\nu}{1-\nu}\alpha\frac{dT}{dr}.$$

The solution of this differential equation is

$$u = Cr + \frac{D}{r^2} + \frac{1+\nu}{1-\nu}\frac{\alpha}{r^2}\int_{R_1}^{r} r^2 T(r)\,dr.$$

Its substitution back into the stress-strain relations gives

$$\sigma_{rr} = \frac{E}{1-2\nu}C - \frac{2E}{1+\nu}\frac{D}{r^3} - \frac{2\alpha E}{1-\nu}\frac{1}{r^3}\int_{R_1}^{r} r^2 T(r)\,dr,$$

$$\sigma_{\theta\theta} = \frac{E}{1-2\nu}C + \frac{E}{1+\nu}\frac{D}{r^3} + \frac{\alpha E}{1-\nu}\frac{1}{r^3}\int_{R_1}^{r} r^2 T(r)\,dr - \frac{\alpha E T(r)}{1-\nu}.$$

The boundary conditions give

$$\sigma_{rr}(R_1) = 0: \quad \frac{E}{1-2\nu}C - \frac{2E}{1+\nu}\frac{D}{R_1^3} = 0,$$

$$\sigma_{rr}(R_2) = 0: \quad \frac{E}{1-2\nu}C - \frac{2E}{1+\nu}\frac{D}{R_2^3} = \frac{2\alpha E}{1-\nu}\frac{1}{R_2^3}\int_{R_1}^{R_2} r^2 T(r)\,dr,$$

which specify the integration constants as

$$C = \frac{2\alpha(1-2\nu)}{1-\nu}\frac{1}{R_2^3 - R_1^3}\int_{R_1}^{R_2} r^2 T(r)\,dr,$$

$$D = \frac{\alpha(1+\nu)}{1-\nu}\frac{R_1^3}{R_2^3 - R_1^3}\int_{R_1}^{R_2} r^2 T(r)\,dr.$$

For the steady-state heat flow, the temperature satisfies the Laplacian equation $\nabla^2 T = 0$, which is, in the spherical coordinates and with radial symmetry,

$$\frac{d^2 T}{dr^2} + \frac{2}{r}\frac{dT}{dr} = 0.$$

Using the boundary conditions $T(R_1) = T_1$ and $T(R_2) = T_2$, the integration gives

$$T(r) = \frac{R_1 R_2}{R_2 - R_1}\frac{T_1 - T_2}{r} + \frac{R_2 T_2 - R_1 T_1}{R_2 - R_1}.$$

Thus,

$$\int_{R_1}^{R_2} r^2 T(r)\,dr = \frac{5}{6}R_1 R_2 (R_1 + R_2)(T_1 - T_2) + \frac{1}{3}(R_1^3 T_1 - R_2^3 T_2).$$

This specifies the integration constants C and D above and thus the required stress distribution within the sphere.

Problem 13.4. Determine the displacement field associated with a doublet (a pair of nearby dilatation and compression centers at small distance h as in Fig. 13-1) in an infinite

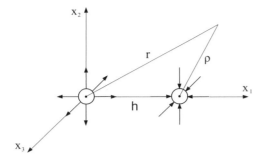

Figure 13-1. A pair of nearby dilatation and compression centers at small distance h.

elastic solid. The displacement components due to dilatation center at the origin is

$$u_i = -A \frac{1-2\nu}{8\pi\mu(1-\nu)} \frac{\partial}{\partial x_i}\left(\frac{1}{r}\right),$$

where $A = Pd$ is the center's strength (the product of a small distance d between a pair of opposite forces and the magnitude of each force P). The shear modulus is μ and Poisson's ratio is ν.

Solution: Let the dilatation center be at the origin, and position the x_1 axis along the axis of the doublet, as shown in the figure. By superposition, the displacement field due to the doublet is

$$u_i = -A \frac{1-2\nu}{8\pi\mu(1-\nu)} \left[\frac{\partial}{\partial x_i}\left(\frac{1}{r}\right) - \frac{\partial}{\partial x_i}\left(\frac{1}{\rho}\right)\right],$$

where

$$\rho^2 = (x_1 - h)^2 + x_2^2 + x_3^2 \approx r^2 - 2hx_1.$$

It readily follows that

$$\rho \approx r\left(1 - h\frac{x_1}{r^2}\right), \quad \frac{1}{\rho} \approx \frac{1}{r}\left(1 + h\frac{x_1}{r^2}\right).$$

Consequently,

$$\frac{\partial}{\partial x_i}\left(\frac{1}{\rho}\right) \approx \frac{\partial}{\partial x_i}\left(\frac{1}{r}\right) + h\frac{\partial}{\partial x_i}\left(\frac{x_1}{r^3}\right),$$

and the displacement components are

$$u_i = Ah \frac{1-2\nu}{8\pi\mu(1-\nu)} \frac{\partial}{\partial x_i}\left(\frac{x_1}{r^3}\right).$$

Problem 13.5. Determine the displacement field due to a couple of magnitude M at the origin. The displacement components due to a concentrated force P at the origin, directed in the negative x_3 direction, is

$$u_i = \frac{P}{4\pi\mu}\left[-\frac{\delta_{i3}}{r} + \frac{1}{4(1-\nu)}\frac{\partial}{\partial x_i}\left(\frac{x_3}{r}\right)\right].$$

Problems 13.5–13.6

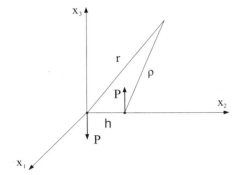

Figure 13-2. A concentrated couple formed by two opposite forces at small distance h.

Solution: Let the couple be formed by two opposite forces at small distance h, as shown in Fig. 13-2, such that $M = Ph$. By superposition, the displacement field due to this couple is then

$$u_i = \frac{P}{4\pi\mu}\left[-\frac{\delta_{i3}}{r} + \frac{\delta_{i3}}{\rho} + \frac{1}{4(1-\nu)}\frac{\partial}{\partial x_i}\left(\frac{x_3}{r} - \frac{x_3}{\rho}\right)\right].$$

Since

$$\rho^2 = x_1^2 + (x_2 - h)^2 + x_3^2 \approx r^2 - 2hx_2,$$

we have

$$\frac{1}{\rho} \approx \frac{1}{r}\left(1 + h\frac{x_2}{r^2}\right).$$

Consequently,

$$\frac{\partial}{\partial x_i}\left(\frac{x_3}{\rho}\right) \approx \frac{\partial}{\partial x_i}\left(\frac{x_3}{r}\right) + h\frac{\partial}{\partial x_i}\left(\frac{x_2 x_3}{r^3}\right),$$

and the displacement components are

$$u_i = \frac{M}{4\pi\mu}\left[\frac{x_2}{r}\delta_{i3} - \frac{1}{4(1-\nu)}\frac{\partial}{\partial x_i}\left(\frac{x_2 x_3}{r^3}\right)\right].$$

Problem 13.6. Consider an axisymmetric 3D elasticity problem. Derive the biharmonic equation for the corresponding Love's function.

Solution: For axisymmetric problems, the field variables are independent of the polar angle θ (Fig. 13-3). The circumferential component of displacement vanishes ($u_\theta = 0$), whereas the radial and longitudinal components depend on the radius r and the longitudinal coordinate z, i.e.,

$$u_r = u_r(r, z), \quad u_z = u_z(r, z).$$

The corresponding infinitesimal strains are

$$e_{rr} = \frac{\partial u_r}{\partial r}, \quad e_{\theta\theta} = \frac{u_r}{r}, \quad e_{zz} = \frac{\partial u_z}{\partial z},$$

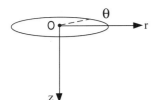

Figure 13-3. The cylindrical coordinates for axisymmetric 3D problems.

and
$$\epsilon_{rz} = \frac{1}{2}\left(\frac{\partial u_r}{\partial z} + \frac{\partial u_z}{\partial r}\right).$$

The equilibrium equations are
$$\frac{\partial \sigma_{rr}}{\partial r} + \frac{\partial \sigma_{rz}}{\partial z} + \frac{1}{r}(\sigma_{rr} - \sigma_{\theta\theta}) = 0,$$

$$\frac{\partial \sigma_{rz}}{\partial r} + \frac{\partial \sigma_{zz}}{\partial z} + \frac{1}{r}\sigma_{rz} = 0.$$

The problems involving purely elastic deformations can be conveniently solved by expressing the displacement components in terms of Love's function $\Omega = \Omega(r, z)$ as
$$u_r = -\frac{1}{2G}\frac{\partial^2 \Omega}{\partial r \partial z}, \quad u_z = \frac{1}{2G}\left[2(1-\nu)\nabla^2 \Omega - \frac{\partial^2 \Omega}{\partial z^2}\right],$$

where
$$\nabla^2 = \frac{\partial^2}{\partial r^2} + \frac{1}{r}\frac{\partial}{\partial r} + \frac{\partial^2}{\partial z^2}.$$

The elastic constants of an isotropic material are the shear modulus G and the Poisson's ratio ν.

By using the above expressions and the generalized Hooke's law, the stress components can be expressed in terms of Love's function as
$$\sigma_{rr} = \frac{\partial}{\partial z}\left(\nu \nabla^2 \Omega - \frac{\partial^2 \Omega}{\partial r^2}\right),$$

$$\sigma_{\theta\theta} = \frac{\partial}{\partial z}\left(\nu \nabla^2 \Omega - \frac{1}{r}\frac{\partial \Omega}{\partial r}\right),$$

$$\sigma_{zz} = \frac{\partial}{\partial z}\left[(2-\nu)\nabla^2 \Omega - \frac{\partial^2 \Omega}{\partial z^2}\right],$$

$$\sigma_{rz} = \frac{\partial}{\partial r}\left[(1-\nu)\nabla^2 \Omega - \frac{\partial^2 \Omega}{\partial z^2}\right].$$

When these expressions are inserted into equilibrium equations, it follows that the first of them is identically satisfied, while the second requires Ω to be a biharmonic function in the (r, z) domain, *i.e.*,
$$\nabla^4 \Omega = \frac{\partial^4 \Omega}{\partial r^4} + 2\frac{\partial^4 \Omega}{\partial r^2 \partial z^2} + \frac{\partial^4 \Omega}{\partial z^4} = 0.$$

Problems 13.7–14.1

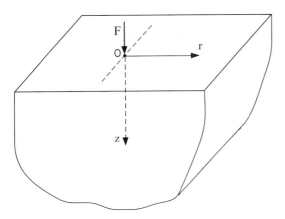

Figure 13-4. The vertical concentrated force on the boundary of the half-space.

Problem 13.7. Derive the displacement components of the points on the plane $z = 0$ due to the vertical concentrated force acting on the boundary of the half-space (Fig. 13-4).

Solution: The Love's function for the Boussinesq problem of concentrated force on a semi-infinite body is

$$\Omega = \frac{F}{2\pi} z \left[1 + 2\nu\sqrt{1 + r^2/z^2} + (1 - 2\nu) \ln\left(1 + \sqrt{1 + r^2/z^2}\right) - \ln z \right].$$

The corresponding displacement field is determined from

$$u_r = -\frac{1}{2G} \frac{\partial^2 \Omega}{\partial r \partial z}, \quad u_z = \frac{1}{2G}\left[2(1 - \nu)\nabla^2 \Omega - \frac{\partial^2 \Omega}{\partial z^2}\right],$$

which gives

$$u_r = \frac{F(1 - 2\nu)}{4\pi G} \frac{1}{r} \left[\frac{1}{1 - 2\nu} \frac{r^2/z^2}{(1 + r^2/z^2)^{3/2}} + \frac{1}{(1 + r^2/z^2)^{1/2}} - 1 \right],$$

$$u_z = \frac{F}{4\pi G} \frac{1}{z} \left[\frac{1}{(1 + r^2/z^2)^{3/2}} + 2(1 - \nu) \frac{1}{(1 + r^2/z^2)^{1/2}} \right].$$

The displacements at the points of the plane $z = 0$ are, consequently,

$$u = -\frac{F(1 - 2\nu)}{4\pi G} \frac{1}{r}, \quad w = \frac{F(1 - \nu)}{2\pi G} \frac{1}{r}.$$

CHAPTER 14

Problem 14.1. Derive an expression for the Young's modulus E_n in an arbitrary direction **n** in terms of elastic compliances of a linearly elastic fully anisotropic material. Derive also the expressions for the Poisson's ratio ν_{mn} and the shear modulus G_{mn}. Specialize the results in the case of a transversely isotropic material.

Solution: The strain in the direction **n** can be expressed in terms of the strain tensor components as $e_n = n_i e_{ij} n_j$. If the strain state is produced by tension σ_n applied along the direction **n**, the stress state is

$$\sigma = \sigma_{kl} \mathbf{e}_k \mathbf{e}_l = \sigma_n \mathbf{n} \, \mathbf{n},$$

which gives $\sigma_{kl} = \sigma_n n_k n_l$. For a generally anisotropic linearly elastic material, the stress-strain relation is $e_{ij} = S_{ijkl}\sigma_{kl}$, where $S_{ijkl} = C_{ijkl}^{-1}$ are the components of the elastic compliance tensor (the inverse of the elastic moduli tensor). Thus, we have

$$e_n = n_i e_{ij} n_j = \sigma_n n_i n_j S_{ijkl} n_k n_l.$$

Since the Young's modulus in the direction **n** is defined by $E_n = \sigma_n/e_n$, we recognize from the above equation that

$$\frac{1}{E_n} = n_i n_j S_{ijkl} n_k n_l.$$

Similarly, the Poisson's ratio in the direction **m** due to longitudinal stress in the orthogonal direction **n** is

$$\nu_{mn} = -\frac{e_m}{e_n} = -\frac{m_i m_j S_{ijkl} n_k n_l}{n_\alpha n_\beta S_{\alpha\beta\gamma\delta} n_\gamma n_\delta}.$$

The shear modulus between the orthogonal directions **m** and **n** is

$$\frac{1}{G_{mn}} = 4 m_i n_j S_{ijkl} m_k n_l.$$

Problem 14.2. Specialize the results from previous problem in the case of a transversely isotropic material, whose axis of transverse isotropy is in the coordinate direction $\mathbf{e}_3$. Use $\mathbf{m} = \{0, -\sin\theta, \cos\theta\}$ and $\mathbf{n} = \{0, \cos\theta, \sin\theta\}$, where θ is the angle from the plane of isotropy.

Solution: The compliance matrix of a transversely isotropic material (around the axis $\mathbf{e}_3$) is

$$\mathbf{S} = \begin{bmatrix} s_{11} & s_{12} & s_{13} & 0 & 0 & 0 \\ s_{12} & s_{11} & s_{13} & 0 & 0 & 0 \\ s_{13} & s_{13} & s_{33} & 0 & 0 & 0 \\ 0 & 0 & 0 & s_{44} & 0 & 0 \\ 0 & 0 & 0 & 0 & s_{44} & 0 \\ 0 & 0 & 0 & 0 & 0 & 2(s_{11}-s_{12}) \end{bmatrix}.$$

The Voigt notation is used for the compliance components, so that $S_{1111} = s_{11}$, $S_{2222} = s_{22}$, $S_{3333} = s_{33}$, $S_{1122} = s_{12}$, $S_{2233} = S_{3311} = s_{13}$, $4S_{2323} = 4S_{3131} = s_{44}$, and $4S_{1212} = 2(s_{11}-s_{12})$. Consequently, we obtain

$$\frac{1}{E_n} = s_{11}\cos^4\theta + s_{33}\sin^4\theta + (2s_{13}+s_{44})\sin^2\theta\cos^2\theta,$$

$$\nu_{mn} = -\frac{s_{13} + (s_{11}+s_{33}-s_{44}-2s_{13})\sin^2\theta\cos^2\theta}{s_{11}\cos^4\theta - s_{33}\sin^4\theta + (s_{44}+2s_{13})\sin^2\theta\cos^2\theta},$$

$$\frac{1}{G_n} = (s_{22}+s_{33}-2s_{13})\sin^2 2\theta + s_{44}\cos^2 2\theta.$$

Problem 14.3

Problem 14.3. A homogeneous rectangular block of a fully anisotropic elastic material is subject to shear stress $\sigma_{12} = \tau$. Determine the length changes of the edges a, b, c of the block, and the volume change of the block. Derive the expressions for the displacement components.

Solution: The stress-strain relations for a fully anisotropic linearly elastic material

$$\begin{bmatrix} e_{11} \\ e_{22} \\ e_{33} \\ 2e_{23} \\ 2e_{31} \\ 2e_{12} \end{bmatrix} = \begin{bmatrix} s_{11} & s_{12} & s_{13} & s_{14} & s_{15} & s_{16} \\ s_{12} & s_{22} & s_{23} & s_{24} & s_{25} & s_{26} \\ s_{13} & s_{23} & s_{33} & s_{34} & s_{35} & s_{36} \\ s_{14} & s_{24} & s_{34} & s_{44} & s_{45} & s_{46} \\ s_{15} & s_{25} & s_{35} & s_{45} & s_{55} & s_{56} \\ s_{16} & s_{26} & s_{36} & s_{46} & s_{56} & s_{66} \end{bmatrix} \begin{bmatrix} \sigma_{11} \\ \sigma_{22} \\ \sigma_{33} \\ \sigma_{23} \\ \sigma_{31} \\ \sigma_{12} \end{bmatrix}.$$

In the case of pure shear $\sigma_{12} = \tau$ this gives

$$e_{11} = s_{16}\tau, \quad e_{22} = s_{26}\tau, \quad e_{33} = s_{36}\tau,$$

$$2e_{23} = s_{46}\tau, \quad 2e_{31} = s_{56}\tau, \quad 2e_{12} = s_{66}\tau.$$

The lengths of the edges of the block are thus changed by $\Delta a = as_{16}\tau$, $\Delta b = bs_{26}\tau$ and $\Delta c = cs_{36}\tau$, with the resulting volume change $\Delta V = abc(s_{16} + s_{26} + s_{36})\tau$.

Since the strain-displacement relations are

$$e_{11} = \frac{\partial u_1}{\partial x_1}, \quad e_{22} = \frac{\partial u_2}{\partial x_2}, \quad e_{33} = \frac{\partial u_3}{\partial x_3},$$

$$2e_{23} = \frac{\partial u_2}{\partial x_3} + \frac{\partial u_3}{\partial x_2}, \quad 2e_{31} = \frac{\partial u_3}{\partial x_1} + \frac{\partial u_1}{\partial x_3}, \quad 2e_{12} = \frac{\partial u_1}{\partial x_2} + \frac{\partial u_2}{\partial x_1},$$

their integration gives

$$u_1 = s_{16}\tau x_1 + \frac{1}{2}s_{66}\tau x_2 + \frac{1}{2}s_{56}\tau x_3 - \omega_3 x_2 + \omega_2 x_3 + u_1^0,$$

$$u_2 = \frac{1}{2}s_{66}\tau x_1 + s_{26}\tau x_2 + \frac{1}{2}s_{46}\tau x_3 + \omega_3 x_1 - \omega_1 x_3 + u_2^0,$$

$$u_3 = \frac{1}{2}s_{56}\tau x_1 + \frac{1}{2}s_{46}\tau x_2 + s_{36}\tau x_3 - \omega_2 x_1 + \omega_1 x_2 + u_3^0.$$

The integration constants can be specified by imposing, for example, the following boundary conditions at the origin

$$u_1 = u_2 = u_3 = 0, \quad \frac{\partial u_2}{\partial x_1} = \frac{\partial u_3}{\partial x_1} = 0, \quad \frac{\partial u_2}{\partial x_3} - \frac{\partial u_3}{\partial x_2} = 0.$$

This gives $u_1^0 = u_2^0 = u_3^0 = 0$, $\omega_1 = 0$, $\omega_2 = s_{56}\tau/2$, and $\omega_3 = -s_{66}\tau/2$. Thus, the resulting displacements are

$$u_1 = \tau(s_{16}x_1 + s_{66}x_2 + s_{56}x_3),$$

$$u_2 = \tau\left(s_{26}x_2 + \frac{1}{2}s_{46}x_3\right),$$

$$u_3 = \tau\left(\frac{1}{2}s_{46}x_2 + s_{36}\tau x_3\right).$$

Problem 14.4. A hollow sphere of inner radius R_1 and outer radius R_2 is made from a spherically uniform material, which has the same elastic constants at every point relative to a local spherical coordinate system at that point. If the material is at each point transversely isotropic about its radial direction r, determine the stresses in the sphere due to internal pressure p_1 and external pressure p_2.

Solution: Due to spherical symmetry, the only displacement component is the radial displacement $u = u(r)$. The corresponding strain components are

$$e_{rr} = \frac{du}{dr}, \quad e_{\theta\theta} = e_{\phi\phi} = \frac{u}{r}.$$

The nonvanishing stress components are the radial stress σ_{rr} and the hoop stresses $\sigma_{\theta\theta} = \sigma_{\phi\phi}$. In the absence of body forces, the equilibrium equation is

$$\frac{d\sigma_{rr}}{dr} + \frac{2}{r}(\sigma_{rr} - \sigma_{\theta\theta}) = 0.$$

For a transversely isotropic material about the radial direction, we have at each point

$$\begin{bmatrix} \sigma_{\theta\theta} \\ \sigma_{\phi\phi} \\ \sigma_{rr} \\ \sigma_{\phi r} \\ \sigma_{r\theta} \\ \sigma_{\theta\phi} \end{bmatrix} = \begin{bmatrix} c_{11} & c_{12} & c_{13} & 0 & 0 & 0 \\ c_{12} & c_{11} & c_{13} & 0 & 0 & 0 \\ c_{13} & c_{13} & c_{33} & 0 & 0 & 0 \\ 0 & 0 & 0 & c_{44} & 0 & 0 \\ 0 & 0 & 0 & 0 & c_{44} & 0 \\ 0 & 0 & 0 & 0 & 0 & \frac{1}{2}(c_{11} - c_{12}) \end{bmatrix} \cdot \begin{bmatrix} e_{\theta\theta} \\ e_{\phi\phi} \\ e_{rr} \\ 2e_{\phi r} \\ 2e_{r\theta} \\ 2e_{\theta\phi} \end{bmatrix}.$$

Thus,

$$\sigma_{rr} = c_{33}e_{rr} + 2c_{13}e_{\theta\theta}, \quad \sigma_{\theta\theta} = (c_{11} + c_{12})e_{\theta\theta} + c_{13}e_{rr}.$$

When this is introduced into the equilibrium equation, there follows

$$\frac{d^2u}{dr^2} + \frac{2}{r}\frac{du}{dr} - 2k_0\frac{u}{r^2} = 0, \quad k_0 = \frac{c_{11} + c_{12} - c_{13}}{c_{33}}.$$

The solution of this differential equation is

$$u = Ar^{-(1-3k)/2} + Br^{-(1+3k)/2}, \quad k = \frac{1}{3}\sqrt{1 + 8k_0}.$$

Problems 14.4–14.5

The corresponding stresses are

$$\sigma_{rr} = A\left[2c_{13} - \frac{1}{2}c_{33}(1-3k)\right]r^{-3(1-k)/2}$$
$$+ B\left[2c_{13} - \frac{1}{2}c_{33}(1+3k)\right]r^{-3(1+k)/2},$$

$$\sigma_{\theta\theta} = A\left[c_{11} + c_{12} - \frac{1}{2}c_{13}(1-3k)\right]r^{-3(1-k)/2}$$
$$+ B\left[c_{11} + c_{12} - \frac{1}{2}c_{13}(1+3k)\right]r^{-3(1+k)/2}.$$

The boundary conditions for determination of the constants A and B are $\sigma_{rr}(R_1) = -p_1$ and $\sigma_{rr}(R_2) = -p_2$. This gives

$$A\left[2c_{13} - \frac{1}{2}c_{33}(1-3k)\right] = \frac{p_1 R_1^{3/2} R_2^{-3k/2} - p_2 R_2^{3/2} R_1^{-3k/2}}{(R_2/R_1)^{3k/2} - (R_1/R_2)^{3k/2}},$$

$$B\left[c_{11} + c_{12} - \frac{1}{2}c_{13}(1+3k)\right] = \frac{p_2 R_2^{3/2} R_1^{3k/2} - p_1 R_1^{3/2} R_2^{3k/2}}{(R_2/R_1)^{3k/2} - (R_1/R_2)^{3k/2}}.$$

Problem 14.5. Derive the expressions for the elastic compliances of a transversely isotropic elastic material in terms its elastic moduli.

Solution: Assuming that the axis of transverse isotropy is along the x_3 direction, the normal stresses are given in terms of normal strains by

$$\sigma_{11} = c_{11}e_{11} + c_{12}e_{22} + c_{13}e_{33},$$

$$\sigma_{22} = c_{12}e_{11} + c_{11}e_{22} + c_{13}e_{33},$$

$$\sigma_{33} = c_{13}e_{11} + c_{13}e_{22} + c_{33}e_{33}.$$

Thus,

$$\sigma_{11} - \sigma_{22} = (c_{11} - c_{12})(e_{11} - e_{22}),$$

$$\sigma_{11} + \sigma_{22} = (c_{11} + c_{12})(e_{11} + e_{22}) + 2c_{13}e_{33},$$

and

$$e_{33} = \frac{1}{c_{33}}[\sigma_{33} - c_{13}(e_{11} + e_{22})].$$

By appropriate combination of these expressions, we obtain

$$e_{11} - e_{22} = \frac{1}{c_{11} - c_{12}}(\sigma_{11} - \sigma_{22}),$$

$$e_{11} + e_{22} = \frac{1}{c_{11} + c_{12} - \dfrac{2c_{13}^2}{c_{33}}}\left(\sigma_{11} + \sigma_{22} - 2\frac{c_{13}}{c_{33}}\sigma_{33}\right).$$

These can be solved for the two strain components to give

$$e_{11} = s_{11}\sigma_{11} + s_{12}\sigma_{22} + s_{13}\sigma_{33},$$

$$e_{22} = s_{12}\sigma_{11} + s_{11}\sigma_{22} + s_{13}\sigma_{33},$$

where

$$s_{11} = \frac{c_{11} - \dfrac{c_{13}^2}{c_{33}}}{c_{11} - c_{12}} s, \quad s_{12} = -\frac{c_{12} - \dfrac{c_{13}^2}{c_{33}}}{c_{11} - c_{12}} s,$$

$$s_{13} = -\frac{c_{13}}{c_{33}} s, \quad s = -\left(c_{11} + c_{12} - 2\frac{c_{13}^2}{c_{33}}\right)^{-1}.$$

The substitution in the earlier expression for e_{33} also gives

$$e_{33} = s_{13}\sigma_{11} + s_{13}\sigma_{22} + s_{33}\sigma_{33},$$

where

$$s_{33} = \frac{c_{11} + c_{12}}{c_{33}} s.$$

Finally, since

$$\sigma_{12} = (c_{11} - c_{12})e_{12}, \quad \sigma_{23} = 2c_{44}e_{23}, \quad \sigma_{31} = 2c_{44}e_{31},$$

we have

$$s_{44} = s_{55} = \frac{1}{c_{44}}, \quad s_{66} = 2(s_{11} - s_{12}) = \frac{2}{c_{11} - c_{12}}.$$

Problem 14.6. The components of the elastic moduli tensor of a cubic crystal, with respect to its cubic axes, are

$$C_{ijkl} = c_{12}\delta_{ij}\delta_{kl} + 2c_{44}I_{ijkl} + (c_{11} - c_{12} - 2c_{44})A_{ijkl},$$

where

$$I_{ijkl} = \frac{1}{2}(\delta_{ik}\delta_{jl} + \delta_{il}\delta_{jk}),$$

$$A_{ijkl} = a_i a_j a_k a_l + b_i b_j b_k b_l + c_i c_j c_k c_l.$$

The components of the unit vectors $\mathbf{a}$, $\mathbf{b}$, and $\mathbf{c}$ along the cubic axes are a_i, b_i, and c_i, respectively. Evidently, all the components A_{ijkl} are equal to zero, except $A_{1111} = A_{2222} = A_{3333} = 1$. Derive the expressions for the elastic compliance components S_{ijkl} of a cubic crystal in terms of its elastic moduli.

Solution: The elastic compliance tensor is the inverse of the elastic moduli tensor, so that

$$S_{ijmn}C_{mnkl} = I_{ijkl}.$$

Problems 14.6–15.1

Assume that the components of elastic compliance tensor can be expressed as
$$S_{ijkl} = s_1 \delta_{ij} \delta_{kl} + s_2 I_{ijkl} + s_3 A_{ijkl}.$$
Substituting this in the above condition for the inverse, there follows
$$[(c_{11} + 2c_{12})s_1 + c_{12}s_2 + c_{12}s_3]\delta_{ij}\delta_{kl} + 2c_{44}s_2 I_{ijkl}$$
$$+ [(c_{11} - c_{12} - 2c_{44})s_2 + (c_{11} - c_{12})s_3] A_{ijkl} = I_{ijkl}.$$
In the derivation, note that $A_{ijmn} A_{mnkl} = A_{ijkl}$ and $A_{ijkk} = \delta_{ij}$. Consequently, we must have
$$(c_{11} + 2c_{12})s_1 + c_{12}s_2 + c_{12}s_3 = 0,$$
$$2c_{44}s_2 = 1,$$
$$(c_{11} - c_{12} - 2c_{44})s_2 + (c_{11} - c_{12})s_3 = 0.$$
Upon solving for s_1, s_2, and s_3, we obtain
$$s_1 = -\frac{c_{12}}{(c_{11} - c_{12})(c_{11} + 2c_{12})},$$
$$s_2 = \frac{1}{2c_{44}},$$
$$s_3 = \frac{1}{c_{11} - c_{12}} - \frac{1}{2c_{44}}.$$
Consequently, the components of the elastic compliance tensor of a cubic crystal, expressed in terms of its elastic moduli, are
$$S_{ijkl} = -\frac{c_{12}}{(c_{11} - c_{12})(c_{11} + 2c_{12})} \delta_{ij}\delta_{kl} + \frac{1}{2c_{44}} I_{ijkl} + \left(\frac{1}{c_{11} - c_{12}} - \frac{1}{2c_{44}}\right) A_{ijkl}.$$

CHAPTER 15

Problem 15.1. Consider the loading on a circular region of radius a of the general form
$$p(r) = p_0[1 - (r/a)^2]^n.$$
Calculate the displacements under the circle of loading (see Fig. 15-1) for $n = 0$ and $n = -1/2$.

Solution: The general analysis for arbitrary n has been described by Johnson (1985). Here, we consider the cases where $n = 0$, *i.e.*, uniform loading, and $n = -1/2$, which produces uniform indentation.

Case $n = 0$: Recall the point force solutions for the surface displacements, given in (13.121) and (13.122). Now Fig. 15-1 shows a polar coordinate system in the surface plane, $z = 0$, where $q(x, y)$ is a typical point at which the displacement is to be computed. The

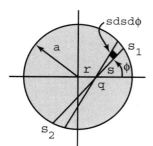

Figure 15-1. Polar coordinate system for a point $q(x, y)$ interior to the circle of loading. s is the polar distance from the typical field point, q.

radial symmetry makes it clear that such a point suffices for our analysis. A typical surface element, $s\,ds\,d\phi$, is at distance s from q and, thus, (13.122) becomes

$$\bar{u}_z(r) = \frac{1-\nu}{2\pi G} \int_S p(s, \phi)\,ds\,d\phi.$$

In the case of uniform loading, this is simply

$$\bar{u}(r) = \frac{1-\nu}{2\pi G} p_0 \int_S d\phi\,ds.$$

For points interior to the circle of loading, the limits on the radial coordinate s are

$$s_{1,2} = -r\cos\phi \pm [r^2\cos^2\phi + (a^2 - r^2)]^{1/2},$$

and thus

$$\bar{u}_z(r) = \frac{1-\nu}{2\pi G} p_0 \int_0^\pi 2[r^2\cos^2\phi + (a^2 - r^2)]^{1/2}\,d\phi$$

$$= \frac{2(1-\nu)}{\pi G} p_0 a \int_0^{\pi/2} [1 - (r/a)^2 \sin^2\phi]^{1/2}\,d\phi$$

$$= \frac{2(1-\nu)}{\pi G} p_0 a\, E(r/a), \quad r \le a,$$

where $E(r/a)$ is the complete elliptic integral of the second kind with modulus r/a. At the center, the maximum depth of indentation is, with $E(0) = \pi/2$,

$$\bar{u}(0) = \delta = \frac{1-\nu}{G} p_0 a.$$

At the perimeter $r = a$, where $E(1) = 1$, we have

$$\bar{u}(a) = \frac{2(1-\nu)}{\pi G} p_0 a,$$

Case $n = -1/2$: In this case an integration procedure, similar to that just described, yields

$$\bar{u}_z(r) = \frac{\pi(1-\nu)p_0 a}{2G}, \quad r \le a.$$

The total load is

$$P = \int_0^a 2\pi r p_0 [1 - (r/a)^2]^{-1/2}\,dr = 2\pi a^2 p_0,$$

Problems 15.1–15.2

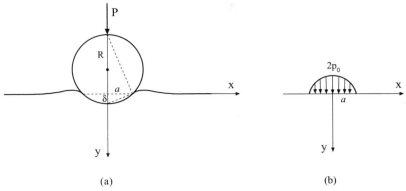

Figure 15-2. (a) A circular cylinder of radius R pressed by the vertical force P into the half-space. The depth of the indentation is δ and the width of the contact region is $2a$. (b) The corresponding pressure distribution $p = p(x)$.

so that the load vs. indentation relation becomes

$$\bar{u}_z = \delta = \frac{P(1-\nu)}{4Ga}.$$

The other components of displacement are, of course, readily calculated from the expressions given for the point force solutions.

Problem 15.2. Derive the pressure distribution under a circular rigid indenter of radius R pressed into an isotropic elastic half space by the vertical force P (per unit length in the z direction), as shown in Fig. 15-2. Derive also the expressions for the width of the contact region $2a$ and the depth of the indentation δ.

Solution: In the absence of tangential load, (15.23) gives

$$\frac{\partial \bar{u}_y}{\partial x} = -\frac{2(1-\nu^2)}{\pi E} \int_{-a}^{a} \frac{p(s)ds}{x-s}, \quad |x| \leq a,$$

where p is the pressure distribution in the contact region of width $2a$. Letting

$$x = a\cos\phi, \quad s = a\cos\vartheta,$$

we have

$$\frac{\partial}{\partial \phi} = -a\sin\phi \frac{\partial}{\partial x}, \quad \frac{\partial}{\partial x} = -\frac{1}{a\sin\phi}\frac{\partial}{\partial \phi}.$$

Thus,

$$\frac{\partial \bar{u}_y}{\partial \phi} = a\sin\phi \frac{2(1-\nu^2)}{\pi E} \int_0^\pi \frac{p(\vartheta)\sin\vartheta}{\cos\phi - \cos\vartheta} d\vartheta, \quad 0 \leq \phi \leq \pi.$$

Having in mind the result

$$\int_0^\pi \frac{\cos n\vartheta}{\cos\phi - \cos\vartheta} d\vartheta = -\frac{\pi \sin n\phi}{\sin\phi},$$

we search the solution by using the expansions for $p(\vartheta)$ and $\partial \bar{u}_y/\partial \phi$ of the following type

$$p(\vartheta) = \sum_{n=0}^{\infty} p_n \frac{\cos n\vartheta}{\sin \vartheta},$$

$$\frac{\partial \bar{u}_y}{\partial \phi} = \sum_{n=1}^{\infty} u_n \sin n\phi.$$

The substitution into the integral equation gives

$$\sum_{n=1}^{\infty} u_n \sin n\phi = -\frac{2(1-\nu^2)}{E} a \sum_{n=0}^{\infty} p_n \sin n\phi,$$

i.e.,

$$p_n = -\frac{E u_n}{2(1-\nu^2) a}, \quad n \geq 1.$$

However, for a symmetric punch and loading, $\bar{u}_y$ is symmetric in x, so that $u_1 = 0$ (otherwise, $u_1 \cos \phi \sim u_1 x$ would give an antisymmetric contribution to $\bar{u}_y$). This implies that $p_1 = 0$ as well. The coefficient p_0 is obtained from the equilibrium condition

$$P = \int_{-a}^{a} p(s) ds = \int_0^{\pi} p(\vartheta) a \sin \vartheta \, d\vartheta = a \sum_{n=0}^{\infty} \int_0^{\pi} \cos n\vartheta \, d\vartheta = \pi a p_0,$$

which gives

$$p_0 = \frac{P}{\pi a}.$$

For a sufficiently shallow indentation, the circular indentation profile is reproduced by using $n = 2$ term, i.e.,

$$\frac{\partial \bar{u}_y}{\partial \phi} = u_2 \sin 2\phi, \quad \bar{u}_y = \frac{u_2}{2}(1 - \cos 2\phi),$$

so that $\bar{u}_y(\phi = 0, \pi) = 0$. If the indentation depth at $x = 0$ ($\phi = \pi/2$) is δ, we have

$$u_2 = \delta \approx \frac{a^2}{2R}.$$

A simple geometry reveals from Fig. 15-2 that a is a geometric mean between δ and $2R - \delta$, so that for $\delta \ll R$,

$$a^2 = \delta(2R - \delta) \approx 2R\delta.$$

which was used in the previous expression. Alternatively, to derive $u_2 \approx a^2/2R$ one can proceed from the curvature expression $\partial^2 \bar{u}_y/\partial x^2 \approx -1/R$. Consequently,

$$p_2 = -\frac{E u_2}{2(1-\nu^2) a} = -\frac{E a}{4(1-\nu^2) R},$$

and

$$p(\vartheta) = \frac{1}{\sin \vartheta} \left[\frac{P}{\pi a} - \frac{E a}{4(1-\nu^2) R} \cos 2\vartheta \right].$$

Problems 15.2–16.1

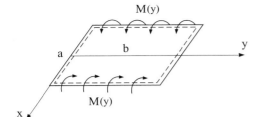

Figure 16-1. A rectangular simply supported plate under distributed edge moments.

If $p(\vartheta)$ is not to be singular at $\vartheta = 0$ and $\vartheta = \pi$ (i.e., $x = \pm a$),

$$\frac{P}{\pi a} - \frac{Ea}{4(1-\nu^2)R} = 0 \quad \Rightarrow \quad a = 2\left[\frac{(1-\nu^2)PR}{\pi E}\right]^{1/2}.$$

The substitution into the expression for $p(\vartheta)$ gives

$$p(\vartheta) = \frac{Ea}{2(1-\nu^2)R}\sin\vartheta,$$

or, in terms of x,

$$p(x) = \frac{2P}{\pi a}\left(1 - \frac{x^2}{a^2}\right)^{1/2}.$$

The width and the depth of the indentation, expressed in terms of P, R, and the elastic properties E, ν are

$$2a = 4\left[\frac{(1-\nu^2)PR}{\pi E}\right]^{1/2}, \quad \delta = 2\frac{(1-\nu^2)P}{\pi E}.$$

CHAPTER 16

Problem 16.1. A rectangular plate is loaded along its edges by uniform bending moments M_x and M_y. Determine the deflected shape of the plate (see Fig. 16-1).

Solution: The bending and twisting moments in the plate are related to the deflection w of the plate by

$$M_x = D\left(\frac{\partial^2 w}{\partial x^2} + \nu\frac{\partial^2 w}{\partial y^2}\right),$$

$$M_y = D\left(\frac{\partial^2 w}{\partial y^2} + \nu\frac{\partial^2 w}{\partial x^2}\right),$$

$$M_{xy} = D(1-\nu)\frac{\partial^2 w}{\partial x \partial y} = 0.$$

Consequently,

$$\frac{\partial^2 w}{\partial x^2} = \frac{M_x - \nu M_y}{D(1-\nu^2)},$$

$$\frac{\partial^2 w}{\partial y^2} = \frac{M_y - \nu M_x}{D(1-\nu^2)},$$

$$\frac{\partial^2 w}{\partial x \partial y} = 0.$$

The integration yields

$$w = \frac{M_x - \nu M_y}{2D(1-\nu^2)} x^2 + \frac{M_y - \nu M_x}{2D(1-\nu^2)} y^2 + C_1 x + C_2 y + C_3.$$

By placing the coordinate origin in the center of the midsurface of the deformed plate, we obtain $C_1 = C_2 = C_3 = 0$, and

$$w = \frac{1}{2D(1-\nu^2)} \left[(M_x - \nu M_y) x^2 + (M_y - \nu M_x) y^2 \right].$$

Problem 16.2. Determine the deflected shape of the simply supported rectangular plate loaded by a distributed moment $M(y)$ along its edges $x = \pm a/2$.

Solution: Expand the edge moment $M(y)$ in a Fourier series

$$M(y) = \sum_{n=1}^{\infty} M_n \sin \frac{n\pi y}{b}, \quad M_n = \frac{2}{b} \int_0^b M(y) \sin \frac{n\pi y}{b} dy.$$

Since no load is applied over the lateral surface of the plate, the governing equation for deflection is a homogeneous biharmonic equation $\nabla^4 w = 0$. In view of the symmetry with respect to $x = 0$, we take

$$w = \sum_{n=1}^{\infty} \left(B_n \cosh \frac{n\pi x}{b} + C_n x \sinh \frac{n\pi x}{b} \right) \sin \frac{n\pi y}{b}.$$

The boundary conditions of vanishing deflection and bending moment along $y = 0$ and $y = b$ are automatically satisfied with this representation of w. The remaining boundary conditions, along $x = \pm a/2$, are

$$w = 0, \quad D \frac{\partial^2 w}{\partial x^2} = M(y).$$

These are satisfied provided that

$$B_n = -C_n \frac{a}{2} \tanh \frac{n\pi a}{2b}, \quad C_n = \frac{b M_n}{2n\pi D \cosh(n\pi a/2b)}.$$

Thus, the deflected shape of the plate is

$$w = -\frac{b}{2\pi D} \sum_{n=1}^{\infty} \frac{\sin \frac{n\pi y}{b}}{n \cosh \frac{n\pi a}{2b}} M_n \left(\frac{a}{2} \tanh \frac{n\pi a}{2b} \cosh \frac{n\pi x}{b} - x \sinh \frac{n\pi x}{b} \right).$$

Problems 16.2–16.3

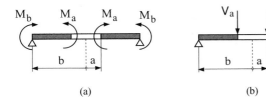

Figure 16-2. Simply supported annular plate of inner radius a and outer radius b under (a) edge moments M_a and M_b and (b) shear force V_a.

For example, if the edge moment is constant, i.e., $M(y) = M_0$, we have

$$M_n = \frac{4M_0}{n\pi}, \quad n = 1, 3, 5, \ldots,$$

and

$$w = -\frac{2M_0 b}{\pi^2 D} \sum_{n=1,3}^{\infty} \frac{\sin\frac{n\pi y}{b}}{n^2 \cosh\frac{n\pi a}{2b}} \left(\frac{a}{2}\tanh\frac{n\pi a}{2b}\cosh\frac{n\pi x}{b} - x\sinh\frac{n\pi x}{b}\right).$$

Problem 16.3. Derive the deflected shape of the annular circular plate shown in Fig. 16-2 due to (a) edge moments, and (b) shear force around the inner edge.

Solution: (a) Since there is no shear force Q_r in the annular plate under axisymmetric edge moments, we have

$$Q_r = D\frac{d}{dr}\left(\frac{dw}{dr^2} + \frac{1}{r}\frac{dw}{dr}\right) = D\frac{d}{dr}\left[\frac{1}{r}\frac{d}{dr}\left(r\frac{dw}{dr}\right)\right] = 0.$$

Upon three successive integrations, there follows

$$w = \frac{1}{4}Ar^2 + B\ln\frac{r}{b} + C.$$

The integration constants A, B, C are specified from the boundary conditions $w = 0$ at $r = b$, and

$$M_r = D\left(\frac{d^2 w}{dr^2} + \nu\frac{1}{r}\frac{dw}{dr}\right) = \begin{cases} M_a, & r = a, \\ M_b, & r = b. \end{cases}$$

This gives

$$w = \frac{1}{2(1+\nu)D}(b^2 M_b - a^2 M_a)\frac{r^2 - b^2}{b^2 - a^2} - \frac{1}{(1-\nu)D}\frac{a^2 b^2}{b^2 - a^2}(M_b - M_a)\ln\frac{r}{b}.$$

Recall that w in Chapter 16 is measured positive when upwards, so that downward deflections are negative.

(b) By vertical equilibrium, we must have

$$2\pi r Q_r(r) + 2\pi a V_a = 0,$$

which gives $Q_r(r) = -aV_a/r$. Thus,

$$Q_r = D\frac{d}{dr}\left[\frac{1}{r}\frac{d}{dr}\left(r\frac{dw}{dr}\right)\right] = -\frac{aV_a}{Dr}.$$

After three successive integrations, there follows

$$w = -\frac{aV_a}{4D}r^2\left(\ln\frac{r}{b} - 1\right) + \frac{1}{4}Ar^2 + B\ln\frac{r}{b} + C.$$

The integration constants A, B, C are specified from the boundary conditions $w = 0$ at $r = b$, and $M_r = 0$ at both $r = a$ and $r = b$. This leads to the following expression for the deflected shape of the plate

$$w = -\frac{ab^2 V_a}{4D}\left\{\left(1 - \frac{r^2}{b^2}\right)\left[\frac{3+\nu}{2(1+\nu)} - \frac{a^2}{b^2 - a^2}\ln\frac{a}{b}\right]\right.$$
$$\left. + \left(\frac{r^2}{b^2} + \frac{1+\nu}{1-\nu}\frac{2a^2}{b^2 - a^2}\ln\frac{a}{b}\right)\ln\frac{r}{b}\right\}.$$

CHAPTER 17

Problem 17.1. Derive the stress and displacement fields for screw dislocation in an infinite isotropic medium.

Solution: This is an antiplane strain problem of linear elasticity, for which the displacement components are

$$u_x = u_y = 0, \quad u_z = u_z(x, y).$$

The corresponding nonvanishing strain components are

$$e_{zx} = \frac{1}{2}\frac{\partial u_z}{\partial x}, \quad e_{zy} = \frac{1}{2}\frac{\partial u_z}{\partial y},$$

with the resulting stresses

$$\sigma_{zx} = 2\mu e_{zx} = \mu\frac{\partial u_z}{\partial x}, \quad \sigma_{zy} = 2\mu e_{zy} = \mu\frac{\partial u_z}{\partial y}.$$

Substituting these into the equilibrium equation

$$\frac{\partial \sigma_{zx}}{\partial x} + \frac{\partial \sigma_{zy}}{\partial y} = 0,$$

gives the Laplacian equation for the out-of-plane displacement

$$\frac{\partial^2 u_z}{\partial x^2} + \frac{\partial^2 u_z}{\partial y^2} = 0.$$

The displacement u_z is thus a harmonic function, which satisfies a discontinuity condition

$$\oint_C du_z = b,$$

where b is the magnitude of the Burgers vector, and C is any closed contour around the core of the dislocation. Evidently, this is satisfied if we take

$$u_z = \frac{b}{2\pi}\theta = \frac{b}{2\pi}\tan^{-1}\frac{y}{x}.$$

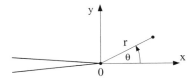

Figure 17-1. Semi-infinite crack in an infinite medium.

The resulting stresses are

$$\sigma_{zx} = \mu \frac{\partial u_z}{\partial x} = -\frac{\mu b}{2\pi} \frac{y}{x^2 + y^2}, \quad \sigma_{zy} = \mu \frac{\partial u_z}{\partial y} = \frac{\mu b}{2\pi} \frac{x}{x^2 + y^2}.$$

The polar coordinates counterparts are

$$\sigma_{zr} = 0, \quad \sigma_{z\theta} = \frac{\mu b}{2\pi} \frac{1}{r}.$$

Note the order of the stress singularity due to r^{-1} term at the center of the dislocation core.

Problem 17.2. Derive the asymptotic (near crack-tip) stress field for a semi-infinite crack in an infinite isotropic medium under mode III (antiplane shear) loading (Fig. 17-1).

Solution: As in the previous problem, the only nonvanishing displacement component is out-of-plane displacement $u_z = u_z(x, y)$, which satisfies the Laplacian equation

$$\frac{\partial^2 u_z}{\partial x^2} + \frac{\partial^2 u_z}{\partial y^2} = 0.$$

When this is rewritten in polar coordinates (see figure), we have

$$\frac{\partial^2 u_z}{\partial r^2} + \frac{1}{r} \frac{\partial u_z}{\partial r} + \frac{1}{r^2} \frac{\partial^2 u_z}{\partial \theta^2} = 0.$$

For antiplane shear loading, u_z should be an odd function of θ, i.e., $u_z(r, \theta) = -u_z(r, -\theta)$. This is satisfied by taking

$$u_z = A r^n \sin(n\theta),$$

where n is a real number and A is an arbitrary constant. The boundary conditions on the traction free faces of the crack are $\sigma_{\theta z}(r, \theta = \pm \pi) = 0$. This gives

$$\sigma_{\theta z} = \mu \frac{1}{r} \frac{\partial u_z}{\partial \theta}\bigg|_{\theta=\pm\pi} = \mu A n r^{n-1} \cos(n\pi) = 0,$$

which is satisfied for $n = 1/2, 3/2, 5/2, \ldots$ Taking $n = 1/2$ (which gives the dominating stress contribution near the crack tip, and u_z that monotonically increases from $\theta = 0$ to $\theta = \pi$), we have

$$u_z = A r^{1/2} \sin \frac{\theta}{2}.$$

The corresponding stresses are

$$\sigma_{\theta z} = \mu \frac{1}{r} \frac{\partial u_z}{\partial \theta} = \frac{\mu A}{2\sqrt{r}} \cos \frac{\theta}{2},$$

$$\sigma_{rz} = \mu \frac{\partial u_z}{\partial r} = \frac{\mu A}{2\sqrt{r}} \sin \frac{\theta}{2}.$$

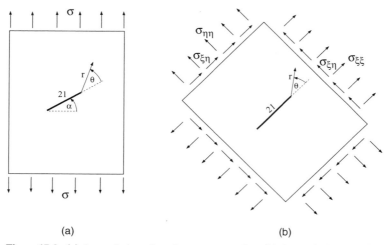

Figure 17-2. (a) An angled crack under remote tension. (b) An equivalent loading configuration.

Note the order of the stress singularity due to $r^{-1/2}$ term at the crack tip.

Problem 17.3. Determine the near crack tip stress field for an angled crack in an infinite plate under remote tension.

Solution: The loading shown in Fig. 17-2(a) is equivalent to the loading shown in Fig. 17-2(b). By the transformation of stress formulas, we have

$$\sigma_{\xi\xi} = \sigma \sin^2 \alpha, \quad \sigma_{\eta\eta} = \sigma \cos^2 \alpha, \quad \sigma_{\xi\eta} = \sigma \sin \alpha \cos \alpha.$$

This gives rise to a combined (mixed) mode I and II loading, with the corresponding stress intensity factors

$$K_I = \sigma_{\eta\eta} \sqrt{\pi l} = \sigma \cos^2 \alpha \sqrt{\pi l},$$

$$K_{II} = \sigma_{\xi\eta} \sqrt{\pi l} = \sigma \sin \alpha \cos \alpha \sqrt{\pi l}.$$

The near crack tip stress field for the mode I portion of the loading (expressed in polar coordinates) is

$$\sigma_{rr} = \frac{K_I}{\sqrt{2\pi r}} \left(\frac{5}{4} \cos \frac{\theta}{2} - \frac{1}{4} \cos \frac{3\theta}{2} \right),$$

$$\sigma_{\theta\theta} = \frac{K_I}{\sqrt{2\pi r}} \left(\frac{3}{4} \cos \frac{\theta}{2} + \frac{1}{4} \cos \frac{3\theta}{2} \right),$$

$$\sigma_{r\theta} = \frac{K_I}{\sqrt{2\pi r}} \left(\frac{1}{4} \sin \frac{\theta}{2} + \frac{1}{4} \sin \frac{3\theta}{2} \right).$$

Problems 17.3–17.4

Similarly, for the mode II portion of the loading, we have

$$\sigma_{rr} = \frac{K_{II}}{\sqrt{2\pi r}} \left(-\frac{5}{4} \sin \frac{\theta}{2} + \frac{3}{4} \sin \frac{3\theta}{2} \right),$$

$$\sigma_{\theta\theta} = \frac{K_{II}}{\sqrt{2\pi r}} \left(-\frac{3}{4} \sin \frac{\theta}{2} - \frac{3}{4} \sin \frac{3\theta}{2} \right),$$

$$\sigma_{r\theta} = \frac{K_{II}}{\sqrt{2\pi r}} \left(\frac{1}{4} \cos \frac{\theta}{2} + \frac{3}{4} \cos \frac{3\theta}{2} \right).$$

The total stress field near the crack tip is the sum of the above two fields.

Problem 17.4. The Peierls–Nabarro dislocation model accounts for the discreteness of the crystalline lattice. For an edge dislocation, the displacement discontinuity along the slip plane $y = 0$ is

$$\Delta u_x = \frac{b_x}{\pi} \tan^{-1} \frac{2x}{w}, \quad w = \frac{d}{1-\nu},$$

where b_x is the Burgers vector, d is the interplanar spacing across the slip plane, and w is the so-called width of the dislocation. It can be shown that the Airy stress function for this dislocation model is

$$\Psi = -\frac{\mu b_x}{4\pi (1-\nu)} y \ln \left[x^2 + \left(y \pm \frac{w}{2} \right)^2 \right].$$

The plus sign in $(y \pm w/2)$ holds for $y > 0$, and the minus sign for $y < 0$. Derive the corresponding stress field and verify that there is no stress divergence at the center of the dislocation core.

Solution: The stress components are derived from

$$\sigma_{xx} = \frac{\partial^2 \Psi}{\partial y^2}, \quad \sigma_{yy} = \frac{\partial^2 \Psi}{\partial x^2}, \quad \sigma_{xy} = -\frac{\partial^2 \Psi}{\partial x \partial y}.$$

This gives

$$\sigma_{xx} = -\frac{\mu b_x}{2\pi (1-\nu)} \left\{ \frac{y \pm w}{x^2 + (y \pm w/2)^2} + \frac{2x^2 y}{[x^2 + (y \pm w/2)^2]^2} \right\},$$

$$\sigma_{yy} = -\frac{\mu b_x}{2\pi (1-\nu)} \left\{ \frac{y}{x^2 + (y \pm w/2)^2} - \frac{2x^2 y}{[x^2 + (y \pm w/2)^2]^2} \right\},$$

$$\sigma_{xy} = +\frac{\mu b_x}{2\pi (1-\nu)} \left\{ \frac{x}{x^2 + (y \pm w/2)^2} - \frac{2xy(y \pm w/2)}{[x^2 + (y \pm w/2)^2]^2} \right\},$$

$$\sigma_{zz} = \nu(\sigma_{xx} + \sigma_{yy}) = -\frac{\nu \mu b_x}{\pi (1-\nu)} \frac{y \pm w/2}{x^2 + (y \pm w/2)^2}.$$

Clearly, the stresses remain bounded at the center of the dislocation core ($x = y = 0$), although the strains become large in the core region (so that the applicability of Hooke's law may actually be exceeded there).

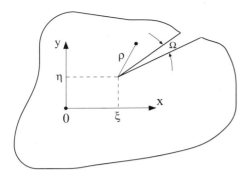

Figure 18-1. A disclination of an angle Ω in an infinite medium.

Problem 17.5. The displacement field for the Peierls–Nabarro screw dislocation is

$$u_z = -\frac{b_z}{2\pi} \tan^{-1} \frac{x}{y \pm d/2},$$

where b_z is the Burgers vector, and d is the interplanar spacing. The plus sign in $(y \pm d/2)$ holds for $y > 0$, and the minus sign for $y < 0$. Derive the corresponding stress field.

Solution: Since

$$\sigma_{zx} = \mu \frac{\partial u_z}{\partial x}, \quad \sigma_{zy} = \mu \frac{\partial u_z}{\partial y},$$

it readily follows that

$$\sigma_{zx} = -\frac{\mu b_z}{2\pi} \frac{y \pm d/2}{x^2 + (y \pm d/2)^2},$$

$$\sigma_{zy} = \frac{\mu b_z}{2\pi} \frac{x}{x^2 + (y \pm d/2)^2}.$$

The polar counterparts are

$$\sigma_{z\theta} = \frac{\mu b_z}{2\pi} \frac{1}{r} \left(1 - \frac{d}{2r} \sin\theta\right),$$

$$\sigma_{zr} = -\frac{\mu b_z}{2\pi} \frac{1}{r} \frac{d}{2r} \cos\theta.$$

For $r \gg d/2$, the above reduces to stress components of the corresponding Volterra dislocation

$$\sigma_{z\theta} = \frac{\mu b_z}{2\pi} \frac{1}{r}, \quad \sigma_{zr} = 0.$$

CHAPTER 18

Problem 18.1. The Airy stress function for a wedge dislocation (also called disclination) is

$$\Phi^\Omega = \frac{1}{4} k\Omega \, \rho^2 \ln \rho^2, \quad \rho^2 = (x - \xi)^2 + (y - \eta)^2,$$

Problems 18.1–18.2

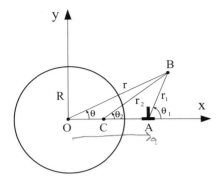

Figure 18-2. The radii r, r_1, r_2 and the angles $\theta, \theta_1, \theta_2$ appearing in the expressions for the stress components at point B due to edge dislocation at point A. The radius of the void is R, and the lengths $\overline{OA} = a$ and $\overline{OC} = R^2/a$.

where Ω is the disclination angle (see Fig. 18-1), and $k = \mu/2\pi(1-\nu)$. Show that the Airy stress functions for the edge dislocations with the Burgers vectors b_x and b_y can be deduced from Φ^Ω as

$$\Phi^{b_x} = \frac{b_x}{\Omega} \frac{\partial \Phi^\Omega}{\partial \eta}, \quad \Phi^{b_y} = -\frac{b_y}{\Omega} \frac{\partial \Phi^\Omega}{\partial \xi}.$$

Solution: It readily follows that

$$\frac{\partial \Phi^\Omega}{\partial \eta} = -\frac{k\Omega}{2}(y-\eta)(1+\ln \rho^2).$$

Since linear terms in x, y are immaterial to the Airy stress function (stresses being defined by its second derivatives), we obtain

$$\Phi^{b_x} = -kb_x(y-\eta)\ln \rho,$$

which is the Airy stress function for the dislocation b_x at the point (ξ, η). A similar derivation proceeds for Φ^{b_y}.

Problem 18.2. Derive the stress field due to an edge dislocation with the Burgers vector b_x near the circular void of radius R. The Airy stress function for the problem was derived by Dundurs and Mura (1964) as

$$\Phi = -\frac{Gb_x}{2\pi(1-\nu)} \Big[r_1 \ln r_1 \sin \theta_1 - r_2 \ln r_2 \sin \theta_2 + r \ln r \sin \theta$$

$$+ \frac{\zeta^2 - 1}{2\zeta^3} R \sin 2\theta_2 - \frac{(\zeta^2-1)^2}{\zeta^4} \frac{R^2}{2r_2} \sin \theta_2 + \frac{R^2}{2r} \sin \theta \Big].$$

Solution: Consider the stress state at point B due to edge dislocation at point A, at the distance $\overline{OA} = a$ from the center of circular void shown in Fig. 18-2. A nondimensional parameter $\zeta = a/R$ defines the position of dislocation relative to the void. The stresses are derived from the stress function as

$$\sigma_{xx} = \frac{\partial^2 \Phi}{\partial y^2}, \quad \sigma_{yy} = \frac{\partial^2 \Phi}{\partial x^2}, \quad \sigma_{xy} = -\frac{\partial^2 \Phi}{\partial x \partial y}.$$

This gives

$$\sigma_{xx} = -\frac{Gb_x y}{2\pi(1-\nu)} \left[\frac{1}{r_1^2}\left(1+2\frac{x_1^2}{r_1^2}\right) - \frac{1}{r_2^2}\left(1+2\frac{x_2^2}{r_2^2}\right) + \frac{1}{r^2}\left(1+2\frac{x^2}{r^2}\right) \right.$$
$$+ 2\frac{\zeta^2-1}{\zeta^3}\frac{Rx_2}{r_2^4}\left(1-4\frac{x_2^2}{r_2^2}\right) - \frac{(\zeta^2-1)^2}{\zeta^4}\frac{R^2}{r_2^4}\left(1-4\frac{x_2^2}{r_2^2}\right)$$
$$\left. + \frac{R^2}{r^4}\left(1-4\frac{x^2}{r^2}\right) \right],$$

$$\sigma_{yy} = -\frac{Gb_x y}{2\pi(1-\nu)} \left[\frac{1}{r_1^2}\left(1-2\frac{x_1^2}{r_1^2}\right) - \frac{1}{r_2^2}\left(1-2\frac{x_2^2}{r_2^2}\right) + \frac{1}{r^2}\left(1-2\frac{x^2}{r^2}\right) \right.$$
$$- 2\frac{\zeta^2-1}{\zeta^3}\frac{Rx_2}{r_2^4}\left(3-4\frac{x_2^2}{r_2^2}\right) + \frac{(\zeta^2-1)^2}{\zeta^4}\frac{R^2}{r_2^4}\left(1-4\frac{x_2^2}{r_2^2}\right)$$
$$\left. - \frac{R^2}{r^4}\left(1-4\frac{x^2}{r^2}\right) \right],$$

$$\sigma_{xy} = -\frac{Gb_x}{2\pi(1-\nu)} \left[\frac{x_1}{r_1^2}\left(1-2\frac{x_1^2}{r_1^2}\right) - \frac{x_2}{r_2^2}\left(1-2\frac{x_2^2}{r_2^2}\right) + \frac{x}{r^2}\left(1-2\frac{x^2}{r^2}\right) \right.$$
$$+ \frac{\zeta^2-1}{\zeta^3}\frac{R}{r_2^2}\left(1-8\frac{x_2^2}{r_2^2}+8\frac{x_2^4}{r_2^4}\right) + \frac{(\zeta^2-1)^2}{\zeta^4}\frac{R^2 x_2}{r_2^4}\left(3-4\frac{x_2^2}{r_2^2}\right)$$
$$\left. - \frac{R^2 x}{r^4}\left(3-4\frac{x^2}{r^2}\right) \right].$$

It these expressions $x = r\cos\theta$, $x_1 = r_1 \cos\theta_1 = x - a$, and $x_2 = r_2\cos\theta_2 = x - R^2/a$. The surface of the void exerts the force on the dislocation that is equal to

$$F_x = b_x \sigma_{xy}(a,0) = -\frac{Gb_x^2}{2\pi(1-\nu)}\frac{R^2}{a^3}\frac{2a^2-R^2}{a^2-R^2}.$$

The minus sign indicates the attractive nature of the force toward the stress free surface of the void. The singular self-stress of the dislocation at $x = a$ is excluded from $\sigma_{xy}(a,0)$ in evaluating the force on the dislocation due to void in the above expression.

CHAPTER 19

Problem 19.1. Consider a Griffith crack of length $2c$ in an elastically anisotropic material under uniform remote loading σ_{ij}^0. Show that near the crack tip the traction components along the crack plane are

$$\sigma_{2i} = \frac{\sigma_{2i}^0}{(2r/c)^{1/2}},$$

where r is the distance from the crack tip.

Problems 19.1–19.2

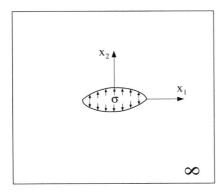

Figure 19-1. A nonsymmetric profile of the deformed crack faces under uniform pressure σ in an anisotropic materials.

Solution: By superimposing a uniform state of stress to (19.149), to achieve the traction-free crack faces, we obtain

$$t_2(x_1, 0) = \frac{|x_1|}{(x_1^2 - c^2)^{1/2}} \mathbf{t}_2^0.$$

By writing $x_1 = c + r$, and letting $r \ll c$, there follows

$$\mathbf{t}_2(r, 0) = \frac{\mathbf{t}_2^0}{(2r/c)^{1/2}},$$

i.e.,

$$\sigma_{2i}(r, 0) = \frac{\sigma_{2i}^0}{(2r/c)^{1/2}},$$

independently of the type of elastic anisotropy or the values of the elastic moduli. This was previously demonstrated in the main text by (19.47).

Problem 19.2. Consider a Griffith crack of length $2c$ in an elastically anisotropic material. The crack is loaded by uniform pressure σ over its faces. Show that the opening of the crack faces is not symmetrical with respect to the x_2 axis. Also, calculate the maximum vertical displacement along the crack faces and the horizontal displacements of the two crack tips.

Solution: For this loading we have

$$\mathbf{t}_2^0 = \begin{bmatrix} 0 \\ \sigma \\ 0 \end{bmatrix}.$$

Since $\mathbf{S} \cdot \mathbf{L}^{-1}$ is always antisymmetric, its diagonal elements vanish and from (19.150) we obtain along the crack faces ($|x_1| < c$, $x_2 = \pm 0$) the following displacement components

$$u_1(x_1, x_2 = \pm 0) = \left[\pm (c^2 - x_1^2)^{1/2} L_{12}^{-1} + x_1 (SL^{-1})_{12} \right] \sigma,$$

$$u_2(x_1, x_2 = \pm 0) = \pm (c^2 - x_1^2)^{1/2} L_{22}^{-1} \sigma.$$

Thus, unless $L_{12}^{-1} = 0$, the horizontal displacement component along the crack faces is not symmetric with respect to x_2 axis, although the crack tips move horizontally in a symmetric manner by

$$u_1(x_1 = \pm c, x_2 = 0) = \pm c(SL^{-1})_{12} \sigma.$$

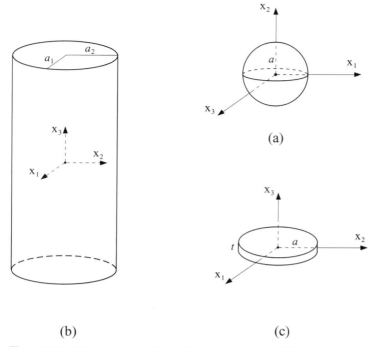

Figure 20-1. (a) A spherical, (b) an elliptical cylinder, and (b) a penny-shape inclusion.

The maximum vertical displacement of the points along the crack faces is

$$u_2(x_1 = 0, x_2 = \pm 0) = \pm c L_{22}^{-1} \sigma.$$

CHAPTER 20

Problem 20.1. By appropriately specializing the general results for an ellipsoidal inclusion, listed in Section 20.4, write down the components of the Eshelby tensor S_{ijkl} for: (a) a spherical inclusion, (b) an elliptical cylinder, and (c) a penny-shape inclusion.

Solution: (a) For a spherical inclusion $a_1 = a_2 = a_3 = a$ (Fig. 20-1a), we obtain

$$I_1 = I_2 = I_3 = \frac{4\pi}{3},$$

and

$$I_{11} = I_{22} = I_{33} = I_{12} = I_{23} = I_{31} = \frac{4\pi}{5a^2}.$$

This leads to

$$S_{1111} = S_{2222} = S_{3333} = \frac{7 - 5\nu}{15(1 - \nu)},$$

$$S_{1122} = S_{2233} = S_{3311} = S_{1133} = S_{2211} = S_{3322} = \frac{5\nu - 1}{15(1 - \nu)},$$

$$S_{1212} = S_{2323} = S_{3131} = \frac{4 - 5\nu}{15(1 - \nu)}.$$

Problem 20.1

The tensor components S_{ijkl} can be compactly expressed as

$$S_{ijkl} = \frac{4-5\nu}{15(1-\nu)}(\delta_{ik}\delta_{jl} + \delta_{il}\delta_{jk}) + \frac{5\nu-1}{15(1-\nu)}\delta_{ij}\delta_{kl}.$$

(b) For an elliptical cylinder with $a_3 \to \infty$ (Fig. 20-1b), we obtain

$$I_1 = \frac{4\pi a_2}{a_1 + a_2}, \quad I_2 = \frac{4\pi a_1}{a_1 + a_2}, \quad I_3 = 0,$$

$$I_{12} = \frac{4\pi}{(a_1 + a_2)^2}, \quad I_{11} = \frac{4\pi}{3a_1^2} - \frac{1}{3}I_{12}, \quad I_{22} = \frac{4\pi}{3a_2^2} - \frac{1}{3}I_{12},$$

$$I_{13} = I_{23} = I_{33} = 0,$$

$$a_3^2 I_{13} = I_1, \quad a_3^2 I_{23} = I_2, \quad a_3^2 I_{33} = 0.$$

The Eshelby stress components are accordingly

$$S_{1111} = \frac{1}{2(1-\nu)}\left[\frac{a_2(2a_1 + a_2)}{(a_1 + a_2)^2} + (1-2\nu)\frac{a_2}{a_1 + a_2}\right],$$

$$S_{2222} = \frac{1}{2(1-\nu)}\left[\frac{a_1(2a_2 + a_1)}{(a_1 + a_2)^2} + (1-2\nu)\frac{a_1}{a_1 + a_2}\right],$$

$$S_{1122} = \frac{1}{2(1-\nu)}\left[\frac{a_2^2}{(a_1 + a_2)^2} - (1-2\nu)\frac{a_2}{a_1 + a_2}\right],$$

$$S_{2211} = \frac{1}{2(1-\nu)}\left[\frac{a_1^2}{(a_1 + a_2)^2} - (1-2\nu)\frac{a_1}{a_1 + a_2}\right],$$

$$S_{1133} = \frac{\nu}{1-\nu}\frac{a_2}{a_1 + a_2},$$

$$S_{2233} = \frac{\nu}{1-\nu}\frac{a_1}{a_1 + a_2},$$

$$S_{1212} = \frac{1}{4(1-\nu)}\left[\frac{a_1^2 + a_2^2}{(a_1 + a_2)^2} + 1 - 2\nu\right],$$

$$S_{2323} = \frac{a_1}{2(a_1 + a_2)},$$

$$S_{3131} = \frac{a_2}{2(a_1 + a_2)},$$

$$S_{3333} = S_{3311} = S_{3322} = 0.$$

(c) For a penny-shape inclusion, with $a_1 = a_2 = a \gg a_3 = t$ (Fig. 20-1c), we have

$$I_1 = I_2 = \frac{\pi^2 t}{a}, \quad I_3 = 4\pi - \frac{2\pi^2 t}{a},$$

$$I_{11} = I_{22} = I_{12} = I_{21} = \frac{3\pi^2 t}{4a^3}, \quad I_{33} = \frac{4\pi}{3a^2},$$

$$I_{13} = I_{23} = I_{31} = I_{32} = \frac{4\pi}{a^2} - \frac{3\pi^2 t}{a^3}.$$

The Eshelby stress components are consequently

$$S_{1111} = S_{2222} = \frac{\pi(13 - 8\nu)}{32(1 - \nu)} \frac{t}{a},$$

$$S_{3333} = 1 - \frac{\pi(1 - 2\nu)}{4(1 - \nu)} \frac{t}{a},$$

$$S_{1122} = S_{2211} = \frac{\pi(8\nu - 1)}{32(1 - \nu)} \frac{t}{a},$$

$$S_{1133} = S_{2233} = \frac{\pi(2\nu - 1)}{8(1 - \nu)} \frac{t}{a},$$

$$S_{3311} = S_{3322} = \frac{\nu}{1 - \nu} - \frac{\pi(4\nu + 1)}{8(1 - \nu)} \frac{t}{a},$$

$$S_{1212} = \frac{\pi(7 - 8\nu)}{32(1 - \nu)} \frac{t}{a},$$

$$S_{1313} = S_{2323} = \frac{1}{2} + \frac{\pi(\nu - 2)}{8(1 - \nu)} \frac{t}{a}.$$

The remaining nonzero components satisfy the connections

$$S_{kk11} = S_{kk22} = \frac{\nu}{1 - \nu} + \frac{\pi(1 - 2\nu)}{4(1 - \nu)} \frac{t}{a},$$

$$S_{kk33} = 1 - \frac{\pi(1 - 2\nu)}{2(1 - \nu)} \frac{t}{a},$$

with sum on k.

Problem 20.2. By appropriately specializing the stress expressions for an ellipsoidal inclusion, listed in Section 20.4, write down the stress components for: (a) a spherical inclusion, (b) an elliptical cylinder, and (c) a penny-shape inclusion.

Solution: (a) From the general formulas given in Section 20.4, we obtain

$$\sigma_{11} = -\frac{2\mu}{15(1 - \nu)} [8e_{11}^T + (5\nu + 1)(e_{22}^T + e_{33}^T)],$$

Problems 20.2–20.3

$$\sigma_{12} = -\frac{2\mu}{15(1-\nu)}(7-5\nu)e_{12}^T.$$

The other stress components are obtained by cyclic permutations of $(1, 2, 3)$.

(b) In the case of an inclusion of the elliptical cylinder shape, the stresses are

$$\sigma_{11} = -\frac{\mu}{1-\nu}\frac{a_1}{(a_1+a_2)^2}\left[(2a_1+a_2)e_{11}^T + a_2 e_{22}^T + 2\nu(a_1+a_2)e_{33}^T\right],$$

$$\sigma_{22} = -\frac{\mu}{1-\nu}\frac{a_2}{(a_1+a_2)^2}\left[(2a_2+a_1)e_{22}^T + a_1 e_{11}^T + 2\nu(a_1+a_2)e_{33}^T\right],$$

$$\sigma_{33} = -\frac{2\mu}{1-\nu}\frac{1}{a_1+a_2}\left[\nu a_1 e_{11}^T + \nu a_2 e_{22}^T + (a_1+a_2)e_{33}^T\right],$$

$$\sigma_{12} = -\frac{2\mu}{1-\nu}\frac{a_1 a_2}{(a_1+a_2)^2}e_{12}^T,$$

$$\sigma_{23} = -2\mu\frac{a_2}{a_1+a_2}e_{23}^T,$$

$$\sigma_{31} = -2\mu\frac{a_1}{a_1+a_2}e_{31}^T.$$

(c) In the case of a penny-shape inclusion, the stresses are found to be

$$\sigma_{11} = -\frac{2\mu}{1-\nu}\left[e_{11}^T + \nu e_{22}^T - \frac{13\pi t}{32a}e_{11}^T - \frac{(16\nu-1)\pi t}{32a}e_{22}^T + \frac{(2\nu+1)\pi t}{8a}e_{33}^T\right],$$

$$\sigma_{22} = -\frac{2\mu}{1-\nu}\left[e_{22}^T + \nu e_{11}^T - \frac{13\pi t}{32a}e_{22}^T - \frac{(16\nu-1)\pi t}{32a}e_{11}^T + \frac{(2\nu+1)\pi t}{8a}e_{33}^T\right],$$

$$\sigma_{33} = -\frac{\mu(2\nu+1)}{4(1-\nu)}\frac{\pi t}{a}\left(e_{11}^T + e_{22}^T + \frac{2}{2\nu+1}e_{33}^T\right),$$

$$\sigma_{12} = -2\mu\left[1 - \frac{7-8\nu}{16(1-\nu)}\frac{\pi t}{a}\right]e_{12}^T,$$

$$\sigma_{23} = -\frac{\mu(2-\nu)}{2(1-\nu)}\frac{\pi t}{a}e_{23}^T,$$

$$\sigma_{31} = -\frac{\mu(2-\nu)}{2(1-\nu)}\frac{\pi t}{a}e_{31}^T.$$

Problem 20.3. Using equations of linear isotropic elasticity, derive the stress field in an infinite solvent matrix due to substitution of the solute atom of radius $R_2 = R_1 + \Delta$, where R_1 is the radius of the solvent atom.

Solution: Because of spherical symmetry the displacement field is radial, i.e., the only displacement component is $u_r = u_r(r)$. The corresponding nonvanishing strains are

$$e_{rr} = \frac{du_r}{dr}, \qquad e_{\theta\theta} = e_{\phi\phi} = \frac{u_r}{r},$$

with the stresses

$$\sigma_{rr} = \frac{2\mu}{1-2\nu}[(1-\nu)e_{rr} + 2\nu e_{\theta\theta}], \qquad \sigma_{\theta\theta} = \sigma_{\phi\phi} = \frac{2\mu}{1-2\nu}(\nu e_{rr} + e_{\theta\theta}).$$

Substituting these into the equilibrium equation

$$\frac{d\sigma_{rr}}{dr} + \frac{2}{r}(\sigma_{rr} - \sigma_{\theta\theta}) = 0$$

gives a differential equation for u_r,

$$\frac{d^2 u_r}{dr^2} + \frac{2}{r}\frac{du_r}{dr} - 2\frac{u_r}{r^2} = 0.$$

Its solution is

$$u_r = \frac{C_1}{r^2} + C_2 r.$$

For the external problem (solvent matrix), $C_2 = 0$ because $u_r \to 0$ as $r \to \infty$. If p is the pressure required for the substitution of the solvent atom, the condition $\sigma_{rr}(R_1) = -p$ specifies

$$C_1 = \frac{pR_1^3}{2\mu}.$$

For the internal problem (solute matrix), $C_1 = 0$ because $u_r = 0$ at $r = 0$. Thus the radial stress is uniform in the solute and equal to p, which gives

$$C_2 = -\frac{1-2\nu}{2(1+\nu)}\frac{p}{\mu}.$$

We assumed the same elastic constants in both solute and solvent.

To determine the unknown pressure p, we next use the misfit condition

$$u_r^{\text{ext}}(R_1) - u_r^{\text{int}}(R_2) = \Delta.$$

Neglecting the small quantities of higher order, this gives

$$p = 2\mu \frac{1+\nu}{2-\nu}\frac{\Delta}{R_1}.$$

Consequently, the stress field in the solvent matrix is found to be

$$\sigma_{rr} = -2\mu \frac{1+\nu}{2-\nu}\frac{R_1^2 \Delta}{r^3}, \qquad \sigma_{\theta\theta} = \sigma_{\phi\phi} = -\sigma_{rr}.$$

Problem 20.4. Derive the stress and displacement fields for sliding circular inclusion under biaxial transformation strain.

Problem 20.4

Solution: The Papkovich–Neuber potentials for the displacement in the sliding inclusion associated with the transformation strain e_{xx}^T and e_{yy}^T are

$$\Phi_0 = A_1 r^2 \cos 2\theta, \quad \Phi_1 = A_2 r^3 \cos 3\theta + A_3 r \cos\theta,$$

$$\Phi_2 = A_2 r^3 \sin 3\theta + A_3 r \sin\theta,$$

where r and θ denote the polar coordinates. The displacement components are derived from

$$u_x = \frac{\partial}{\partial x}(\Phi_0 + x\Phi_1 + y\Phi_2) - 4(1-\nu)\Phi_1 \;|\; + e_{xx}^T x,$$

$$u_y = \frac{\partial}{\partial y}(\Phi_0 + x\Phi_1 + y\Phi_2) - 4(1-\nu)\Phi_2 \;|\; + e_{yy}^T y.$$

The terms $e_{xx}^T x$ and $e_{yy}^T y$, appearing to the right of the vertical (|) line, correspond to stress-free eigenstrain, and should not be taken into account when calculating the stresses.

The Papkovich–Neuber potentials for displacements in the matrix are

$$\Phi_0 = B_1 r^{-2} \cos 2\theta, \quad \Phi_1 = B_2 r^{-1} \cos\theta, \quad \Phi_2 = B_3 r^{-1} \sin\theta.$$

The corresponding displacement components are obtained from the above expressions for u_x and u_y by excluding $e_{xx}^T x$ and $e_{yy}^T y$ terms on their right-hand side. The boundary conditions for the sliding inclusion are the vanishing of the shear traction at the interface between the inclusion and the matrix, and the continuity of normal traction and normal displacement at the interface. Thus, at $r = a$,

$$\sigma_{r\theta}^I = 0, \quad \sigma_{r\theta}^M = 0, \quad \sigma_{rr}^I = \sigma_{rr}^M, \quad u_r^I = u_r^M.$$

The superscript I designates the inclusion, and M the matrix. Upon calculation, we obtain

$$A_1 = -\frac{3k}{8}(e_{xx}^T - e_{yy}^T), \quad A_2 = \frac{k}{8}(e_{xx}^T - e_{yy}^T)a^{-2}, \quad A_3 = \frac{k}{2}(e_{xx}^T + e_{yy}^T),$$

and

$$B_1 = \frac{k}{8}(e_{xx}^T - e_{yy}^T)a^4, \quad B_{2,3} = -\frac{k}{8}[8k(e_{xx}^T + e_{yy}^T) \pm 3(e_{xx}^T - e_{yy}^T)]a^2,$$

where $k = 1/4(1-\nu)$. The following displacement and stress components result in polar coordinates. The displacements in the inclusion are

$$u_r = \frac{k}{4}\left[4(e_{xx}^T + e_{yy}^T)r + (e_{xx}^T - e_{yy}^T)\left(5 - 8\nu + 2\nu\frac{r^2}{a^2}\right)r\cos 2\theta\right],$$

$$u_\theta = -\frac{k}{4}(e_{xx}^T - e_{yy}^T)\left[5 - 8\nu + (3 - 2\nu)\frac{r^2}{a^2}\right]r\sin 2\theta,$$

and the stresses

$$\sigma_{rr} = -\frac{k\mu}{2}\left[4(e_{xx}^T + e_{yy}^T) + 3(e_{xx}^T - e_{yy}^T)\cos 2\theta\right],$$

$$\sigma_\theta = -\frac{k\mu}{2}\left[4(e_{xx}^T + e_{yy}^T) - 3(e_{xx}^T - e_{yy}^T)\left(1 - 2\frac{r^2}{a^2}\right)\cos 2\theta\right],$$

$$\sigma_{r\theta} = \frac{3k\mu}{2}(e_{xx}^T - e_{yy}^T)\left(1 - \frac{r^2}{a^2}\right)\sin 2\theta.$$

The displacement components in the matrix are similarly

$$u_r = \frac{k}{4}\frac{a}{r}\left\{4(e^T_{xx}+e^T_{yy})a + (e^T_{xx}-e^T_{yy})\left[6(1-\nu) - \frac{a^2}{r^2}\right]a\cos 2\theta\right\},$$

$$u_\theta = -\frac{k}{4}(e^T_{xx}-e^T_{yy})\frac{a}{r}\left[3(1-2\nu) + \frac{a^2}{r^2}\right]a\sin 2\theta,$$

with the corresponding stresses

$$\sigma_{rr} = -\frac{k\mu}{2}\frac{a^2}{r^2}\left[4(e^T_{xx}+e^T_{yy}) + 3(e^T_{xx}-e^T_{yy})\left(2 - \frac{a^2}{r^2}\right)\cos 2\theta\right],$$

$$\sigma_{\theta\theta} = \frac{k\mu}{2}\frac{a^2}{r^2}\left[4(e^T_{xx}+e^T_{yy}) - 3(e^T_{xx}-e^T_{yy})\frac{a^2}{r^2}\cos 2\theta\right],$$

$$\sigma_{r\theta} = -\frac{3k\mu}{2}(e^T_{xx}-e^T_{yy})\frac{a^2}{r^2}\left(1 - \frac{a^2}{r^2}\right)\sin 2\theta.$$

The stress state at all points of the inclusion at the interface $r = a$ is purely dilatational in the sense $\sigma_{rr} = \sigma_\theta$. A discontinuity in the tangential displacement at the boundary of the inclusion is

$$\Delta u_\theta = u_\theta^M(a,\theta) - u_\theta^I(a,\theta) = \frac{1}{4}(e^T_{xx}-e^T_{yy})a\sin 2\theta.$$

A discontinuity in the hoop stress across the interface of the sliding inclusion is constant and equal to $\Delta\sigma_\theta = 4k\mu(e^T_{xx}+e^T_{yy})$.

CHAPTER 21

Problem 21.1. The asymptotic stress and displacement fields near the crack tip of a semi-infinite crack in an infinite medium under mode III (antiplane shear) loading are

$$\sigma_{rz} = \frac{K_{III}}{\sqrt{2\pi r}}\sin\frac{\theta}{2}, \quad \sigma_{\theta z} = \frac{K_{III}}{\sqrt{2\pi r}}\cos\frac{\theta}{2},$$

$$u_z = \frac{2K_{III}}{\mu}\sqrt{\frac{r}{2\pi}}\sin\frac{\theta}{2}.$$

Evaluate the J integral around the crack tip.

Solution: The J integral (for the x direction) around the circle with the center at the coordinate origin (see Fig. 21-1) in the case of antiplane shear is

$$J = \int\left(Wdy - \sigma_{rz}\frac{\partial u_z}{\partial x}ds\right),$$

where $dy = r\cos\theta d\theta$ and $ds = rd\theta$. Since the strain energy of linearly elastic material (per unit volume) is

$$W = \frac{1}{2\mu}(\sigma_{rz}^2 + \sigma_{\theta z}^2) = \frac{K_{III}^2}{4\pi\mu r},$$

Problem 21.1

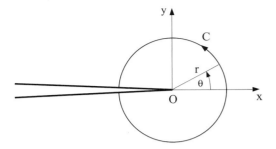

Figure 21-1. A circular integration path around the crack tip used to evaluate the J integral.

and since

$$\frac{\partial u_z}{\partial x} = \frac{\partial u_z}{\partial r}\frac{\partial r}{\partial x} + \frac{\partial u_z}{\partial \theta}\frac{\partial \theta}{\partial x} = -\frac{K_{III}}{\mu\sqrt{2\pi r}}\sin\frac{\theta}{2},$$

the substitution in the above expression for the J integral, and integration from $\theta = -\pi$ to $\theta = \pi$, gives

$$J = \frac{K_{III}^2}{2\mu}.$$

The above analysis can be extended to combined mode I and II loading. It follows that

$$J = \frac{1-v^2}{E}\left(K_I^2 + K_{II}^2\right), \quad \text{plane strain,}$$

$$J = \frac{1}{E}\left(K_I^2 + K_{II}^2\right), \quad \text{plane stress.}$$

In the derivation, the following stress and displacement fields apply for the respective modes.

Mode I:

$$\begin{pmatrix}\sigma_{xx}\\ \sigma_{yy}\\ \sigma_{xy}\end{pmatrix} = \frac{K_I}{\sqrt{2\pi r}}\cos(\theta/2)\begin{pmatrix}1 - \sin(\theta/2)\sin(3\theta/2)\\ \sin(\theta/2)\cos(3\theta/2)\\ 1 + \sin(\theta/2)\sin(3\theta/2)\end{pmatrix},$$

$$\begin{pmatrix}u_x\\ u_y\end{pmatrix} = \frac{K_I}{2\mu}\sqrt{\frac{r}{2\pi}}\begin{pmatrix}\cos(\theta/2)\left[\upsilon - 1 + 2\sin^2(\theta/2)\right]\\ \sin(\theta/2)\left[\upsilon + 1 - 2\cos^2(\theta/2)\right]\end{pmatrix}.$$

where $\upsilon = 3 - 4v$ for plane strain and $\upsilon = (3-v)/(1+v)$ for plane stress.

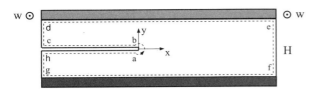

Figure 21-2. An infinitely long strip of thickness H with a semi-infinite crack. The lower side of the strip is fixed and the upper is given a uniform out-of-plane displacement w.

Mode II:

$$\begin{pmatrix} \sigma_{xx} \\ \sigma_{yy} \\ \sigma_{xy} \end{pmatrix} = \frac{K_{II}}{\sqrt{2\pi r}} \begin{pmatrix} -\sin(\theta/2)\left[2 + \cos(\theta/2)\cos(3\theta/2)\right] \\ \cos(\theta/2)\left[1 - \sin(\theta/2)\sin(3\theta/2)\right] \\ \sin(\theta/2)\cos(\theta/2)\cos(3\theta/2) \end{pmatrix},$$

$$\begin{pmatrix} u_x \\ u_y \end{pmatrix} = \frac{K_{II}}{2\mu}\sqrt{\frac{r}{2\pi}} \begin{pmatrix} \sin(\theta/2)\left[\upsilon + 1 + 2\cos^2(\theta/2)\right] \\ -\cos(\theta/2)\left[\upsilon - 1 - 2\sin^2(\theta/2)\right] \end{pmatrix}.$$

The J integral is calculated from

$$J = \int_{-\pi}^{\pi} \left(W\cos\theta - T_k \frac{\partial u_k}{\partial x}\right) r\,d\theta,$$

where T_k ($k = 1, 2$) are the traction components along the circular integration path.

Problem 21.2. Determine the value of the J integral around the crack tip for a cracked strip shown in Fig. 21-2. The upper side of the strip is given a uniform out-of-plane displacement w, while the lower side is fixed.

Solution: As in the similar problem in the text, the J integral for the closed path *abc...ha* vanishes, i.e.,

$$J_{abc...ha} = J_{ab} + J_{bc} + J_{cd} + J_{de} + J_{ef} + J_{fg} + J_{gh} + J_{ha} = 0.$$

The stresses along cd and gh (infinitely remote from the crack tip) vanish, thus $W = 0$ there, whereas $\sigma_{yz} = \mu w/H$ along infinitely remote ef. Imposing the obvious boundary conditions along horizontal parts of the integration contours, we obtain

$$J_{abc...ha} = J_{ab} + J_{ef} = J_{ab} + \int_e^f W\,dy = 0.$$

Since $W = \sigma_{yz}^2/2\mu = \mu w^2/2H^2$ along the segment ef, the above gives

$$J_{ab} = \frac{\mu w^2}{2H}.$$

Problems 21.3–21.4

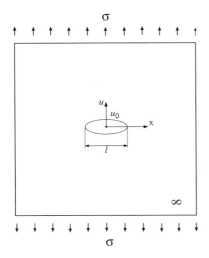

Figure 21-3. A crack of length l in an infinite medium under remote tension (Griffith crack). The deformed crack faces are elliptical in shape.

Problem 21.3. Consider a crack of length l in an infinite medium under remote uniform tension σ (see Fig. 21-3). The crack faces deform into an elliptical shape defined by the displacement function

$$u(x) = u_0\left(1 - 4\frac{x^2}{l^2}\right), \quad u_0 = \frac{\sigma}{E_*}l,$$

where $E_* = E$ for plane stress and $E_* = E/(1-\nu^2)$ for plane strain. Derive an expression for the energy release rate G associated with the extension of the crack length.

Solution: The potential energy P of the cracked medium differs from the potential energy of the uncracked medium under remote tension σ by an amount that is equal to the work done by σ applied to the crack faces to close the crack (and restore the uniform stress state throughout the medium). Therefore, we can write

$$P = P_0 - 2\int_{-l/2}^{l/2}\sigma u(x)\,dx = P_0 - 2\int_{-l/2}^{l/2}\sigma u_0\left(1 - 4\frac{x^2}{l^2}\right)dx = P_0 - \frac{\sigma^2 l^2 \pi}{4E_*}.$$

The energy release rate is defined as the negative gradient of the potential energy with respect to the crack length. Thus,

$$G = -\frac{\partial P}{\partial l} = \frac{\sigma^2 l \pi}{2E_*}.$$

Problem 21.4. Evaluate the M integral around an edge dislocation with the Burgers vector b_x.

Solution: The M integral of two-dimensional elasticity is defined by

$$M = \int_C \left(Wn_i x_i - T_k\frac{\partial u_k}{\partial x_i}x_i\right)dC.$$

In a homogeneous linearly elastic material within a simply connected region this integral vanishes along any closed path C that does not embrace a singularity or defect.

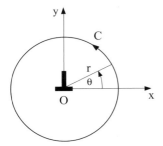

Figure 21-4. A circular path C around an edge dislocation used to evaluate the M integral.

Along a circular path around the dislocation (with the coordinate origin at the center of dislocation, Fig. 21-4), we have

$$x_i n_i = r, \quad \frac{\partial u_k}{\partial x_i} x_i = r \frac{\partial u_k}{\partial r},$$

so that

$$T_k \frac{\partial u_k}{\partial x_i} x_i = r \left(\sigma_{rr} \frac{\partial u_r}{\partial r} + \sigma_{r\theta} \frac{\partial u_\theta}{\partial r} \right).$$

The elastic strain energy density for plane strain is

$$W = \frac{1}{4\mu} \left[(1-\nu)(\sigma_{rr}^2 + \sigma_{\theta\theta}^2) - 2\nu \sigma_{rr} \sigma_{\theta\theta} + 2\sigma_{r\theta}^2 \right].$$

For an edge dislocation with the Burgers vector b_x, we have

$$\sigma_{rr} = \sigma_{\theta\theta} = -\frac{\mu b_x}{2\pi(1-\nu)} \frac{\sin\theta}{r}, \quad \sigma_{r\theta} = \frac{\mu b_x}{2\pi(1-\nu)} \frac{\cos\theta}{r},$$

with the displacements

$$u_r = -\frac{b_x}{4\pi(1-\nu)} \left[(1-2\nu) \ln\frac{r}{b_x} \sin\theta - 2(1-\nu)\theta \cos\theta - \frac{1}{2} \sin\theta \right],$$

$$u_\theta = -\frac{b_x}{4\pi(1-\nu)} \left[(1-2\nu) \ln\frac{r}{b_x} \cos\theta + 2(1-\nu)\theta \sin\theta + \frac{1}{2} \cos\theta \right].$$

It readily follows that

$$W x_i n_i - T_k \frac{\partial u_k}{\partial x_i} x_i = \frac{\mu b_x^2}{4\pi^2(1-\nu)} \frac{\cos^2\theta}{r},$$

and, upon integration from $\theta = 0$ to $\theta = 2\pi$,

$$M = \frac{\mu b_x^2}{4\pi(1-\nu)}.$$

Problem 21.5. Evaluate the M integral around the crack tip of a semi-infinite crack under asymptotic stress field given in Problem 21.1. The coordinate origin is the the crack tip.

Problems 21.5–21.6

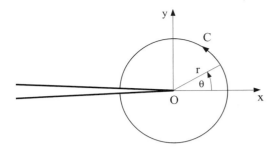

Figure 21-5. A circular path C around the crack tip of a semi-infinite crack used to evaluate the M integral.

Solution: Along a circular path around the crack tip (Fig. 21-5), we have

$$\frac{\partial u_k}{\partial x_i} x_i = r \frac{\partial u_k}{\partial r} = \frac{1}{2} u_k.$$

The last identity follows because u_k are homogeneous functions of r of degree $1/2$. Since $x_i n_i = r$, we obtain

$$M = \int_C \left(W n_i x_i - T_k \frac{\partial u_k}{\partial x_i} x_i \right) dC = \int_0^{2\pi} \left(W r - \frac{1}{2} T_k u_k \right) r \, d\theta.$$

Now, from the asymptotic fields around the crack tip, listed in Problem 21.1, we recall that $T_k \sim r^{-1/2}$, $u_k \sim r^{1/2}$ and $W \sim r^{-1}$. Consequently, without any further calculation, by taking the limit as $r \to 0$, we obtain $M = 0$. Because along the traction free crack faces $x_i n_i = 0$ and $T_k = 0$, we conclude that $M = 0$ around the crack tip regardless or r.

Problem 21.6. If the value of M integral, defined with the coordinate origin placed at the point O, is M_0, show that the value of the M integral, defined with the coordinate origin at the point A (see Fig. 21-6), is

$$M_A = M_0 - x_i^A J_i.$$

Solution: The M integral with the coordinate origin at $A(x_1^A, x_2^A)$ is

$$M_A = \int_C \left(W n_i \xi_i - T_k \frac{\partial u_k}{\partial \xi_i} \xi_i \right) dC.$$

Since

$$\xi_i = x_i - x_i^A,$$

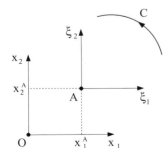

Figure 21-6. A path C used to evaluate the M integral with respect to coordinates with the origin at O and A.

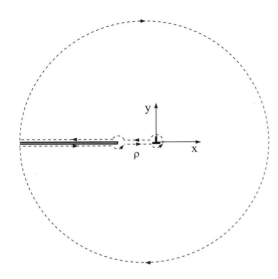

Figure 21-7. A path used to evaluate the M integral and derive an expression for the dislocation force due to a nearby semi-infinite crack at distance ρ.

the substitution in the above integral gives

$$M_A = \int_C \left(W n_i x_i - T_k \frac{\partial u_k}{\partial x_i} x_i \right) dC - x_i^A \int_C \left(W n_i - T_k \frac{\partial u_k}{\partial x_i} \right) dC.$$

Thus,

$$M_A = M_0 - x_i^A J_i,$$

where

$$J_i = \int_C \left(W n_i - T_k \frac{\partial u_k}{\partial x_i} \right) dC, \quad (i = 1, 2).$$

Problem 21.7. Derive an expression for the force on an edge dislocation due to traction free crack faces of a nearby semi-infinite crack. The dislocation is at the distance ρ from the crack tip, as shown in Fig. 21-7.

Solution: The M integral along the closed contour around the crack tip and the dislocation, shown in the figure, vanishes. By placing the origin at the center of the dislocation, the only nonvanishing contributions to the M integral are along the small circles around the dislocation and the crack tip (because of the traction free crack faces, and because along large circle the stresses due to crack tip dominate over those due to nearby dislocation. Also, along remote circle the distance ρ is not observed, and from Problem 21.5 we know that $M = 0$ around the crack tip with the origin at the crack tip). Therefore, recalling the results from Problems 21.4 and 21.6,

$$M_O = \frac{\mu b_x^2}{4\pi(1-\nu)} - \rho J_x = 0,$$

i.e.,

$$J_x = \frac{\mu b_x^2}{4\pi(1-\nu)} \frac{1}{\rho}.$$

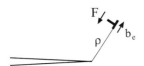

Figure 21-8. An edge dislocation with the Burgers vector b_e at distance ρ along an inclined slip plane from the crack tip of a semi-infinite crack.

This is the energy release rate of the crack associated with its extension toward the dislocation (configurational force on the crack tip due to dislocation). The opposite force of the same magnitude is the force on the dislocation due to the traction free crack faces.

If the slip plane of the dislocation is at an angle to the crack faces (see Fig. 21-8), a similar analysis reveals that the attractive force from the crack tip on the edge dislocation with the Burgers vector b_e, is

$$F = \frac{\mu b_e^2}{4\pi(1-\nu)} \frac{1}{\rho}.$$

If the dislocation is of a screw character, the dislocation force is

$$F = \frac{\mu b_s^2}{4\pi} \frac{1}{\rho}.$$

Problem 21.8. Consider a double cantilever beam shown in Fig. 21-9.

(a) Calculate the load to propagate the crack of length l, if the surface energy of the crack faces is γ. Estimate the elastic stiffness of the system by using the simple beam theory.
(b) Derive the force-displacement relationship if the crack propagates in a neutrally stable mode (transition between stable and unstable mode).
(c) Independently of part (b), examine the stability of crack growth under constant force and constant displacement conditions.

Solution: (a) The potential energy of the system is the sum of the elastic strain energy and the load potential, *i.e.*,

$$\Pi = \frac{1}{2}Fu - Fu.$$

If, at some state of loading, the crack extends its length from l to $l + dl$, while the force changes from F to $F + dF$ and displacement from u to $u + du$, the potential energy of the system changes by

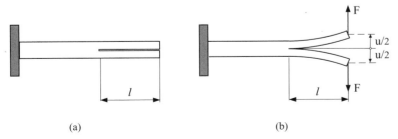

(a)   (b)

Figure 21-9. A double-cantilever specimen: (a) before and (b) after load application.

$$d\Pi = \frac{1}{2}Fdu + \frac{1}{2}dFu - Fdu.$$

There is no load potential of dF on the previously applied displacement u. The crack extends its length by dl if the energy release rate of the system is sufficient to supply the surface energy increase $2\gamma\,dl$ (times the unit thickness), that is if $-d\Pi = 2\gamma\,dl$. This gives the propagation condition $G = 2\gamma$, where

$$G = -\frac{\partial \Pi}{\partial l} = \frac{1}{2}\left(F\frac{du}{dl} - u\frac{dF}{dl}\right)$$

is the energy release rate.

If the elastic compliance of the system is $s = s(l)$, we have

$$s = \frac{u}{F} \quad \Rightarrow \quad \frac{ds}{dl} = \frac{1}{F^2}\left(F\frac{du}{dl} - u\frac{dF}{dl}\right).$$

Therefore, the energy release rate is

$$G = \frac{1}{2}F^2\frac{ds}{dl}.$$

If $k = k(l)$ is the elastic stiffness of the system, such that $F = ku$ and $ks = 1$, we have

$$\frac{ds}{dl} = -\frac{1}{k^2}\frac{dk}{dl}.$$

Consequently, the energy release rate can be expressed as either of

$$G = \frac{1}{2}F^2\frac{ds}{dl} = -\frac{1}{2}u^2\frac{dk}{dl}.$$

For the double cantilever beam we can adopt from elementary beam-bending theory the simple relationship $u/2 = Fl^3/(3EI)$, where EI is the cantilever bending stiffness. Thus,

$$s(l) = \frac{2l^3}{3EI}, \quad k(l) = \frac{3EI}{2l^3},$$

and

$$\frac{ds}{dl} = \frac{2l^2}{EI}, \quad \frac{dk}{dl} = -\frac{9EI}{2l^4},$$

The crack propagation condition then becomes

$$\frac{1}{2}F^2\frac{2l^2}{EI} = \frac{1}{2}u^2\frac{9EI}{2l^4} = 2\gamma.$$

This gives the critical force for the crack propagation, and the corresponding displacement,

$$F^2 = \frac{2\gamma EI}{l^2}, \quad u^2 = \frac{8\gamma l^4}{9EI}.$$

(b) By forming the product $F^4 u^2$ from the above two expressions, we have

$$F^4 u^2 = \frac{32}{9}\gamma^3 EI \quad \Rightarrow \quad F^2 = \frac{4\gamma}{3}\sqrt{2\gamma EI}\,\frac{1}{u}.$$

Problem 21.8

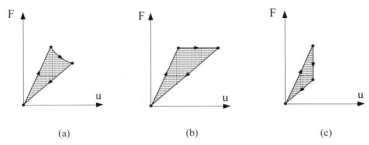

Figure 21-10. (a) The force-displacement relation during: (a) metastable crack propagation, (b) unstable crack propagation under constant force, and (c) stable crack propagation under constant displacement.

This defines the force displacement relationship $F = F(u)$ during the neutrally stable crack growth (Fig. 21-10a). During this growth

$$\left(\frac{\partial^2 \Pi}{\partial l^2}\right)_{G=2\gamma} = 0.$$

This can be easily verified from the expressions

$$\frac{\partial^2 \Pi}{\partial l^2} = \frac{1}{2}\left(u\frac{d^2 F}{dl^2} - F\frac{d^2 u}{dl^2}\right),$$

and

$$F^2 = \frac{2\gamma EI}{l^2} \quad \Rightarrow \quad \frac{d^2 F}{dl^2} = 2\sqrt{2\gamma EI}\, l^{-3},$$

$$u^2 = \frac{8\gamma l^4}{9EI} \quad \Rightarrow \quad \frac{d^2 u}{dl^2} = \frac{4\sqrt{2\gamma}}{3\sqrt{EI}} l.$$

(c) The crack will propagate in a stable mode if

$$\left(\frac{\partial^2 \Pi}{\partial l^2}\right)_{G=2\gamma} > 0,$$

and in an unstable mode if the reverse inequality applies.

If the crack propagation is taking place in a test under constant force (Fig. 21-10b), then

$$\Pi = -\frac{1}{2}Fu = -\frac{1}{2}F^2 s(l), \quad s(l) = \frac{2l^3}{3EI},$$

$$G = -\left(\frac{\partial \Pi}{\partial l}\right)_F = \frac{1}{2}F^2 \frac{ds}{dl} = 2\gamma \quad \Rightarrow \quad F^2 = \frac{2\gamma EI}{l^2},$$

$$\left(\frac{\partial^2 \Pi}{\partial l^2}\right)_{G=2\gamma} = -\frac{1}{2}F^2\frac{d^2 s}{dl^2} = -\frac{4\gamma}{l} < 0.$$

Thus, under the constant force, the crack propagates in an unstable mode.

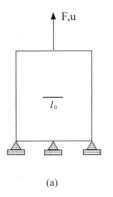
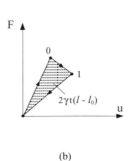

Figure 21-11. (a) The cracked specimen under applied force F. The initial crack length is l_0. (b) The force vs. displacement, loading–unloading path of the cracked specimen.

(a)   (b)

If the crack propagation is taking place in a test under constant displacement (Fig. 21-10c), then

$$\Pi = \frac{1}{2} F u = \frac{1}{2} k(l) u^2, \quad k(l) = \frac{3EI}{2l^3},$$

$$G = -\left(\frac{\partial \Pi}{\partial l}\right)_u = -\frac{1}{2} u^2 \frac{dk}{dl} = 2\gamma \quad \Rightarrow \quad u^2 = -\frac{8\gamma l^4}{9EI},$$

$$\left(\frac{\partial^2 \Pi}{\partial l^2}\right)_{G=2\gamma} = \frac{1}{2} u^2 \frac{d^2 k}{dl^2} = \frac{8\gamma}{l} > 0.$$

Thus, under the constant displacement, the crack propagates in a stable mode. This is also obvious from the result in part (b), because for any force-displacement curve, corresponding to crack growth, below the curve in Fig. 21-10a the crack will propagate in a stable mode, and for any force-displacement curve above the curve in Fig. 21-10a the crack will propagate in an unstable mode.

Problem 21.9. In a fracture test of cracked specimen with initial crack length l_0 (Fig. 21-11a), the crack began to propagate when the force F was F_0 and the corresponding displacement of its point of application u_0. As the crack propagated to length l_1, the force had fallen to F_1, with the corresponding displacement u_1. The specimen was then unloaded, and the elastic unloading went back to the origin (Fig. 21-11b). Assuming that the force-displacement curve from (F_0, u_0) to (F_1, u_1) is a straight line, evaluate the surface area γ of the crack faces.

Solution: The work done by the force from the initial state $(0, 0)$ to state (F_1, u_1) is

$$\frac{1}{2} F_0 u_0 + \int_0^1 F(u) du = \frac{1}{2} F_0 u_0 + \frac{1}{2} (F_0 + F_1)(u_1 - u_0),$$

the integral being evaluated as the area of the trapezoid under the force-displacement line $0 \to 1$. The difference of this work and the unloading work

$$\frac{1}{2} F_1 u_1$$

Problems 21.9–22.1

is equal to the increase of the surface energy of the crack faces, associated with crack extension from length l_0 to l_1. This is

$$2\gamma t(l_1 - l_0),$$

where t is the thickness of the specimen. Thus,

$$\frac{1}{2}F_0 u_0 + \frac{1}{2}(F_0 + F_1)(u_1 - u_0) - \frac{1}{2}F_1 u_1 = 2\gamma t(l_1 - l_0).$$

After solving for γ, we obtain

$$\gamma = \frac{F_0 u_1 - F_1 u_0}{4t(l_1 - l_0)}.$$

CHAPTER 22

Problem 22.1. Evaluate the effect of the couple stresses on the stress concentration at the surface of the circular hole in an infinite medium under remote shear loading.

Solution: Consider a stress-free circular void of radius R in an infinite medium under remote shear stresses σ_{13}^∞ and σ_{23}^∞. The displacement field is $w = w^0 + w^*$, where

$$w^0 = \left(\frac{\sigma_{13}^\infty}{\mu}r + A\frac{R^2}{r}\right)\cos\theta + \left(\frac{\sigma_{23}^\infty}{\mu}r + B\frac{R^2}{r}\right)\sin\theta,$$

$$w^* = RK_1\left(\frac{r}{l}\right)(C\cos\theta + D\sin\theta).$$

The constants A, B, C, and D are determined from the boundary conditions of vanishing reduced stress tractions along the surface of the hole $r = R$, which are

$$\bar{t}_{r3} = t_{r3} - \frac{1}{2R}\frac{\partial m_{rr}}{\partial \theta} = 0, \quad m_{r\theta} = 0.$$

The boundary conditions giving rise to uniform shear stresses σ_{13}^∞ and σ_{23}^∞ at $r \to \infty$ are identically satisfied by the selected form of the displacement function. The first boundary condition gives

$$[\mu R^2 + 2(\alpha + \beta)]A + (\alpha + \beta)\left[K_1\left(\frac{R}{l}\right) - \frac{R}{l}K_1'\left(\frac{R}{l}\right)\right]C = R^2\sigma_{13}^\infty,$$

$$[\mu R^2 + 2(\alpha + \beta)]B + (\alpha + \beta)\left[K_1\left(\frac{R}{l}\right) - \frac{R}{l}K_1'\left(\frac{R}{l}\right)\right]D = R^2\sigma_{23}^\infty,$$

and the second

$$2(\alpha + \beta)A + \frac{R^2}{l^2}\left[(\alpha + \beta)K_1''\left(\frac{R}{l}\right) - \beta K_1\left(\frac{R}{l}\right)\right]C = 0,$$

$$2(\alpha + \beta)B + \frac{R^2}{l^2}\left[(\alpha + \beta)K_1''\left(\frac{R}{l}\right) - \beta K_1\left(\frac{R}{l}\right)\right]D = 0.$$

It readily follows that

$$A = \frac{a}{b}\frac{\sigma_{13}^\infty}{\mu}, \quad B = \frac{a}{b}\frac{\sigma_{23}^\infty}{\mu}, \quad C = -\frac{2}{b}\frac{\sigma_{13}^\infty}{\mu}, \quad D = -\frac{2}{b}\frac{\sigma_{23}^\infty}{\mu},$$

with the parameters

$$a = a_0 + 2K_1, \quad b = a_0 + 4K_1, \quad a_0 = \frac{R}{l}K_0 + \frac{\alpha}{\alpha+\beta}\frac{R^2}{l^2}K_1.$$

The values of the modified Bessel functions at $r = R$ are denoted by K_0 and K_1. The resulting displacement field is

$$w = \left[r + \frac{a}{b}\frac{R^2}{r} - \frac{2R}{b}K_1\left(\frac{r}{l}\right)\right]\frac{\sigma_{r3}^\infty}{\mu},$$

where

$$\sigma_{r3}^\infty = \sigma_{13}^\infty \cos\theta + \sigma_{23}^\infty \sin\theta.$$

In the limit as $R/l \to \infty$, the ratio $a/b \to 1$ and we recover the classical elasticity result

$$w = \left(r + \frac{R^2}{r}\right)\frac{\sigma_{r3}^\infty}{\mu}.$$

To evaluate the effect of the couple stresses on the stress concentration at the points on the surface of the hole, consider the shear stress components $t_{\theta 3}$ and $t_{3\theta}$ at $r = R$. It is found that

$$t_{\theta 3} = \frac{2c}{b}\sigma_{\theta 3}^\infty, \quad t_{3\theta} = \left(1 + \frac{d}{b}\right)\sigma_{\theta 3}^\infty,$$

with $c = a_0 + 3K_1$, $d = a_0 - 2K_1$ and

$$\sigma_{\theta 3}^\infty = -\sigma_{13}^\infty \sin\theta + \sigma_{23}^\infty \cos\theta.$$

The stress magnification factor for the shear stress $t_{\theta 3} = 2\zeta \sigma_{\theta 3}^\infty$ due to couple stress effects is

$$\zeta = \frac{c}{b} = \frac{\dfrac{\alpha}{\alpha+\beta} + \dfrac{l}{R}\left(3\dfrac{l}{R} + \dfrac{K_0}{K_1}\right)}{\dfrac{\alpha}{\alpha+\beta} + \dfrac{l}{R}\left(4\dfrac{l}{R} + \dfrac{K_0}{K_1}\right)}.$$

For example, for a small hole with the radius $R = 3l$ and with $\beta = 0$, this gives $\zeta = 0.936$ (indicating a decrease of the maximum stress due to couple stress effects).

CHAPTER 23

Problem 23.1. Determine the stress field for a screw dislocation array parallel to a bimaterial interface, and at distance h from it. The dislocation spacing in the array is p as shown in Fig. 23-1.

Problems 23.1–23.2

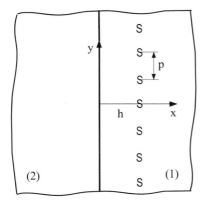

Figure 23-1. A screw dislocation array with the Burgers vector at a distance h from a bimaterial interface. The dislocation spacing is p.

Solution: By the summation of the stresses due to individual dislocations, located within the material (1), it is readily found that the shear stresses in two materials are

$$\sigma_{zx}^{(1)} = -\frac{\mu_1 b_z}{2p} \sin\psi \left(\frac{1}{C} - c\frac{1}{A}\right),$$

$$\sigma_{zy}^{(1)} = \frac{\mu_1 b_z}{2p} \left(\frac{\sinh\vartheta}{C} - c\frac{\sinh\varphi}{A}\right),$$

and

$$\sigma_{zx}^{(2)} = -\frac{\mu_2 b_z}{2p}(1+c)\frac{\sin\psi}{C},$$

$$\sigma_{zy}^{(2)} = \frac{\mu_2 b_z}{2p}(1+c)\frac{\sinh\vartheta}{C}.$$

The introduced parameters are

$$A = \cosh\varphi - \cos\psi, \quad C = \cosh\vartheta - \cos\psi,$$

and

$$\xi = \frac{x}{p}, \quad \eta = \frac{y}{p}, \quad h_0 = \frac{h}{p}, \quad \vartheta = 2\pi(\xi - h_0), \quad \varphi = 2\pi(\xi + h_0), \quad \psi = 2\pi\eta.$$

Problem 23.2. Determine the stress field for an edge dislocation array parallel to a bimaterial interface, and at distance h from it. The dislocations Burgers vector is b_x, and the dislocation spacing in the array is p as shown in Fig. 23-2.

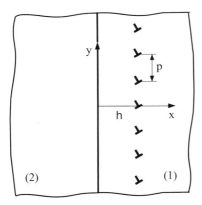

Figure 23-2. An edge dislocation array with the Burgers vector $\{b_x, b_y\}$ at a distance h from a bimaterial interface. The dislocation spacing is p.

Solution: By the summation of the stresses due to individual dislocations, it is readily found that the stresses in two materials are

$$\sigma_{xx}^{(1)} = -\frac{\pi k_1 b_x \sin \psi}{A^2 p} (X_1 + q X_2 + a\beta X_3),$$

$$\sigma_{yy}^{(1)} = \frac{\pi k_1 b_x \sin \psi}{A^2 p} (Y_1 + q Y_2 + a\beta Y_3),$$

$$\sigma_{xy}^{(1)} = \frac{\pi k_1 b_x}{A^2 p} (Z_1 + q Z_2 + a\beta Z_3),$$

and

$$\sigma_{xx}^{(2)} = -\frac{\pi k_1 a b_x \sin \psi}{C^2 p} (\vartheta \sinh \vartheta + C + \beta\varphi \sinh \vartheta),$$

$$\sigma_{yy}^{(2)} = \frac{\pi k_1 a b_x \sin \psi}{C^2 p} [\vartheta \sinh \vartheta - C + \beta(4\pi\xi \sinh \vartheta - \vartheta \sinh \vartheta - 2C)],$$

$$\sigma_{xy}^{(2)} = \frac{\pi k_1 a b_x}{C^2 p} [D\vartheta - \beta(C \sinh \vartheta - D\varphi)].$$

The parameters k_1, β, a and q are introduced in the text. The utilized abbreviations are

$$X_1 = \frac{A^2}{C^2} (\vartheta \sinh \vartheta + C),$$

$$Y_1 = \frac{A^2}{C^2} (\vartheta \sinh \vartheta - C),$$

$$Z_1 = \frac{A^2}{C^2} D\vartheta,$$

and

$$X_2 = \varphi \sinh \varphi + A + \frac{8\pi^2 h_0 \xi}{A} (B + \sinh^2 \varphi),$$

$$Y_2 = (\vartheta - 4\pi h_0) \sinh \varphi - A + \frac{8\pi^2 h_0 \xi}{A} (B + \sinh^2 \varphi),$$

$$Z_2 = B\vartheta + \frac{8\pi^2 h_0 \xi}{A} \sinh \varphi (B - \sin^2 \psi).$$

In addition, $X_3 = Y_3 = A$ and $Z_3 = A \sinh \varphi$. The variables A and C are as in Problem 23.1, and

$$B = \cosh \varphi \cos \psi - 1, \quad D = \cosh \vartheta \cos \psi - 1.$$

Problem 23.3. Determine the stress field for an edge dislocation array parallel to a bimaterial interface, and at distance h from it. The dislocations Burgers vector is b_y, and the dislocation spacing in the array is p (Fig. 23-2).

Solution: By the summation of the stresses due to individual dislocations, it is readily found that the stresses in two materials are

$$\sigma_{xx}^{(1)} = \frac{\pi k_1 b_y}{A^2 p}(X_1 + qX_2 + a\beta X_3),$$

$$\sigma_{yy}^{(1)} = \frac{\pi k_1 b_y}{A^2 p}(Y_1 + qY_2 + a\beta Y_3),$$

$$\sigma_{xy}^{(1)} = \frac{\pi k_1 b_y \sin\psi}{A^2 p}(Z_1 + qZ_2 + a\beta Z_3),$$

and

$$\sigma_{xx}^{(2)} = \frac{\pi k_1 a b_y}{C^2 p}[D\vartheta + \beta(C\sinh\vartheta + D\varphi)],$$

$$\sigma_{yy}^{(2)} = \frac{\pi k_1 a b_y}{C^2 p}[2C\sinh\vartheta - D\vartheta + \beta(C\sinh\vartheta - D\varphi)],$$

$$\sigma_{xy}^{(2)} = \frac{\pi k_1 a b_y \sin\psi}{C^2 p}(\vartheta\sinh\vartheta - C + \beta\varphi\sinh\vartheta).$$

The introduced abbreviations are in this case

$$X_1 = \frac{A^2}{C^2} D\vartheta,$$

$$Y_1 = \frac{A^2}{C^2}(2C\sinh\vartheta - D\vartheta),$$

$$Z_1 = \frac{A^2}{C^2}(\vartheta\sinh\vartheta - C),$$

and

$$X_2 = B\vartheta - \frac{8\pi^2 h_0 \xi}{A}\sinh\varphi\,(B - \sin^2\psi),$$

$$Y_2 = 2A\sinh\varphi - (\varphi + 4\pi h_0)B + \frac{8\pi^2 h_0 \xi}{A}\sinh\varphi\,(B - \sin^2\psi),$$

$$Z_2 = \varphi\sinh\varphi - A - \frac{8\pi^2 h_0 \xi}{A}(B + \sinh^2\varphi).$$

In addition, $X_3 = -Y_3 = -A\sinh\varphi$ and $Z_3 = -A$. The variables A, B, C, and D are as in Problem 23.2.

CHAPTER 24

Problem 24.1. A thin circular film of thickness h_f is bonded to a circular substrate of thickness $h_s \gg h_f$ (see Fig. 24-1). If the elastic mismatch strain between the film and substrate is

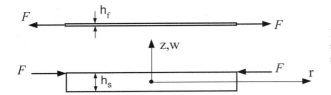

Figure 24-1. A thin film and its thick substrate. The force due to elastic mismatch strain is F.

e_m, derive the expression for the curvature of the substrate produced by the bonded film. The elastic constants of the film and substrate are (E_f, ν_f) and (E_s, ν_s), respectively.

Solution: The inplane radial force (per unit length) in the film needed to produce the elastic mismatch strain e_m is

$$F = \sigma_f h_f = \frac{E_f}{1 - \nu_f} e_m h_f.$$

Upon bonding of the film, the substrate can be considered to be strained by opposite F, as indicated in the figure. This is statically equivalent to radial force F within the midplane of the substrate and the bending moment $M = F h_s/2$. The deflected shape of the circular plate due to this distributed edge moment (see Chapter 15) is

$$w_s = \frac{M}{2(1 + \nu_s) D_s} r^2, \quad D_s = \frac{E_s h_s^3}{12(1 - \nu_s^2)},$$

where r is the radial distance from the center of plate and D_s is the bending stiffness of the plate (substrate). Thus, the curvature of the substrate is

$$\kappa_s = \frac{\partial^2 w}{\partial r^2} = \frac{M}{(1 + \nu_s) D_s} = \frac{12(1 - \nu_s) M}{E_s h_s^3},$$

i.e.,

$$\kappa_s = \frac{6F(1 - \nu_s)}{E_s h_s^2} \quad \text{(generalized Stoney's formula)}.$$

Using the expression for F, this gives

$$\kappa_s = \frac{E_f}{E_s} \frac{1 - \nu_s}{1 - \nu_f} \frac{6 h_f e_m}{h_s^2}.$$

Note that the neutral plane of the substrate is defined by

$$\sigma_{rr} = -\frac{12 M}{h_s^3} z - \frac{F}{h_s} = 0, \quad M = F \frac{h_s}{2},$$

which gives $z = -h_s/6$, regardless of the sign or magnitude of F.

Problem 24.2. Consider two orthogonal, but otherwise identical, dislocation arrays at the interface between the layer and its substrate. The Burgers vector of dislocations in the first array is $\{b_x, b_y, b_z\}$, with the screw component b_z. The Burgers vector of dislocations in the second array is $\{\hat{b}_x, \hat{b}_y, \hat{b}_z\}$, with the screw component $\hat{b}_y$, relative to the same (x, y, z) coordinate system. In the presence of uniform initial misfit strain e^m, derive an expression for the total elastic strain energy per unit area of the free surface.

Problems 24.2–24.3

Solution: The total elastic strain energy per unit area of the free surface is

$$E = E^{\text{d}} + E^{\hat{\text{d}}} + E^{\text{m}} + E^{\text{d,m}} + E^{\hat{\text{d}},\text{m}} + E^{\text{d},\hat{\text{d}}}.$$

The energy per unit length associated with the first dislocation array is derived in the text as

$$E^{\text{d}} = \frac{k}{2} \left\{ (b_x^2 + b_y^2) \left[\ln \frac{\sinh \varphi_0}{\rho_0} - \frac{\varphi_0^2}{2 \sinh^2 \varphi_0} + \frac{1}{4(1-\nu)} \right] \right.$$
$$\left. + (b_x^2 - b_y^2)\left(\frac{1}{2} - \varphi_0 \coth \varphi_0\right) + (1-\nu) b_z^2 \ln \frac{\sinh \varphi_0}{\sinh \rho_0} \right\}.$$

The energy $E^{\hat{\text{d}}}$ of the second dislocation array is given by the above equation in which b_x^2 is replaced by $\hat{b}_x^2$, b_y^2 by $\hat{b}_z^2$, and b_z^2 by $\hat{b}_y^2$. The misfit energy E^{m} is given by

$$E^{\text{m}} = \frac{1}{2}\left(\sigma_{yy}^{\text{m}} e_{yy}^{\text{m}} + \sigma_{zz}^{\text{m}} e_{zz}^{\text{m}} + 2\sigma_{zy}^{\text{m}} e_{zy}^{\text{m}}\right) hp,$$

where h is the film thickness and p the dislocation spacing. The interaction energy between the first dislocation array and the misfit strain $E^{\text{d,m}}$ is given by either one of the expressions

$$E^{\text{d,m}} = -(\sigma_{yy}^{\text{m}} b_y + \sigma_{zy}^{\text{m}} b_z) h = (\sigma_y^0 e_{yy}^{\text{m}} + \sigma_z^0 e_{zz}^{\text{m}} + 2\sigma_{zy}^0 e_{zy}^{\text{m}}) hp.$$

The second of these is the work of the average dislocation stresses in the layer on the misfit strains. The nonvanishing average dislocation stresses in the layer segment of dimensions $h \times p$ are

$$\sigma_y^0 = -4\pi k b_y/p, \quad \sigma_z^0 = \nu \sigma_y^0, \quad \sigma_{zy}^0 = -\mu b_z/p.$$

The interaction energy between the second dislocation array and the misfit strain is

$$E^{\hat{\text{d}},\text{m}} = -(\sigma_{zz}^{\text{m}} \hat{b}_z + \sigma_{zy}^{\text{m}} \hat{b}_y) h,$$

which is the work of uniform misfit stresses on dislocation jump displacements, associated with the second array. In view of the Betti reciprocal work theorem, the interaction energy between two dislocation arrays is

$$E^{\text{d},\hat{\text{d}}} = -2(\sigma_{zz}^0 \hat{b}_z + \sigma_{zy}^0 \hat{b}_y) h = 4\pi k \frac{h}{p}\left[2\nu b_y \hat{b}_z + (1-\nu) b_z \hat{b}_y\right].$$

Problem 24.3. Consider a thin film and a substrate which share the same cubic lattice and orientation, with the interface parallel to their (001) crystallographic planes. The fractional mismatch of a lattice parameter is e^{m}. Let the first dislocation array consist of dislocations on (111) planes, with dislocation Burgers vector along $[0\bar{1}1]$ crystallographic direction, so that relative to the (xyz) coordinate system $b_x = -b/\sqrt{2}$, $b_y = -b/2$, and $b_z = b/2$, where the magnitude of b is equal to $a_l/\sqrt{2}$. Let the second dislocation array consist of dislocations on $(1\bar{1}1)$ planes, with dislocation Burgers vector along $[011]$ crystallographic direction, so that relative to the same (xyz) coordinate system $\hat{b}_x = -\hat{b}/\sqrt{2}$, $\hat{b}_y = \hat{b}/2$, and $\hat{b}_z = -\hat{b}/2$, and the magnitude of $\hat{b}$ is equal to the magnitude of b. Derive the necessary condition for the formation of arrays by using the Matthews criterion.

Solution: The critical condition for the formation of array is $E/p = E^{\text{m}}/p$. By using the expression for E from the previous problem, we then have

$$E^{\text{d}} + E^{\hat{\text{d}}} + E^{\text{d,m}} + E^{\hat{\text{d}},\text{m}} + E^{\text{d},\hat{\text{d}}} = 0.$$

Since the fractional mismatch of a lattice parameter is e^m, the corresponding misfit strain and stress components are

$$e_{yy}^m = e_{zz}^m = e^m, \quad e_{zy}^m = 0,$$

and

$$\sigma_{yy}^m = \sigma_{zz}^m = \sigma^m = 4\pi k(1+\nu)e^m, \quad \sigma_{zy}^m = 0.$$

Thus, from the expressions given in Problem 24.2, we obtain

$$E^{d,m} + E^{\hat{d},m} = \frac{1}{2}h\sigma^m(b+\hat{b}) = 2\pi k(1+\nu)e^m h(b+\hat{b}),$$

and

$$E^{d,\hat{d}} = \pi k(1+\nu)\frac{h}{p}b\hat{b}.$$

To have the energy relaxation, $\hat{b}$ should be of the same sign as b, otherwise the dislocation/misfit interaction energy would vanish. Also, for the relaxation to occur, $e^m b$ must be negative. The dislocation energy is

$$E^d + E^{\hat{d}} = 2E^d = \frac{kb^2}{4}\left[(4-\nu)\ln\frac{\sinh\varphi_0}{\rho_0} - \frac{3\varphi_0^2}{2\sinh^2\varphi_0} - \varphi_0\coth\varphi_0 \right.$$
$$\left. + \frac{5-2\nu}{4(1-\nu)}\right].$$

Substituting these in the criterion for the formation of the arrays gives

$$(4-\nu)\ln\frac{\sinh\varphi_0}{\rho_0} - \frac{3\varphi_0^2}{2\sinh^2\varphi_0} - \varphi_0\coth\varphi_0 + 4\pi(1+\nu)\left(4e^m\frac{h}{b} + \frac{h}{p}\right)$$
$$+ \frac{5-2\nu}{4(1-\nu)} = 0.$$

This condition differs from the condition corresponding to a single dislocation array,

$$(4-\nu)\ln\frac{\sinh\varphi_0}{\rho_0} - \frac{3\varphi_0^2}{2\sinh^2\varphi_0} - \varphi_0\coth\varphi_0 + 16\pi(1+\nu)e^m\frac{h}{b}$$
$$+ \frac{5-2\nu}{4(1-\nu)} = 0,$$

by the presence of an extra term $4\pi(1+\nu)h/p$. If $p \to \infty$, i.e., $\varphi_0 \to 0$, both conditions reduce to

$$(4-\nu)\ln\frac{2h}{\rho} + 16\pi(1+\nu)e^m\frac{h}{b} - \frac{5-8\nu}{4(1-\nu)} = 0,$$

which is the criterion for the deposition of an isolated misfit dislocation. When the layer thickness becomes large, the dislocation spacing tends to a constant value $p = -(b/8e^m)(5+\nu)/(1+\nu)$. This is slightly higher than the spacing $p_0 = b/2e^m$ at which the array would completely relax biaxial initial mismatch strain e^m.

Problem 24.4. Derive the necessary condition for the formation of intersecting dislocation arrays from the previous problem by using the Frank and van der Merwe criterion.

Problems 24.4–25.1

Solution: If the Frank and van der Merwe criterion is used, an analogous procedure to that described in the text leads to the condition

$$p \frac{dE^d}{dp} = E^d + E^{d,m} + E^{d,\hat{d}},$$

which gives

$$(4-\nu)\ln\frac{\sinh\varphi_0}{\varphi_0} - \frac{\varphi_0^2}{2\sinh^2\varphi_0}(7 - 6\varphi_0 \coth\varphi_0) + (2-\nu)(\varphi_0 \coth\varphi_0 - 1)$$

$$+ 8\pi(1+\nu)\left(2e^m\frac{h}{b} + \frac{h}{p}\right) - \frac{3(1-2\nu)}{4(1-\nu)} = 0.$$

If $p \to \infty$, this gives the same critical thickness as that following from the analysis of a single dislocation array. If h becomes very large, the limiting spacing $p = -(b/4e^m)(5 + \nu)/(1+\nu)$ is obtained, which is two times as large as the limiting spacing predicted by the Matthews criterion. According to the model, therefore, in the considered case no complete strain relaxation by misfit dislocations is possible during the film growth.

CHAPTER 25

Problem 25.1. Consider a layer/substrate system $Si_75Ge_{0.25}/Si$, which share the same cubic lattice and orientation. The interface is parallel to their (001) crystallographic planes, as in Problem 24.3. Derive the critical value of the parameter λ^* for the nominally flat surface of the film to be stable with respect to a shallow doubly periodic perturbation

$$u_x = a \cos\frac{2\pi y}{\lambda_y} \cos\frac{2\pi z}{\lambda_z}.$$

Assume that the surface energy is $\gamma = 1.2$ J/m², and that the elastic properties of the layer are $E = 123$ GPa and $\nu = 0.278$. The lattice parameters of the Si and Ge are $a_{Si} = 0.54305$ nm and $a_{Ge} = 0.56576$ nm.

Solution: If the lattice parameters of the layer and substrate are a_l and a_s, the fractional mismatch of the lattice parameter is

$$e^m = \frac{a_s - a_l}{a_l}.$$

The associated misfit strain components are

$$e^m_{yy} = e^m_{zz} = e^m,$$

and the biaxial stress state is

$$\sigma^m_{yy} = \sigma^m_{zz} = \bar{E}(1+\nu)e^m, \quad \bar{E} = \frac{E}{1-\nu^2}.$$

The lattice parameter of the layer is approximately, by Vegard's rule,

$$a_l = 0.25 a_{Ge} + 0.75 a_{Si},$$

and of the substrate $a_s = a_{Si}$. Since $a_{Si} = 0.54305$ nm and $a_{Ge} = 0.56576$ nm, we obtain

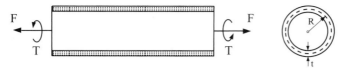

Figure 26-1. A thin-wall tube subject to tension–torsion test. The midradius of the tube is R, and its thickness t.

$a_l = 0.54873$ nm. The misfit strain is thus $e^m = (a_s - a_l)/a_l = -0.01$. The corresponding biaxial stress is

$$\sigma_0 = \bar{E}(1+\nu)e^m = \frac{E}{1-\nu}e^m = -1.7\,\text{GPa}.$$

Thus, the critical value of the wavelength parameter is

$$\lambda_*^{\text{cr}} = \left(\frac{1}{\lambda_y^2} + \frac{1}{\lambda_z^2}\right)_{\text{cr}}^{-1/2} = \frac{\pi\gamma\bar{E}}{\sigma_0^2} = \frac{\pi\gamma E}{\sigma_0^2(1-\nu^2)} = 174\,\text{nm}.$$

CHAPTER 26

Problem 26.1. A thin-wall tube shown in Fig. 26-1 is subject to tension–torsion test, in which the applied torque is related to the applied tensile force F by $T = FR/2$, where R the midradius of the tube.

(a) At what value of F will the tube yield according to von Mises criterion, if the yield stress in simple tension test is $Y = 200$ MPa, and $R = 10$ cm?

(b) Find the ratios of plastic components of the rate of deformation tensor $D^p_{zz}/D^p_{z\theta}$ and $D^p_{rr}/D^p_{z\theta}$ at the onset of yield.

The longitudinal direction of the tube is z, and (r, θ) are the polar coordinates in the plane orthogonal to z. The initial thickness of the tube is $t = R/10$.

Solution: (a) The longitudinal stress is

$$\sigma_{zz} = \frac{F}{A} = \frac{F}{2R\pi t} = \frac{5F}{R^2\pi},$$

while the circumferential shear stress is

$$\sigma_{z\theta} = \frac{T}{I_0}R = \frac{FR/2}{2R^3\pi t}R = \frac{5F}{2R^2\pi} = \frac{\sigma_{zz}}{2}.$$

The von Mises yield criterion is $f = 0$, where

$$f = \frac{1}{2}\left[(\sigma_{zz}-\sigma_{rr})^2+(\sigma_{rr}-\sigma_{\theta\theta})^2+(\sigma_{\theta\theta}-\sigma_{zz})^2\right]+3\left(\sigma_{zr}^2+\sigma_{r\theta}^2+\sigma_{\theta z}^2\right)-Y^2.$$

This gives

$$\sigma_{zz}^2 + 3\sigma_{z\theta}^2 = Y^2.$$

Since $\sigma_{z\theta} = \sigma_{zz}/2$ in the considered case, we obtain

$$\sigma_{zz} = \frac{2}{\sqrt{7}}Y.$$

Problems 26.1–26.2

Thus, the force $F = F_Y$ at the onset of plastic yield is

$$\frac{5F_Y}{R^2\pi} = \frac{2}{\sqrt{7}}Y \Rightarrow F_Y = \frac{2}{5\sqrt{7}} R^2 \pi Y = 9.5 \text{ kN}.$$

(b) The nonvanishing components of the plastic rate of deformation are

$$D^p_{zz} = \dot\lambda \frac{\partial f}{\partial \sigma_{zz}} = 2\dot\lambda\, \sigma_{zz},$$

$$D^p_{rr} = \dot\lambda \frac{\partial f}{\partial \sigma_{rr}} = -\dot\lambda\, \sigma_{zz},$$

$$D^p_{\theta\theta} = \dot\lambda \frac{\partial f}{\partial \sigma_{\theta\theta}} = -\dot\lambda\, \sigma_{zz},$$

$$D^p_{z\theta} = \dot\lambda \frac{\partial f}{\partial \sigma_{z\theta}} = 6\dot\lambda\, \sigma_{z\theta}.$$

Since $\sigma_{z\theta} = \sigma_{zz}/2$ and $\sigma_{zz} = 2Y/\sqrt{7}$ at the onset of yield, we obtain

$$D^p_{zz} = \dot\lambda \frac{4Y}{\sqrt{7}},$$

$$D^p_{rr} = D^p_{\theta\theta} = -\dot\lambda \frac{2Y}{\sqrt{7}},$$

$$D^p_{z\theta} = \dot\lambda \frac{6Y}{\sqrt{7}}.$$

Thus, the required ratios are

$$\frac{D^p_{zz}}{D^p_{z\theta}} = \frac{2}{3}, \quad \frac{D^p_{rr}}{D^p_{z\theta}} = -\frac{1}{3}.$$

Note the plastic incompressibility

$$D^p_{zz} + D^p_{rr} + D^p_{\theta\theta} = 0,$$

in accord with the pressure-independence of the von Mises yield criterion ($\partial f/\partial \sigma_{kk} = 0$), and the normality rule $D^p_{ij} \sim \partial f/\partial \sigma_{ij}$.

Problem 26.2. A thin-walled tube of initial length l_0, thickness t_0 and midradius R_0 is subject to axial force F and torque T. Assuming an isotropic rigid-plastic material obeying the von Mises (J_2) yield condition and isotropic hardening, determine the stress state in the tube during the plastic loading path specified by a given relationship $l = l(\phi)$, where l is the current length of the tube, and ϕ is the angle of rotation of one end of the tube relative to the other end.

Solution: The velocity field at an arbitrary stage of plastic loading (see Problem 7.13) is

$$v_r = -\frac{1}{2} r \frac{\dot l}{l}, \quad v_\theta = r \frac{z}{l} \dot\phi, \quad v_z = \frac{\dot l}{l} z.$$

The corresponding components of the rate of deformation tensor are

$$D_{rr} = D_{\theta\theta} = -\frac{1}{2}\frac{\dot{l}}{l}, \quad D_{zz} = \frac{\dot{l}}{l}, \quad D_{z\theta} = \frac{1}{2}\frac{r}{l}\dot{\phi}.$$

The current thickness of the tube is t and its current midradius is r. The nonvanishing stress components are σ_{zz} and $\sigma_{z\theta}$, with their deviatoric components

$$\sigma'_{rr} = \sigma'_{\theta\theta} = -\frac{1}{3}\sigma_{zz}, \quad \sigma'_{zz} = \frac{2}{3}\sigma_{zz}, \quad \sigma'_{z\theta} = \sigma_{z\theta}.$$

The von Mises yield condition is

$$J_2 = k^2, \quad J_2 = \frac{1}{2}\sigma' : \sigma' = \frac{1}{3}\left(\sigma_{zz}^2 + 3\sigma_{z\theta}^2\right).$$

The current size of the yield surface k (the yield stress in pure shear test), is a function of the generalized or equivalent (plastic) strain, i.e.,

$$k = k(\vartheta), \quad \vartheta = \int_0^t (2\,\mathbf{D} : \mathbf{D})^{1/2}\, dt.$$

The constitutive equation of the J_2 isotropic hardening plasticity is

$$\mathbf{D} = \frac{1}{4h}\frac{\dot{J}_2}{J_2}\sigma', \quad \frac{\dot{J}_2}{J_2} = \frac{2\sigma' : \dot{\sigma}}{\sigma' : \sigma'} = \frac{2\sigma' : \overset{\triangledown}{\sigma}}{\sigma' : \sigma'}.$$

Rigid-plastic model is used, so that $\mathbf{D}^e = 0$ and $\mathbf{D} = \mathbf{D}^p$. Since $J_2 = k^2$, the above can be rewritten as

$$\mathbf{D} = \frac{1}{2h}\frac{\dot{k}}{k}\sigma'.$$

The hardening modulus is $h = dk/d\vartheta$. Thus, we have

$$D_{rr} = D_{\theta\theta} = -\frac{1}{2}\frac{\dot{l}}{l} = \frac{1}{2h}\frac{\dot{k}}{k}\left(-\frac{1}{3}\sigma_{zz}\right).$$

$$D_{zz} = \frac{\dot{l}}{l} = \frac{1}{2h}\frac{\dot{k}}{k}\left(\frac{2}{3}\sigma_{zz}\right),$$

$$D_{z\theta} = \frac{1}{2}\frac{r}{l}\dot{\phi} = \frac{1}{2h}\frac{\dot{k}}{k}\sigma_{z\theta}.$$

Evidently,

$$\frac{D_{zz}}{D_{z\theta}} = \frac{2}{r}\frac{dl}{d\phi} = \frac{2}{3}\frac{\sigma_{zz}}{\sigma_{z\theta}} \quad \Rightarrow \quad \frac{\sigma_{z\theta}}{\sigma_{zz}} = \frac{r}{3}\frac{1}{dl/d\phi},$$

and

$$k = J_2^{1/2} = \frac{\sigma_{zz}}{\sqrt{3}}\left[1 + 3\left(\frac{\sigma_{z\theta}}{\sigma_{zz}}\right)^2\right]^{1/2} = \frac{\sigma_{zz}}{\sqrt{3}}\left[1 + \frac{r^2}{3}\frac{1}{(dl/d\phi)^2}\right]^{1/2}.$$

But,

$$v_r = \dot{r} = -\frac{1}{2}r\frac{\dot{l}}{l},$$

Problems 26.2–26.3

gives

$$2\frac{\dot{r}}{r} = -\frac{\dot{l}}{l} \quad \Rightarrow \quad \left(\frac{r}{r_0}\right)^2 \frac{l}{l_0} = 1,$$

so that

$$k = \frac{\sigma_{zz}}{\sqrt{3}}\left[1 + \frac{r_0^2}{3}\frac{l_0}{l}\frac{1}{(dl/d\phi)^2}\right]^{1/2}.$$

If this is substituted into the expression for D_{zz}, we obtain

$$\frac{\dot{l}}{l} = \frac{k}{\sqrt{3}h}\left[1 + \frac{r_0^2}{3}\frac{l_0}{l}\frac{1}{(dl/d\phi)^2}\right]^{-1/2}.$$

Since $dl = (dl/d\phi)d\phi$, the above can be rewritten as

$$\frac{dk}{h(k)} = \left[\frac{r_0^2 l_0}{l(\phi)} + 3\left(\frac{dl}{d\phi}\right)^2\right]^{1/2}\frac{d\phi}{l(\phi)}.$$

For a given loading path $l = l(\phi)$, the numerical integration

$$\int_{k_0}^{k}\frac{dk}{h(k)} = \int_0^{\phi}\left[\frac{r_0^2 l_0}{l(\phi)} + 3\left(\frac{dl}{d\phi}\right)^2\right]^{1/2}\frac{d\phi}{l(\phi)}$$

gives

$$k = k(\phi).$$

The normal stress, corresponding to the angle of rotation ϕ and the path $l = l(\phi)$, is then

$$\sigma_{zz} = \sqrt{3}k\left[1 + \frac{r_0^2}{3}\frac{l_0}{l}\frac{1}{(dl/d\phi)^2}\right]^{-1/2},$$

whereas the shear stress is

$$\sigma_{z\theta} = \frac{r_0}{3}\left(\frac{l_0}{l}\right)^{1/2}\frac{\sigma_{zz}}{dl/d\phi}.$$

Problem 26.3. Prove that in the previous problem

$$\sigma' : \dot{\sigma} = \sigma' : \overset{\triangledown}{\sigma}.$$

Solution: This was proven in a general context in Problem 5.7. We here verify the result explicitly by using the expressions for the Jaumann rates of stress, derived in Problem 5.10, i.e.,

$$\overset{\triangledown}{\sigma}_{zz} = \dot{\sigma}_{zz} + \sigma_{z\theta}\frac{r}{l}\dot{\phi}, \quad \overset{\triangledown}{\sigma}_{z\theta} = \dot{\sigma}_{z\theta} - \frac{1}{2}\sigma_{zz}\frac{r}{l}\dot{\phi}, \quad \overset{\triangledown}{\sigma}_{\theta\theta} = -\sigma_{z\theta}\frac{r}{l}\dot{\phi},$$

the remaining Jaumann rates being equal to zero. Consequently,

$$\sigma' : \dot{\sigma} = -\frac{1}{3}\sigma_{zz}\overset{\triangledown}{\sigma}_{\theta\theta} + \frac{2}{3}\sigma_{zz}\overset{\triangledown}{\sigma}_{zz} + 2\sigma_{z\theta}\overset{\triangledown}{\sigma}_{z\theta}$$

$$= \frac{2}{3}\sigma_{zz}\dot{\sigma}_{zz} + 2\sigma_{z\theta}\dot{\sigma}_{z\theta} = \sigma' : \dot{\sigma} = \dot{J}_2.$$

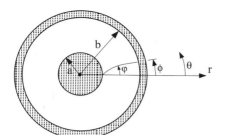

Figure 26-2. Simple shear of a hollow cylinder.

Problem 26.4. Consider a hollow circular cylinder bounded by two rigid casings as shown in Fig. 26-2. If the inner casing is fixed, whereas the outer casing is rotated by a small angle ϕ, determine the deformation and stress field within the cylinder in the case of (a) elastic and (b) elastoplastic response. In the latter case assume that material is linearly hardening.

Solution: Using the cylindrical coordinates, the displacement components are

$$u_\theta = r\varphi(r), \quad u_r = u_z = 0,$$

where $\varphi(a) = 0$ and $\varphi(b) = \phi$. The nonvanishing strain component is

$$e_{r\theta} = \frac{1}{2}\left(\frac{du_\theta}{dr} - \frac{u_\theta}{r}\right) = \frac{r}{2}\frac{d\varphi}{dr}.$$

Since normal strains are zero, the normal stresses also vanish (assuming small deformations). Thus, the only nonvanishing stress is $\sigma_{r\theta}$, satisfying the equilibrium equation

$$\frac{d\sigma_{r\theta}}{dr} + \frac{2}{r}\sigma_{r\theta} = 0.$$

This has a general solution

$$\sigma_{r\theta} = \frac{C_1}{r^2},$$

where the integration constant C_1 depends on the material properties.

(a) If deformation is purely elastic,

$$\sigma_{r\theta} = \frac{C_1}{r^2} = 2\mu e_{r\theta} = \mu r \frac{d\varphi}{dr}.$$

Upon integration, using the boundary conditions $\varphi(a) = 0$ and $\varphi(b) = \phi$, there follows

$$\varphi(r) = \frac{\phi}{1 - a^2/b^2}\left(1 - \frac{a^2}{r^2}\right).$$

The corresponding stress is

$$\sigma_{r\theta} = \frac{2\mu\phi}{1 - a^2/b^2}\frac{a^2}{r^2}.$$

(b) If the yield stress in pure shear is $\sigma_{r\theta}^0$, the plastic deformation begins at $r = a$ when the angle ϕ is equal to

$$\phi^0 = \left(1 - \frac{a^2}{b^2}\right)\frac{\sigma_{r\theta}^0}{2\mu}.$$

Problem 26.4

For $\phi > \phi^0$, plasticity spreads outward and at an arbitrary stage of deformation before reaching $r = b$ the elastic-plastic interface is at $r = c$. Assuming linear hardening,

$$\sigma_{r\theta} = \sigma_{r\theta}^0 + 2h e_{r\theta}^p,$$

where plastic part of the shear strain is

$$e_{r\theta}^p = e_{r\theta} - \frac{1}{2\mu}\sigma_{r\theta} = \frac{r}{2}\frac{d\varphi}{dr} - \frac{1}{2\mu}\sigma_{r\theta}.$$

Thus,

$$\left(1 + \frac{h}{\mu}\right)\sigma_{r\theta} = \sigma_{r\theta}^0 + hr\frac{d\varphi}{dr}.$$

Recalling that $\sigma_{r\theta} = C_1/r^2$, we obtain a differential equation

$$h\frac{d\varphi}{dr} = \left(1 + \frac{h}{\mu}\right)\frac{C_1}{r^3} - \sigma_{r\theta}^0 \frac{1}{r},$$

whose solution, satisfying $\varphi(a) = 0$, is

$$h\varphi(r) = \frac{C_1}{2a^2}\left(1 + \frac{h}{\mu}\right)\left(1 - \frac{a^2}{r^2}\right) - \sigma_{r\theta}^0 \ln\frac{r}{a}, \quad a \le r \le c.$$

On the other hand, in the elastic region we have

$$\varphi(r) = -\frac{C_1}{2\mu}\frac{1}{r^2} + C_2, \quad c \le r \le b.$$

Imposing the boundary condition at $r = b$ and the continuity condition at the elastic-plastic interface $r = c$,

$$\varphi(b) = \phi, \quad \varphi(c - 0) = \varphi(c + 0),$$

it readily follows that

$$\frac{C_1}{2a^2} = \frac{h\phi + \sigma_{r\theta}^0 \ln\frac{c}{a}}{1 - \frac{a^2}{c^2} + \frac{h}{\mu}\left(1 - \frac{a^2}{b^2}\right)},$$

$$hC_2 = \left(1 - \frac{a^2}{c^2} + \frac{h}{\mu}\right)\frac{C_1}{2a^2} - \sigma_{r\theta}^0 \ln\frac{c}{a}.$$

The relationship between the angle ϕ and the radius of the elastic-plastic interface $r = c$ can be obtained from the condition

$$\sigma_{r\theta}(c) = \frac{C_1}{c^2} = \sigma_{r\theta}^0.$$

This gives

$$\frac{2h}{\sigma_{r\theta}^0}\phi = \frac{c^2}{a^2} - 1 - 2\ln\frac{c}{a} + \frac{h}{\mu}\left(\frac{c^2}{a^2} - \frac{c^2}{b^2}\right).$$

Plasticity spreads throughout the cylinder ($c = b$) when $\phi = \phi^*$, where

$$\frac{2h}{\sigma_{r\theta}^0} \phi^* = \left(1 + \frac{h}{\mu}\right)\left(\frac{b^2}{a^2} - 1\right) - 2\ln\frac{b}{a}.$$

At this instant

$$\frac{2h}{\sigma_{r\theta}^0} \varphi(r) = \left(1 + \frac{h}{\mu}\right)\left(\frac{b^2}{a^2} - \frac{b^2}{r^2}\right) - 2\ln\frac{r}{a}, \quad a \leq r \leq b.$$

The corresponding shear stress is

$$\sigma_{r\theta} = \sigma_{r\theta}^0 \frac{b^2}{r^2}, \quad a \leq r \leq b.$$

Problem 26.5. Solve the previous problem by using the rigid-plastic J_2 flow theory of plasticity with a linear hardening.

Solution: The deformation mapping in cylindrical coordinates is

$$r = R,$$

$$\theta = \Theta + \varphi(r, t),$$

$$z = Z.$$

The coordinates in the undeformed configuration are (R, Θ, Z) and in the deformed configuration (r, θ, z); t is a monotonically increasing time like parameter, e.g., the prescribed outer angle of rotation ϕ during a continuous loading. The corresponding velocity field is

$$v_r = \dot{r} = 0, \quad v_\theta = r\dot{\theta} = r\frac{\partial \varphi}{\partial t}, \quad v_z = \dot{z} = 0,$$

with the components of the rate of deformation tensor

$$D_{rr} = D_{\theta\theta} = D_{zz} = 0, \quad D_{r\theta} = \frac{1}{2}r\frac{\partial^2 \varphi}{\partial r \partial t}, \quad D_{rz} = D_{\theta z} = 0.$$

The constitutive equation of the rigid-plastic J_2 flow theory is

$$\mathbf{D} = \frac{1}{4h}\frac{\dot{J}_2}{J_2}\boldsymbol{\sigma}', \quad J_2 = \frac{1}{2}\boldsymbol{\sigma}' : \boldsymbol{\sigma}'.$$

Assuming that the stress tensor is

$$\boldsymbol{\sigma} = \begin{bmatrix} \sigma_{rr} & \sigma_{r\theta} & 0 \\ \sigma_{\theta r} & \sigma_{\theta\theta} & 0 \\ 0 & 0 & \sigma_{zz} \end{bmatrix},$$

Problem 26.5

the deviatoric stress components are

$$\sigma'_{rr} = \frac{1}{3}(2\sigma_{rr} - \sigma_{\theta\theta} - \sigma_{zz}),$$

$$\sigma'_{\theta\theta} = \frac{1}{3}(2\sigma_{\theta\theta} - \sigma_{zz} - \sigma_{rr}),$$

$$\sigma'_{zz} = \frac{1}{3}(2\sigma_{zz} - \sigma_{rr} - \sigma_{\theta\theta}),$$

$$\sigma'_{r\theta} = \sigma_{r\theta}.$$

Thus,

$$D_{rr} = \frac{1}{4h}\frac{\dot{J}_2}{J_2}\frac{1}{3}(2\sigma_{rr} - \sigma_{\theta\theta} - \sigma_{zz}) = 0,$$

$$D_{\theta\theta} = \frac{1}{4h}\frac{\dot{J}_2}{J_2}\frac{1}{3}(2\sigma_{\theta\theta} - \sigma_{zz} - \sigma_{rr}) = 0,$$

$$D_{zz} = \frac{1}{4h}\frac{\dot{J}_2}{J_2}\frac{1}{3}(2\sigma_{zz} - \sigma_{rr} - \sigma_{\theta\theta}) = 0.$$

These are satisfied if

$$\sigma_{rr} = \sigma_{\theta\theta} = \sigma_{zz} = -p,$$

where p is an arbitrary pressure. In view of one of the equilibrium equations, $i.e.$,

$$\frac{\partial \sigma_{rr}}{\partial r} + \frac{\sigma_{rr} - \sigma_{\theta\theta}}{r} = 0,$$

there follows $\partial p/\partial r = 0$, so that $p = p(t)$ only. The deviatoric normal stresses are zero,

$$\sigma'_{rr} = \sigma'_{\theta\theta} = \sigma'_{zz} = 0,$$

and

$$J_2 = \frac{1}{2}\sigma':\sigma' = \sigma_{r\theta}^2, \quad \dot{J}_2 = 2\sigma_{r\theta}\frac{\partial \sigma_{r\theta}}{\partial t}.$$

Since

$$D_{r\theta} = \frac{1}{4h}\frac{\dot{J}_2}{J_2}\sigma_{r\theta} = \frac{1}{2}r\frac{\partial^2 \varphi}{\partial r \partial t},$$

there follows

$$\frac{\partial \sigma_{r\theta}}{\partial t} = hr\frac{\partial^2 \varphi}{\partial r \partial t}.$$

The integration gives

$$\sigma_{r\theta} = hr\frac{\partial \varphi}{\partial r} + f(r).$$

On the other hand, using the assumption of linear hardening, we can write
$$J_2^{1/2} = k = \sigma_{r\theta}^0 + h\vartheta,$$
where
$$\vartheta = \int_0^t (2\mathbf{D}:\mathbf{D})^{1/2}\, dt = \int_0^t 2 D_{r\theta}\, dt = r\,\frac{\partial \varphi}{\partial r}.$$

This gives
$$\sigma_{r\theta} = \sigma_{r\theta}^0 + hr\,\frac{\partial \varphi}{\partial r}.$$

The comparison with the earlier expression reveals that $f(r) = \sigma_{r\theta}^0$. Note also that at the interface $r = c$ between plastic and rigid regions, $\partial \varphi / \partial r = 0$, so that $\sigma_{r\theta}(c, t) = \sigma_{r\theta}^0$.

But, from the remaining equilibrium equation
$$\frac{d\sigma_{r\theta}}{dr} + \frac{2}{r}\sigma_{r\theta} = 0,$$
we obtain
$$\sigma_{r\theta} = \frac{C(t)}{r^2}.$$

Combining this expression with the previous expression gives
$$\frac{C(t)}{r^2} = hr\,\frac{\partial \varphi}{\partial r} + \sigma_{r\theta}^0,$$
or, upon integration using the boundary condition $\varphi(a, t) = 0$,
$$h\varphi(r, t) = \frac{C(t)}{2a^2}\left(1 - \frac{a^2}{r^2}\right) - \sigma_{r\theta}^0 \ln\frac{r}{a}.$$

Since at $r = c \leq b$, we have
$$\varphi(c, t) = \phi(t),$$
it readily follows that
$$\frac{C(t)}{2a^2} = \frac{1}{1 - a^2/c^2}\left[h\phi(t) + \sigma_{r\theta}^0 \ln\frac{c}{a}\right].$$

Consequently,
$$\varphi(r, t) = \left[\phi(t) + \frac{\sigma_{r\theta}^0}{h}\ln\frac{c}{a}\right]\frac{1 - a^2/r^2}{1 - a^2/c^2} - \frac{\sigma_{r\theta}^0}{h}\ln\frac{r}{a}.$$

The angle ϕ^* at which plasticity first reaches the radius $r = b$ is obtained from the condition $\sigma_{r\theta}(b) = \sigma_{r\theta}^0$, i.e.,
$$\frac{2a^2}{b^2 - a^2}\left[h\phi^* + \sigma_{r\theta}^0 \ln\frac{b}{a}\right] = \sigma_{r\theta}^0.$$

This gives
$$\phi^* = \left(\frac{b^2 - a^2}{2a^2} - \ln\frac{b}{a}\right)\frac{\sigma_{r\theta}^0}{h}.$$

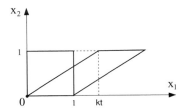

Figure 26-3. Simple shear of a rectangular block.

Problem 26.6. Consider a simple shear deformation of of rectangular block (see Fig. 26-3), for which the velocity field is

$$v_1 = kx_2, \quad v_2 = v_3 = 0 \quad (k = \text{const.}).$$

Assuming the rigid-plastic material, which hardens in a kinematic hardening mode with the von Mises type yield condition,

$$(\sigma'_{ij} - \alpha_{ij})(\sigma'_{ij} - \alpha_{ij}) = \frac{2}{3}\sigma_0^2, \quad (\sigma_0 = \text{const.}),$$

$$\overset{\triangledown}{\alpha}_{ij} = c D_{ij}, \quad (c = \text{const.}),$$

determine the stress components at an arbitrary stage of deformation.

Solution: The only nonvanishing component of the velocity gradient is $L_{12} = k$, with the corresponding nonvanishing components of the rate of deformation and spin tensors

$$D_{12} = D_{21} = \frac{k}{2}, \quad W_{12} = -W_{21} = \frac{k}{2}.$$

Thus, since

$$\overset{\triangledown}{\alpha}_{ij} = \dot{\alpha}_{ij} - W_{ik}\alpha_{kj} + \alpha_{ik}W_{kj} = c D_{ij},$$

we obtain

$$\dot{\alpha}_{11} = -\dot{\alpha}_{22} = k\alpha_{12},$$

$$\dot{\alpha}_{12} = \dot{\alpha}_{21} = k\left(\frac{c}{2} - \alpha_{11}\right).$$

Consequently, by combining these two equations,

$$\ddot{\alpha}_{12} + k^2\alpha_{12} = 0,$$

which has the general solution

$$\dot{\alpha}_{12} = a_1 \cos kt + a_2 \sin kt.$$

Since the components of the back stress are initially ($t = 0$) equal to zero, we have $\alpha_{12}(0) = 0$ and $\dot{\alpha}_{12}(0) = kc/2$, which specifies the integration constants $a_1 = 0$ and $a_2 = c/2$. Thus,

$$\alpha_{12} = \frac{c}{2} \sin kt.$$

From the differential equation for α_{11} it then readily follows that

$$\alpha_{11} = \frac{c}{2}(1 - \cos kt), \quad \alpha_{22} = -\alpha_{11}.$$

Now, from the normality rule for the rigid-plastic material giving $D_{ij} \sim (\sigma'_{ij} - \alpha_{ij})$, we have

$$\frac{D_{11}}{\sigma'_{11} - \alpha_{11}} = \frac{D_{22}}{\sigma'_{22} - \alpha_{22}} = \frac{D_{12}}{\sigma_{12} - \alpha_{12}}.$$

Since $D_{11} = D_{22} = 0$, we conclude that $\sigma'_{11} = \alpha_{11}$ and $\sigma'_{22} = \alpha_{22} = -\alpha_{11}$. Thus, from the yield condition

$$(\sigma'_{11} - \alpha_{11})^2 + (\sigma'_{22} - \alpha_{22})^2 + 2(\sigma_{12} - \alpha_{12})^2 = \frac{2}{3}\sigma_0^2,$$

we obtain $\sigma_{12} - \alpha_{12} = \sigma_0/\sqrt{3}$. The nonvanishing stress components at an arbitrary stage of deformation are therefore

$$\sigma'_{11} = -\sigma'_{22} = \frac{c}{2}(1 - \cos kt), \quad \sigma_{12} = \frac{\sigma_0}{\sqrt{3}} + \frac{c}{2}\sin kt.$$

An arbitrary hydrostatic pressure can be superposed to this without affecting the considered simple shear deformation. The obtained spurious oscillation of the stress response corresponding to monotonically increasing shear deformation ($\gamma = kt$) is an indication of the limitation of the utilized simple evolution equation for the back stress. More involved representation of this evolution is required to eliminate such oscillations.

Problem 26.7. Determine the critical stress for a cylindrical cavitation instability in an incompressible elastic-plastic linearly hardening material.

Solution: Consider a cylindrical void of initial radius R_0 in an isotropic infinite medium under remote biaxial tension σ. Assume the plane strain conditions. The stress state at an arbitrary point consists of the radial stress σ_{rr}, the hoop stress $\sigma_{\theta\theta}$, and the longitudinal stress σ_{zz}. Because of cylindrical symmetry, the stress components depend only on the radial distance r. In the absence of body forces the equilibrium equation is

$$\frac{d\sigma_{rr}}{dr} - \frac{1}{r}(\sigma_{\theta\theta} - \sigma_{rr}) = 0.$$

If the material response is within the infinitesimal elastic range, the compatibility equation reduces to

$$\frac{d}{dr}(\sigma_{rr} + \sigma_{\theta\theta}) = 0, \quad \text{i.e.,} \quad \sigma_{rr} + \sigma_{\theta\theta} = 2\sigma.$$

Combining above equations, it follows that

$$\sigma_{rr} = \sigma\left(1 - \frac{R_0^2}{r^2}\right), \quad \sigma_{\theta\theta} = \sigma\left(1 + \frac{R_0^2}{r^2}\right).$$

The longitudinal stress is $\sigma_{zz} = 2\nu\sigma$, and the radial displacement is

$$u = \frac{\sigma}{2G}\left[(1 - 2\nu)r + \frac{R_0^2}{r}\right].$$

Problem 26.7

According to the von Mises yield criterion, the plastic deformation commences when

$$\frac{3}{4}(\sigma_{\theta\theta} - \sigma_{rr})^2 + \left(\sigma_{zz} - \frac{\sigma_{rr} + \sigma_{\theta\theta}}{2}\right)^2 = Y^2.$$

Thus, the threshold value of applied stress σ for the onset of plastic deformation at the surface of the void is

$$\sigma_{yy} = \frac{Y}{[3 + (1 - 2\nu)^2]^{1/2}}.$$

With an increase of stress beyond this value, the plastic zone expands outward and at an arbitrary instant of deformation the elastic-plastic boundary has reached the radius ρ. The deformation gradient tensor, associated with a cylindrical expansion of the void and expressed in cylindrical coordinates, becomes

$$\mathbf{F} = \begin{bmatrix} \dfrac{dr}{dr_0} & 0 & 0 \\ 0 & \dfrac{r}{r_0} & 0 \\ 0 & 0 & 1 \end{bmatrix}.$$

The corresponding volume change of an infinitesimal material element is

$$\frac{dV}{dV_0} = \det \mathbf{F} = \frac{r}{r_0}\frac{dr}{dr_0}.$$

If plastic deformation is assumed to be isochoric, the volume change is entirely due to elastic deformation, so that

$$\frac{dV - dV_0}{dV} = \frac{1}{E}[(1-\nu)(\sigma_{rr} + \sigma_{\theta\theta}) - 2\nu\sigma_{zz}].$$

The closed form analysis can be pursued further for an elastically incompressible material ($\nu = 1/2$). The longitudinal stress σ_{zz} is then an arithmetic mean of the other two stress components throughout the medium, and

$$r^2 - r_0^2 = R^2 - R_0^2 = \rho^2 - \rho_0^2.$$

In the elastic zone $r \geq \rho$, the stress components are

$$\sigma_{rr} = \sigma - \frac{Y}{\sqrt{3}}\frac{\rho^2}{r^2}, \quad \sigma_{\theta\theta} = \sigma + \frac{Y}{\sqrt{3}}\frac{\rho^2}{r^2}, \quad \sigma_{zz} = \sigma.$$

They satisfy the yield condition at the elastic-plastic interface $r = \rho$, and the condition $\sigma_{rr} + \sigma_{\theta\theta} = 2\sigma$ everywhere in the elastic region. Because the strain is elastic and infinitesimal at the elastic-plastic interface, we can write

$$\frac{\rho^2 - \rho_0^2}{\rho^2} \approx 2e_{\theta\theta}(\rho) = \frac{\sqrt{3}Y}{E},$$

and

$$R^2 = R_0^2 + \frac{\sqrt{3}Y}{E}\rho^2.$$

At the advanced stages of plastic deformation, when R becomes much greater than R_0, the ratio ρ/R approaches a constant value

$$\frac{\rho}{R} = \left(\frac{E}{\sqrt{3}Y}\right)^{1/2}.$$

The last two equations hold regardless of the type of material hardening in the plastic zone.

The deformation gradient tensor for the void expansion in an incompressible material is

$$\mathbf{F} = \begin{bmatrix} \dfrac{r_0}{r} & 0 & 0 \\ 0 & \dfrac{r}{r_0} & 0 \\ 0 & 0 & 1 \end{bmatrix}.$$

By using the multiplicative decomposition of the deformation gradient into its elastic and plastic parts ($\mathbf{F} = \mathbf{F}^e \cdot \mathbf{F}^p$), we have

$$\mathbf{F} = \begin{bmatrix} \dfrac{r_p}{r} & 0 & 0 \\ 0 & \dfrac{r}{r_p} & 0 \\ 0 & 0 & 1 \end{bmatrix} \begin{bmatrix} \dfrac{r_0}{r_p} & 0 & 0 \\ 0 & \dfrac{r_p}{r_0} & 0 \\ 0 & 0 & 1 \end{bmatrix}.$$

The corresponding logarithmic strain

$$\mathbf{E} = \begin{bmatrix} -\ln\dfrac{r}{r_0} & 0 & 0 \\ 0 & \ln\dfrac{r}{r_0} & 0 \\ 0 & 0 & 0 \end{bmatrix}$$

can then be additively decomposed as $\mathbf{E} = \mathbf{E}^e + \mathbf{E}^p$, i.e.,

$$\mathbf{E} = \begin{bmatrix} -\ln\dfrac{r}{r_p} & 0 & 0 \\ 0 & \ln\dfrac{r}{r_p} & 0 \\ 0 & 0 & 0 \end{bmatrix} + \begin{bmatrix} -\ln\dfrac{r_p}{r_0} & 0 & 0 \\ 0 & \ln\dfrac{r_p}{r_0} & 0 \\ 0 & 0 & 0 \end{bmatrix}.$$

If the material is linearly hardening with respect to the generalized plastic strain

$$\bar{E}^p = \left(\frac{2}{3}\mathbf{E}^p : \mathbf{E}^p\right)^{1/2} = \frac{2}{\sqrt{3}} \ln\frac{r_p}{r_0},$$

the equivalent yield stress $\sqrt{3}(\sigma_{\theta\theta} - \sigma_{rr})/2$ at an arbitrary stage of deformation is

$$\frac{\sqrt{3}}{2}(\sigma_{\theta\theta} - \sigma_{rr}) = Y + \frac{2k}{\sqrt{3}} \ln\frac{r_p}{r_0} = Y + \frac{2k}{\sqrt{3}}\left(\ln\frac{r}{r_0} - \ln\frac{r}{r_p}\right).$$

By using Hooke's type law for the elastic component of strain,

$$\ln\frac{r}{r_p} = \frac{3}{4E}(\sigma_{\theta\theta} - \sigma_{rr}),$$

Figure 26-4. A thin circular disk under internal pressure p. The current midradius of the ring is R and its thickness t.

the substitution in the previous equation yields

$$\sigma_{\theta\theta} - \sigma_{rr} = \hat{Y} + \hat{k} \ln \frac{r}{r_0}.$$

The abbreviations are here used

$$\hat{Y} = \frac{2Y/\sqrt{3}}{1 + k/E}, \quad \hat{k} = \frac{4k/3}{1 + k/E}.$$

When this is introduced into the equilibrium equation, the integration gives

$$\sigma_{rr} = \hat{Y} \ln \frac{r}{R} + \frac{\hat{k}}{4} \sum_{n=1}^{\infty} \frac{1}{n^2} \left[\left(\frac{R^2 - R_0^2}{R^2} \right)^n - \left(\frac{R^2 - R_0^2}{r^2} \right)^n \right].$$

Evaluating this at $r = \rho$, and equating the result with $\sigma_{rr}(\rho) = \sigma - Y/\sqrt{3}$, provides an expression for the applied stress σ in terms of the plastic zone radius ρ. This is

$$\sigma = \frac{Y}{\sqrt{3}} + \hat{Y} \ln \frac{\rho}{R} + \frac{\hat{k}}{4} \sum_{n=1}^{\infty} \frac{1}{n^2} \left[\left(\frac{R^2 - R_0^2}{R^2} \right)^n - \left(\frac{\sqrt{3}Y}{E} \right)^n \right].$$

If the plastic zone around the void has spread to a large extent, we have

$$\sigma_{cr} = \frac{Y/\sqrt{3}}{1 + k/E} \left(1 + \ln \frac{E}{\sqrt{3}Y} + \frac{\pi^2}{18} \frac{\sqrt{3}k}{Y} \right).$$

This is the critical stress for a cylindrical cavitation instability in an incompressible elastic linearly hardening material. If $k \ll E$, the result reduces to that for a pressurized cylindrical void in an infinitely extended material (Hill, 1950).

Problem 26.8. Determine the pressure at the onset of necking of a thin circular ring assuming a rigid-plastic J_2 flow theory. The initial midradius of the ring is R_0 and its square cross section has initial size t_0.

Solution: The hoop stress in the ring at an arbitrary stage of deformation before necking is $\sigma = pR/t$, where R and t are the current dimensions of the ring (Fig. 26-4). Thus, $p = \sigma t/R$. At the onset of necking, p reaches its maximum, so that

$$dp = 0 \quad \Rightarrow \quad \frac{d\sigma}{\sigma} + \frac{dt}{t} - \frac{dR}{R} = 0.$$

Since the deviatoric stress components are

$$\sigma'_{\theta\theta} = \frac{2}{3}\sigma \quad \sigma'_{rr} = \sigma'_{zz} = -\frac{1}{3}\sigma$$

we have
$$D_{rr} = D_{zz} = \frac{\dot{t}}{t} = \frac{1}{4h}\frac{\dot{J}_2}{J_2}\left(-\frac{1}{3}\sigma\right),$$

$$D_{\theta\theta} = \frac{\dot{R}}{R} = \frac{1}{4h}\frac{\dot{J}_2}{J_2}\left(\frac{2}{3}\sigma\right),$$

where
$$J_2 = \frac{1}{3}\sigma^2, \quad \frac{\dot{J}_2}{J_2} = \frac{2\dot{\sigma}}{\sigma}.$$

Evidently,
$$\frac{dt}{t} = -\frac{1}{2}\frac{dR}{R},$$

which is a consequence of plastic incompressibility ($Rt^2 = R_0 t_0^2$). Thus, at the onset of necking
$$\frac{d\sigma}{\sigma} = \frac{3}{2}\frac{dR}{R}.$$

Since from the flow rule
$$\frac{dR}{R} = \frac{1}{4h}\frac{dJ_2}{J_2}\left(\frac{2}{3}\sigma\right) = \frac{d\sigma}{3h},$$

we obtain that the hoop stress at the necking is
$$\sigma = 2h.$$

For the linearly hardening material
$$J_2^{1/2} = k = k_0 + h\vartheta \quad \Rightarrow \quad \sigma = \sigma_0 + \sqrt{3}h\vartheta,$$

where σ_0 is the hoop stress at the beginning of plastic deformation, and
$$\vartheta = \int_0^t (2\mathbf{D}:\mathbf{D})^{1/2}\, dt = \sqrt{3}\ln\frac{R}{R_0}.$$

Thus, the necking condition becomes
$$\sigma = \sigma_0 + 3h\ln\frac{R}{R_0} = 2h,$$

which gives the radius of the ring at the onset of necking
$$R = R_0 \exp\left(\frac{2}{3} - \frac{\sigma_0}{3h}\right).$$

The corresponding pressure is
$$p_{\max} = \frac{2ht_0}{R_0}\exp\left(\frac{\sigma_0}{2h} - 1\right).$$

An alternative derivation proceeds by writing the function $p = p(R)$ directly from $p = \sigma t/R$, and
$$\sigma = \sigma_0 + 3h\ln\frac{R}{R_0}, \quad Rt^2 = R_0 t_0^2.$$

Problems 26.8–26.9

This gives

$$p(R) = \sqrt{R_0}\, t_0 \left(\sigma_0 + 3h \ln \frac{R}{R_0}\right) R^{-3/2}.$$

The necking condition is simply $dp/dR = 0$, i.e.,

$$2h - \left(\sigma_0 + 3h \ln \frac{R}{R_0}\right) = 0,$$

in accord with the previously derived result.

Problem 26.9. Consider a thin cylindrical shell under internal pressure p and axial force N, such that

$$\sigma_{\theta\theta} = \frac{Rp}{t}, \quad \sigma_{zz} = \frac{Rp}{2t} + \frac{N}{2R\pi t}, \quad \sigma_{rr} \approx 0,$$

where R is the current radius of the shell, and t is its thickness. Suppose that $N = cpR^2\pi$, where $c =$const. Assuming a rigid-plastic J_2 material model and isotropic hardening with constant hardening rate h, determine c such that $R = R_0 =$const. throughout the loading process. Determine in that case the relationship between the pressure p and the thickness t of the deformed shell, and discuss the onset of instability.

Solution: For the considered rigid-plastic material model, the components of the rate of deformation tensor are

$$D_{ij} = \frac{1}{2h} \frac{\sigma'_{kl} \dot{\sigma}'_{kl}}{\sigma'_{mn} \sigma'_{mn}} \sigma'_{ij}.$$

The stress components are

$$\sigma_{\theta\theta} = 2\sigma, \quad \sigma_{zz} = (1+c)\sigma, \quad \sigma_{rr} \approx 0,$$

where $\sigma = Rp/2t$. The corresponding deviatoric components

$$\sigma'_\theta = \left(1 - \frac{c}{3}\right)\sigma, \quad \sigma'_z = \frac{2c}{3}\sigma, \quad \sigma'_r = -\left(1 + \frac{c}{3}\right)\sigma.$$

It readily follows that

$$\sigma'_{mn}\sigma'_{mn} = 2\left(1 + \frac{c^2}{3}\right)\sigma^2, \quad \sigma'_{kl}\dot{\sigma}'_{kl} = 2\left(1 + \frac{c^2}{3}\right)\sigma\dot{\sigma},$$

so that

$$D_{ij} = \frac{1}{2h} \frac{\dot{\sigma}}{\sigma} \sigma'_{ij}.$$

Thus,

$$D_r = \frac{\dot{t}}{t} = \frac{1}{2h}\frac{\dot{\sigma}}{\sigma}\sigma'_r = -\frac{1}{2h}\left(1 + \frac{c}{3}\right)\dot{\sigma},$$

$$D_\theta = \frac{\dot{R}}{R} = \frac{1}{2h}\frac{\dot{\sigma}}{\sigma}\sigma'_\theta = \frac{1}{2h}\left(1 - \frac{c}{3}\right)\dot{\sigma},$$

$$D_z = \frac{\dot{l}}{l} = \frac{1}{2h}\frac{\dot{\sigma}}{\sigma}\sigma'_z = \frac{c}{3h}\dot{\sigma}.$$

Clearly, there is no change in radius of the shell ($\dot{R} = 0$) if $c = 3$. In this case, then,

$$\frac{\dot{t}}{t} = -\frac{1}{h}\dot{\sigma} \quad \Rightarrow \quad \ln\frac{t_0}{t} = \frac{1}{h}\left(\frac{pR_0}{2t} - \frac{p_0 R_0}{2t_0}\right).$$

The plastic yielding begins when

$$\left(\frac{1}{2}\sigma'_{mn}\sigma'_{mn}\right)^{1/2} = \frac{\sigma^0_{yy}}{\sqrt{3}} \quad \Rightarrow \quad p_0 = \frac{\sigma^0_{yy}}{\sqrt{3}}\frac{t_0}{R_0},$$

where σ^0_{yy} is the initial yield stress in simple tension. Consequently, the pressure-thickness relationship is

$$p = \frac{t}{R_0}\left(\frac{\sigma^0_{yy}}{\sqrt{3}} + 2h\ln\frac{t_0}{t}\right).$$

This is a monotonically increasing function of the ratio t_0/t up to the instability condition

$$\frac{dp}{d(t_0/t)} = 0 \quad \Rightarrow \quad \frac{t_0}{t} = \exp\left(1 - \frac{\sigma^0_{yy}}{2\sqrt{3}h}\right).$$

The corresponding (maximum) pressure is

$$p_{\max} = 2h\frac{t_0}{R_0}\exp\left(\frac{\sigma^0_{yy}}{2\sqrt{3}h} - 1\right).$$

The hardening rate has to be such that $p_{\max} > p_0$, which is satisfied for any $h > 0$. For $h = 0$ (ideal plasticity), we must require $\dot{\sigma} = 0$, which gives $p = p_0 t/t_0$. This implies instability from the onset of deformation ($t < t_0 \Rightarrow p < p_0$).

Problem 26.10. Consider an elastoplastic deformation process which is characterized by elastic isotropy and plastic incompressibility. Let the elastic strain energy per unit initial volume be

$$\Phi = \Phi(\mathbf{E}^e), \quad \mathbf{E}^e = \frac{1}{2}\left(\mathbf{F}^{eT}\cdot\mathbf{F}^e - \mathbf{I}\right),$$

where $\mathbf{F}^e$ is the elastic part of the elastoplastic deformation gradient from the multiplicative decomposition $\mathbf{F} = \mathbf{F}^e \cdot \mathbf{F}^p$. The corresponding symmetric Piola–Kirchhoff stress tensor, corresponding to elastic deformation $\mathbf{F}^e$, is

$$\mathbf{S}^e = (\det \mathbf{F}^e)\,\mathbf{F}^{e-1}\cdot\boldsymbol{\sigma}\cdot\mathbf{F}^{e-T} = \frac{\partial \Phi}{\partial \mathbf{E}^e},$$

where $\boldsymbol{\sigma}$ is the Cauchy stress. Prove that the symmetric Piola–Kirchhoff stress tensor, corresponding to elastoplastic deformation $\mathbf{F}$, is

$$\mathbf{S} = \frac{\partial \Phi}{\partial \mathbf{E}}.$$

Solution: The elastoplastic Lagrangian strain is

$$\mathbf{E} = \frac{1}{2}\left(\mathbf{F}^T\cdot\mathbf{F} - \mathbf{I}\right) = \mathbf{F}^{pT}\cdot\mathbf{E}^e\cdot\mathbf{F}^p + \mathbf{E}^p,$$

where the Lagrangian strain corresponding to plastic deformation gradient $\mathbf{F}^p$ is defined by

$$\mathbf{E}^p = \frac{1}{2}\left(\mathbf{F}^{pT} \cdot \mathbf{F}^p - \mathbf{I}\right).$$

Thus,

$$\mathbf{E}^e = \mathbf{F}^{p-T} \cdot (\mathbf{E} - \mathbf{E}^p) \cdot \mathbf{F}^{p-1}.$$

The elastic strain energy per unit unstressed volume can now be expressed as

$$\Phi = \Phi(\mathbf{E}^e) = \Phi\left[\mathbf{F}^{p-T} \cdot (\mathbf{E} - \mathbf{E}^p) \cdot \mathbf{F}^{p-1}\right].$$

Thus, by partial differentiation

$$\frac{\partial \Phi}{\partial E_{ij}} = \frac{\partial \Phi}{\partial E^e_{kl}} \frac{\partial E^e_{kl}}{\partial E_{ij}}.$$

Since

$$\frac{\partial E^e_{kl}}{\partial E_{ij}} = F^{p-T}_{ki} F^{p-1}_{jl},$$

we obtain

$$\frac{\partial \Phi}{\partial \mathbf{E}} = \mathbf{F}^{p-1} \cdot \frac{\partial \Phi}{\partial \mathbf{E}^e} \cdot \mathbf{F}^{p-T}.$$

By substituting into this the expression

$$\frac{\partial \Phi}{\partial \mathbf{E}^e} = (\det \mathbf{F}^e)\, \mathbf{F}^{e-1} \cdot \boldsymbol{\sigma} \cdot \mathbf{F}^{e-T},$$

and by using $\mathbf{F}^{p-1} \cdot \mathbf{F}^{e-1} = \mathbf{F}^{-1}$ and the plastic incompressibility constraint giving $\det \mathbf{F}^e = \det \mathbf{F}$, we finally obtain

$$\frac{\partial \Phi}{\partial \mathbf{E}} = (\det \mathbf{F})\mathbf{F}^{-1} \cdot \boldsymbol{\sigma} \cdot \mathbf{F}^{-T} = \mathbf{S}.$$

Note also that

$$\mathbf{S}^e = \mathbf{F}^p \cdot \mathbf{S} \cdot \mathbf{F}^{pT}.$$

Problem 26.11. Prove equation (26.189), *i.e.*, when the current state is taken as the reference state, show that the Jaumann rate of the Kirchhoff stress is

$$\overset{\triangledown}{\boldsymbol{\tau}} = \overset{\triangledown}{\boldsymbol{\sigma}} + \boldsymbol{\sigma}\, \mathrm{tr}\mathbf{D}.$$

Solution: Apply the Jaumann rate to $\boldsymbol{\tau} = (\det \mathbf{F})\boldsymbol{\sigma}$ to obtain

$$\overset{\triangledown}{\boldsymbol{\tau}} = \frac{d}{dt}(\det \mathbf{F}) + (\det \mathbf{F})\overset{\triangledown}{\boldsymbol{\sigma}}.$$

Since

$$\frac{d}{dt}(\det \mathbf{F}) = (\det \mathbf{F})\, \mathrm{tr}\mathbf{D},$$

we have
$$\overset{\triangledown}{\tau} = (\det \mathbf{F})(\overset{\triangledown}{\sigma} + \sigma\, \mathrm{tr}\mathbf{D}).$$

When the current state is taken to be the reference state, we have $\mathbf{F} = \mathbf{I}$ and $\det \mathbf{F} = 1$, so that
$$\overset{\triangledown}{\tau} = \overset{\triangledown}{\sigma} + \sigma\, \mathrm{tr}\mathbf{D} = \dot{\sigma} + \sigma\, \mathrm{tr}\mathbf{D} - \mathbf{W} \cdot \sigma + \sigma \cdot \mathbf{W}.$$

Problem 26.12. The elastoplastic partitions of the rate of deformation tensor and the Jaumann rate of the Kirchhoff stress are defined such that
$$\mathbf{D} = \mathbf{D}^e + \mathbf{D}^p, \quad \mathbf{D}^e = \Lambda^{e\,-1} : \overset{\triangledown}{\tau},$$
and
$$\overset{\triangledown}{\tau} = \overset{\triangledown}{\tau}{}^e + \overset{\triangledown}{\tau}{}^p, \quad \overset{\triangledown}{\tau}{}^e = \Lambda^e : \mathbf{D},$$
where Λ^e is the instantaneous elastic moduli tensor. Derive the relationship between the plastic parts $\overset{\triangledown}{\tau}{}^p$ and $\mathbf{D}^p$.

Solution: By applying a trace product with Λ^e to
$$\mathbf{D} = \Lambda^{e\,-1} : \overset{\triangledown}{\tau} + \mathbf{D}^p,$$
we obtain
$$\overset{\triangledown}{\tau} = \Lambda^e : \mathbf{D} - \Lambda^e : \mathbf{D}^p.$$
Thus
$$\overset{\triangledown}{\tau}{}^p = -\Lambda^e : \mathbf{D}^p.$$
Clearly, there is an identity
$$\overset{\triangledown}{\tau}{}^p : \mathbf{D}^e = -\overset{\triangledown}{\tau} : \mathbf{D}^p = -\mathbf{D}^p : \Lambda^e : \mathbf{D}^e.$$

Problem 26.13. If the material obeys the Ilyushin's postulate of nonnegative net work in an isothermal cycle of strain, then
$$\overset{\triangledown}{\tau}{}^p : \mathbf{D} < 0.$$
Show that
$$\overset{\triangledown}{\tau}{}^e : \mathbf{D}^p > 0, \quad \overset{\triangledown}{\tau}{}^p : \mathbf{D}^p < 0,$$
$$\overset{\triangledown}{\tau}{}^e : \mathbf{D} > 0, \quad \overset{\triangledown}{\tau} : \mathbf{D}^e > 0,$$
and
$$\overset{\triangledown}{\tau} : \mathbf{D}^p > -\mathbf{D}^p : \Lambda^e : \mathbf{D}^p.$$

Problems 26.13–26.14

Solution: Since from Problem 26.13, $\overset{\triangledown}{\boldsymbol{\tau}}{}^{\mathrm{p}} = -\boldsymbol{\Lambda}^{\mathrm{e}} : \mathbf{D}^{\mathrm{p}}$, we have

$$\overset{\triangledown}{\boldsymbol{\tau}}{}^{\mathrm{p}} : \mathbf{D} = -\mathbf{D}^{\mathrm{p}} : \boldsymbol{\Lambda}^{\mathrm{e}} : \mathbf{D} = -\mathbf{D}^{\mathrm{p}} : \overset{\triangledown}{\boldsymbol{\tau}}{}^{\mathrm{e}} < 0.$$

This proves the first inequality. The second inequality follows immediately, because

$$\overset{\triangledown}{\boldsymbol{\tau}}{}^{\mathrm{p}} : \mathbf{D}^{\mathrm{p}} = -\mathbf{D}^{\mathrm{p}} : \boldsymbol{\Lambda}^{\mathrm{e}} : \mathbf{D}^{\mathrm{p}} < 0,$$

the elastic moduli tensor being positive definite. Similarly, the third and fourth inequality follow from

$$\overset{\triangledown}{\boldsymbol{\tau}}{}^{\mathrm{e}} : \mathbf{D} = \mathbf{D} : \boldsymbol{\Lambda}^{\mathrm{e}} : \mathbf{D} > 0,$$

and

$$\overset{\triangledown}{\boldsymbol{\tau}} : \mathbf{D}^{\mathrm{e}} = \mathbf{D}^{\mathrm{e}} : \boldsymbol{\Lambda}^{\mathrm{e}} : \mathbf{D}^{\mathrm{e}} > 0.$$

The last inequality is deduced from

$$\overset{\triangledown}{\boldsymbol{\tau}}{}^{\mathrm{e}} : \mathbf{D}^{\mathrm{p}} = \left(\overset{\triangledown}{\boldsymbol{\tau}} - \overset{\triangledown}{\boldsymbol{\tau}}{}^{\mathrm{p}}\right) : \mathbf{D}^{\mathrm{p}} > 0.$$

This gives

$$\overset{\triangledown}{\boldsymbol{\tau}} : \mathbf{D}^{\mathrm{p}} + \mathbf{D}^{\mathrm{p}} : \boldsymbol{\Lambda}^{\mathrm{e}} : \mathbf{D}^{\mathrm{p}} > 0,$$

i.e.,

$$\overset{\triangledown}{\boldsymbol{\tau}} : \mathbf{D}^{\mathrm{p}} > -\mathbf{D}^{\mathrm{p}} : \boldsymbol{\Lambda}^{\mathrm{e}} : \mathbf{D}^{\mathrm{p}}.$$

If elastoplastic material is in the hardening range, then $\overset{\triangledown}{\boldsymbol{\tau}} : \mathbf{D}^{\mathrm{p}} > 0$.

Problem 26.14. Consider a pressurized hollow sphere. At an instant when the applied pressure over its internal surface is p, and its inner and outer radii are a and b, determine the stress field within the sphere, assuming that it is made from elastically rigid, power-law plastic material, governed by the constitutive expression

$$D_{ij} = \dot{\gamma}_0 \left(\frac{J_2^{1/2}}{K_0}\right)^m \frac{\sigma'_{ij}}{J_2^{1/2}}, \quad m \geq 1, \quad J_2^{1/2} < K_0.$$

Note that for $m = 1$ the model corresponds to linearly viscous material, while rigid ideally plastic material is obtained in the limit $m \to \infty$.

Solution: The nonvanishing velocity component is the radial velocity $v_r = v_r(r)$. The rate of deformation components are

$$D_{rr} = \frac{dv_r}{dr}, \quad D_{\theta\theta} = D_{\phi\phi} = \frac{v_r}{r}.$$

The incompressibility constraint is

$$D_{rr} + D_{\theta\theta} + D_{\phi\phi} = 0 \quad \Rightarrow \quad \frac{dv_r}{dr} + 2\frac{v_r}{r} = 0,$$

which can be integrated to give
$$v_r = \frac{A}{r^2}.$$

Thus,
$$D_{rr} = -\frac{2A}{r^3}, \quad D_{\theta\theta} = D_{\phi\phi} = \frac{A}{r^3}.$$

The nonvanishing stress components are σ_{rr} and $\sigma_{\theta\theta} = \sigma_{\phi\phi}$, with the corresponding deviatoric parts
$$\sigma'_{rr} = -\frac{2}{3}(\sigma_{\theta\theta} - \sigma_{rr}), \quad \sigma'_{\theta\theta} = \sigma'_{\phi\phi} = \frac{1}{3}(\sigma_{\theta\theta} - \sigma_{rr}).$$

In the considered problem the radial stress is compressive and the hoop stress tensile, so that
$$J_2^{1/2} = \frac{1}{\sqrt{3}}(\sigma_{\theta\theta} - \sigma_{rr}).$$

Thus, from the constitutive expression we can write
$$\frac{1}{\sqrt{3}}(\sigma_{\theta\theta} - \sigma_{rr}) = K_0 \left(\frac{\sqrt{3}\,A}{\dot{\gamma}_0\,r^3}\right)^{1/m}.$$

On the other hand, the equilibrium equation is
$$\frac{d\sigma_{rr}}{dr} + \frac{2}{r}(\sigma_{\theta\theta} - \sigma_{rr}) = 0 \quad \Rightarrow \quad \sigma_{\theta\theta} - \sigma_{rr} = \frac{r}{2}\frac{d\sigma_{rr}}{dr}.$$

Combining this with the previous expression gives
$$\frac{1}{\sqrt{3}}\frac{r}{2}\frac{d\sigma_{rr}}{dr} = K_0\left(\frac{\sqrt{3}\,A}{\dot{\gamma}_0\,r^3}\right)^{1/m},$$

and, upon integration,
$$\sigma_{rr} = -\frac{2m}{\sqrt{3}}K_0\left(\frac{\sqrt{3}A}{\dot{\gamma}_0}\right)^{1/m} r^{-3/m} + B.$$

The boundary conditions at the considered instant of deformation process are
$$\sigma_{rr}(r = a) = -p, \quad \sigma_{rr}(r = b) = 0.$$

They give
$$B = \frac{2m}{\sqrt{3}}K_0\left(\frac{\sqrt{3}A}{\dot{\gamma}_0}\right)^{1/m} b^{-3/m} + B,$$

$$\frac{2m}{\sqrt{3}}K_0\left(\frac{\sqrt{3}A}{\dot{\gamma}_0}\right)^{1/m} = -\frac{p}{b^{-3/m} - a^{-3/m}}.$$

The stresses are consequently

$$\sigma_{rr} = -p\,\frac{b^{-3/m} - r^{-3/m}}{b^{-3/m} - a^{-3/m}}, \qquad \sigma_{\theta\theta} = \sigma_{\phi\phi} = -p\,\frac{b^{-3/m} - (1 - \frac{3}{2m})r^{-3/m}}{b^{-3/m} - a^{-3/m}}.$$

The radial velocity is

$$v_r = \frac{\dot{\gamma}_0}{\sqrt{3}}\left[\frac{\sqrt{3}}{2m}\,\frac{p}{K_0(b^{-3/m} - a^{-3/m})}\right]^m \frac{1}{r^2}.$$

CHAPTER 27

Problem 27.1. Consider a uniaxial tension of a single crystal rod of fcc nickel, oriented with the [100] direction parallel to the rod axis. Calculate the Schmid factor for the slip system {111}, <110> involved in the plastic flow of nickel.

Solution: The angle between the loading axis [100] and the slip plane normal [111] is determined from

$$\cos\phi = \{0, 0, 1\} \cdot \left\{\frac{1}{\sqrt{3}}, \frac{1}{\sqrt{3}}, \frac{1}{\sqrt{3}}\right\} = \frac{1}{\sqrt{3}}, \quad \phi = 54.7°.$$

The angle between the loading axis [100] and the slip direction [$\bar{1}$01] is determined from

$$\cos\psi = \{0, 0, 1\} \cdot \left\{-\frac{1}{\sqrt{2}}, 0, \frac{1}{\sqrt{2}}\right\} = \frac{1}{\sqrt{2}}, \quad \psi = 45°.$$

Thus, the Schmid factor is

$$\text{S.F.} = \cos\phi\,\cos\psi = \frac{1}{\sqrt{3}}\,\frac{1}{\sqrt{2}} = \frac{1}{\sqrt{6}} = 0.408.$$

Problem 27.2. It was found that a cylindrical rod of the cross sectional area $A = 5\,\text{cm}^2$, made from zinc, yields at the axial load 2.1 kN, if the slip system (0001), [11$\bar{2}$0] is oriented such that $\phi = 83.5°$ and $\psi = 18°$. Calculate at what load will the rod yield if the slip system is oriented relative to the axial load such that $\phi = 13°$ and $\psi = 78°$.

Solution: The first information specifies the critical resolved shear stress of the slip system

$$\tau_{cr} = \frac{2.1}{5 \times 10^{-4}}\,\cos 83.5°\,\cos 18° = 452.2\,\text{kPa}.$$

The crystal will therefore yield in the second case if the load is

$$F = \frac{A\tau_{cr}}{\cos 13°\,\cos 78°} = \frac{5 \times 10^{-4} \times 452.2}{\cos 13°\,\cos 78°} = 1.12\,\text{kN}.$$

Problem 27.3. Derive the necking condition in a uniaxial tension test. Assuming material incompressibility, and the stress-strain relationship $\sigma = K\epsilon^n$, determine the strain at the onset of necking, and the corresponding maximum force.

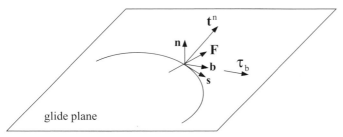

Figure 27-1. A curved segment of dislocation with the Burgers vector **b** within the glide plane whose unit normal is **n**. The traction vector at a point of the dislocation is $\mathbf{t}_n$.

Solution: The applied force is $F = \sigma A$, where σ is the true stress and A the current cross-sectional area. Thus,

$$dF = d\sigma A + \sigma dA.$$

Necking occurs at the maximum load, where $dF = 0$ and

$$\frac{d\sigma}{\sigma} = -\frac{dA}{A}.$$

This is the necking condition. If the plastic strain is large and dominates the elastic strain component, one can assume the overall incompressibility, *i.e.*, $AL =$ const., where L is the current length of the specimen. Consequently,

$$dAL + AdL = 0 \quad \Rightarrow \quad -\frac{dA}{A} = \frac{dL}{L} = d\epsilon.$$

Combining this with the previously derived necking condition, we obtain

$$\frac{d\sigma}{d\epsilon} = \sigma.$$

If $\sigma = K\epsilon^n$, this gives the necking strain $\epsilon = n$. The corresponding maximum force is

$$F_{\max} = KA_0 \left(\frac{n}{e}\right)^n.$$

Problem 27.4. A glide force on an infinitesimal dislocation segment, which is normal to the dislocation line, is defined as $F = b\tau_b$, where b is the magnitude of the Burgers vector **b** of the curved dislocation, and τ_b the resolved shear stress over the glide plane and in the direction of **b**. Derive the expression for this force in terms of the stress tensor σ_{ij}.

Solution: By the Cauchy relation, the components of the traction vector and the stress tensor at an arbitrary point of the dislocation line (Fig. 27-1) are related by

$$t_i^n = \sigma_{ij} n_j.$$

Thus, the Peach–Koehler glide force is

$$F = b\tau_b = b_i t_i^n = b_i \sigma_{ij} n_j.$$

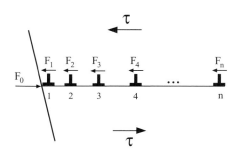

Figure 27-2. A dislocation pileup against a grain boundary. There are n dislocations in the pileup, which is under remote shear stress τ.

If **s** is the unit vector along the dislocation line, the vectorial expression for the glide force is

$$\mathbf{F} = b\tau_b(\mathbf{n} \times \mathbf{s}) = (\mathbf{b} \cdot \boldsymbol{\sigma}) \times \mathbf{s}.$$

Problem 27.5. Consider a dislocation pileup at a strong obstacle, such as grain boundary (Fig. 27-2). The mutual repulsion of the dislocations tends to spread out the dislocation arrangement, while the applied shear stress drives the dislocations closer together against the obstacle. If there are n dislocations in the pileup, calculate the force F_0 exerted on the obstacle.

Solution: Let the Burgers vector of the edge dislocations be b, and the applied shear stress τ. The dislocation forces on the dislocations from the pileup are

$$F_1 = \tau b + F_{1,2} + F_{1,3} + F_{1,4} + \ldots + F_{1,n},$$

$$F_2 = \tau b - F_{2,1} + F_{2,3} + F_{2,4} + \ldots + F_{2,n},$$

$$F_3 = \tau b - F_{3,1} - F_{3,2} + F_{3,4} + \ldots + F_{3,n},$$

$$\ldots\ldots\ldots\ldots\ldots\ldots \text{etc.} \ldots\ldots\ldots\ldots\ldots\ldots,$$

$$F_n = \tau b - F_{n,1} - F_{n,2} - F_{n,3} - \ldots - F_{n,n-1},$$

where $F_{i,j}$ is the interaction force on the i-th dislocation due to the j-th dislocation. Clearly, $F_{i,j} = F_{j,i}$. In the equilibrium pileup configuration, each dislocation is in the equilibrium, so that

$$F_1 - F_0 = 0, \quad F_2 = F_3 = \ldots = F_n = 0.$$

Thus, by summing up the expressions for the dislocation forces, we obtain

$$\sum_{i=1}^{n} F_i = F_1 = n\tau b,$$

because the interaction forces cancel each other,

$$\sum_{i=1}^{n}\sum_{j=1}^{n} F_{i,j} = 0.$$

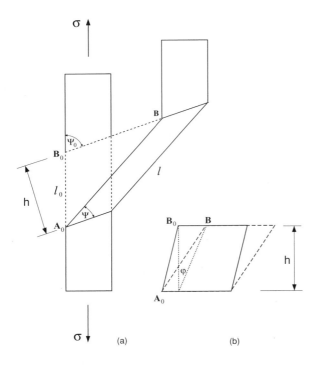

Figure 28-1. Simple shearing along the slip direction, which is at an angle ψ_0 in the undeformed configuration, and ψ in the deformed configuration (relative to longitudinal directions of the rod l_0 and l). The resulting shear strain is $\gamma = \overline{B_0 B}/h$.

Therefore, the force exerted on the obstacle is

$$F_0 = F_1 = n\tau b.$$

CHAPTER 28

Problem 28.1. Consider a plastic extension of the rod produced by a single slip of amount γ as depicted in Fig. 28-1. Assume that the slip plane normal, the slip direction and the loading axis of the rod are all in one plane. Derive the relationship between the longitudinal stretch ratio l/l_0 and the slip induced shear strain γ.

Solution: The shear strain in an element of the rod of initial length l_0 because of slip of amount $\overline{B_0 B}$ over the slip plane and in the slip direction is

$$\gamma = \tan\varphi = \frac{\overline{B_0 B}}{h}.$$

From the geometry of the triangle $A_0 B_0 B$ we have

$$h = l_0 \sin\psi_0 = l \sin\psi,$$

$$\overline{B_0 B} = l\cos\psi - l_0 \cos\psi_0.$$

Since

$$l\cos\psi = l\left(1 - \sin^2\psi\right)^{1/2} = l_0\left[\left(\frac{l}{l_0}\right)^2 - \sin^2\psi_0\right]^{1/2},$$

Problems 28.1–28.2

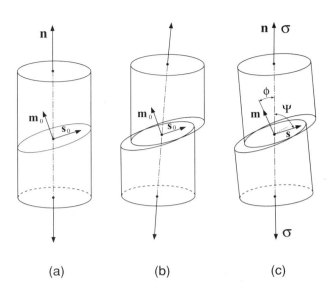

Figure 28-2. Single crystal under uniaxial tension oriented for single slip along the slip direction s_0 in the slip plane with the normal m_0; parts (a) and (b). The lattice rotates during deformation so that the slip direction s in the deformed configuration makes an angle ψ with the longitudinal direction n; part (c). The angle between the slip plane normal m and the longitudinal direction is ϕ. The vectors m_0, s_0, and n, in general, do not belong to one plane.

we can write

$$\overline{B_0 B} = l_0 \left\{ \left[\left(\frac{l}{l_0} \right)^2 - \sin^2 \psi_0 \right]^{1/2} - \cos \psi_0 \right\}.$$

Therefore, the shear strain γ is related to the stretch ratio l/l_0 by

$$\gamma = \frac{1}{\sin \psi_0} \left\{ \left[\left(\frac{l}{l_0} \right)^2 - \sin^2 \psi_0 \right]^{1/2} - \cos \psi_0 \right\}.$$

If the slip plane normal does not belong to the plane formed by the loading axis and the slip direction, the above expression generalizes to

$$\gamma = \frac{1}{\cos \phi_0} \left\{ \left[\left(\frac{l}{l_0} \right)^2 - \sin^2 \psi_0 \right]^{1/2} - \cos \psi_0 \right\},$$

as originally derived by Schmid and Boas.

Problem 28.2. Derive an expression for the instantaneous modulus $H = d\sigma/de$ of a rigid-plastic crystal oriented for single slip under uniaxial tension. Show that the lattice rotation can cause an apparent softening of the crystal, even when the slip direction is still hardening.

Solution: Consider a specimen under uniaxial tension oriented for single slip along the direction s_0, on the slip plane with the normal m_0 (see Fig. 28-2). The corresponding rate of deformation and the spin tensors can be expressed from the equations derived in text as

$$\mathbf{D} = \frac{1}{a} (\mathbf{P}\,\mathbf{P}) : \overset{\triangledown}{\sigma},$$

$$\mathbf{W} = \mathbf{W}^* + \frac{1}{a} (\mathbf{Q}\,\mathbf{P}) : \overset{\triangledown}{\sigma},$$

where
$$a = h - \mathbf{P} : (\mathbf{Q} \cdot \boldsymbol{\sigma} - \boldsymbol{\sigma} \cdot \mathbf{Q}),$$

and
$$\mathbf{P} = \frac{1}{2}(\mathbf{sm} + \mathbf{ms}), \quad \mathbf{Q} = \frac{1}{2}(\mathbf{sm} - \mathbf{ms}).$$

Suppose that the specimen is under uniaxial tension in the direction $\mathbf{n}$, which is fixed by the grips of the loading machine. The Cauchy stress tensor is then
$$\boldsymbol{\sigma} = \sigma \, \mathbf{n}\mathbf{n},$$

the material spin is $\mathbf{W} = \mathbf{0}$, and
$$\overset{\triangledown}{\boldsymbol{\sigma}} = \dot{\sigma} \, \mathbf{n}\mathbf{n}.$$

It follows that
$$\mathbf{P} : \overset{\triangledown}{\boldsymbol{\sigma}} = \dot{\sigma} \, (\mathbf{m} \cdot \mathbf{n})(\mathbf{s} \cdot \mathbf{n}) = \dot{\sigma} \, \cos \phi \, \cos \psi,$$

where ϕ is the angle between the current slip plane normal $\mathbf{m}$ and the loading direction $\mathbf{n}$, while ψ is the angle between the current slip direction $\mathbf{s}$ and the loading direction $\mathbf{n}$, as shown in Fig. 28-2. It is easily found that
$$\mathbf{P} : (\mathbf{Q} \cdot \boldsymbol{\sigma} - \boldsymbol{\sigma} \cdot \mathbf{Q}) = \frac{1}{2} \sigma \left[(\mathbf{m} \cdot \mathbf{n})^2 - (\mathbf{s} \cdot \mathbf{n})^2 \right] = \frac{1}{2} \sigma \left(\cos^2 \phi - \cos^2 \psi \right).$$

Therefore, upon substitution into the expression for the rate of deformation,
$$\mathbf{D} = \frac{\dot{\sigma} \, \cos \phi \, \cos \psi}{h - \frac{1}{2} (\cos^2 \phi - \cos^2 \psi)} \, \mathbf{P}.$$

Denoting by e the longitudinal strain in the direction of the specimen axis $\mathbf{n}$, we can write
$$\dot{e} = \mathbf{n} \cdot \mathbf{D} \cdot \mathbf{n},$$

and therefore
$$\dot{e} = \frac{\dot{\sigma} \, \cos^2 \phi \, \cos^2 \psi}{h - \frac{1}{2} (\cos^2 \phi - \cos^2 \psi)},$$

i.e.,
$$\dot{\sigma} = H \dot{e}, \quad H = \frac{h}{\cos^2 \phi \, \cos^2 \psi} - \frac{\sigma (\cos^2 \phi - \cos^2 \psi)}{2 \cos^2 \phi \, \cos^2 \psi}.$$

Depending on the current orientation of the active slip system, the instantaneous modulus H can be positive, zero, or negative. If the lattice has rotated such that
$$\cos^2 \phi - \cos^2 \psi > \frac{2h}{\sigma},$$

the current modulus is negative, although the slip direction may still be hardening ($h > 0$). The resulting apparent softening is purely geometrical effect, due to rotation of the lattice caused by crystallographic slip, and is referred to as geometric softening. In the derivation it was assumed that the lattice rotation does not activate the slip on another slip system.

Problem 28.3. Derive the expression (28.78), *i.e.*, show that

$$\overset{\triangledown}{\boldsymbol{\tau}}{}^* = \mathbf{F}^* \cdot (\boldsymbol{\Lambda}^* : \dot{\mathbf{E}}^*) \cdot \mathbf{F}^{*T} + \mathbf{D}^* \cdot \boldsymbol{\tau}^* + \boldsymbol{\tau}^* \cdot \mathbf{D}^*.$$

Solution: We start from the relationship between the symmetric Piola–Kirchhoff stress relative to the lattice and the corresponding Kirchhoff stress $\tau^* = (\det \mathbf{F}^*)\sigma$,

$$\mathbf{S}^* = \mathbf{F}^{*-1} \cdot \boldsymbol{\tau}^* \cdot \mathbf{F}^{*-T}.$$

Applying the time like derivative, we obtain

$$\dot{\mathbf{S}}^* = \left(\mathbf{F}^{*-1}\right)^{\cdot} \cdot \boldsymbol{\tau}^* \cdot \mathbf{F}^{*-T} + \mathbf{F}^{*-1} \cdot \dot{\boldsymbol{\tau}}^* \cdot \mathbf{F}^{*-T} + \mathbf{F}^{*-1} \cdot \boldsymbol{\tau}^* \cdot \left(\mathbf{F}^{*-T}\right)^{\cdot}.$$

Since

$$\left(\mathbf{F}^{*-1}\right)^{\cdot} = -\mathbf{F}^{*-1} \cdot \dot{\mathbf{F}} \cdot \mathbf{F}^{*-1} = -\mathbf{F}^{*-1} \cdot \mathbf{L}^*,$$

the previous expression can be rewritten as

$$\dot{\mathbf{S}}^* = \mathbf{F}^{*-1} \cdot \left(\dot{\boldsymbol{\tau}}^* - \mathbf{L}^* \cdot \boldsymbol{\tau}^* - \boldsymbol{\tau}^* \cdot \mathbf{L}^{*T}\right) \cdot \mathbf{F}^{*-T}.$$

The bracketed term is the convected rate of the Kirchhoff stress relative to the lattice,

$$\overset{\diamond}{\boldsymbol{\tau}}{}^* = \dot{\boldsymbol{\tau}}^* - \mathbf{L}^* \cdot \boldsymbol{\tau}^* - \boldsymbol{\tau}^* \cdot \mathbf{L}^{*T} = \overset{\triangledown}{\boldsymbol{\tau}}{}^* - \mathbf{D}^* \cdot \boldsymbol{\tau}^* - \boldsymbol{\tau}^* \cdot \mathbf{D}^*,$$

where

$$\overset{\triangledown}{\boldsymbol{\tau}}{}^* = \dot{\boldsymbol{\tau}}^* - \mathbf{W}^* \cdot \boldsymbol{\tau}^* + \boldsymbol{\tau}^* \cdot \mathbf{W}^*$$

is the Jaumann rate corotational with the lattice. Thus, we obtain

$$\overset{\triangledown}{\boldsymbol{\tau}}{}^* = \mathbf{F}^* \cdot \dot{\mathbf{S}}^* \cdot \mathbf{F}^{*T} + \mathbf{D}^* \cdot \boldsymbol{\tau} + \boldsymbol{\tau} \cdot \mathbf{D}^*.$$

Finally, since

$$\dot{\mathbf{S}}^* = \boldsymbol{\Lambda}^* : \dot{\mathbf{E}}^*,$$

we arrive at the desired result. If plastic deformation is assumed to be volume preserving, then $\det \mathbf{F}^* = \det \mathbf{F}$ and $\tau^* = \tau$.

CHAPTER 29

Problem 29.1. Consider a plane strain bifurcation of an incompressible nonlinear elastic material from the state of biaxial stretch ($\lambda_1 = \lambda$, $\lambda_2 = \lambda^{-1}$). Show that

$$\overset{\triangledown}{\sigma}_{11} - \overset{\triangledown}{\sigma}_{22} = 2\mu^*(D_{11} - D_{22}), \quad \overset{\triangledown}{\sigma}_{12} = 2\mu D_{12},$$

where

$$2\mu = \frac{\lambda^4 + 1}{\lambda^4 - 1}(\sigma_1 - \sigma_2), \quad 4\mu^* = \lambda \frac{d}{d\lambda}(\sigma_1 - \sigma_2).$$

Solution: Let the bifurcation strain rate

$$\mathbf{D} = \begin{bmatrix} D_{11} & D_{12} \\ D_{12} & D_{22} \end{bmatrix}.$$

be superposed to the biaxially stretched state with the deformation gradient and corresponding Cauchy stress

$$\mathbf{F} = \begin{bmatrix} \lambda_1 & 0 \\ 0 & \lambda_2 \end{bmatrix} = \begin{bmatrix} \lambda & 0 \\ 0 & \lambda^{-1} \end{bmatrix}, \quad \boldsymbol{\sigma} = \begin{bmatrix} \sigma_1 & 0 \\ 0 & \sigma_2 \end{bmatrix}.$$

The rate of Lagrangian strain is

$$\dot{\mathbf{E}} = \mathbf{F}^T \cdot \mathbf{D} \cdot \mathbf{F} = \begin{bmatrix} \lambda^2 D_{11} & D_{12} \\ D_{12} & \lambda^{-2} D_{22} \end{bmatrix}.$$

For an incompressible material, the rate of the symmetric Piola–Kirchhoff stress is related to the convected rate of the Cauchy stress by

$$\dot{\mathbf{S}} = \mathbf{F}^{-1} \left(\stackrel{\triangledown}{\boldsymbol{\sigma}} - \mathbf{D} \cdot \boldsymbol{\sigma} - \boldsymbol{\sigma} \cdot \mathbf{D} \right) \cdot \mathbf{F}^{-T},$$

i.e.,

$$\begin{bmatrix} \dot{S}_{11} & \dot{S}_{12} \\ \dot{S}_{12} & \dot{S}_{22} \end{bmatrix} = \begin{bmatrix} \lambda^{-2}(\stackrel{\triangledown}{\sigma}_{11} - 2\sigma_1 D_{11}) & \stackrel{\triangledown}{\sigma}_{12} - (\sigma_1 + \sigma_2) D_{12} \\ \stackrel{\triangledown}{\sigma}_{12} - (\sigma_1 + \sigma_2) D_{12} & \lambda^2(\stackrel{\triangledown}{\sigma}_{22} - 2\sigma_2 D_{22}) \end{bmatrix}.$$

The rates of stress and strain are related by $\dot{S}_{ij} = \Lambda_{ijkl} \dot{E}_{kl}$, which gives

$$\dot{S}_{11} = \Lambda_{1111} \dot{E}_{11} + \Lambda_{1122} \dot{E}_{22},$$

$$\dot{S}_{22} = \Lambda_{2211} \dot{E}_{11} + \Lambda_{2222} \dot{E}_{22},$$

$$\dot{S}_{12} = 2\Lambda_{1212} \dot{E}_{12}.$$

The (symmetric) instantaneous elastic moduli are from (7.100) given by

$$\Lambda_{1111} = \frac{1}{\lambda_1} \frac{\partial S_1}{\partial \lambda_1}, \quad \Lambda_{1122} = \frac{1}{\lambda_2} \frac{\partial S_1}{\partial \lambda_2} + \frac{S_1 - S_2}{\lambda_1^2 - \lambda_2^2},$$

$$\Lambda_{2222} = \frac{1}{\lambda_2} \frac{\partial S_2}{\partial \lambda_2}, \quad \Lambda_{2211} = \frac{1}{\lambda_1} \frac{\partial S_2}{\partial \lambda_1} + \frac{S_1 - S_2}{\lambda_1^2 - \lambda_2^2},$$

$$\Lambda_{1212} = \frac{S_1 - S_2}{\lambda_1^2 - \lambda_2^2}, \quad \Lambda_{1112} = \Lambda_{2212} = 0.$$

Upon combining of the above results, we obtain

$$\stackrel{\triangledown}{\sigma}_{11} = \left(2\sigma_1 + \lambda_1^3 \frac{\partial S_1}{\partial \lambda_1} - \lambda_1 \frac{\partial S_1}{\partial \lambda_2} - \frac{S_1 - S_2}{\lambda_1^2 - \lambda_2^2} \right) D_{11},$$

$$\stackrel{\triangledown}{\sigma}_{22} = \left(2\sigma_2 + \lambda_2^3 \frac{\partial S_2}{\partial \lambda_2} - \lambda_2 \frac{\partial S_2}{\partial \lambda_1} - \frac{S_1 - S_2}{\lambda_1^2 - \lambda_2^2} \right) D_{22},$$

$$\stackrel{\triangledown}{\sigma}_{12} = \left(\sigma_1 + \sigma_2 + 2 \frac{S_1 - S_2}{\lambda_1^2 - \lambda_2^2} \right) D_{12}.$$

To derive the desired results, we recall that

$$\sigma_1 = \lambda_1^2 S_1, \quad \sigma_2 = \lambda_2^2 S_2, \quad (\lambda_1 = \lambda, \; \lambda_2 = \lambda^{-1}),$$

Problems 29.1–29.2

so that

$$\frac{d\sigma_1}{d\lambda} = \frac{\partial(\lambda_1^2 S_1)}{\partial \lambda_1}\frac{\partial \lambda_1}{\partial \lambda} + \frac{\partial(\lambda_1^2 S_1)}{\partial \lambda_2}\frac{\partial \lambda_2}{\partial \lambda}.$$

This gives

$$\lambda\frac{d\sigma_1}{d\lambda} = 2\sigma_1 + \lambda_1^3 \frac{\partial S_1}{\partial \lambda_1} - \lambda_1 \frac{\partial S_1}{\partial \lambda_2},$$

and similarly

$$\lambda\frac{d\sigma_2}{d\lambda} = -2\sigma_2 - \lambda_2^3 \frac{\partial S_2}{\partial \lambda_2} + \lambda_2 \frac{\partial S_2}{\partial \lambda_1}.$$

Recalling that $D_{11} + D_{22} = 0$ by the incompressibility constraint, we finally obtain

$$\overset{\triangledown}{\sigma}_{11} - \overset{\triangledown}{\sigma}_{22} = \left[\frac{1}{2}\lambda \frac{d}{d\lambda}(\sigma_1 - \sigma_2)\right](D_{11} - D_{22}),$$

and

$$\overset{\triangledown}{\sigma}_{12} = \left[\frac{\lambda^4 + 1}{\lambda^4 - 1}(\sigma_1 - \sigma_2)\right] D_{12}.$$

Problem 29.2. Derive the expressions for the moduli μ and μ^* of the previous problem if the strain energy function is of the Mooney–Rivlin type

$$\Phi = \frac{1}{2}\mu_0 \left[\left(\frac{1}{2} + \beta\right)(I_B - 3) - \left(\frac{1}{2} - \beta\right)(II_B + 3)\right].$$

Solution: Since, in the three-dimensional setting,

$$\mathbf{F} = \begin{bmatrix} \lambda & 0 & 0 \\ 0 & \lambda^{-1} & 0 \\ 0 & 0 & 1 \end{bmatrix}, \quad \mathbf{B} = \begin{bmatrix} \lambda^2 & 0 & 0 \\ 0 & \lambda^{-2} & 0 \\ 0 & 0 & 1 \end{bmatrix},$$

we obtain

$$I_B = \operatorname{tr} \mathbf{B} = \lambda^2 + \lambda^{-2} + 1, \quad II_B = \frac{1}{2}[\operatorname{tr}(\mathbf{B}^2) - (\operatorname{tr} \mathbf{B})^2] = -(\lambda^2 + \lambda^{-2} + 1),$$

so that

$$\Phi = \frac{1}{2}\mu_0(\lambda^2 + \lambda^{-2} - 2).$$

Thus,

$$\sigma_1 = -p_0 + \lambda_1 \frac{\partial \Phi}{\partial \lambda_1} = -p_0 + \mu_0 \lambda_1^2, \quad \sigma_2 = -p_0 + \lambda_2 \frac{\partial \Phi}{\partial \lambda_2} = -p_0 + \mu_0 \lambda_2^2.$$

Consequently,

$$\sigma_1 - \sigma_2 = \mu_0(\lambda^2 - \lambda^{-2}), \quad \lambda\frac{\partial}{\partial \lambda}(\sigma_1 - \sigma_2) = 2\mu_0(\lambda^2 + \lambda^{-2}),$$

and, therefore,

$$\mu = \mu^* = \frac{1}{2}\mu_0(\lambda^2 + \lambda^{-2}).$$

CHAPTER 30

A polycrystalline aggregate is considered to be macroscopically homogeneous by assuming that local microscopic heterogeneities (because of different orientation and state of hardening of individual crystal grains) are distributed in such a way that the material elements beyond some minimum scale have essentially the same overall macroscopic properties. This minimum scale defines the size of the representative macroelement or representative cell. The representative macroelement can be viewed as a material point in the continuum mechanics of macroscopic aggregate behavior. To be statistically representative of the local properties of its microconstituents, the representative macroelement must include a sufficiently large number of microelements. For example, for relatively fine-grained metals, a representative macroelement of volume 1 mm^3 contains a minimum of 1000 crystal grains. The concept of the representative macroelement is used in various branches of the mechanics of heterogeneous materials and is also referred to as the representative volume element.

Experimental determination of the mechanical behavior of an aggregate is commonly based on the measured loads and displacements over its external surface. Consequently, the macrovariables introduced in the constitutive analysis should be expressible in terms of this surface data alone.

Problem 30.1. Let

$$\mathbf{F}(\mathbf{X}, t) = \frac{\partial \mathbf{x}}{\partial \mathbf{X}}, \quad \det \mathbf{F} > 0,$$

be the deformation gradient at the microlevel of description, associated with a (continuous and piecewise continuously differentiable) microdeformation within a crystalline grain, $\mathbf{x} = \mathbf{x}(\mathbf{X}, t)$. The reference position of the particle is $\mathbf{X}$, and its current position at time t is $\mathbf{x}$. Express the volume averages of the deformation gradient, rate of deformation gradient, velocity gradient, and Cauchy and Kirchhoff stress, all in terms of the surface data.

Solution: The volume average of the deformation gradient over the reference volume V_0 of the macroelement is

$$\langle \mathbf{F} \rangle = \frac{1}{V_0} \int_{V_0} \mathbf{F} \, dV_0 = \frac{1}{V_0} \int_{S_0} \mathbf{x} \, \mathbf{n}_0 \, dS_0,$$

by the Gauss divergence theorem. The unit outward normal to the bounding surface S_0 of the macroelement volume is $\mathbf{n}_0$.

The volume average of the rate of deformation gradient,

$$\dot{\mathbf{F}}(\mathbf{X}, t) = \frac{\partial \mathbf{v}}{\partial \mathbf{X}}, \quad \mathbf{v} = \dot{\mathbf{x}}(\mathbf{X}, t),$$

where $\mathbf{v}$ is the velocity field, is

$$\langle \dot{\mathbf{F}} \rangle = \frac{1}{V_0} \int_{V_0} \dot{\mathbf{F}} \, dV_0 = \frac{1}{V_0} \int_{S_0} \mathbf{v} \, \mathbf{n}_0 \, dS_0.$$

Problems 30.1–30.2

Table 34.1. *Designation of slip systems in fcc crystals*

Plane	(111)			($\bar{1}\bar{1}1$)			($\bar{1}11$)			($1\bar{1}1$)		
Slip Rate	a_1	a_2	a_3	b_1	b_2	b_3	c_1	c_2	c_3	d_1	d_2	d_3
Slip Direction	$[0\bar{1}1]$	$[10\bar{1}]$	$[\bar{1}10]$	$[011]$	$[\bar{1}0\bar{1}]$	$[1\bar{1}0]$	$[0\bar{1}1]$	$[\bar{1}0\bar{1}]$	$[110]$	$[011]$	$[10\bar{1}]$	$[\bar{1}\bar{1}0]$

If the current configuration is taken as the reference configuration ($\mathbf{x} = \mathbf{X}$, $\mathbf{F} = \mathbf{I}$, $\dot{\mathbf{F}} = \mathbf{L} = \partial \mathbf{v}/\partial \mathbf{x}$), the above equation gives the volume average of the velocity gradient

$$\{\mathbf{L}\} = \frac{1}{V} \int_V \mathbf{L} \, dV = \frac{1}{V} \int_S \mathbf{v} \mathbf{n} \, dS.$$

The current volume of the deformed macroelement is V, and S is its bounding surface with the unit outward normal $\mathbf{n}$. Enclosure within $\{\}$ brackets is used to indicate that the average is taken over the deformed volume of the macroelement.

Let $\mathbf{P} = \mathbf{P}(\mathbf{X}, t)$ be a nonsymmetric nominal stress field within the macroelement. The nominal traction $\mathbf{p}_n$ is related to the true traction $\mathbf{t}_n$ by $\mathbf{p}_n \, dS_0 = \mathbf{t}_n \, dS$. The volume average of the nominal stress is

$$\langle \mathbf{P} \rangle = \frac{1}{V_0} \int_{V_0} \mathbf{P} \, dV_0 = \frac{1}{V_0} \int_{S_0} \mathbf{X} \mathbf{p}_n \, dS_0.$$

If current configuration is chosen as the reference, ($\mathbf{P} = \boldsymbol{\sigma}$, $\mathbf{p}_n = \mathbf{t}_n$), and we have

$$\{\boldsymbol{\sigma}\} = \frac{1}{V} \int_V \boldsymbol{\sigma} \, dV = \frac{1}{V} \int_S \mathbf{x} \mathbf{t}_n \, dS.$$

Note that

$$\int_{V_0} \boldsymbol{\tau} \, dV_0 = \int_V \boldsymbol{\sigma} \, dV = \int_S \mathbf{x} \mathbf{t}_n \, dS = \int_{S_0} \mathbf{x} \mathbf{p}_n \, dS_0,$$

so that

$$\langle \boldsymbol{\tau} \rangle = \frac{1}{V_0} \int_{V_0} \boldsymbol{\tau} \, dV_0 = \frac{1}{V_0} \int_{S_0} \mathbf{x} \mathbf{p}_n \, dS_0.$$

Problem 30.2. The slip in an fcc crystal occurs on the octahedral planes in the directions of the octahedron edges (see Fig. 30-1). There are three possible slip directions in each of the four distinct slip planes, making a total of 12 slip systems (if counting both senses of a slip direction as one) or 24 (if counting opposite directions separately). The positive senses of the slip directions are chosen as indicated in Table 34.1. The letters a, b, c, and d refer to four slip planes. With attached indices 1, 2, and 3, they designate the slip rates in the respective positive slip directions. If elastic (lattice) strains are disregarded, calculate the components of the rate of deformation tensor, expressed on the cubic axes, due to simultaneous slip rates in 12 slip directions.

Solution: The components of the rate of deformation can be calculated from

$$\mathbf{D} = \sum_{\alpha=1}^{12} \mathbf{P}^\alpha \dot{\gamma}^\alpha = \sum_{\alpha=1}^{12} \frac{1}{2} (\mathbf{s}^\alpha \mathbf{m}^\alpha + \mathbf{m}^\alpha \mathbf{s}^\alpha) \dot{\gamma}^\alpha,$$

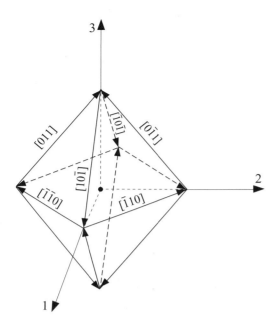

Figure 30-1. Twelve different slip directions in fcc crystals (counting opposite directions as different) are the edges of the octahedron shown relative to principal cubic axes. Each slip direction is shared by two intersecting slip planes so that there is a total of 24 independent slip systems (12 if counting opposite slip directions as one).

where $\mathbf{m}^\alpha$ is the unit slip plane normal and $\mathbf{s}^\alpha$ is the slip direction. For example, the contribution from the slip rate $\dot{\gamma} = a_1$ is obtained by using

$$\mathbf{m} = \frac{1}{\sqrt{3}}(1,1,1), \quad \mathbf{s} = \frac{1}{\sqrt{2}}(0,-1,1),$$

which gives

$$\frac{1}{2}(\mathbf{s}^\alpha \mathbf{m}^\alpha + \mathbf{m}^\alpha \mathbf{s}^\alpha) a_1 = \frac{a_1}{2\sqrt{6}} \begin{pmatrix} 0 & -1 & 1 \\ -1 & -2 & 0 \\ 1 & 0 & 2 \end{pmatrix}.$$

An analogous calculation proceeds for other slip rates, which yields the following expressions for the overall rate of deformation components

$$\sqrt{6}\, D_{11} = a_2 - a_3 + b_2 - b_3 + c_2 - c_3 + d_2 - d_3,$$

$$\sqrt{6}\, D_{22} = a_3 - a_1 + b_3 - b_1 + c_3 - c_1 + d_3 - d_1,$$

$$\sqrt{6}\, D_{33} = a_1 - a_2 + b_1 - b_2 + c_1 - c_2 + d_1 - d_2,$$

$$2\sqrt{6}\, D_{23} = -a_2 + a_3 + b_2 - b_3 - c_2 + c_3 + d_2 - d_3,$$

$$2\sqrt{6}\, D_{31} = -a_3 + a_1 + b_3 - b_1 + c_3 - c_1 - d_3 + d_1,$$

$$2\sqrt{6}\, D_{12} = -a_1 + a_2 - b_1 + b_2 + c_1 - c_2 + d_1 - d_2.$$

Problem 30.3. Determine the number of independent sets of five slips in an fcc crystal.

Solution: An arbitrary rate of deformation tensor has five independent components ($\text{tr}\,\mathbf{D} = 0$ for a rigid-plastic crystal), and therefore can only be produced by multiple slip

Problem 30.3 823

over a group of slip systems containing an independent set of five. Of the $C_5^{12} = 792$ sets of five slips, only 384 are independent. The 408 dependent sets are identified as follows. First, only two of three slip systems in the same slip plane are independent. The unit slip rates along a_1, a_2, and a_3 directions together produce the zero resultant rate of deformation. The same applies to three slip directions in b, c, and d slip planes. We can write this symbolically as

$$a_1 + a_2 + a_3 = 0, \quad b_1 + b_2 + b_3 = 0, \quad c_1 + c_2 + c_3 = 0, \quad d_1 + d_2 + d_3 = 0.$$

Thus, if the set of five slip systems contains a_1, a_2, and a_3, there are $C_2^9 = 36$ possible combinations with the remaining nine slip systems. These 36 sets of five slips cannot produce an arbitrary **D**, with five independent components, and are thus eliminated from 792 sets of five slips. Additional $3 \times 36 = 108$ sets, associated with dependent sets of three slips in b, c, and d planes, can also be eliminated. This makes a total of 144 dependent sets corresponding to the above constraints.

Of the remaining 648 sets of five slips, 324 involve two slips in each of two slip planes with one in a third ($6 \times 3^2 \times 6 = 324$), whereas 324 involve two slips in one slip plane and one in each of the other three slip planes ($4 \times 3^4 = 324$). In the latter group, there are $3 \times 8 = 24$ sets involving the combinations

$$a_1 - b_1 + c_1 - d_1 = 0, \quad a_2 - b_2 + c_2 - d_2 = 0, \quad a_3 - b_3 + c_3 - d_3 = 0.$$

These expressions can also be interpreted as meaning that such combinations of unit slips produce zero resultant rate of deformation. The 24 sets of five slips, involving four slip rates according to above, can thus be eliminated (these sets necessarily consists of two slips in one plane and one slip in each of the remaining three slip planes). Additional 12 sets are eliminated, which correspond to conditions obtained from the above equations by adding or subtracting $\sum a_i, \ldots, \sum d_i$, one at a time, to each of the equation. A representative of these is $a_1 - b_1 + c_1 + d_2 + d_3 = 0$.

There are $4 \times 33 = 132$ dependent sets associated with

$$a_1 + b_2 + d_3 = 0, \quad a_2 + b_1 + c_3 = 0, \quad a_3 + c_2 + d_1 = 0, \quad b_3 + c_1 + d_2 = 0.$$

Each group of 33 sets consists of 21 sets involving two slips in one plane and one slip in each of other three planes and 12 sets involving two slips in two planes and one slip in one plane. Additional 84 sets can be eliminated by subtracting $\sum a_i, \sum b_i$, and $\sum d_i$, one at a time, from the first of the above set of equation, and similarly for the other three. This makes 12 groups of 7 sets. A representative group is associated with $a_1 + b_2 - d_1 - d_2 = 0$. Four of the 7 sets consist of two slips in two planes and one slip in one plane, whereas 3 sets consist of two slips in one plane and one slip in each of the other three planes. Finally, 12 more sets (making total of 228 dependent sets associated with the above set of equations and their equivalents) can be eliminated by subtracting appropriate one of $\sum a_i, \ldots, \sum d_i$ from each of the 12 previous group equations. An example is $-a_1 + b_1 + b_3 + d_1 + d_2 = 0$. They all involve two slips in each of two planes and one slip in another plane.

In summary, there is a total of 408 dependent sets of five slips: 144 sets with three slips in the same plane, 108 sets with two slips in each of two planes and one in a third, and 156 sets with two slips in one plane and one slip in each of the other three planes. The total number of independent sets of five slips is $792 - 408 = 384$. Taylor (1938a) originally considered only 216 sets as geometrically admissible (involving double slip in each of two planes) and did not observe 168 admissible sets with double slip in only one plane. These were originally identified by Bishop and Hill (1951b).

Problem 30.4. According to the Voigt assumption, when a polycrystalline aggregate is subjected to the overall uniform strain, the individual crystals will all be in the same state of applied strain. Based on this assumption, calculate the effective elastic constants of an isotropic aggregate of cubic crystals.

Solution: For a cubic crystal the elastic moduli can be expressed as

$$C^c_{ijkl} = c_{12}\delta_{ij}\delta_{kl} + 2c_{44}I_{ijkl} + (c_{11} - c_{12} - 2c_{44})A_{ijkl},$$

where

$$I_{ijkl} = \frac{1}{2}(\delta_{ik}\delta_{jl} + \delta_{il}\delta_{jk}),$$

and

$$A_{ijkl} = a_ia_ja_ka_l + b_ib_jb_kb_l + c_ic_jc_kc_l.$$

The vectors **a**, **b**, and **c** are the orthogonal unit vectors along the principal cubic axes, and the usual notation for the elastic constants c_{11}, c_{12}, and c_{44} is employed. Two independent linear invariants of C^c_{ijkl} are

$$C^c_{iijj} = 3(c_{11} + 2c_{12}), \quad C^c_{ijij} = 3(c_{11} + 2c_{44}).$$

Denote by κ and μ the overall (effective) bulk and shear moduli of an isotropic aggregate of cubic crystals. The corresponding elastic moduli are

$$C_{ijkl} = 2\mu I_{ijkl} + \left(\kappa - \frac{2}{3}\mu\right)\delta_{ij}\delta_{kl}.$$

According to the Voigt assumption, when a polycrystalline aggregate is subjected to the overall uniform strain, the individual crystals will all be in the same state of applied strain (which gives rise to stress discontinuities across the grain boundaries). Thus, by requiring that the overall stress is the average of the local stresses, there follows

$$C_{ijkl} = \{C^c_{ijkl}\}.$$

Instead of performing the integration

$$C_{ijkl} = \frac{1}{8\pi^2}\int_\Omega C^c_{ijkl}\,d\Omega,$$

the effective polycrystalline constants can be obtained directly by observing that the linear invariants of C_{ijkl} and C^c_{ijkl} must be equal. Thus, equating the earlier expressions for C^c_{iijj} and C^c_{ijij} to

$$C_{iijj} = 9\kappa, \quad C_{ijij} = 3\kappa + 10\mu,$$

we obtain the Voigt estimates

$$\kappa = \frac{1}{3}(c_{11} + 2c_{12}), \quad \mu^V = \frac{1}{5}(c_{11} - c_{12} + 3c_{44}).$$

Problem 30.5. According to the Reuss assumption, when a polycrystalline aggregate is subjected to the overall uniform stress, the individual crystals will all be in the same state of stress. Based on this assumption, calculate the effective elastic constants of an isotropic aggregate of cubic crystals.

Solution: By requiring that the overall strain in the aggregate is the average of the local crystalline strains, there follows

$$S_{ijkl} = \{S^c_{ijkl}\}.$$

The components of elastic compliance tensor of cubic crystals (see Problem 14.6) are

$$S^c_{ijkl} = -\frac{c_{12}}{(c_{11} - c_{12})(c_{11} + 2c_{12})} \delta_{ij}\delta_{kl} + \frac{1}{2c_{44}} I_{ijkl}$$
$$+ \left(\frac{1}{c_{11} - c_{12}} - \frac{1}{2c_{44}}\right) A_{ijkl}.$$

Two independent linear invariants of S^c_{ijkl} are

$$S^c_{iijj} = \frac{3}{c_{11} + 2c_{12}}, \quad S^c_{ijij} = \frac{3(c_{11} + c_{12})}{(c_{11} - c_{12})(c_{11} + 2c_{12})} + \frac{3}{2c_{44}}.$$

The components of the elastic compliance tensor of an isotropic aggregate of cubic crystals are

$$S_{ijkl} = \frac{1}{2\mu} I_{ijkl} + \frac{1}{3}\left(\frac{1}{3\kappa} - \frac{1}{2\mu}\right) \delta_{ij}\delta_{kl},$$

with the corresponding invariants

$$S_{iijj} = \frac{1}{\kappa}, \quad S_{ijij} = \frac{5}{2\mu} + \frac{1}{3\kappa}.$$

Consequently, the Reuss estimates are

$$\kappa = \frac{1}{3}(c_{11} + 2c_{12}), \quad \mu^R = \frac{5}{\dfrac{4}{c_{11} - c_{12}} + \dfrac{3}{c_{44}}}.$$

It can be shown that μ^V is the upper bound and that μ^R is the lower bound on the true value of the effective shear modulus, *i.e.*,

$$\mu^R \leq \mu \leq \mu^V.$$

It can also be shown that the effective Lamé constant is bounded such that $\lambda^V \leq \lambda \leq \lambda^R$.

CHAPTER 31

Problem 31.1. Derive the rate-type constitutive equation given by (31.16) of the main text.

Solution: The deformation gradient is decomposed as

$$\mathbf{F} = \mathbf{F}^* \cdot \mathbf{F}^p,$$

with
$$\dot{\mathbf{F}}^p \cdot \mathbf{F}^{p-1} = \sum_{\alpha=1}^{2} \dot{\gamma}_\alpha \, \mathbf{s}_\alpha \, \mathbf{m}_\alpha.$$

The two Lagrangian strain measures, relative to the initial reference configuration, are
$$\mathbf{E} = \frac{1}{2}\left(\mathbf{F}^T \cdot \mathbf{F} - \mathbf{I}\right), \quad \mathbf{E}^p = \frac{1}{2}\left(\mathbf{F}^{pT} \cdot \mathbf{F}^p - \mathbf{I}\right),$$

whereas
$$\mathbf{E}^* = \frac{1}{2}\left(\mathbf{F}^{*T} \cdot \mathbf{F}^* - \mathbf{I}\right).$$

These strain measures are related by
$$\mathbf{E} = \mathbf{F}^{pT} \cdot \mathbf{E}^* \cdot \mathbf{F}^p + \mathbf{E}^p.$$

By differentiating, we obtain the relationship
$$\dot{\mathbf{E}} = \mathbf{F}^{pT} \cdot \dot{\mathbf{E}}^* \cdot \mathbf{F}^p + \mathbf{F}^{pT} \cdot \left[\mathbf{C}^* \cdot \left(\dot{\mathbf{F}}^p \cdot \mathbf{F}^{p-1}\right)\right]_{\mathrm{sym}} \cdot \mathbf{F}^p,$$

where $\mathbf{C}^* = \mathbf{F}^{*T} \cdot \mathbf{F}^*$. The above can be rewritten as
$$\dot{\mathbf{E}}^* = \mathbf{F}^{p-T} \cdot \dot{\mathbf{E}} \cdot \mathbf{F}^{p-1} - \left[\mathbf{C}^* \cdot \left(\dot{\mathbf{F}}^p \cdot \mathbf{F}^{p-1}\right)\right]_{\mathrm{sym}}$$

The two symmetric Piola–Kirchhoff stress tensors are derived from the strain energy $\Phi(\mathbf{E}^*)$ by the gradient operations
$$\mathbf{S}^* = \frac{\partial \Phi}{\partial \mathbf{E}^*}, \quad \mathbf{S} = \frac{\partial \Phi}{\partial \mathbf{E}},$$

with the connection
$$\mathbf{S}^* = \mathbf{F}^p \cdot \mathbf{S} \cdot \mathbf{F}^{pT}.$$

The relationship between the rates of the stress tensors $\mathbf{S}^*$ and $\mathbf{S}$ is obtained by simple differentiation, which gives
$$\dot{\mathbf{S}}^* = \mathbf{F}^p \cdot \dot{\mathbf{S}} \cdot \mathbf{F}^{pT} + 2\left[\left(\dot{\mathbf{F}}^p \cdot \mathbf{F}^{p-1}\right) \cdot \mathbf{S}^*\right]_{\mathrm{sym}}.$$

Finally, by taking the time derivative of $\mathbf{S}^* = \partial \Phi / \partial \mathbf{E}^*$, there follows
$$\dot{\mathbf{S}}^* = \mathbf{\Lambda}^* : \dot{\mathbf{E}}^*, \quad \mathbf{\Lambda}^* = \frac{\partial^2 \Phi}{\partial \mathbf{E}^* \, \partial \mathbf{E}^*}.$$

When the previously derived expressions for $\dot{\mathbf{S}}^*$ and $\dot{\mathbf{E}}^*$ are substituted in this relationship, we obtain
$$\dot{\mathbf{S}} = \mathbf{\Lambda} : \dot{\mathbf{E}} - \sum_{\alpha=1}^{2} \dot{\gamma}_\alpha \mathbf{X}_\alpha,$$

where
$$\Lambda_{ijkl} = F^{p\,-1}_{im} F^{p\,-1}_{jn} \Lambda^*_{mnpq} F^{p\,-1}_{pk} F^{p\,-1}_{ql},$$

and
$$\mathbf{X}_\alpha = \mathbf{F}^{p-1} \cdot \left(\mathbf{\Lambda}^* : \mathbf{A}_\alpha + 2\mathbf{B}_\alpha\right) \cdot \mathbf{F}^{p-1}.$$

Problems 31.1–32.2

The second-order symmetric tensors $\mathbf{A}_\alpha$ and $\mathbf{B}_\alpha$ are defined by

$$\mathbf{A}_\alpha = [\mathbf{C}^* \cdot (\mathbf{s}_\alpha \mathbf{m}_\alpha)]_{\text{sym}}, \quad \mathbf{B}_\alpha = [(\mathbf{s}_\alpha \mathbf{m}_\alpha) \cdot \mathbf{S}^*]_{\text{sym}}.$$

CHAPTER 32

Problem 32.1. Consider the total body force acting on the body of variable mass, whose current volume is V and mass density ρ, i.e.,

$$\mathbf{F}_b = \int_V \mathbf{b}\, \rho\, dV,$$

where $\mathbf{b}$ is the body force per unit current mass. Derive an expression for the time rate of change of this force.

Solution: Applying the time derivative to $\mathbf{F}_b$, we have

$$\frac{d\mathbf{F}_b}{dt} = \frac{d}{dt}\int_V \mathbf{b}\, \rho\, dV = \int_V \frac{d}{dt}(\rho\, \mathbf{b}\, dV) = \int_V \rho \frac{d\mathbf{b}}{dt} dV + \int_V \mathbf{b}\, \frac{d}{dt}(\rho\, dV).$$

Since

$$\frac{d(\rho\, dV)}{dt} = r^g dV,$$

there follows

$$\frac{d\mathbf{F}_b}{dt} = \int_V \left(\rho \frac{d\mathbf{b}}{dt} + r^g\, \mathbf{b}\right) dV.$$

In view of the continuity equation

$$\frac{d\rho}{dt} + \rho \nabla \cdot \mathbf{v} = r^g,$$

we can also express the rate of the total body force acting on the body of variable mass as

$$\frac{d\mathbf{F}_b}{dt} = \int_V \left[\rho \frac{d\mathbf{b}}{dt} + \left(\frac{d\rho}{dt} + \rho \nabla \cdot \mathbf{v}\right)\mathbf{b}\right] dV.$$

If there is no mass growth,

$$r^g = 0, \quad \frac{d\rho}{dt} + \rho \nabla \cdot \mathbf{v} = 0,$$

the previous expression reduces to classical expression of the mechanics of constant mass

$$\frac{d\mathbf{F}_b}{dt} = \int_V \rho \frac{d\mathbf{b}}{dt} dV.$$

Problem 32.2. Consider a circumferential growth of a cylindrical blood vessel (Fig. 32-1). If the opening angle Ω ($\Omega < 0$ for growth; $\Omega > 0$ for resorption) is assumed to be constant along the thickness of the vessel, and if the plain strain conditions prevail, write down the components of the deformation gradients $\mathbf{F}$, $\mathbf{F}^g$, and $\mathbf{F}^e$.

Solution: The deformation mapping from the initial to current state is

$$r = r(R), \quad \theta = \Theta, \quad z = Z,$$

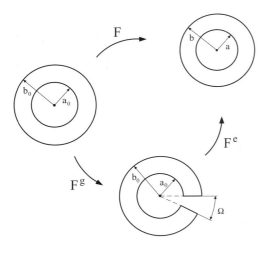

Figure 32-1. Circumferential growth of a cylindrical blood vessel. The opening angle is Ω.

where (R, Θ, Z) are the cylindrical coordinates in the initial configuration and (r, θ, z) are those in the current configuration. The corresponding deformation gradient at an arbitrary point of the vessel is

$$\mathbf{F} = \begin{bmatrix} dr/dR & 0 & 0 \\ 0 & r/R & 0 \\ 0 & 0 & 1 \end{bmatrix}.$$

A discontinuous mapping giving rise to an opening angle Ω is

$$\rho = R, \quad \vartheta = \left(1 - \frac{\Omega}{2\pi}\right)\Theta, \quad \zeta = Z,$$

where (ρ, ϑ, ζ) are the cylindrical coordinates in the intermediate (stress free) configuration. Thus,

$$\mathbf{F}^{g} = \begin{bmatrix} 1 & 0 & 0 \\ 0 & \omega & 0 \\ 0 & 0 & 1 \end{bmatrix}, \quad \omega = 1 - \frac{\Omega}{2\pi}.$$

The elastic part of the deformation gradient $\mathbf{F}^e$ now follows from the multiplicative decomposition $\mathbf{F} = \mathbf{F}^e \cdot \mathbf{F}^g$ as

$$\mathbf{F}^e = \mathbf{F} \cdot \mathbf{F}^{g-1} = \begin{bmatrix} dr/dR & 0 & 0 \\ 0 & r/(\omega R) & 0 \\ 0 & 0 & 1 \end{bmatrix}.$$

Problem 32.3. Referring to previous problem, derive the expressions for the stress components in terms of the elastic principal stretches $\lambda_r^e, \lambda_\theta^e, \lambda_z^e$. Assume that the blood vessel is an incompressible elastic material with the strain energy function

$$W = \frac{1}{2}\alpha[\exp(Q) - 1] - p(\lambda_r^e \lambda_\theta^e \lambda_z^e - 1),$$

where α is a material parameter and p plays a role of the Lagrangian multiplier corresponding to the incompressibility constraint $\lambda_r^e \lambda_\theta^e \lambda_z^e = 1$. The function Q depends on the strain components according to

Problem 32.3

$$Q = \beta[(E_r^e)^2 + (E_\theta^e)^2 + (E_z^e)^2],$$

where β is another material parameter. The pressures on the inner and outer surface of the vessel, p_0 and p_1, are assumed to be known.

Solution: The elastic stretches in the previous problem are

$$\lambda_r^e = \frac{dr}{dR}, \quad \lambda_\theta^e = \frac{1}{\omega}\frac{r}{R}, \quad \lambda_z^e = 1.$$

In view of the incompressibility constraint $\lambda_r^e \lambda_\theta^e \lambda_z^e = 1$, we have

$$\frac{dr}{dR}\frac{r}{R} = \omega, \quad \text{i.e.,} \quad r^2 = \omega R^2 + C_1.$$

The integration constant is to be determined from the boundary condition at the inner (or outer) surface of the blood vessel.

The Cauchy stress components can be expressed in terms of the principal stretches as

$$\sigma_{rr} = \lambda_r^e \frac{\partial W}{\partial \lambda_r^e} - p, \quad \sigma_{\theta\theta} = \lambda_\theta^e \frac{\partial W}{\partial \lambda_\theta^e} - p, \quad \sigma_{zz} = \lambda_z^e \frac{\partial W}{\partial \lambda_z^e} - p.$$

Since

$$e_{rr}^e = \frac{1}{2}[(\lambda_r^e)^2 - 1], \quad e_{\theta\theta}^e = \frac{1}{2}[(\lambda_\theta^e)^2 - 1], \quad E_z^e = \frac{1}{2}[(\lambda_z^e)^2 - 1],$$

we have

$$Q = \frac{1}{2}\beta\left\{[(\lambda_r^e)^2 - 1]^2 + [(\lambda_\theta^e)^2 - 1]^2 + [(\lambda_z^e)^2 - 1]^2\right\},$$

and

$$\frac{\partial Q}{\partial \lambda_r^e} = 2\beta\lambda_r^e[(\lambda_r^e)^2 - 1], \quad \frac{\partial Q}{\partial \lambda_\theta^e} = 2\beta\lambda_\theta^e[(\lambda_\theta^e)^2 - 1], \quad \frac{\partial Q}{\partial \lambda_z^e} = 0.$$

Consequently,

$$\frac{\partial W}{\partial \lambda_r^e} = \alpha\beta\lambda_r^e[(\lambda_r^e)^2 - 1]\exp(Q) - \frac{p}{\lambda_r^e},$$

$$\frac{\partial W}{\partial \lambda_\theta^e} = \alpha\beta\lambda_\theta^e[(\lambda_\theta^e)^2 - 1]\exp(Q) - \frac{p}{\lambda_\theta^e}.$$

Therefore, the stress components become

$$\sigma_{rr} = \alpha\beta(\lambda_\theta^e)^2[(\lambda_\theta^e)^2 - 1]\exp(Q) - p,$$

$$\sigma_{\theta\theta} = \alpha\beta(\lambda_\theta^e)^2[(\lambda_\theta^e)^2 - 1]\exp(Q) - p,$$

$$\sigma_{zz} = -p.$$

The explicit determination of the function $p = p(r)$ requires a numerical solution of the differential equation of equilibrium

$$\frac{d\sigma_{rr}}{dr} + \frac{\sigma_{rr} - \sigma_{\theta\theta}}{r} = 0.$$

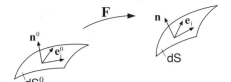

Figure 33-1. The membrane element of area dS^0 is mapped by the deformation gradient $\mathbf{F}$ into an element with area dS. The corresponding unit normal vectors are $\mathbf{n}^0$ and $\mathbf{n}$. The inplane membrane vectors $\mathbf{e}_i^0$ are transformed into $\mathbf{e}_i = \mathbf{F} \cdot \mathbf{e}_i^0$ ($i = 1, 2$).

The integration constant is specified by imposing the remaining boundary condition on the inner or outer surface of the blood vessel.

Problem 32.4. Show that in the case of isotropic uniform growth of a spherical blood vessel, there are no stresses due to growth alone.

Solution: The deformation mapping due to growth alone is

$$\rho = \lambda^g R, \quad \varphi = \Phi, \quad \vartheta = \Theta,$$

where (R, Φ, Θ) are the spherical coordinates in the initial configuration, and $(\rho, \varphi, \vartheta)$ in the intermediate configuration. The growth stretch ratio is λ^g. The corresponding deformation gradient at an arbitrary point of the vessel is

$$\mathbf{F} = \begin{bmatrix} d\rho/dR & 0 & 0 \\ 0 & \rho/R & 0 \\ 0 & 0 & \rho/R \end{bmatrix} = \begin{bmatrix} \lambda^g & 0 & 0 \\ 0 & \lambda^g & 0 \\ 0 & 0 & \lambda^g \end{bmatrix}.$$

This is clearly a pure dilatation. If there is no pressure applied to inner or outer surface of the vessel, we have

$$r = \rho = \lambda^g R, \quad \phi = \varphi = \Phi, \quad \theta = \vartheta = \Theta,$$

where (r, ϕ, θ) are the spherical coordinates in the current configuration. Thus, $\mathbf{F} = \mathbf{F}^g$ and $\mathbf{F}^e = \mathbf{I}$. Since $\mathbf{F}^e = \mathbf{I}$, there are no stresses in the blood vessel during its isotropic uniform growth alone.

CHAPTER 33

Problem 33.1. Derive the relationship for the fractional change in reference area of the membrane $\alpha = \lambda_1 \lambda_2 - 1$, where λ_1 and λ_2 are the principal stretches in the tangent plane of the membrane.

Solution: Let $\mathbf{n}^0 = \mathbf{e}_1^0 \times \mathbf{e}_2^0$ be the unit normal vector to the membrane element in the undeformed configuration, where $\mathbf{e}_1^0$ and $\mathbf{e}_2^0$ are the orthogonal unit vectors in the tangent plane of the membrane in the directions of its principal membrane stretches (Fig. 33-1). Then, the right Cauchy–Green stretch tensor is

$$\mathbf{U} = \lambda_1 \, \mathbf{e}_1^0 \, \mathbf{e}_1^0 + \lambda_2 \, \mathbf{e}_2^0 \, \mathbf{e}_2^0 + \lambda_3 \, \mathbf{n}^0 \, \mathbf{n}^0 \,.$$

The Nanson's relation (4.74) provides an expression for the area change of the surface element according to

$$\mathbf{n} dS = (\det \mathbf{F}) \mathbf{F}^{-T} \cdot \mathbf{n}^0 dS^0 \,,$$

Problem 33.1

where **n** is the unit normal to the deformed surface element of area dS. The deformation gradient that maps dS^0 into dS is **F**. Consequently,
$$dS^2 = (\det \mathbf{F})^2 \left(\mathbf{n}^0 \cdot \mathbf{C}^{-1} \cdot \mathbf{n}^0\right) (dS^0)^2 \,.$$

The inverse of the right Cauchy–Green deformation tensor is
$$\mathbf{C}^{-1} = \mathbf{U}^{-2} = \lambda_1^{-2}\,\mathbf{e}_1^0\,\mathbf{e}_1^0 + \lambda_2^{-2}\,\mathbf{e}_2^0\,\mathbf{e}_2^0 + \lambda_3^{-2}\,\mathbf{n}^0\,\mathbf{n}^0 \,,$$

and $\det \mathbf{F} = \det \mathbf{U} = \lambda_1 \lambda_2 \lambda_3$. Since
$$\mathbf{n}^0 \cdot \mathbf{C}^{-1} \cdot \mathbf{n}^0 = \lambda_3^{-2} \,,$$

we obtain
$$dS^2 = \lambda_1^2 \lambda_2^2 (dS^0)^2 \,,$$

which gives the fractional area change
$$\alpha = \frac{dS - dS^0}{dS^0} = \lambda_1 \lambda_2 - 1 \,.$$

Bibliography

[1] Adkins, J. E. (1961), Large Elastic Deformations. In *Progress in Solid Mechanics* (R. Hill and I. N. Sneddon, eds.), Vol. 2, pp. 1–60, North-Holland, Amsterdam.

[2] Aero, E. L., and Kuvshinskii, E. V. (1960), Fundamental Equations of the Theory of Elastic Media with Rotationally Interacting Particles, *Fiz. Tverd. Tela*, Vol. 2, pp. 1399–1409.

[3] Anand, L. (1986), Moderate Deformations in Extension-Torsion of Incompressible Isotropic Elastic Materials, *J. Mech. Phys. Solids*, Vol. 34, pp. 293–304.

[4] Anand, L., and Kothari, M. (1996), A Computational Procedure for Rate-Independent Crystal Plasticity, *J. Mech. Phys. Solids*, Vol. 44, pp. 525–558.

[5] Antman, S. S. (1995), *Nonlinear Problems of Elasticity*, Springer-Verlag, New York.

[6] Armstrong, P. J., and Frederick, C. O. (1966), A Mathematical Representation of the Multi-axial Bauschinger Effect, *G.E.G.B. Report RD/B/N*, 731.

[7] Arruda, E. M., and Boyce, M. C. (1993), A Three-Dimensional Constitutive Model for the Large Stretch Behavior of Rubber Elastic Materials, *J. Mech. Phys. Solids*, Vol. 41, pp. 389–412.

[8] Asaro, R. J. (1979), Geometrical Effects in the Inhomogeneous Deformation of Ductile Single Crystals, *Acta Met.*, Vol. 27, pp. 445–453.

[9] Asaro, R. J. (1983a), Crystal Plasticity, *J. Appl. Mech.*, Vol. 50, pp. 921–934.

[10] Asaro, R. J. (1983b), Micromechanics of Crystals and Polycrystals, *Adv. Appl. Mech.*, Vol. 23, pp. 1–115.

[11] Asaro, R. J. (1985), Rate Dependent Modeling of Polycrystals, *J. Metals*, Vol. 37, p. A42.

[12] Asaro, R. J., and Barnett, D. M. (1974), The Non-Uniform Transformation Strain Problem for an Anisotropic Ellipsoidal Inclusion, *J. Mech. Phys. Solids*, Vol. 23, pp. 77–83.

[13] Asaro, R. J., and Barnett, D. M. (1976), Applications of the Geometrical Theorems for Dislocations in Anisotropic Elastic Media. In *Computer Simulation for Materials Applications* (R. J. Arsenault, J. R. Beeler, Jr., and J. A. Simmons, eds.), p. 313, NBS, Gaithersburg, Maryland.

[14] Asaro, R. J., and Hirth, J. P. (1974), Planar Dislocation Interactions in Anisotropic Media with Applications to Nodes, *J. Phys. F*, Vol. 3, pp. 1659–1671.

[15] Asaro, R. J., Hirth, J. P., Barnett, D. M., and Lothe, J. (1973), A Further Synthesis of Sextic and Integral Theories for Dislocations and Line Forces in Anisotropic Media, *Phys. Status Solidi B*, Vol. 60, pp. 261–271.

[16] Asaro, R. J., Krysl, P., and Kad, B. (2003), Deformation Mechanism Transitions in Nanoscale FCC Metals, *Phil. Mag. Letters*, Vol. 83, pp. 733–743.

[17] Asaro, R. J., and Needleman, A. (1985), Texture Development and Strain Hardening in Rate Dependent Polycrystals, *Acta Metall.*, Vol. 33, pp. 923–953.

[18] Asaro, R. J., and Rice, J. R. (1977), Strain Localization in Ductile Single Crystals, *J. Mech. Phys. Solids*, Vol. 25, pp. 309–338.

[19] Asaro, R. J., and Suresh, S. (2005), Mechanistic Models for the Activation Volume and Rate Sensitivity in Metals with Nanocrystallne Grains and Nano-Scale Twins, *Acta Mater.*, Vol. 53, pp. 3369–3382.

[20] Asaro, R. J., and Tiller, W. A. (1972), Interface Morphology Development during Stress Corrosion Cracking. Part I: Via Surface Motion, *Metall. Trans.*, Vol. 3, pp. 1789–1796.

[21] Ashby, M. F. (1971), The Deformation of Plastically Non-Homogeneous Alloys. In *Strengthening Methods in Crystals* (A. Kelly and R. B. Nicholson, eds.), p. 137, Appl. Science Publ., London.

[22] Atkinson, C., and Leppington, F. G. (1974), Some Calculations of the Energy-Release Rate G for Cracks in Micropolar and Couple-Stress Elastic Media, *Int. J. Fracture*, Vol. 10, pp. 599–602.

[23] Atkinson, C., and Leppington, F. G. (1977), The Effect of Couple Stresses on the Tip of A Crack, *Int. J. Solids Struct.*, Vol. 13, pp. 1103–1122.

[24] Bailey, R., and Asaro, R. J. (2004), Deformation Rate Sensitivity in Nanocrystalline Ni, *UCSD Report SE-Nano-4*, University of California at San Diego, La Jolla, CA.

[25] Barber, J. R. (2002), *Elasticity*, 2nd ed., Kluwer, Dordrecht, The Netherlands.

[26] Barendareght, J. A., and Sharpe, W. N., Jr. (1973), The Effect of Biaxial Loading on the Critical Resolved Shear Stress of Zinc Single Crystals, *J. Mech. Phys. Solids*, Vol. 21, pp. 113–123.

[27] Barnett, D. M., and Asaro, R. J. (1972), The Fracture Mechanics of Slit-Like Cracks in Anistropic Elastic Media, *J. Mech. Phys. Solids*, Vol. 20, pp. 353–366.

[28] Barnett, D. M., and Lothe, J. (1973), Synthesis of the Sextic and the Integral Formalism for Dislocation, Green's Functions and Surface Waves in Anisotropic Elastic Solids, *Phys. Norv.*, Vol. 7, pp. 13–19.

[29] Basinski, S. J., and Basinski, Z. S. (1979), Plastic Deformation and Work Hardening. In *Dislocations in Solids: Dislocations in Metallurgy*, Vol. 4 (F. R. N. Nabarro, ed.), pp. 261–362, North-Holland, Amsterdam.

[30] Basinski, Z. S., and Jackson, P. I. (1965), The Instability of the Work-Hardened State – I. Slip in Extraneously Deformed Crystals, *Phys. Status Solidi*, Vol. 9, pp. 805–823.

[31] Bassani, J. L. (1990), Single Crystal Hardening, *Appl. Mech. Rev.*, Vol. 43, pp. S320–S327.

[32] Bassani, J. L. (1993), Plastic Flow of Crystals, *Adv. Appl. Mech.*, Vol. 30, pp. 191–258.

[33] Batdorf, S. S., and Budiansky, B. (1949), A Mathematical Theory of Plasticity Based on the Concept of Slip, *N.A.C.A. TN*, 1871.

[34] Batdorf, S. S., and Budiansky, B. (1954), Polyaxial Stress-Strain Relations of a Strain-Hardening Meter, *J. Appl. Mech.*, Vol. 21, pp. 323–326.

[35] Bažant, Z. P., and Cedolin, L. (1991), *Stability of Structures: Elastic, Inelastic, Fracture, and Damage Theories*, Oxford University Press, New York.

[36] Beatty, M. F. (1987), Topics in Finite Elasticity: Hyperelasticity of Rubber, Elastomers, and Biological Tissues – with Examples, *Appl. Mech. Rev.*, Vol. 40, pp. 1699–1734.

[37] Beatty, M. F. (1996), Introduction to Nonlinear Elasticity. In *Nonlinear Effects in Fluids and Solids* (M. M. Carroll and M. A. Hayes, eds.), pp. 13–112, Plenum, New York.

[38] Beevers, C. J., and Honeycombe, R. W. (1962), Deformation and Fracture of Aluminum 5.5% Copper Crystals, *Acta Metall.*, Vol. 10, pp. 17–24.

[39] Besseling, J. F., and Van der Giessen, E. (1994), *Mathematical Modeling of Inelastic Deformation*, Chapman & Hall, London.

[40] Biot, M. A. (1965), *Mechanics of Incremental Deformations*, Wiley, New York.

[41] Bishop, J. F. W., and Hill, R. (1951a), A Theory of the Plastic Distortion of a Polysrystalline Aggregate Under Combined Stresses, *Phil. Mag.*, Vol. 42, pp. 414–427.

[42] Bishop, J. F. W., and Hill, R. (1951b), A Theoretical Derivation of the Plastic Properties of a Polycrystalline Face-Centered Metal, *Phil. Mag.*, Vol. 42, pp. 1298–1307.

[43] Bishop, J. F. W., Hill, R., and Mott, N. F. (1945), The Theory of Indentation and Hardness Tests, *Proc. Phys. Soc.*, Vol. 57, pp. 147–159.

[44] Blatz, P. J., and Ko, W. L. (1962), Application of Finite Elasticity Theory to the Deformation of Rubbery Materials, *Trans. Soc. Rheol.*, Vol. 6, pp. 223–251.

Bibliography

[45] Boal, D. (2002), *Mechanics of the Cell*, Cambridge University Press, New York.
[46] Boas, W., and Ogilvie, G. J. (1954), The Plastic Deformation of a Crystal in a Polycrystalling Aggregate, *Acta Metall.*, Vol. 2, pp. 655–659.
[47] Boehler, J. P. (1987), *Application of Tensor Functions in Solid Mechanics*, Springer-Verlag, Wien.
[48] Boley, B. A., and Weiner, J. H. (1960), *Theory of Thermal Stresses*, Wiley, New York.
[49] Bragg, W. H., and Bragg, W. L. (1933), *The Crystalline State*, Bell, London.
[50] Brillouin, L. (1964), *Tensors in Mechanics and Elasticity*, Academic Press, New York.
[51] Broek, D. (1987), *Elementary Engineering Fracture Mechanics*, 4th ed., Martinus Nijhoff, Dordrecht, The Netherlands.
[52] Bronkhorst, C. A., Kalindindi, S. R., and Anand, L. (1992), Polycrystalline Plasticity and the Evolution of Crystallographic Texture in FCC Metals, *Phil. Trans. Roy. Soc. London, Ser. A.*, Vol. 341, pp. 443–477.
[53] Brown, J. W., and Churchill, R. V. (2001), *Fourier Series and Boundary Value Problems*, 6th ed., McGraw-Hill, Boston.
[54] Brulin, O., and Hsieh, R. K. T., eds. (1982), *Mechanics of Micropolar Media*, World Scientific, Singapore.
[55] Budiansky, B. (1959), A Reassessment of Deformation Theories of Plasticity, *J. Appl. Mech.*, Vol. 26, pp. 259–2664.
[56] Budiansky, B., Hutchinson, J. W., and Slutsky, S. (1982), Void Growth and Collapse in Viscous Solids. In *Mechanics of Solids: The Rodney Hill 60th Anniversary Volume* (H. G. Hopkins and M. J. Sewell, eds.), pp. 13–45, Pergamon, Oxford, UK.
[57] Budiansky, B., and Rice, J. R. (1973), Conservation Laws and Energy-Release Rates, *J. Appl. Mech.*, Vol. 40, pp. 201–203.
[58] Bunge, H.-J. (1982), *Texture Analysis in Materials Science – Mathematical Methods*, Butterworths, London.
[59] Callen, H. B. (1960), *Thermodynamics*, Wiley, New York.
[60] Cammarata, R. C. (1994), Surface and Interface Stress Effects in Thin Films, *Progress in Surface Science*, Vol. 46, pp. 1–38.
[61] Carlson, D. E., and Shield, R. T., eds. (1982), *Finite Elasticity*, Martinus Nijhoff, The Hague.
[62] Carreker, R. P., and Hibbard, W. R. (1953), Tensile Deformation of High-Purity Copper as a Function of Temperature, Strain Rate, and Grain Size, *Acta Metall.*, Vol. 1, pp. 654–655.
[63] Chadwick, P. (1960), Thermoelasticity. The Dynamical Theory. In *Progress in Solid Mechanics*, Vol. I (I. N. Sneddon and R. Hill, eds.), pp. 263–328, North-Holland, Amsterdam.
[64] Chadwick, P. (1974), Thermo-Mechanics of Rubberlike Materials, *Phil. Trans. Roy. Soc. Lond. A*, Vol. 276, pp. 371–403.
[65] Chadwick, P. (1999) *Continuum Mechanics: Concise Theory and Problems*, Dover, New York.
[66] Chadwick, P., and Smith, G. D. (1982), Surface Waves in Cubic Elastic Materials. In *Mechanics of Solids: The Rodney Hill 60th Anniversary Volume* (H. G. Hopkins and M. J. Sewell, eds.), pp. 47–100, Pergamon, Oxford, UK.
[67] Chakrabarty, J. (1987), *Theory of Plasticity*, McGraw-Hill, New York.
[68] Chakrabarty, J. (2000), *Applied Plasticity*, Springer-Verlag, New York.
[69] Champion, Y., Langlois, C., Guerin-Mailly, S., Langlois, P., Bonnentien, J. L., and Hytch, M. J. (2003), Near-Perfect Elastoplasticity in Pure Nanocrystalline Copper, *Science*, Vol. 300, pp. 310–311.
[70] Chang, Y. W., and Asaro, R. J. (1981), An Experimental Study of Shear Localization in Aluminum-Copper Single Crystals, *Acta Metall.*, Vol. 29, pp. 241–257.
[71] Chen, W. F., and Han, D. J. (1988), *Plasticity for Structural Engineers*, Springer-Verlag, New York.
[72] Cherepanov, G. P. (1979), *Mechanics of Brittle Fracture*, McGraw-Hill, New York.
[73] Chiem, C. Y., and Duffy, J. D. (1981), Strain Rate History Effects and Observations of Dislocation Substructure in Aluminum Single Crystals Following Dynamic Deformation, *Brown University Report*, MRL E-137, Providence, Rhode Island.

[74] Chokshi, A. H., Rosen, A., Karch, J., and Gleiter, H. (1989), On the Validity of the Hall-Petch Relationship in Nanocrystalline Materials, *Scripta Metall.*, Vol. 23, pp. 1679–1684.
[75] Christoffersen, J., and Hutchinson, J. W. (1979), A Class of Phenomenological Corner Theories of Plasticity, *J. Mech. Phys. Solids*, Vol. 27, pp. 465–487.
[76] Chung, T. J. (1996), *Applied Continuum Mechanics*, Cambridge University Press, Cambridge.
[77] Coleman, B. D., and Gurtin, M. (1967), Thermodynamics with Internal Variables, *J. Chem. Phys.*, Vol. 47, pp. 597–613.
[78] Conrad, H. (1965), In *High Strength Materials* (ed. V. F. Zackay), Wiley, New York.
[79] Conrad, H., and Narayan, J. (2000), On the Grain Size Softening in Nanocrystalline Materials, *Scripta Mater.*, Vol. 42, pp. 1025–1030.
[80] Conrad, H. (2003), Grain Size Dependence of the Plastic Deformation Kinetics in Cu, *Mater. Sci. Eng. A*, Vol. 341, pp. 216–228.
[81] Cook, R. D., and Young, W. C. (1999), *Advanced Mechanics of Materials*, 2nd ed., Prentice Hall, Upper Saddle River, New Jersey.
[82] Cosserat, E., and F. (1909), *Théorie des Corps Deformables*, Hermann, Paris.
[83] Cottrell, A. H. (1961), *Dislocations and Plastic Flow in Crystals*, Oxford University Press, London.
[84] Cottrell, A. H. (1964), *The Mechanical Properties of Matter*, Wiley, New York.
[85] Cowin, S. C. (1990), Structural Adaptation of Bones, *Appl. Mech. Rev.*, Vol. 43, pp. S126–S133.
[86] Cowin, S. C. (2004), Tissue Growth and Remodeling, *Annu. Rev. Biomed. Eng.*, Vol. 6, pp. 77–107.
[87] Cowin, S. C., and Hegedus, D. H. (1976a), Bone Remodeling I: Theory of Adaptive Elasticity, *J. of Elasticity*, Vol. 6, pp. 313–326.
[88] Cowin, S. C., and Hegedus, D. H. (1976b), Bone Remodeling II: Small Strain Elasticity, *J. of Elasticity*, Vol. 6, pp. 337–352.
[89] Cristescu, N. (1967), *Dynamic Plasticity*, North-Holland, Amsterdam.
[90] Cristescu, N., and Suliciu, I. (1982), *Viscoplasticity*, Martinus Nijhoff, The Hague.
[91] Dafalias, Y. F. (1983), Corotational Rates for Kinematic Hardening at Large Plastic Deformations, *J. Appl. Mech.*, Vol. 50, pp. 561–565.
[92] Dafalias, Y. F., and Popov, E. P. (1975), A Model of Nonlinearly Hardening Materials for Complex Loading, *Acta Mech.*, Vol. 21, pp. 173–192.
[93] Dafalias, Y. F., and Popov, E. P. (1976), Plastic Internal Variables Formalism of Cyclic Plasticity, *J. Appl. Mech.*, Vol. 43, pp. 645–651.
[94] Dalla Torre, F., Van Swygenhoven, H., and Victoria, M. (2002), Nanocrystalline Electrodeposited Ni: Microstructure and Tensile Properties, *Acta Mater.*, Vol. 50, pp. 3957–3970.
[95] Dao, M., and Asaro, R. J. (1993), Non-Schmid Effects and Localized Plastic-Flow in Intermetallic Alloys, *Mat. Sci. Eng. A*, Vol. 170, pp. 143–160.
[96] Dao, M., and Asaro, R. J. (1994), Coarse Slip Bands and the Transition to Macroscopic Shear Bands, *Scripta Metall.*, Vol. 30, pp. 791–796.
[97] Dao, M., and Asaro, R. J. (1996), Localized Deformation Modes and non-Schmid Effects in Crystalline Solids. Part I: Critical Conditions for Localization, *Mech. Mater.*, Vol. 23, pp. 71–102.
[98] D'Arcy Thomson, W. (1942), *On Growth and Form*, Cambridge University Press, Cambridge.
[99] Datta, A. K. (2002), *Biological and Bioenvironmental Heat and Mass Transfer*, Marcel Dekker, New York.
[100] De Borst, R., and Van der Giessen, E., eds. (1998), *Material Instabilities in Solids*, Wiley, Chichester, UK.
[101] DeHoff, R. T. (1993), *Thermodynamics in Materials Science*, McGraw-Hill, New York.
[102] Denbigh, K. (1981), *The Principles of Chemical Equilibrium*, 4th ed., Cambridge University Press, Cambridge.
[103] Desai, C. S., and Siriwardane, H. J. (1984), *Constitutive Laws for Engineering Materials – with Emphasis on Geological Materials*, Prentice Hall, Englewood Cliffs, New Jersey.

Bibliography

[104] Dhaliwal, R. S., and Singh, A. (1987), Micropolar Thermoelasticity. In *Thermal Stresses II* (R. B. Hetnarski, ed.), pp. 269–328, Elsevier, Amsterdam.

[105] Dickerson, R. E. (1969), *Molecular Thermodynamics*, The Benjamin/Cummings Publ., Menlo Park, California.

[106] Dillamore, I. L., Roberts, J. G., and Bush, A. C. (1979), Occurence of Shear Bands in Heavily Rolled Cubic Metals, *Metal Sci.*, Vol. 13, pp. 73–77.

[107] DiMaggio, F. L., and Sandler, I. S. (1971), Material Model for Granular Soils, *J. Engry. Mech.*, ASCE, Vol. 97, pp. 935–950.

[108] Doghri, I. (2000), *Mechanics of Deformable Solids*, Springer-Verlag, Berlin.

[109] Doyle, T. C., and Ericksen, J. L. (1956), Nonlinear Elasticity, *Adv. Appl. Mech.*, Vol. 4, pp. 53–115.

[110] Drucker, D. C. (1951), A More Fundamental Approach to Plastic Stress-Strain Relations. In *Proc. 1st Natl Congr. Appl. Mech.* (E. Sternberg, ed.), ASME, New York, pp. 487–491.

[111] Drucker, D. C. (1959), A Definition of Stable Inelastic Material, *J. Appl. Mech.*, Vol. 26, pp. 101–106.

[112] Drucker, D. C. (1960), Plasticity. In *Structural Mechanics – Proc. 1st Symp. Naval Struct. Mechanics* (J. N. Goodier and N. J. Hoff, eds.), Pergamon, New York, pp. 407–455.

[113] Drucker, D. C., and Prager, W. (1952), Soil Mechanics and Plastic Analysis or Limit Design, *Q. Appl. Math.*, Vol. 10, pp. 157–165.

[114] Duffy, J. D. (1983), Strain Rate History Effects and Dislocation Structure. In *Material Behavior under High Stress and Ultrahigh Loading Rates* (J. Mescall and V. Weiss, eds.), pp. 21–37, Plenum, New York.

[115] Dundurs, J. (1969), Elastic Interactions of Dislocations with Inhomogeneities. In *Mathematical Theory of Dislocations* (T. Mura, ed.), ASME, New York, pp. 70–115.

[116] Dundurs, J., and Markenscoff, X. (1989), The Sternberg–Koiter Conclusion and other Anomalies of the Concentrated Couple, *J. Appl. Mech.*, Vol. 56, pp. 240–245.

[117] Dundurs, J., and Markenscoff, X. (1993), Invariance of Stresses under a Change in Elastic Compliances, *Proc. Roy. Soc. Lond. A*, Vol. 443, pp. 289–300.

[118] Dundurs, J., and Mura, T. (1964), Interaction Between Edge Dislocation and a Circular Inclusion, *J. Mech. Phys. Solids*, Vol. 12, pp. 177–189.

[119] Duszek, M. K., and Perzyna, P. (1991), On Combined Isotropic and Kinematic Hardening Effects in Plastic Flor Processes, *Int. J. Plasticity*, Vol. 9, pp. 351–363.

[120] Ebrahimi, F., Bourne, G. R., Kelly, M. S., and Matthews, T. E. (1999), Mechanical Properties of Nanocrystalline Nickel Produced by Electrodeposition, *Nanostr. Mater.*, Vol. 11, pp. 343–350.

[121] Elam, C. F. (1926), Tensile Tests of Large Gold, Silver, and Copper Crystals, *Proc. Roy. Soc. Lond., Ser. A*, Vol. 112, pp. 289–294.

[122] Elam, C.F. (1927a), Tensile Tests on Alloy Crystals. Part II: Solid Solution Alloys of Copper and Zinc, *Proc. Roy. Soc. Lond., Ser. A*, Vol. 115, pp. 148–169.

[123] Elam, C. F. (1927b), Tensile Tests on Crystals. Part IV: A Copper Alloy Containing Five Per Cent Aluminum, *Proc. Roy. Soc. London, Ser. A*, Vol. 13, pp. 73–77.

[124] Elam, C. F. (1936), The Distortion of β-Brass and Iron Crystals, *Proc. Roy. Soc. London, Ser. A*, Vol. 153, pp. 273–301.

[125] Elsherik, A. M., Erb, U., Palumbo, G., and Aust, K. T. (1992), Deviations from Hall-Petch Behavior in as-prepared Nanocrystalline Nickel, *Scripta Metall.*, Vol. 27, pp. 1185–1188.

[126] Embury, J. D. (1971), Strengthening by Dislocation Substructures. In *Strengthening Methods in Crystals* (A. Kelly and R. B. Nicholson, eds.), p. 331, Appl. Science Publ., London.

[127] Ericksen, J. L. (1960), Tensor Fields. In *Hanbuch der Physik* (S. Flugge, ed.), Band III/1, Springer-Verlag, Berlin.

[128] Ericksen, J. L. (1977), Special Topics in Elastostatics, *Adv. Appl. Mech.*, Vol. 17, pp. 189–244.

[129] Ericksen, J. L. (1991), *Introduction to the Thermodynamics of Solids*, Chapman and Hall, London.

[130] Eringen, A. C. (1967), *Mechanics of Continua*, Wiley, New York.

[131] Eringen, A. C. (1971), Tensor Analysis. In *Continuum Physics* (A. C. Eringen, ed.), Academic Press, New York.

[132] Eringen, A. C. (1999), *Microcontinuum Field Theories*, Springer-Verlag, New York.

[133] Eringen, A. C., and Suhubi, E. S. (1964), Nonlinear Theory of Simple Microelastic Solids, *Int. J. Eng. Sci.*, Vol. 2, I: pp. 189–203, II: pp. 389–404.

[134] Eshelby, J. D. (1951), The Force on an Elastic Singularity, *Phil. Trans. Roy. Soc. A*, Vol. 244, pp. 87–112.

[135] Eshelby, J. D. (1956), The Continuum Theory of Lattice Defects. In *Solid State Physics* (F. Seitz and D. Turnbull, eds.), Vol. 3, Academic Press, New York, pp. 79–144.

[136] Eshelby, J. D. (1957), The Determination of the Elastic Field of an Ellipsoidal Inclusion, and Related Problems, *Proc. Roy. Soc. Lond. A*, Vol. 241, pp. 376–396.

[137] Eshelby, J. D. (1961), Elastic Inclusions and Inhomogeneities. In *Progress in Solid Mechanics*, Vol. 2 (I. N. Sneddon and R. Hill, eds.), pp. 87–140.

[138] Eshelby, J. D. (1970), Energy Relations and the Energy-Momentum Tensor in Continuum Mechanics. In *Inelastic Behavior of Solids* (M. F. Kanninen, W. F. Adier, A. R. Rosenfield, and R. I. Janee, eds.), pp. 77–115, McGraw-Hul, New York.

[139] Eshelby, J. D. (1975), The Elastic Energy-Momentum Tensor, *J. of Elasticity*, Vol. 5, pp. 321–335.

[140] Eshelby, J. D., Read, W. T., and Shockley, W. (1953), Anisotropic Elasticity with Applications to Dislocation Theory, *Acta Metall.*, Vol. 1, pp. 251–259.

[141] Evans, E. A., and Skalak, R. (1980), *Mechanics and Thermodynamics of Biomembranes*, CRC Press, Boca Raton, Florida.

[142] Ewing, J. A., and Rosenhain, W. (1899), Experiments in Micro-Metallurgy: Effects of Strain, Preliminary Notice, *Proc. Roy. Soc. Lond.*, Vol. 65, pp. 85–90.

[143] Ewing, J. A., and Rosenhain, W. (1900), The Crystalline Structure of Metals, *Proc. Roy. Soc. Lond.*, Vol. 193, pp. 353–375.

[144] Filonenko-Borodich, M. (1964), *Theory of Elasticity*, P. Noordhoff, Groningen, Netherlands.

[145] Fischer-Cripps, A. C. (2000), *Introduction to Contact Mechanics*, Springer-Verlag, New York.

[146] Fitzgerald, E. A. (1991), Dislocations in Strained-Layer Epitaxy: Theory, Experiment, and Applications, *Mater. Sci. Reports*, Vol. 7, pp. 87–142.

[147] Fleck, N. A., and Hutchinson, J. W. (1997), Strain Gradient Plasticity, *Adv. Appl. Mech.*, Vol. 33, pp. 295–361.

[148] Follansbee, P. S., and Kocks, U. F. (1988), A Constitutive Description of the Deformation of Copper based on the Use of the Mechanical Threshold Stress as an Internal State Variable, *Acta Metall.*, Vol. 36, pp. 81–93.

[149] Follansbee, P. S., Regazzoni, G., and Kocks, U. F. (1984), Mechanical Properties of Materials at High Rates of Strain, *Inst. Phys.*, Conf. Ser. 70, pp. 71–81.

[150] Franciosi, P., Berveiller, M., and Zaoui, A. (1980), Latent Hardening in Copper and Aluminum Singel Crystals, *Acta Metall.*, Vol. 28, pp. 273–283.

[151] Franciosi, P., and Zaoui, A (1982), Multislip Tests on Copper Crystals: A Junctions Hardening Effect, *Acta Met.*, Vol. 30, pp. 2141–2151.

[152] Frank, F. C., and Read, W. T., Jr. (1950), Multiplication Processes for Slow Moving Dislocations, *Phys. Rev.*, Vol. 79, pp. 722–724.

[153] Frenkel, J. (1926), Zur Theorie der Elastizitätsgrenze und der Festigheit kristallinischer Körper, *Z. Phys.*, Vol. 37, pp. 572–609.

[154] Freund, L. B. (1978), Stress Intensity Factor Calculations Based on a Conservation Integral, *Int. J. Solids Struct.*, Vol. 14, pp. 241–250.

[155] Freund, L. B. (1987), The Stability of a Dislocation Threading a Strained Layer On A Substrate, *J. Appl. Mech.*, Vol. 54, pp. 553–557.

[156] Freund, L. B. (1990), The Driving Force for Glide of A Threading Dislocation in a Strained Epitaxial Layer on a Substrate, *J. Mech. Phys. Solids*, Vol. 38, pp. 657–679.

[157] Freund, L. B. (1993), The Mechanics of Dislocations in Strained-Layer Semiconductor Materials, *Adv. Appl. Mech.*, Vol. 30, pp. 1–66.

[158] Freund, L. B. (1990), *Dynamic Fracture Mechanics*, Cambridge University Press, Cambridge.

Bibliography

[159] Freund, L. B., and Suresh, S. (2003), *Thin Film Materials: Stress, Defect Formation and Surface Evolution*, Cambridge University Press, New York.

[160] Friedel, J. (1964), *Dislocations*, Pergamon, New York.

[161] Fung, Y.-C. (1965), *Foundations of Solid Mechanics*, Prentice Hall, Englewood Cliffs, New Jersey.

[162] Fung, Y.-C. (1973), Biorheology of Soft Tissues, *Biorheology*, Vol. 10, pp. 139–155.

[163] Fung, Y.-C. (1990), *Biomechanics: Motion, Flow, Stress, and Growth*, Springer-Verlag, New York.

[164] Fung, Y.-C. (1993), *Biomechanics: Mechanical Properties of Living Tissues*, 2nd ed., Springer-Verlag, New York.

[165] Fung, Y.-C. (1995), Stress, Strain, Growth, and Remodeling of Living Organisms, *Z. Angew. Math. Phys.*, Vol. 46, pp. S469–S482.

[166] Galin, L. A. (1961), *Contact Problems in the Theory of Elasticity*, School of Physical Sciences and Applied Mathematics, North Carolina State College.

[167] Gao, H., and Nix, W. D. (1999), Surface Roughening of Heteroepitaxial Thin Films, *Ann. Rev. Mater. Sci.*, Vol. 29, pp. 173–209.

[168] Gaskell, D. R. (2003), *Introduction to the Thermodynamics of Materials*, 4th ed., Taylor & Francis, New York.

[169] Gdoutos, E. E. (1990), *Fracture Mechanics Criteria and Applications*, Kluwer, Dordrecht, The Netherlands.

[170] Gdoutos, E. E. (1993), *Fracture Mechanics: An Introduction*, Kluwer, Dordrecht, The Netherlands.

[171] Germain, P., Nguyen, Q. S., and Suquet, P. (1983), Continuum Thermodynamics, *J. Appl. Mech.*, Vol. 50, pp. 1010–1020.

[172] Gertsman, V. Y., Hoffman, M., Gleiter, H., and Birringer, R. (1994), The Study of Grain-Size Dependence of Yield Stress of Copper for a Wide Grain-Size Range, *Acta Metall.*, Vol. 42, pp. 3539–3544

[173] Gladwell, G. M. L. (1980), *Contact Problems in the Classical Theory of Elasticity*, Alphen aan den Rijn, The Netherlands.

[174] Gleiter, H. (1989), Nanocrystalline Materials, *Progr. Mater. Sci.*, Vol. 33, pp. 223–315.

[175] Goldberg, R. R. (1961), *Fourier Transforms*, Cambridge University Press, Cambridge, UK.

[176] Gosling, T. J., Jain, S. C., Willis, J. R., Atkinson, A., and Bullough, R. (1992), Stable Configurations in Strained Epitaxial Layers, *Phil. Mag. A*, Vol. 66, pp. 119–132.

[177] Gray G. T., *et al.* (1997), Influence of Strain Rate & Temperature on the Mechanical Response of Ultrafine-Grained Cu, Ni, and Al-4Cu-0.5Zr, *Nanostructured Materials*, Vol. 9, pp. 477–480.

[178] Green, A. E., and Adkins, J. E. (1960), *Large Elastic Deformations*, Oxford University Press, Oxford, UK.

[179] Green, A. E., and Zerna, W. (1954), *Theoretical Elasticity*, Oxford University Press, London.

[180] Green, A. E., and Naghdi, P. M. (1965), A Dynamical Theory of Interacting Continua, *Int. J. Eng.*, Vol. 3, pp. 231–241.

[181] Grioli, G. (1960), Elasticity Asimmetrica, *Ann. Mat. Pura Appl., Ser. IV*, Vol. 50, pp. 389–417.

[182] Günther, W. (1958), Zur Statik und Kinematik des Cosseratschen Kontinuums, *Abh. Braunschw. Wiss. Ges.*, Vol. 10, pp. 195–213.

[183] Günther, W. (1962), Uber einige Randintegrale der Elastomechanik, *Abh. Braunschw. Wiss. Ges.*, Vol. 14, pp. 53–72.

[184] Gurson, A. L. (1977), Continuum Theory of Ductile Rapture by Void Nucleation and Growth. Part I: Yield Criteria and Flow Rules for Porous Ductile Media, *J. Eng. Mater. Tech.*, Vol. 99, pp. 2–15.

[185] Gurtin, M. E. (1981a), *Topics in Finite Elasticity*, SIAM, Philadelphia.

[186] Gurtin, M. E. (1981b) *An Introduction to Continuum Mechanics*, Academic Press, New York.

[187] Hadamar, J. (1903), *Leçons Sur la Propagation des Ondes et les Équations de la L'Hydrodinamique*, Hermann, Paris.

[188] Haddow, J. B., and Ogden, R. W. (1990), Thermoelasticity of Rubber-like Solids at Small Strain. In *Elasticity – Mathematical Methods and Applications* (G. Eason and R. W. Ogden, eds.), pp. 165–179, Ellis Horwood, Chichester, UK.

[189] Hall, E. O. (1951), The Deformation and Aging of Mild Steel: III. Discusion of Results, *Proc. Phys. Soc. London, Sec. B*, Vol. 64, pp. 747–753.

[190] Han, W., and Reddy, B. D. (1999), *Plasticity – Mathematical Theory and Numerical Analysis*, Springer-Verlag, New York.

[191] Harren, S. V. (1989), The Finite Deformation of Rate-Dependent Polycrystals – I and II, *J. Mech. Phys. Solids*, Vol. 39, pp. 345–383.

[192] Harren, S. V., and Asaro, R. J. (1989), Nonuniform Deformation in Polycrystals and Aspects of the Validity of the Taylor Model, *J. Mech. Phys. Solids*, Vol. 37, pp. 191–232.

[193] Harren, S. V., Deve, H. E., and Asaro, R. J. (1988), Shear Band Formation in Plane Strain Compression, *Acta Metall.*, Vol. 36, pp. 2435–2480.

[194] Harren, S. V., Lowe, T. C., Asaro, R. J., and Needlemann, A. (1989), Analysis of Large-Strain Shear in Rate-Dependent Face-Centered Cubic Polycrystals: Correlation of Micro- and Macromechanics, *Phil. Trans. Roy. Soc. London, Ser. A*, Vol. 328, pp. 442–500.

[195] Haupt, P., and Kurth, J. A. (2002), *Continuum Mechanics and Theory of Materials*, Springer-Verlag, Wien.

[196] Havner, K. S. (1971), A Discrete Model for the Prediction of Subsequent Yield Surfaces in Polycrystalling Plasticity, *Int. J. Solids Struct.*, Vol. 7, pp. 179–730.

[197] Havner, K. S. (1992), *Finite Plastic Deformation of Crystalline Solids*, Cambridge University Press, Cambridge, UK.

[198] Head, A. K. (1953a), The Interaction of Dislocations and Boundaries, *Phil. Mag.*, Vol. 44, pp. 92–64.

[199] Head, A. K. (1953b), Edge Dislocations in Inhomogeneous Media, *Proc. Phys. Soc. Lond.*, Vol. B66, pp. 793–801.

[200] Hencky, H. (1924), Zur Theorie Plastischer Deformationen und der Hierdurch in Material Hervorgerufenen Nachspannungen, *Z. Angew. Math. Mech.*, Vol. 4, pp. 323–334.

[201] Heredia, F., and Pope, D. P. (1988), Improving the Ductility of Single Crystalline Ni_3Al with Baron Additions, *J. Metals*, Vol. 40, p. 107.

[202] Hill, R. (1948), A Theory of the Yielding and Plastic Flow of Anisotropic Metals, *Proc. Roy. Soc. Lond. A*, Vol. 193, pp. 281–297.

[203] Hill, R. (1950), *The Mathematical Theory of Plasticity*, Oxford University Press, London.

[204] Hill, R. (1958), A General Theory of Uniqueness and Stability in Elastic-Plastic Solids, *J. Mech. Phys. Solids*, Vol. 6, pp. 236–249.

[205] Hill, R. (1962), Acceleration Waves in Solids, *J. Mech. Phys. Solids*, Vol. 10, pp. 1–16.

[206] Hill, R. (1965a), Continuum Micro-Mechanics of Composite Materials, *J. Mech. Phys. Solids*, Vol. 13, pp. 89–101.

[207] Hill, R. (1965b), A Self-Consistent Michanics of Composite Materials, *J. Mech. Phys. Solids*, Vol. 13, pp. 213–222.

[208] Hill, R. (1967), The Essential Structures of Constitutive Laws for Metal Composites and Polycrystals, *J. Mech. Phys. Solids*, Vol. 15, pp. 79–95.

[209] Hill, R. (1972), On Constitutive Macro-Variables for Heterogeneous Solids at Finite Strain, *Proc. Roy. Soc. Lond. A*, Vol. 326, pp. 131–147.

[210] Hill, R. (1978), Aspects of Invariance in Solid Mechanics, *Adv. Appl. Mech.*, Vol. 18, pp. 1–75.

[211] Hill, R. (1981), Invariance Relations in Thermoelasticity with Generalized Variables, *Math. Proc. Camb. Phil. Soc.*, Vol. 90, pp. 373–384.

[212] Hill, R., and Havner, K. (1982), Perspectives in the Mechanics of Elastoplastic Crystals, *J. Mech. Phys. Solids*, Vol. 30, pp. 5–22.

[213] Hill, R., and Hutchinson, J. W. (1975), Bifurcation Phenomena in the Plane Tension Test, *J. Mech. Phys. Solids*, Vol. 23, pp. 239–264.

[214] Hill, R., and Rice, J. R. (1972), Constitutive Analysis of Elastic-Plastic Crystals at Arbitrary Strains, *J. Mech. Phys. Solids*, Vol. 20, pp. 401–413.

Bibliography

[215] Hill, R., and Rice, J. R. (1973), Elastic Potentials and the Structure of Inelastic Constitutive Laws, *SIAM J. Appl. Math.*, Vol. 25, pp. 448–461.

[216] Hirth, J. P., and Lothe, J. (1982), *Theory of Dislocations*, 2nd ed., Wiley, New York.

[217] Hoffman, K., and Knuze, R. (1961), *Linear Algebra*, Prentice Hall, Englewood Cliffs, New Jersey.

[218] Hoger, A. (1987), The Stress Conjugate to the Logarithmic Strain, *Int. J. Solids Struct.*, Vol. 23, pp. 1645–1656.

[219] Holzapfel, G. A. (2000), *Nonlinear Solid Mechanics*, Wiley, Chichester, UK.

[220] Holzapfel, G. A., Gasser, T. C., and Ogden, R. W. (2000), A New Constitutive Framework for Arterial Wall Mechanics and a Comparative Study of Material Models, *J. of Elasticity*, Vol. 61, pp. 1–48.

[221] Holzapfel, G. A., and Weizsacker, H. W. (1998), Biomechanical Behavior of the Arterial Wall and its Numerical Characterization, *Computers in Biology and Medicine*, Vol. 28, pp. 377–392.

[222] Hsu, F. (1968), The Influences of Mechanical Loads on the Form of a Growing Elastic Body, *Biomechanics*, Vol. 1, pp. 303–311.

[223] Huang, Y., Hutchinson, J. W., and Tvergaard, V. (1991), Cavitation Instabilities in Elastic-Plastic Solids, *J. Mech. Phys. Solids*, Vol. 39, pp. 223–241.

[224] Hughes, G. D., Smith, S. D., Pande, C. S., Johnson, H. R., and Armstrong, R. W. (1986), Hall-Petch Strengthening for the Microhardness of 12 Nanometer Grain Diameter Electrodeposited Nickel, *Scripta Metall.*, Vol. 20, pp. 93–97.

[225] Hull, A. W. (1919), The Positions of Atoms in Metals, *Proc. Am. Inst. Elec. Eng.*, Vol. 38, pp. 1171–1192.

[226] Hull, D., and Bacon, D. J. (1999), *Introduction to Dislocations*, 3rd ed., Butterworth-Heinemann, Boston.

[227] Humphrey, J. D. (1995), Mechanics of the Arterial Wall: Review and Directions, *Critical Reviews in Biomed. Eng.*, Vol. 23, pp. 1–162.

[228] Humphrey, J. D. (2002), *Cardiovascular Solid Mechanics: Cells, Tissues, and Organs*, Springer-Verlag, New York.

[229] Humphrey, J. D. (2003), Continuum Thermomechanics and the Clinical Treatment of Disease and Injury, *Appl. Mech. Rev.*, Vol. 56, pp. 231–260.

[230] Hunter, S. C. (1983), *Mechanics of Continuous Media*, Ellis Horwood, Chichester, UK.

[231] Hutchinson, J. W. (1964a), Plastic Stress-Strain Relations of F.C.C. Polycrystalline Metals Hardening According to Taylor Rule, *J. Mech. Phys. Solids*, Vol. 12, pp. 11–24.

[232] Hutchinson, J. W. (1964b), Plastic Deformation of B.C.C. Polycrystals, *J. Mech. Phys. Solids*, Vol. 12, pp. 25–33.

[233] Hutchinson, J. W. (1970), Elastic-Plastic Behavior of Polycrystalline Metals and Composites, *Proc. Roy. Soc. London*, Vol. A319, pp. 247–275.

[234] Hutchinson, J. W. (1976), Bounds and Self-Consistent Estimates for Creep of Polycrystalline Materials, *Proc. Roy. Soc. London*, Vol. A319, pp. 347–272.

[235] Ilyushin, A. A. (1947), Theory of Plasticity at Simple Loading of the Bodies Exhibiting Plastic Hardening, *Prikl. Mat. Mekh.*, Vol. 11, pp. 291–296 (in Russian).

[236] Ilyushin, A. A. (1961), On the Postulate of Plasticity, *Prikl. Mat. Mekh.*, Vol. 25, pp. 503–507 (in Russian).

[237] Ilyushin, A. A. (1963), Plasticity, Foundations of the Grand Mathematicl Theory, *Izd. Akd. Nauk SSSR*, Moscow (in Russian).

[238] Indenbom, V. L., and Lothe, J. (1992), *Elastic Strain Fields and Dislocation Mobility*, North-Holland, Amsterdam.

[239] Inui, H., Oh, M. H., Nakamura, A., and Yamaguchi, M. (1992), Room-Temperature Tensile Deformation of Polycrystalline Twinned (PST) Crystals of TiAl, *Acta Metall.*, Vol. 40, pp. 3095–3104.

[240] Iyer, R. S., Frey, C. A., Sastry, S. M. L., Waller, B. E., and Buhro, W. E. (1999), Plastic Deformation of Nanoscale Cu and Cu-Co-0.2wt%B, *Mater. Sci. Eng. A*, Vol. 264, pp. 210–214.

[241] Jackson, P. J., and Basinski, Z. S. (1967), Latent Hardening and the Flow Stress in Copper Single Crystals, *Can. J. Phys.*, Vol. 45, pp. 707–735.

[242] Jaeger, J. C., and Cook, N. G. W. (1976), *Fundamentals of Rock Mechanics*, Chapman and Hall, London.

[243] Jain, S. C., Gosling, T. J., Willis, J. R., Totterdell, D. H. J., and Bullough, R. (1992), A New Study of Critical Layer Thickness, Stability and Strain Relaxation in Pseudomorphic Ge$_z$Si$_{1-x}$ Strained Epilayers, *Phil. Mag. A.*, Vol. 65, pp. 1151–1167.

[244] Jasiuk, I., and Ostoja-Starzewski, M. (1995), Planar Cosserat Elasticity of Materials with Holes and Intrusions, *Appl. Mech. Rev.*, Vol. 48, pp. S11–S18.

[245] Jauzemis, W. (1967), *Continuum Mechanics*, The Macmillan, New York.

[246] Johnson, G. R., and Cook, W. H. (1983), A Constitutive Model and Data for Metals Subjected to Large Strain Rates, and High Temperatures. In *Proceedings of the 7th International Sysposium on Ballistics*, pp. 1–7, ADPA, The Hague.

[247] Johnson, K. L. (1985), *Contact Mechanics*, Cambridge University Press, New York.

[248] Johnson, W., and Mellor, P. B. (1973), *Engineering Plasticity*, Van Nostrand Reinhold, London.

[249] Joshi, N. R., and Green, R. E., Jr. (1980), Continuous X-Ray Diffraction Measurement of Lattice Rotation during Tensile Deformation of Aluminum Crystals, *J. Mat. Sci.*, Vol. 15, pp. 729–738.

[250] Kachanov, L. M. (1971), *Foundation of Theory of Plasticity*, North-Holland, Amsterdam.

[251] Kanninen, M. F., and Popelar, C. H. (1985), *Advanced Fracture Mechanics*, Oxford University Press, New York.

[252] Kawabata, T., Kanai, T., and Izumi, O. (1985), Positive Temperature-Dependence of the Yield Stress in TiAl(L1$_0$) Type Superlattice Intermetallic Compound Single-Crystals at 293-1273 K, *Acta Metall.*, Vol. 33, pp. 1355–1366.

[253] Kelly, P. D. (1964), A Reacting Continuum, *Int. J. Eng. Sci.*, Vol. 2, pp. 129–153.

[254] Kelly, A., Groves, G. W., and Kidd, P. (2000), *Crystallography and Crystal Defects*, Wiley, New York.

[255] Kelly, A., and Nicholson, R. B., eds. (1971), *Strengthening Methods in Crystals*, Appl. Science Publ., London.

[256] Kestin, J. (1979), *A Course in Thermodynamics*, McGraw-Hill, New York.

[257] Kestin, J., and Rice, J. R. (1970), Paradoxes in the Application of Thermodynamics to Strained Solids. In *A Critical Review of Thermodynamics* (E. B. Stuart, B. Gal-Or, and A. J. Brainard, eds.), pp. 275–298, Mono-Book, Baltimore.

[258] Khan, A. S., and Huang, S. (1995), *Continuum Theory of Plasticity*, Wiley, New York.

[259] Kienzler, R., and Maugin, G. A., eds. (2001), *Configurational Mechanics of Materials*, Sringer-Verlag, Wien.

[260] Kittel, C. (1996), *Introduction to Solid State Physics*, 7th ed., Wiley, New York.

[261] Knapp, J. A., and Follstaedt, D. M. (2004), Hall-Petch Relationship in Pulsed-Laser Deposited Nickel Films, *J. Mater. Res.*, Vol. 19, pp. 218–227.

[262] Knowles, J. K., and Sternberg, E. (1972), On a Class of Conservation Laws in Linearized and Finite Elastostatics, *Arch. Ration. Mech. Anal.*, Vol. 44, pp. 187–211.

[263] Kochendörfer, A. (1941), *Reine und angewandte Metallkunde*, Springer-Verlag, Berlin.

[264] Kocks, U. F. (1964), Latent Hardening and Secondary Slip in Aluminum and Silver, *Trans. Met. Soc.*, AIME, Vol. 230, pp. 1160–1167.

[265] Kocks, U. F. (1970), The Relation Between Polycrystal Deformation and Single-Crystal Deformation, *Metall. Trans.*, Vol. 1, pp. 1121–1142.

[266] Kocks, U. F. (1987), Constitutive Behavior Based on Crystal Plasticity. In *Unified Constitutive Equations for Creep and Plasticity* (A. K. Miller, ed.), pp. 1–88, Elsevier, London.

[267] Kocks, U. F., and Brown, T. J. (1966), Latent Hardening in Aluminum, *Acta Metall.*, Vol. 14, pp. 87–98.

Bibliography

[268] Kocks, U. F., Tomé, C. N., and Wenk, H.-R. (1998), *Texture and Anisotropy: Preferred Orientations in Polycrystals and Their Effect on Materials Properties*, Cambridge University Press, Cambridge, UK.

[269] Koehler, J. S. (1952), The Nature of Work-Hardening, *Phys. Rev.*, Vol. 86, pp. 52–59.

[270] Koiter, W. T. (1964), Couple-Stresses in the Theory of Elasticity, *Proc. Ned. Akad. Wet. (B)*, Vol. 67, I: pp. 17–29, II: pp. 30–44.

[271] Kovalenko, A. D. (1969), *Thermoelasticity*, Wolters–Noordhoff, Groningen, The Netherlands.

[272] Krajcinovic, D. (1996), *Damage Mechanics*, Elsevier, New York.

[273] Krausz, A. S., and Eyring, H. (1975), *Deformation Kinetics*, Wiley, New York.

[274] Krieg, R. D. (1975), A Practical Two Surface Plascitiy Theory, *J. Appl. Mech.*, Vol. 42, pp. 641–646.

[275] Kuhlman-Wilsdorf, D. (1975), Recent Progress in Understanding of Pure Metal and Alloy Hardening. In *Work Hardening in Tension and Fatigue* (A. W. Thompson, ed.), p. 1, AIME, New York.

[276] Kunin, I. A., and Sosnina, E. G. (1972), Ellipsoidal Inhomogeneity in an Elastic Medium, *Sov. Phys. Doklady*, Vol. 16, p. 534–536.

[277] Kumar, K. S., Van Swygenhoven, H., and Suresh, S. (2003), Mechanical Behavior of Nanocrystalline Metals and Alloys, *Acta Mater.*, Vol. 51, pp. 5743–5774.

[278] Lai, W. M., Rubin, D., and Krempl, E. (1993), *Introduction to Continuum Mechanics*, Pergamon, New York.

[279] Landau, L. D., and Lifshitz, E. M. (1986), *Theory of Elasticity*, 3rd English ed., Pergamon, New York.

[280] Lee, E. H. (1969), Elastic-Plastic Deformation at Finite Strains, *J. Appl. Mech.*, Vol. 36, pp. 1–6.

[281] Legros, M., Elliott, B. R., Rittner, M. N., Weertman, J. R., and Hemker, K. J. (2000), Microsample Tensile Testing of Nanocrystalline Metals, *Phil. Mag. A*, Vol. 80, pp. 1017–1026.

[282] Lehmann, Th., and Liang, H. Y. (1993), The Stress Conjugate to the Logarithmic Strain, *Z. angnew. Math. Mech.*, Vol. 73, pp. 357–363.

[283] Leigh, D. C. (1968), *Nonlinear Continuum Mechanics*, McGraw-Hill, New York.

[284] Lekhnitskii, S. G. (1981), *Theory of Elasticity of an Anisotropic Elastic Body*, Mir Publishers, Moscow.

[285] Lighthill, M. J. (1955), *Introduction to Fourier Analysis and Generalized Functions*, Cambridge University Press, Cambridge.

[286] Lin, T. H. (1964), Slip and Stress Fields of a Polycrystalline Aggregate at Different Stages of Loading, *J. Mech. Phys. Solids*, Vol. 12, pp. 391–408.

[287] Little, R. W. (1973), *Elasticity*, Prentice Hall, Englewood Cliffs, New Jersey.

[288] Liu, S. Q., and Fung, Y.-C. (1988), Zero-Stress States of Arteries, *J. Biomech. Eng.*, Vol. 110, pp. 82–84.

[289] Liu, S. Q., and Fung, Y.-C. (1989), Relationaship Betwen Hypertension, Hypertrophy, and Opening Angle of Zero-Stress State of Arteries Following Aortic Constriction, *J. Biomech. Eng.*, Vol. 111, pp. 325–335.

[290] Lomer, W. M. (1951), A Dislocation Reaction in the Face-Centered Cubic Lattice, *Phil. Mag.*, Vol. 41, pp. 1327–1331.

[291] Love, A. E. H. (1944), *A Treatise on the Mathematical Theory of Elasticity*, Dover, New York.

[292] Lu, L., Li, S. X., and Lu, K. (2001), An Abnormal Strain Rate Effect on Tensile Behavior in Nanocrystalline Copper, *Scripta Mater.*, Vol. 45, pp. 1163–1169.

[293] Lu, L., Schwaiger, R., Shan, Z. W., Dao, M., Lu, K., and Suresh, S. (2005), Nano-Sized Twins Induce High Rate Sensitivity of Flow Stress in Pure Copper, *Acta Mater.*, Vol. 53, pp. 2169–2179.

[294] Lu, L., Shen, Y., Chen, X., and Lu, K. (2004), Ultrahigh Strength and High Electrical Conductivity in Copper, *Science*, Vol. 304, pp. 422–426.

[295] Lubarda, V. A. (1997), Energy Analysis of Dislocation Arrays Near Bimaterial Interfaces, *Int. J. Solids Struct.*, Vol. 34, pp. 1053–1073.

[296] Lubarda, V. A. (1999), Dislocations Arrrays at the Interface between an Epitaxial Layer and Its Substrate, *Math. Mech. Solids*, Vol. 4, pp. 411–431.

[297] Lubarda, V. A. (2002), *Elastoplasticity Theory*, CRC Press, Boca Raton, Florida.

[298] Lubarda, V. A. (2003a), The Effects of Couple Stresses on Dislocation Strain Energy, *Int. J. Solids Struct.*, Vol. 40, pp. 3807–3826.

[299] Lubarda, V. A. (2003b), Circular Inclusions in Anti-Plane Strain Couple Stress Elasticity, *Int. J. Solids Struct.*, Vol. 40, pp. 3827–3851.

[300] Lubarda, V. A. (2004), Constitutive Theories Based on the Multiplicative Decomposition of Deformation Gradient: Thermoelasticity, Elastoplasticity and Biomechanics, *Appl. Mech. Rev.*, Vol. 57, pp. 95–108.

[301] Lubarda, V. A., and Hoger, A. (2002), On the Mechanics of Solids with A Growing Mass, *Int. J. Solids Struct.*, Vol. 39, pp. 4627–4664.

[302] Lubarda, V. A., and Lee, E. H. (1981), A Correct Definition of Elastic and Plastic Deformation and its Computational Significance, *J. Appl. Mech.*, Vol. 48, pp. 35–40.

[303] Lubarda, V. A., and Markenscoff, X. (2003), On Conservation Integrals in Micropolar Elasticity, *Phil. Mag. A*, Vol. 83, pp. 1365–1377.

[304] Lubarda, V. A., Mastilovic, S., and Knap, J. (1996), Brittle-Ductile Transition in Porous Rocks by a Cap Model, *ASCE J. Engr. Mech.*, Vol. 122, pp. 633–642.

[305] Lubarda, V. A., and Shih, C. F. (1994), Plastic Spin and Related Issues in Phenomenological Plasticity, *J. Appl. Mech.*, Vol. 61, pp. 524–529.

[306] Lubliner, J. (1990), *Plasticity Theory*, Macmillan, New York.

[307] Lupis, C. H. P. (1983), *Chemical Thermodynamics of Materials*, Prentice Hall, Englewood Cliffs, New Jersey.

[308] Lurie, A. I. (1990), *Nonlinear Theory of Elasticity*, North-Holland, Amsterdam.

[309] Ma, E. (2004), Watching the Nanograins Roll, *Science*, Vol. 305, pp. 623–624.

[310] Malvern, L. E. (1969), *Introduction to the Mechanics of a Continuous Medium*, Prentice Hall, Englewood Cliffs, New Jersey.

[311] Marsden, J. E., and Hughes T. J. R. (1983), *Mathematical Foundations of Elasticty*, Prentice Hall, Englewood Cliffs, New Jersey.

[312] Marsden, J. E., and Tromba, A. J. (2003), *Vector Calculus*, 5th ed., W. H. Freeman and Company, New York.

[313] Markenscoff, X. (1994), Some Remarks on the Wedge Paradox and Saint-Venant's Principle, *J. Appl. Mech.*, Vol. 61, pp. 519–523.

[314] Martin, J. B. (1975), *Plasticity: Fundamentals and General Results*, MIT Press, Cambridge, Massachusetts.

[315] Matthews, J. W. (1979), Misfit Dislocations. In *Dislocations in Solids* (F. R. N. Nabarro, ed.), Vol. 2, pp. 461–545, North-Holland, Amsterdam.

[316] Matthews, J. W., and Blakeslee, A. E., (1974), Defects in Epitaxial Multilayers I. Misfit Dislocatons, *J. Cryst. Growth*, Vol. 27, pp. 118–125.

[317] Maugin, G. A. (1992), *The Thermomechanics of Plasticity and Fracture*, Cambridge University Press, Cambridge.

[318] Maugin, G. A. (1993), *Material Inhomogeneities in Elasticity*, Chapman & Hall, New York.

[319] McClintock, F. A., and Argon, A. S. (1966), *Mechanical Behavior of Materials*, Addison-Wesley, Reading, Massachusetts.

[320] McHugh, P. E., Varias, A. G., Asaro, R. J., and Shih, C. F. (1989), Computational Modeling of Microstructures, *Future Generation Computer Systems*, Vol. 5, pp. 295–318.

[321] McLellan, A. G. (1980), *The Classical Thermodynamics of Deformable Materials*, Cambridge University Press, Cambridge.

[322] Mear, M. E., and Hutchinson, J. W. (1985), Influence of Yield Surface Curvature on Flow Localization in Dilatant Plasticity, *Mech. Mater.*, Vol. 4, pp. 395–407.

[323] Melan, E. (1938), Zur Plastizität des räumlichen Kontinuums, *Ing. Arch.*, Vol. 9, pp. 116–126.

Bibliography

[324] Mindlin, R. D., (1964), Micro-Structure in Linear Elasticity, *Arch. Ration. Mech. Anal.*, Vol. 16, pp. 51–78.

[325] Mindlin, R. D., and Tiersten, H. F. (1962), Effects of Couple-Stresses in Linear Elasticity, *Arch. Ration. Mech. Anal.*, Vol. 11, pp. 415–448.

[326] Minonishi, Y. (1991), Plastic Deformation of Single-Crystals of Ti_3Al with $DO19$ Structure, *Phil. Mag. A*, Vol. 63, pp. 1085–1093.

[327] Mirsky, L. (1955), *An Introduction to Linear Algebra*, Clarendon Press, Oxford, UK.

[328] Mitchell, T. E. (1964), Dislocations and Plasticity in Single Crystals of Face-Centered Cubic Metals and Alloys. In *Progress in Applied Materials Research* (E. G. Stanford, J. H. Fearon, and W. J. McGonnagle, eds.), Vol. 6, pp. 119–237, A Heywood Books, London.

[329] Mitchell, T. E., and Thornton, P. R. (1964), The Detection of Secondary Slip During the Deformation of Copper and α-Brass Single Crystals, *Phil. Mag.*, Vol. 10, pp. 315–323.

[330] Mitra, R., Hoffman R. A., Madan, A., and Weertman, J. R. (2001), Effect of Process Variables on the Structure, Residual Stress, and Hardness of Sputtered Nanocrystalline Nickel Films, *J. Mater. Res.*, Vol. 16, pp. 1010–1027.

[331] Miura, S., and Saeki, Y. (1978), Plastic Deformation of Aluminum Bicrystals 100 Oriented, *Acta Metall.*, Vol. 26, pp. 93–101.

[332] Molinari, A. (1997), Self-Consistent Modelling of Plastic and Viscoplastic Polycrystalline Materials. In *Large Plastic Deformation of Crystalline Aggregates* (C. Teodosiu, ed.), pp. 173–246, Springer-Verlag, Wien.

[333] Molinari, A., Ahzi, S., and Koddane, R. (1997), On the Self-Consistent Modeling of Elastic-Plastic Behavior of Polycrystals, *Mech. Mater.*, Vol. 26, pp. 43–62.

[334] Molinari, A., Canova, G. R., and Ahzi, S. (1987), A Self-Consistent Approach of the Large Deformation Polycrystal Viscoplasticity, *Acta Metall.*, Vol. 35, pp. 2983–2994.

[335] Moran, B., Ortiz, M., and Shih, C. F. (1990), Formulation of Implicit Finite Element Methods for Multiplicative Finite Deformation Plasticity, *Int. J. Numer. Methods Eng.*, Vol. 29, pp. 483–514.

[336] Mori, K., and Nakayama, Y. (1981), Shear Bands in Rolled Copper Single Crystals, *Japan Inst. Metals*, Vol. 22, pp. 857–864.

[337] Morrison, W. B. (1966), The Effect of Grain Size on the Stress-Strain Relationship in Law-Carbon Steel, *Trans. ASM*, Vol. 59, p. 824–846.

[338] Mróz, Z. (1967), On the Description of Anisptropic Work-Hardening, *J. Mech. Phys. Solids*, Vol. 15, pp. 163–175.

[339] Mróz, Z. (1976), A Non-Linear Hardening Model and Its Application to Cyclic Plasticity, *Acta Mech.*, Vol. 25, pp. 51–61.

[340] Müller, I. (1985), *Thermodynamics*, Pitman, Boston.

[341] Mura, T. (1987), *Micromechanics of Defects in Solids*, Martinus Nijhoff, Dordrecht, The Netherlands.

[342] Murnaghan, F. D. (1951), *Finite Deformation of an Elastic Solid*, Wiley, New York.

[343] Musgrave, M. J. P. (1970), *Crystal Acoustics: Introduction to the Study of Elastic Waves and Vibrations in Crystals*, Holden-day, San Francisco.

[344] Muskhelishvili, N. I. (1963), *Some Basic Problems of the Mathematical Theory of Elasticity*, Noordhoff, Groningen, Netherlands.

[345] Naghdi, P. M. (1960), Stress-Strain Relations in Placticity and Thermoplasticity. In *Plascitity – Proc. 2nd Symp. Naval Struct. Mechanics* (E. H. Lee and P. Symonds, eds.), pp. 121–167, Pergamon, New York.

[346] Nabarro, F. R. N., ed. (1979), *Dislocations in Solids*, North-Holland, Amsterdam.

[347] Nabarro, F. R. N. (1987), *Theory of Crystal Dislocations*, Dover, New York.

[348] Needleman, A. (1972), Void Growth in an Elastic-Plastic Medium, *J. Appl. Mech.*, Vol. 39, pp. 964–970.

[349] Needleman, A. (1979), Non-Normality and Bifurcation in Plane Strain Tension and Compression, *J. Mech. Phys. Solids*, Vol. 27, pp. 231–254.

[350] Needleman, A. (1989), Dynamic Shear Band Development in Plane Strain, *J. Appl. Mech.*, Vol. 56, pp. 1–9.

[351] Needleman, A., Tvergaard, V., and Hutchinson, J. W. (1992), Void Growth in Plastic Solids. In *Topics in Fracture and Fatigue* (A. S. Argon, ed.), pp. 145–178, Springer-Verlag, New York.

[352] Nemat-Nasser, S. (1983), On Finite Plastic Flow of Crystalline Solids and Geomaterials, *J. Appl. Mech.*, Vol. 50, pp. 1114–1126.

[353] Nemat-Nasser, S. (2004), *Plasticity: A Treatise on Finite Deformation of Heterogeneous Inelastic Materials*, Cambridge University Press, New York.

[354] Nemat-Nasser, S., and Hori, M. (1999), *Micromechanics: Overall Properties of Heterogeneous Materials*, 2nd ed., Elsevier, Amsterdam.

[355] Nemat-Nasser, S., and Shokooh, A. (1980), On Finite Plastic Flows of Compressible Materials with Internal Friction, *Int. J. Solids Struct.*, Vol. 16, pp. 495–514.

[356] Nieman, G. W., Weertman, J. R., and Siegal, R. W. (1992), Mechanical Behavior of Nanostructured Metals, *Nanostructured Materials*, Vol. 1, pp. 185–190.

[357] Nix, W. D. (1969), Mechanical Properties of Thin Films, *Metall. Trans. A*, Vol. 20, pp. 2217–2245.

[358] Noda, N., Hetnarski, R. B., and Tanigawa, Y. (2003), *Thermal Stresses*, Taylor & Francis, New York.

[359] Noether, E. (1918), Nachr. König. Gessel. Wissen, *Gottingen Math. Phys. Klasse*, Vol. 2, p. 235; Translated in *Transport Theory and Stat. Phys.*, Vol. 1, p. 186, (1971).

[360] Novozhilov, V. V. (1961), *Theory of Elasticity*, Pergamon, New York.

[361] Nowacki, W. (1986), *Theory of Asymmetric Elasticity*, Pergamon, Oxford, UK.

[362] Ogden, R. W. (1982), Elastic Deformations of Rubberlike Solids. In *Mechanics of Solids: The Rodney Hill 60th Anniversary Volume* (H. G. Hopkins and M. J. Sewell, eds.), pp. 499–537, Pergamon, Oxford, UK.

[363] Ogden, R. W. (1984), *Non-Linear Elastic Deformations*, Ellis-Horwood, Chichester, UK.

[364] Olver, P. J. (1986), *Applications of Lie Groups to Differential Equations*, Springer-Verlag, New York.

[365] Orowan, E. (1934), Zur Kristallplastizität. III. Über den Mechanismus des Gleituorganges, *Z. Phys.*, Vol. 89, pp. 634–659.

[366] Orowan, E. (1963), Dislocations in Plasticity. In *The Sorby Centennial Symposium on the History of Metallurgy* (C. S. Smith, ed.), p. 359, Gordon and Breach, New York.

[367] Ortiz, M. (1985), A Constitutive Theory for the Inelastic Behavior of Concrete, *Mech. Mater.*, Vol. 4, pp. 67–93.

[368] Paidar, V., Pope, D. P., and Vitek, V. (1984), A Theory of the Anomalous Yield Behavior in Ll2 Ordered Alloys, *Acta Metall.*, Vol. 32, pp. 435–448.

[369] Palsson, B. Ø., and Bhatia, S. N. (2004), *Tissue Engineering*, Pearson – Prentice Hall, Upper Saddle River, New Jersey.

[370] Pan, J., and Rice, J. R. (1983), Rate Sensitivity of Plastic Flow and Implications for Yield-Surface Vertices, *Int. J. Solids Struct.*, Vol. 19, pp. 973–987.

[371] Papoulis, A. (1962), *The Fourier Integral and Its Application*, McGraw-Hill, New York.

[372] Peach, M., and Koehler, J. S. (1950), The Forces Exerted on Dislocations and the Stress Fields Produced by Them, *Phys. Rev.*, Vol. 80, pp. 436–439.

[373] Peirce, D., Asaro, R. J., and Needleman, A. (1982), An Analysis of Nonuniform and Localized Deformation in Ductile Single Crystals, *Acta Metall.*, Vol. 30, pp. 1087–1119.

[374] Peirce, D., Asaro, R. J., and Needleman, A. (1983), Material Rate Dependency and Localized Deformation in Crystalline Solids, *Acta Metall.*, Vol. 31, pp. 1951–1976.

[375] Petch, N. J. (1953), The Cleavage Strength of Polycrystals, *J. Iron Steel Inst.*, Vol. 174, pp. 25–28.

[376] Piercy, G. R., Cahn, R. W., and Cottrell, A. H. (1955), A Study of Primary and Conjugate Slip in Crystals of Alpha-Brass, *Acta Metall.*, Vol. 3, pp. 331–338.

[377] Pinkus, A., and Samy, Z. (1997), *Fourier Series and Integral Transforms*, Cambridge University Press, New York.

Bibliography

[378] Polanyi Von, M. (1922), Röntgenographische Bestimmung von Kristallanodnungen, *Natwrwissenschaften*, Vol. 10, pp. 411–416.

[379] Polanyi Von, M. (1934), Über eine Art Gitterstorung, die einen Kristall Plastisch Machen Kinnte, *Z. Phys.*, Vol. 80, pp. 660–664.

[380] Prager, W. (1955), The Theory of Plasticity: A Survey of Recent Achievements, (James Clayton Lecture), *Proc. Inst. Mech. Eng.*, Vol. 169, pp. 41–67.

[381] Prager, W. (1956), A New Method of Analyzing Stresses and Strains in Work-Hardening Plastic Solids, *J. Appl. Mech.*, Vol. 23, pp. 493–496.

[382] Prager, W. (1961), *Introduction to Mechanics of Continua*, Ginn and Co., Boston, MA.

[383] Price, R. J., and Kelly, A. (1964a), Deformation of Age-Hardened Aluminum Alloy Crystals. I: Plastic Flow, *Acta Metall.*, Vol. 12, p. 159–169.

[384] Price, R. J., and Kely, A. (1964b), Deformation of Age-Hardened Aluminum Alloys Crystals. II: Fracture, *Acta Metall.*, Vol. 12, pp. 979–991.

[385] Ragone, D. V. (1995), *Thermodynamics of Materials*, Wiley, New York.

[386] Ramaswami, B., Kocks, U. F., and Chalmers, B. (1965), Latent Hardening in Silver and an Ag-Au Alloy, *Trans. AIME*, Vol. 233, pp. 927–931.

[387] Ratner, B. D., Hoffman, A. S., Schoen, F. J., and Lemons, J. E., eds. (2004), *Biomaterials Science: An Introduction to Materials in Medicine*, 2nd ed., Elsevier, San Diego.

[388] Reinhardt, W. D., and Dubey, R. N. (1996), Application of Objective Rates in Mechanical Modeling of Solids, *J. Appl. Mech.*, Vol. 118, pp. 692–698.

[389] Rice, J. R. (1968a), A Path Independent Integral and Approximate Analysis of Strain Concentration by Notches and Cracks, *J. Appl. Mech.*, Vol. 38, pp. 379–386.

[390] Rice, J. R. (1968b), Mathematical Analysis in the Mechanics of Fracture. In *Fracture – An Advanced Treatise* (H. Liebowitz, ed.), Vol. II, pp. 191–311, Academic Press, New York.

[391] Rice, J. R. (1970), On the Structure of Stress-Strain Relations for Time Dependent Plastic Deformation in Metals, *J. Appl. Mech.*, Vol. 37, pp. 728–737.

[392] Rice, J. R. (1971), Inelastic Constitutive Relations for Solids: An Internal Variable Theory and Its Application to Metal Plasticity, *J. Mech. Phys. Solids*, Vol. 19, pp. 433–455.

[393] Rice, J. R. (1975), Continuum Mechanics and Thermodynamics of Plasticity in Relation to Microscale Deformation Mechanisms. In *Constitutive Equations in Plasticity* (A. S. Argon, ed.), pp. 23–79, MIT Press, Cambridge, MA.

[394] Rice, J. R. (1977), The Localization of Plastic Deformation. In *Theoretical and Applied Mechanics* (W. T. Koiter, ed.), pp. 207–220, North-Holland, Amsterdam.

[395] Rice, J. R. (1985), Conserved Integrals and Energetic Forces. In *Fundamentals of Deformation and Fracture* (B. A. Bilby, K. J. Miller, and J. R. Willis, eds.), Cambridge University Press, Cambridge.

[396] Rice, J. R., and Tracey, D. M. (1969), On the Ductile Enlargement of Voids in Triaxial Stress Fields, *J. Mech. Phys. Solids*, Vol. 17, pp. 201–217.

[397] Rivlin, R. S. (1960), Some Topics in Finite Elasticity. In *Structural Mechanics* (J. N. Goodier and N. Hoff, eds.), pp. 169–198, Pergamon, New York.

[398] Rivlin, R. S., ed. (1977), *Finite Elasticity*, ASME, AMD, Vol. 27, New York.

[399] Rivlin, R. S., and Sawyers, K. N. (1976), The Strain-Energy Function for Elastomers, *Trans. Soc. Rheol.*, Vol. 20, pp. 545–557.

[400] Rodrigez, E. K., Hoger, A., and McCulloch, A. D. (1994), Stress-Dependent Finite Growth in Soft Elastic Tissues, *J. Biomechanics*, Vol. 27, pp. 455–467.

[401] Rudnicki, J. W., and Rice, J. R. (1975), Conditions for the Localization of Deformation in Pressure-Sensitive Dilatant Materials, *J. Mech. Phys. Solids*, Vol. 23, pp. 371–394.

[402] Sachs, G. (1928), Zur Ableitung einer Fleissbedingung, *Z. d. Ver. deut. Ing.*, Vol. 72, pp. 734–736.

[403] Sacks, M. S. (2000), Biaxial Mechanical Evaluation of Planar Biological Materials, *J. of Elasticity*, Vol. 61, pp. 199–246.

[404] Saiomoto, S., Hasford, W. F., and Backofen, W. A. (1965), Ductile Fracture in Copper Single Crystals, *Phil. Mag.*, Vol. 12, pp. 319–333.

[405] Salençon, J. (1977), *Application of the Theory of Plasticity in Soil Mechanics*, Wiley, Chichester, UK.

[406] Sanders, J. L. (1955), Plastic Stress-Strain Relations Based on Infinitely Many Plane Loading Surfaces. In *Proc. 2nd U.S. Nat. Congr. Appl. Mech.* (P. M. Naghdi, ed.), pp. 445–460, ASME, New York.

[407] Sanders, P. G., Eastman, J. A., and Weertman, J. R. (1997), Elastic and Tensile Behavior of Nanocrystalline Copper and Palladium, *Acta Mater.*, Vol. 45, pp. 4019–4025.

[408] Schiotz, J., Di Tolla, F. D., and Jacobsen, K. W. (1998), Softening of Nanocrystalline Metals at Very Small Grain Sizes, *Nature*, Vol. 391, pp. 561–563.

[409] Schmid, E. (1924), Neuere Untersuchungen an Metallkristallen. In *Proc. 1st Intl. Cong. Appl. Mech.* (C. B. Biezeno and J. M. Burgers, eds.), Delft, Waltman, pp. 342–353.

[410] Schmid, E. (1926), Ueber dis Schubverfestigung von Einkristallen bei Plasticher Deformation, *Z. Phys.*, Vol. 40, pp. 54–74.

[411] Schmid, E., and Boas, W. (1968), *Plasticity of Crystals*, Chapman and Hall, London.

[412] Schoenfield, S. E., Ahzi, S., and Asaro, R. J. (1995), Elastic-Plastic Crystal Mechanics for Low Symmetry Crystals, *J. Mech. Phys. Solids*, Vol. 43, pp. 415–446.

[413] Schuh, C. A., Nieh, T. G., and Yamasaki, T. (2002), Hall-Petch Breakdown Manifested in Abrasive Wear Resistance of Nanocrystalline Nickel, *Scripta Mater.*, Vol. 46, pp. 735–740.

[414] Schwaiger, R., Moser, B., Chollacoop, N., Dao, M., and Suresh, S. (2003), Some Critical Experiments on the Strain-Rate Sensitivity of Nanocrystalline Nickel, *Acta Mater.*, Vol. 51, pp. 5159–5172.

[415] Sedov, L. I. (1966), *Foundations of the Non-Linear Mechanics of Continua*, Pergamon, Oxford, UK.

[416] Sewell, M. J. (1987), *Maximum and Minimum Principles – A Unified Approach with Applications*, Cambridge University Press, Cambridge.

[417] Shames, I. H., and Cozzarelli, F. A. (1997), *Elastic and Inelastic Stress Analysis*, Taylor & Francis, London.

[418] Shapery, R. A. (1968), On a Thermodynamic Constitutive Theory and Its Application to Various Nonlinear Materials. In *Irreversible Aspects of Continuum Mechanics* (H. Parkus and L. I. Sedov, eds.), pp. 259–285, Springer-Verlag, Berlin.

[419] Skalak, R., Dasgupta, G., Moss, M., Otten, E., Dullemeijer, P., and Vilmann, H. (1982), Analytical Description of Growth, *J. Theor. Biol.*, Vol. 94, pp. 555–577.

[420] Simo, J. C., and Hughes, T. J. R. (1998), *Computational Inelasticity*, Springer-Verlag, New York.

[421] Sneddon, I. N. (1951), *Fourier Transforms*, McGraw-Hill, New York.

[422] Sneddon, I. N. (1961), *Fourier Series*, Routledge and Paul, London.

[423] Sneddon, I. N. (1972), *The Use of Integral Transforms*, McGraw-Hill, New York.

[424] Sneddon, I. N. (1974), *The Linear Theory of Thermoelasticity*, CISM Udine, Springer–Verlag, Wien.

[425] Sokolnikoff, I. S. (1956), *Mathematical Theory of Elasticity*, 2nd ed., McGraw-Hill, New York.

[426] Spencer, A. J. M. (1971), Theory of Invariants. In *Continuum Physics* (A. C. Eringen, ed.), Vol. 1, Academic Press, New York.

[427] Spencer, A. J. M. (1992), *Continuum Mechanics*, Longman Scientific & Technical, London.

[428] Spitzig, W. A., Sober, R. J., and Richmond, O. (1975), Pressure Dependence of Yielding and Associated Volume Expansion in Tempered Martensite, *Acta Metall.*, Vol. 23, pp. 885–893.

[429] Stojanović, R. (1970), *Recent Developments in the Theory of Polar Continua, CISM Lecture Notes*, Springer-Verlag, Wien.

[430] Stören, S., and Rice, J. R. (1975), Localized Necking in Thin Sheets, *J. Mech. Phys. Solids*, Vol. 23, pp. 421–441.

[431] Stroh, A. N. (1958), Dislocations and Cracks in Anisotropic Elasticity, *Phil Mag.*, Vol. 3, pp. 625–646.

[432] Suo, Z. (1997), Motions of Microscopic Surfaces in Materials, *Adv. Appl. Mech.*, Vol. 33, pp. 193–294.

Bibliography

[433] Suresh, S. (2003), The Golden Jubilee Issue – Selected Topics in Materials Science and Engineering: Past, Present and Future – Preface, *Acta Mater.*, Vol. 51, p. 5647.

[434] Suresh, S., Giannakopoulos, A. E., and Alcala, J. (1997), Spherical Indentation of Compositionally Graded Materials: Theory and Experiments, *Acta Mater.*, Vol. 45, pp. 1307–1321.

[435] Swalin, R. A. (1972), *Thermodynamics of Solids*, Wiley, New York.

[436] Swearengen, J. C., and Taggart, R. (1971), Low Amplitude Cyclic Deformation and Crack Nucleation in Copper and Copper-Aluminum Bicrystals, *Acta Metall.*, Vol. 19, pp. 543–559.

[437] Synge, J. L., and Schild, A. (1949), *Tensor Calculus*, University Press, Toronto.

[438] Szilard, R. (2004), *Theories and Applications of Plate Analysis: Classical, Numerical, and Engineering Methods*, Wiley, Hoboken, New Jersey.

[439] Taber, L. A. (1995), Biomechanics of Growth, Remodeling, and Morphogenesis, *Appl. Mech. Rev.*, Vol. 48, pp. 487–545.

[440] Taber, L. A. (2004), *Nonlinear Theory of Elasticity – Application to Biomechanics*, World Scientific, New Jersey.

[441] Taber, L. A., and Eggers, D. W. (1996), Theoretical Study of Stress-Modulated Growth in the Aorta, *J. Theor. Biol.*, Vol. 180, pp. 343–357.

[442] Taber, L. A., and Perucchio, R. (2000), Modeling Heart Development, *J. Elasticity*, Vol. 61, pp. 165–197.

[443] Takeuchi, S., and Kuramoto, E. (1973), Temperature and Orientation Dependence of Yield Stress in Ni_3Ga Single-Crystals, *Acta Metall.*, Vol. 21, pp. 415–425.

[444] Taylor, G. I. (1926), The Distortion of Single Crystals of Metals. In *Proc. 2nd Intl. Cong. Appl. Mech.* (E. Meissner, ed.), pp. 46–52, Orell Füssli, Zurich.

[445] Taylor, G. I. (1927), The Distortion of Crystals of Aluminum Under Compression. Part II: Distortion by Double Slipping and Changes in Orientation of Crystal Axes During Compression, *Proc. Roy. Soc. London*, Vol. A116, pp. 16–38.

[446] Taylor, G. I. (1927), The Distortion of Crystals of Aluminum Under Compression. Part III: Measurements of Stress, *Proc. Roy. Soc. London, Sec A*, Vol. 116, pp. 39–60.

[447] Taylor, G. I. (1934), The Mechanism of Plastic Deformation of Crystals. Part I: Theoretical, *Proc. Roy. Soc. London, Sec. A*, Vol. 145, pp. 362–387.

[448] Taylor, G. I. (1938a), Pleastic Strain in Metals, *J. Inst. Met.*, Vol. 62, pp. 307–325.

[449] Taylor, G. I. (1938b), Analysis of Plastic Strain in a Cubic Crystal. In *Stephen Timoshenko 60th Anniversary Volume* (J. M. Lessels, ed.), McMillian, New York.

[450] Taylor, G. I., and Elam, C. F. (1923), The Distortion of an Aluminum Crystal During a Tensile Test, *Proc. Roy. Soc. London, Sec. A*, Vol. 102, pp. 634–667.

[451] Taylor, G. I., and Elam, C. F. (1925), The Plastic Extension and Fracture of Aluminum Crystals, *Proc. Roy. Soc. London, Sec. A*, Vol. 108, pp. 28–51.

[452] Taylor, G. I., and Quinney, H. (1931), The Plastic Distortion of Metals, *Phil. Trans. Roy. Soc. A*, Vol. 230, pp. 232–262.

[453] Temam, R. (1985), *Mathematical Problems in Plasticity*, Gauthier-Villars, Paris.

[454] Teodosiu, C. (1982), *Elastic Models of Crystal Defects*, Springer-Verlag, Berlin.

[455] Thompson, A. W. (1975), Polycrystal Hardening. In *Work Hardening in Tension and Fatigue* (A. W. Thompson, ed.), p. 89, AIME, New York.

[456] Thompson, A. W., and Baskes, M. I. (1973), The Influence of Grain Size on the Work Hardening of Face-Centered-Cubic Polycrystals, *Phil. Mag.*, Vol. 28, pp. 301–308.

[457] Timoshenko, S. P., and Goodier, J. N. (1970), *Theory of Elasticity*, 3rd ed., McGraw-Hill, New York.

[458] Timoshenko, S. P., and Woinowsky-Krieger, S. (1987), *Theory of Plates and Shells*, 2nd ed., McGraw-Hill, New York.

[459] Timpe, A. (1905), *Probleme der Spannungsverteilung in ebenen Systemen einfach gelost mit Hilfe der airyschen Funktion*, Göttingen Disseration, Leipzig.

[460] Ting, T. C. T. (1991), The Stroh Formalism and Certain Invariances in Two-Dimensional Anisotropic Elasticity. In *Modern Theory of Anisotropic Elasticity and Applications* (J. J. Wu, T. C. T. Ting, and D. M. Barnett, eds.), SIAM, Philadelphia.

[461] Ting, T. C. T. (1996), *Anisotropic Elasticity: Theory and Applications*, Oxford University Press, New York.
[462] Toupin, R. A. (1962), Perfectly Elastic Materials with Couple Stresses, *Arch. Ration. Mech. Anal.*, Vol. 11, pp. 385–414.
[463] Treloar, L. R. G. (1975), *The Physics of Rubber Elasticity*, Clarendon Press, Oxford, UK.
[464] Triantafyllidis, N. (1980), Bifurcation Phenomena in Pure Bending, *J. Mech. Phys. Solids*, Vol. 28, pp. 221–245.
[465] Triantafyllidis, N. (1983), On the Bifurcation and Postbifurcation Analysis of Elastic-Plastic Solids under General Prebifurcation Conditions, *J. Mech. Phys. Solids*, Vol. 31, pp. 499–510.
[466] Truesdell, C. (1985), *The Elements of Continuum Mechanics*, Springer-Verlag, New York.
[467] Truesdell, C., and Noll, W. (1965), The Non-Linear Field Theories of Mechanics. In *Handbuch der Physik* (S. Flugge, ed.), Band III/3, Springer-Verlag, Berlin.
[468] Truesdell, C., and Toupin, R. (1960), The Classical Field Theories. In *Handbuch der Physik* (S. Flügge, ed.), Band III/l, Springer-Verlag, Berlin.
[469] Tvergaard, V. (1982), On Locatization in Ductile Metarials Containing Spherical Voids, *Int. J. Fracture*, Vol. 18, pp. 237–252.
[470] Tvergaard, V., and Needleman, A. (1984), Analysis of the Cup-Cone Fracture in a Round Tensile Bar, *Acta Metall.*, Vol. 32, pp. 157–169.
[471] Ugural, A. C. (1999), *Stresses in Plates and Shells*, 2nd ed., McGraw-Hill, Boston.
[472] Ugural, A. C., and Fenster, S. K. (2003), *Advanced Strength and Applied Elasticity*, 4th ed., Prentice Hall, Upper Saddle River, New Jersey.
[473] Valanis, K. C., and Landel, R. F. (1967), The Strain-Engery Function of a Hyperelastic Material in Terms of the Extension Ratios, *J. Appl. Phys.*, Vol. 38, pp. 2997–3002.
[474] Valiev, R. Z., Alexandrov, I. V., Zhu, Y. T., and Lowe, T. L. C. (2002), Paradox of Strength and Ductility in Metals Processed by Severe Plastic Deformation, *J. Mater. Res.*, Vol. 17, pp. 5–8.
[475] Van der Merwe, J. H. (1991), Strain Relaxation in Epitaxial Overlayers, *J. Electr. Mat.*, Vol. 20, pp. 793–803.
[476] Van Swygenhoven, H. (2003), Plastic Deformation in Metals with Nanosized Grains: Atomistic Simulations and Experiments, *Mater. Sci. Forum*, Vol. 447, pp. 3–10.
[477] Van Swygenhoven, H., Caro, A., and Farkas, D. (2001), Grain Boundary Structure and its Influence on Plastic Deformation of Polycrystalline FCC Metals at the Nanoscale: A Molecular Dynamics Study, *Scripta Mater.*, Vol. 44, pp. 1513–1516.
[478] Van Swygenhoven, H., Derlet, P. M., and Hasnaoui, A. (2002), Atomic Mechanism for Dislocation Emission from Nanosized Grain Boundaries, *Phys. Rev. B*, Vol. 66, Art. No. 024101.
[479] Van Vliet, K. J., Tsikata, E., and Suresh, S. (2003), Model Experiments for Direct Visualization of Grain Boundary Deformation in Nanostructured Metals, *Appl. Phys. Lett.*, Vol. 83, pp. 1441–1443.
[480] Voigt, W. (1887), Theoritiscke Studienüber die Elastizitatsverhältnisse der Krystalle, *Abhandl. Ges. Wiss. Gottingen*, Vol. 34, pp. 3–51.
[481] Volterra, V. (1907), Sur l'Équilibre des Corps Élastiques Multiplement Convexes, *Ann. Ec. Norm., Ser. 3*, Vol. 24, pp. 401–517.
[482] Wang, C.-C., and Truesdell, C. (1973), *Introduction to Rational Elasticity*, Noordhoff, Leyden, The Netherlands.
[483] Wang, Y. M., Ma, E., and Chen, M. W. (2002), Enhanced Tensile Ductility and Toughness in Nanostructured Cu, *Appl. Phys. Lett.*, Vol. 80, pp. 2395–2397.
[484] Wang, Y. M., and Ma, E. (2003), Temperature and Strain Rate Effects on the Strength and Ductility of Nanostructured Copper, *Appl. Phys. Lett.*, Vol. 83, pp. 3165–3167.
[485] Wang, Y. M., and Ma, E. (2004), Strain Hardening, Strain Rate Sensitivity, and Ductility of Nanostructured Metals, *Mater. Sci. Eng. A*, Vol. 375-77, pp. 46–52.
[486] Wasilewski, R. J., Butler, S. R., and Hanlon, J. E. (1967), Plastic Deformation of Single-Crystal NiAl, *Trans. Metall. Soc. AIME*, Vol. 239, p. 1357.
[487] Watson, G. N. (1995), *A Treatise on the Theory of Bessel Punctiohs*, 2nd ed., Cambridge University Press, Cambridge.

Bibliography

[488] Wei, Q., Cheng, S., Ramesh, K. T., and Ma, E. (2004), Effect of Nanocrystalline and Ultrafine Grain Sizes on the Strain Rate Sensitivity and Activation Volume: FCC versus BCC Metals, *Mater. Sci. Eng. A*, Vol. 381, pp. 71–79.

[489] Williams, F. A. (1985) *Combustion Theory – The Fundamental Theory of Chemically Reacting Flow Systems*, 2nd ed., Addison-Wesley, New York.

[490] Willis, J. R. (1964), Anisotropic Inclusion Problems, *Q. J. Mech. Appl. Math.*, Vol. 17, pp. 157–174.

[491] Willis, J. R., Jain, S. C., and Bullough, R. (1990), The Energy of an Array of Dislocations: Implications for Strain Relaxation in Semiconductor Heterostructures, *Phil. Mag.*, Vol. 62, pp. 115–129.

[492] Willis, J. R., Jain, S. C., and Bullough, R. (1991), The Energy of an Array of Dislocations II: Consideration of a Capped Expitaxial Layer, *Phil. Mag.*, Vol. 64, pp. 629–640.

[493] Wolf, K. B. (1979), *Integral Transforms in Science and Engineering*, Plenum, New York.

[494] Worthington, P. J., and Smith, E. (1964), The Formation of Slip Bands in Polycrystalline 3% Silicon Iron in the Pre-Yield Microstrain Region, *Acta Metall.*, Vol. 12, pp. 1277–1281.

[495] Wrede, R. C. (1972), *Introduction to Vector and Tensor Analysis*, Dover Pub., New York.

[496] Wu, J. J., Ting, T. C. T., and Barnett, D. M., eds. (1991), *Modern Theory of Anisotropic Elasticity and Applications*, SIAM, Philadelphia.

[497] Xia, Z. C., and Hutchinson, J. W. (1996), Crack Tip Fields in Strain Gradient Plasticity, *J. Mech. Phys. Solids*, Vol. 44, pp. 1621–1648.

[498] Yamakov, V., Wolf, D., Phillpot, S. R., and Gleiter, H. (2002), Grain Boundary Diffusion Creep in Nanocrystalline Palladium by Molecular Dynamics Simulation, *Acta Mater.*, Vol. 50, pp. 61–73.

[499] Yang, W., and Lee, W. B. (1993), *Mesoplasticity and Its Applications*, Springer-Verlag, Berlin.

[500] Youssef, K. M., Scattergood, R. O., Murty, K. L., and Koch, C. C. (2004), Ultratough Nanocrystalline Copper with a Narrow Grain Size Distribution, *Appl. Phys. Lett.*, Vol. 85, pp. 929–931.

[501] Ziegler, H. (1959), A Modicification of Prager's Hardening Rule, *Q. Appl. Math.*, Vol. 17, pp. 55–65.

[502] Ziegler, H. (1983), *An Introduction to Thermomechanics*, 2nd revised ed., North-Holland, Amsterdam.

[503] Zyczkowski, M. (1981), *Combined Loadings in the Theory of Plasticity*, PWN, Polish Scientific Publishers, Warszawa.

Index

absolute temperature, 115
acoustic tensor, 266
adaptive elasticity, 609
adiabatic loading, 119
affinity, 127, 129, 618
Airy stress function, 179, 410, 415, 419
angle of twist, 214
angular change
    between fibers, 76
    of principal directions, 76
anisotropic
    elastic solid, 264, 345, 604
    elasticity, 332
    hardening, 475
antiplane
    shear, 390
    strain, 386
aorta, 630
apposition, 609
arc length, 59
area change, 64
    for a membrane, 636
Asaro–Tiller field, 459
associative flow rule, 480
asymptotic stress fields, 459
axial vector, 17, 87
axisymmetric problems, 211

back stress, 475
balance of angular momentum, 95
Barnett–Lothe tensors, 333
Bauschinger effect, 475, 595
Beltrami–Michell equations, 172
bending of beams, 225
Bessel functions, 385
Betti's reciprocal theorem, 177, 318
    couple-stress elasticity, 393
Bianchi conditions, 171

bifurcation, 498
biharmonic equation, 179, 184
    polynomial solutions, 185
    with body forces, 188
bimaterial interface, 407, 410, 419
binary alloy, 145
biomaterials, 609
Blatz–Ko strain energy, 700
blood vessel, 630
body force potential, 178
Boussinesq–Papkovitch solutions, 248
Burgers vector, 294, 407, 410, 419

Cauchy elasticity, 149
Cauchy stress, 92
    nonsymmetric, 398
    principal values of, 96
    symmetry of, 95
Cauchy tetrahedron, 92
Cauchy–Green deformation tensor, 58
cavitation instability, 500, 501
Cayley–Hamilton theorem, 22
Cesàro integrals, 172
characteristic equation, 13, 23
chemical potential, 139, 370, 449
Christoffel stiffness tensor, 266
circular plate, 288
Clapeyron's formula, 111
Clausius–Duhem inequality, 115
coarse slip bands, 558
coefficient
    of compressibility, 122
    of thermal expansion, 122
compatibility equations, 169, 377
    Beltrami–Michell, 172
    Saint-Venant, 171
compliance tensor, 118
configurational
    entropy, 144
    force, 400

conservation
    integrals, 403
    laws, 404
    of mass, 34
consistency condition, 472, 481
constitutive equations
    linear elasticity, 161
    nonlinear elasticity, 149
    plasticity
        deformation theory, 485
        isotropic hardening, 475
        kinematic hardening, 477
        nonassociative, 480
        pressure-dependent, 480
        rate-dependent, 482
constrained
    equilibrium, 127
    field, 338
contact problems, 271
continuity equation, 34, 610
convected
    lattice vectors, 540
    stress rate, 105, 680
convolution integrals, 48
coordinate transformation, 3
core energy, 424, 426
corner theory of plasticity, 487
corotational stress rate, 106
correspondence theorem, 378, 389
Cotter–Rivlin convected rate, 680
Coulomb's law, 27
couple-stress, 375, 376, 398
coupled heat equation, 120
crack extension force, 323
crack opening displacement, 325
creep, 483
critical
    conditions for localization, 574
    film thickness, 432
    hardening rate, 567
    nucleus size, 374
    resolved shear stress, 505
crystal plasticity, 538, 601
    laminate model, 601
    single slip, 605
curl, 55, 56
curvature tensor, 376, 399
cylindrical
    coordinates, 668
    void, 500

Debye's temperature, 124
deformation gradient, 56
    cylindrical coordinates, 670
    multiplicative decomposition, 490, 622
deformation rate, 69, 73, 74
deformation theory of plasticity, 484, 486, 596
degenerate solutions, 204
densification, 611

determinant, 6
deviatoric
    plane, 469
    work, 464
dilatancy factor, 481
dilatant materials, 480
Dirac delta function, 51, 233, 264
Dirichlet conditions, 42
disclination, 760
dislocation, 293, 299
    core, 424
    density, 511
    distribution, 296
    driving force, 439, 441
    edge, 419, 509
    energy, 424, 426
    forces on, 365
    forest, 515
    injection, 444
    interactions, 514
    line, 299
    misfit, 432, 433
    multiplication, 517
    near free surface, 426
    partial, 513
    perturbed array, 443
    recession, 439
    spacing, 432
    threading, 432, 437
dislocation array
    energy, 430
    formation, 432, 438
    stress field, 428
dislocation force
    array, 431
    edge dislocation, 415, 418, 421, 423
    screw dislocation, 408, 410
    straight, 427
displacement discontinuity, 421, 424, 430
displacements
    determination of, 196
    half space, 238
    in beam, 187
    nonsingle valued, 293
distributed contact loading, 274
div, 56
divergence theorem, 26, 33, 34
double Fourier series, 37
double slip, 576
Drucker–Prager yield criterion, 468
Duhamel–Neumann expression, 132
Dulong–Petit limit, 124
dyadic notation, 13

eccentricity, 636
edge dislocation, 299, 410
    couple-stress elasticity, 381
    near free surface, 417, 422

Index

eigenvalues, 12, 63
  symmetric tensors, 14
eigenvectors, 12, 63
  orthogonality of, 63
  stretch of, 63
elastic
  compliance tensor, 118
  constants, 169
  deformation gradient, 490
  moduli tensor, 158, 627
  pseudomoduli, 155
  stiffness tensor, 118
  unloading, 489
elastic-plastic interface, 497
elasticity
  3D, 264
  Cauchy, 149
  crystal elasticity, 547
  Green, 148
  isotropic, 150
  nonlinear, 148
energy
  Gibbs, 117, 128
  Helmholtz free, 116, 128
  internal, 113
  kinetic, 113
  of a dislocation line, 366
  total, 113
energy equation, 114
  with mass growth, 616
energy factor
  matrix, 511
  tensor, 315
energy momentum tensor, 358
  finite deformations, 361
  micropolar elasticity, 404
energy release rate, 315, 323
enthalpy, 117, 128
  of mixing, 145
entropic elasticity, 116
entropy, 114
  of mixing, 144
epitaxial
  growth, 432
  layer, 432
equilibrium equations, 95
Eshelby
  inclusion problem, 335
  tensor, 341
Euler's laws of motion, 614
Eulerian
  strain, 676
    rate of, 76
  triad, 80
evolution equation, 129, 479
  for mass growth, 631
  for stretch ratio, 630
evolution of back stress, 476, 477

extended
  dislocation, 513
  Taylor model, 587
fiber
  rotation of, 19
  stretch, 60
  stretching rate, 73
film thickness, 432
first law of thermodynamics, 113
first Piola–Kirchhoff stress, 99
Flamant solution, 723
flat punch, 278
flexural rigidity, 284
flow potential, 482
flow rule
  associative, 480
  nonassociative, 480
flux, 31, 127, 129, 618
force
  generalized, 357
  on a dislocation, 365
  on a precipitate interface, 373
  on an interface, 359
forest dislocations, 515
Fourier
  double series, 37
  integral theorem, 46
  kernel, 41
  law of conduction, 119
  loading, 191, 193
  series, 36
  transform, 39, 48, 265
frame indifference, 102
Frank and van der Merwe criterion, 434
Frank–Read source, 530
free energy, 356

Galerkin vector, 256
Gauss
  divergence theorem, 26
  law, 28
generalized
  force, 357
  plastic strain, 474
geometrical
  hardening, 607
  softening, 508, 580, 607, 816
Gibbs
  conditions of equilibrium, 129
  energy, 117, 128, 141, 356
Gibbs–Duhem equation, 142
grad, 56
Green
  elasticity, 148
  function, 264
  isotropic elasticity, 150
  lattice strain, 603
  strain, 59
  stretch tensor, 86

Griffith
    crack, 332, 773
    criterion, 320
Gurson yield criterion, 470

half-plane loading, 232
half-space solutions, 229
Hall–Petch relation, 517
hard tissue, 609
hardening, 473
    isotropic, 474
    kinematic, 475
heat
    capacity, 124
    conduction, 113
        Fourier law, 119
    equation, 120
    flow, 113
Helmholtz equation, 378
Helmholtz free energy, 116, 128
    with mass growth, 620
Hertz problem, 259
hollow dislocation, 384
Hooke's law, 150, 161
hyperelasticity, 149
hypertension, 630
hypertrophy, 609

ideal
    plasticity, 473
    solution, 144
ideally plastic material, 497
identity tensor, 6, 10
image force on a defect, 371
inclusion
    elastic energy of, 343
    field at the interface, 352
    field in the matrix, 352
    inhomogeneous, 344
    isotropic spherical, 353
    problem statement, 335
incompressibility constraint, 153
incompressible elasticity, 153
infinite strip, 242
inflation of balloon, 702
inhomogeneous inclusion, 344
instantaneous elastic moduli, 155
integral
    derivative, 83
    transform, 39
interaction energy, 362, 432, 442, 444
    between dislocations, 318
interface dislocation
    edge, 415, 420
    screw, 409
interface force, 359
intermediate configuration, 489
internal energy, 113
    with mass growth, 619
internal variables, 127

invariant functional, 401
invariants, 6, 12, 14
inverse
    Fourier transform, 48
    of a tensor, 6
    pole figures, 593
irreversible thermodynamics, 127
isochoric plastic deformation, 474
isotropic
    Green elasticity, 150
    hardening, 474
    inclusion, 350
    material, 150

J integral, 358, 404
J_2 deformation theory of plasticity, 486
Jaumann stress rate, 106, 107, 541
    cylindrical coordinates, 685
    on crystal spin, 541
Johnson–Cook model, 483
Joule's effect, 137

kernel, 41
kinematic hardening, 475
    linear, 477
    nonlinear, 477
    Prager, 477
    Ziegler, 477
kinetic energy, 113
Kirchhoff stress, 101
Kolosov constant, 180, 411
Kronecker delta, 2

L integral, 404
Lagrangian
    multiplier, 153
    strain, 59
    triad, 79
Lamé
    constants, 168
    problem, 251
laminate plasticity, 601
Laplace's equation, 172, 378
latent
    hardening, 519, 547
    heat, 118
lattice
    base vectors, 540
    parameter, 433
lattice rotation, 506
    geometrical softening, 580, 582
    shear bands, 580, 582
left stretch tensor, 21
line integrals
    material derivative, 85
linear dependence, 13
linearly hardening material, 499
loading index
    kinematic hardening, 476
    pressure-dependent plasticity, 479

localized plastic deformation, 557
Love's potential, 257

M integral, 404, 773
macroscopic shear bands, 558
mass
   conservation, 34
   flow, 31
   growth, 609, 611, 630
   resorption, 631
material derivative, 71
   line integrals, 85
   surface area, 84
   surface integrals, 84
   volume integrals, 84
material length, 383
Matthews–Blakeslee criterion, 433
maximum shear stress, 97
Maxwell relations, 128
mechanical power input, 108
Mellin transforms, 40
Michell's solution, 724
micropolar
   continuum, 375
   elasticity, 398
Mindlin's stress functions, 380
misfit dislocation, 432, 433
mismatch strain, 433, 434
mixed dislocations, 304
Mohr's envelope, 467
Mohr–Coulomb yield criterion, 467
molar Gibbs energy, 142
Mooney–Rivlin material, 153, 698
morphogenesis, 610
multiple slip, 576
multiplicative decomposition, 490
   in biomechanics, 622
multiply connected
   cross section, 222
   regions, 172
Murnaghan's constants, 150

nanocrystalline grains, 530
Nanson's relation, 65
necking, 487
neo-Hookean material, 153
Newtonian fluid, 701
Noether's theorem, 400
nominal
   stress, 98, 99, 101
   traction, 678
non-Schmid
   effect, 571
   stress, 563
nonassociative flow rule, 480
nonlinear elasticity, 148
norm of a function, 37

objective rate, 90, 105
octahedral
   plane, 464, 468
   shear stress, 464
Oldroyd rate, 154, 679, 680
Onsager reciprocity relations, 129
open thermodynamic system, 139
opening angle, 631
ordered crystals
   localized deformation, 560
orthogonal tensors, 17
   geometrical interpretation, 18
   specific forms of, 88
orthonormal basis, 2, 11

Papkovich–Neuber potentials, 769
partial dislocations, 513, 531
Peach–Koehler force, 129, 365, 512, 812
Peierls–Nabarro dislocation, 759
permutation tensor, 3
phenomenological plasticity, 461
plane
   strain, 178
   stress, 179
plane stress modulus, 567
plastic
   potential, 480
   deformation gradient, 490
   potential surface, 480
   rate potential, 488
   secant modulus, 486
   strain, 484
   tangent modulus, 474, 486
   void growth, 495
plasticity, 461
   associative, 480
   corner theory, 487
   deformation theory, 484, 486
   ideal, 473
   nonassociative, 480
   strain hardening, 461
plates
   equilibrium, 282
   flexural rigidity, 284
point force, 237, 257, 261
Poisson's equation, 172
polar decomposition, 20, 63
polynomial solutions, 184
porosity, 470
positive definite tensors, 14
potential energy, 175, 355
   bent plate, 284
   couple-stress elasticity, 396
Poynting effect, 700
Prager's hardening, 477
Prandtl stress function, 217
precipitation, 373

pressure-dependent plasticity, 478
pressure-sensitive plastic flow, 572
pressurized
    cylinder, 209
    sphere, 250
principal
    directions, 76
    stresses, 96
    stretch, 21, 62, 78, 152
principle of virtual work, 109
proportional loading, 484, 486
pseudotraction, 678
pseudomoduli, 155

quadratic forms, 22

rate
    of deformation, 69, 627
    of working, 101
    potential, 488
    sensitivity, 602
    tangent modulus, 548
rate-dependent
    plasticity, 482
    slip, 547, 592
recall term, 478
reciprocal symmetry, 480
regular solution, 145
Reiner–Rivlin fluid, 700
remodeling, 610
residual strain, 631
resolved shear stress, 482
Reuss estimates, 825
reversible thermodynamics, 116
Reynolds transport theorem, 613
right stretch tensor, 21
rigid
    inclusion, 252
    indenters, 271
rigid body motion, 88
Rivlin–Ericksen tensors, 657
rocks, 480
rotation tensor, 376
rubber model, 153

Saint-Venant
    compatibility equations, 170, 171
    principle, 187
Saint-Venant–Kirchhoff assumption, 150, 696
scalar field, 55
    gradient, 70
scalar product, 1
Schmid
    rule, 505, 560
    stress, 561
screw dislocation, 299, 302, 407
    couple-stress elasticity, 391
    near free surface, 409

secant modulus, 484
second law of thermodynamics, 114
second Piola–Kirchhoff stress, 102
semi-inverse method, 225
shear
    center, 227
    modulus, 161
    strain, 60
shear stress
    maximum, 97
    resolved, 482
simple beam
    Fourier loading, 191
simple shear, 58, 85
single valued displacements, 208
singular integral equation, 296
slip
    steps, 502
    system, 482
    traces, 502
slip-plane hardening, 547
small strain, 61
soft tissue, 609, 629
softening, 473
    geometric, 816
    vertex, 487
soil mechanics, 468
specific heat, 118
sphere
    subject to temperature gradient, 254
spherical
    coordinates, 248
    indentation, 255
    void, 495
spin rate, 74
    plastic part, 604
spin tensor, 79
stability
    array, 441
    bounds, 439, 445
    dislocation array, 439
stacking faults, 513
    emission from grain boundaries, 531
state variables, 116
static equilibrium, 92, 96
stationary discontinuity, 563
stiffness tensor, 118
Stokes theorem, 26
Stoney's formula, 786
strain
    definitions, 56
    Eulerian, 56
    integration of, 187
    Lagrangian, 56
    logarithmic, 56
    natural, 56
    nominal, 56
    potentials, 256

Index

rate, 76
relaxation, 437
shear, 60
small, 61
strain energy, 148
   for soft tissues, 629
   isotropic elasticity, 151
strain hardening, 461
   crystal plasticity, 544
   localization, 574
   origins of, 505
   single crystal, 603
strength-differential effect, 480
stress
   Cauchy, 92
   first Piola–Kirchoff, 99
   function, 330
   invariants, 462
   Kirchhoff, 101
   nominal, 98, 101
   power, 109
   rate, 106
   second Piola–Kirchhoff, 102
   work conjugate, 99
stretch, 56
   principal, 62
   tensor, 59
Stroh formalism, 329
structural rearrangements, 127
substrate, 432
surface
   energy, 449
   instability, 455
surface integrals, 30
   material derivative, 84
symmetric tensor, 14

tangent modulus, 484
   plastic, 486
   secant, 486
Taylor
   lattice, 510
   model, 587
temperature, 115, 483
tensile crack, 296
tensors
   antisymmetric, 14
   axial vector, 17
   characteristic equation, 13
   conductivities, 119
   determinant, 6
   field, 27
   invariants, 6, 14
   inverse, 6
   orthogonal, 17
   positive definite, 14
   product, 10
   spectral forms, 14
   symmetric, 14
   trace, 6
   transpose, 5
texture determination, 593
thermal strains, 178
thermodynamic
   force, 127
   potential, 127, 128
   system, 113
thermoelastic effect, 136
thermoelasticity, 127, 131, 178
thin
   films, 432
   plates, 280
thin-walled section, 727
third law of thermodynamics, 126
threading dislocation, 432, 437
time-independent behavior, 483
tissue, 609
torsion, 214
   displacements, 214
   elliptical cross section, 221
   energy of, 216
   function, 214
   multiply connected cross sections, 222
   rectangular cross section, 217
   rigidity, 216
total energy, 355
total strain theories, 596
trace, 6, 10
traction vector, 92
transformation strain, 335, 336
   polynomial, 345
transmural cut, 631
transport formulae, 83
Tresca yield criterion, 465
triple product, 1, 2, 7, 13

uniform contact pressure, 276
uniqueness of solution, 174
unit cells, 505

vector field, 27
   curl of, 55
   differentiable, 55
   divergence, 55
   gradient, 70
vector product, 1
Vegard's rule, 433
velocity, 71
velocity gradient, 71
   antisymmetric part, 74
   crystal plasticity, 539
   elastoplastic deformation, 490
   symmetric part, 74
velocity strain, 73
vertex softening, 487
virtual work, 107

void growth, 479, 495
Voigt
   estimates, 824
   notation, 164
Volterra's integral, 301
volume
   change, 63
   integrals, 84
   rate of change, 84
volumetric
   strain, 161
   strain rate, 479
von Mises yield criterion, 463

wedge problem, 271
width of dislocation, 759
work conjugate stress, 99

yield
   cone, 469
   surface, 475
   vertex, 487
yield criterion
   Drucker–Prager, 468, 479
   Gurson, 470, 479
   Mohr–Coulomb, 467
   pressure-dependent, 468, 478
   Tresca, 465
   von Mises, 463
yield surface, 474
Young's modulus, 161

Ziegler's hardening, 477